ORGANIC SYNTHESIS ENGINEERING

TOPICS IN CHEMICAL ENGINEERING
A Series of Textbooks and Monographs

SERIES EDITOR
Keith E. Gubbins, North Carolina State University

ASSOCIATE EDITORS
Mark A. Barteau, University of Delaware
Edward L. Cussler, University of Minnesota
Douglas A. Lauffenburger, MIT
Manfred Morari, ETH

W. Harmon Ray, University of Wisconsin
William B. Russel, Princeton University
Matthew V. Tirrell, University of Minnesota

Receptors: Models for Binding, Trafficking, and Signalling   D. Lauffenburger and J. Linderman
Process Dynamics, Modeling, and Control   B. Ogunnaike and W. H. Ray
Microstructures in Elastic Media   N. Phan-Thien and S. Kim
Optical Rheometry of Complex Fluids   G. Fuller
Nonlinear and Mixed Integer Optimization: Fundamentals and Applications   C. A. Floudas
An Introduction to Theoretical and Computational Fluid Dynamics   C. Pozrikidis
Mathematical Methods in Chemical Engineering   A. Varma and M. Morbidelli
The Engineering of Chemical Reactions   L. D. Schmidt
Analysis of Transport Phenomena   W. M. Deen
The Structure and Rheology and Complex Fluids   R. Larson
Discrete-Time Dynamic Models   R. Pearson
Smoke, Dust, and Haze, Second Edition   S. Friedlander
Organic Synthesis Engineering   L. K. Doraiswamy

Organic Synthesis Engineering

L. K. DORAISWAMY

Anson Marston Distinguished Professor in Engineering

Department of Chemical Engineering
Iowa State University

2001

OXFORD
UNIVERSITY PRESS

Oxford   New York
Athens   Auckland   Bangkok   Bogotá   Buenos Aires   Calcutta
Cape Town   Chennai   Dar es Salaam   Delhi   Florence   Hong Kong   Istanbul
Karachi   Kuala Lumpur   Madrid   Melbourne   Mexico City   Mumbai
Nairobi   Paris   São Paulo   Shanghai   Singapore   Taipei   Tokyo   Toronto   Warsaw

and associated companies in
Berlin   Ibadan

Copyright © 2001 by Oxford University Press, Inc.

Published by Oxford University Press, Inc.,
198 Madison Avenue, New York, New York 10016
http://www.oup-usa.org

Oxford is a registered trademark of Oxford University Press.

All rights reserved. No part of this publication may be reproduced,
stored in a retrieval system, or transmitted, in any form or by any means,
electronic, mechanical, photocopying, recording, or otherwise,
without the prior permission of Oxford University Press.

Library of Congress Cataloging-in-Publication Data
Doraiswamy, L. K. (Laxmangudi Krishnamurthy)
Organic synthesis engineering / L.K. Doraiswamy.
    p. cm.—(Topics in chemical engineering)
Includes index.
ISBN 0-19-509689-4
1. Chemical engineering.   2. Chemical reactors.   3. Catalysis.
4. Organic compounds—Synthesis.   I. Title.
II. Topics in chemical engineering (Oxford University Press)

TP155.7.D65   2000
660'.2844—dc21   00-052427

1 3 5 7 9 8 6 4 2

Printed in the United States of America
on acid-free paper

I dedicate this book to

my daughter   *my son*
  Sandhya   **Deepak**

When one sees Eternity in things that pass away and Infinity in finite things, then one has pure knowledge.

But if one merely sees the diversity of things, with their divisions and limitations, then one has impure knowledge.

And if one selfishly sees a thing as if it were everything, independent of the ONE and the many, then one is in the darkness of ignorance.

Bhagvad Gita XVIII, 20–22

PREFACE

The last twenty years have witnessed an explosion in the number of books dealing with chemical reaction engineering in one form or another. The subject has undergone a sea change since its beginnings as applied kinetics in Wisconsin more than a half century ago under the inspiring leadership of Olaf A. Hougen, and the process of change continues. To put it simply at the very outset, this book attempts to capture an emerging shift in the context of the growing need (and a hesitant willingness) to integrate chemistry with engineering.

The growth of chemical reaction engineering (CRE) has been marked by an unrestrained expansion of its scope. Thus what started out as an engineering science approach to the design of reactors for conventional organic and inorganic chemical reactions quickly began to encompass reactors for such diverse reactions as biochemical, electrochemical, photochemical, sonochemical, and a variety of other chemically oriented transformations. Lost in this seamless expansion of CRE has been the convergence of certain areas toward an objective focus. More specifically, these areas are increasingly being used for the common purpose of enhancing the rates and selectivities of organic reactions.

A quick perusal of the introduction makes the scope and content of the book obvious enough, but it does not tell the story of the title—and thus misses a personal point. The title captures more tellingly the main spirit of the attempt than perhaps any other: integrating the role of the synthetic organic chemist with that of the reaction engineer—something I had nurtured for a long time, particularly since I took over the stewardship of the National Chemical Laboratory in India. I realized how difficult it was to bring chemist and chemical engineer together to promote a common endeavor. Each talked a different language. Thus I had a list prepared of the questions (thirty in all as a starter) that must be answered, mainly by chemists, before processes developed by them could be

taken up for more detailed scale-up studies. This seems to have led to some informal conversational nuggets. The chemical engineers, I was told, called it the *essential* thirty, whereas the chemists dismissed it as the *dirty thirty*. The point is that the task has never been an easy one. If the present book can help reaffirm the need for more chemistry in engineering, can provide a broader perspective to practicing chemist and engineer alike and expose them to the principles and tools of the game delivered in a common kit, one can derive some satisfaction that the *game's afoot*.

The book consists of five parts. Part I is written as an introduction addressed equally to chemist and chemical engineer to bring them to a common level of appreciation of the approaches involved. Part II, dealing with catalysis in its various forms as applied to organic synthesis, is basically for the chemist and the technologist, but the chemical engineer should not be far behind in understanding it. In fact, in one of the chapters (concerned with the engineering aspects of catalysts), the roles might be reversed. Parts III and IV are concerned with the design of reactors for the various types of systems encountered in the manufacture of organic chemicals: bulk, intermediate, and fine. Part V describes various strategies for enhancing the rates of organic reactions. It was conceived as a basic element of organic synthesis engineering (OSE), and therefore was written with the elaboration it deserves. Anything less might have been diminishing.

A feature of this book is the preponderance of examples, qualitative and quantitative, to illustrate an application or justify a strategy. The motivation for this is the simple adage that nothing convinces more than an example or, in the more engaging words of Mark Twain: "Few things are harder to put up with than the annoyance of a good example."

OSE was not a topic that one day suddenly erupted in my mind. It made its embryonic beginnings in the creative vigor and discipline of my associations at NCL and is being nurtured to this day in the equally friendly and stimulating atmosphere of Iowa State University. I had the advantage of associating with a number of outstanding students and colleagues at NCL. B. D. Kulkarni, R. V. Chaudhari, V. R. Choudhary, (the late) R. A. Rajadhyaksha, S. D. Prasad, S. Ravikumar, A. N. Gokarn, and S. S. Tambe, among several other students, have contributed greatly to the development of chemical reaction engineering at NCL (and in India). This advantage fully spilled over to my stay at ISU. Leigh Hagenson, Ore Sofekun, Sridhar Desikan, Sanjeev Naik, Jennifer Anderson, Holger Glatzer, and Justinus Satrio have all helped, often without asking, and have added new dimensions to OSE, mainly through their efforts in the areas of phase transfer catalysis, sonochemistry, and microphase engineering. Three other graduate students, Meiyu Shen, R. Venkataraman, Kristine Bendixen, and Chandrika Mulakala have also been helpful. If the truth be told, I am not sure to this day whether I learned more from my students at NCL and ISU or they from me.

In addition to the help rendered to me in a general way by many of my students and some colleagues as indicated above through their long associations with me, I have also received invaluable help on specific aspects of the book from several chemists and engineers from India and USA (including some of my

students). I would like specially to mention R. V. Chaudhari (homogeneous catalysis, and three-phase reactors), K. Jayaraman (electroorganic synthesis engineering), A. Basha (electroorganic synthesis engineering), V. K. Gaikar (multifunctional reactors and other strategies), Leigh Hagenson (biphasing, use of ultrasound, and preparation of figures and tables), S. Desikan (phase-transfer catalysis, miscellaneous strategies, and solution of some problems), S. Naik (phase transfer catalysis and miscellaneous strategies), P. Ratnasami (catalysis), A. V. Rama Rao (asymmetric synthesis and some general aspects of organic synthesis), P. J. Reilly (bioorganic engineering), S. Rajappa (certain aspects of catalysis), V. K. Jayaraman (certain aspects of bioorganic engineering), A. B. Pandit (use of ultrasound), V. S. Moholkar (use of ultrasound), and H. Glatzer and J. Satrio (reading and criticism of the final manuscript and help in many other ways). I am grateful to all of them for their assistance, without which I should have been hard put to write some of the crucial chapters of the book.

I would also like to acknowledge the inspiration I derived in a general way from M. M. Sharma and R. A. Mashelkar through several impromptu sessions of stimulating discussions—which, together with many other circumstances of similar quality, has been an important factor in providing an integrated form to this book.

I would like to express my appreciation to Terry King, chair of Chemical Engineering at ISU for the larger part of my stay here so far, for his constant courtesy, continued encouragement, and for not overloading me with administrative responsibilities. The same courtesy and help are being extended by the present chair, Charles Glatz, and I am equally grateful to him. I must mention here that it was Peter Reilly who first suggested that I come to ISU after retiring from NCL, and Richard Seagrave made this possible. I am deeply appreciative of their interest. I would like specially to thank the late Maurice Larson for his professional collaboration (the results of which are presented in one of the chapters of the book) and for his many personal kindnesses. My grateful thanks are also due to Ed Lightfoot, a true friend of over three decades, who has helped me in many ways, particularly since my arrival in this country over ten years ago, and to Rutherford Aris from whom I have learned much.

The typing of this manuscript must have been a nightmare for Linda Edson, what with my ceaseless corrections, changes of format, and endless additions, modifications, and deletions of figures and tables. I am grateful to her for her patience, and most of all for making sense out of my scribblings, often unintelligible even to me.

The writing of this book has been plagued by delays arising largely from bouts of temporary ill health. I am sincerely grateful to Bob Rogers, senior editor of Oxford University Press during that period, for putting up with these delays and encouraging me with expressions of continued interest in the completion of this book. I am also thankful to Cynthia Garver for her patience and help during the production of the book.

My biggest source of strength has been my family: son Deepak, daughter Sandhya, son-in-law Sankar, and grandchildren Rahul and Priya. Their care and affection have sustained me through the protracted evolution of this book.

More chemistry in CRE? Well, what's new about that? It is being done without much fuss over the years. In the words of Stephen Jay Gould: "If these paeans and effusions were invariably true, I could compose my own lyrical version of the consensus and end this book forthwith." The essential truth is not in question, however. It has been expressed and occasionally illustrated in articles and books, but there has always been a gulf between the expression and the thematic demonstration of many of the strategies in a single consolidated effort. And it is in the strength of such a connected demonstration of these strategies superimposed on a canvas of CRE principles that this book will, one hopes, find its greatest appeal.

Had this book been a hastily written one, the inevitable errors might have been brazenly attributed to time and condoned by an understanding readership. As things are, time and those who assisted me are blameless. I bear the full responsibility for any errors and would be glad to have them brought to my attention.

Ames, Iowa L. K. D.
April 1999

CONTENTS

Chapter 1 Introduction and Structure of the Book, 3
   Why Organic Synthesis Engineering, 3
   The Organic Chemicals Ladder and the Role of Catalysis, 4
   Process Intensification, 8
   Structure of the Book, 9
   Internal Organization of Chapters, 13

PART I REACTIONS AND REACTORS IN ORGANIC SYNTHESIS: BASIC CONCEPTS

Chapter 2 Rates and Equilibria in Organic Reactions: The Thermodynamic and Extrathermodynamic Approaches, 17
   Basic Thermodynamic Relationships and Properties, 17
   Thermodynamics of Reactions in Solution, 26
   The Extrathermodynamic Approach, 28
   Extrathermodynamic Relationships between Rate and Equilibrium Parameters, 30

Chapter 3 Estimation of Properties of Organic Compounds, 36
   Correlation Strategy, 37
   Basic Properties, 37
   Thermodynamic and Equilibrium Properties, 38
   Reaction Properties, 43
   Transport Properties, 46
   Surface Tension and Ultrasonic Velocity, 51

Chapter 4 Reactions and Reactors: Basic Concepts, 52
  Reaction Rates, 52
  Stoichiometry of the Rate Equation, 56
  Reactors, 58
  Transport between Phases, 77
  Laboratory Reactors, 81

Chapter 5 Complex Reactions, 85
  Reduction of Complex Reactions, 86
  Rate Equations, 92
  Selectivity and Yield, 95
  Simultaneous Homogeneous and Catalytic (or Autocatalytic) Reactions, 110

Notation to Part I, 113

References to Part I, 118

PART II CATALYSIS IN ORGANIC SYNTHESIS AND TECHNOLOGY

Chapter 6 Catalysis by Solids, 1: Organic Intermediates and Fine Chemicals, 125
  Modified Forms of the More Common Catalysts, 127
  Zeolites, 129
  Heteropolyacid Catalysts, 143
  Clays in Catalysis: A Class of Supported Reagents, 145
  Solid Superacids, 149
  Solid Base Catalysts, 149
  Metallic Glass Catalysts, 150
  Ion-Exchange Resins, 152
  Catalysis by Inclusion: Cyclodextrins, 154
  Titanates, 155
  Catalysts for Selected Classes of Reactions, 156
  "Heterogenized" Homogeneous Catalysts, 163
  Role of Solvent in Catalysis by Solids, 168

Chapter 7 Catalysis by Solids, 2: The Catalyst and Its Microenvironment, 171
  Modeling of Solid Catalyzed Reactions, 171
  Role of Diffusion in Pellets: Catalyst Effectiveness, 183
  Effect of External Mass and Heat Transfer, 201
  Combined Effects of Internal and External Diffusion, 204
  Relative Roles of Mass and Heat Transfer in Internal and External Diffusion, 205
  Regimes of Control, 206
  Eliminating or Accounting for Transport Disguises, 207
  Experimental Reactors, 210

Chapter 8 Homogeneous Catalysis, 213
    Homogeneous versus Heterogeneous Catalysis, 214
    Formalisms in Transition-Metal Catalysis, 214
    The Operational Scheme of Homogeneous Catalysis, 222
    The Basic Reactions of Homogeneous Catalysis, 223
    Main Features of Transition-Metal Catalysis in Organic Synthesis:
      A Summary, 226
    Important Classes of Reactions with Industrial Examples, 228
    General Kinetic Analysis, 238

Chapter 9 Asymmetric Synthesis, 243
    Basic Definitions/Concepts, 244
    Methods of Preparing Pure Enantiomers, 248
    Asymmetric Synthesis, 258
    Homogeneous Asymmetric Catalysis, 260
    Heterogeneous Asymmetric Catalysis, 276

Notation to Part II, 280

References to Part II, 285

PART III REACTOR DESIGN FOR HOMOGENEOUS AND
         FLUID-SOLID (CATALYTIC) REACTIONS

Chapter 10 Reactor Design for Simple Reactions, 305
    Plug-Flow Reactor with Recycle, 305
    Mixed-Flow Reactors in Series, 309
    Semibatch Reactors, 315
    Varying Volume Reactors, 326
    Measures of Mixing: Comparison of the Series, Recycle, and Variable
      Volume Reactors, 330
    Comments, 331

Chapter 11 Reactor Design for Complex Reactions, 333
    Batch Reactor, 334
    Plug-Flow Reactor, 339
    Continuous Stirred Tank Reactor, 340
    Reactor Choice for Maximizing Yields and Selectivities, 343
    Optimum Temperatures or Temperature Profiles for Maximizing
      Yields and Selectivities, 348

Chapter 12 Reactor Design for Solid-Catalyzed Fluid Phase Reactions, 357
    Fixed-Bed Reactor, 357
    Fluidized-Bed Reactor, 377
    Reactor Choice for a Deactivating Catalyst, 390

Chapter 13 Mixing, Multiple Solutions, and Forced Unsteady-State
   Operation, 396
   Role of Mixing, 396
   Multiple Solutions, 409

Notation to Part III, 417

References to Part III, 423

PART IV FLUID-FLUID AND FLUID-FLUID-SOLID REACTIONS
   AND REACTORS

Chapter 14 Gas-Liquid Reactions, 431
   The Basis, 431
   Diffusion Accompanied by an Irreversible Reaction of General Order, 433
   Effect of Temperature, 441
   Discerning the Controlling Regime in Simple Reactions, 443
   Diffusion Accompanied by a Complex Reaction, 443
   Simultaneous Absorption and Reaction of Two Gases, 448
   Absorption of a Gas Followed by Reaction with Two Liquid-Phase
      Components, 457
   Extension to Complex Rate Models: Homogeneous Catalysis, 458
   Measurement of Mass Transfer Coefficients, 464
   Examples of Gas-Liquid Reactions, 465

Chapter 15 Liquid-Liquid and Solid-Liquid Reactions, 468
   Liquid-Liquid Reactions, 468
   Solid-Liquid Reactions, 477

Chapter 16 Gas-Liquid and Liquid-Liquid Reactor Design, 490
   A Generalized Form of Equation for All Regimes, 490
   Classification of Gas-Liquid Contactors, 493
   Reactor Design for Gas-Liquid Reactions, 495
   Reactor Choice for Gas-Liquid Reactions, 504
   Liquid-Liquid Contactors, 510
   Stirred Tank Reactor: Some Practical Considerations, 515

Chapter 17 Multiphase Reactions and Reactors, 517
   Analysis of Three-Phase Catalytic Reactions, 518
   Design of Three-Phase Catalytic Reactors, 526
   Types of Three-Phase Reactors, 533
   Collection and Interpretation of Laboratory Data for Three-Phase
      Catalytic Reactions, 547

Examples of Three-Phase Catalytic Reactions in Organic Synthesis
and Technology, 548
Three-Phase Noncatalytic Reactions, 549

Notation to Part IV, 559

References to Part IV, 565

PART V STRATEGIES FOR ENHANCING THE RATES OF ORGANIC REACTIONS

Chapter 18 Biphasic Reaction Engineering, 575
  Equilibria in Biphasic Reactions, 576
  Solvent Selection, 585
  Role of Aqueous Phase Volume, 587
  Matching the pH Optima for Reaction Equilibrium and Catalyst Activity, 588
  Modeling of Biphasic Biocatalytic Systems, 589
  Use of Biphasing in Organic Synthesis, 595
  Examples of Biphasing, 601

Chapter 19 Phase-Transfer Reaction Engineering, 606
  General Features, 607
  Kinetics and Modeling of Soluble PTC, 612
  Immobilized PTC, 625
  Industrial Applications of PTC, 641

Chapter 20 Bioorganic Synthesis Engineering, 647
  Microbes and Enzymes, 648
  Kinetics and Modeling of Bioreactions, 653
  Bioreactors, 662
  Examples of the Use of Biocatalysts in Organic Synthesis, 674

Chapter 21 Electroorganic Synthesis Engineering, 682
  Basic Concepts and Definitions, 683
  Modeling of Electroorganic Reactions and Reactors, 693
  Reaction Modeling, 693
  Reactor Modeling, 695
  Scale-Up of Electrochemical Reactors, 704
  Industrial Electrochemical Cell Components and Configurations, 706
  Selected Examples of Industrial Electroorganic Synthesis, 707

Chapter 22 Sonoorganic Synthesis Engineering, 711
  Devices for Producing Ultrasound, 712
  Homogeneous Reactions, 714
  Heterogeneous Reactions, 722

xviii Contents

    Ultrasound in Organic Synthesis, 726
    Scale-Up of Sonochemical Reactors, 733
    Design of Cavitation Reactors, 737
    Sonochemical Effect without Ultrasound: Hydrodynamic Cavitation, 739

Chapter 23 Microphase-Assisted Reaction Engineering, 744
    Classification of Microphases, 744
    Analysis of Microphase Action, 746
    Microphase as a Chemical (Catalytic) or Physical Sink, 749
    Microphase as a Physical Carrier, 757
    Microphase as a Source—Reactant as Microphase, 758
    Solid Product as a Microphase—Microphase-Assisted Autocatalysis, 759

Chapter 24 Membrane-Assisted Reactor Engineering, 765
    General Considerations, 766
    Modeling of Membrane Reactors, 773
    Operational Features, 783
    Comparison of Reactors, 788
    Examples of the Use of Membrane Reactors in Organic Technology and Synthesis, 789

Chapter 25 Multifunctional Reactor Engineering, 792
    Reaction-Extraction, 793
    Distillation-Reaction, 802

Chapter 26 Other Important (and Some Lesser Known) Strategies of Rate Enhancement, 815
    Photochemical Enhancement, 815
    Enhancement by Micelles, 827
    Microwave-Assisted Organic Synthesis, 831
    Organic Synthesis in Supercritical Fluids, 839
    Use of Hydrotropes (Hydrotropic Solubilization), 843
    Combinatorial Strategies, 847
    Use of Micromixing and Forced Unsteady-State Operation, 855

Notation to Part V, 856

References to Part V, 874

A Process Overview, 899

Epilogue, 903

Acknowledgments for Figures and Tables, 905

Index, 909

ORGANIC SYNTHESIS ENGINEERING

Chapter 1

Introduction and Structure of the Book

> The game's afoot!
> William Shakespeare: Henry V

WHY ORGANIC SYNTHESIS ENGINEERING?

A large part of the chemical industry is concerned with organic chemicals from simple to highly complex structures. In dealing with relatively simple structures, there does not appear to be any need usually for a deeper understanding of chemistry than that to which an engineer is normally exposed. Most reaction engineering texts are written with this basic assumption. Catalysis, which is invariably an integral part of the reaction engineer's arsenal, has been limited to the production of large volume chemicals which are often relatively simple in structure. Increasing attempts by chemists today to extend the use of catalysis to the production of medium and small volume chemicals has triggered a change in perspective that augers well for a closer liaison between chemists and engineers. We examine this a little further below by defining an *organic chemicals ladder*, and the merging roles of the two in exploiting this ladder, particularly for chemicals stacked on its intermediate rungs.

Another change that is taking place is the increasing role of *process intensification*, nowhere more evident than in the production of organic chemicals. Process intensification means improvement of a process, mainly the reaction, by any possible means, to increase the overall productivity. This usually takes the form of reaction rate enhancement by extending known or emerging laboratory techniques to industrial scale production. These techniques can be engineering intensive, chemistry intensive, or both. Examples are the use of ultrasound (sonochemistry), light (photochemistry), electrons (electrochemistry), enzymes (biotechnology), agents for facilitating a reaction between immiscible phases (phase-transfer catalysis), microparticles (microphase engineering), membranes (membrane reactor engineering), a second phase (biphasing), combinations of

reactions with different techniques of separation (multifunctional or *combo* reactor engineering), and mixing. Their use in the production of medium and small volume chemicals like pesticides, drugs, pharmaceuticals, perfumery chemicals, and other consumer products is being increasingly explored both by industry and academe. Some of these techniques have progressed little beyond the laboratory stage, although they have been a part of the synthetic organic chemist's repertoire for a number of years. Thus, in addition to the use of catalysis in its various forms, this book will also explore different techniques of reaction rate and/or selectivity enhancement. All of these techniques and operational innovations will be overlaid on a canvas of reaction engineering principles for both homogeneous and heterogeneous systems.

THE ORGANIC CHEMICALS LADDER AND THE ROLE OF CATALYSIS

All organic chemicals are produced from nature's three basic resources, petroleum, coal, and biomass. Of these, petroleum continues to be the dominant resource, followed by biomass. Coal-to-chemicals has always been a subject of debate, marked by sporadic justifications of neglect to perceptions of possible use, depending on such imponderable factors as international politics, economics, and the energy crunch. Taking petroleum as the source, the first step is to convert it to aromatics, alkenes, and paraffins. Their production usually involves the use of solid catalysts in vapor-phase reactions producing several hundred tons of each per day. They are chemically simple molecules such as ethylene, propylene, butylene, benzene, toluene, and naphthalene. These chemicals are then converted to next level chemicals such as ethylene oxide, propylene oxide, chlorobenzene, nitrobenzene, and benzenesulfonic acid. These second level chemicals are converted to the next (third) level chemicals, lower in volume of production but even more complex in structure. This is repeated over a few more stages to obtain a series of higher order chemicals known as *organic intermediates*, till finally the penultimate intermediate in each case is converted to a consumer product. Thus an organic chemicals ladder is established.

Two such ladders leading to a number of drugs and pharmaceuticals are shown in Tables 1.1 and 1.2. Two main features of these ladders are noteworthy. As already noted, the molecules become progressively more complex as one goes down the ladder, and except in rare cases such as thermal cracking, processes for first level chemicals of the ladder are usually catalytic. A large fraction of the second level processes are also catalytic, but those at the lower end tend to be largely noncatalytic. A great change is taking place now in which the chemicals in the lower rungs of the ladder are also sought to be made catalytically because of the many advantages of catalytic processes such as the avoidance of corrosive reagents, easy separation of product, greater safety, and often greater selectivity (particularly where zeolites or asymmetric catalysts are used). Several serial publications such as *Heterogeneous Catalysis and Fine Chemicals* and *Catalysis of Organic Compounds* regularly report on the development of catalysts for making known or (occasionally) new organic molecules. There is, indeed, a

Table 1.1 Examples of a chemical ladder starting from aromatics (obtained from crude oil cracking) and ending in a variety of pharmaceuticals

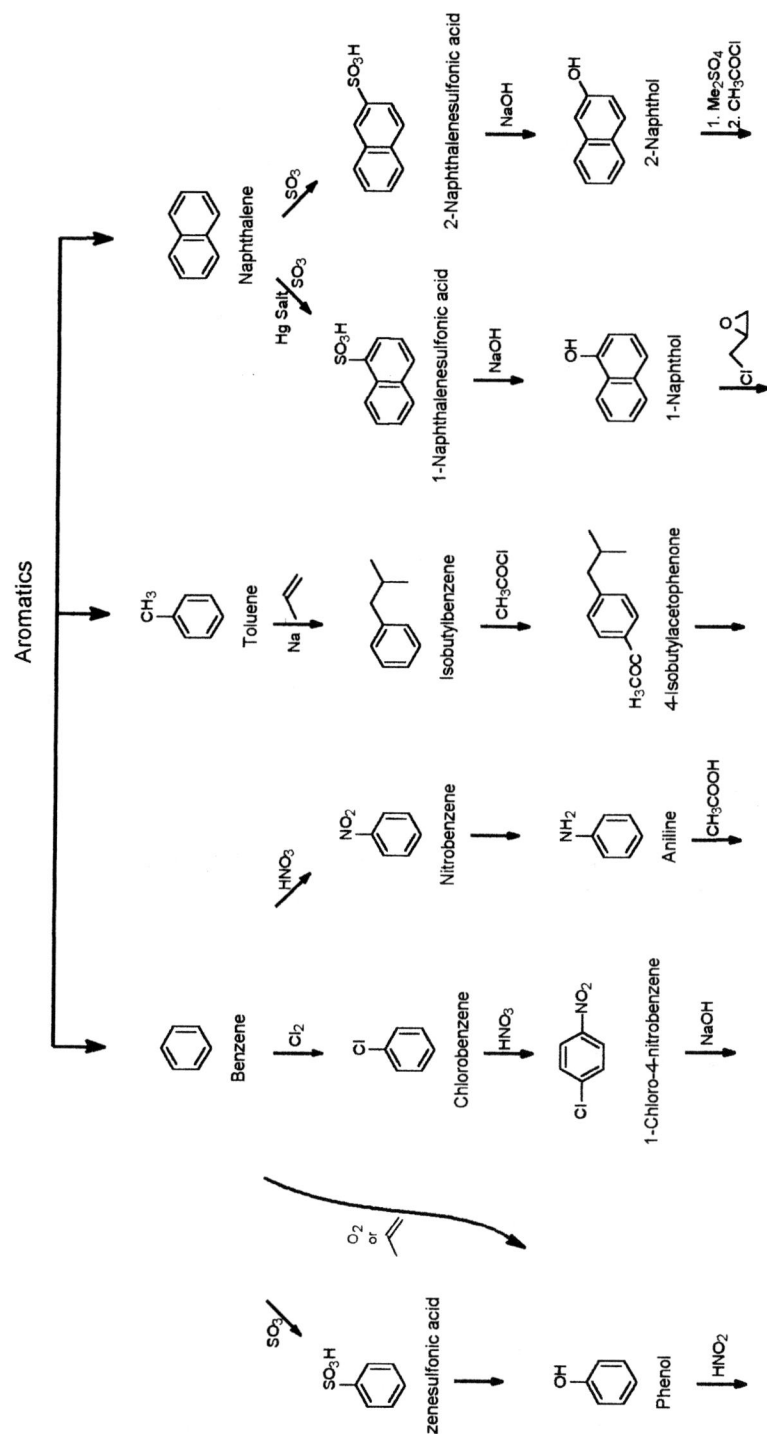

(continued)

Table 1.1 (continued)

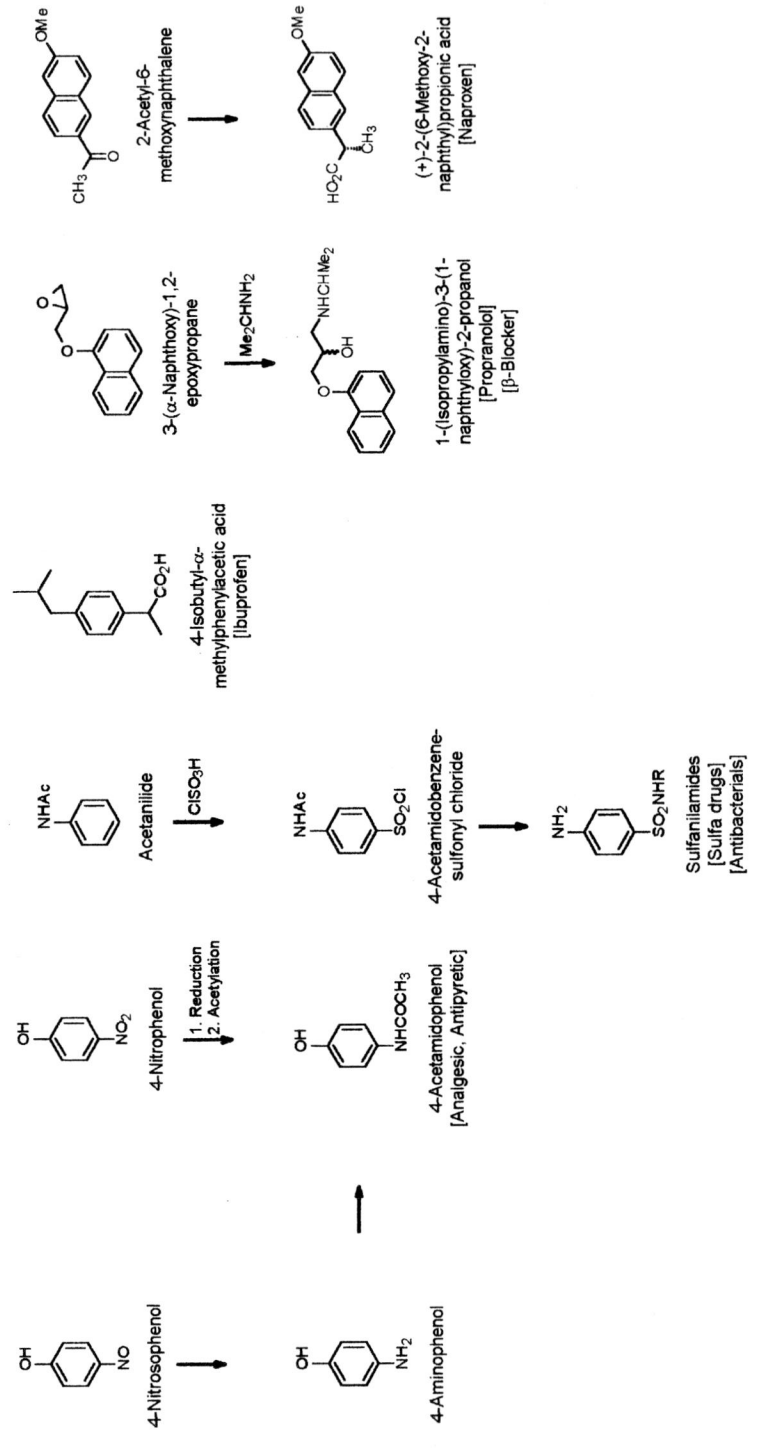

Table 1.2 Examples of a chemical ladder starting from alkenes (obtained from crude oil cracking) and ending in a variety of pharmaceuticals

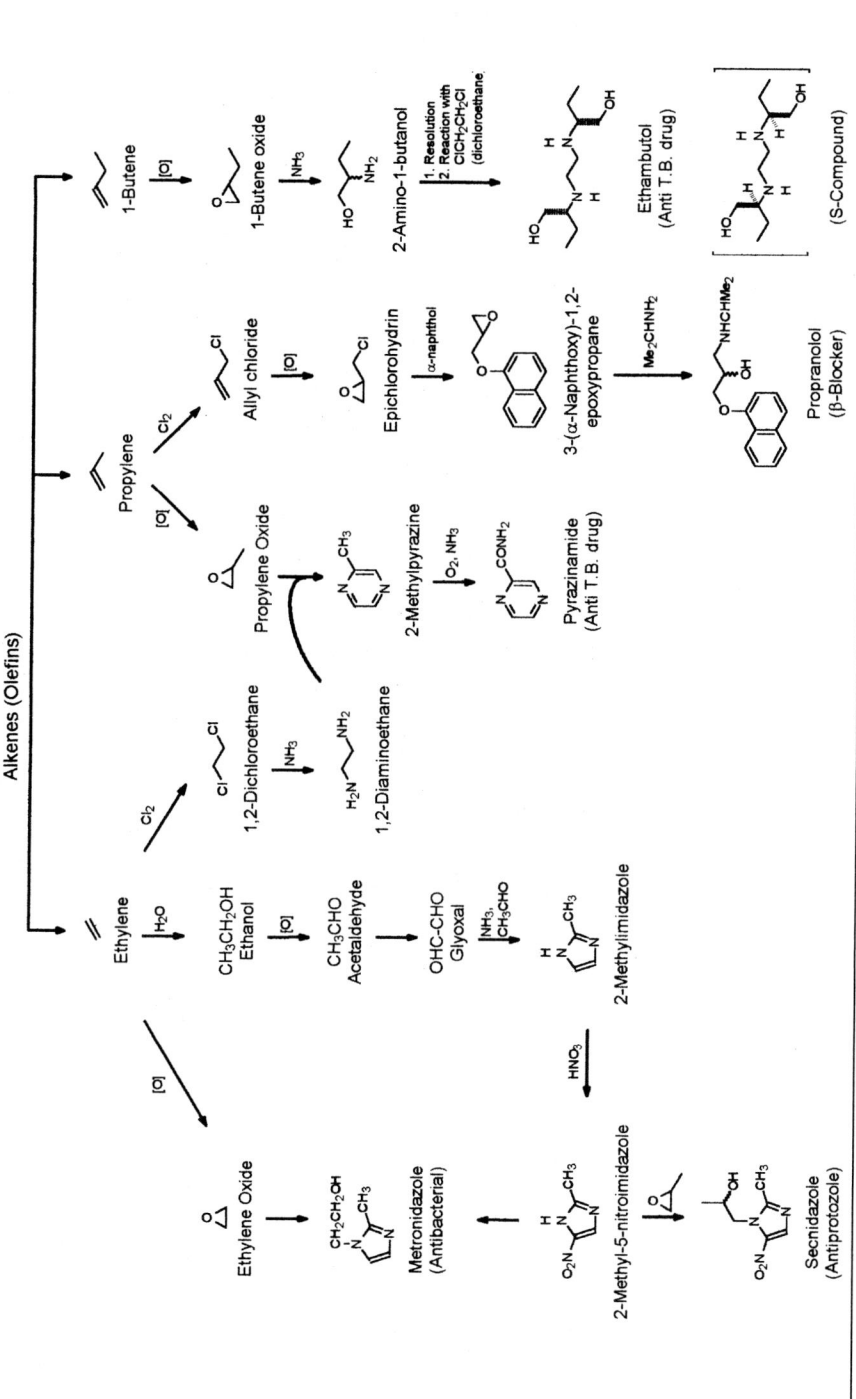

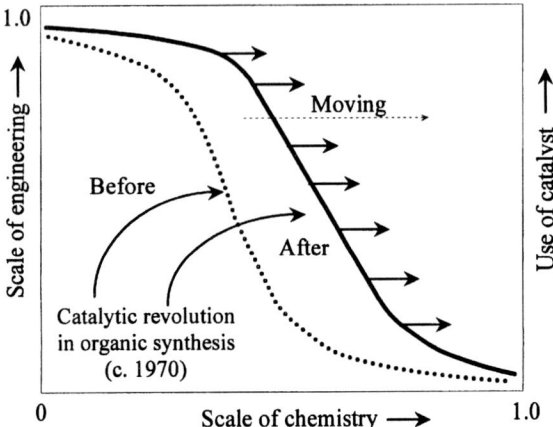

Figure 1.1 Qualitative representation of the increasing role of catalysis in organic synthesis.

spurt of activity in this area in many chemistry and chemical engineering laboratories throughout the world. This trend has spurred research on the engineering aspects of catalytic processes for intermediate and small volume chemicals. Thus the earlier concept of catalysis as the bedrock of large volume productions is slowly but definitely giving place to a much broader role for catalysts in organic synthesis. This trend is qualitatively depicted in Figure 1.1.

PROCESS INTENSIFICATION

Rates of organic reactions can be enhanced and selectivities can frequently be improved by using many techniques besides catalysis. In some cases reactions can be initiated, in some just facilitated, and in many enhanced. They can be used individually, in various combinations (usually binary), or as supplements to catalysis. A list of these techniques was mentioned in the introduction. Unfortunately, although almost all of them are known to perform very well in the chemist's lab, their scale-up to industrial size remains a daunting issue. As an example, the use of light for enhancing or initiating a reaction is well known, but the rational scale-up of reactors for carrying out such reactions has, ironically, reentered the zone of uncertainty from one of fair certainty, thanks mainly to a better understanding of the principles of photoreaction engineering. In the case of ultrasound, the uncertainty has never left, so that today we are still groping for reasonable methods of scale-up. The situation is more encouraging in some of the other areas of enhancement, such as the use of enzymes, membranes, or a second liquid phase. Intensive research in the area of multifunctional reactors has brought their design from first principles within reach.

In this book we attempt to cover the use of many of these techniques for facilitating organic reactions or enhancing their rates. The structural details of the treatment are given in the next section.

STRUCTURE OF THE BOOK

The basic philosophy of this book can be described in terms of the twofold approach depicted in Figure 1.2: establish a broad framework (the three unshaded boxes) and then superimpose on this framework a number of strategies for reaction rate enhancement (the central shaded box). The details are covered in five parts:

Part I   Reactions and reactors in organic synthesis: Basic concepts (Chapters 2–5)
Part II   Catalysis in organic synthesis and technology (Chapters 6–9)
Part III Reactor design for homogeneous and fluid-solid catalytic reactions (Chapters 10–13)
Part IV   Fluid-fluid and fluid-fluid-solid reactions and reactors (Chapters 14–17).
Part V   Strategies for enhancing the rates of organic reactions (Chapters 18–26)

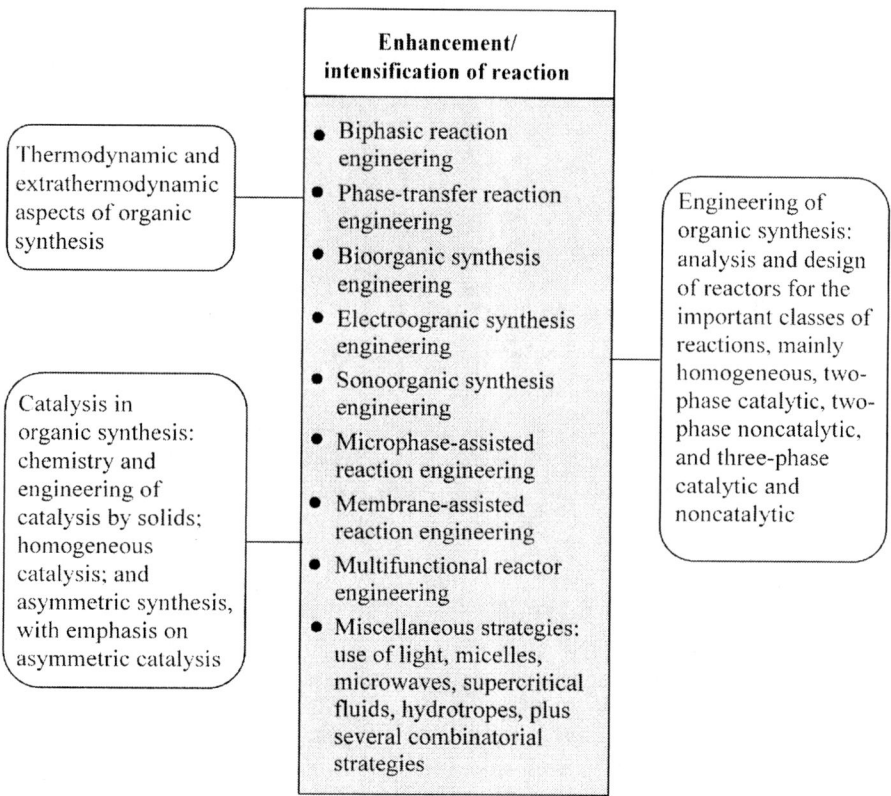

Figure 1.2 Structure of the book (indicative also of a broad morphology of organic synthesis engineering).

Reactions and Reactors in Organic Synthesis: Basic Concepts

This part starts with Chapter 2, which covers the basic aspects of thermodynamics with an introduction to the concept of extrathermodynamics. The latter is necessary because many organic reactions occur in the liquid phase and our ability to predict the behavior of liquid-phase reactions is severely limited. In this approach, enough microscopic detail is added to the macroscopic approach of conventional thermodynamics, thus enabling a certain degree of prediction. The chemical engineer is usually not exposed to these methods. Following logically from the extrathermodynamic approach, we move on to Chapter 3 in which the most useful feature of this approach, the method of group contributions, is considered; we construct procedures for estimating the thermodynamic properties of a variety of organic compounds. The need for knowledge of these properties in reactor design is obvious enough and beyond question, but the need for methods to estimate them in light of their direct availability from computer programs and databases can be a matter of debate. I firmly believe that inclusion of these methods in a book of this kind is neither a superfluous nor an outdated exercise but a necessary one, and hence a brief description of these is presented in Chapter 3. The properties include the thermodynamic properties of relevance to design and also the major transport properties of fluids.

Chapter 4 outlines the elements of stoichiometry, rates, and reaction and reactor analysis. The reader is introduced to the concept of ideal reactors and the principles of their design for simple reactions. Extensions of these ideal reactors form the subject matter of Part III but are anticipated at this stage as an introductory setting for that part. Chapter 5 extends the analysis to complex reactions, but the design of reactors for complex reactions is deferred to Part III.

Catalysis in Organic Synthesis and Technology

This part constitutes a significant and rather extensive departure from other books on chemical reaction engineering. Because catalysis is emerging as an important component of organic synthesis, several aspects of catalysis are described in this part: catalysis by solids in the production of organic chemicals in general (Chapter 6); homogeneous (liquid-phase) catalysis (Chapter 8); and asymmetric catalysis used specifically for the selective production of biologically active isomers (Chapter 9).

In addition, it is important to recognize that catalysis by solids involves diffusion of reactants from fluid bulk to the catalyst surface, as well as diffusion within the solid matrix. The latter invariably occurs simultaneously with the reaction, and the former is usually (but not necessarily rigorously) treated as an independent precursor to it. Thus any analysis of catalysis by solids is based on understanding its action under the physical influence of the microenvironment in which it functions. Catalysis by solids actually occurs on the catalyst surface and constitutes the *surface field problem*, which is the core of its action. Diffusion of reactant within its matrix is an *internal or intraparticle field problem*, and its transport from bulk to surface is an *external or interphase field problem*.

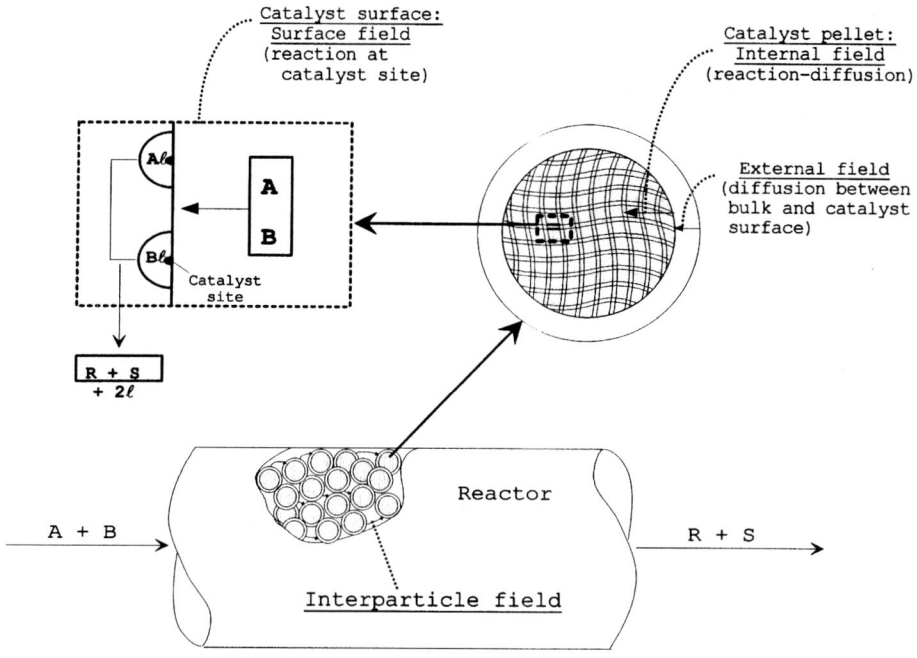

Figure 1.3 The four main components of catalytic reactor analysis and design.

All of these occur within the microenvironment of a single particle or pellet and constitute the contents of Chapter 7.

Consideration of the various field problems just described provides the groundwork for the design of catalytic reactors in Part III. Within the reactor, the happenings in the microenvironment of a pellet translate into the overall behavior of a *bed of pellets*, fixed or fluidized. In other words, the consequences of reaction in a given microenvironment are transferred to a neighboring microenvironment and the process is continued till an integrated effect is realized in a flowing stream of reactant(s) at the exit of the bed. This is basically an *interparticle field problem* and constitutes the core of reactor design.

Thus the complete design of a catalytic reactor involves all of these considerations which are schematically shown in Figure 1.3. Chapter 7 of Part II is concerned with the surface, internal, and external field problems, and the interparticle field problem is considered in Part III devoted to reactor design. Note that these observations are restricted to the use of solid catalysts, which constitutes only a part of the much wider scope of the book.

Reactor Design for Homogeneous and Fluid-Solid Catalytic Reactions

This part is basically an extension of the methods of reactor analysis and design introduced in Chapter 4 and covers the design of homogeneous reactors for both simple and complex reactions (Chapters 10 and 11). Considerable attention is

given to semibatch reactors because of their importance in industrial organic synthesis. Further, these reactors are not normally dealt with in adequate detail in many formal texts on reaction engineering.

In view of the growing importance of catalysis in organic synthesis, the design of catalytic reactors assumes considerable significance. Although published work and industrial practice seem to indicate that catalytic reactors are largely restricted to large scale production, they can also be used for medium and small volume production. The design of fluidized bed reactors, in particular, is aimed at very large scale production, but my personal experience has shown that the procedures apply equally to intermediate-scale production. Therefore, the treatment of catalytic reactors in Chapter 12 follows that of many existing books but is structured differently to fit into the pattern of presentation adopted in this book.

The treatment of reactors in the preceding chapters was based on the assumption of ideality with respect to mixing and the existence of a single solution for a given set of operating conditions (i.e., of a single steady state). Although deviations from these assumptions are usually not severe, they can be, and hence a separate chapter (Chapter 13) is devoted to a brief description of mixing and multiple steady states. Another kind of unsteady-state operation, one in which the direction of reactant flow within a reactor is reversed in cycles, is referred to as forced unsteady-state operation (FUSO). Although such forced cycling increases the conversion and therefore can be regarded as a strategy for process intensification, it is included in Chapter 13 because it is also a form of unsteady-state operation. The analysis of these situations tends to be highly mathematical. Thus only a qualitative treatment is attempted with minimal resort to mathematics.

Fluid-Fluid and Fluid-Fluid-Solid Reactions and Reactors

Heterogeneous reactions involving more than one phase are a common feature of organic synthesis. Among them are gas-liquid reactions, liquid-liquid reactions, solid-liquid reactions, and gas-liquid-solid reactions (e.g., the slurry reaction). The main features of various two-phase reactions are covered in Chapters 14 and 15, and Chapter 16 discusses *reactors* for such systems. Chapter 17 covers the analysis of three-phase *reactions and reactors*.

Strategies for Enhancing the Rates of Organic Reactions

Parts I and II provide material with which both chemist and chemical engineer should be comfortable. Parts III and IV are aimed more toward the chemical engineer, although the text itself (excluding the many equation-laden tables provided) should offer no great hurdles to the chemist. Part V is unique in the sense that it is never a part of any conventional reaction engineering book, except by way of brief excursions to illustrate applications.

In this part of the book we cover several strategies for reaction rate enhancement. Many of these have reached a stage where they merit consideration as

special chapters. Chapters 18 to 25 belong to this category. Chapter 26 includes a number of strategies that are being increasingly explored, but seem currently to lack the potential of the others.

The strategies covered in these chapters are all written in practically the same format: a brief introduction followed by references to major works and texts; an outline of principles; methods of reactor design with illustrative problems where considered necessary; and a listing of examples of their use, accompanied sometimes by brief descriptions of a few important ones. The lists are quite extensive in many cases; consequently, the literature cited is also extensive.

INTERNAL ORGANIZATION OF CHAPTERS

This book is an attempt to combine chemistry with engineering, in the hope that someday soon there will be a real integration of the two. With this objective in mind, the various chapters are organized so that some of them have a chemistry focus and the rest an engineering focus. The organization of material in each is such that specialized portions, mostly equations, conditions, and criteria, are compressed into tables (or figures), so that the nonspecialist has the option to skip them altogether. Some of the chapters might serve as refreshers to the chemical engineer and others to the chemist. The central theme of the book is reaction engineering but with adequate chemistry and catalysis included to present a total picture of organic synthesis engineering.

A few words on Notation and References complete our introduction to this book. Each of Parts I–IV has a theme common enough for including a consolidated list of references and notation at the end. Part V consists of chapters that are quite distinctive, and hence each chapter contains its own notation and list of references. However, in the interest of uniformity, these are listed (chapterwise) at the end of the part.

PART I

REACTIONS AND REACTORS IN ORGANIC SYNTHESIS

Basic Concepts

It is good to make things simple, but not simpler.
Albert Einstein

Chapter 2

Rates and Equilibria in Organic Reactions

The Thermodynamic and Extrathermodynamic Approaches

In any reversible reaction such as

$$\nu_A A + \nu_B B \leftrightarrow \nu_R R + \nu_S S \qquad [2.1]$$

the system inevitably moves toward a state of equilibrium, or maximum probability. This equilibrium state is very important in analyzing chemical reactions because it defines the limit to which any reaction can proceed.

Organic reactions, particularly those constituting a synthetic scheme for a fine chemical, usually involve molecules reacting in the liquid phase. The effects of reactant structure and of the solvent (medium) in which the reaction occurs (the solvation effects) are not included in the conventional macroscopic approach to thermodynamics. Therefore, the treatment of liquid-phase reactions tends to be less exact than that of gas-phase reactions involving simpler molecules without these influences.

A convenient way of approaching this problem is to start with the conventional macroscopic or thermodynamic approach and add enough microscopic detail to allow for the effects of solute (reactant) structure and the medium. This approach is called the *extrathermodynamic* approach and may be regarded as bridging the gap between the two rather disparate fields of rates and equilibria represented by kinetics and thermodynamics, respectively. Such an approach is particularly useful in organic synthesis and forms the subject matter of this chapter.

BASIC THERMODYNAMIC RELATIONSHIPS AND PROPERTIES

Basic Relationships

An important consideration in process calculations is the change that results in the basic thermodynamic properties, internal energy (U), enthalpy (H), Helmholtz

work function (A), and Gibbs free energy (G) when a closed system of constant mass moves from one macroscopic state to another. For a homogeneous fluid, these change equations can be expressed in terms of four exact differential equations, which then can be written in difference form by employing the operator Δ to represent the change from state 1 to state 2:

$$dU = dA + PdS, \quad \Delta U = \Delta A + P\Delta S \tag{2.1}$$

$$dH = dU + PdV, \quad \Delta H = \Delta U + P\Delta V \tag{2.2}$$

$$dA = dU - TdS, \quad \Delta A = \Delta U - T\Delta S \tag{2.3}$$

$$dG = dH - TdS, \quad \Delta G = \Delta H - T\Delta S \tag{2.4}$$

Of these, the enthalpy and free energy change equations are the most frequently used in the analysis of reactions.

Heats of Reaction, Formation, and Combustion

Consider the reaction

$$A + B \rightarrow R + S \tag{2.2}$$

The heat of reaction is given by

$$\Delta H_r = \Delta H_{fR} + \Delta H_{fS} - \Delta H_{fA} - \Delta H_{fB} \tag{2.5}$$

where the terms on the right refer to the enthalpies of formation of R, S, A and B, respectively. The standard heat of reaction, denoted by ΔH_r^0, is defined as the difference between the enthalpies of the products in their standard states and of the reactants in their standard states, all at the same temperature.

The enthalpy of formation, usually known as the *heat of formation*, of a compound is the heat evolved during its formation from its constituent elements. The enthalpies of formation of elements are assumed to be zero. Thus, for the special case of heat of formation, each of the terms on the right in Equation 2.5 is zero, and the heat of reaction becomes equal to the heat of formation.

Usually all of the enthalpies in reaction 2.2 refer to the reactants and products in the ideal gaseous state, and appropriate corrections must be made for a change of state if some of the components are in the liquid or solid state. To appreciate this fact fully, let us consider the combustion of a typical organic compound. Depending on the atoms in the molecule, the final products of combustion would be H_2O, CO_2, SO_2, N_2, and HX (where X is a halogen). The heat evolved in such a reaction is called the *heat of combustion*. Standard heats of combustion are often listed with H_2O in the liquid state (i.e., as water); thus a suitable correction must be made to get the values with all of the products in the gaseous state. This is illustrated in Example 2.1.

Errata Sheet

Organic Synthesis Engineering

L. K. Doraiswamy

On page 18, Equation 2.1 should be replaced by

$dU = TdS - PdV, \quad \Delta U = T\Delta S - P\Delta V$

On page 189, the equation in Table 7.4, column 6, row 2, should be

$$\varepsilon = \frac{1}{\varphi_1}\left(\frac{1}{\tanh(3\varphi_1)} - \frac{1}{3\varphi_1}\right)$$

On page 363, the second line of reaction E12.1.1 should be (A) → (B) where (A) is below nitrobenzene and (B) is below aniline.

On page 698, the third paragraph should read "Equations 21.40 and 21.39" and Equation 21.42 should be

$$i_A = nFk'[A]_0(1 - X_A)$$

EXAMPLE 2.1

Calculating the heat of reaction from the heats of combustion

The heats of combustion of gaseous methyl alcohol and dimethyl ether (with H_2O as liquid water) are 182.6 and 347.6 kcal/mol, respectively. Calculate the heat of reaction for the dehydration of methyl alcohol to methyl ether when the reactants and products are all in the gaseous state and when they are all in the liquid state.

The combustion of methyl alcohol and dimethyl ether is represented by the reactions

$$2CH_3OH(g) + 3O_2(g) \rightarrow 2CO_2(g) + 4H_2O(l) + 365.2 \qquad [E2.1.1]$$

$$(\text{i.e.}, 2 \times 182.6) \text{ kcal}$$

$$CH_3OCH_3(g) + 3O_2(g) \rightarrow 2CO_2(g) + 3H_2O(l) + 347.6 \text{ kcal} \qquad [E2.1.2]$$

Combining these,

$$2CH_3OH(g) \rightarrow CH_3OCH_3(g) + H_2O(l) + 17.6 \text{ kcal} \qquad [E2.1.3]$$

Thus the heat of reaction is $17.6/2 = 8.8$ kcal/mol of methy alcohol undergoing reaction, when all the reactants and products are in the gaseous state except water. If even water is to be in the gaseous state, we must subtract the heat required to vaporize one mole of water, its heat of vaporization. Thus we get

$$2CH_3OH(g) \rightarrow CH_3OCH_3(g) + H_2O(g) + 7.6 \text{ kcal} \qquad [E2.1.4]$$

Therefore, the heat of reaction with all the products in the gaseous state is 7.6/2 or 3.8 kcal mol of methyl alcohol reacting.

When the reactants and products are all in the liquid state, the heats of vaporization of methyl alcohol and methyl ether should also be considered. Thus we should subtract the heat absorbed in vaporizing two moles of methyl alcohol and add the heats evolved in condensing one mole each of dimethyl ether and H_2O:

$$2(\Delta H_r^0) = 7.6 - 2(8.4) + 4.8 + 10 = 5.6$$

$$(\Delta H_r^0) = 2.8 \text{ kcal/mol}$$

Implications of Liquid-Phase Reactions

The thermodynamic implications of reactions in the liquid state are important. Let us consider the case where a gas, liquid, or solid is dissolved in a solvent, and the products also remain in solution (i.e., the reaction occurs in the liquid phase). The method illustrated in the previous example is applicable to such cases. Because all of these involve energy changes associated with condensation, as well as dissolution and mixing with solvents, they are far more complicated than reactions in the gaseous state. As will be emphasized in the following discussion, the formal thermodynamic approach fails to give predictive correlations for such cases, and resort to empirical combinations of the microscopic effects of solvents with the formal macroscopic approach becomes necessary.

Free Energy Change and Equilibrium Constant

Standard free energy change and equilibrium constant

The concept of free energy is useful in defining the possibility of a reaction and in determining its limiting or equilibrium conversion. The formal definition of the equilibrium state of a chemical reaction is the state for which the total free energy is a minimum. Thus the well-known rule: reaction can occur if ΔG is negative; it cannot occur if ΔG is positive. Now we shall present the main features of the equilibrium state for ideal and nonideal gases.

IDEAL GASES

Consider reaction 2.1. The free energy change accompanying this reaction is given by the well-known equation

$$\Delta G_r^0 = -R_g T \ln K \tag{2.6}$$

where

$$K = \frac{k_{\text{forward}}}{k_{\text{reverse}}} = \frac{p_R^{\nu_R} p_S^{\nu_S}}{p_A^{\nu_A} p_B^{\nu_B}} \tag{2.7}$$

is the *thermodynamic equilibrium constant* of the reaction. It also represents a state where the two rates (not rate constants) are equal, so that the slightest parametric disturbance will drive it in any one direction.

NONIDEAL GASES

In applying the concepts just presented to nonideal gases, it is necessary to introduce quantities that may be regarded as nonideal gas equivalents of ideal gases. These are *fugacity, fugacity coefficient, partial fugacity, and partial fugacity coefficient*. The first is defined by the equation

$$f_j = \phi_j P \tag{2.8}$$

where f_j is the *pure-component fugacity* (the nonideal gas equivalent of pressure) and ϕ_j is the *pure-component fugacity coefficient* which is independent of composition and is a measure of nonideality. Pure-component fugacities can be estimated from generalized charts of f/P as a function of reduced pressure and temperature (see Sandler, 1977).

When nonideal gases form an ideal solution, the equation for K becomes

$$K = \frac{p_R^{\nu_R} p_S^{\nu_S}}{p_A^{\nu_A} p_B^{\nu_B}} \left(\frac{\phi_R \phi_S}{\phi_A \phi_B} \right) \tag{2.9}$$

On the other hand, when they form a nonideal solution, the fugacity coefficient is replaced by the partial fugacity coefficient $\bar{\phi}_j$ defined as

$$\bar{f}_j = \bar{\phi}_j y_j P \tag{2.10}$$

Notice that $\bar{\phi}_j$ is not independent of composition, and hence it is not possible to estimate it with any degree of certainty.

TEMPERATURE DEPENDENCE OF K

The following thermodynamic expression can be derived for K as a function of T:

$$R_g \ln K = -\frac{I_1}{T} + I_2 + \sum_{j=1}^{N} v_j \left(a_j \ln T + \frac{b_j T}{2} + \frac{c_j T^2}{6} + \frac{d_j T^3}{12} \right) \quad (2.11)$$

where I_1 is the constant of the equation

$$\Delta H^0 = I_1 + \sum_{j=1}^{N} v_j \left(a_j T + \frac{b_j T^2}{2} + \frac{c_j T^3}{3} + \frac{d_j T^4}{4} \right) \quad (2.12)$$

and can be estimated from a known value of ΔH^0 at any one temperature, usually 298 K, and those of a_j, b_j, c_j, and d_j from the method described in Chapter 3. I_2 can similarly be obtained from Equation 2.11 from the value of K at a single temperature, also usually 298 K. These two equations can be used to calculate ΔH^0 and K as functions of T.

Equilibrium compositions in gas phase reactions

Of primary importance in conducting any reaction is knowledge of the equilibrium conversion and the composition of the reaction mixture at equilibrium. Noddings and Mullet (1965) considered the most general reaction

$$v_A A + v_B B + v_C C + v_D D \leftrightarrow v_F F + v_G G + v_H H + v_I I + v_L L \quad [2.3]$$

and give extensive tables of equilibrium composition versus equilibrium constant for several simplified forms of this general stoichiometry. Listed in Table 2.1 are

Table 2.1 Expressions for equilibrium conversion for a few common types of reactions

Reaction	Expression for K^a	Pressure dependence
1. $A \rightleftarrows R$	$\dfrac{X}{1-X}$	No (no vol. change)
2. $A \rightleftarrows R + S$	$\dfrac{X^2 P}{(1-X)(1+X)}$	Direct (vol. increase)
3. $A + B \rightleftarrows R$	$\dfrac{X(2-X)}{(1-X)^2 P}$	Inverse (vol. decrease)
4. $A + B \rightleftarrows R + S$	$\dfrac{X^2}{(1-X)^2}$	No (no vol. change)
5. $A \rightleftarrows 2R$	$\dfrac{2XP}{(1-X)(1+X)}$	Direct (vol. increase)

^{a} *Example:* Reaction 2. $K = K_N \left(\dfrac{P}{N_t}\right)^{\Delta v}$ where $K_N = \dfrac{N_R N_S}{N_A}$.
For 1 mol A and a conversion of X, $K = \dfrac{X^2 P}{(1-X)(1+X)}$.

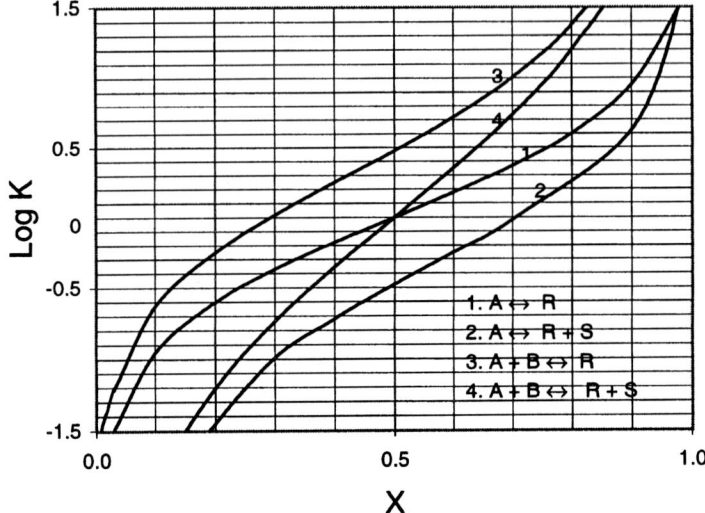

Figure 2.1 Equilibrium conversion as a function of the equilibrium constant for four common types of reactions.

the conversion-equilibrium constant relationships for five commonly encountered reaction types in organic synthesis/technology. Plots of K versus X_A for four of these reactions are shown in Figure 2.1. Because there is a volume change in reactions 2, 3, and 5 of the table, the pressure dependence of the X_A–K relationship for one of these reactions is illustrated in Figure 2.2.

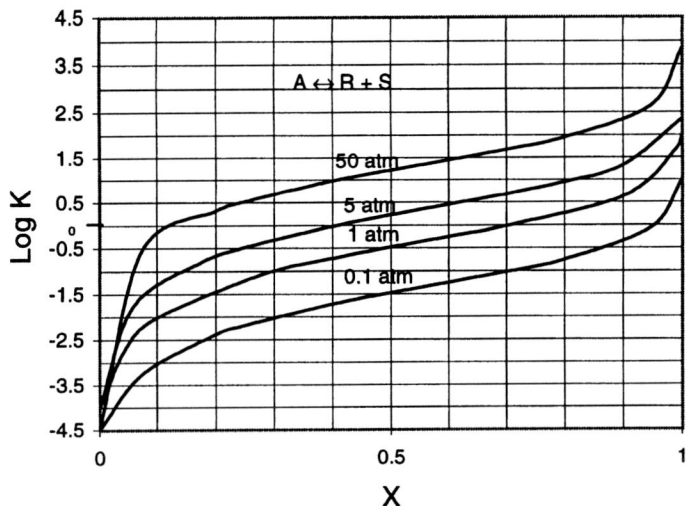

Figure 2.2 Equilibrium conversion as a function of the equilibrium constant and pressure for a reaction with volume change.

Accounting for condensed phase(s)

In the reactions considered in Table 2.1, all of the components are in the ideal gaseous state. To treat reactions where all of the components are liquid, we assume that the system is ideal and its pressure is not much different from the component $p_\nu s$. Hence (mole fraction × vapor pressure) may be used in place of partial pressure, and then the equilibrium composition can be calculated from equations similar to those in Table 2.1. However, where one of the components is a liquid, its activity with respect to a standard state of an ideal gas can be taken as the partial pressure at the temperature of the system. Thus, if R is in the liquid phase in reaction 3 of Table 2.1, the equilibrium constant is given by

$$K = \frac{p_{\nu R}}{p_A p_B} \qquad (2.13)$$

Alternatively, its free energy of formation can be calculated from

$$\Delta G_f^0(l) = \Delta G_f^0(g) + R_g T \ln(P_\nu) \qquad (2.14)$$

and then this value can be used in calculating the modified ΔG_r^0 and the corresponding K.

Thermodynamics can also be usefully employed in analyzing the role of a catalyst in organic synthesis. A typical example is the Gatterman–Koch reaction.

EXAMPLE 2.2

Thermodynamics of the Gattermann–Koch reaction

In this reaction, a −CHO group is introduced into a molecule such as benzene with the assistance of a catalyst of the type (HCl + AlCl$_3$) according to the reaction

$$C_6H_6(l) + CO(g) \rightarrow C_6H_5CHO(l) \qquad [E2.2.1]$$

It has been postulated that formyl chloride is a transient intermediate in this reaction. The free energy change of the reaction leading to its formation

$$HCl(g) + CO(g) = ClCHO(g) \qquad [E2.2.2]$$

may be calculated as $(-43.17) - (-22.78 - 32.82) = 12.43$ kcal/mol (Dilkey and Ely, 1949). In view of the positive value of ΔG_{298}^0, this compound clearly could not be isolated, but the fact that it is formed is indirectly substantiated by the negative value of ΔG_{298}^0 for the following reaction:

$$ClCHO(g) + C_6H_6(g) = C_6H_5CHO(g) + HCl(g) \qquad [E2.2.3]$$

| ΔG_{f298}^0 | -43.17 | 30.99 | -1.07 | -22.78 |

$\Delta G_{r298}^0 = -11.67$ kcal/mol

The calculations presented are not conclusive enough. Hence another line of investigation was pursued by Dilkey and Eley (1949). The thermodynamics of formation of the catalytic complex with different chlorides such as those of Al, Sn, Fe, and Sb was studied to see if complex formation does occur and if so whether a consistent ranking of the

catalysts might be discerned through increments in the estimated values of the enthalpy, entropy, and free energy, or equilibrium constant. For this purpose, two quantities were measured: (1) the calorimetric enthalpy of mixing of the solid halide with liquid benzaldehyde, that is, the enthalpy of complex formation and (2) the concentration of the species at equilibrium. This yielded the equilibrium constant K in the mixing experiments in which complex formation occurs according to the reaction

$$C_6H_5CHO(l) + \tfrac{1}{2}M_2Cl_6(s) = C_6H_5CHO \cdot MCl_3(s) \qquad [E2.2.4]$$

where $M = $ Al, Sn, Fe, Sb. The results (Stull et al., 1969) clearly show that complex formation does occur. Further, the increments in the enthalpy, free energy (or equilibrium constant), and entropy all point to the same order of performance of the different catalysts.

Complex Equilibria

Voluminous literature exists on the calculation of reaction equilibria in complex networks. The following two procedures are particularly useful: simultaneous solution of the equilibrium equations and minimization of the free energy.

Simultaneous solution of equilibrium equations

A simple and direct method of determining the equilibrium composition of a complex reaction is to solve simultaneously all of the equations comprising the complex network. The actual number of equations to be solved is equal to the number of independent reactions of the network. Chapter 5 deals formally with the treatment of complex reactions, and the method outlined therein can be applied to this problem. Where the number of reactions is relatively small, say 2 or 3, simple, less formal methods can be used as illustrated in the following example.

EXAMPLE 2.3

Equilibrium composition in the complex reaction
$CH_3Cl(g) + H_2O(g) \leftrightarrow CH_3OH(g) + HCl(g)$,
$2CH_3OH(g) \leftrightarrow (CH_3)_2O(g) + H_2O(g)$ (see Stull et al., 1969)

The equilibrium constant of the first reaction is given as 0.00154 and of the second as 10.6 at 600 K, and it is desired to calculate the equilibrium composition of the mixture produced by reacting methyl chloride with water.

Let us start with one mole each of methyl chloride and water. Assuming that X moles of HCl and Y of dimethyl ether are formed, the amounts of the different constituents at equilibrium would be $CH_3Cl = (1 - X)$, $HCl = X$, $CH_3OH = X - 2Y$, $(CH_3)_2O = Y$, and $H_2O = (1 - X + Y)$. Thus the equilibrium compositions of the two reactions can

be expressed as

$$0.00154 = \frac{(X-2Y)X}{(1-X)(1-X+Y)} \quad \text{(E2.3.1)}$$

$$10.6 = \frac{Y(1-X+Y)}{(X-2Y)^2} \quad \text{(E2.3.2)}$$

Solution of these equations gives

$X = 0.048$

$Y = 0.009$

Then the equilibrium composition can be readily calculated from these values.

EXTENSION TO A NONIDEAL SYSTEM

One can extend the treatment to a nonideal reaction by using Equation 2.9. Further, for any complex scheme such as

$$A + B \leftrightarrow 2R \quad [2.4.1]$$

$$3R + B \leftrightarrow 3S \quad [2.4.2]$$

it is more convenient to use the concept of *extent of reaction* (moles converted by a given reaction) than conversion (see Chapter 5). Thus we define

ξ_1 = moles of A or B converted by reaction 2.4.1

ξ_2 = moles of B converted by reaction 2.4.2

based on which the number of moles of each component at a given extent of each reaction (starting with N_{A0} moles of A and N_{B0} moles of B) can be written and added up to give the total number of moles $N_t = N_{A0} + N_{B0} - \xi_2$. Then the equilibrium constants of the two reactions can be written as

$$K_1 = \left(\frac{N_R^2}{N_A N_B}\right) K_{\phi 1}, \quad K_2 = \left(\frac{N_S^3}{N_B N_R^3}\right)\left(\frac{N_t}{P}\right) K_{\phi 2} \quad (2.15)$$

where

$$K_{\phi 1} = \frac{\phi_R^2}{\phi_A \phi_B}, \quad K_{\phi 2} = \frac{\phi_S^3}{\phi_B \phi_R^3} \quad (2.16)$$

and the ϕs represent the pure component fugacity coefficients. Equations 2.15 and 2.16 can be readily solved by using any good nonlinear equation solver.

Minimization of free energy

The total free energy G of a reacting system reaches a minimum at equilibrium. For an ideal gaseous system, the component partial pressures of any reaction,

simple or complex, are related to this free energy by the equation

$$\frac{G}{R_g T} = \sum_{j=1}^{N_t} N_j \left[\left(\frac{\Delta G_{fj}^0}{R_g T} \right) + \ln P + \ln y_j \right] \tag{2.17}$$

where y_j is the mole fraction of j given by $p_j = y_j P$, that is, N_j/N_t. Now we find the number of moles of each component at *equilibrium* by requiring that they produce a minimum in $G/R_g T$ in the above equation and simultaneously satisfy the elemental balance

$$\sum_{j=1}^{c} n_{ij} N_j - a_i = 0, \quad i = 1, 2, 3, \ldots, m \tag{2.18}$$

where n_{ij} is the number of atoms of element i in molecular species j, a_i is the total moles of element i in the system, c is the total number of molecular species, and m is the total number of elemental species. Mathematical techniques for accomplishing this are discussed in several books (Prausnitz, 1969; Walas, 1985) and will not be considered here.

THERMODYNAMICS OF REACTIONS IN SOLUTION

Most treatments of reactor design focus on the gaseous state. Many organic reactions are carried out in the liquid state, often in solvents, and hence in this section we consider the thermodynamics of reactions in solution.

Partial Molal Properties

A number of chemical reactions involve at least two chemical species, sometimes more. Therefore, any thermodynamic property M of a system of, say, N components $(N_1, N_2, \ldots, N_N)$ must be written in terms of the contributions of the individual species. The total property value is given by NM (or simply by M) whose functional dependence on other parameters can be defined as

$$M = f(T, P, N_1, N_2, \ldots, N_N) \tag{2.19}$$

Now we formally define a *partial molal quantity* represented by $\bar{M}_i$ as

$$\bar{M}_i = \left[\frac{\partial M}{\partial N_i} \right]_{P,T,N \neq N_i} \tag{2.20}$$

where M is the molal thermodynamic property of a solution of constant composition. Based on this definition, the following equation, named after Gibbs and Duhem, can be derived:

$$M = N_1 \bar{M}_1 + N_2 \bar{M}_2 + \cdots + N_i \bar{M}_i + \cdots \tag{2.21}$$

Physically, the quantity $\bar{M}_i$ is equal to the limit of the change in property M of the solution per mole of i added at constant temperature and pressure, as the amount of i added approaches zero. This is a very useful equation in dealing with dilute solutions.

Medium and Substituent Effects on Standard Free Energy Change, Equilibrium Constant, and Activity Coefficient

Where reactions are carried out in solution, two situations are possible: the solution can be dilute or concentrated. For dilute solutions, the concentration can be used directly as an exact measure of its activity. This is often justified in organic synthesis because normally solvents are used in large excess. But where the reactant concentration is high, its activity in solution cannot be replaced by concentration without an appropriate correction factor. Note that in both cases we would be dealing with partial molal quantities.

For dilute solutions, the dependence of the partial molal free energy of any component i on its concentration is expressed by the equation

$$\bar{G}_i = \bar{G}_i^0 + R_g T \ln [i] \tag{2.22}$$

In analogy with Equation 2.6, the following relationship can be derived:

$$\Delta \bar{G}_r^0 = -R_g T \ln K_c \tag{2.23}$$

where $\Delta \bar{G}^0$ is the partial molal standard free energy change for the reaction and K_c is the concentration based equilibrium constant. For any reaction, such as 2.1, K_c is given by

$$K_c = \frac{[R]^{\nu_R}[S]^{\nu_S}}{[A]^{\nu_A}[B]^{\nu_B}} \tag{2.24}$$

Note that the constant K_c is different from the K of Equation 2.7 based on partial pressures, but one can be converted to the other by using the gas law, $P = R_g T[i]$. Clearly, where all the reaction orders are unity, the $R_g T$ terms cancel out, and the two equilibrium constants are equal.

Equation 2.23 applied to concentrated solutions becomes

$$\Delta \bar{G}_r^0 = -R_g T \ln K_a \tag{2.25}$$

with

$$K_a = K_\gamma K_c \tag{2.26}$$

where K_γ is a correction factor based on the activity coefficients γ of the different components and for reaction 2.1 is given by

$$K_\gamma = \frac{\gamma_R^{\nu_R} \gamma_S^{\nu_S}}{\gamma_A^{\nu_A} \gamma_B^{\nu_B}} \tag{2.27}$$

It is useful to correlate any thermodynamic property M as the difference δ_M between the value for a given solute–solvent combination and that for a selected "standard" combination. The *operator* δ can represent the effect on the reaction of changing the solvent structure for a given solute (the *solvent operator*) or of changing the solute structure for a given solvent (the *solute* or *substituent operator*).

Thus, considering ΔG, equations for the solvent and solute operators are

SOLVENT OPERATOR

$$\delta_S \Delta \bar{G}_i^0 = \Delta \bar{G}_{i,\text{solvent S}}^0 - \Delta \bar{G}_{i,\text{standard solvent}}^0 \qquad (2.28)$$

Recall that the operator Δ refers to the effect of a chemical reaction, or the change in free energy accompanying a reaction, and should be distinguished from the newly defined solvent operator δ_S.

SUBSTITUENT OPERATOR

$$\delta_R \Delta \bar{G}^0 = \Delta \bar{G}_R^0 - \Delta \bar{G}_{R_0}^0 \qquad (2.29)$$

where R and R_0 represent the two substituent groups. R_0 is usually but not necessarily the hydrogen atom.

A striking example of the solvent effect is revealed in the keto–enol tautomerization of benzoyl camphor (Hammett, 1940). A similar substituent effect is seen in the acid dissociation constants in various solvents (Davis and Hetzer, 1958).

Comments

Despite extensive theoretical and experimental studies on solvent and solute effects based on the operators just defined, there are no general theoretical models available yet that can predict these effects with any certainty. From a practical point of view, therefore, one must look to methods other than those based on the application of formal thermodynamics. Thus we turn to the extrathermodynamic approach.

THE EXTRATHERMODYNAMIC APPROACH

Basic Principles

The basis of the extrathermodynamic approach is simple. Consider any property such as the heat of formation. The effect, for instance, of an amino group in benzene is assumed to be the same as its effect in any aromatic compound (e.g. toluene or any of the xylenes). The principle is best illustrated for the ionization constant of carboxylic acids in a number of solvents:

$$\text{RCOOH} \xleftrightarrow{\text{water}} \text{RCOO}^- + \text{H}^+ \qquad [2.5]$$

$$\text{RCOOH} \xleftrightarrow{\text{alcohol}} \text{RCOO}^- + \text{H}^+ \qquad [2.6]$$

The extrathermodynamic relationship is

$$\log K_{a,w} = m \log K_{a,a} \qquad (2.30)$$

where $K_{a,w}$ and $K_{a,a}$ are the ionization constants of a given acid in water and alcohol, respectively. Or the relationship can be

$$\log K_{a,R} = n \log K_{a,R'} \qquad (2.31)$$

in a given solvent such as water, where R and R′ represent different substituent groups.

In its most primitive form, the structure of the parent molecule is inconsequential. Thus if the value of the substituent group is known, it can be used in any molecule regardless of its structure. Linear relationships based on such similarity of effects are referred to as extrathermodynamic relationships, because the approach does not call for information on the microscopic nature of the structures. To that extent it retains the fundamental character of the thermodynamic approach. A typical linear relationship with respect to the substituent effect in the rearrangement of aminotriazoles, i.e.,

$$\begin{array}{c}\text{Ar}\\|\\ \text{N}\\ \diagup\,\diagdown\\ \text{N}\quad\quad\text{C—NH}_2\\ \|\quad\quad\|\\ \text{N———C—}\phi\end{array}\ \xrightleftharpoons[\text{glycol}]{\text{in ethylene}}\ \begin{array}{c}\text{H}\\|\quad\quad\text{H}\\ \text{N}\quad\quad|\\ \diagup\,\diagdown\\ \text{N}\quad\quad\text{C—N—Ar}\\ \|\quad\quad\|\\ \text{N———C—}\phi\end{array} \qquad [2.7]$$

is

$$\Delta_R \Delta \bar{H}^c = n \delta_R \Delta \bar{S}^0 \qquad (2.32)$$

The results (Lieber et al., 1957) provide remarkable proof of this relationship.

The Group Contributions or Additivity Principle

As a result of the fortuitous simplifying circumstance mentioned earlier, a molecule can be divided into "action" and "neutral" zones. The latter zone usually occupies the bulk of the molecule's size. The word "action" is more appropriate than "reaction" as used by many authors, because the concept is applicable even where no reaction occurs. Changes in properties are assumed to occur only as a result of changes in the action zone. Rules can be formulated for the effect of different substituent groups in the action zone. These are the *additivity rules*, or the *rules of group contributions*. It must be noted, however, that in the interest of greater accuracy in properties estimation, it may often be necessary to introduce higher order approximations that violate the neutrality of the neutral zone; but one pays a price for this: an increase in the number of empirical parameters.

Two general aspects of additivity methods are noteworthy:

1. Consider a compound with two functional groups, such as succinic acid $HOOC(CH_2)_2COOH$. If we are interested in only one of the COOH groups acting as the reactive site and the other as substituent, we must divide the observed rate constant by a *statistical factor* of 2. For m reactive sites, the statistical factor is m, and if there are two reagents with m and n reactive sites, the statistical factor is mn.
2. The same order of approximation should be used for all *part-structures* (generally called *groups*). Use of even a single lower order approximation would tend to subvert the accuracy of the higher order approximations.

Let us take, for instance, the value of N in NO_2 and NH_2. If different values are used for N in the two groups, then a similar higher order approximation should also be used for other atoms, such as O in COOH and CO.

A simple way to address the question of the order of approximation is to assume that the contributions from any two groups are exactly additive if they are sufficiently apart within a given structure (Benson and Buss, 1958). Thus for a molecule CH_3XCH_3 where X is the intervening structure, the zeroth-order approximation would be CH_3CH_3 (i.e., no intervening structure). For a first-order approximation, X could be a simple atom such as N or O or a CH_2 group. Higher order approximations would involve correspondingly more CH_2 groups and even branched chains.

EXTRATHERMODYNAMIC RELATIONSHIPS BETWEEN RATE AND EQUILIBRIUM PARAMETERS

The Polanyi and Brönsted Relationships

Consider any two reactions in a family of reactions such as isomerization of hydrocarbons or dehydration of alcohols. The difference in the activation energies of the two reactions is assumed to be directly proportional to the difference in their heats of reaction:

$$\delta_P E = -\alpha \delta_P(-\Delta H_r^0) \tag{2.33}$$

where δ_P may be regarded as a Polanyi operator. Using this relationship, the following important expression can be derived relating the rate constant of a reaction to its thermodynamic equilibrium constant:

$$k = (\text{constant})K^\alpha \tag{2.34}$$

This is referred to as the Brönsted relationship. The proportionality constant and the constant α are characteristic of a given family of reactions.

When applied to reactions catalyzed by acids or bases, the Brönsted relationship has a slightly different connotation. Examples of these are the base-catalyzed halogenation of ketones and esters and the acid-catalyzed dehydration of acetaldehyde hydrate. For an acid-catalyzed reaction,

$$k_a = (\text{constant})K_a^\alpha \tag{2.35}$$

where K_a is the dissociation equilibrium constant of the acid (say HA) given by

$$HA \xleftrightarrow{K_a} A^- + H^+ \qquad [2.8]$$

Equation 2.35, which is the Brönsted relationship for acid-catalyzed reactions, states that the rate constant of a reaction catalyzed by an acid is proportional to some power of the dissociation equilibrium constant of the acid used as the catalyst.

Similarly, for reactions catalyzed by a base,

$$k_b = (\text{constant})K_b^\beta \tag{2.36}$$

where k_b is the rate constant of the reaction and K_b is the dissociation constant of the base.

A striking example of the application of Equation 2.35 is the isomerization of substituted 5-amino-triazoles in ethylene glycol shown in reaction 2.7. Figure 2.3 is a log-log plot of the rate constant versus equilibrium constant for this reaction (Leffler and Grunwald, 1963).

The Hammett Relationship for Dissociation Constants

The Hammett linear free energy relationship is a widely used extrathermodynamic relationship for organic reactions. Although developed specifically for dissociation constants, in principle the method is applicable to any organic reaction.

Consider the family of reactions involving the ionization of benzoic acid in water followed by reaction with ethyl alcohol and those of *para* or *meta* substituted benzoic acids in water also followed by reaction with ethyl alcohol:

$$C_6H_5COO^- + C_2H_5OH \xrightleftharpoons{K_{a,0}} C_6H_5COOC_2H_5 + H_3O^+ \qquad [2.9]$$

$$m\text{- or } p\text{-}XC_6H_4COO^- + C_2H_5OH \xrightleftharpoons{K_a} m\text{- or } p\text{-}XC_6H_4COOC_2H_5 + H_3O^+ \qquad [2.10]$$

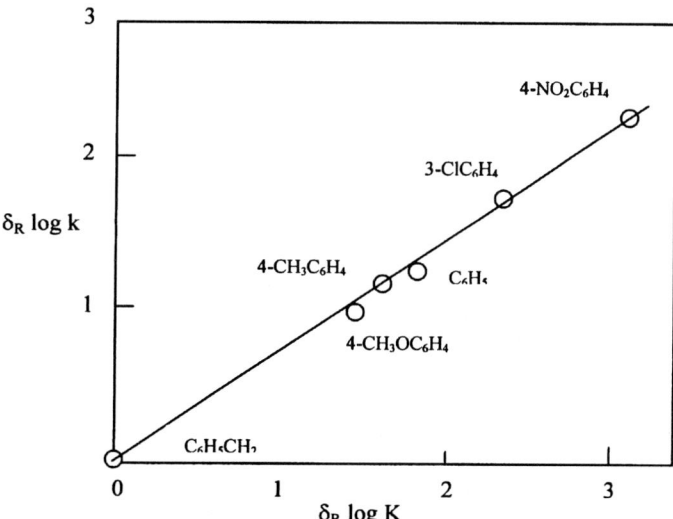

Figure 2.3 The rate–equilibrium relationship in 5-aminotriazole rearrangement (Leffler and Grunwald, 1963).

where $K_{a,0}$ and K_a are the acid dissociation constants for the two reactions. The basic feature of the method is that the ratio σ of the dissociation constants of the two reactions is used as the correlating parameter for other members of the family, provided that the reaction conditions are the same. Thus we can write

$$\log \frac{k}{k_0} = \log \frac{K_a}{K_{a,0}} = \rho\sigma \tag{2.37}$$

where the constant ρ is a function of the reaction and the reaction conditions used. A detailed compilation of σ values for various reactions is given by McDaniels and Brown (1958) and Shorter (1982).

Extrathermodynamic Approach to Selectivity

The selectivity for a given product in a complex reaction is an important practical consideration in carrying out a reaction. Let us first define the *reactivity* of a reagent. Clearly, for any bimolecular reaction, it has any meaning only in relation to the second reactant, the substrate. However, as long as the substrates are reasonably similar, each reagent exhibits a characteristic substrate-independent reactivity that enables a broad ordering of reagents. Therefore, we may choose (or postulate) a "standard substrate" and define reactivity as

$$\begin{bmatrix} \text{Reactivity of} \\ \text{a reagent} \end{bmatrix} \equiv \begin{bmatrix} \text{rate constant } k_0 \\ \text{for reaction with a} \\ \text{standard substrate} \end{bmatrix} \tag{2.38}$$

Consider the reaction

$$A + B \xrightarrow{k_I} R \qquad [2.11a]$$

$$A + C \xrightarrow{k_{II}} S \qquad [2.11b]$$

From a thermodynamic point of view, it is convenient in a complex reaction of this type to view selectivity (say for R) as the preference of A to reactant with B over C.

Theoretical analysis

If we designate the selectivity of A to B as S_1 and of A to C as S_2, then we can define S_1 as

$$S_1 = R_g T \ln\left(\frac{k_I}{k_{II}}\right) \tag{2.39}$$

Now we seek to obtain a relationship between reactivity represented by the overall reaction of A with B and C, its conversion to R and S, and its selectivity for B given by Equation 2.39. Clearly, as the overall reactivity increases, every encounter between A and B and between A and C would be successful leading to reaction with a selectivity of zero. On the other hand, continuous lowering of reactivity to zero can enhance selectivity to unity. Although not a universal law,

this is the basis of the commonly observed increase in selectivity with a decrease in conversion.

The reactivity–selectivity relationship can be analyzed by postulating the following with respect to the reactants: reactant A is designated as A_H if it is highly reactive (hot) and as A_C if it is less reactive (cold), and the second reactant (which is also referred to as the substrate) is designated as B_H if it is more reactive and as B_C if less. Thus the selectivity will vary depending on different combinations of A and B.

Consider the reactions

$$A_H + B_H \rightarrow \text{Products} \qquad [2.12.1]$$

$$A_H + B_C \rightarrow \text{Products} \qquad [2.12.2]$$

and

$$A_C + B_H \rightarrow \text{Products} \qquad [2.13.1]$$

$$A_C + B_C \rightarrow \text{Products} \qquad [2.13.2]$$

where the less reactive substrate B_C can be, say, benzene, and the more reactive one B_H can be a substituted benzene. Reactions 2.12.1 and 2 represent the effect of changing the substrate structure from hot to cold for a highly reactive (hot) A. Similarly, reactions 2.13.1 and 2 denote the effect of substrate structure for a less reactive (cold) A. The use of these reactions is illustrated in Example 2.4.

A prelude to reactive separation (Chapter 25)

An important situation arises in organic reactions when two bases compete for a proton. If the proton-donating compound is the reagent and the proton recipient is the substrate,

$$\underset{\text{Reagent}}{B_1 H^+} + \underset{\text{Substrate}}{B_2} \leftrightarrow B_1 + B_2 H^+ \qquad [2.14]$$

it is easy to see that because the transition state has the maximum free energy, the transition complex with the weaker base will be more stable than that with the stronger base. Thus the stronger acid reacts preferentially over the weaker acid. This is the basis for the separation of two acidic organic compounds (or bases) by a process of reactive separation. Such processes are considered in Chapter 25.

EXAMPLE 2.4

Grading the selectivity of a reagent (a carbonyl compound) for reaction between two competing bases: Base-catalyzed halogenation of ketones and esters

Consider the base-catalyzed halogenation of ketones and esters. In this reaction there is competition for a proton between a carbonyl compound and a catalyzing base.

$$\underset{\text{Reagent}}{\text{H}-\overset{\overset{\text{O}}{|}}{\underset{|}{\text{C}}}-\overset{\overset{\text{O}}{\|}}{\text{C}} + \text{RCO}^{\ominus}} \xrightleftharpoons{K} \underset{\text{Substrate}}{\overset{\text{O}}{\|}{\text{RCOH}} + \text{C}=\text{C}\overset{\text{O}^{\ominus}}{}}$$ [E2.4.1]

For a given carbonyl compound (reagent) reacting with a series of carboxylate ions (substrates), the rate constant k is related to the dissociation constant K_b of the base by the equation

$$\delta_R \log k = \beta' \delta_R \log K_b \tag{E2.4.1}$$

Because the equilibrium constant K of reaction E2.4.1 is proportional to K_b, we may write

$$\delta_R \log k = \beta \delta_R \log K \tag{E2.4.2}$$

which may be immediately recognized as a rate–equilibrium relationship.

Bell and co-workers (1949, 1952) reported extensive studies on the base-catalyzed halogenation of a series of ketones and esters and related the reactivity as measured by k_0 with the parameter β. We know that k_0 has to be defined for a standard substrate. In this particular case, it is defined as the rate constant for the halogenation of a hypothetical catalyzing base with a dissociation equilibrium constant K_b of 10^{-10}. Specifically, let us demonstrate the procedure for the reagent acetone. Equation (E2.4.2) can be written as

$$\log k_{B_1} - \log k_{B_2} = \beta(\log K_{B_1} - \log K_{B_2}) \tag{E2.4.3}$$

where B_1 and B_2 refer to the two bases (or substrates) used for the halogenation of acetone. Now, if we take B_2 as the standard base, we can write

$$\log k_{B_1} - \log k_{\text{standard}} = \beta(\log K_{B_1} - \log 10^{-10}) \tag{E2.4.4}$$

and calculate k_{standard} or k_0 directly from it. By repeating the procedure for a number of reagents, β can be obtained as a function of the reactivity k_0.

We recall that our objective is to relate reactivity to selectivity. For this purpose we have to establish a relationship between β and selectivity as follows. Consider two bases R_1COO^- and R_2COO^- represented by B_1 and B_2, respectively, that compete for reaction

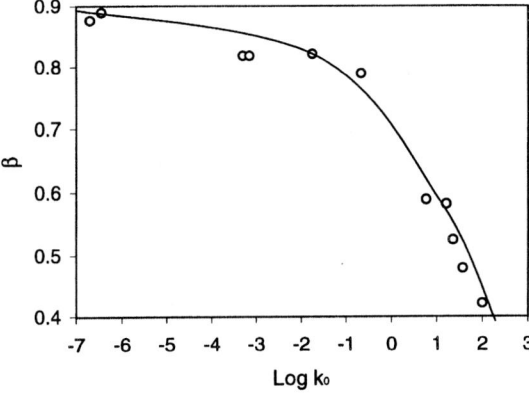

Figure 2.4 Effect of increasing reactivity k_0 of the carbonyl compound on the parameter β (redrawn from Leffler and Grunwald, 1963).

with a given carbonyl compound A. Reaction 2.12 or 2.13 may be written specifically for this case as

$$A + B_1 = R + S \tag{E2.4.2a}$$
$$A + B_2 = U + V \tag{E2.4.2b}$$

From Equation 2.39, the selectivity for substrate 1 can be written as

$$S_1 = R_g T \ln \frac{k_I}{k_{II}} = \beta R_g T \ln \frac{K_{B_1}}{K_{B_2}} \tag{E2.4.5}$$

Now, if the carbonyl compound is changed but B_1 and B_2 are kept the same, the selectivity of the carbonyl compound will vary in proportion to β, and the proportionality constant is $R_g T \ln(K_{B_1}/K_{B_2})$. Thus β can be regarded as a direct measure of selectivity.

A plot of β as a function of $\log k_0$ is shown in Figure 2.4. Note that the selectivity approaches the limits of unity or zero asymptotically as the reactivity becomes very small or very large. This confirms the previous theoretical observation.

Chapter 3

Estimation of Properties of Organic Compounds

A primary requirement of any reactor design or process development computation is knowledge of the major properties of the compounds involved. Although most of these can be obtained from the literature, there is still a need to estimate them from correlations.[1] The main difficulty is the large number of correlations proposed for a given property and the need to select the best from among them. No single correlation works with equally high precision under all conditions. On the other hand, correlations that can be used with acceptable levels of precision over a wide range of conditions are also available for a number of properties. The slight sacrifice of accuracy is often more than compensated for by the ease and generality of application of these methods. Our emphasis here will be on such correlations.

For a detailed treatment, reference should be made to books devoted exclusively to properties estimation. The book by Reid, Prausnitz, and Poling (1987), along with its earlier versions by Reid and Sherwood (1958, 1966) and Reid, Prausnitz, and Sherwood (1977), and the works of Janz (1958), Hansch and Leo (1979), and Lyman, Reehl, and Rosenblatt (1982) are noteworthy.[2] The following methods selected for a few properties are based in part on the recommendations contained in these treatises.

---

[1] As already noted in Chapter 1, it is often argued that, because most thermodynamic properties can be computed from readily available computer programs and databases, there is no longer any need for procedures to estimate them. Although this may be true from a practical point of view. I firmly believe that such a purely utilitarian view would produce a real gap in the armory of the technologist/engineer. Hence no apology is offered here for including a short chapter on this subject.

[2] The 1977 book of Reid, Prausnitz, and Sherwood is particularly useful.

CORRELATION STRATEGY

The two most important bases for formulating correlations for estimating the properties of organic compounds (indeed of any compound) are *the law of corresponding states* (LCS), and the *method of group contributions* (GC).

LCS is based on the concept that all substances exhibit identical properties under conditions equally removed from their critical states. The "equally removed" state for any property is usually expressed as the ratio of its value at that state to the value at the critical state and is referred to as the *reduced property*. Thus $T_r = T/T_c$, $P_r = P/P_c$, $V_r = V/V_c$ and $\eta_r = \eta/\eta_c$ are the reduced temperature, pressure, volume, and viscosity, respectively. If the simple ideal gas law $PV/R_g T = Z$, where Z is the compressibility, can be recast in terms of reduced properties as $P_r V_r / R_g T_r = Z/Z_c$, then the PVT behavior of all fluids can be represented as P_r versus T_r plots for different values of Z_r. Because Z_c is constant at 0.27, a very close approximation for many nonpolar compounds, this is equivalent to correlating Z as a function of T_r and P_r. The method can be extended to other reduced properties such as reduced viscosity.

The second method (GC), based on structure dependence, is an outgrowth of the extrathermodynamic approach outlined in Chapter 2 and encompasses a whole range of properties. Starting from zero or first-order approximations, these methods have been developed over the years to remarkably high levels of precision.

Methods for the following general classes of properties are presented here: basic properties, thermodynamic properties, reaction properties, and transport properties, but the treatment is restricted in each class to properties of direct relevance to our subject. Also included is a short section on estimating the velocity of ultrasound in single and mixed liquid media because of its importance in a subsequent chapter. It must be emphasized that, wherever possible, reported experimental values should be used because estimation methods can never be as precise.

BASIC PROPERTIES

The properties most commonly used as correlating parameters for most other properties are the critical properties T_c, P_c and V_c, and the boiling point T_b. Therefore, we begin by presenting procedures for estimating these basic properties.

Critical Properties

Of the many procedures available for estimating critical properties, those based on GC are the most accurate. The following relationships proposed by Joback (1984) give accurate predictions:

$$T_c = T_b \left[0.584 + 0.965 \sum \Delta_T - \left(\sum \Delta_T \right)^2 \right]^{-1} \quad (3.1a)$$

or

$$T_{rb} = \frac{T_b}{T_c} = \left[0.584 + 0.965 \sum \Delta_T - \left(\sum \Delta_T\right)^2\right] \quad (3.1b)$$

$$P_c = \left(0.113 + 0.0032 N_a - \sum \Delta_P\right)^{-2} \quad (3.2)$$

$$V_c = 17.5 + \sum \Delta_V \quad (3.3)$$

In these equations Δ_T, Δ_P, and Δ_V represent the incremental contributions of the various structural components of a molecule to T_c (K), P_c (bar), and V_c (cm^3/mol), respectively; and N_a is the number of atoms in the molecule. The structural components can be molecular fragments, bonds, rings, etc. The values of Δ_T, Δ_P and Δ_V for different part-structures are listed by Jovak (1984).

Boiling Point

The following equation of Miller (1980), based on the earlier correlations of Rackett (1970) and Tyn and Calus (1975), is recommended in preference to a few others that have been proposed, for example, Myers, 1979; More and Capparelli, 1980; Joback, 1984; Ambrose, 1976 (specific to alkanols):

$$T_c = \frac{\exp \beta}{R_g} \quad (3.4)$$

where

$$\beta = \frac{[(1 - T_{rb})^{2/7} - 0.048] \ln V_c + (1 - T_{rb})^{2/7} \ln P_c + 1.255}{(1 - T_{rb})^{2/7}} \quad (3.5)$$

P_c is in bar, V_c is in cm^3/mol, and $R_g = 82.054$ cm^3 atm/mol K. Using the value of T_{rb} estimated from Equation 3.1b, β can be obtained from Equation 3.5, then T_c from Equation 3.4, and finally T_b from $T_b = T_c T_{rb}$. Errors of the order of 5–8% result, however. Hence it should be used only when the boiling point cannot be determined experimentally, a rather rare situation.

THERMODYNAMIC AND EQUILIBRIUM PROPERTIES

Vapor Pressure

Numerous generalized correlations are available for estimating vapor pressure (e.g., Klein, 1949; Fishtine 1963; Chen, 1965; Miller, 1965; Narasimhan, 1965, 1967; Thek and Stiel, 1967; Ambrose, 1972; Wagner, 1973a,b, 1977; Lee and Kesler, 1975; Gomez-Nieto and Thodos, 1977a,b, 1978, Kratzke, 1980; McGarry, 1983; Gupte and Daubert 1985). Although no single equation can be recommended as the best, the Riedel–Plank–Miller equation appears very attractive due to its simplicity relative to others of comparable accuracy. In its final form, this equation is

$$\ln P_{r,vp} = -\frac{A}{T_r}[1 - T_r^2 + B(T_r + 3)(1 - T_r)^3] \quad (3.6)$$

where $P_{r,vp}$ is the reduced vapor pressure,

$$A = 0.4835 + 0.4605\, T_{rb} \frac{\ln P_c}{(1 - T_{rb})}, \tag{3.7}$$

$$B = \frac{\dfrac{T_{rb} \ln P_c}{A(1 - T_{rb})} - (1 + T_{rb})}{(3 + T_{rb})(1 - T_{rb})^2} \tag{3.8}$$

where P_c is in atm. It enables estimating the vapor pressure as a function of temperature purely from T_c, P_c, and T_b. It is very accurate for nonpolar fluids and less so for polar fluids.

Heat of Vaporization

A simple but approximate equation is that of Reid et al. (1977) based on the earlier studies of Pitzer et al. (1955) and Carruth and Kobayashi (1972):

$$-\frac{\Delta H_v}{R_g T_c} = 7.08(1 - T_r)^{0.354} + 10.95\omega(1 - T_r)^{0.456},$$

$$0.6 < T_r \leq 1.0 \tag{3.9}$$

where ω is the *acentric factor*, which is a measure of the complexity of the molecule, essentially its polarity and deviation from sphericity (being zero for monatomic gases). Values of ω for a number of organic compounds are listed by Pitzer et al. (1955). They can also be estimated from Edmister's relationship

$$\omega = \frac{3}{7}\left(\frac{T_{rb}}{1 - T_{rb}}\right)\log P_c - 1 \tag{3.10}$$

(cited by Reid et al., 1977). Thus the heat of vaporization can be calculated from Equation 3.9 purely from knowledge of T_c, P_c, and T_b.

Two other methods, both due to Doraiswamy and co-workers (Kunte and Doraiswamy, 1960; Sastri et al., 1969), are more accurate. The first of these, based on the rigorous Clausius–Clapeyron equation, gives plots of the "reduced latent heat" ($\Delta H_V/T_c$) versus T_r for different values of P_r, assuming an average value of $Z_c = 0.276$. These plots (see original reference) predict heats of vaporization to within about 2%. The second method was developed specifically for the heat of vaporization ΔH_{vb} at the normal boiling point, which is usually adequate for most purposes (see also Lyman et al., 1982).

ΔH_{vb} can also be estimated to within about 3–5% from the following analytical expression proposed by Vetere (1973):

$$\frac{\Delta H_{Vb}}{T_{rb}} = R_g T_c \frac{0.4343 \ln P_c - 0.6886 + 0.89584\, T_{rb}}{0.37691 - 0.37306\, T_{rb} + 0.1488 P_c^{-1} T_{rb}^{-2}} \tag{3.11}$$

where P_c is in atm. Other methods (see Lyman et al., 1982) include those of Watson (1943), Klein (1949), Giacalone (1951), Fishtine (1963, 1966), Chen (1965), Narasimhan (1965, 1967).

A simple rule of thumb (to within 15%) for ΔH_{vb} is the classical Trouton rule:

$$\frac{\Delta H_{Vb}}{T_b} \approx 21 \tag{3.12}$$

Ideal Gas Heat Capacities

The most accurate method is that of Benson (1968; see also Benson et al., 1969), who gives an elaborate listing of group values at six temperatures (from 300 to 1000 K), for estimating the ideal gas heat capacity C_p^0 and also the reaction properties considered later on. Instead of focusing on individual groups within a molecule, this method deals with each atom, as differentiated by the number and type of bonds connecting it to neighboring atoms. Thus the listing of contributions runs into several pages with explanations, rules, correction terms, etc. This method is particularly accurate for hydrocarbons. The values have been updated several times (O'Neal and Benson, 1970; Shaw, 1971; Eigenmann et al., 1973; Stein et al., 1977; Majer et al., 1979, Bures et al., 1981) with marginal enhancements in accuracy.

On the other hand, a much more general though slightly less accurate method, proposed earlier by Rihani and Doraiswamy (1965), is applicable to a wide variety of compounds and enables estimating C_p as a *continuous function* of temperature (thus eliminating the need for interpolation). Here the constants of the regression equation

$$C_p = a + bT + cT^2 + dT^3 \tag{3.13}$$

are expressed in terms of the simpler and more common group contributions (not atomic contributions as in the Benson method), thus enabling the formulation of an equation for C_p as a function of temperature. The equation is accurate to within about 2.5%.

Similar methods that involve more elaborate equations have been proposed by Thinh (1971), Yoneda (1979), and Joback (1984). Thinh's method is of limited use because of its restriction to hydrocarbons, whereas Joback's method is more general and therefore more useful in organic synthesis. A few other methods have also been proposed (e.g., Kothari and Doraiswamy, 1964) that are easy to use, but are less accurate.

Considering the fact that the Rihani–Doraiswamy and Joback methods are applicable to a wide range of organic compounds and provide C_p as a continuous function of T (a very useful feature in reactor design), they are clearly the recommended methods.

Heat Capacities of Liquids

To estimate C_{pL} as a relatively precise continuous function of temperature, the following equation based on LCS may be used (Bondi, 1966; Rowlinson, 1969):

$$\frac{C_{pL} - C_p}{R_g} = 1.45 + 0.45(1 - T_r)^{-1}$$
$$+ 0.25\omega[17.11 + 25.2(1 - T_r)^{1/3}T_r^{-1} + 1.742(1 - T_r)^{-1}] \tag{3.14}$$

where C_p is estimated from the Rihani–Doraiswamy or Joback method, and ω from Equation 3.10 if not found in the list of Pitzer et al. Other methods based on LCS are those of Lyman and Danner (1976) which uses a new parameter, *radius of gyration*, and Yuan and Stiel (1970) which uses another new parameter *polar factor*.

To estimate C_{PL} at or around 20 °C (near room temperature), the group contributions method of Chueh and Swanson (1973) may be used, as it is claimed to be more accurate than other methods at room temperature (Reid et al., 1977). Group contributions at other temperatures are also available (Missenard, 1965).

Gas/Vapor Density

The gas-phase density is the reciprocal of the molar volume, $\rho_V = 1/V$, and hence can be estimated from any equation of state, for example, the Redlich–Kwong equation (1949). Kunte and Doraiswamy's (1958) plots of reduced density as a function of P_r and T_r provide a quick and reasonably accurate method of estimating vapor density. From our point of view, the most useful are the charts developed by Nelson and Obert (1954) in which the compressibility factor Z defined as $Z = PV/R_g T$ is plotted as a function of P_r for various values of T_r based on experimental data for a number of compounds. Then the density may be calculated from

$$\rho_V = \frac{PM}{ZR_g T} \tag{3.15}$$

where M is the molecular weight. The ideal gas law ($Z = 1$ in the above equation), however, is usually adequate and is at least as accurate as the Redlich–Kwong equation (Lyman et al., 1982).

Liquid Density

The following equation of Yen and Woods (1966) based on LCS is recommended:

$$\frac{\rho_L}{\rho_c} = 1 + \sum_{j=1}^{4} K'_j (1 - T_r)^{j/3} \tag{3.16}$$

where

$$K'_1 = 17.4425 - 214.578 Z_c + 989.625 Z_c^2 - 1522.06 Z_c^3$$
$$K'_2 = -3.28257 + 13.6377 Z_c + 107.4844 Z_c^2 - 384.211 Z_c^3 \quad \text{if } Z_c \leq 0.26$$
$$K'_2 = 60.2091 - 402.063 Z_c + 501.0 Z_c^2 + 641.0 Z_c^3 \quad \text{if } Z_c > 0.26 \tag{3.17}$$
$$K'_3 = 0$$
$$K'_4 = 0.93 - K'_2$$

It predicts liquid densities usually with errors less than 3–5%.

Other methods are also available such as those of Gambill (1979), Bhirud (1978), Chueh and Prausnitz (1967, 1969), Gunn and Yamada (1971), Goyal and Doraiswamy (1966), and Lyckman et al. (1965), but Equation 3.16 of Yen and Woods remains the method of choice.

Distribution Coefficient

The distribution coefficient m of a solute distributed between two solvents, usually between an organic phase and an aqueous phase, is the ratio of its concentrations in the two phases:

$$m = \frac{\text{concentration in phase 1}}{\text{concentration in phase 2}} \qquad (3.18)$$

This property is obviously of great importance in two-phase liquid-liquid systems such as those involving simple physical extraction, liquid-liquid reaction (Chapter 14), or reactive extraction described in Chapter 25. There is only one major correlation available for estimating this property, and this involves an aqueous phase. Even this is an indirect correlation in that it is based on an empirical logarithmic relationship between m for a specific system (designated m_s) and for octanol-water (designated as m_0). Therefore, we shall first describe the method for estimating m_0 and then extend it to m_s.

Estimation of m_0

Hansch and Leo (1979) proposed a GC correlation based on two distinct constituents, fragments, representing atoms or groups, and factors, representing structural contributions. Thus

$$\log m_0 = \text{sum of fragments } (f) + \text{sum of factors } (F) \qquad (3.19)$$

One way to use Equation 3.19 is to select a structurally similar compound for which an experimental value is available and make adjustments by adding or subtracting appropriate fragment and factor values:

$$m_0(\text{new chemical}) = m_0 \text{ (similar chemical)}$$
$$\pm \text{ fragments } (f) \pm \text{ factors } (F) \qquad (3.20)$$

Thus, if we need an estimate of m_0 for $C_6H_5NH_2$ and an experimental value is available for $C_6H_5NO_2$, then we merely subtract the value for NO_2 and add that for NH_2. The second method is to start from scratch and build up the value for the chemical directly from Equation 3.19.

The fragment and factor contributions have been worked out by Hansch and Leo in very fine detail, covering a wide spectrum of structural combinations available so far; the tables and explanations that run into several pages are not reproduced here.

Estimation of m_s

To extend this method to a combination of any of a variety of solvents and water, we reverse the procedure of Leo and Hansch (1971) and Leo et al. (1971)

who developed it essentially to obtain the octanol–water value from the value for any other solvent and water. The regression equation to be used is

$$\log m_s = (1/a)m_0 - b/a \tag{3.21}$$

A set of tables along with rules for selecting the values of a and b for 31 solvent–water systems is available in the original reference. These include such common solvents as benzene, toluene, xylene, carbon tetrachloride, ethyl ether, chloroform, methyl isobutyl ketone, cyclohexanol, hexane, and cyclohexane.

REACTION PROPERTIES

The most important reaction property is the rate constant. Unfortunately, there is no method known yet that can predict this property, except for an approximate method for the dissociation constant. The other properties directly relevant to reaction are the heat of combustion, and the heat, entropy, and free energy of formation.

Heat of Combustion

The importance of the heat of combustion ΔH_c^0 in calculating the heat of formation of an organic molecule was pointed out in Chapter 2. A method of estimating it has been proposed (Cardozo, 1986) in which an *equivalent chain length* of a molecule is defined as

$$N = N_a + \sum_i \Delta N_i \tag{3.22}$$

where N_a is the total number of atoms in the molecule and ΔN_i are the corrections for various structures and phases. A long list of these corrections has been provided by Cardozo. First, the heat of combustion (in kJ/mol) is determined from

$$\Delta H_c^0(g) = -198.42 - 615.14\,N \tag{3.23}$$

$$\Delta H_c^0(l) = -196.98 - 610.13\,N \tag{3.24}$$

$$\Delta H_c^0(s) = -206.21 - 606.56\,N \tag{3.25}$$

Then, the standard heat of formation can be calculated as follows.

Standard Heat of Formation

The standard heat of formation of a compound at 298 K can be obtained from the extensive tables published in the literature (see Chapter 2). If the heat of formation is required at any other temperature, it can be obtained from the relationship

$$\Delta H_{fT}^0 = \Delta H_{f298}^0 + \int_{298}^{T} \Delta C_p \, dT \tag{3.26}$$

where ΔC_p is the difference between the ideal gas heat capacities of the chemical and of the constituent elements. Several methods of estimating the heat of formation from GC have been developed. As in the case of heat capacity, the most elaborate are those of Benson (Benson, 1968; Benson et al., 1969; Eigenmann et al., 1973; and Yoneda, 1979), which give the values at specific temperatures. A more general though slightly less accurate method for certain structures is the method of Franklin (1949, 1953) and Verma and Doraiswamy (1965). Verma and Doraiswamy used the regression equation

$$\Delta H^0_{fT} = a + bT \tag{3.27}$$

and give group contributions for the constants a and b for two temperature ranges. The error in this method is generally less than 3%. Another method, that of Anderson et al. (1944), is less accurate than either the Verma–Doraiswamy or the Franklin methods.

Note that Joback's method estimates the values only at 298 K, and hence is not useful as a general method for estimating ΔH^0_{fT}. This is also true of the method based on the heat of combustion calculated from Equation 3.28 in kJ/mol (Cardozo, 1986):

$$\Delta H^0_f(298\,\text{K}) = -393.78 N_c - 121.00(N_H - N_X)$$
$$- 271.81 N_F - 92.37 N_{Cl}$$
$$- 36.26 N_{Br} + 24.81 N_I - 297.26 N_S - \Delta H^0_c \tag{3.28}$$

where N_C, N_H, N_F, N_{Cl}, N_{Br}, N_I, and N_S are, respectively, the numbers of atoms of carbon, hydrogen, fluorine, chlorine, bromine, iodine, and sulfur in the compound; N_x is the total number of hydrogen atoms; and ΔH^0_c is the heat of combustion estimated from Equation 3.23, 3.24, or 3.25. Note that all of the products of combustion are assumed to be in the ideal gaseous state in this equation. To account for H_2O in the liquid state (as in many reported tables), a suitable correction must be made, as illustrated in Example 2.1.

Ideal Gas Entropy

The entropy of a molecule is a measure of its disorder. Thus it is customary to express the entropy at any temperature in the ideal gaseous state S^0_T relative to the most ordered state, namely, the solid state at 0 K. With this definition, the entropy change for any reaction can be readily obtained by subtracting the sum of the values of S^0_T for the reactants from the sum for the products. In theory, it is also possible to estimate the entropy of formation ΔS^0_f by using the values of S^0_T for the constituent elements, but no method is yet available for the purpose.

Here again, the method of group contributions is the most useful, and within this class the elaborate tables of Benson (1968) and Benson et al. (1969) are the most accurate. Anderson et al. (1944) were the first to propose such contributions for evaluating the ideal gas entropy at 298 K. Essentially, this method involves identifying a base structure and estimating the value for the compound

of interest by (algebraically) adding the contributions from the other groups that comprise the structure.

Standard Free Energy of Formation

As with the heat of formation, ΔG_T^0 can also be expressed as a linear function of temperature:

$$\Delta G_T^0 = a + bT \tag{3.29}$$

where the constants a and b can be estimated from the group contributions given by Van Krevelen and Chermin (1951, 1952). Again, as with the heat of formation, incremental group values have been tabulated for two temperature ranges. According to Reid et al. (1977), this method gives values to within 5 kcal/g mole. Such a level of accuracy is adequate for predicting the feasibility of a reaction but not for calculating equilibrium compositions.

To calculate equilibrium compositions, the more elaborate tabulations of Benson (1968) and Benson et al. (1969) at specific temperatures are recommended.

Acid and Base Dissociation Constants

The properties of primary importance in acid-base chemistry are the acid dissociation constant K_a and the base dissociation constant K_b. Consider first the dissociation of a weak organic acid HA:

$$HA + H_2O \leftrightarrow H_3O^+ + A^{-1} \qquad [3.1]$$

In principle, the method of estimating K_a parallels that for the distribution coefficient, but there is a significant difference. Unlike the procedure for the distribution coefficient, we can only add appropriate substituent values but not subtract any. Further, there are two distinct contributions here: the reaction constant ρ characteristic of the parent group, and the substituent constant σ characteristic of the substituent group(s). Using this principle, Equation 2.37 can be written as

$$\log \frac{K_a}{K_{a,0}} = \log \frac{K_a \text{ (substituted benzoic acid)}}{K_{a,0} \text{(benzoic acid)}} = \sigma\rho \tag{3.30}$$

where $K_{a,0}$ is the dissociation constant for the parent acid. Two other equivalent forms of this equation are

$$K_a = K_{a,0} 10^{\sigma\rho}$$
$$pK_a = pK_{a,0} - \sigma\rho \tag{3.31}$$

The following equation can be developed for the dissociation constant of a base in terms of the acid constant:

$$K_b = \frac{10^{-14}}{K_a} \tag{3.32}$$

$$pK_b = 14 - pK_a$$

Hammett's reaction constants ρ for various aromatic parent compounds were obtained by letting $\rho = 1$ for the ionization constant of benzoic acid in an aqueous solution. Based on this assumption, values of ρ have been extensively tabulated (Jaffe, 1953; McDaniels and Brown, 1958; Wells, 1968; Shorter, 1982). His constants σ for different substituents to these aromatic parent compounds are also included in these references, along with exact procedures for finding the correct values.

A similar procedure can be used for aliphatic acids. The reaction constants for different parent aliphatic substances and for different substituents are included in the same references.

TRANSPORT PROPERTIES

All three transport properties, viscosity, diffusivity, and thermal conductivity, are important in reactor design. Viscosity is a measure of momentum transfer, diffusivity of mass transfer, and thermal conductivity of heat transfer.

Viscosity

Viscosity is a rather unique property in that an increase in temperature raises it for a gas but lowers it for a liquid. The following correlations may be used for gases and liquids which conform to the criteria set out earlier.

Gas viscosities

First we consider viscosity at low pressures and define a dimensionless group $\mu^*_{c,\text{org}}$ specifically applicable to organic compounds, which may be regarded as a *pseudocritical viscosity* that represents the viscosity at the critical temperature but at a low pressure (Reid et al., 1977):

$$\mu^*_{c,\text{org}} = \frac{M^{1/2} T_c}{\sum_i n_i C_i}, \quad \mu P \tag{3.33}$$

In this expression n_i is the number of atomic groups C_i of type i. Because pressure has no effect at low pressures, this group can be used to correlate μ as a function exclusively of temperature expressed as T_r. A very useful correlation is that of Reichenberg (1973, 1975), which in its simpler form is

$$\mu = \frac{\mu^*_{c,\text{org}} T_r}{[1 + 0.36 T_r (T_r - 1)]^{1/6}} \left[\frac{T_r (1 + 270 \mu_r^4)}{T_r + 270 \mu_r^4} \right], \quad \mu P \tag{3.34}$$

A generalized method of estimating μ at any given T and P is by using reduced plots where the reduced viscosity

$$\mu_r = \frac{\mu}{\mu_c} \qquad (3.35)$$

is plotted as a function of T_r and P_r (Uyehara and Watson, 1944). The critical viscosity μ_c can be obtained from the pseudocritical viscosity $\mu^*_{c,org}$ using the following relationship of Licht and Stechert (1944):

$$\mu_c = 7.7\mu^*_{c,org}, \quad \mu^*_{c,org} = \frac{M^{1/2}P_c^{2/3}}{T_c^{1/6}}, \quad \mu P \qquad (3.36)$$

Babu Rao and Doraiswamy (1966) presented plots for different values of Z_c that predict viscosities to within about 10%. A number of other more accurate but correspondingly more elaborate procedures based on LCS have been proposed, of which the methods of Lucas (1980, 1981) and Reichenberg (1971, 1973, 1975a,b) are noteworthy.

Care must be exercised in using these correlations for gaseous mixtures. Although methods are available (e.g., Dean and Stiel, 1965), they are not reliable enough to warrant the additional computation involved. We recommend the use of the additivity rule after ensuring that no maxima or minima result (a distinct possibility in polar–nonpolar mixtures).

Liquid viscosities

Liquid viscosities at 20 °C may be obtained from the following equation of Orrick and Erbar (1974):

$$\ln \frac{\mu_L}{\rho_L M} = A + \frac{B}{T} \qquad (3.37)$$

where μ_L is in cP and ρ_L is in g/cm^3. For liquids whose boiling points are below 20 °C, ρ_L at the boiling point is used, whereas for liquids with freezing points above 20 °C, ρ_L at the freezing point is used. The constants A and B are estimated from group contributions. Compounds containing nitrogen and sulfur cannot be treated by this method. Several other methods are also available, for example, Morris (1964), Hildebrand (1971), Van Velzen et al. (1972a,b), Lyman et al. (1982). The method of Lyman et al. is the most general and can be applied to almost any organic liquid.

To estimate μ_L at higher temperatures, a quick and reasonably accurate procedure is to use the generalized chart of Lewis and Squires (1934). This can be expressed in equation form as (Reid et al., 1987)

$$\mu_{L,T}^{-0.266} = \mu_{L,T(\text{ref})}^{-0.266} + \frac{T - T(\text{ref})}{233}, \quad cP \qquad (3.38)$$

where $\mu_{L,T(\text{ref})}$ is the known viscosity at $T(\text{ref})$ and $\mu_{L,T}$ is the required viscosity at T. The temperature can be in °C or K. This method has the advantage that it requires a value of viscosity at just one temperature, whereas the other available

methods (e.g., Cornellisen and Waterman, 1955; Eversteijn et al., 1960; Medani and Hasan, 1977) require two datum points.

Diffusion Coefficient

Diffusivities in gases

Binary gaseous diffusion coefficients are important parameters in the design of reactors for two-phase reactions that involve a gas and a liquid or a solid (either as catalyst or reactant). The recommended equation for low pressures is a modified form of the theoretical Chapman–Enskog equation, but a more readily usable equation is the following modification (Fuller et al., 1966) of the original equation proposed by Gilliland (1934):

$$D_{AB} = 10^{-3} \frac{T^{1.75}}{P(V_A^{1/3} + V_B^{1.3})^2} \sqrt{\frac{1}{M_A} + \frac{1}{M_B}}, \quad cm^2/s \tag{3.39}$$

where P is in atm, V_i is in cm/mol, and M_i is in g/mol. No reliable method is available for estimating D_{AB} at elevated pressures. A rule of thumb is to assume that PD_{AB} is constant provided that $\rho_r < 1$.

In a multicomponent system, the diffusivity of A is estimated by assuming that it diffuses through a stagnant film of the other gases. Then the overall diffusivity, say, of component j can be calculated from the various constituent binary diffusivities using one of the following two equations (the second is slightly more accurate):

$$\frac{1}{\bar{D}_j} = \frac{1}{1-y_j} \sum_{i \ne j}^{N} \frac{y_i}{D_{ji}} \tag{3.40}$$

In the presence of a chemical reaction, the stoichiometric coefficients of the various components come into the picture, giving

$$\frac{1}{\bar{D}_j} = \frac{\sum_{i \ne j}^{N} \frac{1}{D_{ji}} [y_i - y_j(\nu_i/\nu_j)]}{1 - y_j \sum_{i \ne j}^{N} \frac{\nu_i}{\nu_j}} \tag{3.41}$$

In these equations $\bar{D}_j$ is the diffusion coefficient of j in a mixture of $1 + 2 + \cdots + i + \cdots + N$ components, D_{ji} are the binary diffusion coefficients of j and i $(i \ne j)$, ν_i is the stoichiometric coefficient of i in the reaction, and y_i is its mole fraction.

Diffusivities in liquids

We restrict the treatment here to binary mixtures of, say, A and B, in which A diffuses in B at infinite dilution, that is, at very low concentrations of A in B. Reference may be made to Reid et al. (1987) for correlations at high concentrations.

All correlations proposed for estimating the diffusivity at infinite dilution are modifications of the original Stokes–Einstein equation

$$D_{AB} = \frac{R_g T}{6\pi \mu r_a} \tag{3.42}$$

where μ is the solvent viscosity and r_a is the radius of the solute assumed to be spherical. Of these modifications, the following have been most widely used: those of Scheibel (1954), Wilke and Chang (1955), Reddy and Doraiswamy (1967), Hayduk and Laudie (1974), Tyn and Calus (1975), Nakanishi (1978), Hayduk and Minhas (1982). These relate the diffusivity to viscosity (as in the Stokes–Einstein equation), but impart greater generality by including the molar volume of A or those of A and B. The following simplified form of the Tyn–Calus correlation (1975) may be used:

$$D_{AB} = 8.93 \times 10^{-8} \frac{V_B^{0.267}}{V_A^{0.433}} \frac{T}{\mu_B} \tag{3.43}$$

where the molar volume is obtained from

$$V_i = 0.285(V_c)_i^{1.048}, \quad i = A, B \tag{3.44}$$

and V_i and $(V_c)_i$ are the molar and critical volumes, respectively, of species i.

EXAMPLE 3.1

Estimation of liquid diffusivities

Estimate the diffusion coefficient at 85 °C of (a) benzyl chloride in a mixture of 1 mole benzyl chloride and 100 ml toluene and of (b) octyl bromide in a mixture of 2 moles of the bromide and 100 ml toluene (using alternatives to Equation 3.43).

SOLUTION

All diffusivity correlations require values of the corresponding viscosities. Hence, we first estimate the viscosities of pure liquids by the group contributions method of Van Velzen (1972a,b). The values at 85 °C are:

  benzyl chloride 5.16×10^{-4} Pa s
  octyl bromide 6.61×10^{-4} Pa s
  toluene 3.16×10^{-4} Pa s

The values of the two mixtures are then obtained from

$$\ln \mu_{mix} = \sum_i v_i \ln \mu_i \tag{E3.1.1}$$

where v_i is the volume fraction of component i in the mixture. The values obtained are

  benzyl chloride-toluene   0.3324 cP
  octyl bromide-toluene   0.3930 cP

The diffusion coefficients is then estimated from the following three equations:

WILKE-CHANGE CORRELATION

$$D_{AB} = 7.4 \times 10^{-8} \frac{(\phi M_B)^{1/2} T}{\mu_B V_A^{0.6}}$$

$(A = \text{solute}, B = \text{solvent})$ \hfill (E3.1.2)

where ϕ is the association factor, which usually is equal to 1 for nonpolar organic solvents.

REDDY-DORAISWAMY CORRELATION

$$D_{AB} = \frac{K_{RD} M_B^{1/2} T}{\mu_B (V_A V_B)^{1/3}} \tag{E3.1.3}$$

where $K_{RD} = 10 \times 10^{-8}$ for $V_B/V_A \leq 1.5$
$= 8.5 \times 10^{-8}$ for $V_B/V_A > 1.5$

SCHEIBEL CORRELATION

$$D_{AB} = \frac{K_S T}{\mu_B V_A^{1/3}} \tag{E3.1.4}$$

where $K_S = 8.2 \times 10^{-8} \left[1 + \left(\frac{3 V_B}{V_A} \right)^{2/3} \right]$

The molar volumes of the individual liquids were estimated by the method of Le Bas (1915), which after over 80 years is still surprisingly valid.

The values obtained from the three correlations are listed in the following table:

	$D_{AB} \times 10^5$ cm² s	
Correlation	Benzyl chloride-toluene	Octyl bromide-toluene
Wilke-Change	3.98	2.64
Reddy-Doraiswamy	4.03	3.02
Scheibel	4.89	3.06

It will be noticed that the predictions from all the three correlations are reasonably close to each other. Thus any of these equations can be used.

Thermal Conductivity

Thermal conductivities of gases

A number of studies of thermal conductivity have been reported (e.g., Roy and Thodos, 1968). Based on these, the following modification of the well-known Euken correlation is still surprisingly accurate:

$$\frac{\lambda M}{\mu} = 3.52 + \frac{1.32 C_p}{\gamma} \tag{3.45}$$

where λ is the thermal conductivity and γ is the ratio of the constant pressure to constant volume heat capacities C_p/C_v. For an ideal gas $(C_p - C_V) = R_g$ giving $\gamma = R_g + 1$.

Thermal conductivities of liquids

A useful (and perhaps the most accurate) correlation based on elaborate sets of group contributions is that of Baroncini et al. (1980, 1981, 1983, 1984), but a simpler and reasonably accurate correlation (Reid et al., 1987; see also Riedel, 1951) is

$$\lambda_L = \frac{1.11}{M^{1/2}} \left[\frac{3 + 20(1 - T_r)^{2/3}}{3 + 20(1 - T_{rb})^{2/3}} \right], \quad W/m\,K \tag{3.46}$$

This is a simple equation needing only T_b, T_c and M to estimate λ_L at any temperature. To get the values in cal/cm s K, replace 1.11 by 2.64×10^{-3} in the equation. Several other methods have also been reported, for example Robbins and Kingrea (1962), Missenard (1973), Mathur et al. (1978), Teja and Rice (1981, 1982), and Ogiwara et al. (1982), but Equation 3.46 is recommended.

SURFACE TENSION AND ULTRASONIC VELOCITY

Surface tension is an important property that defines the behavior of liquids at the interface. Using this property, the velocity of sound in liquids (pure or mixtures) can be found. This last property is of particular importance in equations relating to rate enhancement through the use of ultrasound, considered in Chapter 22. Hence we give the equations for estimating these properties.

Brock and Bird (1955) suggest the following correlation between critical constants and surface tension:

$$\frac{\sigma}{(T_c P_c^2)^{1/3}} = (0.133\alpha_c - 0.28)(1 - T_r)^{11/9}, \quad \alpha_c = 0.91\left(1 + \frac{T_{rb} \ln P_c}{1 - T_{rb}}\right) \tag{3.47}$$

In the case of binary mixtures pseudocritical properties T_{cm}, P_{cm}, σ_{cm} (estimated as mole fraction averages) can be used.

Using the Brock-Bird relation (Toxvaerd, 1975), the Flory theory (see Pandey et al., 1999, for a discussion), and the well-known Auerbach relation (Auerbach, 1948), the following equation can be formulated to predict ultrasonic velocity:

$$u_{us} = \left(\frac{\sigma_m}{6.3 \times 10^{-2} \rho_m}\right)^{2/3} \tag{3.48}$$

where ρ_m and σ_m are, respectively, the mole fraction average density and surface tension of a mixture. The equation can also be used for single components.

Another method used to evaluate the ultrasonic velocity is the Flory statistical theory, which, as it happens, has no direct link with ultrasonic velocity. Patterson and Rastogi (1970) have used this theory to calculate a characteristic surface tension using appropriately defined characteristic pressure and temperature. Equation 3.47 can then be used to calculate u_{us}.

Chapter 4

Reactions and Reactors

Basic Concepts

Organic synthesis is replete with countless classes of reactions, including several that are named after their discoverers (the *name reactions*), but fortunately they can all be conducted in less than a half-dozen broad types of reactors. Choosing a reactor for a given reaction is based on several considerations and combines *reaction analysis* with *reactor analysis*. Thus in this chapter we consider the following aspects of reactions and reactors, much of which should serve as an introduction to chemists and a refresher to chemical engineers: reaction rates, stoichiometry, and rate equations; the basic reactor types, as a prelude to a more rigorous treatment of these in Parts III and IV; transport of mass (represented by reactant and product molecules) and heat across phase boundaries for heterogeneous reactions; and types of laboratory reactors used by chemists and chemical engineers for their specific objectives.

REACTION RATES

The first step in any consideration of reaction rates is the definition of reaction time. This depends on the mode of reactor operation, batch or continuous (see Figure 4.1). For the batch reactor, the reaction time is the elapsed time; whereas for the continuous reactor, it is given by the time the reactant spends in the reactor, called the *residence time*, that is measured by the ratio of reactor volume to flow rate (volume/volume per unit time with units of time). An equally important consideration is the concept of *reaction space* (which can have units of volume, surface, or weight), leading to different definitions of the reaction rate. We begin this section by considering different ways of defining the reaction rate based on different definitions of reaction time and space.

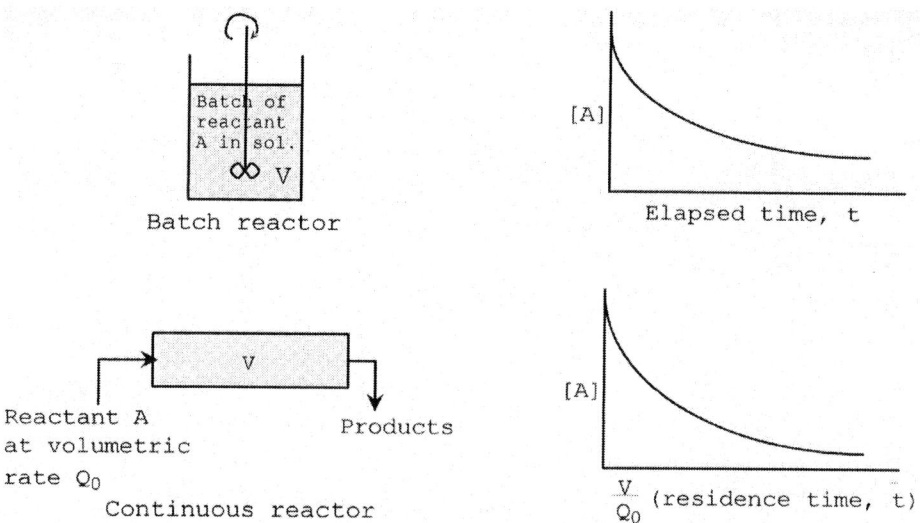

Figure 4.1 Batch and continuous operation concepts.

Different Definitions of the Rate

The basis of all reactor design is an equation for the reaction rate. The rate is expressed as

$$\begin{bmatrix} \text{Rate of consumption} \\ \text{of a reactant or} \\ \text{formation of a product} \end{bmatrix} = \frac{\text{moles consumed or formed}}{(\text{time})(\text{reaction space})} \quad (4.1)$$

Different definitions are possible, depending on the definition of reaction space. This in turn depends on whether the reaction is homogeneous or heterogeneous.

Consider a reactor that contains a fixed amount of reactant in a single phase. If the reactant is a gas, it occupies the entire reactor space. If a liquid, the reaction space is less than the reactor volume, usually about two-thirds. This volume is generally indicated by the symbol V (or sometimes specifically by V_r). Thus for any component i the rate is defined as

$$\text{Rate}, r_i = \frac{1}{V}\left(\frac{dN_i}{dt}\right) \quad (4.2)$$

In the case of a catalytic reactor, the reaction space can be the weight, volume, or surface of the catalyst. It can also be the total reactor volume (catalyst plus voids). Thus four definitions are normally used:

$$r_{Vi} \text{ (or } r_i) = \frac{1}{V_r}\left(\frac{dN_i}{dt}\right) \quad \begin{array}{l}\text{moles per unit time per unit volume}\\ \text{of reactor (catalyst plus voids in}\\ \text{a catalytic reactor)}\end{array} \quad (4.3a)$$

$$r_{wi} = \frac{1}{W}\left(\frac{dN_i}{dt}\right) \quad \begin{array}{l}\text{moles per unit time per unit weight}\\ \text{of catalyst}\end{array} \quad (4.3b)$$

$$r_{vi} = \frac{1}{v}\left(\frac{dN_i}{dt}\right) \quad \begin{array}{l}\text{moles per unit time per unit volume}\\ \text{of catalyst}\end{array} \quad (4.3c)$$

$$r_{Si} = \frac{1}{S}\left(\frac{dN_i}{dt}\right) \quad \begin{array}{l}\text{moles per unit time per unit}\\ \text{surface of catalyst}\end{array} \quad (4.3d)$$

In gas-liquid or liquid-liquid reactions, Equation 4.3a based on total reactor volume, which is identical to Equation 4.2 for homogeneous reactions, may be used to give r_i. Alternatively, the rate can be expressed as r_{Si} given by Equation 4.3d for catalytic reactions but based on gas-liquid or liquid-liquid (instead of the catalyst) interfacial area. In this case (see Part IV) we generally use the symbol r'_i.

There could be situations where all the three phases are present: gas, liquid, and solid (as a catalyst or reactant). The most common example of this is the slurry reactor used in reactions such as hydrogenation. Here the rate is sometimes expressed as it is for catalytic reactions (i.e., per unit catalyst volume, weight, or surface), but more commonly in terms of the total reactor volume (liquid plus gas plus catalyst).

The various rates are summarized in Table 4.1 along with interrelationships among them. Depending on the nature of the reaction, any of these definitions may be used in handling organic synthesis. Because this chapter is largely restricted to homogeneous reactions, the rate based on total volume (r_i) is used throughout the chapter.

The Basic Rate Equation

Any rate equation can be written in its most general form as

$$\text{Rate} = f\begin{bmatrix}\text{temperature, concentration, history of catalyst in the}\\ \text{case of catalytic reactions, and mode of contact}\end{bmatrix} \quad (4.4)$$

In this equation, no parameter is independent of another, and hence a single functionality is assigned. This is referred to as a *nonseparable kinetic equation*. For homogeneous reactions, this usually reduces to

$$\text{Rate} = f(\text{temperature, concentration}) \quad (4.5)$$

which can be further simplified to

$$\text{Rate} = f_1(\text{temperature}) f_2(\text{concentration}) \quad (4.6)$$

In this *separable rate equation*, temperature and concentration are treated as independent functions, and the effects of the two are separated. Our treatment will be based on separable kinetics.

Table 4.1 Units of rates/rate constants (for a first-order reaction) for different classes of reactions

Reaction	Reaction rate	Rate constant
Any reaction	r_{VA} (or r_A), $\dfrac{\text{mol}}{\text{m}^3 \text{ reactor s}}$	k (or k_V), $\dfrac{1}{s}$
Gas-liq, liq-liq reactions	r'_{VA} (or r'_A), $\dfrac{\text{mol}}{\text{m}^2 \text{ interfacial area s}}$	k', $\dfrac{m}{s}$
Catalytic reactions	r_{vA}, $\dfrac{\text{mol}}{\text{m}^3 \text{ cat s}}$	k_v, $\dfrac{1}{s}$
	r_{SA}, $\dfrac{\text{mol}}{\text{m}^2 \text{ surface area s}}$	k_S, $\dfrac{m}{s}$
	r_{wA}, $\dfrac{\text{mol}}{\text{kg cat s}}$	k_W, $\dfrac{\text{m}^3}{\text{kg cat s}}$

Interrelationships

k_V (or k) $= (1 - f_B)k_v = (1 - f_B)S_g \rho_c k_s$
k_V (or k) $= ak'$, $k_w = k_v/\rho_c$

The temperature dependence is expressed in terms of the Arrhenius equation,

$$k = k^0 e^{-E/R_g T} \tag{4.7}$$

where E is the activation energy for the reaction, k^0 is the preexponential term which may be regarded as the temperature-independent rate constant and R_g is the gas-law constant.

The concentration dependence is usually expressed as a power of the concentration, and this power is called the *reaction order*. Thus for the reactions

$$A \rightarrow R \qquad [4.1]$$
$$A + B \rightarrow R \qquad [4.2]$$

the rate equations, referred to as *power law models*, are

$$-r_A = -\frac{d[A]}{dt} = k[A]^n \tag{4.8}$$

$$-r_A = -\frac{d[A]}{dt} = k[A]^m[B]^n \tag{4.9}$$

where m and n are the reaction orders with respect to A and B, respectively. If the reaction order is equal to the *stoichiometric coefficient* of the concerned reactant (i.e., the number of molecules of the reactant taking part in the reaction), then it is called an *elementary reaction*.

Reactions 4.1 and 4.2 involve only a single step and are called simple reactions. Rate equations similar to Equations 4.8 and 4.9 can be written for a

STOICHIOMETRY OF THE RATE EQUATION

Basic Relationships

It is clear from the previous section that the rate of reaction depends on the concentrations of reactants. In reversible reactions, it depends additionally on the concentrations of products. Often in a laboratory experiment, one of the reactants is selected, and its concentration is monitored as a function of time or flow rate, depending on whether the reactor is batch or continuous. Then the results are expressed in terms of the conversion of that reactant (say A) defined as

$$X_A = \frac{\text{moles } A \text{ converted}}{\text{moles } A \text{ fed}} \tag{4.10}$$

It is often necessary to convert the rate of reaction of any component to that of another. Thus a stoichiometric relationship between the rates is required. To obtain such a relationship, consider the general reaction

$$\nu_A A + \nu_B B \rightarrow \nu_R R + \nu_S S \tag{4.3}$$

From its stoichiometry, we can easily write

$$\frac{-r_A}{\nu_A} = \frac{-r_B}{\nu_B} = \frac{r_R}{\nu_R} = \frac{r_S}{\nu_S} \tag{4.11}$$

Conversion-Concentration Relationships

Consider reaction 4.3. No stoichiometric relationship is needed for calculating the conversion which is given simply by Equation 4.10. But the rate equation contains other concentrations as well, and these can be related to X_A through the stoichiometry of the reaction. These relationships will depend on whether a volume change accompanies the reaction. Volume change can occur due to a change in number of moles ($\nu_A + \nu_B \neq \nu_R + \nu_S$), temperature, or pressure, or combinations of these factors. Reactions in which no volume change occurs due to any of these factors are called *constant density (constant volume)* reactions. Those in which one or more of these occur are called *variable density (variable volume)* reactions. The effect of volume change is important in gas-phase reactions but is negligible in homogeneous liquid-phase reactions.

For a batch reactor of volume V, the concentration of A is given by

$$\text{Batch reactor:} \quad [A] = \frac{N_A}{V} \tag{4.12}$$

where the volume V is constant and equal to the initial volume V_0. For a continuous flow reactor with a volumetric feed rate of Q_A liters per second and a molar rate of F_{A0} moles per second, the concentration is given by

$$\text{Continuous reactor:} \quad [A] = \frac{F_{A0}}{Q_A} \tag{4.13}$$

where Q_A is constant and equal to the initial value Q_{A0}. We also define the following two parameters:

$$\psi_i = \frac{[i]_0}{[A]_0} = \frac{F_{i0}}{F_{A0}} \quad (4.14)$$

and the volume change parameter δ_i

$$\delta_i = \frac{\text{change in total number of moles at complete conversion of } i}{\text{Number of moles of } i \text{ reacted}} \quad (4.15)$$

which is equal to zero for a constant density reaction. Considering reaction 4.3 with A as the key component, now we can write the following expressions for the number of moles of each component that remain after a certain conversion X_A:

$$N_A = N_{A0}(1 - X_A) \quad (4.16)$$

$$N_i = N_{A0}\left(\psi_i \pm \frac{\nu_i}{\nu_A} X_A\right), i \neq A \quad (4.17)$$

$$N_t = N_{t0} + \delta_A N_{A0} X_A \quad (4.18)$$

where

$$\delta_A = \frac{\nu_R + \nu_S - \nu_A - \nu_B}{\nu_A} \quad (4.19)$$

Note that $i \neq A$, and the sign is negative for reactant and positive for product. Because there is no volume change, we can combine Equations 4.16–4.18 with Equation 4.12 to give

$$[A] = [A]_0(1 - X_A) \quad (4.20)$$

$$[i] = [A]_0\left(\psi_i \mp \frac{\nu_i}{\nu_A} X_A\right) \quad (4.21)$$

Identical concentration equations can be derived for a flow system, using F in place of N in Equations 4.16–4.18 and Equation 4.13 for the concentration.

Variable density reactions

The general case of variable density reactions is applicable mostly to gas-phase reactions and seldom to liquid-phase reactions. Because the gas law gives the precise relationship between P, V, T, and N, we start with that equation. Based on the ideal gas law ($P_0 V_0 / N_0 = PV/N$), we can write

$$V = V_0\left(\frac{P_0}{P}\right)\left(\frac{T}{T_0}\right)\left(\frac{N_t}{N_{t0}}\right) = V_0\left(\frac{P_0}{P}\right)\left(\frac{T}{T_0}\right)(1 + \delta_A y_{A0} X_A) \quad (4.22)$$

where

$$y_{A0} = \frac{N_{A0}}{N_{t0}} \quad (4.23)$$

is the initial mole fraction of A. Our concern here is with the change in total volume, when A has reacted completely, in relation to the total volume present initially. Thus we define a volume change parameter ε_A

$$\varepsilon_A = \frac{(V)_{X_A=1} - (V)_{X_A=0}}{(V)_{X_A=0}} \qquad (4.24)$$

Because δ_A represents the volume change at complete reaction per mole of A reacted (see Equation 4.19),

$$\varepsilon_A = y_{A0}\delta_A \qquad (4.25)$$

Combining Equations 4.22 and 4.25, we get

$$V = V_0 \left(\frac{P_0}{P}\right)\left(\frac{T}{T_0}\right)(1 + \varepsilon_A X_A) \qquad (4.26)$$

which is the basic equation to account for volume/density change. Now the concentration equations readily follow:

$$[A] = [A]_0 \left(\frac{1 - X_A}{1 + \varepsilon_A X_A}\right)\left(\frac{T_0}{T}\right)\left(\frac{P}{P_0}\right) \qquad (4.27)$$

$$[i] = [A]_0 \left[\frac{\psi_i \mp \left(\frac{\nu_i}{\nu_A}\right) X_A}{(1 + \varepsilon_A X_A)}\right]\left(\frac{T_0}{T}\right)\left(\frac{P}{P_0}\right) \qquad (4.28)$$

The analysis can be extended to a flow system by expressing the volume change in terms of the change in Q_A. The following expression similar to Equation 4.26 results:

$$Q_A = Q_{A0}\left(\frac{P_0}{P}\right)\left(\frac{T}{T_0}\right)(1 + \varepsilon_A X_A) \qquad (4.29)$$

Combining this equation with 4.13, equations similar to 4.27 and 4.28 can be written.

REACTORS

A variety of reactors can be used to conduct organic reactions. These are classified as homogeneous or heterogeneous reactors, depending on whether the reaction itself is homogeneous or heterogeneous. Essentially these involve: a single homogeneous liquid or gas phase, or two or more phases that are heterogeneous combinations of gas, liquid, and solid (as catalyst or reactant).

To understand the underlying principles of these reactors, which are treated in Parts III and IV, we provide the groundwork in this section by considering the basic reactor types. Thus we first describe the *batch reactor* (BR) most commonly used in organic synthesis and its continuous counterpart, the *plug-flow reactor* (PFR). These represent one extreme characterized by total absence of backmixing from fluid elements "downstream" in time or space. On the other extreme, we have the fully (or *perfectly*) *mixed-flow reactor* (MFR), also called the

continuous stirred tank reactor (CSTR), which, as the name implies, is perfectly mixed. The three reactors mentioned above are called *ideal reactors*. Effects due to departures from these limits of perfect and no mixing lead to *partially mixed reactors*.

The Batch Reactor

The batch reactor[1] is essentially a reactor in which a batch of reactants placed in it is allowed to react under predetermined conditions. The reactor is continuously stirred to maintain uniform composition at any time during the reaction. This composition will of course change with time.

Batch reactors are most common in the pharmaceutical, perfumery, essential oil, and other fine chemicals industries. They are also extensively used in the pesticides industry.

For relatively small scale productions, it is customary not to have a batch reactor exclusively for a single reaction. The scheduling of its use can be done so as to ensure maximum utilization.

Consider the simple irreversible reaction 4.1, $A \rightarrow R$, carried out in a batch reactor. The following material balance can be written:

$$(\text{Input}) - (\text{output}) = \begin{pmatrix} \text{disappearance} \\ \text{by reaction} \end{pmatrix} + (\text{accumulation}) \quad (4.30)$$

As there is no continuous input or output of material, this can be written as

$$-\frac{dN_A}{dt}\frac{1}{V} = (-r_A) \quad (4.31)$$

Because $N_A = N_{A0}(1 - X_A)$,

$$t = N_{A0} \int_{X_{A0}}^{X_{Af}} \frac{dX_A}{V(-r_A)} \quad (4.32)$$

Equation 4.32 is written in terms of the number of moles; hence it is independent of volume and holds equally for reactions with and without volume change. If the moles are expressed as concentrations, however, volume comes into the picture, and different expressions result for reactions with and without volume change.

Reactions with no volume change

Because V is constant, $[A] = N_A/V$, and Equation 4.31 becomes

$$-\frac{d[A]}{dt} = (-r_A) \quad (4.33)$$

---

[1] The organic chemistry literature usually mentions, or almost always implies, the use of batch reactors with very little reference to other reactor types. An exception is Tundo's *Continuous Flow Methods in Organic Synthesis* (1991).

Table 4.2 Analytical solutions (design equations) for simple reactions in a batch reactor (also valid for PFR with t replaced by $\bar{t}$)

Reaction	Rate equation	Analytical solution[a]
1. $A \to R$	$-r_A = k[A]$	$kt = \ln\left[\dfrac{1}{1-X_A}\right]$, $\dfrac{[A]}{[A]_0} = e^{-kt}$
2. $2A \to R$	$-r_A = k[A]^2$	$k[A]_0 t = \left[\dfrac{X_A}{1-X_A}\right]$, $\dfrac{[A]}{[A]_0} = \dfrac{1}{1+k[A]_0 t}$
3. $3A \to R$	$-r_A = k[A]^3$	$k[A]_0^2 t = \dfrac{1}{2}\left[\dfrac{1}{(1-X_A)^2} - 1\right]$, $2kt = \dfrac{1}{[A]^2} - \dfrac{1}{[A]_0^2}$
4. $A \to R$	$-r_A = k[A]^n$	$k[A]_0^{n-1} t = \dfrac{1}{(n-1)}[(1-X_A)^{1-n} - 1], n \neq 1$
5. $A + B \to R$ $\psi_B = 1$	$-r_A = k[A][B]$	$k[A]_0 t = \left[\dfrac{X_A}{1-X_A}\right]$, $t = \dfrac{1}{k}\left(\dfrac{1}{[A]} - \dfrac{1}{[A]_0}\right)$
6. $A + B \to R$ $\psi_B \neq 1$	$-r_A = k[A][B]$	$k[A]_0 t = \dfrac{1}{(\psi_B - 1)}\ln\left[\dfrac{\psi_B - X_A}{\psi_B(1-X_A)}\right]$
7. $\nu_A A + \nu_B B \to R$ $\psi_B = \nu_B/\nu_A$	$-r_A = k[A][B]$	$k[A]_0 t = \dfrac{1}{\psi_B}\left[\dfrac{X_A}{1-X_A}\right]$
8. $A + 2B \to R$ $\psi_B = 2$	$-r_A = k[A][B]^2$	$k[A]_0^2 t = \dfrac{1}{8}\left[\dfrac{1}{(1-X_A)^2} - 1\right]$
9. $A + B \to R$ $\psi_B = 1$	$-r_A = k[A][B]^2$	$k[A]_0^2 t = \dfrac{1}{2}\left[\dfrac{1}{(1-X_A)^2} - 1\right]$
10. $A + 2B \to R$ $\psi_B \neq 2$	$-r_A = k[A][B]^2$	$k[A]_0^2 t = \dfrac{1}{(2-\psi_B)^2}\left[\ln\dfrac{\psi_B - 2X_A}{\psi_B(1-X_A)} + \dfrac{2X_A(2-\psi_B)}{\psi_B(\psi_B - 2X_A)}\right]$
11. $\nu_A A + \nu_B B \to R$ $\psi_B \neq \nu_B/\nu_A$	$-r_A = k[A][B]$	$k[A]_0 t = \dfrac{1}{\left[\psi_B - \left(\dfrac{\nu_B}{\nu_A}\right)\right]}\ln\left[\dfrac{\psi_B - \left(\dfrac{\nu_B}{\nu_A}\right)X_A}{\psi_B(1-X_A)}\right]$
12. $\nu_A A + \nu_B B \to R$ $\psi_B = \nu_B/\nu_A$	$-r_A = k[A]^n[B]^m$	$k[A_0]^{m+n-1} t = \dfrac{1}{\psi_B(m+n-1)}\left[\dfrac{1}{(1-X_A)^{m+n-1}} - 1\right]$

[a] LHS = $k[A_0]^{n-1} t$ where k has the units of an nth [or $(m+n)$th]-order reaction, $\left(\dfrac{m^3}{mol}\right)^{n-1}\dfrac{1}{s}$; $\psi_B = \dfrac{[B]_0}{[A]_0}$.

Integrating between the limits of initial and final concentrations gives

$$t = -\int_{[A]_0}^{[A]} \frac{d[A]}{(-r_A)} = \int_{[A]}^{[A]_0} \frac{d[A]}{(-r_A)} = [A]_0 \int_0^{X_A} \frac{dX_A}{(-r_A)} \qquad (4.34)$$

The time t for achieving a stated conversion can be obtained by introducing the appropriate equation for the rate in Equation 4.34 and solving it. For the simple first-order case ($-r_A = k[A]$), the solution is

$$\frac{[A]}{[A]_0} = e^{-kt}, \quad \text{or} \quad \ln\frac{[A]_0}{[A]} = kt \qquad (4.35)$$

Solutions to a number of rate forms corresponding to different reaction schemes are given in Table 4.2.

A common type of reaction in organic technology and synthesis is the second-order reversible reaction represented by

$$A + B \leftrightarrow R + S \qquad [4.4]$$

Esterification reactions such as

$$C_2H_5OH + CH_3COOH \leftrightarrow CH_3COOC_2H_5 + H_2O \qquad [4.5]$$

are typical of this class. The rate equation for this second-order reversible reaction is given by

$$-r_A = k_+[A][B] - k_-[R][S] \qquad (4.36)$$

Analytical solution to this reaction is given in Table 4.3, along with solutions to a few other reversible reactions.

Equation 4.34 can also be solved by graphical integration of a plot of $(-1/r_A)$ versus X_A or $[A]$ between the limits 0 and X_{Af} or $[A]_0$ and $[A]_f$, as shown in Figure 4.2. However, there are situations where the rate passes through a maximum, as in adiabatic or autocatalytic reactions. Such curves are considered in Chapter 10.

Reactions with volume change

First let us consider the volume change resulting from a change in the number of moles, such as in the reaction

$$A \rightarrow R + S \qquad [4.6]$$

To account for volume change, we express V in terms of Equation 4.26 and modify Equation 4.32 to give

$$t = [A]_0 \int_0^{X_A} \frac{dX_A}{(-r_A)(1 + \varepsilon_A X_A)} = \int_{[A]}^{[A]_0} \frac{d[A]}{[1 + \varepsilon_A([A]/[A]_0)](-r_A)} \qquad (4.37)$$

This can be graphically integrated as shown in Figure 4.3 to give the time for accomplishing a given conversion.

Rate constants from batch reactor data

The main use of the equations just presented is in determining the reaction time for a given volume of production. They are equally useful in the reverse problem of determining the rate constant and reaction order for a given reaction from concentration-time data. In gas-phase reactions, pressure can be advantageously used in place of concentration. The rate equations can also be used in their differential or integrated forms.

FROM CONCENTRATION DATA

Let us take the nth order irreversible reaction 4.1 with rate Equation 4.8 written in the form

Table 4.3 Analytical solutions (design equations) for some reversible reactions at constant density in a batch reactor in terms of a, b, c of the equation

$$t = \frac{1}{k_+} \int_0^{X_A} \frac{dX_A}{aX_A^2 + bX_A + c}{}^a$$

Reaction	Constants		
	a	b	c
1. $A \underset{k_-}{\overset{k_+}{\rightleftharpoons}} 2R$	$-4\dfrac{[A]_0}{K_c}$	$-\left(1 + 4\psi_R \dfrac{[A]_0}{K_c}\right)$	$\left(1 - \dfrac{[A]_0 \psi_R^2}{K_c}\right)$
2. $2A \underset{k_-}{\overset{k_+}{\rightleftharpoons}} R$	$[A]_0$	$-\left(2[A]_0 + \dfrac{1}{2K_c}\right)$	$[A]_0 - \dfrac{\psi_R}{K_c}$
3. $A + B \underset{k_-}{\overset{k_+}{\rightleftharpoons}} R$ $\psi_B \neq 1$	$[A]_0$	$-\left([A]_0(1 + \psi_B) + \dfrac{1}{K_c}\right)$	$[A]_0 \psi_B - \dfrac{\psi_R}{K_c}$
4. $A + B \underset{k_-}{\overset{k_+}{\rightleftharpoons}} R + S$ $\psi_B \neq 1, \psi_R \neq \psi_S$	$[A]_0 \left(1 - \dfrac{1}{K_c}\right)$	$-[A]_0 \left[1 + \psi_B + \dfrac{1}{K_c}(\psi_R + \psi_S)\right]$	$[A]_0 \left(\psi_B - \dfrac{\psi_R \psi_S}{K_c}\right)$
5. $2A \underset{k_-}{\overset{k_+}{\rightleftharpoons}} R + S$ $\psi_R \neq \psi_S$	$[A]_0 \left(1 - \dfrac{1}{4K_c}\right)$	$-[A]_0 \left[2 + \dfrac{1}{2K_c}(\psi_R + \psi_S)\right]$	$[A]_0 \left(1 - \dfrac{\psi_R \psi_S}{K_c}\right)$
6. $2A \underset{k_-}{\overset{k_+}{\rightleftharpoons}} 2R$	$[A]_0 \left(1 - \dfrac{1}{K_c}\right)$	$-2[A]_0 \left(1 + \dfrac{\psi_R}{K_c}\right)$	$[A]_0 \left(1 - \dfrac{\psi_R^2}{K_c}\right)$

The integrated forms are

$$t = \frac{1}{k_+} \int_0^{X_A} \frac{dX_A}{aX_A^2 + bX_A + c} = \frac{1}{k_+ \sqrt{b^2 - 4ac}} \ln \frac{(2aX_A + b - \sqrt{b^2 - 4ac})(b + \sqrt{b^2 - 4ac})}{(2aX_A + b + \sqrt{b^2 - 4ac})(b - \sqrt{b^2 - 4ac})}, \quad b^2 > 4ac$$

$$t = \frac{1}{k_+} \int_0^{X_A} \frac{dX_A}{aX_A^2 + bX_A + c} = \frac{2}{k_+ \sqrt{4ac - b^2}} \left[\tan^{-1} \frac{2aX_A + b}{\sqrt{4ac - b^2}} - \tan^{-1} \frac{b}{\sqrt{4ac - b^2}}\right], \quad b^2 < 4ac$$

$$t = \frac{1}{k_+} \left[\frac{4aX_A}{(2aX_A + b)b}\right], \quad b^2 = 4ac$$

^{a} From Holland and Anthony (1989). K_c = equilibrium constant.

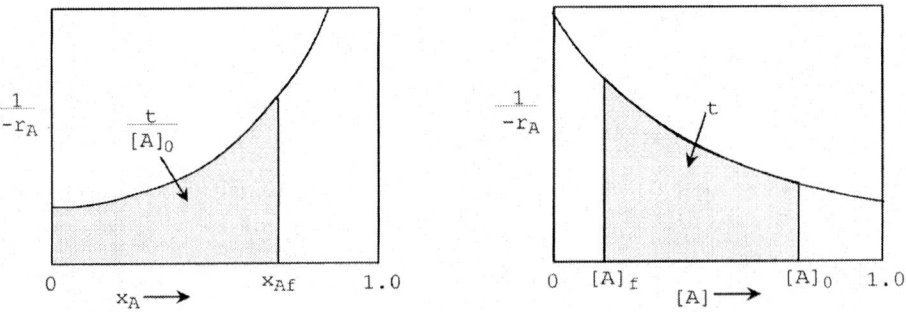

Figure 4.2 Integration of Equation 4.34 to give reaction time t.

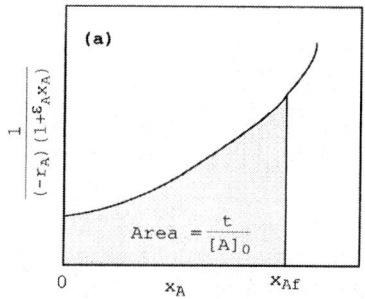 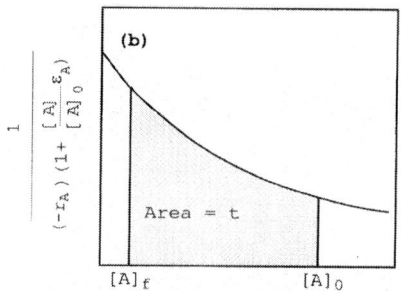

Figure 4.3 Determination of reaction time for a reaction with volume change.

$$\ln(-r_A) = \ln k + n \ln [A] \tag{4.38}$$

In the differential method of estimating the rate parameters, first we plot $[A]$ as a function of time, and differentiate it either graphically or by curve fitting. The slope thus obtained gives the rate directly for a reaction with no volume change, based on which the kinetic parameters n and k can be determined as shown in Figure 4.4. But for a reaction with volume change, the rate is obtained from the following modified form of Equation 4.33:

$$-\frac{d[A]}{dt} = (-r_A) \frac{[A]_0}{[A]_0 + \varepsilon_A [A]} \tag{4.39}$$

Now, let us consider a bimolecular reaction

$$A + B \rightarrow R + S \tag{4.7}$$

In its most general form, the rate may be expressed as

$$-r_A = k[A]^m [B]^n \tag{4.40}$$

and may be measured as described for a unimolecular reaction. Equation 4.40 can be recast as

$$\ln(-r_A) = \ln k + m \ln [A] + n \ln [B] \tag{4.41}$$

and the parameters k, m, and n determined by regression analysis.

In the integral method, a suitable function of $[A]$ or x_A, which can be found in Table 4.2 for several reactions, is plotted as a function of time. This is illustrated in Figure 4.5 for four types of reactions.

FROM PRESSURE DATA

Typically, a liquid-phase organic reaction is carried out in a closed vessel, and the progress of reaction monitored by changes in concentration (by analysis of samples withdrawn periodically). Where gas-phase reactions are concerned or a liquid-phase reaction in which one of the products is a gas, the progress of

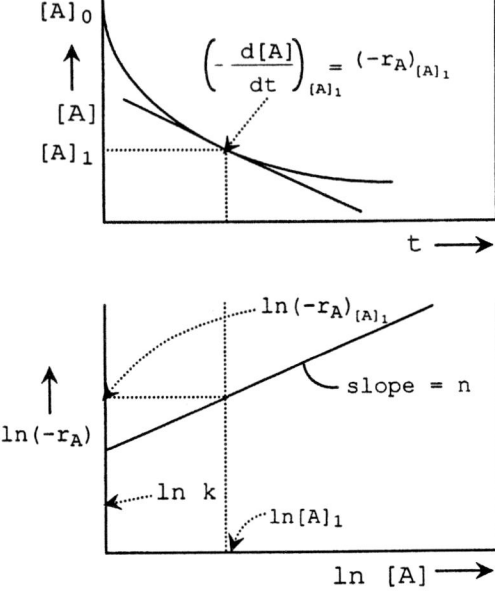

Figure 4.4 Rate constant from concentration-time data for $r_A = k[A]^n$.

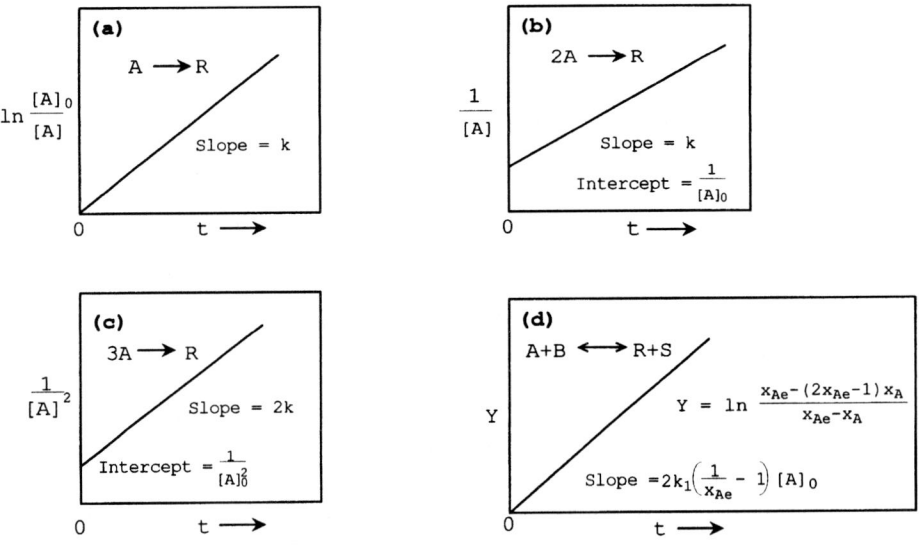

Figure 4.5 Use of integrated forms for obtaining rate constants.

reaction can be more easily monitored by recording the pressure as a function of time. This method is particularly useful for reactions with volume change.

Thus for a typical (and common) gas-phase reaction with volume change

$$A \rightarrow R + S \tag{4.8}$$

the following expression can be derived:

$$\frac{dP_A}{dt} = \bar{k}(2P_{A0} - P)^n \tag{4.42}$$

where

$$\bar{k} = k(R_g T)^{1-n} \tag{4.43}$$

Now, we can use either the differential or the integral method to get the rate parameters.

EXAMPLE 4.1

Decomposition of di-t-butyl peroxide to acetone and ethane in the vapor phase

The reaction

$$\underset{(A)}{(CH_3)_3COOC(CH_3)_3} \rightarrow \underset{(R)}{2(CH_3)_2CO} + \underset{(S)}{C_2H_6} \tag{E4.1.1}$$

was carried out in a constant-volume batch reactor (Peters and Skorpinski, 1965), and its progress was monitored by the increase in pressure. However, as the P–t data were not recorded in the original paper, they have been generated from the data given, and the values at two temperatures are given in Table 4.4.

Table 4.4 Pressure–temperature data for the decomposition of di-t-butyl peroxide

\multicolumn{2}{c}{$T = 154.6\,°C$}		$T = 147.2\,°C$	
Time (min)	P (mmHg)	Time (min)	P (mmHg)
0	173.5	0	182.6
2	187.3	2	190.5
3	193.4	6	201.7
5	205.3	10	213.6
6	211.3	14	224.3
8	222.9	18	235.0
9	228.6	20	240.4
11	239.8	22	245.4
12	244.4	26	255.6
14	254.5	30	265.2
15	259.2	34	274.4
17	268.7	38	283.3
18	273.9	40	288.0
20	282.0	42	292.0
21	286.8	46	300.2

Nitrogen was used as the diluent in the reaction. Its partial pressure remained approximately constant throughout the reaction (P_{N_2} at $154.6\,°C = 8.1\,mm$ and at $147.2\,°C = 4.5\,mmHg$). The initial partial pressures of the peroxide (A) were 168 and 179 mm Hg at $154.6\,°C$ and $147.2\,°C$, respectively. Obtain a suitable first-order rate equation for the reaction, assuming an ideal gas.

SOLUTION

The material balance for A gives

$$-\frac{1}{V_t}\frac{dN_A}{dt} = -\frac{d[A]}{dt} = k_v \frac{N_A}{V_t} \quad (E4.1.1)$$

For constant-volume conditions,

$$-\frac{d[A]}{dt} = k_v[A] \quad (E4.1.2)$$

Also,

$$N_t = N_A + N_R + N_S + N_{N_2}$$
$$N_R = 2N_S = 2(N_{A_0} - N_A)$$
$$N_t = N_A + 2(N_{A_0} - N_A) + (N_{A_0} - N_A) + N_{N_2}$$

giving

$$N_A = \frac{3N_{A0} + N_{N_2} - N_t}{2} \quad (E4.1.3)$$

Assuming the ideal gas law applies, $[A] = p_A/R_g T$,

$$-\ln\left(\frac{[A]}{[A]_0}\right) = k_v t \quad (E4.1.4)$$

or

$$\ln\left[\frac{p_{A0}}{p_A}\right] = \ln\left[\frac{3p_{A0} + p_{N_2} - p_{t0}}{3p_{A0} + p_{N_2} - p_t}\right] = k_v t \quad (E4.1.5)$$

Thus a plot of $\ln[3p_{A0} + p_{N_2} - p_t]$ versus t gives a slope of k_v. From such a plot we get

$$k_v = 0.0086\,min^{-1} \text{ at } 147.2\,°C$$
$$= 0.0193\,min^{-1} \text{ at } 154.6\,°C$$

Thus the rate model is

$$-r_A = k_v[A] \quad (E4.1.6)$$

where A represents butyl peroxide and k_v is the first-order rate constant with the values given above.

Nonisothermal operation

Most reactions are characterized by reasonable heats of reaction, and hence it is not always possible to operate them under isothermal conditions with a constant

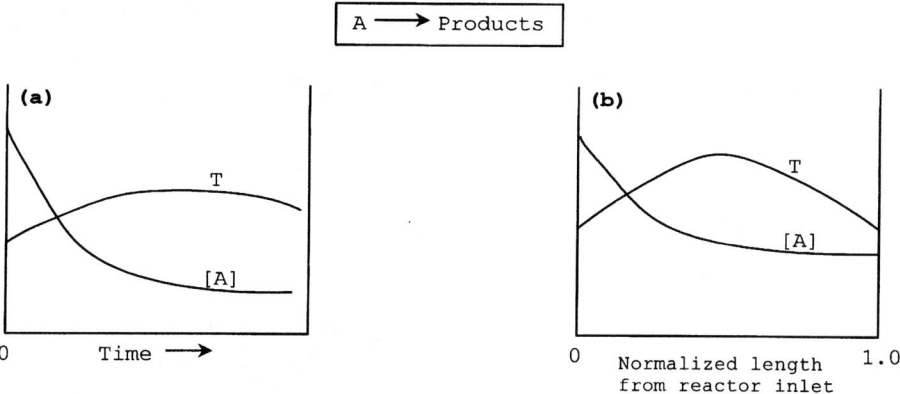

Figure 4.6 Temperature (T) and concentration ([A]) profiles (a) over time for the batch reactor and (b) along reactor length for the PFR.

heat exchange rate. A temperature profile is established (along with concentration profiles) with progress in time, as shown in Figure 4.6(a). It is necessary in such cases to write the energy balance in addition to the mass balance and solve them simultaneously.

Some kind of heat exchange is necessary to control the temperature, and this is usually done by circulating a heat exchange fluid through the reactor jacket or an immersed coil as the case may be. Three major parameters are involved in accounting for heat exchange between the fluids: the heat transfer area, the temperature difference or driving force beween the phases, and the rate of heat tranfer expressed through a heat transfer coefficient. We shall anticipate the concept of such a coefficient at this stage and explain it more fully later on.

The mass and energy balances are

$$\frac{dX_A}{dt} = \frac{V}{N_{A0}}(-r_A)(X_A, T) \tag{4.44}$$

$$N_{t0}C_{pm}\frac{dT}{dt} = V(-\Delta H_r)(-r_A)(X_A, T) + UA_h(T_w - T) \tag{4.45}$$

where U is the overall heat transfer coefficient at the heat exchange surface, A_h is the heat transfer surface, $(T_w - T)$ is the temperature driving force between the wall and the reaction mixture, and C_{pm} is the molar heat capacity of the total entering feed. They can be combined to give

$$N_{t0}C_{pm}\frac{dT}{dt} - (-\Delta H_r)N_{A0}\frac{dX_A}{dt} = UA_h(T_w - T) \tag{4.46}$$

Integration of this equation between the inlet (X_{A0}, T_0) and outlet (X_A, T) gives

$$N_{t0}C_{pm}(T - T_0) - N_{A0}(-\Delta H_r)(X_A - X_{A0}) = U(T_w - T)A_h t \tag{4.47}$$

For the special case of adiabatic operation, there is no abstraction or addition of heat. Hence the last term of Equation 4.47 vanishes, and the following unique relationship between temperature and conversion results:

$$(X_A - X_{A0}) = \alpha(T - T_0) \tag{4.48}$$

where

$$\alpha = \frac{C_{PA}}{(-\Delta H_r)} \tag{4.49}$$

and $C_{PA} = C_{pm}(N_{t0}/N_{A0})$ = heat capacity of the total feed stream per mole of entering A. Note that no such relationship exists for nonadiabatic operation. Therefore, solutions for nonisothermal operation depend on whether it is adiabatic or nonadiabatic.

ADIABATIC OPERATION

We can recast Equations 4.44 and 4.48 as

$$\Delta t = \beta \Delta X_A \tag{4.50}$$

$$\Delta X_A = \alpha \Delta T \tag{4.51}$$

where

$$\beta = \frac{N_{A0}}{V(-r_A)} \tag{4.52}$$

and solve them numerically.

NONADIABATIC OPERATION

Here, we solve Equations 4.44 and 4.45 simultaneously using the following relationship for the overall heat transfer coefficient (assuming that the inner and outer diameters of the reactor are approximately equal):[2]

$$\frac{1}{U} = \frac{1}{h_r} + \frac{t_w}{\lambda_w} + \frac{1}{h_c} \tag{4.53}$$

where h_r is the heat transfer coefficient on the reaction mixture side of the reactor and is given by (Chilton et al., 1944)

$$\frac{h_r d_T}{\lambda}\left(\frac{\mu_w}{\mu}\right) = 0.36 \left[\frac{d_s^2 (rps)\rho}{\mu}\right]^{0.66} \left(\frac{C_p \mu}{\lambda}\right)^{0.33} \tag{4.54}$$

μ_w is the viscosity of the reaction mixture at the wall temperature, t_w is the thickness of the reactor wall, rps is the number of stirrer revolutions per second, d_s is the propeller diameter, and all other properties are for the reaction mixture.

---

[2] The more rigorous version is given in Equation 4.87.

More elaborate equations for estimating it have also been proposed (e.g., Strek, 1963; Chapman et al., 1964). h_c is the heat transfer coefficient on the control fluid side of the reactor, for which the following correlation for turbulent conditions may be used:

$$\frac{h_c d_i}{\lambda} = 0.023 \left(\frac{d_i u \rho}{\mu}\right)^{0.8} \left(\frac{C_{pc}\mu_c}{\lambda}\right)^{0.4} \tag{4.55}$$

where d_i is the inner diameter of the heat transfer tube and the subscript c stands for the control fluid.

Optimal operating policies

Although the design of the batch reactor already considered gives the batch time for a given duty, this time is not necessarily the optimum reaction time for maximum profit. Aris (1965, 1969) suggests a method for calculating the optimum time for maximizing profit at a given temperature. It is more important, however, to compute a time–temperature policy for maximizing performance. Thus, because the reaction rate for a simple reaction always increases with temperature, the optimum temperature policy for a simple reaction is merely the maximum temperature possible. This is fixed by considerations such as the material of construction of the reactor, catalyst deactivation (in catalytic reactions), etc.

On the other hand, different time–temperature policies are optimal for different classes of complex reactions. These are considered in Chapter 11 for complex reactions. Although the reversible reaction is also a complex reaction in the sense that two reactions occur, it is equally true that no additional species are involved in the second (reverse) reaction. Hence the reversible reaction can also be regarded as a simple reaction. If the reaction is endothermic, its reversible nature makes no difference because both the reaction rate constant and the equilibrium constant increase with temperature, and the maximum practicable temperature continues to be the optimal temperature. But if the reaction is exothermic, an increase in temperature has opposite effects: it lowers the equilibrium constant but raises the rate constant. Hence an optimum temperature exists. For any reaction such as $A \leftrightarrow R$ whose rate equation is $-r_A = k([A] - [R]/K)$, this optimum can be found by integrating the expression

$$t = [A]_0 \int_0^{X_A} \frac{dX_A}{[-r_A(X_A, T)]} \tag{4.56}$$

for different constant values of T (and hence of k, K, and the rate) and finding the optimum temperature for minimum reaction time.

The Plug-Flow Reactor

The characteristic feature of the plug-flow reactor (PFR) is that there is no feedback from downstream to upstream. This kind of ideal behavior eliminates

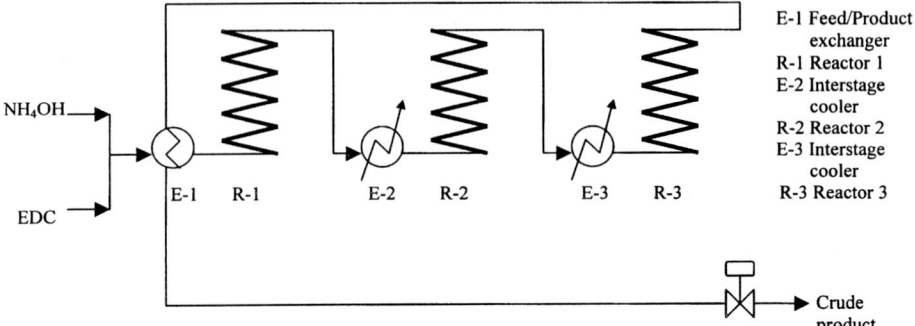

Figure 4.7 A typical PFR for the production of ethylenediamine (Venkitakrishnan and Doraiswamy, 1982).

many complications associated with fluid flow and leads to relatively simple reactor equations.

The plug-flow reactor is usually a long tube, straight or coiled, a set of straight tubes connected in series at their ends, or a bank of independent tubes. The diameter of the tube is usually not more than 3 cm. For the PFR assumption to be valid, the length-to-diameter ratio should be very high, at least 30. These reactors are common for the solid catalyzed vapor-phase reactions considered in Chapter 12. Where no solids are present, the reactor tube is sometimes coiled to accommodate high residence times. An example of this is the coiled reactor for the production of ethylenediamine by reaction between ethylene dichloride (EDC) and aqueous ammonia sketched in Figure 4.7 (Venkitakrishnan and Doraiswamy, 1982).

A serious drawback of PFR is the variation of temperature with length (Figure 4.6b), leading to significant temperature gradients within the reactor. If a reaction is highly temperature sensitive, has a large heat of reaction, and operation at an optimum temperature (determined from laboratory experiments) is essential, a PFR cannot be used. Either a batch reactor or a mixed-flow reactor (to be described later in this chapter) is a more suitable choice. However, a PFR with controlled heat exchange to give a favorable temperature profile can also be used and may often be the preferred candidate. Other advantages are high selectivity in complex reactions and low reactor volume. In general, liquid-phase reactions are carried out in batch or semibatch reactors, and large-volume vapor-phase catalytic reactions in plug-flow reactors of the heat exchanger type, to be described in Chapter 12.

The basic PFR equation

A sketch of the reactor along with the inlet and outlet concentrations and flow rates is shown in Figure 4.8. The material balance of Equation 4.1 holds equally for a differential element dV of this reactor over which the rate is assumed to be constant. However, there is a finite flow into the element and a finite flow out of it, and as a result of the steady-state assumption, there is no accumulation in the

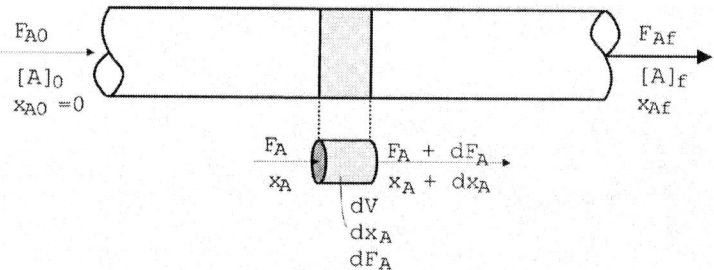

Figure 4.8 Sketch of a plug-flow reactor.

reactor. Therefore, in terms of the nomenclature of Figure 4.8, this equation can be written as

$$F_A - (F_A + dF_A) = 0 + (-r_A)dV \tag{4.57}$$

or

$$-dF_A = (-r_A)dV$$

Because $F_A = F_{A0}(1 - X_A)$,

$$F_{A0}dX_A = (-r_A)dV \tag{4.58}$$

Solution of this equation with boundary conditions $V = 0$, $V = V$, and $X_A = 0$, $X_A = X_{Af}$ gives

$$\frac{V}{F_{A0}} = \int_{X_{A0}}^{X_{Af}} \frac{dX_A}{(-r_A)}, \quad X_{A0} = 0 \text{ (usually)} \tag{4.59}$$

This is the basic equation for a PFR.

It is useful to modify Equation 4.59 so that it becomes equivalent to Equation 4.34 for a BR, particularly because the two are identical if time is replaced by residence time V/Q or l/u. For this purpose, we define a parameter

$$\bar{t} = \frac{V[A]_0}{F_{A0}} = \frac{V}{Q_0} \tag{4.60}$$

which represents the time needed to treat one reactor volume of the feed stream. This is equal to the residence time for a constant-density system. Thus Equation 4.59 becomes

$$\bar{t} = [A]_0 \int_0^{X_{Af}} \frac{dX_A}{(-r_A)}, \quad \varepsilon_A = \text{any value} \tag{4.61}$$

$$= \int_{[A]_f}^{[A]_0} \frac{d[A]}{(-r_A)}, \quad \varepsilon_A = 0 \tag{4.62}$$

Design equations

Note that Equations 4.59 and 4.61 are general and valid irrespective of volume change. Integrated forms for constant-density systems are identical to those for a

Table 4.5 Analytical solutions (design equations) for some simple variable density reactions in a PFR

Reaction	Rate equation	Design equation
1. $A \rightarrow R$	$-r_A = k[A]$	$kt = (1 + \varepsilon_A) \ln\left(\dfrac{1}{1 - X_A}\right) - \varepsilon_A X_A$
2. $A \rightleftarrows \nu_R R$	$-r_A = k_+[A] - k_-[R]$	$k\bar{t} = \dfrac{\psi_R + \nu_R X_{Ae}}{\psi_R + \nu_R}\left[(1 + \varepsilon_A X_{Ae}) \ln\left(\dfrac{X_{Ae}}{X_{Ae} - X_A}\right) - \varepsilon_A X_A\right]$
3. $2A \rightarrow R$	$-r_A = k[A]^2$	$k[A]_0 \bar{t} = 2\varepsilon_A(1 + \varepsilon_A)\ln(1 - X_A) + \varepsilon_A^2 X_A + (\varepsilon_A + 1)^2\left(\dfrac{X_A}{1 - X_A}\right)$
4. $A + B \rightarrow R$ $\psi_B = 1$	$-r_A = k[A][B]$	

BR (Table 4.2). Solutions to a few variable-density reactions are presented in Table 4.5. Graphical integration is straightforward and gives the reactor volume directly, as shown in Figure 4.9. The reciprocal rate is plotted as a function of either X_A (Equation 4.61) or $[A]$ (Equation 4.62). Alternatively, any of several numerical methods can be used, and this is perhaps the most attractive.

Rate parameters from PFR data

The rate constant and order of reaction can be determined from PFR data by using the methods described in the section on BR. The only difference is that now we use the ratio (V/F_{A0}) instead of time. Note that BR operation enables measuring the concentration continuously as a function of time. Thus a single run is often sufficient to construct a concentration–time plot. On the other hand, because a PFR is a steady-state operation, every run at a given feed rate F_{A0} (and therefore a given value of V/F_{A0}, because V is constant) gives a single point on the $[A]$–(V/F_{A0}) plot. Hence a number of runs are required to prepare such a single plot. Then the rate and kinetic parameters are found by the procedure described for the BR.

Nonisothermal operation

In a PFR, where time is not usually a parameter, nonisothermicity is reflected in a temperature profile in the reactor from inlet to outlet, as shown in Figure 4.6b.

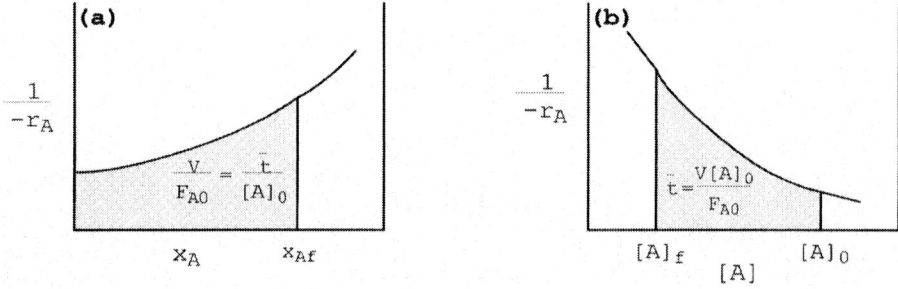

Figure 4.9 Graphical integration of (a) Equation 4.61 and (b) Equation 4.62.

Therefore, nonisothermal operation of a PFR is an inherent feature of the reactor.

Consider reaction 4.1 (with first-order kinetics) conducted in a nonisothermal PFR. Now, the mass balance for the PFR given by Equation 4.58 must be supplemented by the energy balance. Restricting our attention to adiabatic operation, the energy balance may be written as

$$F_{A0}\,dX_A(-\Delta H_r) = F_t C_{pm}\,dT \tag{4.63}$$

or

$$dX_A = \frac{F_t C_{pm}}{F_{A0}(-\Delta H_r)}\,dT \tag{4.64}$$

Integration of Equation 4.64 leads to Equation 4.48 with α modified as

$$\alpha = \frac{F_t C_{pm}}{F_{A0}(-\Delta H_r)} \tag{4.65}$$

Then, assuming first-order kinetics with Arrhenius dependence of k on T, Equation 4.59 can be written as

$$\frac{V}{F_{A0}} = \frac{1}{k^0[A]_0} \int_{T_0}^{T} \frac{e^{E/R_g T}\alpha}{\{1 - [X_{A0} + \alpha(T - T_0)]\}}\,dT \tag{4.66}$$

A fully analytical solution to this equation is not possible, but a semianalytical solution has been given by Douglas and Eagleton (1962). However, numerical solution is quite straightforward, as shown here, and appears to be the method of choice. The objective is to establish the temperature and concentration profiles in the reactor. For this purpose, we express V as the product $A_c l$ (where A_c is the cross-sectional area of the reactor) and use Equations 4.58, 4.64, and 4.65 to write the following equivalent forms of Equations 4.50 and 4.51 of the batch reactor:

$$\Delta X_A = \beta' \Delta l \tag{4.67}$$

$$\Delta T = \alpha' \Delta l \tag{4.68}$$

where

$$\alpha' = \frac{(-\Delta H_r) A_c (-r_A)}{C_{pm} F_t}, \quad \beta' = \frac{A_c(-r_A)}{F_{A0}} \tag{4.69}$$

The use of Equations 4.67 and 4.68 in computing the profiles in a packed-bed tubular reactor is illustrated in Example 12.3.

The Mixed-Flow Reactor

The mixed-flow reactor is a continuous reactor with a constant volumetric inflow of reactants and outflow of products, but the fluid (usually liquid) within the reactor is in a state of perfect mixing. As a result, the composition of this liquid is spatially uniform within the reactor. Thus the outflow from the reactor will be at the same composition as the liquid within.

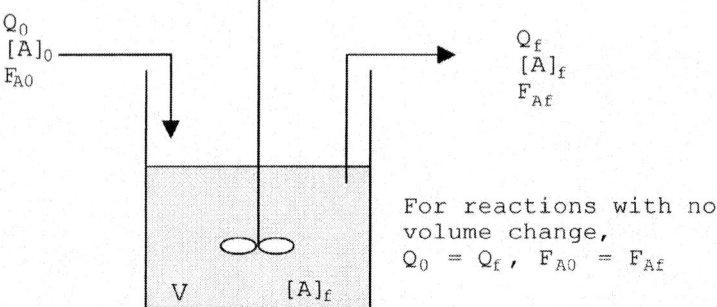

Figure 4.10 Mixed-flow reactor (MFR), also called continuous stirred tank reactor (CSTR).

The basic MFR equation

A sketch of a mixed-flow reactor is shown in Figure 4.10. The inflow and outflow rates and compositions in the two streams, as well as within the reactor, are clearly marked. Again, the material balance of Equation 4.30 holds, leading to

$$Q_0[A]_0 - (-r_{Af})V = Q_f[A]_f \tag{4.70}$$

for component A. Note that the rate has been written in terms of the outlet conditions. This is a distinctive feature of an MFR, one that enables algebraic equations to be written instead of differential equations as in a PFR in which the rate and concentrations change from inlet to outlet.

For liquid-phase systems, we can assume that $Q_0 = Q_f$, and Equation 4.70 simplifies to

$$[A]_0 - [A]_f = (-r_{Af})\bar{t}, \quad \text{or} \quad \bar{t} = \frac{[A]_0 - [A]}{-r_{Af}} \tag{4.71}$$

From the relationship $X_A = [([A]_0 - [A])/[A]_0]$ and the definition of residence time given by Equation 4.60, this equation can be recast as

$$\frac{V}{F_{A0}} = \frac{X_{Af}}{-r_{Af}}, \quad \text{or} \quad \bar{t} = \frac{[A]_0 X_{Af}}{-r_{Af}} \tag{4.72}$$

which for a first-order reaction becomes

$$X_{Af} = \frac{k\bar{t}}{1 + k\bar{t}} \tag{4.73}$$

Design equations

Equation 4.72 is the basic equation for a constant-volume MFR and can be solved for any rate form to give the corresponding design equations. Table 4.6 lists these equations for a few common reactions.

Table 4.6 Design equations for some common reactions in an MFR

		Design equation	
Reaction	Rate equation	Volume change	No volume change[a]
1. $A \to R$	$-r_A = k[A]$	$k\bar{t} = \dfrac{X_A(1+\varepsilon_A X_A)}{1-X_A}$	$\dfrac{[A]}{[A]_0} = \dfrac{1}{1+k\bar{t}}$
2. $2A \to R$	$-r_A = k[A]^2$	$k[A]_0 \bar{t} = \dfrac{X_A(1+\varepsilon_A X_A)^2}{(1-X_A)^2}$	$k\bar{t} = \dfrac{[A]_0 - [A]}{[A]^2}$
3. $A + \nu_B B \to R$	$-r_A = k[A][B]$	$k[A]_0 \bar{t} = \dfrac{X_A(1+\varepsilon_A X_A)^2}{(1-X_A)(\psi_B - \nu_B X_A)}$	$k\bar{t} = \dfrac{[A]_0 - [A]}{[A]([B]_0 - \nu_B[A]_0 + \nu_B[A])}$
4. $3A \to R + S$	$-r_A = k[A]^3$	$k[A]_0^2 \bar{t} = \dfrac{X_A(1+\varepsilon_A X_A)^3}{(1-X_A)^3}$	$k\bar{t} = \dfrac{[A]_0 - [A]}{[A]^3}$
5. $A \underset{k_-}{\overset{k_+}{\rightleftharpoons}} \nu_R R$	$-r_A = k_+[A] - k_-[R]$	$k_+ \bar{t} = \dfrac{\psi_R + \nu_R X_{Ae}}{\psi_R + \nu_R} \dfrac{X_A(1+\varepsilon_A X_A)}{(X_{Ae} - X_A)}$	$k_+ \bar{t} = \dfrac{[R]_e}{[A]_0(\psi_R + \nu_R)} \dfrac{([A]_0 - [A])}{([A] - [A]_e)}$

[a] These expressions are given in terms of concentration because a simple relationship $[A] = [A]_0(1 - X_A)$ exists between conversion and concentration for $\varepsilon_A = 0$.

76 Reactions and Reactors

The procedure for variable-volume reactions is the same as described for a PFR. Thus we account for a volume change through the term $(1 + \varepsilon_A X_A)$. With this modification, the design equation for reaction 4.1, with a first-order rate law, for instance, is

$$k\bar{t} = \frac{X_A(1 + \varepsilon_A X_A)}{1 - X_A} = \frac{[A]_0([A]_0 - [A])}{[A]([A]_0 + \varepsilon_A[A])} \tag{4.74}$$

Equations for a few other variable-volume reactions are included in Table 4.6.

Note that these are simple algebraic equations. All one has to do is write the rate equation for a reaction at hand and compute the residence time. However, in the production of small volume organic chemicals, one is often faced with the reverse problem, finding the rate of production in an available reactor of known size. This is so because separate reactors are seldom allotted for each reaction. The only change in procedure in this case is that Q is obtained for a fixed value of V from Equation 4.72.

Rate parameters from MFR data

Equation 4.72 shows that the rate can be obtained directly by operating an MFR at a fixed value of F_{A0}. Because the reactor is fully mixed, the rate corresponds to the conditions in the outlet stream, which are the same as those within the reactor. Different flow rates corresponding to different values of F_{A0} give different outlet compositions and therefore different conversions.

Thus the rate can be obtained directly from Equation 4.72 as a function of $[A]$ (or X_A) without the need for differentiation of $[A]$ (or X_A) versus t or V/F_{A0} data (as in a BR or a PFR). Obtaining rates by this method is experimentally very attractive and should be used wherever possible.

Nonisothermal Operation

The heat effect in a stirred reactor is quite different from that in a PFR. Unlike in a PFR, there is no spatial variation of temperature or concentration in a stirred reactor because it is fully mixed. Hence the effect of the energy balance is restricted to raising or lowering the single temperature at which the reactor is operating. As a result, the analysis is simpler. To the mass balance given by Equation 4.70, we now add the energy balance

$$Q_0 C_p \rho (T_0 - T) + k(T)[A]V(-\Delta H_r) - UA_h(T_0 - T_c) = 0 \tag{4.75}$$

Combining the two equations leads to

$$Q_0 \rho C_p (T_0 - T) - UA_h(T - T_c) = \frac{-V(-\Delta H_r)k(T))Q_0[A]_0}{Q_0 + Vk(T)} \tag{4.76}$$

or

$$J[(T - T_0) + \kappa(T - T_c)] = \frac{\bar{t}k(T)}{1 + \bar{t}k(T)} \tag{4.77}$$

where

$$J = \frac{\rho C_p}{(-\Delta H_r)[A]_0} \text{ (units of 1/K)}, \quad \kappa = \frac{UA_h}{Q_0 \rho C_p} \text{ (dimensionless)} \quad (4.78)$$

and $k(T)$ is given by the Arrhenius equation $k^0 \exp(-E/R_g T)$. Note that ρC_p can be replaced by the molar product $\rho_m C_{pm}$. Equation 4.77 gives the temperature at which the reactor would operate at steady state. For adiabatic operation, the heat transfer term (second in the brackets on the LHS) vanishes.

This equation is important in the context of establishing the temperature of the reactor and also because it can have more than one solution, leading to operation at more than one steady state. This is part of a more general phenomenon and is considered in Chapter 13.

Comparison of Batch, Plug-Flow, and Mixed-Flow Reactors

We already derived the performance equations for the three ideal reactors, batch, plug-flow, and mixed-flow. The batch and plug-flow reactors are exactly comparable, and the reaction time t in a BR is related to the residence time $\bar{t}$ at the corresponding axial position in a PFR by

$$t = \left(\frac{l}{L}\right)\bar{t} \quad (4.79)$$

An MFR, on the other hand, operates on a different principle, that of complete backmixing, and therefore no direct relationship between an MFR and a BR or PFR is possible.

TRANSPORT BETWEEN PHASES

The Chemist's and Chemical Engineer's Understandings of Mass Transport

Before fully appreciating any concept of mass transport, an understanding of the frequently used term *flux* is necessary. It is simply the amount of fluid transported per unit time per unit area (perpendicular to the direction of transport).

According to the well-known Fick's law, the flux is proportional to the concentration difference of a "moving" gas between two points divided by the distance separating the points, and the proportionality constant is known as the *diffusion coefficient* or *diffusivity* [with units of (distance)^2/time]. This is the way the chemist normally looks at mass transport and is essentially tantamount to a distributed parameter model in which the concentration is viewed as varying with time and position within a phase.

When transport occurs between phases, we usually express the flux as a function only of the concentration difference of the "diffusing gas." The proportionality constant here is known as the *mass transfer coefficient* (with units of [distance/time]). More puristically, it is referred to as the *phenomenological mass transfer coefficient*, to distinguish it from other, often more useful,

definitions. This is usually the chemical engineer's way of looking at mass transport and is essentially a lumped parameter model in which the transport is "localized" and distance does not appear as a dimension.

The concept of diffusion is used whenever one is dealing with transport within a phase as a function of time and position. For example, when a chemical reaction occurs in a catalyst pellet, the reactant has to diffuse through the catalyst and react while it is still diffusing. Thus, in any rational analysis of such a situation, we (chemists or chemical engineers) are concerned with diffusion. As we shall see in Chapter 7, the Thiele modulus, which is central to the analysis of catalytic reactions, is based on the joint use of diffusion and reaction coefficients in a single dimensionless group.

On the other hand, the synthetic organic chemist, who has often to deal with immiscible or sparingly miscible fluids, has to contend with the problem of transport of a desired species across an interface such as gas-liquid, liquid–liquid, or fluid-solid. In such a situation, one has to fall back on the concept of mass transfer coefficient by defining a hypothetical *film* across which transport occurs—an approach to which the chemist is unaccustomed. Although the two models (one based on diffusion and the other on mass transfer coefficient) are related, we shall largely be concerned (more as chemical engineers now) with the latter in dealing with interfacial phenomena (e.g., in Chapters 7 and 14–17).

Theories of Mass Transfer

When the reactants are present in two different phases, one of them must diffuse from its phase into the other for reaction to occur there. If the distribution coefficients of the two reactants do not favor any particular phase, the reaction can occur in both phases, particularly if both are liquid. Clearly, therefore, in general the rates of *mass transfer* of reactants between phases becomes an important consideration in heterogeneous systems.

Three major theories of mass transfer have been in vogue to explain interphase transport: Whitman's film theory, and the Higbie and Danckwerts versions of the penetration theory. The *film theory* in essence asserts that adjacent to any interface, there is a stagnant *film* of thickness δ through which transport of any species occurs by molecular conduction (there is no convection). Conditions in the rest of the fluid, called the *bulk*, are assumed to remain constant, so that the driving force for transport is consumed entirely by the film, as depicted in Figure 4.11a. If, after or during transport, the diffusing molecule participates in a reaction in this phase, the reaction can occur either in the film, the bulk, or in both. Hence one has to combine the theories of mass transfer and reaction in situations of this kind, which are more fully considered in Part IV.

Another theory of mass transfer is based on the postulate that elements of the fluid impinge on the interface where they remain for a specified period of time during which they shed their load of reactant and then return to the body of the fluid. The contact time of an element with the interface can be constant for all elements (Higbie, 1935) or vary from element to element (Danckwerts, 1953). Such a postulate, sketched in Figure 4.11b, is the basis of the *penetration theory*.

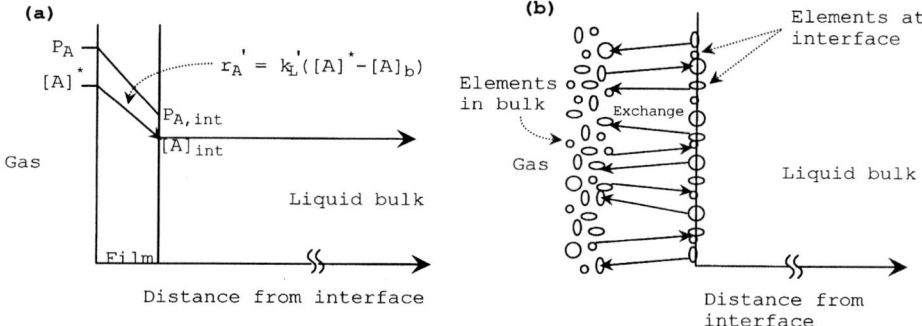

Figure 4.11 Schematic representations of (a) film and (b) penetration theories.

The basic equation (for any theory) for mass transfer of species A can be written as

$$r'_A = k'_L([A]^* - [A]_b) \tag{4.80}$$

where $[A]^*$ and $[A]_b$ are the concentrations of A at saturation and in the bulk, respectively, and k'_L is the *specific mass transfer coefficient* (with units of m/s). It is in the definition of k'_L that the film and penetration theories differ from each other. Note that $[A]_{int}$ is assumed equal to $[A]_b$.

FILM THEORY

$$k'_L = \frac{D_A}{\delta} \tag{4.81}$$

The advantage of using Equation 4.81 for the mass transfer coefficient lies in the assumption that under similar hydrodynamic conditions, the film thickness δ is constant, so that k'_L is directly proportional to D_A.

PENETRATION THEORIES

Astarita (1967) has combined the Higbie and Danckwerts approaches to give the following general equation for the mass transfer coefficient:

$$k'_L = \sqrt{\frac{D_A}{t_D}} \tag{4.82}$$

where t_D is an *equivalent diffusion time* that replaces t^* (the time each element spends at the surface in the Higbie model), and $1/s$ (the average life of surface elements, where s is the rate of surface renewal of the Danckwerts model).

Thus we have two models represented by Equations 4.81 and 4.82 for the film and penetration theories, respectively, to choose from. In making a choice, note that one is often concerned with the *ratio* of diffusivities of the two reactants of a two-phase system (see Chapter 14), and not the diffusivity of just one diffusing component. Because the diffusion coefficients of organic compounds in many of the solvents normally used do not greatly differ from one another, the difference

between D_B/D_A and $\sqrt{D_B/D_A}$ tends to be negligibly small. Hence one is often justified in using the film model except when unsteady-state behavior is clearly indicated. As a result, our treatment of interphase transport will largely be based on the film theory; the penetration theory is invoked only in some special cases.

Two-Film Theory of Mass and Heat Transfer for Fluid-Fluid Reactions in General

Mass transfer

To provide a basis to account for the influence of phase heterogeneity in reaction analysis, the theories presented were based on a single film associated with a single phase. In applying them to real sytems, two films must be considered, one on either side of the interface, as shown in Figure 4.12. Now, because two mass transfer coefficients are involved, an overall mass transfer coefficient is defined as

$$\frac{1}{k'_{GL}} = \frac{1}{k'_G} + \frac{1}{k'_L} \quad \text{(gas-liquid)} \tag{4.83}$$

$$\frac{1}{k'_{L_1L_2}} = \frac{1}{k'_{L_1}} + \frac{1}{k'_{L_2}} \quad \text{liquid-liquid} \tag{4.84}$$

where the mass transfer coefficient k' has units of m/s.

Heat transfer

The basic concepts of mass transfer can be readily extended to heat transfer by writing the rate of heat transfer (from phase 1 to phase 2) across a surface as

$$Q = UA_h(T_1 - T_2) \tag{4.85}$$

where U is *an overall heat transfer coefficient* defined as

$$\frac{1}{U} = \frac{1}{h_1} + \frac{1}{h_2} \tag{4.86}$$

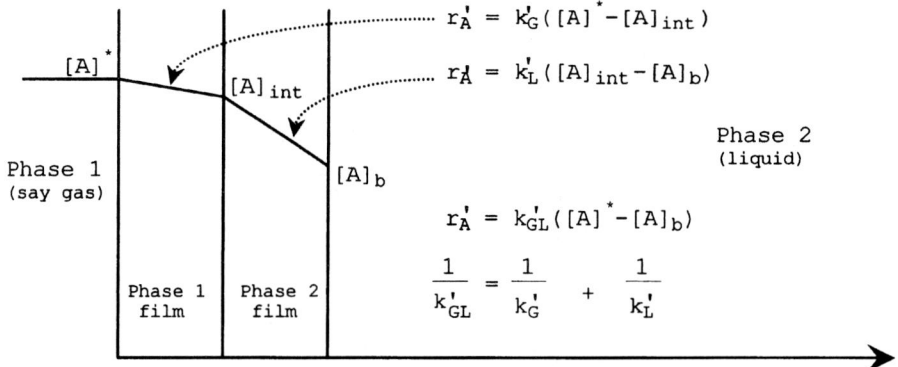

Figure 4.12 Mass transfer coefficients and reaction rates per the two-film theory of mass transfer.

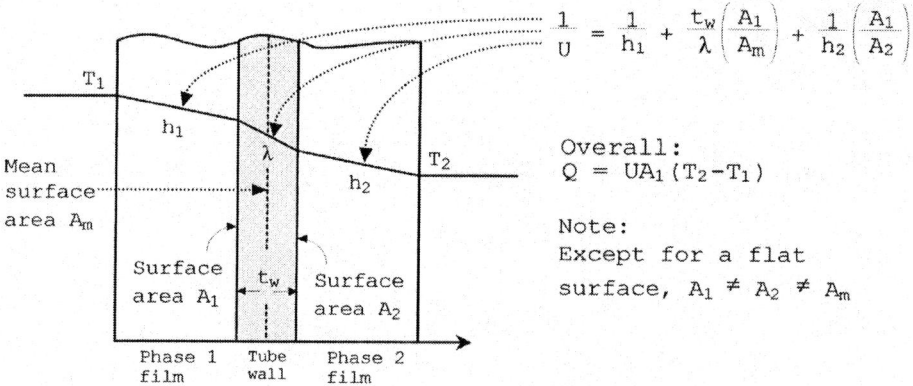

Figure 4.13 Two-film representation of heat transfer in the cooling (or heating) of reactor fluids by passing a control fluid through the reactor jacket or an internal coil.

and h_1 and h_2 are the individual coefficients for the two films (see Figure 4.13). Note, however, that, unlike in mass transfer, a third mode of transport is also involved in heat transfer, conduction through the wall (of thickness t_w). Thus Equation 4.86 should be modified as

$$\frac{1}{U} = \frac{1}{h_1} + \frac{t_w}{\lambda}\frac{A_1}{A_m} + \frac{1}{h_2}\frac{A_1}{A_2} \qquad (4.87)$$

where A_1 and A_2 are the surface areas of the two sides of the wall (one containing the reaction mixture and the other the control fluid), and A_m is the logarithmic mean of A_1 and A_2.

LABORATORY REACTORS

The type of reactor to be used in laboratory experiments to obtain data for process evaluation or kinetic modeling depends on the nature of the reaction: homogeneous gas or liquid phase reaction, gas-liquid or liquid-liquid reaction, gas or liquid phase reaction on a solid catalyst, or three-phase reaction. Solid phase reactions are also possible, but they are quite rare in organic synthesis and technology. Laboratory reactors for organic synthesis can roughly be divided into two categories: reactors for gathering data in a chemist's laboratory with the object of developing a feasible synthetic route for a chemical, and those used to obtain precise kinetic data under isothermal conditions, which also take into account the mass and heat transfer features of the reaction. Figure 4.14 lists the main laboratory reactors used for different reaction systems along with an indication of the chapters in which they are considered subsequently in the book. In these reactors, one either fully eliminates mass transfer resistances or accounts for them wherever they are inevitably present. Designs that eliminate mass and heat transfer effects are called *gradientless reactors*. They may be regarded as heterogeneous reactor equivalents of homogeneous reactors which, by their very

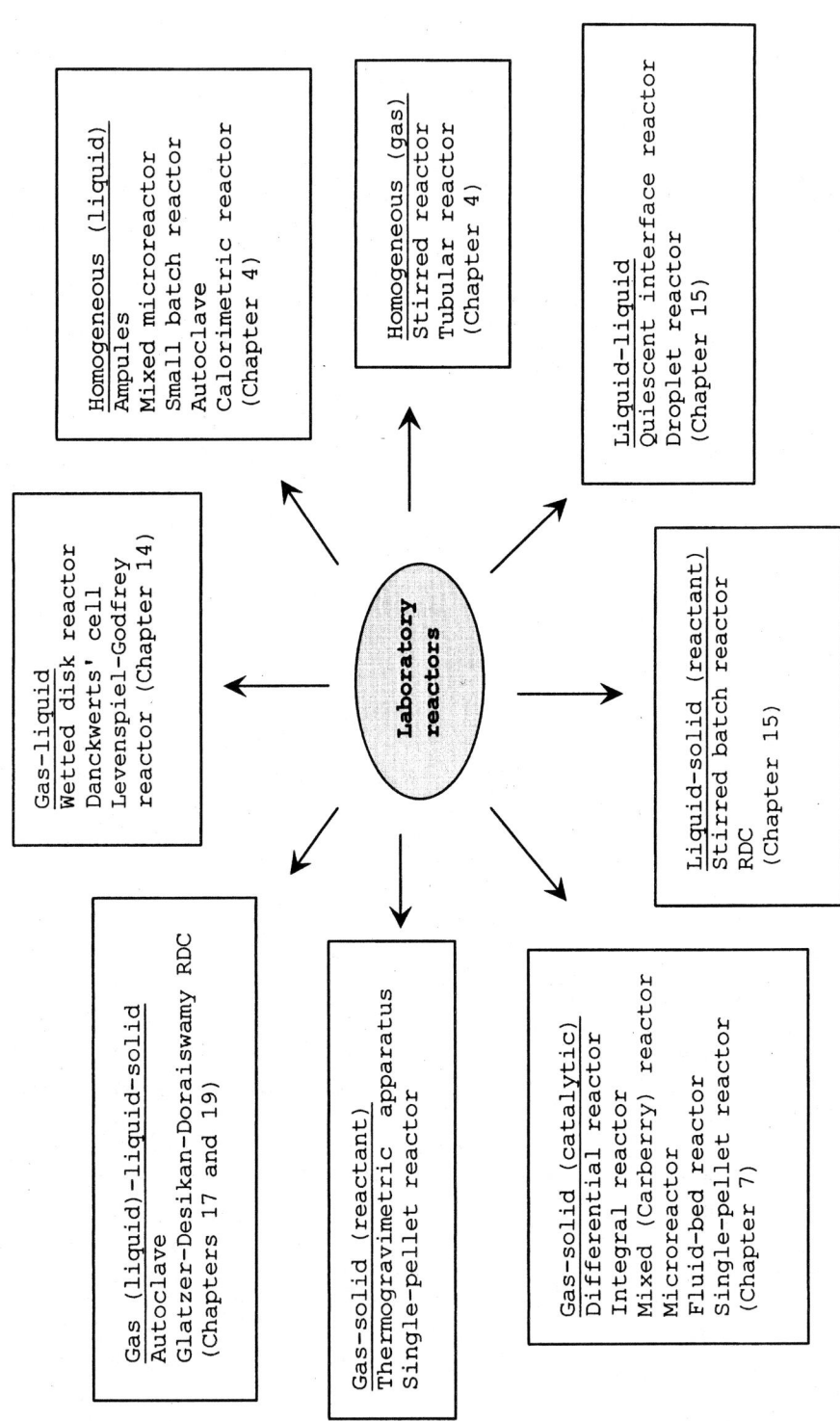

Figure 4.14 Main types of laboratory reactors for studying various homogeneous and heterogeneous reactions. RDC = rotating disk contactor.

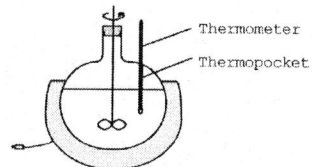
a. Round bottomed flask in a heating mantle.

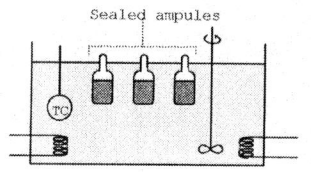
b. Ampules in a thermostat

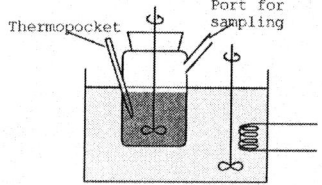
c. Small bench scale reactor in a thermostat

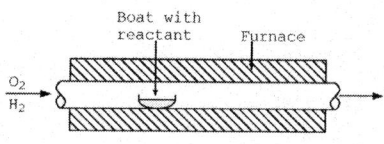
d. Boat containing liquid reactant in a furnace with or without a flowing gaseous reactant.

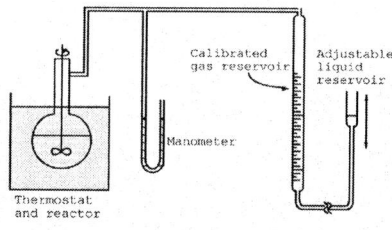
e. Reactor with provision for measuring evolving gas.

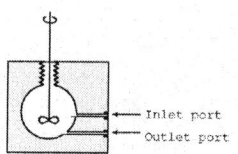
f. Mixed microreactor

Figure 4.15   Laboratory reactors for homogeneous reactions.

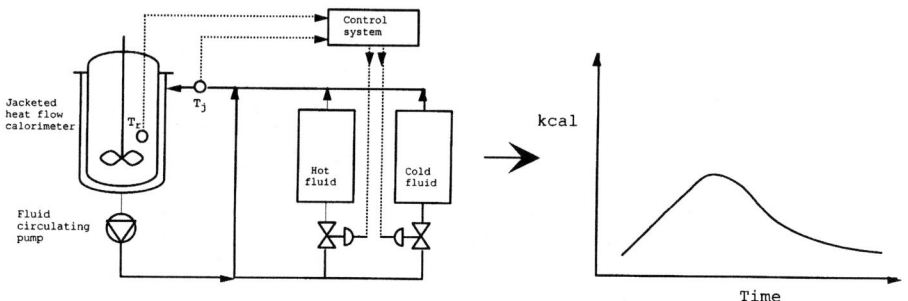

a. Calorimetric reactor
   T_r = reactor temperature,
   T_j = jacket temperature

b. Output from calorimetric reactor.

Figure 4.16   Reactor with heat generation rate as output (no chemical analysis).

nature, are gradientless (except for very fast reactions where local gradients in the vicinity of reaction spots may exist).

For the homogeneous reactions considered in this chapter, a number of laboratory designs are in use, some mainly in the chemist's laboratory. Several designs are sketched in Figures 4.15 and 4.16, which are all self-explanatory. High-pressure batch reactors such as the Parr autoclave are also extensively used. The heat flow calorimetric reactor (Figure 4.16) is a particularly interesting design because it eliminates chemical analysis and depends only on the rate of heat evolution as a function of time to determine the conversion as a function of time (see Regenass, 1978, for details).

Chapter 5

Complex Reactions

When a reactant or a set of reactants undergoes several reactions (at least two) simultaneously, the reaction is said to be a complex reaction. The total conversion of the key reactant, which is used as a measure of reaction in simple reactions, has little meaning in complex reactions, and what is of primary interest is the fraction of reactant converted to the desired product. Thus the more pertinent quantity is *product distribution* from which the conversion to the desired product can be calculated. This is usually expressed in terms of the *yield* or *selectivity* of the reaction with respect to the desired product.

From the design point of view, an equally important consideration is the analysis and quantitative treatment of complex reactions, a common example of which is the dehydration of alcohol represented by

$$2C_2H_5OH \rightarrow C_2H_5OC_2H_5 + H_2O \qquad [5.1.1]$$

$$C_2H_5OC_2H_5 \rightarrow 2C_2H_4 + H_2O \qquad [5.1.2]$$

$$C_2H_5OH + C_2H_4 \rightarrow C_2H_5OC_2H_5 \qquad [5.1.3]$$

$$C_2H_5OH \rightarrow C_2H_4 + H_2O \qquad [5.1.4]$$

We call such a set of simultaneous reactions a complex *multiple* reaction.

It is also important to note that many organic syntheses involve a number of steps, each carried out under different conditions (and sometimes in different reactors), leading to what we designate as *multistep* reactions (normally called a synthetic scheme by organic chemists). This could, for example, be a sequence of reactions like dehydration, oxidation, Diels–Alder, and hydrogenation.

This chapter outlines simple procedures for the treatment of complex multiple and multistep reactions and explains the concepts of selectivity and yield. For a more detailed treatment of multiple reactions, the following books may be

consulted: Aris (1969) and Nauman (1987). We conclude the chapter by considering a reaction with both catalytic and noncatalytic steps, which also constitutes a kind of complex reaction.

REDUCTION OF COMPLEX REACTIONS

Stoichiometry of Simple and Complex Reactions

Because both chemists and chemical engineers are involved in formulating a practical strategy for accomplishing an organic synthesis, it is important to appreciate the roles of each. We do so by considering a "simple" reaction such as the chlorination of methane to methyl chloride:

$$CH_4 + Cl_2 \rightarrow CH_3Cl + HCl \qquad [5.2]$$

Understanding the kinetics and mechanism of this reaction involves understanding the elementary steps leading to the ultimate reaction represented by 5.2. These steps are

$$Cl_2 \rightarrow 2Cl\cdot \qquad [5.2.1]$$

$$Cl\cdot + CH_4 \rightarrow HCl + CH_3\cdot \qquad [5.2.2]$$

$$CH_3\cdot + Cl_2 \rightarrow CH_3Cl + Cl\cdot \qquad [5.2.3]$$

Thus, although reaction 5.2 is apparently a single-step reaction with its own stoichiometry, it is in reality complex. The breakup of this reaction into these mechanistic steps and the study of the controlling mechanism fall in the province of the chemist. The chemical engineer, on the other hand, is concerned primarily with reaction 5.2 per se, and the evaluation of the rate of that reaction in terms of the concentrations of the reacting species, as determined by the stoichiometry of the reaction. This chapter will largely be concerned with the latter approach.

Now consider the set of reactions

$$CH_4 + Cl_2 \rightarrow CH_3Cl + HCl \qquad [5.3.1]$$

$$CH_3Cl + Cl_2 \rightarrow CH_2Cl_2 + HCl \qquad [5.3.2]$$

$$CH_2Cl_2 + Cl_2 \rightarrow CHCl_3 + HCl \qquad [5.3.3]$$

$$CHCl_3 + Cl_2 \rightarrow CCl_4 + HCl \qquad [5.3.4]$$

The difference between reactions 5.2 and 5.3 is that the intermediates in reaction 5.2 are transitory, whereas those in reaction 5.3 are stable compounds present in finite quantities in the final product. Thus, although reaction 5.2 can be treated as a "single" reaction, reaction 5.3 cannot. It is a multiple reaction. Our primary concern in this chapter is with reactions of the type represented by 5.3.

An important point concerning the set of reactions 5.2 is worth noting. As already mentioned, it is generally assumed in such reactions that the net rate of generation of an intermediate can be set to zero. In another approach, if an equilibrium step is involved, we use the quasi-equilibrium approximation, in which the ratio of the forward to backward rates of a rapid elementary step is set equal to unity. In either case, it must be remembered that the fundamental

stoichiometric relationships between the compounds cannot be violated. Only then would the simpler equation be truly valid. This can be achieved by a systematic algorithmic approach, which ensures that the simplified equation coincides with the asymptotics of the original complex mechanism that does not ignore the transient species (see also Vasil'eva et al., 1995; Haario et al., 1999). However, as a general rule, chemical engineers assume the correctness of the simplified equation, without any such rigorous validation, purely from intuitive considerations.

EXAMPLE 5.1

A typical Friedel–Crafts reaction to illustrate the difference between the chemist's and chemical engineer's definitions of the reaction rate

The Friedel–Crafts reaction is an important reaction in organic synthesis and is widely used for the substitution of a side chain in an aromatic ring with $AlCl_3$ as catalyst. The reaction between benzene and phthalic anhydride, which constitutes a step in the preparation of anthraquinone, is one such reaction. We give below the chemist's and the chemical engineer's ways of looking at this reaction (see Rose, 1981, for a detailed discussion).[1]

THE CHEMIST'S DETAILED MECHANISM

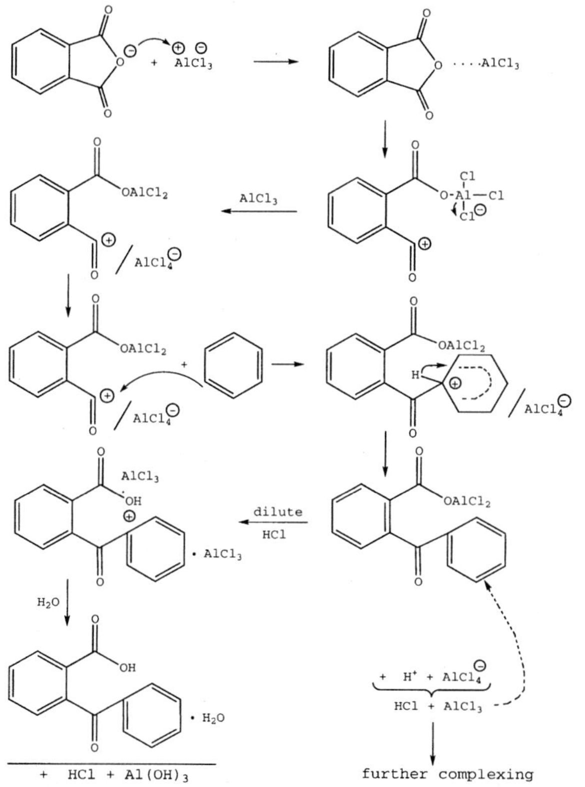

[E5.1.1]

[1] An important point to note in this particular reaction is that 2 moles of $AlCl_3$ are required per mole of the product. This leads to messy disposal and environmental problems. Therefore, industry would like to avoid this reaction altogether and devise other schemes for anthraquinone.

THE CHEMICAL ENGINEER'S "UTILITARIAN" SCHEME

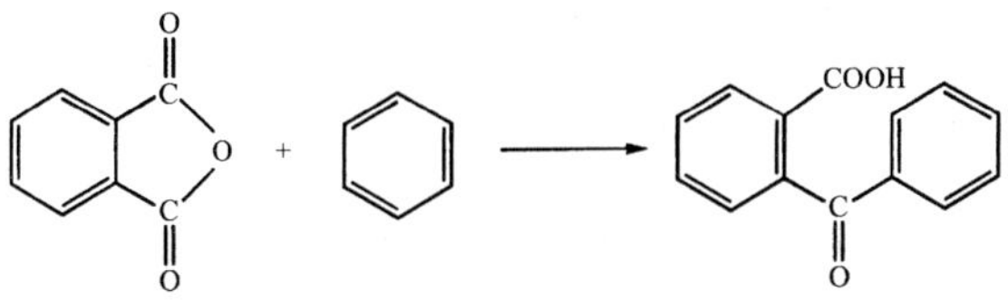

[E5.1.2]

Mathematical Representation of Simple and Complex Reactions

To the chemist, a simple reaction such as

$$A + 2B \to R \tag{5.4}$$

is fully defined when its stoichiometry—complete with its mechanistic antecedents—is fully understood. To the chemical engineer, however, a mathematical representation such as

$$-A - 2B + R = 0 \tag{5.5}$$

is often more convenient—and adequate. In this equation, we bring all the constituents to one side and set the other side equal to zero. Note that the reactants are denoted by a negative sign and the products by a positive sign. Then the equation is further modified to read as

$$\sum_{i=A,B,R}^{N} \nu_i A_i = 0 \tag{5.1}$$

where the term within the summation sign represents $(-A - 2B + R)$. For the most general case of a single reaction involving a large number of components $A_1, A_2, \ldots, A_j, \ldots, A_N$, we can write

$$\sum_{i=1}^{N} \nu_i A_i = 0 \tag{5.2}$$

where i represents any species from 1 to N.

If a reaction system consists of a number of components reacting with one another in more than one reaction, the result is a complex reaction network. Mathematically, a complex reaction consisting of N components and M reactions can be represented as

$$\sum_{i=1}^{N} \nu_{ij} A_i = 0, \quad j = 1, 2, \ldots, M \tag{5.3}$$

where ν_{ij} is the stoichiometric coefficient of A_i in the jth reaction.

Now we shall see how a complex reaction network can be conveniently represented in matrix form. Thus consider a simple reaction

$$A \to R \tag{5.6}$$

with the rate equation

$$-\frac{d([A]V)}{dt} = -r_A V \tag{5.4}$$

Clearly, a single rate equation is all that is needed to describe the system kinetically. But when extended to a complex reaction represented by Equation 3, a set of N ordinary differential equations, one for each component, must be written to describe the system. These may be expressed concisely as

$$\frac{d(\mathbf{c}V)}{dt} = \mathbf{vr}V \tag{5.5}$$

where $\mathbf{c}$ is a vector ($N \times 1$ matrix) of component concentrations, $\mathbf{v}$ is an ($N \times M$) matrix of stoichiometric coefficients, and $\mathbf{r}$ is a vector ($M \times 1$ matrix) of reaction rates.

Independent Reactions

A typical complex organic reaction usually consists of a number of reactions, some of which can be obtained by algebraic addition of two or more reactions of the network. Thus in scheme 5.1 describing the dehydration of alcohol, reaction 5.1.4 can be obtained by the addition of reactions 5.1.1 and 5.1.2 and hence is not an independent reaction. Similarly, reaction 5.1.3 can be obtained from reactions 5.1.1 and 5.1.4 and is hence not an independent reaction. This can be stated more formally as follows: for a set of reactions to be *independent*, no reaction from the set shall be obtainable by algebraic additions of other reactions (as such or in multiples thereof), and each member shall contain one new species exclusively.

Mathematically, if a set of complex reactions is represented in matrix form (Equation 5.5), then the number of independent reactions is given by the rank of the matrix, as illustrated in the following example. It can also be found by a simple stepwise manipulation of the matrix (see Aris, 1969).

EXAMPLE 5.2.

The number of independent reactions in the reactions of propylene glycol in the cyclization of ethylenediamine and propylene glycol and in the ethylation of aniline

REACTIONS OF PROPYLENE GLYCOL

The cyclization of ethylenediamine (EDA) and propylene glycol (PG) over a mixture of zinc and chromium oxides to 2-methylpyrazine (MP) is a basic step in the synthesis of 2-amidopyrazine, a well-known antitubercular drug. This is a highly complex reaction in which EDA and PG each react independently to give a variety of products, as shown in Table 5.1. (Forni and Miglio, 1993). We desire to find the number of independent reactions from this set.

Table 5.1 Possible stoichiometric relationships in the formation of products from propylene glycol (from Forni and Miglio, 1993)

CH_3—CHOH—CH_2—OH + H_2 → CH_3—CH_2—OH + CH_3—OH
CH_3—CH_2—OH ⇔ CH_3—CHO + H_2
CH_3—CHOH—CH_2—OH → H_2O + CH_3—CO—CH_3
CH_3—CHOH—CH_2—OH → H_2O + CH_3—CH_2—CHO
CH_3—CH_2—CHO + H_2 ⇔ CH_3—CH_2—CH_2—OH
CH_3—CHOH—CH_2—OH → CH_2=CH—CH_2—OH + H_2O
CH_3—CHOH—CH_2—OH → CH_2=CH—CHO + H_2O + H_2
CH_3—CH_2—OH + CH_3—CH_2—CH_2—OH → CH_3—CO—CH_2—CH_3 + CO + 3H_2
2CH_3—CH_2—OH → CH_3—CO—CH_3 + CO + 3H_2
2CH_3—CH_2—CH_2—OH → CH_3—CH_2—CO—CH_2—CH_3 + CO + 3H_2
2CH_3—CO—CH_3 → $(CH_3)_2$C=CH—CO—CH_3 + H_2O
2CH_3—CH_2—CHO → CH_3—CH_2—CH=C(CH_3)—CHO + H_2O

SOLUTION

The reaction scheme shown in Table 5.1 consists of 12 reactions involving 16 species. Thus a (12 × 16) matrix of stoichiometric coefficients can be written as shown in Table 5.2. The species are marked S1, S2, ..., S16, and the reactions R1, R2, ..., R12. The full nomenclature is included in the table.

Now we use the fact that the rank of the matrix is the number of independent rows in the matrix, which in turn is the number of independent reactions. We use a software called MATLAB (of MathWorks Inc.). After loading the matrix, a single command gives the rank directly. Thus we find

rank = number of independent reactions = 12

In other words, all of the reactions in the set are independent. A useful conclusion from this illustration is that it is not always possible to reduce the number of reactions to be considered from a given complex reaction sequence.

Table 5.2 Matrix representation of the reactions of propylene glycol[a]

	S1	S2	S3	S4	S5	S6	S7	S8	S9	S10	S11	S12	S13	S14	S15	S16
R1	1	1	−1	−1	0	0	0	0	0	0	0	0	0	0	0	0
R2	0	−1	1	0	−1	0	0	0	0	0	0	0	0	0	0	0
R3	1	0	0	0	0	−1	−1	0	0	0	0	0	0	0	0	0
R4	1	0	0	0	0	−1	0	−1	0	0	0	0	0	0	0	0
R5	0	1	0	0	0	0	0	1	−1	0	0	0	0	0	0	0
R6	1	0	0	0	0	−1	0	0	0	−1	0	0	0	0	0	0
R7	1	−1	0	0	0	−1	0	0	0	0	−1	0	0	0	0	0
R8	0	−3	1	0	0	0	0	0	1	0	0	−1	−1	0	0	0
R9	0	−3	2	0	0	0	−1	0	0	0	0	0	−1	0	0	0
R10	0	−3	0	0	0	0	0	0	2	0	0	0	−1	−1	0	0
R11	0	0	0	0	0	−1	2	0	0	0	0	0	0	0	−1	0
R12	0	0	0	0	0	−1	0	2	0	0	0	0	0	0	0	−1

[a] Species: S1 = $CH_3CHOHCH_2OH$; S2 = H_2; S3 = CH_3CH_2OH; S4 = CH_3OH; S5 = CH_3CHO; S6 = H_2O; S7 = CH_3COCH_3; S8 = CH_3CH_2CHO; S9 = $CH_3CH_2CH_2OH$; S10 = CH_2=$CHCH_2OH$; S11 = CH_2=$CHCHO$; S12 = $CH_3COCH_2CH_3$; S13 = CO; S14 = $CH_3CH_2COCH_2CH_3$; S15 = $(CH_3)_2$=$CHCOCH_3$; S16 = CH_3CH_2CH=$C(CH_3)CHO$.

Table 5.3 Reactions in the ethylation of aniline (from Goyal and Doraiswamy, 1970)

1	$C_6H_5NH_2 \;(A_1) + C_2H_5OH \;(A_2) \rightleftharpoons C_6H_5NH(C_2H_5) \;(A_3) + H_2O \;(A_4)$
2	$C_6H_5NH(C_2H_5) \;(A_3) + C_2H_5OH \;(A_2) \rightleftharpoons C_6H_5N(C_2H_5)_2 \;(A_5) + H_2O \;(A_4)$
3	$C_6H_5NH_2 \;(A_1) + 2\,C_2H_5OH \;(2A_2) \rightleftharpoons C_6H_5N(C_2H_5)_2 \;(A_5) + 2\,H_2O \;(2A_4)$
4	$2\,C_2H_5OH \;(2A_2) \rightleftharpoons (C_2H_5)_2O \;(A_6) + H_2O \;(A_4)$
5	$C_6H_5NH_2 \;(A_1) + \tfrac{1}{2}(C_2H_5)_2O \;(\tfrac{1}{2}A_6) \rightleftharpoons C_6H_5NH(C_2H_5) \;(A_3) + \tfrac{1}{2}H_2O \;(\tfrac{1}{2}A_4)$
6	$C_6H_5NH(C_2H_5) \;(A_3) + \tfrac{1}{2}(C_2H_5)_2O \;(\tfrac{1}{2}A_6) \rightleftharpoons C_6H_5N(C_2H_5)_2 \;(A_5) + \tfrac{1}{2}H_2O \;(\tfrac{1}{2}A_4)$
7	$C_2H_5OH \;(A_2) \rightleftharpoons C_2H_4 \;(A_7) + H_2O \;(A_4)$
8	$C_6H_5NH_2 \;(A_1) + C_2H_4 \;(A_7) \rightleftharpoons C_6H_5NH(C_2H_5) \;(A_3)$
9	$C_6H_5NH(C_2H_5) \;(A_3) + C_2H_4 \;(A_7) \rightleftharpoons C_6H_5N(C_2H_5)_2 \;(A_5)$
10	$C_6H_5NH_2 \;(A_1) + 2\,C_2H_4 \;(2A_7) \rightleftharpoons C_6H_5N(C_2H_5)_2 \;(A_5)$

ETHYLATION OF ANILINE

In studying the kinetics of this reaction, Goyal and Doraiswamy (1970) considered all possible reactions (Table 5.3). Let us represent the different components of the reaction in accordance with the nomenclature of Equation 5.3: A_1 = aniline, A_2 = alcohol, A_3 = monoethylaniline, A_4 = water, A_5 = diethylaniline, A_6 = diethyl ether, and A_7 = olefin. Then, using the method illustrated for the reactions of propylene glycol, it can be shown that there are only 4 independent reactions from a total of 11 reactions.

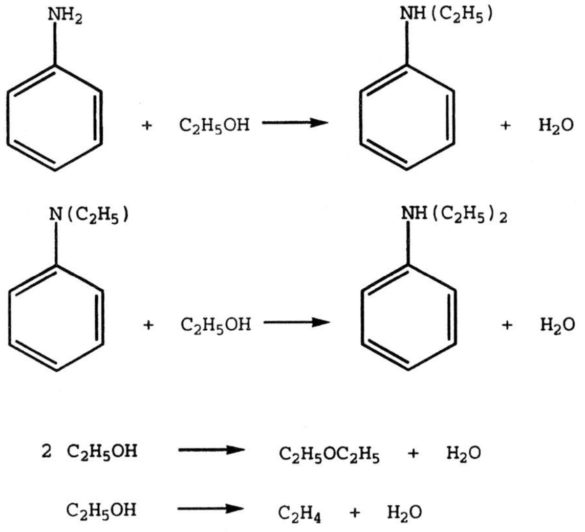

$$\text{C}_6\text{H}_5\text{NH}_2 + \text{C}_2\text{H}_5\text{OH} \longrightarrow \text{C}_6\text{H}_5\text{NH}(\text{C}_2\text{H}_5) + \text{H}_2\text{O}$$

$$\text{C}_6\text{H}_5\text{N}(\text{C}_2\text{H}_5) + \text{C}_2\text{H}_5\text{OH} \longrightarrow \text{C}_6\text{H}_5\text{NH}(\text{C}_2\text{H}_5)_2 + \text{H}_2\text{O}$$

$$2\,\text{C}_2\text{H}_5\text{OH} \longrightarrow \text{C}_2\text{H}_5\text{OC}_2\text{H}_5 + \text{H}_2\text{O}$$

$$\text{C}_2\text{H}_5\text{OH} \longrightarrow \text{C}_2\text{H}_4 + \text{H}_2\text{O}$$

RATE EQUATIONS

There are two aspects to a rate equation, its formulation from laboratory kinetic data and its use in reactor design. In this section we shall consider a procedure for formulating rate expressions for the independent reactions of a complex set and defer the question of reactor design to Part III, Chapter 11.

The Concept of Extent of Reaction

Consider a simple reaction

$$\nu_A A + \nu_B B \rightarrow \nu_R R \qquad [5.7]$$

with no restriction regarding volume change. In such a situation, the amounts of A and B converted and R formed can be expressed in terms of the actual number of moles before and after reaction, because these are independent of volume change:

$$-\frac{N_A - N_{A0}}{\nu_A} = -\frac{N_B - N_{B0}}{\nu_B} = \frac{N_R - N_{R0}}{\nu_R} = \xi \qquad (5.6)$$

where ξ is the *extent of reaction* or *reaction coordinate*. Note that ξ has the units of moles, whereas the conversion X_A is dimensionless. The rate and extent of reaction are obviously related.

Thus let us consider the reaction

$$2A + 3B = R + 2S \qquad [5.8]$$

The rate of this reaction can be understood only in terms of the rates of formation or disappearance of the various components. The rate of a reaction as such is difficult to define unless it is postulated that it is based on the rate of formation of a product or disappearance of a reactant with a specified stoichiometric coefficient. *Usually it is the rate of formation of a product with a stoichiometric coefficient of unity*. Thus in this case we choose the rate of formation of R. Then the rate of disappearance of A is twice this rate, that of B is three times this rate, and the rate of formation of S is twice this rate.

Because one mole of R is formed in reaction 5.8, its rate of formation has been conveniently used as the basis for expressing the rates of formation/disappearance of the other components. On the other hand, in a reaction of the type

$$2A + 3B = 2R \qquad [5.9]$$

no such basis exists. In such a case, we postulate a fictitious product F with a stoichiometric coefficient of $+1$:

$$2A + 3B = 2R + F \qquad [5.10]$$

and use this as the basis for expressing the rates of formation/disappearance of the others. Thus,

rate of formation of $A = -2r_F$

rate of formation of $B = -3r_F$

rate of formation of $R = 2r_F$

Now let us extend the concept to each of the reactions comprising a complex set, such as

$$2A + 3B = 2R$$
$$2C + D = S \qquad [5.11]$$

Considering the first reaction, the condition that a fictitious product F_1 have a stoichiometric coefficient of unity, though necessary, is no longer sufficient. We have to postulate additionally that F_1 is not present in the second reaction, that is $\nu_{F_1,2} = 0$. Similarly, the extent of reaction of the second reaction is obtained by postulating a hypothetical product F_2 for that reaction with the same conditions as before. However, because S happens to be such a product for this reaction, that is, $\nu_{S,2} = 1$, there is no need to postulate F_2.

Thus, now it is possible to express the rate of each reaction in terms of a single, indivisible parameter ξ. Then the rates of formation or disappearance of the different components of the reaction can be expressed as multiples of this reaction coordinate. For example, for reaction 5.11 we can write

94   Reactions and Reactors

$$-\frac{N_A - N_{A0}}{2} = -\frac{N_B - N_{B0}}{3} = \frac{N_R - N_{R0}}{2} = \xi_1 \qquad (5.7)$$

$$-\frac{N_C - N_{C0}}{2} = -\frac{N_D - N_{D0}}{1} = \frac{N_S - N_{S0}}{1} = \xi_2 \qquad (5.8)$$

Then the compositions of A, B, C, D, R, and S can be expressed in terms of the reaction coordinates ξ_1 and ξ_2 defined by Equations 5.7 and 5.8, respectively. Hence the number of equations would be: (a) six if written in terms of the rates of formation/disappearance of the individual components or (b) two if written in terms of the extent of reaction in each step. This concept is applied in Chapter 11 to the actual design of reactors for complex reactions

Determination of the Individual Rates in a Complex Reaction

Whether the system of equations is written in terms of (a) or (b), it is clearly useful to have the rates of formation and disappearance of the individual components as a function of some measure of time (t for batch, $\bar{t}$ for MFR and PFR) to formulate rate equations. A procedure for doing this is illustrated here.

The method depends on whether the reactor is operated as a PFR or an MFR. If operated as an MFR, the experimentally determined product composition gives the rates directly. Thus, considering A, the rate is given by

$$-r_{Af} = \frac{[A]_0 X_{Af}}{\bar{t}} \qquad (5.9)$$

where $\bar{t} = [A]_0 V_r / F_{A0}$. Note that r_{Af} is the rate corresponding to the final composition. By varying the initial composition and flow rate, the rates corresponding to different compositions can be obtained.

On the other hand, for a PFR, the outlet conversion (or concentration) of any component i must be plotted as a function of $\bar{t}$, and the rate at any value of $\bar{t}$ determined by measuring the slope of the curve at that value (Figure 5.1), for example,

$$(\text{Measured slope}) = r_R = \frac{d[R]}{d\bar{t}} \qquad (5.10)$$

This rate corresponds to the composition at that value of $\bar{t}$. Reactions carried out on a catalyst in a tubular reactor (which conforms to plug flow) can also be treated similarly (see Chapters 4 and 12), with this difference that we now plot conversion to i as a function of W/F_{A0} where W is the weight of the catalyst (Figure 5.2). The dimensions of the rate in this case are moles per unit weight of catalyst per unit time.

An instructive application of the methods previously outlined is the determination of reaction rates in the complex network describing the deamination of diethylenetriamine. Nitrogen-containing heterocyclic compounds with two nitrogen atoms such as pyrazine, piperazine, and their derivatives find a wide range of applications in the synthesis of organic intermediates. One of the methods of synthesizing them is by vapor-phase deamination of diethylenetriamine. Although this method lacks selectivity for any one compound, a

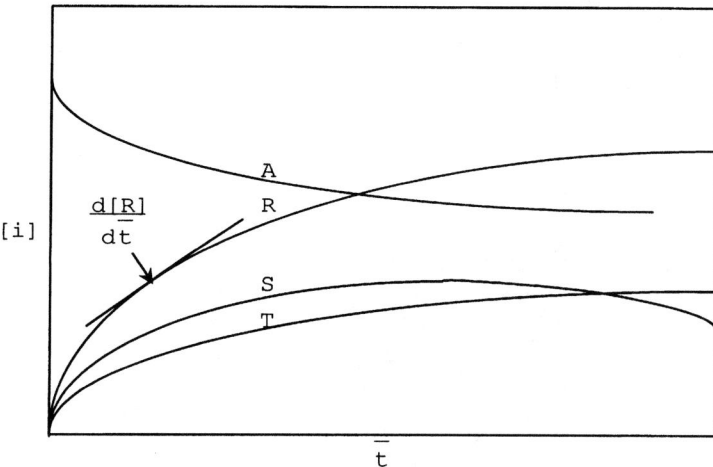

Figure 5.1 Product distribution plot in terms of concentration vs. $\bar{t} = [A]_0 V_r / F_{A0}$.

number of them can be prepared according to the reaction network shown in Figure 5.3, and the required components can be separated by distillation or other techniques. Bhat et al. (1985) determined the kinetic parameters (activation energies and preexponential factors) of all nine reactions of the network by setting up ODEs for all of the components involved and solving them.

SELECTIVITY AND YIELD

We consider in this section the concepts of yield and selectivity as applied to multiple, as well as multistep reactions. The reaction engineering literature uses the two terms "multiple" and "multistep" interchangeably. However, when

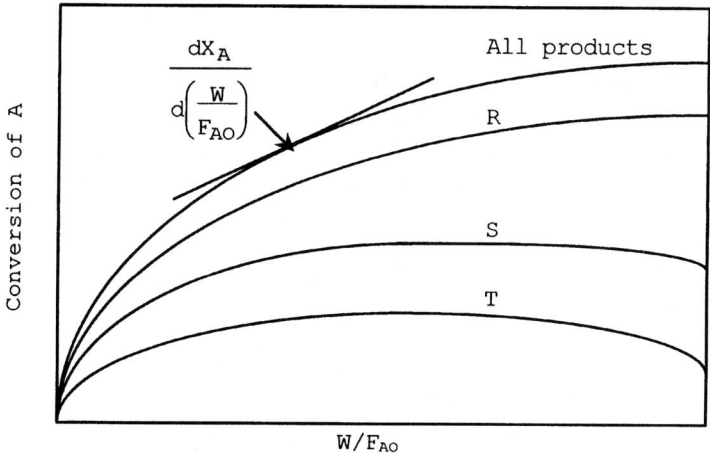

Figure 5.2 Product distribution plot in terms of total conversion (X_A) of A to products or to a specific product $i(X_i)$ vs. W/F_{A0}.

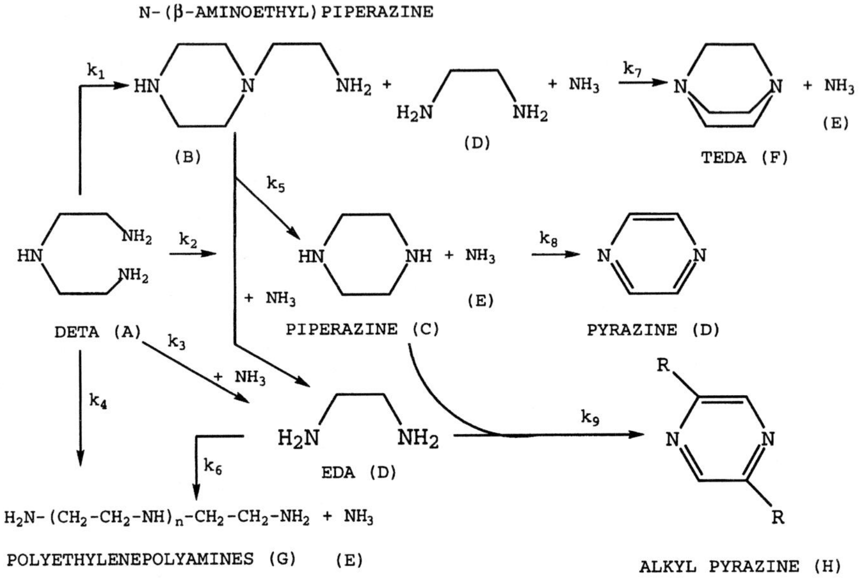

Figure 5.3 Plausible reaction network in the deamination of diethylenetriamine (DETA) (Bhat et al., 1985).

dealing with an overall organic synthesis comprising many separate steps, it is desirable to distinguish between the two.

Multiple Reactions

Definitions

Consider the reaction

$$A + B \rightarrow R$$
$$A + B \rightarrow S \qquad [5.12]$$

with arbitrary stoichiometry. Suppose we start with 100 moles of A and 150 moles of B. Let the final product contain 60 moles of R, 20 moles of S, 20 moles of A, and 70 moles of B. Let us also suppose that R is the desired product. The conversion, yield, and selectivity are defined as

$$\text{conversion, } X_A = \frac{\text{moles A converted}}{\text{moles A fed}} = \frac{[A]_0 - [A]}{[A]_0} = \frac{80}{100} = 80\%$$

$$\text{yield, } \qquad Y_R = \frac{\text{moles R formed}}{\text{moles A fed}} = \frac{[R] - [R]_0}{[A]_0} = \frac{60}{100} = 60\%$$

$$\text{selectivity, } S_R = \frac{\text{moles R formed}}{\text{moles A converted}} = \frac{[R] - [R]_0}{[A]_0 - [A]} = \frac{60}{80} = 75\%$$

Note: yield = (conversion) (selectivity) = $0.80 \times 0.75 = 60\%$.

These definitions can be generalized for any multiple reaction consisting of one or more reactants and a number of products. It is useful to express these in terms of the cost-determining reactant, or the *key reactant*.

Simplifying the kinetic structure: lumping

Multiple reactions can involve a large number of reactions and species (sometimes exceeding 100), for example thermal and catalytic cracking of petroleum feedstocks. The difficulty in the mathematical analysis of such reactions is often compounded by the imprecise composition of the single feed used. Because both experimental characterization and mathematical analysis of these reactions are highly complicated, a technique known as *lumping* is commonly used, in which the numbers of reactions and species are kept to a reasonable level by lumping like species together.

Lumping has been a subject of considerable interest in the petrochemical industry, and many studies have been reported, for example, Aris and Gavalas (1966), Luss and collaborators (Hutchinson and Luss, 1970; Luss and Hutchinson, 1971; Luss and Golikeri, 1975), Lee (1985), Nace et al. (1971), Jacob et al. (1976), Akella and Lee (1981). The methods described in these studies can be useful in the analysis of complex organic transformations involved in the treatment of waste streams of uncertain composition from organic process plants. Such a treatment can in theory be extended to the recovery of products (like resorcinol) from the waste stream of a typical organic intermediates plant. However, in view of the very limited use of this technique in organic technology and synthesis where the feed is almost always precisely defined, we will not consider it further here.

Analytical solutions

To illustrate the procedures used in the analysis of multiple reactions, we considered relatively complex cases in the earlier sections. Such complex schemes do not normally admit analytical solutions. Many industrially important schemes, however, are less complex, for which analytical solutions can be found (see Frost and Pearson, 1961; Rodigin and Rodigina, 1964, Levenspiel, 1993). The simplest are the *parallel* and the *consecutive* (or series) reactions:

$$A \begin{array}{c} \nearrow R \\ \searrow S \end{array} \qquad A \longrightarrow R \longrightarrow S \qquad [5.13]$$

Parallel　　　Series (consecutive)

Three other important schemes commonly encountered in organic technology are

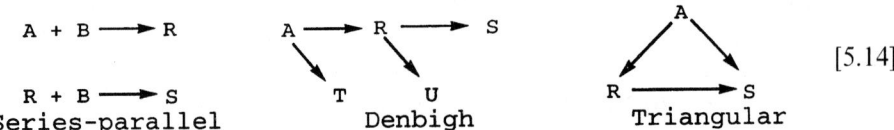

$$\begin{array}{c} A + B \longrightarrow R \\ R + B \longrightarrow S \\ \text{Series-parallel} \end{array} \qquad \begin{array}{c} A \longrightarrow R \longrightarrow S \\ \searrow \quad \searrow \\ T \quad U \\ \text{Denbigh} \end{array} \qquad \begin{array}{c} A \\ \swarrow \quad \searrow \\ R \longrightarrow S \\ \text{Triangular} \end{array} \qquad [5.14]$$

The rate equations and solutions for all five schemes are summarized in Table 5.4. The solutions will depend on whether the reactions are carried out under plug-flow or fully mixed conditions. Expressions for both cases are included in the table.

The evolution of concentration with time clearly indicates whether a particular component is an end product or an intermediate product that undergoes further reaction even as it is formed. The concentration-time (t or $\bar{t}$) profiles for the series, series-parallel, Denbigh, and triangular schemes mentioned above are sketched in Figure 5.4. An important feature of an intermediate product is that it has a maximum at a specific time t_{max}. If that compound happens to be the desired product, it is best to operate the reactor at t_{max}. Relationships for $[R]_{max}$ in terms of the kinetic parameters of the reactions are important in maximizing production. They can be derived from the solutions given in Table 5.4, and are summarized in Table 5.5 for a few selected schemes in which R is produced as an intermediate.

Another method of interpreting experimental data in formulating the kinetics of a complex reaction is by stoichiometric plots in which the selectivity for a product is plotted as a function of time. Thus for the parallel scheme

$$A \xrightarrow{k_1} 2R$$
$$A \xrightarrow{k_2} S \qquad [5.15]$$

the plots for R and S (moles formed per mole A reacted) are both parallel to the time axis, as shown in Figure 5.5a. On the other hand, for the series-parallel scheme

$$A \xrightarrow{k_1} 2R$$
$$A \xrightarrow{k_2} S \qquad [5.16]$$
$$R \xrightarrow{k_3} T$$

the plots would be as shown in Figure 5.5b. Reactions that produce curves such as those for R and S are called *primary reactions*, and reactions in which the products of the primary reaction react further (producing curves such as that for T) are called *secondary reactions* (see Holland and Anthony, 1989, for further details). However, the interpretation becomes more involved if *tertiary reactions* also occur in which the products of the secondary reactions react further. Thus this procedure can be used only as a preliminary step for the evaluation of certain simple schemes.

Table 5.4 Integrated forms of rate equations for the more common complex schemes

Reaction	Rate equations	Solutions for BR/PFR	Solutions for MFR
$A \underset{2}{\overset{1}{\nearrow}} \begin{matrix} R \\ S \end{matrix}$ (Parallel)	$-\dfrac{d[A]}{dt} = (k_1 + k_2)[A]$ $\dfrac{d[R]}{dt} = k_1[A]$ $\dfrac{d[S]}{dt} = k_2[A]$	$\ln\dfrac{[A]_0}{[A]} = (k_1 + k_2)t$ $[R] = [R]_0 + \dfrac{k_1[A]_0}{k_1 + k_2}\{1 - \exp[-(k_1 + k_2)t]\}$ $[S] = [S]_0 + \dfrac{k_2[A]_0}{k_1 + k_2}\{1 - \exp[-(k_1 + k_2)t]\}$	$\dfrac{[A]}{[A]_0} = \dfrac{1}{1 + (k_1 + k_2)\bar{t}}$ $\dfrac{[R]}{[A]_0} = \dfrac{[R]_0}{[A]_0} - \dfrac{k_1\bar{t}}{1 + (k_1 + k_2)\bar{t}}$ $\dfrac{[S]}{[A]_0} = \dfrac{[R]_0}{[A]_0} - \dfrac{k_2\bar{t}}{1 + (k_1 + k_2)\bar{t}}$
$A \xrightarrow{1} R \xrightarrow{2} S$ (Series)	$-\dfrac{d[A]}{dt} = k_1[A]$ $\dfrac{d[R]}{dt} = k_1[A] - k_2[R]$ $\dfrac{d[S]}{dt} = k_2[R]$	$\dfrac{[A]}{[A]_0} = \exp(-k_1 t)$ $\dfrac{[R]}{[A]_0} = \dfrac{k_1}{k_2 - k_1}[\exp(-k_1 t) - \exp(-k_2 t)] + \dfrac{[R]_0}{[A]_0}\exp(-k_2 t)$ $[S] = [A]_0 + [R]_0 + [S]_0 - [A] - [R]$	$\dfrac{[A]}{[A]_0} = \dfrac{1}{1 + k_1 \bar{t}}$ $\dfrac{[R]}{[A]_0} = \dfrac{k_1 \bar{t}}{(1 + k_1 \bar{t})(1 + k_2 \bar{t})} + \dfrac{[R]_0}{[A]_0}\dfrac{1}{1 + k_2 \bar{t}}$ $[S] = [A]_0 + [R]_0 + [S]_0 - [A] - [R]$
$A + B \xrightarrow{1} R$ $R + B \xrightarrow{2} S$ (Series-parallel)	$-\dfrac{d[A]}{dt} = k_1[A][B]$ $\dfrac{d[R]}{dt} = k_1[A][B] - k_2[R][B]$ $\dfrac{d[S]}{dt} = k_2[R][B]$ $-\dfrac{d[B]}{dt} = k_1[A][B] + k_2[R][B]$	$\dfrac{d[A]}{dt} = k_1[A][A]_0$ $\times \left\{\dfrac{[B]_0}{[A]_0} - \dfrac{k_2}{k_1 - k_2}\left[\dfrac{[A]}{[A]_0} - 1\right] + \dfrac{k_2}{k_1 - k_2}\left[\left(\dfrac{[A]}{[A]_0}\right)^{k_2/k_1} - 1\right] + \dfrac{[A]}{[A]_0} - 1\right\}$; $[A] = f(t)$ (numerical soln.) $\dfrac{[B]}{[A]_0} = \dfrac{[B]_0}{[A]_0} - \dfrac{k_2}{k_1 - k_2}\left(\dfrac{[A]}{[A]_0} - 1\right) + \dfrac{k_1}{k_1 - k_2}\left[\left(\dfrac{[A]}{[A]_0}\right)^{k_2/k_1} - 1\right] + \dfrac{[A]}{[A]_0} - 1$ $\dfrac{[R]}{[A]_0} = -\dfrac{[A]}{[A]_0}\left(\dfrac{k_1}{k_1 - k_2}\right) + \dfrac{k_1}{k_1 - k_2}\left(\dfrac{[A]}{[A]_0}\right)^{k_2/k_1}$ $\dfrac{[S]}{[A]_0} = \dfrac{[S]_0}{[A]_0} + \dfrac{k_2}{k_1 - k_2}\left(\dfrac{[A]}{[A]_0} - 1\right) - \dfrac{k_1}{k_1 - k_2}\left[\left(\dfrac{[A]}{[A]_0}\right)^{k_2/k_1} - 1\right]$	$\dfrac{[A]}{[A]_0} = 1 - \dfrac{\bar{t}}{[A]_0}k_1[A][B]$ $\dfrac{[R]}{[A]_0} = \dfrac{[R]_0}{[A]_0} + \dfrac{\bar{t}}{[A]_0}(k_1[A][B] - k_2[R][B])$ $\dfrac{[B]}{[A]_0} = \dfrac{[B]_0}{[A]_0} - \dfrac{\bar{t}}{[A]_0}(k_1[A][B] + k_2[R][B])$ $[S] = [A]_0 + [R]_0 + [S]_0 - [A]$

(continued)

Table 5.4 (continued)

Reaction	Rate equations	Solutions for BR/PFR	Solutions for MFR
$A \underset{3}{\overset{1}{\nearrow}} \underset{2}{\searrow} S$ R (Triangular)	$-\dfrac{d[A]}{dt} = (k_1 + k_3)[A]$ $\dfrac{d[R]}{dt} = k_1[A] - k_2[R]$ $\dfrac{d[S]}{dt} = k_3[A] + k_2[R]$	$\dfrac{[A]}{[A]_0} = \exp[-(k_1+k_3)t]$ $\dfrac{[R]}{[A]_0} = \dfrac{k_1}{k_2 - k_1 - k_3}\{\exp[-(k_1+k_3)t] - \exp(-k_2 t)\}$ $\dfrac{[S]}{[A]_0} = 1 - \dfrac{k_3}{k_1 + k_3}\exp[-(k_1+k_3)t] - \dfrac{k_2 k_1}{(k_1+k_3)(k_2-k_1-k_3)}$ $\times \exp[-(k_1+k_3)t] + \dfrac{k_1}{k_2-k_1-k_3}\exp(-k_2 t)$	$\dfrac{[A]}{[A]_0} = \dfrac{1}{1+(k_1+k_3)\bar{t}}$ $\dfrac{[R]}{[A]_0} = \dfrac{[R]_0/[A]_0 + k_1\bar{t}/[1+(k_1+k_3)\bar{t}]}{1+k_2\bar{t}}$ $[S] = [A]_0 + [R]_0 + [S]_0 - [A] - [R]$
$A \overset{1}{\to} R \overset{2}{\underset{4}{\to}} S$ $\underset{3}{\downarrow}$ $T \quad V$ (Denbigh)	$-\dfrac{d[A]}{dt} = (k_1+k_3)[A]$ $\dfrac{d[R]}{dt} = k_1[A] - (k_2+k_4)[R]$ $\dfrac{d[S]}{dt} = k_2[R]$ $\dfrac{d[T]}{dt} = k_3[A]$ $\dfrac{d[V]}{dt} = k_4[R]$	$\dfrac{[A]}{[A]_0} = \exp[-(k_1+k_3)t]$ $\dfrac{[R]}{[A]_0} = \dfrac{k_1}{k_2+k_4-k_1-k_3}\{\exp[-(k_1+k_3)t] - \exp[-(k_2+k_4)t]\} + \dfrac{[R]_0}{[A]_0}\exp[-(k_2+k_4)t]$ $\dfrac{[S]}{[A]_0} = \dfrac{k_1 k_2}{k_2+k_4-k_1-k_3}\left\{\dfrac{\exp[-(k_2+k_4)t]}{k_2+k_4} - \dfrac{\exp[-(k_1+k_3)t]}{k_1+k_3}\right\}$ $+ \dfrac{k_1 k_2}{(k_1+k_3)(k_2+k_4)} + \dfrac{[R]_0}{[A]_0}\dfrac{k_2}{k_2+k_4}\{1-\exp[-(k_2+k_4)t]\} + \dfrac{[S]_0}{[A]_0}$ $\dfrac{[T]}{[A]_0} = \dfrac{k_3}{k_1+k_3}\{1-\exp[-(k_1+k_3)t]\} + \dfrac{[T]_0}{[A]_0}$ $\dfrac{[U]}{[A]_0}$, same as for $\dfrac{[S]}{[A]_0}$ but with $k_2 \leftrightarrow k_4$ and $[S]_0 \leftrightarrow [V]_0$	$\dfrac{[A]}{[A]_0} = \dfrac{1}{1+(k_1+k_3)\bar{t}}$ $\dfrac{[R]}{[A]_0} = \dfrac{k_1\bar{t}}{[1+(k_1+k_3)\bar{t}][1+(k_2+k_4)\bar{t}]}$ $+ \dfrac{[R]_0}{[A]_0}\dfrac{k_2\bar{t}}{[1+(k_2+k_4)\bar{t}]}$ $\dfrac{[S]}{[A]_0} = \dfrac{k_1 k_2 \bar{t}^2}{[1+(k_1+k_3)\bar{t}][1+(k_2+k_4)\bar{t}]}$ $+ \dfrac{[R]_0}{[A]_0}\dfrac{k_2\bar{t}}{[1+(k_1+k_3)\bar{t}]} + \dfrac{[T]_0}{[A]_0}$ $\dfrac{[T]}{[A]_0} = \dfrac{k_3\bar{t}}{[1+(k_1+k_3)\bar{t}]} + \dfrac{[T]_0}{[A]_0}$ $\dfrac{[U]}{[A]_0}$, same as for $\dfrac{[S]}{[A]_0}$ but with $k_2 \leftrightarrow k_4$ and $[S]_0 \leftrightarrow [V]_0$

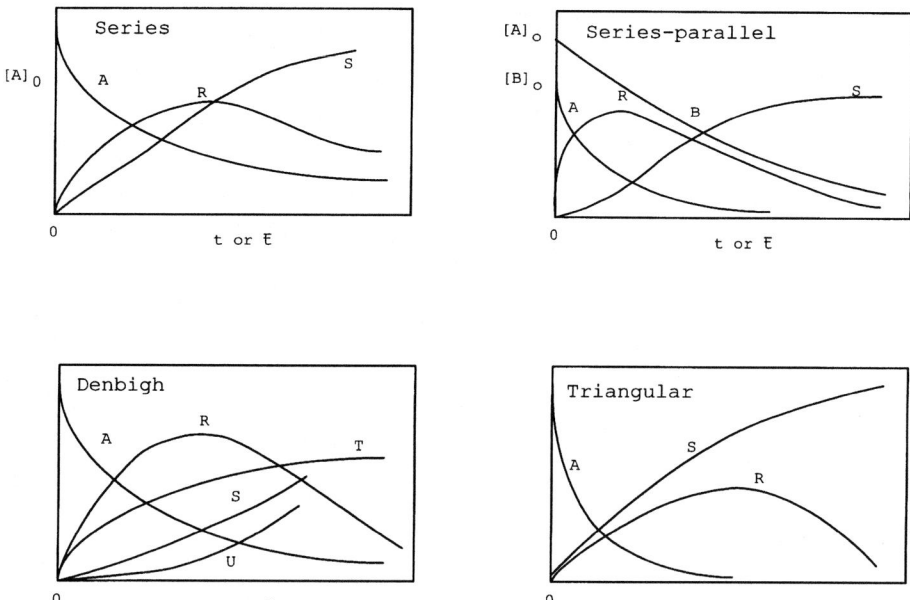

Figure 5.4 Concentration–time profiles of the various components in series, series-parallel, Denbigh, and triangular schemes.

Table 5.5 Equations for $[R]_{max}/[A]_0$ for different complex reaction schemes

Class of reaction	Batch reactor (or PFR)[a]	Mixed reactor[a]
Series reaction $A \xrightarrow{k_1} R \xrightarrow{k_2} S$	$\left(\dfrac{k_1}{k_2}\right)^{\frac{k_2}{k_2-k_1}}$	$\dfrac{1}{\left[\left(\dfrac{k_2}{k_1}\right)^{1/2}+1\right]^2}$
Series-parallel reaction $A+B \xrightarrow[n=2]{k_1} R$ $R+B \xrightarrow[n=2]{k_2} S$	$\left(\dfrac{k_1}{k_2}\right)^{\frac{k_2}{k_2-k_1}}$	$\dfrac{1}{\left[\left(\dfrac{k_2}{k_1}\right)^{1/2}+1\right]^2}$
Denbigh reaction $A \xrightarrow{k_1} R \xrightarrow{k_2} S$, $A\xrightarrow{k_3} T$, $R\xrightarrow{k_4} U$	$\left(\dfrac{k_1}{k_{13}}\right)\left(\dfrac{k_{13}}{k_{24}}\right)^{\frac{k_{24}}{k_{24}-k_{13}}}$	$\left(\dfrac{k_1}{k_{13}}\right)\dfrac{1}{\left[\left(\dfrac{k_{24}}{k_{13}}\right)^{1/2}+1\right]^2}$

[a] $k_{13} = k_1 + k_3$, $k_{24} = k_2 + k_4$. No product in feed. Note: All steps are first order except where noted otherwise.

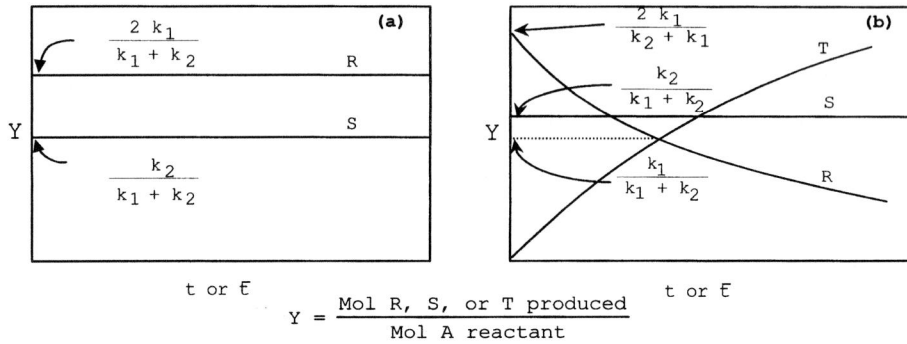

Figure 5.5 Selectivity vs. time plots of (a) $A \to 2R$, $A \to S$, and (b) $A \to 2R$, $A \to S$, and $R \to T$ (redrawn from Holland and Anthony, 1989).

Maximizing selectivity in a complex reaction: important considerations

Choosing a reaction pathway for any reaction, simple or complex, is always a difficult task, more so for a new, untried product. Even so, several considerations are common to both new and old products. Among them are cost and economic recovery of solvents, minimization of side reactions, use of relatively mild reaction conditions (although extreme conditions are not always precluded), and minimization and efficient disposal of wastes. In a multistep (as opposed to multiple) process, considered in the next section, minimization of the number of steps is particularly important.

Many of these considerations can be factored into a statistical program, with selectivity for the desired product as the objective function. On the other hand, a package of simple kinetic "rules" and other chemical considerations can often be invoked as an effective substitute or additional input for any such purely mathematical study. We illustrate such a procedure in Example 5.3 for the manufacture of β-hydroxy-β-methyl butyrate (see Barac, 1998).

EXAMPLE 5.3

Reaction optimization in the production of β-hydroxy-β-methyl butyrate (HMB)

HMB is used as a nutritional supplement for human beings and also as an animal feed additive, and acts by suppressing protein breakdown. It is manufactured by Metabolic Technologies Inc. (Ames, Iowa). The reactions comprising the synthesis of HMB are the main reaction involving the decomposition of diacetone alcohol, known as the haloform reaction, accompanied by several side reactions.

MAIN (HALOFORM) REACTION

HMB is produced by reacting 4-hydroxy-4-methylpentan-2-one (diacetone alcohol) with a mixture of NaOH and Cl_2 known as bleach:

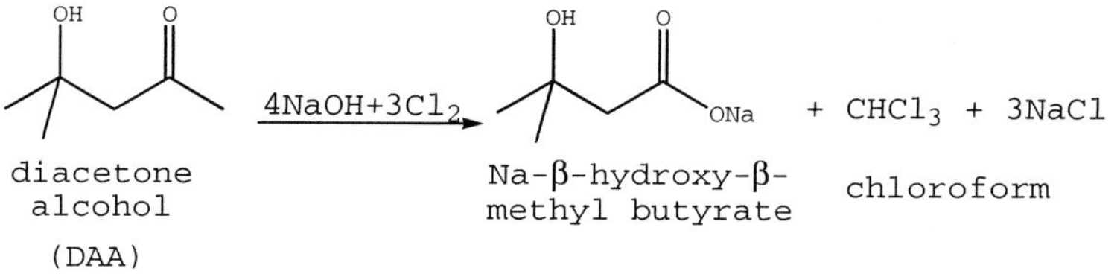

Complex Reactions

$$\text{diacetone alcohol (DAA)} \xrightarrow{4NaOH + 3Cl_2} \text{Na-}\beta\text{-hydroxy-}\beta\text{-methyl butyrate} + \underbrace{CHCl_3}_{\text{chloroform}} + 3NaCl + 3H_2O \quad [E5.3.1]$$

SIDE REACTIONS

In the presence of NaOH (hydroxyl ions), diacetone alcohol can also undergo the following reactions:

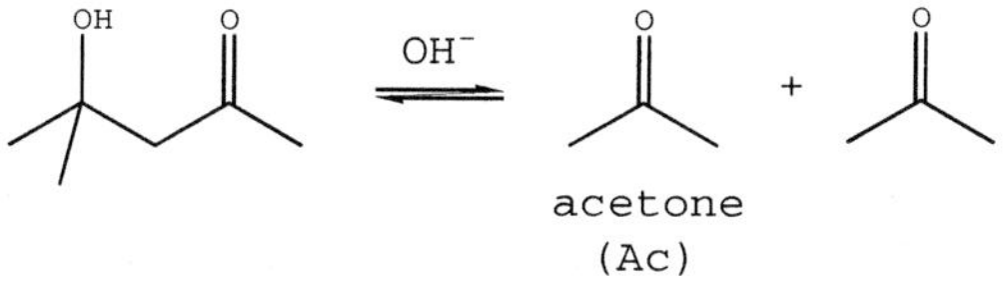

$$\text{DAA} \xrightleftharpoons{OH^-} \text{acetone (Ac)} + \text{acetone} \quad [E5.3.2]$$

Acetone from reaction E5.3.2 reacts with the bleach to form sodium acetate:

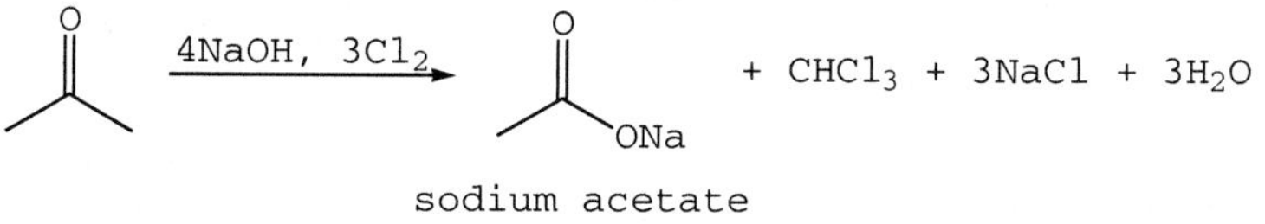

$$\text{acetone} \xrightarrow{4NaOH,\ 3Cl_2} \text{sodium acetate} + CHCl_3 + 3NaCl + 3H_2O \quad [E5.3.3]$$

Diacetone alcohol (DAA) is also halogenated at the α-position of the non-methyl group, followed by oxidation, halogenation of the methyl group, and finally chain cleavage. The overall reaction yielding the diol (DHMB) is

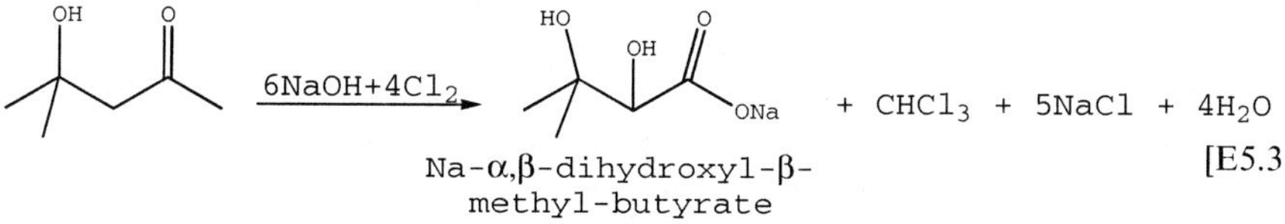

$$\text{DAA} \xrightarrow{6NaOH + 4Cl_2} \text{Na-}\alpha,\beta\text{-dihydroxyl-}\beta\text{-methyl-butyrate} + CHCl_3 + 5NaCl + 4H_2O \quad [E5.3.4]$$

Another reaction that occurs is the dissolution of chlorine in NaOH resulting in the well-known reaction

$$Cl_2 + 2NaOH \leftrightarrow NaOCl + NaCl + H_2O \quad [E5.3.5]$$

Yet another reaction that can occur but probably does not is

$$CHCl_3 + NaOCl \leftrightarrow CCl_4 + NaOH \quad [E5.3.6]$$

Considering all the reactions (except E5.3.6), the overall scheme for the complex haloform reaction can be represented by the following network:

```
           k₁, r₁
              ↗ HMB
          k₂, r₂
    DAA ⇌ DHMB                                    [E5.3.7]
          k₃, r₃
          k₄, r₄ ↘        k₅, r₅
                    Ac ─────────→ NaAc
```

Employing this network, it is desired to outline a procedure for maximizing the selectivity for HMB (S_{HMB}) defined as

$$S_{HMB} = \frac{r_1}{r_2 + r_3 - r_4} \tag{E5.3.1}$$

Optimization can be done by proper reactor choice followed by a suitable temperature progression in the case of a batch or semibatch reactor, or by temperature profiling in the case of a tubular reactor. An even more effective way is to optimize reactant concentrations, pressure, and/or temperature by applying certain simple rules of kinetics and manipulation of the chemistry (wherever possible). Hence the combined efforts of chemist and chemical engineer are needed to optimize selectivity in a given complex reaction.

Based on several preliminary laboratory experiments, the following useful information was obtained.

1. All the haloform reactions take place simultaneously. Hence, a considerable excess of hypochlorite is needed to achieve complete oxidation of DAA.
2. All the three haloform reactions are first order in OCl^-. They are also first order in OH^-, except reaction 2, which is less than first order (0.9). Thus the bleach concentration in the reactor must be maintained as high as possible.
3. The use of solvents is not a viable option. Chloroform, being a product of the reaction, is a good a priori choice, but its use did not result in increased HMB yield.
4. The chloride ion has a strong positive influence on the net rate of production of HMB. High yields can be obtained by reacting DAA with bleach solutions of NaCl. This can be practically achieved by using a semibatch reactor containing an initial charge of the bleach solution into which DAA is continuously introduced. The main limiting factor is the solubility of NaCl at the reaction temperature.
5. All the three main reactions are exothermic. The heat of reaction at 25 °C for HMB, diol, and acetic acid formations are 210, 313, and 278 kcal/mol, respectively. Thus all are favored by low-temperature operation, and temperature selectivity is difficult to achieve unless elaborate calculations are carried out to impose temperature progression (see Chapter 11 for an elementary discussion).
6. The role of heat of reaction is much less critical than that of the activation energy. To obtain the activation energies for the individual steps, the analysis is repeated at different temperatures for scheme E5.3.7. These values are then substituted in Equation E5.3.1 and the optimum temperature determined by plotting S_{HMB} versus T. An important qualitative rule (see Figure 11.6) is that higher temperatures favor reactions with higher E, and vice versa.

QUANTITATIVE ANALYSIS

The qualitative conclusions listed can be regarded as preliminary aids to process development. For a quantitative analysis of the problem, rate equations must be written for all the species of this complex reaction and solved simultaneously with the equations for the type of reactor chosen. In accordance with the methods developed in this chapter, the rates of formation of the individual species can be written by using the following designations: r_{ij} for the ith species in the jth reaction where $i = 1$ for DAA, 2 for HMB, 3 for DHMB, 4 for acetone, 5 for sodium acetate (or acetic acid, HAc), 6 for NaOCl, and 7 for NaOH.

Then the rates are (with r_{ij} written as $r_{i,j}$ for clarity):

$$\sum_j r_{1,j} = -r_{1,1} - r_{1,2} - r_{1,3} + r_{1,4}$$

$$= -k_1[\text{DAA}][\text{NaOCl}][\text{NaOH}] - k_2[\text{DAA}][\text{NaOCl}][\text{NaOH}]^{0.9}$$

$$- k_3[\text{DAA}][\text{NaOH}] + k_4[\text{Ac}]^2[\text{NaOH}] \quad (E5.3.2)$$

$$\sum_j r_{2,j} = r_{2,1} = k_1[\text{DAA}][\text{NaOCl}][\text{NaOH}] \quad (E5.3.3)$$

$$\sum_j r_{3,j} = r_{3,2} = k_2[\text{DAA}][\text{NaOCl}][\text{NaOH}]^{0.9} \quad (E5.3.4)$$

$$\sum_j r_{4,j} = 2(r_{4,3} - r_{4,4}) - r_{4,5} = 2(k_3[\text{DAA}][\text{NaOH}] - k_4[\text{Ac}]^2[\text{NaOH}])$$

$$- k_5[\text{Ac}][\text{NaOCl}][\text{NaOH}] \quad (E5.3.5)$$

$$\sum_j r_{5,j} = r_{5,5} = k_5[\text{Ac}][\text{NaOCl}][\text{NaOH}] \quad (E5.3.6)$$

$$\sum_j r_{6,j} = -3r_{6,1} - 4r_{6,2} - 3r_{6,5}$$

$$= -3k_1[\text{DAA}][\text{NaOCl}][\text{NaOH}] - 4k_2[\text{DAA}][\text{NaOCl}][\text{NaOH}]^{0.9}$$

$$- 3k_5[\text{Ac}][\text{NaOCl}][\text{NaOH}] \quad (E5.3.7)$$

$$\sum_j r_{7,j} = 0 \quad (E5.3.8)$$

The last equation is zero because NaOH does not get consumed in the reactions, although it catalyzes all of them.

This set of equations can be solved from experimental concentration versus time plots for the different species. The kinetic parameters thus obtained define these equations completely. By repeating the calculations at different temperatures, the activation energies for all the reactions can be found. Then, as already stated, these equations can be combined with the appropriate reactor equations to obtain the concentration-time history for the chosen reactor. This is illustrated in Example 11.2.

Multistep Reactions

Definitions

Multistep reactions may be classified as *simple multistep* and *complex multistep* reactions. In the single multistep scheme, each step of the synthesis is a simple

reaction with no side products (this can often be assumed if reaction conditions for each step are so chosen that side products, if any, are negligibly small). A general example of such a scheme would be

Step 1:	$A \rightarrow B + C$	
Step 2:	$C + D \rightarrow E$	[5.17]
Step 3:	$E + F \rightarrow G + H$	
Step 4:	$H \rightarrow I + J$	

On the other hand, if each step of a synthetic strategy happens to be complex, then it is a complex multistep reaction, for example

Step 1:	$A + B \rightarrow C$	
	$C + B \rightarrow D$	
Step 2:	$C \rightarrow E$	
	$C \rightarrow F$	
Step 3:	$E + G \rightarrow H$	[5.18]
	$H + G \rightarrow I$	
	$H + E \rightarrow J$	
Step 4:	$H + K \rightarrow L$	
	$L \rightarrow M$	

with L as the final product. For such a scheme, we define the overall conversion, yield, and selectivity as follows:

$$\text{overall conversion,} \quad X_{ov} = \prod_n (\text{conversion in each step})$$

$$\text{overall yield,} \quad Y_{ov} = \prod_n (\text{yield in each step}) \quad (5.11)$$

$$\text{overall selectivity,} \quad S_{ov} = \prod_n (\text{selectivity in each step})$$

where n is the number of steps. The application of these definitions is illustrated in Example 5.4 for two synthetic schemes.

EXAMPLE 5.4

Conversions, yields, and selectivities in the synthesis of (S)-naproxen and (S)-ibuprofen

(S)-Naproxen is produced by a three-step process shown in scheme E5.4.1, and (S)-ibuprofen also by a three-step process shown in scheme E5.4.2.

(E5.4.1)

(E5.4.2)

108   Reactions and Reactors

The third (asymmetric) step in each of these is the critical step involving the production of the desired (S)-enantiomers of the α-arylalkanoic acids. The yields of the first two steps are given as

(S)-Naproxen: 0.90 and 0.95.

(S)-Ibuprofen: 0.95 and 0.95

and for the third step the enantiomeric excesses (see Chapter 9) are given as

(S)-Naproxen: ee = 0.985

(S)-Ibuprofen: ee = 0.950

Calculate the overall yield in each case.

SOLUTION

It will be assumed that the conversion is 100% in each of the steps of the two processes. The two enantiomeric forms R and S are formed by a parallel reaction scheme represented by

$$A \rightarrow (R)\text{-enantiomer}$$
$$A \rightarrow (S)\text{-enantiomer}$$
(E5.4.3)

and the enantiomeric excess is defined as (see Chapter 9)

$$ee = \frac{[R] - [S]}{[R] + [S]}$$
(E5.4.4)

An ee of 100% corresponds to an enantiomerically pure form, and the reaction is said to be enantiospecific. On the other hand, an ee of 0% corresponds to a 1:1 mixture of the enantiomers known as a racemic mixture or racemate. In the case of (S)-naproxen,

$$0.985 = \frac{[S] - [R]}{[S] + [S]}$$

giving

$$\frac{[S]}{[R]} = 132.33$$

Because 100% conversion is assumed, the yield of S is given by

$$\frac{[S]}{[S] + [R]} = 0.9925$$

A similar calculation for (S)-ibuprofen gives the yield of the third step as 0.975.
   The overall yields based on the starting material in each case may now be calculated from Equation (5.11).

Yield of (S)-naproxen = 0.90 × 0.95 × 0.9925 = 0.8486

Yield of (S)-ibuprofen = 0.95 × 0.95 × 0.975 = 0.8799

Table 5.6 summarizes the results. Note that the overall yield decreases with an increase in the number of steps. This is discussed further below. The selectivities will be the same as the yields because the conversion in each step is assumed to be 100%.

Table 5.6 Overall yields in the production of (S)-naproxen and (S)-ibuprofen

Process	Yield		Enantiomeric excess		Final yield
(S)-Naproxen	Step 1:	0.90			
	Step 2:	0.95			0.849
	Step 3:	0.99	Step 3:	0.985	
(S)-Ibuprofen	Step 1:	0.95			
	Step 2:	0.95			0.880
	Step 3:	> 0.975	Step 3:	0.95	

Yield versus number of steps

It would be instructive to elaborate on the relationship between yield and the number of steps. For this we consider the reaction

$$A + B \to C$$
$$C + D \to E$$
$$E + F \to G \qquad (5.19)$$
$$G + H \to I$$
$$I + J \to K$$

Assuming the same yield for each step, the overall yield is plotted in Figure 5.6 as a function of the number of steps for different values of the individual yield. We see that for a five-step reaction, quite common in organic synthesis, the overall yield is only 58% for individual yields of 90% in each step. If the individual yields can be raised to 95% (an increase of just 5%), the overall yield goes up to

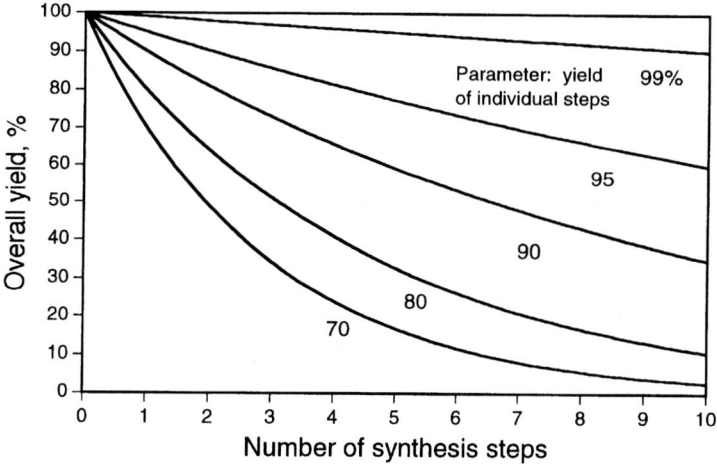

Figure 5.6 Overall final yield as a function of the number of synthesis steps at selected values of the individual yields.

110 Reactions and Reactors

77%, an increase of about 20%. This underscores the importance of maximizing the yield of each step in a multistep reaction.

SIMULTANEOUS HOMOGENEOUS AND CATALYTIC (OR AUTOCATALYTIC) REACTIONS

As mentioned, complexity can also arise by a combination of homogeneous and catalytic steps in a single reaction. The approach to such a reaction need not follow the formal procedure outlined for complex reactions in general. It tends to be system specific, and we illustrate in Example 5.5 an approach that can be modified as required to suit individual systems.

EXAMPLE 5.5

A complex reaction simultaneously involving homogeneous and autocatalytic steps

Methyl chlorosilanes, used in the manufacture of a variety of resins, elastomers, and silicone oils, are manufactured by reacting silicon (a solid) with methyl chloride (a gas) in the presence of an alloy Cu_3Si (called the η-phase) as catalyst. Although strictly a gas-solid reaction, and hence more appropriate to Chapter 15, we illustrate the procedure here by ignoring all mass transfer effects and focusing only on the two reactions involved (one homogeneous and the other autocatalytic).

The overall scheme of reactions may be represented as

$$\begin{array}{c}
Si + CH_3Cl \longrightarrow (CH_3)_nCl_{4-n}Si \\
\uparrow \\
Cu_3Si\ (\eta\text{-phase}),\ \text{catalyst for main reaction} \\
\boxed{\text{Autocatalysis leading to } Cu_3Si\ (\eta\text{-phase})} \\
4CuCl + nSi \longrightarrow SiCl_4 + 4Cu^* + (n-1)\ Si^* \\
(A)\quad (B)\qquad\quad (P)\quad (Q)\qquad (B) \\
3Cu^* + Si^* \longrightarrow Cu_3Si\ (\eta\text{-phase})
\end{array}$$

[E5.1.1]

It will be noted that the first step (starting at the bottom) is the formation of the η-phase, which then catalyzes the main reaction shown at the top (methylchlorosilane formation). The asterisks denote that the corresponding species are in an activated state. Our concern here is with the modeling of step 1 leading to η-phase formation.

Experimental curves of conversion versus time (reproduced in Figure 5.7a) clearly show autocatalytic behavior. However, a simultaneous homogeneous reaction all but overwhelmed by this autocatalytic step cannot be ruled out. Hence, in any rational modeling of this reaction, both steps must be considered. It is desired to find the rate constants for both these steps.

SOLUTION

We use the method of Tamhankar et al. (1981) by first writing the simple homogeneous reaction as

$$A + B \rightarrow P + Q \quad \text{[E5.5.2]}$$

giving

$$R_1 = -\frac{d[A]}{dt} = (k[B])[A] = k_0[A] \quad \text{(E5.5.1)}$$

Since experimental curves (reproduced in Figure 5.7a) show autocatalysis, we write

$$A + B + Q \rightarrow P + 2Q \quad \text{[E5.5.3]}$$

$$R_2 = -\frac{d[A]}{dt} = k_{ac}[A][Q] \quad \text{(E5.5.2)}$$

Total rate is given by

$$R = R_1 + R_2 = k_0[A] + k_{ac}[A][Q] \quad \text{(E5.5.3)}$$

If X_A is the conversion,

$$R = -\frac{d[A]}{dt} = A_0 \frac{dX_A}{dt} \quad \text{(E5.5.4)}$$

giving

$$\frac{dX_A}{dt} = \frac{R}{[A]_0} = \frac{k_0[A]_0(1 - X_A) + k_{ac}[A]_0^2 X_A(1 - X_A)}{[A]_0} = R'$$

or

$$\frac{dX_A}{dt} = k_0(1 - X_A) + k_{ac}[A]_0 X_A(1 - X_A) = \frac{R}{[A]_0} \quad \text{(E5.5.5)}$$

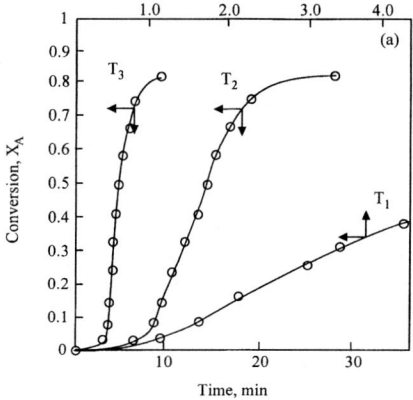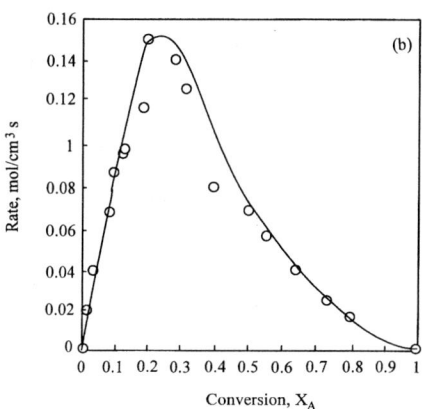

Figure 5.7 Example 5.5: (a) fractional conversion of CuCl as a function of time and (b) rate as a function of conversion for run T_2 of part *a* (redrawn from the data of Tamhankar et al., 1981).

Thus a plot of $R/(1 - X_A)[A]_0$ versus X_A should give a straight line of slope $k_{ac}[A]_0$ and intercept k_0.

Another interesting method is to rewrite Equation E5.5.5 and integrate it to give

$$\ln \frac{X_A + k_1}{1 - X_A} = k_2(t - t_i) + \ln k_1 \tag{E5.5.6}$$

where

$$k_1 = k_0/[A]_0 k_{ac} \tag{E5.5.7}$$

$$k_2 = k_0 + [A]_0 k_{ac} \tag{E5.5.8}$$

$t_i =$ induction period for the reaction.

A plot of LHS versus $(t - t_i)$ should give k_2 if k_1 is known (but it is not).

To get k_1 we differentiate the S-shaped X_A versus t curves (Figure 5.7a). Because of the nature of autocatalysis, the resulting rate curves (see Figure 5.7b) show a maximum. Thus at $X_A = X_{Am}$

$$\frac{d(dX_A/dt)}{dX_A} = \frac{dR}{dt} = [A]_0^2 k_{ac} - k_0[A]_0 - 2[A]_0^2 k_{ac} X_{Am} = 0 \tag{E5.5.9}$$

giving

$$k_0/[A]_0 k_{ac} = k_1 = (1 - 2X_{Am}) \tag{E5.5.10}$$

where $X_{Am} = X_A$ at R_{max}

Using k_1 for getting the LHS in plotting Equation E5.5.6, k_2 can be obtained. Finally,

$$\left. \begin{array}{l} k_0 = \dfrac{k_1 k_2}{(k_1 + 1)} \\ k_{ac} = \dfrac{k_2}{[A]_0 (k_1 + 1)} \end{array} \right\} \tag{E5.5.11}$$

the actual values obtained are (Tamhankar et al., 1981):

$T_1 = 300\,°C \quad k_0 = 2.368 \times 10^{-4}\,s^{-1} \quad k_{ac} = 0.175\,cm^3\,mol^{-1}\,s^{-1}$

$T_2 = 330\,°C \quad k_0 = 1.333 \times 10^{-3}\,s^{-1} \quad k_{ac} = 1.778\,cm^3\,mol^{-1}\,s^{-1}$

$T_3 = 360\,°C \quad k_0 = 2.345 \times 10^{-3}\,s^{-1} \quad k_{ac} = 9.770\,cm^3\,mol^{-1}\,s^{-1}$

(See original reference for details of each run.)

Notation to Part I

A	Reactant; Helmholtz work function; parameters defined by Equations 3.7 and 3.37.
A_c, A_H	Less reactive (cold) and highly reactive (hot) A.
A_c	Cross-sectional area, m².
A_h	Heat transfer surface, m².
A_i	Species i.
a_i	Total moles of element i in a compound.
a_j, b_j, c_j, d_j	Constants of Equation 2.11.
a, b, c, d	Constants of Equation 3.13; any constant.
B	Reactant; parameter defined by Equation 3.8.
B_C, B_H	Less reactive (cold) and highly reactive (hot) B.
C_i	Atomic group.
C_p	Heat capacity at constant pressure, kcal/kg K.
C_{PL}	Liquid heat capacity, kcal/kg K.
C_{pm}	Molar (or mean molar) heat capacity at constant pressure, kcal/mol K.
C_v	Heat capacity at constant volume, kcal/kg K.
c	Total number of molecular species.
c	Concentration matrix, mol/s.
D, D'	Chemical species.
D_{AB}	Binary diffusion coefficient of A in B, m²/s.
D_i	Diffusivity of species i, m²/s.
$\bar{D}_j$	Diffusion coefficient of j in a mixture excluding j, m²/s.
D_{ji}	Binary diffusion coefficient of j in i, m²/s.
d_i	Internal diameter of heat transfer tube, m.
d_s	Diameter of stirrer, m.
d_T	Diameter of tank, m.
E	Activation energy, kcal/mol.
F	Factors of Equation 3.20.

F_i	Molar flow rate of i, mol/s.
F_t	Total feed rate, mol/s.
f	Fragments of Equation 3.20.
f_j	Fugacity of species j.
$\bar{f}_j$	Partial fugacity of species j.
G	Gibb's free energy, kcal/mol.
G_i	Partial molal free energy of i, kcal/mol.
G_i^0	Standard partial molal free energy of i, kcal/mol.
ΔG^0	Standard molal free energy change, kcal/mol.
ΔG_{fj}^0	Standard free energy of formation of component j, kcal/mol.
ΔG_r	Free energy change due to reaction, kcal/mol.
g	Gas phase.
ΔG_r^0	Standard free energy change due to reaction, kcal/mol.
ΔG_T^0	Standard free energy change at any temperature T, K.
H	Enthalpy, kcal/mol.
ΔH^0	Partial molal enthalpy change, kcal/mol.
ΔH_c	Heat of combustion, kcal/mol.
ΔH_c^0	Standard heat of combustion, kcal/mol.
ΔH_f	Heat of formation, kcal/mol.
ΔH_f^0	Standard heat of formation, kcal/mol.
$\Delta H_{fT}^0, \Delta H_{f298}^0$	Heat of formation at T and 298 K.
ΔH_r	Enthalpy change due to reaction, or heat of reaction, kcal/mol.
ΔH_r^0	Standard enthalpy change due to reaction or standard heat of reaction, kcal/mol.
ΔH_v	Heat of vaporization, kcal/mol.
ΔH_{vb}	Heat of vaporization at normal boiling point, kcal/mol.
h_c	Heat transfer coefficient on the control fluid side of reactor, kcal/m s K.
h_r	Heat transfer coefficient on the reaction mixture side of reactor, kcal/m s K.
h_1, h_2	Heat transfer coefficient on the two sides of a surface, kcal/m K s.
I_1, I_2	Constants of integration of Equations 2.11 and 2.12.
K	General symbol for equilibirum constant.
K_a	Equilibrium constant in terms of activities; ionization (dissociation) constant of an acid.
$K_{a,0}$	Acid dissociation constant for a standard reaction.
$K_{a,R}, K_{a,R'}$	Ionization constants in a given solvent for compounds with substituent groups R and R'.
K_b	Ionization (dissociation) constant of a base.
K_c	Equilibrium constant in terms of concentrations, $(\text{mol/m}^3)^{\nu_R+\nu_S+-\nu_A-\nu_B}$.
$K_{0,w}, K_{a,a}$	Ionization constants of a given acid in water and alcohol.
K_γ	Correction factor for nonideality defined by Equation 2.27.
$K_{\phi 1}, K_{\phi 2}$	Fugacity coefficients groups defined by Equation 2.16.
K_1, K_2	Equilibrium constants of reactions 2.4.1 and 2.4.2.
K_1', K_2', K_3', K_4'	Parameters defined by Equation 3.17.
k	General symbol for rate constant, appropriate units.
k_i	Rate constant for reaction i, appropriate units.
k	Rate constant vector, appropriate units.
k_a, k_b	Rate constants for acid- and base-catalyzed reactions.
k_G'	Gas-side mass transfer coefficient, m/s.
k_{GL}'	Overall mass transfer coefficient at gas-liquid interface, m/s.

k'_L	Liquid-side mass transfer coefficient, m/s.
$k'_{L_1 L_2}$	Overall mass transfer coefficient at liquid–liquid interface, m/s.
k'_{L1}, k'_{L2}	Mass transfer coefficient for liquid films 1 and 2, m/s.
k^0	Arrhenius preexponential factor with units of the rate constant.
$\bar{k}$	Modified rate constant given by Equation 4.43.
k_0	Rate constant for reaction with a standard substrate, appropriate units.
k_+, k_-	Forward and reverse rate constants, appropriate units.
L	Total length, m; liquid phase.
M	Any thermodynamic property, such as A, H, G, S; molecular weight; number of reactions.
M_i	Molecular weight of species i, kg/mol.
$\overline{M}_i$	Partial molar quantity of species i (i.e. $\bar{G}_i, \bar{H}_i$) defined by Equation 2.20.
m	Reaction order; distribution coefficient defined by Equation 3.18; total number of elemental species.
m_0	Distribution coefficient for octanol-water system.
m_s	Distribution coefficient for any solvent-water system (other than) octanol-water.
Δ_m	Log mean of the inner and outer areas of tube wall.
N	Species N; number of components; equivalent chain length.
N_a	Number of atoms in a molecule.
N_i, N_j	Number of moles of species i, j.
N_t	Total number.
ΔN_i	Correction for structure in Equation 3.22.
n	Reaction order; exponent.
n_c	Number of components in a reacting system.
n_i	Number of atomic groups C_i of type i.
n_{ij}	Number of atoms of element i in molecular species j.
P	Total pressure, atm (Pa).
P_c	Critical pressure, atm (Pa).
P_r	Reduced pressure, P/P_c.
p_R	Partial pressure of R, atm.
$P_{r,vp}$	Reduced vapor pressure.
p_i	Partial pressure of species i, atm (Pa).
$\Delta_p, \Delta_T, \Delta_V$	Structural increments for P_C, T_C, V_C.
Q_i	Volumetric flow rate of i, m³/s.
R	Product; substituent group.
R_g	Gas constant, m³ atm/mol/K.
$\mathbf{r}$	Matrix of reaction rates.
r_a	Radius of the solute molecule, m.
r_{Si}, r'_i	Reaction rate of species i, mol/m² surface s.
$r_{Vi}; r_i$	Reaction rate of species i, mol/m³ reactor s.
r_{vi}	Reaction rate of species i, mol/m³ cat s.
r_{wi}	Reaction rate of species i, mol/kg cat s.
S	Product; solid; entropy, kcal/mol K; selectivity; surface area, m².
S_i	Selectivity of species i.
S_1, S_2	Selectivity of A to B and to C in Equation 2.39.
S_T^0	Entropy at T K in the ideal gaseous state.
ΔS^0	Partial molal entropy change, kcal/mol.
s	Rate of surface renewal, s⁻¹.
T	Temperature, °C or K.

116 Reactions and Reactors

T_b	Boiling point, K.
T_c	Critical temperature, K.
T_r	Reduced temperature, T/T_c.
T_{rb}	Reduced boiling point.
T_w	Wall temperature, K.
t	Time, s.
t_D	Equivalent diffusion time, s.
t_t	Total time, s.
t_w	Wall thickness, m.
$\bar{t}$	Time parameter defined as V/Q_0.
t^*	Surface renewal time, s.
U	Internal energy, kcal/mol; overall heat transfer coefficient, kcal/m s K.
u	Linear velocity, m/s.
V	Volume, m^3; product.
V_c	Critical volume, m^3 (usually given in cm^3 in tables).
V_i	Molecular volume of species i, cm^3/mol.
V_L	Volume of liquid in reactor, m^3.
V_r	Reduced volume, V/V_c; reactor volume, m^3.
V_t	Total volume, m^3.
V_v	Vapor volume, m^3.
v	Volume of catalyst, m^3.
W	Weight of catalyst, kg.
X_i	Conversion of species i.
Y_i	Yield of species i.
y_j	Mole fraction of species j (i.e., N_j/N_t).
Z	Compressibility factor.
Z_c	Compressibility factor at the critical condition.

Greek

α	Exponent of Brönsted equation Equation 2.35 for acid; parameter defined by Equations 4.49 and 4.65.
α'	Group defined by Equation 4.69.
β	Exponent of Equation 2.36; parameter of Equation 3.5; group defined by Equation 4.52.
β'	Proportionality constant of Equation E2.4.1; group defined by Equation 4.69.
δ	Film thickness, m; operator.
δ_i	Volume change parameter.
δ_M	General representation of solvent and substituent operators.
δ_p	Polanyi operator.
δ_R	Substituent operator.
δ_s	Solvent or medium operator.
ε_A	Volume change parameter.
ϕ_j	Fugacity coefficient of species j.
$\bar{\phi}_j$	Partial fugacity coefficient of species j.
l	Length, m.
γ	Ratio C_p/C_v; activity coefficient.
λ	Thermal conductivity, kcal/m s K; length, m.
λ_L	Liquid thermal conductivity, kcal/m s K.

μ	Viscosity, kg/m s.
μ_c	Critical viscosity, kg/m s.
$\mu_{c,\text{org}}^*$	Pseudocritical viscosity, kg/m s (mP).
μ_j	Chemical potential of species j.
μ_L	Liquid viscosity, kg/m s.
$\mu_{L,T}, \mu_{L,T(\text{ref})}$	Liquid viscosities at T and reference T, kg/m s.
μ_r	Reduced viscosities at T and reference T, $\mu/\mu c$.
$\boldsymbol{\nu}$	Matrix of stoichiometric coefficients.
ν_{ij}	Stoichiometric coefficient of species i in reaction j.
ν_i	Stoichiometric coefficient of species i.
ρ	Function defined by the Hammett relationship; density, kg/m^3.
ρ_c	Critical density, kg/m^3.
ρ_L	Liquid density, kg/m^3.
ρ_r	Reduced density.
ρ_v	Vapor density, kg/m^3.
σ	Ratio of the dissociation constants $K_a/K_{a,o}$ (Hammett's substituent constant).
ω	Acentric factor.
ξ	Extent of reaction, mol.
ξ'	Extent of reaction, mol/s.
ψ_i	Ratio $[i]/[A]_0$, $i \neq A$.

Subscripts/Superscripts

Superscripts

a	Alcohol.
c	Control fluid; critical property.
cm	Pseudocritical.
f	Final (leaving) condition.
int	Interface.
L	Liquid.
m	Mole fraction average; molar; mean
N	Nth stage.
0	Entering condition.
ov	Overall.
s	Surface.
w	Wall; water.

References to Part I

(Note: General references not referred to in the text are marked with an asterisk.)

Akella, L.M., and Lee, H.H. *Chem. Eng. J.,* **22**, 25 (1981).
Ambrose, D. *Appl. Chem. Biotechnology,* **26**, 711 (1976); Natl. Phys. Lab. (U.K.) Rep. Chem. 57 (Dec. 1958). Ambrose, D. Vapor-Pressure Equations, Natl. Phys. Lab. Rep. Chem., Nov. 1972.
Anderson, J.W., Beyer, G.H., and Watson, K.M. *Natl. Pet. News Tech. Sec.,* **36**, R 476 (1944).
Aris, R. *Introduction to the Analysis of Chemical Reactors,* Prentice-Hall, Englewood Cliffs, N.J., 1965; *Elementary Chemical Reactor Analysis,* Prentice-Hall, Englewood Cliffs, N.J., 1969; *Introduction to the Analysis of Chemical Reactors,* Prentice-Hall, Englewood Cliffs, N.J., 1969.
Aris, R., and Gavalas, G.R. *Phil. Trans. R. Soc. London,* **260**, 351 (1966).
Astarita, G. *Mass Transfer with Chemical Reaction,* Elsevier, Amsterdam, 1967.
Auerbach, R. *Experimentia,* **4**, 473 (1948).
Babu Rao, K., and Doraiswamy, L.K. *Ind. J. Tech.,* **4**, 141 (1966).
Barac, G., Ph.D Thesis in chemical engineering, Iowa State University, Ames, Iowa, 1998.
Baroncini, C., Di Filippo, P., and Latini, G. *Intern. J. Refrig.,* **6**, 60 (1983).
Baroncini, C., Di Filippo, P., Latini, G., and Pacelli, M. *Intern. J. Thermophys.,* **1**, 159 (1980); **2**, 21 (1981).
Baroncini, C., Latini, G., and Pierpaoli, P. *Inter. J. Thermophys.,* **5**, 387 (1984).
Bell, R.P., Gelles, E., and Moller, E. *Proc. R. Soc. London,* **A198**, 310 (1949).
Bell, R.P., and Goldsmith, H.L. *Proc. R. Soc. London,* **A210**, 322 (1952).
*Bender, M.L. *Mechanism of Homogeneous Catalysis from Protons to Proteins,* Wiley Interscience, New York, 1971.
Benson, S.W. *Thermochemical Kinetics,* Wiley, New York, 1968.
Benson, S.W., and Buss, J.H. *J. Chem. Phys.,* **29**, 550 (1958).

Benson, S.W., Cruickshank, F.R, Golden, D.M., Haugen, G.H., O'Neal, H.E., Rodgers, A.S., Shaw, R., and Walsh, R. *Chem. Rev.*, **69**, 279 (1969).
Bhat, Y.S., Kulkarni, B.D., and Doraiswamy, L.K. *Ind. Eng. Chem. Proc. Des. Dev.*, **24**, 525 (1985).
Bhirud, V.S. *AIChE J.*, **24**, 1127 (1978).
Bondi, A. *Ind. Chem. Fundam.*, **5**, 443 (1966).
Brock, J.R., and Bird, R. B. *AIChE J.*, **1**, 174 (1955).
Bures, M., Majer, V., and Zabransky, M. *Chem. Eng. Sci.*, **36**, 529 (1981).
Cardozo, R.L. *AIChE J.*, **32**, 844 (1986).
Carruth, G.F. and Kobayashi, R. *Ind. Eng. Chem. Fundam.*, **11**, 509 (1972).
Chapman, F.S., Dellenbach, H., and Holland, F.A. *Trans. Inst. Chem. Eng.*, **42**, T398 (1964).
Chen, N.H. *J. Chem. Eng. Data*, **10**, 207 (1965).
Chilton, T.H., Drew, T.B., and Jebens, R.H. *Ind. Eng. Chem.*, **36**, 510 (1944).
Chueh, P.L., and Prausnitz, J.M. *AIChE J.*, **13**, 1099 (1967); **15**, 471 (1969).
Chueh, C.F., and Swanson, A.C. *Chem. Eng. Prog.*, **69**(7), 83 (1973); *Can. J. Chem. Eng.*, **51**, 596 (1973).
*Cooper, A.R., and Jeffreys, G.V. *Chemical Kinetics and Reactor Design*, Prentice-Hall, Englewood Cliffs, N.J., 1971.
Cornelissen, J., and Waterman, H.I. *Chem. Eng. Sci.*, **4**, 238 (1955).
Danckwerts, P.V. *Chem. Eng. Sci.*, **2**, 1 (1953).
Davis, M.M., and Hetzer, H.B. *J. Res. Natl. Bur. Stand. (U.S.A.)*, **60**, 569 (1958).
Dean, D.E., and Stiel, L.I. *AIChE J.*, **11**, 526 (1965).
Dilke, M.H., and Eley, D.D. *J. Chem. Soc.*, 2601 (1949).
Douglas, J.M., and Eagleton, L.C. *Ind. Eng. Chem. Fundam.*, **1**, 116 (1962).
Eigenmann, H.K., Golden, D.M., and Benson, S.W. *J. Phys. Chem.*, **77**, 1687 (1973).
Eversteijn, F.C., Stevels, J.M., and Waterman, H.I. *Chem. Eng. Sci.*, **11**, 267 (1960).
Fishtine, S.H. *Ind. Eng. Chem.*, **55**(4), 20 (1963); **55**, 49 (1963); **55**(6), 47 (1963); *Hydrocarbon Proc. Petro. Ref.*, **45**, 173 (1966).
Forni, L., and Miglio, R. In *Heterogeneous Catalysis and Fine Chemicals*, (eds., Guisnet, M., Barbier, J., Barrault, J., Bouchoule, C., Dupez, D., Montassier, C., and Pirot, G., Elsevier, Amsterdam, 1993.
Franklin, J.L. *Ind. Eng. Chem.*, **41**, 1070 (1949); *J. Chem. Phys.*, **21**, 2029 (1953).
Frost, A.A., and Pearson, R.G. *Kinetics and Mechanism*, 2nd ed., Wiley, New York, 1961.
Fuller, E.N., Shetler, P.D., and Giddings, J.C. *Ind. Eng. Chem.*, **58**, 19 (1966).
Gambill, W.R. *Chem. Eng. (N.Y.)*, **66**, 193 (1959).
Giacalone, A. *Gazz. Chim. Ital.*, **81**, 180 (1951).
Gilliland, E.R. *Ind. Eng. Chem.*, **26**, 681 (1934).
Gomez-Nieto, M., and Thodos, G. *Can. J. Chem. Eng.*, **55**, 445 (1977a), *Ind. Eng. Chem. Fundam.*, **16**, 254 (1977b); **17**, 45 (1978).
Goyal, P., and Doraiswamy, L.K. *Hydrocarbon Process*, **45**, 200 (1966); *Ind. Eng. Chem. Proc. Des. Dev.*, **9**, 25 (1970).
Gunn, R.D., and Yamada, T. *AIChE J.*, **17**, 1341 (1971).
Gupte, P. A., and Daubert, T.E. *Ind. Eng. Chem. Proc. Des. Dev.*, **24**, 674 (1985).
Haario, H., Kalachev, L., Salmi, T., and Lehtonen, J. *Chem. Eng. Sci.*, **54**, 1131 (1999).
Hammett, L.P. *Physical Organic Chemistry*, McGraw-Hill, New York, 1940.
Hansch, C., and Leo, A.J. *Substituent Constants for Correlation Analysis in Chemistry and Biology*, Wiley, New York, 1979.
Hayduk, W., and Minhas, B.S. *Can. J. Chem. Eng.*, **60**, 295 (1982).
Hayduk, W., and Laudie, H. *AIChE J.*, **20**, 611 (1974).

Higbie, R. *Trans. Am. Inst. Chem. Eng.*, **31**, 365 (1935).
Hildebrand, J.H. *Science*, **174**, 490 (1971).
Holland, C.D., and Anthony, R.G. *Fundamental of Chemical Reaction Engineering*, Prentice-Hall, Englewood Cliffs, N.J., 1989.
*Horak, J., and Pasek, J. *Design of Industrial Chemical Reactors from Laboratory Data*, Heyden, London, 1978.
Hutchinson, P., and Luss, D. *Chem. Eng. J.*, **1**, 129 (1970).
Jacob, S.M., Gross, B., Voltz, S.E., and Weekman, V.W. *AIChE J.*, **22**, 701 (1976).
Jaffe, H.H. *Chem. Rev.*, **53**, 191 (1953).
Janz, G.J. *Thermodynamic Properties of Organic Compounds:* Estimation Methods and Practice, Academic Press, New York, 1958.
Joback, K.G. M.S. Thesis in Chemical Engineering, Massachusetts Institute of Technology, Cambridge, Mass., June 1984.
Klein, V.A. *Chem. Eng. Prog.*, **45**, 675 (1949).
Kothari, M.S., and Doraiswamy, L.K. *Hydrocarbon Proc. and Petro. Ref.*, **43**, 133 (1964).
Kratzke, H. *J. Chem. Thermodyn.*, **12**, 305 (1980).
Kunte, M.V., and Doraiswamy, L.K. *Chem. Proc. Eng.*, 157 (1958); *Chem. Eng. Sci.*, **12**, 1 (1960).
Le Bas, G., in *Molecular Volumes of Liquid Chemical Compounds*, Longmans, Green, London, 1915.
Lee, B.I., and Kesler, M.G. *AIChE J.*, **21**, 510 (1975).
Lee, H.H. *Heterogeneous Reactor Design*, Butterworth, Boston, 1985.
Leffler, J.E., and Grunwald, E. *Rates and Equilibrium of Organic Reactions*, Dover, New York, 1963.
Leo, A., and Hansch, C. *J. Org. Chem.*, **36**, 1539 (1971).
Leo, A., Hansch, C., and Elkins, D. *Chem. Rev.*, **71**, 525 (1971).
Levenspiel, O. *The Chemical Reactor Omnibook*, OSU Bookstore, Corvallis, Or, 1993.
Lewis, W.K., and Squires, L. *Refiner Nat. Gasoline Manuf.*, **13**(12), 448 (1934).
Licht, W., Jr. and Stechert, D.G. *J. Phys. Chem.*, **48**, 23 (1944).
Lieber, E., Rao, C.N.R., and Chao, T.S. *J. Am. Chem. Soc.*, **79**, 5962 (1957).
Lucas, K. *Phase Equilibria and Fluid Properties in the Chemical Industry*, Dechema, Frankfurt, 1980; *Chem. Ing. Tech.*, **53**, 959 (1981).
Luss, D., and Golikeri, S.V. *AIChE J.*, **21**, 865 (1975).
Luss, D., and Hutchinson, P. *Chem. Eng. J.*, **2**, 172 (1971).
Lyckman, E.W., Eckert, C.A., and Prausnitz, J.M. *Chem. Eng. Sci.*, **20**, 703 (1965).
Lyman, T.J., and Danner, R.P. *AIChE J.*, **22**, 759 (1976).
Lyman, W.J., Reehl, W.F., and Rosenblatt, D.H. *Handbook of Chemical Property Estimation Methods*, McGraw-Hill, New York, 1982.
Majer, V., Bures, M., and Zabransky, M. *Chem. Prumysl.*, **29**(9), 462 (1979).
Mathur, V.K., Singh, J.D., and Fitzgerald, W.M. *J. Chem. Eng. Jpn*, **11**, 67 (1978).
McDaniels, D.H., and Brown, H.C. *J. Org. Chem.*, **23**, 420 (1958).
McGarry, J., *Ind. Eng. Chem. Proc. Des. Dev.*, **22**, 313 (1983).
Medani, M.S., and Hasan, M.A. *Can. J. Chem. Eng.*, **55**, 203 (1977).
Miller, D.G., Univ. California Rad. Lab. Rep. 14115-T, Apr. 21, 1965.
Miller, C.O.M., a personal communication in 1980, cited in Lyman, W.J., Reehl, W.F., and Rosenblatt, D.H. McGraw-Hill, New York, 1982.
Missenard, F.A. *C. R.*, **260**, 5521 (1965); *Rev. Gen. Thermodyn.*, **141**, 751 (1973).
More, R., and Capparelli, A.L. *J. Phys. Chem.*, **84**, 1870 (1980).
Morris, P.S. M.S. Thesis, Polytechnic Institute of Brooklyn, Brooklyn, N.Y. 1964.
Myers, R.T. *J. Phys. Chem.*, **83**, 294 (1979).

Nace, D.M., Voltz, S.E., and Weekman, V.W. *Ind. Eng. Chem. Proc. Des. Dev.*, **10**, 530 (1971).
Nakanishi. K. *Ind. Eng. Chem. Fundam.*, **17**, 253 (1978).
Narasimhan, G. *Br. Chem. Eng.*, **10**, 253 (1965); **12**, 897 (1967).
Nauman, E.B. *Chemical Reactor Design*, Wiley, New York, 1987.
Nelson, L.C., and Obert, E.F. *Trans. ASME*, **76**, 1057 (1954).
Noddings, C.R., and Mullet, G.M., *Handbook of Compositions at Thermodynamic Equilibrium*, Wiley Interscience, New York, 1965.
Ogiwara, K., Arai, Y., and Saito, S., *J. Chem. Eng. Jpn*, **15**, 335 (1982).
O'Neal, H.E., and Benson, S.W. *J. Chem. Eng. Data*, **15**, 266 (1970).
Orrick, C., and Erbar, J. H. Private communications to Reid, R. C., Praunitz, J. M., and Poling, B. E., 1987.
Pandey, J.D., Dubey, G.P., and Tripathi, N. *Proceed., Chemal Sciences*, Indian Acad. Sci., **111**, 361 (1999).
Patterson, D., and Rastogi, A.K. *J. Phys. Chem.*, **74**, 1067 (1970).
*Perrin, D.D., *Dissociation Constants of Organic Bases in Aqueous Solution*, Butterworths, London, 1965.
Peters, M.S., and Skorpinski, E.J. *J. Chem. Educ.*, **42**, 329 (1965).
Pitzer, K.S., Lippmann, D.Z., Curl, R.F., Huggins, C.M., and Petersen, D.E. *J. Am. Chem. Soc.*, **77**, 3433 (1955).
Prausnitz, J.M. *Molecular Thermodynamics of Fluid Phase Equilibria*, Prentice-Hall, Englewood Cliffs, N.J., 1969.
Rackett, H.G. *J. Chem. Eng. Data*, **15**, 514 (1970).
*Rase, H.F. *Chemical Reactor Design for Process Plants*, Wiley, New York, 1977; *Fixed Bed Reactor Design and Diagnostics*, Butterworth, Boston, 1990.
Reddy, K.A., and Doraiswamy, L.K. *Ind. Eng. Chem. Fundam.* **6**, 77 (1967).
Redlich, O., and Kwong, J.N.S. *Chem. Rev.*, **44**, 233 (1949).
Regenass, W. *Am. Chem. Soc. Symp. Ser.*, No. 65, Chemical Reaction Engineering, Houston, TX, 1978.
Reichenberg, D. DCS Rep., National Physical Laboratory, Teddington, U.K., 1971; *AIChE J.*, **19**, 854 (1973); *AIChE J.*, **21**, 181 (1975a); *The Viscosity of Pure Gases at High Pressures*, Natl. Eng. Lab. Rep. Chem. 38, East Kilbride, Glasgow, Scotland, 1975b.
Reid, R.C., Prausnitz, J.M., and Poling, B.E. *The Properties of Gases and Liquids*, 4th ed., McGraw-Hill, New York, 1987.
Reid, R.C., Prausnitz, J. M., and Sherwood, T.K. *Properties of Gases and Liquids*, 3rd ed., McGraw-Hill, New York, 1977.
Reid, R.C., and Sherwood, T.K. *The Properties of Gases and Liquids*, 1st ed., McGraw-Hill, New York, 1958; 2nd ed., 1966.
Riedel, L. *Chem. Ing. Tech.*, **23**, 59, 321, 465 (1951).
Rihani, D.N., and Doraiswamy, L.K. *Ind. Eng. Chem. Fundam.*, **4**, 17 (1965).
Robbins, L.A., and Kingrea, C.L. *Hydrocarbon Proc. Petro. Ref.*, **41**(5), 133 (1962).
Rodigin, N.M., and Rodigina, E.N. *Consecutive Chemical Reactions*, Van Nostrand, New York, 1964.
Rose, L.M. *Chemical Reactor Design in Practice*, Elsevier, Amsterdam, 1981.
Rowlinson, J.S. *Liquids and Liquids Mixtures*, 2nd ed., Butterworth, London, 1969.
Roy, D., and Thodos, G. *Ind. Eng. Chem. Fundam.*, **7**, 529 (1968).
Sandler, S.I. *Chemical and Engineering Thermodynamics*, Wiley, New York, 1977.
Sastri, S.R.S., Ramana Rao, M.V., Reddy, K.A., and Doraiswamy, L.K. *Br. Chem. Eng.*, **14**, 959 (1969).

Scheibel, E.G. *Ind. Eng. Chem.*, **46**, 2007 (1954).
Shaw, R. *J. Phys. Chem.*, **75**, 4047 (1971).
Shorter, J.A. *Correlation Analysis of Organic Reactivity*, Research Studies Press (Wiley), New York, 1982.
Stein, S.E., Golden, D.M., and Benson, S.W. *J. Phys. Chem.*, **81**, 314 (1977).
Strek, F. *Intl. Chem. Eng.*, **3**, 33 (1963).
Stull, D.R., Westrum, E.F., Jr., and Sinke, G.C. *The Chemical Thermodynamics of Organic Compounds*, Wiley, New York, 1969.
Tamhankar, S.S., Gokarn, A.N., and Doraiswamy, L.K. *Chem. Eng. Sci.*, **36** (1981).
Teja, A.S., and Rice, P. *Chem. Eng. Sci.*, **36**, 417 (1981); **37**, 790 (1982).
Thek, R.E., and Stiel, L.I. *AIChE J.*, **12**, 599 (1966), **13**, 626 (1967).
Thinh, T.P., Duran, J.L., and Ramlho, R.S. *Ind. Eng. Chem. Proc. Des. Dev.*, **10**, 576 (1971).
Toxvaerd, S. *Statistical mechanics*, Chem. Soc., London, Vol. 2, Chapter 4, 1975.
Tundo, P. *Continuous Flow Methods in Organic Synthesis*, Ellis Horwood, New York, 1991.
Tyn, M.T., and Calus, W.F. *Processing*, **21**(4), 16 (1975).
Uyehara, O.A., and Watson, K.M. *Natl. Pet. News*, **36**, R-714 (1944).
Van Krevelen, D.W. and Chermin, H.A.G. *Chem. Eng. Sci.*, **1**, 66 (1951); **1**, 238 (1952).
Van Velzen, D.R., Cardozo, R.L., and Langenkamp, H. *Ind. Eng. Chem. Fundam.*, **11**, 20 (1972a); *Euratom*, 4735e, Joint Nuclear Research Center, Ispra Establishment, Italy (1972b).
Vasil'eva, A.B., and Butuzov, V. F. *Asymmetric Expansions of Solutions of Singularity Perturbed Equations*, Nauka (in Russiab), Moscow, 1995.
Venkitakrishnan, G.R., and Doraiswamy, L.K. National Chemical Laboratory report, 1982.
Verma, K.K., and Doraiswamy, L.K. *Ind. Eng. Chem. Fundam.*, **4**, 389 (1965).
Vetere, A. *Snam Progetti*, San Donalo, Milanese (1973).
Wagner, W. *Cryogenis*, **13**, 470 (1973a); Bull. Inst. Froid. Annexe, No.4, 65 (1973b); *A New Correlation Method for Thermodynamic Data Applied to the Vapor-Pressure Curve of Argon, Nitrogen, and Water* (trans., and ed., Watson, J.T.R.), IUPAC Thermodynamic Tables Project Center: London, 1977.
Walas, S.M. *Phase Equilibria in Chemical Engineering*, Butterworth, Boston, 1985; *Reaction Kinetics for Chemical Engineers*, Butterworth, Stoneham, Mass., 1989.
Watson, K.M. *Ind. Eng. Chem.*, **35**(4), 398 (1943).
Wells, P.R. *Linear Free Energy Relationships*, Academic Press, New York, 1968.
Wilke, C.R., and Chang, P. *AIChE J.*, **1**, 264 (1955).
Yen, L.C., and Woods, S.S. *AIChE J.*, **12**, 95 (1966).
Yoneda, Y. *Bull. Chem. Soc. Jpn*, **52**, 1297 (1979).
Yuan, T. F., and Stiel, L.I. *Ind. Chem. Fundam.*, **9**, 393 (1970).

PART II

CATALYSIS IN ORGANIC SYNTHESIS AND TECHNOLOGY

Many bodies have the property of exerting on other bodies an action which is very different from chemical affinity. By means of this action they produce decomposition in bodies, and form new compounds into the composition of which they do not enter. This new power, hitherto unknown, is common both in organic and inorganic nature—I shall call it catalytic power. I shall also call catalysis the decomposition of bodies by this force.

J. J. Berzelius,
Edinburgh New Philosophical Journal, **XXI**, 223 (1886)

Chapter 6

Catalysis by Solids, 1

Organic Intermediates and Fine Chemicals

The use of solid catalysts has a number of advantages compared to catalysis in solution. The most important advantage is their discrete state, usually stationary, which enables easy separation of the product from the catalyst. Although catalysis by solids in organic technology was largely restricted till about the mid-1970s to bulk chemicals produced by continuous processes, it has since been extended to organic intermediates and fine chemicals (which are usually medium to small-volume production in batch processes).

We devote this chapter to a brief review of the major types of solid catalysts used in the production of intermediates and fine chemicals. Though these reactions can be carried out in both the vapor and liquid phases, the substrates used in organic synthesis are often relatively complex liquid molecules which tend to decompose under harsh conditions. Hence it is usually desirable to operate under softer conditions, thus preserving the liquid state of the substrate and preventing any likely decomposition to unwanted products.

Because catalysis by solids will almost certainly play a major role in organic syntheses of the future, surface science studies involving complex organic molecules are being increasingly undertaken (see, e.g., Rylander, 1979, 1985; Molnar, 1985; Kim and Barteau, 1989; Joyner, 1990; Idriss et al., 1992; Schulz and Cox, 1992, 1993; Pierce and Barteau, 1995; and the recent review by Smith, 1996). However, this book will not be concerned with such mechanistic considerations.

There are a few classes of catalysts that have acquired a degree of prominence during the last decade in the synthesis of organic intermediates and fine chemicals that marks them as uniquely relevant in the context of industry's irreversible shift to green technology. In addition to the homogeneous catalysts considered subsequently in Chapter 9, they include a wide variety of solid catalysts. These catalysts can be classified in two ways: (1) as distinct classes of catalysts that cut

Table 6.1 Classification of solid catalysts in organic synthesis

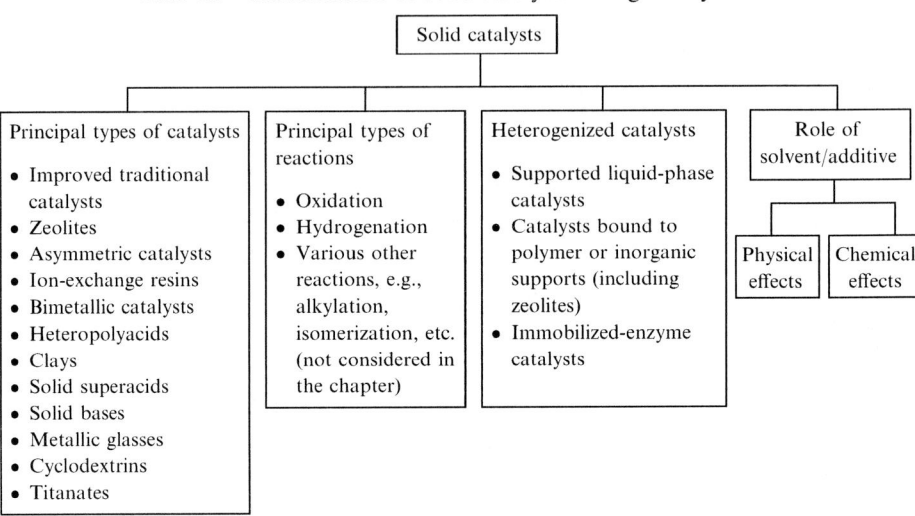

across different types of reactions, including dissolved catalysts supported on solids; and (2) as catalysts specific to different types of reactions. It is also possible to control catalytic action by using appropriate solvents/additives. Table 6.1 summarizes these aspects of solid catalysis in organic synthesis and technology. It is important to note that the two broad classes mentioned are not mutually exclusive. We briefly consider all of the subclasses included in the table with greater emphasis on zeolites because of their rising importance in organic synthesis.

A rapidly developing area of catalysis is the use of combinatorial chemistry and solid-phase synthesis to build up large libraries (of close to a million members) from which the best candidates can be selected for a given reaction (see, e.g., Lam et al., 1991; Liu and Ellman, 1995; Menger et al., 1995; Francis et al., 1996). We will not be concerned with such libraries in this book.

There has been a series of books under the title *Catalysis of Organic Reactions* edited by different authors: Moser (1981), Kosak (1984), Augustine (1985), Rylander et al. (1988), Blackburn (1990), Poscoe (1992), Kosak and Johnson (1994), Scaros and Prunier (1995), and Malz (1996), which together contain a wealth of information on organic reactions catalyzed both by homogeneous and heterogeneous catalysts. Another series that is equally important appears under the title *Heterogeneous Catalysis and Fine Chemicals* edited by Guisnet et al. (1988, 1991, 1993, 1996).

An important consideration in catalysis is selectivity, and several books devoted to this aspect of it have appeared in the recent literature, for example, those of Oyama and Hightower (1993) and Davis and Suib (1993). Two particularly useful books (both to the chemist and the chemical engineer) on the principles and practice of catalysis are those of Satterfield (1991) and Thomas and Thomas (1997). Catalytic processes are being constantly improved, and a good book from which

to obtain this information is *Advances in Catalytic Processes* (1995). Synthetic organic chemists will greatly benefit from Augustine's (1996) treatment of catalysis specially addressed to them. Zeolites, clays, and heteropolyacids are among the catalysts being increasingly explored for use in organic synthesis. A lucid combined treatment of these can be found in Izumi et al. (1992).

MODIFIED FORMS OF THE MORE COMMON CATALYSTS

Modifications of traditional catalysts have been introduced in several ways and may be broadly divided into two categories: general nonspecific methods such as addition of promoters, changing the methods of preparation, and using different types of supports; and (2) subjecting common catalysts to some specific treatments that have developed into well-defined techniques of modification. We give examples of some general methods and then refer to one specific method, sulfidation.

Modifications by Nonspecific Methods

A rewarding line of research has been the use of vapor-phase reactions wherever feasible, usually for intermediates and occasionally for finished fine chemicals. A good example of this is the development of a catalyst consisting of a mixture of Zn and Cr oxides (in the ratio 3:1) impregnated with 1 wt% Pd for the cyclization of ethylenediamine (1) and propylene glycol (2) to 2-methylpyrazine (3) (reaction 6.1) (Forni et al., 1987a,b; Forni, 1988; Forni and Nestori, 1988; Forni and Miglio, 1993).

$$\text{(1)} + \text{(2)} \longrightarrow \text{(3)} + 2H_2O + 3H_2 \quad [6.1]$$

This reaction is a basic step in the synthesis of 2-amidopyrazine, a well-known antitubercular drug, with a selectivity of more than 90%.

A particularly important vapor-phase organic synthesis is the partial oxidation of toluene to benzaldehyde, used in the manufacture of fragrance chemicals. The importance of this chemical has spurred extensive research over the last fifty years (e.g. Parks and Yula, 1941; Downie et al., 1961; Adams, 1968; Reddy and Doraiswamy, 1969; Andersson, 1986; Grzybowska et al., 1987). In all of these studies, a conversion of no more than about 10% was obtained with selectivities of up to 50%. In a more recent study, Ai (1991) used mixed oxides of Mo, U, and Sb on pumice originating from volcanic stone to obtain a selectivity of 60% with a conversion of about 25% (an increase by a factor of 2.5). Benzaldehyde can also be produced (again in the vapor phase) by the hydrogenation of benzoic acid or its ester (methyl benzoate). It has been shown (Aboulayt et al., 1993), for instance, that methyl benzoate can be hydrogenated to the aldehyde with high selectivity using ZrO_2 or CeO_2 as catalysts.

Two other important modifications of vapor-phase reactions in organic synthesis with improved selectivity may be mentioned: monoethylaniline by ethylation of aniline with ethyl alcohol over bauxite (Goyal and Doraiswamy, 1970; Doraiswamy and Venkitakrishnan, 1978) and oxidative dehydrogenation of 3-hydroxy-4-methyl-4-penten-2-one to 4-methyl-4-penten-2,3-dione on CuO/Al_2O_3 and CuO/SiO_2 (Lansink-Rotgerink et al., 1991).

The use of supported copper has been explored in a number of reactions with good commercial potential. Some examples are oxidative dehydrogenation (already mentioned), dehydrogenation of esters (Evans et al., 1974; Agarwal et al., 1987; Chen et al., 1989), hydrolysis of nitriles (Lee et al., 1988), amination of esters (Barrault et al., 1991), and one-step synthesis of dissymmetric amines [R_2NCH_3 or $RN(CH_3)_2$] from the corresponding nitriles, methanol, and hydrogen (Barrault et al., 1993).

Sulfidation of Common Catalysts

Recent studies by Moreau et al. (1988, 1991) have established the effectiveness of sulfided catalysis in the synthesis of organic intermediates and small-volume chemicals. An example is the hydroprocessing of chloronitrobenzenes over a sulfided industrial catalyst HR 306 (3% CoO, 14% MoO_3 on γ-alumina) at $T = 50\text{--}250\,°C$ and $P = 20$ atm.

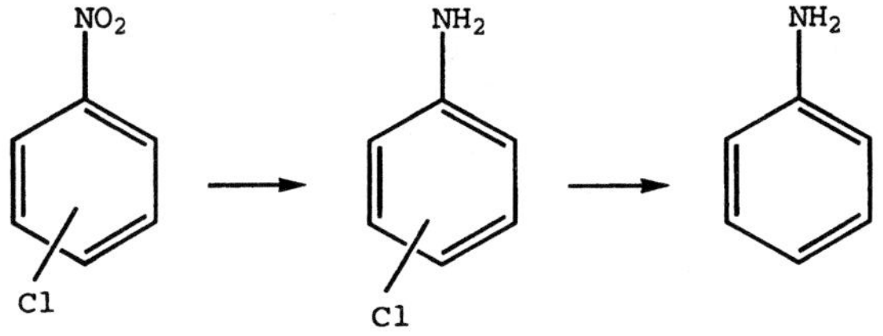

[6.2]

Another unique example of the use of sulfided catalysts is in the production of 6-chloro-2(1H)-quinoxalinone (5) from 6-chloro-2(1H)-quinoxaline-4-oxide (4) on platinum sulfide (Malz et al., 1993).

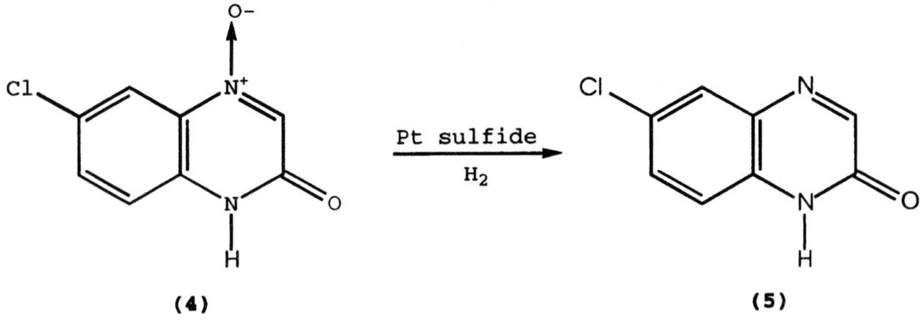

[6.3]

(4)   (5)

Both (4) and (5), as well as substituted quinoxalines, are intermediates in the manufacture of drugs for the central nervous system and agricultural chemicals such as Nissan Chemical's *Assure* and Ciba-Geigy's *Agile*.

Optimization with Common Catalysts

Sometimes, optimization of process conditions using a common catalyst can be more rewarding than searching for new catalysts. A good example of this is the synthesis of isophoronediamine (IPDA) used in paints and varnishes. Processes using Raney cobalt as such or doped with chromium (Dombeck and Wenzel, 1990a,b), and a few other processes (Hirako, 1987; Merger et al., 1991), were found inferior to a process based on the common Raney nickel under optimized conditions (Gillet et al., 1993). This is a new two-step process involving azine synthesis from isophoronenitrile, followed by hydrogenation of the azine to IPDA on Raney nickel in the presence of ammonia.

ZEOLITES

Zeolites first made their appearance as cracking catalysts in the petroleum refining industry and were then quickly taken up by the petrochemicals industry. Chen et al. (1989) and Corma (1991) give excellent reviews of these applications. The extraordinary shape-selective properties of zeolites have since been exploited by many sectors of the chemical industry. The properties of zeolites are so much in tune with the selectivity demands of industry and the environmental regulations of society that one can enthusiastically agree with the statement that "zeolite catalysis and technology will (almost) certainly be the future cornerstone of a clean, environmentally friendly organic chemicals industry" (Hoelderich and van Bekkum, 1991). The factual basis for such optimism will be reviewed in this section.

Structure and Shape Selectivity

Zeolites are crystalline aluminosilicates possessing characteristic pore and cage structures, as shown in Figure 6.1 for a common zeolite (ZSM-5). These structures are responsible for their several remarkable properties: unusually high selectivities based on reactant or product shape and bulk, thermal stability, acidities of both Brönsted and Lewis types, and a wide hydrophobicity/hydrophilicity spectrum. Many of these features are described in several reviews, for example, Hoelderich (1988, 1991a,b), Hoelderich et al. (1988), Hoelderich and van Bekkum (1991), and van Bekkum and Kouwenhoven (1988, 1989). Shape selectivity in particular has been discussed, among others, by Csicsery (1984, 1986), Parton et al. (1989), Hoelderich (1991b), and Davis (1993). Newer zeolites are being continually synthesized and characterized for their various properties (Suib, 1993). The vast experience gained in the exploitation of these properties in the petrochemicals industry has been increasingly applied in recent years to the selective synthesis of high value organic chemicals.

The dimensions of the zeolite pores and cages are determined by the arrangement of the oxygen atoms in the lattice. The channel diameter in various types of zeolites in common use ranges from 0.74 nm for large-pore Y zeolites to 0.41 nm

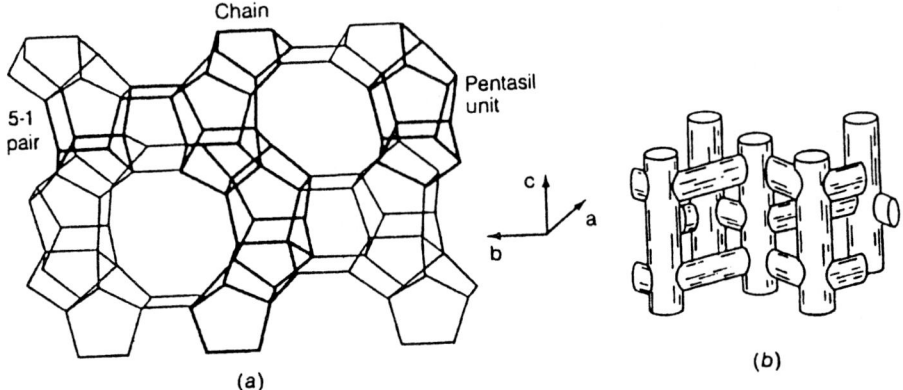

Figure 6.1 Structure of the zeolite ZSM-5. (a) Final stage of evolution by stacking of silicalite layers. (b) Pore structure of ZSM-5.

for small-pore A zeolites. The pentasil zeolites such as ZSM-5 and silicalite are medium-pore zeolites that have a diameter of about 0.55 nm. In some zeolites, there may be larger cavities behind the pore openings; these are important for the location of metal complexes for catalytic activity.

Apart from the zeolites composed of aluminosilicates, there are also other families of microporous materials. One consists of aluminum phosphates which are $AlPO_4$ polymorphs. These materials also exist in a wide range of open tetrahedral network structures. Structural modifications have also been carried out on these materials. An important example is the modification referred to as SAPO, in which silicon atoms replace some of the phosphorus atoms.

The course of zeolite-catalyzed chemical transformations is determined by the types of restrictions imposed by the lattice structure, collectively referred to as *shape selectivity*. Based on these restrictions, three types of shape selectivity can be recognized:

1. *Reactant selectivity* arising from the fact that only certain reactant molecules whose dimensions fall within specific limits can pass through the pores and reach the site of reaction
2. *Product selectivity* where only those products are obtained whose dimensions let them diffuse out of the pores of the zeolite
3. *Transition-state* selectivity imposed by the cavity dimensions on the size of the transition state for the reaction

Hoelderich (1990) has written a thought-provoking review on the shape-selective aspects of zeolite catalysis, and the various features of zeolites have been summarized by Perot and Guisnet (1990). Obviously, shape selectivity can operate only if the reaction occurs within the zeolite pores. Some reactions also take place on the outer surface of the zeolite; shape selectivity effects are not manifested in such reactions.

Applications in Organic Synthesis: General Considerations

The application of zeolites in organic synthesis can be classified in three distinct categories:

1. Noncatalytic uses
2. Reactions catalyzed by zeolites
3. Reactions brought about by hybrid catalysts

Organic chemists are frequently interested in regio- and stereoselective reactions. The use of zeolites as solid acids might result in regioselectivity caused by the spatial demand of fitting the transition state inside the cage and desorbing the product. Furthermore, multifunctional catalysts can be generated by using zeolites as carriers for other organometallic or metal complex catalysts. Finally, it is worth remembering that zeolites are catalysts with very high surface area (typically 400–500 $m^2 g^{-1}$) with more than 99% of the surface area inside the bulk (Thomas, 1994).

From the properties of zeolites mentioned, it is clear that they have always been regarded as solids that promote acid-catalyzed reactions. A spectacular new development has been their use in base-catalyzed reactions (Davis, 1993). Another development that merits equal attention is the synthesis of titanium silicalites (TS-1) with their unique oxidative potential. We shall refer to these applications later following a review of their "more conventional" use as Brönsted and Lewis acid zeolites.

Noncatalytic Uses of Zeolites

The ability of zeolites to adsorb and retain small molecules such as water forms the basis of their use in the noncatalytic synthesis of fine chemicals (Van Bekkum and Kouwenhoven, 1988, 1989). One of the best recent examples is the use of NaA zeolite in the Sharpless asymmetrical epoxidation of allylic alcohols (see Chapter 10) (Gao et al., 1987; Antonioletti et al., 1992). In this Ti(IV)-catalyzed epoxidation by *t*-butyl hydroperoxide in the presence of diethyl tartrate (reaction 6.4), it has been demonstrated that the inclusion of zeolites (3 Å or 4 Å) leads to high conversion (>95%) and high enantioselectivity (90–95% *ee*). The effect of the zeolite is quite dramatic. It is believed that the role of the zeolite is to protect the titanium isopropoxide catalyst from water, perhaps generated during the reaction.

$$R\diagup\!\!\!\diagdown\!\!\!\diagup OH \quad \xrightarrow[\substack{(+)\text{-diethyl tartrate} \\ (\text{in } CH_2Cl_2,\ -20\,^\circ C)}]{Ti(OPr^i)\ t\text{-BuOOH}} \quad R\diagup\!\!\overset{O}{\triangle}\!\!\diagup OH \qquad [6.4]$$

Similarly, zeolites have been used in several other water-producing reactions to shift the equilibrium in the desired direction. Examples include acetalization, esterification, and enamine formation (Van Bekkum and Kouwenhoven, 1988, 1989).

Table 6.2 Values of the Hammett acidity function (H_0)

Acid	Hammett acidity function (H_0)
p-Toluenesulfonic acid	+0.55
Montmorillonite	
Natural	1.5 to −3.0
Cation-exchanged	−5.6 to −8.0
Amberlyst-15	−2.2
Sulfuric acid (40%)	−2.4
Sulfuric acid (100%)	−12.3
Nafion	−11 to −13
HY-zeolites	−13.6 to −12.7
$H_3PW_{12}O_4$ and $Cs_{2.5}H_{0.5}PW_{12}O_{40}$ (HPAs)	−13.16
Lanthanum and cerium-exchanged HY-zeolites	< −14.5
Fluorosulfonic acid	−15.07
Sulfated zirconia	−16
HSO_3F-SbF_5 (superacid)	−20

Zeolites as Brönsted and Lewis Acid Catalysts

Because of their Brönsted acidity, zeolites can, in principle, be used as catalysts for any organic reaction subject to conventional proton catalysis. Table 6.2 lists the acidity levels of several classes of acid catalysts. Note that zeolites with a Hammett acidity function in the range of −13.6 to −12.7 are quite high in the acidity range. The only limitation would be the bulkiness of the molecule to be generated inside the zeolite cavity and desorbed from there.

Alkylation of aromatics

GENERAL COMMENTS

Alkylation by zeolites made a major entry in the field of industrial catalysis through the highly acclaimed Mobil-Badger process for ethylbenzene (Csicsery, 1984, Hoelderich et al., 1988; Hoelderich and Van Bekkum, 1991) by replacing the toxic, eco-unfriendly, corrosive Friedel–Crafts reaction. A modified version developed by the National Chemical Laboratory (the *Albene process*), using a similar class of catalysts known as *Encilites*, is particularly suited to alkylation by ethyl alcohol of any concentration down to 30% (Bhowmik and Ratnasamy, 1991).

Although the examples given relate to important large-volume chemicals, they have been specifically mentioned in view of their significance in smaller volume production in that they provide a basic understanding of alkylation reactions in general.

2,6-DIISOPROPYLNAPHTHALENE

2,6-Diisopropylnaphthalene (DIPN) is an important raw material for monomers for high-performance polymers. The new Catalytica process uses a mordenite with a pore size in the range of 0.6 to 0.7 nm as the catalyst for this

alkylation. The result is very strong selectivity for the 2,6-derivative (6) (2,6-/2,7-ratio = 2.9), unlike in the older process. It is quite likely that this is a consequence of the shape selectivity of the zeolite (Cusumano, 1992). DIPN can in turn be converted to naphthalene-2,6-dicarboxylic acid (7), 2,6-dihydroxynaphthalene (8), or 6-hydroxy-naphthalene-2-carboxylic acid (9) (reaction 6.5). These are useful raw materials for specialty polyesters and liquid crystal polymers (Kikuchi et al., 1994; Sugi et al., 1994).

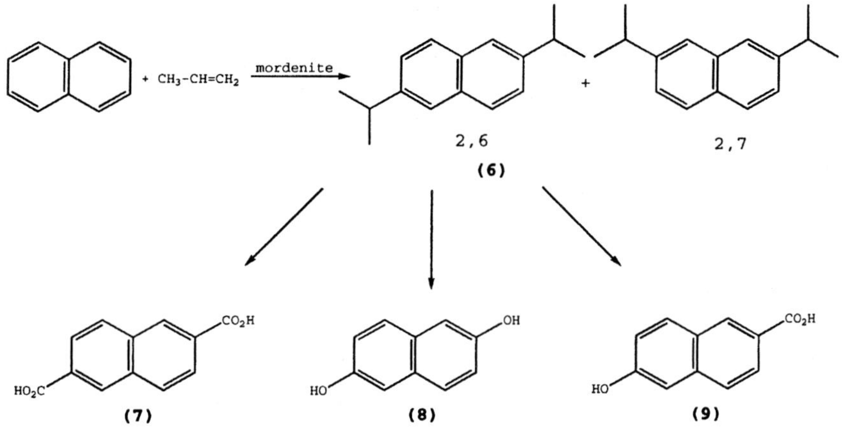

[6.5]

Shape-selective alkylations effected on phenols and their derivatives constitute an important class of reactions (Jacobs et al., 1988; Marczewski et al., 1988; Parton et al., 1989). A particularly useful reaction is the conversion of anisole to *p*-methylanisole by self-alkylation.

Isomerization of aromatics

The use of zeolites in the organic chemical industry took a great step forward with their adoption in the isomerization of *m*-xylene to *p*- and *o*-xylenes used in the manufacture of terephthalic acid and phthalic anhydride, respectively. The ZSM-5 catalysts developed by Mobil Corporation became the prime catalysts for this process all over the world. Remember, however, that catalysts are developed specifically to meet certain criteria and to suit a given feedstock specification. We illustrate this through the following example.

EXAMPLE 6.1

Development of a zeolite catalyst for the isomerization of xylenes: *xylofining*

CRITERIA

The National Chemical Laboratory (NCL) undertook to develop (for a client) a catalyst for the isomerization of xylenes to meet the following specific criteria:

1. A minimum conversion of 58% of the ethylbenzene (EB) in the feedstock.
2. A maximum xylenes loss of 2% at 58% EB conversion.

3. A certain minimum ratio of *o*-to-*p* xylenes—since the prospective client [Indian Petrochemicals Corporation Ltd. (IPCL)] was also a producer of *o*-xylene.
4. The molar ratio of benzene produced (from EB) to EB converted as close to unity as possible.
5. A maximum hydrogen to hydrocarbon ratio of 4.
6. The ability to tolerate up to 5% of nonaromatic aliphatics in the feedstock.
7. A minimum liquid hourly space velocity (LHSV) of 10 mixed xylenes (wt/wt catalyst/h).
8. A minimum cycle length of two years.

Criterion 3 was especially peculiar to IPCL because most other plants produce predominantly *p*-xylene for which a pentasil shape-selective ZSM-5 zeolite is ideally suited because it facilitates the preferential transport of the *para* isomer through its pores.

SCIENTIFIC DISCUSSION OF CATALYST CHOICE

The process involves two main reactions:

1. Isomerization of *m*- to *o*- and *p*-xylenes

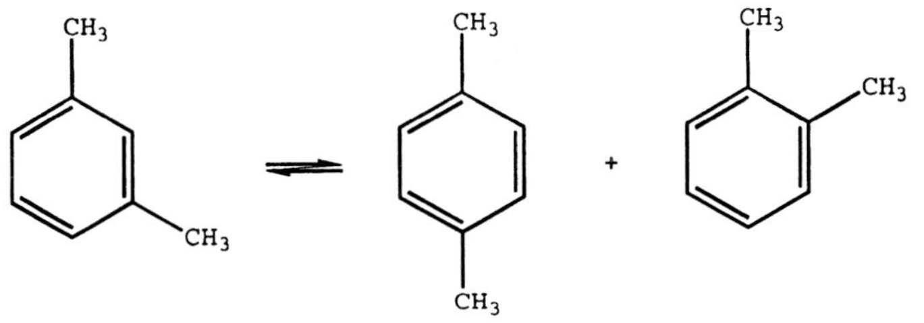

$$\text{m-xylene} \rightleftharpoons \text{p-xylene} + \text{o-xylene} \qquad [E6.1.1]$$

2. Conversion of EB to benzene and xylenes

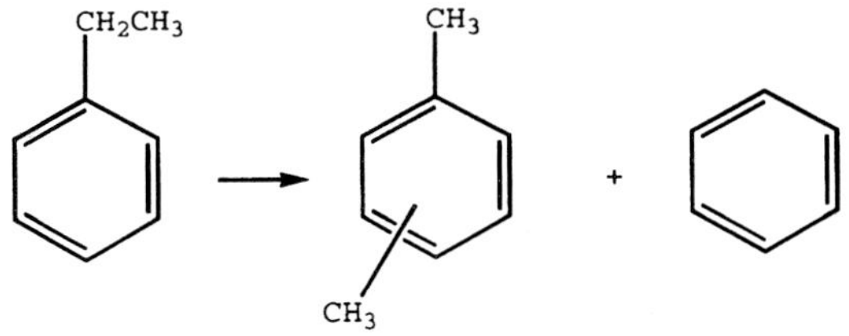

$$\text{ethylbenzene} \rightarrow \text{xylene} + \text{benzene} \qquad [E6.1.2]$$

Reaction E6.1.1 is acid catalyzed; hence a zeolite should be suitable. Reaction E6.1.2 can be achieved by hydrocracking (which needs both an acid site and a metal site), cracking (which needs only an acid site), or hydrogenolysis (which needs only a metal site). The cracking route was ruled out because the ethylene formed in the absence of fast hydrogenation to ethane would lead to fast catalyst deactivation (as in FCC). Thus the choice was restricted to hydrocracking or hydrogenolysis. Both of these use a metal on zeolite type of catalyst. Because the sulfur content of the feedstock was less than 1 ppm, Pt was the preferred catalyst (not Rh which would have been preferred had the sulfur content been higher).

The next question was how to minimize xylene loss by disproportionation to toluene and trimethylbenzenes according to the reaction

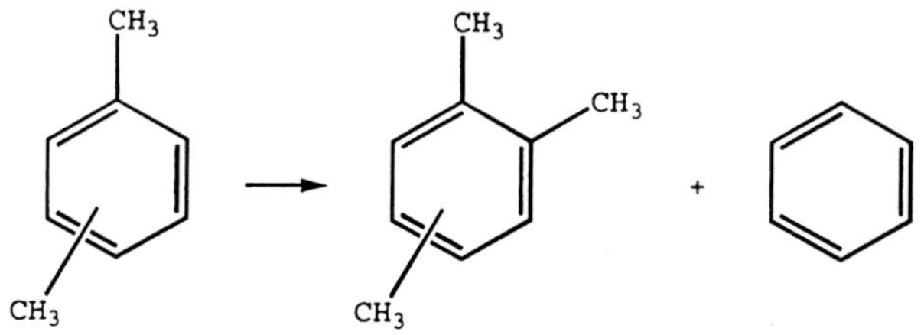

$$\text{2,4-dimethylbenzene} \longrightarrow \text{1,2,4-trimethylbenzene} + \text{benzene} \qquad [\text{E6.1.3}]$$

The only way to have an acid catalyst that would not simultaneously promote this side reaction was to isolate the acid sites and place them in a shape-selective environment. Also, because isomerization is a unimolecular reaction whereas disproportionation is a bimolecular reaction, the former would be favored if the distance between the active sites is increased by increasing the silicon-alumina ratio. But this would lead to a reduction of the acid site density and hence of the acidic activity of the catalyst. One way to compensate for this activity loss is to decrease the diffusional resistance of the zeolite by reducing its particle size. Based on this reasoning, several runs were carried out with reduced particle sizes, but an unexpected complication arose. Because the proportion of active sites at the external surface in the smaller particles is larger, the extent of disproportionation, a non-shape-selective reaction, increased considerably (corresponding to more than 2% xylene loss). This was countered by the simple strategy of deactivating the external sites by steaming.

THE FINAL CHOICE

With the main features of an acceptable catalyst fixed as summarized above, several candidate catalysts were prepared. The variables studied were:

Basic catalyst type, zeolite, mordenite, pentasil ZSM-5
Silica-alumina ratio
Platinum loading
Steaming severity
Pretreatment procedure

The final choice (Balakrishnan et al., 1982) was a ZSM-5 zeolite of silica-alumina ratio 100:1, particle size 0.5 microns, and platinum content 0.1%. This was formulated into extrudates of one-sixteenth inch. Pt impregnation of the calcined extrudate was done by soaking in chloroplatinic acid. The final step following platinum loading was steaming.

Acylation of aromatics

Acylation reactions of arenes using zeolites may become industrially important because of their high selectivity. The Friedel-Crafts acylation of aromatic hydrocarbons by carboxylic acids has been studied by Chiche et al. (1986) [see also Brunel et al. (1989)].

Acylation of phenol with $PhCCl_3$ or benzoyl chloride is effectively catalyzed by HZSM-5 (reaction 6.6). The product *p*-hydroxybenzophenone is used in polymers and as an intermediate in the production of tamoxifen, an anti-breast-cancer drug. High shape selectivity of the catalyst is responsible for the high *para*-selectivity.

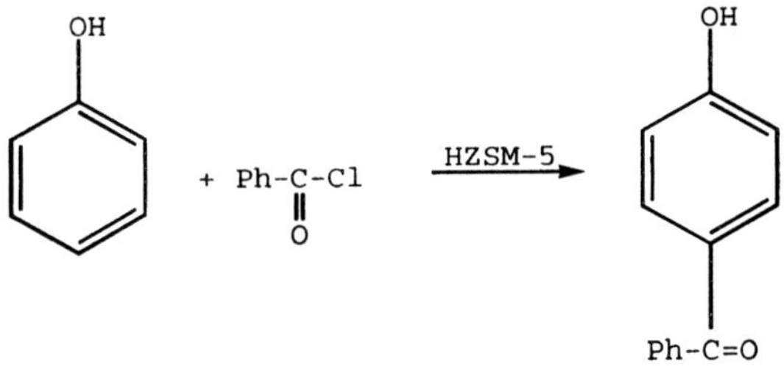

[Scheme 6.6: Phenol + Ph-C(=O)-Cl → 4-hydroxybenzophenone (para-Ph-C(=O)-C₆H₄-OH), catalyzed by HZSM-5] [6.6]

Another reaction that is likely to become industrially important is the acylation of benzene with phthalic anhydride to form anthraquinone, a very useful intermediate in the dye industry. The catalyst most suitable for this is NaX zeolite loaded with 0.1% Pt. The reaction temperature is 350 to 550 °C (Hoelderich et al., 1988). The product mixture contains 98% anthraquinone and 2% benzophenone.

Fries rearrangement

The Fries rearrangement of phenol esters can lead to the attachment of the carbonyl group to either the *o*- or the *p*-position with respect to the phenol. Geneste and co-workers (Chiche et al., 1986; Brunel et al., 1989) have found that rearrangement of phenyl benzoate is best catalyzed by the acidic zeolite HY (Si/Al ratio 15:1), giving a conversion of 38% with an *o/p* ratio of 15:1 (not possible by conventional AlCl$_3$ catalysis).

Hydrogen shift and methyl migration

The rearrangement of 3,5,5-trimethylcyclohex-2-ene-1,4-dione to trimethylhydroquinone, an intermediate in vitamin E synthesis, can be accomplished over HY-zeolite in the presence of hydrogen (Hoelderich et al., 1988).

Condensation reactions

Several condensation reactions are catalyzed by zeolites. For example, ranitidine, an H$_2$-receptor antagonist, is a powerful inhibitor of gastric acid secretion and is extensively used in peptic ulcer therapy (Daly and Price, 1983). 1-Methyl-amino-1-methylthio-2-nitroethene [(13) in reaction 6.7] is a crucial intermediate in the synthesis of ranitidine. To date, two processes are known for the preparation of this compound (White, 1976), one environmentally unacceptable and the other hazardous.

In a new, environmentally acceptable NCL process developed by Rajappa and collaborates (Deshmukh et al., 1990a,b,c,d; Reddy et al., 1993), nitromethane (10) is condensed with *N*-methylcarbonimidodithioic ester (11) in the presence of rare-earth-exchanged NaY zeolite to give (12) which then is easily converted to (13). The yields are good, and the catalyst can be recycled after calcination. A plausible mechanism for reaction 6.7 has been proposed (Bhawal et al., 1994). It is expected that this new ecofriendly process will soon be taken up by industry.

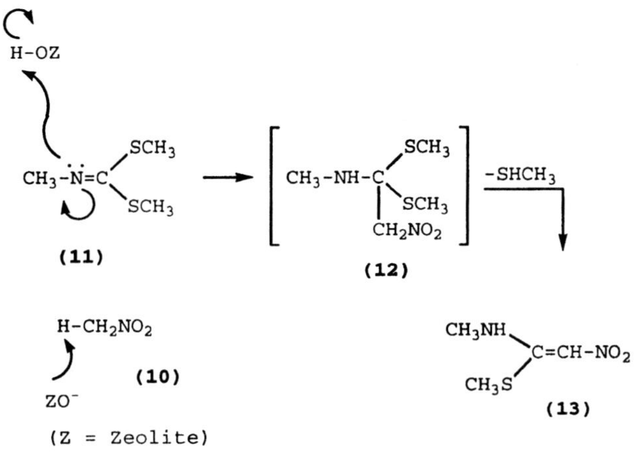

[Scheme showing structures 10, 11, 12, 13 with Z = Zeolite]

$$[6.7]$$

Rearrangement of ketones

Several analgesic and anti-inflammatory drugs such as ibuprofen and naproxen belong to the general class of α-arylalkanoic acids. An often used synthetic route to these molecules is the C–C bond cleavage by Lewis acid catalyzed rearrangement of the ketals of aryl α-haloalkyl ketones (Rieu et al., 1986). Methods based on asymmetric catalysis have also been developed (see Chapter 9). It has recently been shown (Baldovi et al., 1992) that zinc-exchanged Y-zeolites can catalyze the transformation of the ethylene ketal (14) of 2-bromopropiophenone to the α-phenylpropanoic ester (15) (reaction 6.8).

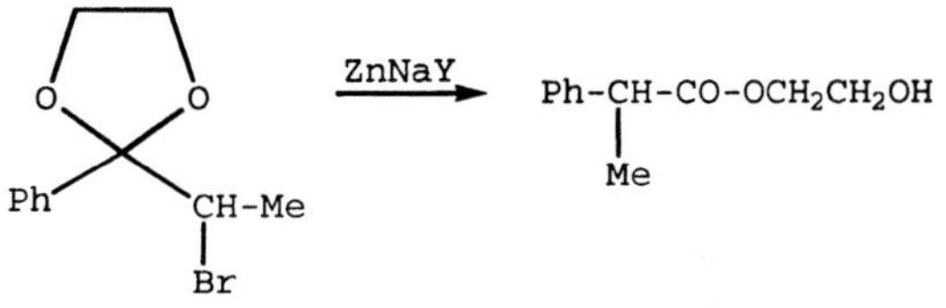

[Reaction scheme: (14) ethylene ketal of 2-bromopropiophenone → ZnNaY → Ph-CH(Me)-CO-OCH$_2$CH$_2$OH (15)]

$$[6.8]$$

Beckmann rearrangement

Another C–C bond cleavage that can be catalyzed by zeolites is the Beckmann rearrangement. The most important industrial example of this rearrangement is the conversion of cyclohexanone oxime to caprolactam, a large volume chemical (Hoelderich, 1990, 1993). In the existing process, cyclohexanone is first reacted with hydroxylamine sulfate to form the oxime (16). Then this is subjected to H_2SO_4-catalyzed Beckmann rearrangement to form caprolactam (17) (reaction 6.9). Disposal of the enormous amounts of ammonium sulfate formed (4 to 5 tons per ton of caprolactam) and the need to handle fuming sulfuric acid are major drawbacks of the process.

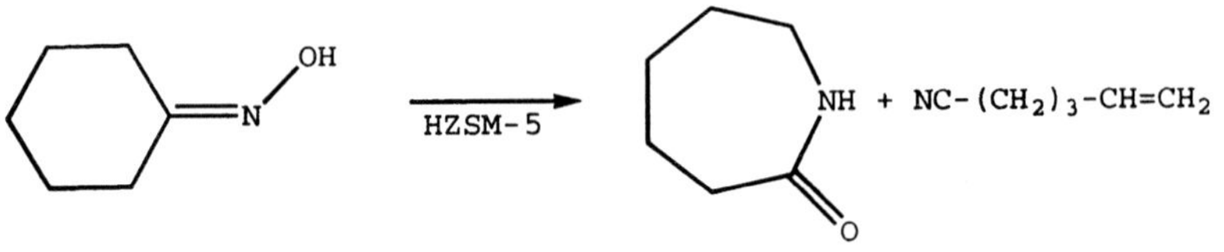

138   Catalysis

$$\text{(16)} \xrightarrow{\text{HZSM-5}} \text{(17)} + NC\text{-}(CH_2)_3\text{-}CH\text{=}CH_2 \quad [6.9]$$

(16) is cyclohexanone oxime; (17) is caprolactam.

Therefore, attempts are being made to use zeolites in both steps. We discuss the first step later. The best catalyst for the second step is the silicoaluminophosphate SAPO-11, according to the recent review by Thomas (1994). A process developed by Union Carbide was at the pilot plant stage at this writing.

Halogenation reactions

HALOGENATION OF ARENES

Using L-type zeolites, chlorination of benzene has been claimed to give *p*-dichlorobenzene with 95% selectivity, a remarkably high value compared to the conventional chlorination with homogeneous iron chloride catalysts (Saeki and Taniguchi, 1986). The *p*-selective chlorination of chlorobenzene and anisole has been studied with Y-type zeolites, and it has been suggested that the chlorination occurs in the pores of the zeolite (Miyake et al., 1989). It has also been reported that bromination of toluene can be catalyzed by NaY zeolite in the presence of an HBr scavenger, propylene oxide. This led to a dramatically improved 98% *p*-selectivity at 10% conversion (de la Vega and Sasson, 1989).

Selectivity in the chlorination of benzenoid compounds by sulfuryl chloride to give either nuclear or side-chain chlorinated products can be controlled at will by employing zeolites as catalysts (Delaude and Laszlo, 1990). Obviously, such selective chlorinations are of great interest to industry.

HYDROLYSIS AND HALOGEN EXCHANGE

Copper-containing zeolites have been used for the hydrolysis of halobenzenes to phenols (Gubelmann et al., 1992). The following order of reactivity in this reaction has been determined: ArCl > ArBr > ArF. Investigation of the mechanism of this reaction has led to the discovery of a new halogen-exchange reaction brought about in the presence of Cu-HZSM-5 (Imhaoulene et al., 1994), for example,

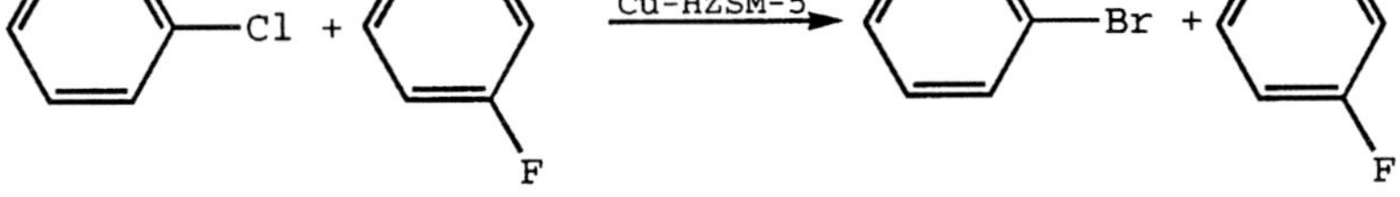

$$\text{Ph-Cl} + \text{Br-C}_6\text{H}_4\text{-F} \xrightarrow{\text{Cu-HZSM-5}} \text{Ph-Br} + \text{Cl-C}_6\text{H}_4\text{-F} \quad [6.10]$$

HALOTHIOPHENES

3-Halothiophenes are crucial starting materials for various industrially important thiophene derivatives, including thiophene oligomers and polymers.

Direct mono-halogenation of thiophene gives only 2-halothiophene. The preparation of 3-halothiophenes involves quite a few steps. In a significant new development, it has been shown that the best way to obtain these 3-halothiophenes is by zeolite-catalyzed isomerization of the 2-halo isomers (Dettmeier et al., 1987). The zeolite used is the strongly acidic silicon-rich HZSM-5.

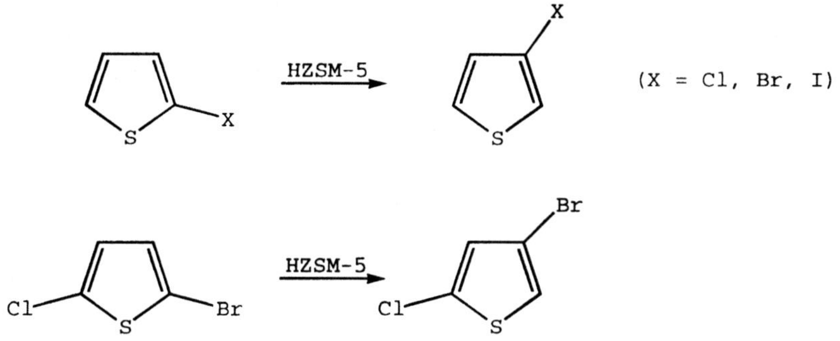

$$\text{2-halothiophene} \xrightarrow{\text{HZSM-5}} \text{3-halothiophene} \quad (X = Cl, Br, I)$$

[6.11]

$$\text{2-Cl-5-Br-thiophene} \xrightarrow{\text{HZSM-5}} \text{2-Cl-4-Br-thiophene}$$

Zeolites as Basic Catalysts

All of the reactions discussed till now have been catalyzed by zeolites functioning as Brönsted or Lewis acids. In a major new development, zeolites have also been tailored to act as base catalysts (Hathaway and Davis, 1988a,b,c; Tsuji et al., 1991). A superbase catalyst of this type has now found application in the synthesis of the key intermediate, 4-methylthiazole, used in the preparation of the anthelmintic, thiabendazole. The existing industrial route for 4-methylthiazole (18) involves using several hazardous chemicals such as chloroacetone and carbon disulfide. The new zeolite-based route uses the base-catalyzed reaction of SO_2 with the imine from acetone (reaction 6.12).

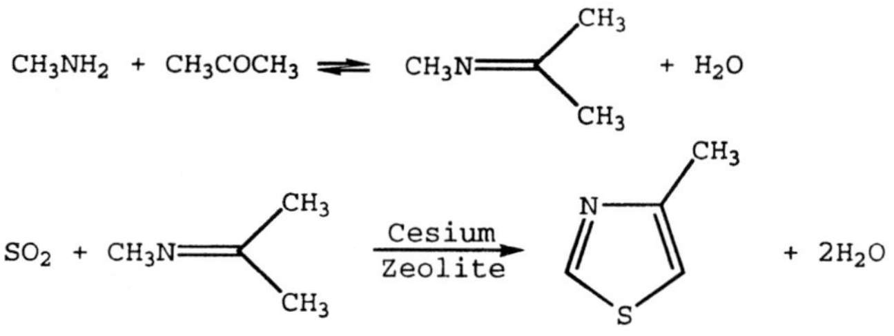

$$CH_3NH_2 + CH_3COCH_3 \rightleftharpoons CH_3N{=}C(CH_3)_2 + H_2O$$

$$SO_2 + CH_3N{=}C(CH_3)_2 \xrightarrow{\text{Cesium Zeolite}} \text{4-methylthiazole (18)} + 2H_2O$$

[6.12]

Zeolite Catalysis for Oxidation Reactions

One of the most important advances in zeolite catalysis has been the introduction of titanium silicalite (TS-1) for the oxidation of organic substrates by hydrogen peroxide (see Bellussi and Fattore, 1991, for a review). Only those cations that meet certain specific steric requirements can fit in the tetrahedral positions of the zeolite lattice. One such cation is titanium. The replacement of silicon in silicalite-1 by titanium results in the formation of titanium silicalite (TS-1).

Hydroquinone from phenol

TS-1 has already found industrial application in the liquid-phase oxidation of phenol to hydroquinone and catechol (Esposito et al., 1985; Notari, 1987, 1991), which are important intermediates used in the manufacture of agrochemicals, fine chemicals, antioxidants, and photographic chemicals. The existing processes for these dihydroxybenzenes are the following: (1) the Eastman Kodak process starting from aniline; (2) the Signal Goodyear process starting from p-diisopropylbenzene; (3) the Rhone-Poulenc process in which phenol is oxidized by H_2O_2 with perchloric acid and phosphoric acid as catalysts; and (4) the Brichima process in which phenol is oxidized by H_2O_2 in the presence of Fe^{2+} and Co^{2+} (Fenton's reagent).

In contrast to these, the Enichem process using TS-1 zeolite for the oxidation of phenol by H_2O_2 leads to a conversion of 20–25%. Even at this high level of conversion, the formation of polyhydroxylated benzene and tars is inhibited by transition-state shape selectivity.

EXAMPLE 6.2

Development of a titanosilicate (TS-1) catalyst for the production of catechol and hydroquinone from phenol

The different routes to these intermediates were outlined in the previous section. In this example we present the catalyst development strategy (Ratnasamy and Sivasankar, 1996) for a process based on the use of extrudates of TS-1 to produce catechol and hydroquinone in a continuous fixed-bed reactor.

CRITERIA

When the developmental work was undertaken at NCL, Enichem had a patented process using TS-1 zeolite powder as a catalyst and hydrogen peroxide as an oxidant. The reactor was a slurry reactor operated in the batch mode. NCL undertook the development of a new process to meet the following criteria:

1. Hydrogen peroxide selectivity should be 80% as against the reported operating selectivity of 70–75%.
2. The catalyst should be capable of being used in a fixed-bed reactor for continuous operation.
3. Energy costs and phenol/hydrogen peroxide ratio should be lower than the figures available at that time.
4. An efficient scheme for regenerating the catalyst should be worked out with a cycle length of at least one month.

CATALYST DESIGN AND PROCESS INNOVATION

The most important variable in TS catalysts is the Si/Ti ratio. At high values, the activity is low, but because there would be no extra lattice Ti ions, the selectivity would be higher. However, the LHSV would have to be very low (say below 35). Hence one has to be

careful to avoid extra lattice octahedral Ti ions which would lead to nonselective oxidation like quinone formation.

Another requirement is that the particle size should be minimum, for it is known that the catalytic activity of TS-1 is drastically reduced above a particle size of 0.2 micron. Preparing catalyst particles of this size requires special filtration techniques. It was found that the usual methods of filtration had to be replaced by centrifugation to obtain satisfactory results. Hence centrifugation must be used at the manufacturing level.

Because the catalyst has to be used in a fixed-bed reactor and the particle size must be very small, the TS-1 particles had to be formulated into extrudates or spheres. This was the most challenging part of the development. Binders like alumina, graphite, etc. could not be used because they seriously interfere with catalytic activity and selectivity. Silica is the only binder that is inert, but formulating spheres or extrudates using silica as binder is extremely difficult, and consumed the major part of the development effort. After several trials, extrudates were produced for use in fixed-bed reactors.

Several important innovations were also introduced on the process side, such as multipoint injection of hydrogen peroxide and the use of cost-reducing water as solvent for phenol. The process is expected to be commercialized in India by the year 2000.

Epoxidation of olefinic compounds

The TS-1 catalyst has also been found effective in the epoxidation of olefinic compounds. It is of particular interest in the preparation of propylene oxide by oxidation of propylene with hydrogen peroxide (Hoelderich, 1988). Ethylene has also been epoxidized to ethylene oxide with 30% H_2O_2 in aqueous t-butanol to obtain 96% selectivity at 97% H_2O_2 conversion (Sheldon, 1991).

Ammoxidation of cyclohexanone to the oxime

The second TS-1 based process which is likely to go into commercial production is the synthesis of cyclohexanone oxime from cyclohexanone, ammonia, and hydrogen peroxide (Roffia et al., 1990) (reaction 6.13), which is the first step in the preparation of ε-caprolactam (Montedipe, pilot plant). Oxime selectivity is >98% at a conversion of 99.9%.

$$\text{cyclohexanone} + NH_3 + H_2O_2 \xrightarrow{\text{TS-1}} \text{cyclohexanone oxime} + 2H_2O \quad [6.13]$$

A more recent study (Montegazza et al., 1996) proposes a comprehensive network for this reaction.

Reactions Brought about by Hybrid Catalysts Consisting of Metal Complexes inside Zeolite Cages

Zeolites can also be used as carriers for other metal catalysts or organometallic complexes. In such cases, hybrid catalysts result in which the properties of the

deposited active components are superimposed on the intrinsic zeolite characteristics (see Ozin and Gil, 1989, for a review).[1] This kind of encapsulation is a logical extension of the concept of immobilizing homogeneous catalysts on solid supports such as silica, alumina, and various polymers. To successfully develop such a hybrid catalyst, it is essential that the active metal complex be located inside the channels or cavities of the zeolite host, not just deposited on the external surface.

The encapsulation of metal carbonyls in cation-exchanged zeolites results in a material that has high catalytic activity. For example, a 45% Li^+-exchanged NaY zeolite incorporating $Mo(CO)_6$ is 96% selective in the hydrogenation of 1,3-butadiene, leading to cis-2-butene (reaction 6.14) (Okamoto et al., 1988).

$$\text{CH}_2=\text{CH-CH}=\text{CH}_2 \xrightarrow[\text{LiNaY-Mo(CO)}_6]{H_2} \text{CH}_3\text{-CH}=\text{CH-CH}_3 \qquad [6.14]$$

BIMETALLIC CATALYSTS

When a second metallic element is added to the common single metal-supported catalyst, bimetallic clusters of particles in the size range 1–50 nm are formed. About 50–95% of the metal atoms in the range of 1–3 nm are exposed to the surface. A particularly important effect of such microparticle clusters supported on silica or alumina is on selectivity, which can often be enhanced to significantly higher levels.

A class of reactions extensively studied is the selective hydrogenation of α, β-unsaturated aldehydes, that is of the C=O bond, to the corresponding unsaturated alcohols, without affecting the C=C double bond. The hydrogenation of crotonaldehyde on Pt/SiO_2, for example, is a complex reaction involving the following scheme:

$$CH_3CH=CHCHO \begin{array}{c} \xrightarrow{H_2} CH_3CH_2CH_2CHO \\ \xrightarrow{H_2} CH_3CH=CHCH_2OH \end{array} \longrightarrow CH_3CH_2CH_2CH_2OH \qquad [6.15]$$

(19)

or an even more complex scheme given by Augustine and Meng (1996). The most desired step is the formation of crotyl alcohol (19) which is also the least thermodynamically favored. Studies have been carried out by King and co-workers (e.g., Wu et al., 1990; Schoeb et al., 1992; Bhatia et al., 1992; Uner et al., 1994) and by Galvagno et al. (1986), Poltarzewski et al. (1986), Raab et al. (1993), and Marinelli et al. (1993, 1995) to elucidate the mechanism of bimetallic catalysis in general. Although no general conclusions are possible, it appears that in the system $Pt-Sn/SiO_2$, for example, the role of the second metal (Sn) is to deactivate the metal sites for C=C bond hydrogenation and activate the C=O group.

---

[1] These fall under the general category of heterogenized or "bottled" catalysts which are considered later as a separate group, but are treated in this section because of their special relationship to zeolites.

Examples

Because bimetallic catalysts are commonly used in hydrogenation reactions, selected examples of their use are given in that section, but we give two interesting examples here.

Pyrrolidones are organic intermediates used in the manufacture of pharmaceuticals and as specialty solvents. An intermediate in the manufacture of these chemicals is γ-butyrolactone (22) for which Phillips Petroleum (Bjornson and Stark, 1992) recently patented a two-step process (scheme 6.16) starting from maleic anhydride (20) and proceeding via succinic anhydride (21). It uses a bimetallic catalyst Pd-Co/SiO$_2$ in the first step and NiO/SiO$_2$ in the second.

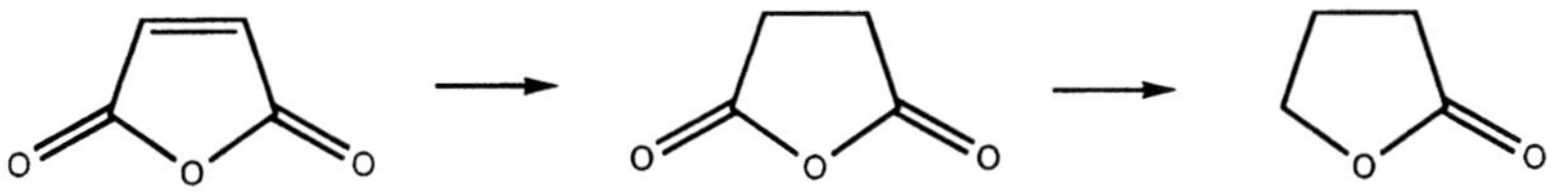

$$\text{(20)} \longrightarrow \text{(21)} \longrightarrow \text{(22)} \qquad [6.16]$$

Another process (Dallons et al., 1989) uses the same bimetallic catalyst (23% Ni, 2% Pd on silica for both steps) with a selectivity of 97% to the lactone.

Bimetallic catalysis does not always mean using two non-precious metals or a combination of precious and nonprecious metals. There are situations where only precious metals give acceptable levels of selectivity. In such cases, it is prudent to "dilute" a "more precious" metal such as Rh with a considerably "less precious" metal like Pt or Pd (less than 5–6% of the cost of Rh). A good example of this is the hydrogenation of a substituted pyridine.

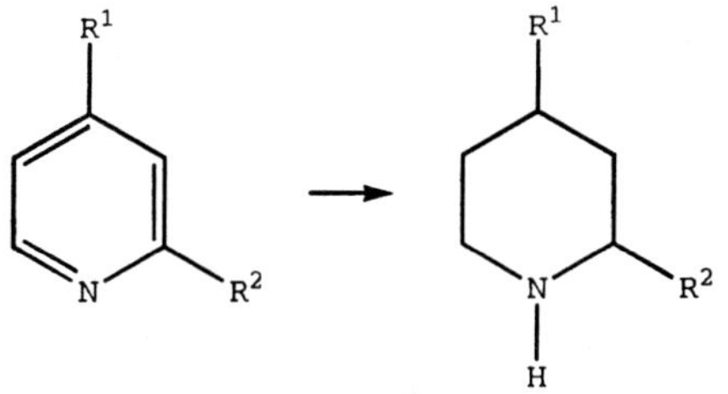

$$\qquad [6.17]$$

Results show (Dodgson, 1993) that 5% Rh on Al$_2$O$_3$ can be thrifted down to 0.5% Rh and 4.5% Pd, with a surprising increase in selectivity from 94% to more than 96%.

HETEROPOLYACID CATALYSTS

A class of catalysts that is being increasingly used in organic synthesis is the strongly acidic *heteropolyacids* (HPA). These catalysts are *polymeric oxometalates* or *metal oxide clusters* incorporating *heteropolyanions* formed by the condensation of different oxoanions, for example,

$$12\text{WO}_4^{2-} + \text{HPO}_4^{2-} + 23\text{H}^+ \rightarrow (\text{PW}_{12}\text{O}_{40})^{3-} + 12\text{H}_2\text{O} \qquad [6.18]$$

Acid forms of these polyanions $(PW_{12}O_{40})^{-3}$ are called *heteropolyacids*. Among a wide range of heteropolyanions, the most important is the so-called *Keggin series* that contains heteropolyanions of the type $XM_{12}O_{40}^{3-}$ where M = W(VI) or Mo(VI), or $XM_n^1M_{12-n}^2O_{40}^{x-}$ where M^1 = W(VI) and M^2 = Mo(VI). The acid forms are strongly acidic with H_0 as high as -13.2 (Table 6.2). Both the acid forms and the anions are usually abbreviated to XM_{12}. The strong acidity of HPAs, which is responsible for their unique catalytic activity, is probably the result of the large size of the polyanion combined with its weak interaction with the proton. For a typical structure of heteropolyacids, see Misono (1987).

Heteropolyacids are soluble in water, as well as in oxygenated organic solvents, which means that they can be regarded as "soluble oxides." Hence they can be used in both homogeneous and heterogeneous catalysis in oxidation, reduction, and other reactions. Liquid-phase reactions can occur in a single homogeneous phase or in an aqueous–organic biphasic medium. The remarkable feature of HPAs is that they are catalytically more active and selective (in solution) than the ordinary protonic acids used in traditional solution catalysis. Another important feature is that the strong acidity of HPAs is often supplemented by a remarkable oxidizing ability (see Misono et al., 1980). Thus, although we treat oxidation separately, oxidation with HPAs will be covered in this section because of their special features.

Examples

Hydration reactions

The remarkable activity of these catalysts in aqueous solution has been put to commercial use for reactions such as the hydration of propene, *n*-butene, and isobutene (Misono, 1987; Onoue et al., 1978).

Alkylation reactions

An environmentally appealing feature of these catalysts is that they can replace the conventional $AlCl_3$ in homogeneous Friedel–Crafts reactions. For example, cesium-exchanged tungstophosphoric acid $Cs_xH_{3-x}PW_{12}O_{40}$ has been found highly active in alkylation reactions. Its activity in the alkylation of 1,3,5-trimethylbenzene with cyclohexene is reported to be four times the activity with the acid alone (Misono, 1993). *t*-Butyl *p*-xylene, which is an important precursor to liquid crystalline polyesters and to polyamides with good solubilities and low melting points, can be obtained by alkylation of *p*-xylene with isobutylene using 12-tungstophosphoric acid. The selectivity achieved with this new catalyst is on the order of 75% compared to 11 and 7%, respectively, with $AlCl_3$-CH_3NO_2 and sulfuric acid.

Oxidation reactions

Oxidation of organic compounds using an HPA as a catalyst is usually carried out with O_2, H_2O_2, or alkyl peroxides as oxidants, thus replacing traditional

stoichiometric oxidants such as lead tetraacetate with their attendant problems of environmental pollution and large salt formation.

Polyoxymetalates such as $H_5Mo_{10}V_2O_{40}$ have been used as catalysts in the oxidative cleavage of cyclic ketones (El Ali et al., 1989a,b) and of vicinal diols (Brégeault et al., 1989) in the presence of oxygen under mild conditions.

$$\text{2-methylcyclohexanone} \xrightarrow[20\ ^\circ C,\ 1\ atm]{O_2,\ PMoV-2} \text{5-oxohexanoic acid (CHO-COOH)} \quad [6.19]$$

96% conversion
90% selectivity

$$\text{1,2-cyclohexanediol} \xrightarrow[EtOH,\ 75\ ^\circ C]{O_2,\ PMoV-2} \text{diethyl adipate (COOEt, COOEt)} \quad [6.20]$$

62% conversion
90% selectivity

Another example is the oxidative cleavage of 2-methylcyclohexanone at room temperature by oxygen to form 5-formylpentanoic acid with a conversion of 90% and selectivity of 90% (Hoelderich, 1993).

Epoxidation of olefins

The scope of HPAs as oxidation catalysts can be considerably widened by applying the technique described previously of incorporating a redox metal ion in a zeolite lattice framework. Thus ions such as Mn(II) or Co(II) can be incorporated in the heteropolyion matrix to give heteropolytungstates of the general formula $(R_4N)_4HMPW_{11}O_{39}$ where M = Mn(II) or Co(II). These catalysts have been used in the epoxidation of olefins with C_6H_5IO (Hill and Brown, 1986) and the hydroxylation of alkanes (Faraj and Hill, 1987).

CLAYS IN CATALYSIS: A CLASS OF SUPPORTED REAGENTS

Clay-supported catalysts belong to the general category of *solid–supported reagents* in which the "reagent" can be a catalyst or just what the term implies—a reagent—and the solid need not necessarily be a clay. Their use is motivated as much by environmental concern as by the need for economy through enhanced regioselectivity. Unfortunately, the vast amount of literature that has accumulated over the last two decades (McKillop and Young, 1979; Laszlo, 1984; Clark et al., 1992; Smith, 1992; Balogh and Laszlo, 1993) does not always distinguish between a supported reagent and a supported catalyst. The situation can be even more confusing. For example, in the chlorination of anisole

by clay-supported Fe(III) chloride, the supported reagent apparently serves as both catalyst and a source of reagent. This is probably because of our poor understanding of the mechanism of molecular adsorption, which is far more important in fine chemicals synthesis than in bulk chemicals production, which practically thrives on catalysis in one form or another. The latter depends, to a large extent, on atomic adsorption which is considerably better understood than molecular adsorption.

Layered versus Porous Catalysts: A Difference in Dimensionality

The dimensionality (d) of three-dimensional space (i.e., any volume) is 3, that of a surface is 2, and of a point is 1. A value of 2 is obviously the best from the point of view of catalysis. The dimensionality of a porous solid is between 2 and 3, which should be regarded as quite poor compared to, say, an enzyme that has a value close to 2.

Clays are known to have a highly organized structure, with platelets of clay stacked vertically on top of one another. Clearly, the thin regions between platelets (1 nm wide) have a dimensionality approaching 2. Further, a very high level of reactant concentration can be built up between layers, leading to enhanced reaction rates.

Structure and Types of Clays

Clays can generally be used up to about 200 °C without any damage to their structure. Beyond this temperature, the interlamellar solvent that keeps the platelets separated from one another is expelled, and the entire structure collapses. It is possible, however, to prevent this collapse by using molecular props or *pillars* in place of the expelled solvent and thus maintain structural integrity. Because of the use of pillaring agents, these clays are referred to as *pillared clays* or *catalysts*. Polyoxycations such as Si, Al, Cr, Zr, Fe, and Mg have been used as pillars to give structural stability even beyond 600 °C. The most common are Al_{13} and Zr_4. Clays impregnated with zinc chloride are being increasingly used.

A number of treated clays have been produced in laboratories, for example, *Clayfen* (clay-supported ferric nitrate) and *Clayzic* (clay-supported zinc chloride). Some of these have been improved and marketed by different companies, for example, *Tixogel* Tonsil K 306, the Englehardt series (F-25, F-34, F-44, F-54, F-124, F-224, G-62), and the *Envirocat* (EPZG, EPG10, EPGE, EPAD) series (made by Contract Catalysts). Most of these belong to the K-catalyst series derived from montmorillonite (Figure 6.2). They are characterized by high acidity (Table 6.2) and are generally superior to the commonly used Brönsted and Lewis acids.

Examples

The use of clays in organic synthesis has been growing. We cite below some selected examples to illustrate the scope of these catalysts in the production of organic intermediates and fine chemicals.

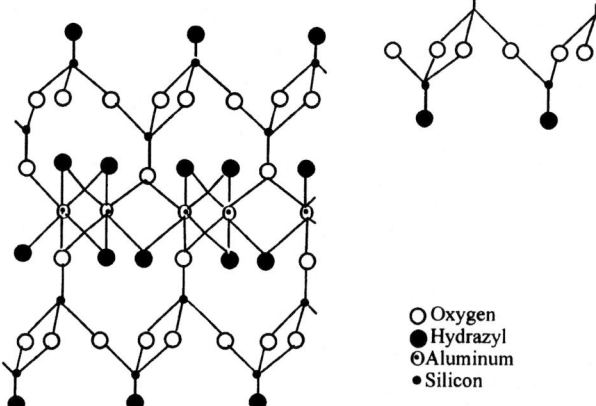

Figure 6.2 Structure of montmorillonite (2:1) clay.

Friedel–Crafts reactions

K-10 montmorillonite, upon exchange of the interlamellar ions with transition-metal ions, functions as a very efficient Friedel–Craft catalyst. It is claimed (Cornelis and Laszlo, 1985; Laszlo and Mathy, 1987) that Clayzec gives 100% regioselectivity to the *para* product. For instance, benzoylation of anisole by benzoyl chloride gives the *para* isomer with almost 100% selectivity at a temperature of 160 °C and a reaction time as low as 5 minutes (Cornelis et al., 1990). Several other benzoylations have been accomplished by using treated clays belonging to the Envirocat EPZG series. These have been claimed to be similar to Clayzic (Cornelis and Laszlo, 1985).

Nitration of phenols

Nitration is classically done with mixtures of nitric and sulfuric acids. The yield in the case of phenol is just over 60%, and the normal *ortho*- and *para*-isomers are accompanied by the *meta*-isomer and a mixture of polynitration products. The use of Clayfen produces the *ortho*- and *para*-isomers exclusively (Laszlo, 1986).

An industrially important extension of this technique is the nitration of estrone by Clayfen to give the desired isomer according to the reaction

$$\text{estrone} \xrightarrow{\text{Clayfen}} \text{2-nitroestrone} \quad [6.21]$$

The conventional route using concentrated nitric acid in glacial acetic acid lacks regioselectivity and also leads to polynitrated products. This is an important step in the manufacture of estrogenic drugs and is claimed to have lowered the cost of production by a factor of 6 (Laszlo, 1984; Cornelis and Laszlo, 1985).

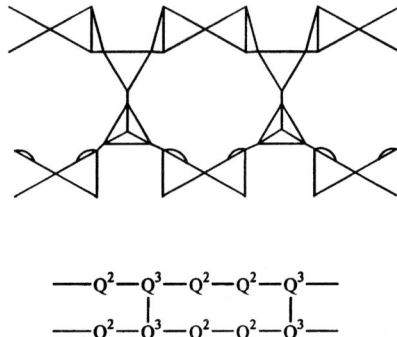

$$—Q^2—Q^3—Q^2—Q^2—Q^3—$$
$$—Q^2—Q^3—Q^2—Q^2—Q^3—$$

Figure 6.3   Structure of the basic clay Xonotlite.

Use of Other Supports

A number of supports other than clay have also been used. In terms of dimensionality, d is between 2 and 3 in these cases as against 2 for clays. Even so, there are many other properties associated with these solids that make them good support matrices. A large number of reactions have been conducted with a variety of supported reagents (McKillop and Young, 1979; Smith, 1992; Clark et al., 1992). Among the major supports used are Celite, silica, graphite, alumina, carbon, Xonotlite, and kiesulguhr. We restrict the treatment below to an interesting basic support: Xonotlite.

Xonotlite: a support with basic sites

The most promising of this less common class of supports are silicates and aluminosilicates with their good basicity and poor nucleophilicity and with negative charges dispersed over all of the oxygen atoms. Xonotlite, a fibrous calcium silicate $[Ca_6Si_6O_{17}(OH)_2]$ whose structure resembles a railroad track cross-linked at every fourth silica, combines the dual requirements of basicity and low dimensionality (Figure 6.3). The space between the parallel tracks restricts the dimensionality to two. Further, its basicity can be boosted by adding t-butoxide.

Xonotlite-t-butoxide has been used to catalyze the Knoevenagel condensation in which a carbonyl double bond (23) is converted to an olefinic bond (24) (reaction 6.22). The yields are generally 20–30% higher than with Al_2O_3 (Cabello et al., 1984; Texier-Boullet and Foucaud, 1982).

$$\underset{(23)}{\overset{R^1}{\underset{R^2}{>}}C=O} + H_2C\overset{CN}{\underset{R^3}{<}} \longrightarrow \underset{(24)}{\overset{R^1}{\underset{R^2}{>}}C=C\overset{CN}{\underset{R^3}{<}}} + H_2O \qquad [6.22]$$

SOLID SUPERACIDS

General Description

Liquid superacids such as HSO_3F and SbS_5 with acid strengths greater than 100% sulfuric acid are quite well known. The use of *solid superacids* in industrial organic synthesis has become a strong possibility in the last decade. Sulfated zirconia is one of the strongest and best known solid superacids. Other common superacids are TiO_2-1 and TiO_2-2. A reference to Table 6.2 shows that their acidity is indeed very high ($H_0 = -16$). Important information has been reported (Hino and Arata, 1980, 1985; Yamaguchi et al., 1986, 1987) on the preparation and structure of superacids.

Examples

A number of reactions relevant to fine chemicals manufacture can be catalyzed by superacids, for example, dehydration of carboxamides to nitriles, Friedel–Crafts alkylation and acylation, Fisher indole synthesis of coumarins, isomerization of limonene, condensation of hydroquinone with aniline, and esterification in general (Hino and Arata, 1980, 1985; Joshi and Rajadhyaksha, 1980; Rajadhyaksha and Chaudhari, 1987; Kumbhar and Yadav, 1989; Rajadhyaksha and Joshi, 1991; Kumbhar et al., 1994). Two particularly important reactions are described here.

Nitriles, which are important intermediates in the manufacture of a wide variety of organic compounds such as amines, aldehydes, amidines, and heterocycles, are manufactured either from alkali cyanides or from carboxamides. The cyanides route is obviously highly toxic, whereas the carboxamide consumes the reagent in stoichiometric quantities. Solid superacids offer a clear alternative to the traditional catalysts. Thus the use of sulfated zirconia enables the dehydration to be accomplished under much milder conditions in the liquid state (Joshi and Rajadhyaksha, 1986; Rajadhyaksha and Joshi, 1991).

Condensation reactions of hydroquinone with aniline and substituted anilines are carried out industrially for the production of N,N'-diaryl-p-phenylenediamine. As with nitriles, conventional reagents like PTSA and chlorides of Al and Zn lead to stoichiometric reactions. Sulfated zirconia, used in the form of a slurry, on the other hand, is needed only in catalytic amounts, and has been shown (Kumbhar and Yadav, 1989) to be a very effective catalyst.

SOLID BASE CATALYSTS

In contrast to solid acid catalysts (including solid superacids) described earlier, solid base catalysts (including superbases) have received much less attention. A common basic catalyst used in organic synthesis is $Ba(OH)_2$. The use of alumina-supported sodium in the double bond isomerization of olefins, first studied by Pines et al. (1955), is also widely known. More recent developments include the use of K, Na/K, MgO, ZrO_2, ThO_2, and $K/KOH/Al_2O_3$ zeolites containing different metal ions to control their acid-base properties, solid superbases, and

nonmetal oxides. The use of zeolites has already been covered in a previous section as part of zeolite catalysis in general. Therefore, we shall restrict the treatment in this section to the other types of catalysts.

Alkaline-earth oxides are perhaps the most extensively used solid base catalysts today, and double bond isomerization is perhaps the most extensively studied reaction. A distinguishing feature of solid base catalysts is that they hydrogenate conjugated dienes exclusively to mono-olefins, which do not react any further under the softer conditions normally employed in such catalyses (see Hattori et al., 1976; Imizu et al., 1979; Tanaka et al., 1980; Nakano et al., 1983).

Examples

Direct hydrogenation of aromatic carboxylic acids to the corresponding aldehydes using modified ZrO_2 (associated with basic properties) has been commercialized by Mitsubishi Chemical Company (Maki et al., 1993), and is an important example of the industrial application of solid base catalysts.

$$ArCOOH + H_2 \rightarrow ArCHO + H_2O \qquad [6.23]$$

Dehydration on basic solid catalysts has also been accomplished industrially. An example is the intramolecular dehydration of monoethylamine to ethyleneimine commercialized by Nippon Shokubai Ltd. using a Si-alkali metal-P-O mixed oxide (Hattori, 1993). Another example of base-catalyzed dehydration is the dehydration of 1-cyclohexylethanol to vinyl cyclohexane (Takahashi et al., 1993), commercialized by Sumitomo Chemicals Ltd. The catalyst used is zirconia modified with bases such as NaOH.

METALLIC GLASS CATALYSTS

The introduction by Klement et al. in 1960 of a technique for the rapid quenching of metals paved the way for the preparation and use as catalysts of a new class of solid materials known as *bimetallic glasses* or *amorphous metal* alloys.

Metallic glasses are metastable materials that are commonly regarded as amorphous but are not entirely so: although they completely lack long-range order, a basic feature of crystals, they do possess some short-range ordering. However, for all practical purposes, they can be considered amorphous. They are nonporous, and hence any reaction is confined to the external surface. Consequently, problems of diffusion associated with porous catalysts (see Chapter 7) are not encountered here. There are reports to indicate (e.g., Cocke and Yoon, 1984; Jefimov et al., 1986; Shibata and Masumoto, 1987) that metallic glasses can give higher activities and selectivities in certain reactions than the more common crystalline solids.

Metallic alloys are produced by a method in which a melted mixture of metals is quenched at an extremely rapid rate, on the order of 10^5–10^6 K/s. Details of the method can be found in Liebermann's review (1983). Note, however, that regardless of its structure the alloy has to be pretreated or "activated" for use as catalyst (Yamashita et al., 1985a,b; Molnár et al., 1989; see also Molnár, 1985).

The technique used, especially the rate of quenching, influences the catalytic activity of the alloy. A commonly used method is melt spinning or strip casting, in which a rotating wheel or disk is brought into contact with the molten alloy. The catalyst comes out in the form of ribbons no more than a few nanometers in thickness. Another technique that is likely to be increasingly practiced is ion sputtering, which has already been used to prepare Fe–Co–Si, Ni–Al–Cr, and Ni–Co (Otsuka Chemical Company, 1982).

Examples

The use of metallic glasses as catalysts has been studied mainly for hydrogenation, dehydrogenation, and hydrogenolysis reactions. It has been found, for instance, that the metallic glass $Ni_{28}T_{72}$ can give 100% selectivity in the hydrogenation of acrolein to propanal (Funakoshi et al., 1985).

$$CH_2=CHCH\!=\!\!O \xrightarrow[H_2]{Ni_{28}Ti_{72}} CH_3CH_2C\!\!=\!\!O\,H \qquad [6.24]$$

Recall that selective hydrogenation of C=C in a molecule containing C=O and C=C (such as acrolein) is difficult. Thus with the usual Ni/alumina catalyst, 1-propanol is also formed. Another important example of the selective hydrogenation of bifunctional substrates over amorphous alloys such as Ni–B, Ni–P, Ni/SO_2, etc. is the hydrogenation of (–)-verbenone (25) over amorphous Ni–P and Ni–B (Torok et al., 1993).

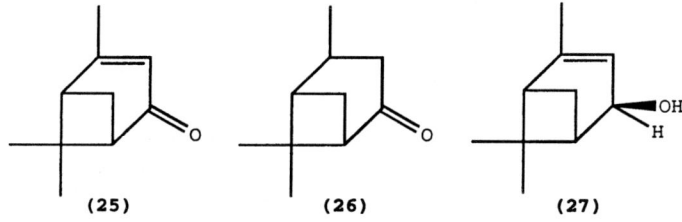

(25)   (26)   (27)

Verbenone is an industrially important molecule as it is the principal building block in the synthesis of a sex pheromone (Silverstein et al., 1966). At a temperature of 373 K, selective saturation of the C=C bond occurs and yields verbanone (26), whereas at a higher temperature (398 K), the C=O bond is hydrogenated and yields the alcohol, verbenol (27).

Another industrially exploitable feature of these alloy catalysts is their restrictive hydrogenation of acetylenic bonds to olefinic bonds. A number of studies using rapidly cooled $Pd_{77}Ge_{23}$ (Carturan et al., 1982; Boitiaux et al., 1983; Molnár et al., 1986) clearly indicate remarkably high selectivities for the semi-hydrogenated double bond product, almost 98%.

The activity of Raney nickel, the most commonly used hydrogenation catalyst, can be improved by enrichment of a glassy phase through appropriate treatment of precursors in Raney nickel preparation (Stiles, 1971; Sinfelt et al.,

1972; Brooks et al., 1982). This so-called *heat soak treatment* involves cooling at the rate of 10^5 K/s followed by treatment in a fluidized bed at 1120 K for 10 hours. The result is an enrichment of the Al_3Ni phase that enhances the catalytic activity of the Raney nickel.

ION-EXCHANGE RESINS

There are numerous applications of ion-exchange resins as catalysts in organic technology and synthesis (Bhagade and Nageshwar, 1977, 1978a,b, 1980, 1981; Olah et al., 1986; Widdecke, 1988; Neier, 1991; Chakrabarti and Sharma, 1993), and the number is growing. An ion-exchange resin (IER) is essentially an insoluble solid matrix containing ions that can be exchanged with ions in the surrounding liquid medium. It is usually an addition copolymer prepared from vinyl monomers. The most common example is a poly(styrene-divinylbenzene) resin with a sulfonic acid ($-SO_3H$) group that can be replaced to any extent by simple ion exchange.

Ion-exchange resins can be divided into two classes, gel and macroreticular. The gel variety constitutes a homogeneous phase with no discontinuities. Its essential quality is that it swells in a solvent. On the other hand, a resin made up of agglomerates of small microporous particles, the macroreticular resin, behaves very differently (Corte and Meyer, 1961; Rohm and Haas, 1963). It does not depend on the ability of the solvent to cause swelling but on its ability to diffuse through the particles and promote reaction inside the pores. An important corollary of this action is that reaction can occur even in a nonswelling medium, which immediately widens the scope of these resins (Bortnick, 1962; Litteral, 1972; Reinicker and Gates, 1974; Wesley and Gates, 1974).

Catalysis by IER: Selectivity Enhancement

Catalysis by IER is a strong function of the nature of the solvent (Pitochelli, 1980). Based on whether the solvent is aqueous or nonaqueous and whether water is formed in the reaction, two broad classes of IER-catalyzed reactions can be identified with two subcategories under each class (Table 6.3).

IERs give selectivities higher than those in solution. Thus in a mixture of small and large molecules, they promote selective catalysis of small molecules by denying access to the large molecules (Pitochelli, 1980). This sieving action can be likened to that of enzymes or zeolites but is less efficient. Even when the molecular sizes are not significantly different, selectivity can be enhanced by preferential sorption of the targeted reactant.

Examples

The important classes of reactions catalyzed by IERs are alkylation, etherification, esterification, and acetylization.

Methyl salicylate has been alkylated with isobutylene and isoamylene on Indion 130, a macroporous sulfonated styrene-divinylbenzene resin with cross-

Table 6.3 Classification of resin-catalyzed reactions

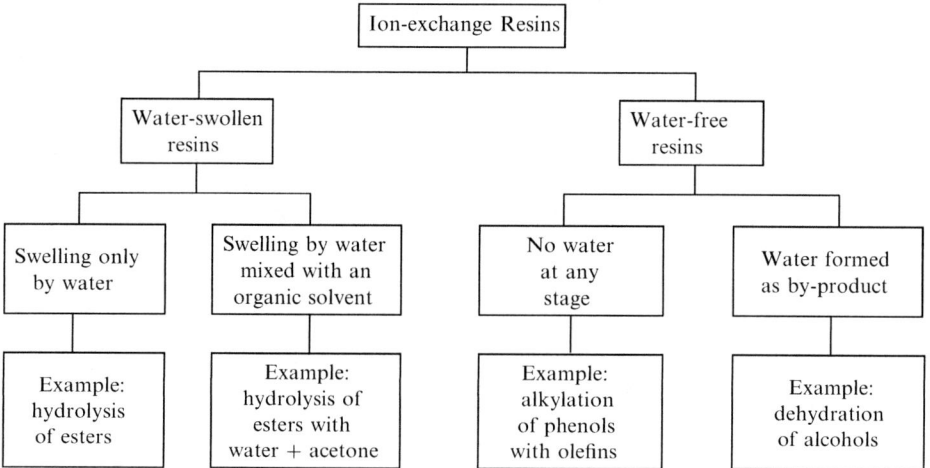

linking of 14%. Generally, two regioisomers are formed in this reaction. However, in resin-catalyzed experiments with isobutylene as the alkylating agent, only one isomer, methyl (5-t-butyl) salicylate, is obtained to the complete exclusion of the other isomer, methyl (3-t-butyl) salicylate. It is even more remarkable that no reaction occurs with the common homogeneous catalyst PTSA.

Another recent example (Ford and Conner, 1994) is the alkylation of aniline or a substituted aniline with p-isopropenylaniline (28) to give bisaniline A (29) with potential applications in polyamides, polyimides, and as a component in epoxy systems.

[6.25]

An example of etherification is the reaction between a mixture of *cis-* and *trans* 2-*t*-butylcyclohexanol and isoamylene on Indion 130. A selectivity of 94% for the *cis isomer* is obtained with a feed consisting of about 68.5% *cis* and 31.5% *trans* isomers.

An interesting aspect of IER catalysis is that the course of reaction with a homogeneous catalyst can be altered in the presence of IER. An example of this is the mono-acylation of diols ranging from 1,3-ethanediol to 1,16-hexa-decanediol by transesterification in ester-solvent mixtures catalyzed by an acidic IER (reaction 6.26) (Takeshi et al., 1994).

$$HO(CH_2)_nOH \xrightarrow[RCOOR'-\text{alkane}]{IER} RCOO(CH_2)_nOH + RCOO(CH_2)_nOCOR \quad [6.26]$$

The symmetric diols are selectively monoacetylated, probably due to the higher rate of monoacetylation compared to the rate of diester formation from the monoester. Thus in the reaction

$$\text{Diol} \xrightarrow{k_1} \text{monoester} \xrightarrow{k_2} \text{diester} \quad [6.27]$$

rate$_1$ is always much higher than rate$_2$ even at long times when the diol concentration is low.

Several hundred reactions are reported to be catalyzed by IERs with good selectivities. Many of these have been commercialized, and many more hold promise of future commercialization. Those commercialized are mainly medium to large volume chemicals, for example, methyl-*t*-butyl ether, *t*-butanol, bisphenol A, isopropyl esters, alkylphenols, and *t*-amyl methyl ether.

A particularly attractive feature of IERs is their use in distillation column reactors, in which reaction and distillation are carried out simultaneously to increase the selectivity of a desired product. This is treated in Chapter 25.

CATALYSIS BY INCLUSION: CYCLODEXTRINS

The remarkable selectivity of enzymes has spurred research in various directions to mimic their specificity. The emergence of the induced-fit model to explain their specificity has brought into focus the remarkable happenings within the cavity of a *host molecule* when a *guest molecule* of suitable dimensions fits itself into this cavity. A practical manifestation of this host–guest interaction is the catalytic transformation of the guest into a new product. This is variously referred to as *catalysis in cavities, host–guest catalysis*, and *inclusion catalysis*. Examples of host molecules are Dianin's compound, urea, hydroquinones, and cyclodextrins. The fitting of the guest molecule within the cavities of these host molecules occurs largely by hydrogen bonding and Van der Waal's interactions, not by formation of covalent bonds. This highly selective fitting process can obviously be exploited as a tool in separation.

The other important host molecule is cyclodextrin. This molecule is especially unique, for it can form inclusion complexes with a guest molecule in both solution and solid phases (unlike the other host molecules which operate exclusively as solids). If a substrate molecule finds itself in the cavity of a cyclodextrin, it is in

the right position with respect to the active centers. This can lead to remarkable catalytic effects (Griffiths and Bender, 1973; Breslow, 1982; Tabushi, 1982).

There are three classes of cyclodextrins, α, β, and γ, each with a cavity in the center which can accommodate guests of suitable dimensions (Saenger, 1980; Tabushi and Kuroda, 1983). The structure of cyclodextrin along with values of cavity dimensions for the three classes can be found in many articles (see, e.g., Syamala et al., 1986).

Examples of inclusion catalysis with cyclodextrin playing host are the chlorination of anisole (Breslow and Campbell, 1971), the carboxylation of phenols (Komiyama and Hirai, 1984), and the one-step synthesis of vitamin K_1 (and K_2) analogs (Tabushi et al., 1979). In the chlorination of anisole by HOCl, use of α-cyclodextrin gives a ratio of *para* to *ortho* chloro derivatives as high as 21:1. In the carboxylation of phenols, β-cyclodextrin steers the attack almost exclusively to the *para* position. For vitamin K_1, which is a factor in blood clotting, the starting material is 2-methylnaphtha-1,4-quinone. The earlier synthetic schemes based on the Friedel–Crafts reaction (e.g. Fieser, 1939; Tishler et al., 1940; Hirschmann et al., 1954) can be replaced by an elegant one-step synthesis using β-cyclodextrin (Tabushi et al., 1977, 1979).

[6.28]

R: a, $-CH_2CH=CH_2$;   b, $-CH_2CH=CH-CH_3$;
    c, $-CH_2\underset{\underset{CH_3}{|}}{C}=CH_2$;   d, $-CH_2CH=\underset{\underset{CH_3}{|}}{C}-CH_3$

The K_1 (or K_2) analogs (a,b) are obtained by using allyl or crotyl bromide in the presence of β-cyclodextrin as the electrophilic reagent to introduce the allyl or crotyl group regiospecifically at C_3.

TITANATES

Esterification and transesterification reactions are commonly used in constructing a synthetic scheme for a product. Because the temperatures employed are usually higher than 200 °C, ion-exchange resins cannot be used, as they are for

alkylation reactions. The common practice has been to use PTSA, but this is environmentally unacceptable. Use of tetrabutyl (or propyl or isopropyl) titanate overcomes this serious deficiency. A striking example of this is the production of dialkyl phthalates such as dioctyl, diisononyl, and diisodecyl because the catalyst is easily decomposed to butanol and TiO_2 which precipitates. Immobilized versions of titanate catalysts can be even more attractive, and are currently under development.

CATALYSTS FOR SELECTED CLASSES OF REACTIONS

Partial Oxidation Catalysts

More than 50% of the world's production of chemicals is based on the presence of an oxidation step. A primary requirement of oxidation catalysts is that their activity be controlled so that the organic reactants are not oxidized to the final products of oxidation, water and carbon dioxide. Thus one is always concerned with *partial oxidation* in such cases.

Pollution abatement pressures in oxidation processes in the production of bulk chemicals are practically nonexistent today because almost all of them have switched to air as the oxidizing agent. On the other hand, many organic intermediates and fine chemicals, mainly value-added products, are produced using stoichiometric oxidants such as potassium permanganate, chromic acid, organic peroxides, metal salts, and chlorates. Hence there has been a continuing effort to replace these environmentally unacceptable processes by cleaner catalytic processes.

Types of oxidants

A wide range of oxidizing agents has been used in laboratory experiments. The more common ones are listed in Table 6.4. An important factor in determining

Table 6.4 The more important oxygen donors with their active oxygen contents (Sheldon, 1991)

Donor	Active oxygen (%)	Coproduct	Environ friendly
H_2O_2	47.0[a]	H_2O	yes
O_3	33.3	O_2	yes
t-BuO_2H	17.8	t-BuOH	yes
NaClO	21.6	NaCl	no
$NaClO_2$	19.9[b]	NaCl	no
NaBrO	13.4	NaBr	no
HNO_3	25.4	NO_x	no
$C_5H_{11}NO_2$[c]	13.7	$C_5H_{11}NO$	no
$KHSO_5$	10.5	$KHSO_4$	no
$NaIO_4$	7.0[b]	NaI	no
C_6H_5IO	7.3	C_6H_5I	

[a] Based on 100% H_2O_2.
[b] Assuming that only one oxygen atom is utilized.
[c] N-methylmorpholine-N-oxide (NMO).

the effectiveness of an oxidation is its active oxygen content. This, of course, varies with the oxidant and is included in the table. The use of stoichiometric oxidants has been completely abandoned now because of the large quantities required and more so because the ratio of the undesirable coproduct formed (listed in the table) to the desired product can be very high. A few other donors, besides those listed in the table, have recently been introduced, for example, sodium perborate (McKillop and Tarbin, 1983, 1987) and sodium percarbonate (Ando et al., 1986). The introduction of these new oxidants notwithstanding, H_2O_2, with water as the coproduct, is obviously the "Mr. Clean" of all of the oxidants and is hence the oxidant of choice in organic synthesis.

Types of catalysts

Oxidation catalysts used in organic synthesis may essentially be divided into four classes: homogeneous catalysts, supported metal catalysts, supported metal ion catalysts, and metal ion-incorporated solid matrices such as zeolites. Among metal-supported catalysts are the various transition metals supported on carbon, silica, etc. The third category consists essentially of ion-exchange resins which were considered separately in a previous section. The fourth category was also considered separately in another section.

A particularly important aspect of oxidation reactions is the combination of catalyst and oxidant. Hence we consider here some selected classes of reactions for which different oxidant-catalyst combinations have been used, irrespective of the type of catalyst.

Examples

Most oxidations occur by transfer of oxygen from the oxidant to the reactant according to the reaction

$$\underset{\text{(Catalyst)}}{M} + \underset{\text{(O-donor)}}{X-O-Y} \xrightarrow{\text{Reactant A}} M + AO + XY \qquad [6.29]$$

Clearly, the scope of oxidation by this method is quite wide because innumerable combinations of metals (including practically all transition metals) and oxidants are possible.

Epoxidations of ethylene and propylene are among the most important oxidation reactions today, and a considerable amount of literature has accumulated on catalysts for these bulk chemicals (Sheldon, 1981; Jorgensen, 1989), but they will not be considered here.

OXIDATION OF ALCOHOLS

The traditional method of using Cr(IV)-based oxidants is obviously unacceptable today. A number of new environmentally benign catalyst systems have recently been developed which can be used with H_2O_2 or TBHP as oxidant. These promote chemoselective oxidation of secondary alcohols and diols to ketones, an important class of reactions in organic synthesis. The metals used are molybdenum (Kurusu and Masuyama, 1986) and titanium.

With Zr as the metal, primary alcohols are selectively oxidized to the aldehydes (with no serial oxidation to the carboxylic acids), whereas the allylic alcohols are chemoselectively oxidized to the α,β-unsaturated aldehydes.

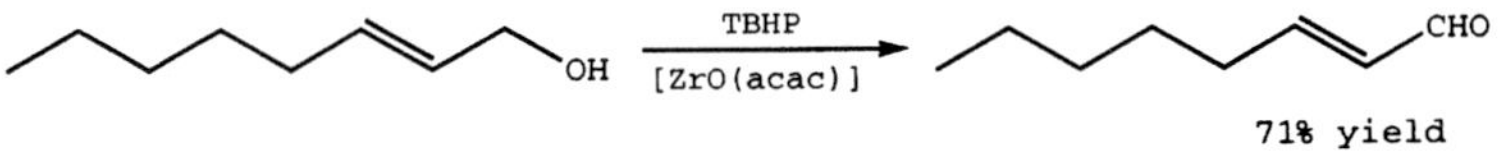

$$\text{allylic alcohol-OH} \xrightarrow[\text{[ZrO(acac)]}]{\text{TBHP}} \text{allylic aldehyde-CHO} \quad 71\% \text{ yield} \qquad [6.30]$$

Ruthenium compounds are perhaps the most effective catalysts for the chemoselective oxidation of primary and secondary alcohols, for they function in conjunction with almost any oxidant. A good example is the ruthenium compound tetrapropyl-ammonium perruthenate (TRAP) with oxidant NMO (Griffith, 1989). An important illustration of the use of TRAP as catalyst is in the synthesis of an intermediate in the production of the antiparasitic agent milbemycin B.

OXIDATIVE CLEAVAGE OF VICINAL DIOLS

Vicinal diol cleavage is of great importance in fine chemicals production. It can be achieved by using oxidant-catalyst combinations involving H_2O_2 with W(VI) and $RuCl_3$ (Mitsui Petrochemical Co., Japan Patent, 1980; Venturello and Ricci, 1986). Oxidations with molecular oxygen in combination with catalysts of the general formula $A_{2+x}Ru_{2-x}O_{7-y}$ (A = Pb, Bi; $0 < x < 1$; $0 < y \leq 0.5$) have also been reported (Felthouse, 1987), for example, the oxidation of cyclohexane-1,3-diol to the adipate.

EXAMPLE 6.3

Comparison of catalysts for the production of Vitamin K_3: menadione

2-Methyl-1,4-naphthoquinone, or vitamin K_3, also known as menadione, is a member of the family of K vitamins and is used in preventing hemorrhages. The conventional method of producing menadione is by stoichiometric oxidation by Cr_2O_3 of 2-methylnaphthalene in coke-oven tar and certain petroleum fractions. An alternative two-step method (scheme E6.3.1) has been developed (Kozhevnikov, 1995) in which 1-naphthol is alkylated by methanol in the vapor phase to 2-methyl-1-naphthol (30), which then is oxidized in the liquid phase to menandione (31) by O_2 in the presence of a heteropolyanion as catalyst.

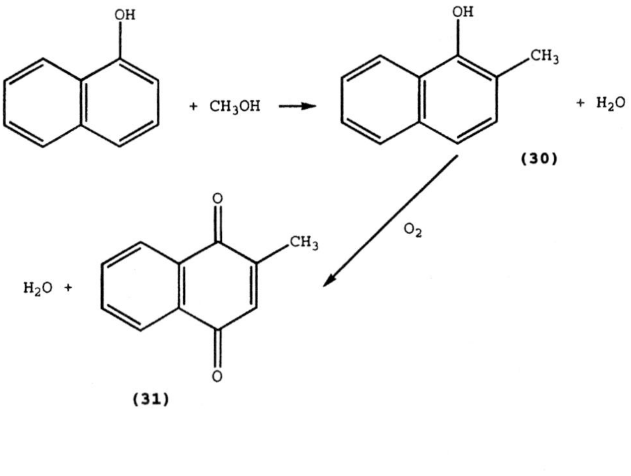

$$\text{1-naphthol-OH} + CH_3OH \longrightarrow \text{2-methyl-1-naphthol-OH, CH}_3 \text{ (30)} + H_2O$$

$$\text{(30)} \xrightarrow{O_2} \text{menadione (31)} + H_2O$$

[E6.3.1]

Table 6.5 Competing catlytic processes for menadione

Process	Reference
Liquid-phase oxidation by H_2O_2 in presence of acetic acid	Takanobu et al. (1977)
Liquid-phase oxidation by H_2O_2 in presence of an ion-exchange resin, Pd(II)–polystyrene sulfonic acid resin—Dowex 50W (H form)	Yamaguchi et al. (1985)
Oxidation by ammonium persulfate in presence of a cerium salt as an emulsified solution	Skarzewski (1984)
Vapor-phase oxidation by O_2 over a variety of catalysts	Ray et al. (1972); Morita and Ohta (1973); Minami et al. (1986); Shapovalov et al. (1989)
Oxidation by H_2O_2 in presence of an HPA, $H_{3+n}PMo_{12-n}V_nO_{40}$ (n = 3, 4, 6)	Neuman and de la Vega (1993)

EVALUATION OF COMPETING CATALYTIC ROUTES

Menadione is a good example of the development of several competing catalytic processes for a fine chemical. The processes include many of the catalysts included in the previous sections, and are summarized in Table 6.5. Based on a survey of the available information on all of these processes, the process using an HPA is the best because it gives the highest selectivity (82%) at a reasonably high conversion (78%).

A process has also been developed for menadione in which the reaction is carried out in a membrane reactor. This is briefly discussed in Chapter 24.

Redox zeolites such as TS-1, and supported metal ion catalysts such as those that use ion-exchange resins as supports are among the other major oxidation catalysts. These were considered earlier in this chapter.

Hydrogenation Catalysts

Hydrogenation is a common reaction in organic technology. The conventionally used catalysts are Fe-hydrochloric acid (as in the earlier processes for the reduction of nitrobenzene to aniline), Zn/NaOH, $NaBH_4$, Na_2S_x, etc. Today, these processes have been almost completely abandoned because they involve expensive disposal of harmful sludge. A conventional catalyst that continues to be used is Raney nickel in a slurry reactor. Nickel supported on silica or alumina is also used but to a lesser extent. These nickel-based slurry hydrogenations involve separation of product from fine slurry particles, which is done by filters of various designs (e.g., magnetic separators and cross-flow filters).

Selectivity problems are not particularly serious where a single functional group is to be hydrogenated such as $-NO_2$ to $-NH_2$, or a double bond is to be saturated as in the production of saturated oils. However, where more than one group is present or a double bond is to be produced from a triple bond, selectivity to the desired product becomes important. An example of the latter is

the production of butenediol (33) from butynediol (32), in which the butenediol can undergo further hydrogenation to butanediol (34) according to reaction 6.31:

$$HO-CH_2C{\equiv}CCH_2OH \rightarrow HO-CH_2CH{=}CHCH_2OH$$
$$(32) \qquad\qquad\qquad (33)$$

$$\rightarrow HO-(CH_2)_4-OH \qquad [6.31]$$
$$(34)$$

This is an important scheme in organic synthesis and technology. As will be shown in Chapter 7, a change of support can enhance the rate of the desired step (and therefore the selectivity) by a favorable alteration of the effect of reactant diffusion on the different steps. Thus it has recently been found (Bautista et al., 1991) that, by using $AlPO_4$ as the support component in a tailored bimetallic (NiCu) catalyst, more than 98% selectivity to (33) can be realized. This suppresses by-product formation, an unavoidable feature of other catalysts (Del Rosso et al., 1984).

HYDROGEN DONORS

Molecular hydrogen is commonly used as the direct H-donor. Use of other donors has also been explored, for example, formic, phosphinic and phosphorous acids, and their salts (Johnstone et al., 1985). Unsaturated hydrocarbons (e.g., cyclohexene) and hydrazine are particularly effective in reducing nitro compounds to amines. Alcohols can also be used in selective hydrogenation of aromatic and aliphatic compounds such as aldehydes, epoxides, nitriles, and nitro groups in general (Kijenski et al., 1988, 1991). Magnesium oxide is a highly effective catalyst with an alcohol as the donor. The reaction occurs in the vapor phase by hydrogen transfer between the nitro compound and the alcohol (reactions 6.32 and 6.33).

PRIMARY ALCOHOL

$$R^1NO_2 + 3R^2CH_2OH \rightarrow R^1NH_2 + 3R^2C\overset{\displaystyle O}{\underset{\displaystyle H}{\diagdown}} + 2H_2O \qquad [6.32]$$

SECONDARY ALCOHOL

$$R^1NO_2 + 3R^2R^3CHOH \rightarrow R^1NH_2 + 3R^2R^3C{=}O + 2H_2O \qquad [6.33]$$

Examples

HYDROGENATION OF α, β-UNSATURATED ALDEHYDES TO ALCOHOLS

As mentioned previously, a particularly difficult hydrogenation is that of a compound with two double bonds in which only one is to be hydrogenated, for example, preferential hydrogenation of C=O over C=C in a given molecule. Among the important factors that influence this selective hydrogenation are the nature of the support, the crystal size and morphology, and the presence of additive species. Industrially useful hydrogenations of this type are the selective hydrogenation of cinnamaldehyde (35) to cinnamyl alcohol (36)

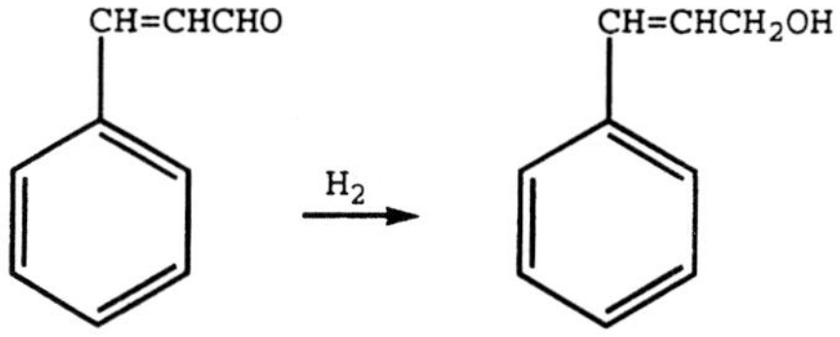

$$\text{(35)} \xrightarrow{H_2} \text{(36)} \qquad [6.34]$$

cinnamaldehyde (35) to cinnamyl alcohol (36)

and that of citral (37) to citronellol (38).

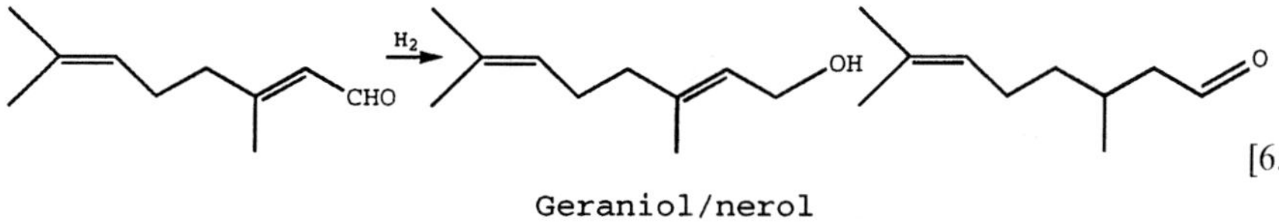

$$\text{(37)} \xrightarrow{H_2} \text{Geraniol/nerol} \quad \text{(38)} \qquad [6.35]$$

Use of monometallic catalysts with metals such as Ni, Pt, and Rh on traditional supports gives less than 5% conversion. By changing the support to ZrO_2, Pt-encapsulated zeolite-β, or zeolite-Y, the selectivity to cinnamyl alcohol could be increased to 80% (Gallezot et al., 1992; Kumbhar et al., 1994), and by using carbon-supported Ru, it could be increased to more than 90% (Coq et al., 1994). Even higher selectivities can be obtained (Didillon et al., 1991) by using a bimetallic Ru-Sn/Al_2O_3 catalyst prepared by a technique known as controlled surface reaction.

HYDROGENATION OF HALOGENONITROBENZENES

Aromatic haloamines such as 4-chloronitrobenzene and 3,4-dichloronitrobenzene are useful intermediates in the production of a variety of drugs and agrochemicals. They are produced by catalytic hydrogenation of the corresponding nitro compounds, for example

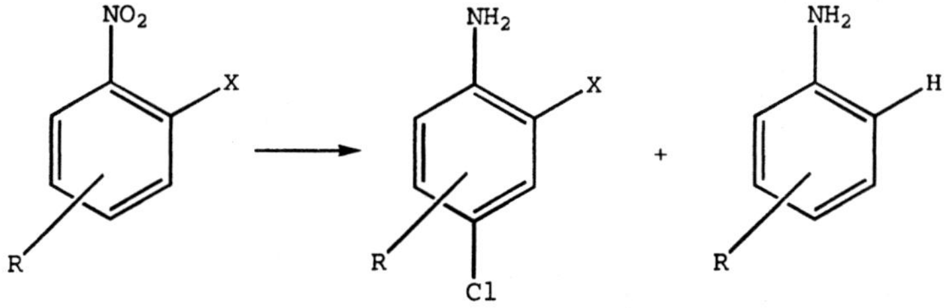

$$[6.36]$$

The traditional catalyst used in such hydrogenations is Raney nickel. This catalyst, however, causes extensive dehalogenation in these reactions, as shown in the previous reaction, resulting in very low selectivity. Thus Raney nickel catalysts have been modified by adding a base (Bayer, 1970), phosphorus, amines (Burge et al., 1980; Wisniak and Klein, 1984), various sulfided additives (Bayer, 1970), and several other modifiers. The catalyst developed by Baumeister et al. (1991) uses Raney nickel modified with formamidine acetate to give a selectivity as high as 97%. This catalyst is in commercial use today at Ciba-Geigy.

A catalyst has recently been developed (Cordier et al., 1994) based on the use of thiourea as a new selectivity promoter, which gives almost 100% hydrogenation selectivity. A few other promoters such as thiophene and sodium sulfide also give good selectivities, but thiourea is by far the best. The azo and azoxy compounds that are usually the main side products are hardly formed at all (<10 ppm).

HYDROGENATION OF AROMATIC CARBOXYLIC ACIDS TO ALDEHYDES

We noted previously that aldehydes such as benzaldehyde can be prepared by direct oxidation of the side chain in the reactant. It is also possible to go from the opposite direction and reduce a carboxylic acid to the corresponding aldehyde. In both cases, side-chain halogenation using hazardous chemicals is avoided. A one-step direct hydrogenation process has been developed by Mitsubishi using a Cr^{3+} modified zirconia catalyst (Yokoyama et al., 1992), which is reported to give aldehyde selectivities of 80–97% in the hydrogenation of benzoic acid, o-methylbenzoic acid, m-phenoxybenzoic acid, dimethyl terephthalate, etc.

PARTIAL HYDROGENATION OF M-PHENYLENEDIAMINE: A NEW ROUTE TO M-AMINOPHENOL

m-Aminophenol (MAP) is an important organic intermediate used in the synthesis of thermal- and pressure-sensitive dyes used in imaging and in the synthesis of unsymmetric diaminodiphenyl ethers that are key intermediates in the production of new classes of polymers. The conventional method of production involves three steps: sulfonation of nitrobenzene, reduction to m-aminobenzenesulfonic acid, and caustic fusion. This process is not environmentally acceptable today because of large by-product salt formation and the extremely hazardous nature of the alkali fusion step. A new process recently commercialized by Sumitomo is based on the aminolysis of resorcinol (Franck and Stadelhafer, 1988). One can also think of a route based on the relatively cheap m-phenylenediamine (MPD) as a commercially viable alternative. Jacobson of Dupont (1994) recently proposed such a route.

The process involves three steps: partial hydrogenation of MPD to 3-amino-2-cyclohexene-1-imine (3-ACI), hydrolysis of the imine to 3-amino-2-cyclohexene-1-one (3-ACO), and dehydrogenation of 3-ACO to MAP (scheme 6.37).

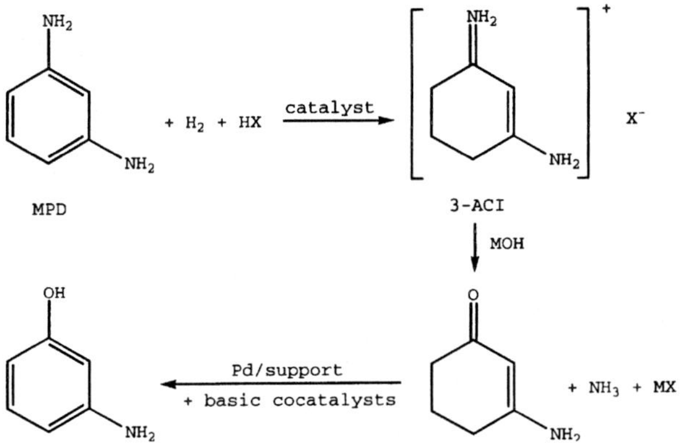

[6.37]

Table 6.6 Classification of methods for immobilizing (or heterogenizing) homogeneous catalysts

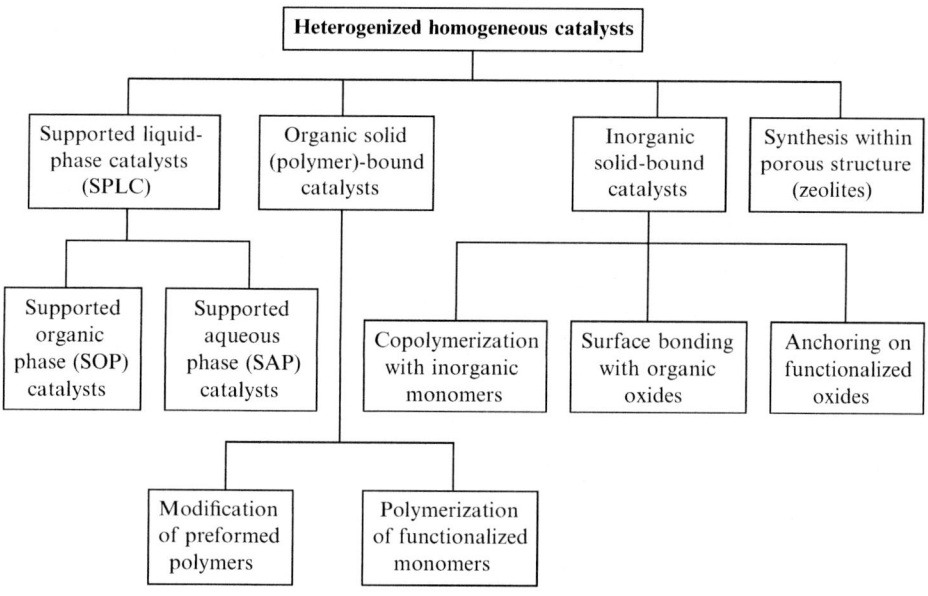

The partial hydrogenation step was carried out using Pd on carbon in a solution of MPD in aqueous acetic acid, and a conversion of 100% with a 3-ACI selectivity of 99% was obtained. Platinum on carbon gave 74% conversion but no 3-ACI at all.

"HETEROGENIZED" HOMOGENEOUS CATALYSTS

A particularly desirable catalyst would be one in which the advantages of homogeneous and heterogeneous catalysis are combined. This can be accomplished by "bottling" a homogeneous catalyst within the pores of a solid support or by binding or anchoring it in some way to it, and it is often referred to as "heterogenizing" a homogeneous catalyst. A catalyst of this type inherits the solid's property of insolubility without in any way sacrificing its homogeneous analog's reactivity and selectivity.

There are four major classes of heterogenized homogeneous catalysts, and these are listed in Table 6.6 along with subcategories of each class.[2] The most important and recent is the supported aqueous phase (SAP) catalyst.

---

[2] It is often the practice to include SLPC under inorganic solid-bound catalysts, but we prefer to consider it a separate class.

Supported Liquid-Phase Catalysts (SLPC)

Supported organic-phase (SOP) catalysts

An SLPC consists of a homogeneous solution of a catalyst dispersed in the pore volume of the solid support and is a free-flowing solid that can be used as any conventional solid catalyst (Acres et al., 1966). Extensive studies have been reported by Rony (1969), Rony and Roth (1975), and Chen and Rinker (1978) on the preparation and properties of these catalysts. Catalysts in these systems are usually dissolved in apolar organic liquids. Therefore, we call them *supported organic-phase* (SOP) catalysts.

Applications of SOP catalysts are few. A list of reactions carried out with these catalysts was tabulated by Doraiswamy and Sharma (1984), and the number has not significantly increased since then. Although all of these applications are for the production of bulk organic chemicals, for example, the hydrogenation of propylene (Rony and Roth, 1975) and the isomerization of quadricyclene (Wilson and Rinker, 1976), they also have good potential as fixed-bed or slurry catalysts for smaller volume production. Several studies on modeling these systems have been reported (Livbjerg et al., 1974, 1976; Rony, 1969; Villadsen and Livbjerg, 1978).

Supported aqueous-phase (SAP) catalysts

The preparation and applications of a novel class of complexes, water-soluble organometallic complexes, are described in Chapter 9. In what promises to be an even more important application, it is now possible to immobilize these water-soluble catalysts on solid supports (Arhancet et al., 1989, 1990, 1991; Wan and Davis, 1994a,b, 1995) to give *supported aqueous-phase* (SAP) catalysts. Figure 6.4 illustrates the general SAP catalysis scheme described by Arhancet et al. (1989). A high surface area hydrophilic support such as silica or porous glass is impregnated with an aqueous solution of the selected organometallic complex. Then water is evacuated from the pores, leaving the water-free complex distributed over the entire surface. Next, the surface is exposed to water vapor for a predetermined length of time to allow a precise amount of water to condense on it. This results in a thin coating of the aqueous film containing the catalyst, giving a supported liquid-phase catalyst which is as free flowing as the parent silica itself. Now this catalyst can be placed in contact with an organic liquid containing the reactant for reaction to occur at the water-organic phase surface, with the product(s) diffusing back to the organic phase.

As described later in Chapter 9, it is necessary to sulfonate an asymmetric catalyst to produce a water-soluble version. In supporting this on a solid as described, the water is first evacuated. However, an exact amount is reintroduced by contact with water vapor because the presence of water is necessary and the extent of enantioselectivity depends on the amount of water. A useful application of SAP catalysis is the use of SAP-Ru-BINAP-4SO$_3$Na supported on controlled pore glass GPG-240 (supplied by CPG Inc.) in the enantioselective hydrogenation of 2-(6'-methoxy-2'-naphthyl)acrylic acid to naproxen (Wan and Davis,

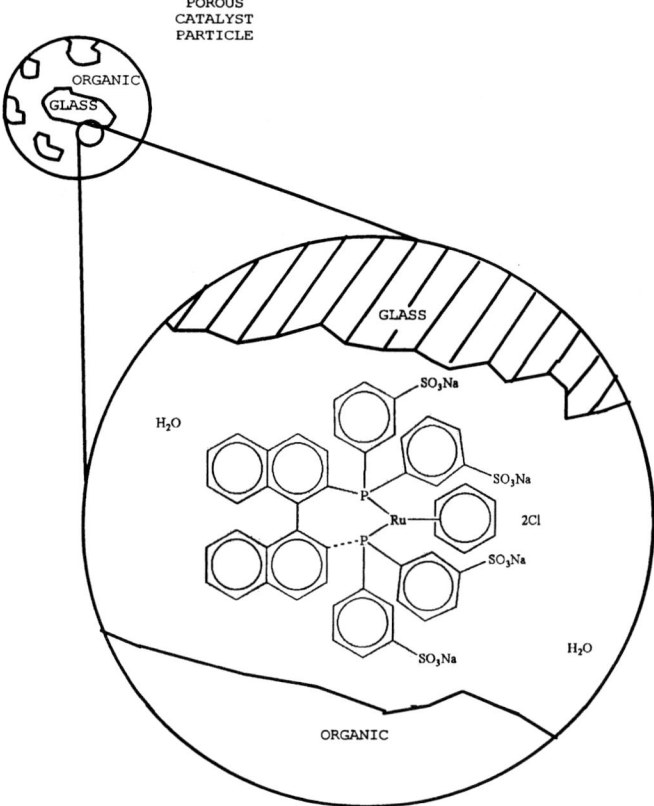

Figure 6.4 Supported aqueous-phase (SAP) catalysts (redrawn from Wan and Davis, 1994b).

1994b). Here, the water is introduced by impregnation with an organic phase (ethyl acetate) saturated with water.

Polymer-Bound Catalysts

The preparation and uses of these catalysts have been extensively reviewed, for example, Heineman (1971), Pittman and Evans (1973), Burwell (1974), Michalska and Webster (1974), Bailar (1974), Delmon and Jannes (1975), Grubbs (1977), Hodge and Sherrington (1980), Bailey and Langer (1981), Sherrington and Hodge (1988), and Pittman (in particular: 1978, 1980, 1977–1992, 1994). Based on these reviews, polymer-bound catalysts can be divided into two categories, depending on the method of preparation: modification of preformed polymers, or polymerization of functionalized monomers. We illustrate these methods through a few selected examples.

Modification of preformed polymers

The basic requirement of any organic support is the presence of functional groups which could serve as potential ligands. Certain polymers such as polystyrenes and polyamides contain such groups. Where functional groups do not exist, the preformed polymer is functionalized so that a catalytic complex can be attached. An example is the introduction of cyclopentadienyl ligands by treating preformed chloromethylated polystyrene with $SnCl_4$ followed by reaction with sodium cyclopentadienide in THF to produce the functionalized polymer (39) (Bonds et al., 1975)

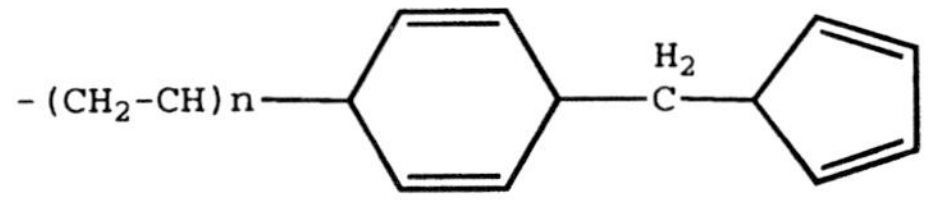

(39)

Phosphorus, the most important ligand element, can be introduced via many routes. For example, (1) halogenated polystyrene is reacted with butyllithium followed by reaction with $ClPPh_2$ (Evans et al., 1974) and (2) the resin is treated with PPh_2Li (Nonaka et al., 1974) to produce the same functionalized polymer. The supports thus produced have been used to immobilize carbonyls such as $Ni(CO)_4$ (Nonaka et al., 1974), $(PPh_3)_3RhH(CO)$ (Pittman and Felis, 1974), and $Rh_6(CO)_{16}$ (Collman et al., 1972).

Whether a polymer is used as such or is modified by functionalization, immobilization of the catalyst is quite straightforward. It can be achieved by dissolving the carbonyl in a suitable solvent, equilibrating the solution with the resin, and then removing the solvent or by refluxing the mixture and recovering the catalyst by filtration.

Polymerization of functionalized monomers

In this method, one starts with a monomer containing the ligand or the ligand-metal complex and polymerizes it to the appropriate resin. Examples are monomers containing metal carbonyls along with groups that can undergo vinyl polymerization (Pittman, 1977, 1978), for example,

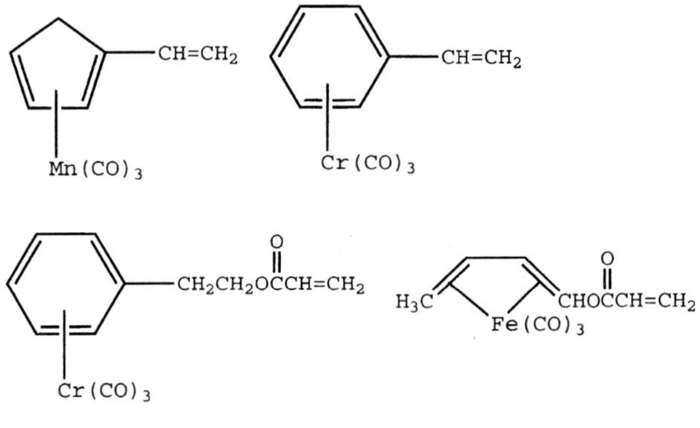

Another example is the preparation of vinyl derivatives of phosphines, which represent the most useful class of ligands in homogeneous catalysis. Specifically, p-diphenylphosphinostyrene (40) can be copolymerized with styrene and divinylbenzene to give a polymer (41) that can be used as a solid support (Nonaka et al., 1974).

$$\text{(40)} + \text{styrene} + \text{divinylbenzene} \longrightarrow \text{(41)} \quad [6.38]$$

Ligand groups containing nitrogen are also used to anchor metal carbonyls. They can be prepared by polymerizing nitrogen-containing monomers in the presence of the metal carbonyl. The immobilization is accomplished by the same procedure as described before (mostly for phosphine ligands).

As a sequel to what was mentioned previously, it is useful to note that a method has recently been developed for the rapid and simultaneous evaluation of each member of a large library of polymer-bound catalysts (Taylor and Morken, 1989). Because every catalyst will cause a temperature change in the reaction (when initiated at the same temperature), this change is used as a measure to grade the members of a library.

Immobilization on Inorganic Supports

The use of inorganic supports in catalysis is quite common and has been extensively reviewed, for example, Bailey and Langer (1981), Clark et al. (1992), Smith (1992). The most frequently used supports are γ-Al_2O_3 and SiO_2, and those less common are η-Al_2O_3, MgO, chloromethylated SiO_2, graphite, controlled pore glass, NaY and HY zeolites, Al_2O_3 impregnated with Pt, La_2O_3, ZnO, phosphated SiO_2, TiO_2, BeO, CaO, ZrO_2, and combinations of them. The inorganic solid-supported transition metal complexes exhibit three principal advantages over their polymeric counterparts:

1. They have much higher thermal stability (300 °C) compared to polymer-supported complexes (160 °C). Surprisingly, complexing increases the thermal stability in relation to the homogeneous (unanchored) solution.
2. The deactivation that results from intermolecular condensation due to polymer flexibility is avoided in the rigid inorganic support.
3. The diffusion characteristics of inorganic supports are fixed and predictable, unlike those of polymer supports that vary with swelling and, therefore, can neither be predicted nor controlled. Thus the rate and course of a reaction as influenced by diffusion (see Chapter 7) can be more easily predicted for inorganic supports.

ROLE OF SOLVENT IN CATALYSIS BY SOLIDS

The linear free energy relationship was used in Chapter 2 to include microscopic effects such as those of substrate and solvent structures in liquid-phase reactions. The Hammett relationship is the most commonly used empirical expression to predict these effects. When a catalyst is present in solid form, generalizations are less tenable, and it is best to analyze each reaction separately for any solvent effect. Even so, some generalizations are available which, though of limited value, merit brief mention. We confine the treatment in this section to specific examples of reactions in which solvents have been used to good purpose.

The solvent effect can be physical or chemical. The physical effect is usually manifested in the form of solubility or the dielectric constant. Here, we consider a new feature known as *haptophilicity*. The chemical effect depends on the acid/base character or reactivity of the solvent. We shall demonstrate these effects through selected examples (see Gilbert and Mercier, 1993).

The Effect of Polarity: Haptophilicity

The effect of the reactant is basically a kinetic effect and is hence part of the kinetic analysis considered in Chapter 7. We confine our discussion in this section to the effect of solvent polarity, which has a direct influence on the performance of the catalyst. It is known (Thompson and Naipawer, 1973; Sehgal et al., 1975) that certain functional groups in a substrate can bind to the catalyst surface during a reaction in such a way as to enforce the addition of hydrogen from its own side of the molecule, which is opposite to that expected on the basis of steric hindrance, namely binding occurs from the opposite side. This effect, termed haptophilicity, can be exploited in selecting the right kind of solvent to enhance the selectivity to a desired product.

Thus consider compound (42). The nature of the solvent determines the type of isomer formed, *trans* (43) or *cis* (44), when this compound is hydrogenated on a catalyst such as Pd/C (Thompson et al., 1976).

$$\text{(42)} \xrightarrow[\text{solvent}]{H_2,\ Pd/C} \text{(43)} + \text{(44)} \qquad [6.39]$$

If a polar solvent such as alcohol or DMF is used, then the polar group in the substrate (CH_2OH) interacts with the solvent group thus "engaging" it. Then hydrogenation occurs on the opposite face of the molecule leading to the *trans* isomer (43). On the other hand, when a nonpolar solvent such as hexane is used, the polar group on the substrate tends to chemisorb on the catalyst making it more polar. Now hydrogenation occurs by addition to the same face, leading to the *cis* isomer (44). Thus the polarity of the solvent can have a significant influence on isomer distribution. In the case under consideration, the *trans/cis* ratio decreases from 94/6 with DMF to 39/61 with hexane. It is possible, however, that the selectivity changes will not always be as clear.

Chemical Effects

Solvents can be effectively used to increase the selectivity of an intermediate in a consecutive reaction by forcing the desorption of the desired compound from the catalyst surface and thus preventing further reaction. A good example of this is the formation of benzene hydroxylamine (45), an often undetected intermediate in the production of aniline from nitrobenzene (reaction 6.40).

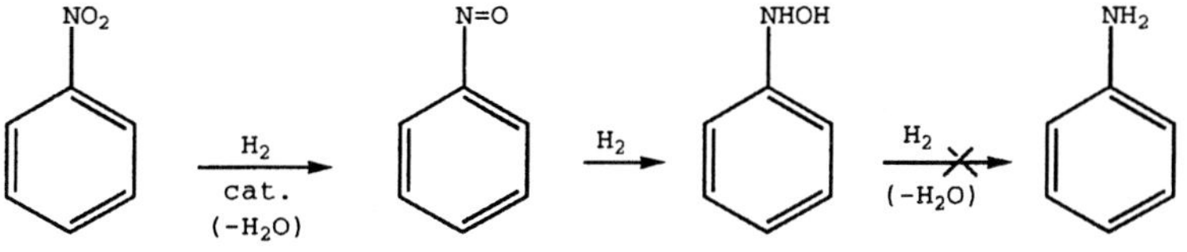

[6.40]

The use of pyridine or piperidine promotes desorption of the hydroxylamine, thus enhancing its selectivity (Leludec, 1972). Such a strategy is useful in producing intermediates for biologically active molecules (Von Pierre, 1989), for example,

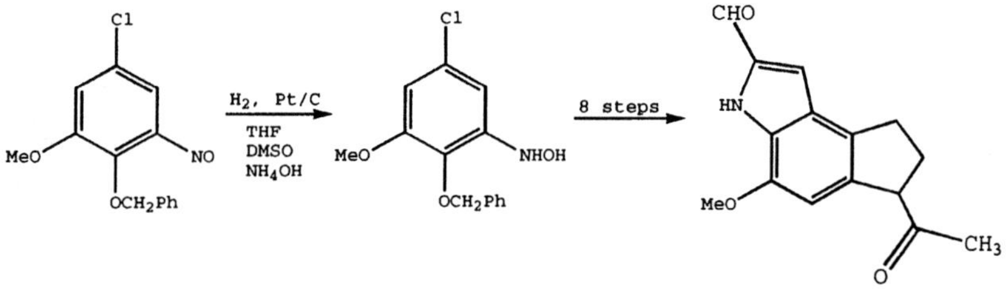

[6.41]

We have already seen how different types of catalysts can be used to arrest the hydrogenation of a triple bond at the double-bond stage. A solvent or additive can also accomplish this by competitive chemisorption on the catalyst. An example (reaction 6.42) is the hydrogenation of dehydrolinalyl acetate (46) to linalyl acetate (47), a perfumery chemical, without further hydrogenation to the unwanted dihydrolinalyl acetate (48). Compound (48) should be present in less than 2% to preserve the perfumery quality of (47). This can be accomplished by using Pd/C as catalyst with pyridine as the additive or solvent (Gilbert and Mercier, 1993). A similar application can be found in the synthesis of ethylenic diol (Baillard et al., 1988) used in agrochemicals.

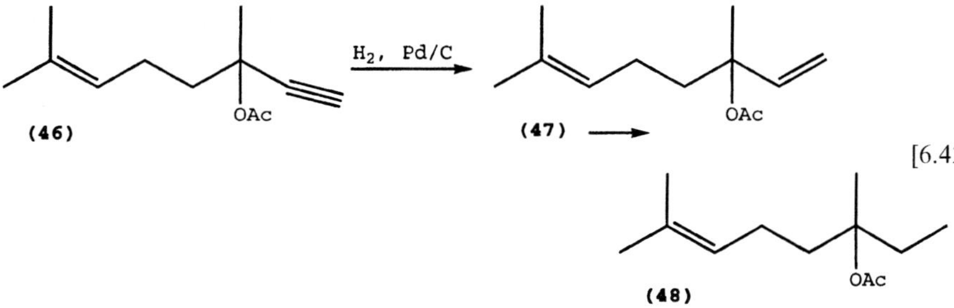

[6.42]

A striking use of a solvent or additive is in the manufacture of dimethylaniline (DMA) (49), an intermediate in the production of drugs and dyes, by a continuous catalytic process (Doraiswamy et al., 1981), compared to the usual high-pressure batch process from aniline, methanol, and sulfuric acid. In this process (reaction 6.43)

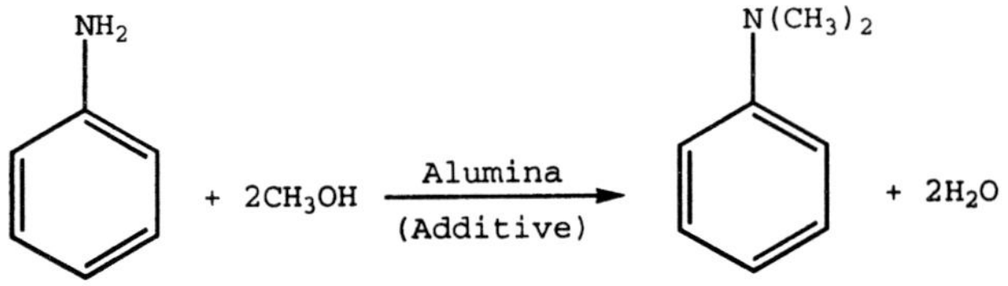

$$C_6H_5NH_2 + 2CH_3OH \xrightarrow[\text{(Additive)}]{\text{Alumina}} C_6H_5N(CH_3)_2 + 2H_2O \qquad [6.43]$$

$$(49)$$

aniline and methanol are passed in the vapor phase over a fixed bed of specially treated alumina to give more than 99% DMA. The catalyst tends to be deactivated very rapidly. This is counteracted by feeding, along with the reactants, an additive that continuously dissolves the impurities that adhere to the catalyst surface and block its catalytic action. Thus the catalyst operates over a prolonged period of time without any loss of activity.

Note

The synthesis of a catalytic moiety within the porous structure of a solid (mainly zeolites) constitutes another class of immobilized catalysts listed in Table 6.1. Because of their special connection to zeolites, they were considered in the section on reactions by hybrid catalysts, but I must emphasize that they also belong to the general class of immobilized catalysts.

Chapter 7

Catalysis by Solids, 2

The Catalyst and Its Microenvironment

Solid catalysts by their very nature involve diffusion of reactant fluids within their matrix. These fluids react even as they diffuse. Thus the problem of *internal diffusion* accompanied by reaction becomes important. Another problem of equal importance is the transport of the reactant from the fluid bulk to the catalyst surface—often referred to as *external diffusion*. Together these constitute the *microenvironment* of the catalyst pellet and form the subject matter of this chapter.

MODELING OF SOLID CATALYZED REACTIONS

The Overall Scheme

Consider the reaction between aniline and methanol on pellets of alumina to give dimethylaniline:

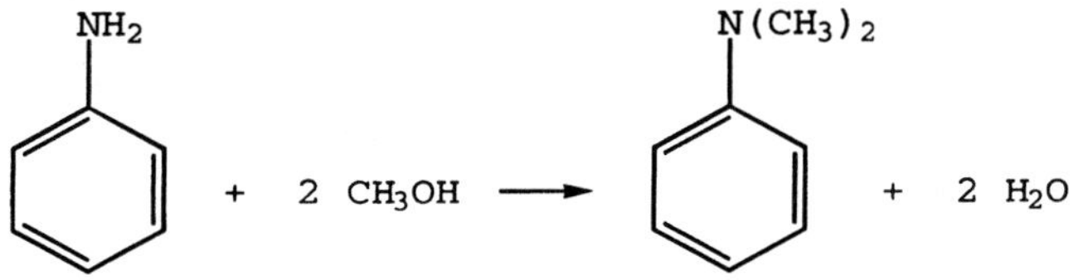

$$C_6H_5NH_2 + 2\ CH_3OH \longrightarrow C_6H_5N(CH_3)_2 + 2\ H_2O \qquad [7.1]$$

Let the pellets be placed in a flowing stream of reactants inside a tubular reactor. Restricting our attention now to a single pellet and its immediate environment, the various steps involved in the overall process are shown in Figure 7.1. This physical-chemical circuit is built in analogy with the electrical circuit shown at the bottom of the figure. Clearly, the overall process is a complex combination

172 Catalysis

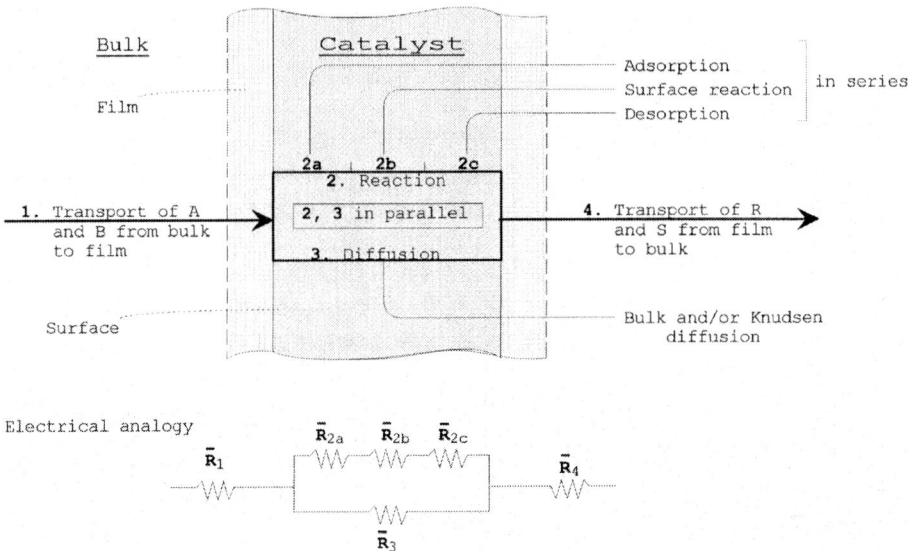

Figure 7.1 Major steps in the solid-catalyzed reaction $A + B \rightarrow R + S$. $\bar{R}$ is electrical resistance.

of chemical and physical steps. Note, however, that the mathematical analysis of the parallel pathways (diffusion and reaction) is not always based on the addition of reciprocal resistances as in parallel electrical circuits but on the fact that the two occur simultaneously on a single pathway, that is, the molecule reacts even as it diffuses.

The construction of a model based on one of the adsorption-reaction-desorption steps (shown in the figure) being the limiting step constitutes the core of the semiempirical approach considered in this chapter. In this approach, the microscopic origins of the observed macroscopic effects of catalysts (as described by many authors, e.g., Plath, 1989) are ignored. The models thus developed are commonly known as Langmuir–Hinshelwood models among chemists and as Hougen–Watson models among chemical engineers. We choose to call them *Langmuir–Hinshelwood–Hougen–Watson (LHHW)* models. For a more basic approach to kinetic modeling of catalytic reactions in general, reference may be made, among others, to Compton (1991) and Van Santen and Niemantsverdriet (1995).

Langmuir–Hinshelwood–Hougen–Watson (LHHW) Models

In the interest of generality, we consider hypothetical reactions and derive rate equations for a few typical *LHHW* models (Hougen and Watson, 1947; Yang and Hougen, 1950; Satterfield, 1980, Butt, 1980; Doraiswamy and Sharma, 1984; Boudart and Djéga-Mariadassou, 1984). As the Langmuir isotherm is the basis of all LHHW models, we begin by a simple derivation of this isotherm.

The Langmuir isotherm

Unlike in homogeneous reactions where the rate is proportional to the reactant concentration (say, [A]), in catalytic reactions it is proportional to the surface concentration $[A]_s$. Because $[A]_s$ is not usually known, it is convenient to express it in terms of [A] by equating the rates of adsorption and desorption:

$$k_{Aa}[A][\ell]_v = k_{Ad}[A]_s \tag{7.1}$$

giving

$$\theta_A = K_A[A]\theta_v = K_A[A](1 - \theta_A) \tag{7.2}$$

where $K_A = k_{Aa}/k_{Ad}$, and θ_A and θ_v are the fractions of adsorbed A and vacant site (ℓ) on the surface, respectively, given by

$$\theta_A = \frac{[A]_s}{[\ell]_t}, \quad \theta_v = \frac{[\ell]_v}{[\ell]_t} \tag{7.3}$$

Thus

$$\theta_A = \frac{K_A[A]}{1 + K_A[A]}, \quad \theta_v = \frac{1}{1 + K_A[A]} \tag{7.4}$$

When other components besides A are also adsorbed, such as B, R, I (inert), Equation 7.4 becomes

$$\theta_A = \frac{K_A[A]}{1 + K_A[A] + K_B[B] + K_R[R] + K_I[I]}$$
$$\theta_v = \frac{1}{1 + K_A[A] + K_B[B] + K_R[R] + K_I[I]} \tag{7.5}$$

To use these expressions in formulating LHHW rate models, an understanding of the "slowest" or the rate-determining step is necessary.

The rate-determining step

The basic assumption of the LHHW models is that the slowest of many possible steps involving adsorption, surface reaction, and desorption is rate controlling. Because the rates of all of the steps must be equal at steady state, this statement tends to be misleading. To dispel any such confusion, it is helpful to regard the rate-limiting step as the one that consumes most of the available driving force. A clarifying analogy is that of a current passing through a set of resistances in series: although the current (corresponding to the rate) is the same, the conductivity of any one of the resistors can be lower than that of the others, making it the rate-limiting resistor.

Now we develop rate equations by assuming any one of several steps involved in a given reaction as the rate-determining step.

The basic procedure

Consider the reaction

$$A + B \leftrightarrow R + S \tag{7.2}$$

The various steps comprising the reaction are

Adsorption of A and B:

$$A + \ell \to A\ell, \quad B + \ell \to B\ell \qquad [7.3a]$$

Surface reaction:

$$A\ell + B\ell \to R\ell + S\ell \qquad [7.3b]$$

Desorption of R and S:

$$R\ell \to R + \ell, S\ell \to S + \ell \qquad [7.3c]$$

According to the concept of the rate-determining step, any one of these steps can be controlling.

Because all the steps are reversible, some further clarification is needed with respect to the rate-determining step. It can be imagined that the forward and reverse rates of four of the five steps are much larger than those of the fifth step. In fact, they are so much larger that the forward and reverse rates of each step may be assumed to be nearly equal. In other words, these four steps may be assumed to be in quasi-equilibrium. Note, however, that the net rate is still the same for all the steps, that is, $r(\text{net}) \cong (r_{1+} - r_{1-}) \cong (r_{2+} - r_{2-}) \ldots \cong (r_{5+} - r_{5-})$. For the fifth step, we have the additional fact that $r_{5+} \neq r_{5-}$, while for the other steps $r_+ \cong r_-$. Therefore, it is the rate-determining step.

Focusing on the case where surface reaction is controlling, the basic procedure in developing a LHHW model is to write the rate equation in terms of the surface coverage θ_A of reactant A rather than its concentration $[A]$. Sometimes, as in reactions requiring a second (vacant) site for adsorbing a product (e.g., $A \to R + S$), the rate will also depend directly on the fraction of surface covered by vacant sites θ_v, and when there is dissociation of a reactant, a pair of adjacent vacant sites should be available, so that now the rate of adsorption would be proportional to θ_v^2 rather than to θ_v. For example

$A \to R$, both adsorbed $\quad -r_{wA} = k_w \theta_A$

$A \to R + S$, all adsorbed $\quad -r_{wA} = k_w \theta_A \theta_v$

$A_2 \to 2A \to P$, A_2 dissociates $\quad -r_{wA} = k_w \theta_A$ (or $k_w \theta_A^2$)

where, for this case,

$$\theta_A = \frac{(K_A[A])^{1/2}}{1 + (K_A[A])^{1/2}}$$

$A + B \to R + S$, all adsorbed $\quad -r_{wA} = k_w \theta_A \theta_B$

Now let us turn to the case where adsorption of one of the components is controlling. Thus consider the reaction

$$A \leftrightarrow R \qquad [7.4]$$

where adsorption of A is controlling. Clearly, adsorption equilibrium does not exist, and we cannot use Equation 7.2 to get θ_A. Instead, because now the reaction would be in equilibrium, we use

$$-r_{wA} = k_{Aa}\left(\theta_A - \frac{\theta_R}{K'}\right) \tag{7.6}$$

where K' is the equilibrium constant for the surface reaction. This constant must be distinguished from the true thermodynamic equilibrium constant K given by $[R]/[A]$. Then simple algebraic manipulations lead to expressions for $\theta_A, \theta_B, \theta_R, \theta_S$, and θ_v. Using these, the final expression for the rate can be developed.

All LHHW models can be consolidated into a single general form

$$\text{Rate} = \frac{(\text{kinetic term})(\text{potential term})}{(\text{adsorption term})^n} \tag{7.7}$$

in which the exponent n in the adsorption term denotes the number of sites participating in the catalysis. Yang and Hougen (1950) list the various terms for several classes of reactions. It is a relatively simple matter to construct a full LHHW model from this table. A more elaborate method of accounting for all possible models (often more than one hundred) for a given reaction has been proposed by Barnard and Mitchell (1968). Table 7.1 lists the model equations for a few selected types of reactions and controlling mechanisms.

In some bimolecular reactions like disproportionation of propylene to butylene and ethylene and hydrogenation of ethylene to ethane to ethane, a modified form of LHHW models has to be used. Here, the reaction is assumed to occur by a molecule of one of the reactants (say A) striking an obsorbed molecule of B (or another A). Thus the rate equation would be: $r_A = kp_A\theta_B$ (or $\theta_A p_B$ if B reacts from the gas phase). Such models are generally referred to as Eley-Rideal models.

Selection of the Most Plausible Model

Several methods are available for model selection. We shall be concerned only with statistical methods, which in their simplest form are based on the selection of the most plausible model from among a number of candidates. It must be emphasized that these methods do not uniquely select a candidate as the only acceptable model but converge to one that is the most acceptable statistically.

Preliminary short listing of models

The first step is to write the rate equations for all possible controlling steps. Then preliminary shortlisting can be done by reducing the models to intial conditions when no product would have formed. Under these conditions, each model exhibits a specific behavior with variation in total pressure. Table 7.2 lists the complete rate equations, the initial rate equations (expressed in terms of the total

176   Catalysis

Table 7.1   LHHW models for a few selected reaction types[a]

Reaction	Controlling mechanism	LHHW equation for $(-r_{wA})$
$A \leftrightarrow R$	Surface reaction	$\dfrac{k_w K_A \left([A] - \dfrac{[R]}{K}\right)}{1 + K_A[A] + K_R[R]}$
$A + B \leftrightarrow R + S$	Surface reaction	$\dfrac{k_w K_A K_B \left([A][B] - \dfrac{[R][S]}{K}\right)}{(1 + K_A[A] + K_B[B] + K_R[R] + K_S[S])^2}$
$A + B \leftrightarrow R + S$	Adsorption of B	$\dfrac{k_{Ba}\left([B] - \dfrac{[R][S]}{[A]K}\right)}{1 + K_A[A] + K_B \dfrac{[R][S]}{K[A]} + K_R[R] + K_S[S]}$
$A + B \leftrightarrow R + S$	Desorption of R	$\dfrac{k_{Rd} K \left(\dfrac{[A][B]}{[S]} - \dfrac{[R]}{K}\right)}{\left(1 + K_A[A] + K_B[B] + K K_R \dfrac{[A][B]}{[S]} + K_S[S]\right)}$
$A + B \leftrightarrow R + S$	Adsorption of A with dissociation	$\dfrac{k_{Aa}\left([A] - \dfrac{[R][S]}{K[B]}\right)}{\left(1 + \sqrt{\dfrac{K_A[R][S]}{K[B]}} + K_B[B] + K_R[R] + K_S[S]\right)^2}$
$A + B \leftrightarrow R$	Surface reaction with dissociation of A, B not adsorbed	$\dfrac{k_w K_A \left([A][B] - \dfrac{[R]}{K}\right)}{(1 + \sqrt{K_A[A]} + K_R[R])^2}$

[a] Constructed from Yang and Hougen's table (1950).

pressure P), and the corresponding r_{wA0}–P plots for three representative models. A more complete treatment can be found in Yang and Hougen (1950). Similar effects can be produced by changing the ratio of reactants in a bimolecular reaction.

While the initial rate is an attractive preliminary tool for model elimination, three important considerations must be noted:

1. Not all models may be amenable to the kind of simplification seen in the three models shown in Table 7.2.
2. Even though the product partial pressures may be low, if the adsorption is strong enough, all the product terms in the denominator may not vanish. This leaves only the numerator to be simplified.
3. Impurities and other components may compete for adsorption and occupy a fraction of the sites. Hence a variation of the pressure would probably have less effect than when all the sites are occupied only by reactants and products. This may not significantly change the trends as

Table 7.2 Initial rate–pressure relationships for different rate forms

Reaction	Model equation for $-r_{wA}$	Equation for the initial rate, $-r_{wA0}$	Behavior of $(-r_{A0})$ with variation in total pressure P
1. $A \leftrightarrow R + S$ (Surface reaction controlling)	$\dfrac{k_{wp} K_A \left(p_A - \dfrac{p_R p_S}{K}\right)}{\left(1 + K_A p_A + K_R p_R + K_S p_S\right)^2}$	$\dfrac{aP}{(1+bP)^2}$	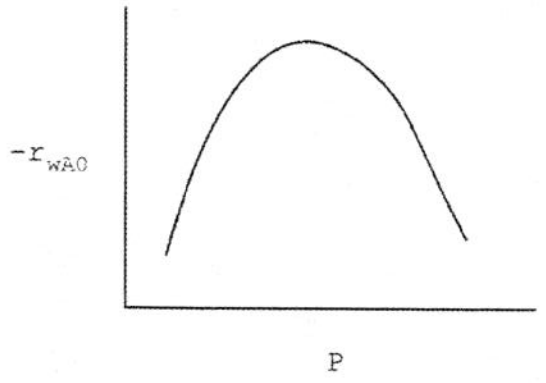
2. $A + B \leftrightarrow R$ (Surface reaction controlling, B not adsorbed)	$\dfrac{k_{wp} K_A \left(p_A p_B - \dfrac{p_R}{K}\right)}{\left(1 + K_A p_A + K_R p_R\right)}$	$\dfrac{aP^2}{(1+bP)}$	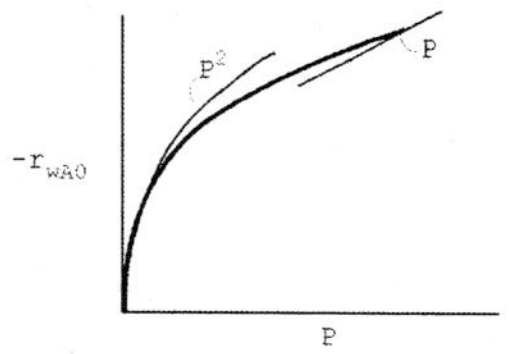
3. $A \leftrightarrow B + S$ (Desorption of R controlling)	$\dfrac{k_{Rd} K \left(\dfrac{p_A}{p_S} - \dfrac{p_R}{K}\right)}{1 + K_A p_A + K K_R \dfrac{p_A}{p_S} + K_S p_S}$	$\dfrac{k_R}{K_R} = \text{const.}$	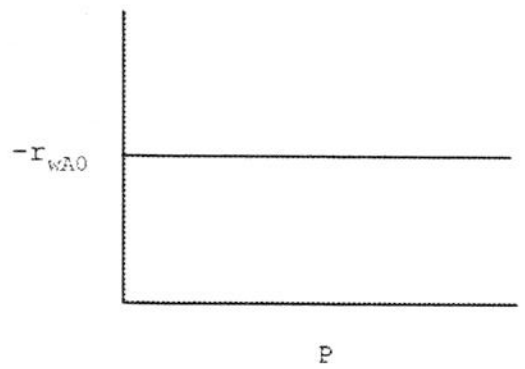

shown in Table 7.2 for many reactions, but it does inject an element of uncertainty.

The basic steps in model selection

Having narrowed down the field by the initial rate method, the following procedure can then be used to select the most probable model from the surviving list of contenders.

1. Write all possible mechanisms and the corresponding rate equations.
2. Linearize the equations as illustrated below for a typical case.

$$-r_{wA} = \frac{k_w K_A}{(1 + K_A[A] + K_R[R])}\left([A] - \frac{[R]}{K}\right) \tag{7.8}$$

Recast this as

$$\bar{r} = a + b[A] + c[R] \tag{7.9}$$

where

$$\bar{r} = \frac{\left([A] - \dfrac{[R]}{K}\right)}{(-r_{wA})}, \quad k_w = \frac{1}{b}, \quad K_A = \frac{b}{a}, \quad K_R = \frac{c}{a} \tag{7.10}$$

Then minimize the sum of squares of the residuals

$$s^2 = \frac{\sum_{i=1}^{n}(y_i - \bar{y})^2}{n - 1} \tag{7.11}$$

where $\bar{y}$ is the arithmetic mean of n measurements and y_i is an individual measurement, and determine the rate and adsorption equilibrium constants at each temperature. This is the well-known linear least squares adaptation of the regression theory.

3. However, the regression theory requires that the errors be normally distributed around $(-r_{wA})$, and not around $\bar{r}$ as in the linearized version just described. Hence use the values determined as initial estimates to obtain more accurate values of the constants by minimizing the sum of squares of the residuals of the rates directly from the "raw rate equation" by nonlinear least squares analysis.
4. The analysis described under step 3 can be carried out by the differential method in which the rates to be used in the equations are obtained directly in a differential reactor or by appropriate manipulation of integral data obtained in an integral reactor.
5. From standard statistical t-tests, make sure that all constants are significantly different from zero, and discard models for which even one of these constants is significantly negative because none of these constants can reasonably be negative.
6. From the models surviving step 5, reject those for which the rate constant k_w decreases with temperature or the adsorption equilibrium

constant K_A or K_R increases with temperature. This is because the rate constant for a single reaction step must always increase with temperature, and adsorption being exothermic, K_A and K_R must decrease with temperature.
7. A simple way to check roughly for the consistency of the selected LHHW model is to apply a test proposed by Everett (1950) and Boudart et al. (1967). This is not a powerful test.
8. From the models surviving step 6, choose the one that best fits the experimental data.

Refinements in nonlinear estimation of parameters

Himmelblau (1970) and Kittrell (1970) give very useful treatments of nonlinear estimation of kinetic parameters, and a good recent book is that of Huet et al. (1996). A simple procedure is illustrated in detail for the isomerization of *n*-butene to *i*-butene by Raghavan and Doraiswamy (1977) (see also Huet et al., 1996). Among other useful illustrations are those of Franckaerts and Froment (1964), Dumez and Froment (1976), and Dumez et al. (1977). In addition, several refinements in statistical procedures have been proposed over the years for model discrimination and parameter optimization, for example, Box and Lucas (1959), Marquardt (1963), Box and Draper (1965), Kittrell and Mezaki (1967), Hoffmann and Hofmann (1971, 1976), Hofmann (1972, 1985), Seinfeld (1970), Froment (1974, 1975), Kim et al. (1991). The sequential discrimination method (Box and Hill, 1967; Box and Henson, 1969; Froment and Mezaki, 1970; Hosten, 1974; Van Parijs et al., 1986) in which successive experiments are planned to sequentially update the probability of several competing models is particularly useful. In this method, if properly planned, only one model with probability close to one will finally emerge as the most plausible. [Sequential design can also be used for updating the parameter values of the simpler power law model with a high correlation between preexponential factor and activation energy, often a vexing problem (Dovi et al., 1994)]. More recently, some good strategies have been formulated (Watts, 1994) in which the original LHHW models are reformulated and the parameters are transformed to produce "well-behaved" estimates. All these refinements notwithstanding, there is still some uncertainty regarding the LHHW models, mainly because different models can give the same degree of fit to experimental data (Knözinger et al., 1973; Boag et al., 1978; Corma et al., 1988).

Comments

Some of the criteria just outlined for model selection may not always be valid. There are many theoretical and experimental pitfalls.

THEORETICAL PITFALLS

1. More than one step can be rate controlling, as in the case of dehydrogenation of *sec*-butyl alcohol (Thaller and Thodos, 1960; Bischoff and Froment, 1962; Shah, 1965; Choudhary and Doraiswamy, 1972).

It is also possible that the effect of temperature is much more severe on one or more of the other, faster steps than on the slowest, or the rate-determining, step. Thus the actual rate-determining step can change with temperature. Pressure changes can also exert a similar influence (Boudart and Djéga-Mariadassou, 1984).

2. There are a few instances of endothermic adsorption, and in such cases the adsorption equilibrium constants will increase with temperature (see Doraiswamy and Sharma, 1984, for examples).
3. In the presence of attractive interactions between adsorbed molecules on the surface, the adsorption equilibrium constant can be negative, but such situations are rare.
4. Even if all of the criteria are satisfied, there is still the possibility that the reaction may proceed according to a different mechanism. Dehydrogenation of methylcyclohexane is an interesting example of this (Sinfelt, 1964).

EXPERIMENTAL PITFALLS

The observed (i.e., experimental) rate constant or activation energy may be quite different from the true (i.e., intrinsic) value. Thus:

1. Consider the first reaction in Table 7.1. If A and R are not strongly adsorbed, or if R is not adsorbed and the partial pressure of A is low, then the denominator vanishes and a simple first order equation would result. Experiment would then give the observed rate constant (k_{obs}) as the product $k_w K_A$. Applying the Arrhenius equation to k_w, and assuming the adsorption to be exothermal (as it usually is), we get $k_w = k_w^0 \exp(-E/R_g T) \exp(\Delta H_{ads}/R_g T)$, or $E_{obs} = (E - \Delta H_{ads})$. In other words, the observed value would be less than the true value.
2. Consider now the second reaction and assume that the products are not adsorbed. If A and B are weakly adsorbed, all the adsorption terms in the denominator would be zero, so that only the numerator must be considered. This gives $k_{obs} = k_w K_A K_B$ and $E_{obs} = E - \Delta H_A - \Delta H_B$.
3. If in case 2, A is weakly adsorbed and B is strongly adsorbed, then the rate equation reduces to $k_w K_A p_A/(K_B p_B)^2$, and we have $E_{obs} = E - \Delta H_A + \Delta H_B$.

Clearly, the observed values can be quite different from the true values if simple power law models are force fitted to the data. Such pitfalls should be avoided.

Another pitfall, of a slightly different kind, is also worth noting. Consider the rate equation $-r_{wA} = k_w K_A p_A/(1 + K_A p_A)$. For very high values of p_A, $K_A p_A \gg 1$, and we have $-r_A = k_w$. By contrast, for very low values of p_A, $K_A p_A \ll 1$, and we have $-r_A = k_w K_A p_A$. In other words, the reaction can change from zero to first order with change in pressure and assume values between 0 and 1 at intermediate pressures. This can lead to erroneous conclusions about the mechanism.

Use of isothermal rate equations for scale-up: A cautionary note

We have so far considered only procedures for obtaining rate equations from isothermal data. It is also possible to obtain such data from nonisothermal experiments, especially those carried out adiabatically. Since, as will be shown later in Chapter 10, there is a unique relationship between conversion and temperature in an adiabatic reactor, a single equation can be used to obtain concentration and temperature profiles in such a reactor (in fact, one can do away with concentrations altogether). Then, by considering several constant temperature zones (which can be as narrow as possible) of the reactor, equations can be written for each section in which the rate constant is expressed in terms of an Arrhenius equation. This equation can then be solved to obtain the kinetic parameters k and E from nonisothermal data.

In a recent study Wang and Hofmann (1999) have stressed the importance of nonisothermal rate data. From a simple theoretical analysis they conclude that kinetic and transport data obtained under isothermal conditions in a laboratory reactor cannot logically be used to simulate any other type of reactor. This is because of the behavior of the Lipschitz constant **L**, which is a measure of the sensitivity of the reaction to different models. It tells us how any two models would diverge at the end of a reactor under different thermal conditions of operation. It is therefore a useful criterion for selecting the best model. It has been shown that **L** is different for different reactor models:

$$\mathbf{L}(\text{adia}) > \mathbf{L}(\text{noniso–nonadia}) > \mathbf{L}(\text{iso})$$

This means that for the same initial conditions, the hot spot temperature in a nonadiabatic-nonisothermal reactor or the exit temperature in an adiabatic reactor is always higher than that in an isothermal reactor. Wang and Hofmann (1999) also give correlations for calculating **L** for different reactor types.

The general conclusion from these studies is that the methods suggested earlier in this chapter for obtaining the best LHHW model for a given reaction using isothermal data are quite valid. However, in a strict sense, they cannot be used to describe transient phenomena (Simon and Vortmeyer, 1978; Graham and Lynch, 1987), or extrapolated to nonisothermal reactors without testing them in laboratory reactors with different values of **L** operated under nonisothermal conditions.

A General Approach to Modeling

The procedures just outlined rely exclusively on statistical analysis of rate data. Such an exercise converges to the most plausible model, not necessarily the only acceptable one. It is possible to combine the statistical methods with other experimental methods to obtain a more unique model. One such method is to perturb the system by changing any of the parameters (such as the reactant concentration, flow rate, temperature, pressure, etc.), and determine the time taken to regain steady state. This gives important information on the rates of

the individual steps and therefore narrows the number of possible rate-determining steps (see, e.g., Kobayashi and Kobayashi, 1974; Lee and Agnew, 1977; Burghardt and Smith, 1979). Statistical analysis can then be employed to determine the best from among the remaining candidates.

Another method is to independently measure the adsorption equilibrium constants of the components involved, and use them in determining the rate equation. Since the adsorption equilibrium values are very different under actual reaction conditions than under the conditions they are normally determined, care must be taken to ensure their measurement under reaction conditions. A chromatographic method based on the use of central moments appears to be particularly useful (Kubin, 1965; Kucera, 1965; Schneider and Smith, 1968). A combination of this method with statistical analysis is a sound strategy for determining the rate equation for a given reaction. This method has been successfully employed by Raghavan and Doraiswamy (1977) for the isomerization of butenes.

Influence of surface nonideality

In addition to the limitations specifically applicable to the procedure just outlined, the models themselves suffer from a few genetic defects and therefore have been the subject of some criticism and much commentary. The chief limitations are these:

1. All sites are equally active with equal heats of adsorption; this is not true because there is usually a distribution of activity on the surface that is ignored in the LHHW models.
2. Interaction between adsorbed molecules is negligible, again not true.
3. Molecules are always adsorbed at random on the surface, also not true; like molecules may tend to adsorb in contiguity forming their own islands and this leads to a completely different mechanism of surface reaction.

A number of isotherms that dispense with assumptions (1) and (2) have been proposed (see Doraiswamy, 1991). Some studies (e.g., Kiperman et al., 1989) indicate that it may not be possible to model certain reactions without invoking the role of surface nonideality. Fortuitously, the use of more rigorous isotherms does not materially affect the companion problem of the diffusion-reaction behavior of systems (Shendye et al., 1993)—a topic considered in the next section.

The paradox of heterogeneous kinetics: The "placebo effect"

An interesting feature of LHHW kinetics is worth noting. Many reactions on surfaces known to be nonideal surprisingly follow the ideal LHHW models, a situation that can only be described as the "placebo effect" or the paradox of heterogeneous kinetics (Boudart et al., 1967; Boudart, 1986). In the same vein but with less justification, it has also been argued for more than four decades—

for example, from Weller (1956) to Bouzek and Rousár (1996)—that rate data for a given reaction can be correlated equally well by simple power law kinetics (thus dispensing with the surface science approach altogether). In general, LHHW models supplemented by rigorous methods of parameter estimation do represent a valid mechanistic approach that can be accepted as a reasonably sound basis for reactor design. This is particularly true of *facile* or surface-insensitive reactions—reactions largely uninfluenced by the crystal structure of the catalyst. Further, such reactions require only one or two adjacent atoms, unlike surface-sensitive reactions that would need clusters, multiplets, or ensembles of more than two atoms. Thus, for the latter class of reactions, a more puristic approach would be required with a firmer anchoring to the more rigorous theories of surface science.

It is a fortuitously helpful circumstance that many common organic reactions are reasonably structure-insensitive and therefore amenable to LHHW type of modeling. Some examples are hydrogenation of alkenes and arenes, dehydrogenation of *sec*-alcohols, ethylation of aniline, many hydrogenations on nickel, and a number of isomerizations.

Another explanation of the paradox of heterogeneous kinetics may be the possibility that the more active sites on a distributed site activity surface (i.e., a nonideal surface) tend to compensate for the less active sites, thus creating the placebo effect.

ROLE OF DIFFUSION IN PELLETS: CATALYST EFFECTIVENESS

Catalysts are normally used in the form of pellets, except in fluidized-bed reactors where powders are used. Thus problems of resistance to diffusion within the pellets are common. These have been quite extensively studied and many texts written (see in particular Aris, 1975). We present some basic equations here and outline methods for a quick evaluation of these effects.

Modes of Diffusion

The basic requirement in any study of internal diffusion is an understanding of the various modes of transport in a straight capillary: bulk, Knudsen, configurational, and surface. Then this knowledge is extended to diffusion in the porous matrix of a pellet to formulate expressions for an *effective diffusion coefficient*. We confine our treatment in this book to listing in Table 7.3 the more important equations for direct use in estimation. Given in this table are the equations for bulk (for macropores with $r_p > 200 \text{ Å}$), Knudsen (for micropores with $r_p < 50 \text{ Å}$), and combined (for all pore sizes) diffusion in a straight capillary. The last is also referred to as diffusion in the *transition regime*. These are followed by equations for an effective diffusion coefficient in a pellet in these regimes. Two models are considered for the transition regime: (1) the parallel path model (Johnson and Stewart, 1965; Feng and Stewart, 1973) that accounts for a single overall pore size distribution; and (2) the *micro-macro* or *random pore model* (Wakao and Smith, 1962, 1964) that assumes a *bimodal distribution* of the

Table 7.3 Equations for estimating diffusivities (of A) in pellets in various regimes[a]

Serial number	Flow regime/pore structure	Parameter to be estimated	Equation	Main input data
1.	Bulk diffusion (independent of pore structure)	Bulk diffusivity D_{bA} (cm²/s)	$D_{bA} = \dfrac{10^{-3} T^{1.75}}{P(V_A^{1/3} + V_B^{1/3})^2} \left(\dfrac{1}{M_A} + \dfrac{1}{M_B}\right)^{1/2}$	Molecular weights M_i, temperature, and molar volumes V_i
2.	Bulk diffusion in a porous pellet	Effective bulk diffusivity D_{ebA}	$D_{ebA} = \dfrac{f_c}{\tau} D_{bA}$	Tortuosity factor τ, pellet porosity f_c
3.	Knudsen diffusion in a straight capillary	Knudsen diffusivity (cm²/s), $\bar{r}_p$ in Angstroms	$D_{KA} = 9.7 \times 10^{-5}\, \bar{r}_p \left(\dfrac{T}{M_A}\right)^{1/2}$	Capillary radius, temperature, and molecular weight
4.	Knudsen diffusion in a porous pellet	Effective Knudsen diffusivity D_{eKA}	$D_{eKA} = \dfrac{f_c}{\tau} D_{KA}$	Tortuosity factor τ, pellet porosity f_c
5.	Transition regime in a straight capillary	Combined diffusivity D_{cA}	$D_{cA} = D_{bA}\, \dfrac{\ln\left(\dfrac{1 - \alpha y_{AL} + R_D}{1 - \alpha y_{A0} + R_D}\right)}{\alpha(y_{A0} - y_{AL})}$	Bulk and Knudsen diffusivities from 1 and 3, and α from experiment or the Hoogschagen relation

6. Transition regime in a porous pellet	Effective diffusivity D_{eA}	$D_{eA} = D_{ebA} \dfrac{\ln\left(\dfrac{1-\alpha y_{AL}+R_{D_e}}{1-\alpha y_{A0}+R_{D_e}}\right)}{\alpha(y_{A0}-y_{AL})}$	Effective bulk and Knudsen diffusivities from 2 and 4, and α from experiment or the Hoogschagen relationship
7. Computational models			
a. Macropore–micropore model	Effective diffusivity D_{eA}	See Wakao and Smith (1962)	Bulk diffusivity, Knudsen diffusivities in micro- and macropores, molecular weights, macro- and micro porosities, and mean micro- and macropore radii
b. Rigorous equation accounting for pore size distribution	Effective diffusivity D_{eA}	$D_{eA} = \dfrac{k^p D_{bA}}{\alpha(y_{A0}-y_{AL})}\displaystyle\int_0^\infty \ln\left(\dfrac{1-\alpha y_{A0}+R_D}{1-\alpha y_{AL}+R_D}\right) f(r_p)\,dr_p$	Bulk and Knudsen diffusivities and pore size distribution; assume $k^p \approx 1/3$ (see Johnson and Stewart, 1965)

[a] $R_D = D_{bA}/D_{KA}$; $R_{D_e} = D_{ebA}/D_{eKA}$; $f(r_p)$ is volume of pores between r_p and $(r_p + dr_p)$ per unit volume of pellet; $\bar{r}_p$ = average pore radius; $\alpha = [1+(N_{bA}/N_{bB})] = [1-(\sqrt{M_B/M_A})]$ – Hoogschagen's relation; y_{A0}, y_{AL} = mole fractions of A at $\ell = 0$ and $\ell = L$, respectively; D_{eA} and D_{eA} are combined diffusivities in a straight capillary and pellet, respectively, but since various other factors are also involved in diffusion in a pellet, the diffusivity in this case is called effective diffusivity; V_i may be estimated from the tables of Le Bas or Fuller et al. (see Chapter 3).

pore structure, one for the space between particles in a pellet (usually the macropores) and the other for the pores within a particle (usually the micropores). *Multimodal structures* involving more than one macropore distribution are also possible (Cunningham and Geankopolis, 1968), but such complexity is almost never consistent with the quality of basic data that can be generated. Other pore structures have also been proposed, for example, the *pore network model* of Beekman and Froment (1982). Extensive testing of these models has shown that the parallel path model is slightly superior (Satterfield and Cadle, 1968a,b; Brown et al., 1969; Patel and Butt, 1974).

Effectiveness Factor

The resistance to diffusion (expressed in terms of an effective diffusion coefficient as defined before) has the effect of progressively reducing the concentration of the reactant molecule from the catalyst surface to the center. This leads to a lower reaction rate and to a lower value of the rate constant, that is,

$$k_a = \varepsilon k \tag{7.12}$$

where k_a is the *actual rate constant*, k is the *true* or *intrinsic rate constant*, and ε is commonly referred to as the catalyst *effectiveness factor* (or *utilization factor*). Note that because it is a codeterminant of the apparent rate, it is as important a factor as the true rate constant itself in the analysis and design of catalytic reactors. Several detailed treatments of the subject are available, for example, Petersen (1965), Aris (1975), Carberry (1976), Luss (1977), Doraiswamy and Sharma (1984), Froment and Bischoff (1990), and we restrict the treatment to a brief outline of the approaches and equations.

Isothermal Effectiveness Factors

First-order reaction in a sphere

Consider a simple first-order reaction

$$A \rightarrow \text{products}, \tag{7.5}$$

$$-r_A = k_v[A] \tag{7.13}$$

occurring in a spherical catalyst pellet, as shown in Figure 7.2. (Note that, for convenience, we have switched from rates based on catalyst weight to rates based on catalyst volume.) A mass balance on A at steady state gives

$$\frac{d^2[A]}{dr^2} + \frac{2}{r}\frac{d[A]}{dr} = \frac{(-r_{vA})}{D_{eA}} \tag{7.14a}$$

$$= \frac{k_v[A]}{D_{eA}} \text{ (first order)} \tag{7.14b}$$

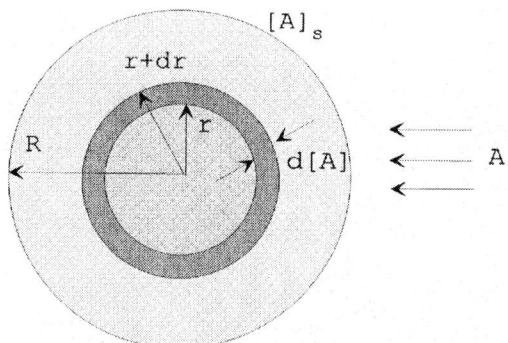

Figure 7.2 A differential element of a spherical pellet.

This can be recast in dimensionless form as

$$\frac{d^2[\hat{A}]}{d\hat{R}^2} + \frac{2}{\hat{R}}\frac{d[\hat{A}]}{d\hat{R}} = \phi_{s1}[\hat{A}] \tag{7.15}$$

where $[\hat{A}] = [A]/[A]_s$, $\hat{R} = r/R$, and

$$\phi_{s1} = R\sqrt{\frac{k_v}{D_{eA}}} \tag{7.16}$$

is known as the Thiele modulus for a first-order reaction in a sphere. The symbol ϕ represents the modulus in general and the subscripts specify pellet shape and reaction order. It is a measure of the relative rates of reaction and diffusion: low values denote chemical control, and high values diffusion control.

Equation 7.15 can be solved by specifying the boundary conditions at the surface and center. These are

Surface:

$$[\hat{A}] = 1, \tag{7.17a}$$

Center:

$$\frac{d[\hat{A}]}{d\hat{R}} = 0 \tag{7.17b}$$

and the solution is

$$[\hat{A}] = \frac{\sinh(\phi_{s1}\hat{R})}{\hat{R}\sinh\phi_{s1}} \tag{7.18}$$

from which the concentration can be computed as a function of the radial position for various values of the Thiele modulus.

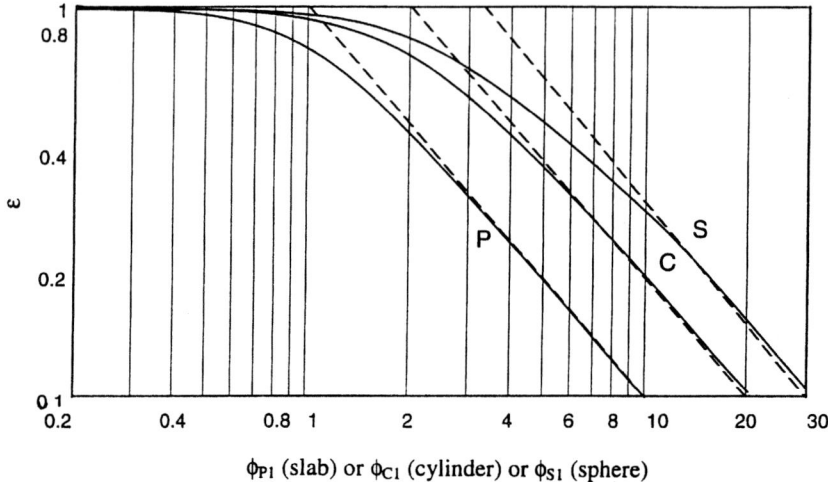

Figure 7.3 Effectiveness factors for a first-order reaction in a slab (P), cylinder (C), and sphere (S) as functions of the Thiele modulii for the three shapes (Aris, 1969).

We define an effectiveness factor as

$$\varepsilon = \frac{\text{(actual rate within the pellet based on average concentration)}}{\text{(rate based on surface conditions throughout the pellet)}}$$

$$= \frac{\frac{1}{R}\int_0^R [A]\,dr}{[A]_s} \tag{7.19}$$

Thus we merely combine Equations 7.18 and 7.19 and integrate between the limits $r = 0$ and $r = R$ to give

$$\varepsilon = \frac{3}{\phi_{s1}^2}(\phi_{s1}\coth\phi_{s1} - 1) \tag{7.20}$$

This is graphically displayed as curve S in Figure 7.3.

Generalization with respect to catalyst shape and reaction order

By procedures similar to that already presented for a sphere, we can derive expressions for other shapes as well. We consider two other shapes: a flat plate (or slab) and a cylinder. The equations for all the three shapes are given in Table 7.4, and the corresponding plots are included in Figure 7.3. Equations for a single pore (item 4 in the table) are similar to those for the flat plate.

Because the equations for the three shapes produce three different curves, it is desirable to formulate a single Thiele modulus which will be applicable to these and other shapes. It is also desirable to generalize the modulus to include any

Table 7.4 Intraphase diffusion parameters and equations for a first-order reaction ($A \rightarrow$ products) for different catalyst shapes

Shape and definition	Thiele modulus	Generalized Thiele modulus ϕ_1 based on Λ_0	Modified distance parameter Λ_0	Equation for ε	Equation for ε based on generalized modulus ϕ_1
1. Infinite slab a. One end open 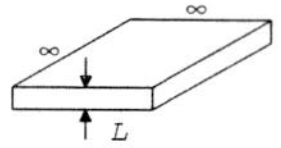	$\phi_{p1} = L\left(\dfrac{k_v}{D_e}\right)^{1/2}$	$\Lambda_0\left(\dfrac{k_v}{D_e}\right)^{1/2}$	L	$\varepsilon = \dfrac{1}{\phi_{p1}}(\tanh \phi_{p1})$	$\varepsilon = \dfrac{1}{\phi_1}(\tanh \phi_1)$
b. Both ends open			$\dfrac{L}{2}$		
2. Sphere 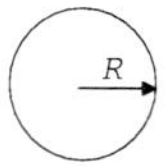	$\phi_{s1} = R\left(\dfrac{k_v}{D_e}\right)^{1/2}$	$\Lambda_0\left(\dfrac{k_v}{D_e}\right)^{1/2}$	$\dfrac{R}{3}$	$\varepsilon = \dfrac{3}{\phi_{s1}}\left(\dfrac{1}{\tanh \phi_{s1}} - \dfrac{1}{\phi_{s1}}\right)$	$\varepsilon = \dfrac{\phi_{s1}}{\phi_1}\left(\dfrac{1}{\tanh(3\phi_1)} - \dfrac{1}{3\phi_1}\right)$
3. Infinite cylindrical rod (sealed ends) 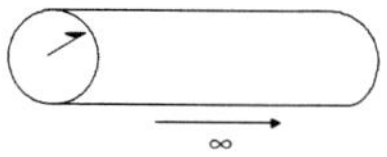	$\phi_{c1} = R\left(\dfrac{k_v}{D_e}\right)^{1/2}$	$\Lambda_0\left(\dfrac{k_v}{D_c}\right)^{1/2}$	$\dfrac{R}{2}$	$\varepsilon = \dfrac{2}{\phi_{c1}}\left(\dfrac{I_1(\phi_{c1})}{I_0(\phi_{c1})}\right)$ where I_0 and I_1 are modified Bessel functions of the first kind of order 0 and 1, respectively	$\varepsilon = \dfrac{1}{\phi_1}\left(\dfrac{I_1(2\phi_1)}{I_0(2\phi_1)}\right)$
4. Single pore (open ends) 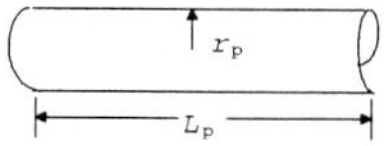	$\phi_{p1} = L_p\left(\dfrac{k_v}{D_e}\right)^{1/2}$ $= \left[L_p\left(\dfrac{2k_s}{r_p D_e}\right)^{1/2}\right]$			Same as for infinite slab with L replaced by L_p	

order n. The equation obtained for a first-order reaction is

$$\frac{d^2[\hat{A}]}{d\hat{x}^2} + \frac{s}{\hat{x}}\frac{d[\hat{A}]}{d\hat{x}} = \phi_1^2[\hat{A}] \qquad (7.21)$$

where $\hat{x} = x/x_0$ is the *normalized length cordinate*; x is a *generalized length coordinate* (r for the sphere or cylinder, and ℓ for the flat plate); ϕ_1 is a modulus expressed in terms of a *new* shape parameter,

$$\Lambda_0 = \frac{\text{volume of shape}}{\text{area of shape}} \qquad (7.22)$$

with $\Lambda_0 = L/2$ for the plate, $R/2$ for the cylinder, and $R/3$ for the sphere; and s is a shape constant, with values of 2 for the sphere, 1 for the cylinder, and 0 for the flat plate. We now expand the definition of the modulus to include the effect of reaction order as follows:

$$\phi = \Lambda_0\sqrt{\frac{(n+1)k_v[A]_s^{n-1}}{2D_{eA}}} \qquad (7.23)$$

The effectiveness factor equations obtained by solving the shape-generalized Equation 7.21 for a first-order reaction ($\phi = \phi_1$) are included in Table 7.4.

Delineating the control regions

THE WEISZ MODULUS (A PRACTICALLY USEFUL QUANTITY)

Recall that for calculating the Thiele modulus a knowledge of the rate constant is needed, which requires elaborate kinetic studies under conditions free of diffusional effects. A practically more useful modulus based on *observable quantities* can be obtained by recasting Equation 7.16 in the form

$$\phi^2 = R^2\frac{(\text{true rate})}{D_{eA}[A]_s} \qquad (7.24)$$

and defining the new modulus as

$$\phi_a = R^2\frac{(\text{actual rate})}{D_{eA}[A]_s} = R^2\left[\frac{(\text{true rate})}{D_{eA}[A]_s}\right]\varepsilon = \phi^2\varepsilon \qquad (7.25)$$

This modulus, named for Weisz who first proposed it along with Prater in 1954, can easily be prepared from the more common $\varepsilon - \phi$ plot.

DELINEATION OF REGIMES

Figure 7.4 shows a plot of the effectiveness factor against the Weisz modulus ϕ_a using the generalized length parameter Λ_0 and is hence valid for all shapes. The ε versus ϕ plot is also shown in the figure. Notice that the two curves coincide with each other except for a small range in the shaded region. Three regions can be identified: chemical control, diffusion control, and mixed control (shaded). These are clearly marked on the figure with corresponding values of ϕ_1 and ϕ_a.

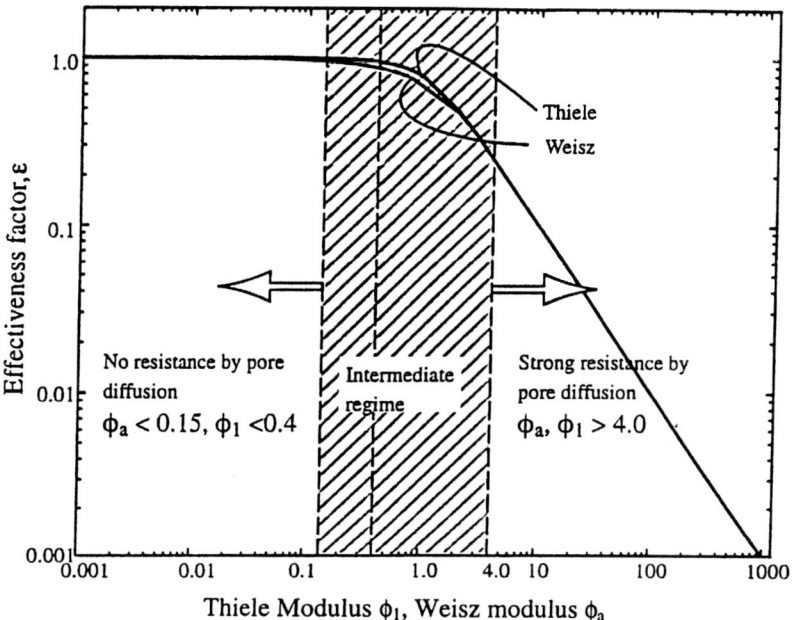

Figure 7.4 Effectiveness factor as a function of the Weisz (observable) modulus ϕ_a (and also the Thiele modulus ϕ_1).

Extension to bimolecular reactions

The analysis just presented was restricted to reactions with a single diffusing reactant. Reactions with more than one diffusing component have also been considered (see, e.g., Schneider and Mitschka, 1965, 1966a,b; Maymó and Cunningham, 1966; Schneider, 1975; Xu and Chuang, 1997). A class of reactions of particular importance in organic technology is the bimolecular reaction

$$A + B \leftrightarrow R + S \quad [7.6]$$

with the rate equation

$$-r_{vA} = k_v\{[A][B] - ([R][S])/K\} \quad (7.26)$$

The effect of diffusion in such a reaction can be determined by writing the concentrations of B, R, and S in terms of the concentration of A. This allows us to use the mass balance of Equation 7.14a with the rate term written exclusively in terms of the concentration of A within the pellet (Schneider and Mitschka, 1966b; Xu and Chuang, 1997):

$$-r_{vA} = (\alpha[A]^2 + \beta[A] - \gamma) \quad (7.27)$$

where

$$\alpha = k_v(\delta_B - \delta_R \delta_S/K)$$

$$\beta = k_v\left([B]_s - \delta_B[A]_s - \frac{\delta_R[S]_s}{K} + \frac{2\delta_R\delta_S[A]_s}{K} - \frac{\delta_S[R]_s}{K}\right)$$

$$\gamma = \frac{k_v}{K}[[A]_s(\delta_R\delta_S[A]_s - \delta_S[R]_s - \delta_R[S]_s) + [R]_s[S]_s]$$

$$\delta_j = \frac{v_j}{v_A}\frac{D_{eA}}{D_{ej}}, \quad j = B, R, \text{ or } S$$

(7.28)

Note that the diffusivities involved here are multicomponent diffusivities, which can be calculated by the method described in Chapter 3.

Using the above expressions, Equation 7.14a can be nondimensionalized to give

$$\frac{d^2y}{dx^2} - \frac{2}{1-x}\frac{dy}{dx} = \phi'^2 \frac{y(y+M)}{(1+M)}$$

(7.29)

where

$$y = \frac{[A] - [A]^*}{[A]_s - [A]^*}$$

$$x = \frac{(R-r)}{R}$$

$$M = \frac{[2[A]^* + (\beta/\alpha)]}{([A]_s - [A]^*)}$$

(7.30)

$$N = \alpha(1+M)([A]_s - [A]^*)^2 = -r_{vA}(y=1)$$

$$\phi' = R\sqrt{\frac{N}{[D_{eA}([A]_s - [A]^*)]}} \quad \text{(equivalent to the Weisz modulus)}$$

and [A]* is the equilibrium concentration of A. Now Equation 7.29 can be solved for the boundary conditions $x = 0, y = 1$, and $x = 1, dy/dx = 0$, and the effectiveness factor can be obtained from Equation 7.19 by using the flux at the pellet surface as the numerator and the surface reaction as the denominator.

The results of the numerical solution are plotted as ε versus $\phi'/3$ with M (defined in Equation 7.30) as parameter in Figure 7.5. This figure is particularly useful for esterification reactions carried out over catalysts like ion-exchange resins such as Amberlyst 15. Several experimental studies on such systems have been reported, for example, Kaiser et al. (1962), Beasley and Jakovac (1984), Leung et al. (1986), and Patwardhan and Sharma (1990). In a more recent study (Ihm et al., 1996), the macroreticular resin (see Chapter 6) is regarded as a two-phase system, one consisting of microspheres of uniform size and the other of pores formed by the space between the microspheres. Then the reaction is analyzed in terms of the micro- and macroeffectiveness

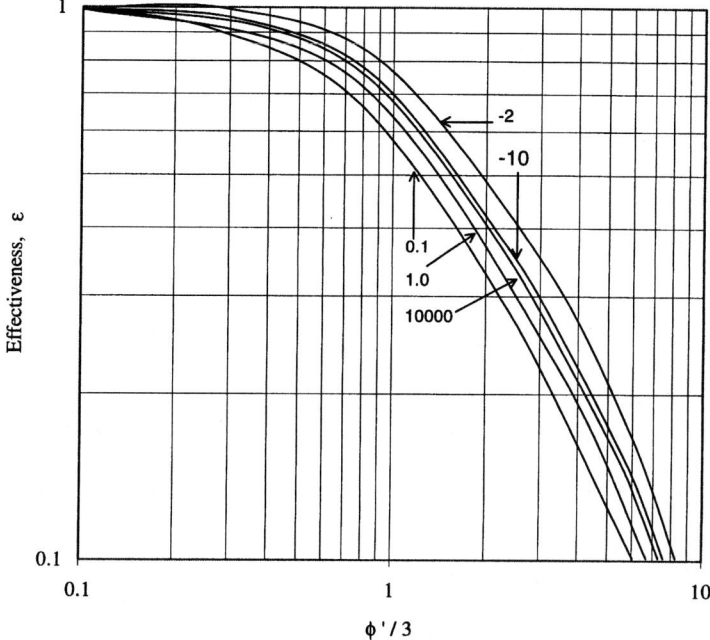

Figure 7.5 Effectiveness factor plots for a biomolecular reaction (redrawn from Xu and Chuang, 1997).

factors corresponding to these two phases and the fraction of active sites on the surface layer.

Nonisothermal Effectiveness Factors

The generation of heat inside a pellet due to reaction and its transport through the pellet can greatly affect the reaction rate. For endothermic reactions, there is a fall in temperature within the pellet. As a result, the rate falls, thus augmenting the retarding effect of mass diffusion. On the other hand, for exothermic reactions, there is a rise in temperature within the pellet. This leads to an increase in rate which can more than offset the decrease due to lowered concentration. Thus the effectiveness factor can actually be greater than one.

In analyzing the effect of thermal diffusion within the pellet, we use the following fundamental heat balance:

$$D_{eA}(-\Delta H_r)([A]_s - [A]) = \lambda(T - T_s) \tag{7.31}$$

where $(-\Delta H_r)$ is the heat of reaction and λ the thermal conductivity of the pellet. This can be expressed in dimensionless form as

$$\frac{\Delta T}{T_s} = \beta_m(1 - [\hat{A}]) \tag{7.32}$$

where

$$\beta_m = \frac{(-\Delta H_r) D_{eA}[A]_s}{\lambda T_s} \tag{7.33}$$

represents the maximum temperature rise, that is, the rise when the inside concentration $[\hat{A}](=[A]/[A]_s)$ is zero. Another commonly used group is the Arrhenius parameter expressed at the surface temperature

$$\alpha_s = \frac{E}{R_g T_s} \tag{7.34}$$

Now we consider a differential section of a pellet of any shape and write the following continuity equation:

$$\frac{d^2[A]}{dx^2} + \frac{s}{x}\frac{d[A]}{dx} = [-r_{vA}([A], T_s)] \tag{7.35}$$

In dimensionless form, this becomes

$$\frac{d^2[\hat{A}]}{d\hat{x}^2} + \frac{s}{\hat{x}}\frac{d[\hat{A}]}{d\hat{x}} = \phi_1^2[\hat{A}]$$

$$= (\phi_1^2)_s \exp\left[\alpha_s \beta_m \frac{1-[\hat{A}]}{1+\beta_m(1-[\hat{A}])}\right][\hat{A}] \tag{7.36}$$

where $(\phi_1)_s$ represents the Thiele modulus for a first-order reaction at the surface temperature. Solutions can be obtained as effectiveness factor plots with β_m and α_s as parameters. A large number of studies have been reported on various aspects of the solutions including the occurrence of multiple steady states (see Aris, 1975).

The plots of Weisz and Hicks (1962) are reproduced in Figure 7.6. The nature of the curves at high values of β_m suggests multiple solutions. In other words, the reaction can occur at three steady states, two stable and one unstable. We shall not be concerned with this aspect of effectiveness factors, but it is instructive to note that ε given by one of the solutions in the multiple steady-state region can be orders of magnitude higher than unity. Instabilities of this kind are essentially local in nature, and are briefly considered in Chapter 12. The reactor as a whole can also exhibit multiple steady states, a feature that is briefly treated in Chapter 13.

Multicomponent Diffusion

The mathematical description of diffusion of more than one component is a complex problem and not germane to our subject. However, with the present trend toward increasing use of solid catalysts in organic synthesis, many situations arise in which multicomponent diffusion is involved. This problem was considered in Chapter 3 and two relatively simple expressions were given (Equations 3.40 and 3.41) of that chapter] for estimating the diffusivity $\bar{D}_j$ of a

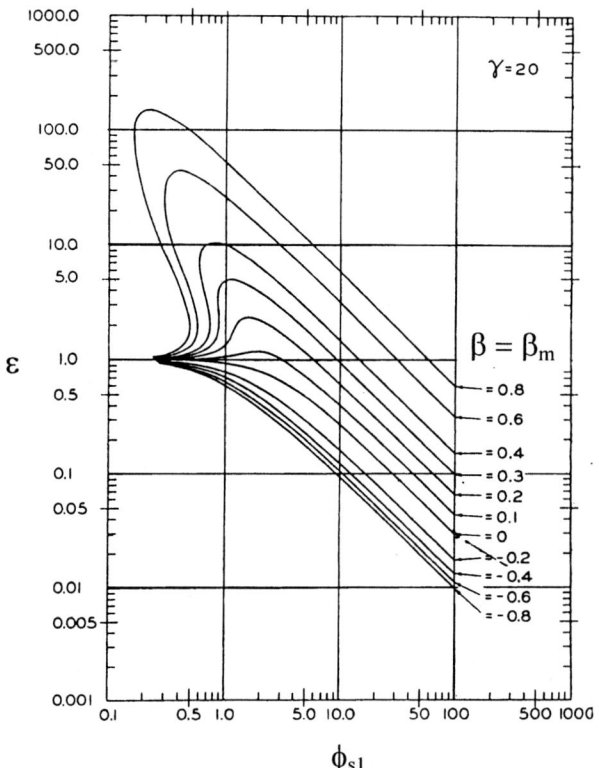

Figure 7.6 Effectiveness factor with a first-order reaction in a spherical nonisothermal catalyst pellet (Weisz and Hicks, 1962).

component j in a mixture of N components. Then the effective diffusivity can be found from $D_{eA} = \rho_c \bar{D}_j / \tau$

Miscellaneous Effects

A number of factors can influence the effectiveness factor. Some of them are particle size distribution in a mixture of particles/pellets, change in volume upon reaction (Weekman and Gorring, 1965; Kehoe and Aris, 1973; Haynes, 1978; Jayaraman et al., 1983), pore shape and constriction (such as ink-bottle type pores) (Chu and Chon, 1970; Ferraiolo et al., 1973), radial and length dispersion of pores (Schmalzer, 1969), micro-macro pore structure (Tartarelli et al., 1970; Kulkarni et al., 1981; Jayaraman et al., 1983), flow regime (such as bulk or Knudsen) (Scott, 1962; Otani et al., 1965), surface diffusion (Aris, 1971; Jayaraman and Kulkarni, 1981), nonuniform environment around a pellet (Bischoff, 1968; Copelowitz and Aris, 1970), dilution of catalyst bed or pellet (Ruckenstein, 1970; Varghese and Wolf, 1980), distribution of catalyst activity in a pellet (see, in particular, the review by Gavriilidis and Varma, 1993), transverse diffusion (Bischoff, 1965), the external surface of the catalyst (Goldstein and

Extension to LHHW Kinetics

In the developments already presented, the rate equations used were simple power law expressions. But, as discussed previously, catalytic rate equations are much more complex and often require the use of LHHW models. Many attempts have been made to incorporate these models in the analysis (e.g., Chu and Hougen, 1962; Krasuk and Smith, 1965; Roberts and Satterfield, 1965, 1966; Hutchings and Carberry, 1966; Schneider and Mitschka, 1966a,b; Kao and Satterfield, 1968; Rajadhyaksha et al., 1976; see in particular: Aris, 1975; Luss, 1977). Clearly, graphical representation becomes cumbersome when a large number of adsorbed species is involved. However, the problem is quite tractable where only one species is adsorbed.

Thus consider the simple reaction A → R with the rate equation

$$-r_{wA} = \frac{k_w K_A[A]}{(1 + K_A[A])} \tag{7.37}$$

where only A is adsorbed. If there is an external film resistance, the surface concentration $[A]_s$ should be used (see section on the effect of external mass and heat transfer). By incorporating this expression recast in terms of r_{vA} and k_v for the rate in Equation 7.14a, curves for different values of $K_A[A]_s$ can be produced, as shown in Figure 7.7. Notice that the lower bound of these curves corresponding to decreasing values of $K_A[A]_s$ is the first-order curve (i.e., for $K_A[A]_s = 0$), and the upper bound corresponding to increasing values is the zero-order curve (i.e., for $K_A[A] = \infty$).

The treatment can be extended to reactions involving more than one adsorbed species provided that the number of additional parameters (i.e., $K_B[B]$, $K_R[R]$, etc.) can be reduced, preferably to one. An elegant method has been proposed by Roberts and Satterfield (1965, 1966) in which precisely this is accomplished: all of the adsorption parameters are combined into a single parameter Kp_{As}. This method has been further generalized (Rajadhyaksha et al., 1976) by extending it to nonisothermal pellets, but graphical representation in this case becomes unwieldy because now two additional (thermal) parameters α_s and β_m must be accommodated.

Extension to Complex Reactions

In Chapter 5 we considered the mathematical treatment of complex reactions. It was also shown how some simpler reaction schemes such as parallel, series, and series-parallel reactions are amenable to analytical solution. In this section we consider the role of pore diffusion in these complex reactions. We omit the mathematical details and in Table 7.5 present the salient features of the effect of pore structure, monomodal or bimodal distribution, on yield and conversion

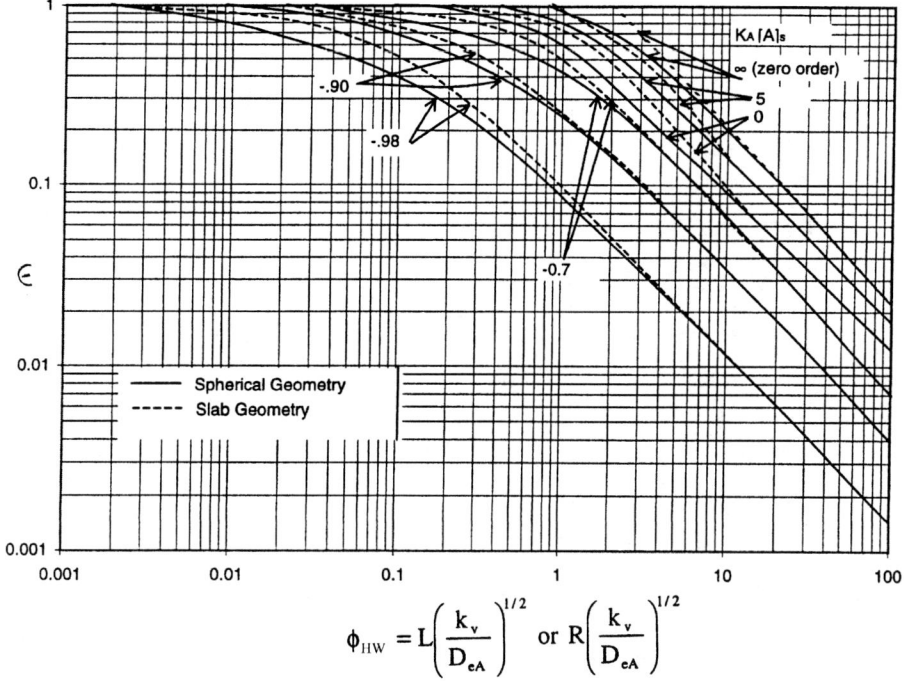

$$\phi_{HW} = L\left(\frac{k_v}{D_{eA}}\right)^{1/2} \text{ or } R\left(\frac{k_v}{D_{eA}}\right)^{1/2}$$

Figure 7.7 Effectiveness factor plots for LHHW kinetics. Negative values represent desorption of product with $K_A p_{As} = K p_{As}$ and $k_v = k_m$, a new parameter (adapted from Roberts and Satterfield, 1965).

in a few selected types of complex reactions. Product R (bold in the table) is considered the desired product.

Effectiveness Factors for Liquid-Phase Reactants

Because several organic reactions, particularly those used in medium and small volume chemicals manufacture, are carried out in the liquid phase, diffusion of liquids in solid catalysts becomes an important consideration. Although the procedures for estimating effectiveness factors given in the preceding sections apply equally to liquid-phase reactants, the actual values of the diffusion coefficients to be used often pose considerable uncertainty.

The diffusivity of a liquid-phase reactant depends on its concentration (see Ruthven and Loughlin, 1971) and on the presence of other sorbate species. Although the concentration effect can be ignored in dilute solutions (as in gas-phase diffusion at all concentrations), it needs to be accounted for in concentrated solutions. Once the diffusivity is determined (usually by ignoring the effect of concentration), the effective diffusivity is best calculated from the relationship (Ternan, 1987, 1996)

$$\frac{D_{eA}}{D_{bA}} = \frac{(1-\lambda)^2}{(1+a\lambda)} \tag{7.38}$$

Table 7.5 Effect of pore diffusion on selectivities and yields in the more common classes of complex organic reactions

Reaction **R** = desired product	Pore structure	Intrinsic selectivity and yield	Actual selectivity and yield	Main features
1. Independent $A \xrightarrow{1} R$ $B \xrightarrow{2} S$	Monodispersed	$s = \dfrac{k_{v1}}{k_{v2}}$	$s_a = \left(\dfrac{s}{\alpha_D}\right)^{1/2}$	Greater diffusional resistance for reaction 2 enhances Y_{Ra}
2. Parallel $A \diagup^{1,m \ \ R}_{2,n \ \ S}$	Monodispersed	$s = \dfrac{k_{v1}}{k_{v2}}$ $Y_R = \dfrac{1}{1 + p_{nm}}$ where $p_{nm} = \dfrac{k_{v2}}{k_{v1}[A]_s^{m-n}}$ $(m > n)$	$s_a = s^{1/2}$ $\dfrac{Y_{Ra}}{Y_R} = \dfrac{n+1}{2m - n + 1}$ (Pawlowski, 1961; Roberts, 1972)	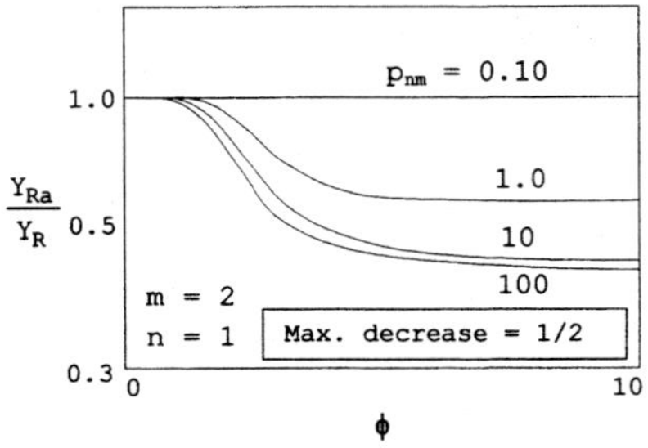
3. Consecutive $A \xrightarrow{1} R \xrightarrow{2} S$	Monodispersed	$s = \dfrac{k_{v1}}{k_{v2}}$, $\mathfrak{I} = \dfrac{1}{s}$	$Y_R = \dfrac{s}{1-s}[(1-X_A)^s - (1-X_A)]$ $\mathfrak{I} = \dfrac{1}{s^{1/2}}$	
	Bidispersed	$s = \dfrac{k_{v1}}{k_{v2}}$, $\mathfrak{I} = \dfrac{1}{s}$	$\mathfrak{I} = \dfrac{1}{s^{1/4}}$ (Carberry, 1962; Doraiswamy and Sharma, 1984)	

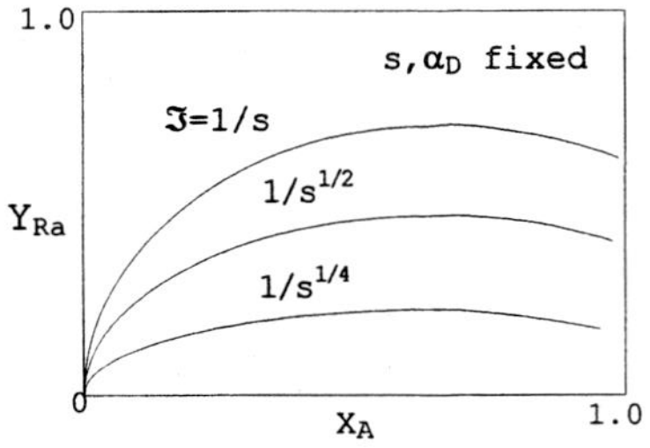

4. Parallel-Consecutive				
(a)  A →₁ R →₂ S, A →₃ (R branch)	Monodispersed	$s_1 = \dfrac{k_{v1}}{k_{v2}}$	$Y_{Ra} = f(\phi, s_n)$ (see Wirges and Rähse, 1975, for full equation).	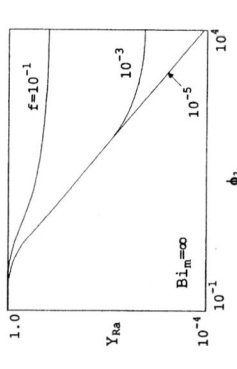 (plot of Y_{Ra} vs X_A, s_n fixed, curves labeled 0.2, 0.1, arrow showing ϕ)
(b)  A →₁ R, with branches to S and T	Monodispersed	$s_{13} = \dfrac{k_{v1} + k_{v3}}{k_{v2}}$	$s_n = s_1$ for $k_{v3} = 0$ $s_n = s_{13}$ for $k_{v3} \neq 0$	
(c) A + B →₁ R; R + B →₂ S	Monodispersed	$Y_{RB} = \dfrac{[R]}{[B]_0}$ $Y_{RA} = \dfrac{[R]}{[A]_0}$ $= Y_{RB}\left(\dfrac{[B]_0}{[A]_0}\right)$	No analytical solution for $(Y_{RB})_a$ or $(Y_{RA})_a$. Numerical solution by Wirges and Rähse (1975)	(plot of $(Y_{RB})_a$ vs X_A, curves labeled 0.1, 10, axis $\frac{[B]_0}{[A]_0}$)
(d) Same as (a)	Monodispersed with the external surface constituting a finite fraction (f) of the total surface area of the catalyst	Same as for (a) or (b)	No analytical solution. Numerical solution by Varghese et al. (1978)	(plot of Y_{Ra} vs ϕ_1, curves labeled $f = 10^{-1}$, 10^{-3}, 10^{-5}, $Bi_m = \infty$) Y_{Ra} is a strong function of f. It increases with Bi_m. Also

where $\lambda = d_m/d_p$, the ratio of molecular to particle diameters, and a is an adjustable parameter obtained from the calculated values of λ and D_{bA}.

For zeolites, the following relationship can be used to directly estimate the concentration dependence of the effectiveness factor for a first-order reaction (involving single-component diffusion) in a flat plate (Ruthven, 1972):

$$\varepsilon = \left(\frac{2}{[\hat{A}]^*\phi_{p1}}\right)[-\hat{A} - \ln(1 - \hat{A})^{1/2}] \qquad (7.39)$$

where $[\hat{A}]^*$ is the ratio of sorbate concentration at the surface to that at saturation and ϕ_{p1} is the Thiele modulus based on the diffusivity at the surface. In general, analytical solutions obtained for mildly concentration-dependent diffusivities have shown that the effect is minor except in the strongly diffusion-controlled regime corresponding to $\phi \to$ infinity (Pereira and Varma, 1978).

Interference by other sorbate species in multicomponent diffusion can be far more severe for zeolites than for other catalysts such as ion-exchange resins. Thus it is unfortunate that, although a number of studies have been reported on single-component diffusion in zeolites, for example, Satterfield and Cheng (1971), Choudhary and Singh (1986), Choudhary and Mamman (1990), Choudhary et al. (1989, 1992), those on multicomponent diffusion are relatively few. A particularly important feature of multicomponent diffusion appears to be that the sorption of a slow diffusing component in a multicomponent system is drastically reduced in the presence of a fast diffusing species. For example, the sorption of isopropyl benzene, a slow diffusing component, is further reduced (in fact, quite drastically) in the presence of p-xylene or n-isopropyl benzene (Choudhary et al., 1989).

Effectiveness Factors in Supported Liquid-Phase/Immobilized Catalysts

The use of supported liquid-phase catalysts (SLPCs) in organic reactions was mentioned in Chapter 6. The diffusion-reaction problem in such catalysts is considerably more complicated than that for solid catalysts (Rony, 1969; Abed and Rinker, 1973; Livbjerg et al., 1974, 1976). In a more recent analysis of SLPC systems, it has been shown (Datta and Rinker, 1985; Datta et al., 1985) that a critical parameter is the ratio of the effective diffusivity of a liquid-loaded pellet to that of a dry pellet $(D_{eA,L}/D_{eA})$. This ratio is a strong function of the liquid loading q and the gas-liquid partition coefficient H_A of reactant A, as shown in Figure 7.8. Thus the effectiveness factor is a function of liquid loading.

An emerging system similar to the preceding employs what has come to be known as triphase catalysis, in which a phase-transfer catalyst is immobilized on a solid support for use in a liquid-liquid reacting system. In view of the potential importance of such a system, it is considered at greater length in Chapter 19 on phase-transfer catalysis.

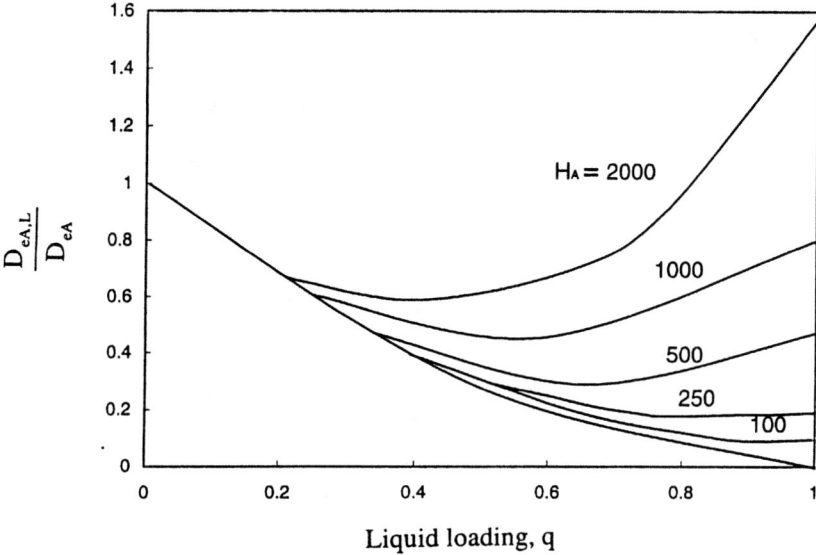

Figure 7.8 The ratio of the effective diffusivity in a liquid-loaded pellet to the effective diffusivity in the dry pellet vs. the liquid loading (Datta et al., 1985).

EFFECT OF EXTERNAL MASS AND HEAT TRANSFER

Uniform and Nonuniform Films

A uniform surface is one that is *equiaccessible* to fluid elements all around it. Common examples are a rotating disk in which fluid in laminar flow is uniformly distributed over the disk surface due to its rotation and the internal surface of a tube. A less common example is the region of an obstruction placed in a tube such that it is impinged upon by the central layers of a flowing fluid in laminar flow with a parabolic velocity profile (see Balaraman et al., 1980). Extensive theoretical studies have been carried out on these configurations, but we shall be concerned only with surfaces of packed catalyst pellets, which are clearly nonuniform. Fortunately, by assuming that such a surface is uniform, errors of less than 10% are introduced (Carra, 1972).

Another assumption made in all analyses of external transport is that the catalyst surface is smooth without any kinks or crevices. If such imperfections are present, it is likely that there would be pockets of stagnant fluid in those regions that would adversely affect mass transfer. Correlations for the mass transfer coefficient do not account for this nonideality, leading sometimes to erroneous conclusions regarding the role of external mass transfer (Pignet and Schmidt, 1974; Loffler and Schmidt, 1975; Hori and Schmidt, 1975).

Notwithstanding the wealth of theory about such external transport processes, the most accurate equations available for estimating transport coefficients are empirical.

Empirical Correlations

All empirical expressions for estimating mass and heat transfer coefficients are based on experimental data correlated in terms of dimensionless numbers. The chief experimental variables are fluid flow rate, particle dimension, and fluid properties. Based on these, the following groups are normally used:

$$\text{Sherwood number} \quad Sh = \frac{k'_G R}{D} = \frac{\text{mass diffusivity}}{\text{molecular diffusivity}}$$

$$\text{Nusselt number} \quad Nu = \frac{hR}{\lambda} = \frac{\text{total heat transfer}}{\text{conductive heat transfer}}$$

$$\text{Prandtl number} \quad Pr = \frac{C_p \mu}{\lambda} = \frac{\text{kinematic viscosity}}{\text{thermal diffusivity}} \quad (7.40)$$

$$\text{Schmidt number} \quad Sc = \frac{\mu}{\rho D} = \frac{\text{kinematic viscosity}}{\text{molecular diffusivity}}$$

$$\text{Reynolds number} \quad Re' = \frac{d_p G}{\mu} = \frac{\text{inertial forces}}{\text{viscous forces}}$$

where k'_G and h are the required mass and heat transfer coefficients and R represents the characteristic dimension of the shape (can be d_T, d_p, or L). The Sherwood and Nusselt numbers (which may also be regarded as the ratio of the time scale of molecular or thermal diffusion to the time scale of global mass or heat transfer) are also known as Biot numbers for mass and heat transfer, respectively.

Two approaches have usually been taken based on (1) correlating the mass transfer number Sh with Re' and Sc, and the heat transfer number Nu with Re' and Pr and (2) defining a composite dimensionless factor j (j_d for mass and j_h for heat transfer) and correlating it with Re':

$$j = \frac{(\text{mass or heat transfer group})}{(\text{flow group})(\text{property group})} \quad (7.41)$$

giving

$$j_d = \frac{Sh}{Re' Sc}, \quad j_h = \frac{Nu}{Re' Sc} \quad (7.42)$$

based on which correlations of the form

$$j_d \text{ (or } j_h\text{)} = a Re'^b \quad (7.43)$$

have been developed. Note that b is always negative in this form of equation and that $j_h = 0.8 j_d$.

Numerous correlations have been proposed over the years. It is difficult to pick certain correlations as the best, even for limited ranges of variables. The following correlations have been found useful.

Medium and high Re'

$$j_d = 1.24 \left(\frac{Re'}{1-f_B}\right)^{-0.39} \quad (7.44)$$

$$50 < Re' < 5000, \ 0.3 < f_B < 0.5, \ 0.6 < Sc < 2000$$

$$Nu = 2.0 + 1.8 Re' Pr^{-1/3}, \ Re' > 100 \quad (7.45)$$

Note: Equation 7.44 can also be used for heat transfer through the conversion factor $j_h = 0.8 j_d$, but Equation 7.45 is preferred.

Low Re'

$$j_d = 0.41 (Re')^{-0.06}, \ 2 < Re' < 100 \quad (7.46)$$

$$Sh = \frac{Nu}{0.8} = 0.7 Re', \ Re' < 2.0 \quad (7.47)$$

External Effectiveness Factor

It is useful to define an *external effectiveness factor* along the lines of the effectiveness factor ε for diffusion within the solid, which may now be more appropriately called the "internal effectiveness factor."

$$\begin{pmatrix} \text{external} \\ \text{effectiveness} \\ \text{factor } \eta \end{pmatrix} = \frac{\begin{pmatrix} \text{actual rate in the} \\ \text{presence of external} \\ \text{diffusional resistance} \end{pmatrix}}{\begin{pmatrix} \text{rate under conditions} \\ \text{where the surface and} \\ \text{bulk concentrations} \\ \text{are the same} \end{pmatrix}} \quad (7.48)$$

A simple mathematical formulation of this definition for a first-order isothermal reaction can be obtained by the addition of resistances, that is,

$$\frac{1}{k_a} = \frac{1}{k_v \eta} = \frac{1}{k_v} + \frac{1}{k'_G a} \quad (7.49)$$

giving

$$\eta = \frac{1}{1 + (k_v/k'_G a)} = \frac{1}{1 + Da_1} \quad (7.50)$$

where k_a is the observed rate constant ηk_v and $k_v/k'_G a$ is the Damköhler number for a first-order reaction (Da_1).

As will be shown later, heat transfer is considerably more important than mass transfer in the external film. Detailed theoretical analyses of the external film problem are available (Carberry and Kulkarni, 1973; Carberry, 1975) and will not be considered here.

COMBINED EFFECTS OF INTERNAL AND EXTERNAL DIFFUSION

Thus far we have considered the two effects of internal and external diffusion separately. The combined effects can be accounted for as follows.

Solve Equation 7.21 with a modified surface boundary condition which accounts for the diffusional resistance across the fluid film on the surface and hence gives the true surface concentration $[A]_s$. Thus, the boundary condition will not be $[A]_b = [A]_s$ but must be modified to (for a flat plate)

$$\frac{D_{eA}}{L}\left(\frac{d[A]}{d\ell}\right) = k'_G([A]_b - [A])$$

or

$$[\hat{A}] = [\hat{A}]_b - \frac{1}{Bi_m}\frac{d[\hat{A}]}{d\ell} \tag{7.51}$$

The solution (for $s = 0$) is then

$$\varepsilon = \frac{\tanh \phi_{p1}}{\phi_{p1}\left(1 + \frac{\phi_{p1}\tanh \phi_{p1}}{Bi_m}\right)} \tag{7.52}$$

A practical way of plotting this equation is to use the observable quantity $\phi_a = \varepsilon \phi_1^2$ instead of ϕ_1, where ϕ_1 is independent of shape. Such a plot is shown in Figure 7.9.

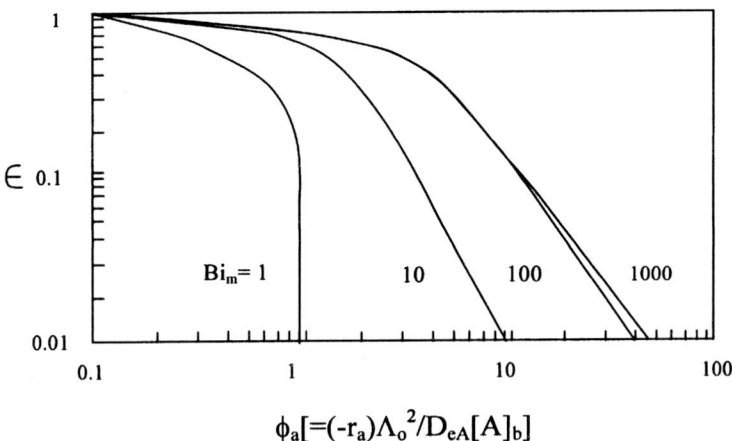

Figure 7.9 Effectiveness factor vs. Weisz modulus for different Biot numbers (Carberry, 1976).

RELATIVE ROLES OF MASS AND HEAT TRANSFER IN INTERNAL AND EXTERNAL DIFFUSION

Gas-Phase Reactants

An appreciation of the relative magnitudes of heat and mass transfer effects in internal and external diffusion is useful. A measure of the relative magnitudes is the ratio of the mass to heat Biot numbers:

$$B = \frac{Bi_m}{Bi_h} = \frac{k'_G L / D}{hL/\lambda} = \begin{bmatrix} \frac{\Delta C_{\text{int}}}{\Delta C_{\text{ext}}} \\ \frac{\Delta T_{\text{int}}}{\Delta T_{\text{ext}}} \end{bmatrix} = \begin{bmatrix} \left(\frac{\Delta C}{\Delta T}\right)_{\text{int}} \\ \left(\frac{\Delta C}{\Delta T}\right)_{\text{ext}} \end{bmatrix} \quad (7.53)$$

Because this ratio usually has values in the range of 10–500 for gas-solid systems, it may be concluded that ΔC_{int} and ΔT_{ext} must be very high. In other words, heat transfer would be the controlling resistance externally and mass transfer internally. This can be understood by considering a highly exothermic reaction for which the heat generated within a pellet is transported quickly enough to the surface, but further dissipation across the film is slow, leading to increased surface temperatures and to enhanced reaction rates.

A quantitative analysis of the relative magnitudes of the temperature gradients across the external film and within the pellet leads to a very useful relationship based only on observable quantities (Carberry, 1975):

$$\frac{\Delta T_{\text{ext}}}{\Delta T_{\text{ov}}} = \frac{B\eta Da}{1 + \eta Da(B-1)} \quad (7.54)$$

where ηDa for a first-order reaction is $k_v a / k'_G a = (-r_a)/k'_G a[A]_b$ (an observable quantity). Figure 7.10, which is a graphical representation of this equation,

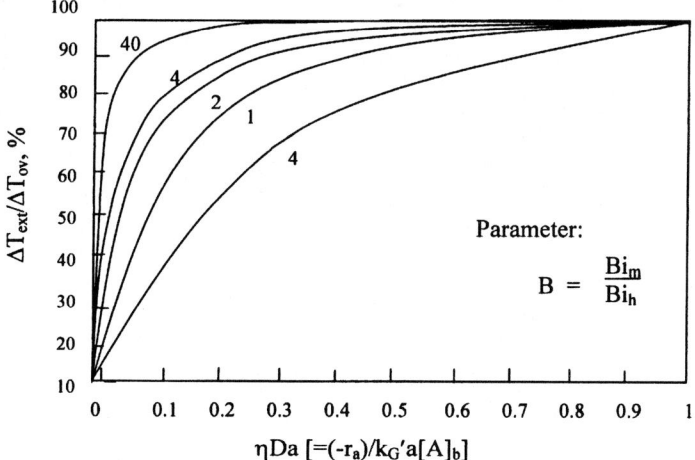

Figure 7.10 Ratio of external to total ΔT vs. the observable quantity εDa for solid-catalyzed reactions, with the ratio of mass to heat transfer Biot numbers B as parameter (Carberry, 1975).

clearly shows a marked rise in the external gradient with an increase in the ratio of Biot numbers.

Liquid-Phase Reactants

When the reactants are liquids (as in many organic reactions), the values of B are much less than one, and hence the conclusions are quite the converse of those for gas-phase reactants: the major fraction of the temperature gradient resides within the solid, whereas the concentration gradient is largely confined to the external film. Indicative value ranges for gas-solid and liquid-solid systems are given in Table 7.6. This is an important table in analyzing liquid-solid catalytic reactions (common in small and medium scale organic syntheses).

REGIMES OF CONTROL

As the controlling regime changes, the values of the kinetic parameters E and n also change. The limiting values of E for different regimes are indicated in Figure 7.11 which is a representation of these changes on an Arrhenius plot for reactions with nonnegligible heat effects. It is an expanded version of one of the earliest plots of this kind for reactions with negligible heat effects (Vaidyanathan and Doraiswamy, 1968).

Note that E can change from a high positive value for chemical control to a negligible value for external diffusion control. There can also be a region of negative activation energy corresponding to surface diffusion control, but this is almost never observed and is not considered here. For highly exothermic reactions, the activation energy can rise to almost infinity as the temperature is raised. Also, the value in the limit of pore diffusion control depends on the thermal nature of the reaction and is equal to $E_{true}/2$ for reactions with no heat effect. The limiting values of E and n for different conditions are discussed by Languasco et al., (1972), Rajadhyaksha and Doraiswamy (1976), and Doraiswamy and Sharma (1984), and are included in Figure 7.11.

Table 7.6 Ranges of important intra- and interphase parameter values for gas-solid and liquid-solid reactions (Carberry, 1976)

Parameter	Gas-solid	Liquid-solid
$\alpha_b = \dfrac{E}{R_g T_b}$	5–40	5–40
$\beta_{m,int} \left[= \dfrac{\Delta T_{int}}{T_0} \right]$	0.001–0.250	0.001–0.100
$\beta_{m,ext} \left[= \dfrac{\Delta T_{ext}}{T_0} \right]$	0.01–2.0	0.001–0.05
$B = \dfrac{Bi_m}{Bi_h}$	10–10^4	10^{-4}–10^{-1}

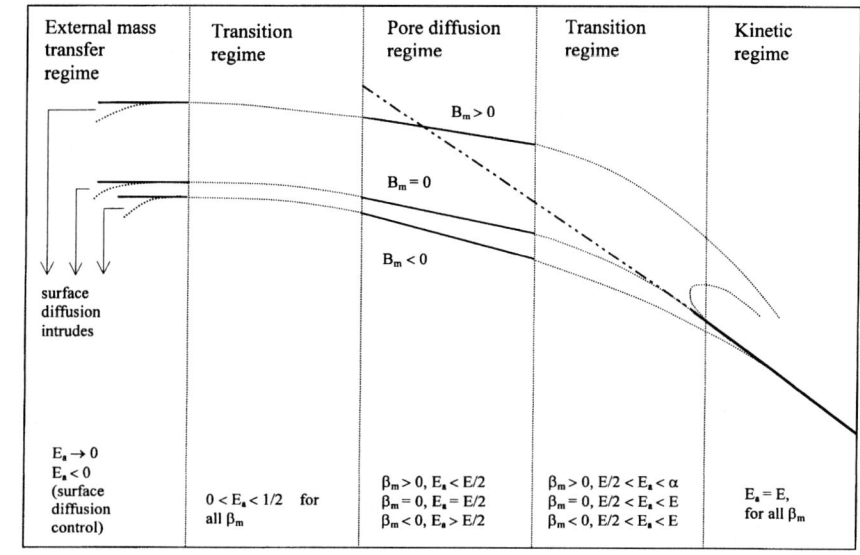

Figure 7.11 Schematic representation of regimes of operation on an Arrhenius diagram.

ELIMINATING OR ACCOUNTING FOR TRANSPORT DISGUISES

It is necessary to eliminate transport disguises to obtain precise kinetic data. The methods of accomplishing this have been discussed in detail by Doraiswamy and Sharma (1984). When they cannot be eliminated, they should be accounted for, and the true values extracted from the "contaminated data." The following procedures/comments should be useful:

1. To eliminate external transport effects, run the reaction at a constant value of W/F_{A0}, but change the velocity. In other words, change the hydrodynamic conditions but keep the kinetic factor (W/F_{A0}) constant. The conversion increases and then levels off at a velocity beyond which external mass transfer would have no effect (Figure 7.12). This can be done by changing the reactor diameter or by changing both W and F (higher F corresponds to higher velocity). One should be cautious in the very low velocity region where the curve might show a deceptive peak before resuming its normal course (Doraiswamy and Sharma, 1984).
2. Carry out runs at different particle sizes. The rate will increase with decrease in particle size, reaching a constant value that indicates absence of pore diffusional effect. On an Arrhenius plot, the effect of pore diffusion for a given particle size will be as shown in Figure 7.13. This can be divided into two asymptotic regions, one corresponding to chemical control and the other to diffusion control. We are concerned only with the chemical control line.

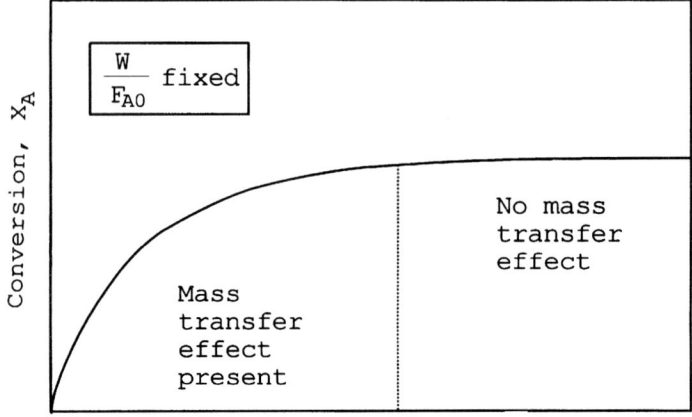

Figure 7.12 Experimental evaluation of the effect of external mass transfer. Velocity is varied by varying F_{A0} and W simultaneousy so that W/F_{A0} (contact time) remains constant.

3. Use a reactor that is sufficiently long to ensure absence of axial diffusion. A length of at least 40–50 pellet diameters is recommended (see Mears, 1971b, 1976, for further details).
4. If experimental tests for diffusional effects as described under 1–4 cannot be carried out, use the criteria assembled in Table 7.7 to confirm the absence of these effects in the runs that were carried out.
5. Ultimately, the only positive *diagnostic* test for the absence of these effects is to *verify directly that the reaction is in the kinetic regime*. For this purpose, the catalyst powder is diluted with an inert powder of the

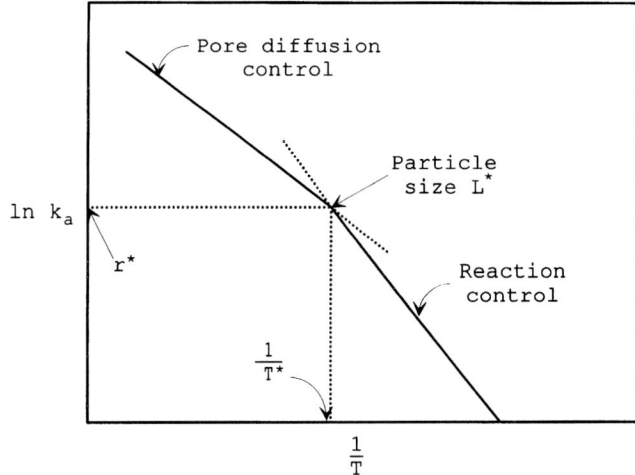

Figure 7.13 Arrhenius plot for formulating rate equations in chemical/pore diffusion regimes.

Table 7.7 Criteria for the absence of mass and heat transfer effects[a]

Interphase	Intraparticle	Interphase and intraparticle
Mass transfer	**Isothermal**	**Isothermal**
1. $\eta = 1 - \dfrac{k_a d_p^{1.5}}{11\sqrt{D_b u}}$;	1. $\dfrac{d_p^2 r_a}{4 D_{eA}[A]_s} < 1,\ n = 1$	1. $\alpha_b \beta_{mb} + 0.3 n \alpha_b \left(\dfrac{(-\Delta H)_r r_a d_p}{2 h_{fp} T_b}\right) < 0.05n$
Ruthven (1968)	Weisz and Prater (1954); Weisz (1957)	Mears (1971a); see also Guha and Narasimhan (1972)
2. $\dfrac{r_a d_p}{2[A]_b k'_G} < \dfrac{0.15}{n}$	2. $\dfrac{d_p^2 r_a}{4 D_{eA}[A]_s} < 1|n|,\ n \neq 0$,	**Nonisothermal**
	Stewart and Villadsen (1969); Mears (1971a)	2. $\dfrac{r_a d_p^2}{4[A]_b D_{eA}} < \dfrac{1 + 0.33\alpha_b \dfrac{(-\Delta H_r) r_a d_p}{2 h_{fp} T_b}}{|n - \alpha_b \beta_{mb}| \left[1 + 0.33 n \left(\dfrac{r_a d_p}{2[A]_b k_G}\right)\right]}$
	With volume change:	Mears (1971a)
Heat transfer	3. $\dfrac{d_p^2 r_a}{4[A]_s D_{eA}} < \dfrac{1}{(1+\theta)n},\ n \neq 0$,	
$\left|\dfrac{(-\Delta H_r) r_a d_p}{2 T_b h_{fp}}\right| < \dfrac{0.15}{\alpha_b}$,	Kubota et al. (1969); Mears (1971a)	
Mears (1971a)	**Nonisothermal**	
	4. $\dfrac{d_p^2 r_a}{4[A]_s D_{eA}} < \dfrac{1}{|n - \alpha_s \beta_m|}$,	
	Kubota and Yamanaka (1969); Mears (1971a)	

[a] r_a = observed rate (mol/m^3 cat s); k_a = observed rate constant (1/s); $\alpha_b = \alpha$ at bulk conditions; η = external effectiveness factor; θ = volume change parameter given by $(\nu_B - 1) y_{As}$, where ν_B is the stoichiometric coefficient in the reaction $A \to \nu_B B$ and y_{As} is the mole fraction of A at the catalyst surface. See Mears (1971b, 1976) for interparticle criteria (i.e., criteria for the reactor as a whole) such as that for the absence of axial diffusion.

same size and characteristics (such as the unimpregnated support). If it is found that

$$\frac{\text{Rate (diluted pellet)}}{\text{Rate (undiluted pellet)}} = \frac{\text{weight of pure catalyst in mixture}}{\text{weight of catalyst in undiluted catalyst}} \tag{7.55}$$

then it is clear evidence that the reaction is in the kinetic regime (Koros and Nowak, 1967).

EXPERIMENTAL REACTORS

Integral, Differential, and Mixed Reactors

There are two fundamental types of experimental reactors for measuring solid-catalyzed reaction rates, integral and differential. The integral reactor consists essentially of a tube of diameter less than 3 cm filled with, say, W g of catalyst. Each run comprises steady-state operation at a given feed rate, and based on several such runs, a plot of the conversion X_A versus W/F_{A0} is prepared. Differentiation of this curve gives the rate at any given X_A (i.e., concentration) as

$$-r_{wA} = \frac{dX_A}{d(W/F_{A0})}, \quad \text{mol/g cat s} \tag{7.56}$$

This is illustrated in the top half of Figure 7.14. A differential reactor, on the other hand, uses a differential amount of catalyst (usually less than 1 g) in which a differential conversion (less than 1–2%) occurs, so that the rate may be obtained directly as

$$-r_{wA} = \frac{\Delta X_A}{\Delta W/F_{A0}} \tag{7.57}$$

at the average concentration in the bed. This is illustrated in the lower portion of the figure.

A convenient way of operating a differential reactor at integral conversions is to use a fully mixed reactor in which a constant concentration within the reactor is imposed by efficient mixing. Several innovative designs to achieve mixing in a fixed bed of pellets have been proposed. Some of them are the spinning basket reactor (Carberry, 1976), the internal circulation reactor (Berty et al., 1969), the rotating pot reactor (Choudhary and Doraiswamy, 1972), the recycle reactor (see, e.g., Satterfield and Roberts, 1968; Carberry, 1976), etc. Detailed treatments of a number of experimental reactors are given by Doraiswamy and Tajbl (1974), Weekman (1974), Doraiswamy and Sharma (1984), Hofmann (1985), and Pratt (1987), and useful introductory treatments by Rose (1981) and Carberry (1976). A few important designs are sketched in Figure 7.15. In a recent design (Borman et al., 1994), perfect mixing is achieved by circulating the gas in the reactor using an axial flow impeller in a well-streamlined enclosure. The mass transfer coefficients in this variation of the internal recycle reactor (reactor (c) in Figure 7.15)

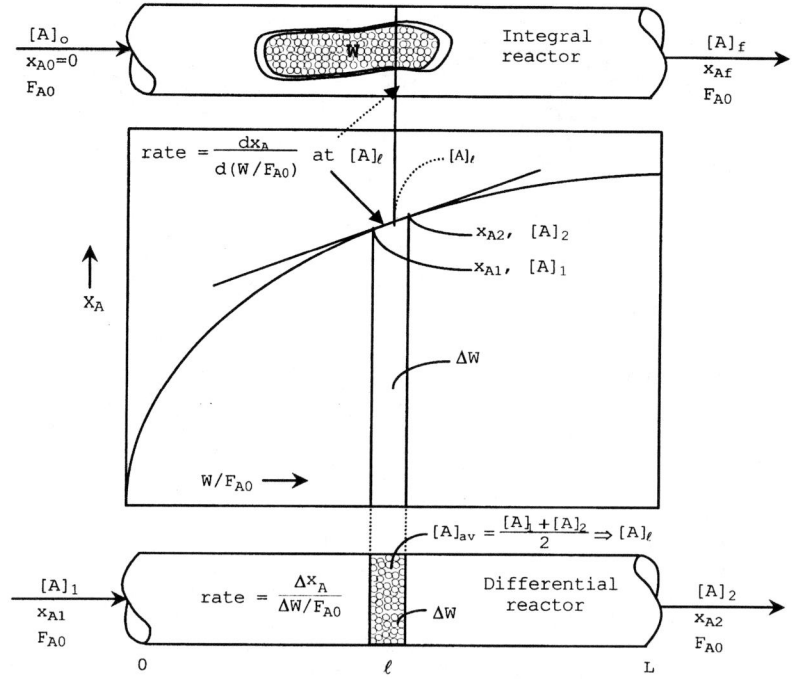

Figure 7.14 Integral and differential reactors for experimental rate determination.

are stated to be 4–6 times higher. In another design (Gullett et al., 1990), a reactor system for short-time rate measurements, particularly suited for rapidly deactivating catalysts, was described. In this design, a catalyst bed is allowed to slide into a process gas stream and then is automatically retracted after a preset exposure time (which can be as low as 0.3 s).

It is always best to operate an experimental reactor under conditions where all diffusional disguises are lifted (by using the criteria listed in the previous section). A less acceptable alternative is to account for them through appropriate effectiveness factors and external transport coefficients. A number of highly sophisticated computer-controlled reactor systems such as the Berty recycle reactor are commercially available. Many of them are available with software and appropriate interfacing that can set and implement the experiments for each of a series of sequential runs (see Mandler et al., 1983), resulting in the emergence of the most acceptable model at the end of the exercise.

Catalyst Dilution as a Means of Obtaining Precise Kinetic Data

A basic requirement for obtaining precise kinetic data is isothermicity of the catalyst pellet. One commonly used method of achieving this is to dilute the bed with inert solids (see Rihani et al., 1965). However, the exit product composition can be influenced to the extent of 3–5% by the manner of inert solids distribution in the bed and can become important in experiments of high precision. Van den

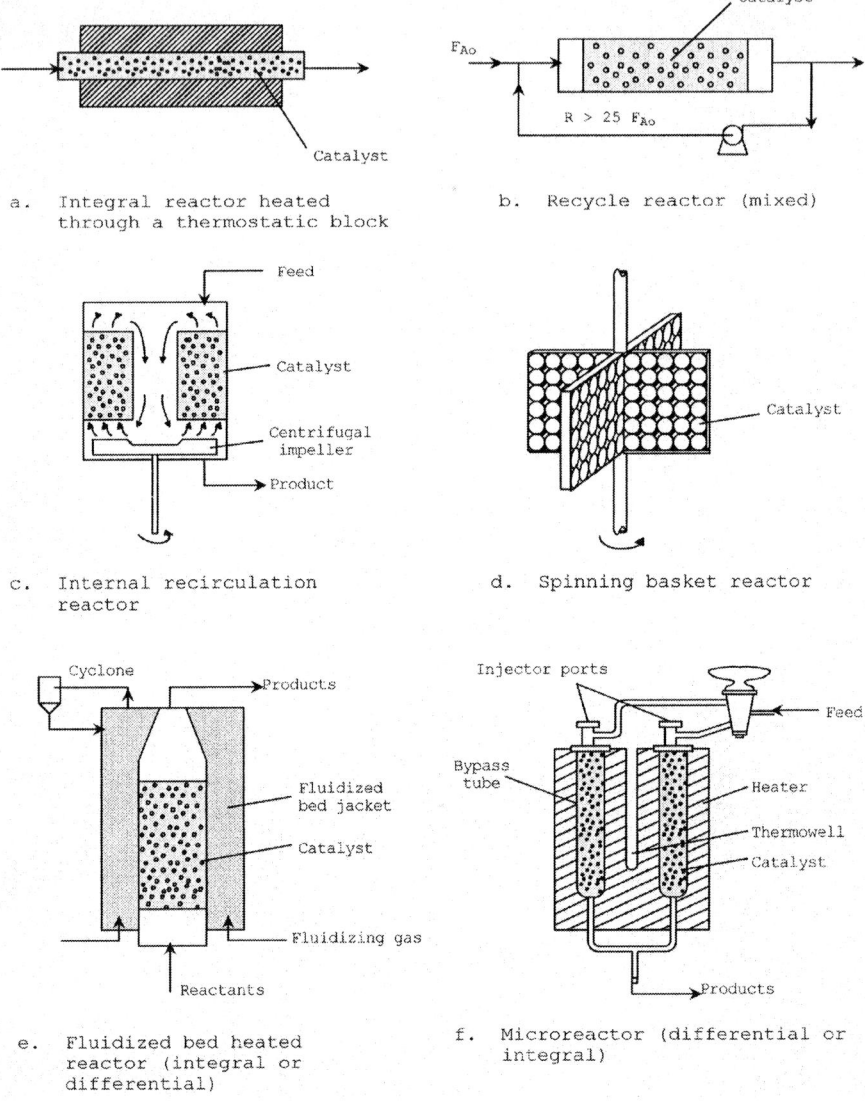

Figure 7.15 Some common experimental reactors for solid-catalyzed reactions.

Bleek et al. (1969) proposed the following criterion for neglecting this *dilution effect*:

$$\frac{bd_p}{\ell\delta} < 4 \times 10^{-3} \tag{7.58}$$

where b is the dilution ratio (weight of total solid particles/weight of catalyst particles), ℓ is the undiluted height, and δ is the experimental error. Sofekun et al. (1994) further examined this problem and proposed a more rigorous statistical criterion for the absence of the dilution effect.

Chapter 8

Homogeneous Catalysis

Catalysis by soluble complexes of transition metals is a rapidly gaining mode of catalysis in organic synthesis. These metals form bonds with one or more carbons in an organic reactant resulting in complexes that are known as *organometallic complexes*. Catalysis by these complexes is often referred to as homogeneous catalysis. Among the important applications of *homogeneous catalysis* in organic synthesis are isomerization of olefins; hydrogenation of olefins (carried out using Wilkinson type catalysts); oligomerization; hydroformylation of olefins to aldehydes with CO and H_2 (the *oxo* process); carbonylation of unsaturated hydrocarbons and alcohols with CO (and coreactants such as water); oxidation of olefins to aldehydes, ketones, and alkenyl esters (Wacker process); and metathesis of olefins (a novel kind of disproportionation).

Enantioselective catalysis that rivals enzymes in selectivity is a major development in homogeneous catalysis. As a result, many earlier processes in the pharmaceutical and perfumery industries are being replaced by more elegant syntheses using soluble catalysts in which "handedness" is introduced in the critical step of the process, thus avoiding the costly separation of racemic mixtures. In view of its importance in organic synthesis, enantioselective (or asymmetric) catalysis was briefly introduced in Chapter 6 and is again considered as a powerful synthetic tool in Chapter 9.

This chapter is concerned with the use in general of homogeneous catalysis in organic synthesis (including asymmetric synthesis). Among the several books and reviews written on the subject, the following may be mentioned: Halpern (1975, 1982), Bau et al. (1978), Parshall (1980), Masters (1981), Collman and Hegedus (1980), Eby and Singleton (1983), Chaudhari (1984), Davidson (1984), Kegley and Pinhas (1986), Collman et al. (1987), Parshall and Nugent (1988),

Noyori and Kitamura (1989), Parshall and Ittel (1992), Gates (1992), Chan (1993), Akutagawa (1995).

HOMOGENEOUS VERSUS HETEROGENEOUS CATALYSIS

Gas (or liquid)-phase reactions on solid catalysts are among the most common industrial reactions. However, homogeneous catalysis is rapidly catching up. Excluding applications in petroleum refining, the dollar value of organic chemicals produced worldwide by homogeneous catalysis (more than $35 billion) is quite impressive compared to that by heterogeneous catalysis (more than $45 billion). Attempts are now under way to find an integrated approach to homogeneous and heterogeneous catalyses (Moulijn et al., 1993).

Many of the processing problems of homogeneous catalysts arise from the "liquid state" of the catalysts. One way of overcoming them is to heterogenize the catalyst by bonding or anchoring it to a solid support, thus combining the advantages of homogeneous and heterogeneous catalyses. We discussed this in Chapter 6 as part of the general strategy of heterogenizing liquid-phase catalysts.

FORMALISMS IN TRANSITION-METAL CATALYSIS

The Uniqueness of Transition Metals

Transition metals have the distinctive property (not shared by other metals) that their d shells are only partially filled with electrons. This gives them the unique ability to exist in several oxidation states. As a result, the use of organo-transition-metal complexes can provide pathways for an extraordinary range of reactions.

A typical transition metal atom has nine valence shell orbitals: one s, three p, and five d, in which it accommodates valence electrons that bond with other moieties known as *ligands* (usually represented by the letter L) to form two types of bonds, *covalent* and *coordinated*. In a typical covalent bond, one or more electrons are shared between two atoms. However, the electrons that constitute a bond need not be equally shared; in fact, all of the electrons can come from just one of the atoms. Although this kind of bond is also a covalent bond, it is usually regarded as a subcategory of the covalent bond and is referred to as a coordinate bond. This ability to form these two types of bonds with a number of ligands is responsible for the unique catalytic properties of the transition metals and their complexes.

EXAMPLE 8.1

The hydride complex $ReH_7(PMe_2Ph)_2$ is known to be catalytically very active. Why is this so?

The hydride complex contains phosphine as the ligand (Figure 8.1) (see Bau et al., 1978). There are seven Re–H bonds and two P → Re bonds. The seven metal–hydrogen bonds are covalent bonds formed by pairing of the lone electrons from seven hydrogen atoms

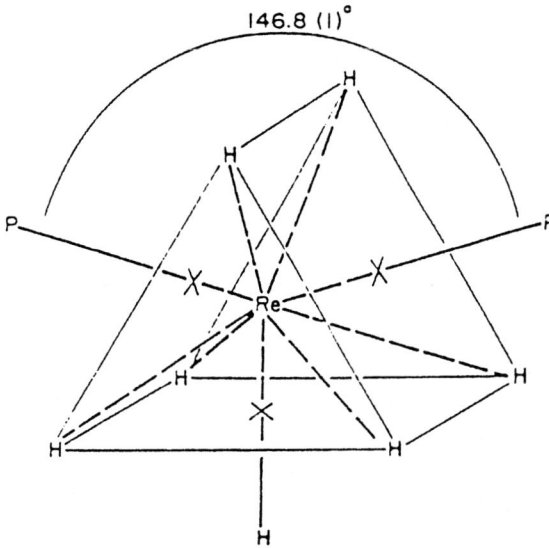

Figure 8.1   A postulated structure of H$_7$Re (P-Me$_2$Ph)$_2$ (Bau et al., 1978).

with valence electrons of rhenium, one from each of its seven orbitals. They are metal–ligand bonds similar to any C–H bond of an organic molecule. On the other hand, the two P → Re bonds are coordinate bonds formed by donation of a complete electron pair from just one partner, which then is shared by the other. The donor component in this case is the phosphine ligand.

The metal bonds with nine ligands in two different ways, seven covalently and two coordinatively. This ability of the transition metal to bond with such a large number of ligands makes it unique. The changes that this coordination introduces in the electron distribution in the complex greatly enhance the reactivity of the ligand molecule, leading to its unique catalytic action.

Another concept that is fundamental to an understanding of homogeneous catalysis is that of the oxidation state of a complex. Although commonly used in general chemistry, a less puristic definition of this state, which may be regarded as a formalism more suited to transition-metal complexes, has become the cornerstone of homogeneous catalysis.

Indeed, the application of transition-metal catalysis in organic synthesis is built around many such formalisms. In addition to the formal oxidation state, these include *coordinative unsaturation, coordination number,* and *coordination geometry*; *hydride formalism*; and the *18-electron* and *16-18-electron rules*, also referred to as *electron bookkeeping*.

The Oxidation State of a Metal

According to Collman et al. (1987), the *oxidation state* is defined as "the charge left on the central atom when the ligands are removed in their normal closed-

Table 8.1 Relationships between oxidation state and d^n in transition metals (Collman et al., 1987)

Group number		IVa	Va	VIa	VIIa	VIIIa			Ib	
First row	3d	Ti	V	Cr	Mn	Fe	Co	Ni	Cu	
Second row	4d	Zr	Nb	Mo	Tc	Ru	Rh	Pd	Ag	
Third row	5d	Hf	Ta	W	Re	Os	Ir	Pt	Au	
Oxidation state	0	4	5	6	7	8	9	10	—	d^n
	I	3	4	5	6	7	8	9	10	
	II	2	3	4	5	6	7	8	9	
	III	1	2	3	4	5	6	7	8	
	IV	0	1	2	3	4	5	6	7	

shell configurations." A particularly important fact in organometallic chemistry is that both hydrogen and carbon are more electronegative than the transition metal (recall that H is electropositive in acids and C in organic molecules). Thus the M-H group is considered as M^+H^-. This is commonly referred to as the *hydride formalism*. Oxidation states can be similarly assigned to any ligand if its normal closed-shell configuration is known.

The formal oxidation state of a metal is related to its d electron configuration as shown in Table 8.1 (Collman et al., 1987). This table is constructed on the assumption that the outer electrons are all d electrons. This is usually a good approximation. Thus Co(0) is d^9, Re(IV) is d^3, Ir(II) is d^7, and so on. A metal complex consists of the metal and a ligand. Thus the formal ligand charge should be known before we can assign a formal oxidation state to the metal. Collman et al.'s table of ligand charges (1987) is reproduced in Table 8.2. Example 8.2 illustrates the procedure for obtaining the oxidation states of metals in various complexes (see Kegley and Pinhas, 1986; Collman et al., 1987, for more examples).

EXAMPLE 8.2

Determine the oxidation state of the metal in each of the following complexes: $[Fe(CO)_4]^{2-}$, $[Ni(CO)_4]$, $[Cu(NH_3)_4]^{2+}$, $[(\eta^5\text{-}C_5Me_5)IrMe_4]$, and $(PMe_3)_2Pd(\eta^3\text{-}C_3H_3)]^+$

Note from Table 8.2 that the carbonyl ligands are formally neutral. Because the charge on the iron carbonyl is 2^-, the charge on iron is $(-2-0)=-2$. Hence the oxidation state of iron in this complex is -2, which is represented as Fe(-II).

In $[Ni(CO)_4]$, because the charge on the molecule is zero and the carbonyl ligands are also neutral, the charge on the nickel is also zero. Thus the oxidation state of nickel in nickel carbonyl is Ni(0).

Considering the complex $[Cu(NH_3)_4]^{2+}$, note from Table 8.2 that the ammonia ligand is also neutral. Hence the charge on Cu is $(2-0)=2$, that is, the oxidation state of copper in the complex is Cu(+II).

In the iridium complex $[(\eta^5-C_5Me_5)IrMe_4]$, the charges on CH_3 and $\eta^5-C_5Me_5$ are -1. Hence the oxidation state of Ir in the complex is $[0-(-5)]=5$, that is, Ir(+V).

Table 8.2 Charges and corresponding coordination numbers for typical ligands (Collman et al., 1987)

Ligand	Charge[a]	Coordination number[a]
X (Cl, Br, I)	−1	1(2)
H	−1	1(2, 3)
CH_3	−1	1(2)
Ar	−1	1
RCO	−1	1(2)
Cl_3Sn	−1	1
R_3Z (Z = N, P, As, Sb)	0	1
R_2E (E = S, Se, Te)	0	1
CO	0	1(2, 3)
RNC:	0(−2)	1(2)
RN:	0(−2)	1(2)
R_2C:	0(−2)	1(2)
N_2	0	1(2)
$R_2C=CR_2$	0(−2)	1(2)
$RC\equiv CR$	0(−2)	1(2)
CN	−1	1
η^4-Cyclobutadiene	0	2
$CH_2=CHCH_2$-	−1	1
η^3-Allyl	−1	2
η^6-Benzene (C_6H_6)	0	3(2, 1)
η^5-Cyclopentadienyl (C_5H_5)	−1	3
η^7-Cycloheptadienyl (C_7H_7)	+1	3
NO[b]	+1(−1)	1(2)
ArN_2 [b]	+1(−1)	1(2)
O	−2	2
O_2	−2(−1)	2(1)

[a] The less common or alternative formulation is in parentheses.
[b] Noninnocent ligand (two or more discrete bonding modes).
Note: The superscript on η implies that all carbon atoms interact with the metal.

The ligand ($\eta^3 - C_3H_3$) in the complex $[(PMe_3)_2Pd(\eta^3 - C_3H_3)]^+$ has a charge of −1, and PMe_3 has a charge of 0. Because the complex itself has a charge of 1, the oxidation state of Pd is $[1 - (-1)] = 2$, that is, Pd(II).

Coordinative Unsaturation, Coordination Number, and Coordination Geometry

The presence of a vacant coordination site (analogous to an active site in heterogeneous catalysis) is an important prerequisite for homogeneous catalysis and is termed coordinative unsaturation. When the total number of electrons in metal–ligand binding is 18, the complex is considered coordinatively saturated. If the electron count is 16 or less, the metal ion possesses at least one vacant coordination site and is said to be coordinatively unsaturated. For example, the dissociation of PPh_3 (triphenylphosphine) from $RhCl(PPh_3)_3$ (Wilkinson's complex) is an important step in forming the coordinatively unsaturated reactive species

RhCl(PPh$_3$)$_2$ in the hydrogenation of olefins. This kind of vacant coordination site can either be built into a catalyst or created in situ by appropriate design of the metal ion–ligand–solvent system.

A single metallic ion is usually surrounded by a variable number of ligands resulting in different structures. The number of ligands involved in each structure is known as the *coordination number* n_c of the complex, and is an important parameter of transition-metal catalysis.

The structures resulting from the bonding can be linear, square planar, trigonal planar, tetrahedral, octahedral, trigonal bipyramidal, or square pyramidal as shown in Figure 8.2.

Ligands and Their Role in Transition-Metal Catalysis

Another important formalism implicit in the definition of the oxidation state considered previously is with respect to the classification of ligands. We formally define a ligand as "an element or a combination of elements which form(s) chemical bond(s) with a transition element" (Masters, 1981). Essentially, two types of ligands have been formally recognized: ionic, such as Cl$^-$, H$^-$, OH$^-$, CN$^-$, alkyl$^-$, aryl-, and COCH$_3^-$; and neutral, such as primary-, secondary- and tertiary-phosphines, CO, alkene, amine, etc. Such a clear-cut distinction can be misleading to the purist in coordination chemistry. Thus the "ionic" ligands such as H$^-$ and CH$_3^-$ are nearly neutral in terms of metal–ligand charge separation in the classical sense, and the "neutral" ligands such as tertiary phosphines are known to violate their neutrality (Chatt and Leigh, 1978).

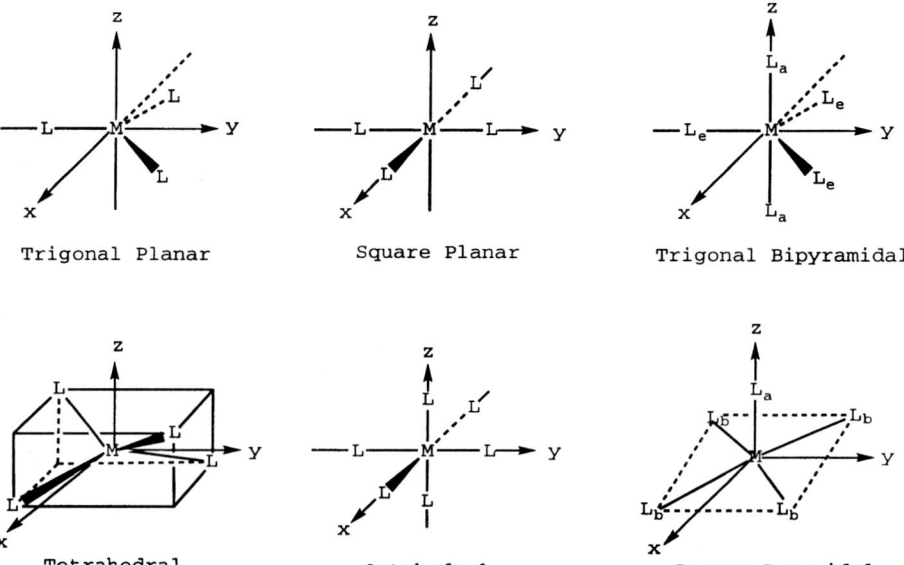

Figure 8.2 Coordination geometries of metal complex catalysts (redrawn from Collman et al., 1987).

Ligands can be bonded to metal through single coordination sites or through more than one coordination site. These are known as *unidentate* and *multidentate ligands*, respectively. Examples of unidentate ligands are the halides, NH_3, and H_2O. CO can form both unidentate and multidentate bonds as shown below.

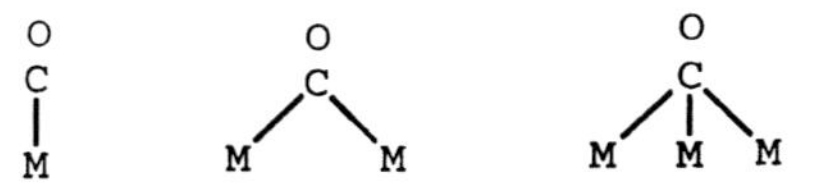

```
    O           O              O
    ‖           ‖              ‖
    C           C              C
    |          / \            /|\
    M         M   M          M M M

  linear    bidentate      tridentate
```

The iodide ligand is particularly important in homogeneous catalysis. The phosphine ligands are even more important; typical examples are unidentate phosphine, unidentate triphenylphosphine (TPP), bidentate diphosphine (DIPHOS), and bidentate chiral diphosphine (DIPAMP), with the following structures

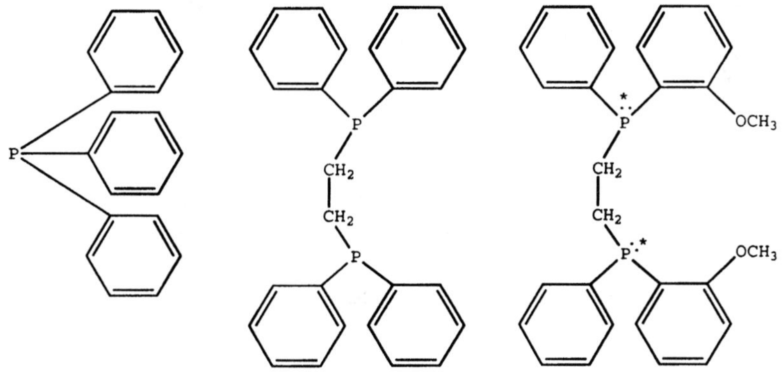

TPP   DIPHOS   DIPAMP

Several more are given in Figure 9.5.

Multidentate ligands are not commonly used in industry because of the heavy costs associated with their manufacture. It is also important to note that hydride (H^-) and alkyl (R^-) are two of the most commonly used ligands in homogeneous catalysis.

A practically more useful classification can be made in terms of the actual presence or otherwise of a ligand in the final product(s) of reaction (see Masters, 1981). In some cases the ligands are carried over intact in the products of catalytic cycles, such as CO in a catalytic carbonylation or hydroformylation, for example,

$$RCH = CH_2 + CO + H_2 \xrightarrow{CO_2(CO)_8} RCH_2CH_2CHO + RCH(CHO)CH_3$$

[8.1]

Such ligands are termed *participative ligands*. On the other hand, there are ligands that promote a catalytic cycle but do not end up in the product(s), such as in the preparation of dimethyl maleate from acetylene, CO, and

MeOH, catalyzed by palladium chloride in the presence of thiourea (NH_2CSNH_2) and a trace of oxygen:

$$CH\equiv CH + 2CO + 2MeOH \xrightarrow{PdCl_2/NH_2CSNH_2 + O_2 (trace)} MeO_2CCH=CHCO_2Me \quad [8.2]$$

Note that chlorine and thiourea do not appear in the product. Such ligands are known as *nonparticipative* ligands and can be usefully employed by exploiting their structural properties (Tolman, 1977).

The Electron Rules ("Electron Bookkeeping")

It should be clear from the earlier sections that the basic requirement in applying the principles of organometallic chemistry is counting electrons associated with a complex. The formalisms arising out of this have been combined into two powerful rules of electron bookkeeping: the *18-electron rule* and its corollary the *16-18-electron rule*.

The 18-electron rule

This rule arises from the assumption that the valence shell electrons of the metals are all in the Nd shell, where N is the principal quantum number. From quantum theory considerations, the d levels are usually associated with the highest energies and are hence the most amenable to the exchange of electrons. Because the number of electrons is related to the oxidation state of a metal, it is clear that the number of d electrons, denoted by d^n, determines this state. This bookkeeping function of the d orbitals in determining the oxidation state of the complex is best illustrated by its usefulness in formulating a rule for the maximum allowable number of ligands for each d^n. The rule in its final form, known as the *18-electron rule*, is

$$d^n + 2n_{C,max} = 18 \quad (8.1)$$

where $n_{c,max}$ is the maximum coordination number. This rule is limited to cases where the complex has only one metal atom (i.e., is mononuclear) and d^n is an even number with all electrons paired (diamagnetic). Stated simply, the rule asserts that for a metal complex to be stable, the nine outer orbitals of a transition metal must accommodate 18 electrons.

Exceptions to this rule are not many but do exist. Thus more than 18 valence electrons can be accommodated in some cases. The more important cases are those where the number is less than 18, leading to what are known as *coordinatively unsaturated complexes*. Some of the latter (the 14- and 16-electron complexes) are particularly useful in catalysis, resulting in the important rule to be described later. A procedure for determining the stabilities of complexes based on this rule is illustrated below for two complexes selected from a large list considered by Kegley and Pinhas (1986).

EXAMPLE 8.3

Determining the stability of complexes

Are the complexes $Ru(PPh_3)_2(CO)_2$ and $Cp_2NbH(C_2H_4)$ stable and observable?

To answer this question, it is convenient to construct a table of ligand coordination numbers and electrons for the different components of each complex, as shown here.

Complex	Ligand(s)	Total coordination no.	No. of electrons	Charge
$Ru(PPh_3)_2(CO)_2$	2 CO	$1 \times 2 = 2$	4	Neutral
	2 PPh_3	$1 \times 2 = 2$	4	Neutral
	Ru(0)	–	8	Neutral
		4	16	
$Cp_2NbH(C_2H_4)$	Cp^-	$3 \times 2 = 6$	12	-2
($Cp = \eta^5-C_5H_5$)	H^-	$1 \times 1 = 1$	2	-1
	C_2H_4	$1 \times 1 = 1$	2	Neutral
	Nb(III)	–	2	$+3$
		8	18	

In preparing the table, the total coordination number for any ligand (column 3) is calculated as (n_C) (number of ligand units), and n_C is obtained from Table 8.2. Thus for CO in complex 1, the value is $(1 \times 2) = 2$. The number of electrons for any ligand (column 4) is calculated as (the number of electrons for each unit of a ligand) (the number of ligand units). Thus for CO the value is $(2 \times 2) = 4$. When all of the d electrons are paired, the number of electrons is twice the coordination number. Recall that our treatment is restricted to such systems. Column 4 gives directly the number of d electrons in outer shell.

The conclusions from column 4 for the two compounds are that compound 1 is an unstable 16-electron complex and compound 2 is a stable and observable 18-electron complex.

The 16-18-electron rule

When a complex is coordinatively saturated, any subsequent ligand substitution by nucleophilic attack can occur only by a mechanism in which the nucleophile does not appear in the rate-determining step. In other words, the mechanism is governed solely by transformations within the complex in which the 18-electron compound would dissociate to give an unsaturated complex which then can bind (i.e., associate) with other potential ligands. Such a mechanism corresponds to the S_N1 mechanism, and has given rise to the *dissociation-association* or *16-18-electron rule*, first proposed by Tolman (1972). We explain this rule below by an example.

EXAMPLE 8.4

Illustration of the 16-18-electron rule

We desire to replace two carbonyl groups in the chromium-carbonyl complex $Cr(CO)_6$ considered in the previous section by triphenylphosphine ligands. Illustrate the 16-18-electron rule for this replacement.

Parshall (1980) suggests the following steps in the replacement scheme:

$$Cr(CO)_6 \xrightarrow{-CO \text{ (slow)}} Cr(CO)_5 \xrightarrow{PPh_3} Cr(CO)_5(PPh_3) \qquad [E8.4.1]$$
$$18\,e \qquad\qquad\qquad 16\,e \qquad\qquad\qquad 18\,e$$

$$\xrightarrow{-CO \text{ (slow)}} Cr(CO)_4(PPh_3) \xrightarrow{PPh_3} Cr(CO)_4(PPh_3)_2$$
$$\qquad\qquad\qquad 16\,e \qquad\qquad\qquad 18\,e$$

Because the initial reaction of $Cr(CO)_6$ follows the S_N1 mechanism, it is unimolecular and hence independent of the triphenyl-phosphine concentration.[1] The first step in the reaction scheme is dissociation of the stable 18-electron $Cr(CO)_6$ to the more reactive 16-electron $Cr(CO)_5$ that has been characterized spectroscopically (Perutz and Turner, 1975). Then this latter moiety undergoes a fast associative reaction with a triphenylphosphine molecule to form another stable 18-electron compound. This is followed by a second slow dissociation to give the 16-electron compound $Cr(CO)_4(PPh_3)$, which on further rapid association with another triphenylphosphine gives the final stable 18-electron compound, bis(triphenylphosphine).

A useful interpretation of this rule is that in a series of steps in a reaction, no step is possible in which the number of valence electrons changes by more than 2.

THE OPERATIONAL SCHEME OF HOMOGENEOUS CATALYSIS

It is clear from the basic principles (or, more correctly, the formalisms) of homogeneous catalysis just outlined that the essence of homogeneous catalysis lies in the formation of a transition-metal complex with a coordination sphere that offers an environment conducive to chemical change. These transformations may involve rearrangement or migration of ligands already present and their elimination as product(s) or insertion of an external ligand within the coordination sphere to form a product that then is eliminated from the complex.

The second essential feature of any homogeneous catalytic process is that the catalyst must be regenerable. Clearly, this involves a *catalytic cycle* as shown in Figure 8.3 for a simple hypothetical case. Each step constitutes a fundamental reaction of organometallic chemistry. Justifiably, therefore, the critical economic issue that determines the success of a technology is its ability to regenerate the costly catalyst (say Pd) at the expense of cheap raw materials (such as oxygen in oxidation). Based on these observations, the complete methodology for developing a process using homogeneous catalysis is described in the chart shown in Figure 8.4.

---

[1] This is not necessarily true of all such complexes. For example, $Mo(CO)_6$ and $W(CO)_6$ follow the S_N2 mechanism, that is, the rates of substitution in these cases also depend on the substituent ligand concentration (Graham and Angelici, 1967).

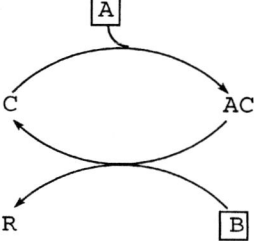

Figure 8.3 A typical catalytic cycle. Reaction $A + B \to R$.

THE BASIC REACTIONS OF HOMOGENEOUS CATALYSIS

The chief basic reactions of transition-metal chemistry are reactions of ligands, coordination and addition reactions (sometimes called the *elementary* or *activation* steps), and insertion and elimination reactions (the main reactions). We consider each of these here with examples.

Reactions of Ligands (Mainly Replacement)

The replacement of one ligand by another in a transition-metal complex is a common basic reaction in homogeneous catalysis. In a reaction of this kind where two reactants are involved, the kinetics can be described by the $S_N 1$ or $S_N 2$ mechanism (see Parshall, 1980). Thus for $Cr(CO)_6$, the replacement of one CO by a triphenylphosphine ligand occurs by the $S_N 1$ mechanism. In other words, it is influenced only by the local environment within the coordination sphere of the complex. Similarly, one can also think of ligands that conform to the $S_N 2$ mechanism by actively participating in the displacement of an existing

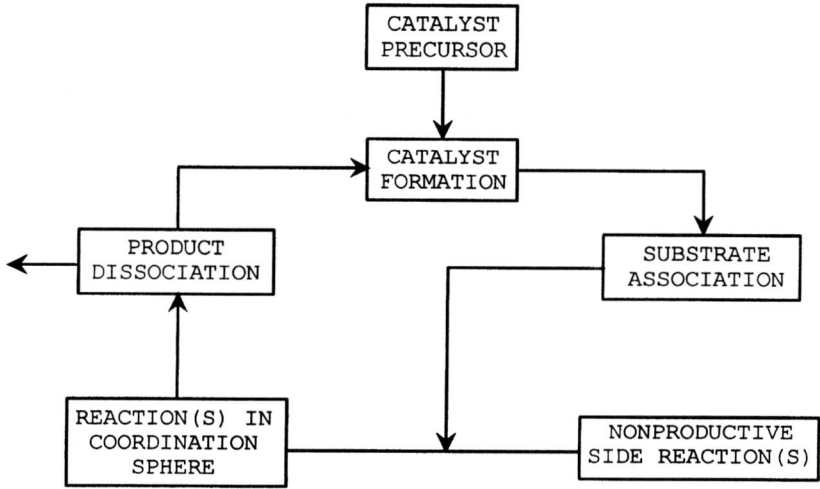

Figure 8.4 Overall methodology of homogeneous catalysis.

ligand. Examples of these are the substitution reactions of $Mo(CO)_6$ and $W(CO)_6$. All of these examples belong to even-electron-count systems, that is, diamagnetic complexes.

Examples of odd-electron-count (i.e., paramagnetic) intermediates are also known but are very rare and much less understood. Therefore, our focus will continue to be the even-electron-count complexes considered so far.

The Elementary Reactions (or Activation Steps)

Essentially, two broad types of elementary reactions occur in homogeneous catalysis: coordination and addition. Although these are reactions and we designate them so, they should more appropriately be viewed as steps in a catalytic cycle that "activate" the substrate prior to reaction.

Coordination reactions

When CO or an alkene coordinates with a metal center, it appears in the final product without losing its integrity. Consider, for example, the reaction (Masters, 1981):

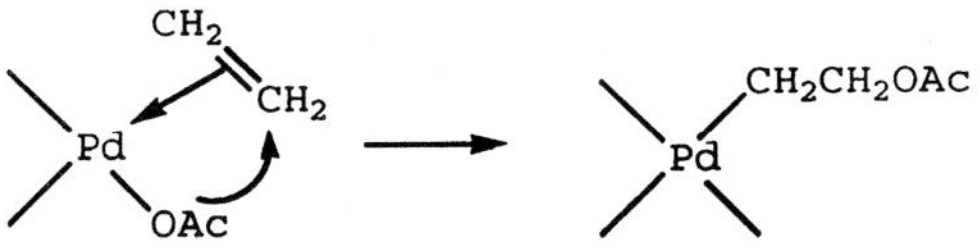

$$\text{[Pd complex with CH}_2\text{=CH}_2 \text{ and OAc]} \longrightarrow \text{[Pd complex with CH}_2\text{CH}_2\text{OAc]} \qquad [8.3]$$

Here, the ethylene does not lose its integrity in the reaction with acetate in the coordination sphere of Pd. Reactions of this type are known as *coordination reactions* (comparable to nondissociative adsorption in heterogeneous catalysis).

Addition reactions

In *addition reactions*, the substrate splits, and each fraction bonds separately with the catalyst (metal) center. These reactions can be classified as *heterolytic addition*, *homolytic addition*, and *oxidative addition*. They are distinguished by the nature of the change in the oxidation state upon the addition of a substrate to a metal center.

HETEROLYTIC ADDITION

$$M^n L_y + XY \rightarrow M^n L_{y-1} X + Y^+ + L^- \qquad [8.4]$$

This is characterized by the addition of a substrate XY to a single metal center (through X or Y) with no overall change in the oxidation state or coordination number of the metal.

HOMOLYTIC ADDITION

$$2M^n L_y + XY \rightarrow 2M^{n+1}(X)(Y)L_y \qquad [8.5]$$

Here a substrate XY adds on to two metal centers so that the formal oxidation state of each metal center increases by one.

OXIDATIVE ADDITION

$$M^n L_y + XY \rightarrow M^{n+2}(X)(Y)L_y \qquad [8.6]$$

In this addition, the formal oxidation state of the metal is increased by two. The reverse of this reaction is known as *reductive elimination*. An example is the following oxidative addition in a Rh complex (Forster, 1979):

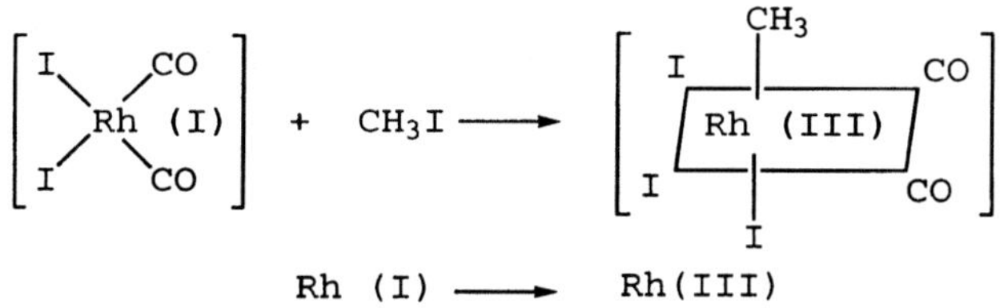

$$\begin{bmatrix} I & CO \\ & Rh\ (I) \\ I & CO \end{bmatrix} + CH_3I \longrightarrow \begin{bmatrix} & CH_3 & \\ I & | & CO \\ & Rh\ (III) & \\ I & | & CO \\ & I & \end{bmatrix} \qquad [8.7]$$

Rh (I) $\longrightarrow$ Rh(III)

The Main Reactions

After the complex has been formed, that is, the reactant molecule has been activated, the next step is the main reaction leading to the formation of the products. These reactions generally occur by either insertion or elimination.

Insertion

SINGLE METAL CENTER

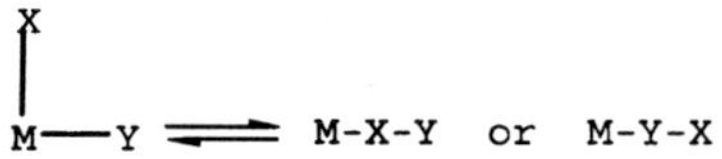

$$\overset{X}{\underset{M-Y}{|}} \rightleftharpoons M-X-Y \quad \text{or} \quad M-Y-X \qquad [8.8]$$

MULTIPLE METAL CENTERS

$$\overset{X}{\underset{M_1}{|}} \overset{Y}{\underset{M_2}{|}} \longrightarrow \overset{XY}{\underset{M_1-M_2}{|}} \qquad [8.9]$$

As can be seen from both of these general schemes, what is really involved is migration of a ligand within the molecule. Insertion is the key step in several catalytic cycles such as those associated with hydrogenation, oligomerization, and hydroformylation. For example, in hydroformylation reactions using $HRh(CO)(PPh_3)_3$ as the catalytic complex, insertion of precoordinated CO into the alkyl group by ligand migration within the coordination sphere of the complex is an important step in the catalytic cycle (Evans et al., 1968a,b):

$$(CH_3CH_2)Rh(CO)_2(PPh_3)_2 \rightarrow (CH_3CH_2CO)Rh(CO)(PPh_3)_2 \qquad [8.10]$$

Elimination

REDUCTIVE ELIMINATION

When $Cp(CO)_2Fe\text{-}CH_3$ is reacted with HBF_4, a complex is formed from which CH_4 is eliminated:

$$[Cp(CO)_2Fe(CH_3)H]^+[BF_4]^- \xrightarrow[\text{elimination}]{\text{Reductive}} [Cp(CO)_2Fe]^+[BF_4]^- + CH_4$$

$$Fe(IV) \longrightarrow Fe(II) \qquad [8.11]$$

In this step, the two ligands H and CH_3 combine to form the product CH_4 which then leaves the coordination sphere of the metal. The formal oxidation state of the metal in this reductive elimination step is reduced by two units, which may be regarded as the reverse of oxidative addition.

β-HYDROGEN ELIMINATION

This is the reverse of insertion in which the β-hydrogen of an alkyl group migrates to the metal atom and further strengthens the description of the elimination reaction as a ligand migration reaction. α-Elimination is also possible but is much less understood.

An example of β-H elimination is shown in Figure 8.5 as one step in a catalytic cycle (Bergman, 1980). This is an instructive cycle from our point of view because it includes two other major elementary steps besides β-elimination: insertion and reductive elimination.

MAIN FEATURES OF TRANSITION-METAL CATALYSIS IN ORGANIC SYNTHESIS: A SUMMARY

The special features of transition metal complexes that are responsible for their remarkable attractiveness in organic synthesis may be summarized as follows:

1. General readiness to bond with a large number of metals in the periodic table and with just about any organic molecule.
2. Ability to activate (through coordination) a variety of industrially available feedstocks such as CO, H_2, olefins, and alcohols or their derivatives.
3. Ability to stabilize unstable intermediates such as metal hydrides and metal alkyls in relatively stable but kinetically reactive complexes.
4. Accessibility to different oxidation states and ability to move from one oxidation state to another during the course of a reaction, as exemplified by the hydrogenation of an olefin

$$RhCl(PPh_3)_3 + H_2 \rightarrow RhH_2Cl(PPh_3)_3 \leftrightarrow RhH_2Cl(PPh_3)_2 + PPh_3 \qquad [8.12]$$

$$RhH_2Cl(PPh_3)_2 + RCH=CH_2 \rightarrow RhCl(PPh_3)_2 + RCH_2CH_3 \qquad [8.13]$$

in which the rhodium undergoes the following I-III-I oxidation/reduction cycle:

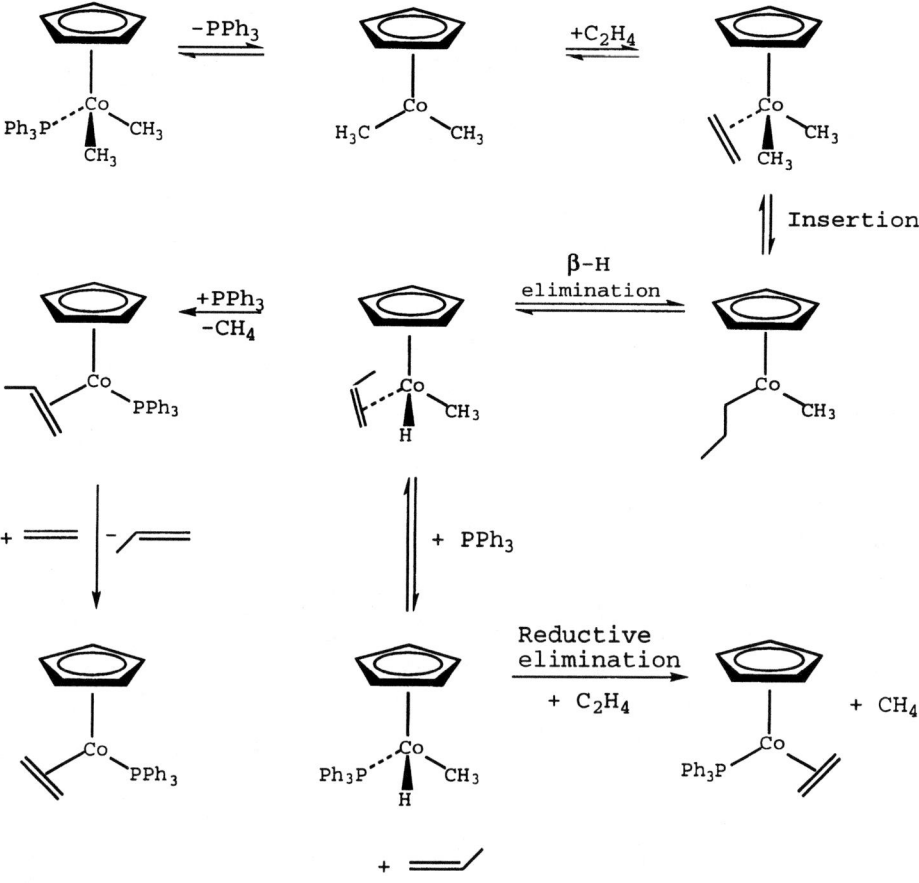

Figure 8.5 An example of β-H elimination (including also insertion and reductive elimination) (Bergman, 1980)

$$Rh^{I}Cl(PPh_3)_2 \xrightarrow{H_2} Rh^{III}H_2Cl(PPh_3)_2$$
$$\xrightarrow{RCH=CH_2} Rh^{III}H(RCH_2CH_2)Cl(PPh_3)_2$$
$$Rh^{I}Cl(PPh_3)_2 + RCH_2CH_3 \qquad [8.14]$$

5. Ability to assemble and orient various reactive components within the coordination sphere (the *template effect*).
6. Ability to accommodate both participative and nonparticipative ligands within the coordination sphere; this is useful in modifying the steric and electronic properties important in determining catalytic activity and selectivity.

IMPORTANT CLASSES OF REACTIONS WITH INDUSTRIAL EXAMPLES

Based on the activation and elementary steps outlined, a variety of catalytic reactions can be better understood and catalytic cycles defined. We consider a few major classes of reactions: isomerization, hydrogenation, carbonylation, hydroformylation, oxidation, and metathesis.

Isomerization

Many isomerization reactions occur by double-bond migration. Two common mechanisms of double-bond migration are (Parshall and Ittel, 1992) (1) equilibration between internal and terminal olefins by hydrogen migration and (2) interconversion between branched and linear chains of an olefin. Another mechanism of olefin isomerization is the shift of an allylic hydrogen from the 3- to the 1-position. We confine our discussion to the more important double-bond migration mechanism.

The most prominent example of equilibration between terminal and internal olefins is the Shell Higher Olefins Process (SHOP) for linear α-olefins (Gum and Freitas, 1979). Here the internal olefin, 2-decene, is converted to the corresponding terminal olefin which then is hydroformylated to the linear aldehyde

$$C_7H_{15}CH=CHCH_3 \leftrightarrow C_8H_{17}CH=CH_2 \xrightarrow{H_2, CO} C_8H_{17}CH_2CH_2CHO$$
$$\text{Isomerization} \qquad \text{Hydroformylation} \qquad [8.15]$$

The aldehyde, undecanal, is an intermediate in perfumery manufacture (Bauer and Garbe, 1985), as well as in detergent manufacture. This process is a unique example of a homogeneous catalyst that is simultaneously active in two sequential reactions, isomerization and hydroformylation. The catalyst used is $HCo(CO)_3(PBu_3)$ (Slaugh and Mullineaux, 1966), which further hydrogenates the aldehyde to the detergent alcohol (the end product of the process). Other useful (or potentially useful) isomerization catalysts are $Fe(CO)_5$, $Fe_3(CO)_{12}$, $PdCl_2$ (Bingham et al., 1974), and $HCo(CO)_4$ (Orchin, 1966). For example, $Fe(CO)_5$ can be used in the manufacture of anethole, a licorice fragrance.

DuPont's process for adiponitrile, used in the manufacture of Nylon from butadiene and HCN, contains a step in which a branched chain (2-methyl-3-butenenitrile) is converted to a straight chain olefin

$$\underset{\underset{CH_3}{|}}{CH_2=CH-CH-CN} \leftrightarrow CH_3CH=CHCH_2CN \leftrightarrow CH_2=CH-CH_2-CH_2CN$$
$$[8.16]$$

followed by equilibration of the internal olefin with its terminal isomer $CH_2=CHCH_2CH_2CN$.

Hydrogenation

The three types of activation characterized by reactions 8.4, 8.5, and 8.6 can be written specifically for hydrogen activation by Ru, Co, and Rh as follows:

$$Ru^{2+} + H_2 = Ru^{2+}H + H^+ \quad \text{(heterolytic splitting)} \quad [8.17]$$

$$2Co^0 + H_2 = 2Co^{1+}H \quad \text{(homolytic splitting)} \quad [8.18]$$

$$Rh^+ + H_2 = Rh^{3+}H_2 \quad \text{(oxidative addition)} \quad [8.19]$$

Of these, oxidative addition is a particularly important class, with several examples of reactions catalyzed by complexes of iridium (I) and rhodium (I). The following reaction scheme for iridium (I) is illustrative:

$$Ir^I(CO)(PPh_3)_2Cl + XY \rightarrow Ir^{III}(Cl)(X)(Y)(CO)(PPh_3)_2 \quad [8.20]$$

where the substrate XY can be

$$X - Y \Rightarrow H - H, \quad H - Cl, \quad CH_3 - I, \quad CH_3CO - Cl, \quad Cl - Cl$$

Because the oxidation state increases, activation by oxidative addition is more effective for metals in a low state of oxidation. Therefore, noble metals of Group VIII that satisfy this requirement are the most suited for this purpose. (See Chaloner et al., 1994, for a detailed treatment of homogeneous hydrogenation.)

The Wilkinson catalyst

Hydrogenation by homogeneous catalysis has been widely studied over the years both with respect to mechanism and application. Ruthenium, iridium, and rhodium are the most frequently used metals. The best known is the rhodium catalyst $RhCl(PPh_3)_3$ that was first discovered by Wilkinson[2] and bears his name (see Jardine et al., 1967; Osborn et al., 1966; James, 1973).

THE CATALYTIC CYCLE

The mechanism of Wilkinson hydrogenation, extensively investigated by Halpern and co-workers (1973, 1976, 1978), is briefly outlined here, mainly to demonstrate a procedure for formulating appropriate catalytic cycles and formulating rate models. The complete cycle for Wilkinson hydrogenation is given in Figure 8.6 and, based on it, the following conclusions are important:

Step 6 in the catalytic cycle (boxed) represents substrate insertion and is invariably the rate-controlling step. This is a rather fortunate circumstance because the substrate and complex concentrations are directly measurable and based on them the kinetics can be readily established. Thus let us consider the hydrogenation of an olefin by Wilkinson's catalyst (Henri-Olive and Olive, 1977). The catalytic cycle in Figure 8.6 shows that all steps other than step 6 are

---

[2] G. Wilkinson and E.O. Fischer were awarded the Nobel prize for this discovery which started a burst of activity in homogeneous catalysis.

230   Catalysis

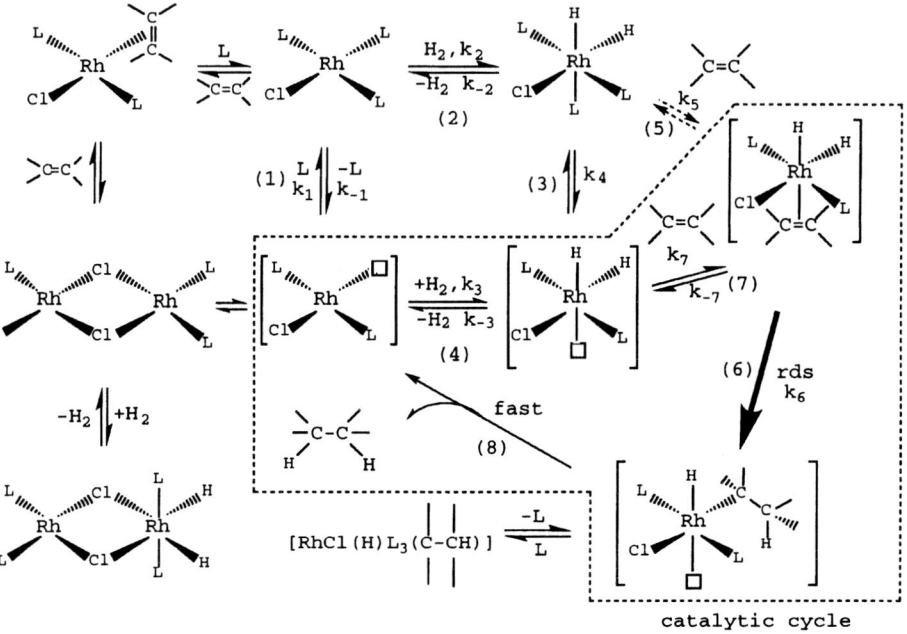

Figure 8.6 Catalytic cycle for hydrogenation of olefins using a Wilkinson catalyst (redrawn from Halpern et al., 1976). L = PPh$_3$, ☐ = vacant coordination site, rds = rate-determining step

characterized as fast or have reached equilibrium. Hence the rate-determining step of the cycle is step 6.

KINETICS AND MODELING

The kinetics of this reaction can be formulated by the method outlined in Chapter 7 (see also Gates, 1992, for a specific reaction). Let A = olefin, B = PPh$_3$, C = RhCl(PPh$_3$)$_2$H$_2$A, and D = RhCl(PPh$_3$)$_3$H$_2$. Then the rate of disappearance of olefin is given by the rate of step 6:

$$-r_A = -\frac{d[A]_b}{dt} = \frac{d[C]_b}{dt} = k_6[C]_b \tag{8.2}$$

where subscript b represents the liquid bulk. The equilibrium of the previous step is

$$K_5 = \frac{[B]_b[C]_b}{[A]_b[D]_b} \tag{8.3}$$

Because both the complexes C and D are saturated octahedral complexes, we can use the total concentration of the rhodium complex

$$[T]_b = [C]_b + [D]_b \tag{8.4}$$

as a readily measurable correlating parameter to eliminate the unknown concentration [C]. Thus, from Equations 8.3 and 8.4

$$[C]_b = \frac{k_5[A]_b[T]_b}{K_5[A]_b + [B]_b} \tag{8.5}$$

Combining this with Equation 8.3 leads to

$$-r_A = \frac{k_6 K_5[A]_b[T]_b}{K_5[A]_b + [B]_b} \tag{8.6}$$

Note that the final rate equation given by Equation 8.6 contains only quantities that are readily measurable (with the possible exception of K_5).

A general hydrogenation model

A similar analysis can also be made for other reactions. However, because detailed information on the catalytic cycle for a reaction at hand would not normally be available, the value of the equilibrium constant corresponding to K_5 would not be known. From a practical (reaction engineering) point of view, therefore, Equation 8.6 may be recast as a more general two-parameter rate equation of the form

$$-r_A = \frac{k_a[A]_b[T]_b}{K_5[A]_b + [B]_b} \tag{8.7}$$

and the constants k_a and K_5 determined from simple statistical methods, as described in Chapter 7. Then the constant k_6 can be obtained from $k_6 = k_1/K_5$. The equation can be generalized by letting K_5 be any constant k_b.

It is interesting to note that the nature of the ligand that is bonded to the metal and that is not a reactant is important in determining the reactivity of the complex. The data of O'Connor and Wilkinson (1968) show that the reactivity of a Wilkinson hydrogenation catalyst can be enhanced by a factor exceeding 50 merely by changing the ligand.

Reactions with Carbon Monoxide

Reactions with CO constitute a very important class of reactions catalyzed by transition-metal complexes. When CO alone is added to the substrate, such as in the reaction

$$\text{ROH} + \text{CO} \rightarrow \text{RCOOH} \tag{8.21}$$

or is added along with other reactant(s), there is an addition of CO either by itself or along with other moieties. Such reactions are generally referred to as *carbonylation*.

Specifically, when hydrogen is a coreactant along with CO, such as in the following reaction of an alkene:

$$\text{CH}_3\text{CH}=\text{CH}_2 + \text{H}_2 + \text{CO} \rightarrow \text{CH}_3\text{CH}_2\text{CH}_2\text{CHO} \tag{8.22}$$

we are basically adding a formaldehyde (H-CHO) molecule to the substrate. Therefore, such a reaction is termed as hydroformylation. Reaction 8.22 for the manufacture of butyraldehyde from propylene is perhaps the most prominent example of this class of reactions because it is the penultimate step in the process for the manufacture of the important plasticizer alcohol, 2-ethylhexanol.

The coreactant can also be water, an alcohol, or an amine, in which case the product will be a carboxylic acid or its corresponding derivative, such as in the reactions

$$RCH=CH_2 + CO + H_2O \rightarrow RCH_2CH_2COOH \qquad [8.23]$$

$$RCH=CH_2 + CO + R'OH \rightarrow RCH_2CH_2COOR' \qquad [8.24]$$

$$RCH=CH_2 + CO + R'NH_2 \rightarrow RCH_2CH_2CONHR' \qquad [8.25]$$

Similar reactions are possible with acetylenic and alcoholic substrates, but the former (an industrial component of Reppe chemistry which held sway for a number of years) are no longer of any practical importance. These specific types of carbonylation are referred to as *hydroxycarboxylation*. We shall very briefly consider some important reactions belonging to these classes.

Carbonylation

Basic reactions

Carbonylation substrates are usually alkynes, alkenes, alcohols, and esters, for example,

$$RC\equiv CH + CO + H_2O \rightarrow RCH=CH-COOH \qquad [8.26]$$

$$RCH=CH_2 + CO + H_2O \rightarrow RCH_2-CH_2-COOH \qquad [8.27]$$

$$ROH + CO \rightarrow RCOOH \qquad [8.28]$$

$$RCOOR' + CO \rightarrow \begin{matrix} ROC \\ \searrow \\ O \\ \nearrow \\ R'OC \end{matrix} \qquad [8.29]$$

Of these, the carbonylation of methanol to acetic acid:

$$CH_3OH + CO \rightarrow CH_3COOH \qquad [8.30]$$

is among the most important applications of homogeneous catalysis.

Carbonylation of methanol: kinetics and modeling

The kinetics and mechanism of the carbonylation of methanol to acetic acid using Monsanto's rhodium complex catalyst has been extensively studied. The reaction is first order in both rhodium and CH_3I promoter but zero order in CO pressure. It is believed that oxidative addition of CH_3I is the rate-controlling step in this process. This is a unique example of designing a catalyst system with commercial viability in which the substrate (methanol) is first converted to CH_3I

by a stoichiometric reaction with HI. Then the principal carbonylation reaction occurs between CH_3I and CO to give CH_3COI, which in the presence of water is converted to acetic acid, regenerating HI.

$$CH_3OH + HI \rightarrow CH_3I + H_2O \qquad [8.31]$$

$$CH_3OH + CH_3COOH \rightarrow CH_3COOCH_3 + H_2O \qquad [8.32]$$

$$CH_3I + CO \rightarrow CH_3COI \qquad [8.33]$$

$$CH_3COI + H_2O \rightarrow CH_3COOH + HI \qquad [8.34]$$

$$\text{Overall reaction: } CH_3OH + CO \rightarrow CH_3COOH \qquad [8.35]$$

In this reaction, the presence of water is essential although it is not consumed. The catalytic cycle describing the carbonylation step is shown in Figure 8.7. The nature of the active catalytic species has been extensively studied by Forster (1976, 1979), and it has been shown that a species of the type $[Rh(CO)_2I_2]^-$ is formed as the active species. Oxidative addition of this species to CH_3I to give $[MeRh(CO)_2I_3]^-$ is well accepted as the rate-determining step (Hjortkjaer and Jensen, 1977). Thus the kinetics of the overall reaction is

$$-r_A = k[B]_b[P]_b \qquad (8.8)$$

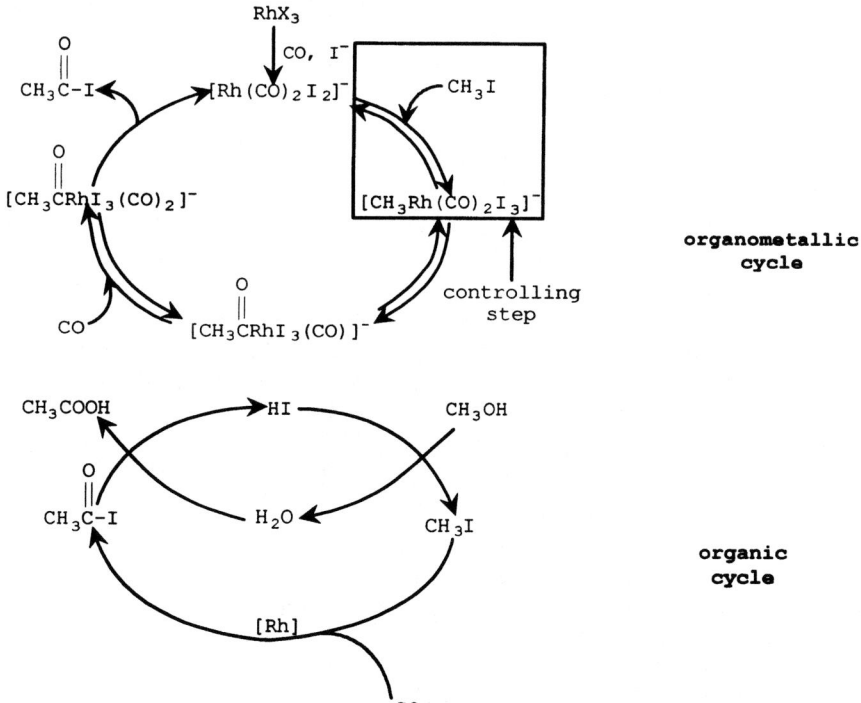

Figure 8.7 Proposed mechanism of rhodium complex-catalyzed carbonylation of methanol (Forster, 1976).

where B and P represent the complex $[Rh(CO)_2I_2]^-$ and promoter CH_3I, respectively. The rate constant k has a value of 3.5×10^6 ℓ/mol s.

In a more recent study, Dake et al. (1989) showed that the kinetics of the reaction changes at CO pressures less than 10 atm. They observed that below a pressure of 10 atm, the rate of carbonylation also depends on CO concentration. The following rate equation is proposed:

$$-r_A = \frac{kp_{co}[B]_b[P]_b}{(1 + Kp_{co})} \qquad (8.9)$$

In another study, direct experimental observation of the complex $[MeRh(CO)_2I_3]^-$ was reported (Haynes et al., 1993), providing strong evidence for the catalytic cycle of Figure 8.7.

In another study using a Ni complex as catalyst and N-containing ligands such as isoquinoline, a significant rate enhancement was reported (Example 8.5). This catalytic system offers high activity and selectivity like Rh and, in addition, is very cheap. The rate model for the reaction is given by (Kelkar et al., 1995)

$$-r_A = \frac{kp_{co}[C]_b^{0.89}[A]^*[P]_b}{(1 + K_1 p_{co})(1 + K_2[A]^*)(1 + K_3[W]_b)^2} \qquad (8.10)$$

where A is the reactant, and W is water. Water is seen to seriously inhibit the reaction—probably because of its interaction with the active species to give inactive species. In other words, water is not preferentially adsorbed. Hence it is surprising that water should inhibit the reaction.

Hydroformylation

Hydroformylation of higher olefins is important in the manufacture of plasticizer alcohols and perfumery chemicals. One of the most important applications is in the production of butyraldehyde, an intermediate in the manufacture of 2-ethylhexanol. Upon reaction with phthalic anhydride, the alcohol gives 2-ethylhexyl phthalate, which is extensively used as a plasticizer in polymer applications, especially for PVC. Another use is its application in exploiting 4-carbon feedstocks as an alternative to the more common 2- and 3-carbon feedstocks in the manufacture of organic chemicals. An example of the latter is the alkoxycarbonylation of butadiene to adipic esters which are subsequently hydrolyzed to the acid (see Kummer et al., 1979). Based largely on extensive studies on these two technologies, especially the production of butyraldehyde, certain general models for hydroformylation have emerged.

A major factor in modeling hydroformylation reactions is that two gases (H_2 and CO) are present compared to one gas (CO) in carbonylation. The most commonly used complexes are those of Co and Rh. A typical kinetic model for hydroformylation has the following hyperbolic (LHHW) form (Martin, 1954):

$$-r_A = \frac{kp_H[C]_b[P]_b}{p_{co} + K_H p_H} \qquad (8.11)$$

A more representative expression based on a detailed study of a number of olefinic substrates with different substituents (e.g., 1-hexene, allyl alcohol, and vinyl acetate) has a slightly more complicated form (Deshpande and Chaudhari, 1988, 1989a,b).

$$-r_A = \frac{k p_H^m p_{co}[B]_b[C]_b}{(1 + K_{co}p_{co})^n(1 + K_b[B]_b)^p} \tag{8.12}$$

with appropriate units for k.

An important observation from both forms is the negative order dependence of the rate on CO pressure. This behavior is consistent with the mechanism shown in Figure 8.8 wherein the formation of a dicarbonyl Rh complex is postulated. This species, which is inactive in hydroformylation, is formed at the expense of the active species. As the amount formed increases with CO pressure, there is a corresponding decrease in the active species concentration, leading to a progressive inhibition of the rate. A comprehensive rate equation

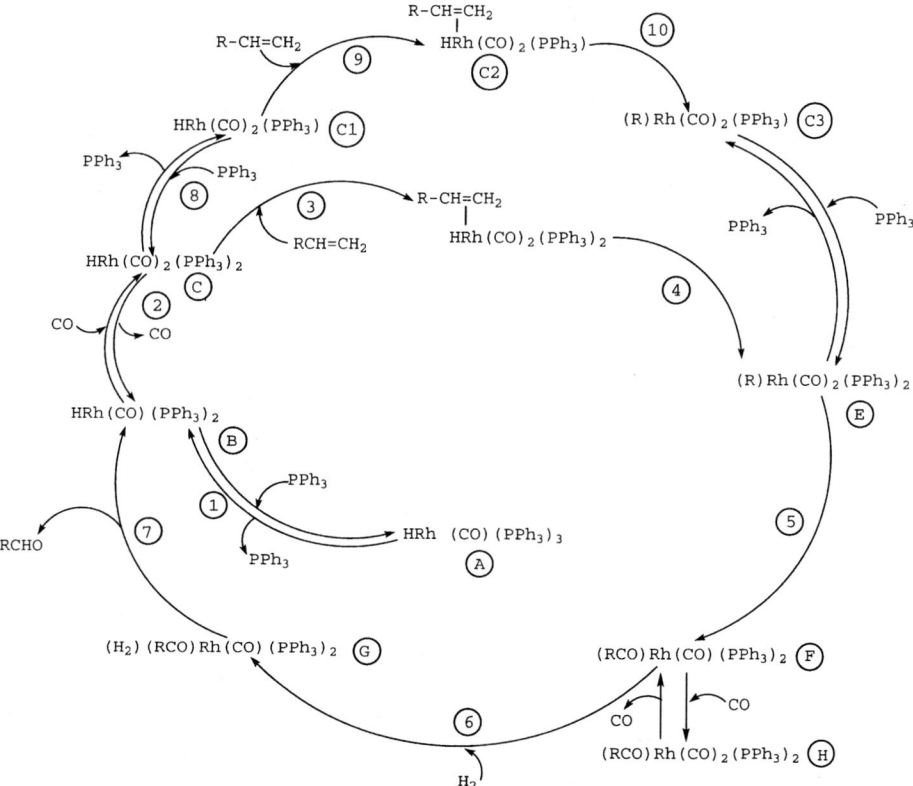

Figure 8.8 Mechanism of HRh(CO)(PPh$_3$)-catalyzed hydroformylation reactions (redrawn from Evans et al., 1968a; see also Brown and Wilkinson, 1970). R' represents RCH$_2$CH$_2$ or RCHCH$_3$

derived on the basis of the catalytic cycle of Figure 8.8 is given below assuming that step 6 is controlling (Divekar et al., 1993).

$$-r_A = \frac{kp_H p_{co}[A]^*[C]_b}{1 + K_1 p_{co} + K_1 K_2 p_{co}[A]^* + K_1 K_2 K_3 p_{co}^2[A]^* + K_1 K_2 K_3 K_4 p_{co}^3[A]^*} \quad (8.13)$$

This equation quantitatively predicts the negative order effect of CO concentration.

Oxidation/Epoxidation

Among the major initial impacts of homogeneous catalysis more than 30 years ago was the industrial production of acetaldehyde by the oxidation of ethylene. More recently, enantioselective epoxidation of olefins has been attracting increasing attention in the synthesis of pharmaceutical and specialty chemicals.

OXIDATION: WACKER PROCESS FOR ACETALDEHYDE FROM ETHYLENE

Known as the Wacker synthesis, this oxidation is accomplished with a mixture of $PdCl_2$ and $CuCl_2$ in aqueous solution as a homogeneous catalyst. The kinetics and mechanism of the reaction have been fully discussed in the literature, for example, Smidt et al. (1959). The main conclusion is that no single step is controlling and that reactor design should be based on all of the reactions being equally controlling [see the next section for a discussion of multistep control].

EPOXIDATION

Direct air epoxidation of propylene to propylene oxide suffers from selectivity problems. Epoxidation by alkyl hydroperoxide, as practiced by Arco, is based on the use of $Mo(CO)_6$ as a homogeneous catalyst. The most impressive use of homogeneous catalysis in epoxidation, however, is in the Sharpless asymmetric oxidation of allylic alcohols. In view of its importance, this enantioselective reaction is included in Chapter 9 which is devoted mainly to asymmetric catalysis.

Homogeneous Catalysis in Asymmetric Synthesis: Applications in the Pharmaceutical Industry

A number of drugs and pharmaceuticals are enantiomeric isomers and hence are prepared by any of the methods of asymmetric synthesis described in Chapter 9. Homogeneous catalysis is one such method and is a highly effective one that has been widely used by industry in the manufacture of bioactive chemicals such as drugs. A list of selected drugs prepared by asymmetric synthesis using homogeneous catalysts is given in Table 8.3 [see also Tables 9.6 and 9.7]. Hydrogenation and Sharpless synthesis are by far the most common reactions used.

Table 8.3 Examples of the application of homogeneous catalysis in the pharmaceutical industry (adapted from Mills and Chaudhari, 1997)

Reaction	Catalyst	Product (application)	Reference
Selective hydrogenation of C-22/C-23 double bond in synthesis of Ivermectin	$Rh(I)Cl(PPh_3)_3$	Ivermectin—an antiparastic drug	Chabala and Fischer (1980)
Hydrogenation of acetamido cinnamic acid (3-methyl-4-acetoxy derivative)	Rh-DIAMP chiral complex	L-Dopa	Knowles (1986)
Asymmetric hydrogenation of 2-naphthyl-4-methoxy acrylic acid	Ru-BINAP chiral complex	S-Naproxen	Ohta et al. (1987)
Synthesis of α-tocopherol by asymmetric hydrogenation	Ru-BINAP complex	Vitamin E	Akutagawa (1995)
Hydrogenation of acetoxyzentidine-2-one derivative	Ru-BINAP complex	β-Lactam	Akutagawa (1995)
Asymmetric hydrogenation of α-acetamido cinnamic acid	$[Rh(PNNP)(NBD)]$, a chiral complex	L-Phenylalanine	Fiorini and Giongo (1979)
Hydroformylation of 1,4-diacetoxybutene	$HRh(CO)(PPh_3)_3$ or Rh complex without PPh_3	4-Acetoxy-2-methyl crotonaldehyde—an intermediate in vitamin A synthesis	Stinson (1986)
Hydroformylation of acrylonitrile	Co-carbonyl complex	$NC-(CH_2)-CHO$—an intermediate in sodium glutamate synthesis	Botteghi et al. (1987)
Oxidation of 2-methyl naphthalene	Metal salt and heteropolyacid	2-Methyl 1,4 naphthoquinone	Kozhevnikov (1995)

GENERAL KINETIC ANALYSIS

Intrinsic Kinetics

Recall that in the majority of reactions using homogeneous catalysts in the liquid phase, a gas phase is also present, mainly hydrogen and/or carbon monoxide. This diffusion of gas in liquid can falsify the kinetics. In this section we consider the modeling of gas-liquid reactions in the absence of diffusional effects.

The development of a mechanistic model for a homogeneous reaction requires constructing a catalytic cycle, which is quite difficult. On the other hand, simple kinetic expressions of both the power law and hyperbolic types can be readily derived. These are usually adequate for reactor design. Thus in the analysis of homogeneous catalysis involving a gas-liquid reaction, the following general hyperbolic form of the rate equation may be used:

$$-r_A = \frac{k[A]^*[B]_b[C]_b}{(1 + K_A[A]^* + K_B[B]_b)^n} \qquad (8.14)$$

or

$$-r_A = \frac{k[A]^*[B]_b[C]_b}{(1 + K_E[E]^*)^n} \qquad (8.15)$$

where A, B, and C represent, respectively, the gas and liquid phase reactants and the catalyst. The second form, where E represents carbon monoxide, is particularly useful for hydroformylation reactions. The value of n varies depending on the substrate used. For hydroformylation of hexene, for example, $n = 2$, and for that of allyl alcohol $n = 3$. Table 8.4 summarizes the hyperbolic models for selected reactions. These models are clearly reminiscent of the LHHW models discussed in Chapter 7.

Multistep Control

In the foregoing examples, rate equations were developed on the basis of a single rate-determining step. It is possible that many steps of a cycle are simultaneously controlling, as in the Wacker process. The rate equation for such a reaction tends to be more complicated but can be developed by the methods discussed in Chapter 7. Thus for the oxidation of triphenylphosphine with a Pt complex, a rate equation can be developed based on the catalytic cycle shown in Figure 8.9 (Halpern and Pickard, 1970; Birk et al., 1968a,b):

$$\text{rate} = \frac{k_1 k_2 [\text{Pt complex}][\text{PPh}_3][O_2]}{k_1[O_2] + k_2[\text{PPh}_3]} \qquad (8.16)$$

where k_1 and k_2 are the rate constants for the steps marked 1 and 2 in the cycle:

$$\text{Pt}(\text{PPh}_3)_3 + O_2 \xrightarrow{k_1} \text{Pt}(\text{PPh}_3)_2 O_2 + \text{PPh}_3 \qquad [8.36]$$

$$\text{Pt}(\text{PPh}_3)_2 O_2 + \text{PPh}_3 \xrightarrow{k_2} \text{Pt}(\text{PPh}_3)_3 O_2 \qquad [8.37]$$

Table 8.4 Summary of kinetic models used in homogeneous catalysis (adapted from Mills et al., 1992)

S. No.	Reaction system	Catalyst	Rate equation, $-r_A$ (mol/cm^3 s)[a]	Reference
1.	Hydrogenation of cyclohexane	RhCl(PPh$_3$)$_3$	$\dfrac{k[A]^*[B]_b[C]_b}{1+K_A[A]^*+K_B[B]_b}$	Osborn et al. (1966)
2.	Hydrogenation of allyl alcohol	RhCl(PPh$_3$)$_3$	$\dfrac{k[A]^*[B]_b[C]_b}{1+K_A[A]^*+K_B[B]_b{}^2}$	Wadkar and Chaudhari (1983)
3.	Oxidation of cyclohexane	Mn(OAc)$_2$	$\dfrac{k[B]_b[C]_b}{(k_1+k_2[C]_b)}$	Kamiya and Kotake (1973)
4.	Oxidation of ethylene	RCl(PPh$_3$)$_3$	$\dfrac{k[A]^*[B]_b[C]_b}{(1+K_A[B]^*)}$	Chaudhari (1984)
5.	Carbonylation of methanol	RhCl$_3$/HI solution	$k[P]_b[C]_b$	Roth et al. (1971)
6.	Hydroformylation of propylene	HCo(CO)$_4$	$\dfrac{k[A]^*[B]^*[C]_b}{[E]^*}$	Natta et al. (1954)
7.	Hydroformylation of diisobutylene	Co$_2$(CO)$_8$	$\dfrac{k[A]^*[P]_b[C]_b}{E^*+K_A[A]^*}$	Martin (1954)
8.	Hydroformylation of hexene and allyl alcohol	HRhCO(PPh$_3$)$_3$	$\dfrac{k[A]^*[B]_b[C]_b}{(1+K_E[E]^*)^n}$, $n=2$ for hexene, 3 for allyl alc.	Deshpande and Chaudhari (1983)
9.	Hydroformylation of allyl alcohol	HRhCO(PPh$_3$)$_3$	$\dfrac{k[A]^{*1.52}[E]^*[C]_b[B]_b}{(1+K_E[E]^*)^3(1+K_B[B]_b)^2}$	Deshpande and Chaudhari (1989a)
10.	Hydroformylation of 1-decene	HRh(CO)(PPh$_3$)$_3$	Equation (8.13) of text	Divekar et al. (1993)

[a] $[A]^*$ = interfacial concentration of O$_2$, H$_2$, CO; $[B]^*$ = interfacial concentration of gaseous substrate; $[B]_b$ = concentration of liquid-phase substrate; $[C]_b$ = concentration of catalyst; $[E]^*$ = interfacial concentration of CO; $[P]_b$ = concentration of promoter.

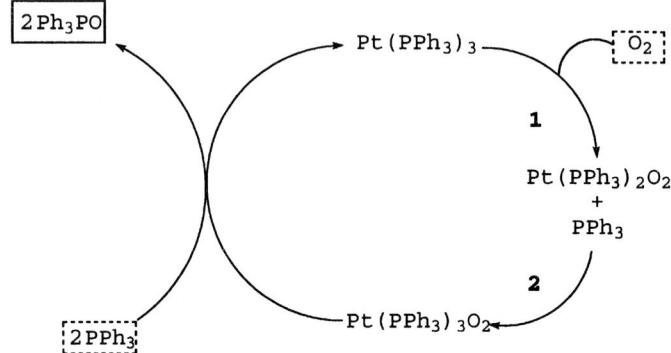

Figure 8.9 Catalytic cycle for the oxidation of triphenylphosphine (drawn from reactions given by Halpern and Pickard, 1970).

Role of Diffusion

Recall that there are a number of reactions where homogeneous catalysis involves two phases, liquid and gas, for example, hydrogenation, oxidation, carbonylation, and hydroformylation. The role of diffusion becomes important in such cases. In Chapter 7 we considered the role of diffusion in solid-catalyzed fluid-phase reactions. The treatment of gas-liquid reactions is based on a different approach, however, and uses an "enhancement factor" to express the enhancement in the rate of absorption due to reaction, rather than an effectiveness factor to express the effect of diffusion on reaction. A catalyst may or may not be present. If there is no catalyst, we have a simple noncatalytic gas-liquid heterogeneous reaction in which the reaction rate is expressed by simple power law kinetics. On the other hand, when a dissolved catalyst is present, as in homogeneous catalysis, the rate equations acquire a hyperbolic form (similar to LHHW models discussed in Chapter 7). Therefore, the mathematical analysis of such reactions becomes more complex. Because this is a logical extension of noncatalytic gas-liquid reactions, we defer a consideration of the role of diffusion in homogeneous catalysis to Part IV concerned with heterogeneous reactions involving a liquid phase (noncatalytic and catalytic).

The situation is quite different for supported homogeneous catalysts. They were briefly considered in Chapter 7, and will be treated at greater length in the section on triphase catalysis in Chapter 19.

EXAMPLE 8.5

Development of a new catalyst for the carbonylation of methanol to acetic acid

Acetic acid is commercially produced via the homogeneously catalyzed carbonylation of methanol in a high pressure gas-liquid reactor. Because it is one of the largest scale operating processes, the catalyst for this reaction has been undergoing continual improvement. This example describes the development of a new catalyst in this continuing effort.

EXISTING CATALYSTS

Carbonylation of methanol to acetic acid is described by reactions 8.31–8.35. The main reactants are methanol in the liquid phase and carbon monoxide in the gas phase, and the heat of reaction is 33 kcal/mol. The present state of the art of the catalysts used for this process is as follows:

1. *Cobalt complex catalyst:* A homogeneous cobalt carbonyl complex was used in the first commercial process (developed by BASF) which requires pressures as high as 600–700 atm and a temperature of 250°C.
2. *Low pressure rhodium complex catalyst:* Due to the need for high activity and selectivity under milder operating conditions, a new catalyst system consisting of a soluble rhodium complex catalyst was developed by Monsanto. The carbonylation occurs at milder operating conditions (150–200°C and 50–70 atm pressure).

CATALYST DEVELOPMENT

Although the Monsanto catalyst is a great improvement over the original BASF catalyst, it suffers from the following drawbacks: Rh is very expensive and any losses even in trace amounts have a direct impact on process economics, and Rh reserves are limited and available only in certain regions. The latter is a matter of great concern in view of the continuous growth in acetic acid capacity.

Any alternative catalyst should preferably be based on cheaper metals like Ni or Fe. These catalysts are known to require high pressures with efficiencies even poorer than for the Co-complexes. Therefore, an alternative catalyst can be designed only by using cocatalysts/promoters and or new ligands. The catalyst should be designed so that the rate of oxidative addition of methyl iodide is increased. Such a catalyst was developed at the NCL (Kelkar et al., 1990, 1992) by stabilizing the nickel carbonyl iodide complex of the type $[Ni(CO)_3I]^-$ under reaction conditions. This was achieved by using isoquinoline as the ligand. Stabilization of the complex by quaternization was thus facilitated, resulting in a significant increase in activity and selectivity. The possible structure of the complex is

$[Ni(CO)_3I]^-$ [isoquinolinium structure]$^+$, R = H or CH_3

The overall performances of the three catalysts are compared in Table 8.5. It is obvious that the Rh and Ni catalysts are superior to the Co catalyst in all respects. However, unlike Rh, Ni is not yet proven on a commercial scale.

Table 8.5 Comparison of catalysts for methanol carbonylation

Catalyst	Temperature (°C)	Pressure (atm)	Selectivity (%) Methanol based	Selectivity (%) CO based
Co Complex (BASF)	210–250	600–700	85	60
Rh Complex (Monsanto)	150–200	50–70	99	90
Ni Complex (NCL catalyst)	180–220	40–50	98	95

EXAMPLE 8.6

Illustrating the use of homogeneous catalysis in developing a cleaner technology for an available product: Ibuprofen

A striking example of this is the manufacture of ibuprofen using the Boots Hoechst-Celanese process based on homogeneous catalytic carbonylation of *p*-isobutylphenyl-ethanol as a key step.

The conventional process for the manufacture of ibuprofen consists of a six-step synthesis involving the use of stoichiometric reagents and large volumes of toxic solvents. A new catalytic route, which exploits carbonylation chemistry in one of the steps, is compared with the conventional route in Figure 8.10. The carbonylation step is critical and has many advantages. Further, this new catalytic route gives almost 99% atom utilization efficiency.

Note: Other competing technologies, such as that based on asymmetric catalysis (see Chapter 9) are also available for ibuprofen.

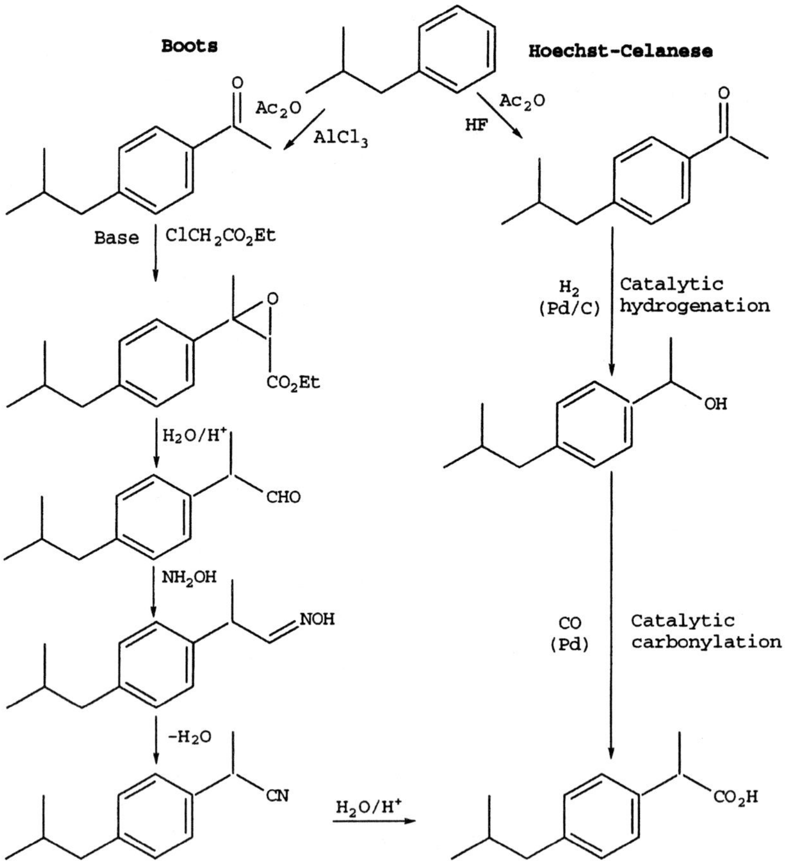

Figure 8.10 Conventional and catalytic routes for ibuprofen.

Chapter 9

Asymmetric Synthesis

The manufacture of fine chemicals, particularly drugs, fragrances, and flavors, is undergoing a major revolution now as a result of the capability of chemists to prepare these chemicals, mainly drugs, in their purest isomeric forms (as stereoisomers). This shift to pure forms has been described by Brown in the following words (1990) (see also Deutsch, 1991): "A mixture of stereoisomers in a medicine will (now) need to be justified just the same way as any other mixture of compounds." Indeed, in the United States today (as in many other advanced countries), the use of pure enantiomeric forms is practically a requirement since extensive justification is needed to continue with racemates (FDA, 1992). As a consequence, the combined sales of the "chiral top ten" drugs (ammoxycillin, enalapril, ampicillin, captopril, pravastatine, diltiazem, ibuprofen, lovastatin, naproxen, and fluoxetine) in 1994 amounted to more than 16 billion dollars (Sheldon, 1996). (Of these, ibuprofen and fluoxetine are still sold as racemates.)

Because patent expiry for a racemate tends to proliferate the drug as generics, product line extension for an existing racemate technology can be obtained by switching to a single stereoisomeric form (the *racemic switch*). This is an incentive to produce the drugs in their pure enantiomeric forms.

The use of homogeneous catalysis in organic synthesis and technology was considered in the previous chapter. A particularly useful application of homogeneous catalysis is in the production of stereoisomers in pure forms by a rapidly expanding technique known as *chiral (asymmetric) synthesis* or *chirotechnology*. Note, however, that the term asymmetric synthesis is much broader in scope, and the use of homogeneous catalysis to achieve *chirality* is only one of the methods of doing so. When applied to this specific case, it is generally referred to as *chiral* or *asymmetric catalysis*. This again is a much broader term and includes both homogeneous and heterogeneous catalyses.

Although chirality is not a prerequisite for biological activity, the presence of stereogenic centers in bioactive molecules gives rise to large differences in the activities of the individual enantiomers. It is this fact that is exploited in the production of the desired pure enantiomers. Among the many recent books and reviews on the subject, the following are particularly noteworthy: Bartok (1985), Morrison (1985), Brunner (1988), Ojima et al. (1989), Blaser and Müller (1991), Collins et al. (1992), Aitken and Kilěnyi (1992), Federsel (1993), Ojima (1993), Chan (1993), Chan et al. (1994), Brunner and Zettlmeier (1993), Sheldon (1990a,b, 1993, 1996).

We begin our treatment with a brief introduction to the basic principles of chirality and then describe its use as a powerful tool in the economic manufacture of a variety of stereospecific chemicals (thus all but eliminating the untoward effects of admixture with their stereoisomers).

Note

Although Part II is devoted to catalysis in organic synthesis, a distinction has been made in the case of asymmetric catalysis. Because the induction of asymmetry in a molecule is one of the most important aspects of organic synthesis today and because the dividing line between asymmetric catalysis and other means of chiral induction is rather fuzzy, the scope of this chapter has been expanded to asymmetric synthesis in general with emphasis on asymmetric catalysis.

BASIC DEFINITIONS/CONCEPTS

The basic concepts and other important features of asymmetric synthesis are explained next.

Enantiomers, Stereoisomers, and Chirality

These are explained in a number of books and reviews (e.g., Aitken and Kilěnyi, 1992; Wade, 1987; Federsel, 1993). Every molecule, indeed every object, has a mirror image. Most molecules such as bromochloromethane, however, can be superimposed on their mirror images. Hence the mirror images of such molecules are not isomers. On the other hand, if the mirror images are not superimposable on each other, they represent different entities and are regarded as isomers. These molecular structures that are constitutionally identical but differ only in the three-dimensional arrangement of their atoms, for example, alanine, constitute a special class of isomers known as enantiomers.

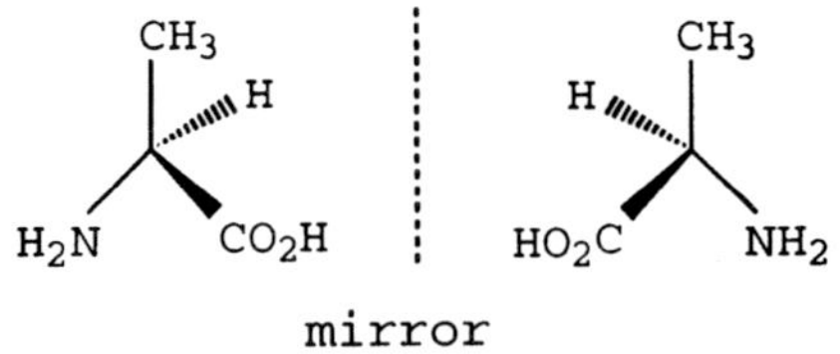

Two enantiomeric forms of alanine

Note, however, that there can be three-dimensional arrangements of atoms in a molecule that are not mirror images of each other. The resulting isomers are known as *diastereomers*. Both these space-oriented forms of isomers, one representing a nonsuperimposable set of mirror images and the other that are not mirror images at all, belong to a general class of three-dimensional isomers known as *stereoisomers*.

Molecules that are nonsuperimposable mirror images of each other are said to be *chiral* (Greek word for "handed") and the property is referred to as *chirality*. The central carbon C* in any organic compound is referred to as the *stereogenic center* if it is attached to four different groups. A well-known example is

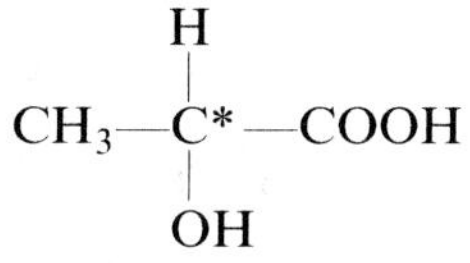

$$CH_3 - \overset{H}{\underset{OH}{C^*}} - COOH$$

Lactic acid

A single stereogenic center in a molecule always makes it chiral. On the other hand, the presence of more than one stereogenic center can destroy chirality, that is, make the molecule *achiral*, and result in the formation of diastereomers. If there are n stereogenic centers, anywhere up to 2^n diastereomers can be formed.

If an achiral molecule contains two groups (or atoms), one of which leads to enantiomeric products whereas the other does not, these groups are said to be *enantiotopic* and the molecule itself is *prochiral*. Similarly, if one of the groups leads to diastereomeric products, then the groups are said to be *diastereotopic*. Also note that there are three types of chiral molecules, *central*, *axial*, and *planar* (Table 9.1). The *central configuration* in which the groups or atoms are arranged about a stereogenic center is the most common.

Table 9.1 Types of chirality

Type of stereogenic unit	Example	Remarks
Central	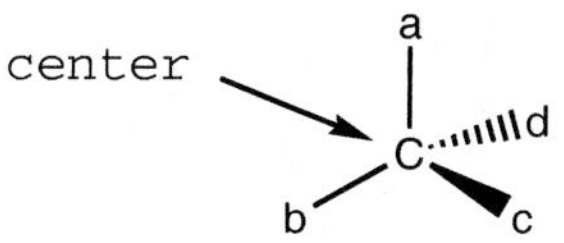 center → C with a, b, c, d	This tetrahedral structure with a *stereogenic center* is the most common.
Axial	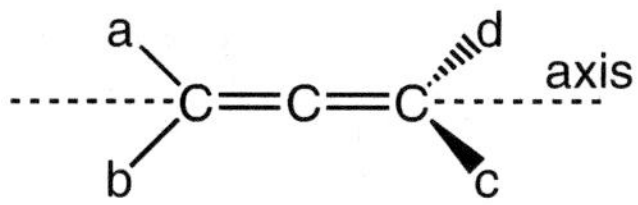 a, b — C=C=C — d, c axis	Chirality is the result of arrangement of molecules about a *stereogenic axis*.
Planar	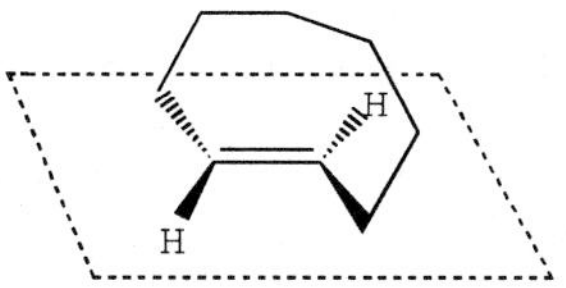 (ring with H substituents in a plane)	Chirality is the result of arrangement of molecules about a *stereogenic plane*.

The nonsuperimposable character of mirror images that causes enantiomerism is also responsible for their optical activity. This is the only feature that distinguishes enantiomers from each other, for they are identical in every other way, including chromatographic retention time and IR and NMR spectra. The enantiomer that rotates a plane of plane-polarized light in the clockwise direction is denoted (+) or d, and that which shows equal and opposite rotation under the same conditions is noted (−) or l.

Absolute Configuration: Assignment of Priorities

The classification just described does not give any information about the neighborhood of the stereogenic center. In other words, its absolute configuration, that is, the nature of the groups attached to its four bonds, remains unclear. A detailed methodology for determining and specifying this neighborhood has been provided by Cahn et al. (1966) (see also Prelog and Helmchen, 1982). This methodology is based on placing the four groups attached to a tetrahedral center in a predetermined *order of priority*, or simply *priority*. The priority, which varies from 1 to 4 (4 is the lowest), is based on the atomic numbers of the groups. Thus for a molecule Cabcd in which the ranking in terms of atomic numbers is $a > b > c > d$, the enantiomers (R) and (S) are identified as follows:

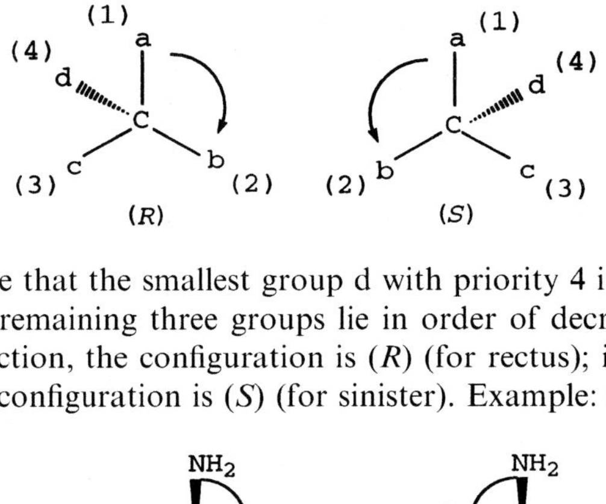

Note that the smallest group d with priority 4 is at the back of the molecule. If the remaining three groups lie in order of decreasing priority in the clockwise direction, the configuration is (R) (for rectus); if in the anticlockwise direction, the configuration is (S) (for sinister). Example: (R)- and (S)-phenylglycine.

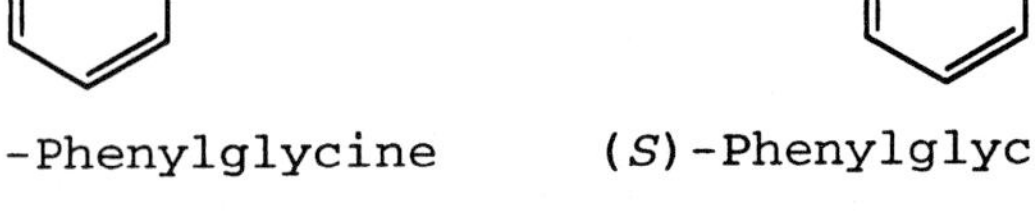

(R)-Phenylglycine   (S)-Phenylglycine

This is the basic concept behind asymmetric synthesis.

Racemic Mixtures, Resolution, and Enantiomeric Excess

When a compound prepared from a molecule that contains a chiral center is optically inactive, it does not mean that it has lost its chirality. What actually happens is that even a small sample of a compound contains a number of

different molecules, and for every molecule oriented one way there is another oriented as its mirror image. The net result is that there is no polarization of light. Therefore, the optical inactivity is the result of the mutual cancellation of equal and opposite optical effects. Such molecules are known as *racemic mixtures* or *racemates*. Because only one of the enantiomeric forms of this mixture is the desired bioactive molecule, it has to be separated from this mixture. The process of separation is known as *resolution*.

The resolution of a racemic mixture can give the desired enantiomer in pure form. Often, however, complete resolution cannot be achieved, and one should therefore have a measure of the degree of resolution, that is, of the amount of the desired enantiomer that would be present (after the resolution) over and above the ratio of 1:1 in the racemic mixture. Such a measure is also needed in characterizing the efficiency of preparing the desired isomer from a substrate by any of the methods described later for asymmetric synthesis. This measure is the *enantiomeric excess* (ee), also called *enantioselectivity* or *optical yield*, defined as

$$\text{Enantiomeric excess (ee)} = \frac{[R] - [S]}{[R] + [S]} \quad (9.1)$$

Enantioselectivity is a kinetic phenomenon caused by the difference in activation energies (ΔE) between the reactions

$$A \to R$$

$$A \to S$$

whose rate constants are given by

$$k_R = k_R^0 e^{-E_R/R_g T}, \quad k_S = k_S^0 e^{-E_S/R_g T} \quad (9.2)$$

Because the products have almost identical physical properties, it can be assumed that frequency factors k_R^0 and k_S^0 are equal, so that

$$\frac{p_R}{p_S} = e^{-(E_R - E_S)/R_g T} = e^{-\Delta E/R_g T} \quad (9.3)$$

Thus one can prepare a plot of the enantiomeric ratio p_R/p_S as a function of ΔE at different temperatures. A more useful way of respresenting the results (Sheldon, 1996) is to plot the ee of the remaining substrate, say (S), as a function of conversion for different values of k_R/k_S assuming that (R) reacts selectively (Figure 9.1). It is clear from the figure that high values of this ratio (also called the enantiomeric ratio) are needed to obtain high values of ee for the remaining substrate (usually over 100) (Sheldon, 1996). This is a vital factor in choosing between kinetic resolution and asymmetric synthesis, considered later.

Importance of Chirality

It is a crucial fact of nature that most of the important building blocks that constitute the biological macromolecules of living systems are constructed only from one enantiomeric form. Thus when a biologically active compound such as

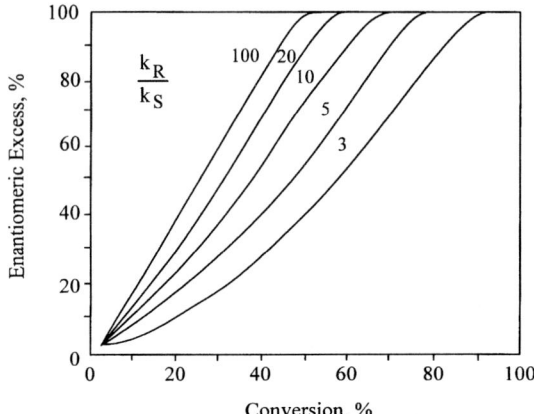

Figure 9.1 Dependence of ee of remaining substrate (S) on the enantiomeric ratio k_R/k_S and total conversion.

a drug interacts with its receptor site which is chiral, it is clear that the nature of the interaction will depend on the enantiomeric form of the drug. This has generated an immense amount of research in identifying the enantiomer responsible for a desired activity and also in manufacturing this enantiomer in the most economical way possible.

Ever since the thalidomide debacle[1] (Roskies, 1972) one-quarter century ago, it has become increasingly clear that administration of racemic mixtures to a patient can have dangerous consequences. Usually, only one of the enantiomers has the desired medicinal effect. The other can be neutral or toxic. The relationship between the form of the enantiomer [i.e., its absolute configuration, (R) or (S)] and biological effect is shown in Table 9.2. It may be argued that the racemic form can be used directly when the unwanted enantiomer is not toxic, as in the case of ibuprofen; but this is not an economically sound argument in view of the high cost of letting the unwanted isomer function as a natural diluent. Also, developing methods for preparing pure enantiomers is a clinical necessity.

It is important to note, however, that economic constraints may often make it too expensive to replace an existing technology that consists of racemate preparation followed by separation, by a process based on chiral synthesis.

METHODS OF PREPARING PURE ENANTIOMERS

Several methods of isolating an enantiomer in its pure form or of synthesizing it directly from a substrate have been in use. A classification of these is given in Table 9.3.

[1] This drug, sold as a racemic mixture, was used for its sedative action in pregnant women. It was found later that this action was caused by its (+)-enantiomer, whereas the (−)-enantiomer caused birth deformities in children. This finding is regarded as historical because it revolutionized thinking about the role of enantiomeric impurities in drugs and has set the tone for legislation requiring the use of pure enantiomeric forms only.

Table 9.2 Relationship between absolute configuration and biological activity of enantiomerically related compounds

Serial no.	Name of compound	Absolute configuration	Biological effect
1	Asparagine	R	Sweet taste
		S	Bitter taste
2	Carvone	R	Spearmint flavor
		S	Caraway flavor
3	1-Chloropropane-2,3-diol	R	Toxic
		S	Antifertility activity
4	Propranolol	R	Contraceptive
		S	Antihypertensive, antiarrhythmic
5	Thalidomide	R	Sedative, hypnotic
		S	Extreme teratogen
6	Levodopa	R	Toxic
		S	Treatment for Parkinsonism
7	Naproxen	R	Liver toxin
		S	Anti-inflammatory
8	Ibuprofen	R	Inactive
		S	Anti-inflammatory
9	Chloramphenicol	R, R	Antibacterial
		S, S	Inactive
7	Ethambutol	R, R	Causes blindness
		S, S	Tuberculostatic
10	Paclobutrazol	R, R	Fungicide
		S, S	Plant growth regulator
11	Deltamethrin	R, R, S	Potent insecticide
		S, S, R	Inactive
6	Fluazifop butyl	R	Herbicide
		S	Inactive
12	Penicillamin	R	Toxic
		S	Antiarthritic
13	Limonene	R	Orange odor
		S	Lemon odor
14	Warfarin		Hypoprothrombinaemic; (S) 5–6 times more potent than (R)

Traditional Synthesis Followed by Resolution

The traditional method of preparing a pure enantiomer is by conventional organic synthesis that results in a racemic mixture, followed by its resolution using processes that are different from the usual physical separation techniques (Knowles, 1986). In general, the disadvantages of this traditional approach may be summarized as follows: (1) the highest possible yield is only 50%, but much lower in actual practice; (2) although sometimes the undesirable enantiomer can be racemized and the racemic mixture subjected to resolution again, such a process would evidently be tedious and expensive; and (3) all resolution processes require at least stoichiometric amounts of resolving agents, which again adds to the cost of production.

Table 9.3 Classification of routes for pure enantiomer synthesis

```
                          Pure enantiomer
                             synthesis
                    ┌────────────┴────────────┐
              From racemates            Asymmetric
                                         synthesis
      ┌──────────┬──────────┐         ┌─────┴─────┐
Diastereomer  Direct    Kinetic   From achiral  From
crystallization crystallization resolution  or prochiral  chiral
                                          substrates    pool
                          ┌────┴────┐    ┌─────┴─────┐
                      Enzymatic  Chemical  Asymmetric  Enzyme
                                            catalysis  catalysis
```

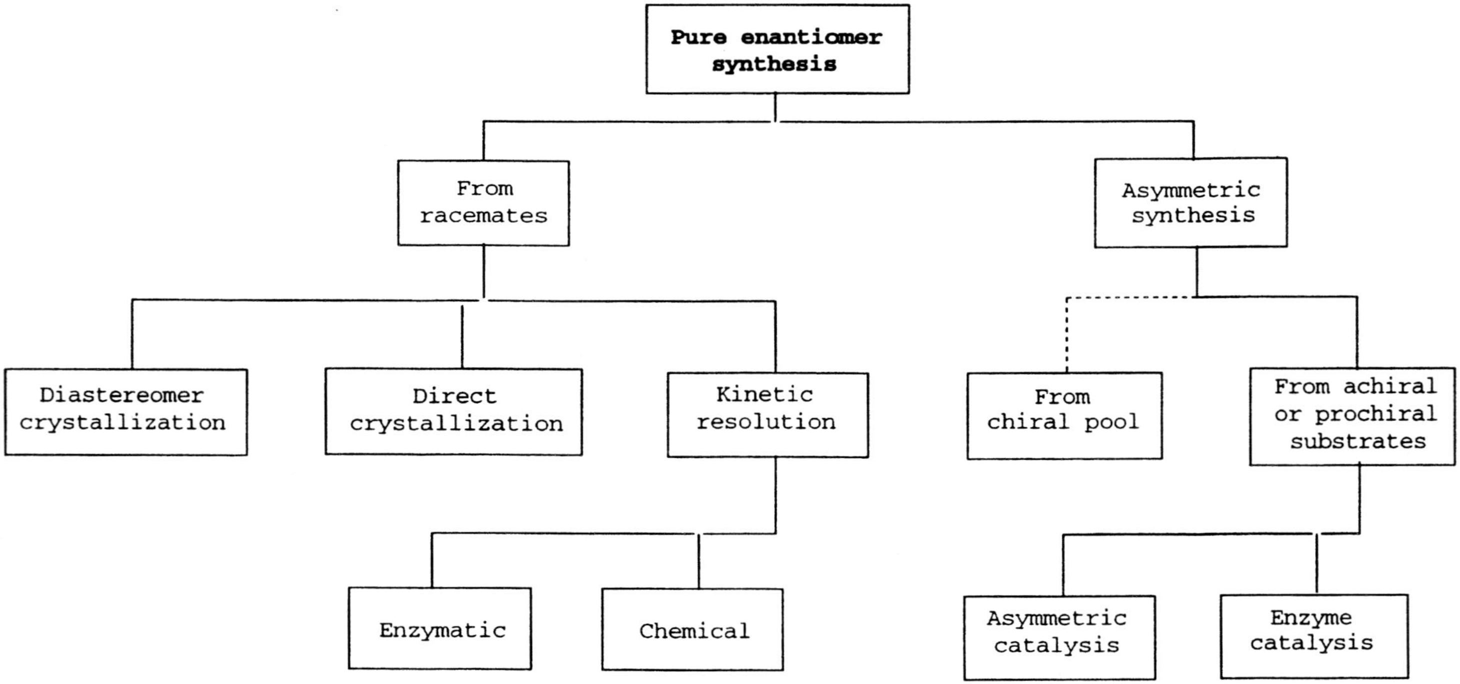

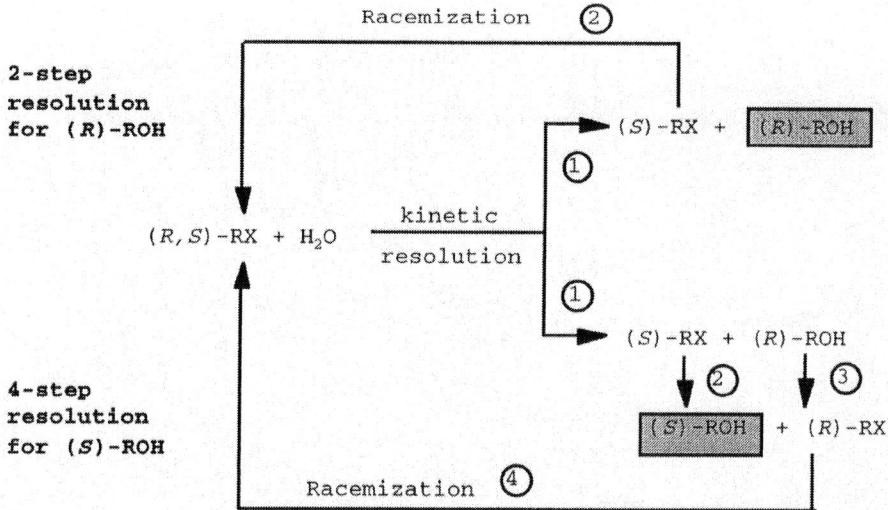

Figure 9.2 Direct (two-step) and indirect (four-step) resolution of the desired entantiomer (boxed) (redrawn from Sheldon, 1996).

A general point to note is that the number of steps involved in any resolution process must be kept to the minimum. The actual number of steps depends on whether the resolution leads directly to the desired enantiomer or some intermediate steps are present. Figure 9.2 illustrates schemes for direct and indirect resolution involving two and four steps, respectively (Sheldon, 1996).

There are essentially three important variations of this traditional method, as described next.

Diastereomer crystallization

The most common method of resolving a racemic mixture is to use an enantiomerically pure organic acid or base as the *resolving agent* that bonds with the enantiomers. Such bonding results in a pair of diastereomers which, because of their differing physical properties, can be more easily separated. In spite of the introduction of newer processes based on asymmetric catalysis, this classical technique of preparing pure enantiomers still remains the most widely used in industry. Some important examples are ethambutol (a tuberculostat), (+) or D-phenylglycine (a drug intermediate), diltiazem (an antihypertensive), and deltamethrin (an insecticide). The method is best illustrated by the following two processes:

1. For the drug diltiazem, using (S)-α-methylbenzylamine as the resolving agent:

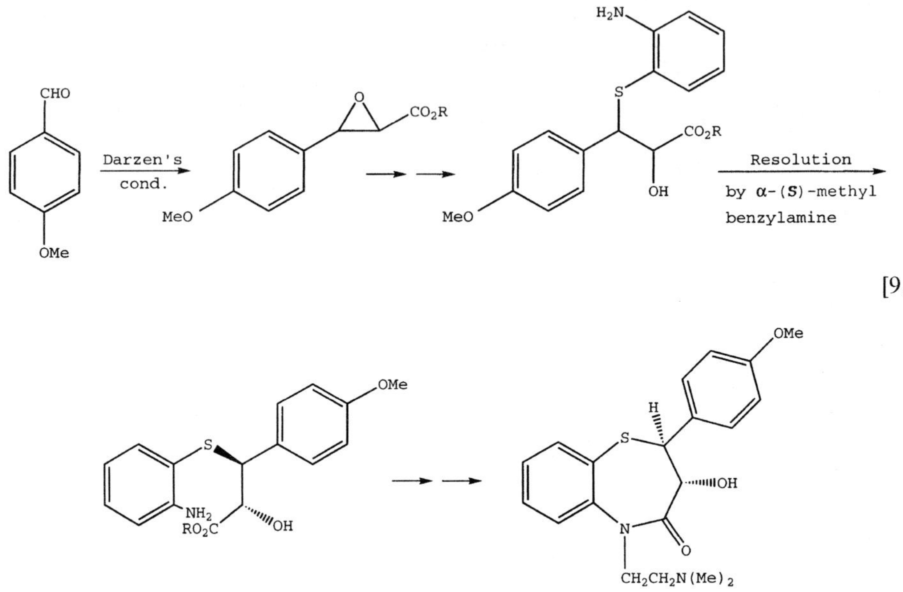

[9.1]

This process is currently in industrial production in India (Rama Rao, 1993).
2. For the insecticide deltamethrin (Roussel-Uclaf).

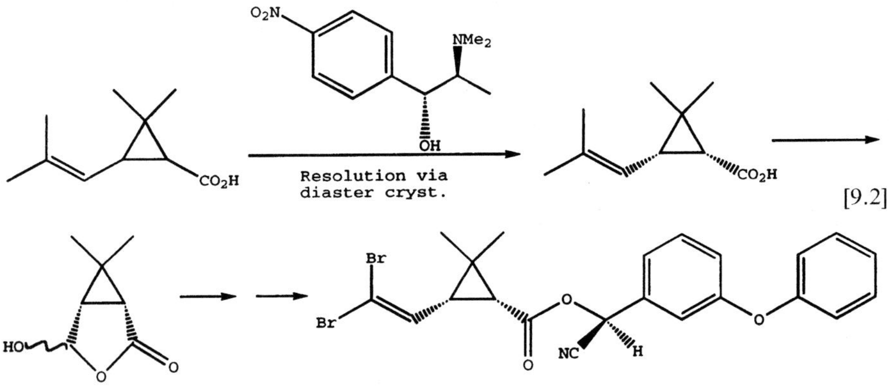

[9.2]

As stated earlier, resolution processes give a maximum yield of 50% of the desired enantiomer, but in practice it is usually less than 40%. The yield can be raised beyond 50% if the diastereomer that remains in solution can be made to undergo spontaneous epimerization. An example is the synthesis of dextropropoxyphene by asymmetric transformation of the diastereomeric salt formed from the Mannich reaction product using dibenzoyl-L-tartaric acid (reaction 9.3).

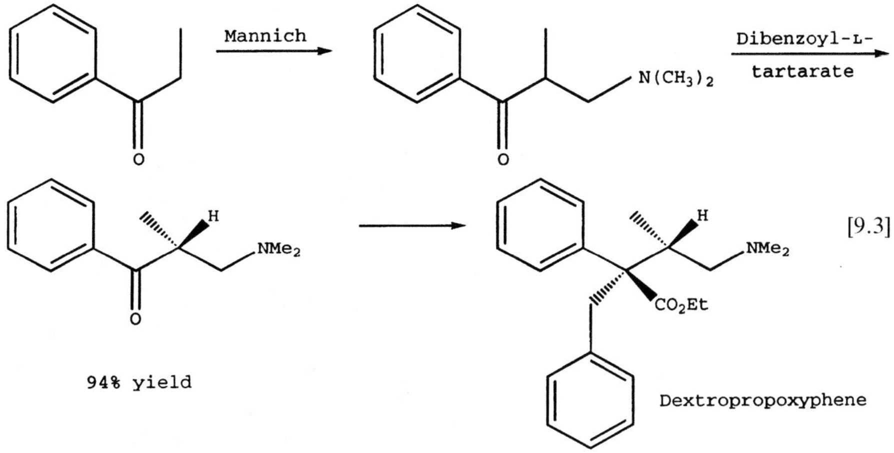

[Scheme showing synthesis leading to Dextropropoxyphene, labeled 9.3, with 94% yield noted]

An important factor in the resolution methods is the need to introduce the resolution step much earlier in the synthetic scheme to avoid carrying the isomeric ballast through a number of steps. Figure 9.3 for the synthesis of captopril illustrates the obvious difference between late and early introduction of the resolution step.

CHOICE OF RESOLVING AGENT

There are no set guidelines for selecting a suitable resolving agent to form a diastereomeric salt, nor is it possible to predict the solubilities of such complexes. The selection is mostly by trial and error with the following broad guidelines in mind: (1) the resolving agent should be easily available at low cost; (2) it should be readily recoverable from the crystallization step; (3) the chiral center should be close to the functional group responsible for the diastereomeric salt formation; (4) the resolving agent should preferably be a strong acid or base in order to ensure formation of a stable salt under the conditions of resolution, that is, elevated temperatures of crystallization; and (5) preferably, both of its enantiomers should be available. Some examples are amino acids, alkaloids, and D-glucosamine for resolving acids, and tartaric acid, camphor, sulfonic acid, and camphor sulfonic acid for resolving bases. Further, several synthetic chiral intermediates are available on a commercial scale, which can also be used as resolving agents.

Kinetic resolution

Another method is by *kinetic resolution*, in which one of the enantiomers reacts faster with a reagent than the other. An example is the resolution of (R, S)-mandelic acid. The (R)-isomer reacts faster than the (S)-isomer with (S)-menthol. Thus by taking a stoichiometric deficiency of (S)-menthol, a product mixture can be obtained containing mainly (S)-menthyl and (R)-mandelate and a very small quantity of (S)-menthyl and (S)-mandelate (Greenstein and Winitz, 1961).

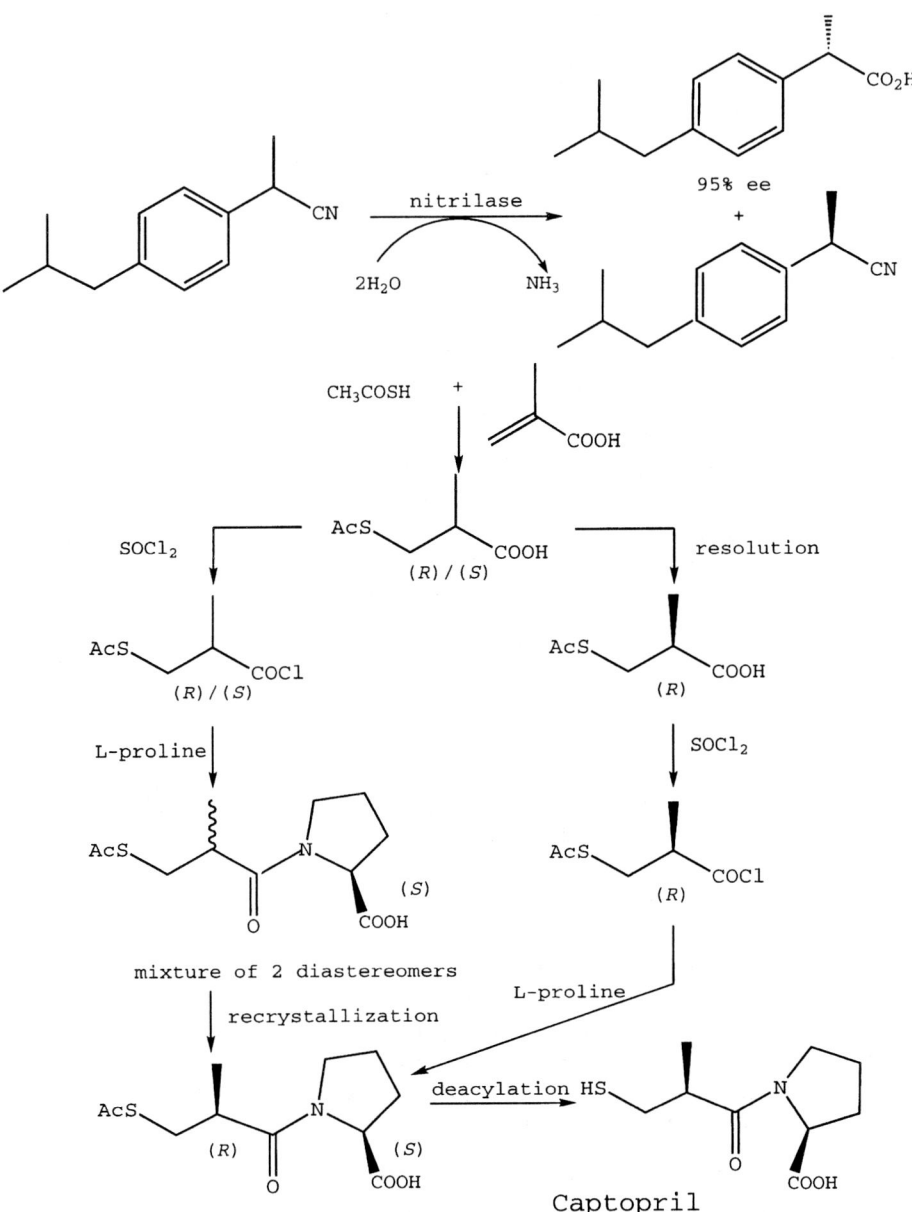

Figure 9.3 Early vs. late resolution in captopril synthesis (from Sheldon, 1996).

An advantage of this process is that it is possible to circumvent the racemization step if the unwanted isomer can be made to undergo spontaneous in situ racemization under the reaction conditions. In such a *dynamic kinetic resolution* process, a theoretical yield of 100% is possible as in asymmetric synthesis. Another example is the production of (*R*)-phenylglycine.

Because kinetic resolution depends on the relative rates of reaction of the two enantiomers, several studies of the analysis of kinetic resolution have been reported. However, as kinetic resolution is being increasingly restricted to enzymatic reactions, we consider the kinetics of resolution later in the section devoted to it.

Directed crystallization

In this method, referred to also as *resolution by entrainment*, the desired enantiomer is selectively seeded out of the racemate (Knowles, 1986). The method, which does not use a resolving agent, is applicable only to *R–S* mixtures that form conglomerate crystals (i.e., mechanical mixtures consisting of equal amounts of *R* and *S* crystals). It is not commonly used, because less than 20% of all racemates exist as conglomerates. Whether a given compound is a true racemate or a conglomerate is determined by using a differential scanning colorimeter. A conglomerate has a minimum melting point compared to its enantiomers, whereas a true racemate has a maximum melting point which is often equal to that of the pure enantiomers (Figure 9.4). The method is described in considerable detail by Reinhold et al. (1968) and Jacques et al. (1981).

A particularly important application of this method is in the synthesis of chloramphenicol (by Roussell-Uclaf), an antityphoid drug, in which resolution of the DL-threo base (reaction 9.4) is accomplished by its dissolution in aqueous HCl from which the less soluble D-chloramphenicol HCl preferentially crystallizes. To the mother liquor which now contains excess L-compound, an equal quantity of DL-base is added when the L-chloramphenicol HCl preferentially crystallizes. This process is repeated several times to obtain the D and L forms alternately in pure forms.

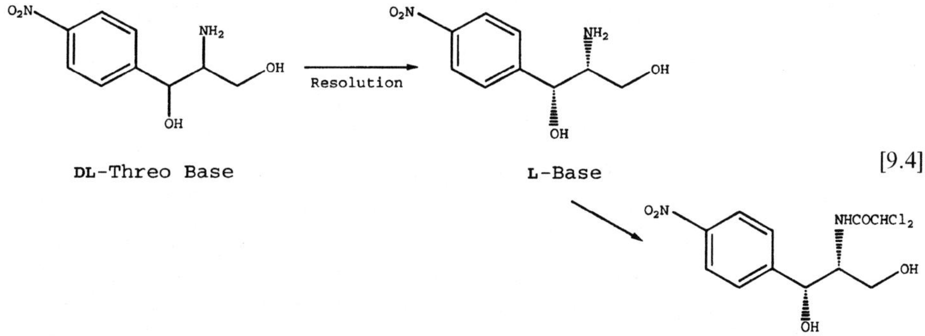

[9.4]

Another important application is in the separation of 1-α-methyl-dopa (Krieger et al., 1968).

Chromatographic Resolution

This method uses chromatographic columns containing particles whose surfaces are coated with chiral molecules. The basis of the separation is that the enantio-

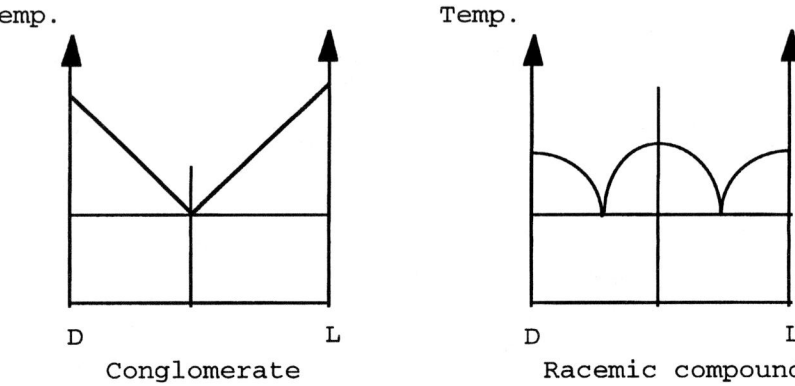

Figure 9.4 Melting point diagrams for racemates.

mers are converted into diastereomers that have different retention times (recall that enantiomers have the same retention time) and therefore can be more easily separated. An important example is the production of (S)-Praziquantel, an anthelmintic drug. The process is a complicated one, in which a continuous chromatographic process is coupled with multistage crystallization (Lim et al., 1995).

Biocatalysis

Biocatalysts can be used either in synthesizing the desired enatiomer directly with very high enantioselectivity from a given achiral substrate or in resolving a racemate by reacting with only one of the enantiomers. Both of these aspects of biocatalysis are briefly considered here.

Direct use in asymmetric synthesis

This method, based on the use of microbial cells or enzymes, exploits three important features of biocatalysts: (1) directing a reaction exclusively toward one enantiomer, (2) transforming a prochiral center to a chiral product, and (3) carrying out transformations on nonfunctionalized centers. The traditional problem associated with the enzymatic method is the presumption that these reactions should, of necessity, be carried out in dilute aqueous solutions to mimic biological systems. This leads to problems such as expensive separations and sensitivity of fermentations to deactivating influences. Despite these limitations, the biochemical route offers an attractive alternative for synthesizing an enantiomer directly (Knowles, 1986; Sheldon, 1996).

A particularly important application is in the synthesis of diltiazem, for which two routes have been proposed. The original route developed by Tanabe (see Sheldon, 1996) involving late resolution via diastereomeric salt crystallization

has been replaced by the more economic enzymatic resolution (Hulshof and Roskam, 1989; Matsumae et al., 1993, 1994).

In attempts to overcome the limitations of aqueous phase enzymatic reactions, organic media such as ethyl acetate, cyclohexane, and benzene are being tried as solvents. An industrially viable application of this variation is the lipase-catalyzed preparation of (R)-glycidyl butyrate (1), used in the production of (S)-(S)-β-blockers (2) (Ladner and Whitesides, 1984).

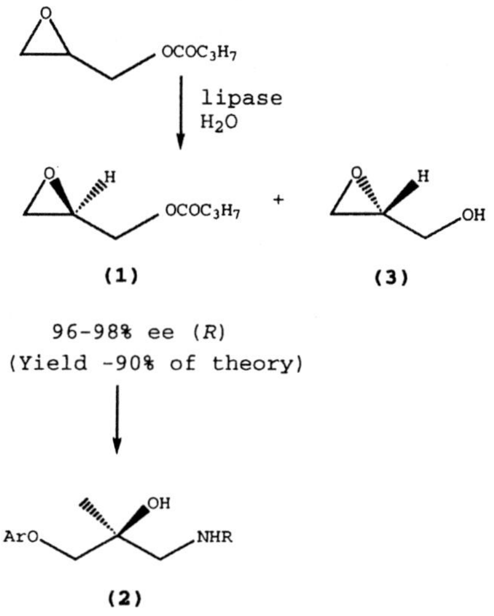

[9.5]

The by-product (R)-glycidol (3), which is also known for its potential as a drug intermediate, is not pure enough for use without expensive workup.

Another example of the use of organic solvents in enzymatic asymmetric synthesis is the development of a new scheme for ibuprofen. Although sound methods for the synthesis of (S)-ibuprofen (the active form) exist, a synthetic scheme based on lipase-catalyzed hydrolysis of organic soluble esters has also been developed in which the (S)-isomer is preferentially produced and leaves the (R) form untouched. In fact, lipase-catalyzed hydrolysis of esters to the desired acids or alcohols is likely to be an increasingly useful technique in asymmetric synthesis.

In what appears to be a significant discovery, de Zoete et al. (1993, 1994) found that in the presence of certain lipases (e.g., *Candida antarctica*), esters undergo ammoniolysis under very mild conditions using a saturated solution of ammonia in *t*-butanol (reaction 9.6), and for chiral esters the reaction is highly enantioselective (reaction 9.7).

$$R^1CO_2R^2 + NH_3 \xrightarrow{\text{NH}_3 \atop \text{lipase}} R^1CONH_2 + R^2OH \qquad [9.6]$$

258   Catalysis

[Reaction scheme 9.7: 2-chloroethyl ester of ibuprofen + NH₃:t-BuOH with C. antarctica lipase → ibuprofen amide + recovered ester, 56% conversion, 96% ee]

[9.7]

Kinetic analysis

Because resolution is a kinetic phenomenon, an expression for the relative rates of conversion of the two enantiomers is desirable for a quantitative analysis of resolution. As the development and use of such an expression requires an understanding of the kinetics of enzymatic reactions, we defer considering this to Chapter 20 which deals with biochemical methods of enhancing reaction rates and selectivities.

Asymmetric Catalysis

Most of the methods already described, particularly those involving resolution, tend to be tedious and expensive. An attractive alternative is *chiral* or *asymmetric synthesis*. This method of asymmetric induction (asymmetric catalysis in particular) constitutes the main thrust of this chapter and is considered in greater detail below.

ASYMMETRIC SYNTHESIS

In its broadest sense, asymmetric synthesis signifies the induction of asymmetry in prochiral or achiral substrates by using a chiral reagent or catalyst. Asymmetry can also reside in the substrate itself, in which case any synthetic scheme (with or without a catalyst) must carry the asymmetry through to the final product.

Methods of Generating Chirality; Chiral Pools

There are essentially four methods of generating chirality. These are frequently referred to as *first, second, third,* and *fourth generation* methods:

1. Synthesis starting with an enantiomerically pure compound, almost always of natural origin, that is incorporated into the final product (e.g., synthetic penicillins and cephalosporins)
2. Synthesis starting with a nonchiral substrate and a chiral auxiliary that is attached to the substrate and is removed after serving its purpose

3. Synthesis starting with a nonchiral substrate that is directly converted to the chiral product by using a chiral reagent
4. Synthesis using a chiral catalyst to direct the conversion of a nonchiral substrate to the desired chiral product.

Note that in the first and second generation methods, asymmetry derives from the starting material or a chiral auxiliary, whereas in the third and fourth generation methods, it derives from a reagent or a catalyst. All of these methods require inducing chirality (by the substrate, reagent, or catalyst) either at the start or at some point during the synthetic sequence. Enantioselectivity is possible in all of these cases.

Thus a collection of chiral compounds must be available from which the best choice can be made for a proposed synthesis. Such a collection, usually known as a *chiral pool*, consists of compounds obtained from the following three sources:

1. Natural chiral molecules such as carbohydrates, amino acids, carboxylic acids, terpenes, and alkaloids
2. Fermentation of carbohydrate feedstocks such as sucrose and molasses that give simple chiral molecules such as L-amino acids, lactic acid, tartaric acid, and complex compounds such as hormones and vitamins
3. Nonnatural, man-made structures different from those isolated from natural compounds, for example, (R)-phenylglycine.

A list of natural products available in high enantiomeric purity has been compiled by Scott (1984). A few examples of all the classes are given in Table 9.4.

Classification of Asymmetric Synthesis

A classification of the different synthetic routes for chiral induction was presented in Table 9.3. Included in the table is *asymmetric catalysis* which forms the subject matter of the rest of the chapter.

Asymmetric catalysis can be homogeneous or heterogeneous and can take place either in the organic phase or (as recently found) in the aqueous phase. Table 9.5 presents a broad classification of the major categories of asymmetric catalysis. The most common is homogeneous catalysis in the aqueous phase, and the most recent and novel is the use of aqueous phase catalysts immobilized on solid supports. The various categories are considered in the following sections.

Table 9.4 Representative substances from the chiral pool

Compound
Ascorbic acid; (+)-calcium pantothenate; (−)-carvone; anhydrous dextrose; ephedrine hydrochloride; (+)-limonene; L-lysine; mannitol; monosodium glutamate; norephedrine hydrochloride; quinidine sulphate; quinine sulphate; sorbitol; L-threonine; L-tryptophan; tartaric acid; all amino acids; most sugars, e.g., glucose, xylose, ribose.

Table 9.5 Classification of asymmetric catalysis and catalysts

```
                        Asymmetric catalysis/catalysts
                                    |
                    ┌───────────────┴───────────────┐
               Homogeneous                     Heterogeneous
                    |                               |
          ┌─────────┴─────────┐     ┌───────────────┼───────────────┐
    Catalysis in      Catalysis in   Solid catalysts  Homogeneous    Footprint
    organic phase     aqueous phase  with modifiers   catalysis      catalysts
                                                      supported on
                                                      solids
                            |                               |
                ┌───────────┴───────────┐       ┌───────────┴───────────┐
          Chiral solids          Nonchiral      Catalysts in      Supported
          with nonchiral         solids with    organic phase     aqueous phase
          modifiers              chiral         encapsulated      (SAP) catalysts
                                 modifiers      in zeolite
                                                cages
```

HOMOGENEOUS ASYMMETRIC CATALYSIS

Note at the outset that asymmetric catalysis in the synthesis of fine chemicals is rarely a single-step process that converts a reactant directly to the final product. It is usually one of the steps in a total synthesis but is often the key step. Hence the analysis of the overall yield will be based on the methods described in Chapter 5. There are many types of reactions where asymmetric catalysis can be applied. The most important of these are C–C bond-forming reactions such as alkylation or nucleophilic addition, oxidation, reduction, isomerization, Diels–Alder reaction, Michael addition, deracemization, and Sharpless expoxidation (of allyl alcohols). A few representative examples (homogeneous and heterogeneous) are given in Table 9.6.

Hydrogenation

Commercially, the most extensively used class of reactions in asymmetric synthesis is catalytic hydrogenation with chiral transition metal complexes. The industrial importance of homogeneous chiral hydrogenations stems from the fact that almost quantitative yields can be obtained with practically no by-products. Chirality in homogeneous catalysts may lie either at the metal bonding site or at a site removed from the metal, but good enantiomeric selectivity is obtained irrespective of its location (Rylander, 1973; Schrock and Osborn, 1976; Nogradi, 1987). The most widely used metal is rhodium complexed to a wide range of chiral ligands, generally biphosphines. Other metals such as cobalt, ruthenium, nickel, molybdenum, platinum, vanadium, and iridium have also been used.

Figure 9.5 lists the more important P-, N-, and S-containing chiral ligands used in asymmetric synthesis. Of these, Noyori's BINAP (axially dissymmetric phosphine-substituted binaphthyl) is probably the most effective and is used, for instance, as a Ru complex in the synthesis of (S)-naproxen (Noyori and Takaya,

Table 9.6 Selected chiral syntheses belonging to different classes of reactions using homogeneous catalysts

No.	Reaction		Comments
1	Diels–Alder reaction 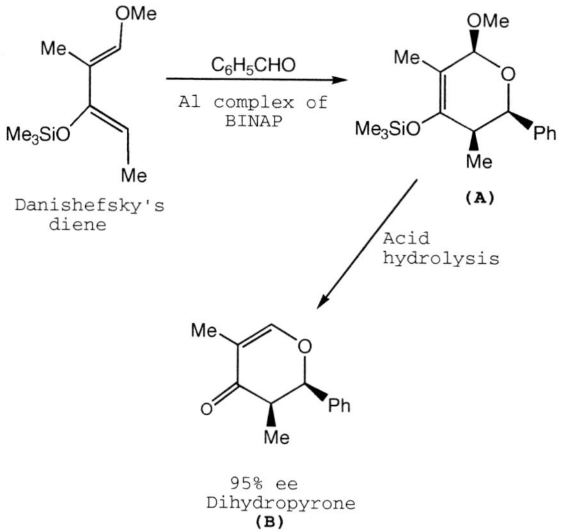	Danishefsky's diene + C_6H_5CHO, Al complex of BINAP → adduct (A); acid hydrolysis → 95% ee Dihydropyrone (B)	The initial adduct (A) is hydrolyzed to the final pyrone (B) (Maruoka et al., 1998).
2	Reduction 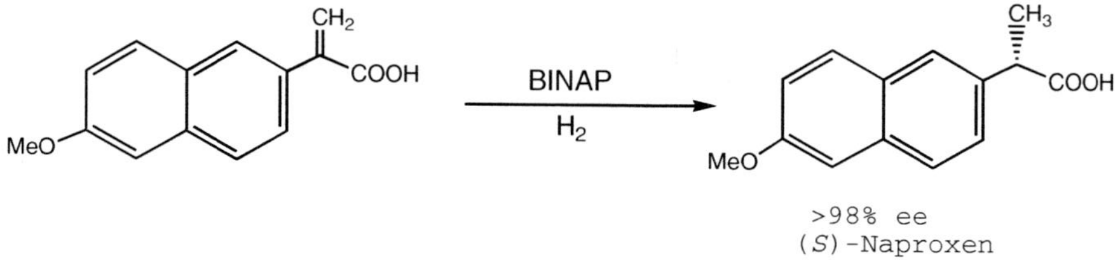	methoxynaphthalene acrylic acid + H_2, BINAP → >98% ee (S)-Naproxen	Final step in the manufacture of (S)-naproxen (see Table 9.4).

(Continued)

Table 9.6 (Continued)

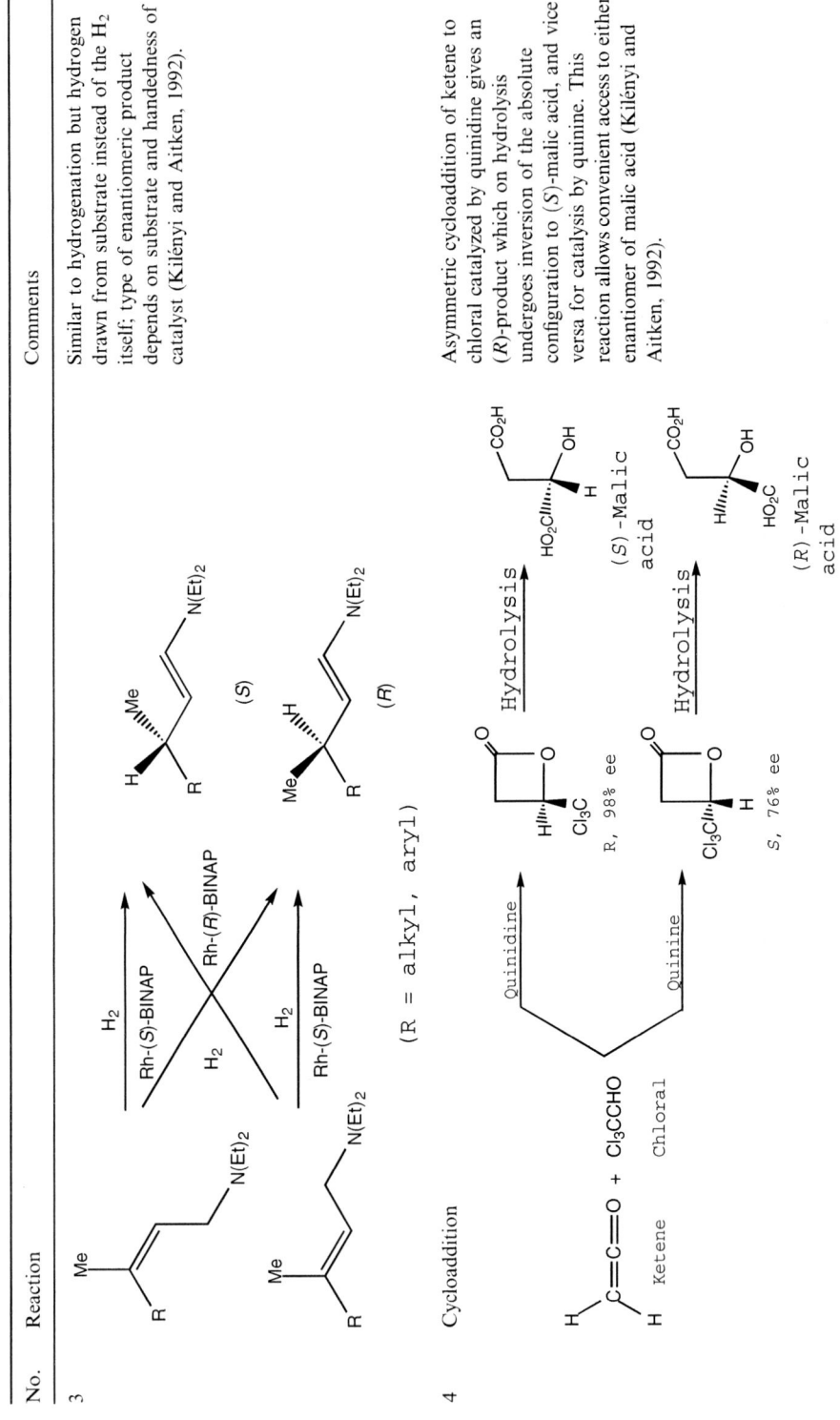

No.	Reaction	Comments
3		Similar to hydrogenation but hydrogen drawn from substrate instead of the H_2 itself; type of enantiomeric product depends on substrate and handedness of catalyst (Kilényi and Aitken, 1992).
4	Cycloaddition	Asymmetric cycloaddition of ketene to chloral catalyzed by quinidine gives an (R)-product which on hydrolysis undergoes inversion of the absolute configuration to (S)-malic acid, and vice versa for catalysis by quinine. This reaction allows convenient access to either enantiomer of malic acid (Kilényi and Aitken, 1992).

5. Sharpless epoxidation

[reaction: Allyl alcohol + Bu^tOOH →(Titanium complex of (+) or (−) enantiomers of diethyl tartrate)→ glycidol + Bu^tOOH; 90% ee (R) or (S) Glycidol]

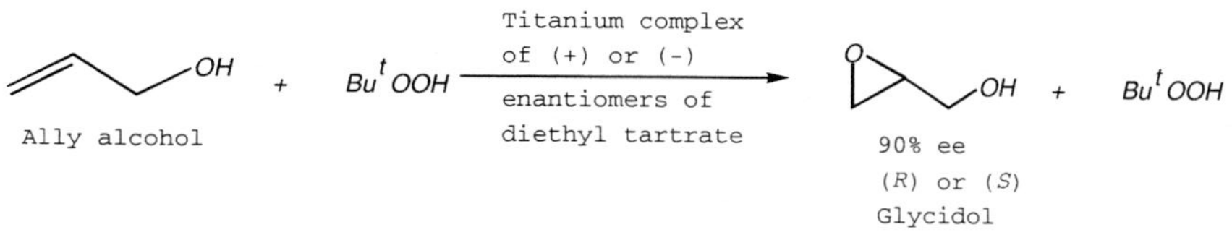

Glycidol is used as intermediate in specialty chemicals (Chem. Week, 1990; Chem. Eng. News, 1989)

6. Sulfide oxidation

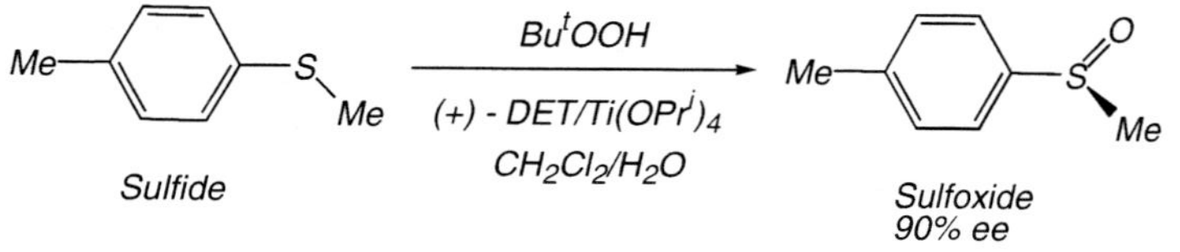

[reaction: Sulfide (4-MeC6H4-S-Me) →(Bu^tOOH, (+)-DET/Ti(OPr^i)4, CH2Cl2/H2O)→ Sulfoxide 90% ee]

ee drops to zero in the absence of water which is the opposite of Sharpless epoxidation; hence the nature of the Ti complex is different (Kagan and Rebiere, 1990)

7. Deracemization

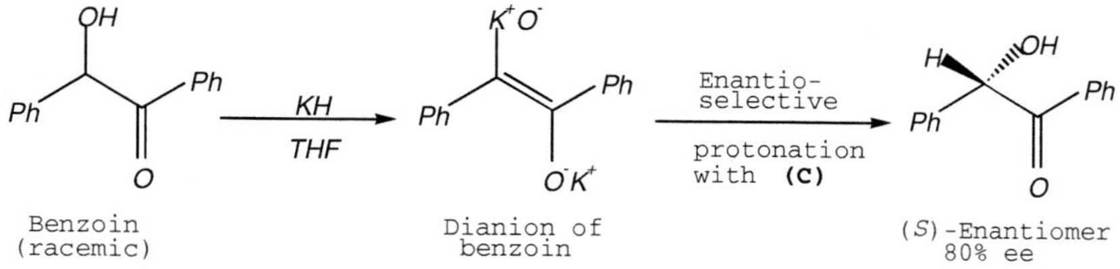

[reaction: Benzoin (racemic) →(KH, THF)→ Dianion of benzoin →(Enantioselective protonation with (C))→ (S)-Enantiomer 80% ee]

Racemic benzoin is planar and therefore achiral; (2R,3R)-o,o-dipivaloyl-tartaric acid (C) converts the (R)- to the (S)-enantiomer (Duhamal et al., 1984). Note: this is not resolution but conversion of one enantiomer to the other

(*Continued*)

Table 9.6 (Continued)

No.	Reaction	Comments
8	Alkylation 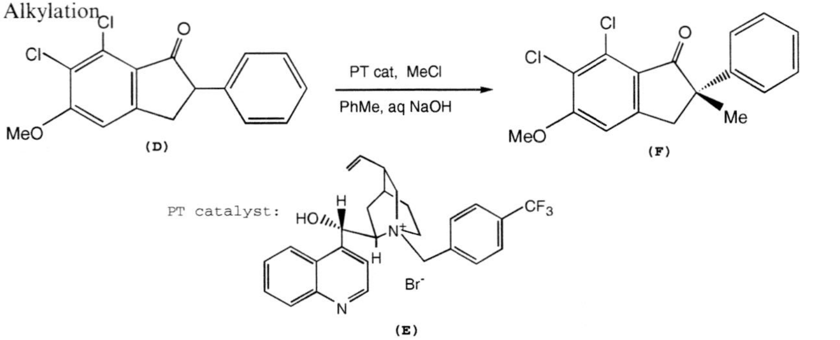	This reaction is a step in the synthesis of the diuretic indacrinone (Hughes et al., 1987) and uses a quaternary salt of a cinchona alkaloid (E) as an asymmetric phase-transfer catalyst (Wynberg, 1986). The product enolate (F) formed from the ketone is a precursor of indacrinone
9.	Michael addition 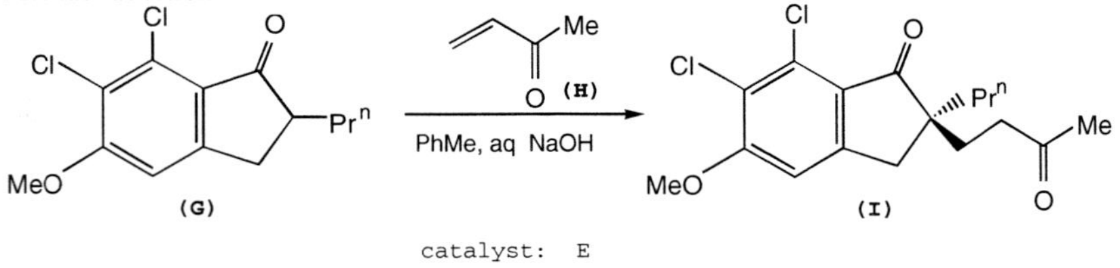	Michael addition of indanone (G) to methyl vinyl ketone (H) in the presence of (E) gives product (I) in 80% ee and 95% yield (Conn et al., 1986)

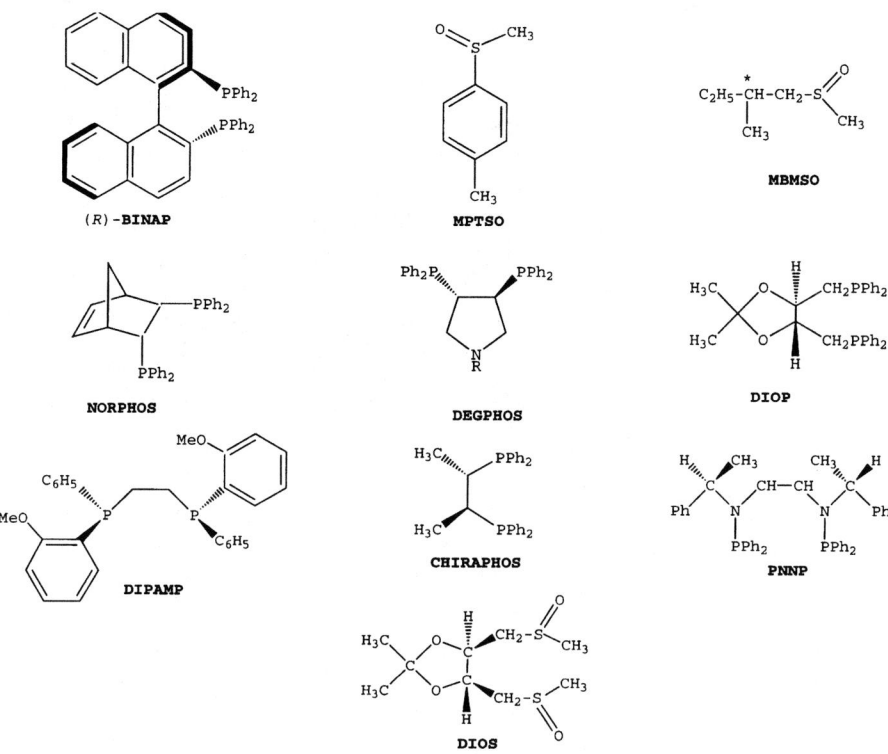

Figure 9.5 Some P-, and S-containing ligands

1990; Taber and Silverberg, 1991; Noyori, 1992). The (*R*) and (*S*) Ru(II) complexes of BINAP [2,2′-bis(diphenyl-phosphino)-1,1′-binaphthyl] are shown in Figure 9.6. Another important ligand is (*R,R*)-1,2-ethanediylbis[(*o*-methoxyphenyl)phenylphosphine], known as DIPAMP (Vineyard et al., 1977). Complexed to rhodium (Figure 9.7), this ligand is used in the production of L-dopa.

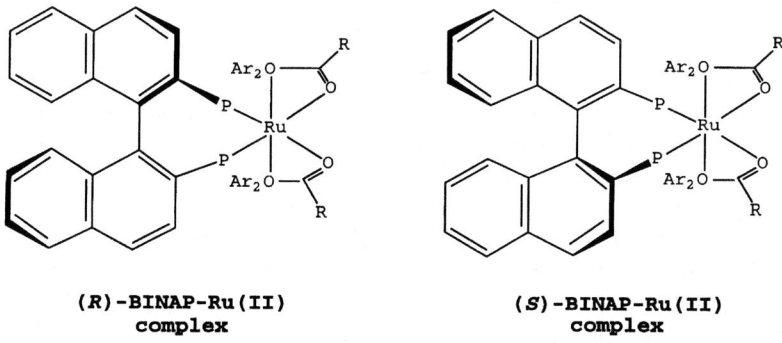

Figure 9.6 Ruthenium complexes of BINAP

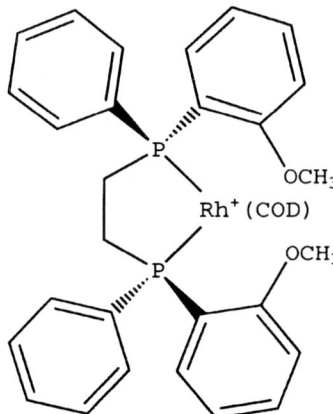

Figure 9.7 Rhodium complex of the chiral ligand DIPAMP: [Rh(1,5-cyclo-octadiene = COD)DIPAMP] (tetrafluoroborate).

Sharpless Synthesis

The basic methodology

Sharpless epoxidation of allyl alcohols (Sharpless, 1985, 1988; Pfenninger, 1986; Rossiter, 1985; Woodard et al., 1991; Finn and Sharpless, 1991; Corey, 1990a,b), an example of which is included in Table 9.6, is perhaps the most recent and one of the most remarkable applications of asymmetric catalysis. The reaction is normally performed at low temperatures (-30 to $0\,°C$) in methylene chloride with a titanium complex consisting of a chiral component [diethyl tartrate (DET) or diisopropyl tartrate (DIPT)] and a titanium salt (titanium tetraisopropoxide) as the catalyst. The beauty of the synthesis is that both enantiomers of the tartrate are available so that either form of the product can be prepared in more than 90% ee.

Extension to substrates with no functional groups

The main drawback in Sharpless epoxidation is that the substrate must bear a functional group to achieve the precoordination required for high enantioselectivity (as in the case of allyl alcohol). This restriction is not applicable to the epoxidation of alkyl- and aryl-substituted olefins with manganese complexes of chiral Schiffs bases as catalysts. Very high enantioselectivities can be obtained in these reactions (Jacobsen, 1993). The most widely used catalysts that give high enantioselectivity are those derived from the Schiff bases of chiral diamines such as [SS] and [RR] 1,2-diphenylethylenediamine and [SS] and [RR] cyclohexyl-1,2-diamine. An example is the synthesis of cromakalim.

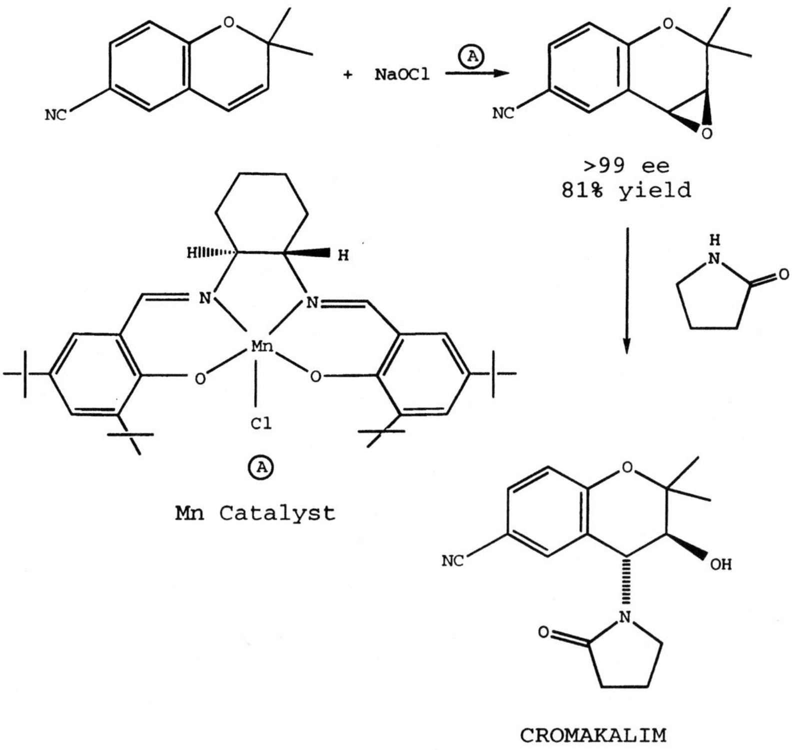

[9.8]

CROMAKALIM

This methodology is open to improvement because several substrates, particularly terminal olefins, do not undergo epoxidation with the desired selectivity. The potential is enormous.

Kinetic resolution of terminal epoxides by catalytic hydrolysis with water

Despite several advances made in asymmetric epoxidation, there is no general practical method of synthesizing enantiomerically pure compounds. Recent studies by Jacobsen and his colleagues (e.g., Tokunaga et al., 1997) show that asymmetric hydrolysis of terminal epoxides using chiral cobalt-based salen complexes can give highly valuable terminal epoxides and 1,2-diols in excellent yields with high enantiomeric enrichment. Water is the only reagent used in this unique reaction.

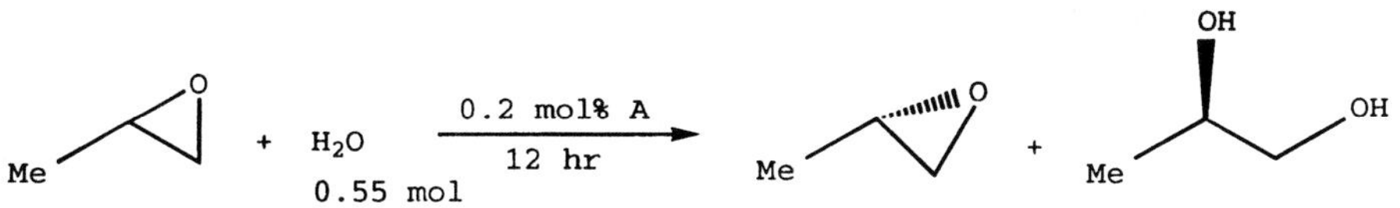

[9.9]

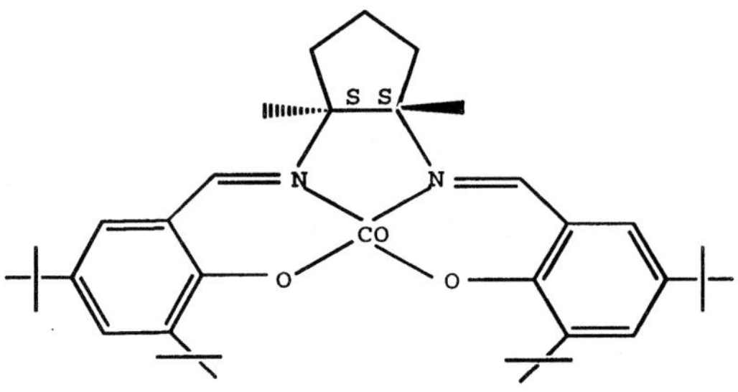

A

This work may be regarded as a major breakthrough in asymmetric synthesis and should pave the way for the synthesis of a variety of biologically active chiral compounds because epoxides are versatile building blocks that are commercially produced by the relatively cheap direct air oxidation of unfunctional olefins.

Reductive Catalysis by *Chemzymes*

One of the outstanding discoveries of recent years is the enantioselective reduction of prochiral ketones with borane in the presence of catalytic amounts of oxazaborolidenes obtained from L-proline (reaction 9.10), and termed *chemzymes* (Corey, 1990a,b).

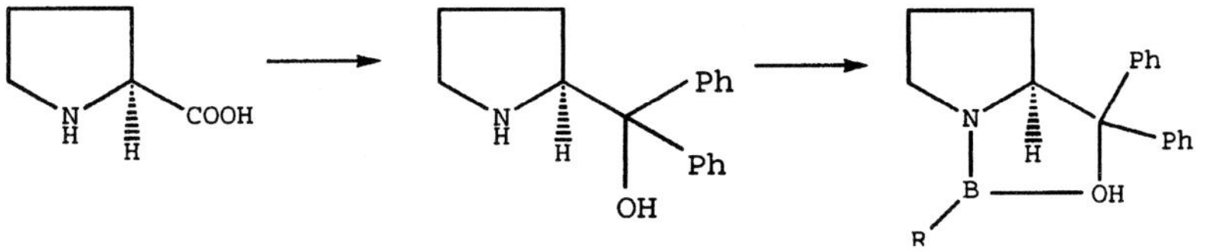

[9.10]

This reaction, represented by the general scheme

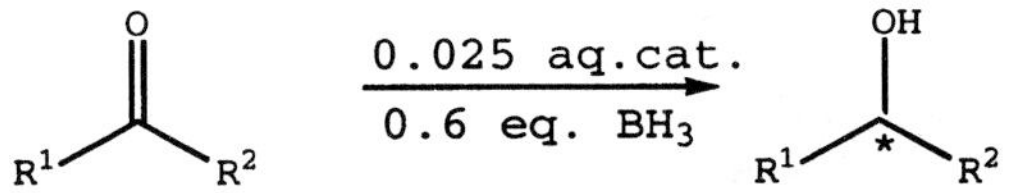

[9.11]

>97% ee

is almost certain to be exploited as a powerful tool in the commercial synthesis of a variety of chiral compounds. Rama Rao et al. (1990, 1992) extended the methodology and developed new (*R*) and (*S*) oxaborolidenes. Of these, the four-membered oxazaborolidines are capable of reducing ketones with 1 M borane to give chiral alcohols with 95–97% ee.

Chiral Compounds as Starting Materials

Table 9.3 shows that syntheses using chiral reactants also fall under the broad category of chiral synthesis. Most natural products such as amino acids, carbo-

hydrates, hydroxyacids, and monoterpenes are chiral and abundantly available as pure enantiomers. They can be used as chiral templates and transformed into a variety of optically pure end products. Thus simple sugars such as D-glucose, D-mannose, and D-mannitol, D-tartaric acid, and L-amino acids can be converted into expensive chiral compounds. A good example is the manufacture of vitamin C. Other examples are the transformation of D-mannitol to (S)-timolol and the production of semisynthetic penicillins and cephalosporins from 7-aminopenicillanic acid (7-APA) and 7-aminocephalosporanic acid (7-ACA), respectively.

Industrial Examples

Besides the more common reactions such as hydrogenation, isomerization, alkylation, and the Diels–Alder reaction, Sharpless epoxidation and dihydroxylation by asymmetrical catalysis are rapidly emerging as reactions with immense industrial potential. Table 9.7 lists some important syntheses based on asymmetric catalysis. These include processes for the pharmaceutical drugs (S)-naproxen, (S)-ibuprofen, (S)-propranolol, L-dopa, and cilastatin, a fragrance chemical, L-menthol, and an insecticide (R)-disparlure. Deltamethrin, an insecticide, is another very good example of industrial asymmetric synthesis. The total synthetic scheme is also given for each product. In general, the asymmetric step is the key step in the total synthesis, but this is not always so, as in the production of ibuprofen. Many of the processes listed in the table are in industrial production.

Asymmetric catalysis is not necessarily the cheapest method of making an enantiomer. Sometimes the number of steps involved in such a synthesis is so large that it becomes uneconomical (see Chapter 5). An example is the synthesis of diltiazem by asymmetric catalysis (Watson et al., 1990) which compares unfavorably with the enzymatic route.

Enantioselective hydrogenation

The family of anti-inflammatory drugs listed in Figure 9.8, all made from α-arylalkanoic acid, provides perhaps the strongest evidence of the role of asymmetrical synthesis in drug manufacture. For various reasons, however, all of them, except naproxen, continue to be sold as racemates. Ibuprofen is the largest selling of them, with sales an order of magnitude higher than the others. All of them can be made by asymmetric hydrogenation. The artificial sweetener aspartame is also produced by a process involving one asymmetric reduction step (see Table 9.7).

Enantioselective isomerization: L-menthol

The synthesis of L-menthol from β-pinene by homogeneous catalysis is an outstanding example of the application of enantioselective catalysis in the perfumery industry (Noyori and Takaya, 1990; Parshall and Ittel, 1992). The synthetic scheme for this process is included in Table 9.7. The key step is isomerization

Table 9.7 Some industrial syntheses based on asymmetric catalysis

Desired molecule	Total synthesis	Steps	Reference(s)
1. (S)-Naproxen		Step 1: Electrocarboxylation of 2-acetyl-6-methoxynaphthalene Step 2: Acid catalyzed dehydration of 2-(6-methoxy-2-naphthyl) lactic acid Step 3: KEY STEP Catalytic asymmetric hydrogenation of 2-(6-methoxy-2-naphthyl) acrylic acid	Parshall and Ittel (1992) Collins et al. (1992) Federsel (1993) Chan et al. (1994)
2. Levodopa	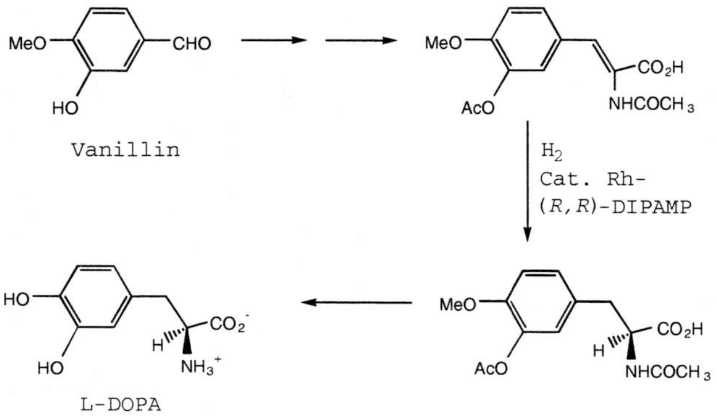	Step 1: Reaction of vanillin with N-acetylglycine in presence of sodium acetate and acetic anhydride Step 2: Hydration of the oxazolidinone Step 3: KEY STEP Catalytic asymmetric hydrogenation of the dehydro derivative	Schrock and Osborn (1976) Knowles (1986) Parshall and Ittel (1992) Collins et al. (1992) Chan et al. (1994) Aitken and Kilényi (1992)

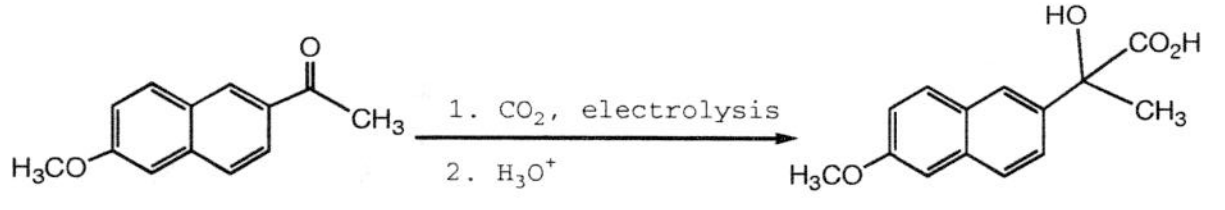

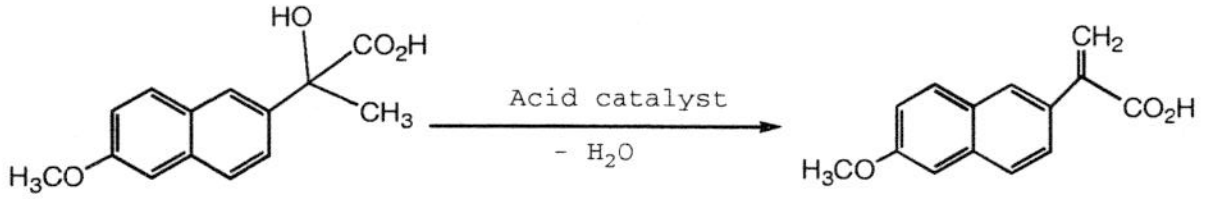

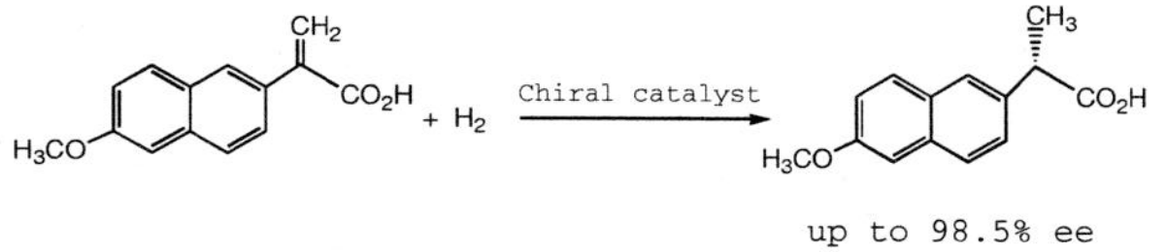

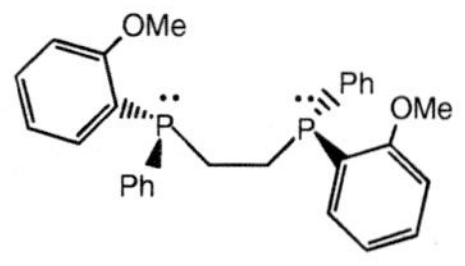

3. (*S*)-Ibuprofen

(*R*,*R*)-DIPAMP

Step 1: Electrocarboxylation — Parshall and Ittel (1992); Collins et al. (1992)
Step 2: Acid-catalyzed dehydration — Federsel (1993)
Step 3: KEY STEP — Catalytic asymmetric hydrogenation — Chan et al. (1994)
Step 4: Acid hydrolysis of the hydrogenation product

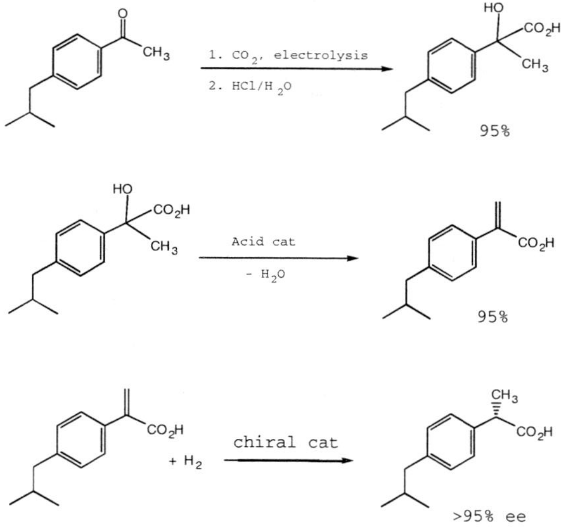

1. CO_2, electrolysis
2. HCl/H_2O

95%

Acid cat, $-H_2O$

95%

chiral cat, $+H_2$

>95% ee

(*S*)-Ibuprofen

(*Continued*)

Table 9.7 (Continued)

Desired molecule	Total synthesis	Steps	Reference(s)
4. (S)-Propranolol	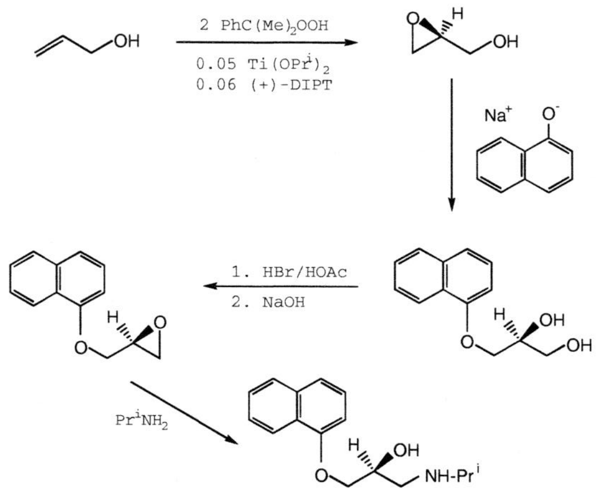	Step 1: KEY STEP: Catalytic Sharpless Epoxidation of allyl alcohol using (R)-diisopropyl tartrate as the source of asymmetry Step 2: Reaction of glycidol with naphthoxide anion Step 3: Dehydration Step 4: Conversion of the epoxide with isopropylamine to (S)-propranolol	Aitken and Kilényi (1992)
5. Cilastatin	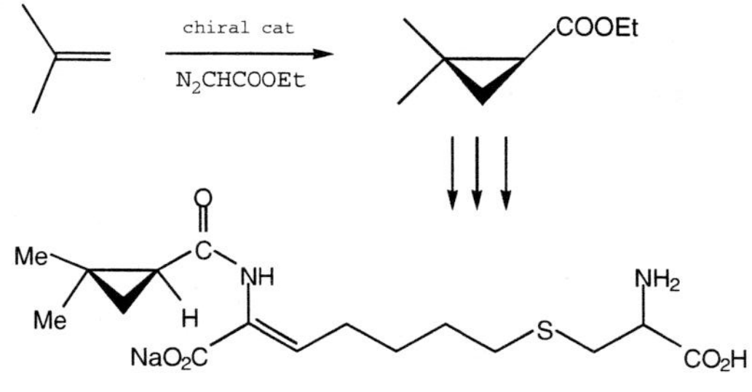	KEY STEP: Catalytic cyclopropanation Ethyl diazoacetate and 2-methylpropene are converted into (+)-(1S)-2,2-dimethylcyclopropanecarboxylate Several conventional organic steps to the final product Cilastatin	Ojima et al. (1989) Parshall and Ittel (1992)

6. Demethoxy aranciamycinone

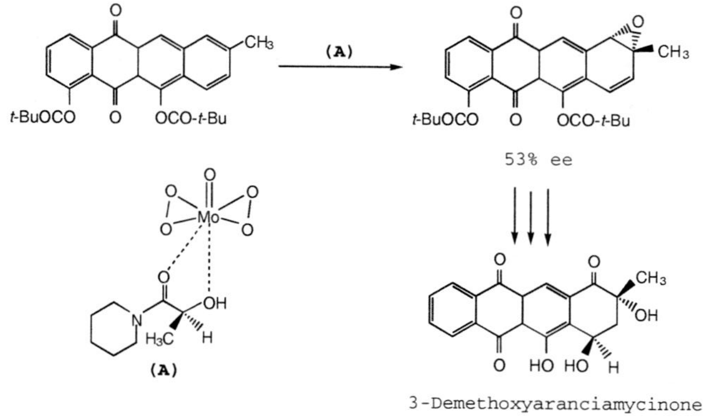

KEY STEP:
Asymmetric epoxidation of the olefin
Catalyst: chiral oxodiperoxo-Mo complex

Several conventional steps to the final product 3-demethoxyaranciamycinone

Ojima et al. (1989)

7. (R)-Disparlure

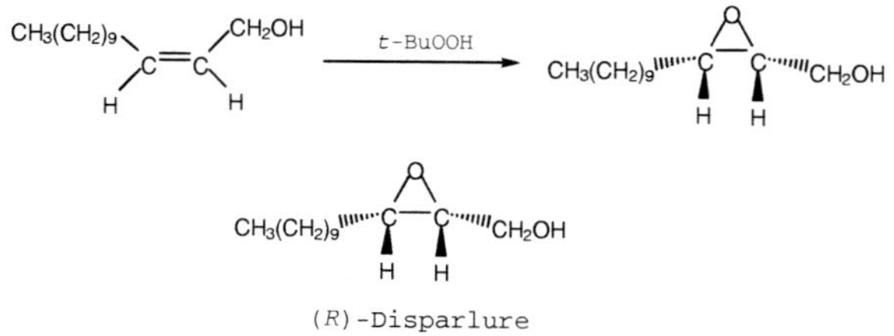

Step 1:
KEY STEP:
Enantioselective oxidation of Z-2-tridecenol to an epoxy alcohol in 90–95% enantiomeric purity before recrystallization

Step 2, 3, 4:
Three conventional organic steps to final product (R)-disparlure

Parshall and Ittel (1992)

(Continued)

Table 9.7 (Continued)

Desired molecule	Total synthesis	Steps	Reference(s)
8. (S)-Menthol	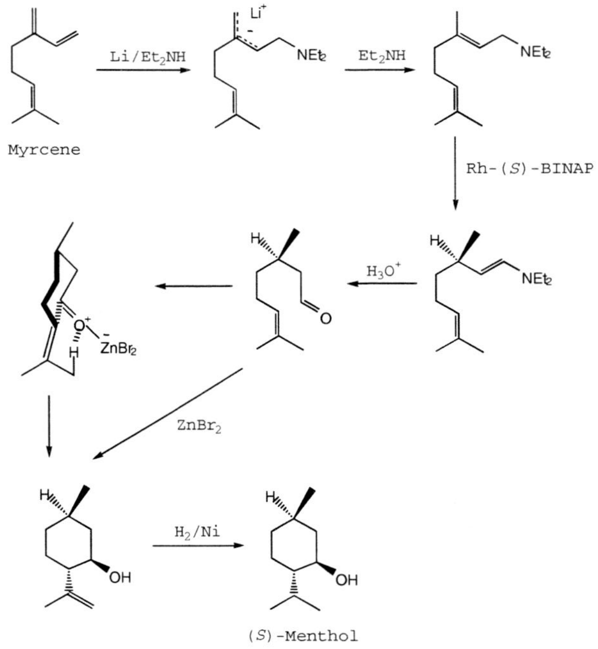 Myrcene → (via Li/Et$_2$NH, then Et$_2$NH) → Z-diene with NEt$_2$ → (Rh-(S)-BINAP) → enamine with NEt$_2$ → (H$_3$O$^+$) → aldehyde → (ZnBr$_2$) → cyclic ZnBr$_2$-ene intermediate → cyclohexenol with OH and isopropenyl → (H$_2$/Ni) → (S)-Menthol	Step 1: Conversion of myrcene with lithium diethylamide to diene giving Z-allylamine Step 2: **KEY STEP** Asymmetric isomerization with rhodium complex of (S)-BINAP to an enamine (almost 100% enantiomeric purity) Step 3: Hydrolysis to an aldehyde Step 4: Formation of a cyclic product; intramolecular ene reaction with Lewis acid ZnBr$_2$; formation of two new stereogenic centers Step 5: Hydrogenation of the double bond with nickel as catalyst	Aitken and Kilényi (1992) Ojima et al. (1989)

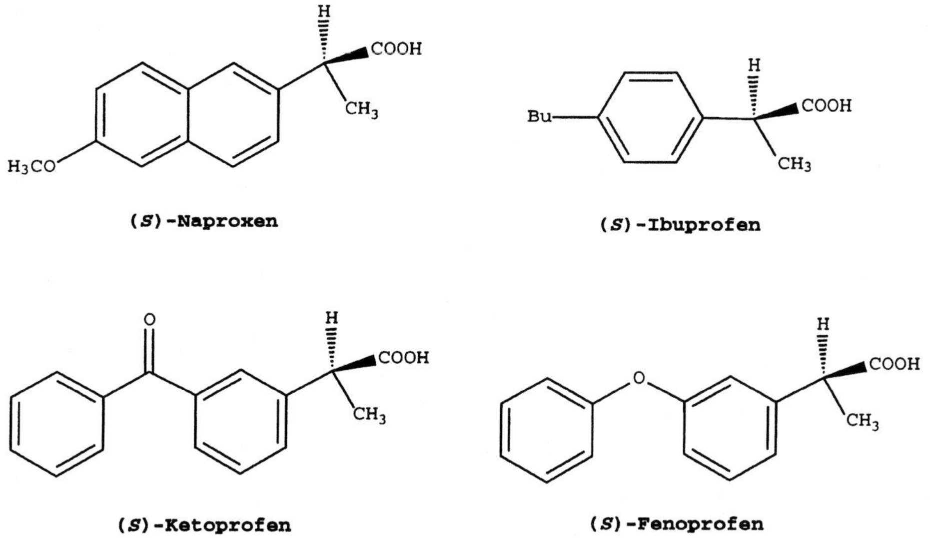
H
COOH
CH3
H3CO
(S)-Naproxen
H
Bu
COOH
CH3
(S)-Ibuprofen
O
H
COOH
CH3
(S)-Ketoprofen
O
H
COOH
CH3
(S)-Fenoprofen

Figure 9.8 Important examples of α-arylalkanoic acids.

of diethylgeranylamine to (*R*)(−)-diethyl-*E*-citronellalenamine which introduces C_2-chirality. This is accomplished by using a rhodium complex containing the ligand 2,2′-bis(diphenylphosphino)-1,1′-binaphthyl (BINAP) (see Figure 9.7). This is the first major use of enantioselective catalysis on a commercial scale (see Nozaki et al., 1966; Kagan, 1982; Noyori and Kitamura, 1989).

Sharpless asymmetric synthesis: disparlure

Enantioselective epoxidation of olefins is unique in that it creates two stereogenic centers. The catalyst is prepared quite easily by treating a titanium (+4) alkoxide with an (*R*)- or (*S*)-diethyl tartrate (Gao et al., 1987). The peroxidation agent is usually t-butyl hydroperoxide.

A chiral epoxide that is now finding increasing use in pest management is the sex attractant (*R*)-disparlure emitted by the female gypsy moth (see Table 9.7). This sex hormone is prepared from an intermediate produced by enantioselective epoxidation of *Z*-2-tridecenol.

Water-Soluble Catalysts

Homogeneous catalysts have traditionally been used in solution in organic solvents. Hence the specification "organic-soluble catalysts" is not commonly used to describe them. However, in view of the novelty of water-soluble homogeneous catalysts, they are described more fully by using the full term. The overriding advantage of these catalysts is that water is environmentally more benign than any other solvent. Additionally, they can also be used in two-phase aqueous–organic and supported configurations. The simplest way of preparing

these catalysts is to incorporate highly polar functional groups such as amino, hydroxide, or sulfonate into phosphine ligands (Kalck and Monteil, 1992).

Another way is to water solubilize an available ligand such as BINAP by sulfonating it. A practical problem is the formation of phosphine oxides during sulfonation. This can be minimized by the simple preparatory technique used by Wan and Davis (1993a): dissolve (*R*)-BINAP in concentrated sulfuric acid, add fuming sulfuric acid (dropwise) over 1–2 hours, keep for 3 days under stirring in an argon atmosphere, quench the reaction, neutralize with caustic solution, add methanol to precipitate sodium sulfate, and recover the sulfonated ligand BINAP(SO$_3$Na)$_4$ by vacuum drying and filtration.

The ligand prepared as indicated was used to prepare the rhodium complex [Rh{(*R*)-BINAP(SO$_3$Na)$_4$}(H$_2$O)$_2$]$^+$, which hydrogenates 2-acetamidoacrylic acid and its methyl ester with the same enantiomeric efficiency as the traditional catalyst in a nonaqueous solvent. A subsequent study (Wan and Davis, 1993b) showed that a ruthenium-sulfonated-BINAP complex can also be an effective hydrogenation catalyst in both methanolic and neat water solvent systems.

HETEROGENEOUS ASYMMETRIC CATALYSIS

Heterogeneous catalysis is the most extensively used form of catalysis in industry. On the other hand, the use of homogeneous catalysis for fine chemicals is more prevalent and is on the rise. However, homogeneous asymmetric catalysts frequently combine very expensive ligands with precious metals. Thus the use of this catalyst in solid form can exploit the main advantage of solid catalysts, that of freedom from separation problems, and make them economically more attractive. In practical terms, this means that if the enantiomeric yields can be made to match those of homogeneous catalysts, they will certainly be the preferred candidates for industrial use. We describe here the four categories of heterogeneous catalysts listed in Table 9.5.

Solid-Supported Chiral Catalysts

Chiral solid catalysts usually have two functions, activation and control. The *activating function* ensures that the solid actually catalyzes a reaction (chemical catalysis), and the *control function* provides the stereochemical direction that yields the required enantiomer. Early studies were carried out with metallic catalysts supported on inherently chiral solids such as quartz, cellulose (Harada and Yoshida, 1970), and polypeptides (Akabori et al., 1956; Beamer et al., 1967), in which the metal provided the activating function and the support provided the control function. More recent emphasis has been on binding chiral molecules to nonchiral supports.

Substances that provide the control function are also referred to as *stereochemical modifiers*. These modifiers are essential because there is no enantiomeric selectivity without them. An example is the hydrogenation of methyl pyruvate (MeCO˙COOMe) to methyl lactate [MeCH(OH)COOMe] (Webb and Wells,

Table 9.8 Combinations of activity and control functions for different classes of reactions (Blaser and Müller, 1991)

Activating function	Controlling function	Reaction type
Metallic surface	Modifier or polymer	Hydrogenation Hydrogenolysis Electrochemistry
Metallic surface	Chiral support	Hydrogenation Dehydrogenation Isomerization
Metal salt or oxide	Modifier or polymer	Polymerization Isomerization
Chiral metal salt	Chiral metal salt	Polymerization Carbene addition SN_2 reaction
Chiral polymer	Chiral polymer	Nucleophilic addition Oxidation
None	Crystal	Bromination Dimerization

1992; Vermerer and Wells, quoted in Dodgson, 1993) on cinchona alkaloid-modified Pt/SiO_2. The theoretical yield rises from zero with no modifier to 60% with modifier. Sometimes, the same substance can act both as catalyst and modifier. Based on a survey of many available systems, combinations of the two functions for different types of reactions are listed in Table 9.8 (Blaser and Müller, 1991).

Despite the many studies reported, only two combinations of catalytic and modifying functions have emerged as commercial possibilities, tartrate-modified catalysts and cinchona-modified catalysts. The substrates that have been most studied are the α, β-ketoesters, and the polymer support most frequently used is preformed styrene–DVB(2%) resin. Asymmetric ligand monomers that contain asymmetric ligand sites have also been synthesized.

Catalysts modified by tartaric acid (commonly referred to as tartrate-modified catalysts) constitute the most extensively studied class of modified catalysts (Izuma, 1983; Bartok, 1985; Sachtler, 1985; Tai and Harada, 1986). The catalyst usually consists of Raney nickel or a noble metal-containing bimetallic system, modified by tartaric acid with NaBr as a comodifier in a mildly polar aprotic solvent such as methyl propionate (Izuma, 1983; Tai and Harada, 1986). The reaction is usually carried out in the temperature range 60–100 °C at 80–120 atm.

The second catalyst system, the cinchona-modified platinum catalysts, has not been as extensively studied but appears to be gaining in importance (see Orito et al., 1979a,b, 1980, 1982; Blaser et al., 1988). The catalysts used are platinum supported on a variety of solids and modified with naturally occurring cinchona alkaloids. Rhodium can also be used as a catalyst and ephedrine derivatives as modifiers, but these are less effective. Apolar solvents with dielectric constants of 2–6 give the best results, but alcohols are also known to give good results in some

cases. It is important to note that an optimum modifier concentration exists which depends on solvent, modifier, and substrate type (Blaser and Müller, 1991). Reaction conditions are less severe than for tartrate-modified catalysts, 20–50 °C at pressures usually in excess of 10 atm.

Several studies have been reported on the mechanism of chiral catalysis by modified solid catalysts. Both modifier and substrate structures play important roles. Refer to reviews by Fish and Ollis (1978), Izuma (1983), Sachtler (1985), Tai and Harada (1986), and Blaser et al. (1988) for details of various postulated mechanisms, but one particular conclusion is significant. For the cinchona-modified platinum catalyst, the presence of nitrogen is considered essential, and the configuration at C_8 of the alkaloid determines which enantiomer of the product is formed.

Examples

Studies have been reported of a large number of model reactions, and these have been systematically classified and analyzed by Blaser and Müller (1991). Industrial applications are relatively few, but with the availability of commercial catalysts, more applications are expected. Two such catalyst systems are reported to be available: Raney nickel/tartrate/NaBr system from Degussa (Blaser and Müller, 1991), and nickel powder/tartrate/NaBr (Brunner et al., 1980).

Examples of industrial production based largely on these modified solid catalysts are a sex hormone (Tai et al., 1978, 1979), ligands for homogeneous enantioselective hydrogenation (Bakos et al., 1981; Tai and Harada, 1986), and an intermediate in the production of the angiotensin-converting enzyme inhibitor, benazepril (Morrison and Mosher, 1971).

Zeolite-Supported Catalysts

One way of creating a heterogeneous catalyst is to immobilize a liquid-phase catalyst on a solid support either by dispersing it within its pores or by binding it to a chemical moiety on its surface. Various types of immobilization have been described by Doraiswamy and Sharma (1984). In view of the general importance of this method of heterogenizing a liquid-phase catalyst, including homogeneous chiral catalysts, it was treated as one of the techniques for *heterogenizing a homogeneous catalyst* in Chapter 6. We merely note here that such immobilized systems consisting of Rh(I), Ru(II), Ni(II), or Co(II) anchored to modified USY zeolites have been developed by Corma and his collaborators (Corma et al., 1991, 1992; Sanchez et al., 1991).

Supported Aqueous Phase (SAP) Catalysts

Immobilization on zeolites mentioned before has the obvious limitation that the small size of their pores restricts the reaction to small substrate molecules. This has been overcome by a combination of two developments: (1) removing the need for necessarily using an organic solvent for the catalyst by replacing it with water-soluble chiral rhodium- or ruthenium-sulfonated-BINAP (see section on

water-soluble catalysts) and (2) supporting this new class of catalysts on a solid by a technique developed earlier to produce *supported aqueous phase (SAP) catalysts* (see Chapter 6). Clearly, this is also a heterogenized liquid-phase catalyst but specially designed to heterogenize an aqueous solution of the sulfonated-BINAP complex.

Footprint Catalysis

A unique class of catalysts with enzyme-like specificity has recently been developed. They can be tailor-made for a desired selective conversion, including enantioselective conversion. The basis of their unique action is the creation in aluminum ion-doped silica gel of cavities with molecular recognition properties with a Lewis acid site at the bottom. The Lewis sites are created during the preparation of the gel, and the molecular recognition capability is imprinted during pretreatment with molecular species that are basically transition-state analogs of the substrate molecule. These molecular species may be regarded as templates that stabilize the substrate molecules in their transition state. In other words, they *create molecular footprints* that lower the activation energy for the reaction specific to a substrate on that footprint. The cavity as a whole functions by the complementary actions of the acidity of the Lewis sites and the molecular recognition capability of the footprints left by the expelled template molecules. Because of the highly specific and "individualized" nature of the cavities, the resulting catalysis is referred to as *footprint catalysis*. These catalysts show substrate specificity, enantioselectivity, and feedback regulation. The art of footprint catalysis lies in choosing the right kind of template for a given substrate under nucleophilic attack.

Most studies reported so far on footprint catalysis have come from Morihara and his collaborators in a series of papers starting from 1988 (Morihara et al., 1988a,b, 1989, 1992, 1993a,b,c; Shimada et al., 1992, 1993, 1994; Matsuishi et al., 1992, 1994). Refer to them for further details.

Notation to Part II

A	Component A.
$[A]^*$	Concentration of A at saturation, mol/m^3.
$[\hat{A}]$	Defined by $[A]/[A]_s$.
$[\hat{A}]_b$	Defined by $[A]/[A]_b$.
$[\hat{A}]^*$	Ratio of sorbate (A) concentration at the surface to that at saturation, mol/m^3.
a	Constants of Equations 7.9 and 7.43; interfacial area, m^2/m^3.
a'	External surface area, m^2/g.
B	Component B; ratio of mass to heat Biot numbers.
Bi_m, Bi_h	Mass and heat transfer Biot numbers, $k'_G R/D, hR/\lambda$.
b	Constant of Equation 7.9; exponent of Equation 7.43; catalyst dilution ratio of Equation 7.58.
C_p	Heat capacity of fluid, kcal/K kg.
c	Constant of Equation 7.9.
D_{bA}	Bulk duffision coefficient of A, m^2/s.
D_{cA}	Combined diffusity of A, m^2/s.
D_{eA}	Effective diffusivity of A, m^2/s.
$D_{eA,L}$	Effective diffusivity of a liquid-loaded pellet, m^2/s.
D_{ji}	Bulk diffusion coefficient of j in i, m^2/s.
$\bar{D}_j$	Bulk diffusion coefficient of j in a mixture of N components, m^2/s.
D_{KA}	Knudsen diffusivity of A, m^2/s.
D_{eKA}	Effective Knudsen diffusivity of A, m^2/s.
Da_n	Damköhler number for an nth order reaction, $k_v/[A]^{n-1} k'_G a$.
d_m	Molecular diameter, m.
d_p	Particle (pellet) diameter, m.

Notation to Part II

E	Activation energy, kcal/mol.
E_i	Activation energy of species i, kcal/mol.
ee	Enantiomeric excess.
F_{A0}	Feed rate of A, mol/s.
f	Fraction of total surface on the exterior of the pellet; porosity, m^3/m^3.
f_B	Bed porosity.
H_A	Gas-liquid partition coefficient of A, (mol A/m^3 gas)/(mol A/m^3 liquid).
$-\Delta H_r$	Heat of reaction, kcal/mol.
h	Heat transfer coefficient, kcal/m^2 K s; Table 7.7.
h_{fp}	Heat transfer coefficient specifically for fluid-particle systems, kcal/m^2 K s.
$[i]$	Concentration of i within a pellet, mol/m^3.
$[i]_s$	Concentration of i at pellet surface, mol/m^3.
$[\hat{i}]$	Normalized concentration of i, $[i]/[i]_s$.
j, j_d, j_d	Transfer factors defined by Equations 7.41 and 7.42.
K	Thermodynamic equilibrium constant.
K_i	Adsorption equilibrium constant of component i, m^3/mol (or 1/atm).
K'	Equilibrium constant of the surface reaction.
k	General symbol for reaction rate constant (for an nth order reaction, (m^3/mol)$^{n-1}$ 1/s, 1/s for a first-order reaction).
k^o	Arrhenius preexponential factor, units of rate.
k_i^0	Arrhenius preexponential factor for the formation of i, units of rate.
k_a	Actual (observed) reaction rate constant, appropriate units.
k_{ia}	Adsorption rate constant of i, units depending on the units of the rate.
k_{id}	Desorption rate constant of i, units depending on the units of the rate.
k_i	Reaction rate constant for the formation of i, 1/s.
k_s, k'	Reaction rate constant based on catalyst surface, m/s.
k_v	Reaction rate constant based on catalyst volume, (m^3/mol)$^{n-1}$ (1/s).
k_w	Reaction rate constant based on catalyst weight (mol/kg cat s)/(mol/m^3)^n.
k_{wp}	Reaction rate constant, (mol/kg cat s)/(atm)^n.
k'_G	Mass transfer coefficient, m/s.
L	Length, m.
$\hat{L}$	Defined as ℓ/L.
L^*	Defined in Figure 7.13.
L	Lipschitz constant.
M	Parameter defined by Equation 7.30.
M_i	Molecular weight of i, kg/mol.
m	Reaction order.
N	Parameter defined by Equation 7.30.
Nu	Nurselt number, hR/λ.
n	Reaction order; number of measurements.
P	Pressure, atm.
p_i	Partial pressure of i, atm.
Pr	Prandtl number, $C_p\mu/\lambda$.
q	Liquid loading, kg/kg.

R	Component R; radius of pellet, m.
$\hat{R}$	Defined by r/R.
R_D	Ratio of diffusivities, D_{bA}/D_{KA}.
R_{D_e}	Ratio of effective diffusivities, D_{ebA}/D_{eKA}.
R_g	Gas constant, kcal/mol K.
Re	Reynolds number, $d_p u \rho / \mu$.
r	Radial coordinate of spherical pellet, m.
r^*	Defined in Figure 7.13.
r_a	*Actual* rate of reaction, mol/m^3 s.
r_i	Rate of reaction of i, mol/m^3 s.
$\bar{r}$	Parameter defined by Equation 7.10.
r_p	Radius of pore, m.
$\bar{r}_p$	Mean pore radius, m.
$-r_A$	General notation for rate of disappearance of A, mol/m^3 s.
$-r_{vA}$	Rate of disappearance of A, mol/m^3 cat s.
$-r_{wA}$	Rate of disappearance of A, mol/kg cat s.
S	Component S.
S_g	Pore surface, m^2/g.
Sc	Schmidt number, $\mu/\rho D$.
Sh	Sherwood number, $k'_G R/D$.
s	Intrinsic selectivity, k_{v1}/k_{v2}; shape parameter of Equation 7.21.
s_a	Actual selectivity.
s^2	Sum of squares of residuals.
T	Temperature, K.
T^*	Defined in Figure 7.13.
V	Volume, m^3.
u	Velocity, m/s.
V'	Pellet volume per unit weight, m^3/kg.
V_g	Pore volume, m^3/g.
W	Weight of catalyst, kg.
X_i	Conversion of i.
x	Parameter defined by Equation 7.30; length coordinate, m.
x_0	Characteristic shape length, m.
$\hat{x}$	Normalized length, x/x_0.
Y_R	Intrinsic yield of R.
Y_{Ra}	Actual yield of R.
Y_{RA}, Y_{RB}	Parameters defined in Table 7.5.
y	Parameter defined by Equation 7.30.
y_i	Mole fraction of i; an experimental measurement.
$\bar{y}$	Arithmetic mean of n measurements.

GREEK

α	Parameter defined by Equation 7.28; parameter defined in Table 7.3; Arrhenius group defined by $E/R_g T$.
α_b	Arrhenius group at bulk conditions, $E/R_g T_b$.

α_D	Ratio of effective diffusivities of B to A.
α_s	Arrhenius group at surface conditions, $E/R_g T_s$.
β	Parameter defined by Equation 7.28.
β_m	Maximum temperature rise defined by Equation 7.33.
δ	Experimental error.
$\delta_B, \delta_R, \delta_S$	Parameters defined by Equation 7.28.
ε	Effectiveness factor.
γ	Parameter defined by Equation 7.28.
ϕ	Shape and order generalized Thiele modulus; also general symbol for Thiele modulus.
ϕ'	Parameter defined by Equation 7.30.
ϕ_a^2	Observed (Weisz) modulus defined by Equation 7.25.
ϕ_1	Thiele modulus for a first-order reaction in any shape.
ϕ_{c1}	Thiele modulus in a cylindrical pellet for a first-order reaction.
ϕ_{p1}	Thiele modulus in a flat plate for a first-order reaction.
ϕ_{s1}	Thiele modulus in a spherical catalyst for a first-order reaction.
ϕ_{sn}	Thiele modulus in a spherical pellet for an nth-order reaction in a spherical pellet.
$(\phi_{s1})_s$	ϕ_s at pellet surface conditions.
η	External effectiveness factor.
λ	Ratio of molecular to particle diameters, d_m/d_p; thermal conductivity, kcal/m K s.
λ_e	Effective thermal conductivity, kcal/m K s.
μ	Viscosity of fluid, kg/m s.
ν_i	Stoichiometric coefficient of component i.
θ_i	Fraction of sites occupied by i.
θ_v	Fraction of vacant sites.
ρ	Density, kg/m^3.
σ	Force constant.
τ	Tortuosity factor.
Λ	Generalized length coordinate in pellet, r or ℓ, m.
Λ_0	Generalized pellet length, R or L, m.
$\bar{\Lambda}$	Normalized length parameter, Λ/Λ_0.
ℓ	Length parameter, m; active site; undiluted height of a diluted bed, m.
Ω	Collision integral.
$\Im$	Parameter defined in Table 7.5.

Subscripts/Superscripts

Subscripts

a	Actual (observed) adsorption.
b	Bulk diffusion; bulk.
c	Cylinder; catalyst.
d	Desorption.
e	Effective.

ext	External.
f	Fluid; outlet condition.
G	Gas.
int	Internal.
K	Knudsen diffusion.
0	Inlet condition.
ov	Overall.
p	Flat plate.
s	Surface, sphere.
t	Total.
v	Vacant site.

Superscripts
*	Equilibrium.

References to Part II

Abed, R., and Rinker, R.G. *AIChE J.*, **19**, 618 (1973).
Aboulayt, A., Chambellan, A., Marzin, M., Saussey, J., Mauge, F., Lavalle, J.C., Mercier, C., and Jacquot, R. In *Heterogeneous Catalysis and Fine Chemicals* (Eds., Guisnet, M., Barbier, J., Barrualt, J., Bouchoule, C., Duprez, D., Montassier, C., and Pérot, G.), Elsevier, Amsterdam, 1993.
Acres, G.J.K., Bond, G.C., Cooper, B.J., and Dawson, J.A. *J. Catal.*, **6**, 139 (1966).
Adams, C.R. *J. Catal.*, **10**, 355 (1968).
Advances in Catalytic Processes, JAI Press, Greenwich, Conn., 1995.
Agarwal, A.K., Cant, N.W., Wainwright, M.S., and Trimm, D.L. *J. Mol. Catal.*, **43**, 79 (1987).
Ai, M. In *Heterogeneous Catalysis and Fine Chemicals* (Eds., Guisnet, M., Barbier, J., Barrualt, J., Bouchoule, C., Duprez, D., Montassier, C., and Pérot, G.), Elsevier, Amsterdam, 1991.
Aitken, R.A., and Kilênyi, S.N. *Asymmetric Synthesis*, Blackie Academic and Professional, Chapman & Hall, Glasgow, U.K., 1992.
Akabori, S., Izumi, Y., Fuji, Y., and Sakurai, S. *Nature*, **178**, 323 (1956).
Akutagawa, S. In *Catalysis of Organic Reactions*, Dekker, New York, 1995.
Andersson, S.L.T. *J. Catal.*, **98**, 138 (1986).
Ando, T., Cork, D.G., and Kimura, T. *Chem. Lett.*, **665** (1986).
Antonioletti, R., Bonadies, F., Locati, L., and Scettri, A. *Tetrahedron Lett.*, **33**, 3205 (1992).
Arhancet, J.P., Davis, M.E., Merola, J.S., and Hanson, B.E. *Nature* (London), **339**, 454 (1989); *J. Catal.*, **121**, 327 (1990).
Arhancet, J.P., Davis, M.E., and Hanson, B.E. *J. Catal.*, **129**, 94 (1991).
Aris, R. *Elementary Chemical Reactor Analysis*, McGraw-Hill, New York, 1969 (also Butterworth, Boston, MA, 1989); *J. Catal.*, **22**, 232, (1971); *The Mathematical Theory of Diffusion and Reaction in Permeable Catalysts*, Vols. 1 and 2, Clarendon Press, Oxford, U.K., 1975.

Augustine, R.L. *Heterogeneous Catalysis for the synthetic Chemist*, Dekker, New York, 1996.
Augustine, R.L., and Meng, L. In *Catalysis of Organic Reactions* (Ed., Malz, R.E., Jr.), Dekker, New York, 1996.
Bailar, J.C. *Catal. Rev.*, **10**, 17 (1974).
Bailey, D.C., and Langer, S.H. *Chem. Rev.*, **81**, 109 (1981).
Baillard, R.M., Aranda, Y., Mas, J.M., and Jacquot, R. EP 259, 234 to Rhone-Poulenc Agro (1988).
Bakos, J., Toth, I., and Marko, L. *J. Org. Chem.*, **46**, 5427 (1981).
Balakrishnan, I., Rao, B.S., Kavedia, C., Babu, G.P., Kulkarni, S.B., and Ratnasamy, P. *Chem. Ind.*, **410** (19 June), 1982.
Balaraman, K.S., Mashelkar, R.A., and Doraiswamy, L.K. *AIChE J.*, **26**, 635 (1980).
Baldovi, M.V., Corma, A., Fornes, V., Garcia, H., Martinez, A., and Primo, J. *J. Chem. Soc., Chem. Commun.*, **949** (1992).
Balogh, M., and Laszlo, P. *Organic Chemistry Using Clays*, Springer-Verlag, New York, 1993.
Barnard, J.A., and Mitchell, D.S. *J. Catal.*, **12**, 376 (1968).
Barrault, J., Brunet, S., Essayem, N., Piccirilli, A., Guimon, C., and Gamet, J.P. In *Heterogeneous Catalysis and Fine Chemicals* (Eds., Guisnet, M., Barbiar, J., Barrault, J., Bouchoule, C., Dupez, D., Montassier, C., and Pérot, G.), Elsevier, Amsterdam, 1993.
Barrault, J., Delahay, G., Essayem, N., Gaizi, G., Forquy, C., and Brouard, R. In *Heterogeneous Catalysis and Fine Chemicals* (Eds., Guisnet, M., Barbiar, J., Barrault, J., Bouchoule, C., Dupez, D., Montassier, C., and Pérot, G.), Elsevier, Amsterdam, 1991.
Bartok, M. In *Stereochemistry of Heterogeneous Metal Catalysts*, Ch. XI, Wiley, New York, 1985.
Bau, R., Carroll, W.E., Hart, D.W., Teller, R.G., and Koetzle, T.F. In *Transition Metal Hydrides* (Ed., Bau, R.), *Am. Chem. Soc. Ser.*, **167**, Washington, D.C., 1978.
Bauer, K., and Garbe, D. *Common Fragrance and Flavor Materials*, Verlag Chimie, Weinheim, 1985.
Baumeister, P., Blaser, H.U., and Scherrer, W. In *Heterogeneous Catalysis and Fine Chemicals II* (Eds., Guisnet, M., Barrault, J., Bouchoule, C., Duprez, D., Perot, G., Maurel, R., and Montassier, C.), Elsevier, Amsterdam, 1991.
Bautista, F.M., Campelo, J.M., Garcia, A., Guardeno, R., Luna, D., and Marinas, J.M. In *Heterogeneous Catalysis and Fine Chemicals* (Eds., Guisnet, M., Barrault, J., Bouchoule, C., Dupez, D., Montassier, C., and Pérot, G.), Elsevier, Amsterdam, 1991.
Bayer, A.G. British Patent 1191610 (1970).
Beamer, R.L., Belding, R.H., and Fickling, C.S. *J. Pharm. Sci.*, **58**, 1142; 1419 (1967).
Beasley, G.B., and Jakovac, I.J. In *Ion Exchange Resin Catalysts Having Improved Catalytic Activity and Enhanced Thermal Stability, Ion Exchange Technology*, Ellis Horwood, New York, 1984.
Beekman, J.W., and Froment, G.F. *Ind. Eng. Chem. Fundam.*, **21**, 243 (1982).
Bellussi, G., and Fattore, V. In *Zeolite Chemistry and Catalysis* (Eds., Jacobs, P.A., Jaeger, N.I., Kubelkova, L., and Wichterlova, B.), *Stud. Surf. Sci. Catal.*, **69**, 79 (1991).
Bergman, R.G. *Acct. Chem. Res.*, **13**, 113 (1980).
Berty, J.M., Hanibrick, J.O., Malone, T.R., and Ullock, D.S. *64th National Meeting of the AIChE*, New Orleans, Paper no. 42E, 1969.

Bhagade, S.S., and Nageshwar, G.D. *Chem. Petro-Chem. J.*, **8**, 9 (1977); **9**, 3 (1978a); **9**, 21 (1978b); **11**, 23 (1980); **12**, 21 (1981).
Bhatia, S., Wu, X., Sanders, D.K., Gerstein, B.C., Pruski, M., and King, T.S. *Catal. Today*, **12**, 165 (1992).
Bhawal, B.M., Vetrivel, R., Reddy, T.I., Deshmukh, A.R.A.S., and Rajappa, S. *J. Phys. Org. Chem.*, **7**, 377 (1994).
Bhowmik, S.K., and Ratnasamy, P. In *Proceedings 4th World Congr. of Chem. Eng.*, Verlag Chemie, Berlin, 1991.
Bingham, D., Hudson B., Webster, D., and Wells, P.B. *J. Chem. Soc., Dalton Trans.*, 1521 (1974).
Birk, J.P., Halpern, J., and Pickard, A.L. *J. Am. Chem. Soc.*, **90**, 4491 (1968a); *Inorg. Chem.*, **7**, 2672 (1968b).
Bischoff, K.B., and Froment, G.F. *Ind. Eng. Chem. Fundam.*, **1**, 195 (1962).
Bischoff, K.B. *AIChE J.*, **11**, 351 (1965); *Chem. Eng. Sci.*, **23**, 451 (1968).
Bjornson, G., and Stark, J.J. U.S. Patent 5,086,030 1992.
Blaser, H.U., and Müller, M. In *Heterogeneous Catalysis and Fine Chemicals II* (Eds., Guisnet, M., Barrault, J., Bouchoule, C., Duprez, D., Perot, G., Maurel, R., and Montessier, C.), Elsevier, Amsterdam, 1991.
Blaser, H.U., Jalett, H.P., Monti, D.M., Reber, J.F., and Wehrli, J.T. In *Heterogeneous Catalysis and Fine Chemicals II* (Eds., Guisnet, M., Barbier, J., Barrault, J., Bouchoule, C., Duprez, D., Pérot, G.), Elsevier, Amsterdam, 1988.
Boag, I.F., Bacon, D.W., and Downie, J. *Can. J. Chem. Eng.*, **56**, 389 (1978).
Boitiaux, J.P., Cosyns, J., and Vasudevan, S. *Appl. Catal.*, **6**, 41 (1983).
Bonds, W.D., Jr., Brubaker, C.H., Jr., Chandrashekaran, E.S., Gibbons, C., Grubbs, R.H., and Kroll, L.C. *J. Am. Chem. Soc.*, **97**, 2128 (1975).
Borman, P.C., Bos, A.N.R., and Westerterp, K.R. *AIChE J.*, **40**, 862 (1994).
Bortnick, N.M. U.S. Pat. 3,037,052 1962; *Chem. Abstr.*, **57**, 6125 (1962).
Botteghi, C., Ganzerla, R., Lenarada, M., and Moretti, G. *J. Mol. Catal.*, **40**, 129 (1987).
Boudart, M., and Djega-Mariadassou, G. *Kinetics of Heterogeneous Catalytic Reactions*, Princeton University Press, Princeton, N.J., 1984.
Boudart, M. *Ind. Eng. Chem. Fundam.*, **25**, 658 (1986).
Boudart, M., Mears, D.E., and Vannice, M.A. *Ind. Chim. Belge*, **32**, 281 (1967).
Bouzek, K., and Rousăr, I. *J. Chem. Technol. Biotechnol.*, **66**, 131 (1996).
Box, G.E.P., and Draper, N.R. *Biometrika*, **52**, 355 (1965).
Box, G.E.P., and Henson, T.L. M.B.R. Tech. Rept. No. 51, University of Wisconsin, Madison, Wisc., January 1969.
Box, G.E.P., and Hill, W.J. *Technometrics*, **9**, 57 (1967).
Box, G.E.P., and Lucas, H.L. *Biometrika*, **46**, 77 (1959).
Bregault, J.M., El Ali, B., Mercier, J., Martin, J., and Martin, C.C.R. *Acad. Sci. Paris*, **309**(11), 459 (1989).
Breslow, R. and Campbell, P. *Bio-org. Chem.*, **1**, 140 (1971).
Breslow, R. *Science* (Washington, D.C.), **218**, 532 (1982).
Brooks, C.S., Lemkey, F.D., and Golden, G.S. *Proc. Mater. Res. Soc. Symp.*, **8**, 397 (1982).
Brown, C.K., and Wilkinson, G. *J. Chem. Soc. A*, 3133 (1970).
Brown, J.R. *Drug Info J.*, **24**, 117 (1990).
Brown, L.F., Hanyes, H.W., and Manogue, W.H. *J. Catal.*, **14**, 220 (1969).

Brunel, D., Chamoumi, M., Chiche, B., Finiels, A., Gauthier, C., Geneste, P., Marichez, F., and Moreau, P. In *Recent Advances in Zeolite Sciences* (Eds., Klinowski, J., and Barrie, J.), *Stud. Surf. Sci. Catal.*, **52**, 139 (1989).
Brunner, H., and Zettlmeier, W. *Handbook of Enantioselective Catalysis: With Transition Metal Compounds*, VCH, New York, 1993.
Brunner, H., Muschiol, M., Wischert, T., and Wiehl, J. *Tetrahedron Asymm.*, **1**, 159 (1990).
Brunner, H. *Topics in Stereochemistry*, **18**, 129 (1988).
Burge, H.D., Collins, D.J., and Davis, B.H. *Ind. Eng. Chem. Prod. Res. Dev.*, **19**, 389 (1980).
Burghardt, A., and Smith, J.M. *Chem. Eng. Sci.*, **34**, 267 (1979).
Burwell, R.L. *Chemtech.*, **370** (1974).
Butt, J.B. *Reaction Kinetics and Reactor Design*, Prentice-Hall, Englewood Cliffs, N.J., 1980.
Cabello, J.A., Campelo, J.M., Garcia, E., Luna, D., and Marinas, J.M. *J. Org. Chem.*, **49**, 5195 (1984).
Cahn, R.H., Ingold, C.K., and Prelog, V. *Angew. Chem., Int. Ed. Engl.*, **5**, 385 (1966).
Carberry, J.J. *Chem. Eng. Sci.*, **17**, 675 (1962); *Ind. Eng. Chem. Fundam.*, **14**, 129 (1975); *Chemical and Catalytic Reaction Engineering*, McGraw-Hill, New York, 1976.
Carberry, J.J., and Kulkarni, A.A. *J. Catal.*, **31**, 41 (1973).
Carra, S. *Chim. Ind.* (Milan), **54**, 434 (1972).
Carturan, G., Facchin, G., Cocco, G., Enzo, S., and Navazio, G. *J. Catal.*, **76**, 405 (1982).
Catalysis of Organic Reactions, Dekker, New York, Eds., Moser, W.R. (1981), Kosak, J.R. (1984), Augustine, R.L. (1985), Rylander, P.N., Greenfield, H., and Augustine, R.L. (1988), Blackburn, D.W. (1990), Poscoe, W. (1992), Kosak, J.R., and Johnson, T.A. (1994), Scaros, M. and Prunier, M.L. (1995), and Malz, R.E. Jr. (1996).
Chabala, J.C., and Fischer, M.H. U.S. Patent 4,199,569 1980.
Chakrabarti, A., and Sharma, M.M. *Reactive Polym.*, **20**, 1 (1993).
Chaloner, P.A. (et al.), *Homogeneous Hydrogenation*, Dordrecht, Kluwer Academic, 1994.
Chan, A.S.C. *Chemtech.*, 46 (1993).
Chan, A.S.C., Laneman, S.A., Miller, R.E., Wagenknecht, J.H., and Coleman, J.P. In *Catalysis of Organic Reactions* (Eds., Kosak, J.R., and Johnson, T.A.), Dekker, New York, 1994.
Chatt, J., and Leigh, G.J. *Angew. Chem., Int. Ed. Engl.*, **17**, 400 (1978).
Chaudhari, R.V. In *Frontiers in Chemical Reaction Engineering* (Eds. Doraiswamy, L.K., and Mashelkar, R.A.), Wiley Eastern, New Delhi, Vol. I, 1984.
Chem. Eng. News, 50 (16 October 1989).
Chem. Week, **29** (28 February 1990).
Chen, N.Y., Garwood, W.E., and Dwyer, F.G. *Shape Selective Catalysis in Industrial Applications*, Dekker, New York, 1989.
Chen, O.T., and Rinker, R.G. *Chem. Eng. Sci.*, **33**, 1201 (1978).
Chiche, B., Finiels, A., Gauthier, C., Geneste, P., Graille, J., and Pioch, D., *J. Org. Chem.*, **51**, 2128 (1986).
Choudhary, V.R., and Doraiswamy, L.K. *Ind. Eng. Chem. Proc. Des. Dev.*, **11**, 420 (1972).
Choudhary, V.R., and Singh, A.P. *Zeolites*, **6**, 206 (1986).
Choudhary, V.R., Akolekar, D.B., and Singh, A.P. *Chem. Eng. Sci.*, **44**, 1047 (1989).
Choudhary, V.R., and Mamman, A.S. *AIChE J.*, **36**, 1577 (1990).
Choudhary, V.R., Nayak, V.S., and Mamman, A.S. *Ind. Eng. Chem.*, **31**, 624 (1992).
Chu, C., and Chon, K. *J. Catal.*, **17**, 71 (1970).

Chu, C., and Hougen, O.A. *Chem. Eng. Sci.*, **17**, 167 (1962).
Clark, J.H., Kybett, A.P., and Macquarrie, D.J. *Supported Reagents: Preparation, Analysis, and Applications*, VCH, Weinheim, Germany, 1992.
Cocke, D.L. and Yoon, C. In *Proc. 5th Int. Conf. Rapidly Quenched Met.*, Wurzburg, 1984.
Collins, A.N., Sheldrake, G.N., and Crosby, J. *Chirality in Industry, The Commercial Manufacture and Applications of Optically Active Compounds*, Wiley, New York, 1992.
Collman, J.P., Hegedus, L.S., Cooke, M.P., Norton, J.R., Dolcetti, G., and Marquardt, D.N. *J. Am. Chem. Soc.*, **94**, 1789 (1972).
Collman, J. P., and Hegedus, L.S. *Principles and Applications oif Organotransition Metal Chemistry*, University Science Books, Mill Valley, Calif., 1980.
Collman, J.P., Hegedus, L.S., Norton, J.R., and Finke, R.G. *Principles and Applications of Organotransition Metal Chemistry*, 2nd ed, University Science Books, Mill Valley, Calif., 1987.
Compton, R.G., Ed. *Kinetic Models of Catalytic Reactions*, Elsevier, New York, 1991.
Conn, R.S.E., Lovell, A.V., Karady, S., and Weinstock, L.M. *J. Org. Chem.*, **51**, 4710 (1986).
Copelowitz, I., and Aris, R. *Chem. Eng. Sci.*, **25**, 885 (1970).
Coq, B., Plaueix, J.M., Kumbhar, P.S., Geneste, P., Dutartre, R., Conster, N., Breton, V., and Ajayau, P.M. *J. Am. Chem. Soc.*, **116**, 7934, (1994).
Cordier, G., Grosselin, J.M., and Bailliard, R.M. In *Catalysis of Organic Reactions* (Eds., Kosak, J.R., and Johnson, T.A.), Dekker, New York, 1994.
Corey, E.J. *J. Org. Chem.*, **55**(6), 1693 (1990a); *Pure & Appl. Chem.*, **62**, 1209 (1990b).
Corma, A., Llopis, S., Monton, J., and Weller, S. *Chem. Eng. Sci.*, **43**, 785 (1988).
Corma, A. In *Zeolite Microporous Solids: Synthesis, Structure and Reactivity* (Eds., Derouane, E.G., Lemos, F., Naccache, C., and Ribeiro, F.R.), NATO ASI Series, Kluwer Academic, Dordrecht, 1991.
Corma, A., Iglesias, M., del Pinto, C., and Sanchez, F. *J. Chem. Soc., Chem. Commun.*, 1253 (1991); *J. Organomet. Chem.*, 431 (1992).
Cornelis, A., and Laszlo, P. *Synthesis*, 909 (1985).
Cornelis, A., Gerstmans, A., Laszlo, P., Mathy, A., and Zieba, I. *Catal. Lett.*, **6**, 103 (1990).
Corte, H., and Meyer, A., Ger. Pat. 1,045,102 1958; *Chem. Abstr.*, **55**, 1969 (1961).
Csicsery, S.M. *Pure and Appl. Chem.*, **58**, 841 (1986); Zeolites, **4**, 202 (1984).
Cunningham, R.S., and Geankopolis, J. *Ind. Eng. Chem. Fundam.*, **7**, 535 (1968).
Cusumano, J.A. *Chemtech.*, **22**, 482 (1992).
Dake, S.B., Jaganathan, R., and Chaudhari, R.V. *Ind. Eng. Chem. Res.*, **28**, 1107 (1989).
Dallons, J.L., Jacobs, P., Martens, J., Tastenhoye, P., Van den Eynde, I., and Van Gysel, A. EP 339,012 (1989).
Daly, M.J., and Price, B.J. In *Progress in Medicinal Chemistry* (Eds., Ellis, G.P., and West, G.B.), **20**, 337 (1983).
Datta, R., and Rinker, R.G. *J. Catal.*, **95**, 181 (1985).
Datta, R., Savage, W., and Rinker, R. *J. Catal.*, **95**, 193 (1985).
Davidson, J.M. In *Recent Advances in the Engineering Analysis of Chemically Reacting Systems* (Ed., Doraiswamy, L.K.), Wiley Eastern, New Delhi, 1984.
Davis, M.E., and Suib, S.L., Eds. *Selectivity in Catalysis*, Am. Chem. Soc., Washington, D. C., 1993.
Davis, M.E. *Acc. Chem. Res.*, **26**, 111 (1993).
de la Vega, F., and Sasson, Y. *J. Chem. Soc., Chem. Commun.*, 653 (1989).

de Zoete, M.C., Kock-Van Dalen, A.C., Van Rantwijk, F., and Sheldon, R.A. *J. Chem. Soc., Chem. Commun.*, 1831 (1993); *Biocatalysis*, **10**, 307 (1994).
Del Rosso, R., Mazzochia, C., Grouchi, P., and Centola, P. *Appl. Catal.*, **9**, 269 (1984).
Delaude, L., and Laszlo, P. *J. Org. Chem.*, **55**, 5260 (1990).
Delmon, B., and Jannes, G. (Eds. *Catalysis: Heterogeneous and Homogeneous*, Elsevier, Amsterdam, 1975.
Deshmukh, A.R.A.S., Gumaste, V.K., Shiralkar, V.P., and Bapat, B.V. Indian Pat. Appl. 1024/DEL/90, 1990a.
Deshmukh, A.R.A.S., Gumaste, V.K., and Shiralkar, V.P. Indian Pat. Appl. 1321/DEL/90, 1990b.
Deshmukh, A.R.A.S., Reddy, T.I., Bhawal, B.M., Shiralkar, V.P., and Rajappa, S. *J. Chem. Soc., Perkin Trans. I*, 1217 (1990c).
Deshmukh, A.R.A.S., Bhawal, B.M., Shiralkar, V.P., and Rajappa, S. U.S. Patent 4,967,007, 1990d.
Deshpande, R.M., and Chaudhari, R.V. *Ind. J. Tech.*, **21**, 351 (1983); *Ind. Eng. Chem. Res.*, **27**, 1996 (1988); *J. Catal.*, **115**, 326 (1989a); *J. Mol. Catal.*, **57**, 177 (1989b).
Dettmeier, U., Eichler, K., Kuhlein, K., Leupold, E.I., and Litterer, H. *Angew. Chem., Int. Ed.*, **26**, 468 (1987).
Deutsch, D.H., *Chemtech.*, 157 (March 1991).
Didillon, B., Monsour, A., Candy, J.P., Bournville, J.P., and Basset, M. In *Heterogeneous Catalysis and Fine Chemicals* (Eds., Guisnet, M., Barbier, J., Barrault, J., Bouchoule, C., Duprez, D., Montassier, C., and Pérot, G.), Elsevier, Amsterdam, 1991.
Divekar, S.S., Deshpande, R.M., and Chaudhari, R.V. *Catal. Lett.*, **21**, 191 (1993).
Dodgson, I. In *Heterogeneous Catalysis and Fine Chemicals* (Eds., Guisnet, M., Barbier, J., Barrault, J., Bouchoule, C., Dupez, D., Montassier, C., and Pérot, G.), Elsevier, Amsterdam, 1993.
Dombeck, B.D., and Wenzel, T.T. Eur. Patent 394 967, 1990a; Eur. Patent 394 968, 1990b.
Doraiswamy, L.K. *Prog. Sur. Sci.*, **37**, 1 (1991).
Doraiswamy, L.K., and Sharma, M.M. *Heterogeneous Reactions: Analysis, Examples and Reactor Design*, Vol. 1, Wiley, New York, 1984.
Doraiswamy, L.K., and Tajbl, D.G. *Catal. Rev. Sci. Eng.*, **10**, 177 (1974).
Doraiswamy, L.K., and Venkitakrishnan, G.R. National Chemical Laboratory report, 1978.
Doraiswamy, L.K., Krishnan, G.R.V., and Mukherjee, S.P. *Chem. Eng.* (Chemical Technology Section), **88**, July 13 (1988).
Dovi, V.G., Reverberi, A.P., and Acevedo, D.L. *Ind. Eng. Chem. Res.*, **33**, 62 (1994).
Downie, J., Shestad, K.A., and Graydon, W.F. *Can. J. Chem.*, 201 (1961).
Duhamal, L., Duhamal, P., Launay, J.C., and Plaquevent, J.C. *Bull. Soc. Chim. Fr. II*, 421 (1984).
Dumez, F.J., and Froment, G.F. *Ind. Eng. Chem. Proc. Des. Dev.*, **15**, 291 (1976).
Dumez, F.J., Hosten, L.H., and Froment, G.F. *Ind. Eng. Chem. Fundam.*, **16**, 298 (1977).
Eby, R.T., and Singleton, T.C. In *Applied Industrial Catalysis* (Ed., Leach, B.E.), Academic Press, New York, Vol. 1, 1983.
El Ali, B., Bregault, J.M., Martin, J., Martin, C. *New J. Chem.*, **13**, 173 (1989a).
El Ali, B., Bregault, J.M., Mercier, J., Martin, J., Martin, C., and Convert, O. *J. Chem. Soc., Chem. Commun.*, 825 (1989b).
Esposito, A., Taramasso, M., Neri, C., and Buonomo, F. Br. Patent 2,116,974 (1985).
Evans, D., Osborn, J.A., and Wilkinson, G. *J. Chem. Soc. A.*, 3133 (1968a).
Evans, D., Yagupsky, G., and Wilkinson, G. *J. Chem. Soc. A.*, 2660 (1968b).

Evans, G.O., Pittman, C.U., Jr., McMillan, R., Beach, R.T., and Jones, R. *J. Organomet.*, **67**, 295 (1974).
Everett, D.H. *Trans. Faraday Soc.*, **46**, 942 (1950).
Faraj, M., and Hill, C.L. *J. Chem. Soc., Chem. Commun.*, 1487 (1987).
FDA's Policy Statement for the Development of New Stereoisomeric Drugs, *Chirality*, **4**, 338 (1992).
Federsel, H.J. *Chemtech*, 24 (24 December 1993).
Felthouse, T.R. *J. Am. Chem. Soc.*, **109**, 7566 (1987).
Feng, C.F., and Stewart, W.E. *Ind. Eng. Chem. Fundam.*, **12**, 143 (1973).
Ferraiolo, G., Peloso, A., Reverberi, A., Del Borgi, M., and Beruto, D. *Can. J. Chem. Eng.*, **51**, 447 (1973).
Fieser, L.F. *J. Am. Chem. Soc.*, **61**, 2559 (1939).
Finn, M.G., and Sharpless, K.B. *J. Am. Chem. Soc.*, **113**, 113 (1991).
Fiorini, M., and Giongo, G.M. *J. Mol. Cat.*, **5**, 303 (1979).
Fish, M.J., and Ollis, D.F. *Catal. Rev. Sci. Eng.*, **18**, 258 (1978).
Ford, M.E., and Conner, M.D. In *Catalysis of Organic Reactions* (Eds., Kosak, J.R., and Johnson, T.A.), Dekker, New York, 1994.
Forni, L. Intl. Patent Appl. *J. Catal.*, **111**, 199 (1988).
Forni, L., and Miglio, R. In *Heterogeneous Catalysis and Fine Chemicals* (Eds., Guisnet, M., Barbier, J., Barrault, J., Bouchoule, C., Dupez, D., Montassier, C., and Pérot, G.), Elsevier, Amsterdam, 1993.
Forni, L., and Nestori, S. In *Heterogeneous Catalysis and Fine Chemicals* (Eds., Guisnet, M., Barbier, J., Barrault, J., Bouchoule, C., Durprez, D., Montassier, C., and Pérot, G.), Elsevier, Amsterdam, 1988.
Forni, L., Stern, G., and Gatti, M. *Appl. Catal.*, **29**, 161 (1987a).
Forni, L., Oliva. C., and Regina, A. *Proceedings 6th Italian-Czechoslavak Symposium on Catalysis*, Ital. Chem. Soc., 74 (1987b).
Forster, D. *Adv. Organomet. Chem.*, **17**, 255 (1979); *J. Am. Chem. Soc.*, **98**, 846 (1976).
Franck, H.G., and Stadelhafer, J.W. In *Industrial Organic Chemistry*, Springer-Verlag, Berlin, 1988.
Franckaerts, J., and Froment, G.F. *Chem. Eng. Sci.*, **19**, 807 (1964).
Francis, M.B., Finney, N.S., and Jacobsen, E.S. *J. Am. Chem. Soc.*, **118**, 8983 (1996).
Froment, G.F., and Bischoff, K.B. *Chemical Reactor Analysis and Design*, Wiley, New York, 1990.
Froment, G.F., and Mezaki, R. *Chem. Eng. Sci.*, **25**, 293 (1970).
Froment, G.F. In *Proc. 7th Eur. Symp., Computer Application in Process Development*, Erlangen, Dechema, 1974; *AIChE J.*, **21**, 1041 (1975).
Funakoshi, M., Komiyama, H., and Inoue, H. *Chem. Lett.*, 245 (1985).
Gallezot, P., Biroir-Fedler, A., and Rechard, D. In *Catalysis of Organic Reactions* (Ed., Pascoe, W.), Dekker, New York, 1992.
Galvagno, S., Poltarzewski, Z., Donato, A., Neri, G., and Pietropaolo, R. *J. Chem. Soc., Chem. Commun.*, 1729 (1986).
Gao, Y., Hanson, R.M., Klunder, J.M., Ko, S.Y., Masamune, H., and Sharpless, K.B. *J. Am. Chem. Soc.*, **109**, 5765 (1987).
Gates, B.C. *Catalytic Chemistry*, Wiley, New York, 1992.
Gavriilidis, A., and Varma, A. *Catal. Rev. Sci. Eng.*, **35**, 399 (1993).
Gilbert, L., and Mercier, C. In *Heterogeneous Catalysis and Fine Chemicals* (Eds., Guisnet, M., Barbier, J., Barrault, J., Bouchoule, C., Dupez, D., Montassier, C., and Pérot, G.), Elsevier, Amsterdam, 1993.

Gillet, J.P., Kervennal, J., and Pralus, M. In *Heterogeneous Catalysis and Fine Chemicals* (Eds., Guisnet, M., Barbier, J., Barrault, J., Bouchoule, C., Dupez, D., Montassier, C., and Pérot, G.), Elsevier, Amsterdam, 1993.
Goldstein, W., and Carberry, J.J. *J. Catal.*, **28**, 33 (1973).
Goyal, P., and Doraiswamy, L.K. *Ind. Chem. Eng./Process Dev. Design*, **9**, 26 (1970).
Graham, J.R., and Angelici, R.J. *Inorg. Chem.*, **6**, 2082 (1967).
Graham, W.R., and Lynch, D.T. *AIChE. J.* **33**, 792 (1987).
Greenstein, J.P., and Winitz, M. *Chemistry of the Amino Acids*, Wiley, New York, 1961.
Griffith, W.P. *Platinum Met. Rev.*, **33**, 181 (1989).
Griffiths, D.W., and Bender, M.L. *Adv. Catal.*, **23**, 209 (1973).
Grubbs, R.H. *Chemtech*, 512 (1977).
Grzybowska, B., Czerwenka, M., and Sloczynski, J. *Catal. Today*, **1**, 157 (1987).
Gubelmann, M.H., Guisnet, M., Pérot, G., and Pouilloux, Y. *Coll. Czech. Chem. Commun.*, **57**, 809 (1992).
Guha, B.K., and Narasimhan, G., *Chem. Eng. Sci.*, **27**, 703 (1972).
Guisnet, M., Barbier, J., Barrault, J., Bonchoule, C., Duprez, D., Montassier, C., and Pérot, G. Eds. *Heterogeneous Catalysis and Fine Chemicals*, Elsevier, Amsterdam, 1988, 1991, 1993, 1996.
Gullet, B.K., Bruce, K.R., and Machilek, R.M. *Rev. Sci. Instrum.*, **61**(2), 904 (1990).
Gum, E.R., and Freitas, C.R. *Chem. Eng. Prog.*, **75**, 73 (1979).
Halpern, J., and Pickard A.L. *Inorg. Chem.*, **9**, 2798 (1970).
Halpern, J., and Wong, C.S. *J. Chem. Soc., Chem. Commun.*, 629 (1973).
Halpern, J. In *Organotransition Metal Chemistry* (Eds., Ishidu, Y., and Tsutsui, M.), Plenum, 1975; *Trans. Am. Crystallogr. Assoc.*, **14**, 59 (1978); *Science*, **217**, 401 (1982).
Halpern, J., Okamoto, T., and Zakhariev, A. *J. Mol. Catal.*, **2**, 65 (1976).
Harada, K., and Yoshida, T., *Naturwiss.*, **57**, 131 and 306 (1970).
Hathaway, P.E., and Davis, M.E. *J. Catal.*, **116**, 253 (1988a); **116**, 279 (1988b); **117**, 497 (1988c).
Hattori, H. In *Heterogeneous Catalysis and Fine Chemicals* (Eds., Guisnet, M., Barbier, J., Barrault, J., Bouchoule, C., Dupez, D., Montassier, C., and Pérot, G.), Elsevier, Amsterdam, 1993.
Hattori, H., Tanaka, K., and Tanabe, K. *J. Am. Chem. Soc.*, **98**, 4652 (1976).
Haynes, A., Mann, B.E., Morris, G.E., and Maitlis, P.M. *J. Am. Chem. Soc.*, **115**, 4093 (1993).
Haynes, H.W., Jr. *Can. J. Chem. Eng.*, **56**, 582 (1978).
Heinemann, H. *Chemtech*, 286 (1971).
Henri-Olive, G., and Olive, S. *Trans. Met. Chem.*, **1**, 77 (1977).
Hill, C.L., and Brown, R.B. *J. Am. Chem. Soc.*, **108**, 536 (1986).
Himmelblau, D.M. *Process Analysis by Statistical Methods*, Wiley, New York, N.Y., 1970.
Hino, A., and Arata, K. *J. Chem. Soc., Chem. Commun.*, **24**, 1148 (1980); *J. Appl. Catal.*, **18**, 401 (1985).
Hirako, Y. Japanese Patent 62 123 154, 1987.
Hirschmann, R., Miller, R., and Wendler, N.L. *J. Am. Chem. Soc.*, **76**, 4592 (1954).
Hjortkjaer, J., and Jensen, O.R. *Ind. Eng. Chem. Prod. Res. Dev.*, **16**, 281 (1977).
Hodge, P., and Sherrington, D.C., Eds. *Polymer-Supported Reactions in Organic Synthesis*, Wiley, New York, 1980.
Hoelderich, W.F. In *Heterogeneous Catalysis and Fine Chemicals* (Eds., Guisnet, M., Barbier, J., Barrault, J., Bouchoule, C., Duprez, D., Montassier, C., and Pérot, G.), *Stud. Surf. Sci. Catal.*, **41**, 83 (1988); In *Guidelines for Mastering the Properties*

of Molecular Sieves (Eds., Barthomeuf, D., Derouane, E.G., and Holderich, W.), Plenum, New York, 1990; In *Zeolite Microporous Solids: Synthesis, Structure, and Reactivity* (Eds., Derouane, E.G., Lemos, F., Naccache, C., and Ribeiro, F.R.), NATO ASI Series, Kluwer Academic, Dordrecht, 1991a; In *Structure-Activity and Selectivity Relationships in Heterogeneous Catalysis* (Eds., Grasselli, R.K., and Sleight, A.W.), *Stud. Surf. Sci. Catal.*, **67**, 257 (1991b); In *New Frontiers in Catalysis* (Eds., Guczi, L. et al.), *10th Int. Cong. on Catalysis* (19–24 July 1992), Elsevier, Budapest, Hungary, 1993.

Hoelderich, W.F., Hesse, M., and Naumann, F. *Angew. Chem.*, **100**, 232 (1988); *Angew. Chem., Int. Ed.*, **27**, 226 (1988).

Hoelderich, W.F., and Van Bekkum, H. In *Introduction to Zeolite Science and Practice* (Eds., van Bekkum, H., Flanigen, E.M., and Jansen, J.C.), *Stud. Surf. Sci. Catal.*, **58**, 631 (1991).

Hoffmann, U., and Hofmann, H. *Einfuhrung in die Optimierung*, Verlag Chemie, Weinheim, BRD, 1971; *Chem. Ing. Techn.*, **48**, 465 (1976).

Hofmann, H. *Adv. Chem. Ser.*, **109**, 519 (1972).

Hofmann, H. In *Chemical Reactor Design and Technology: Overview of the New Development of Energy and Petrochemical Reactor Technologies, Projections for the 90's* (Ed., de Lasa, H.I.), Proceedings of the NATO Advanced Study Institute on Chemical Reactor Design and Technology, London, Ontario, Canada, 2–12 June, 1985, Kluwer Academic, Dordrecht, Netherlands, 1986.

Hori, G.K., and Schmidt, L.D. *J. Catal.*, **38**, 335 (1975).

Hosten, L.H. *Chem. Eng. Sci.*, **29**, 2247 (1974).

Hougen, O.A., and Watson, K.M. *Chemical Process Principles*, Vol. III, Wiley, New York, 1947.

Huet, S., Bouvier, A., Gruet, M.A., and Jolivet, E. *Statistical Tools for Numerical Regression*, Springer, New York, 1996.

Hughes, D.L., Dolling, U.H., Ryan, K.M., Schoenewaldt, E.F., and Grabowski, E.J. *J. Org. Chem.*, **52**, 4745 (1987).

Hulshof, L., and Roskam, J.H. European Patent Appl. 0343714, 1989 to Stamicarbon.

Hutchings, J., and Carberry, J.J. *AIChE J.*, **12**, 20 (1966).

Idriss, H., Kim, K.S., and Barteau, M.A. *Surf. Sci.*, **262**, 113 (1992).

Ihm, S.K., Ahn, J.H., and Jo, Y.D. *Ind. Eng. Chem. Res.*, **35**, 2946 (1996).

Imhaoulene, S., Vivier, L., Guisnet, M., Pérot, G., and Gubelmann, M. *Tetrahedron*, **50**, 2912 (1994).

Imizu, Y., Hattori, H., and Tanabe, K. *J. Catal.*, **57**, 35 (1979).

Izumi, Y. *Adv. Catal.*, **32**, 215 (1983).

Izumi, Y., Urabe, K., and Onaka, M., *Zeolite, Clay, and Heteropoly Acid in Organic Reactions*, Kodansha, Tokyo; VCH, New York, 1992.

Jacobs, J.M., Parton, R.F., Boden, A.M., and Jacobs, P.A. In *Heterogeneous Catalysis and Fine Chemicals* (Eds., Guisnet, M., Barbier, J., Barrault, J., Bouchoule, C., Duprez, D., Montassier, C., and Pérot, G.), *Stud. Surf. Sci. Catal.*, **41**, 221 (1988).

Jacobsen, E.N. In *Catalysis in Asymmetric Synthesis* (Ed., Ojima, I.), VCH, New York, 1993.

Jacobson, S.E. In *Catalysis of Organic Reactions* (Eds., Kosak, J.R., and Johnson, T.A.), Dekker, New York, 1994.

Jacques, J., Collet, A., and Willen, S.H. *Enantiomers, Racemates and Resolutions*, Wiley, New York, 1981.

James, B.R. *Homogeneous Hydrogenation*, Wiley, New York, 1973.

Jardine, F.H., Osborn, J.A., and Wilkinson, G. *J. Chem. Soc. (A)*, 1574 (1967).

Jayaraman, V.K., and Kulkarni, B.D. *Chem. Eng. J.*, **21**, 261 (1981).
Jayaraman, V.K., Kulkarni, B.D., and Doraiswamy, L.K. *AIChE J.*, **29**, 521 (1983).
Jefimov, J.V., Glasov, M.V., and Vernova, L.I. *Chem. Technol.*, **38**, 70 (1986).
Johnson, M.L.F., and Stewart, W.E. *J. Catal.*, **4**, 248 (1965).
Johnstone, R.A.W., Wilby, A.H., and Entwistle, I.D. *Chem. Rev.*, **85**, 129 (1985).
Jorgensen, K.A. *Chem. Rev.*, **89**, 431 (1989).
Joshi, G.W., and Rajadyaksha, R.A. *Chem. Ind.*, 876 (1980).
Joyner, R.W. *J. Chem. Soc., Faraday Trans.*, **86**, 2675 (1990).
Kagan, H.B., and Rebiere, F. *Synlett.*, 643 (1990).
Kagan, H.B. In *Comprehensive Organometallic Chemistry*, Vol. 4 (Eds., Wilkinson, G., Stone, F.G.A., and Abel, E.W.), Pergamon, Oxford, U.K., 1982.
Kaiser, J.R., Beuther, H., Moore, L.D., and Odioso, R.C. *Ind. Eng. Chem. Prod. Des. Dev.*, **1**, 296 (1962).
Kalck, P., and Monteil, F. *Adv. Organomet. Chem.*, **34**, 219 (1992).
Kamiya, Y., and Kotake, M. *Bull. Chem. Soc. Jpn.*, **46**, 2780 (1973).
Kao, H.S.P., and Satterfield, C.N. *Ind. Eng. Chem. Fundam.*, **7**, 664 (1968).
Kegley, S.E., and Pinhas, A.R. *Problems and Solutions in Organometallic Chemistry*, University Science Books, Mill Valley, Calif., 1986.
Kehoe, J.P.G. and Aris, R. *Chem. Eng. Sci.*, **28**, 2094 (1973).
Kelkar, A.A., Jaganathan, R., Kolhe, D.S., and Chaudhari, R.V. U.S. Patent 4,902,659, 1990.
Kelkar, A.A., Ubale, R.S., Chaudari, R.V. *J. Catal.*, **136**, 605 (1992).
Kelkar, A.A., Ubale, R.S., Deshpande, R.M., and Chaudhari, R.V. *J. Catal.*, **156**, 290 (1995).
Kijenski, J., Glinski, M., Wisniewski, R. and Murghani, S. In *Heterogeneous Catalysis and Fine Chemicals* (Eds. Guisnet, M., Barbier, J., Barrault, J., Bouchoule, C., Dupez, D., Montassier, C., and Pérot, G.), Elsevier, Amsterdam, 1991.
Kijenski, S., Glinski, M., and Reinhercs, J. In *Heterogeneous Catalysis and Fine Chemicals* (Eds., Guisnet, M., Barbier, J., Barrault, J., Bouchoule, C., Dupez, D., Montassier, C., and Pérot, G.), Elsevier, Amsterdam, 1988.
Kikuchi, E., Sawada, K., Maeda, M., and Matsuda, T. In *Acid-Base Catalysis II* (Eds., Hattori, H., Ono, Y., and Misono, M.), Elsevier, Amsterdam, 1994.
Kilényi, S.N., and Aitken, R.A. In *Asymmetric Synthesis* (Eds., Aitken, R.A., and Kilenyi, S.N.), Blackie Academic and Professional, Chapman and Hall, Glasgow, U.K., 1992.
Kim, I.W., Edgar, T.F., and Bell, N.H. *Computers Chem. Eng.*, **15**, 361 (1991).
Kim, K.S., and Barteau, M.A. *Surf. Sci.*, **223**, 13 (1989).
Kiperman, S.L., Kumbilieva, K.E., and Petrov, L.A. *Ind. Eng. Chem. Res.*, **28**, 376 (1989).
Kittrell, J.R., and Mezaki, R. *AIChE J.*, **13**, 389 (1967).
Kittrell, J.R. In *Advances in Chemical Engineering* (Eds., Drew, T.B., and Hoops, J.W.), Vol. 8, Academic Press, New York, 1970.
Klement, K., Jr., Willens, R.H., and Duwez, P. *Nature*, **187**, 869 (1960).
Knozinger, H., Kochloefl, K., and Meye, W. *J. Catal.*, **28**, 69 (1973).
Knowles, W.S. *J. Chem. Educ.*, **63**, 222 (1986).
Kobayashi, M., and Kobayashi, H. *Shokubai* (Tokyo), **16**, 8 (1974).
Komiyama, M., and Hirai, H. *J. Am. Chem. Soc.*, **106**, 174 (1984).
Koros, R.M., and Nowak, E.J. *Chem. Eng. Sci.*, **22**, 470 (1967).
Kozhevnikov, I.V. Catal. Rev. Sci. Eng., **37**, 311 (1995).
Krasuk, J.H., and Smith, J.M. *Ind. Eng. Chem. Fundam.*, **4**, 102 (1965).
Krieger, K.H., Lago, J., and Wantuck, J.A. U.S. Patent 3,405,159 1968.

Kubin, M. *Collect. Czech. Chem. Commun.*, **30**, 1104 (1965); **30**, 2000 (1965).
Kubota, H., and Yamanaka, Y. *J. Chem. Eng. Jpn*, **2**, 238 (1969).
Kubota, H., Yamanaka, Y., and Lana, I.G.D. *J. Chem. Eng. Jpn*, **2**, 71 (1969).
Kucera, E. *J. Chromatogr.*, **19**, 237 (1965).
Kulkarni, B.D., Jayaraman, K., and Doraiswamy, L.K. *Chem. Eng. Sci.*, **36**, 943 (1981).
Kumbhar, P.S., and Yadav, G.D. *Chem. Eng. Sci.*, **44**, 2535 (1989).
Kumbhar, P.S., Coq, B., Moureau, C., Moureau, P., and Figueras, F. *J. Phys. Chem.*, **98**, 10180 (1994).
Kummer, R., Schneider, H.W., Platz, R., Magnussen, P., Weiss, F. U.S. Patent 4,171,451, 1979.
Kurusu, Y., and Masuyama, Y. *Polyhedron*, **5**, 289 (1986).
Ladner, W.E., and Whitesides, G.M. *J. Am. Chem. Soc.*, **106**(23), 7250 (1984).
Lam, K.S., et al. *Nature*, **354**, 82 (1991).
Languasco, J.M., Cunningham, R.E., and Calvelo, A. *Chem. Eng. Sci.*, **27**, 1459 (1972).
Lansink-Rotgerink, H.G.J., Penn, G., Funfschilling, P.C., and Baiker, A. In *Heterogeneous Catalysis and Fine Chemicals* (Eds., Guisnet, M., Barbier, J., Barrault, J., Bouchoule, C., Duprez, D., Montassier, C., and Pérot, G.), Elsevier, Amsterdam, 1991.
Laszlo, P. and Mathy, A. *Helv. Chim. Acta*, **70**, 577 (1987).
Laszlo, P. *Acc. Chem. Res.*, **19**, 121 (1986); *Recherche*, **15**, 1282 (1984).
Lee, H.H. *Heterogeneous Reactor Design*, Butterworth, Boston, 1985.
Lee, J.C., Trimm, D.L., Kohler, M.A., Wainwright, M.S., and Cant, N.W. *Catal. Today*, **2**, 643 (1988).
Lee, R.S.H., and Agnew, J.B. *Ind. Eng. Chem. Proc. Des. Dev.*, **16**, 495 (1977).
Leludec, J. Fr. Patent 72-19130 to Rhone-Poulenc Chimie 1972.
Leung, P., Zorrilla, C., Recasens, F., and Smith, J.M. *AIChE J.*, **32**, 1839 (1986).
Liebermann, H.H. In *Amorphous Metallic Alloys* (Ed., Luborsky, F.E.), Butterworth, London, 1983.
Lim, B.G., Ching, C.B., and Tan, R.B.H. *Chem. Eng. Sci.*, **50**, 2289 (1995).
Litteral, C.J. U.S. Patent 3,694,405, 1972; *Chem. Abstr.*, **78**, 17047 (1973).
Liu, G., and Ellman, J.A. *J. Org. Chem.*, **60**, 7712 (1995).
Livbjerg, H., Jensen, K.F., and Villadsen, J. *J. Catal.*, **45**, 216 (1976).
Livbjerg, H., Sorensen, B., and Villadsen, J. *Adv. Chem. Ser.*, **133**, 242 (1974); *J. Catal.*, **45**, 216 (1976).
Loffler, D.G., and Schmidt, L.D. *AIChE J.*, **21**, 786 (1975).
Luss, D. *Chemical Reactor Theory: A Review* (Eds., Lapidus, L., and Amundson, N.R.), Prentice-Hall, Englewood Cliffs, N.J., 1977, Chapter 4.
Maki, T., Yokoyama, T., and Fuji, K. *Shokubai* (Catalyst), **35**, 2 (1993).
Malz, R.E., Jr., Reynolds, M.P., and Fagouri, C.J. In *Heterogeneous Catalysis and Fine Chemicals* (Eds., Guisnet, M., Barbier, J., Barrault, J., Bouchoule, C., Duprez, D., Montassier, C., and Pérot, G.), Elsevier, Amsterdam, 1993.
Mandler, J., Lavie, R., and Sheintuch, M. *Chem. Eng. Sci.*, **38**, 979 (1983).
Marczewski, M., Pérot, G., and Guisnet, M. In *Heterogeneous Catalysis and Fine Chemicals* (Eds., Guisnet, M., Barbier, J., Barrault, J., Bouchoule, C., Duprez, D., Montassier, C., and Pérot, G.), *Stud. Surf. Sci. Catal.*, **41**, 273 (1988).
Marinelli, T.B.L.W., Nabuurs, S., and Ponec, J. *J. Catal.*, **151**, 431 (1995).
Marinelli, T.B.L.W., Vleeming, J.H., and Ponec, V. *Stud. Surf. Sci. Catal.*, **75**, 1211 (1993).
Marquardt, D.W. *J. Soc. Ind. Appl. Math.*, **2**, 431 (1963).
Martin, A.R. *Chem. Ind.* (London), 1536 (1954).

Maruoka, M., Itoh, T., Shirasaka, T., and Yamamoto, H. *J. Am. Chem. Soc.*, **110**, 310 (1988).
Masters, C. *Homogeneous Transition-Metal Catalysis*, Chapman Hall, New York, 1981.
Matsuishi, T., Shimada, T., and Morihara, K. Part IX, *Bull. Chem. Soc. Jpn*, **67**, 748 (1994); *Chem. Lett.*, 1921 (1992).
Matsumae, H., Furui, M., Shibatani, T., and Tosa, T. *J. Ferment. Bioeng.*, **75**, 93 (1993); **78**, 59 (1994).
Maymó, J.A., and Cunningham, R.E. *J. Catal.*, **6**, 186 (1966).
McKillop, A., and Tarbin, J.A. *Tetrahedron Lett.*, **24**, 1505 (1983); *Tetrahedron*, **43**, 1753 (1987).
McKillop, A., and Young, D.W. *Synthesis*, **401**, 481 (1979).
Mears, D.E. *Ind. Eng. Chem. Proc. Des. Dev.*, **10**, 541 (1971a); *Chem. Eng. Sci.* **26**, 1361 (1971b); *Ind. Eng. Chem. Fundam.*
Menger, F.M., Eliseev, A.V., and Migulin, V.A. *J. Org. Chem.*, **60**, 6666 (1995).
Merger, F., Priester, C.U., Witzel, T., Koppenhoeffer, G., and Harder, W. Eur. Patent 449 089, 1991.
Michalska, Z.M., and Webster, D.E. *Platinum Met. Rev.*, **18**, 65 (1974).
Mills, P.L., and Chaudhari, R.V., *Catl. Today*, **37**, 367 (1997).
Mills, P.L., Ramachandran, P.A., and Chaudhari, R.V. *Rev. Chem. Eng.*, **8**, 1(1992).
Minami, R., Nishizaki, T., and Kumagai, Y. Jpn. Kokai JP 61,221,148 (1986).
Misono, M. *Catal. Rev. Sci. Eng.*, **29**, 269 (1987).
Misono, M. In *New Frontiers in Catalysis* (Eds., Guczi, L., Solymosi, F., and Tetenyi, P.), Elsevier, Amsterdam, and Akademiai Kiado, Budapest, 1993.
Misono M., Sakata, K., Yoneda, Y., and Lee, W.Y. *Proc. 7th Intl. Cong. Catal.*, Tokyo (1980) (Eds., Seiyama, T. and Tanabe, K.), Kodansha-Tokyo-Elsevier Amsterdam, 1981.
Mitsui Petrochemical Ltd. Japanese Patent 80, 1980, 102,527; *Chem. Abstr.*, **94**, 46783b (1981).
Miyake, T., Sekizawa, K., Hironaka, T., and Tsutsumi, Y., U.S. Patent 4,861,929, to Tohos Corp., 1989.
Molnar, A. In *Stereochemistry of Heterogeneous Metal Catalysis* (Ed., Bartok, M.), Wiley, New York, 1985.
Molnar, A., Smith, G.V., and Bartok, M. *Adv. Catal.*, **36**, 329 (1989); *J. Catal.*, **101**, 67 (1986).
Montegazza, M.A., Cesana, A., and Pastori, M. In *Catalysis of Organic Reactions* (Ed., Malz, R.E., Jr.), Dekker, New York, 1996.
Moreau, C., Durand, R., Graffin, P., and Geneste, P., *Stud. Surf. Sci. Catal.*, **41**, 139 (1988).
Moreau, C., Saenz, C., Geneste, P., Breysse, M., and Lacroix, M. In *Heterogeneous Catalysis and Fine Chemicals* (Eds., Guisnet, M., Barbier, J., Barrault, J., Bouchoule, C., Dupez, D., Montassier, C., and Pérot, G.), Elsevier, Amsterdam, 1991.
Morihara, K., Kurihara, S., and Suzuka, J. Part I, *Bull. Chem. Soc. Jpn.*, **61**, 3991 (1988a).
Morihara, K., Nishihara, E., Kojima, M., and Miyake, S., Part II, *Bull. Chem. Soc. Jpn.*, **61**, 3999 (1988b).
Morihara, K., Tanaka, E., Takeuchi, Y., Miyazaki, K., Yamamoto, N., Sayaway, Y., Kawamoto, E., and Shimada, T. Part III, *Bull. Chem. Soc. Jpn.*, **62**, 499 (1989).
Morihara, K., Kawasaki, S., Kofuji, M., and Shimada, T. Part VI, *Bull. Chem. Soc. Jpn.*, **66**, 906 (1993a).
Morihara, K., Doi, S., Takiguchi, M., and Shimada, T., Part VII, *Bull. Chem. Soc. Jpn.*, **66**, 2977 (1993b).

Morihara, K., Iijima, T., Usui, H., and Shimada, T., Part VIII, *Bull. Chem. Soc. Jpn.*, **66**, 3047 (1993c).
Morihara, K., Kurakawa, M., Kamata, Y., and Shimada, T. *J. Chem. Soc., Chem. Commun.*, 358 (1992).
Morita, M., and Ohta, K. Ger. Offen. DE 2234597 730201, 1973.
Morrison, J.D., Ed. *Asymmetric Synthesis*, Vol. 5, Academic Press, London, 1985.
Morrison, J.D., and Mosher, H.S. *Asymmetric Organic Reactions*, Prentice-Hall, Englewood Cliffs, N.J., 1971.
Moulijn, J.A., Van Leeuwen, P.W.N.M., and Van Santen, R.A. Catalysis: *An Integrated Approach to Homogeneous, Heterogeneous and Industrial Catalysis*, Elsevier, Amsterdam, (1993).
Nakano, Y., Yamaguchi, T., and Tanabe, K. *J. Catal.*, **80**, 307 (1983).
Natta, G., Ercoli, R., Castellano, S., and Barbieri, F.H. *J. Am. Chem. Soc.*, **76**, 4049 (1954).
Neier, W. In *Ion Exchangers* (Ed., Dorfner, K.), Walter de Gruyter, Berlin, 1991.
Neumann, R., and de la Vega, M. *J. Mol. Catal.*, **84**, 93 (1993).
Nogradi, M., *Stereoselective Synthesis*, VCH, Weinheim, Germany, 1987.
Nonaka, Y., Takahashi, S., and Hagihara, N. *Mem. Inst. Sci. Ind. Res.*, Osaka Univ., **31**, 23 (1974).
Notari, B. In *Innovation in Zeolite Materials Science* (Eds., Grobet, P.J., Mortier, W.J., Vansant, E.F., and Schulz-Ekloff, G.), *Stud. Surf. Sci. Catal.*, **47**, 413 (1987).
Notari, B. In *Structure-Activity and Selectivity Relationships in Heterogeneous Catalysis* (Eds., Grasselli, R.K., and Sleight, A.W.), *Stud. Surf. Sci. Catal.*, **67**, 243 (1991).
Noyori, R., *Chemtech.*, 360 (1992).
Noyori, R., and Kitamura, M. In *Modern Synthesis Methods* (Ed., Scheffold, R.), Springer Verlag, Amsterdam, 1989.
Noyori, R., and Takaya, H. *Acc. Chem. Res.*, **23**, 345 (1990).
Nozaki, H., Moriuti, S., Takaya, H., and Noyori, R. *Tetrahedron Lett.*, 5239 (1966).
O'Connor, C., and Wilkinson, G. *J. Chem. Soc. (A)*, 2665 (1968).
Ohta, H., Takaya, H., Kitamura, M., Nagai, K., and Noyori, R. *J. Org. Chem.*, **52**, 3174 (1987).
Ojima, I. (Ed.), *Catalytic Asymmetric Synthesis*, VCH, New York, (1993).
Ojima, I., Clos, N., and Bastos, C. *Tetrahedron*, **45**, 6901 (1989).
Okamoto, Y., Maezawa, A., Kane, H., and Imanaka, T. *J. Chem. Soc., Chem. Commun.*, 380 (1988).
Olah, G.A., Iyer, P.S., and Prakash, G.K.S. *Synthesis*, 513 (1986).
Onoue, Y., Mizutani, Y., Akiyama, S., and Izumi, Y. *Chemtech*, **8**, 432 (1978).
Orchin, M. *Adv. Catal.*, **16**, 1 (1966).
Orito, Y., Imai, S., Niwa, S., and Nguyen, G.H. *J. Synth. Org. Chem. Jpn.*, **37**, 173 (1979a); *J. Chem. Soc. Jpn.*, 1118 (1979b); 670 (1980); 137 (1982).
Osborn, J.A., Jardine, F.H., Young, J.F., and Wilkinson, G. *J. Chem. Soc. (A)*, 1711 (1966).
Otani, S., Wakao, N., and Smith, J.M. *AIChE J.*, **11**, 446 (1965).
Otsuka Chemical Company Ltd., Japanese Kokai Tokkyo Koho, 57, 184, 442 (82, 184, 442), 1982; *Chem. Abstr.*, **99**, 96476g (1983).
Oyama, S.T., and Hightower, J.W., Eds. *Catalytic Selective Oxidation*, Am. Chem. Soc., Washington, D.C., 1993.
Ozin, G.A., and Gil, C. *Chem. Rev.*, **89**, 1749 (1989).
Parks, W.G., and Yula, R.W. *Ind. Eng. Chem.*, **33**, 891 (1941).

Parshall, G.W., and Ittel, S.D. *Homogeneous Catalysis: The Applications and Chemistry of Catalysis by Soluble Transition Metal Complexes*, 2nd ed., Wiley, New York, 1992.
Parshall, G.W., and Nugent, W.A. *Chemtech*, in three parts: Part 1 (p. 184), Part 2 (p. 314), Part 3 (p. 388), 1988.
Parshall, G.W. *Homogeneous Catalysis*, Wiley, New York, 1980.
Parton, R.F., Jacobs, J.M., Huybrechts, D.R., and Jacobs, P.A. In *Zeolites as Catalysts, Sorbents and Detergent Builders* (Eds., Karge, H.G., and Weitkamp, J.), *Stud. Surf. Sci. Catal.*, **46**, 163 (1989).
Patel, P.V., and Butt, J.B. *Ind. Eng. Chem. Proc. Des. Dev.*, **14**, 298 (1974).
Patwardhan, A.A., and Sharma, M.M. *Reactive Polymer*, **13**, 161 (1990).
Pawlowski, J. *Chem. Eng. Tech.*, **33**, 492 (1961).
Pereira, C.J., and Varma, A. *Chem. Eng. Sci.*, **33**, 396 (1978).
Pérot, G., and Guisnet, M. *J. Mol. Catal.*, **61**, 173 (1990).
Perutz, R.N., and Turner, J.J. *J. Am. Chem. Soc.*, **97**, 4791, 4805 (1975).
Petersen, E.E. *Chemical Reaction Analysis*, Prentice-Hall, Englewood Cliffs, N. J., 1965.
Pfenninger, A. *Synthesis*, **2**, 89 (1986).
Pierce, K.G., and Barteau, M.A. *Surf. Sci. Lett.*, **326**, 473 (1995).
Pignet, T., and Schmidt, L.D. *Chem. Eng. Sci.*, **29**, 1123 (1974).
Pines, H., Veseley, J.A., and Ipatieff, V.N. *J. Am. Chem. Soc.*, **77**, 6314 (1955).
Pitochelli, A.R. *Ion Exchange Catalysis and Matrix Effects*, Rohm and Haas, Philadelphia, 1980.
Pittman, C.U., Jr., and Felis, R.F. *J. Organomet. Chem.*, **72**, 389 (1974).
Pittman, C.U., Jr. In *Organometallic Polymers* (Eds., Carraher, C.E., Jr., Sheats, J.E., and Pittman, C.U., Jr.), Academic Press, New York, 1978.
Pittman, C.U., Jr. In *Organometallic Reactions and Synthesis* (Eds., Becker, E.I., and Tsutsui, M.), Plenum, New York, 1977, Vol. 6.
Pittman, C.U., Jr. In *Polymer-Supported Reactions in Organic Synthesis* (Eds., Hodge, P., and Sherrington, D.C.), Wiley, New York, 1980.
Pittman, C.U., Jr. *Polymer Supports in Organic Synthesis, Polymer News*, regular contributions, 1977–1992, 1994, **19**, No. 1 (1994).
Pittman, C.U., Jr., and Evans G.O. *Chemtech*, 560 (1973).
Plath, P., Ed. *Optimal Structures in Heterogeneous Reaction Systems*, Springer-Verlag, Berlin, 1989.
Poltarzewski, Z., Galvagno, S., Pietropaolo, R., and Staiti, P. *J. Catal.*, **102**, 190 (1986).
Pratt, K.C. In *Catalysis: Science and Technology*, Vol. 8 (Eds., Anderson, J.R., and Boudart, M.), Springer-Verlag, 1987.
Prelog, V., and Helmchen, G. *Angew. Chem., Int. Ed. Engl.*, **21**, 567 (1982).
Raab, C.G., Englisch, M., Marenelli, T.B.L.W., and Lercher, J.A. *Stud. Surf. Sci. Catal.*, **78**, 211 (1993).
Raghavan, N.S., and Doraiswamy, L.K. *J. Catal.*, **48**, 21 (1977).
Rajadhyaksha, R.A., and Chaudhari, D.D. *Ind. Eng. Chem. Res.*, **26**, 1743 (1987).
Rajadhyaksha, R.A., and Doraiswamy, L.K. *Catal. Rev. Sci. Eng.*, **13**, 209 (1976).
Rajadhyaksha, R.A., and Joshi, G.W. In *Heterogeneous Catalysis and Fine Chemicals* (Eds., Guisnet, M., Barbier, J., Barrault, J., Bouchoule, C., Dupez, D., Montassier, C., and Pérot, G.), Elsevier, Amsterdam, 1991.
Rajadhyaksha, R.A., Vasudeva, K., and Doraiswamy, L.K. *J. Catal.*, **41**, 61 (1976).
Rama Rao, A.V., Gurjar, H.K., and Kaiwar, V. *Tetrahedron Asymmetry*, **3**, 839 (1992).
Rama Rao, A.V., Gurjar, H.K., Sharma, P.A.S., and Kaiwar, V. *Tetrahedron Lett.*, **31**, 2341 (1990).

Rama Rao, A.V. private communication (1993).
Ratnasamy, P., and Sivasankar, S. U.S. Patent 5,493,061 1996.
Ray, S.K., Murty, G.S., and Rao, H.S. *Proc. Symp. Chem. Oil Coal*, **372**, 9 (1972).
Reddy, K.A., and Doraiswamy, L.K. *Chem. Eng. Sci.*, **24**, 1415 (1969).
Reddy, T.I., Bhawal, B.M., and Rajappa, S. *Tetrahedron*, **49**, 2101 (1993).
Reinhold, D.F., Firestone, R.A., Gaines, W.A., Chemerde, J.M., and Sletzinger, M. *J. Org. Chem.*, **33**, 1209 (1968).
Reinicker, R.A., and Gates, B.C. *AIChE J.*, **20**, 933 (1974).
Rieu, J.-P., Boucherle, A., Cousse, M., and Mouzin, G. *Tetrahedron*, **42**, 4095 (1986).
Rihani, D.N., Narayanan, T.K., and Doraiswamy, L.K. *Ind. Eng. Chem. Proc. Des. Dev.*, **4**, 403 (1965).
Roberts, G.W., and Satterfield, C.N. *Ind. Eng. Chem. Fundam.*, **4**, 288 (1965); **5**, 317 (1966).
Roberts, G.W. *Chem. Eng. Sci.*, **27**, 1409 (1972).
Roffia, P., Leofanti, G., Cesana, A., Mantegazza, M., Padovan, M., Petrini, G., Tonti, S., and Gervasutti, P. In *New Developments in Selective Oxidation* (Eds., Centi, G., and Trifiro, F.), *Stud. Surf. Sci. Catal.*, **55**, 43 (1990).
Rohm and Haas Co., Br. Pat. 932,125 and 932,126, 1963; *Chem. Abstr.*, **59**, 11731 (1963).
Rony, P.R., and Roth, J.F. In *Catalysis: Heterogeneous and Homogeneous* (Eds., Delmon, B., and Jannes, G.), Elsevier, Amsterdam, 1975.
Rony, P.R. *J. Catal.*, **14**, 142 (1969).
Rose, L.M. *Chemical Reactor Design in Practice*, Elsevier, Amsterdam, 1981.
Roskies, E. In *Abnormality and Normality: The Mothering of Thalidomide Children*, Cornell University Press, London, U.K., 1972.
Rossiter, B.E. In *Asymmetric Synthesis*, Vol. 5 (Ed., Morrison, J.D.), Academic Press, Orlando, Fla., 1985, Chap. 7.
Roth, J.F., Craddock, J.M., Hershman, A., and Paulik, F.E. *Chemtech*, 600 (1971).
Ruckenstein, E. *AIChE J.*, **16**, 151 (1970).
Ruthven, D.M., and Loughlin, K.F. *Trans. Faraday Soc.*, **67**, 1661 (1971).
Ruthven, D.M. *Chem. Eng. Sci.*, **23**, 759 (1968); *J. Catal.*, 25, 259 (1972).
Rylander, P.N. *Organic Synthesis with Noble Metal Catalysts*, Academic Press, New York, 1973; In *Catalytic Hydrogenation in Organic Syntheses*, Academic Press, New York, 1979; *Hydrogenation Methods*, Academic Press, London, 1985.
Sachtler, W.M.H. *Chem. Ind.*, **22**, 189 (1985).
Saeki, K., and Taniguchi, K. *Chem. Econ. Eng. Rev.*, **12**, 5 (1986).
Saenger, W. *Angew. Chem., Intl. Ed. Engl.*, **19**, 344 (1980).
Sanchez, F., Iglesias, M., del Pinto, C., and Corma, A. *J. Catal.*, **70**, 369 (1991).
Satterfield, C.N., and Cadle, P.J. *Ind. Eng. Chem. Fundam.*, **7**, 202 (1968a); 256 (1968b).
Satterfield, C.N., and Cheng, C.S. *AIChE Symp. Ser.*, **67**(117), 43 (1971).
Satterfield, C.N., and Roberts, G.W. *AIChE J.*, **14**, 159 (1968).
Satterfield, C.N. *Mass Transfer in Heterogeneous Catalysis*, MIT Press, Cambridge, Mass., 1970; *Heterogeneous Catalysis in Practice*, McGraw-Hill, New York, 1980.
Satterfield, C.N. *Heterogeneous Catalysis in Industrial Practice*, McGraw-Hill, New York, 1991.
Schmalzer, D.K. *Chem. Eng. Sci.*, **24**, 615 (1969).
Schneider, P., and Mitschka, P. *Coll. Czech.. Chem. Commun.*, **30**, 146 (1965); **31**, 3677 (1966a); *Chem. Eng. Sci.*, **21**, 455 (1966b).
Schneider, P., and Smith, J.M. *AIChE J.*, **14**, 762 (1968).
Schneider, P. *Catal. Rev. Sci. Eng.*, **12**, 201 (1975).

Schoeb, A.M., Raeker, T.J., Yang, L., Wu, X., King, T.S., and DePristo, A.E. *Surf. Sci. Lett.*, **278**, L125 (1992).
Schrock, R.R., and Osborn, J.A. *J. Am. Chem. Soc.*, **98**, 2134 (1976).
Schulz, K.H., and Cox, D.F. *J. Phys. Chem.*, **96**, 7394 (1992); **97**, 3555 (1993).
Scott, D.S. *Can. J. Chem. Eng.*, **40**, 173 (1962).
Scott, J.W. In *Asymmetric Synthesis*, 4 (Eds., Morrison, J.D., and Scott, J.W.), Academic Press, New York, 1984.
Sehgal, R.K., Koenigsberger, R.U., and Howard, T.J. *J. Org. Chem.*, **40**, 3073 (1975).
Seinfeld, J.H. *Ind. Eng. Chem.*, **62**, 32 (1970).
Shah, M.J. *Ind. Eng. Chem.*, **57**, 18 (1965).
Shapovalov, A.A., Sembaev, D.K., Suvorov, B.V., and MarÚyasova, G.A. *Izv. Akad. Nauk Kaz. SSR, Ser. Khim.*, **4**, 17 (1989).
Sharpless, K.B. *Chemtech*, 692 (1985); *Janssen Chim. Acta*, **6**(1), 3 (1988).
Sheldon, R.A. In *Aspects of Homogeneous Catalysis* (Ed., Ugo, R.), Reidel, Dordrecht, Vol. 4, 1981; *Chem. Ind.* (London), 212 (1990a); *Drug Inf. J.*, **24**, 129 (1990b); *Chemtech*, 566 (1991); *Chirotechnology*, Dekker, New York, 1993; *J. Chem. Technol. Biotechnol.*, **67**, 1 (1996).
Sherrington, D.C., and Hodge, P., Eds. *Syntheses and Separations Using Functional Polymers*, Wiley, New York, 1988.
Shendye, R.V., Dowd, M.K., and Doraiswamy, L.K. *Chem. Eng. Sci.*, **48**, 1995 (1993).
Shibata, M., and Masumoto, T. In *Preparation of Catalysts IV*, (Eds., Delmon, B., Grange, P., Jacobs, P.A., and Poncelet, G.), Elsevier, Amsterdam, 1987.
Shimada, T., Hirose, R., and Morihara, K. Part X, *Bull. Chem. Soc. Jpn.*, **67**, 227 (1994).
Shimada, T., Kurazono, R., and Morihara, K. Part V, *Bull. Chem. Soc. Jpn*, **66**, 836 (1993).
Shimada, T., Nakanishi, K., and Morihara, K. Part IV, *Bull. Chem. Soc. Jpn.*, **65**, 954 (1992).
Silverstein, R.M., Rodin, J.O., and Wood, D.L. *Science*, **154**, 509 (1966).
Simon, B., and Vortmeyer, D. *Chem. Eng. Sci.*, **33**, 109 (1978).
Sinfelt, J.H. *Adv. Chem. Eng.*, **5**, 37 (1964).
Sinfelt, J.H., Carter, J.L., and Yates, D.J.C. *J. Catal.*, **24**, 283 (1972).
Skarzewski, J. *Tetrahedron*, **40**, 4997 (1984).
Slaugh, L.H., and Mullineaux, R.D. U.S. Patents 3,239,596 and 3,239,570, 1966.
Smidt, J., Hafner, W., Jira, R., Sedlmeier, J., Sieber, R., Rutlinger, R., and Kojer, H. *Angew Chem.*, **71**(5), 176 (1959).
Smith, G.V., 1995 Paul N. Ryder Award Address In *Catalysis of Organic Reactions* (Ed., Malz, R.E.), Dekker, New York, 1996.
Smith, K. *Solid Supports and Catalysts in Organic Synthesis*, Ellis Horwood PTR Prentice-Hall, New York, 1992.
Sofekun, O.A., Rollins, D.K., and Doraiswamy, L.K. *Chem. Eng. Sci.*, **49**, 2611 (1994).
Stewart, W.E., and Villadsen, J. *AIChE J.*, **15**, 28 (1969).
Stiles, A.B. U.S. Patent 3,627,790, 1971.
Stinson, S.C. *Chem. Eng. News*, June & September (1986).
Sugi, Y., Matsuzaki, T., Honaoka, T., Kubota, Y., Kim., J.H., Tu, X., and Matsumoto, M. *Catal. Lett.*, **27**, 315 (1994).
Suib, S.L. *Chem. Rev.*, **93**, 803 (1993).
Syamala, M.S., Reddy, G.D., Rao, B.N., and Ramamurthy, V. *Curr. Sci.*, **55**, 875 (1986).
Taber, D.F., and Silverberg, L.J. *Tetrahedron Lett.*, **32**(34), 4227 (1991).
Tabushi, I., and Kuroda, Y. *Adv. Catal.*, **32**, 417 (1983).
Tabushi, I. *Acc. Chem. Res.*, **15**, 66 (1982).

Tabushi, I., Fujita, K., and Kawakubo, H. *J. Am. Chem. Soc.*, **99**, 6456 (1977).
Tabushi, I., Yamamura, K., Fujita, K., and Kawakubo, H. *J. Am. Chem. Soc.*, *101-4*, 1019 (1979).
Tai, A., and Harada, T. In *Tailored Metal Catalysts* (Ed., Iwasawa, Y.), Reidel, Dordrecht, 1986.
Tai, A., Imaida, M., Oda, T., and Watanabe, H. *Chem. Lett.*, **61** (1978).
Tai, A., Watanabe, H., and Harada, T. *Bull. Chem. Soc. Jpn*, **52**, 1468 (1979).
Takahashi, K., Hibi, T., Higashio, Y., and Araki, M. *Shokubai Catalyst*, **35**, 12 (1993).
Takanobu, E., Baba, R., Saito, Y., and Yokoyama, S. *Japanese Kokai*, **77**, 108959 (1977).
Takeshi, N., Shizuo, F., Yasuhiro, Y., and Akiko, N. *J. Org. Chem.*, **59**, 1191 (1994).
Tanaka, Y., Imizu, Y., Hattori, H., and Tanabe, K. *Proc. 7th Intl. Cong. Catal.*, Tokyo, 1980.
Tartarelli, R., Cionis, S., and Caporani, M. *J. Catal.*, **18**, 212 (1970).
Taylor, S.J., and Morken, J.P. *Science*, **280**, 267 (1989).
Ternan, M. *Can. J. Chem. Eng.*, **65**, 244 (1987); In Pore Diffusion of Vacuum Residue Molecules and Hydrogen Dissociation on Reaction Sites: Essential Steps in Hydrocracking Catalysis, *14th Canadian Symposium on Catalysis*, Whistler, B.C., 1996.
Texier-Boullet, F., and Foucaud, A. *Tetrahedron Lett.*, **23**, 4927 (1982).
Thaller, L.H., and Thodos, G. *AIChE J.*, **6**, 369 (1960).
Thomas, J.M. *Angew. Chem. Int. Ed. Eng.*, **33**, 913 (1994).
Thomas, J.M., and Thomas, W.J. *Principles and Practice of Heterogeneous Catalysis*. VCH, New York, 1997.
Thompson, H.W., and Naipawer, R.E. *J. Am. Chem. Soc.*, **95**, 6379 (1973).
Thompson, H.W., McPherson, E., and Lences, B.L. *J. Org. Chem.*, **41**, 2903 (1976).
Tishler, M., Fieser, L.F., and Wendler, N.L. *J. Am. Chem. Soc.*, **62**, 1982 (1940).
Tokunaga, M., Larrow, J.F., Kakiuchi, F., and Jacobsen, E.N. *Science*, **277**, 936 (1997).
Tolman, C.A., *Chem. Soc. Rev.*, **1**, 337 (1972); *Chem. Rev.*, **77**, 313 (1977).
Torok, B., Molnar, A., Borszeky, K., Toth-Kadar, E., and Bakonyi, I. In *Heterogeneous Catalysis and Fine Chemicals* (Eds., Guisnet, M., Barbier, J., Barrault, J., Bouchoule, C., Dupez, D., Montassier, C., and Pérot, G.), Elsevier, Amsterdam, 1993.
Tsuji, H., Yagi, F., and Hattori, H. *Chem. Lett.*, 1881 (1991).
Uner, D.O., Pruski, M., Gerstein, B.C., and King, T.S. *J. Catal.*, **146**, 530 (1994).
Vaidyanathan, K., and Doraiswamy, L.K. *Chem. Eng. Sci.*, **23**, 537 (1968).
Van Bekkum, H., and Kouwenhoven, H.W. In *Heterogeneous Catalysis and Fine Chemicals* (Eds., Guisnet, M., Barbier, J., Barrault, J., Bouchoule, C., Duprez, D., Montassier, C., and Pérot, G.), *Stud. Surf. Sci. Catal*, 1988; *Rec. Trav. Chim. Pays-Bas*, **108**, 283 (1989).
Van den Bleek, C.M., Van der Wiele, K., and Van den Berg, P.J. *Chem. Eng. Sci.*, **24**, 681 (1969).
Van Parijs, I.A., Hosten, L.H., and Froment, G.F. *Ind. Eng. Chem. Prod. Des. Dev.*, **25**, 437 (1986).
Van Santen, R.A., and Niemantsverdriet, J.W. *Chemical Kinetics and Catalysis*, Plenum Press, New York, 1995.
Varghese, P., and Wolf, E.E. *AIChE J.*, **26**, 55 (1980).
Varghese, P., Varma, A., and Carberry, J.J. *Ind. Eng. Chem. Fundam.*, **17**, 195 (1978).
Venturello, C., and Ricci, M. *J. Org. Chem.*, **51**, 1599 (1986).

Vermerer, W., and Wells, P.B., quoted in Dodgson, I. In *Heterogeneous Catalysis and Fine Chemicals* (Eds., Guisnet, M., Barbier, J., Barrualt, J., Bouchoule, C., Duprez, D., and Pérot, G.), Elsevier, New York, 1993.

Villadsen, J., and Livbjerg, H. *Catal. Rev. Sci. Eng.*, **17**, 203 (1978).

Vineyard, B.D., Knowles, W.S., Sabacky, M.J., Bachman, G.L., and Weinhauff, D.J. *J. Am. Chem. Soc.*, **99**, 5946 (1977).

Von Pierre, M. *Helv. Chim. Acta*, **72**, 1554 (1989).

Wade, L.G., Jr. *Organic Chemistry*, Prentice-Hall, Englewood Cliffs, N. J., 1987.

Wadkar, J.G., and Chaudhari, R.V. *J. Mol. Catal.*, **22**, 105 (1983); *Ind. J. Tech.*, **21**, 351 (1983).

Wakao, N., and Smith, J.M. *Chem. Eng. Sci.*, **17**, 825 (1962); *Ind. Eng. Chem. Fundam.*, **3**, 123 (1964).

Wan, K., and Davis, M.E. *J. Chem. Soc., Chem. Commun.*, 1262 (1993a); *Tetrahedron: Asymmetry*, **4**(12), 2461 (1993b); *Nature*, **370**, 449 (1994a); *J. Catal.*, **148**, 1 (1994b); *J. Catal.*, **152**, 25 (1995).

Wang, S., and Hofmann, H. *Chem. Eng. Sci.*, **54**, 1639 (1999).

Watson, K.G., Fung, Y.M., Gredley, M., Bird, G.J., Jackson, R., Gountzos, H., and Mathews, B.R. *J. Chem. Soc., Chem. Commun.*, 1018 (1990).

Watts, D. *Can. J. Chem. Eng.*, **72**, 701 (1994).

Webb, G., and Wells, P.B. *Catal. Today*, **12**, 319 (1992).

Weekman, V.W., Jr., and Gorring, R.L. *J. Catal.*, **4**, 260 (1965).

Weekman, V.W., Jr. *AIChE J.*, **20**, 833 (1974).

Weisz, P.B. *Z. Physik. Chem.* (Frankfurt), **11**, 1 (1957).

Weisz, P.B., and Hicks, J.S. *Chem. Eng. Sci.*, **17**, 265 (1962).

Weisz, P.B., and Prater, C.D. *Adv. Catal.*, **6**, 143 (1954).

Weller, S. *AIChE J.*, **2**, 59 (1956).

Wesley, R.B., and Gates, B.C. *J. Catal.*, **34**, 288 (1974).

White, G.R. *Ger. Offen.*, 2,621,092, 1976.

Widdecke, H. In *Synthesis and Separations using Functional Polymers* (Eds., Sherrington, D.C., and Hodge, P.), Wiley, Chichester, 1988.

Wilson, H.D., and Rinker, R.G. *J. Catal.*, **42**, 268 (1976).

Wirges, H.P., and Rahse, W. *Chem. Eng. Sci.*, **30**, 647 (1975).

Wisniak, J., and Klein, M. *Ind. Eng. Chem. Prod. Res. Dev.*, **23**, 44 (1984).

Woodard, S.S., Finn, M.G., and Sharpless, K.B. *J. Am. Chem. Soc.*, **113**(1), 106 (1991).

Wu, X., Gerstein, B.C., and King, T.S. *J. Catal.*, **123**, 43 (1990).

Wynberg, H. *Top. Stereochem.*, **16**, 87 (1986).

Xu, Z.P., and Chuang, K.T. *Chem. Eng. Sci.*, **52**, 3011 (1997).

Yamaguchi, S., Inoue, M., and Enomoto, S. *Chem. Lett.*, 827 (1985).

Yamaguchi, T., Jin, T., Ishida, T., and Tanabe, K. *Mat. Chem. Phys.*, **17**, 3 (1987).

Yamaguchi, T., Tanabe, K., and Kung, Y.C. *Mater. Chem. Phys.*, **16**, 67 (1986).

Yamashita, H., Yoshikawa, M., Funabiki, T., and Yoshida, N. *J. Mater. Sci. Lett.*, **4**, 1241 (1985a); *J. C. S. Faraday*, **1**, 81, 2485 (1985b).

Yang, K.H., and Hougen, O.A. *Chem. Eng. Prog.*, **46**, 14 (1950).

Yokoyama, T., Setooyama, T., Fujita, N., Nakajima, M., Maki, T., and Fuji, K. *Appl. Catal.*, **A88**, 149 (1992).

PART III

REACTOR DESIGN FOR HOMOGENEOUS AND FLUID-SOLID (CATALYTIC) REACTIONS

Originality does not consist in saying what no one has ever said before, but in saying exactly what you think yourself.
                                    J. F. Stephen, Horae Sabbaticae

Chapter 10

Reactor Design for Simple Reactions

Ideal reactors and their design principles were discussed in Chapter 4. In addition to these ideal reactors, there are certain reactors in which a reasonably well-defined[1] measure of mixing can be introduced. These are the recycle plug-flow reactor and a sequence of fully mixed reactors. Many organic reactions are conducted in a stirred reactor containing a batch of the same or a second reactant, and continuously feeding, or withdrawing, or feeding and withdrawing one or more of the reactants and/or products. These are referred to as semibatch reactors. They belong to a more general class of reactors known as variable volume reactors. The design of all of these types of reactors is briefly considered in this chapter.

PLUG-FLOW REACTOR WITH RECYCLE

The Basic Design Equation

The principle of the recycle-flow reactor (RFR) is sketched in Figure 10.1. The single parameter that distinguishes it from a PFR is the recycle flow ratio R

$$R = \frac{\text{volume of product recycled}}{\text{volume of product leaving the reactor}} \qquad (10.1)$$

---

[1] These are not truly well defined in terms of the theories of turbulent mixing, but we regard them as well defined in the sense that a predetermined level of macromixing (see Chapter 13) can be introduced by controlling the recycle or the number of reactors. Even so, it is by no means certain that the same effect of mixing will hold for different scales of operation. But this is the assumption we make.

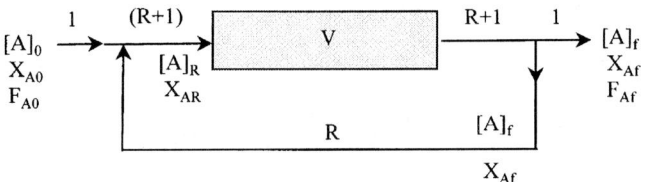

Figure 10.1 Recycle reactor.

With this recycle, the conditions at R corresponding to the reactor inlet are $[A] = [A]_R, F_A = F_{AR}$. Thus the PFR equation given in Chapter 4 becomes

$$\frac{V}{F_{A0}} = (1 + R) \int_{X_{AR}}^{X_{Af}} \frac{dX_A}{-r_A} \tag{10.2}$$

A material balance across the reactor gives

$$X_{AR} = \frac{X_{A0} + RX_{Af}}{1 + R} \tag{10.3}$$

Because there is usually no conversion initially,

$$X_{AR} = \frac{RX_{Af}}{1 + R} \tag{10.4}$$

Equation 10.3 can also be written in terms of concentration as

$$[A]_R = \frac{[A]_0 + R[A]_f}{1 + R} \quad \text{for } \varepsilon_A = 0 \tag{10.5}$$

In terms of $\bar{t}$, Equation 10.2 becomes

$$\bar{t} = \frac{[A]_0 V}{F_{A0}} = (1 + R)[A]_0 \int_{X_{AR}}^{X_{Af}} \frac{dX_A}{(-r_A)}, \quad \varepsilon_A \neq 0 \tag{10.6}$$

or

$$\bar{t} = \frac{[A]_0 V}{F_{A0}} = (1 + R) \int_{[A]_f}^{[A]_R} \frac{d[A]}{-r_A}, \quad \varepsilon_A = 0 \tag{10.7}$$

Integration of Equation 10.7 for a first-order reaction

$$A \rightarrow \text{Products} \tag{10.1}$$

gives

$$\tau = k\bar{t} = (1 + R) \ln \left[\frac{[A]_0 + R[A]_f}{(1 + R)[A]_f} \right] \tag{10.8}$$

Optimal Design of RFR

Note from Equations 10.3 and 10.5 that the recycle ratio R is given by

$$R = \frac{X_{AR} - X_{A0}}{X_{Af} - X_{AR}} \quad \text{for any } \varepsilon_A \tag{10.9a}$$

$$= \frac{[A]_0 - [A]_R}{[A]_R - [A]_f} \quad \text{for } \varepsilon_A = 0 \tag{10.9b}$$

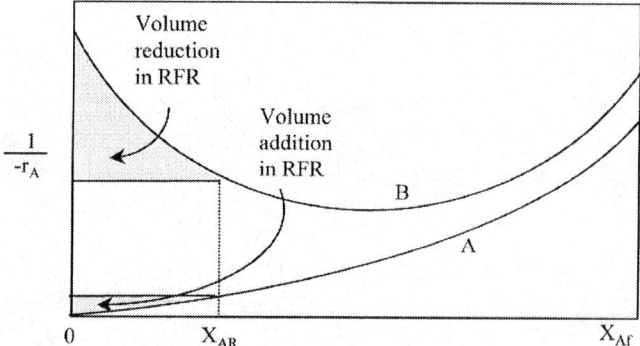

Figure 10.2 Plots of $1/-r_A$ vs. X_A for common reactions (curve A) and for autocatalytic or adiabatic reactions (curve B).

Also, for the common shape of curve shown as A in Figure 10.2, RFR can never have a volume less than PFR. The recycle-flow reactor is at its best when the $(1/-r_A)$ versus X_A curve is continuously falling, but that seldom happens. It can, however, exhibit a minimum (curve B of Figure 10.2), in an autocatalytic or adiabatic reactor. Clearly, in such a case, the recycle reactor can be superior to the plug-flow reactor under certain conditions, as shown by the lower reactor volume for curve B in the figure.

The central problem in the design of an RFR is determining the optimum value of X_{AR} to minimize the reactor volume. This can be obtained by setting

$$\frac{d[\bar{t}/[A]_0]}{dR} = 0,$$

that is,

$$\frac{d[\bar{t}/[A]_0]}{dR} = \frac{d \int_{X_{AR}}^{X_{Af}} (1+R) \frac{dX_A}{-r_A}}{dR} = 0 \tag{10.10}$$

with the result (Levenspiel, 1993)

$$\left|\frac{1}{-r_A}\right|_{X_{AR}} = \frac{\int_{X_{AR}}^{X_{Af}} \frac{dX_A}{-r_A}}{X_{Af} - X_{AR}} \tag{10.11}$$

Expressed in words, this means that

$$(1/\text{rate}) \text{ at } X_{AR} = (1/\text{rate})_{\text{ave}} \text{ (or } 1/\bar{r}_A\text{) in the reactor} \tag{10.12}$$

Equation 10.11 can be solved by iteration to find that value of X_{AR} which satisfies both sides of it. It can also be solved graphically as illustrated in Figure 10.3. The recycle is introduced in such a way that the rate corresponding to the value of X_{AR} is exactly equal to the *average rate* in the reactor. Figures 10.3a and 10.3b show the situation when the two rates are not equal, and Figure

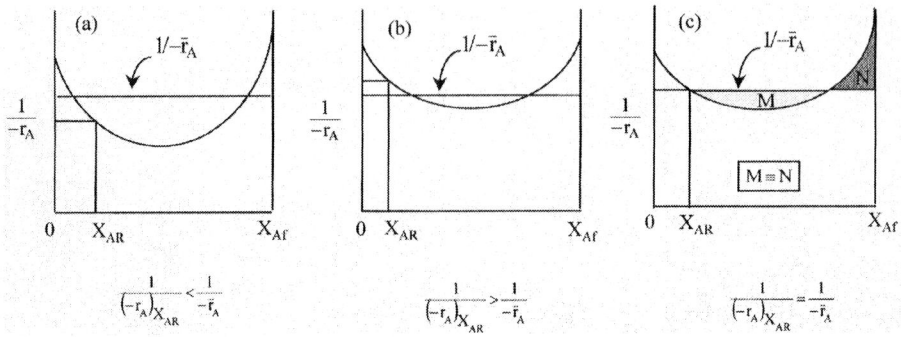

Figure 10.3 Minimization of reactor volume in a recycle-flow reactor.

10.3c is for the case when the two rates are exactly equal and therefore is the optimal condition. This is easily found by locating X_{AR} so that the areas M and N are equal.

Use of an RFR to Resolve a Selectivity Dilemma

The effect of mixing on the yield or selectivity of a desired product is an important consideration in reactor choice (see Chapter 13). An RFR is well suited for imposing a controlled level of mixing to maximize selectivity. In this connection, we note the following two important effects of mixing in an isothermal reaction:

1. Mixing is detrimental to the yield of an intermediate product and favors the formation of the final product.
2. Mixing favors the lowest order reaction in a system that involves reactions of varying orders and has no effect when all orders are the same.

These effects, which can be used to advantage in many complex reactions, lead to contradictory reactor choices when applied to the following scheme:

$$A \xrightarrow{k_1} R \xrightarrow{k_2} S \quad \text{(main reaction)} \quad [10.2a]$$

$$A + A \xrightarrow{k_3} T \quad \text{(side reaction)} \quad [10.2b]$$

where R is the desired product. Thus, although conclusion 1 calls for a PFR to maximize R in the main reaction, conclusion 2 would require a CSTR to minimize the loss of A via the side reaction. It can be shown (Van de Vusse, 1964; Gillespie and Carberry, 1966) that neither of the extremes, PFR or CSTR, is the best for this reaction and that a reactor with an intermediate level of mixing determined by the value of the recycle ratio R is optimal.

MIXED-FLOW REACTORS IN SERIES

The concept of using two or more MFRs in series stems from the circumstance that mixed-flow performance can be made to approach plug-flow performance by increasing the number of MFRs. In fact, the number of reactors (or tanks) can be regarded as a measure of the degree of mixing, and this description of partial mixing is commonly referred to as the *tanks-in-series model*.

Design Equations

Considering a sequence of stirred reactors in series, the residence time for reactor 1 is given by

$$\bar{t}_1 = \frac{V_1 [A]_0}{F_{A0}} \qquad (10.13)$$

Combining this with the performance equation (see Chapter 4) for a first-order reaction,

$$\bar{t}_1 = \frac{V_1 [A]_0}{F_{A0}} = \frac{[A]_0 - [A]_1}{k [A]_1} \qquad \text{for } \varepsilon_A = 0 \qquad (10.14)$$

By writing similar equations for $\bar{t}_2, \ldots, \bar{t}_N$, the following expression for conversion at the end of N reactors results:

$$\frac{[A]_N}{[A]_0} = (1 - X_A) = \frac{1}{\prod_{i=1}^{N} (1 + k \bar{t}_i)} \qquad (10.15)$$

This equation has been validated by the extensive experimental results of Elridge and Piret (1950).

For a reversible reaction, first order in both directions,

$$A \rightleftarrows R \qquad [10.3]$$

the following final expression for $[A]_N$ can be derived (Butt, 1980):

$$[A]_N = \alpha^N [A]_0 + k_- \bar{t}([A]_0 + [R]_0) \left(\frac{\alpha}{1-\alpha} - \frac{\alpha^{N+1}}{1-\alpha} \right) \qquad (10.16)$$

where

$$\alpha = (1 + k_+ \bar{t} + k_- \bar{t})^{-1} \qquad (10.17)$$

Now if we assume that $\bar{t}$ is constant—that is, that all of the reactors are of the same volume for a given volumetric flow rate—Equation 10.15 for a first-order irreversible reaction becomes

$$N\bar{t} = \frac{N}{k} \left[\left(\frac{[A]_0}{[A]_N} \right)^{1/N} - 1 \right] \qquad (10.18)$$

It is evident that PFR operation is approached as N approaches infinity (see Chapter 13 for a further discussion). Under this condition, the PFR equation $(k\bar{t} = \ln[A]_0/[A])$ is recovered. This fact is a clear incentive to replace a single MFR by an MFR sequence.

A similar analysis of non-first-order reactions leads to quite cumbersome equations for $[A]_N/[A_0]$. As the order increases, the telescoping functions involved in these equations become progressively more unwieldy, so that a simple expression for $[A]_N/[A_0]$ for a series of N reactors similar to Equation 10.15 for a first-order reaction becomes impossible. It is necessary in such cases to resort to step-by-step algebraic calculations. The equation for a single step in any such sequence is known as the *recursion equation*. Such equations for $[A]_{i-1}/[A]_i$ (for two consecutive stages) for different types of reactions are listed in Table 10.1.

Minimization of Reactor Volume

For the simple case of a first-order isothermal reaction, the concentration of reactant A in the Nth reactor of a CSTR sequence is minimum (i.e., the conversion is maximum) when the residence times in all N reactors are equal, that is, the reactors are all of equal volume. The general problem of optimizing a CSTR sequence with respect to both temperature and residence time in each reactor is

Table 10.1 Recursion equations for a CSTR sequence (from Butt, 1980)

Reaction	Rate equation	Recursion equation
1. $A \to R$	$-r_A = k$	$\dfrac{[A]_{i-1}}{[A]_i} = 1 + \dfrac{k\bar{t}_i}{[A]_i}$
2. $A \to R$	$-r_A = k[A]$	$\dfrac{[A]_{i-1}}{[A]_i} = 1 + k\bar{t}_i$
3. $2A \to R$ (or $A + B \to R$)	$-r_A = k[A]^2$	$\dfrac{[A]_{i-1}}{[A]_i} = 1 + k\bar{t}_i[A]_i$
4. $A + B \to R$ (E = stoichiometric excess)	$-r_A = [A][B]$	$\dfrac{[A]_{i-1}}{[A]_i} = 1 + k(\mathrm{E} + [A]_i)\bar{t}_i$ $[B]_i = (E + [A]_i),\ E = [B]_0 - [A]_0$
5. $A \leftrightarrow R$	$-r_A = k_+[A] - k_-[R]$	$\dfrac{[A]_{i-1}}{[A]_i} = 1 + \left[k_+ + k_-\left(1 - \dfrac{[A]_0}{[A]_i}\right)\right]\bar{t}_i$
6. $2A \leftrightarrow R$	$-r_A = k_+[A]^2 - k_-[R]$	$\dfrac{[A]_{i-1}}{[A]_i} = 1 + \left[k_+[A]_i + \dfrac{k_-}{2}\left(1 - \dfrac{[A]_0}{[A]_i}\right)\right]\bar{t}_i$
7. $A + B \to R$	$-r_A = k_+[A][B] - k_-[R]$	$\dfrac{[A]_{i-1}}{[A]_i} = 1 + \left[k_+(\mathrm{E} + [A]_i) + k_-\left(1 - \dfrac{[A]_0}{[A]_i}\right)\right]\bar{t}_i$ $[B]_i = (E + [A]_i)$
8. $A + B \to R + S$	$-r_A = k_+[A][B] - k_-[R][S]$	$\dfrac{[A]_{i-1}}{[A]_i} = 1 + \left[k_+(\mathrm{E} + [A]_i) - k_-\dfrac{([A]_0 - [A]_i)^2}{[A]_i}\right]\bar{t}_i$

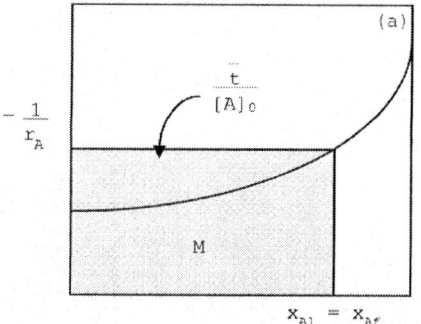 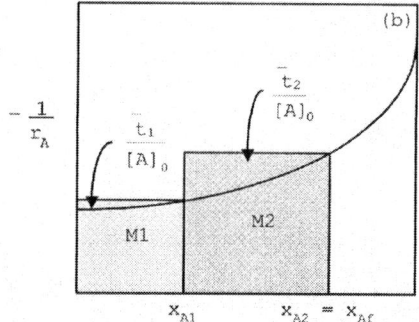

Figure 10.4 Two stirred reactors in series.

considerably more complex and can be handled by the method of dynamic programming (Aris, 1961, 1964) or the discrete maximum principle (Horn, 1961). The former is the preferred method for optimizing any staged operation. Simpler (approximate) methods have also been developed (Denbigh, 1944, 1947; Denbigh et al., 1948; Jones, 1951; Levenspiel, 1972, 1993), and we describe a particularly useful graphical procedure later on.

Equation 10.14 can be written in the more general form

$$\frac{X_{Af}}{-r_A} = \frac{\bar{t}}{[A]_0} \qquad (10.19)$$

Then, a plot of $(1/-r_A)$ versus X_A, as sketched in Figure 10.4a, can be used to represent the volume of the reactor as given by the shaded area M. For two reactors, the volumes will be represented by areas $M1$ and $M2$ in Figure 10.4b, with $(M1 + M2) < M$. The same conclusion can be reached by plotting $(1/-r_A)$ as a function of $[A]$. In an optimal design, however, it is necessary to minimize the total volume needed to achieve the desired conversion. A simple graphical procedure recommended by Levenspiel (1993) is illustrated in Figure 10.5.

Note that reactors of different sizes may be more expensive than the same number of reactors of a single average size. The choice of equal size reactors also has the practical advantage that in a multiproduct pharmaceutical plant, for instance, the reactors can be used interchangeably for a number of reactions. Thus, after obtaining the reactor sizes for minimum volume, it may often be necessary to use personal judgment to select an appropriate common size.

EXAMPLE 10.1

Comparison of different reactors: Reduction of an organic compound by hydrogen

The following rate data are given for the hydrogenation of an organic compound (B) by hydrogen (A) at a temperature of 225 °C and pressure of pure $H_2 = 1.0$ atm:

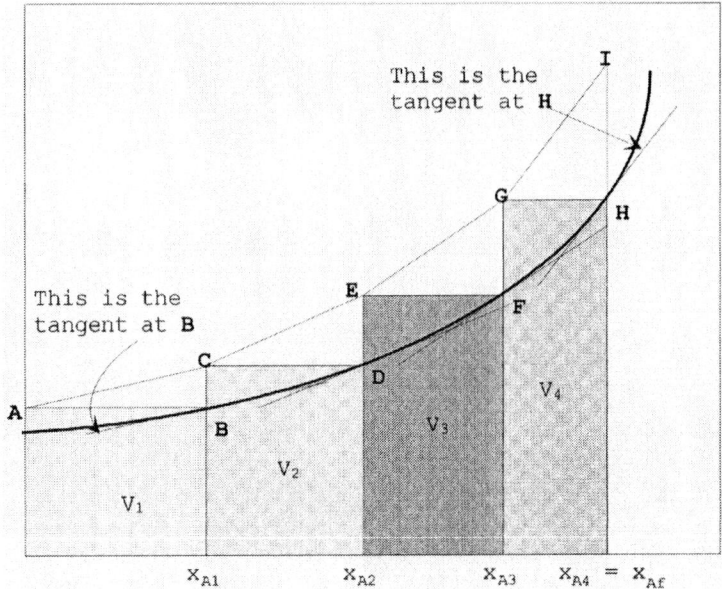

Figure 10.5 Graphical procedure for minimizing the total reactor volume in a series of MFRs (adapted from Levenspiel, 1993). *Steps:* Assume a value of X_{A1}; draw a line from A parallel to the tangent at B; find C and locate D; draw CE parallel to the tangent at D; and repeat the steps till the desired conversion is reached. If the number of reactors is more than anticipated, guess another value of X_{A1} and repeat till the desired number of steps is obtained.

$$[A], \frac{\text{mmol}}{\ell} \quad 24.6 \quad 24.16 \quad 23.91 \quad 23.42 \quad 22.54 \quad 20.06 \quad 17.83 \quad 15.60 \quad 13.29$$

$$-r_A, \frac{\text{mmol}}{\ell \, \text{min}} \quad 0.31 \quad 0.55 \quad 0.74 \quad 1.07 \quad 1.13 \quad 1.22 \quad 0.806 \quad 0.458 \quad 0.218$$

It is desired to treat 100 ℓ/min of the feed and obtain a conversion of 45%. Examine the following alternative reactor arrangements with the object of recommending an arrangement with the least volume:

1. a long tubular reactor ($L/d_T > 25.0$), that is, plug flow is nearly attained.
2. a PFR with optimum recycle; the optimum recycle-flow rate is to be determined.
3. two stirred tanks in series (assuming ideal behavior); comparison of this with the one-tank arrangement.
4. an arrangement of a PFR and an MFR to minimize the total volume of reactors needed.

SOLUTION

1. Assuming that H_2 is an ideal gas, which is a valid assumption, the initial concentration of hydrogen (denoted by A) can be calculated as

$$[A]_0 = \frac{p_{A0}}{R_g T} = \frac{1000}{0.08206 \times 498} = 24.47 \text{ mmol}/\ell$$

Because we are dealing with a constant-volume reaction,

$$[A]_f = [A]_0[1 - X_{Af}] = 24.47[1 - 0.45] = 13.46 \frac{\text{mmol}}{\ell}$$

A plot of $(-1/r_A)$ versus $[A]$ is shown in Figure 10.6a. For a PFR in a constant volume reaction, the residence time $\bar{t}$ is obtained from

$$\bar{t} = \int_{[A]}^{[A]_0} \frac{d[A]}{-r_A} \tag{E10.1.1}$$

Therefore,

$$V_r = \bar{t} \times Q = 17.27 \times 100 \ \ell = 1.727 \text{ m}^3$$

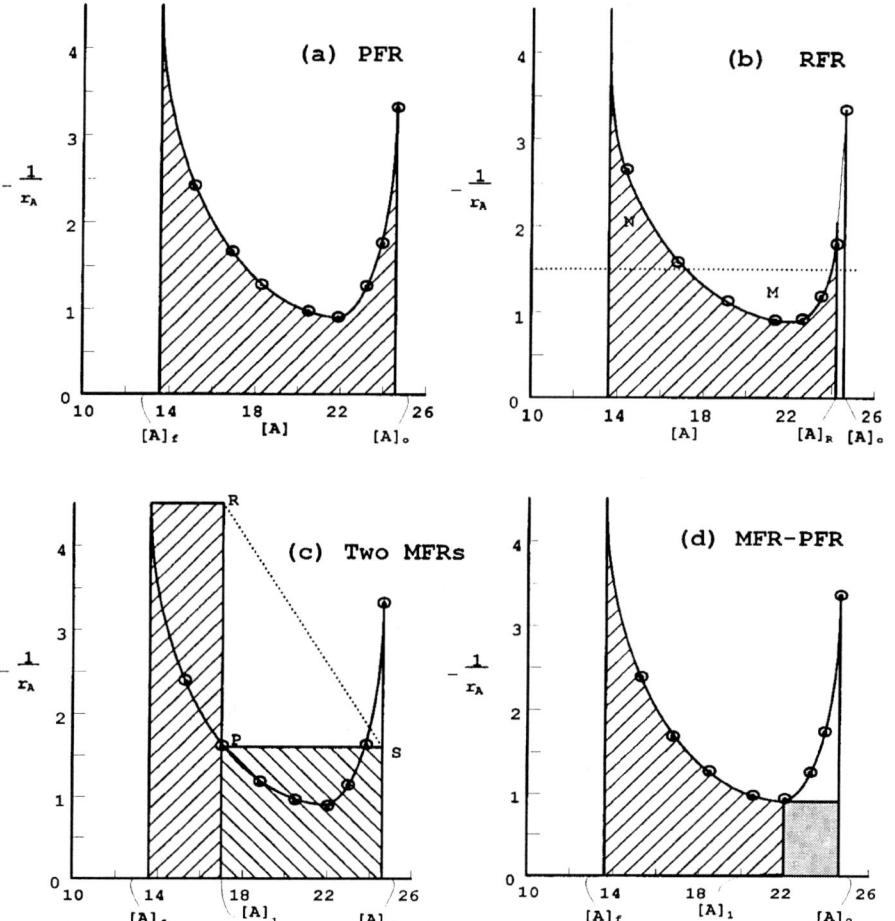

Figure 10.6 Example 10.1: Determination of residence times through plots of $1/-r_A$ versus $[A]$.

2. For a reactor with recycle, $[A]_R$ is located in Figure 10.6b such that areas M and N are equal. Then, from Equation 10.9,

$$R = \frac{24.47 - 24.0}{24.0 - 13.46} = 0.0452$$

Hence 4.52 ℓ/min are recycled. The volume of the RFR is obtained as

$$V_r = \bar{t} \times Q = 16.52 \times 100 = 1.652 \text{ m}^3$$

3. For a mixed-flow reactor, the governing equation is

$$\bar{t} = \frac{([A]_0 - [A]_f)}{[-r_A]_f} \tag{E10.1.2}$$

$$= (24.47 - 13.46) \times 4.25 = 46.79 \text{ min}$$

$$\therefore V_r = 46.79 \times 0.1 = 4.68 \text{ m}^3$$

For two mixed-flow reactors in series, the optimum size ratio of the two reactors is achieved when the slope of the rate curve at P equals the slope of the diagonal RS as shown in Figure 10.6c. By trial and error, we get the intermediate concentration as

$$[A]_1 = 17.0 \frac{\text{mmol}}{\ell}$$

Thus

$$\bar{t}_1 = (24.47 - 17.0)(1.5) = 11.2 \text{ min}, \quad V_{r1} = 1.12 \text{ m}^3$$

and

$$\bar{t}_2 = (17.0 - 13.46)(4.25) = 15.045 \text{ min}, \quad V_{r2} = 1.50 \text{ m}^3$$

giving

$$V_r(\text{total}) = 2.62 \text{ m}^3$$

Because two equally sized units are preferable from overall economic considerations and as because the advantage of the minimum size system over the equally sized system is quite small, equally sized units of 1.50 m³ each are recommended.

4. Here we employ a combination of an MFR and a PFR in series. It can be noted from the $1/-r_A$ versus $[A]$ plots that the portion of the curve up to the minimum is best operated under fully mixed conditions. For the portion beyond this point, a PFR is preferred.

From Figure 10.6d,

$$\left[-\frac{1}{r_A}\right]_{\min} = 0.80 \frac{\ell \text{ min}}{\text{mmol}}$$

Therefore,

$$[A]_1 = 22.0 \frac{\text{mmol}}{\ell}$$

$$\bar{t}_1|_{\text{MFR}} = (0.90)(24.47 - 22.0) = 2.22 \text{ min}$$

$$V_{r1} = 0.222 \text{ m}^3$$

and

$$\bar{t}_2|_{PFR} = -\int_{[A]_i=21.0}^{[A]_f=13.49} \frac{d[A]}{-r_A}$$

By graphical integration,

$$\bar{t}_2 = 13.36 \text{ min}; \qquad V_{r2} = 1.336 \text{ m}^3$$

$$\therefore \quad \text{Total volume of reactors } V_{r1} + V_{r2} = 1.558 \text{ m}^3$$

A comparison of reactor volumes for all four cases considered shows that a combination of an MFR and a PFR in series gives the minimum reactor volume. However, economic considerations may point to a single recycle reactor as the best alternative.

Nonisothermal Operation

Equation 4.77 for a nonisothermal CSTR establishes the temperature at which a stirred reactor operates for a given set of parameter values. This is also true for adiabatic operation. The only difference is that the heat exchange term $UA_h(T - T_c)$ vanishes. In either case, the equation is transcendental and not amenable to extension to a CSTR sequence as a single generalized equation for N reactors. On the other hand, for a first-order reaction, a general recursion formula can be written for N reactors in series. This requires that the temperature of each stage is known to enable calculation of the rate constant.

Thus, expressing the rate constant in Arrhenius form in both the concentration and temperature equations for each stage, the following recursion formulas for the Nth stage can be developed:

$$[A]_N = \frac{[A]_0}{\sum_{i=1}^{N}\left[1 + \left(k^o e^{-E/R_g T_i}\right)\bar{t}\right]} \tag{10.20}$$

$$T_N = T_{N-1} - \frac{k^o(e^{-ER_g/T_N})[A]_0 \bar{t}(-\Delta H_r)}{\rho C_p \prod_{i=1}^{N}\left[1 + \left(k^o e^{-ER_g/T_i}\right)\bar{t}\right]} \tag{10.21}$$

Note that the denominators of these equations involve the sum or product of terms for all N stages, and each stage contains the temperature of that stage. Thus the equations can only be solved stage by stage all the way from stage 1 to N.

SEMIBATCH REACTORS

Semibatch reactors (SBRs) are very common in industrial organic synthesis in general. The basic principle is that a reactant is placed in the reactor and the same or a second reactant, usually the latter, is added continuously. The product

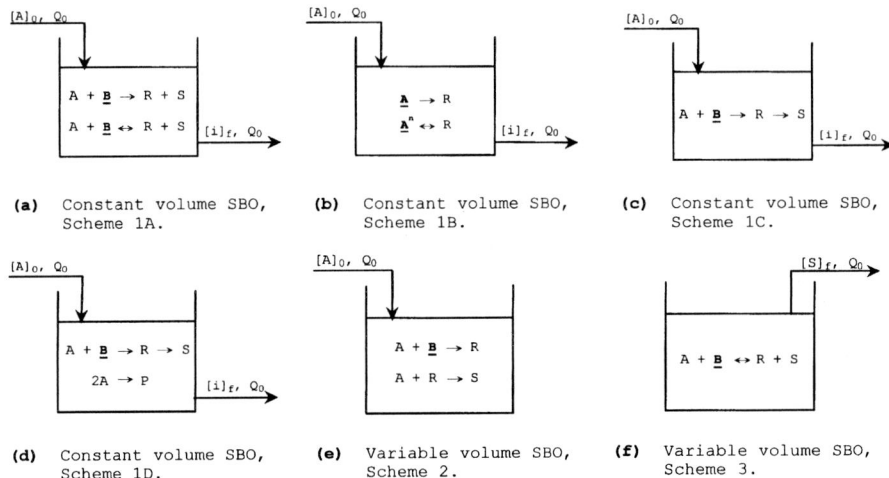

Figure 10.7 Representative modes of semibatch operation (SBO) for different reaction schemes. Bold underlined letters indicate components already present in the reactor.

may or may not be withdrawn. Clearly, several modes of such semibatch operation can be envisaged. The more common of them are sketched in Figure 10.7.

Constant-Volume Reactions with Constant Rates of Addition and Removal: Schemes 1A, B, C, D

In this constant volume semibatch operation, a stirred-tank reactor is charged initially with a reactant or reactants (these are denoted in the figures and text by boldfaced underlined letters). One reactant is fed to the reactor at a constant volumetric flow rate Q_0, and the product is removed at the same rate. This is a general description of the reactions of scheme 1 (i.e., 1A–1D) shown as (a), (b), (c), (d) in Figure 10.7, which differ mainly in reaction kinetics. In analyzing these schemes, we make the following assumptions to simplify the equations: there is no volume change upon mixing of the two liquids or upon reaction, and the reactor is perfectly mixed.

First consider the general case of a simple irreversible second-order reaction

$$A + \underline{\mathbf{B}} \xrightarrow{k_1} R + S \qquad \text{(Scheme 1A-1)} \qquad [10.4]$$

As shown in Figure 10.7a, the tank is initially charged with reactant B, and reactant A is added at rate Q_0. The outlet stream contains both reactants A and B and products R and S. A general material balance can be written as

$$\text{In} - \text{out} - \text{disappearance} + \text{generation} = \text{accumulation} \qquad (10.22)$$

For reactant A, this becomes

$$Q_0[A]_0 - Q_0[A] - (-r_A)V = \frac{d(V[A])}{dt}$$

$$= V\frac{d[A]}{dt} \quad \text{(constant } V\text{)} \qquad (10.23)$$

which can be recast as

$$\frac{d[A]}{dt} = -(-r_A) + \frac{[A]_0 - [A]}{\bar{t}} \tag{10.24}$$

where $\bar{t} = V/Q_0$ and solved for $[A]$. Because generally $[A] \ll [B]$ in this mode of operation, the rate can be written as $(-r_A) = k_1[A]$, leading to the general solution

$$[A] = [A]_0 \left(\frac{1}{k_1\bar{t}+1}\right) + \left([A]_i - \frac{[A]_0}{k_1\bar{t}+1}\right) \exp\left[-\left(\frac{k_1\bar{t}+1}{\bar{t}}\right)t\right] \tag{10.25}$$

where $[A]_i$ is the concentration of A initially charged in the tank. Note, however, that $[A]_i = 0$ in the present case. Similar equations can be written for B, R, and S and are given in Table 10.2.

The analysis can be readily extended to the reverse reaction

$$A + \underline{B} \leftrightarrow R + S \qquad \text{(Scheme 1A-2)} \qquad [10.5]$$

The following other schemes are also important:

$$\underline{A}^n \rightarrow R \qquad \text{(Scheme 1B-1)}$$

$$\underline{A}^n \leftrightarrow R \qquad \text{(Scheme 1B-2)} \qquad [10.6]$$

in which species A is charged to the tank initially at a concentration of $[A]_i$ (Figure 10.7b), the series reaction (Figure 10.7c)

$$A + \underline{B} \xrightarrow{k_1} R \xrightarrow{k_2} S \qquad \text{(Scheme 1C)} \qquad [10.7]$$

and the Van de Vusse reaction (Figure 10.7d)

$$A + \underline{B} \xrightarrow{k_1} R \xrightarrow{k_2} S$$

$$2A \xrightarrow{k_3} P \qquad \text{(Scheme 1D)} \qquad [10.8]$$

The differential equations for these reactions and analytical solutions wherever possible are included in Table 10.2. The semibatch mode of operation of the Van de Vusse scheme gives results similar to those of the recycle reactor. Thus higher yields and selectivities for product R can be realized than in a PFR or an MFR when $k_3[A]_0 \gg k_2$.

Variable Volume Reactor with Constant Rate of Inflow: Scheme 2

Here a tank is charged with an initial volume V_0 of reactant B. Beginning at time zero, reactant A is added at rate Q_0 to the reactor, and no tank products are withdrawn (Figure 10.7e). This SBO scheme is useful in several situations. If the reaction is highly exothermic, the rate of addition of one reactant can be manipulated to control the amount of heat evolved. Furthermore, if undesirable side reactions occur, it is possible to control the selectivity for the desired product. For example, consider the scheme

Table 10.2 List of terms that can be methodically combined to give design equations for several schemes of semibatch operation[a]

Scheme	Reaction[b]	$\frac{d[A]}{dt}, r$	$\frac{d[B]}{dt}, r$	$\frac{d[R]}{dt}, r$	$\frac{d[S]}{dt}, r$	$\frac{d[P]}{dt}, r$	Exact solutions for [A], [R], [R], [S] (see below)
Part 1.	Matrix for identification of terms for constructing design equations						
1A-1	$A + \mathbf{B} \xrightarrow{k_1} R + S$	$\alpha 1, r1$	$\beta 1, r1$	$\gamma 1, r1$	$\delta 1, r1$	—	1a, 1b, 1r, 1s
1A-2	$A + \mathbf{B} \xleftrightarrow{k_1} R + S$	$\alpha 1, r2$	$\beta 1, r2$	$\gamma 1, r2$	$\delta 1, r2$	—	No
1B-1	$\mathbf{A}^n \xrightarrow{k_1} R$	$\alpha 1, r3$	—	$\gamma 1, r3$	—	—	1a, 1r if $n = 1$
1B-2	$\mathbf{A}^n \leftrightarrow R$	$\alpha 1, r4$	—	$\gamma 1, r4$	—	—	No
1C	$A + \mathbf{B} \xrightarrow{k_1} R \xrightarrow{k_2} S$	$\alpha 1, r1$	$\beta 1, r1$	$\gamma 1, r5$	$\delta 1, r6$	—	1a, 1b, eqns. get messy for R & S
1D	$A + \mathbf{B} \xrightarrow{k_1} R \xrightarrow{k_2} S$ $2A \xrightarrow{k_3} P$	$\alpha 1, r7$	$\beta 1, r1$	$\gamma 1, r5$	$\delta 1, r6$	$\rho 1, r8$	No
2	$\left.\begin{array}{c} A + \mathbf{B} \xrightarrow{k_1} R \\ A + R \xrightarrow{k_2} S \end{array}\right\}$	$\alpha 2, r9$	$\beta 2, r1$	$\gamma 2, r10]$	$\delta 2, r11$	—	No
3	$A + \mathbf{B} \leftrightarrow R + S$	$\alpha 3, r12$	$\beta 3, r12$	$\gamma 3, r12$	$\delta 3, r12$	—	No

Part 2. Exact solutions[c]

1a $[A] = \left(\frac{[A]_0}{1 + k_1 \bar{t}}\right) + \left([A]_i - \frac{[A]_0}{1 + k_1 \bar{t}}\right) \exp\left[-\left(\frac{1 + k_1 \bar{t}}{\bar{t}}\right)t\right]$

1b $[B] = -\frac{k_1 \bar{t}[A]_0}{1 + k_1 \bar{t}} + ([A]_0 - [A]_i + [B]_i) \exp\left(-\frac{t}{\bar{t}}\right) + \left([A]_i - \frac{[A]_0}{1 + k_1 \bar{t}}\right) \exp\left[-\left(\frac{1 + k_1 \bar{t}}{\bar{t}}\right)t\right]$

1r $[R] = \frac{k_1 \bar{t}[A]_0}{1 + k_1 \bar{t}} + \left(\frac{[A]_0}{1 + k_1 \bar{t}} - [A]_i\right) \exp\left[-\left(\frac{1 + k_1 \bar{t}}{\bar{t}}\right)t\right] + ([A]_i - [A]_0 + [R]_i) \exp\left(-\frac{t}{\bar{t}}\right)$

1s $[S] = \frac{k_1 \bar{t}[A]_0}{1 + k_1 \bar{t}} + \left(\frac{[A]_0}{1 + k_1 \bar{t}} - [A]_i\right) \exp\left[-\left(\frac{1 + k_1 \bar{t}}{\bar{t}}\right)t\right] + ([A]_i - [A]_0 + [S]_i) \exp\left(-\frac{t}{\bar{t}}\right)$

Part 3. Equations for $\alpha, \beta, \ldots, \rho$

$\alpha 1 = \frac{d[A]}{dt} = -r + \frac{([A]_0 - [A])}{\bar{t}}$

$\beta 1 = \frac{d[B]}{dt} = -r - \frac{[B]}{\bar{t}}$

$\gamma 1 = \frac{d[R]}{dt} = r - \frac{[R]}{\bar{t}}$

$\delta 1 = \frac{d[S]}{dt} = r - \frac{[S]}{\bar{t}}$

$\rho 1 = \frac{d[P]}{dt} = \frac{r}{2} - \frac{[P]}{\bar{t}}$

$\alpha 2 = \frac{d[A]}{dt} = -r + \frac{([A]_0 - [A])}{(\bar{t}_0 + t)}$

$\beta 2 = \frac{d[B]}{dt} = -r - \frac{[B]}{(\bar{t}_0 + t)}$

$\gamma 2 = \frac{d[R]}{dt} = r - \frac{[R]}{(\bar{t}_0 + t)}$

$\delta 2 = \frac{d[S]}{dt} = r - \frac{[S]}{(\bar{t}_0 + t)}$

$\alpha 3 = \frac{d[A]}{dt} = -r - \frac{[A]}{(\bar{t}_0 - t)}$

$\beta 3 = \frac{d[B]}{dt} = -r - \frac{B}{(\bar{t}_0 - t)}$

$\gamma 3 = \frac{d[R]}{dt} = r + \frac{[R]}{(\bar{t}_0 - t)}$

$\delta 3 = \frac{d[S]}{dt} = r + \frac{([S] - [S]_f)}{(\bar{t}_0 - t)}$

Scheme	Reaction[b]	$\frac{d[A]}{dt}, r$	$\frac{d[B]}{dt}, r$	$\frac{d[R]}{dt}, r$	$\frac{d[S]}{dt}, r$	$\frac{d[P]}{dt}, r$	Exact solutions for [A], [R], [R], [S] (see below)

Part 4. Equations for reaction rates, $r_1, r_2, \ldots, r_{12}$.[d]

$r1 = k_1[A]$
$r3 = k_1[A]^n$
$r5 = k_1[A] - k_2[R]$
$r7 = k_1[A] + k_3[A]^2$
$r9 = k_1[A] + k_2[A][R]$
$r11 = k_2[A][R]$

$r2 = k_+[A] - k_-[R][S]$
$r4 = k_+[A]^n - k_-[R]$
$r6 = k_2[R]$
$r8 = k_3[A]^2$
$r10 = k_1[A] - k_2[A][R]$
$r12 = k_+[A][B] - k_-[R][S]$

[a] Locate scheme in Part 1 and read exact solution, if available, in Part 2. If not available, construct ODE for numerical solution from Parts 3 and 4.
[b] **A**, **B** indicate that A, B are initially present in the reactor.
[c] Finite values of $[A]_i$, $[B]_i$, $[R]_i$, $[S]_i$ are assumed in Part 2 solutions; each is zero in the absence of initial charge.
[d] All rates are denoted by r_i and not $(-r_i)$ for disappearance; thus $(-r_A)$ in the text is equivalent to r_A in the table.

$$A + \underline{B} \xrightarrow{k_1} R \quad [10.9a]$$

$$A + R \xrightarrow{k_2} S \quad [10.9b]$$

We can assume that $[A] \ll [B]$ but not that $[R] \ll [A]$, giving the following rate equations: $(-r_A)_1 = k_1[A]$ and $(-r_A)_2 = k_2[A][R]$. The resulting mass balance for A is

$$\frac{d(V[A])}{dt} = V\frac{d[A]}{dt} + [A]\frac{dV}{dt} = Q_0[A]_0 - (-r_A)_1 V - (-r_A)_2 V \quad (10.26)$$

Assuming a linear variation of V with time,

$$V = V_0 + Q_0 t \quad (10.27)$$

or

$$\frac{dV}{dt} = Q_0 \quad (10.28)$$

we get

$$\begin{aligned}\frac{d[A]}{dt} &= \frac{Q_0([A]_0 - [A])}{V_0 + Q_0 t} - k_1[A] - k_2[A][R] \\ &= \frac{([A]_0 - [A])}{\bar{t}_0 + t} - k_1[A] - k_2[A][R]\end{aligned} \quad (10.29)$$

where

$$\bar{t}_0 = \frac{V_0}{Q_0} \quad (10.30)$$

Similar expressions can be written for B, R, and S, and are included in Table 10.2.

Variable Volume Reactor with Constant Rate of Outflow of One of the Products: Scheme 3

In this scheme (Figure 10.7f),

$$A + \mathbf{B} \leftrightarrow R + S \qquad [10.10]$$

the tank is initially charged with one of the reactant species as shown in reaction 10.10 (or with both), and one of the products is withdrawn at a constant rate. This operating mode is particularly useful when the reaction is highly reversible, so that the removal of one of the products causes a favorable shift in equilibrium. For example, in esterification reactions, water (say S) can easily be removed by boiling or under vacuum to shift the equilibrium. The following material balance can then be written for this situation:

$$\frac{d(V[S])}{dt} = V\frac{d[S]}{dt} + [S]\frac{dV}{dt} = V\frac{d[S]}{dt} - Q_0[S] = r_S V - Q_0[S]_f \qquad (10.31)$$

where V is given by Equation 10.27, and $[S_f]$ is the concentration of S in the outlet stream. We assume that the exit stream is pure S with no loss or gain of A, B, or R except by reaction. Equation 10.31 can be recast in the form

$$\frac{d[S]}{dt} = \frac{([S] - [S]_f)}{\bar{t}_0 - t} + k_+[A][B] - k_-[R][S] \qquad (10.32)$$

A mass balance on A gives

$$\frac{d[A]}{dt} = \frac{[A]}{\bar{t}_0 - t} - k_+[A][B] + k_-[R][S] \qquad (10.33)$$

Similar relationships can be written for species B and R. Equations for all of the components are listed in Table 10.2.

General Expression for a Semibatch Reactor for Multiple Reactions with Inflow of Liquid and Outflow of Liquid and Vapor: Scheme 4

The schemes considered so far were all single or relatively simple two-step reactions with a clear specification of inlet and outlet flows of liquid wherever such flows were present. To generalize the approach, we remove these restrictions and write equations for the most general isothermal case that would include multiple reactions, inflow of liquid, and outflow of liquid and vapor. Thus consider N species ($i = 1, 2, \ldots, N$) undergoing M reactions ($j = 1, 2, \ldots, M$) with the following continuity equation:

$$\frac{d([i]V)}{dt} = Q_0[i]_0 - Q_f[i] - Q_{vf}[i]_V - V\sum_{j=1}^{M} \nu_{ij} r_j \qquad (10.34)$$

where Q_{vf} is the volumetric vapor removal rate and V is the total volume of the liquid in the reactor at any time. In analogy with Equation 10.28 for a single

(inlet) flow stream, we can write

$$\frac{dV}{dt} = Q_0 - Q_f - Q_{Lf} = Q_n \quad (10.35)$$

Note that because this equation is for liquid flow, we use the liquid flow rate equivalent Q_{Lf} of the vapor flow rate Q_{Vf}. The two are related by the expression

$$Q_{Lf} = Q_{Vf}\left(\frac{273}{T}\right)P\left(\frac{1}{22.4}\right)\left(\frac{1}{[L]_t}\right) \quad (10.36)$$

where $[L]_t$ is the total liquid-phase concentration of all constituents.

Assuming linear variation of V with time, the following equivalent form of Equation 10.27 can be written:

$$V = V_0 + Q_n t \quad (10.37)$$

Now Equation 10.34 can be solved by substituting Equation 10.37 for V (see Froment and Bischoff, 1990, for an illustrative example).

EXAMPLE 10.2

Design of a semibatch reactor for the production of menadione (2-methyl-1,4-naphthoquinone), commonly known as vitamin K_3

One of the methods of producing menadione is by the oxidation of 2-methylnaphthalene by hydrogen peroxide. Although the complete mechanism of the reaction is not clear, the following three reactions, including the desired one leading to menadione, occur:[2]

OXYGENATION

$$2H_2O_2 \rightarrow 2H_2O + 2O$$
$$3H_2O_2 \rightarrow 3H_2O + 3O \quad \text{[E10.2.1]}$$

OXIDATION

$$\text{A} + 2O \rightarrow \text{P} \quad \text{[E10.2.2]}$$

(A = 2-methylnaphthalene, 2B = 2O, P = 2-methyl-1,4-naphthoquinone)

[2] Note that this is a complex reaction and hence belongs strictly in the next chapter. However, it is included here because it is a very instructive illustration of semibatch reactor design and the basics of complex reactions have already been explained in Chapter 5. A formal treatment of complex reactor design is attempted in Chapter 11.

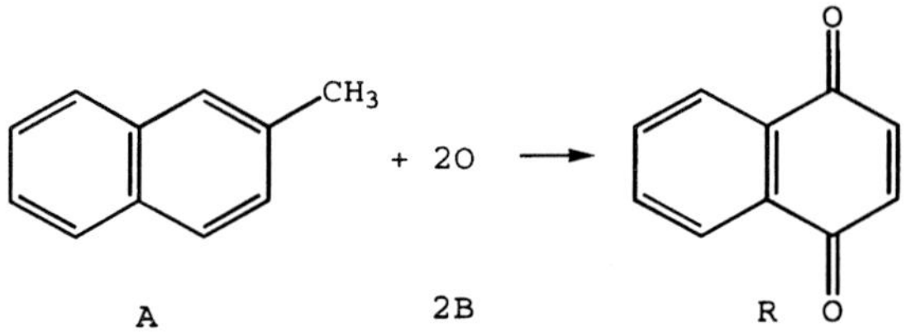

$$A + 2B \longrightarrow R \quad \text{[E10.2.3]}$$

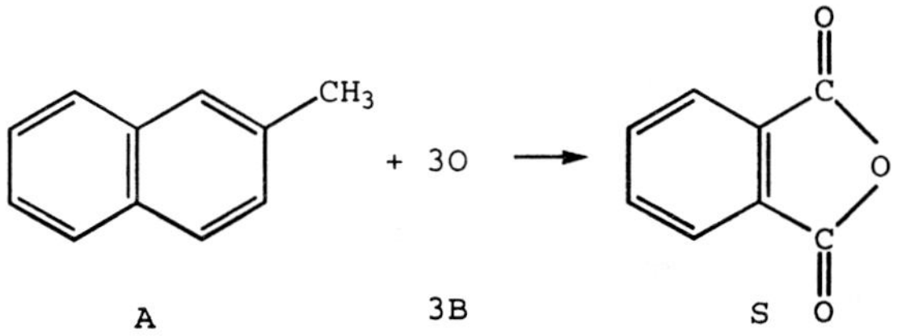

$$A + 3B \longrightarrow S \quad \text{[E10.2.4]}$$

The rate equations along with values of the rate constants are[3]

$$r_1 = k_1[A][B]^2 \quad \text{(E10.2.1)}$$
$$r_2 = k_2[A][B]^2 \quad \text{(E10.2.2)}$$
$$r_3 = k_3[A][B]^3 \quad \text{(E10.2.3)}$$

with

$$k_1 = 2.8 \times 10^{-5} \; \ell^2/\text{mol}^2 \, \text{s}$$
$$k_2 = 5.0 \times 10^{-6} \; \ell^2/\text{mol}^2 \, \text{s}$$
$$k_3 = 1.5 \times 10^{-5} \; \ell^3/\text{mol}^3 \, \text{s}$$

We are required to conduct the reaction in a semibatch reactor and, for a few selected modes of operation, to determine the effect of operating variables on the rate of production of vitamin K_3 and on the selectivity of the reaction for the vitamin.

SOLUTION

Modes of Semibatch Operation

The following two modes of operation will be considered (Figure 10.8):

1. An initial charge of A (methylnaphthalene) is taken in the reactor at a concentration of $[A]_i$, B (hydrogen peroxide solution) at a concentration of $[B]_0$ is added, and the products are withdrawn, both continuously, at the same rate. This mode of operation is continued for a certain length of time corresponding to a fraction f_{SB} of the total time, after which the flow of B is stopped, and the reaction is continued in the batch mode for the remaining fraction of time $(1 - f_{SB})$. This may be regarded as constant volume, semibatch operation.

---

[3] These values were empirically generated from the experimental product distribution and selectivity data of Takanobu et al. (1977). They are representative enough to illustrate the procedure for estimating the selectivity as a function of the rate of production for any mode of semibatch operation.

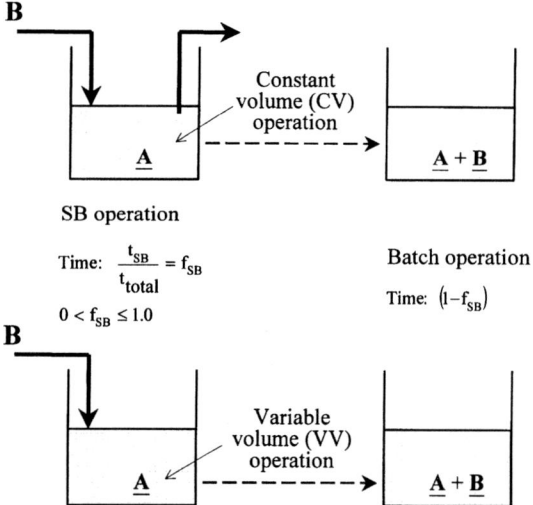

Figure 10.8 The two modes of operation considered in menadione production.

2. Same as in (1) but the product is not withdrawn during the run. This corresponds to variable volume, semibatch operation.

The following data may be used: $[A]_i = 0.86$ mol/ℓ, $[B]_0 = 8.83$ mol/ℓ of water (30 wt%), initial volume of $A = 300\ell$.

Yield and Selectivity Equations

The yield, selectivity, and rate of production of P may be calculated from

$$\text{Yield} = \frac{N_P}{N_{A0}} \tag{E10.2.4}$$

$$\text{Selectivity} = \frac{N_P}{N_P + N_R + N_S} \tag{E10.2.5}$$

$$\text{Rate of production} = \frac{N_P}{t_{\text{total}}} \tag{E10.2.6}$$

where N_{Ai}, the initial number of moles of A in the charge, is obtained as

$$N_{Ai} = [A]_i V_0 = 0.86 \times 300$$
$$= 258 \text{ mol}$$

Equations for Constant-Volume Operation

Expressions for the rate of change of concentration of each species during the hydrogen peroxide addition phase, that is the semibatch mode of operation, may be written as follows:

$$\frac{d[A]}{dt} = -\frac{[A]}{\bar{t}} - (k_1 + k_2 + k_3[B])[A][B]^2 \qquad (E10.2.7)$$

$$\frac{d[B]}{dt} = \frac{([B]_0 - [B])}{\bar{t}} - (2k_1 + 2k_2 + 3k_3[B])[A][B]^2 \qquad (E10.2.8)$$

$$\frac{d[P]}{dt} = -\frac{[P]}{\bar{t}} + k_1[A][B]^2 \qquad (E10.2.9)$$

$$\frac{d[R]}{dt} = -\frac{[R]}{\bar{t}} + k_2[A][B]^2 \qquad (E10.2.10)$$

$$\frac{d[S]}{dt} = -\frac{[S]}{\bar{t}} + k_3[A][B]^3 \qquad (E10.2.11)$$

During the batch mode of operation, input and output flows are both absent ($dV/dt = Q_0 = 0$). Thus the first term on the RHS of each of the above equations vanishes. For example, for species B,

$$\frac{d[B]}{dt} = -(k_1 + k_2 + k_3[B])[A][B]^2 \qquad (E10.2.12)$$

The total number of moles of a particular species (say B, again) that is formed is the sum of the moles formed during the semibatch and batch modes of operation. This may be written as

$$N_B = \int_0^{t_{SB}} Q_0[B]dt + V[B]_{batch} \qquad (E10.2.13)$$

Now we solve Equations E10.2.12 and E10.2.13 simultaneously for [B], and this can be done only via numerical methods. The procedure is repeated for all other species.

Equations for Variable-Volume Operation

In this case, volume becomes a function of time

$$V = V_0 + Q_0 t \qquad (E10.2.14)$$

and hence Equations E10.2.7–E10.2.12 are changed as follows:

$$\frac{d[A]}{dt} = -\frac{[A]}{\bar{t} + t} - (k_1 + k_2 + k_3[B])[A][B]^2 \qquad (E10.2.15)$$

$$\frac{d[B]}{dt} = \frac{[B]_0 - [B]}{\bar{t} + t} - (2k_1 + 2k_2 + 3k_3[B])[A][B]^2 \qquad (E10.2.16)$$

$$\frac{d[P]}{dt} = -\frac{[P]}{\bar{t} + t} + k_1[A][B]^2 \qquad (E10.2.17)$$

$$\frac{d[R]}{dt} = -\frac{[R]}{\bar{t} + t} + k_2[A][B]^2 \qquad (E10.2.18)$$

$$\frac{d[S]}{dt} = -\frac{[S]}{\bar{t} + t} + k_3[A][B]^3 \qquad (E10.2.19)$$

Because no product is removed until the reactor is emptied at $t = t_{total}$, the total number of moles of each species is calculated as shown below for species B.

$$N_B = (V_0 + Q_0 t_{SB})[B]_{t_{total}} \qquad (E10.2.20)$$

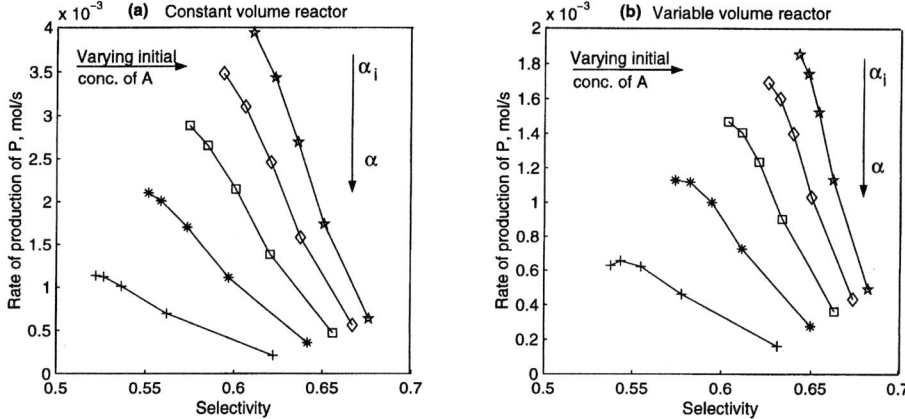

Figure 10.9 Rate of production of menadione as a function of selectivity for (a) constant volume operation and (b) variable volume operation. Total time = residence time = 12.5 h. Entire production is semibatch, i.e. $f_{SB} = 1$. Each line represents a different initial ratio of reactants α_i.

Now we solve Equations E10.2.16 and E10.2.20 numerically. The procedure is repeated for all other species.

Computation of Results

Equations E10.2.7–E10.2.11 and E10.2.15–E10.2.19 are the governing equations for the constant and variable volume reactors, respectively. In each case, the reactor is run in the SB mode for a fraction f_{SB} of the total time t_{total} and in the batch mode for the rest of the time. Defining the mole ratio of the reactants as the ratio α_i of the total number of moles of A in the initial charge to the total number of moles of B fed into the reactor in both the constant and variable volume modes of operation, the following variables can be recognized as important in determining the performance of the reactor: mole ratio of reactants α_i, total reaction time t_{total}, fraction f_{SB} of the total time for which the reactor is run in the SB mode, and the initial concentration $[A]_i$ of methylnaphthalene (A). Using different combinations of these variables, the governing equations are solved by a straightforward interative procedure for the total rate of production as a function of selectivity.

The results are plotted in Figure 10.9 as the rate of production of menadione versus the selectivity for different values of α_i both for (a) CV and (b) VV operations with $t_{total} = 12.5$ hr, $E = 12.5$ hr, and $f_{SB} = 1$ (i.e., the entire operation is semibatch with zero batch time).

Nonisothermal Operation

As mentioned earlier in this section, semibatch operation can be very advantageous for highly exothermic reactions because the rate of heat generation can be controlled by the rate of reactant addition. Specifically, what needs to be controlled is the evolving temperature progression with time. To do this for any of the schemes already considered we must write the energy balance corresponding

to the mass balance of that scheme and solve the two equations simultaneously. Because the mass balance given by Equation 10.34 is general, we shall write the energy balance corresponding to this equation. Thus

$$\rho C_P \frac{d(VT)}{dt} = Q_0 C_p \rho T_0 - Q_f C_p \rho T - Q_{Vf} \sum_{i=1}^{N} [i]_v \Delta H_v$$

$$+ \Delta H_{ex} + V \sum_{j=1}^{M} r_j (-\Delta H_j) \qquad (10.38)$$

where ΔH_v is the heat of vaporization and ΔH_{ex} is the amount of heat exchanged by the control fluid.

The term ΔH_{ex} can be estimated from

$$\Delta H_{ex} = W_c C_{pc}(T_{c0} - T_{cf}) = UA_h \frac{(T_{c0} - T) - (T_{cf} - T)}{\ln \dfrac{T_{cf} - T}{T_{cf} - T}} \qquad (10.39)$$

where T_{c0} and T_{cf} are the inlet and outlet temperatures of the control fluid, respectively, W_c is the flow rate of the control fluid (kg/s), C_{pc} is the heat capacity of the control fluid, and U is the overall heat transfer coefficient whose value depends on whether the reactor is heated by jacket or coil. If by jacket, Equations 4.53–4.55 can be used for estimating it; if by coil, the appropriate equation must be found from the heat transfer literature.

Because all of the parameters of Equations 10.34 and 10.38 are known now or can be estimated, the two equations can be solved by the Runge–Kutta–Gill routine to obtain the various concentrations and temperatures as functions of time.

VARYING VOLUME REACTORS

In the previous section, we considered semibatch operation as it is normally practiced. In this section we attempt to place this operation in a more general setting—as part of a broader class of reactors. In its most general form, this class of reactors is best called the *variable volume reactor* (VVR) (Lund and Seagrave, 1971a,b; Ridlehoover and Seagrave, 1973), and we shall largely be using this wider (inclusive) designation of Seagrave's.

Mathematical Description of VVR

Because our concern here is mainly with liquid-phase reactions, any volume change due to reaction can be neglected, so that the material balance of Equation 10.22 continues to be valid. It is convenient, however, to recast this

equation in dimensionless form by using the following dimensionless variables:

$$a' = \frac{[A]}{[A]_0}, \quad \bar{t} = \frac{V_R}{Q_R}, \quad t^* = \frac{t}{\bar{t}} = \frac{tQ_R}{V_R},$$

$$V^* = \frac{V}{V_R}, \quad Q^* = \frac{Q}{Q_R}, \quad Q_0^* = \frac{Q_0}{Q_R}, \quad \tau^* = k\bar{t} = \frac{kV_R}{Q_R} \tag{10.40}$$

In these definitions, Q_R and V_R have been used as arbitrary normalizing parameters. They can be readily identified for a constant volume reactor: $Q_R = Q_0 = Q_f = Q$, and $V_R = V = V_m$ where V_m represents the maximum volume of the reactor (in other words, the volume at which the reactor would function as a constant volume CSTR). Clearly, in this case $\bar{t} = V/Q = V_0/Q_0$ is the true residence time. However, for the variable volume reactor, Q_R and V_R have to be selected carefully on a case by case basis, and the residence time $\bar{t} = V_R/Q_R$ should be regarded as no more than a parameter whose dimension is time or simply as *pseudoresidence time*. Using these dimensionless parameters (with normalizing variables different from those for an MFR), the general material balance of Equation 10.22 can be expressed as

$$\frac{da'}{dt^*} = \frac{Q_0^*}{V^*}(1 - a') - a'\tau^* \tag{10.41}$$

For the simple steady-state, mixed-flow reactor (CSTR), $da'/dt^* = 0$, $V^* = 1$, $Q_0^* = 1$, and Equation 4.71 (Chapter 4) is recovered. Written in terms of conversion, this equation becomes

$$X_A = \left(1 - \frac{[A]}{[A]_0}\right) = \frac{\tau}{1 + \tau} \tag{10.42}$$

where now $\tau = kV/Q$. Clearly, X_A is maximum when τ is infinity, that is when the reactor operates in the batch mode.

It is possible to break up the time for continuous volume variation into discrete time zones that can be manipulated according to requirement. We describe first a complete policy which considers all possible time zones, and then describe two simpler ones. These policies are essentially semibatch operating policies.

Semibatch Operation

Semibatch operation in its complete description consists of four time zones: filling, reaction, emptying, and downtime (Figure 10.10). During the filling mode, the reactant stream is fed at a constant flow rate of Q_0 into the reactor that contains a volume V_0 of liquid. The liquid volume increases linearly from V_0 to a maximum of V_m. This is followed by batch reaction, during which there is no flow into or out of the reactor. In the next step, the reactor contents are emptied at a constant rate of Q_f (not necessarily equal to the filling rate) till the volume decreases to the initial level V_0. The final step is the downtime period, which has to be taken into account in the overall scheduling of the reactor before

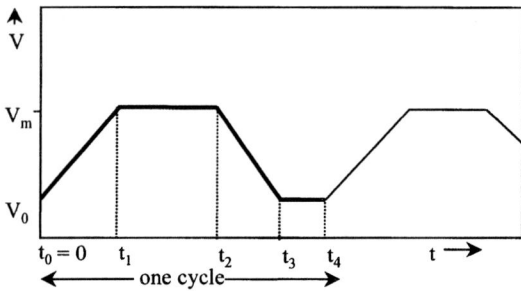

Figure 10.10 Time cycle for semibatch operation: SBO policy-1.

the next batch is loaded. After a few cycles, the reactor reaches a steady state. Thus this procedure essentially involves alternating the reactor volume between a minimum and a maximum by adjusting the input and output flow rates.

Let us denote the times for filling, batch reaction, emptying, and downtime by $t_1, (t_2 - t_1), (t_3 - t_2)$, and $(t_4 - t_3)$, respectively. Normalized with respect to the total cycle time t_4,

$$\theta_1 + \theta_2 + \theta_3 + \theta_4 = 1 \tag{10.43}$$

where

$$\theta_1 = t_1/t_4, \quad \theta_2 = (t_2 - t_1)/t_4,$$
$$\theta_3 = (t_3 - t_2)/t_4, \quad \theta_4 = (t_4 - t_3)/t_4 \tag{10.44}$$

Now we can solve this general policy for the relative conversion X_A^* defined as

$$X_A^* = \frac{X_{A,SB}}{X_{A,MF}} \tag{10.45}$$

which then can be simplified for specific situations of practical importance. Thus we consider two policies: SBO policy-1 and SBO policy-2.

SBO policy-1

In this policy we use the intuitive hypothesis that the filling and emptying times should be equal for optimal operation. Further, we restrict the analysis to a single cycle, which has the advantage that downtime, the time between cycles, need not be considered. We also assume that the reactor is completely emptied, that is, it alternates between $V = V_0 = 0$ and $V = V_m$. These lead to the following modified definitions:

$$V_0 = 0, t_0 = 0, (t_4 - t_3) = 0 \text{ (i.e., downtime} = 0), \theta_1 = \theta_3$$
$$= t_1/t_3 = (t_3 - t_2)/t_3 \text{ (filling time} = \text{emptying time)} \tag{10.46}$$

Note that now the total time is t_3, not t_4. Hence the different time zones are normalized with respect to t_3 and not t_4. With these modifications, Seagrave and collaborators derived an expression for the relative conversion X_A^*, based on

which we note that (1) there is an optimum value of τ_{MF} for every filling time at which SBO-1 gives a maximum value of X_A^* and (2) this maximum has the highest value when the filling is instantaneous, that is, batch operation is approached.

SBO policy-2 (partially emptying reactor)

It will be recalled that the lower limit of V^* is zero in policy-1. In other words, the reactor is emptied fully after each batch. Now if we remove this restriction and also assume instantaneous filling and emptying, then we have a much simpler mode of operation (Figure 10.11). A reactor operating in this semibatch mode is often referred to as a partially *emptying batch reactor* (PEBR) (Levenspiel, 1993), and we designate this mode of operation as SBO policy-2.

As already explained, this is a simplified version of VVR but very useful in organic technology. Because the filling and emptying times in this case are each zero, it would be more logical to compare the performance of this reactor with that of a batch reactor. The main features of this reactor policy may be restated as

$$V_R = V_m, \quad 0 \leq V_0 < V_m, \quad t_1 - t_0 = t_3 - t_2 = 0, \quad t_4 \text{ does not arise} \quad (10.47)$$

The operation of the reactor is sketched in Figure 10.12. If it is to be run as a batch reactor, the performance equation is

$$t = -\int_{[A]_0}^{[A]_f} \frac{d[A]}{(-r_A)} \quad (10.48)$$

If operated in the PEBR mode, the inlet boundary condition would be different, as shown in Figure 10.12, and the following modified performance equation results:

$$t = \int_{[A]_R}^{[A]_f} \frac{d[A]}{(-r_A)}, \quad [A]_R = (1 - V^*)[A]_0 + V^*[A]_f \quad (10.49)$$

For a first-order reaction, $-r_A = k[A]$, and Equation 10.49 can be integrated to give

$$kt = \ln\left[(1 - V^*)\frac{[A]_0}{[A]_f} + V^*\right] \quad (10.50)$$

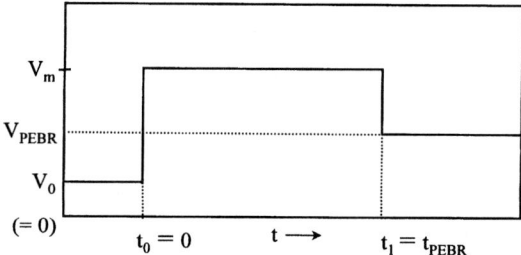

Figure 10.11 Time cycle for the special case of negligible filling and emptying times and no downtime: SBO policy-2.

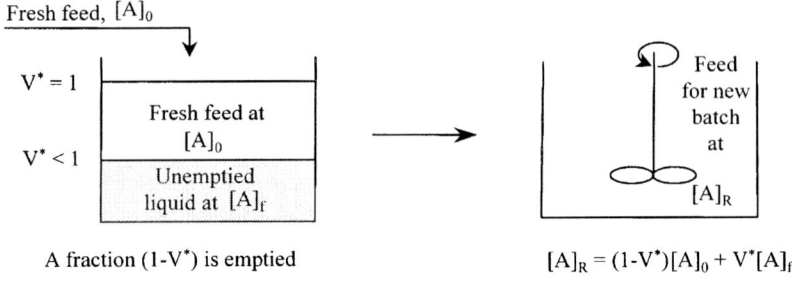

Figure 10.12 The partially emptying reactor: SBO policy-2. $V^* = V/V_m$.

MEASURES OF MIXING: COMPARISON OF THE SERIES, RECYCLE, AND VARIABLE VOLUME REACTORS

The three reactor configurations, tanks-in-series, recycle, and variable volume, can be regarded as different methods or models for characterizing mixing. The parameters of these models are

Tanks-in-series model: Number of tanks, N
Recycle model: Ratio of volume recycled to volume leaving, R
Variable volume model: Ratio of initial to final volumes, V^*

Now let us compare the variable volume and recycle models described, respectively, by Equations 10.50 and 10.8. The following identities can be written straightaway:

$$V^* = \frac{R}{(1+R)}, \quad R = \frac{V^*}{(1-V^*)} \tag{10.51}$$

We also note that

when $V^* = 0$, batch operation prevails, and $\theta_2 > 0$;

when $V^* = 1$, $\theta_2 = 0$ and mixed-flow operation prevails.

The parameter of the recycle reactor R is directly related to the parameter of the partially emptying reactor V^*. Thus for design purposes, one can treat this as a recycle reactor, determine R for optimal operation by the procedure described earlier, and then determine V^* from Equation 10.51 to operate the reactor in the VV or PEBR mode.

It is also possible to derive the following relationship between the recycle and tanks-in-series models:

$$\frac{R}{1+R} = \frac{1}{N} \tag{10.52}$$

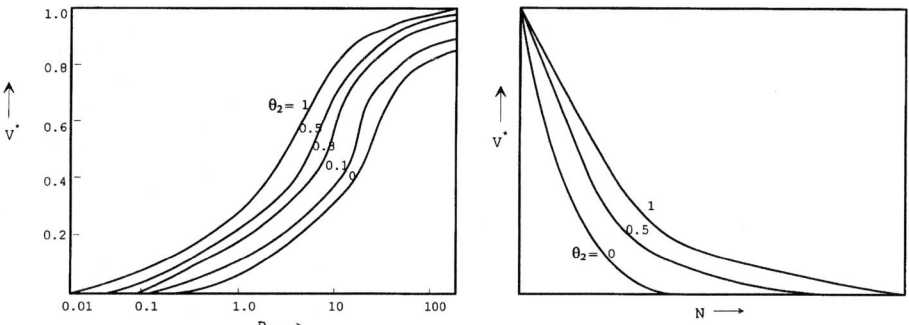

Figure 10.13 Conditions for equal macromixing in RFR, VVR, and CSTRs-in-series (redrawn from Roth et al., 1979).

Thus the relationship among the three models can be expressed in a single equation as

$$V^* = \frac{R}{1+R} = \frac{1}{N} \tag{10.53}$$

In the comparisons just made, the variable volume, semibatch mode of operation was assumed to proceed with negligible filling and emptying times in accordance with SBO policy-2. If we consider SBO policy-1 with equal but finite filling and emptying times, a second parameter, the filling time, has to be introduced. Although this loses the simplicity of the three one-parameter models considered above, it is more general. Without going into the details of derivations, we reproduce in Figure 10.13 the results of Seagrave and collaborators (see Roth et al., 1979). This figure shows plots of the two VVR model parameters, V^* and θ_2 (reaction time), as functions of R and N (the parameters of the other two models) for the same degree of mixing. These are useful plots in comparing the performances of the three important reactors considered in this chapter.

COMMENTS

A large number of liquid-phase organic reactions are carried out in batch or semibatch reactors. For large volume, liquid-phase reactions, the use of a series of CSTRs is quite common. For large volume, vapor-phase catalytic reactions, tubular reactors are often the reactors of choice.

We made a reference to combined reactors in the introduction to this chapter. In these reactors, instead of a series of MFRs, different combinations of a PFR and an MFR are used to minimize the total reactor volume. The theory of these reactors was developed by Douglas (1964), Baburao and Doraiswamy (1965) and Aris (1969). Because they are rarely if ever used industrially, they are not considered in this chapter. But a simple method of calculating the volume of an MFR-PFR combination is included in Example 10.1.

Semibatch operation is involved during the startup or shutdown of a CSTR. This initial period is usually a matter of concern in large-volume production

where shutdowns and start-ups can lead to a reduction in the overall rate of production. During those unsteady periods, the selectivity might also be affected in complex reactions. Another instance of semibatch operation is when a dissolved component from a liquid in a reactor has to be purged (i.e., prevented from accumulating) by continuous introduction of an inert gas such as nitrogen. This often becomes necessary in processes where recycle streams with their unavoidable tendency to accumulate impurities are used. The general problem of unsteady-state operation is briefly considered in Chapter 13.

Chapter 11

Reactor Design for Complex Reactions

Procedures were formulated in Chapter 5 for treating complex reactions. We now turn to the design of reactors for such reactions. Continuing with the ethylation reaction, we consider the following reactor types for which design procedures were formulated earlier in Chapter 4 for simple reactions: batch reactors, continuous stirred reactors (or mixed-flow reactors), and plug-flow reactors. However, we use the following less formal nomenclature: A = aniline, B = ethanol, C = monoethyaniline, D = water, E = diethylaniline, F = diethyl ether, and G = ethylene. The four independent reactions then become

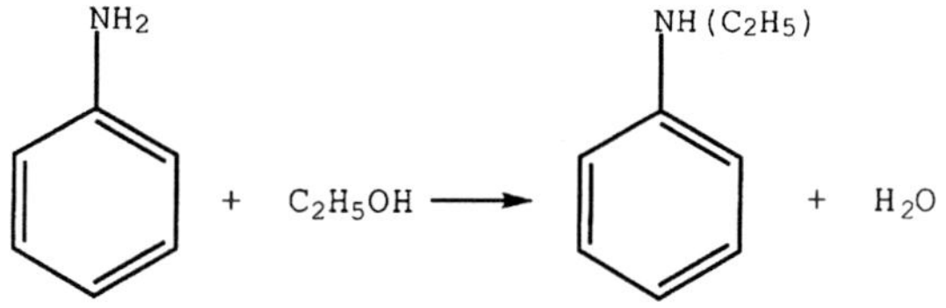

$$C_6H_5NH_2 + C_2H_5OH \longrightarrow C_6H_5NH(C_2H_5) + H_2O$$
$$(A) \qquad (B) \qquad\qquad (C) \qquad\qquad (D) \qquad [11.1.1]$$

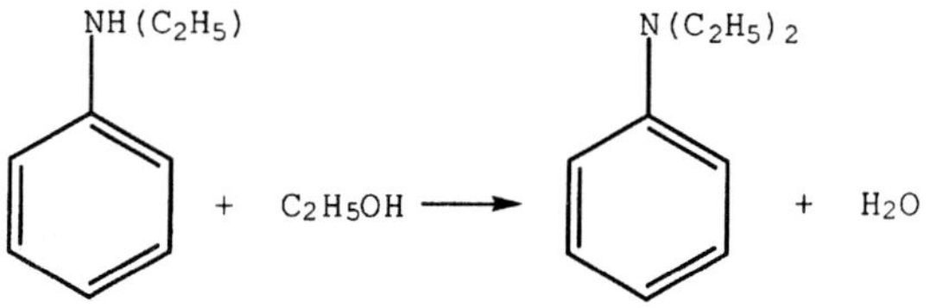

$$C_6H_5NH(C_2H_5) + C_2H_5OH \longrightarrow C_6H_5N(C_2H_5)_2 + H_2O$$
$$(C) \qquad\qquad (B) \qquad\qquad (E) \qquad\qquad (D) \qquad [11.1.2]$$

$$2C_2H_5OH \rightarrow C_2H_5OC_2H_5 + H_2O \qquad [11.1.3]$$
$$(2B) \qquad\quad (F) \qquad\quad (D)$$

$$C_2H_5OH \rightarrow C_2H_4 + H_2O \qquad [11.1.4]$$
$$(B) \qquad\quad (G) \quad (D)$$

Using this set of equations as the basis, we now formulate design equations for various reactor types.[1] Detailed expositions of the theory are presented in a number of books, in particular Aris (1965, 1969) and Nauman (1987).

BATCH REACTOR

Design Based on Number of Components

Consider a reaction network consisting of N components and M reactions. A set of N ordinary differential equations, one for each component, would be necessary to mathematically describe this system. They may be concisely expressed in the form of Equation 5.5 (Chapter 5), or

$$\frac{d(\mathbf{c}V)}{dt} = \mathbf{vr}V \tag{11.1}$$

The use of this equation in developing batch reactor equations for a typical complex reaction is illustrated in Example 11.1.

EXAMPLE 11.1

Batch reactor equations based on number of components: ethylation of aniline

Applying Equation 11.1 in more explicit form to reaction 11.1, and assuming constant volume and first-order dependence of the rate on each concentration, we get

$$\frac{d}{dt}\begin{pmatrix}[A]\\[B]\\[C]\\[D]\\[E]\\[F]\\[G]\end{pmatrix} = \begin{pmatrix}-1 & 0 & 0 & 0\\-1 & -1 & -2 & -1\\1 & -1 & 0 & 0\\1 & 1 & 1 & 1\\0 & 1 & 0 & 0\\0 & 0 & 1 & 0\\0 & 0 & 0 & 1\end{pmatrix}\begin{pmatrix}r_1\\r_2\\r_3\\r_4\end{pmatrix} \tag{E11.1.1}$$

where $r_1 = k_1[A][B]$, $r_2 = k_2[B][C]$, $r_3 = k_3[B]^2$, and $r_4 = k_4[B]$.

We now express the rate of formation/disappearance of each component by accounting for its rates of formation and disappearance by the different steps comprising the reaction. For example, A is formed by reaction 11.1 with a stoichiometric coefficient of -1, and is

---

[1] The reaction scheme considered here is strictly valid in the vapor phase over a solid catalyst. However, the same reactions can also occur in the liquid phase in a batch or continuous stirred reactor; but at the lower temperatures now involved, ethylene may not be formed. Further, the ether may be in the gaseous phase. Even so, because the reaction is reasonably complex, we use it as a model scheme for demonstrating the procedures for different reactor types. We also employ other, simpler schemes in some cases.

not involved in any of the other three reactions. A similar analysis of all components leads to the following set of ordinary differential equations:

$$\frac{d[A]}{dt} = -k_1[A][B]$$

$$\frac{d[B]}{dt} = -k_1[A][B] - k_2[B][C] - 2k_3[B]^2 - k_4[B]$$

$$\frac{d[C]}{dt} = k_1[A][B] - k_2[B][C]$$

$$\frac{d[D]}{dt} = k_1[A][B] + k_2[B][C] + k_3[B]^2 + k_4[B] \tag{E11.1.2}$$

$$\frac{d[E]}{dt} = k_2[B][C]$$

$$\frac{d[F]}{dt} = k_3[B]^2$$

$$\frac{d[G]}{dt} = k_4[B]$$

Note that there are as many equations as number of components. The rate constants would also be known. Hence, these equations can be solved to obtain the product distribution. It is convenient to use the Runge-Kutta-Gill numerical technique, but any other simultaneous differential equation solving routine can also be used.

EXAMPLE 11.2

Design of a semibatch reactor for the Haloform oxidation of diacetone alcohol (a continuation of Example 5.3).

A qualitative analysis of the complex network involved in the Haloform oxidation of DAA was presented in Example 5.3. We shall make use of the conclusions therefrom to formulate a procedure for designing an appropriate reactor for this scheme.

The following data are given, and it is intended to generate concentration profiles for all the components for the selected reactor. Using the same equations, productivity versus selectivity plots can also be generated, but because the procedure for this has already been illustrated for menadione production in Example 5.2, it will not be attempted here.

DATA

Subscript i denotes initial presence and subscript 0 denotes initial or entrance conditions.
$N_{NaOCl}/N_{DAA} = 3$ (stoichiometric), $[DAA]_0 = 8.09\,M$, $[DAA]_i = [HMB]_i = [DHMB]_i = [Ac]_i = [HAc]_i = 0$, $[NaOCl]_i = 2\,M$, $[NaOH]_i = 0.15\,M$, $[HMB]_f = 0.25\,M$, $[HAc]_f = 0.3\,M$, $V_0 = 0.25\,\ell$, $Q = 8.75 \times 10^{-5}\,\ell/min$, t = 240 min, T = 25 °C, $k_1 = 1.0\,(\ell/mol)^2$, 1/min (Barac, 1998), $k_2 = 0.175\,(\ell/mol)^2\,1/min$ (Barac, 1998), $k_3 = 0.477\,(\ell/mol)$ 1/min (Frost and Pearson, 1961), $k_4 = 0.018\,(\ell/mol)^2\,1/min$ (Frost and Pearson, 1961), $k_5 = 5.262\,(\ell/mol^2)^2\,1/min$. (Guthrie et al., 1984).

SOLUTION

From the discussion in Example 5.2, semibatch operation with continuous addition of DAA to hypochlorite appears to be the preferred mode of reactor operation. Using the same nomenclature as in that example, and employing the equations summarized in Table 10.2 for different variable volume semibatch reactions, the following equation can be written for component i in reaction j:

$$\frac{d[i]}{dt} = \frac{[i]_i - [i]}{\bar{t} + t} + \sum_j r_{ij} \tag{E11.2.1}$$

where $\bar{t} = V_0/Q$ and $[i]_i$ is the initial concentration of i.

The rate terms for the different reactions comprising $\Sigma_j r_{ij}$ are given by Equations E5.3.2–E5.3.8. Using these equations, Equation E11.2.1 can be resolved into the following set of variable volume semibatch reactor equations:

$$\frac{d[\text{DAA}]}{dt} = \frac{8.09 - [\text{DAA}]}{2857.14 + t} - [\text{DAA}][\text{NaOCl}][\text{NaOH}]$$

$$- 0.175[\text{DAA}][\text{NaOCl}][\text{NaOH}]^{0.9}$$

$$- 0.477[\text{DAA}][\text{NaOH}] + 0.018[\text{Ac}]^2[\text{NaOH}]$$

$$\frac{d[\text{HMB}]}{dt} = \frac{-[\text{HMB}]}{2857.14 + t} + [\text{DAA}][\text{NaOCl}][\text{NaOH}]$$

$$\frac{d[\text{DHMB}]}{dt} = \frac{-[\text{DHMB}]}{2857.14 + t} + 0.175[\text{DAA}][\text{NaOCl}][\text{NaOH}]^{0.9}$$

$$\frac{d[\text{Ac}]}{dt} = \frac{-[\text{Ac}]}{2857.14 + t} + 0.954[\text{DAA}][\text{NaOH}] - 0.036[\text{Ac}]^2[\text{NaOH}] \qquad [\text{E11.2.2}]$$

$$- 5.262[\text{Ac}][\text{NaOCl}][\text{NaOH}]$$

$$\frac{d[\text{HAc}]}{dt} = \frac{-[\text{HAc}]}{2857.14 + t} + 5.262[\text{Ac}][\text{NaOCl}][\text{NaOH}]$$

$$\frac{d[\text{NaOCl}]}{dt} = \frac{-[\text{NaOCl}]}{2857.14 + 5} - 3[\text{DAA}][\text{NaOCl}][\text{NaOH}] - 0.7[\text{DAA}][\text{NaOCl}][\text{NaOH}]^{0.9}$$

$$- 15.786[\text{Ac}][\text{NaOCl}][\text{NaOH}]$$

$$\frac{d[\text{NaOH}]}{dt} = \frac{-[\text{NaOH}]}{2857.14 + t}$$

These equations can be solved by any differential equation solver, such as the appropriate subroutine in MATLAB. The concentration profiles thus generated are shown in Figure 11.1 (Barac, 1998). Some important observations may be noted:

1. HMB formation stops well before all of the DAA has been added; this is because the concentration of NaOCl becomes so small that $r_{\text{DAA}} < r_{\text{DHMB}}$.
2. Experimental results uphold the model prediction for the final concentration of HMB.

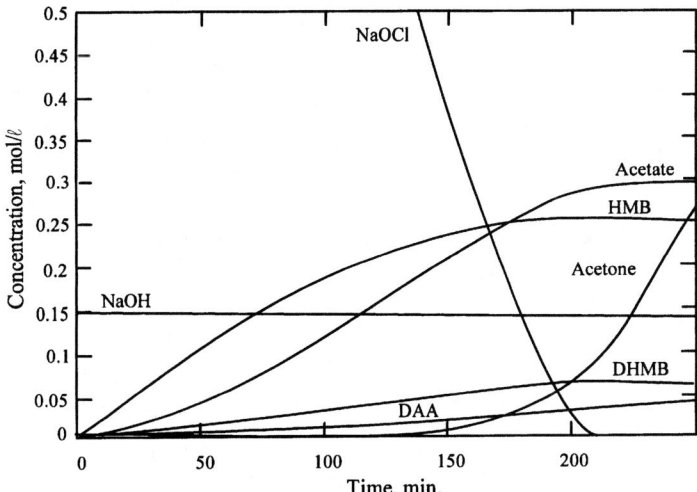

Figure 11.1 Concentration-time profiles for the various products (including unreacted reactants) of the haloform reaction (from Barac, 1998).

Use of Extent of Reaction or Reaction Coordinate

Consider a simple reaction

$$\nu_A A + \nu_B B \to \nu_R R \qquad [11.2]$$

with no restriction regarding volume change. Referring to Equation 5.6, the material balance relations for this reaction can be written as

$$N_A - N_{A0} = \nu_A \xi$$
$$N_B - N_{B0} = \nu_B \xi \qquad (11.2)$$
$$N_R - N_{R0} = \nu_R \xi$$

These can be recast in the form

$$\begin{pmatrix} N_A \\ N_B \\ N_R \end{pmatrix} - \begin{pmatrix} N_{A0} \\ N_{B0} \\ N_{R0} \end{pmatrix} = \begin{pmatrix} \nu_A \\ \nu_B \\ \nu_R \end{pmatrix} \xi \qquad (11.3)$$

When extended to a complex reaction of N components and M reactions, this becomes

$$\begin{pmatrix} N_A \\ N_B \\ N_R \\ \cdot \\ \cdot \end{pmatrix} - \begin{pmatrix} N_{A0} \\ N_{B0} \\ N_{R0} \\ \cdot \\ \cdot \end{pmatrix} = \begin{pmatrix} \nu_{A1} & \nu_{A2} & \cdots & \nu_{AM} \\ \nu_{B1} & \nu_{B2} & \cdots & \nu_{BM} \\ \cdot & \cdot & & \\ \cdot & \cdot & & \end{pmatrix} \begin{pmatrix} \xi_1 \\ \xi_2 \\ \xi_3 \\ \cdot \end{pmatrix} \qquad (11.4)$$

or

$$\mathbf{N} - \mathbf{N}_0 = \mathbf{v}\boldsymbol{\xi} \tag{11.5}$$

where $\mathbf{N}$ and $\mathbf{N}_0$ are $(N \times 1)$ matrices, respectively, of the final and initial moles of each component, $\mathbf{v}$ is the $(M \times N)$ matrix of stoichiometric coefficients, and $\boldsymbol{\xi}$ is the $(M \times 1)$ reaction coordinate matrix.

Since the units of ξ and r are moles and moles/(time) (volume), respectively, we can write

$$\frac{d\boldsymbol{\xi}}{dt} = V\mathbf{r} \tag{11.6}$$

This is the basic equation for a complex reaction and expresses the rates of the individual reactions in terms of the corresponding reaction coordinates. It can be solved by expressing the rates r of the individual elements in terms of the number of moles according to Equation 11.2. Thus, for the complex reaction

$$\begin{aligned} 2A + 3B &\rightarrow R, \quad r_1 = k_1[A][B] \\ B + D &\rightarrow S, \quad r_2 = k_2[B][D] \end{aligned} \tag{11.3}$$

the following relationships are obtained:

$$\begin{aligned} N_A - N_{A0} &= -2\xi_1 \\ N_B - N_{B0} &= -3\xi_1 - \xi_2 \\ N_D - N_{D0} &= -\xi_2 \\ N_R - N_{R0} &= \xi_1 \\ N_S - N_{S0} &= \xi_2 \end{aligned} \tag{11.7}$$

Equation 11.6 can now be expanded to give

$$\frac{d}{dt}\begin{pmatrix} \xi_1 \\ \xi_2 \end{pmatrix} = V \begin{pmatrix} k_1[A][B] \\ k_2[B][D] \end{pmatrix} = \begin{pmatrix} k_1 N_A N_B / V \\ k_2 N_B N_D / V \end{pmatrix} \tag{11.8}$$

and the Ns can then be eliminated by combining with Equations 11.7 to give expressions for $d\xi/dt$ in terms of ξ and N_0.

The method is demonstrated in Example 11.3 for the same complex reaction considered earlier: ethylation of aniline.

EXAMPLE 11.3

Design of a batch reactor based on extent of reaction for the ethylation of aniline

The four independent reactions of this network are given by reaction 11.1. Applying Equation 11.6 to these reactions and resolving the matrices into their elements gives

$$\frac{d}{dt}\begin{pmatrix}\xi_1\\\xi_2\\\xi_3\\\xi_4\end{pmatrix} = V\begin{pmatrix}k_1[A][B]\\k_2[B][C]\\k_3[B]^2\\k_4[B]\end{pmatrix} = \begin{pmatrix}k_1 N_A N_B/V\\k_2 N_B N_C/V\\k_3 N_B^2/V\\k_4 N_B\end{pmatrix} \quad \text{(E11.3.1)}$$

We now express the individual Ns in terms of the corresponding N_0s and ξs by writing Equation 11.7 for the individual components:

$$\begin{aligned}N_A - N_{A0} &= -\xi_1\\N_B - N_{B0} &= -\xi_1 - \xi_2 - 2\xi_3 - \xi_4\\N_C - N_{C0} &= \xi_1 - \xi_2\\N_D - N_{D0} &= \xi_1 + \xi_2 + \xi_3 + \xi_4\\N_E - N_{E0} &= \xi_2\\N_F - N_{F0} &= \xi_3\\N_G - N_{G0} &= \xi_4\end{aligned} \quad \text{(E11.3.2)}$$

Combining Equations E11.3.1 and E11.3.2 leads to

$$\begin{aligned}\frac{d\xi_1}{dt} &= \frac{k_1}{V}[(N_{A0} - \xi_1)(N_{B0} - \xi_1 - \xi_2 - 2\xi_3 - \xi_4)]\\\frac{d\xi_2}{dt} &= \frac{k_2}{V}[(N_{B0} - \xi_1 - \xi_2 - 2\xi_3 - \xi_4)(N_{C0} + \xi_1 - \xi_2)]\\\frac{d\xi_3}{dt} &= \frac{k_3}{V}(N_{B0} - \xi_1 - \xi_2 - 2\xi_3 - \xi_4)^2\\\frac{d\xi_4}{dt} &= k_4(N_{B0} - \xi_1 - \xi_2 - 2\xi_3 - \xi_4)\end{aligned} \quad \text{(E11.3.3)}$$

Since the N_0s and ks are expected to be known in a design calculation, the set of ordinary differential Equations E11.3.3 can be solved to give the individual reaction coordinates (ξ) as functions of time. Thus the product distribution at the end of a stipulated time for a reactor of known volume can be found. Note that only four equations are required (although the number of components is seven).

PLUG-FLOW REACTOR

As explained in Chapter 4, the plug-flow reactor differs from the batch reactor only with respect to the time coordinate. While for the batch reactor time elapsed since the commencement of reaction is directly used as a measure of this coordinate, in the plug-flow reactor it is replaced by the time required to traverse a given distance in the tubular reactor: $t = \ell/u$, where ℓ is the distance and u the average velocity. Thus the rate equation now becomes

$$-r_A = -u\frac{d[A]}{d\ell} \quad (11.9)$$

Rate equations can be written for all the components of a complex reaction. For the ethylation reaction considered in Example 11.3, for instance, these will

be identical to Equations E11.1.2 except that $(d[i]/dt)$ will be replaced by $(u(d[i]/d\ell))$. This set of equations can be solved by any of the well-known numerical methods.

CONTINUOUS STIRRED TANK REACTOR

The performance equation for a continuous stirred tank reactor (CSTR) was developed in Chapter 4. We use the same equation now but with a complex rate equation replacing the simpler one of the earlier chapter. We describe the method for any complex reaction consisting of N components and M reactions. The following material balances can be written for the different constituents of the complex reaction at hand (see Figure 11.2):

$$Q_0[A]_0 + \nu_A(r_{Af})V = Q_f[A]_f$$
$$\vdots \qquad\qquad\qquad\qquad\qquad\qquad (11.10)$$
$$Q_0[M]_0 + \nu_M(r_{Mf})V = Q_f[M]_f$$

where subscript f represents the final condition. In compact matrix form this becomes

$$Q_0 \mathbf{c}_0 + \mathbf{vr}V = Q_f \mathbf{c}_f \qquad (11.11)$$

where $\mathbf{c}_0$ and $\mathbf{c}_f$ are the initial and final concentration matrices, respectively. Thus we have N simultaneous equations and N unknown concentrations plus one unknown outlet flow rate Q_f (i.e., a total of $N+1$ unknowns). If we assume the volumetric flow rate to be approximately constant (which is true for liquid systems, but only very approximately so for gaseous systems), then $Q_0 = Q_f$. Thus we would have only N unknowns, and an equation of state for the system which would relate Q_f to Q_0 would not be needed. If the reactions are assumed to be first order, further simplification results, and Equation 11.11 can be solved

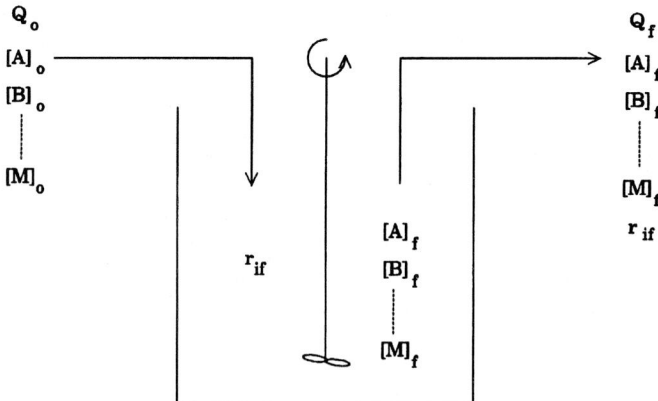

Figure 11.2 Materials flow in an MFR for a complex reaction.

for the N unknown concentrations. Even with these simplifications, the solution becomes quite difficult as the number of reactions increases.

Hence, as in the case of the batch reactor, it is convenient to reduce the number of equations to the number of independent reactions M. For this purpose, we define a reaction coordinate ξ' with units of moles/time, which is equivalent (but not equal) to the reaction coordinate ξ for the batch reactor with units of moles. Using this coordinate we get (without the assumption, $Q_f = Q_0$)

$$Q_0 c_0 + \nu \xi' = Q_f c_f \qquad (11.12)$$

where ξ' is a modified extent of reaction vector whose elements $\xi_1', \xi_2' \ldots \xi_M'$ represent the values for reactions $1, \ldots, M$.

A comparison of Equations 11.11 and 11.12 shows that

$$\nu(\mathbf{r}V - \xi') = 0 \qquad (11.13)$$

which can be recast as

$$\sum \nu_{ij}[r_{jf}V - \xi_j'] = 0 \qquad (11.14)$$

We shall now apply the concept of the basic rate of each reaction as defined in Chapter 5. Thus, for a given reaction of the system, say reaction 1, we postulate a product with a stoichiometric coefficient of $+1$ for that reaction and zero for all other reactions (i.e., $\nu_{i1} = 1, \nu_{i2} = \nu_{i3} = \cdots = 0$). The summation in Equation 11.14 then reduces to the simple equation

$$\xi_1' - r_{1f}V = 0 \qquad (11.15)$$

This equation gives the basic rate of reaction 1, which can be used to express the rates of formation or disappearance of the other components of that reaction as multiples thereof. Similar equations can be written for the other reactions of the system, leading to the following full set of equations:

$$\begin{aligned} \xi_1' - r_{1f}V &= 0 \\ \xi_2' - r_{2f}V &= 0 \\ &\vdots \\ \xi_M' - r_{Mf}V &= 0 \end{aligned} \qquad (11.16)$$

These can now be expressed in compact form as

$$\xi' = \mathbf{r}V \qquad (11.17)$$

Equation 11.17 is the basic design equation for MFR and is applicable to any number of reactions in a complex scheme. To get the rate for use in this equation, recall that in all MFR calculations the rate is based on outlet concentrations. Since these are unknown, we use the material balance Equation 11.10 to express them in terms of the inlet concentrations. Thus for component A we can write

$$Q_f[A]_f - Q_0[A]_0 = \nu_{A1}\xi_1 + \nu_{A2}\xi_2' + \nu_{A3}\xi_3' \ldots \qquad (11.18)$$

or, if $Q_f = Q_0$,

$$Q_0([A]_f - [A]_0) = \nu_{A1}\xi'_1 + \nu_{A2}\xi'_2 + \nu_{A3}\xi'_3 \ldots \tag{11.19}$$

We now illustrate the use of Equation 11.17 in the continuing example of this chapter: ethylation of aniline.

EXAMPLE 11.4

Design of a mixed flow reactor based on extent of reaction for the ethylation of aniline

Considering the four independent reactions of this system (reaction 11.1), and assuming the rates to be first order in each reactant, we get

$$\begin{aligned}\xi'_1 &= r_{1f}V = k_1[A]_f[B]_f V \\ \xi'_2 &= k_2[B]_f[C]_f V \\ \xi'_3 &= k_3[B]_f^2 V \\ \xi'_4 &= k_4[B]_f V\end{aligned} \tag{E11.4.1}$$

The next step is to express the unknown outlet concentrations in terms of the initial concentrations. For this we use the material balance equations for each of the components. For component A, for example, this is given by

$$Q([A]_f - [A]_0) = \nu_{A1}\xi'_1 + \nu_{A2}\xi'_2 \ldots = \sum_{j=1}^{4}\nu_{Aj}\xi'_j \tag{E11.4.2}$$

Thus we get the following equations for the seven components:

$$\begin{aligned}[A]_f &= -\frac{\xi'_1}{Q_0} + [A]_0 = (-\xi'_1 + [A]_0 Q_0)\frac{1}{Q_0} \\ [B]_f &= -\frac{(\xi'_1 + \xi'_2 + 2\xi'_3 + \xi'_4)}{Q_0} + [B]_0 \\ [C]_f &= \frac{(\xi'_1 - \xi'_2)}{Q_0} + [C]_0 \\ [D]_f &= \frac{(\xi'_1 + \xi'_2 + \xi'_3 + \xi'_4)}{Q_0} + [D]_0 \\ [E]_f &= \frac{\xi'_2}{Q_0} + [E]_0 \\ [F]_f &= \frac{\xi'_3}{Q_0} + [F]_0 \\ [G]_f &= \frac{\xi'_4}{Q_0} + [G]_0\end{aligned} \tag{E11.4.3}$$

The final step in the formulation of the design equations is to replace the outlet concentrations appearing in Equation E11.4.1 by the Equations in E11.4.3. This leads

to the following set of four simultaneous equations corresponding to the four independent reactions of the system:

$$\xi'_1 = k_1 \bar{t}(-\xi'_1 + Q_0[A]_0)\{Q_0[B]_0 - (\xi'_1 + \xi'_2 + 2\xi'_3 + \xi'_4)\}\frac{1}{Q_0}$$

$$\xi'_2 = k_2 \bar{t}\{Q_0[B]_0 - (\xi'_1 + \xi'_2 + 2\xi'_3 + \xi'_4)\}(\xi'_1 - \xi'_2 + Q_0[C]_0)\frac{1}{Q_0} \quad \text{(E11.4.4)}$$

$$\xi'_3 = k_3 \bar{t}\{Q_0[B]_0 - (\xi'_1 + \xi'_2 + 2\xi'_3 + \xi'_4)\}^2 \frac{1}{Q_0}$$

$$\xi'_4 = k_4 \bar{t}\{Q[B]_0 - (\xi'_1 + \xi'_2 + 2\xi'_3 + \xi'_4)\}$$

where $\bar{t} = V/Q_0$. Notice that these are algebraic equations as against the differential equations that characterize PFR. The solution can be cumbersome but is relatively straightforward.

REACTOR CHOICE FOR MAXIMIZING YIELDS AND SELECTIVITIES

In reactions where the product does not react further (i.e., parallel reactions), yields and selectivities can be easily calculated from the ratios of the rates. Where a product reacts further, no such simple analysis is possible, and resort to numerical solution is often necessary. As a general rule, however, whenever an intermediate product is the desired product, PFR is the preferred reactor.

Parallel Reactions (Nonreacting Products)

The general case

For a two-step parallel reaction represented by

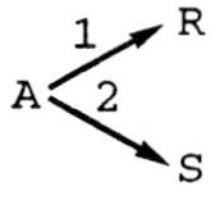

$$A \begin{array}{c} \nearrow^{1} R \\ \searrow_{2} \\ \searrow S \end{array} \qquad [11.4]$$

we can define a point yield as

$$Y_p = \frac{\dfrac{d[R]}{dt}}{-\dfrac{d[A]}{dt}} = -\frac{d[R]}{d[A]} \qquad (11.20)$$

Y_p can be calculated either from knowledge of the rates of formation and disappearance of R and A, respectively, or directly from a plot of $[R]$ versus $[A]$ (see Figure 11.3).

The questions now are: What type of reactor would be best suited for a given parallel scheme? Is the choice to be made on the basis of conversion or yield? The answers to these questions depend largely on the nature of the Y_p versus $[A]$ curve and whether A can be separated from the product and recycled at a

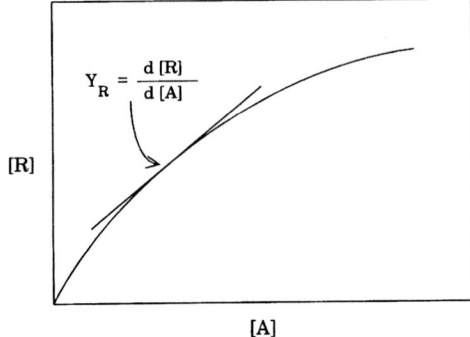

Figure 11.3 Obtaining the yield of R in the parallel scheme $A \to R$, $A \to S$ in a tubular reactor.

relatively low additional cost. This recycle is to be distinguished from the recycle of the recycle flow reactor, where part of the exit stream is recycled as such, without separation of reactant from product.

Let us first examine the question of reactor choice for maximizing [R]. Consider Figure 11.4 which shows a plot of Y_p versus [A]. The values of [R] corresponding to plug and mixed flow modes of operation are given by

$$\text{PFR:} \quad [R] = \int_{[A]_f}^{[A]_0} Y_p d[A] \qquad (11.21)$$

$$\text{MFR:} \quad [R] = Y_{Pf}([A]_0 - [A]_f) \qquad (11.22)$$

These are represented by areas M1 and (M1 + M2), respectively, in the figure. However, different types of yield curves are possible, as shown in Figure 11.5. In part *a* of the figure, Y_p increases with [A]. The value of [R] corresponding to

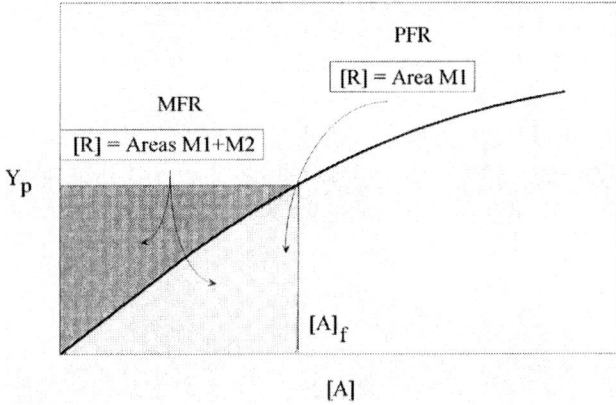

Figure 11.4 Values of final concentration of R in a *PFR* and an *MFR* for the parallel scheme $A \to R$, $A \to S$.

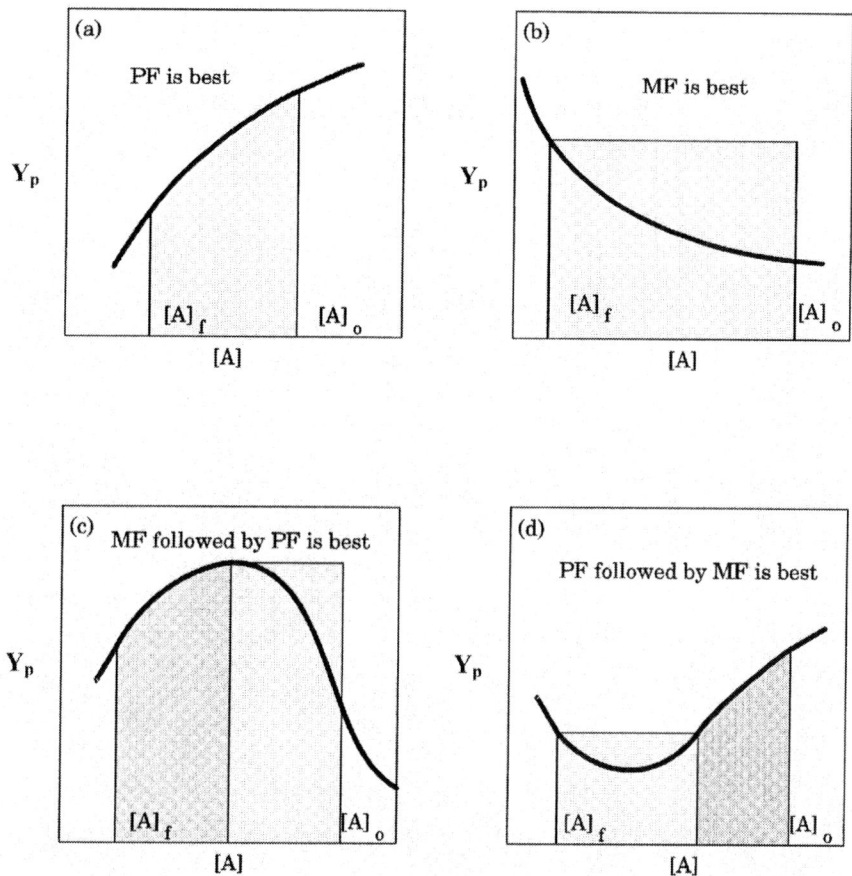

Figure 11.5 Reactor choice for different forms of Y_p–$[A]$ curves for the parallel reaction $A \rightarrow R, A \rightarrow S$.

given initial and final values of $[A]$ is seen to be higher for plug flow than for mixed flow. In other words, for a curve of this type, plug flow is the preferred mode of operation. By the same reasoning, it can be seen that mixed flow is the preferred mode for a curve of the type shown in part b of the figure. Parts c and d indicate combinations of plug- and mixed-flow reactors as the preferred modes. The design principles of these reactors have already been outlined in Chapter 4.

In cases where the reactant is recycled, the extent of conversion is unimportant, and the reactor can be operated at the $[A]$ corresponding to Y_{max}, as shown in Figure 11.6. If the yield curve does not show a maximum, no such clear-cut decision is possible.

Effect of reaction order

Choice of reactor becomes relatively easy if the orders of the two reactions of scheme 11.4 are known. If reaction 1 is first order and 2 is second order in A,

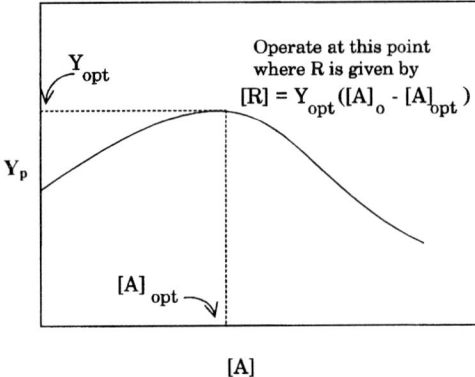

Figure 11.6 Optimum operation strategy for a parallel reaction $A \rightarrow R$, $A \rightarrow S$ where unreacted A is recycled.

then the concentration environment of the reaction will have a major effect on the selectivity of R. Recall that conversion in an MFR occurs at the greatly reduced outlet concentration, while that in a PFR occurs throughout the changing concentration environment of the reactor starting from a high initial value. Thus the selectivity for R will be higher in the MFR. By contrast, if the order of reaction 1 is 2 and of 2 is 1, then the selectivity for R will be lower in the MFR.

Where the orders of the two steps are the same, it is clear that the yield would be the same irrespective of the concentration profiles within the reactor, since any profile would have identical effects on the two reactions. The yield and selectivity for any situation would be given by

$$Y_R = \frac{[R]}{[A]_0}, \quad S_R = \frac{[R]}{[R] + [S]} \tag{11.23}$$

One of the reactants undergoes a second reaction

Let us now consider the scheme

$$A + B \rightarrow R$$
$$B \rightarrow S \tag{11.5}$$

Polymerization of reactants is a common occurrence in organic reactions. Although this is also a parallel scheme, it will be noticed that high concentrations of A combined with low concentrations of B will favor the desired product R. Thus, an SBR would be the preferred candidate since the above condition is met in this reactor. By contrast, the common *BR*, *PFR*, and *MFR* would all give lower selectivities because they all allow the second reaction to proceed without hindrance.

Complex Reactions 347

Parallel-Consecutive Reactions

Consider first the reaction

$$A + B \rightarrow R$$
$$R + B \rightarrow S \qquad [11.6]$$

The selectivity and yield are given by Equations 11.23 and X_A by

$$X_A = \left(1 - \frac{[A]}{[A]_0}\right) \qquad (11.24)$$

The selectivity of R will be high if $[A]$ and $[B]$ are high, and $[B]$ is low at high $[R]$. Since this pattern holds for BR and PFR, these are the preferred reactors for maximizing R. However, for an MRF, $[R]$ within the reactor is uniformly high, being equal always to the exit concentration. Hence the first (desired) reaction is not favored, resulting in a lower selectivity for R than in a PFR or BR.

Scheme 11.6 can be extended to include a number of intermediates, viz.

$$A + B \rightarrow R$$
$$R + B \rightarrow S$$
$$S + B \rightarrow T \qquad [11.7]$$
$$T + B \rightarrow U$$

Industrially important examples are listed in Table 11.1. Based on a detailed study of this scheme (Russell and Buzzelli, 1969), the following observations are important.

1. For a reaction in which the activation energies of the different steps are approximately equal, product distribution is a function only of two variables: mole ratio of reactants, and fraction of product recycled.
2. It is sometimes convenient to use a secondary reactor instead of a recycle to the first (*primary*) reactor. In such a case, while the input ratio (this

Table 11.1 Industrially important examples of reaction 11.7 (a series-parallel scheme)

Reactants		Products		
A	B	R	S	T
Water	Ethylene oxide	Ethylene glycol	Diethylene glycol	Triethylene glycol
Ammonia	Ethylene oxide	Monoethanolamine	Diethanolamine	Triethanolamine
Ammonia	Ethylene dichloride	Ethylenediamine	Diethylenetriamine	Triethylenetriamine
Methyl, ethyl, or butyl alcohol	Ethylene oxide	Monoglycol ether	Diglycol ether	Triglycol ether
Benzene	Chlorine	Monochlorobenzene	Dichlorobenzene	Trichlorobenzene
Methane	Chlorine	Methyl chloride	Dichloromethane	Trichloromethane (chloroform)

time to two reactors) continues to be important, the more critical variable is the allocation of B between the primary and secondary reactors.
3. The choice between recycle to a single (primary) reactor and a secondary reactor is often dictated by cost considerations.

Consider next the scheme

$$A + B \to R$$
$$2A \to S \qquad \qquad [11.8]$$

If R is the desired product, its yield can be maximized by maintaining a low concentration of A throughout the reactor. This can be done by feeding A at various points along a tubular reactor with an inlet feed of (B + some A), or by introducing B in the individual reactors of a CSTR sequence. For further details on this and other similar schemes, reference may be made to Van de Vusse and Voetter (1961) and Denbigh and Turner (1971).

Finally, consider the special consecutive-parallel scheme

$$A \to R \to S$$
$$2A \to T \qquad \qquad [11.9]$$

This poses an interesting problem in that a PFR would favor R by the first reaction while a CSTR would suppress the undesired second reaction (Van de Vusse, 1964). As this problem can be resolved by using a recycle reactor (with its partial mixing), it was considered in Chapter 10.

OPTIMUM TEMPERATURES OR TEMPERATURE PROFILES FOR MAXIMIZING YIELDS AND SELECTIVITIES

In the case of MFR the temperature is uniform within the reactor, and hence one can conceive of a single optimum temperature for maximizing the yield of a product in a complex reaction. In the case of PFR, however, there can be (and often is) a temperature profile within the reactor. The question therefore arises: Can one impose a temperature profile for maximizing the yield in a tubular reactor? We examine both cases next.

Optimum Temperatures

Consider the parallel scheme 11.4 with R as the desired product. If $E_1 > E_2$, the highest practical temperature should be used. For $E_1 < E_2$, an optimum temperature T_{opt} exists, below which the rate would be too low (requiring a huge reactor) and above which the yield of R would be too low. An approximate expression for T_{opt} can be obtained by speculating on the largest allowable reactor size, and therefore τ_{max}, for given $[A]_0$ and F_{A0}—that is, by fixing

$$\tau_{max} = \frac{[A]_0 V_{max}}{F_{A0}} = \text{fixed} \qquad (11.25)$$

This is quite practical since often the reaction is desired to be carried out in an available CSTR. Using this value of τ_{max}, T_{opt} can be obtained from

$$T_{opt} = \frac{E_2}{R_g \ln\left[k_2^0 \tau_{max}\left(\frac{E_2 - E_1}{E_1}\right)\right]} \tag{11.26}$$

The analysis can be readily extended to the scheme

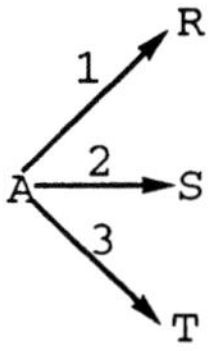

$$A \begin{array}{c} \nearrow^{1} R \\ \xrightarrow{2} S \\ \searrow_{3} T \end{array} \qquad [11.10]$$

where $E_1 < E_2 < E_3$. The relevant equations are (Levenspiel, 1993):

1. S (with intermediate E) is the desired product

$$(E_3 - E_2)k_3^0 \exp\left(\frac{E_3}{R_g T_{opt}}\right)\tau_{max} + (E_1 - E_2)$$
$$k_1^0 \exp\left(\frac{E_1}{R_g T_{opt}}\right)\tau_{max} = E_2 \tag{11.27}$$

2. R (with lowest E) is the desired product: Same as Equation 11.27 with E_1 for E_2
3. T (with highest E) is the desired product: Operate at highest practical T.

Optimum Temperature and Concentration Profiles in a PFR

This problem has been analyzed quite extensively over the years (see Doraiswamy and Sharma, 1984), but we restrict the treatment here to a brief qualitative discussion, followed by a presentation (without derivation) of equations for selected reaction schemes.

Take any two-step scheme in which product R of the first reaction is the desired product, while product S of the second reaction is the undesired product, irrespective of whether the scheme is parallel or consecutive. Let the activation energies of the two reactions be E_1 and E_2, with $E_1 < E_2$. The basic principle used in maximizing the yield of R in this situation is that the rate of reaction 1 is lower than that of reaction 2 at higher temperatures, and higher at lower temperatures. This is illustrated in Figure 11.7.

Parallel reactions

Consider now the parallel scheme $A \rightarrow R$, $A \rightarrow S$, in which $E_1 > E_2$. Clearly, the yield of R is highest at the highest temperature that can be practically used. By contrast, if $E_1 < E_2$, the temperature must be lowered to increase the yield of R. But there is a limit to which the temperature can be reduced consistent with the

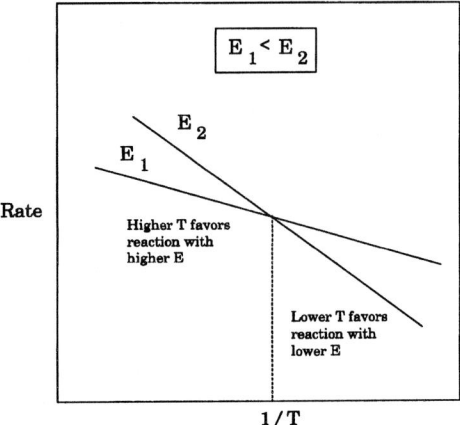

Figure 11.7 Effect of temperature on reactions with different activation energies.

need to maintain a reasonably high rate of reaction. So we employ an increasing temperature profile as shown in Figure 11.8a, in which we start at the minimum feasible temperature (T_{min}) and complete as much reaction as possible at that temperature. In view of the high initial concentration of A, the concentration effect will offset much of the negative temperature effect due to low T. But as reaction progresses and $[A]$ falls, both T and $[A]$ will have a negative effect, and hence the temperature should be progressively raised to increase the overall rate.

In this process an increasing amount of A is lost as S, and hence the basic principle of mathematical optimization is to ensure an overall maximum rate that would still maintain the highest possible overall yield of R (i.e., minimize the loss of A as S). This problem has been extensively studied (e.g., Horn and Troltenier 1961a,b; Skrzypek, 1974). Skrzypek's analysis is based on the

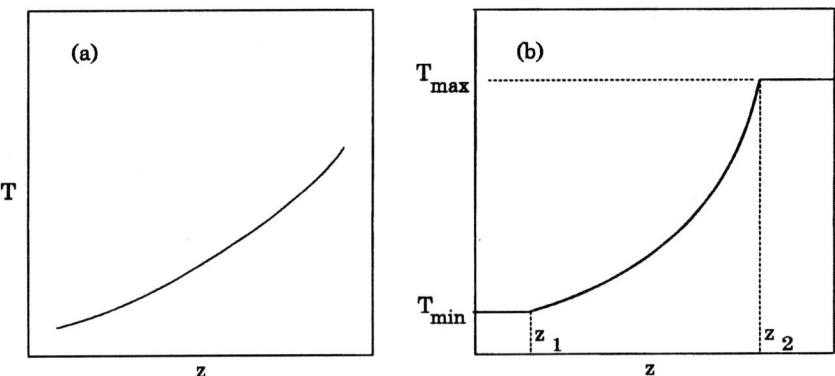

Figure 11.8 Temperature profile in a PFR for maximizing R in the reaction $A \xrightarrow{1} R$, $A \xrightarrow{2} S$ where $E_1 < E_2$.

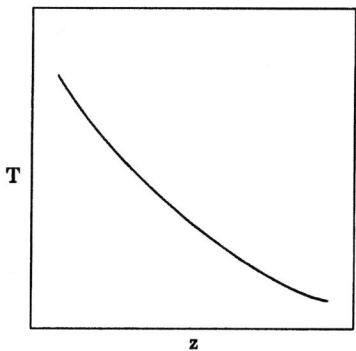 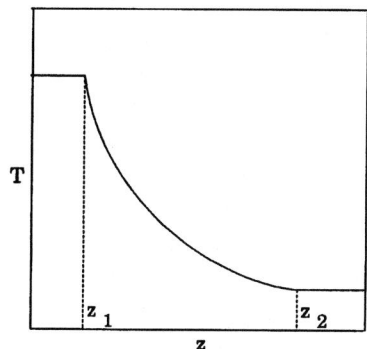

Figure 11.9 Temperature profile in a *PFR* for maximizing R in the reaction $A \to R \to S$ where $E_1 < E_2$.

Pontryagin maximum principle under the constraint $T_{min} < T < T_{max}$. The maximum principle uses a temperature selection for every axial position in a tubular reactor for which the Hamiltonian is maximal with respect to T and remains constant along the temperature profile.

Results show (Figure 11.8b) that the profile actually consists of three sections: (1) the lower horizontal segment ($z = z_1$) where the temperature remains constant at T_{min}, a situation that was anticipated in the preceding qualitative discussion, (2) a section where the temperature rises until $z = z_1$ corresponding to $T = T_{max}$, and (3) the upper horizontal section where T remains constant at T_{max}.

Consecutive reactions

For a consecutive reaction such as $A \to R \to S$ with $E_1 < E_2$, reflection will show that a decreasing temperature profile is optimal (Figure 11.9a). Since R is not present initially, we start with the highest possible temperature and complete as much reaction as possible at that temperature. Since $E_1 < E_2$, the rate of decomposition of R is higher than its rate of formation at higher temperatures, and hence the temperature should be reduced as the reaction progresses in order to maintain a high yield of R.

A rigorous analysis of the temperature profiles in a PFR for such a reaction was made by Bilous and Amundson (1956), Aris (1961, 1965), Coward and Jackson (1965), Denn and Aris (1965), and Fine and Bankhoff (1967). As illustrated in Figure 11.9b, the profile based on these analyses shows a falling temperature region sandwiched between regions of minimum and maximum temperatures.

Reversible reaction

The reversible reaction is a special case of a series reaction in which the reactant is regenerated in the second reaction. Thus a series reaction $A \to R \to S$ becomes

$$A \leftrightarrow R \qquad [11.11]$$

We develop next a simple method of computing the optimum temperature profile for a reversible reaction of the type

$$A + B \underset{k_{-2}}{\overset{k_2}{\rightleftarrows}} R + S \qquad [11.12]$$

carried out in a tubular reactor. The imposition of an optimum temperature profile demands that the rate at every axial position in the reactor be maximum for a given set of concentrations $[A]$, $[B]$, $[R]$, and $[S]$. The mathematical condition for this is

$$\frac{d(-r_A)}{dT} = 0 \qquad (11.28)$$

If the reaction rate is given by

$$-r_A = k_2^0 \exp\left(-\frac{E_1}{R_g T}\right)[A][B] - k_{-2}^0 \exp\left(-\frac{E_2}{R_g T}\right)[R][S] \qquad (11.29)$$

the result is

$$T_{\text{opt}} = \frac{\Delta E}{R_g \ln \dfrac{k_{-2}^0 E_2 [R]^2}{k_2^0 E_1 ([A]_0 - [R])([B]_0 - [R])}} \qquad (11.30)$$

The optimum temperature profile may now be found by the following procedure:

1. Assume a value of X_A, and calculate T_{opt} from Equation 11.30.
2. Calculate $(-r_A)$ at $T = T_{\text{opt}}$ from Equation 11.29.
3. Repeat for various values of X_A and obtain curve A of Figure 11.10.
4. The area under this curve gives $\bar{t}$ from

$$\bar{t} = \frac{V[A]_0}{F_{A0}} = -\int_{[A]_0}^{[A]} \frac{d[A]}{(-r_A)} = [A]_0 \int_0^{X_A} \frac{dX_A}{(-r_A)} \qquad (11.31)$$

5. Compute reactor volume from $V = \bar{t} F_{A0}/[A]_0$ under conditions of optimum temperature at every axial position.

Curve B corresponds to a constant temperature in the reactor maintained at T_0, and curve A is for an optimum temperature profile. The shaded area represents the decrease in volume due to operation with the imposed optimum temperature profile.

Parallel-consecutive reactions

For a scheme such as $A + B \to R$, $R + B \to S$, in which both parallel and consecutive reactions are involved, a profile that combines the features of the two reactions seems to be indicated. Indeed, a decreasing profile followed by an increasing one can be shown to be optimal (Burghardt and Skrzypek, 1974).

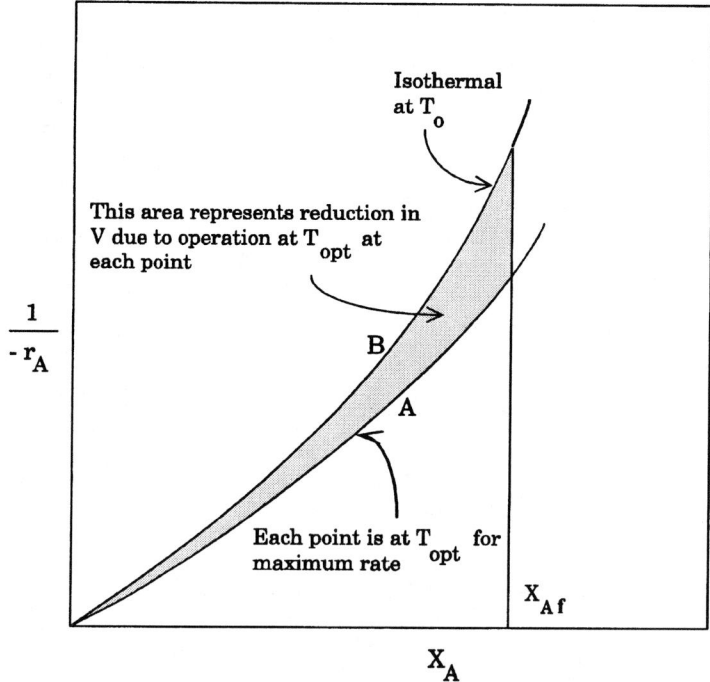

Figure 11.10 Sketch illustrating volume reduction due to operation at optimum temperature at each axial position in a *PFR*.

This is illustrated in Figures 11.11; the second plot shows that the temperature should be held constant at T_{min} for a certain distance in the reactor before the rising profile commences. As already pointed out, these reactions are of considerable importance in organic synthesis.

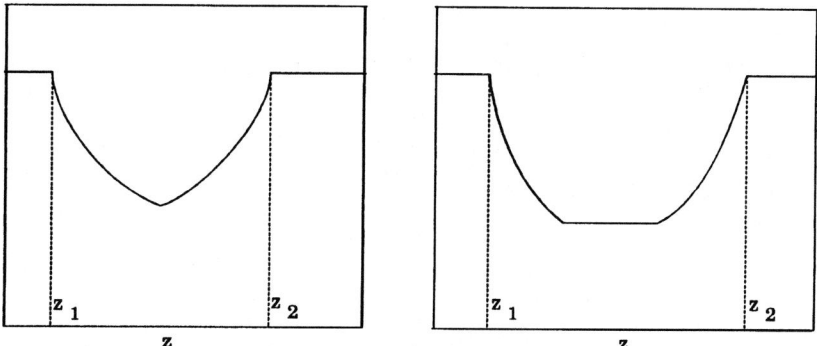

Figure 11.11 Temperature profile for maximizing R in the reaction $A + B \xrightarrow{1} R$, $R + B \xrightarrow{2} S$, where $E_1 < E_2$.

Table 11.2 Temperature–time profiles for various types of complex reactions in a BR (also applicable to PFR with length replacing time).

(1) Reaction	(2) Relative values of E	(3) Optimum profiles
1. $A \underset{2}{\overset{1}{\rightleftharpoons}} R$ (Reversible reaction)	a. $E_1 > E_2$ b. $E_1 < E_2$	$T = T_{max}$ 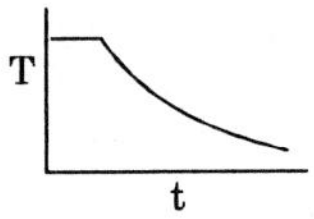 (decreasing T vs t curve)
2. $A \overset{1}{\underset{2}{\swarrow^{R}_{\searrow S}}}$ (Two parallel reactions)	a. $E_1 > E_2$ b. $E_1 > E_2$	$T = T_{max}$ 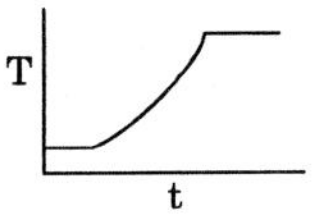 (increasing T vs t curve)
3. $A \overset{1}{\underset{3}{\overset{2}{\rightarrow}}} \begin{matrix}R\\S\\T\end{matrix}$ (Three parallel reactions)	a. $E_1 > E_2, E_3$ or $E_1 < E_2, E_3$ b. E_1 between E_2 and E_3, and $t =$ limited c. E_1 between E_2 and E_3, and $t =$ limited	Same as (2) High t at calculate T (provided high t is acceptable) (increasing T vs t curve)
4. $A \overset{1}{\rightarrow} R \overset{2}{\rightarrow} S$	a. $E_1 > E_2$ b. $E_1 > E_2$	$T = T_{max}$ at calculated t 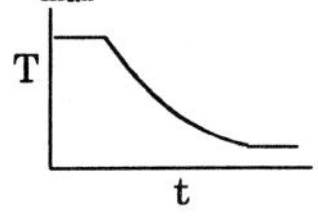 (decreasing T vs t curve)
5. $A \underset{2}{\overset{1}{\rightleftharpoons}} R \overset{3}{\rightarrow} S$	a. $E_1 < E_2, E_3$, and $t =$ unlimited b. $E_1 < E_2, E_3$, and $t =$ limited c. $E_1 > E_2, E_3$ d. $E_2 < E_1 < E_3$ e. $E_3 < E_1 < E_2$	Same as (4) but at $T = T_{min}$ (decreasing T vs t curve) Same as (4a) (increasing T vs t curve) (decreasing T vs t curve)

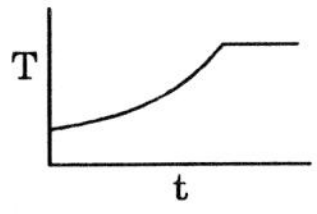

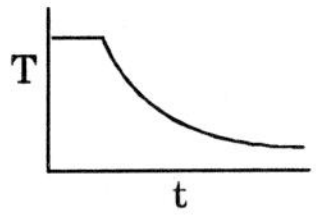

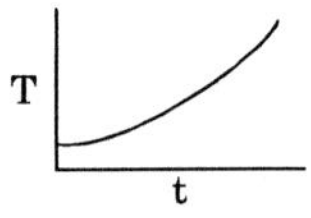

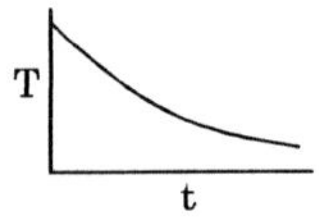

Table 11.2 (Continued)

(1) Reaction	(2) Relative values of E	(3) Optimum profiles
6. $A + B \xrightarrow{1} \text{\textcircled{R}}$ $\text{\textcircled{R}} + B \xrightarrow{2} S$ (Series-parallel reaction)	a. $E_1 > E_2$ b. $E_1 > E_2$	$T = T_{\max}$ at calculated t 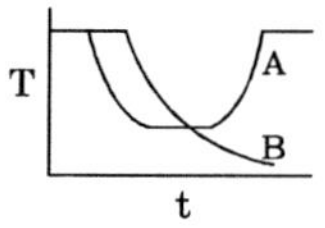
7. $A \xrightarrow{1} S \xrightarrow{2} \text{\textcircled{R}}$ $\searrow_3 \searrow_4$ $T U$ (Denbigh reaction)	a. $E_1 < E_3,\ E_2 > E_4$	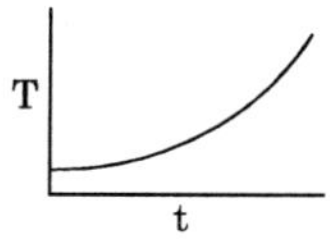
	b. $E_1 > E_3,\ E_2 < E_4$	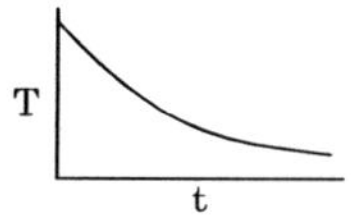
8. $A \xrightarrow{1} \text{\textcircled{R}} \begin{array}{c} \xrightarrow{2} S \\ \searrow_3 T \end{array}$ (Simplified Denbigh reaction)	a. $E_2 < E_1 < E_3$ b. $E_3 < E_1 < E_2$	$T = T_{\max}$ as in (2a) but at $t = t_{\text{calc}}$ $T = T_{\max}$ as in (2b) but at $t = t_{\text{calc}}$

Summary of profiles for common classes of organic reactions

Temperature profiles for eight important classes of reactions commonly encountered in organic synthesis and technology are presented in Table 11.2. These apply equally to plug-flow and batch reactors. For a PFR, the profile represents temperature as a function of axial distance, whereas for a BR, it represents temperature as a function of time (see Rose, 1981). The latter is considered further next.

Extension to a Batch Reactor

It is quite easy to compute the temperature profile for a given reaction by using the methods outlined here. However, from a practical point of view, it is almost impossible to impose such a temperature profile on a tubular reactor. The greatest use of a calculated profile for a PFR is in trying to approach it by experimentally varying the operating parameters and the insight it offers concerning reactor operation.

The situation is quite different, however, for a batch reactor. Here, instead of a temperature profile, we vary the temperature with time. The computational procedure is exactly the same as for a tubular reactor except that we use time t in place of space time V/Q (or ℓ/u). Because batch reactors are very common in

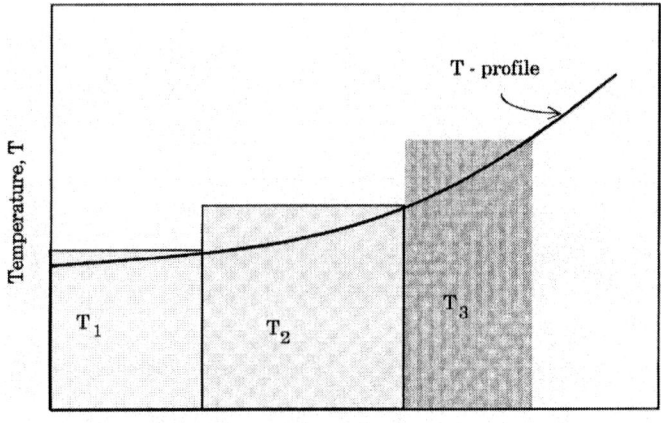

Figure 11.12 A simple way of approximately implementing the theoretical temperature profile for a given complex reaction in a batch reactor.

organic synthesis of small volume chemicals, one does not often face the practical problems of a PFR. It is quite easy to divide the time cycle into different operating time zones, each corresponding to a particular predetermined temperature (Figure 11.12). It is often also sufficient to have qualitative knowledge of the nature of the profile. Operation at three to four discrete temperatures, as indicated by the particular profile in Figure 11.2 can lead to substantial improvements in conversion. The experiments can often be done quite easily in a chemist's laboratory, and the procedure can be implemented on a larger scale without the need for a detailed engineering analysis.

Chapter 12

Reactor Design for Solid-Catalyzed Fluid-Phase Reactions

Catalytic reactions are carried out in reactors with a fixed, fluidized, or moving bed of catalyst. Although the chemical kinetics of the reaction obviously remains the same for all these reactors, the hydrodynamic features vary considerably. Because no complete description of these features is possible, it is convenient to postulate different situations and develop mathematical models to represent these situations for each type of reactor. It is also important to note that wherever solid catalysts are used, the question of catalyst deactivation cannot be ignored. Several books and reviews covering a variety of situations have been written, including those marked with an asterisk in the list of references. They are recommended for general reading.

Our intention in this chapter is limited, however: formulate approaches to the design of two main classes of catalytic reactors, fixed and fluidized bed; briefly describe selected procedures along with a few numerical (or methodological) examples to illustrate their use; and outline a procedure for incorporating the effects of catalyst deactivation in reactor design and operation.

FIXED-BED REACTOR

Approaches to Design

There are basically two types of fixed-bed reactors: (1) multitubular, in which tubes of approximately 1.5 to 4.0 cm in diameter are placed as a bundle within a shell through which a heat exchange fluid is circulated to control the temperature profile within the reactor; and (2) adiabatic, in which the catalyst is placed directly inside a reactor (with no a priori limitation to the diameter), and heat

removal is accomplished by multistaging the bed and removing the heat of reaction by heat exchange between stages.

Four major models have been proposed for describing the behavior of a packed tubular reactor (see Doraiswamy and Sharma, 1984). Of these, the most extensively used is the quasi-continuum model in which the fluid-solid system is assumed to act as a single *pseudohomogeneous* phase with effective properties of its own (as for any true single phase). Thus the procedures developed in Chapters 4 and 10 for the homogeneous model can be used to determine the axial profiles of concentration and temperature. One can also allow for radial transport gradients within each tube [two-dimensional (2-D) models], as opposed to the simpler models in which these gradients are neglected—the one-dimensional (1-D) models. It is also possible to account for deviations from plug flow by allowing for diffusion in the axial direction. Models in which radial and/or axial diffusion is incorporated are referred to as nonideal models. Additionally, it is possible in this quasi-continuum model to explicitly recognize the physical presence of the catalyst and thus allow for mass and heat transfer limitations at the surfaces of the catalyst particles and within them. These are referred to as heterogeneous models.

A broad classification of the various quasi-continuum models is presented in Table 12.1. The simplest is clearly the one-dimensional pseudohomogeneous plug-flow model (A1-a) in which the radial gradients of heat and mass within a tube are neglected. Then complicating factors can be added, one at a time, to allow for increasing reality.

Another useful model is the so-called cell model, which breaks up the reactor into a large number of cells, each cell (or microreactor) corresponding to a single pellet and its immediate neighborhood. By allowing for flow between cells located both in the axial and radial directions, 1-D and 2-D models, as well as different degrees of mixing, can be simulated. The cell model, which involves a set of algebraic equations, is probably preferred for very large volume productions where the effects of nonideality tend to have a more severe economic impact. It is, however, unlikely to match the usefulness of the quasi-continuum models in organic synthesis/technology, considering the scale and diversity of these processes.

In adiabatic reactors, a common situation is one in which a heated reactant is passed over a bed of catalyst placed inside a heat-insulated reactor or a sequence of such reactors (or stages) with interstage cooling. Then the problem is to optimize the number and size of stages for maximum profit. The design methodology is thus different from that for the tubular fixed-bed reactor.

Effective Properties

Recall that the models used in the developments just presented are based on the quasi-continuum approach. As already noted, this means that the gas-solid quasi-continuum is regarded as a single phase with properties of its own. These are not thermodynamic properties but effective properties that can be used to analyze transport phenomena within the continuum, as in any homogeneous system.

The most important of them are the effective thermal conductivity and effective diffusivity. It is important to note that each of these has a radial and an axial

Table 12.1 Classification of models for nonisothermal nonadiabatic reactors

```
                          Nonisothermal
                          nonadiabatic
                               │
                ┌──────────────┴──────────────┐
                A                             B
         Pseudohomogeneous              Heterogeneous
                │                             │
        ┌───────┴───────┐             ┌───────┴───────┐
       A1              A2            B1              B2
  One-dimensional  Two-dimensional  One-dimensional  Two-dimensional
        │               │                │                │
    ┌───┴───┐       ┌───┴───┐       ┌────┴────┐      ┌────┴────┐
   A1-a   A1-b    A2-a    A2-b    B1-a    B1-b    B2-a     B2-b
   PFR   Axial    PFR    Axial   No axial Axial  No axial  Axial
        mixing  (basic   mixing   mixing  mixing  mixing   mixing
                model)

                              │
          ┌───────────────────┼───────────────────┐
       Internal            External         Internal and
       transport           transport        external transport
       B1-a-I              B1-a-E           B1-a-IE
       B1-b-I              B1-b-E           B1-b-IE
       B2-a-I              B2-a-E           B2-a-IE
       B2-b-I              B2-b-E           B2-b-IE
```

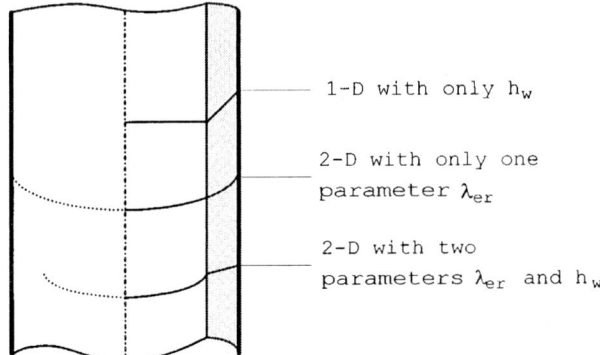

Figure 12.1 Approaches to radial heat transport in a packed bed.

component. A number of correlations have been developed to estimate them (see Kulkarni and Doraiswamy, 1980, for extensive tables; also Doraiswamy and Sharma, 1984; Froment and Bischoff, 1990).

Another equally important property is the heat transfer coefficient for bed-to control fluid heat transfer. If there is no radial heat transfer gradient within the bed (see Figure 12.1), it is assumed that the entire resistance inside the tube is localized within the film adjacent to the wall. This is the value to use in 1-D beds. On the other hand, if there is a radial gradient, a 2-D model must be used, and both heat transfer coefficient and effective radial thermal conductivity are involved, as shown in the figure. Several correlations for these have been listed by Kulkarni and Doraiswamy (1980).

A recent study (Balakotaiah and Dommeti, 1999a) shows that the effective pseudohomogeneous model and the equations based on it (as given subsequently in Table 12.2) are suspect because additional terms are involved besides the dispersion terms based on effective diffusivity and effective thermal conductivity. The effective model is acceptable only if the interphase transfer time is smaller than both the residence time and the characteristic reaction time. Thus it is a reasonable approximation for reactions not greatly influenced by external mass transfer.

Reactor Modeling

We shall first consider the ideal model A1-a of Table 12.1, which is a simple yet reasonable description of an exothermic reaction, and then extend the treatment to model A2-a. Equations for these and a few other models are listed in Table 12.2.

One-dimensional pseudohomogeneous nonisothermal nonadiabatic plug flow (model A1-a)

Consider a simple reaction

$$A \rightarrow R \qquad [12.1]$$

that occurs in a single tube of the multitubular packed-bed reactor (see Figure

4.8). Taking a differential volume dV of the reactor, the material and heat balances can be readily written as

$$u\frac{d[A]}{d\ell} + (-r_A) = 0 \tag{12.1}$$

$$u\rho C_p \frac{dT}{d\ell} + \frac{4U}{d_T}(T - T_w) - (-\Delta H_r)(-r_A) = 0 \tag{12.2}$$

This is an initial value problem with the conditions

$$\left.\begin{array}{l}[A] = [A]_0 \\ T = T_0\end{array}\right\} \quad \text{at } \ell = 0 \tag{12.3}$$

The second term in the heat balance accounts for heat removal by a circulating liquid at temperature T_w from the surface of the volume element dV, assuming an overall heat transfer coefficient U across the tube wall. Note that r_A represents the rate per unit volume of reactor. Hence

$$-r_A \text{ (or } - r_{VA}) = (1 - f_B)(-r_{vA}) \tag{12.4}$$

$$k \text{ (or } k_V) = (1 - f_B)k_v \tag{12.5}$$

where r_{vA} and k_v represent the rate and rate constant based on catalyst volume and f_B is the bed voidage. Now we recast these equations in dimensionless form as

$$\frac{da'}{dz'} + \Re_M(-r_A) = 0 \tag{12.6}$$

$$\frac{d\tau'}{dz'} + \frac{4Ud_p}{u\rho C_p d_T}(\tau' - \tau'_w) - \Re_H(-r_A) = 0 \tag{12.7}$$

where

$$\left.\begin{array}{l}a' = \dfrac{[A]}{[A]_0}, \quad z' = \dfrac{\ell}{d_p}, \quad \tau' = \dfrac{T}{T_0}, \quad \tau'_w = \dfrac{T_w}{T_0} \\ \\ \Re_M = \dfrac{d_p}{u[A]_0}, \quad \Re_H = \dfrac{(-\Delta H_r)d_p}{u\rho C_p T_0}\end{array}\right\} \tag{12.8}$$

The parameters $\Re_M$ and $\Re_H$ are the mass and heat transfer groups (with units of inverse rate), respectively, defined in terms of common physical properties. The initial conditions are

$$\left.\begin{array}{l}a' = 1 \\ \tau' = 1\end{array}\right\} \quad \text{at } z' = 0 \tag{12.9}$$

The solution to these equations is straightforward, but the implications of the solution are important and are discussed later.

REDUCTION TO ISOTHERMAL OPERATION

If the reactor is assumed to be isothermal, Equation 12.2 does not exist, and only Equation 12.1 needs to be solved. An analytical solution can be found, which for a first-order reaction $(-r_{vA} = k_v[A])$ is given by

$$\ln \frac{[A]_0}{[A]} = (1 - f_B) \frac{k_v L}{u} = \frac{k V_T}{Q_T} \tag{12.10}$$

Table 12.2 Some models for interparticle transport: the reactor field equations

Model no. from Table 12.1		Model	Equation[a]	Boundary conditions
A1-a	i.	One-dimensional isothermal plug-flow reactor	$u\dfrac{d[A]}{d\ell} + (-r_A) = 0$	$[A] = [A]_0$ at $\ell = 0$
	ii.	One-dimensional nonisothermal plug-flow reactor	$u\dfrac{d[A]}{d\ell} + (-r_A) = 0$ $u\rho C_p \dfrac{dT}{d\ell} + \dfrac{4U}{d_T}(T - T_w)$ $- (-\Delta H_r)(-r_A) = 0$	$[A] = [A]_0$ at $\ell = 0$ $T = T_0$ at $\ell = 0$
A1-b	i.	One-dimensional isothermal reactor with axial dispersion	$D_{e\ell}\dfrac{d^2[A]}{d\ell^2} - u\dfrac{d[A]}{d\ell} - (-r_A) = 0$	$-D_{e\ell}\dfrac{d[A]}{d\ell} = u([A]_0 - [A])$ at $\ell = 0$ $\dfrac{d[A]}{d\ell} = 0$ at $\ell = L$
	ii.	One-dimensional nonisothermal reactor with axial dispersion	$D_{e\ell}\dfrac{d^2[A]}{d\ell^2} - u\dfrac{d[A]}{d\ell} - (-r_A) = 0$ $\lambda_{e\ell}\dfrac{d^2T}{d\ell^2} - u\rho C_p\dfrac{dT}{d\ell} - \dfrac{4U}{d_T}$ $(T - T_w) + (-\Delta H_r)(-r_A) = 0$	$-D_{e\ell}\dfrac{d[A]}{d\ell} = u([A]_0 - [A])$ at $\ell = 0$ $\dfrac{d[A]}{d\ell} = 0$ at $\ell = L$ $-\lambda_{e\ell}\dfrac{dT}{d\ell} = u\rho C_p(T_0 - T)$ at $\ell = 0$ $\dfrac{dT}{d\ell} = 0$ at $\ell = L$
A2-a		Two-dimensional nonisothermal reactor with radial gradients and no axial dispersion (basic model)	$D_{er}\left(\dfrac{\partial^2[A]}{\partial r^2} + \dfrac{1}{r}\dfrac{\partial[A]}{\partial r}\right)$ $- u\dfrac{\partial[A]}{\partial \ell} - (-r_A) = 0$ $-\lambda_{er}\left(\dfrac{\partial^2 T}{\partial r^2} + \dfrac{1}{r}\dfrac{\partial T}{\partial r}\right)$ $- u\rho C_p\dfrac{\partial T}{\partial \ell} + (-\Delta H_r)(-r_A) = 0$	$\dfrac{\partial[A]}{\partial r} = 0$ at $\ell > 0, r = 0$ $[A] = [A]_0$ at $\ell = 0, r > 0$ $\dfrac{\partial T}{\partial r} = 0$ at $\ell > 0, r = 0$ $-\lambda_{er}\dfrac{\partial T}{\partial r} = h_w(T - T_w)$ at $\ell = 0, r = R$ $T = Y_0$ at $\ell = 0, r > 0$

Table 12.2 (Continued)

Model no. from Table 12.1	Model	Equation[a]	Boundary conditions
A2-b	Two-dimensional nonisothermal reactor with radial gradients and axial dispersion	$D_{er}\left(\dfrac{\partial^2 [A]}{\partial r^2} + \dfrac{1}{r}\dfrac{\partial [A]}{\partial r}\right)$ $+ D_{e\ell}\dfrac{\partial^2 [A]}{\partial \ell^2} - u\dfrac{\partial [A]}{\partial \ell} - (-r_A) = 0$ $\lambda_{er}\left(\dfrac{\partial^2 T}{\partial r^2} + \dfrac{1}{r}\dfrac{\partial T}{\partial r}\right) - u\rho C_p \dfrac{\partial T}{\partial \ell}$ $+ \lambda_{e\ell}\dfrac{\partial^2 T}{\partial \ell^2} + (-\Delta H_r)(-r_A) = 0$	$-D_{e\ell}\dfrac{\partial [A]}{\partial \ell} = u([A]_0 - [A])$ at $\ell = 0, r > 0$ $\dfrac{\partial [A]}{\partial \ell} = 0$ at $\ell = L, r > 0$ $\dfrac{\partial [A]}{\partial r} = 0$ at $\ell > 0, r = 0$ $-\lambda_{e\ell}\dfrac{\partial T}{\partial \ell} = u\rho C_p (T_0 - T)$ at $\ell = 0, r > 0$ $\dfrac{\partial T}{\partial \ell} = 0$ at $\ell = L, r > 0$ $-\lambda_{er}\dfrac{\partial T}{\partial r} = h_w(T - T_w)$ at $\ell = 0, r = R$ $\dfrac{\partial T}{\partial r} = 0$ at $\ell > 0, r = 0$

[a] The equations are given in dimensional form and can be easily recast in nondimensional form.

The reactor length L can be readily found for any given feed velocity u and conversion.

The basic model (model A2-a): two-dimensional pseudohomogeneous nonisothermal nonadiabatic with no axial diffusion

This is perhaps the most useful model and hence we refer to it as the basic model (see Table 12.2 for governing equations). Example 12.1 illustrates the application of the basic model to an industrially important organic reaction.

EXAMPLE 12.1

Catalytic hydrogenation of nitrobenzene to aniline: application of the basic model

Aniline is produced by the catalytic hydrogenation of nitrobenzene in a fixed-bed reactor according to the reaction

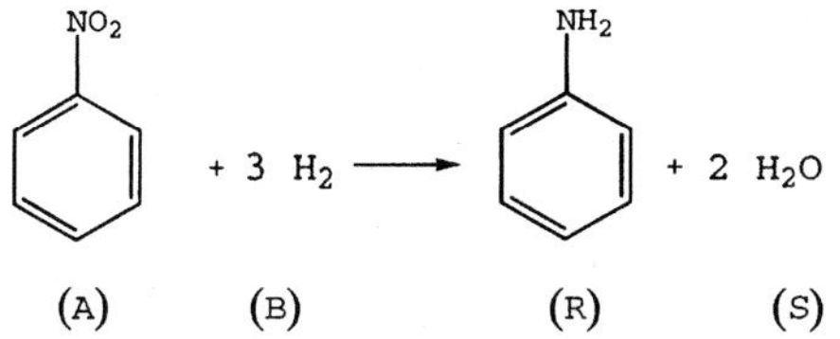

$$\underset{(A)}{C_6H_5NO_2} + 3H_2 \longrightarrow \underset{(R)}{C_6H_5NH_2} + 2H_2O \qquad [E12.1.1]$$
$\quad\;\;(B) \qquad\qquad\qquad\qquad (S)$

The rate equation is given by

$$-r_{wA} = \frac{k_w[A]}{(1 + K_B[B])^2}, \text{ mol/g cat s} \tag{E12.1.1}$$

where

$$k_w = k_w^0 \exp\left(-\frac{E}{R_g T}\right), \text{ cm}^3/\text{g catalyst s} \tag{E12.1.2}$$

$$K_B = K_B^0 \exp\left(\frac{E_{ad}}{R_g T}\right), \text{ cm}^3/\text{mol} \tag{E12.1.3}$$

Assuming a pseudohomogeneous 2-D reactor with plug flow of fluid and constant properties (model A2-a of Table 12.1), calculate the axial concentration and temperature profiles at several radial positions along the axis of a single tube. Use the following property/parameter values: $[A]_0 = 7.1 \times 10^{-4}$ mol/ℓ, $(-\Delta H_r) = 180$ kcal/mol, $\rho_s = 2.18$ g/cm^3, $f_B = 0.312$, $C_p = 0.49$ cal/g (reactant gases) °C at 200 °C, $\lambda = 1.4$ cal/m °C s, $h_w = 9.5 \times 10^{-4}$ cal/cm^2 °C s, $D_{er} = 4.74 \times 10^{-4}$ cm^2/s, $\rho = 0.0944$ g/ℓ, $u = 40$ cm/s, $k_w^0 = 9.46 \times 10^{-3}$ ℓ/g catalyst s, $E = 2631$ cal/mol, $K_B^0 = 10.7$ cm^3/mol, $E_{ad} = 8039$ cal/mol, $T_0 = 160$ °C, $T_w = 100$ °C.

When the equations are solved using these parameter values, the resulting axial profiles of conversion and temperature are as plotted in Figure 12.2. The continuous curve in each case represents the axial profile using the average value of the radial profiles at each axial position. The broken curve represents the one-dimensional model, that is, with no radial gradient (case A1-a, ii of Table 12.2). Note that there is an axial temperature maximum and that this occurs very close to the entrance. Also, there is hardly any difference between the one- and two-dimensional models. The desired conversion of 99.7% is achieved in a length of 8.3 m, whereas a length of 1.9 m is all that is needed for 90% conversion.

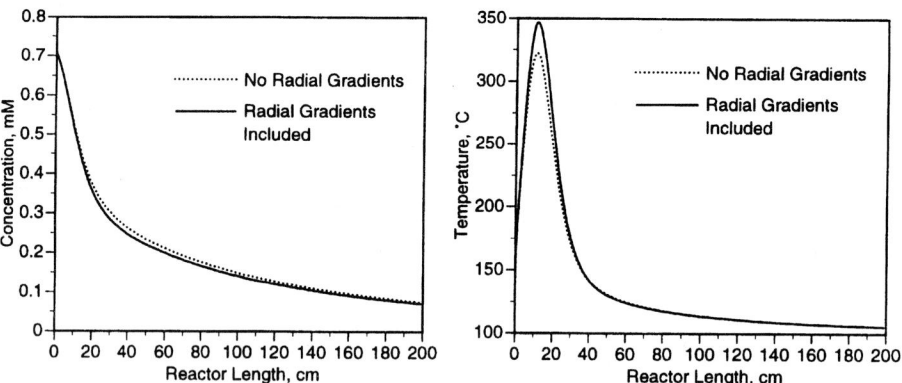

Figure 12.2 Axial and radial conversion and temperature profiles for aniline production in a nonisothermal nonadiabatic reactor (model A2-a).

Extension to Heterogeneous/Nonideal Models

The continuity and energy equations for a number of models listed in Table 12.1 are included in Table 12.2 along with the methods of solution to be used. All of

Table 12.3 Peclet numbers for mass and heat transfer

	Peclet no. for mass transfer	Peclet no. for heat transfer
Axial		
Based on reactor length, L	$Pe_{m\ell} = \dfrac{uL}{D_{e\ell}}$	$Pe_{h\ell} = \dfrac{u\rho_G C_p L}{\lambda_{e\ell}}$
Based on pellet size, d_p	$Pe'_{m\ell} = \dfrac{u d_p}{D_{e\ell}}$	$Pe'_{h\ell} = \dfrac{u\rho_G C_p d_p}{\lambda_{e\ell}}$
Value for a PFR	∞	∞
Value for an MFR	0	0
Radial		
Based on reactor radius, L	$Pe_{mr} = \dfrac{u r_T}{D_{er}}$	$P_{hr} = \dfrac{u\rho_G C_p r_T}{\lambda_{er}}$
Based on pellet size, d_p	$Pe'_{mr} = \dfrac{u d_p}{D_{er}}$	$Pe'_{hr} = \dfrac{u\rho_G C_p d_p}{\lambda_{er}}$
Value for a PFR	0	0
Value for an LFR[a]	Finite	Finite

[a] This represents the laminar-flow reactor which is characterized by a parabolic radial velocity profile and molecular radial diffusion. Hence the radial diffusivity is not exactly zero, and the corresponding Peclet number is finite (see Chapter 13).

the models listed in the table can also be expressed in dimensionless form. For this purpose the effective diffusivities and thermal conductivities in the axial and radial directions are expressed as corresponding Peclet numbers. The definitions of these Peclet numbers along with their values for the extremes of plug flow and full mixing are given in Table 12.3. Despite a recent comparison of the more important diffusion models listed in Table 12.2 (Salmi and Wärmä, 1991), the choice is largely system-specific and depends also on the ultimate reactor size.

Parametric Sensitivity and Runaway

Consider the simple 1-D model A1-a. Solution of the set of Equations 12.1–12.3 that describes this model for an exothermic reaction leads to temperature profiles in the reactor, with a clear maximum T_m in each case, as shown in Figure 12.3. Note that beyond a partial pressure of 0.05 (an illustrative value) referred to as the critical partial pressure p_{cr}, even a slight increase in pressure, say to 0.06, results in an abrupt increase in T_m. The temperature corresponding to p_{cr} is referred to as the critical temperature T_{cr}. At a partial pressure of 0.08, the rise becomes irreversible. The shaded region in which reactor behavior is highly sensitive to operating variables (usually temperature as in this case), is referred to as the region of parametric sensitivity. This is the region where temperature hot spots develop inside the reactor, and as the temperature rises even further ($p = 0.08$ curve), the reaction gets out of control irretrievably. This situation is referred to as temperature runaway.

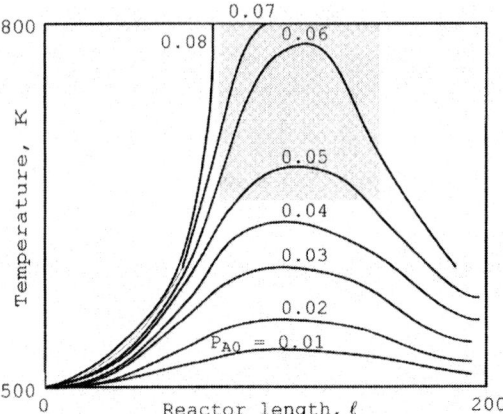

Figure 12.3 Temperature profiles for an exothermic reaction for different initial rectant partial pressures.

Introspection will show that this sensitivity is influenced by heat and mass transfer between pellet and fluid. Therefore the treatment of parametric sensitivity and runaway will be in two parts: runaway in pseudohomogeneous reactors and runaway in heterogeneous reactors.

Runaway in pseudohomogeneous reactors: Van Welsenaere–Froment (VWF) criteria

Clearly, it is important to formulate conditions to avoid these hot spots and any consequent runaway. Studies by Barkelew (1959) followed by those of several other investigators, for example Dente and Collina (1964) and Hlavacek et al. (1969), resulted in a plot of two dimensionless groups to produce a band that demarcates the safe and unsafe regions of operation.

Although Figure 12.3 and the plot just referred to are generally useful in a priori prediction of the region of safe operation, they do not predict the inlet parameter values corresponding to the critical values within the reactor beyond which runaway will occur. Because these are the controllable variables, a method for predicting them would be practically more useful. Van Welsenaere and Froment (1970) developed expressions for predicting the upper and lower bounds of these values. They are given in a slightly modified form below.

LOWER BOUND

$$p_0^\ell = p_{cr} + \frac{A}{B}(T_{cr} - T_0) \tag{12.11}$$

UPPER BOUND

$$p_0^u = \frac{A}{B}(T_{cr} - T_w)\left\{\left[\frac{A}{C}k_{wp}^0 \exp(E/R_g T_{cr})\right]^{-1/2} + 1\right\}^2 \tag{12.12}$$

MEAN

$$(p_0)_{mean} = \frac{1}{2}\left(p_0^\ell + p_0^u\right) \tag{12.13}$$

where

$$A = \frac{MP\rho_B}{\rho}, \quad B = \frac{(-\Delta H_r)\rho_B}{\rho C_p}, \quad C = \frac{4U}{\rho C_p d_T} \tag{12.14}$$

$$T_{cr} = \frac{1}{2}\{b - [b(b - 4T_w)]^{1/2}\}, \quad b = \frac{E}{R_g} \tag{12.15}$$

$$p_{cr} = \frac{(4U/\rho C_p d_T)(T_{cr} - T_w)}{[(-\Delta H_r)\rho_B/\rho C_p]k_{wP}^0 \exp\left(-E/R_g T_{cr}\right)} \tag{12.16}$$

If a second reactant B is present, its partial pressure p_{B0} will appear in the numerators of A and B.

Criteria have also been presented by many other investigators, for example, Thomas (1961), Dente and Collina (1964), Hlavacék et al. (1969), Hlavacék (1970), Oroskar and Stern (1979), and a detailed review has been written by Morbidelli et al. (1987). In general, all other parameters being fixed, the chances of reactor runaway increase as the reaction order increases, the activation energy increases, or the inlet temperature of the reactants increases.

Runaway in heterogeneous reactors: Rajadhyaksha–Vasudeva– Doraiswamy (RVD) criteria

Now we analyze runaway in reactors in the presence of internal concentration gradients within a pellet and external gradients (concentration and temperature) in the film surrounding a pellet. Although the lower and upper bounds of inlet pressure are the same as for the pseudohomogeneous case, now T_{cr} and P_{cr} are found by setting $dp_m/dT_m = 0$ using the following equations for T_m and p_m (for a first-order irreversible reaction in an isothermal pellet—not a reactor) (Rajadhyaksha et al., 1975):

$$T_m = T_p - \frac{\dfrac{(-\Delta H_r)k_{wp}^0}{h_{fp}a_p}\exp\left(-\dfrac{E}{R_g T_p}\right)p_m \varepsilon}{1 + \dfrac{\varepsilon k_{wp}^0 \exp\left(-E/R_g T_p\right)}{k'_{GP}a_p}} \tag{12.17}$$

$$p_m = \frac{(4U/\rho C_p d_T)(T_p - T_w)\left[1 + \dfrac{\varepsilon k_{wp}^0 \exp\left(-E/R_g T_p\right)}{k'_{GP}a_p}\right]}{\varepsilon k_{wp}^0 \exp\left(-E/R_g T_p\right)\left[\dfrac{(-\Delta H_r)\rho_B}{\rho C_p} + \dfrac{4U}{\rho C_p d_T}\left(\dfrac{(-\Delta H_r)}{h_{fp}a_p}\right)\right]} \tag{12.18}$$

where $k'_{GP}a_p$ and k_{wp}^0 have the units of mol/kg cat s atm. These equations are for the extreme case where almost all resistances are operative and become simplified as some of them disappear. Thus it is convenient to divide the parameter space into four regimes. Descriptions of these regimes and the conditions under which

they exist are summarized in Table 12.4. Using these equations, Rajadhyaksha et al. (1975) determined the operating regimes of a number of industrially important reactions.

GLOBAL AND LOCAL SENSITIVITIES

Table 12.4 shows that the terms global sensitivity and local sensitivity have been used. Global sensitivity refers to the behavior of the fluid stream as it passes through the reactor. On the other hand, local sensitivity refers to temperature changes within a pellet due to changes in the fluid temperature in the immediate vicinity of that pellet. These changes are not reflected in the observed temperature profile of the fluid stream but lead to the same deleterious effects such as catalyst deactivation. This local sensitivity can be identified as the sensitivity described in Chapter 7.

EXAMPLE 12.2

Comparison of sensitivity limits in the absence and presence of transport limitations

Using the methods described before, detailed calculations were carried out by Doraiswamy and Sharma (1984) for a hypothetical reaction. It was found that the inlet reactant partial pressure computed by neglecting transport limitations (0.066 atm) was more than 25% higher than the correct value of 0.0523 atm (in the presence of transport limitations) for the onset of sensitivity. This emphasizes the danger of neglecting transport limitations in calculating the maximum inlet partial pressure.

Adiabatic Reactor

The Approach

Because there is no heat removal or addition in an adiabatic reactor, radial transport of heat is absent and that of mass can usually be neglected. Hence, in modeling an adiabatic reactor, a simple one-dimensional model is adequate. What is important, however, is the height of the bed for a given diameter (as determined by the production rate), that is, the allowable conversion within the bed before the product mixture is taken out as the final product or cooled to a predetermined temperature and introduced into a second bed. It may happen that the height of each bed is so small that a number of beds (stages) with interstage cooling are necessary to achieve a desired final conversion. What this means is that two decisions must be taken for each bed: the inlet temperature and the outlet conversion to be achieved in it. In other words, if there are N beds, $2N$ decision variables must be simultaneously varied to arrive at an optimum policy for the reactor that would maximize profit. This leads to an enormous amount of computation. A particularly attractive technique for reducing the amount of computation is dynamic programming introduced by Bellman (1957) (see also Aris, 1961, 1964; Bellman and Dryfus, 1962; Roberts, 1964) in

Table 12.4 Identification of sensitivity regimes (from Rajadhyaksha et al., 1975)[a]

	Regime 1	Regime 2	Regime 3	Regime 4
	Very slow reaction No transport limitations	Slow reaction Significant intraparticle resistance	Fast reaction All resistances	Very fast reaction Interphase mass transfer control
Conditions [Note: the rate is denoted simply by r_w not $(-r_w)$]	$r_w(p_{A0}, T_u) < 0.005 k_a a' p_{A0}$ $\phi^2(T_u) < 1$ $\dfrac{r_w(p_{A0}, T_u)}{r_w(p_{A0}, T_{fu})} - 1 < 0.05$ where $T_{fu} = T_u - \dfrac{r_w(p_{A0}, T_u)(-\Delta H_r)}{h_{fp} a'}$	$\varepsilon r_w(p_{A0}, T_u) < 0.05 k_a a' p_{A0}$ $\phi^2(T_u) > 1$ and/or $\dfrac{r_w(p_{A0}, T_u)}{r_w(p_{A0}, T_{fu})} - 1 > 0.05$	$\varepsilon r_w(p_{A0}, T_u) < 0.05 k_a a' p_{A0}$ $r_w(p_{A0}, T_0) < 20 k_a a' p_{A0}$	$\varepsilon r_w(p_{A0}, T_0) > 20 k_a a' p_{A0}$
Equations for P_0^l and P_0^u	Equations 12.13 and 12.14	P_0^l obtained as noted at (b) and P_0^u from equation at (c) in the footnote	P_0^l obtained as noted at (b) and P_0^u by a method given by Rajadhaksha et al.	
Nature of the maxima curve	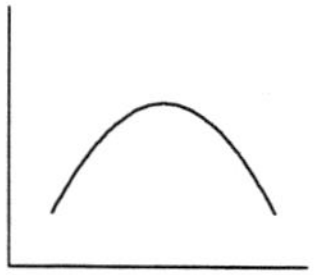	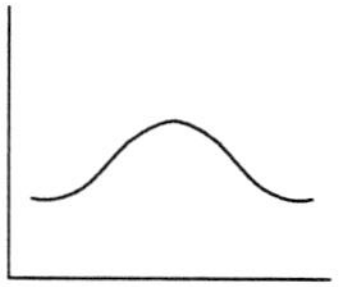	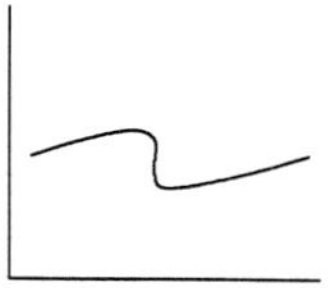	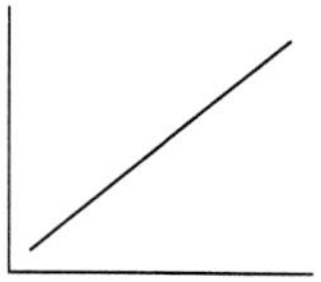
Type of sensitivity	Global	Global	Global and local	No sensitivity

[a] Units: k_a (actual value of k'_{GP}) is in mol/cm² s atm; a' in cm²/g; r in mol/g catalyst s; T_u = upper limit of inlet temperature range
[b] Estimate P_0^l by assuming adiabatic operation between critical point and inlet
[c] $P_0^u = \left(\dfrac{A}{B}\right)(T_{cr} - T_0)\left\{\left[\left(\dfrac{C}{A}\right)k_p^a \varepsilon \exp\left(\dfrac{-E}{R_g}\right)\left(1 + \dfrac{4U}{\rho_B d' h_{fp} a'}\right)^{-1/2}\right] + 1\right\}$, h_{fp} = fluid-particle heat transfer coefficient

which optimization by simultaneous consideration of $2N$ variables is reduced to one of N sequential optimizations of two variables at a time.

The principle of dynamic programming can be exploited in adiabatic reactor design in two ways: (1) by using the rigorous graphical method of Lee and Aris (1963) in which much of the required computation is done on a computer or (2) by using a simple trial-and-error graphical method suggested by Levenspiel (1993). The basis for both methods is the conversion–temperature plot for the adiabatic reactor, and hence we begin by a consideration of this plot.

Trajectories of adiabatic and wall-cooled reactors

Consider the reaction

$$A \underset{k_-}{\overset{k_+}{\rightleftharpoons}} R \qquad [12.2]$$

for which the rate equation can be written as

$$-r_A = [A]_0 \left(\frac{T_0 P}{T P_0}\right) \left[k_+ \left(\frac{1 - X_A}{1 + \varepsilon_A X_A}\right) - k_- \left(\frac{X_A}{1 + \varepsilon_A X_A}\right) \right] \qquad (12.19)$$

If the reaction is exothermic, the equilibrium curve relating the conversion to temperature will have the shape shown in Figure 12.4a, and is designated T_e. The figure also shows a number of constant rate curves, that is, rate contours. The locus of the maxima observed in these contours is designated as the T_m

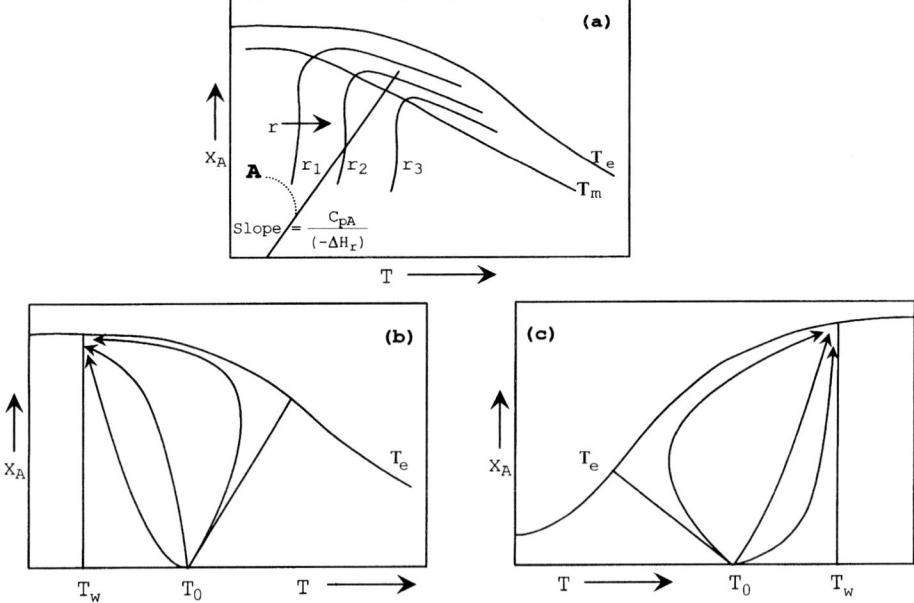

Figure 12.4 Trajectories of adiabatic, wall-cooled (for exothermic), and wall-heated (for endothermic) reactions.

curve. Reverting now to reaction 12.2, the following energy balance can be readily written for an adiabatic reactor:

$$F_{A0}(X_A - X_{A0})(-\Delta H_r) = F_{t0}C_{pm}(T - T_0) \qquad (12.20)$$

giving (because $X_{A0} = 0$)

$$X_A = \left(\frac{F_{t0}C_{pm}}{F_{A0}(-\Delta H_r)}\right)(T - T_0) \qquad (12.21)$$

or (because $\Delta[A] = X_A[A]_0$)

$$\Delta[A] = J\Delta T \qquad (12.22)$$

where

$$J = \frac{C_{pA}[A]_0}{(-\Delta H_r)} \qquad (12.23)$$

and C_{pA} is the total heat capacity per mole of A and is equal to $C_{pm}(F_{t0}/F_{A0})$ and C_{pm} is the mean molar heat capacity of the fluid.

We note from Equation 12.21 that a trajectory of slope $C_{pA}/(-\Delta H_r)$ or $C_{pm}F_{t0}/F_{A0}(-\Delta H_r)$ on an X_A–T plot, as shown by line A in Figure 12.4a, represents the path of an adiabatic reaction. The T_e and T_m curves along with the adiabatic line A and the rate contours shown in the X_A–T plane are central to the design of an adiabatic reactor.

If a constant wall temperature is maintained by cooling, then the reactor is no longer adiabatic and the trajectories will lie between the limits shown in Figure 12.4b where T_w represents the wall temperature. The behavior of an endothermic reaction (where heat has to be supplied) is sketched in Figure 12.4c.

EXAMPLE 12.3

Design of a single-bed adiabatic reactor for the hydrogenation of nitrobenzene to aniline

A fixed-bed adiabatic reactor is to be designed to produce 6000 tonnes per annum (300 working days) of aniline (B). Vertical tubes, 0.8, 1.0, or 1.5 m in diameter, packed with catalyst pellets, are proposed. The desired conversion at the reactor outlet is 99.7%, and there is no heat exchange between the reactor and the surroundings. The rate equation for the disappearance of nitrobenzene (A) is the same as Equation E12.1.1 but given in different unity by the following equation:

$$-r_{wA} = \frac{k_{wP}p_A}{(1 + K_B p_B)^2}, \text{ mol/g cat hr}$$

$$k_{wP} = 8.77 \exp\left(-2631/R_g T\right), \text{ mol/hr g atm}$$

$$K_B = 2.77 \times 10^{-3} \exp\left(8040/R_g T\right), \text{ atm}^{-1}$$

The following data are also given: temperature = 200 °C, pressure = 1 atm, bulk density of catalyst = 2.18 g/cm³, heat of reaction = 180 kcal/mol, ambient temperature = 25 °C, average heat capacity of gases = 7.5 cal/mol °C, bed porosity = 0.312, H₂ : nitrobenzene = 60 : 1 and 80 : 1.

Calculate the reactor length, with conversion and temperature profiles, for each of the tube sizes for both ratios of H_2 to nitrobenzene.

SOLUTION

Production rate of aniline (reaction E12.1.1):

$$\frac{6000 \times 10^3}{300 \times 24 \times 93} = 8961 \text{ mol/hr}$$

From the stoichiometry of the reaction, the feed rate of nitrobenzene (A) is

$$F_{A0} = \frac{8961}{0.997} = 8987 \text{ mol/hr}$$

For the ratio of 60 : 1, for instance, the rate equation can also be recast as

$$-r_{wA} = \frac{8.77 \exp(-2631/2T)\left(\frac{1-X_A}{61-X_A}\right)}{\left[1 + 2.77 \times 10^{-3} \exp(8040/2T)\left(\frac{X_A}{61-X_A}\right)\right]^2} \tag{E12.3.1}$$

A material balance on A gives

$$F_{A0}dX_A = (-r_{wA})dW \tag{E12.3.2}$$

or

$$F_{A0}dX_A = (-r_{wA})\rho_B(1-f_B)A_c d\ell \tag{E12.3.3}$$

and an energy balance gives (as already shown in Equation 12.20)

$$F_{A0}(-\Delta H_r)dX_A = F_{t0}C_{pm}dT \tag{E12.3.4}$$

where F_{t0} is the total flow rate of the inlet gases. In view of the large excess of hydrogen, F_{t0} can be assumed constant.

From Equations E12.3.3 and E12.3.4, we can write

$$dX_A = \left[\frac{\rho_B(1-f_B)A_c(-r_{wA})}{F_{A0}}\right]d\ell \tag{E12.3.5}$$

$$dT = \left[\frac{(-\Delta H_r)\rho_B(1-f_B)A_c(-r_{wA})}{F_{t0}C_{pm}}\right]d\ell \tag{E12.3.6}$$

or

$$dX_A = \bar{\alpha}\, d\ell \tag{E12.3.7}$$

$$dT = \bar{\beta}\, d\ell \tag{E12.3.8}$$

Equations E12.3.7 and E12.3.8 are coupled ordinary differential equations in conversion and temperature. They can be solved for the initial conditions

$$\ell = 0, \quad X_A = 0, \quad T_0 = 473 \text{ K}$$

using the Runge–Kutta fourth-order method for the different tube diameters and ratios given (note that Equation E12.3.1 will be slightly different for the ratio 80 : 1). The results are presented in Figure 12.5.

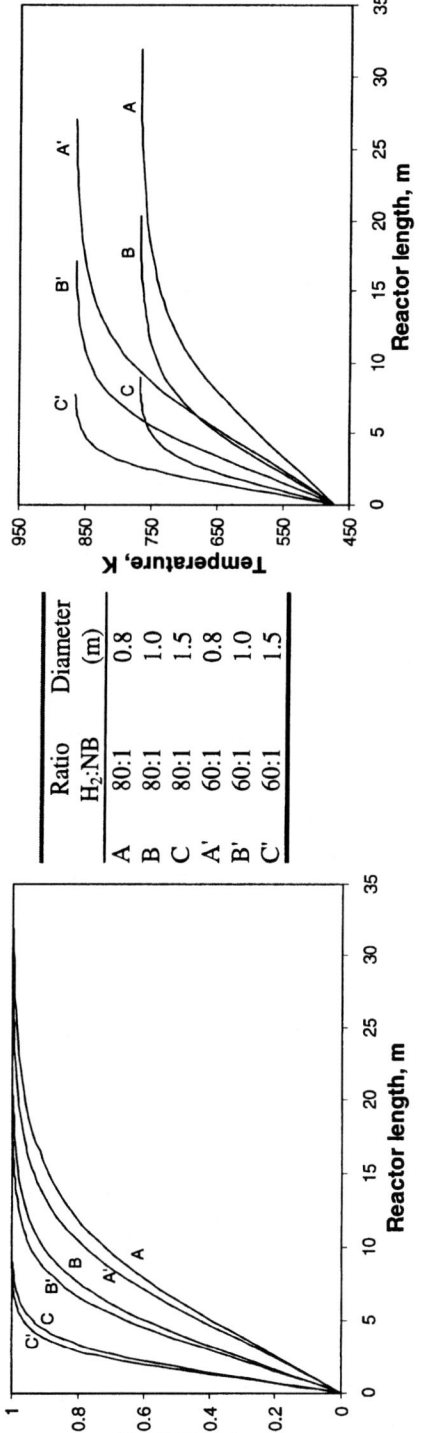

Figure 12.5 Conversion and temperature profiles for an adiabatic fixed-bed reactor for aniline production.

Design by dynamic programming

Consider again the reaction of Example 12.1, hydrogenation of nitrobenzene. Noting the exothermicity of the reaction, it would be desirable to design a multibed reactor. We start with the last bed, designated as stage 1, for which the inlet temperature and conversion are not yet known. The computation begins by optimizing this stage for a whole set of possible inlet conditions. The reasoning is that irrespective of the operating conditions of the preceding stage, the sequence as a whole cannot be optimal unless the last stage is optimal. Then we add one more stage, designated as stage 2, and optimize the two-stage system as a whole. In this new optimal policy, stage 2 will not necessarily operate at the conditions optimal to its performance alone. However, for the combined policy to be optimal, stage 1 has to be optimal for the feed coming from stage 2. It is only necessary to find the inlet conditions for stage 2, which is really the optimal policy for stages 1 and 2 together, because stage 1 has already been optimized.

The same principle can be extended to three stages from the end. Here, we consider stages 1 and 2 as a single pseudostage for which the inlet conditions have already been optimized, and determine the inlet conditions for stage 3 added to this pseudostage. Thus, if there are N stages, the last optimization step would involve optimization of stage N added to an already optimized pseudo single stage consisting of $(N-1)$ stages. This is Bellman's optimization policy and may be formally expressed as

$$\begin{bmatrix} \text{Maximum} \\ \text{profit} \\ \text{from } N \\ \text{stages} \end{bmatrix} = \text{maximum of} \left[\begin{pmatrix} \text{profit} \\ \text{from} \\ \text{stage } N \end{pmatrix} + \begin{pmatrix} \text{maximum} \\ \text{profit from} \\ (N-1) \text{ stages} \\ \text{with feed} \\ \text{from } N\text{th stage} \end{pmatrix} \right] \quad (12.24)$$

Based on this policy, a computer-aided graphical procedure was developed by Lee and Aris (1963).

Strategies for heat exchange

In the development just presented, no special attention was given to heat exchange. It is important from the economic point of view that exchange of heat is accomplished with maximum utilization of the heat generated in the reaction. Thus, consider a multistage plug-flow reactor operating with a cold feed. We can arrange the flows in several ways for maximum heat conservation. One possible scenario is shown in Figure 12.6 with the corresponding X_A-T plane representation. Any number of configurations can be visualized and their feasibility tested. One of those, autothermal operation, was mentioned in Chapter 4 and will be considered in Chapter 13.

A simple graphical procedure

The principle of dynamic programming already outlined can be used, but without much of its rigor, to establish a simple trial-and-error procedure. The various steps in this procedure are outlined in Figure 12.7 for cold shot injection of feed

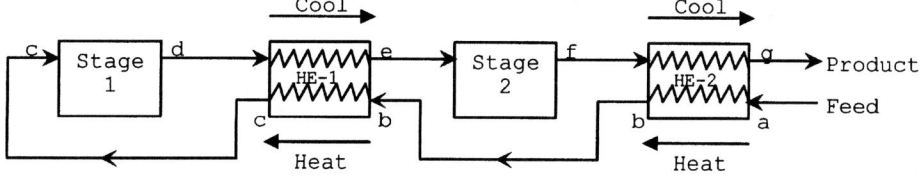

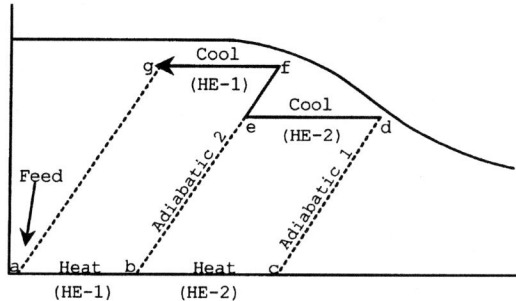

Figure 12.6 Heat exchange between hot product and cold feed by countercurrent flow of the fluids in suitably located heat exchangers.

(Levenspiel, 1993). Other reactors and situations such as recycle-flow reactors, mixed-flow reactors, cooling by inert injection, combination cooling, etc., are also possible.

Choice between Nonisothermal Nonadiabatic Multitubular Reactors and Adiabatic Reactors

An important consideration in reactor choice is the thermal mode of operation: nonadiabatic or adiabatic. Two parameters are useful in making a preliminary choice: adiabatic temperature change at complete conversion (ATC), and temperature sensitivity (TS). These are defined as

$$ATC = \frac{(-\Delta H_r) y_{A0}}{C_{pm}}$$

$$TS = \left[\frac{d(-r_A)}{dT}\right]_{[A]} \frac{1}{(-r_A)} = \frac{E}{R_g T^2}$$

(12.25)

ATC is positive for exothermic reactions and negative for endothermic. The more important is the value of TS. Very high values of the two require cooling during reaction, and hence adiabatic operation is precluded. The following are values for a few representative reactions:

	ATC	TS
Toluene dealkylation	162	5.2 (exo)
Phthalic anhydride from o-xylene	378	12.8 (exo)
Ethane cracking	−979	−29.4 (endo)

376 Reactor Design

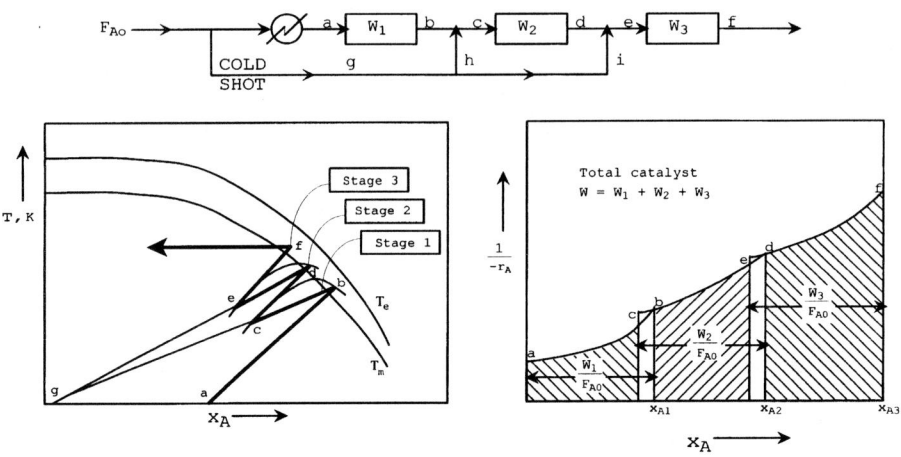

Steps
1. Guess a, b, d
2. Draw adiabatic ab of slope $C_{PA}/(-\Delta H_r)$
3. Locate c at intersection of rate contour from b and line gb where g is the (cold) feed condition
4. Draw cd and locate e at intersection of rate contour from d and gd
5. Repeat for stage 3, and so on
6. Repeat steps 1–5 by guessing new a, b, d, and continue until W/F_{Ao} is minimized

Note
1. b, c do not coincide, neither do e, d, because the conversion based on added cold shot goes down
2. All exchangers but one are eliminated

Flow rates of streams
Representing the flow rate at any position j by $F_{Ao,j}$, we have

$F_{Ao,b} : F_{Ao,c} : F_{Ao,h} = gc : gb : bc$
$F_{Ao,d} : F_{Ao,e} : F_{Ao,i} = ge : gb : de$

Figure 12.7 Graphical design of a *PFR* with cold shot cooling

Many reactions such as hydrogenation of nitrobenzene to aniline, dehydrogenation of ethylbenzene to styrene, dehydrogenation of *n*-butenes to butadiene, amination (by ammonia) of methanol to methylamines, and dehydration of ethanol to ethylene are (or can be) carried out in adiabatic reactors with diameter-to-length (aspect) ratios close to unity and reactor-to-particle diameter ratios of 50 or more. Although such reactions have the potential to generate localized hot spots (Balakotaiah et al., 1999b), transverse temperature gradients are generally ignored in the design (as in the methods described earlier). We believe this is a reasonable approximation for the values of heats of reaction and sizes of production normally encountered in industrial organic synthesis.

Variations in Fixed-Bed Reactor Designs

Most new concepts in fixed-bed reactor design are for very large scale production, and hence we only mention them briefly here. The radial reactor, a relatively well-known design, allows radial flow of gas between the tube and shell sides of the conventional multitubular reactor (through perforations in the tube walls). On the other hand, in the newer spherical fixed-bed reactor (see Hartig and Keil, 1993), flow across the catalyst is accomplished by placing it between two perforated spherical shells inside a solid spherical enclosure. The reactant is introduced between the outer shell and the enclosure and flows through the catalyst between the shells into the inner shell from where it exits the system.

Although the design of fixed-bed reactors is now a well-understood feature of catalytic reactor design in general, it continues to present important challenges that require active collaboration between chemist and chemical engineer. An overall practical methodology for developing a suitable design was suggested by Cropley (1990).

FLUIDIZED-BED REACTOR

In this section we outline the major basic aspects of fluidization and follow it up with a brief description of a few selected models of fluid-bed reactors. Several books and reviews are listed in the References.

Fluidization

Minimum fluidization velocity

The velocity at which the constituent particles of a bed begin to behave as independent entities (not as a single bed) is designated as the minimum fluidization velocity u_{mf}. The pressure drop in the bed remains practically constant thereafter. This velocity can be easily determined in a laboratory experiment. Many correlations are also available for estimating it (see Couderc, 1985, for a review; also Yang et al., 1985; Kunii and Levenspiel, 1991), and the following are recommended:

COARSE PARTICLES (Chitester et al., 1984)

$$Re'_{mf} = [(28.7)^2 + 0.0494 Ar]^{1/2} - 28.7 \tag{12.26}$$

FINE PARTICLES (Wen and Yu, 1966)

$$Re'_{mf} = [(33.7)^2 + 0.0408 Ar]^{1/2} - 33.7 \tag{12.27}$$

where

$$Re'_{mf} = \frac{d_p u_{mf} \rho_G}{\mu}, \; Ar \; (\text{Archimedes no.}) = \frac{d_p^3 \rho_G (\rho_S - \rho_G) g}{\mu^2} \tag{12.28}$$

Two-phase theory of fluidization

The theory of two-phase fluidization postulates that all gas in excess of that required for minimum fluidization passes through the bed as bubbles. Hence the bed consists essentially of two phases, one a mixture of solids and gas at minimum fluidization, called the emulsion phase, and the other comprising the "excess" gas in the form of bubbles, called the bubble phase. Although experimental data suggest that bubbling generally commences at a velocity higher than u_{mf}, the essence of the two-phase theory continues to be the centerpiece of fluid-bed modeling.

Classification of fluidized beds

The fluidization characteristics of powders are largely determined by the density and mean size of the particles. Three broad categories of particles can be identified, as shown in Figure 12.8 (Geldart, 1973). These are clearly marked on the figure.

Velocity limits of a bubbling bed

As already noted, the velocity u_{mb} for the onset of bubbling does not always coincide with that for the onset of fluidization u_{mf}. Depending on the nature of the solid, bubbling can occur at or beyond u_{mf}. The upper limit of velocity for bubbling is the velocity for the onset of slugging. Slugging defines a condition where the bubble size becomes nearly equal to the tube diameter, or (as with group C particles) portions of the bed are bodily lifted resulting in alternate zones of packed and void regions. We are usually concerned with the former, and the velocity for the onset of this condition is given by

$$u_{ms} = u_{mf} + 0.07(gd_T)^{1/2}, \text{ cm/s} \tag{12.29}$$

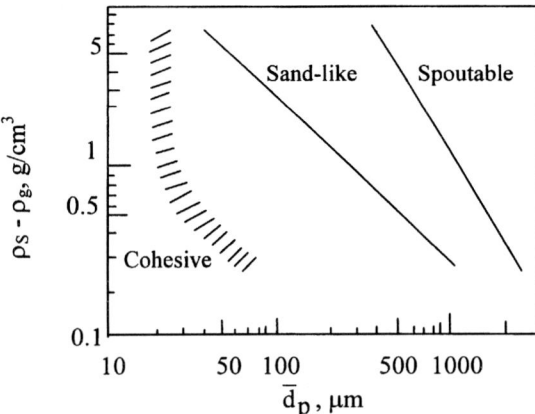

Figure 12.8 Classification of fluidization properties according to the size and density of powders (adapted from Geldart, 1973).

Fluid mechanical models of fluidization

The analysis of fluidized-bed reactors is based largely on the fluid mechanical model first described fully by Davidson and Harrison (1963) and modified later by a number of investigators (e.g., Jackson, 1963; Murray, 1965; Pyle and Rose, 1965; Kunii and Levenspiel, 1968a,b; Rowe, 1971; Orcutt and Carpenter, 1971; Davidson and Harrison, 1971; Davidson et al., 1978; Van Swaaij, 1985). Our description of fluidized-bed reactor modeling will be based on the Kunii–Levenspiel adaptation (see Levenspiel, 1993).

To account for all aspects of the fluidized bed, it is necessary to recognize three regions of the bed, as shown in Figure 12.9: the grid region, the main fluid-bed region, and the dilute-phase region. Much of the conversion occurs in the main fluid-bed region, normally referred to as the bubbling bed, but the reaction in the other regions cannot be ignored.

The Bubbling-Bed Model of Fluidized-Bed Reactors

The bubbling bed

Modeling of the bubbling-bed region is based on several special characteristics, assumptions, and definitions. These are outlined here along with the governing

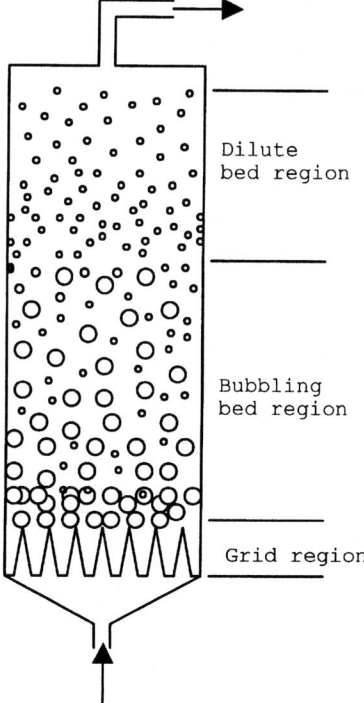

Figure 12.9 The three regions of a complete fluidized bed where reaction can occur.

equations in each case. The equations will depend on the nature of particles used, fine, intermediate size, or large. We restrict our treatment to small particles in view of their greater relevance to organic technology.

It will be assumed that bubble growth occurs close to the distributor, so that a single effective bubble diameter for the entire bed can be assumed. This is a contentious assumption, and several studies accounting for bubble size distribution and other hydrodynamic features of the bed have been reported, for example, Van Deemter (1961, 1967), Partridge and Rowe (1966), Toor and Calderbank (1967), Kato and Wen (1969), Fryer and Potter (1972, 1976), Jayaraman et al. (1980), Chen et al., (1982), Werther (1977), Werther and Schoessler (1984), Grace (1982, 1986). However, we persist with the constant bubble size assumption because it is sound enough for a preliminary design.

Based on this assumption, the distinguishing feature of the bubbling bed is the magnitude of the ratio α of the bubble rise velocity u_b (for a bubble diameter d_b) to the interstitial velocity u_i (equal to u_{mf}/f_{mf})

$$\alpha = \frac{u_b}{u_i} = \frac{u_b}{u_{mf}/f_{mf}} \qquad (12.30)$$

Bubble rise velocity

Now we derive an expression for the bubble rise velocity. This velocity consists of two contributions: (1) the free rise velocity u_{br} which is determined only by the properties of the bed and the gas and is independent of gas velocity and (2) the bulk flow of the bubble phase as a whole which is dependent on gas velocity and is given by $(u_0 - u_{mf})$. Thus

$$u_b = u_{br} + (u_0 - u_{mf}) \qquad (12.31)$$

The free rise velocity u_{br} may be assumed equal to the velocity of a bubble released from a singe nozzle into an inviscid fluid, that is, the rise velocity when the bed is at u_{mf} and is given by

$$u_{br} = 0.711(gd_b)^{1/2}, \text{ cm/s} \qquad (12.32)$$

Equation 12.31 has been empirically modified to provide separate correlations for Geldart A and B particles (Kunii and Levenspiel, 1991), but this simple expression is considered quite adequate for fine particles.

Main features

The bubbling-bed model holds when fast bubbles are rising through a bed of small particles, that is, $u_b \gg u_e$, as depicted in Figure 12.10a. The gas circulation is restricted to the bubble and a small region called the cloud surrounding it. In fact, the bubble gas is completely segregated from the rest of the gas passing through the bed. From simple fluid mechanical concepts, it can be shown that

$$\frac{r_c - r_b}{r_b} \cong \frac{u_i}{u_b} = \frac{1}{\alpha} \qquad (12.33)$$

Fluid Phase Reactions

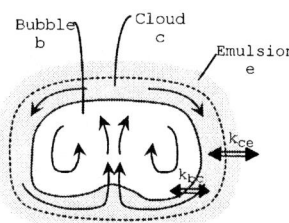
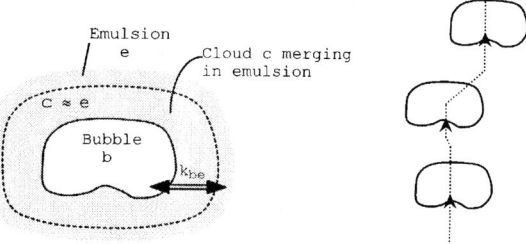

(a) *Fine particles*

$u_b \gg u_e \, (\alpha \gg 1)$

Fast bubble

Thin cloud

Bubble-cloud isolated from emulsion

Two mass transfer steps

Exit gas mainly bubble gas

(b) *Intermediate size particles*

$u_b > u_e \, (\alpha > 1)$

Medium fast bubble

Thick cloud merging with emulsion

Only one mass transfer

Exit gas is bubble and emulsion gas

(c) *Large particles*

$u_b < u_e \, (\alpha < 1)$

Slow bubble

No cloud

No mass transfer step

Exit gain mainly emulsion gas

Figure 12.10 Bubbling-bed models for (a) fine, (b) intermediate, and (c) large particle fluidized beds (see Levenspiel, 1993).

As an example, if the bubble rises ten times as fast as the emulsion gas, the cloud thickness $(r_c - r_b)$ will be just one-tenth the bubble radius (i.e., one-twentieth the bubble diameter).

A more rigorous equation for the ratio of cloud to bubble radii is

$$\frac{r_c}{r_b} = \left(\frac{\alpha + 2}{\alpha - 1}\right)^{1/3} \qquad (12.34)$$

Also, the picture is more complicated than depicted in Figure 12.10a. As the bubble rises, it carries with it a small amount of the solids as wake. Thus a rigorous model should really recognize four regions: emulsion, bubble, cloud, and wake. In the K–L model, it is assumed that the wake solids are evenly distributed in the cloud phase. This simplifies the computations without seriously affecting the accuracy.

Figures 12.10b and c depict the situations for medium and large size particles which, as already noted, will not be considered here.

Mass transfer between bubble and emulsion

An important feature of fluidized-bed reactors is mass transfer between bubble and emulsion. Several models have been proposed for this exchange. The Davidson model assumes no cloud, so that only one mass transfer coefficient k_{be} (for direct bubble–emulsion exchange) is involved. On the other hand, the

K–L model is based on two successive mass transfer steps, leading to the coefficients k_{bc} for bubble–cloud exchange and k_{ce} for cloud–emulsion exchange. The equations for the K–L model are given in Table 12.5.

Solids distribution

Because three phases are present, the extent of reaction in each phase must be computed. And because the reaction occurs only in the presence of solids, the distribution of solids in the three phases, emulsion, cloud, and bubble, must be known. These are expressed as fractions of the bubble phase, s_{bb}, s_{cb}, and s_{eb} for the bubble, cloud, and emulsion phases, respectively. As already mentioned, every bubble carries with it a small amount of solids as wake. The volume fraction of the wake is given by s_{wb} = (volume of wake/volume of bubble phase).

Estimation of bed properties

Knowledge of several properties and parameters of the fluidized bed (including those mentioned above) is necessary in fluidized-bed reactor design. A list of these properties along with the equations to estimate them is included in Table 12.5. Many of these equations will probably have to be revised in the light of the recent observation (Gunn and Hilal, 1997) that u_{mf} is likely to be affected by the scale of equipment and by distributor design. Thus correlations (such as for bed expansion) should more logically not be based on comparisons at the same gas velocity, as has been the practice so far, but at the same excess gas velocity (i.e., the same u_0/u_{mf}).

Table 12.5 Expressions for estimating important fluid-bed properties

Bubble rise velocities	
Single bubble in a quiescent bed at u_{mf}	$u_{br} = 0.711(g\,d_b)^{1/2}$, cm/s
Bubbles in a bubbling bed	$u_b = u_{br} + (u_0 - u_{mf})$, cm/s
Bubble fraction	$\delta = \dfrac{u_0 - u_{mf}}{u_b}$
Solids distribution	$s_{bb} = 0.001\text{–}0.01$
	$s_{cb} = (1 - f_{mf})\left[\dfrac{3u_{mf}/f_{mf}}{u_{br} - u_{mf}/f_{mf}} + s_{wb}\right]$
	$s_{eb} = \left[\dfrac{(1 - f_{mf})(1 - \delta)}{\delta} - s_{cb} - s_{bb}\right]$
	$s_{wb} = 0.2\text{–}2.0$
Mass transfer coefficients	
Bubble cloud	$k_{bc} = 4.50\left(\dfrac{u_{mf}}{d_b}\right) + 5.85\left(\dfrac{D^{1/2}g^{1/4}}{d_b^{5/4}}\right),\,\text{s}^{-1}$
Cloud-emulsion	$k_{ce} = 6.78\left(\dfrac{f_{mf}D u_{br}}{d_b^3}\right)^{1/2},\,\text{s}^{-1}$
Fluid-bed height relationships	$L_{mf} = L_f(1 - \delta)$
	$L_m(1 - f_m) = L_{mf}(1 - f_{mf}) = L_f(1 - f_f)$

Heat transfer

Good heat transfer is one of the most attractive features of the fluidized bed. From the standpoint of its use as a chemical reactor, the most important mode of heat transfer is that from a fluidizing bed to a bank of tubes (with a circulating fluid) immersed within it. The value of the heat transfer coefficient will depend on whether the tube bank is vertical or horizontal. A number of correlations are available for predicting these and other modes of heat transfer in a fluidized bed, and good reviews of these correlations can be found in Botterill (1966), Zabrodsky (1966), Muchi et al. (1984), and Kunii and Levenspiel (1991). Most of them are restricted to relatively narrow ranges of variables. Two useful correlations are listed in Table 12.6. It is important to note that reactions such as the chlorination of methane (Doraiswamy et al., 1972) are entirely heat transfer controlled. The rate of heat removal and design of reactor internals become the crucial considerations in such cases.

Recent studies have made it easier to design reactors with vertical tubular inserts. This arises from the observation (Gunn and Hilal, 1994, 1996, 1997) that the heat transfer coefficients for these systems are almost equal to those for the corresponding open fluidized beds of the same diameter operating with the same particles. Hence, correlations for the latter (which are readily available) can be used for vertical inserts without significant loss of accuracy. Vertical inserts have an additional advantage over horizontal inserts. In horizontal inserts there is accumulation of particles on top of the tubes and depletion of particles at the bottom, a situation that does not exist in the vertical orientation.

Calculation of conversion

Now we come to the main element of the model, calculation of conversion. Here, considering the fact that the amount of emulsion gas is negligible compared to the bubble-phase gas, reaction in the emulsion phase can be neglected. Thus

Table 12.6 Recommended correlations for heat transfer in fluidized beds

Bed-wall heat transfer	$h_w = 35.8 \rho_s^{0.2} \lambda_G^{0.6} d_p^{-0.36}$ (mks units)
Bed-immersed tube heat transfer (vertical tubes)	$\dfrac{h_w d_p}{\lambda_G} = 0.0184(CF)(1-f_f)\left(\dfrac{C_{pG}\rho_G}{\lambda_G}\right)^{0.43}$ $\times \left(\dfrac{d_p \rho_G u}{\mu}\right)^{0.23}\left(\dfrac{C_{p\omega}}{C_{pG}}\right)^{0.8}\left(\dfrac{\rho_s}{\rho_G}\right)^{0.66}$ (for $d_p\rho_G u/\mu = 10^{-2} - 10^2$)
Bed-immersed tube heat transfer (horizontal tubes)	$\dfrac{h_w d_0}{\lambda_G} = 0.66\left(\dfrac{C_{pG}\mu_G}{\lambda_G}\right)^{0.3}\left[\left(\dfrac{d_0 u \rho_G}{\mu}\right)\left(\dfrac{\rho_s}{\rho_G}\right)\left(\dfrac{1-f_f}{f_f}\right)\right]^{0.44}$ (for $d_0 u \rho_G/\mu < 2000$) $\dfrac{h_w d_0}{\lambda_G} = 420\left[\left(\dfrac{C_{pG}\mu_G}{\lambda_G}\right)\left(\dfrac{d_0 u \rho_G}{\mu}\right)\left(\dfrac{\rho_s}{\rho_G}\right)\left(\dfrac{\mu^2}{d_p^3 \rho_s g}\right)\right]^{0.3}$ (for $d_0 u \rho_G/\mu > 2500$)

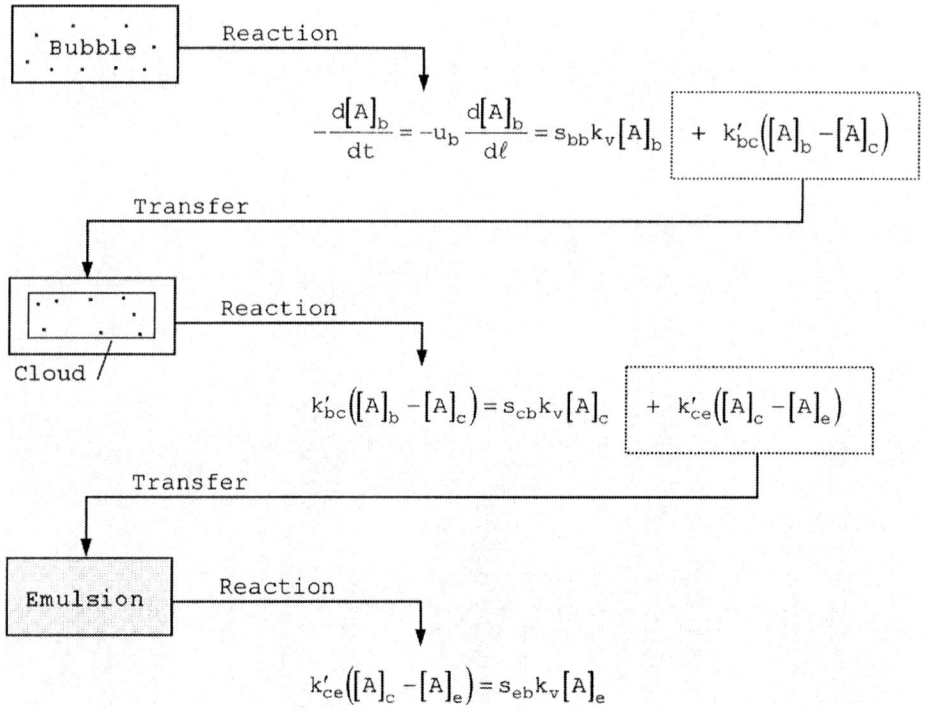

Figure 12.11 The Kunii–Levenspiel model.

conversion is based only on bubble-phase flow. A complete accounting of reactant A can be done as shown in Figure 12.11. Eliminating the intermediate concentration and by simple algebraic manipulations, the following expression can be derived:

$$-u_b \frac{d[A]_b}{d\ell} = K_F [A]_b \tag{12.35}$$

where

$$K_F = \left(s_{bb} k_v + \cfrac{1}{\cfrac{1}{k_{bc}} + \cfrac{1}{s_{cb} k_v + \cfrac{1}{\cfrac{1}{k_{ce}} + \cfrac{1}{s_{eb} k_v}}}} \right) \tag{12.36}$$

Integrating Equation 12.35 results in the following expression for reactant concentration as a function of height:

$$\frac{[A]_b}{[A]_0} = \exp\left(-K_F \frac{\ell}{u_b}\right) \tag{12.37}$$

Expressing this in terms of conversion at the reactor exit,

$$(1 - X_A) = \frac{[A]_{bf}}{[A]_0} = \exp(-\hat{K}_F) = \exp\left(-K_F \frac{L_f}{u_b}\right) \quad (12.38)$$

Note that Equation 12.38 is identical to the plug-flow equation except that K_F is not a true rate constant but a composite constant consisting of the true constant k_v and the various physical parameters of the model.

Extensions of the model

This model has been extended by Doraiswamy and co-workers to the case where there is an increase in volume upon reaction (Irani et al., 1980a), to complex reactions (Irani et al., 1980b, 1981; see also Levenspiel et al., 1978), and to the use of inerts to enhance selectivity (Irani et al., 1979, 1982).

End Region Models

The dilute bed region

In all of the models already developed, it was assumed that the reaction is restricted to the bubbling bed, but the data of Lewis et al. (1962) and Fan et al. (1962) show that an axial distribution of bed density exists. Further, it seems most likely that bubbles carry solid particles along with them through the central region of the bed and enter the dilute phase by a process of bursting on the emulsion surface (Miyauchi, 1974; Miyauchi and Furusaki, 1974). Then a bubble-free emulsion flows down the bed peripherally. This situation clearly leads to some reaction in the dilute phase. An elegant model that accounts for reaction in both the bubbling and dilute regions of the bed has been proposed by Miyauchi (1974) and another by Kunii and Levenspiel (1991) (more in line with their fine particle model).

According to Miyauchi's model, the concentration at the exit of the dilute bed $[A]_t$ is given by (see Figure 12.12)

$$\frac{[A]_t}{[A]_0} = 1 - X_A = \exp[-(K' + K_b + K_d)] = \exp(-K'_R) \quad (12.39)$$

where

$$\frac{1}{K'} = \frac{1}{K_m} + \frac{1}{K_0(1-\delta)} \quad (12.40)$$

The various groups (K_m, K_0, K_b, K_d) are defined in the figure. Note that experimental determination of the bed density distribution $[(1 - \delta)dz_f]$ in the dilute region is necessary to estimate K_d.

The grid or jet region

As the fluidizing gas enters the bed through the openings in the grid plate, it usually issues as fluid jets at velocities in the range of 40–80 m/s (see Behie et al., 1976; Mori and Wen, 1976). These jets penetrate a certain distance into the bed

386 Reactor Design

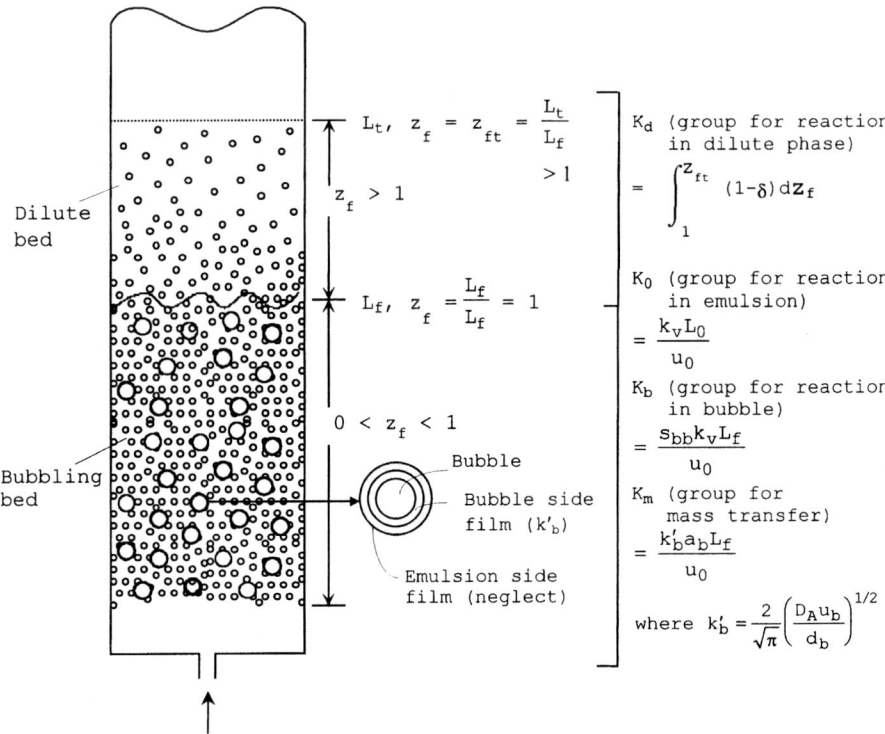

Figure 12.12 Model that accounts for reaction in both the bubbling and dilute regions of the bed.

before they collapse into bubbles. To avoid damage, no reactor internals should be placed in the path of these jets. The jet region is significant in fast reactions in very large scale productions, in which jet velocities tend to be high, but is usually of no particular consequence in the production of organic intermediates.

Practical Considerations

The modeling of fluidized-bed reactors provides a preliminary indication of the size and main characteristics of the reactor for a given reaction. However, there are so many uncertainties in the scale-up of fluidized beds that it is impossible now to avoid pilot plant and semicommercial units before arriving at a suitable commercial design. Some of the problems encountered in scale-up are violent circulating currents (of the order of 20–30 cm/s), known as "gulf streaming," in large beds that result in low bubble residence times in relation to those in smaller laboratory or pilot plant reactors (Davidson, 1973); "caking" of parts of the bed (sometimes the whole bed) due to stagnant pockets in exothermic reactions (Doraiswamy et al., 1968); defluidization of the bed due to sudden changes in temperature or pressure or coke deposition on particles leading to higher u_{mf}; and the location of the inlets of gases and the choice of fluidizing gas for

reactions involving two reactants which cannot be mixed outside the bed due to flammability, as in the chlorination of methane (Doraiswamy et al., 1972). These are only indicative of the many problems involved.

Perhaps the most important problem is the lack of understanding of bubble behavior in beds of different sizes and particle properties. An approximate but useful way of addressing this problem is to operate a pilot plant reactor of a certain size and design the larger reactor so as to simulate the bubble behavior of the pilot reactor. For this purpose, we define an equivalent reactor diameter d_e given by the hydraulic diameter (see Volk et al., 1962)

$$d_e = 4 \left[\frac{\text{free cross-sectional area of the bed}}{\text{wetted perimeter of all vertical surfaces exposed to the bed}} \right] \quad (12.41)$$

and provide vertical internals (usually tubes which can also be used for heat exchange) in the larger reactor so that d_e has the same value as the pilot reactor diameter. This results in bubbles of approximately the same size in the two reactors (corresponding to 1–1.5 times the internal tube pitch in the larger reactor). This simple method will not apply to very high scale-up ratios but should be useful in the scale-up of medium sized organic chemicals production, as demonstrated by Doraiswamy et al. for the chlorination of ethylene to hexachloroethane (1968) and the ethylation of aniline to monoethylaniline (1973).

EXAMPLE 12.4

Design of a fluidized-bed reactor for the hydrogenation of nitrobenzene to aniline

Design a fluid-bed catalytic reactor for the commercial production of 6000 tons per annum of aniline by the hydrogenation of nitrobenzene. A conversion of 97% is to be achieved in the reactor at $u_0 = 36$ cm/s. The following data are given:

molar ratio of hydrogen to nitrobenzene = 20 : 1.
operating temperature = 270 °C.
rate constant $k_v = 1.2$ s^{-1} (based on power law kinetics for the catalyst used in the reactor).
heat of reaction $(-\Delta H_r) = 180$ kcal/mol.
density of hydrogen = 0.0446×10^{-3} g/cm^3 (at 270 °C, 1 atm).
viscosity of reaction mixture = 1.3×10^{-2} cp (at 200°).
diffusivity = 0.25 cm^2/s (at 270 °C).
catalyst particle diameter $d_{ave} = 0.031$ cm.
bulk density of solids $\rho_s = 2.2$ g/cm^3.
void fraction at minimum fluidization $f_{mf} = 0.50$.
wake fraction $s_{wb} = 0.43$.
bubble diameter $d_b = 12$ cm.

SOLUTION

A broad methodology

1. Calculate the minimum fluidization velocity u_{mf}, minimum bubbling velocity u_{mb}, and minimum slugging velocity u_{ms}. Because the calculation of u_{ms} requires a

value of the reactor diameter, which is not known, assume a value, and go through the calculations. This velocity represents the maximum permissible velocity and therefore the minimum diameter. From the given volumetric flow rate, calculate the maximum diameter by assuming a velocity of, say, $3u_{mf}$. Choose a suitable diameter between these limits. Iteration may be needed.

2. Laboratory scale data are best simulated by using a number of vertical internal tubes (which can also be used for heat transfer) in a reactor shell containing the fluidizing solids and ensuring that the equivalent diameter d_e calculated from Equation 10.41 does not exceed 15–20 cm. The effective bubble size should be approximately 1–1.5 times the tube pitch in the larger reactor (which will depend on the size and number of the internal tubes used). Thus design the grid plate so that the maximum and minimum sizes of the bubbles generated by the orifices will straddle the effective bubble size in a narrow range.

3. With the information now available, use Equation 12.35 to calculate the bed height L_f needed to obtain the desired conversion. The present example is primarily concerned with this crucial aspect of design.

4. Because cooling is needed, see if the number of internal tubes calculated in step 2 is adequate. If not, use more tubes, and repeat the pertinent calculations using this new value. A good design is one where the number of tubes calculated from Equation 12.41 is approximately equal to the number based on heat transfer requirements. In any case, an adequate number of tubes must be provided to ensure maintenance of the required temperature.

Bed height for 97% conversion

The rise velocity of bubble is calculated as

$$u_{br} = 0.711[g\, d_b]^{1/2} = 0.711(981 \times 12)^{1/2}$$
$$= 77.14 \text{ cm/s}$$

Then the absolute rise velocity (with $u_{mf} \approx 12$ cm/s) of the bubble is

$$u_b = u_o - u_{mf} + u_{br}$$
$$= 36 - 12 + 77.14 = 101.14 \text{ cm/s}$$

The fraction of the fluidized bed occupied by bubbles is given by

$$\delta = \frac{u_o - u_{mf}}{u_b} = \frac{36 - 12}{101.1} = 0.237$$

The interchange coefficient between bubble and cloud is calculated from the appropriate equation given in Table 12.5.

$$k_{bc} = 4.5\left(\frac{12}{12}\right) + 5.85\left(\frac{0.25^{1/2} 981^{1/4}}{12^{5/4}}\right) = 5.23 \text{ s}^{-1}$$

Similarly, the interchange coefficient between cloud and wake and emulsion phases is calculated as

$$k_{ce} = 6.78\left(\frac{0.5 \times 0.25 \times 77.14}{12^3}\right)^{1/2} = 0.506 \text{ s}^{-1}$$

Now let us calculate the fraction of solids dispersed in the bubble, cloud/wake, and emulsion phases per unit volume of bubble in the bed:

$$\left.\begin{array}{l} s_{bb} \approx 0.003 \\ s_{wb} \approx 0.43 \end{array}\right\} \text{Assume}$$

$s_{cb} = 0.892$ (from Table 12.5)

$s_{eb} = 0.715$ (from Table 12.5)

From Equation 12.38

$$1 - X_A = \exp(-\hat{K}_F)$$

$$1 - 0.97 = \exp(-\hat{K}_F)$$

$$\hat{K}_F = 3.51$$

and from Equation 12.36, $K_F = 1.09$, giving finally:

Height of reactor $L_f = 3.24$ m.

Optimization of Fluid-Bed Reactor Performance

One of the chief drawbacks of the fluid-bed reactor is deviation from plug flow. This is also a feature of other reactor types such as gas-liquid and gas-liquid-solid reactors. However, remedial measures have been introduced in those reactors by providing improved contact through the development of more efficient packings. This is not possible in fluidized beds because particles can lodge in unaerated stagnant zones within a packing's interstices. Such flow maldistribution can be minimized by using structured-grid packings with well-defined uniformly distributed flow areas. Several such packings have been listed by Papa and Zenz (1995) along with an equation for designing them.

An important practical shortcoming of the bubbling-bed model considered in the previous sections is that fluid-bed reactors normally operate at velocities in excess of the limits of bubbling beds, $u_0/u_{mf} > 15$ (Avidan and Edwards, 1986; Bolthrunis, 1989). Reactors of this type give rise to turbulent behavior as opposed to bubbling-bed behavior (Squires et al., 1985). Increasing attention is being paid to these reactors in the recent literature. However, for the relatively small scale production normally encountered in organic technology/synthesis, the bubbling-bed regime is the more likely. A practical overall strategy for fluidized-bed reactor scale-up has also been suggested (Jazayeri, 1995).

REACTOR CHOICE FOR A DEACTIVATING CATALYST

Considering the main features of fixed-bed and fluidized-bed reactors, it appears that the fluidized bed is a better choice for a rapidly deactivating catalyst. In these cases we are usually concerned with activity losses of less than 1% in less than a minute of exposure (due to carbon deposition), and production rates of several hundred tons per day. The catalyst is usually regenerated in a second reactor called the regenerator. This is seldom the situation in organic technology,

more so in organic synthesis. What is important in such reactions is the average conversion that can be obtained over a reasonable length of time after which the catalyst can be regenerated or replaced. Thus both fixed-bed and fluidized-bed reactors become viable candidates. This section is concerned with evaluating the performances of these reactors under such conditions.

The Basic Equation

Restricting the treatment to isothermal plug flow, the continuity equation for a reactor containing a time-decaying catalyst through which reactant A is passing and reacting under diffusion-free conditions may be written as

$$\left(\frac{\partial [A]}{\partial t_p}\right) + u\left(\frac{\partial [A]}{\partial \ell}\right) = -[-r_A([A], t_p)] \quad (12.42)$$

or

$$\frac{f_B \rho_G}{M}\left(\frac{\partial y_A}{\partial t_p}\right) + \frac{G_M}{M}\left(\frac{\partial y_A}{\partial \ell}\right) = -[-r_A(y_A, t_p)] \quad (12.43)$$

where t_p is the production or on-stream time. Note that the rate has been written as a function of both concentration and on-stream time. In other words, at a given concentration, the rate also depends on time (which it would not for a nondecaying catalyst).

Now using the basic equation given by Equation 12.43, we develop the governing equations for fixed- and fluidized-bed reactors and compare their performances.

Fixed-Bed Reactor

The reaction time in a fixed-bed reactor is evidently the same as the total decay (or reaction) time t_{p1}, when viewed from the standpoint of catalyst decay. Hence, the time can be normalized with respect to t_{p1}, and Equation 12.43 recast in dimensionless form as

$$A'\left(\frac{\partial y_A}{\partial \hat{t}}\right) + \frac{\partial y_A}{\partial z} = -B'[-r_A(y_A, \hat{t})] \quad (12.44)$$

where $G_M = \rho_F S_{V,F} L$ (any feed) and $= \rho_L S_{V,L} L$ (liquid feed) and

$$\hat{t} = \frac{t_p}{t_{p1}}, \qquad z = \frac{\ell}{L}$$

$$A' = \frac{f_B \rho_G}{\rho_L S_{V,L} t_{p1}}, \qquad B' = \frac{M}{\rho_L S_{V,F}} = \frac{M}{\rho_L S_{V,L}}, \qquad \frac{m^3 \, s}{mol} \quad (12.45)$$

Equation 12.44 can be simplified if the first term can be neglected. It is entirely reasonable to do so because the constant A' in that term, which represents the ratio of feed transit time to decay time, is usually negligibly small. Thus Equation 12.44 reduces to

$$\frac{dy_A}{dz} = -B'[-r_A(y_A, \hat{t})] \tag{12.46}$$

Now we express the rate as

$$-r_A(y_A, t_p) = k_v(t_p)(1 - f_B)y_A^m \tag{12.47}$$

where k_v has the units of the rate. If we assume that the catalyst decays exponentially with time, that is,

$$k_v(t_p) = k_{v0} \exp(-d_c t_p) \tag{12.48}$$

then

$$\frac{dy_A}{dz} = -B'' \exp(-\lambda \hat{t}) y_A^m \tag{12.49}$$

where d_c is the decay constant (1/s),

$$\lambda = d_c t_{p1}$$

$$B'' = B'(1 - f_B)k_{v0} = \frac{M}{\rho_F S_{V,F}}(1 - f_B)k_{v0} \tag{12.50}$$

Equation 12.49 is the basic nondimensional equation describing the mole fraction of A in a fixed-bed reactor containing an exponentially decaying catalyst as a function of position and time in terms of two dimensionless parameters, B'' and λ. The performance of this reactor can be best judged by solving the equation for the reactor exit, that is, for $z = 1$. The solution for a first-order reaction ($m = 1$) is given in Table 12.7 (Sadana and Doraiswamy, 1971). It is also possible to assume various other forms of catalyst decay. Solutions are included in the table for two other forms, one of them linear.

When a decaying catalyst is used, it is important to estimate the average conversion over a given period of time. From an economic point of view, this quantity is much more important than the conversion at the exit and is given by

$$\bar{X}_A = 1 - \bar{y}_A = 1 - \int_0^1 y_A \, d\hat{t} \tag{12.51}$$

Solutions to this equation (for $z = 1$) are included in the table for all three forms of decay considered.

Fluidized-Bed Reactor

A reasonable assumption for the fluidized-bed reactor is that the fluid is partially mixed, whereas the solid is fully mixed. If we assume that the fluid is in plug flow, the residence time distribution of the solids will be given by $\exp(-\hat{t})$. As for the fixed-bed reactor, we shall first consider exponential decay. Therefore, the average rate constant to be used in solving Equation 12.46 is

$$[k_v(\hat{t})]_{av} = k_{v0} \int_0^\infty \exp(-\hat{t}) \exp(-\lambda \hat{t}) d\hat{t} \tag{12.52}$$

Table 12.7 Expressions for mole fraction and conversion of reactant A for various decay forms (adapted from Sadana and Doraiswamy, 1971)[a]

	Expressions for y_A		Expression for conversion	
	First order	mth order ($m \neq 1$)	First order	mth order ($m \neq 1$)

$k_v = k_{v0} e^{-\lambda \hat{t}}$

Fixed:

(1a) $\exp\{-[B'' \exp(-\lambda \hat{t})]\}$ (1b) $\left[\dfrac{1}{(m-1)B'' e^{-\lambda \hat{t}}+1}\right]^c$ (1c) $1+\dfrac{1}{\lambda}Ei^*(B'')-Ei^*(B''-\lambda)$ (1d) $1-\displaystyle\int_0^1\left[\dfrac{1}{(m-1)B'' e^{-\lambda \hat{t}}+1}\right]^c d\hat{i}$

Fluid:

(2a) $\exp\left(-\dfrac{B''}{\lambda+1}\right)$ (2b) $\left[\dfrac{(\lambda+1)}{(m-1)B''+(\lambda+1)}\right]^c$ (2c) $1-\exp\left(-\dfrac{B''}{\lambda+1}\right)$ (2d)

$k_v = k_{v0} - \lambda \hat{t}$

Fixed:

(3a) $\exp\left[-\left(B'' - \dfrac{B''\lambda \hat{t}}{k_{v0}}\right)\right]$ (3b) $\left[\dfrac{k_{v0}}{B''(m-1)(k_{v0}-\lambda \hat{t})+k_{v0}}\right]^c$ (3c) $1-\displaystyle\int_0^1 \exp\left[-\left(B''-\dfrac{B''\lambda \hat{t}}{k_{v0}}\right)\right]d\hat{i}$ (3d)

Fluid:

(4a) $\exp\left[\left(\dfrac{B''}{k_{v0}}\right)(k_{v0}-\lambda)\right]$ (4b) $\left[\dfrac{k_{v0}}{B''(k_{v0}-\lambda)(m-1)+k_{v0}}\right]^c$ (4c) $1-\exp\left[\left(\dfrac{B''}{k_{v0}}\right)(k_{v0}-\lambda)\right]$ (4d)

$k_v = k_{v0} - \lambda \hat{t}^d$

Fixed:

(5a) $\exp\left(\dfrac{B''}{k_{v0}}\lambda \hat{t}^d - B''\right)$ (5b) $\left[\dfrac{k_{v0}}{(B''k_{v0}-B''\lambda \hat{t}^d)(m-1)+k_{v0}}\right]^c$ (5c) $1-\displaystyle\int_0^1 \exp\left(\dfrac{B''}{k_{v0}}\lambda \hat{t}^d - B''\right)d\hat{i}$ (5d)

Fluid:

(6a) $\exp\left[B''\left(\dfrac{d!\lambda}{k_{v0}}-1\right)\right]$ (6b) $\left[\dfrac{k_{v0}}{k_{v0}+(m-1)B''(k_{v0}-d!\lambda)}\right]^c$ (6c) $1-\exp\left[B''\left(\dfrac{d!\lambda}{k_{v0}}-1\right)\right]$ (6d)

[a] $c = 1/(m-1)$. λ is dimensionless in the first decay form but is a purely empirical quantity with dimensions of mol/m^3 s in the other two.

The solution to this equation is (Weekman, 1968)

$$[k_v(\hat{t})]_{av} = \frac{k_{v0}}{1+\lambda} \tag{12.53}$$

Substituting this equation for the rate constant in Equation 12.47 gives

$$-r_A(y_A, \hat{t}) = \left[\frac{k_{v0}(1-f_B)}{1+\lambda}\right] y_A^m \tag{12.54}$$

for a reaction of order m taking place in a steady-state fluid-bed reactor. Then, upon incorporating this equation in Equation 12.46, we get

$$\frac{dy_A}{dz} = -\left(\frac{B''}{\lambda+1}\right) y_A^m \tag{12.55}$$

Solutions to this equation for $m = 1$ and $m \neq 1$, as well for the other decay forms considered for the fixed-bed reactor (Sadana and Doraiswamy, 1971), are included in Table 12.7. The corresponding expressions for conversion given by

$$X_A = 1 - y_A \tag{12.56}$$

can also be found in the table.

Comparison of Performances

In order to compare the performances of fixed-bed and fluid-bed reactors for a deactivating catalyst, conversion is plotted as a function of the reaction group B'' for two different values of the decay parameter λ in Figure 12.13. It can be clearly noted that the fluid bed is superior and that the difference in conversions between the two reactors increases as the decay parameter increases. Under conditions of no decay ($\lambda = 0$), the equations for both the reactors reduce to

$$X_A = \begin{cases} 1 - \left[\frac{1}{B''(m-1)z+1}\right]^{1/m-1}, & m \neq 1 \\ 1 - \exp(-B''z), & m = 1 \end{cases} \tag{12.57}$$

Thus, for a nondeactivating catalyst, the nature of the solid presence (fixed or fluidized, or even moving as found by Sadana and Doraiswamy, 1971) is of no consequence, as long as the fluid flow patterns are the same, plug flow in this case. It is only when the catalyst is subject to decay that the performances of the reactors differ, and then the nature of the decay equation plays a significant role in determining the conversions achievable.

ILLUSTRATIVE EXAMPLE

Catalyst deactivation in organic synthesis/technology is far more severe in immobilized enzyme catalysts, which are usually more susceptible to deactivation than other solid catalysts. Therefore, the principles outlined here are illu-

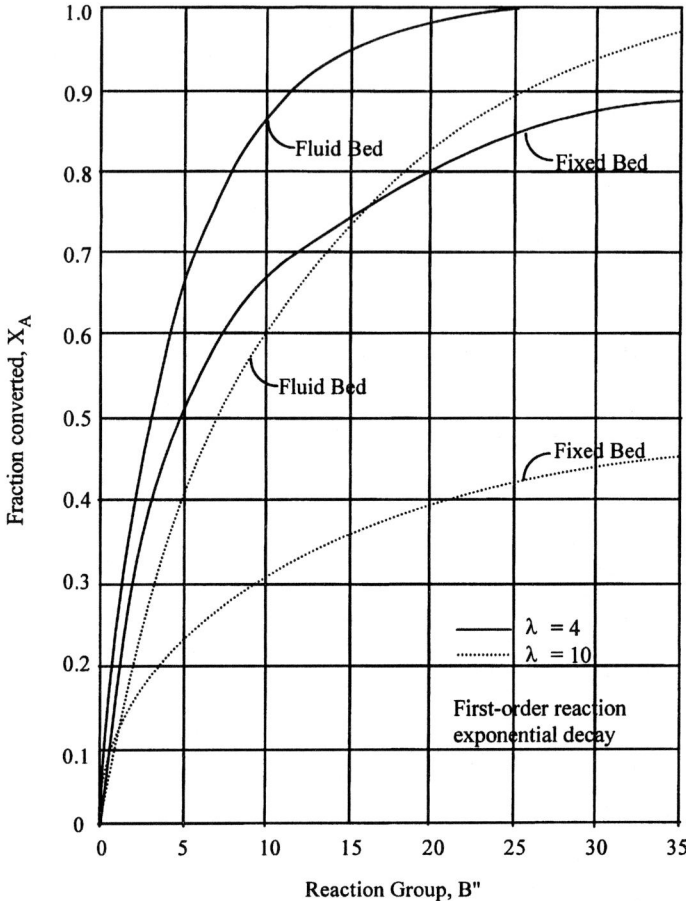

Figure 12.13 Conversions obtained in fixed- and fluid-bed reactors for different values of the decay parameter.

strated through a numerical example in Chapter 20 on biochemical methods in organic synthesis.

The models described have been extended to consecutive reactions by Sadana and Doraiswamy (1971) and Prasad and Doraiswamy (1974).

Comments

The procedures described are relatively simple. More rigorous methods of design for both types of reactors have been documented in several books (e.g., Doraiswamy and Sharma, 1984; Kunii and Levenspiel, 1991).

It is important to note that, unlike the situation previously considered, many catalysts tend to deactivate very rapidly with time. In petroleum cracking, for instance, the deactivation is so rapid that the catalyst must be continuously regenerated in a second reactor. However, the catalysts used in medium- and

small-volume productions are usually required to maintain their activity over a prolonged period of time. Even if they do deactivate, they should do so relatively slowly, for it would be uneconomical to operate reactor–regenerator systems for such reactions. Hence, the problem of reactor–regenerator system design, on which extensive literature is available, is not considered in this book.

Chapter 13

Mixing, Multiple Solutions, and Forced Unsteady-State Operation

Three important (complicating) possibilities were not considered in the treatment of reactors presented in earlier chapters: (1) the residence time of the reactant molecules need not always be fully defined in terms of plug flow or fully mixed flow; (2) the equations describing certain situations can have more than one solution, leading to multiple steady states; and (3) there could be periods of unsteady-state operation with detrimental effects on performance, that is, transients could develop in a reactor.

Actually, reactors can operate under conditions where there is an arbitrary distribution of residence times, leading to different degrees of mixing with consequent effects on reactor performance. Also, multiple solutions do exist for equations describing certain situations, and they can have an important bearing on the choice of operating conditions. And, finally, unsteady-state operation is a known feature of the start-up and shutdown periods of continuous reactor operation; it can also be introduced by intentional cycling of reactants. We briefly review these three important aspects of reactors in this chapter. However, because the subjects are highly mathematical, the treatment will be restricted to simple formulations and qualitative discussions that can act as guidelines in predicting reactor performance.

ROLE OF MIXING

All aspects of mixing in chemical reactors are based on the theory of residence time distribution first enunciated by Danckwerts (1953). Therefore, we begin our discussion of mixing with a brief description of this theory.

Residence Time Distribution

When a steady stream of fluid flows through a vessel, different elements of the fluid spend different amounts of time within it. This distribution of residence times is denoted by a curve which represents, at any given time, the amount of fluid with ages between t and $t + dt$ flowing out in the exit stream. When normalized with respect to the total flow (i.e., expressed as fraction of the total flow), this distribution, known as the residence time distribution (RTD), satisfies the condition

$$\int_0^\infty \delta(t)dt = 1 \tag{13.1}$$

The time spent by each element can vary from zero to infinity. This equation is displayed in Figure 13.1.

The two important characteristics of any distribution are the spread, which is characterized by its mean ($\bar{t}$ in our case), and the shape, which is characterized by its standard deviation σ. The mathematical tool most commonly used to determine these parameters is the analysis of moments, which is fully described in several books, for example, Nauman and Buffham (1983), Nauman (1987), and Levenspiel (1972, 1993). The expression for $\bar{t}$ is

$$\bar{t} = \int_0^\infty t\,\delta(t)dt \tag{13.2}$$

and the standard deviation σ of any function $f(t)$ is given by

$$\sigma^2 = \int_0^\infty [\bar{f} - f(t)]^2 \delta(t)dt \tag{13.3}$$

Cumulative and washout functions

The distribution function is usually derived from different types of experimentally determined curves of concentration (usually of a tracer in an inert stream) versus time. These curves represent responses to different signals introduced at the upstream end of the reactor. Two common signals, pulse and step, are

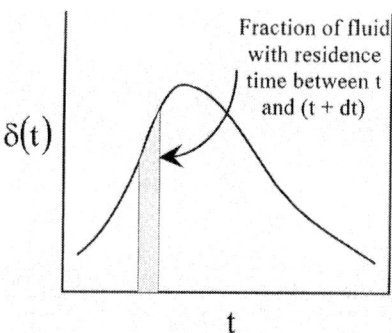

Figure 13.1 A typical residence time distribution (RTD) plot.

sketched in Figure 13.2. The pulse signal, in which a small amount of tracer is instantaneously introduced into a stream of inert, gives the $\delta(t)$ curve directly. The spread of the curve downstream of the reactor represents the degree of mixing, and no spread denotes plug flow. In the step input, the tracer is added at a certain rate into a stream of inert or is cut off from a stream of inert plus tracer at a certain rate. These lead to the cumulative and washout functions of RTD, respectively:

CUMULATIVE FUNCTION

$$F(t) = \int_0^t \delta(t)dt \qquad (13.4)$$

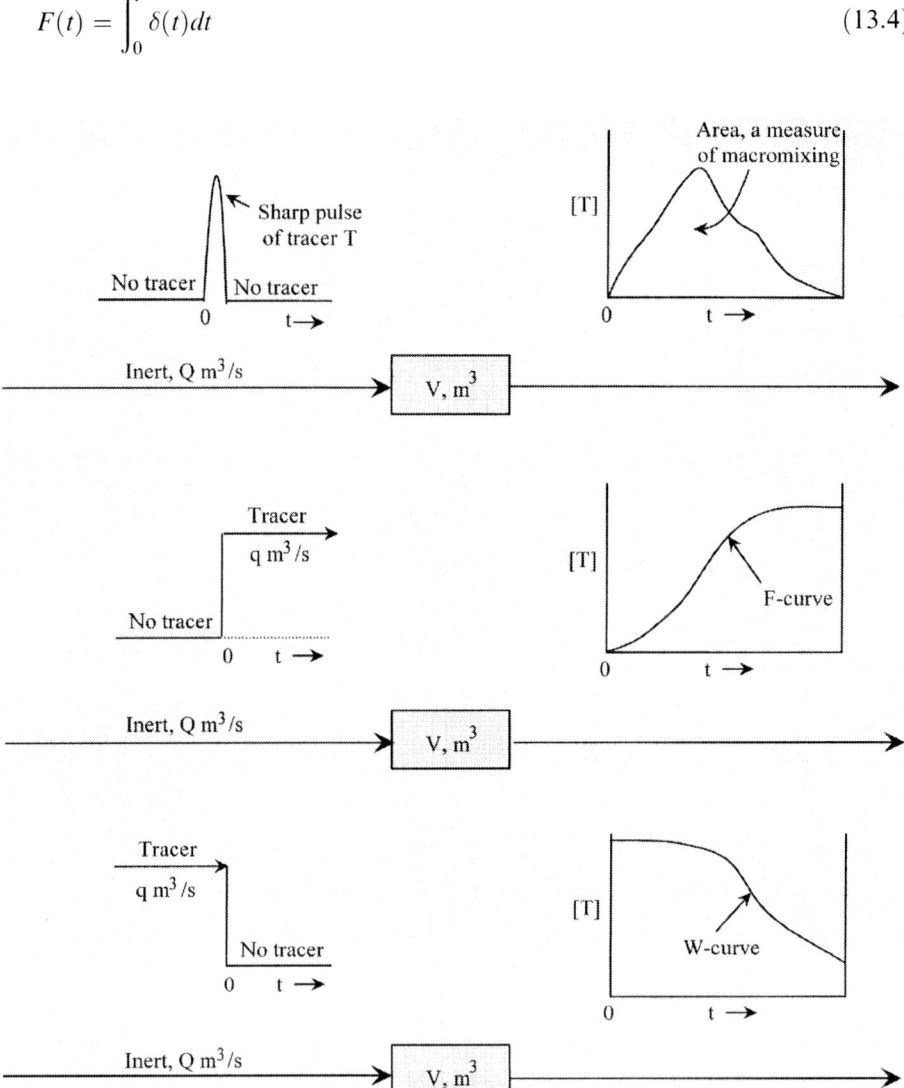

Figure 13.2 Tracer concentrations for instantaneous and step pulse inputs.

This represents the fraction of molecules leaving the reactor that had a residence time of t or less.

WASHOUT FUNCTION

$$W(t) = \int_t^\infty \delta(t)dt \tag{13.5}$$

This represents the fraction of molecules leaving the reactor that had a residence time of t or more.

The following relationships between the various functions can be readily written:

$$F(t) = 1 - W(t) \tag{13.6}$$

$$\delta(t) = \frac{dF(t)}{dt} = -\frac{dW(t)}{dt} \tag{13.7}$$

Types of distribution

The plug-flow limit is represented by the Dirac delta function

$$\delta(t) = g(\bar{t}) \tag{13.8}$$

which shows that $d(t) = 0$ at all times except at $t = \bar{t}$. The fully mixed limit is given by the exponential distribution

$$\delta(t) = \frac{1}{\bar{t}} \exp\left(-\frac{t}{\bar{t}}\right) \tag{13.9}$$

The standard deviation for the delta function is seen to be zero, a consequence of the fact that the pulse is perfectly sharp. On the other hand, it is equal to the residence time for exponential distribution, which denotes that this is the broadest distribution possible. Real (nonideal) RTDs lie between these extremes.

Clearly, then, the effect of RTD should be included in any reactor design, particularly when there is strong reason to believe that the operation would be nonideal (i.e., neither PF nor MF can be assumed). But before this can be done, the nature of the mixing problem must be identified.

Micro- and Macromixing

Whereas mixing is important even where a single reactant is involved in that axial and radial diffusion can influence the reaction in a continuous reactor, it is particularly important where two reactants from two different streams participate. These streams can either be premixed or not premixed (usually referred to as unpremixed). The discussion presented here assumes premixed feed except where specifically stated otherwise (i.e., for fast reactions).

The limits of mixing, as we have understood them so far are plug flow (no mixing) and mixed flow (full mixing). However, there is a region in the vicinity of the fully mixed boundary where different "degrees of full mixing" can exist. In one limit of this region, clumps or aggregates of molecules enter the reactor and move through it without interacting with one another. Within each clump, however, there is complete mixing of the molecules at the molecular level. The

residence time of each molecule within the clump is the same as that of the clump itself. Because the clumps are fully separated from one another, this kind of flow is known as segregated flow. On the other hand, in the other extreme, there are no clumps, and mixing occurs at the molecular level. The kind of mixing considered in the earlier chapters refers to this perfectly mixed condition. Note that now we have introduced a second "fully mixed" condition, segregated flow. Clearly, there can be degrees of mixing among molecules of the clumps, leading to various degrees of mixing or segregation at this level. Thus one can discern two broad regions of mixing. The region between plug flow and fully segregated flow, referred to as the macromixing region, and that between fully segregated and perfectly mixed flows, referred to as the region of micromixing.

In addition to these regions, when the fluids are unpremixed, micromixing can also intrude in the zone now referred to as the macromixing zone. In other words, micromixing can also be associated with plug flow. We shall not be concerned with this aspect of mixing.

One more level of mixing must be specified before we complete the bounds within which real reactors operate. This arises out of the fact that the only RTD that characterizes perfect mixing is an exponential distribution. But each RTD (other than exponential) has its own limit of perfect mixing. This limit is referred to as maximum mixedness and represents the perfect mixing equivalent of any distribution other than exponential.

The four limits of mixing are schematically shown in Figure 13.3. Whether fully segregated or perfectly mixed flow occurs depends on the nature of the fluid. Fluids that tend largely to macromix are called macrofluids, and those that tend largely to micromix are microfluids. In general, fluids with low Schmidt numbers and low viscosities tend to behave as microfluids, whereas the opposite is true for macrofluids. The extent of micromixing is largely controlled by the mixing device used.

Design Equations for Macro- and Micromixing

In view of the various types of mixing already explained, the design equation for a reactor will depend on the region of mixing under consideration. Thus we have

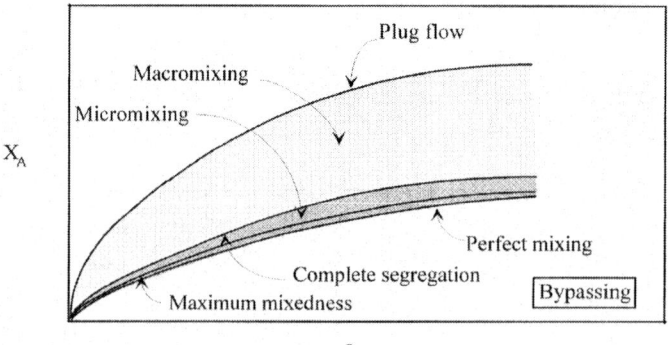

Figure 13.3 Plot showing the different limits of macro- and micromixing.

(1) a plug-flow reactor corresponding to zero macromixing, (2) a mixed-flow reactor corresponding to zero micromixing (fully segregated flow), and (3) a mixed-flow reactor corresponding to perfect (molecular level) mixing and exponential distribution or zero segregation. A fourth limit can be identified for unpremixed feeds: plug flow with zero macromixing and zero micromixing. Cases 1 and 3 correspond, respectively, to PFR and MFR considered in Chapter 4. We develop here the design equation for case (2), fully segregated flow, and then consider situations where there can be partial macromixing or partial micromixing.

Fully segregated flow

We assume that each clump behaves as a batch reactor. Then the total reaction in the reactor is given by the integral

$$\left(\frac{[\bar{A}]}{[A]_0}\right)_f = \int_0^\infty \left(\frac{[A]}{[A]_0}\right)_i \delta(t)\, dt \tag{13.10}$$

where $([A]/[A]_0)_i$ represents the reaction in a batch of fluid of age t and $\delta(t)$ represents the distribution of the little batches or clumps. The equation for $[A]/[A]_0$ in each clump depends on the order of the reaction and can be readily written from the batch reactor equations given in Chapter 4. Thus for an nth-order reaction,

$$\left(\frac{[A]}{[A]_0}\right)_i = \left(\frac{[A]}{[A]_0}\right)_{\text{batch}} = \left[1 + (n+1)[A]_0^{n-1} kt\right]^{1/1-n} \tag{13.11}$$

Inserting this equation in Equation 13.10 and assuming exponential distribution, we get

$$\left[\frac{[\bar{A}]}{[A]_0}\right]_f = \frac{1}{\bar{t}} \int_0^\infty [1 + (n-1)[A]_0^{n-1} kt]^{1/1-n} e^{-t/\bar{t}}\, dt \tag{13.12}$$

When $n = 1$, Equation 13.12 reduces identically to the perfectly mixed CSTR equation $[A]/[A]_0 = 1/(1 + kt)$, that is, micromixing has no effect. Thus, as a general rule, conversion in a first-order reaction is uniquely determined by the residence time distribution. This is true for both premixed and unpremixed feeds. We examine the effect of reaction order more fully in the following section.

Micromixing policy

Any micromixing policy to maximize conversion is irrelevant for a first-order reaction because it is unaffected by the degree of segregation. For non-first-order reactions, micromixing policies depend on two important considerations: whether the reaction order is greater or less than unity and, in bimolecular reactions, whether the feed is premixed or unpremixed.

The rate equations can be classified as concave-up ($n > 1$), linear ($n = 1$), or concave-down ($n < 1$). Because the second derivative of the rate equation is usually continuous, the following postulations can be made:

1. For $n > 1, d^2r/d[A]^2 > 0$, segregation maximizes conversion, and maximum mixedness minimizes it.
2. For $n = 1, d^2r/d[A]^2 = 0$, and the extent of segregation has no effect on conversion.
3. For $n < 1, d^2r/d[A]^2 < 0$, maximum mixedness maximizes conversion, and segregation minimizes it.

This policy also holds for an irreversible bimolecular reaction provided that the two species are premixed. For reversible reactions, again the same policy holds provided that the rate constant of the forward reaction is larger than that of the reverse reaction. No such generalization seems possible for unpremixed feed.

Models for Partial Mixing

So far we have considered design equations for the various mixing limits summarized in Figure 13.3. Clearly, the regions between these limits are equally important, and procedures are necessary for designing reactors operating in these regions. From the discussion of macro- and micromixing already presented, it is important to note that two classes of partial mixing models are possible: for the macromixing region between plug flow and mixed flow and for the micromixing region between full segregation and no segregation. Extensive literature is available on both types of models. However, it is usually adequate to base the design on the closest limiting model. This is particularly true for micromixing because no satisfactory model is yet available. Hence, we limit our treatment to a brief qualitative discussion of the two classes of models.

Models for partial macromixing

The following two models are frequently used to account for partial macromixing: the dispersion model and the tanks-in-series model. In the dispersion model, deviation from plug flow is expressed in terms of a dispersion or effective axial diffusion coefficient. This model was anticipated in Chapter 12, and the governing equations for mass and heat are listed in Table 12.2 of that chapter. The derivation is similar to that for plug flow except that now a term is included for diffusive flow in addition to that for bulk flow. This term appears as $-D_{e\ell}(d[A]/d\ell)$, where $D_{e\ell}$ is the effective axial diffusion coefficient. When the equation is nondimensionalized, the diffusion coefficient appears as part of the Peclet number defined as $Pe = ud/D_{e\ell}$. A number of correlations for predicting the Peclet number for both liquids and gases in fixed and fluidized beds are available and have been reviewed by Wen and Fan (1975).

The equation with axial mixing given in Table 12.2 for a fixed-bed reactor also holds for a homogeneous reactor, except that the effective diffusion coefficient $D_{e\ell}$ is replaced by the real D_ℓ. This can be recast as: $(1/Pe_{m\ell})(d^2a/dz^2) - (da/dz) - (L/[A]_0 u)(-r_{VA}) = 0$, where $Pe_{m\ell}$ is now the axial Peclet number defined as uL/D_ℓ. This equation has been solved (see Danckwerts, 1953; Wehner and Wilhelm, 1956) to obtain an expression for concentration as a

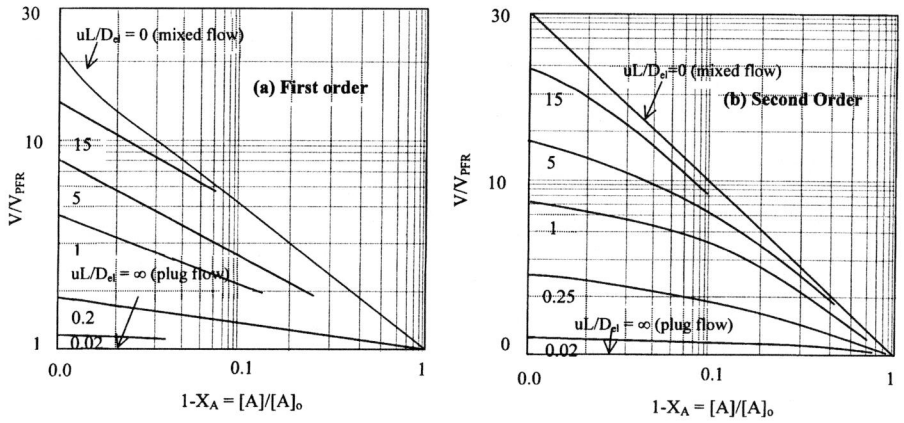

Figure 13.4 Effect of axial mixing (i.e., Peclet number) on reactor volume for (a) a first-order reaction and (b) a second-order reaction (redrawn from Levenspiel, 1993).

function of Pe. The volume ratio V_{axial}/V_{plug} can now be calculated and plotted as a function of concentration, as shown in Figures 13.4a,b for first- and second-order reactions. The effect of Peclet number can be clearly seen, particularly at low reactant concentrations (i.e., high conversions).

The tanks-in-series model, in which a sequence of CSTRs is used to simulate various degrees of partial mixing, was also anticipated and briefly considered in Chapter 10. Equation 10.18 for a first-order reaction shows the approach to PFR of a CSTR as the number of tanks N is increased from one to infinity. This is illustrated in Figure 13.5 (Levenspiel, 1993) as plots of (V_N/V_{PFR}) versus $[A_N]/[A]_0$ for different values of N. Notice that V_N increases by almost a factor

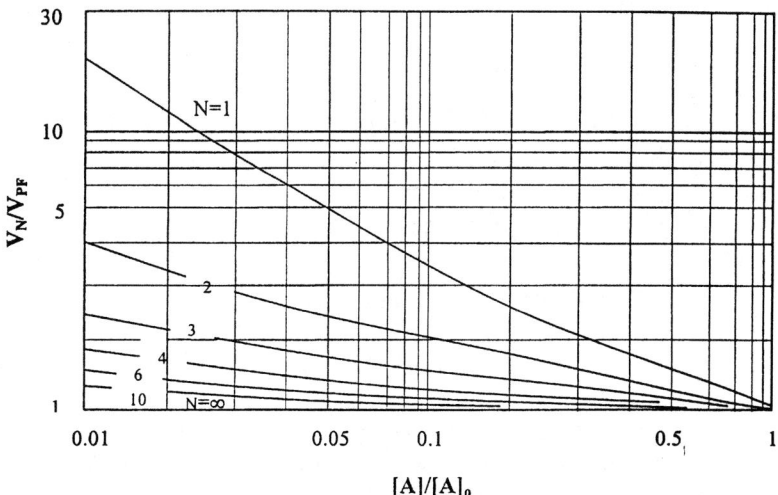

Figure 13.5 Effect of number of tanks on the performance of a CSTR sequence (redrawn from Levenspiel, 1993)

of 3 as the number of tanks is doubled from 1 to 2. Thereafter, the increase is less marked as N approaches infinity (i.e., as PFR performance is reached). Thus N is the tanks-in-series equivalent of Pe of the dispersion model.

Both models assume some sort of symmetry along the direction of flow and are hence unable to account for such common occurrences as shortcircuiting and channeling. A more comprehensive model that accounts for these features is the combined or compartment model in which plug flow, complete mixing, and shortcircuiting are treated as separate components that simultaneously contribute to the flow (see Cholette and Cloutier, 1959). In such a model, conversions can be less than that in a single stirred tank (i.e., fully mixed) reactor, corresponding to $N < 1$. This region is marked as the bypassing zone in Figure 13.3.

Models for partial micromixing

Since the introduction of the parameter J to describe the degree of segregation (with $J = 1$ for complete segregation and $J = 0$ for no segregation) (Danckwerts, 1953; Zwietering, 1959), there has been an explosion of models to describe decay of segregation (i.e., partial micromixing). The basic assumption in all of these models is that they give the same RTD and that differences in performance are attributable to different levels of micromixing sought to be simulated by different physical postulations. The different situations are best analyzed using the general class of population balance models that treat the reaction fluid as a collection of discrete elements, each consisting of a number of molecules—which can be as low as one for the ideal mixer. These models can be divided into three main categories:

1. Two-environment models in which one environment is in a state of complete segregation and the other in a state of maximum mixedness.
2. Fluid element or particle models where the fluid is broken up into small elements and mass transfer occurrs by coalescence and redispersion or diffusion.
3. Fluid flow models, where a simple fluid mechanical model is constructed by dividing the reactor into different zones of macro- and micromixing; this is clearly an extension of the compartment models of macromixing but zones of micromixedness are added.

Laminar flow reactors

In this type of reactor, as the name implies, the flow is laminar. In other words, the radial concentration profile is parabolic and not uniform as in a PFR. This is because there is hardly any mixing and the reactant concentration profile closely matches the velocity profile (lowest near the wall to highest at the center). Thus each element of fluid flowing through the reactor is completely independent of the other elements, so that the fluid as a whole tends to behave as a macrofluid. In essence, therefore, Equation 13.10 would be valid for this case also. Integrated

forms of this equation have been given by Bosworth (1948), Denbigh (1951), and Cheland and Wilhelm (1956).

As may be expected, $V_{laminar}$ is always greater than V_{PFR}. The effect, which is most marked at high conversions, decreases with increasing order. At conversions close to 100%, $V_{laminar}/V_{PFR}$ is around 1.5 for a first-order reaction, decreasing to about 1.3 for a second-order reaction. As laminar flow reactors are not common in industrial organic synthesis, we will not discuss them further here.

Fast Reactions

The basic assumption underlying the mixing models discussed so far is that reaction is slow compared to mixing, so that the ultimate effect of mixing is manifested only through the prevailing state of mixing at the commencement of the reaction. Let us consider a fast bimolecular reaction such as precipitation, neutralization, azo coupling, some substitution and many oxidation reactions, where reactant B is added to A present in the reactor. In other words, we are dealing here with unpremixed feeds, one containing A and the other containing B. If the reaction is very fast, has a half-life of a few seconds, and the residence time is very high, conversion will be independent of residence time and RTD, and the concentration of the limiting reactant would be close to zero. Mass transfer (i.e., mixing) and reaction no longer proceed consecutively but simultaneously, and the type of mixing involved is not macromixing characterized by RTD but turbulent-driven micromixing.

This whole subject is part of what has come to be known as computational fluid dynamics (CFD). Many books and reviews are available, for example, Curl (1963), Hill (1976), Libby and Williams (1980), Brodkey (1975), Nauman and Buffham (1983), Villermaux (1985, 1991), Fox (1996), and Baldyga and Bourne (1999). The general use of CFD in most reactor design calculations is open to question in a number of practical situations. This is particularly true of organic synthesis/technology.

The treatment of such situations tends to be highly mathematical and will not be attempted here. A number of models have been proposed. The two most important classes are the engulfment-deformation-diffusion (EDD) models of Bourne and his colleagues (see Bourne et al., 1981; Bourne, 1984; Baldyga and Bourne, 1989) and the interaction with a mean environment (IEM) model of Villermaux and his colleagues (see Villermaux, 1985). The EDD models basically postulate the formation of deforming eddies (laminated structures) of one fluid in the other followed by reaction and diffusion in the eddies. The latest version of these models suggests that engulfment is the dominating process, so that they are now referred to simply as the E-model.

Among the other models proposed are: 3- and 4-environment models (Ritchie and Togby, 1979; Mehta and Tarbell, 1983), stretch or laminar mixing model (Ou and Ranz, 1983a,b; Chella and Otino, 1984; Ranz, 1985), linear eddy model (Kerstein, 1991), and a generalized IEM model for nonpremixed turbulent reacting flows (Tsai and Fox, 1998).

A complex reaction scheme of great practical importance in organic technology is

$$A + B \to R$$
$$R + B \to S \qquad [13.1]$$

A simple analysis of this scheme can be made by assuming that when a solution of B is added to a solution of A, the rate at which A is transported by general circulation in the vessel to the point of addition of B is much greater than the rate of addition of B. Thus the concentration of A (and of any other component present in the reactor) entering the mixing/reaction zone will be the average value for the vessel. We thus restrict the treatment to micromixing and compare the diffusion time t_D and the reaction time t_R given by the equations

$$t_D = 2(\nu/\varepsilon)^{1/2} \text{arc sinh } (0.05\nu/D) \qquad (13.13)$$

$$t_R = (k_2[AB])^{-1} \qquad (13.14)$$

where [AB] represents the total concentration of A and B after mixing in a stoichiometric ratio of 1:1 (for reaction 13.1), D is the diffusivity, ν is the kinematic viscosity, and ε is the rate of energy dissipation per unit mass of solution. For $t_D \ll t_R$, the reaction is chemically controlled; for $t_D \gg t_R$, it is controlled by mixing; and for $t_D = t_R$, it is controlled by both kinetics and mixing.

Practical Implications of Mixing in Organic Synthesis

General considerations

For a given reaction one can choose the most appropriate type of reactor—such as plug flow, fully mixed, or recycle—to maximize conversion or yield, depending on whether the reaction is simple or complex. The main features of these reactors were considered in Part 1 and in Chapters 10 and 11. In general, where macromixing is the chief mixing phenomenon (as in the preceding cases), PFR is the reactor of choice for maximizing the yield of an intermediate in a complex reaction. However, for a reaction of the type

$$A \to R \to S$$
$$A + A \to T \qquad [13.2]$$

the recycle reactor can be the preferred choice (see Chapter 5).

For fast reactions of the type represented by reaction 13.2, carried out under the conditions stated earlier, micromixing becomes the dominant consideration. However, studies on the effect of micromixing in such reactions are sparse. Some examples are as follows: nitration of aromatic componds in general (Schofield, 1980), potassium metal–provoked reactions of aryl halides with amide and acetone enolate ions (Tremelling and Bunnett, 1980), coupling of 1-naphthol with diazotized sulfanilic acid (Bourne et al., 1981, 1985), reactions of o-(3-butenyl)-halobenzenes and 6-bromo-1-hexene with alkali metals in ammonia/*tert*-butyl alcohol solution (Meijs et al., 1986), and monoacylation of symmetrical diamines (Jacobson et al., 1987). In some fast reactions, hydrogen ions are produced,

which influence the ionic preeqilibria of the reagents and therefore the product distribution. Examples are coupling of 1-hydroxynaphthalene-6-sulfonic acid with benzenediazonium ion (Bourne et al., 1977a) and bromination of resorcinol (Bourne et al., 1977b).

The factors that control the chemical kinetics of a reaction are concentration and stoichiometric ratio. Thus increasing the dilution often enhances the yield, and increasing the ratio $[A]/[B]$ in reaction 13.2 raises the yield of R. The factors that affect mixing are the type of mixer and stirrer speed. These two categories of factors are quantified in terms of t_R and t_D defined by Equations 13.14 and 13.13, respectively. Example 13.1, based on the results of Bourne et al. (1988), clearly bring out the importance of these equations in analyzing the role of micromixing. They also demonstrate, in a general way, the importance of an often ignored fact: the role of addition sequence in reactions involving more than two reagents.

EXAMPLE 13.1

Experiments to illustrate the role of micromixing in determining the yield of a fast reaction and the importance of addition sequence.

A good example of reaction 13.1 is the coupling of 1-naphthol with diazotized sulfanilic acid producing two dyestuffs whose concentrations can be readily measured spectroscopically.[1] The reactions may be represented as

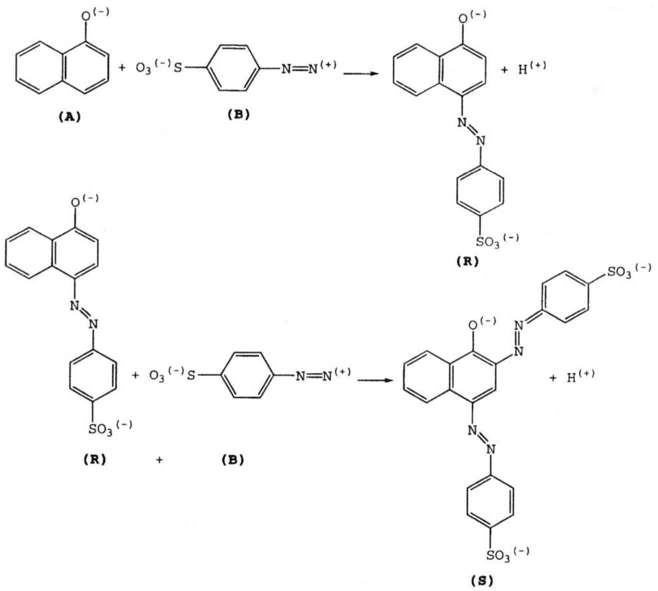

[E13.1.1]

[1] This series-parallel reaction has been extensively used as a model rection for studying the role of micromixing in complex reactions by Bourne and his coworkers (see, e.g., Bourne et al., 1981).

In experiments carried out in a 1-liter beaker stirrred by a 5-cm diameter turbine at a speed of 300 rpm, Bourne et al. (1988) report the following details: $[A] = 1.1071$ mol/m^3, $\nu = 10^{-6}$ m^2/s, $D = 8.5 \times 10^{-10}$m^2/s, $[A]/[B] = 1.05$, volumetric ratio of the two solutions = B added/A in beaker = 1/25, diazonium ion concentration = 2.754 mol/m^3 before mixing, and buffers used to control pH are $P1 = (Na_2CO_3 + NaHCO_3)$ for pH 10 and $P2 = (KH_2PO_4 + Na_2HPO_4)$ for pH 7. Three sets of conditions were studied, in which the rates were calculated from the ionic preequilibria and pK values of the reagents (Bourne et al., 1981).

1. pH 10: A was first buffered with P1 to pH 10, and 20 ml of B (pH $\cong$ 2) were slowly added with stirring. Calculations gave $t_R \cong 8 \times 10^{-4}$ s, $t_D = 0.035$ s, and $e = 7.2 \times 10^{-2}$ W/kg, yield of $R = 98.1\%$.
2. pH 7: Same as above (with the same addition sequence), but with buffer $P2$. Calculations give $t_R = 0.14$ s, $t_D = 0.035$ s (unchanged), yield of $R \cong 99.9\%$.
3. pH 2–10: B (20 ml, pH 2) was added rapidly to 500 ml of unbuffered A in vessel, then 20 ml of buffer P1 were added over 4 min so that coupling could proceed, yield of $R \cong 99.9\%$. Note that the addition seuence was changed in this experiment.

CONCLUSIONS

The following conclusions can be drawn from these experiments.

1. The first experiment, with the addition sequence A-P-B, corresponds to $t_D \gg t_R$. It was therefore controlled by mixing and independent of kinetics.
2. By decreasing the pH to 7 in the second experiment (but with the same sequence A-P-B), the preequilibrium concentrations of the reactive species were changed. This resulted in a drastic reduction of the reaction rate, giving $t_D < t_R$—that is, the reaction was chemically controlled.
3. When the addition sequence was changed (A-P-B to A-B-P), it was found that $t_D = 0$, or $t_D < t_R$, and the reaction was again controlled by kinetics.
4. The controlling mechanism (and frequently the yield) can be affected by the addition sequence of the reagents.

A striking illustration of the role of addition sequence of reagents

The effect of addition sequence was only mildly apparent in Example 13.1. It has been more emphatically illustrated in the esterification reaction between maleianic acid (1) and thionyl chloride (2) (Kumar and Verma, 1984).

[13.3]

When reactants 1 and 2 are mixed and the mixture is poured in any absolute alcohol 3, the same compound 4 results. On the other hand, when 1 is dissolved in absolute methanol or ethanol 3, and 2 is then added dropwise while shaking, compound 5 is formed. Thus the change of sequence from 1-2-3 to 1-3-2 leads to a completely different product. This is ascribed to the mixing effect, but a firmer confirmation of this is needed.

MULTIPLE SOLUTIONS

We noted that Equation 4.77 is very important in the nonisothermal operation of a CSTR. This algebraic equation has more than one solution and leads to the concept of multiple steady states (MSS). On the other hand, the differential equation that characterizes a PFR has only one solution, that is, the PFR operates at a single steady state. Multiple steady states are of particular concern to us because they can occur in the physically realizable range of variables, between zero and infinity, and not at absurd values such as negative concentration or temperature (which would then be no more than a mathematical artifact).

Multiple Steady States in a CSTR

We examine this concept of multiple steady states further by reconsidering Equation 4.77

$$J[(T - T_0) + \kappa(T - T_c)] = \frac{\bar{t}k(T)}{1 + \bar{t}k(T)} \tag{13.15}$$

where

$$J = \frac{\rho C_p}{(-\Delta H_r)[A]_0}(1/K), \quad \kappa = \frac{UA_h}{Q_0 \rho C_p} \text{ (dimensionless)} \tag{13.16}$$

All the terms on the LHS, which we shall collectively designate q_-, represent heat removal (for an exothermic reaction). The term on the RHS, designated q_+, represents heat generation. Thus Equation 13.15 can be more expressively recast as

$$q_-(T) = q_+(T) \tag{13.17}$$

From the nature of the terms, we see that q_- is a linear function of T, whereas q_+ is a nonlinear function of T. Figure 13.6a shows the generation curve specifically for a first-order irreversible reaction, and Figure 13.6b is for a first-order reversible reaction. As can be seen from the figures, the heat generation and removal curves can intersect at one point or at three points. Operation at the temperatures corresponding to these points is referred to as autothermal operation, and the three states are referred to as multiple steady states (see Van Heerden, 1953). Many books, reviews, and leading articles on the general subject of instabilities and oscillations in chemical reactions and reactors have since appeared, for example, Horn (1961), Horn and Troltenier (1961a,b), McGreavy and Adderley (1973, 1974), Luss (1977), McGreavy and Dunbobbin (1984), Aris

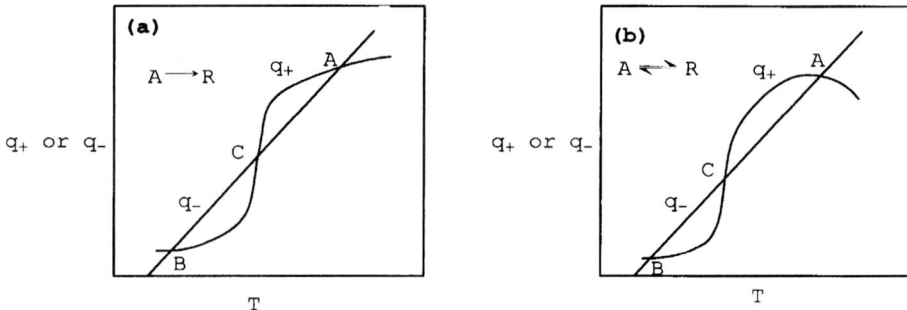

Figure 13.6 Multiple steady states in (a) irreversible and (b) reversible reactions

(1984), Luss and Balakotaiah (1984), Morbidelli et al. (1987), Razon and Schmitz (1987), Gray and Scott (1990), Slin'ko and Jaeger (1994), and will not be discussed further in this book.

Adiabatic CSTR

When a CSTR is operated adiabatically, the heat transfer term in Equation 13.15 vanishes, and

$$J(T - T_0) = \frac{\bar{t}k(T)}{1 + \bar{t}k(T)} \qquad (13.18)$$

Now let us prepare constant $\bar{t}$ plots. For this, we use Equation 4.73 developed in Chapter 4,

$$X_A = \frac{\bar{t}k(T)}{1 + \bar{t}k(T)} \qquad (13.19)$$

This represents the RHS of Equation 13.15 which can be plotted as X_A versus T for different constant values of residence time $\bar{t}$. A typical plot is shown in Figure 13.7. This is the sigmoidal material balance plot. We also plot X_A versus T (the LHS of Equation 13.15) which gives a straight line with slope J.

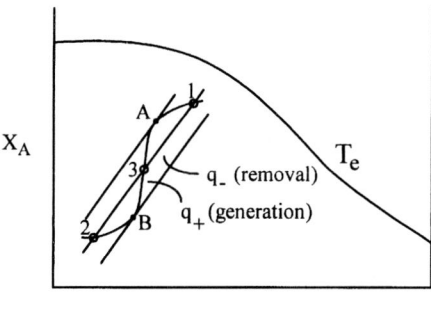

Figure 13.7 Range of multiple steady states in an adiabatic CSTR (on an X–T plot).

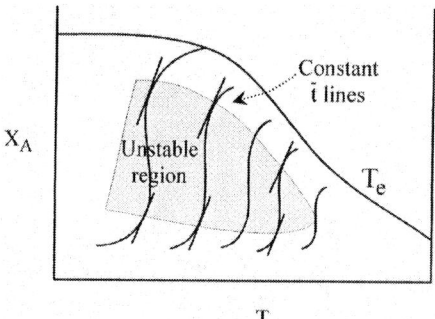

Figure 13.8 The unstable region of adiabatic CSTR operation.

Theoretically, the two curves can intersect at three points. Points (1) and (2) are physically realizable and represent the two steady states of operation, whereas point (3) is not physically realizable and represents the (third) unstable steady state. Note that for such a state to exist, the slope of the material balance line must be greater than the slope J of the energy balance line. This is possible only when the energy balance line lies between the tangential lines at A and B.

When the stability limits represented by A and B are marked on a number of constant $\bar{t}$ lines, as shown in Figure 13.8, the area enclosed by them can be identified as the region of instability of an adiabatic CSTR on the X_A–T plane. Fortunately, however, the practical limits of operation of such a reactor usually lie outside this region, and hence the adiabatic CSTR is almost always in the stable region.

Stability of the steady states

We have seen that there can be two stable steady states, one at a low temperature corresponding to quench conditions and the other at a high temperature corresponding to ignition conditions. An important consideration is the approach to the steady state, for that would determine the fate of a reaction. Thus, if we started at any initial condition, will the reaction approach one of the steady states, and if so how quickly? Alternatively, if we started the reaction at a condition close to a stable steady state, will it approach that steady state? To answer these questions we must modify the steady-state mass and heat balance equations to include a time-dependent component. The resulting transient equations are

$$\bar{t}\frac{dX_A}{dt} = -X_A + \frac{\bar{t}(-r_{Af})}{[A]_0} \tag{13.20}$$

$$\bar{t}\frac{dT}{dt} = T_0 - T - k(T - T_c) + \frac{\bar{t}}{J[A]_0}(-r_{Af}) \tag{13.21}$$

A steady state is said to be stable when the system returns to it after a small perturbation. Alternatively, the perturbation can grow exponentially with time, or the reaction can be quenched to extinction. To determine which of these alternatives will prevail, we linearize the equation about the steady state and examine the behavior of the perturbations in conversion and temperature, that is, of

$$\alpha = X_A - X_{A,SS} \tag{13.22}$$

$$\beta = T - T_{SS} \tag{13.23}$$

These disturbances can be expressed in terms of the simple linear first-order differential equations

$$\bar{t}\frac{d\alpha}{dt} = a\alpha + b\beta \tag{13.24}$$

$$\bar{t}\frac{d\beta}{dt} = c\alpha + d\beta \tag{13.25}$$

which then can be combined to give

$$\frac{d^2\alpha}{dt^2} - (a+d)\frac{d\alpha}{dt} + (ad+bc)\alpha = 0 \tag{13.26}$$

where $a, b, c,$ and d are constant coefficients defined by lengthy expressions involving the steady state values of X_A and T, the activation energy of the reaction, the rate constant, and the residence time. The general solution to this equation is

$$\alpha(t) = C_1 e^{\lambda t/\bar{t}} + C_2 e^{-\lambda t/\bar{t}} \tag{13.27}$$

where the eigenvalues $\pm\lambda$ are related to the coefficients by the expression

$$\lambda = \frac{1}{2}(a+d) \pm \frac{1}{2}[(a-d)^2 + 4bc]^{1/2} \tag{13.28}$$

The solution is stable if the real part of λ is negative, that is, if $[(a-d)^2 + 4bc]^{1/2} > 0$. On evaluating this term using actual values of the coefficients $a, b, c,$ and d, we find that (1) when there is only one steady state, it is a stable one; and (2) when there are three steady states, as shown in Figure 13.8, only the extreme states are stable and the middle one is unstable.

Further analysis of MSS in an adiabatic CSTR reveals some interesting features (see Schmidt, 1998), which we summarize in the form of the following qualitative statements.

1. When the initial temperature T_0 in an adiabatic CSTR is slowly varied, different heat removal lines are obtained, as shown in Figure 13.9a. As expected, either one or three steady states can be obtained. When the corresponding reactor temperatures T are plotted versus T_0, a curve of the type shown in Figure 13.9b results, which clearly reveals a temperature hysteresis.
2. When the initial concentration $[A]_0$ is varied, the resulting situation for an adiabatic CSTR is sketched in Figure 13.10. The number of steady states

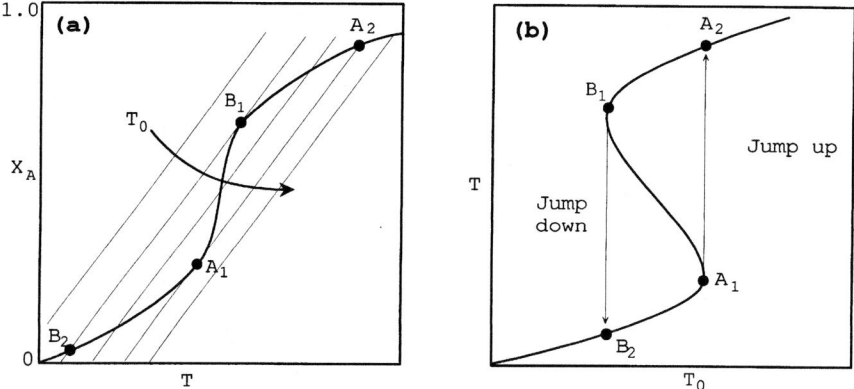

Figure 13.9 The effect of initial temperature T_0 on multiple steady states in an adiabatic CSTR.

varies from a single low conversion steady state at low $[A]_0$ to a single high-conversion steady state at high $[A]_0$ and multiple steady states in between.

3. An important consideration in reactor operation is the time dependence of MSS, that is, the transients that develop during start-up or shutdown. Qualitative curves of X_A versus time for an adiabatic CSTR are shown in Figure 13.11 for a simple first-order irreversible reaction. Note that the system converges to the upper or lower steady state ($X_{A,ss,u}$ or $X_{A,ss,l}$). Any apparent approach to the middle unsteady steady state is deceptive because it quickly turns to either of the extreme steady states. The behavior is identical with respect to temperature.

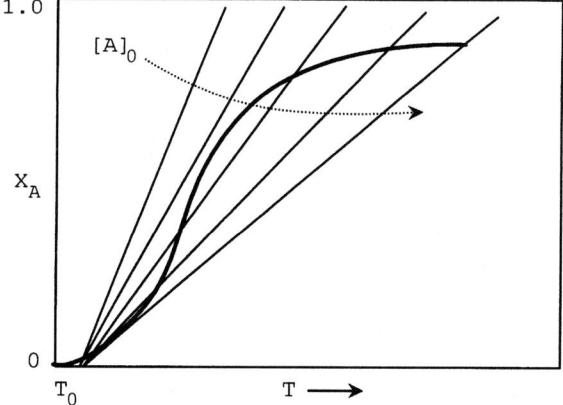

Figure 13.10 The effect of initial concentration on multiple steady states in an adiabatic CSTR.

414   Reactor Design

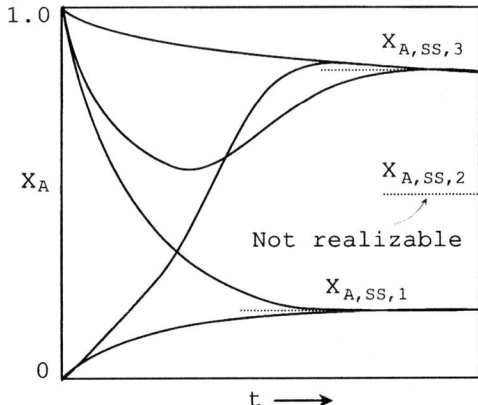

Figure 13.11   Transients in an adiabatic CSTR.

4. When a CSTR is operated nonadiabatically, a heat removal term (for an exothermic reaction) must be added to the system equations. This complicates the situation and increases the likelihood of multiple steady states.

Forced Unsteady-State Operation (FUSO)

Unsteady-state operation of a catalytic reactor can be exploited to enhance the efficiency of a catalytic reaction. A simple way of achieving unsteady-state operation is to force it by periodic reversals of the direction of flow of the reactant mixture. An important parameter in creating this condition is the ratio α_{FUSO} of the time between flow reversals to the time needed for transition to the steady state. For $\alpha_{FUSO} \leq 1$, unsteady operation results, and for $\alpha_{FUSO} > 1$, pseudo-steady-state conditions prevail.

We illustrate the principle of FUSO by considering the oxidation of SO_2, an inorganic reaction that has been extensively studied in this respect (see Unni et al., 1973; Briggs et al., 1977; Boreskov and Matros, 1983; Matros et al., 1984; Matros, 1989, 1996; Matros and Bunimovich, 1996). A bed of catalyst is maintained at 500 °C, and a mixture of SO_2, N_2 and O_2 is introduced at one end of it at a low temperature of 50 °C. Regions of the bed adjacent to the inflowing mixture get gradually cooled as shown in Figure 13.12. As the gases move inward, the heat of reaction causes a rise in the temperature of these regions, and at a certain time before the high temperature wave front leaves the bed, the gas flow is reversed. Now the leaving zone becomes the entering zone and begins to cool because of the low temperature of the entering gas. The temperature and conversion profiles at 4, 15, 30, and 45 minutes following the reversal are shown in Figure 13.13.

At 90 minutes, the flow is reversed again, and this is repeated every 45 minutes. The resulting temperature and conversion profiles settle into a pattern

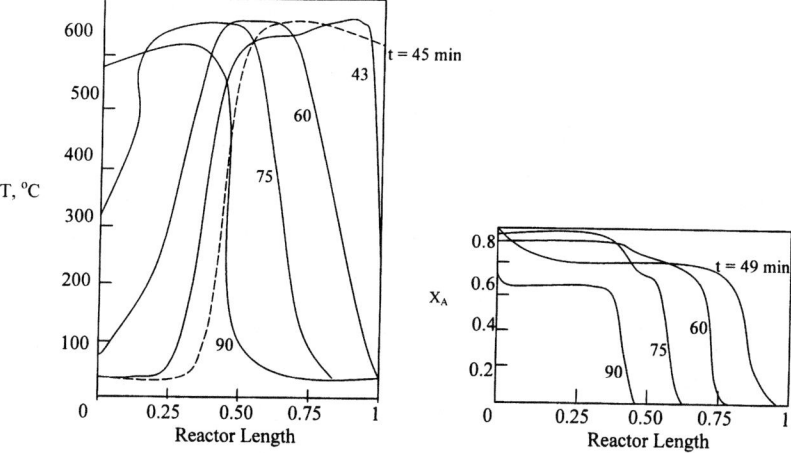

Figure 13.12 The principle of USO: temperature and conversion profiles in an adiabatic catalytic reactor before the first reversal of flow direction (redrawn from Matros et al., 1984).

close to that shown in Figure 13.13. Every temperature profile exhibits a region of steep rise (from T_0 to T_{max}), with the conversion rising to the equilibrium value at T_{max}. This is followed by a stretch of near-constant temperature close to the maximum. Then the heated inlet gas enters the opposite region of the bed and heats it. But the extent of heating progressively decreases, and a falling temperature profile is established. This falling profile is optimal for a reversible exothermic reaction (see Chapter 5), and hence a conversion higher than at T_{max} is obtained. Rigorous mathematical analysis of this situation has been attempted by many investigators and will not be considered here (see Matros, 1996, for a

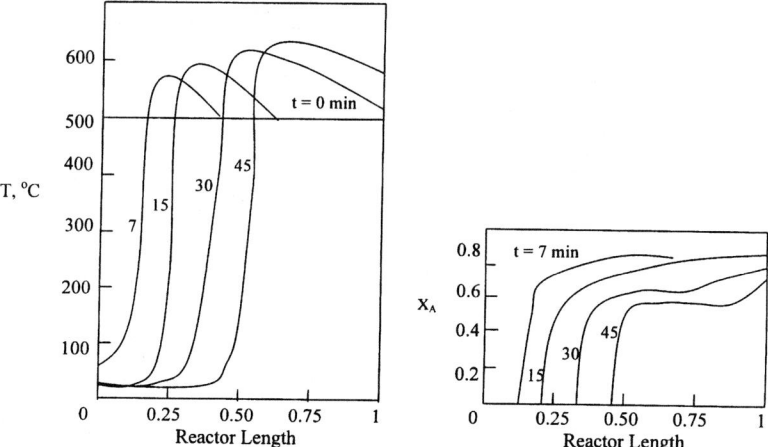

Figure 13.13 Principle of USO (contd.): temperature and conversion profiles after the first reversal of flow direction (redrawn from Matros et al., 1984).

review). It is clear, however, that FUSO enables attainment of higher conversions at low inlet temperatures in a single catalyst bed without the need for any heat exchanger to establish the desired falling profile.

It is important to note that forced cycling is not restricted to the temperature effect. Concentration forcing can also be used to advantage (see Silveston and Hudgins, 1987). Thus FUSO can be considered a potential strategy for rate enhancement. Although a discussion of the various strategies for rate enhancement will be presented in Part V, this particular strategy, as well as the mixing strategy, are covered in this chapter because they are directly related to its subject matter.

Notation to Part III

A	Species A; group defined by Euation 12.14.
A'	Group defined by Equation 12.45.
A_c	Cross-sectional area, m^2.
A_h	Heat transfer area, m^2.
Ar	Archimedes (or Galileo) number defined by Equation 12.28.
a	Constant of Equation 13.24.
a_p	Surface area of pellet, m^2/kg.
a'	Normalized concentration of A, [A]/[A]$_0$.
B	Species B; group defined by Equation 12.14.
B'	Group defined by Equation 12.45.
B''	Group defined by Equation 12.50.
b	Ratio E/R_g, K; constant of Equation 13.24
C	Species C; group defined by Equation 12.14.
C_1, C_2	Parameters of Equation 13.27.
C_p	Heat capacity, kcal/kg K.
C_{pG}	Heat capacity specifically for a gas, kcal/kg K.
C_{pc}	Heat capacity of control fluid, kcal/kg K.
C_{pm}	Heat capacity, kcal/mol K.
c	Constant of Equation 13.25.
c	($N \times 1$) Vector of component concentrations.
D	Product; diffusivity, m^2/s.
D_{eA}	Effective diffusivity of A, m^2/s.
D_{er}	Effective radial diffusivity, m^2/s.
$D_{e\ell}$	Effective axial diffusivity, m^2/s.
d	Constant of Equation 13.25.
d_c	Decay constant, 1/s.
d_b	Bubble diameter, m.
d_e	Equivalent diameter defined by Equation 12.41, m.

d_o	Outer diameter of tube, m.
d_p	Diameter of particle, m.
d_T	Diameter of tube, m.
E	Product; activation energy for reaction, kcal/mol.
E_{ad}	Activation energy for adsorption, kcal/mol.
E_i	Activation energy for reaction i, kcal/mol.
F	Product, molar feed rate, mol/s.
$F(t)$	Cumulative function.
F_i	Molal feed rate of i, mol/s.
F_t	Total molar feed rate, mol/s.
f_B	Bulk voidage of bed.
f_f	Voidage of fluidized bed.
f_{mf}	Bed voidage at minimum fluidization.
f_{SB}	Fraction of total time in semibatch mode.
G	Product.
G_M	Mass flow rate of gas, kg/m^2 s.
g	Acceleration due to gravity, m^2/s.
ΔH_{ex}	Amount of heat exchanged by the control fluid, kcal/s.
H_i	Henry's law constant of i, mol/m^3 atm.
$(-\Delta H_j)$	Exothermic heat of reaction for the jth reaction, kcal/mol.
$(-\Delta H_r)$	Exothermic heat of reaction, kcal/mol.
ΔH_v	Heat of vaporization, kcal/mol.
h_{fp}	Fluid-particle heat transfer coefficient, kcal/m K s.
h_w	Wall heat transfer coefficient, kcal/m K s.
i	Species i.
i	Represents species i initially present in reactor.
$[i]$	Concentration of species i, m^3/s.
$[i]_N$	Concentration of i in the Nth stage.
$[i]_v$	Concentration of i in the vapor phase.
J	Heat removal parameter defined by Equation 12.23 and Equation 13.16.
K_B	Adsorption equilibrium constant of B, m^3/mol.
K_B^0	Temperature independent adsorption equilibrium constant of B, m^3/mol (or 1/atm).
K_F	Overall rate constant defined by Equation 12.36.
K_0, K_m, K_b, K_d	Groups defined in Figure 12.12.
$\hat{K}_F$	Product $K_F(\ell/u_b)$.
K'	Group defined by Equation 12.40.
K_R'	Sum $(K' + K_b + K_d)$.
K_1, K_2	Parameters.
k, k_n	General symbol for reaction rate constant (n = order).
k_i^0	Arrhenius preexponential term for reaction i, units of the rate constant.
k_B	First- or second-order rate constant, 1/s or m^3/mol s.
k_{bc}, k_{ce}, k_{be}	Bubble-to-cloud, cloud-to emulsion, bubble-to-emulsion mass transfer coefficients, 1/s.
k_G	Gas-side mass transfer coefficient, 1/s.
k_i	Reaction rate constant of the ith reaction, appropriate units.
k_p	Reaction rate constant, mol/m^3 s atm^n.
k_v	Reaction rate constant based on catalyst volume, (m^3 cat/mol)$^{n-1}$(1/s).

Notation to Part III

k_{v0}	Reaction rate constant defined specifically for a fresh catalyst, (m^3 cat/mol)$^{n-1}$(1/s).
k_w	Reaction rate constant based on catalyst weight, (m^3)^n/kg cat mol^{n-1} s.
k_{wp}	Reaction rate constant based on catalyst weight and reactant partial pressure, mol/kg cat s atm^n.
$\left.\begin{array}{l} k^0, k_i^0, \\ k_p^0, k_{wp}^0 \end{array}\right\}$	Arrhenius preexponential terms, units of the rate constants.
k_+	Forward reaction rate constant, appropriate units.
k_-	Reverse reaction rate constant, appropriate units.
k'_b	Bubble-side mass transfer coefficient, m/s.
$k'_{bc}, k'_{ce}, k'_{be}$	Bubble-to-cloud, cloud-to emulsion, bubble-to-emulsion mass transfer coefficients, m/s.
k'_G	Gas-side mass transfer coefficient, m/s.
k'_{GP}	Gas-side mass transfer coefficient, mol/m^2 s atm.
L	Reactor length, m; liquid.
L_f	Height of fluidized bed, m.
L_t	Total height of fluidized bed including dilute region, m.
L_{mf}	Bed height at minimum fluidization, m.
$[L]_t$	Total liquid phase concentration, mol/m^3.
M	Number of reactions; Mth reaction; molecular weight.
m	Reaction order.
N	($N \times 1$) Matrix of moles of each component.
N	Number of components or reactors; any total number.
N_j	Number of moles of species j.
n	Reaction order.
P	Product; pressure, atm.
p_{cr}	Critical partial in reactor pressure corresponding to critical temperature, atm.
p_i	Partial pressure of i, atm.
p_m	Partial pressure in reactor corresponding to the maximum of the maxima curve, atm.
p_0^ℓ	Lower bound of initial partial pressure for onset of sensitivity, atm.
p_0^u	Upper bound of initial partial pressure for onset of sensitivity, atm.
Q	Volumetric flow rate, m^3/s.
Q_*	Dimensionless flow rate defined by Q/Q_R.
Q_{Lf}	Outflow rate of liquid, m^3/s.
Q_n	Net flow rate, m^3/s.
Q_0^*	Dimensionless flow rate defined by Q_0/Q_R.
Q_R	Arbitrarily chosen volumetric flow rate parameter of the variable volume reactor, m^3/s.
Q_T	Flow rate in a single tube, m^3/s.
Q_{Vf}	Outflow rate of vapor, m^3/s.
R	Species R; recycle ratio.
q'_+, q_-	Heat generation and heat removal terms.
Re'_{mf}	Reynolds number at minimum fluidization, $d_p u_{mf} \rho_G / \mu$.
R_g	Gas law constant, kcal/mol K.
r	General notation for reaction rate; radial coordinate, m.
r	($M \times 1$) Vector of reaction rates.

r_i'	Specific rate of reaction of i, mol/m² s.
r_b	Radius of bubble, m.
r_c	Radius of cloud, m.
r_i, r_{Vi}	Volumetric rate of reaction of i, mol/m³ reactor s.
r_j	Reaction j in a multiple reaction scheme.
r_T	Radius of tube, m.
r_{wi}	Weight based rate of reaction of i, mol/kg cat s.
S	Species S.
$s_{bb}, s_{cb}, S_{ec}, s_{wb}$	Volume of solid in bubble, cloud, emulsion, or wake/volume of bubble phase.
S_i	Selectivity for species i.
$S_{V,F}(S_{V,L})$	Fluid (liquid) space velocity, 1/s.
T	Species T; temperature, K.
T_c	Temperature of control fluid, K.
T_{cr}	Critical temperature given by Equation 12.15.
T_e	Equilibrium temperature, K.
T_{fU}	Temperature defined in Table 12.4.
T_i	Temperature of the ith stage, K.
T_m	Temperature corresponding to p_m, K.
T_{opt}	Optimum temperature for maximum rate, K.
T_p	Particle temperature, K.
T_U	Temperature defined in Table 12.4.
T_w	Wall temperature, K.
t	Time, s.
$\bar{t}$	Residence time $V/Q_0 (= V[A]_0/F_{A0})$, s.
$\bar{t}_i$	Residence time in reactor i, s.
t_p	Production or on-stream time, s.
t_{pl}	Total reaction (or decay) time, s.
t_{SB}	Time of semibatch reaction, s.
$\hat{t}$	Dimensionless production time, t_p/t_{pl}.
$t*$	Dimensionless time defined as tQ_R/V_R.
t_1, t_2, t_3, t_4	Points in time giving filling time $= t_1$, reaction time $= (t_2 - t_1)$, emptying time $= (t_3 - t_2)$, and down time $= (t_4 - t_3)$.
U	Product; overall heat transfer coefficient, kcal/m² K s.
u	Fluid velocity, m/s.
u_b	Total rise velocity of bubble, m/s.
u_{br}	Free rise velocity of bubble, m/s.
u_e	Emulsion gas velocity, m/s.
u_i	Interstitial velocity of gas, m/s.
u_{mb}	Velocity for onset of bubbling, m/s.
u_{mf}	Velocity for onset of fluidization, m/s.
u_{ms}	Velocity for onset of slugging, m/s.
u_0	Entering velocity, m/s.
V	Volume, m³.
$V*$	Dimensionless volume defined by V/V_R.
V_m	Maximum volume of variable volume reactor, m³.
V_R	Arbitrarily chosen volume parameter of the variable volume reactor, m³.
V_r	Reactor volume, often referred to merely as V, m³.
V_T	Volume of single tube, m³.

W	Catalyst weight, kg.
W_c	Flow rate of the control fluid, kg/s.
$W(t)$	Washout function.
X_A^*	Relative conversion.
X_{AR}	Conversion of A at entry to recycle flow.
X_i	Conversion of species i.
$\bar{X}_A$	Average conversion of A in a fixed-bed reactor.
Y_i	Yield of species i.
Y_p	Point yield.
y_i	Mole fraction of component i.
$\bar{y}_A$	Average mole fraction of A.
z	Dimensionless distance, ℓ/L.
z'	Dimensionless distance, ℓ/d_p.
z_f	Dimensionless height of fluidized bed, L_t/L_f.
z_{ft}	Total height of fluidized bed including the dilute region, $L_t/L_f (> 1)$.

Greek

α	Mole ratio of species A to B in Example 10.2; parameter defined by Equation 10.17; ratio of bubble to emulsion gas velocities, u_b/u_i; concentration perturbation given by Equation 13.22.
$\bar{\alpha}$	Group defined in Equation E12.3.7.
β	Temperature perturbation given by Equation 13.23.
$\bar{\beta}$	Group defined in Equation E12.3.8.
δ	Bubble fraction in a fluidized bed.
$\delta(t)$	Residence time distribution.
ε	Effectiveness factor of catalyst; rate of energy dissipation, m²/s³.
ε_A	Volume change parameter.
κ	Heat removal group defined by Equation 13.16.
ℓ	Length parameter, m.
λ	Decay group, $d_c t_{p1}$; thermal conductivity, kcal/m K.
$\lambda_{e\ell}$	Effective axial thermal conductivity, kcal/m K.
λ_{er}	Effective radial thermal conductivity, kcal/m K.
μ	Viscosity, kg/m s.
ν	Kinematic viscosity, m²/s.
$\mathbf{v}$	$(M \times N)$ Matrix of stoichiometric coefficient.
ν_i	Stoichiometric coefficient of species i.
ν_{ij}	Stoichiometric coefficient of species i in reaction j.
$\theta_1, \theta_2, \theta_3, \theta_4$	Dimensionless filling time, reaction time, emptying time, and down time.
ρ	Density, kg/m³.
ρ_B	Bulk density of catalyst bed, kg/m³.
ρ_G	Density of gas (used specifically for fluidized beds), kg/m³.
ρ_m	Molar density, mol/m³.
ρ_S	Density of solid catalyst, kg/m³.
σ	Standard deviation.
τ	Dimensionless rate constant defined by $\tau = k[A]^{n-1}\bar{t}$, where n is the reaction order (equals $k\bar{t}$ or kV/Q for a first-order reaction), $(m^3/mol)^{n-1}$ 1/s.
τ^*	Dimensionless rate constant for a semibatch reactor defined as kV_R/Q_R for a first-order reaction.

422 Reactor Design

τ'	Dimensionless temperature, T/T_0.
ξ	($M \times 1$) Matrix of extent of reaction, mol.
ξ'	($M \times 1$) Matrix of modified extent of reaction, mol.
ξ_j	Extent of jth reaction, mol.
ξ_j'	Modified extent of jth reaction, mol/s.
T_e	Equilibrium curve in the X_A–T plane.
T_m	Locus of rate maxima in the X_A–T plane.
$\Re_H$	Group $(-\Delta H_r)d_p/u\rho_G C_p T_0$.
$\Re_M$	Group $d_p/u[A]_0$.

Subscripts/Superscripts

Subscripts

a	Actual (observed).
b	Bubble, bulk.
c	Cloud; control fluid; catalyst.
cr	Critical value
e	Effective; emulsion; equilibrium.
F	Any fluid, L or G.
f	Outlet condition; fluidized bed.
ft	Final total height, m.
G	Gas.
i	Initial condition; ith stage.
L	Liquid.
ℓ	Axial coordinate, lower limit of initial partial pressure.
MF	Mixed flow.
m	Maximum.
mf	Minimum fluidization.
ms	Minimum slugging.
0	Inlet condition.
R	Point of mixing of recycle and fresh feed.
r	Radial coordinate, m.
SB	Semibatch.
s	Surface; slugging; solid.
ss	Steady state.
T	Tube.
t	Total.
V	Based on reactor volume; vapor.
v	Based on catalyst volume.
w	Wall, wake; based on catalyst weight.

Superscripts

ℓ	Lower limit of initial partial pressure.
u	Upper limit of initial partial pressure.

References to Part III

(Note: General references not referred to in the text are marked with an asterisk.)

Aris, R., *The Optimal Design of Chemical Reactors—A Study in Dynamic Programming*, Academic Press, New York, 1961; *Discrete Dynamic Programming*, Blaisdell, Mass, 1964; *Introduction to the Analysis of Chemical Reactors*, Prentice-Hall, Englewood Cliffs, N.J., 1965; *Elementary Chemical Reactor Analysis*, McGraw-Hill, New York, 1969; In *Frontiers in Chemical Reaction Engineering* (Eds. Doraiswamy, L.K. and Mashelkar, R.A.), Wiley Eastern, New Delhi, India (1984).
Avidan, A., and Edwards, M. Modeling and Scale-up of Mobil's Fluid Bed MTG Process, *5th Int. Conf. on Fluidization*, Elsinore, Denmark, 18–23 May, 1986.
Baburao, K., and Doraiswamy, L.K. *AIChE J.*, **13**, 397 (1965).
Balakotaiah, V., and Dommeti, S.M.S. *Chem. Eng. Sci.*, **54**, 1621 (1999a).
Baldyga, J., and Bourne, J.R. *Chem. Eng. Commun.*, **28**, 259 (1984a); *Chem. Eng. J. Biochem. Eng.*, **42**, 83 (1989).
Barac, G. Process analysis and yield enhancement for the production of β-hydroxy-β-methyl butyrate (HMB), Ph.D. Thesis, Iowa State University, Ames, IA, 1998.
Barkelew, C.H. *Chem. Eng. Prog. Symp. Ser.*, **55**(25), 38 (1959).
Behie, L.A., Bergougnou, M.A., and Baker, C.G.J. *Fluidization Technology*, Vol. 1 (ed., Keairus, D.L.), Hemisphere, Washington D.C., 1976.
Bellman, R. *Dynamic Programming*, Princeton University Press, Princeton, N.J., 1957.
Bellman, R., and Dryfus, S.E. *Applied Dynamic Programming*, Princeton University Press, Princeton, N.J., 1962.
Bilous, O., and Amundson, N.R. *Chem. Eng. Sci.*, **5**, 81, 115 (1956).
Bolthrunis, C.O. *Chem. Eng. Prog.*, **5**, 51 (1989).
Boreskov, G.K., and Matros, Yu. Sh. *Applied Catal.*, **5**, 337 (1983).
Bosworth, R.C.L. *Phi. Mag.*, **39**, 847 (1948).
Botterill, J.S.M. *Br. Chem. Eng.*, **11**, 122 (1966).
Bourne, J.R., Crivelli, E., and Rys, P. *Helv. Chim. Acta*, **60**, 2944 (1977a).

Bourne, J.R., Hilber, C., and Tovstiga, G. *Chem. Eng. Commun.*, **37**, 293 (1985)
Bourne, J.R., *International Symposium on Chemical Reaction Engineering* (ISCRE) 8, *I. Chem. Symp. Ser.* 87, Edinburgh, 1984.
Bourne, J.R., Kozicki, F., and Rys, P. *Chem. Eng. Sci.*, **36**, 1643 (1981).
Bourne, J.R., Ravindranath, K., and Thoma, S. *J. Org. Chem.*, **53**, 5166 (1988).
Bourne, J.R., Rys, P., and Suter, K. *Chem. Eng. Sci.*, **32**, 711 (1977b).
Briggs, J. P., Hudgins, P. R., and Silveston, P. L. *Chem. Eng. Sci.*, **32**, 1087 (1977).
Brodkey, R.S. Ed. *Turbulence in Mixing Operations*, Academic Press, New York, 1975.
Burghardt, A., and Skryzpek, J. *Chem. Eng. Sci.*, **29**, 1311 (1974).
Butt, J.B. *Reaction Kinetics and Reactor Design*, Prentice-Hall, Englewood Cliffs, N.J., 1980.
* Carberry, J.J. *Chemical and Catalytic Reaction Engineering*, McGraw-Hill, New York, 1976.
* Carberry, J.J., and Varma, A. *Chemical Reaction and Reactor Engineering*, Dekker, New York, 1987.
Cleland, F.A., and Wilhelm, R.H. *AIChE J.*, **2**, 489 (1956).
Chella, R., and Otino, J.M. *Chem. Eng. Sci.*, **39**, 551 (1984).
Chen, G.T., Shang, J.Y., and Wen, C.Y. In *Fluidization, Science and Technology* (Eds., Kwauk, M., and Kunii, D.), Sciences Press, Beijing, 1982.
Chitester, D.C., et al. *Chem. Eng. Sci.*, **39**, 253 (1984).
Cholette, A., and Cloutier, L. *Can. J. Chem. Eng.*, **37**, 105 (1959).
Couderc, J.P. In *Fluidization*, 2nd ed. (eds., Davidson, J.F., Clift, R., and Harrison, D.), Academic Press, New York, 1985.
Coward, I., and Jackson, R. *Chem. Eng. Sci.*, **20**, 911 (1965).
Cropley, J.B. *Chem. Eng. Prog.*, **32** (1990).
Curl, R.L. *AIChE J.*, **9**, 175 (1963).
Danckwerts, P.V. *Chem. Eng. Sci.*, **2**, 1 (1953).
Davidson, J.F. *AIChE Symp. Ser.* (No. 128), **69**, 16 (1973).
* Davidson, J.F., Clift, R., and Harrison, D. *Fluidization*, 2nd ed., Academic Press, London, 1985.
Davidson, J.F., and Harrison, D. *Fluidized Particles*, Cambridge University Press, New York, 1963; *Fluidization*, Academic Press, London, U.K., 1971.
Davidson, J.F., Harrison, D., Darton, R.C., and Lanauze, R.D. *Chemical Reactor Theory—A Review*, Eds., Lapidus, L., and Amundson, N.R.), Prentice-Hall, Englewood Cliffs, N.J., 1978.
Denbigh, K.G. *Trans. Faraday Soc.*, **40**, 352 (1944); **43**, 648 (1947); *J. Appl. Chem.*, **1**, 227 (1951).
Denbigh, K.G., Hicks, M., and Page, F.M. *Trans. Faraday Soc.*, **44**, 479 (1948).
Denbigh, K.G., and Turner, J.C.R. *Chemical Reactor Theory*, Cambridge University Press, New York, 1971.
Denn, M.M., and Aris, R. *Ind. Eng. Chem. Fundam.*, **4**, 213 (1965).
Dente, M., and Collina, A. *Chim. Ind.*, **46**, 752 (1964).
* Doraiswamy L.K. *Recent Advances in the Engineering Analysis of Chemical Reacting Systems*, Wiley Eastern, New Delhi, 1984.
* Doraiswamy, L.K., and Mashelkar, R.A., Eds. *Frontiers in Chemical Reaction Engineering*, Vols. 1 and 2, Wiley Eastern, New Delhi, 1984.
* Doraiswamy, L.K., and Mujumdar, A.S. *Transport in Fluidized Particle Systems*, Elsevier, New York, 1989.
Doraiswamy, L.K., Sadasivan, N., and Venkitakrishnan, G.R. NCL Report, Pune, India (1972); (1973).

Doraiswamy, L.K., Sadasivan, N., Mukherjee, S.P., and Venkitakrishnan, G.R. NCL Report, Pune, India (1968).
Doraiswamy, L.K., and Sharma, M.M. *Heterogeneous Reactions: Analysis, Examples, and Reactor Design*, Vol. 1, Wiley, New York, 1984.
Douglas, J.M. *Chem. Eng. Prog. Symp. Ser.*, **61**(48), 1 (1964).
Elridge, J.W., and Piret, E.L. *Chem Eng. Prog.*, **46**, 290 (1950).
Fan, L.T., Lee, C.J., and Bailie, R.C. *AIChE J.*, **8**, 239 (1962).
Fine, F.A., and Bankhoff, S.G. *Ind. Eng. Chem. Fundam.*, **6**, 288 (1967).
Fox, R.O. *Rev. Inst. Francais du Petrole*, **51**, 215 (1996).
Froment, G.F., and Bischoff, K.B. *Chemical Reactor Analysis and Design*, Wiley, New York, 1990.
Frost, A.A., and Pearson, R.G. *Kinetics and mechanism: A study of homogeneous chemical reactions*, 2nd ed., Wiley, New York, 1961.
Fryer, C., and Potter, O.E. *Ind. Eng. Chem. Fundam.*, **11**, 338, (1972); *AIChE J.*, **22**, 38 (1976).
* Gavalas, G.R., *Non-Linear Differential Equations of Chemically Reacting Systems*, Springer Verlag, New York, 1968.
Geldart, D. *Powder Technology*, **7**, 285 (1973).
* Geldart, D. *Gas Fluidization Technology*, Wiley, New York, 1986.
Gillespie, B.M., and Carberry, J.J. *Ind. Eng. Chem. Fundam.*, **5**, 164 (1966).
Grace, J.R. In *Handbook of Multiphase Systems*, (Eds., Hestroni, G.), Section F., McGraw-Hill, New York, 1982; In *Recent Advances in Engineering Analysis of Chemical Reaction Systems* (Ed., Doraiswamy, L.K.), Wiley Eastern, New Delhi, 1984; In *Gas Fluidization Technology*, (Ed. Geldart, D.), Wiley, New York, 1986.
Gray, P., and Scott, S.K. *Chemical Oscillations and Instabilities*, Oxford Science, New York, 1990.
Gunn, D.J., and Hilal, N. *Int. J. Heat Mass Transfer*, **37**, 2465 (1994); **39**, 3357, (1996); *Chem. Eng. Sci.*, **52**, 2811 (1997).
Guthrie, J.P., Cossar, J., and Clym, A. *J. Am. Chem. Soc.*, **106**, 1351 (1984).
Hartig, F., and Keil, F.J. *Ind. Eng. Chem. Res.*, **32**, 424 (1993).
Hill, J.C. *Annual Rev. Fluid Mech.*, **8**, 135 (1976).
Hlavacek, V. *Ind. Eng. Chem.*, **62**, 8 (1970).
Hlavacek, V., Marek, M., and John, T.M. *Coll. Czech. Chem. Commun.*, **34**, 3868 (1969).
Horn, F. *Chem. Eng. Sci.*, **14**, 77 (1961).
Horn, F., and Troltenier, U. *Chem. Ing. Tech.*, **32**, 383 (1961a); 33, 413 (1961b).
Irani, R.K., Jayaraman, V.K., Kulkarni, B.D., and Doraiswamy, L.K. *Chem. Eng. Sci.*, **36**, 29 (1981).
Irani, R.K., Kulkarni, B.D., and Doraiswamy, L.K. *Ind. Eng. Chem. Proc. Des. Dev.*, **18**, 648 (1979); *Ind. Eng. Chem. Fundam.*, **19**, 424 (1980a); *Ind. Eng. Chem. Proc. Des. Dev.*, **19**, 24 (1980b).
Irani, R.K., Kulkarni, B.D., Hussain, S.Z., and Doraiswamy, L.K. *Ind. Eng. Chem. Proc. Des. Dev.*, **21**, 188 (1982).
Jacobson, A.R., Makris, A.N., and Sayre, L.M. *J. Org. Chem.*, **52**, 2592 (1987).
Jackson, R. *Trans. Inst. Chem. Eng.*, London, **41**, 13 (1963).
Jayaraman, V.K., Kulkarni, B.D., and Doraiswamy, L.K. *ABS PAP A.C.S.*, **180**, 22 (1980).
Jazayeri, B. *Chem. Eng. Prog.*, **91**, 26 (1995).
Jones, R.W. *Chem. Eng. Prog.*, **47**, 46 (1951).
Kato, K., and Wen, C.Y. *Chem. Eng. Sci.*, **24**, 1351 (1969).
Kerstein, A.R. *Phys. Fluids A*, **3**, 1110 (1991).

* Kondelik, P., and Boyarinov, A.I. *Coll. Czech. Chem. Commun.*, **34**, 3852 (1969).
Kulkarni, B.D., and Doraiswamy, L.K. *Catal. Rev. Sci. Eng.*, **22**, 431 (1980).
* Kulkarni, B.D., Mashelkar, R.A., and Sharma, M.M. Eds. *Recent Trends in Chemical Reaction Engineering*, Vols. 1 and 2, Wiley Eastern, New Delhi, 1987.
* Kulkarni, B.D., Ramachandran, P.A., and Doraiswamy, L.K. In *Fluidization* (Eds., Grace, J.R., and Matsen, J.M.), Plenum, New York, 1980.
Kumar, B., and Verma, R.K. *Synth. Commun.*, '**14**(14), 1359 (1984).
* Kumar R., and Mashelkar, R.A. *Reactions and Reaction Engineering*, Festschrift honoring Doraiswamy, L.K., Indian Academy Sciences, Bangalore, 1987.
Kunii, D., and Levenspiel, O. *Ind. Eng. Chem. Fundam.*, **7**, 446, (1968a); *Ind. Eng. Chem. Proc. Des. Dev.*, **7**, 481 (1968b); *Fluidization Engineering*, Butterworth-Heinemann, Boston, 1991.
* Lapidus, L., and Amundson, N.R. *Chemical Reactor Theory – A Review*, Prentice-Hall, Englewood Cliffs, 1977.
* Lee, H.H. *Heterogeneous Reactor Design*, Butterworth, Boston, 1985.
Lee, K.U., and Aris, R. *Ind. Eng. Chem. Proc. Des. Dev.*, **2**, 300, 306 (1963).
* Leva, M. *Fluidization*, McGraw-Hill, New York, 1959.
Levenspiel, O. *Chemical Reaction Engineering*, 2nd ed., Wiley, New York, 1972, 3rd ed., 1999; *Chemical Reactor Omnibook*, Oregon State University Bookstore, Corvallis, Or, 1993.
Levenspiel, O., Baden, N., and Kulkarni, B.D. *Ind. Eng. Chem. Proc. Des. Dev.*, **17**, 478 (1978).
Lewis, W.K., Gilliland, E.R., and Girouard, H. *Chem. Eng. Prog. Symp. Ser. (No. 38)*, **58**, 87 (1962).
Libby, P.A., and Williams, F.a. (Eds.) *Turbulent Reactive Flows*, Springer, Berlin, 1980.
Lund, M.M., and Seagrave, R.C. *AIChE J.*, **17**, 30 (1971a); Ind. Eng. Chem. Fundam., 10, 494 (1971b).
Luss, D. In *Chemical Reactor Theory—A Review* (Eds., Lapidus, L., and Admundson, N.R.), Prentice-Hall, Englewood Cliffs, N.J., Chapter 4, 1977.
Luss, D., and Balakotaiah, V. In *Frontiers in Chemical Reaction Engineering* (Eds., Doraiswamy, L. K., and Mashelkar, R. A., Wiley Eastern, New Delhi, India, 1984.
Matros, Yu. Sh., and Bunimovich, G.A. *Cata. Rev.-Sci. Eng.*, **38**, 1 (1996).
Matros, Yu. Sh. *Catalytic Processes Under Unsteady State Conditions*, Elsevier, Amsterdam, 1989; Can. J. Chem. Eng., **74**, 566 (1996).
Matros, Yu. Sh., Bunimovich, G.A., and Boreskov, G.K. In *Frontiers in Chemical Reaction Engineering* (Eds., Doraiswamy, L.K., and Mashelkar, R.A.), Wiley Eastern, New Delhi, India, 1984.
McGreavy, C., and Adderley, C.I. *Chem. Eng. Sci.*, **28**, 577 (1973); *Adv. Chem. Ser.*, **133**, 519 (1974).
McGreavy, C., and Dunbobbin, B.R. In *Frontiers in Chemical Reaction Engineering* (Eds., Doraiswamy, L.K., and Mashelkar, R.A.), Wiley Eastern, New Delhi, India, 1984.
Mehta, R.V., and Tarbell, J.M. *AIChE J.*, **29**, 320 (1983).
Meijs, G.F., Bunnett, J.F., and Beckwith, A.L.J. *J. Am. Chem. Soc.*, **108**, 4899 (1986).
Miyauchi, T. *Chem. Eng. Jpn*, **7**, 201 (1974).
Miyauchi, T., and Furusaki, S. *AIChE J.*, **20**, 1087 (1974).
Morbidelli, M., Varma, A., and Aris, R. In *Chemical Reaction and Reactor Engineering* (Eds., Carberry, J.J., and Varma, A.), Dekker, New York, 1987.
Mori, S., and Wen, C.Y. *Fluidization Technology* (Ed., Keairus, D.L.), Hemisphere, Washington D.C., 1976.

Muchi, I., Mori, S., and Horio, M. *Chemical Reaction Engineering with Fluidization* (in Japanese), Baifukan, Tokyo, 1984.
Murray, J.D. *J. Fluid Mech.*, **22**, 57 (1965).
Nauman, E.B. *Chemical Reactor Design*, Wiley, New York, 1987.
Nauman, E.B., and Buffham, B.A. *Mixing in Continuous Flow Systems*, Wiley, New York, 1983.
Orcutt, J.C., and Carpenter, B.H. *Chem. Eng. Sci.*, **26**, 1049 (1971).
Oroskar, A., and Stern, S.A. *AIChE J.*, **25**, 903 (1979).
Ou, J., and Ranz, W.E. *Chem. Eng. Sci.*, **38**, 1005 (1983a); 1015 (1983b).
Papa, G., and Zenz, F.A. *Chem. Eng. Prog.*, **91**(4), 32 (1995).
Partridge, B.A., and Rowe, P.N. *Trans. Inst. Chem. Eng.* London, **44**, T335 (1966).
Prasad, K., and Doraiswamy, L.K. *J. Catal.*, **32**, 384 (1974).
Pyle, D.L., and Rose, P.L. *Chem. Eng. Sci.*, **20**, 25 (1965).
Rajadhyaksha, R.A., Vasudeva, K., and Doraiswamy, L.K. *Chem. Eng. Sci.*, **30**, 1399 (1975).
Ranz, W.E. In *Mixing of Liquids by Mechanical Agitation*, (Eds. Ulbrecht, J.J., and Patterson, G.K.), Gordon and Breach, New York, 1985.
Ritchie, B. W., and Togby, A.H. *Chem. Eng. J.*, **17**, 173 (1979).
* Rase, H.F. *Chemical Reactor Design for Process Plants*, Wiley, New York, 1977; *Fixed Bed Reactor Design and Diagnostics*, Butterworth, Boston, 1990.
Razon, L.F., and Schmitz, R.A. *Chem. Eng. Sci.*, **42**, 1005 (1987).
Ridlehoover, G.A., and Seagrave, R. *Ind. Eng. Chem. Fundam.*, **12**, 444 (1973).
Rihani, D.N., Narayanan, T.K., and Doraiswamy, L.K. *Ind. Eng. Proc. Des. Dev.*, **4**, 403 (1965).
Roberts, S.M. *Dynamic Programming in Chemical Engineering and Process Control*, Academic Press, New York, 1964.
Rose, L.M. *Chemical Reactor Design in Practice*, Elsevier, Amsterdam, 1981.
Roth, D.D., Basaran, V., and Seagrave, R.C. *Ind. Eng. Chem. Fundam.*, **18**, 376 (1979).
Rowe, P.N. *Fluidization* (Eds. Davidson, J.F. and Harrison, D.), Academic Press, New York, 1971.
Russell, T.W.F., and Buzzelli, D.T. *Ind. Eng. Chem. Proc. Des. Dev.*, **8**, 2 (1969).
Sadana, A., and Doraiswamy, L.K. *J. Catal.*, **23**, 147 (1971).
Salmi, T., and Wärm(a), J. *Comput. Chem. Eng.*, **15**(10), 715 (1991).
* Schimdt, L.D. *Engineering of Chemical Reactions*, Oxford University Press, New York, 1998.
Schofield, K. *Aromatic Nitration*, Cambridge University Press, New York, 1980.
Skrzypek, J., *Int. Chem. Eng.*, **14**, 214 (1974).
Silveston, P.L., and Hudgins, R.R. In *Recent Trends in Chemical Reaction Engineering* (Eds., Kulkarni, B.D., Mashelkar, R.A., and Sharma, M.M.), Wiley Eastern, New Delhi, India, 1987.
Slin'ko, M.M., and Jaeger, N.I. *Oscillating Heterogeneous Catalytic Systems*. In *Studies in Surface Science and catalysis*, **86**, Elsevier, Amsterdam, 1994.
* Smith, J.M. *Chemical Engineering Kinetics*, McGraw-Hill, New York, 1956.
Squires, A.M., Kwauk, M., and Avidan, A. *Science*, **230**, 1329 (1985).
* Tarhan, M.O. *Catalytic Reactor Design*, McGraw-Hill, New York, 1983.
Thomas, P.H. *Proc. R. Soc. London*, **A262**, 192 (1961).
Takanobu, E.R., Baba, Y., Saito, Y., and Yokoyama, S. *Jap. Kokai*, **77**, 108 (1977).
Toor, F.D., and Calderbank, P.H. *Proc. Int. Symp. Fluidization*, 373 (1967).
Tremelling, M.J., and Bunnett, J.F. *J. Am. Chem. Soc.*, **102**, 7375 (1980).
Tsai, K., and Fox, R.O. *Ind. Eng. Chem. Res.*, **37**, 2131 (1998).

Unni, M.P., Hudgins, R.R., and Silveston, P.L. *Can. J. Chem. Eng.*, **51**, 623 (1973).
* Vaidyanathan, K., and Doraiswamy, L.K. *Chem. Eng. Sci.*, **23**, 537 (1968).
* Valstar, J. A Study of the Fixed Bed Reactor with Application to the Synthesis of Vinyl Acetate, Ph.D. Thesis, Delft University, Netherlands, 1969.
Van de Vusse, J.G. *Chem. Eng. Sci.*, **91**, 994 (1964).
Van de Vusse, J.G., and Voetter, H. *Chem. Eng. Sci.*, **14**, 90 (1961).
Van Deemter, J.J. *Chem. Eng. Sci.*, **16**, 143 (1961); *Proc. Int. Symp. On Fluidization* (Ed., Dirkenburg, A.A.H.), Netherlands University Press, Amsterdam, 1967.
Van Heerden, C. *Ind. Eng. Chem.*, **45**, 1242 (1953).
Van Swaaij, W.P.M. in A.C.S. *Symp. Ser.*, **72**, 193 (1978); In *Fluidization*, 2nd ed. (Ed., Davidson, J.F.), Academic Press, New York, 1985.
Van Welsenaere, R.J., and Froment, G.F. *Chem. Eng. Sci.*, **25**, 1503 (1970).
Villermaux, J. In Chemical Reactor Design and Technology, *Proceedings of the NATO Advanced Study Institute on Chemical Reactor Design and Technology*, London, Ontario, Canada, 1985; *Rev. Chem. Eng.*, **7**, 51 (1991).
Volk, W., Johnson, C.A., and Stotler, H.H. *Chem. Eng. Prog. Symp. Series*, **58** (38), 38 (1962).
Weekman, V.W., Jr. *Ind. Eng. Chem. Proc. Des. Dev.*, **7**, 90 (1968).
Wehner, J.F., and Wilhelm, R.H. *Chem. Eng. sci.*, **6**, 89 (1956)
Wen, C.Y., and Fan, L.T. *Models for Flow Systems and Chemical Reactors*, Dekker, New York, 1975.
Wen, C.Y., and Yu, Y.H. *AIChE J.*, **12**, 610 (1966).
Werther, J. *Chem. Ing. Tech.*, **49**, 777 (1977).
Werther, J., and Schoessler, M. In *Proc. 16th Int. Symp. on Heat and Mass Transfer*, Dubrovnik, 1984.
* Westerterp, K.R., Van Swaaij, W.P.M., and Beenackers, A.A., *Chemical Reactor Design and Operation*, Wiley, New York, 1984.
Yang, W.C., Chitester, D.C., Kornosky, R.M., and Keairns, D.L. *AIChE J.*, **31**, 1085 (1985).
* Yates, J.G. *Fundamentals of Fluidized Bed Chemical Processes*, Butterworth, London, 1983.
Zabrodsky, S.S., *Hydrodynamics and Heat Transfer in Fluidized Beds* (in Russian), translated by Zenz, F.A., MIT Press, Cambridge, Mass, 1966.
* Zenz, F.A., and Othmer, D.F. *Fluidization of Fluid Particle Systems*, Van Nostrand Reinhold, New York, 1960.
Zwietering, T. N. *Chem. Eng. Sci.*, **11**, 1 (1959).

PART IV

FLUID-FLUID AND FLUID-FLUID-SOLID REACTIONS AND REACTORS

A new scientific truth does not triumph by convincing its opponents and making them see the light, but rather because its opponents eventually die, and a new generation grows up that is familiar with it.

Max Planck

Chapter 14

Gas-Liquid Reactions

The general physical picture characterizing fluid-fluid reactions is sketched in Figure 14.1. A is the solute in phase 1 (gas or liquid)[1] and is slightly soluble in phase 2, which is always a liquid. Upon entering phase 2, A reacts with B present in that phase. When phase 1 is a gas, the reaction is almost always restricted to phase 2 (except in the rare case of a desorbing product reacting in the gas phase), but when phase 1 is a liquid or a solid, reaction can occur in both phases. In this chapter we consider the case where reaction is confined to phase 2, gas-liquid reactions.

THE BASIS

Because two phases are present, mass transfer across the interface is clearly an important consideration. Therefore, the basis of the analysis is interaction between mass transfer and reaction, leading to the formulation of conditions and rate expressions for reactions with varying roles in the two processes (i.e., with different controlling regimes).

Consider a reaction of the general form

$$\nu_A A(g) + \nu_B B(l) \to R \qquad [14.1]$$

where $\nu_A = 1$. Our objective is to examine the effect of chemical reaction on mass transfer. Depending on the relative rates of mass transfer and chemical reaction, the following regimes of control can be postulated:

---

[1] Phase 1 can also be a solid, in which case there will be no film inside it. The mechanism of diffusion and reaction in a reactive solid is quite different and is considered in Chapter 15.

very slow reactions: regime 1
slow reactions: regime 2
fast reactions: regime 3
instantaneous reactions: regime 4

Obviously, the distinction among regimes cannot be as sharp as this classification would indicate, but this deficiency can be overcome by accounting for overlaps of regimes and formulating appropriate conditions for different overlaps, such as between regimes 1 and 2, 2 and 3, and 1, 2, and 3.

When the phase 1 reactant is a pure gas, no gas film resistance is involved. On the other hand, when it is a mixture, gas film resistance must also be accounted for. Because no special problems appear, this is dealt with directly in Chapter 16 on the design of fluid-fluid reactors.

The treatment that follows is based essentially on the analysis of gas-liquid reactions by Doraiswamy and Sharma (1984), Danckwerts (1970), Astarita (1967), Hikita and Asai (1964), Shah (1979), Shah et al. (1982), Ramachandran and Chaudhari (1983), Bisio and Kabel (1985), Deckwer (1985), and Joshi et al. (1988). A recent book by Kastanek et al. (1993) provides an extensive review of reactors for gas-liquid reactions. The film theory of mass transfer is the simplest and most extensively used, despite its many limitations. Our treatment too will be based on this theory (see Chapter 4), although reference will be made to the penetration theory in some cases.

It is clear from Figure 14.1 that, in the absence of reaction, pure mass transfer of A occurs across the film. The specific rate of this is given by

$$r'_A = k'_L ([A]^* [A]_b) \tag{14.1}$$

where $[A]^*$ is the concentration of A at the interface of phases 1 and 2 and is equal to the solubility of A in phase 2, and $[A]_b$ is the concentration in the bulk of

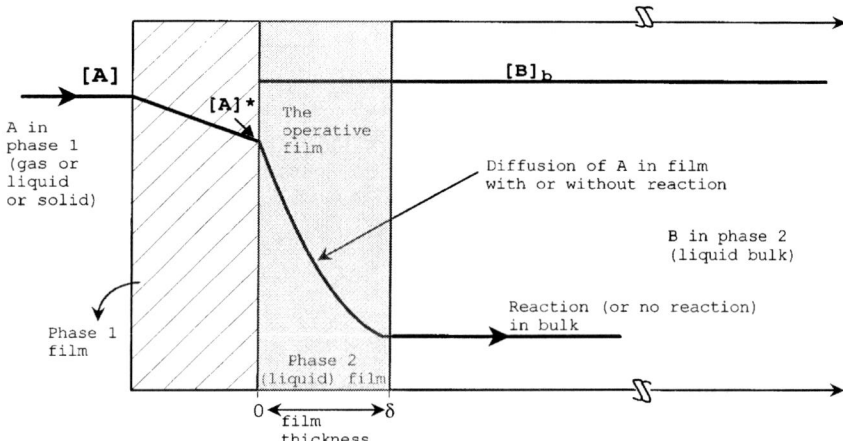

Figure 14.1 Representation of a typical gas-liquid reaction (the actual profile will vary with regime).

phase 2. The rate r'_A has the units of mol/m^2 s (or other consistent units), and k'_L is the conventional phenomenological mass transfer coefficient with units of m/s. If the rate r'_A is multiplied by the interfacial area per unit volume a_L, then

$$r'_A a_L = r_A = k'_L a_L ([A]^* - [A]_b) \tag{14.2}$$

where $r_A = r'_A a_L$ is the volumetric rate of mass transfer with units of mol/m^3 s and $k'_L a_L (= k_L)$ has the units 1/s.

The chemical reaction rate for a reaction of arbitrary order with respect to both the gas- and liquid-phase components is given by

$$-r_A = b k_{mn} [A]_b^m [B]_b^n \tag{14.3}$$

where b is the liquid-phase holdup, that is, the volume fraction of phase 2 in the gas-liquid system, and k_{mn} is the volumetric rate constant for a reaction of orders m and n with respect to A and B, respectively. The entire treatment of gas-liquid reactions, including the determination of the regime of a particular reaction, revolves around the relative values of the rates expressed by Equations 14.2 and 14.3 in one form or another. The analysis can be applied to practically any multicomponent gas-liquid reaction.

The simple case of an irreversible reaction in a gas-liquid system provides the best basis for a clear exposition of the principles underlying the analysis. Thus we begin with this type of reaction. The treatment can also be extended to a number of other reaction types, many of which are listed in Table 14.1. Included in the table is a case involving the desorption of a volatile product and its subsequent reaction. The analysis of all of these systems generally proceeds along similar lines with variations in detail. Thus we give only the final equations for a few such selected systems, following a general qualitative discussion.

DIFFUSION ACCOMPANIED BY AN IRREVERSIBLE REACTION OF GENERAL ORDER

Diffusion and Reaction in Series with No Reaction in Film: Regimes 1 and 2 (Very Slow and Slow Reactions) and Regime between 1 and 2

Here, we are concerned with regimes in which the reaction occurs exclusively in the bulk and is controlled by either chemical reaction (regime 1) or diffusion, that is, mass transfer across the liquid film (regime 2), as well as with the intermediate regime in which there is a mass transfer resistance in the film but the reaction still occurs exclusively in the bulk.

Regimes 1 and 2: Very slow and slow reactions

Consider the case where the reaction is so slow that the liquid bulk is saturated with diffusing A before any measurable reaction occurs between A and B. For such a situation, the condition to be satisfied is given by

$$k'_L a_L [A]^* \gg b k_{mn} [A]^{*m} [B]_b^n \tag{14.4}$$

Table 14.1 Representative types of gas-liquid reactions

Type of reaction	Gas phases(s)	Liquid phase(s)	General representation of reaction	Examples
1. Simple irreversible reaction	A_1	B_1	$A_1(g) + B_1(l) \rightarrow R(l)$	Air oxidation of p-nitrotoluenesulfonic acid to dinitrodibenzyldisulfonic acid, an important step in the manufacture of an intermediate in the production of optical whitening agents 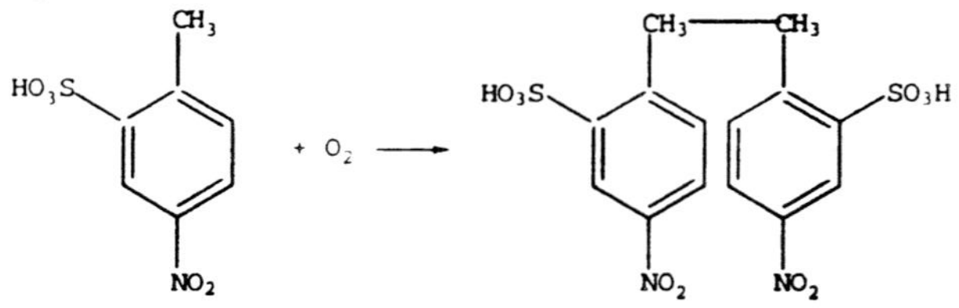
2. Reversible reaction	A_1	B_1	$A_1(g) + B_1(l) \leftrightarrow R(l)$	Absorption of Cl_2 in water in reactions such as the manufacture of chlorohydrins from olefin and water containg HOCl $Cl_2 + H_2O \rightarrow H^+ + Cl^- + HOCl$
3. Consecutive reactio	A_1	B_1	$A_1(g) + B_1(l) \rightarrow R(l) \rightarrow S(l)$	Manufacture of aliphatic carboxylic acids and their corresponding anhydrides by air oxidation of aldehydes in the presence of dissolved catalysts
4. Series-parallel reaction	A_1	B_1	$A_1(g) + B_1(l) \rightarrow R(l)$ $R(l) + A_1(l) \rightarrow S(l)$	Side chain chlorination of toluene 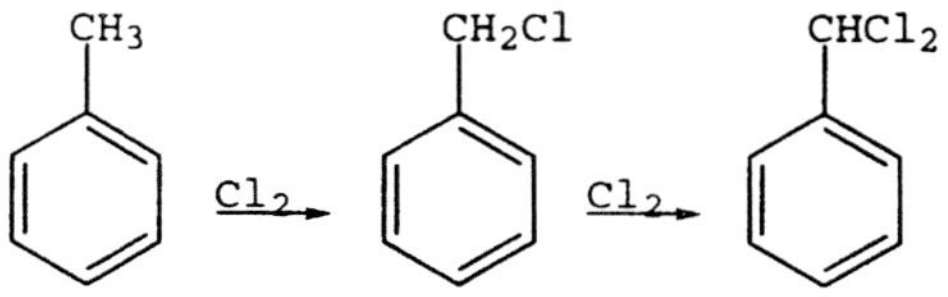

5.	Two gas-phase components reacting with themselves in a nonreactive liquid phase	A_1, A_2	B_1 (nonreactive)	$A_1(g) + A_2(g) \to R(l)$	Manufacture of ethylene dichloride (EDC) from C_2H_2 and Cl_2 dissolved in EDC $C_2H_4 + Cl_2 \to C_2H_4Cl_2$
6.	Two gas-phase components diffusing in a liquid and reacting with a liquid-phase component	A_1, A_2	B_1	$A_1(g) + B_1(l) \to R(l)$ $A_2(g) + B_1(l) \to S(l)$ or $A_1(g) + B_1(l) \to R(l)$ $R(l) + B_1(l) \to S(l)$ $A_2(g) + S(l) \to T(l)$	Absorption of CO_2, H_2S in aqueous diglycolamine solutions
7.	Gas diffusing in a liquid and reacting with two liquid-phase components	A_1	B_1, B_2	$A_1(g) + B_1(l) \to R(l)$ $A_1(g) + B_2(l) \to S(l)$	Alkylation of mixed cresols with isobutylene
8.	Absorption of a gas followed by desorption of a product with further reaction	A_1, A'	B_1	$A_1(g) + B_1(l) \to A'(g)$ $A'(g) + B_1(l) \to R(l)$	Oxidation of organic compounds like toluene and cyclohexane with desorption of water

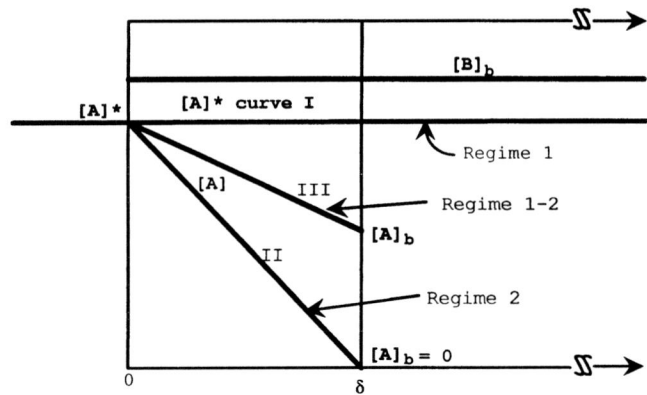

Figure 14.2 Regimes 1 and 2 (very slow and slow reactions) and the regime between 1 and 2.

and the rate by Equation 14.3. This is designated as regime 1, or *very slow reaction regime* (also *kinetic regime*), and is shown as curve I of Figure 14.2.

It can also happen that A first diffuses in the film without reaction and then is fully consumed immediately on reaching the bulk ($[A]_b = 0$). This situation is represented by curve II of Figure 14.2. Thus Equation 14.1 for mass transfer becomes

$$-r_A = k'_L a_L [A]^* \tag{14.5}$$

The reaction is controlled by film diffusion, although it is still quite slow and occurs only in the bulk. This is referred to as regime 2 or the *slow reaction regime*, the condition for which is

$$k'_L a_L [A]^* \ll b k_{mn} [A]^{*m} [B]_b^n \tag{14.6}$$

Another condition for regime 2 is that the amount of A that reacts in the film before reaching the bulk be negligible. This condition may be expressed by defining a parameter $\sqrt{M_{mn}}$ (usually represented simply by $\sqrt{M}$):

$$\sqrt{M} = \sqrt{M_{mn}} = \frac{\text{reaction in film}}{\text{reaction in bulk}}$$

$$= \frac{\sqrt{\frac{2}{m+1} D_A k_{mn} [A]^{*(m-1)} [B]_b^n}}{k'_L} \tag{14.7}$$

Thus the condition is

$$\sqrt{M} \ll 1 \tag{14.8}$$

Sometimes the square M of the above term is used to denote the amount of reaction in the film. Both forms (M and $\sqrt{M}$) are frequently referred to as the *Hatta number* Ha.

Regime between 1 and 2

Now let us visualize a situation where the residual concentration of A in phase 2, $[A]_b$, does not fall to zero, that is, Equation 14.5 does not hold. This situation, shown as curve III in Figure 14.2, corresponds to a regime between 1 and 2, and now the residual concentration of A lies between $[A]^*$ and $[A]_b$. The rate equation for this regime may be obtained by simultaneous solution of Equations 14.2 and 14.3. The nature of the solution will obviously depend on the orders m and n. For a reaction that is first order in both A and B, the solution is given by

$$-r_A = k'_{LR} a_L [A]^* = \frac{[A]^*}{\dfrac{1}{k'_L a_L} + \dfrac{1}{bk[B]_b}} \quad (14.9)$$

where k'_{LR} is an overall constant expressed as the sum of resistances due to mass transfer and reaction:

$$\frac{1}{k'_{LR} a_L} = \frac{1}{k'_L a_L} + \frac{1}{bk[B]_b} \quad (14.10)$$

Diffusion and Reaction in Film, Followed by Negligible or Finite Reaction in Bulk: Regime 3 (Fast Reaction) and Regime Covering 1, 2 and 3

Now we move to a case where the reaction is so fast that it occurs even while A is diffusing through the film and may or may not be completed within the film. If it is not, the rest of the reaction is completed in the bulk. The situation is similar to that considered for gas-solid catalytic reactions in Chapter 7. Here the film may be regarded as the counterpart of the solid, but in addition the reaction can also continue into the bulk. The conceptual equivalence of the two cases has been analyzed by Kulkarni and Doraiswamy (1975).

Reaction entirely in film

The condition for reaction to occur entirely in the film is

$$\sqrt{M} \gg 1 \quad (14.11)$$

Also, if the condition

$$\sqrt{M} \ll \frac{[B]_b}{\nu_B [A]^*} \sqrt{\frac{D_B}{D_A}} \quad (14.12)$$

is satisfied, it can additionally be assumed that the concentration of B is uniform throughout phase 2, that is, there is no depletion of B in the film even though it reacts with A there, that is $[B]^n$ is invariant. This situation is sketched in Figure 14.3. When the reaction occurs in the entire film, it is called a *pseudo-mth-order reaction*, and when it is completed within the film at a distance $x < \delta$, it is usually designated as *fast pseudo-mth-order*.

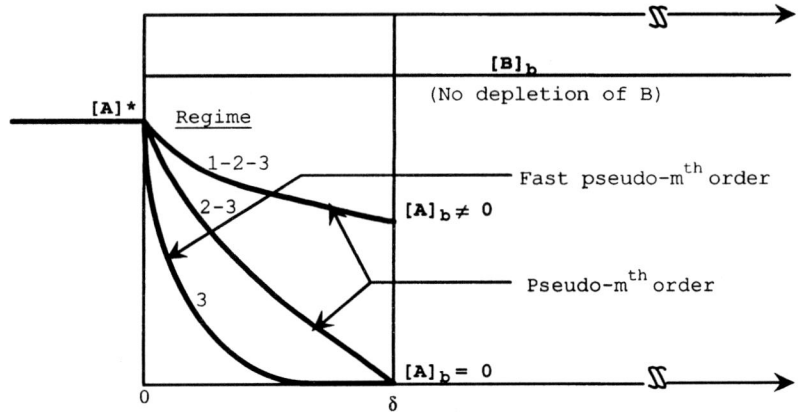

Figure 14.3 Regime 3 (fast reaction) with no depletion of B.

The governing equation for slab geometry is

$$D_A \frac{d^2[A]}{dx^2} = k_{mn}[B]_b^n[A]^m = k_m[A]^m \qquad (14.13)$$

where

$$k_m = k_{mn}[B]_b^n \qquad (14.14)$$

is the pseudo-mth-order rate constant. The boundary conditions are readily seen as

$$\begin{aligned} x = 0, \quad & [A] = [A]^*, \; d[B]/dx = 0 \\ x = \delta, \quad & [A] = 0 \end{aligned} \qquad (14.15)$$

With some manipulation, the solution in its final form is found as

$$-r'_A = k'_L[A]^*\sqrt{M_{mn}} = [A]^*\sqrt{\frac{2}{m+1} D_A k_{mn}[A]^{*(m-1)}[B]_b^n} \qquad (14.16a)$$

$$-r'_A = k'_L[A]^*\sqrt{M_m} = [A]^*\sqrt{\frac{2}{m+1} D_A k_m[A]^{*(m-1)}} \qquad (14.16b)$$

$$-r'_A = k'_L[A]^*\sqrt{M_1} = [A]^*\sqrt{D_A k_1} \qquad (14.16c)$$

where $\sqrt{M_m}$ and $\sqrt{M_1}$ represent the film-to-bulk reaction ratios for pseudo-mth-order and pseudo-first-order reactions, respectively. Equations 14.16b,c may also be written as

$$\sqrt{M_m} = \frac{\sqrt{\frac{2}{m+1} D_A k_m[A]^{*(m-1)}}}{k'_L} \qquad (14.17a)$$

$$\sqrt{M_1} = \frac{\sqrt{D_A k_1}}{k'_L} \qquad (14.17b)$$

Now we define an *enhancement factor* η

$$\eta = \frac{\text{actual rate of reaction}}{\text{rate of mass transfer}} = \frac{(-r_A)}{k'_L a_L [A]^*} \quad (14.18)$$

as a measure of enhancement in mass transfer due to reaction. For a pseudo-first-order reaction, Equation 14.13 can be solved for the rate, $D_A(d[A]/dx)_{x=0}$, to give

$$\eta = \frac{\sqrt{M_1}}{\tanh \sqrt{M_1}} \quad (14.19)$$

Note that, unlike in the definition of effectiveness factor for catalytic reactions (Chapter 7) where the normalizing rate was the rate of reaction, here the normalizing rate is the rate of mass transfer. Thus the reaction is considered the *intruder* (albeit benevolent or *enhancing*), whereas for catalytic reactions, diffusion was the intruder (often, but not always, retarding). For a pseudo-*m*th-order reaction, one can write

$$\eta = \frac{\sqrt{M_m}}{\tanh \sqrt{M_m}} \quad (14.20)$$

Reaction in both film and bulk (regime 1-2-3)

This situation is depicted by the top curve of Figure 14.3. Mathematically, Equation 14.13 continues to be valid, but the boundary conditions will be different:

$$\begin{aligned} x &= 0, \quad [A] = [A]^* \\ x &= \delta, \quad [A] = [A]_b \end{aligned} \quad (14.21)$$

The final rate equation for a pseudo-first-order reaction is given in Table 14.2. With appropriate simplifications, this reduces to the equation for regimes 2 or 3, which are also included in the table.

Instantaneous (or Very Fast) Reaction: Regime 4

Now we move from fast to very fast or instantaneous reaction. Because the reaction is very fast, both A and B are consumed as fast as they are supplied to the locale of the reaction. Thus a reaction plane is established within the film to which B diffuses from the bulk and A from the interface. They react instantaneously and completely at this plane (see Figure 14.4). The condition for such an instantaneous reaction is given by

$$\sqrt{M} \gg \frac{[B]_b}{\nu_B [A]_b} \sqrt{\frac{D_B}{D_A}} \quad (14.22)$$

By equating the rates of diffusion of the components to the reaction plane, $D_A[A]/\lambda$ and $D_B[B]/(\delta - \lambda)$, an expression for the specific rate can be obtained

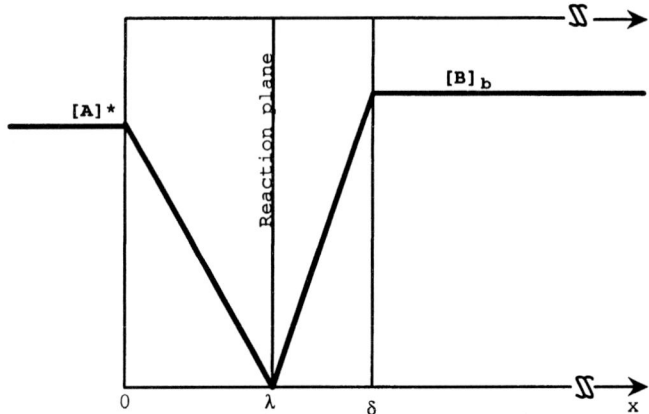

Figure 14.4 Regime 4 (instantaneous reaction).

[Astarita (1967)], and because the reaction is instantaneous, this will not contain any reaction term.

$$-r'_A = k'_L[A]^* \left[1 + \left(\frac{[B]_b D_B}{\nu_B [A]^* D_A}\right)\right] \quad (14.23)$$

Dividing Equation 14.23 by the rate of mass transfer $k'_L[A]^*$ leads to the following expression for the enhancement factor:

$$\eta_a = 1 + \left(\frac{[B]_b D_B}{\nu_B [A]^* D_A}\right) \approx \left(\frac{[B]_b D_B}{\nu_B [A]^* D_A}\right) \quad (14.24)$$

The penetration theory solution is

$$\eta_a = 1 + \left(\frac{[B]_b}{\nu_B [A]^*} \sqrt{\frac{D_B}{D_A}}\right) \approx \frac{[B]_b}{\nu_B [A]^*} \sqrt{\frac{D_B}{D_A}} \quad (14.25)$$

Unlike the enhancement factor of Equation 14.19 or 14.20, that of the above equation refers to a terminal value η_a that corresponds to the enhancement caused by an instantaneous reaction. Therefore, it is referred to as the *asymptotic enhancement factor*. Because this regime involves a reaction plane, it is reasonable to expect that the reaction plane will move with time. It is possible to allow for this transient situation by invoking the penetration theory (Karlsson and Bjerle, 1980). A comparison of the expressions based on the film and penetration theories shows that

$$-r_A \text{ (film theory)} \propto \left(\frac{D_B}{D_A}\right) \quad (14.26)$$

$$-r_A \text{ (penetration theory)} \propto \sqrt{\frac{D_B}{D_A}} \quad (14.27)$$

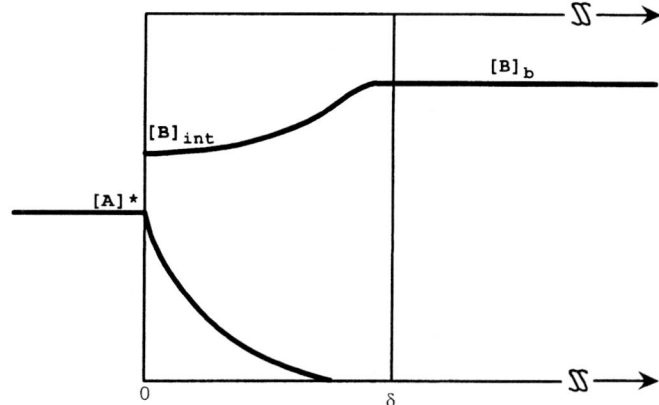

Figure 14.5 Regime between 3 and 4 (depletion regime).

For all practical purposes, this difference is not significant in many systems; and where the diffusivities are equal, there is no difference at all between the two solutions.

Depletion Regime: Regime between 3 and 4

It is conceivable that B will not be fully depleted within the film, as assumed in regime 4. This situation (Figure 14.5) can be analyzed by writing the diffusion equations for A and B in terms of $[A]$ and $[B]_{int}$ instead of $[A]$ and $[B]_b$ (as in regime 4). Then the solution can be obtained as

$$r'_A = [A]^* \sqrt{\frac{2}{m+1} D_A k_{mn} [A]^{*(m-1)} [B]_{int}^n} \qquad (14.28)$$

where $[B]_{int}$ is the interfacial concentration which can be calculated using a procedure given by Doraiswamy and Sharma (1984).

As for other regimes, we can also derive an expression for the enhancement factor for this regime [Hikita and Asai (1964)]:

$$\eta = \sqrt{M} \left(\frac{[B]_{int}}{[B]_b} \right)^{n/2} = \sqrt{M} \left(\frac{\eta_a - \eta}{\eta_a - 1} \right)^{n/2} \qquad (14.29)$$

where η_a is the asymptotic enhancement factor given by Equation 14.24 or 14.25.

EFFECT OF TEMPERATURE

In the developments presented so far, the question of an increase (or decrease) in temperature was not considered. When the heat of reaction is significant (as in most exothermic reactions), the rate of the overall process will increase. We must remember, however, that here we are dealing with a system that is temperature dependent not only through the reaction rate constant but also through other

temperature-sensitive parameters like solubility, diffusivity, mass transfer coefficient, gas-phase partial pressure, and solvent volatility. Thus the overall effect of temperature tends to be far more complex than that in physically homogeneous systems, and the nature of the effect will largely be determined by the controlling regime.

Regimes 1 and 2: Very Slow and Slow Reactions

Clearly, regime 1 is the simplest to analyze because it is kinetically controlled and therefore for all practical purposes can be treated as a homogeneous system. Even here, however, the solubility of A in phase 2 (the liquid), its partial pressure, and solvent vapor pressure can exert a significant influence. Thus, although the rate constant increases with increase in temperature, the solubility generally decreases, but the overall effect would still be positive because the activation energy is almost always higher than the heat of solution. On the other hand, the solubility can also increase with temperature (particularly for liquids), for example, chlorobenzene in aqueous sulfuric acid and esters such as ethyl p-nitrobenzoate and dichloroethyl oxalate in water. This serves only to supplement the effect of temperature on the rate constant.

The effect of temperature on the partial pressure of A and on the vapor pressure of B can also significantly modify the overall effect of temperature on the reaction. Indeed, these effects are so fundamental in the calculations that they are not restricted to any particular regime.

In regime 2, the controlling influence is mass transfer through the liquid side film. Because the enhancing effect of temperature on the mass transfer coefficient k'_L is usually quite small and its retarding effect on the solubility of A is considerably greater, the combined effect of temperature is to decrease the overall rate of reaction. However, in cases where the solubility increases with temperature, the overall effect is positive. An example of this is the nitration of aromatic compounds with mixed acid, in which the solubility of the compound in the acid increases with temperature. An experimental value of $E = 15 \text{ kcal/mol}$ was reported (Albright and Hanson, 1969). This relatively high value of E is due to the combined enhancing effects of temperature on solubility and mass transfer. However, in the absence of data on the abnormal solubility behavior of the system, the usual assumption would be made that the solubility decreases with temperature, and the high value of activation energy would be attributed to the absence of any mass transfer effect. In other words, one might erroneously conclude that the operation was in regime 1. This underscores the need for full information on reaction kinetics and physicochemical properties to discern the controlling regime.

Regimes 3 and 4: Fast and Instantaneous Reactions

As the temperature is raised, the rate constant k_{mn} increases. At the same time, the solubility $[A]^*$ decreases, so that at a given value of $[B]_b$ the ratio $[B]_b/[A]^*$ increases. Noting that k_{mn} appears (embedded in $\sqrt{M}$) as the square root in

condition 14.12 for fast reactions, both sides of the condition tend to be affected roughly to the same extent. Thus the condition remains valid with no change in regime. In general, the effect of temperature in regime 3 can be neglected.

Regime 4 is influenced only by the mass transfer coefficients of A and B through the film. Because these are not greatly temperature-sensitive, the effect of temperature can be neglected in this regime also.

DISCERNING THE CONTROLLING REGIME IN SIMPLE REACTIONS

The exact rate equation to be used for a reaction at hand depends on the regime in which it occurs. It is this equation that must be used in the reactor design methods described in Chapter 16. The equations for various regimes along with the appropriate conditions for their validity, already briefly discussed, are summarized in Table 14.2.

Depending on the regime of a reaction, the controlling rate constant will be either the reaction rate constant or the mass transfer coefficient. The overall effect of temperature on the reaction will be the resultant of its effects on the rate coefficient in the controlling regime of the reaction and on the equilibrium concentration of the gas in the liquid. Table 14.3 summarizes these effects for the different regimes.

DIFFUSION ACCOMPANIED BY A COMPLEX REACTION

There are a number of important gas-liquid organic reactions which are complex. Because more than one reaction is involved in a complex scheme, each with its own value of the rate constant, it is possible that the reactions would be in different regimes. This makes the analysis quite complicated. Hence we restrict the treatment here to a relatively simple but important system, the series-parallel reaction. For more complicated reactions, reference should be made to the original publications (many of which have been listed by Doraiswamy and Sharma, 1984).

Typically, the series-parallel reaction may be represented by (see Chapter 5)

$$A + \nu_B B \rightarrow R$$
$$A + \nu_R R \rightarrow S$$
[14.2]

with the following typical possibilities:

Case 1 Both steps are very slow or slow (regimes 1 and 2).
Case 2 The first step is fast (regime 3) and the second slow (regime 2).
Case 3 Both steps are fast (regime 3).
Case 4 Step 1 is instantaneous (regime 4).

Now we qualitatively discuss the implications of each case.

Table 14.2 Conditions and rate equations for absorption of gas A followed/accompanied by a simple irreversible reaction with B in the liquid in different regimes

Regime	Conditions[a]	Rate equation
1. Regime 1, very slow	$k'_L a_L [A]^* \gg b k_{mn}[A]^{*m}[B]^n_b$	$-r_A = -r'_A a_L = b k_{mn}[A]^{*m}[B]^n_b$ (1)
2. Regime 2, slow	$k'_L a_L [A]^* \ll b k_{mn}[A]^{*m}[B]^n_b$	$-r_A = r'_A a_L = k'_L a_L [A]^*$ (2) $\begin{cases} \text{with } [A]_b = 0 \text{ in} \\ -r'_A a_L = k'_A a_L([A]^* - [A]_b) \end{cases}$ (3)
3. Regime 1–2	Same as for 2 except that $0 < [A]_b < [A]^*$	Simultaneous solution of Equations 1 and 3
4. Regime 3, fast (entire reaction in film, no depletion of B (pseudo-mth-order)	$\sqrt{M} \gg 1$ $\sqrt{M} \ll \dfrac{[B]_b}{\nu_B [A]^*}\sqrt{\dfrac{D_B}{D_A}}$	$-r'_A = k'_L [A]^* \sqrt{M} = [A]^* \sqrt{\dfrac{2}{m+1} D_A k_m [A]^{*(m-1)}}$ or, $-r'_A = k'_L [A]^* \eta$ where, η (enhancement factor) $= \dfrac{\sqrt{M}}{\tanh\sqrt{M}} (\cong \sqrt{M} \text{ for } \sqrt{M} > 3)$
5. Regime 2–3, part of reaction in film, rest in bulk, $[A]_b = 0$, no depletion of B (pseudo-first-order)	$\sqrt{M} \approx 1$ $\sqrt{M} \ll \dfrac{[B]_b}{\nu_B [A]^*}\sqrt{\dfrac{D_B}{D_A}}$	$-r'_A = [A]^* \sqrt{D_A k_1 + k'^2_L}$ where $k_1 = k_n [B]^n_0$ (pseudo first-order)
6. Regime 1–2–3, part of reaction in film, rest in bulk, $[A]_b = 0$, no depletion of B	$\sqrt{M} \leq 1$ $\sqrt{M} \ll \dfrac{[B]_b}{\nu_B [A]^*}\sqrt{\dfrac{D_B}{D_A}}$	$-r'_A a_L = [A]^* \left(\dfrac{1}{bk_1} + \dfrac{1}{a_L \sqrt{D_A k_1 + k'^2_L}} \right)^{-1}$
7. Regime 4, instantaneous	$\sqrt{M} \gg \dfrac{[B]_b}{\nu_B [A]^*}\sqrt{\dfrac{D_B}{D_A}}$	$-r_A = -r'_A a_L = k'_L [A]^* \eta_a$ $\eta_a = 1 + \left(\dfrac{[B]_b D_B}{\nu_B [A]^* D_A} \right)$ (asymptotic enhancement factor)
8. Regime 3–4, depletion regime	$\sqrt{M} \gg \dfrac{[B]_b}{\nu_B [A]^*}\sqrt{\dfrac{D_B}{D_A}}$	For $\sqrt{M}([B]_{int}/[B]_b)^{n/2} > 3$, $-r'_A = [A]^* \sqrt{\dfrac{2}{m+1} D_A k_{mn}[A]^{*(m-1)}[B]^n_{int}}$ (See Doraiswamy and Sharma, 1984, for procedure for finding $[B]_{int}$)

[a] See Equation 14.7 for definition of $\sqrt{M}$.

Table 14.3 Discerning the controlling regimes through temperature effects

No.	Regime	Controlling rate constant	Effect of temperature	Remarks
1	Regime 1	k	Positive on k, negative on $[A]^*$; overall effect: positive	Effect on $[A]^*$ can sometimes be positive, thus supplementing the effect on k
2	Regime 2	k_L	Positive on k_L, negative on $[A]^*$; overall effect: negative	Effect on $[A]^*$ can sometimes be positive; hence overall effect can be positive
3	Regime 3 (no depletion of B)	k	Negligible	Regime remains unchanged
4	Regime 4	k_L for A and B	Slightly positive	Often negligible in the temperature range of practical interest; regime unchanged

Note: (1) k = general representation of rate constant. (2) There can be regime changes in cases 1 and 2.

Profiles and Selectivities in Different Regimes: Qualitative Discussion

Case 1: both steps slow or very slow

When both steps are very slow, mass transfer will be relatively fast, and there will be no gradient in the film. Hence the selectivity will not be affected.

When both steps are slow, first A diffuses through the film with no accompanying reaction, and a pure mass transfer gradient is established. Note that R is formed only in the bulk, with no gradient in the film, and there is no depletion of B. It can be shown that even in this regime, selectivity is not affected.

Case 2: first step is fast, second step slow

For these reactions, diffusion and reaction of A occur simultaneously in the film; A is consumed fully in the film to form R, and no free A diffuses into the bulk for further reaction to occur there. Thus the selectivity for R would be very high. The concentration profiles of A, B and R are shown in Figure 14.6a. However, if part of A diffuses into the bulk (conforming to regime 2–3), it can further react with B to give R or with R to give S. Now, the selectivity of R will be determined by the amounts of A that react with B and R. The concentration profiles of the different species will be as shown in Figure 14.6b.

Clearly, either none or only a small part of A is available in the bulk for the second step to occur there, and hence the selectivity of R would be very high, irrespective of whether the first step is in regime 3 or 2–3 (the second step is in regime 2 in both cases). If the mass transfer coefficient k_L is decreased, the rate of absorption of A will decrease, thus driving the first step more completely into the film. In the resultant absence of A in the bulk, the second step does not occur any further, leading to an increase in R. Overall, therefore, decreasing the mass transfer coefficient increases the selectivity.

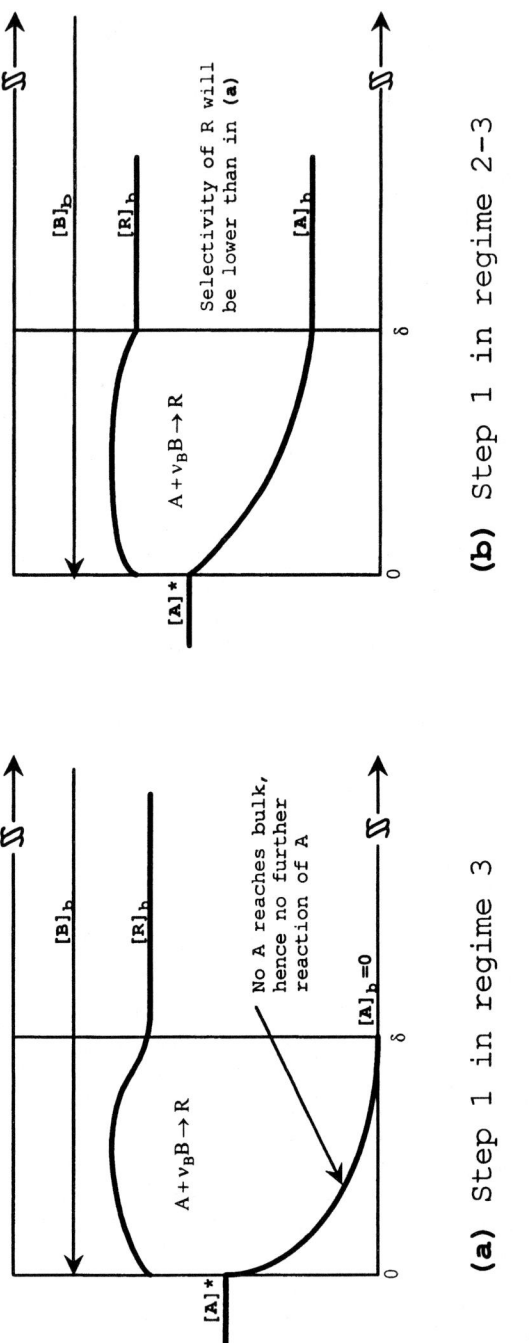

Figure 14.6 Complex reaction, $A + \nu_B B \to R$, $A + \nu_R R \to S$, step 1 in regime 3 or 2–3 and step 2 in regime 2.

Case 3: both steps are fast

Here, both reactions occur in the film with or without depletion of B. Let us first consider the case of no depletion of B. Because R is formed in the film, its concentration in the film would be higher than in the bulk. Also, the second step becomes second order. The situation in the film is similar to that in a plug-flow reactor where the concentration of R is determined by the relative rates of its formation and disappearance and thus exhibits a maximum at a certain position in the film. This is shown in Figure 14.7a. When the bulk concentration of R is high and the partial pressure of A is relatively low, the concentration of R at any point in the film may not be significantly different from that in the bulk (because no reaction occurs in the bulk). Also, both steps become pseudo-first-order. Such a situation would lead to concentration profiles shown in Figure 14.7b.

In either case, mass transfer effects for both steps are restricted to the film and are unlikely to influence selectivity in a significant way. An interesting possibility is that an increase in the value of k_L can lead to a situation where the first step continues in regime 3 but the second shifts to regime 2 and thus can occur only in the bulk. Because most of A would have been consumed in the film, its concentration in the bulk would be very low, thus severely restricting the second step. This will clearly result in an enhancement of selectivity for R.

Now, if we allow for depletion of B (i.e., the regime between 3 and 4), there would still be a maximum in R within the film but there would also be a

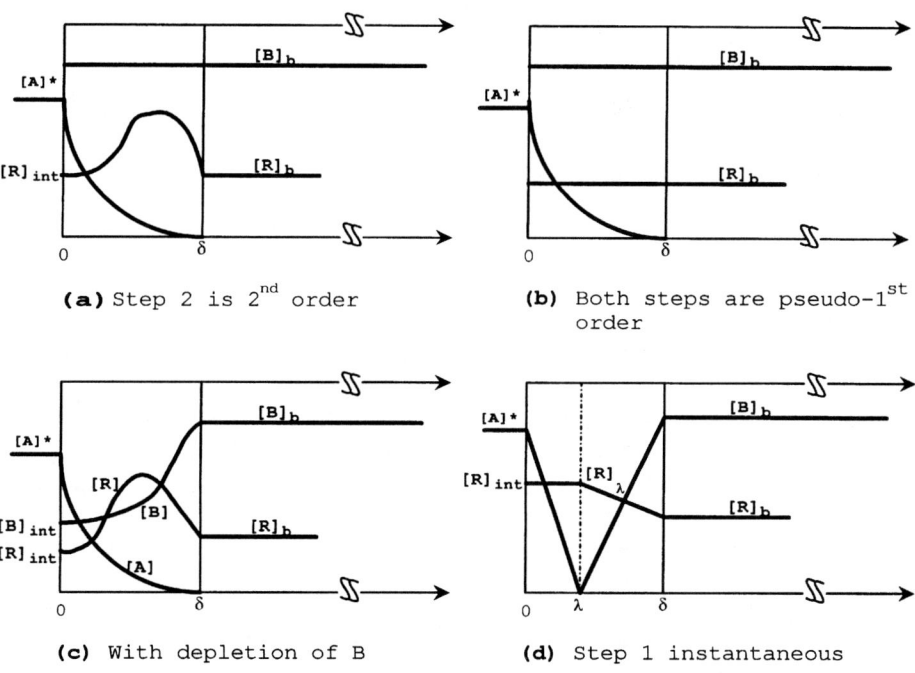

Figure 14.7 Complex reaction, $A + \nu_B B \to R$, $A + \nu_R R \to S$, both steps in film.

decreasing profile of B toward the interface. This is shown in Figure 14.7c. The decrease in the concentration of B in the film lowers the rate of reaction 1, and the increase in the concentration of R accelerates reaction 2. The net result is a decrease in the selectivity for R.

Case 4

Here, the first step is instantaneous and, therefore, occurs at a plane within the film to which A and B diffuse from the interface and bulk, respectively (Figure 14.7d). Reactant A is absent beyond this plane, and hence the second step between A and R must be restricted to the region between this plane and the interface. It can be shown that the concentrations of R at the interface and the reaction plane are almost the same and that an increase in k_L will increase step 1 but leave the second step practically unaffected. Thus the selectivity for R can be enhanced by increasing the mass transfer coefficient.

Equations for the Various Regimes

The criteria for each of the situations discussed above along with the corresponding equations for the rate and selectivity are summarized in Table 14.4.

SIMULTANEOUS ABSORPTION AND REACTION OF TWO GASES

A number of gas-liquid organic reactions involve simultaneous absorption of two gases followed by reaction. Although many such situations can be visualized, the following are more common:

1. The gases react with each other in the liquid phase without participation of the liquid-phase reactant.
2. One or both of the gases react with the liquid-phase reactant but not with themselves.
3. The gases react among themselves and with the liquid-phase reactant.

Absorption and Reaction of Two Gases with Each Other

Two situations can be visualized: (1) the gases are introduced separately and (2) they are premixed and introduced as a single stream. In analyzing these cases here, we assume that the reaction between the gases (A_1 and A_2) is fast or instantaneous, that is, it is confined to the film.

Some examples are as follows: reaction between ethylene and chlorine in a medium of the product, dichloroethane (Balasubramanian et al., 1966); reaction between acetylene and chlorine in a medium of the product, tetrachlorothane; reaction betweeen ethylene and oxygen in a medium of $CuCl/CuCl_2$ with $PdCl_2$ as catalyst to give acetaldehyde (Chandalia, 1968a; Hatch 1970); reaction between ethylene oxide and hydrogen sulfide in a medium of the product, thiodiglycol; simultaneous absorption of NO_2 and dimethyl sulfide in dimethyl sulfoxide (Sittig (1967); simultaneous absorption of vinyl chloride and chlorine in tri-

Table 14.4 Conditions and rate equations for different regimes for the complex (series-parallel) reaction $A + B \xrightarrow{k_I} R$, $A + R \xrightarrow{k_{II}} S$

Regime		Conditions[a]	Rate equations[a]
Step 1	Step 2		
Slow or very slow (regime 1 or 2)	Slow or very slow (regime 1 or 2)	$\sqrt{M_{1,0}} = \dfrac{\sqrt{D_A k_I [B]_b}}{k'_L} \ll 1$ $\sqrt{M_{2,0}} = \dfrac{\sqrt{D_A k_{II} [R]_b}}{k'_L} \ll 1$	$-r'_A a_L = \dfrac{(bk_I[B]_b + bk_{II}[R]_b)k'_L a_L}{(bk_I[B]_b + bk_{II}[R]_b) + k'_L a_L}[A]^*$ $-r'_B a_L = \dfrac{(bk_I[B]_b)k'_L a_L}{(bk_I[B]_b + bk_{II}[R]_b) + k'_L a_L}[A]^*$ $r'_R a_L = \dfrac{(bk_{II}[R]_b)k'_L a_L}{(bk_I[B]_b + bk_{II}[R]_b) + k'_L a_L}[A]^*$
Slow-fast (regime 2–3)	Slow (regime 2)	$\dfrac{\sqrt{D_A k_I [B]_b}}{k'_L} \geq 1, \ll \dfrac{[B]_b}{[A]^*}\sqrt{\dfrac{D_B}{D_A}}$ $\dfrac{\sqrt{D_A k_{II} [R]_b}}{k'_L} \ll 1$	$-r'_A = \dfrac{[A]^* \sqrt{D_A k_I [B]_b}}{\tanh\left(\dfrac{\sqrt{D_A k_I [B]_b}}{k'_L}\right)}$ $-r'_B = \dfrac{[A]^* \sqrt{D_A k_I [B]_b}}{\sinh\left(\dfrac{\sqrt{D_A k_I [B]_b}}{k'_L}\right)}$ $\times \left[\cosh\left(\dfrac{\sqrt{D_A k_I [B]_b}}{k'_L}\right) - \dfrac{k_{II}[R]_b}{k_I[B]_b + k_{II}[R]_b}\right]$ $r'_R = \dfrac{[A]^* \sqrt{D_A k_I [B]_b}}{\sinh\left(\dfrac{\sqrt{D_A k_I [B]_b}}{k'_L}\right)} \cdot \dfrac{k_{II}[R]_b}{k_I[B]_b + k_{II}[R]_b}$
Fast (regime 3)	Slow (regime 2)	$\dfrac{\sqrt{D_A k_I [B]_b}}{k'_L} > 3$	
Fast (regime 3)	Fast (regime 3) (for constant [R] in film)	$1 \ll \dfrac{\sqrt{D_A k_{II} [R]_b}}{k'_L} \ll \dfrac{[R]_b}{[A]^*}\sqrt{\dfrac{D_R}{D_A}}$	$-r'_A = [A]^* \sqrt{D_A (k_I[B]_b + k_{II}[R]_b)}$ $-r'_B = (-r'_A)\dfrac{k_I[B]_b}{k_I[B]_b + k_{II}[R]_b}$ $r'_R = (-r'_A)\dfrac{k_I[R]_b}{k_I[B]_b + k_{II}[R]_b}$
Inst.[b] (regime 4)	In film between $x = \lambda$ and δ	$\dfrac{\sqrt{D_A k_I [B]_b}}{k'_L} \gg \dfrac{[B]_b}{[A]^*}\sqrt{\dfrac{D_B}{D_A}}$	Equations are quite involved (see Doraiswamy and Sharma, 1984)

[a] k_I, k_{II} = rate constants for steps 1 and 2 (both second order), $m^3/mol\,s$.
[b] If step 1 is instantaneous, step 2 must occur in film between $x = \lambda$ and 0 (see Figure 14.7d).

chloroethane (McIver and Ratcliffe, 1974); and sulfoxidation of hydrocarbons. It is difficult to classify them exclusively in terms of unpremixed and premixed operations. Practical consideration often dictate the choice; for example, ethylene and chlorine are introduced separately in the product (dichloroethane) medium. However, the two cases are distinctive in terms of analysis.

The gases are introduced separately

This is a fairly common class of reactions in organic technology. When the two gases are introduced separately, they dissolve at separate locations, diffuse toward each other in the liquid medium, and in most cases react instantaneously to give the product. Because usually pure gases are used, gas side film transfer resistances are absent, and the situation can be depicted as shown in Figure 14.8a for gases A_1 and A_2 (where 1a is phase A and 1b is phase A_2). Note that the solubility of A_1 is assumed to be much lower than that of A_2, so that there is no residual A_1 in the bulk. The method described earlier for reaction with no depletion of the liquid phase reactant will apply.

If the reaction is in the fast regime, again it occurs in the (1a-2) liquid film, but the profiles of A_1 and A_2 will be as shown in Figure 14.8b. The method described earlier for reaction with depletion of B will apply.

The gases are premixed

Consider the case where premixed gases A_1 and A_2 (phase 1) react with each other in a liquid medium (phase 2) according to the simple reaction

$$A_1 + \nu_{A_2} A_2 \to R \tag{14.3}$$

with orders m and n with respect to A_1 and A_2, respectively. If the solubility if A_2 is much greater than that of A_1, $[A_2]^* \gg [A_1]^*$, the methods for diffusion in an irreversible reaction will apply. Because the solubility of A_2 is very high, gas film resistance is possible for A_2, whereas transfer of A_1 occurs under pseudo-mth-order conditions. For $m = n = 1$ (i.e. an overall order of 2), Equation 14.16a of that section can be rewritten for this present case as

$$-r'_{A_1} = [A_1]^* \sqrt{D_{A_2} k_2 [A_2]^*} \tag{14.30}$$

where k_2 is the second-order rate constant, A_2 has been substituted for B, and it has been assumed that $[A_2]_b \approx [A_2]^*$ in view of the high solubility of A_2. Note that this equation is subject to condition 14.12 which requires that there be no depletion of B (A_2 in the present case).

If the solubilities of the two gases are equal, that is, $[A_1]^* \approx [A_2]^*$, and also their diffusivities, then the two gases may be considered identical and the reaction regarded as second order in any one gas, say A_1. The residual concentration of A_2 in the bulk can also be assumed to be zero, that is, $[A_2]_b \approx 0$. Then the rate equation can be written as

$$-r'_A = [A_1]^* \sqrt{(2/3) D_{A_1} k_2 [A_1]^*} \tag{14.31}$$

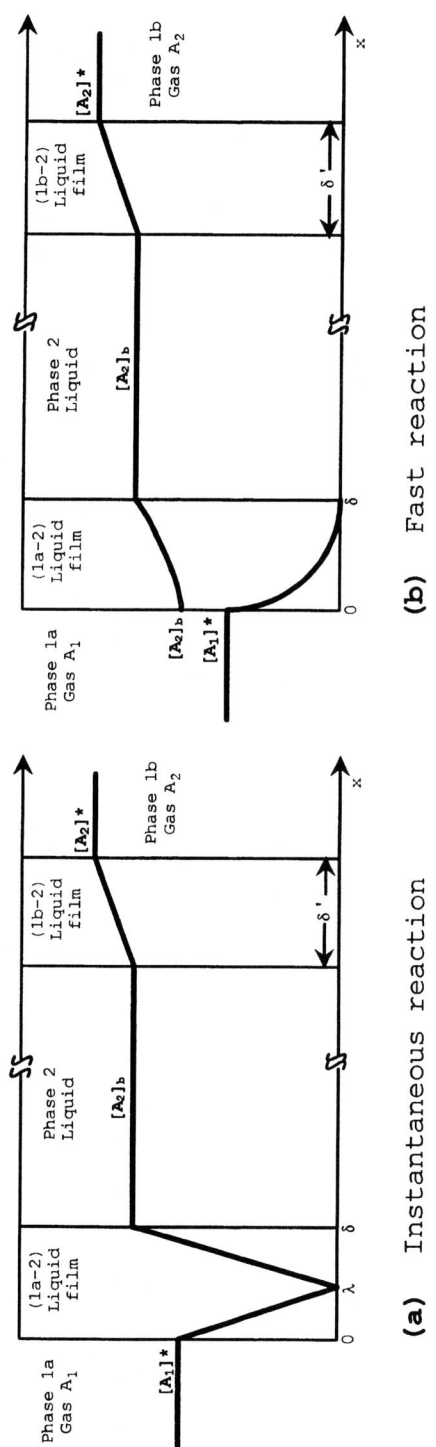

Figure 14.8 Absorption and reaction of two unpremixed gases with each other in a nonreactive solvent.

(a) Instantaneous reaction (b) Fast reaction

Note that in the previous two cases, $[A_2]_b$ was assumed to be either zero or equal to its interfacial concentration (solubility) $[A_2]^*$. The situation gets quite complicated mathematically when $[A_2]_b$ has a finite value between zero and the solubility, that is, $0 < [A_2]_b < [A_2]^*$. To write the absorption equations for this case, knowledge of the concentration profiles in the films is necessary. Ramachandran and Sharma (1971) assumed that the profiles are linear, based on which an approximate solution was obtained. The assumption of a linear profile is only a first-order approximation, because the simultaneous occurrence of diffusion and reaction can lead to a nonlinear profile in the film. This problem was addressed by Sada et al. (1976) and Chaudhari and Doraiswamy (1974) who assumed certain nonlinear profiles and obtained analytical solutions that gave results comparable to those from numerical solution.

Absorption and Reaction of Two Gases with a Liquid-Phase Reactant

This case is similar to that of reaction 1 absorption of A followed by reaction with B in the liquid phase, with the difference that there are now two gases A_1 and A_2. Typical industrial examples are the reactions of CO_2 and H_2S with aqueous solutions of diglycolamine and diisopropanolamine, and of isobutylene and butenes with aqueous sulfuric acid.

This class of reactions can be represented as

$$A_1 + \nu_{A_1} B \rightarrow R \quad [14.4.1]$$

$$A_2 + \nu_{A_2} B \rightarrow S \quad [14.4.2]$$

The dimensionless quantity M defined earlier may be extended to the analysis of these reactions and a few additional quantities defined as follows:

$$\sqrt{M_{A_1}} = \frac{\sqrt{\left(\frac{2}{m+1}\right) D_{A_1} k_{A_1} [B]_b^n [A_1]^{*(m-1)}}}{k_L'}$$

$$\sqrt{M_{A_2}} = \frac{\sqrt{\left(\frac{2}{m'+1}\right) D_{A_2} k_{A_2} [B]_b^{n'} [A_2]^{*(m'-1)}}}{k_L'} \quad (14.32)$$

$$q_{A_1} = \frac{[B]_b D_B}{\nu_{A_1} [A_1]^* D_{A_1}}, \quad q_{A_2} = \frac{[B]_b D_B}{\nu_{A_2} [A_2]^* D_{A_2}}$$

$$q_{A_1 A_2} = \frac{[B]_b}{\nu_{A_1} [A_1]^* (D_{A_1}/D_B) + \nu_{A_2} [A_2]^* (D_{A_2}/D_B)}$$

where m and n denote the orders of A_1 and B in reaction 14.4.1, and m' and n' the orders of A_2 and B in reaction 14.4.2. The Ms represent the fractions reacting in the film in each case.

For pseudo-mth or m'th-order reactions, the $[B]_b$ term is constant and is incorporated in the rate coefficient.

Qualitative analysis

We shall consider three cases: (1) both reactions occur in the bulk (regimes 1 and 2); (2) both reactions occur in the fast reaction regime (regime 3), with no depletion of B (pseudo-mth order) as well as with depletion of B (mnth-order); and (3) one of the reactions is in the instantaneous regime (regime 4), whereas the other is in the slow, fast, or instantaneous regime (regime 2, 3, or 4).

CASE 1

When both are very slow, there are no gradients in the film, and the reactions are chemically controlled. In other words, the heterogeneity of the system has no influence on conversion or selectivity, and the methods of Chapter 5 will apply.

When the reactions are in regime 2 or in regime 1–2, the rate of absorption is controlled by diffusion in the film (Figure 14.9a). The corresponding conditions and rate equations are given in Table 14.5. Note that because the reaction of, say, A_1 occurs in the bulk, its presence does not affect the profile of B in the film and hence the rate of absorption of A_2.

CASE 2

When both reactions occur in the fast reaction regime with no depletion of B, the profiles generated will be as marked in Figure 14.9b. An example of this is the

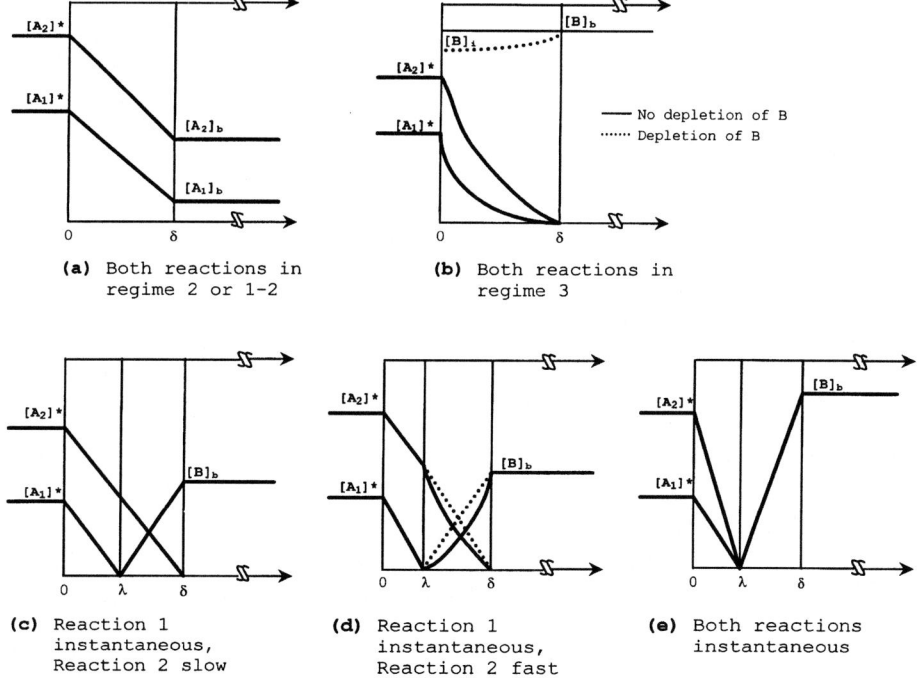

Figure 14.9 Absorption of two gases, A_1 and A_2, followed by reaction with the liquid-phase reactant B.

Table 14.5 Reaction of two absorbing gases A_1 and A_2 with B in liquid, $A_1 + v_{A_1} B \xrightarrow{k_I} R$, $A_2 + v_{A_2} B \xrightarrow{k_{II}} S$

Regime	Conditions[a]	Equations[a]
(1) Both reactions very slow (regime 1)	For A_1: $bk_I[A_1]^{*m}[B]_b^n \ll k'_L a_L [A_1]^*$ For A_2: $bk_{II}[A_2]^{*m}[B]_b^n \ll k'_L a_L [A_2]^*$	For A_1: $-r'_{A_1} = k_I[A_1]^{*m}[B]_b^n$ For A_2: $-r'_2 = k_{II}[A_2]^{*m}[B]_b^n$
(2) Both reactions slow (regime 2)	For A_1: $\sqrt{M_{A_1}} \ll 1$ $bk_I[A_1]^{*m}[B]_b^n \gg k'_L a_L [A_1]^*$ For A_2: $\sqrt{M_{A_2}} \ll 1$ $bk_{II}[A_2]^{*m}[B]_b^n \gg k'_L a_L [A_2]^*$	For A_1: $-r'_{A_1} = k'_L [A_1]^*$ For A_2: $-r'_{A_2} = k'_L [A_2]^*$
(3) Both reactions fast (regime 3)		
(a) No depletion of B, pseudo-mth-order	$\dfrac{\sqrt{M_{A_1}}}{q_{A_1}} + \dfrac{\sqrt{M_{A_2}}}{q_{A_L}} \ll 1 + \dfrac{1}{q_{A_1 A_2}}$	For A_1: $r'_{A_1} = [A_1]^* k'_L \sqrt{\eta_{A_1}}$ $\eta_{A_1} = \dfrac{\sqrt{M_{A_1}}}{\tanh \sqrt{M_{A_1}}}$ $\approx \sqrt{M_{A_1}}$ for $\sqrt{M_{A_1}} > 3$ For A_2: $r'_{A_2} = [A_2]^* k'_L \sqrt{\eta_{A_2}}$ $\eta_{A_2} = \dfrac{\sqrt{M_{A_2}}}{\tanh \sqrt{M_{A_2}}}$ $\approx \sqrt{M_{A_2}}$ for $\sqrt{M_{A_2}} > 3$
(b) Depletion of B	$\dfrac{\sqrt{M_{A_1}}}{q_{A_1}} + \dfrac{\sqrt{M_{A_2}}}{q_{A_2}} \ge \left(1 + \dfrac{1}{q_{A_1 A_2}}\right)$	$\left.\begin{array}{l}-r'_{A_1} = k'_L[A_1]^* \sqrt{M_{A_1}}\, \alpha \\ -r'_{A_2} = k'_L[A_2]^* \sqrt{M_{A_2}}\, \alpha\end{array}\right\}$, $\sqrt{M_{A_1}}, \sqrt{M_{A_2}} > 3$ where $\alpha = \sqrt{\dfrac{[B]_{\text{int}}}{[B]_b}}$ and is obtained from the positive root of $\alpha^2 + \left(\dfrac{\sqrt{M_{A_1}}}{q_{A_1}} + \dfrac{\sqrt{M_{A_2}}}{q_{A_2}}\right)\alpha$ $- \left(1 + \dfrac{1}{q_{A_1 A_2}}\right) = 0$
(4) Reaction 1 instantaneous	$\sqrt{M_{A_1}} \gg q_{A_1}$	
(a) Reaction 2 slow	$\sqrt{M_{A_2}} \ll 1$	$\left.\begin{array}{l}\\ \\ \end{array}\right\}$ Trial-and-error solution (Doraiswamy and Sharma, 1984)
(b) Reaction 2 in film (between λ and δ)	$\sqrt{M_{A_2}} \ge 1$	
(c) Reaction 2 also instantaneous	$\sqrt{M_{A_2}} \gg q_{A_2}$	$-r'_{A_1} = k'_L [A_1]^*$ $\times \left(1 + \dfrac{\dfrac{[B]_b}{v_{A_1}[A_1]^*} \dfrac{D_B}{D_{A_1}}}{1 + \dfrac{v_{A_2}[A_2]^*}{v_{A_1}[A_1]^*} \dfrac{D_{A_2}}{D_{A_1}}}\right)$ Similar for r'_{A_2} with A_2 for A_1 in above equation

[a] k_I and k_{II} are rate constants for steps 1 and 2 (both second order), m^3/mol s; the terms $\sqrt{M_{A_1}}$, $\sqrt{M_{A_2}}$, q_{A_1}, q_{A_2}, $q_{A_1 A_2}$ are defined in Equation 14.32.

simultaneous nitration of benzene and toluene. A mixture of benzene and toluene is introduced into the nitrating liquid, when the following reactions occur:

$$\underset{A_1}{\text{C}_6\text{H}_6} + \underset{A_2}{\text{C}_6\text{H}_5\text{CH}_3} \xrightarrow[\text{liquid}]{\text{Nitrating}} \text{Products} \qquad [14.5]$$

(B)

Note that here the components A_1 and A_2 are liquids, but that does not affect the analysis if it is assumed that the reaction occurs only in the aqueous phase.

Because the two reactions of A_1 and A_2 with B occur independently, the absorption-reaction process for the two gases will also proceed independently of each other. The diffusion equations previously given for the rate and enhancement will therefore apply exactly for each gas.

When there is depletion of B, as shown by the dotted curve in Figure 14.9b, reaction 1 will be (m,n)th-order, and reaction 2 (m',n')th-order, and the equations presented for the depletion regime will apply exactly for each gas diffusing and reacting with B. As in the one-gas case, an important parameter here is the ratio of the concentration of B at the interface to that in the bulk, $[B]_i/[B]_b$.

CASE 3

Interesting situations arise when one of the reactions (say, reaction of A_1 with B) is in the instantaneous regime and the other (reaction of A_2 with B) is in the slow, fast, or instantaneous regime.

In all three cases, A_1 and B diffuse toward each other from the gas and liquid phases, respectively, and react instantaneously at a plane in the film adjacent to A_1, where their concentrations fall to zero. Clearly, B cannot exist in the region between the interface and the reaction plane located at λ. As a result, A_2 cannot react with B in this region. It diffuses from the gas phase without reaction (i.e., pure diffusion) up to the reaction plane, beyond which it reacts either in the bulk, in the film, or in both. The profiles when A_2 reacts with B in the slow, fast, and instantaneous regimes are sketched in Figure 14.9c, d, and e.

Conditions and rate equations for the different cases

The conditions and rate equations for the three cases described are summarized in Table 14.5.

Absorption of Two Gases Followed by a Complex or Reversible Reaction

In the cases just considered, each of the two reactions was simple. Often, however, the reactions are coupled and therefore tend to be much more complex. Many situations of this type can be visualized and appropriate rate equations developed. We consider a few representative cases here.

CASE 1

$$A_1 + B \rightarrow R$$
$$A_2 + R \rightarrow S \quad [14.6]$$

This is a complex reaction common in organic technology. A particularly interesting offshoot of this scheme is when the liquid-phase reactant B is regenerated in the second reaction (Onda et al., 1970a,b; Ramachandran and Sharma, 1971):

$$A_1 + B \rightarrow R$$
$$A_2 + R \rightarrow B + S \quad [14.7]$$

Two important examples of this scheme are the liquid-phase oxychlorination of ethylene to ethylene dichloride:

$$C_2H_4 + 2CuCl_2 \rightarrow 2CuCl + C_2H_4Cl_2$$
$$1/2 O_2 + 2CuCl + 2HCl \rightarrow 2CuCl_2 + H_2O \quad [14.8]$$

and the Wacker process for converting an olefin to the corresponding carbonyl compound (see Chapter 8 on homogeneous catalysis):

$$C_2H_4 + PdCl_2 + H_2O \rightarrow CH_3CHO + 2H^+ + 2Cl^- + Pd$$
$$Pd + 2CuCl_2 \rightarrow PdCl_2 + 2CuCl \quad [14.9]$$
$$1/2 O_2 + 2CuCl + 2HCl \rightarrow 2CuCl_2 + H_2O$$

CASE 2

Another possible situation is absorption of two gases followed by their reactions with B, one or both of which may be reversible:

(a) $\quad A_1 + B \rightarrow R \quad$ (b) $\quad A_1 + B \leftrightarrow R$
$\quad\quad A_2 + B \leftrightarrow S \quad\quad\quad\quad\quad A_2 + B \leftrightarrow S \quad [14.10]$

Let us consider reaction 14.10(a) where the first reaction is instantaneous and the second reversible. The concentration profiles will be determined by the following considerations. Because the first step is instantaneous, there is no B in the region between 0 and λ. This tends to drive the second reaction backward, producing B in this region, which then reacts with more A_1 according to the first reaction. However, A_2 is also formed by reaction 2, which tends to lower the concentration driving force for the transport of A_2. In the region from λ to δ, the full reverse reaction takes place. Clearly, the situation is very complex, but appropriate equations can be set up and solved numerically (Ramachandran, 1971).

ABSORPTION OF A GAS FOLLOWED BY REACTION WITH TWO LIQUID-PHASE COMPONENTS

Alkylation of mixed cresols using isobutylene, co-oxidation of cyclohexane and acetaldehyde, and chlorination of mixed xylenols are examples of gas-liquid organic reactions in which absorption of a gas is followed by reaction with two reactants in the liquid. This class of reactions can be represented as

$$A + \nu_{B_1} B_1 \rightarrow R \qquad [14.11.1]$$

$$A + \nu_{B_2} B_2 \rightarrow S \qquad [14.11.2]$$

Qualitative Discussion

Case 1: Very slow and slow reactions

When the reaction is very slow, all of the components have flat profiles in the film (Figure 14.10a), and the reaction is kinetically controlled. The profiles for regime 2 are shown in Figure 14.10b. Clearly, pure diffusion of A through the film is the controlling resistance.

Case 2: Fast reaction (with and without depletion of B_1 and B_2)

First we consider the case where there is no depletion of either B_1 or B_2 (Figure 14.10c). Therefore, the reaction is pseudo-mth-order in A, and when $m = 1$ it

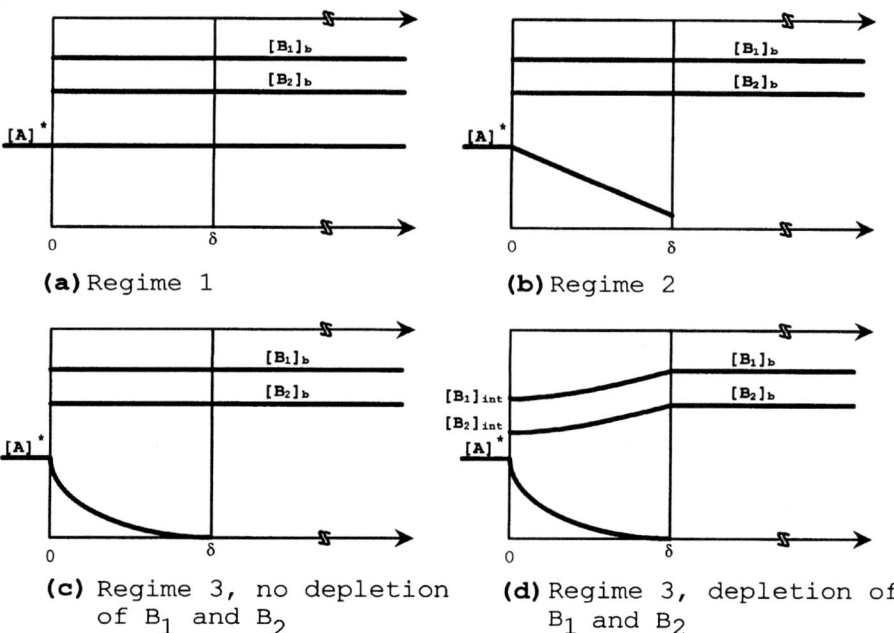

Figure 14.10 Absorption and reaction of a gas with two liquid-phase reactants, both steps in regime 1, 2, or 3.

becomes pseudo-first-order. When there is depletion of the liquid-phase reactants, each will exhibit a concentration profile across the film, as shown in Figure 14.10d.

Case 3: Instantaneous reactions

The profiles here are easy to visualize. As shown in Figure 14.11a, B_1 and B_2 from the liquid phase and A from the gas phase diffuse to the reaction plane at a position λ in the film, where the two reactions occur instantaneously. The concentrations of all three reactants fall to zero in this plane.

An example of this is absorption and reaction of ethylene with Cl_2 and HOCl in an aqueous medium:

$$C_2H_4 + HOCl \rightarrow ClC_2H_4OH$$
$$C_2H_4 + Cl_2 \rightarrow ClC_2H_4Cl$$
[14.12]

Case 4: One reaction instantaneous and the other in regime 2–3

Now we consider an interesting case where reaction between A and B_2 (reaction 14.11.2) is instantaneous and that between A and B_1 (reaction 14.11.1) is in the regime between 2 and 3 (Figure 14.11b). The first question that immediately comes to mind is why not regime 1, 2, or 3 for reaction 1? To answer this question, first we recall that, since reaction 2 is instantaneous, the gas phase reactant cannot exist beyond position λ in the film (i.e., in the region between the reaction plane λ and the bulk). Thus regime 2, where the reaction occurs only in the bulk, is eliminated. Similarly, regime 3 requires that diffusion and reaction occur simultaneously across the entire film, which again is not possible because A would not be present in the portion of the film beyond λ. Thus the reaction can be only in the regime between 2 and 3, where it is complete at some position within the film.

Conditions and Rate Equations

The conditions and rate equations for the various situations considered above are summarized in Table 14.6.

EXTENSION TO COMPLEX RATE MODELS: HOMOGENEOUS CATALYSIS

Hyperbolic equations were used in Chapter 7 to represent reactions catalyzed by solid surfaces. They are referred to as Langmuir–Hinselwood–Hougen–Watson (LHHW) models and they can be empirically extended to homogeneous catalysis in liquid-phase reactions (Chapter 8). The actual rate equation to be used for a given reaction will depend on the regime of that reaction. Methods of discerning the controlling regimes for noncatalytic gas-liquid reactions described in the previous sections were based on simple power law kinetics. Extension of these methods to gas-liquid reactions catalyzed by homogeneous catalysts involves no new principles, but the mathematics becomes more complicated because of the

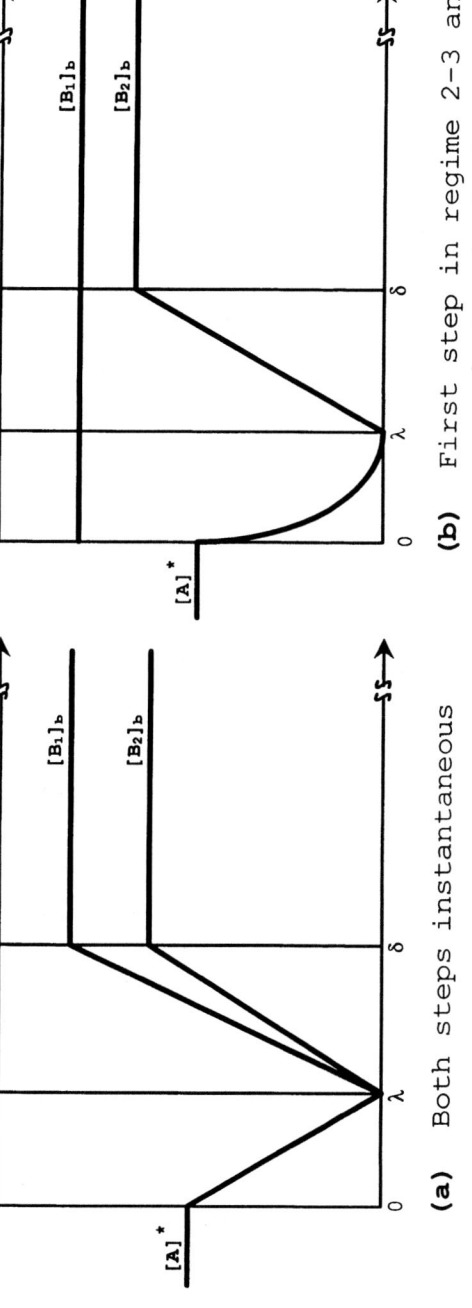

Figure 14.11 Absorption and reaction of a gas with two liquid-phase reactants, second step instantaneous.

Table 14.6 Reaction of absorbing A with two liquid-phase components B_1 and B_2, $A + \nu_{B_1} B_1 \xrightarrow{k_1} R$, $A + \nu_{B_2} B_2 \xrightarrow{k_{II}} S$

Regime	Condition(s)[a]	Rate equations[a]
Both very slow, regime 1	$k'_L a_L \gg bk^*$ where $k^* = k_1 [B_1]_b + k_{II} [B_2]_b$	$-r_A = -r'_A a_L = bk^*[A]^*$ (1) $-r'_{A,1} = -r'_A \left(\dfrac{k_1 [B_1]_b}{k^*} \right)$ $-r_{A,2} = -r'_A \left(\dfrac{k_{II} [B]_b}{k^*} \right)$ (2)
Both slow, regime 2	$k'_L a_L \ll bk^*$ $\dfrac{D_A k^*}{k'^2_L} \ll 1$	$-r'_A = k'_L [A]^*$ ($-r'_{A,1}$ and $-r_{A,II}$ from Equations 1 and 2)
Both fast, regimes 2–3 and 3, no depletion	$1 < \dfrac{\sqrt{D_A k_1 [B_1]_b}}{k'_L} \ll \dfrac{[B_1]_b}{\nu_{B_1} [A]^*} \sqrt{\dfrac{D_{B_1}}{D_A}}$ $1 < \dfrac{\sqrt{D_A k_{II} [B_2]_b}}{k'_L} \ll \dfrac{[B_2]_b}{\nu_{B_2} [A]^*} \sqrt{\dfrac{D_{B_2}}{D_A}}$	$-r'_A = \dfrac{[A]^* \sqrt{D_A k^*}}{\tanh(\sqrt{D_A k^*}/k'_L)}$ (regime 2–3) $-r'_A = [A]^* \sqrt{D_A k^*}$ (regime 3 where $\sqrt{D_A k^*}/k'_L > 3$) ($-r'_{A,I}$ and $-r'_{A,II}$ from Equations 1 and 2)
Both in depletion regime, regime 3–4	Depletion of B_1 and B_2, $\dfrac{\sqrt{D_A k_1 [B_1]_b}}{k'_L} \approx \dfrac{[B_1]_b}{\nu_{B_1} [A]^*} \sqrt{\dfrac{D_{B_1}}{D_A}}$ $\dfrac{\sqrt{D_A k_{II} [B_2]_b}}{k'_L} \approx \dfrac{[B_2]_b}{\nu_{B_2} [A]^*} \sqrt{\dfrac{D_{B_2}}{D_A}}$	$-r'_A = [A]^* \sqrt{(D_A k_1 [B_1]_i + k_{II} [B_2]_{int})}$ $[B_1]_{int}$ and $[B_2]_{int}$ are interface concentrations of B_1 and B_2 and can be calculated by an elaborate procedure given by Doraiswamy and Sharma (1984)

Condition	Rate expression
Both instantaneous, regime 4 $\dfrac{\sqrt{D_A k_I [B_1]_b}}{k'_L} \gg \dfrac{[B_1]_b}{\nu_{B_1}[A]^*}\sqrt{\dfrac{D_{B_1}}{D_A}}$ $\dfrac{\sqrt{D_A k_{II}[B_2]_b}}{k'_L} \gg \dfrac{[B_2]_b}{\nu_{B_2}[A]^*}\sqrt{\dfrac{D_{B_2}}{D_A}}$	$-r'_A = k'_L[A]^*(1+Y_1+Y_2)$ $-r'_{A,1} = r'_A\left(\dfrac{Y_1}{Y_1+Y_2}\right)$ $-r'_{A,2} = r'_A\left(\dfrac{Y_2}{Y_1+Y_2}\right)$ where $Y_1 = \dfrac{[B_1]_b D_{B_1}}{\nu_{B_1}[A]^* D_A}$ $Y_2 = \dfrac{[B_1]_b D_{B_2}}{\nu_{B_2}[A]^* D_A}$
Reaction 2 in instantaneous regime, and reaction 1 in regime 2–3 reaction 2: $\dfrac{\sqrt{D_A k_{II}[B_2]_b}}{k'_L} \gg \dfrac{[B_1]_b}{\nu_{B_2}[A]^*}\sqrt{\dfrac{D_{B_2}}{D_A}}$ reaction 1: $\dfrac{\sqrt{D_A k_I[B_1]_b}}{k'_L} \ll \dfrac{[B_1]_b}{\nu_{B_1}[A]^*}\sqrt{\dfrac{D_{B_1}}{D_A}}$	$-r'_A = \dfrac{[A]^*\sqrt{D_A k_I [B_1]_b}}{\tanh\left(\dfrac{\sqrt{D_A k_I [B_1]_b}}{k'_L}\dfrac{\lambda}{\delta}\right)}$ where λ/δ is the relative distance of the reaction plane from the interface and can be calculated by a procedure given by Doraiswamy and Sharma (1984)

[a] k_I and k_{II} are rate constants for steps 1 and 2 (both second order), m³/mol s.

hyperbolic nature of the rate models. In this section we consider reactions involving one gas and a liquid (e.g., carbonylation, hydrogenation, oxidation) and two gases and a liquid (e.g., hydroformylation).

Reactions Involving One Gas and One Liquid

Regimes 1 and 2

The reaction is controlled by the chemical kinetics in this regime, and hence diffusional limitations are absent. Depending on the reaction, the rate equation can have any of many hyperbolic forms, such as those presented in Chapter 8.

Regime 2 corresponds to film diffusion control and hence is independent of the kinetics. Thus the developments outlined in the previous sections are equally valid for reactions with LHHW kinetics.

Regime between 1 and 2 (reaction in bulk)

In this regime the rate is controlled by both film diffusion and reaction kinetics. Assuming that the reaction is represented by the first equation in Table 8.4 (with $[A]_b$ for $[A]^*$), the following equations must be solved simultaneously for the enhancement factor:

Diffusion: $\quad -r_A = k'_L a_L ([A]^* - [A]_b)$ (14.33)

Reaction: $\quad -r_A = \dfrac{k[A]_b [B]_b [C]_b}{1 + K_A [A]_b + K_B [B]_b}$ (14.34)

where C represents the catalyst. Because the rates are equal at steady state, the unknown $[A]_b$ can be eliminated (as in Table 8.4,) and the following expression obtained for the enhancement factor:

$$\eta = \frac{(-r_{Aa})}{k'_L a_L [A]^*}$$

$$= \frac{(1 + k_a k_b + M_{LH}\gamma) - \sqrt{(1 + k_a + k_b + M_{LH}\gamma)^2 - 4k_a M_{LH}\gamma}}{2k_a} \quad (14.35)$$

where

$$k_a = K_A[A]^*, \quad k_b = K_B[B]_b$$

$$\sqrt{M_{LH}} = \frac{\sqrt{D_A k [B]_b [C]_b}}{k'_L}, \quad \gamma = \frac{k'_L}{D_A a_L} \quad (14.36)$$

The condition to be satisfied is

$$\sqrt{M_{LH}} \leq 0.3\text{--}0.8 \quad (14.37)$$

The value of η is always less than or equal to unity.

Regime 3 (reaction in film)

It was pointed out earlier that regime 3 is controlling (for simple power law kinetics) if $\sqrt{M} \gg 1$. It can similarly be shown (Chaudhari, 1984) that regime 3 is controlling for LHHW models if

$$\sqrt{M_{LH}} \gg 1 \tag{14.38}$$

If there is to be no depletion of B, the following additional condition must be satisfied:

$$\sqrt{M_{LH}} \ll \left(1 + \frac{[A]_b}{[B]_b}\right) \tag{14.39}$$

Then the enhancement factor for such a reaction can be expressed in terms of a generalized Hatta number defined as

$$\sqrt{M'_{LH}} = \frac{\sqrt{2D_A}}{k'_L[A]^*} \left[\int_0^{[A]} (-r_A)\, d[A]\right]$$

$$= \sqrt{M_{LH}} \left\{\frac{2}{k_a^2}\left[k_a + (1+k_b)\ln\frac{1+k_b}{1+k_a+k_b}\right]\right\}^{1/2} \tag{14.40}$$

Equation 14.20 for the enhancement factor is equally valid for the present system, but the definition of the Hatta number is replaced by Equation 14.40:

$$\eta = \frac{\sqrt{M'_{LH}}}{\tanh\sqrt{M'_{LH}}} \tag{14.41}$$

Note that the generalized Hatta number depends on the Hatta number $\sqrt{M_{LH}}$ in addition to the other parameters of the system. Because $\sqrt{M_{LH}}$ depends on the catalyst concentration, the enhancement factor given by Equation 14.41 also depends on the catalyst concentration (and k_a and k_b). This is illustrated in Figure 14.12 which is a solution of this equation for different values of $\sqrt{M_{LH}}$ and k_a for a fixed value of k_b.

An inspection of the figure leads to two major conclusions: (1) an increase in catalyst concentration (i.e., in $\sqrt{M_{LH}}$) leads to a shift in regime from chemical to diffusion control (as expected) and (2) an increase in the diffusion parameter k_a raises the enhancement factor in the kinetic regime but lowers it in the diffusion regime.

Two Gases and a Liquid

The most important class of reactions in this category is hydroformylation. A number of kinetic and modeling studies on the hydroformylation of selected substrates have been reported (see Table 8.4). The role of diffusion in these reactions was also studied and equations for the enhancement factor were proposed (Bhattacharya and Chaudhari, 1987). It was shown that multiple solutions can exist under certain conditions. Also the fact that the reaction is first order in H_2 and negative order in CO (see Table 8.4) leads to results that contradict

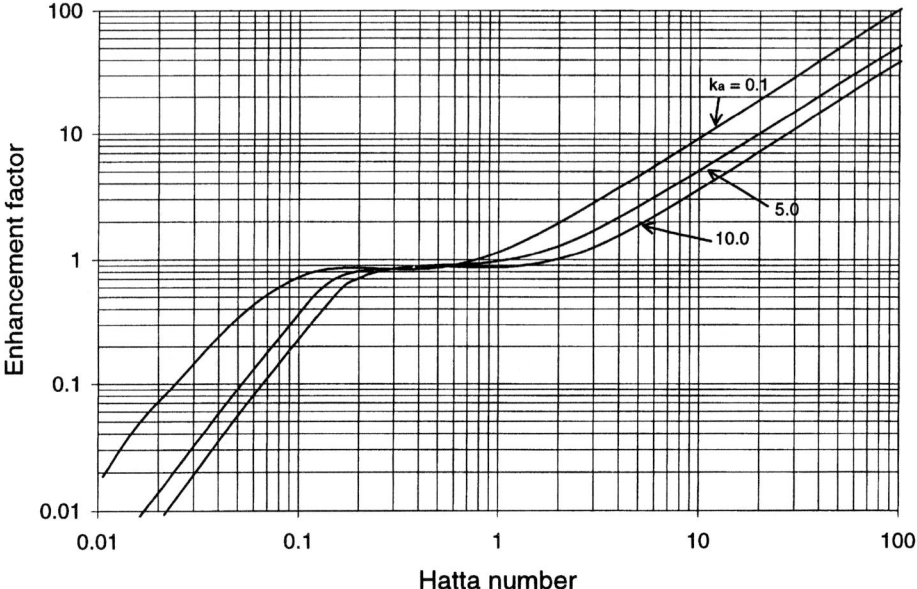

Figure 14.12 Enhancement factor as a function of Hatta number and k_a for fixed k_b (redrawn from Chaudhari, 1984)

accepted trends. Mainly, agitation seems to have no effect under conditions of significant mass transfer limitation.

Liquid-phase oxidation of gaseous substrates with O_2, such as the oxidation of ethylene to acetaldehyde (Wacker process) is another example of this class of reactions. A mathematical model for a bubble column reactor for this reaction, assuming plug flow of gas and mixed flow of liquid, was developed (Rode et al., 1994). It was shown that a critical oxygen concentration in the inlet is necessary to sustain the catalytic cycle, and a model for predicting this was proposed (Bhattacharya and Chaudhari, 1990).

MEASUREMENT OF MASS TRANSFER COEFFICIENTS

It must be clear from the various equations developed above that the gas-liquid interfacial area is a very important parameter in determining the rate of mass transfer. Any precise measurement of the mass transfer coefficient is possible only if the area is correctly known. This is best accomplished by using a stirred cell with a fixed gas-liquid interfacial area, although other experimental reactors such as the wetted wall column, laminar jet, and disk contactor can also be used (see Danckwerts, 1970; Doraiswamy and Sharma, 1984). The two commonly used cell designs are those of Danckwerts (1970) and Levenspiel and Godfrey (1974).

A sketch of the Levenspiel–Godfrey cell is shown in Figure 14.13. It consists of two flanged sections of tubing, one for the upper part (gas phase) and the

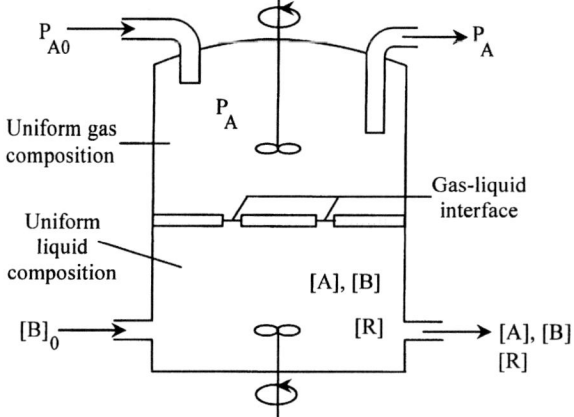

Figure 14.13 A gradientless cell for measuring mass transfer coefficients with provision for independently varying the operating parameters including the interfacial area (Levenspiel and Godfrey, 1974).

other for the lower part (liquid phase), with an interface plate between the two sections. Ports are provided for introducing the gas and liquid phases separately. A distinctive feature of this cell is that the actual interfacial area can be independently varied with a high level of precision by providing holes in the plate and controlling their number. The two phases are independently stirred to produce gradientless conditions in each section (see Chapter 4). As a result, the measured rates give single point values directly.

A number of correlations for estimating mass transfer coefficients for different gas-liquid situations have been reported. Cussler (1997) recommends the following:

PURE GAS BUBBLES IN A STIRRED TANK

$$Sh = \frac{k'_L d_s}{D} = 0.13 \left[\frac{(P/V)d_s^4 \rho^2}{\mu^3} \right]^{1/3} Sc^{1/3} \qquad (14.42)$$

where P/V is the stirrer power per unit volume (kW/m s^3).

PURE GAS BUBBLES IN AN UNSTIRRED TANK

$$Sh = \frac{k'_L d_s}{D} = 0.31 \left[\frac{d_s^3 g(\rho_L - \rho_{G0})\rho_L}{\mu^2} \right] Sc^{1/3} \qquad (14.43)$$

EXAMPLES OF GAS-LIQUID REACTIONS

A number of examples of gas-liquid reactions are listed in Table 14.7. The more recent trend has been to use homogeneous catalysts dissolved in the liquid phase.

Table 14.7 Examples of gas-liquid reactions (including some using homogeneous catalysts)

Reaction	Type of reactor[a]	Reference
1. Ethylation of aniline to 2,6-diethylaniline, a herbicide (Dual^{TM}) intermediate, in the presence of Al-trianilide	MAC	Stroh et al. (1957)
2. Reaction between ketene and acetic acid to make acetic anhydride	PCR	Jeffreys et al. (1961)
3. Reaction between ethylene and chlorine to give dichloroethane	MAC	Balasubramanian et al. (1966)
4. Liquid-phase chlorination of benzene, side-chain chlorination of toluene, etc.	MAC BCR PCR	Miller (1966) Chua and Ratcliffe (1971) Barona and Prengle (1973)
5. Oxychlorination of ethylene in the production of vinyl chloride	BCR	Friend et al. (1968)
6. Wacker process for the conversion of lower olefins to the corresponding carbonyl compounds, for example, ethylene to vinyl acetate or acetaldehyde, propylene to acetone, etc.	MAC BCR	Chandalia (1968a,b)
7. Alkylation of acetic acid with isobutylene to make t-butyl acetate	MAC BCR	Albright and Smith (1968) Gehlawat and Sharma (1971)
8. Conversion of ethylene or propylene to the corresponding chlorohydrins	PCR BCR	Saeki et al. (1972)
9. Ozonolysis of unsaturated fatty acids and esters, for example, of oleic acid in the production of pelargonic acid	MAC BCR	Throckmorton and Pryde (1972) Prengle et al. (1975), Carlson et al. (1977)
10. Oxidation of p-t-butyl toluene to p-t-butyl benzaldehyde, a perfumery chemical, in the presence of a Br promoted Co complex	MAC	Kogami and Kumanotami (1973)
11. Selective hydrogenation of C_{22}/C_{23} double bond in the synthesis of Iverectin, a pharmaceutical, in the presence of Rh(I)Cl(PPh$_3$)$_3$	MAC	Chabala and Fischer (1980)
12. Carbonylation of ethanol to propionic acid, a food preservative/perfumery chemical, in the presence of a Rh-carbonyl or Ni-isoquinoline/Li(I) complex	MAC	Dake et al. (1984), Kelkar et al. (1992)
13. Hydroformylation of 1,4-diacetoxybutene to 4-acetoxy-2-methylcrotonaldehyde, an intermediate in vitamin A production	MAC	Stinson (1986)
14. Hydroformylation of acrylonitrile to NC(CH$_2$)$_2$CHO, a food intermediate	MAC	Botteghi et al. (1987)
15. Hydrogenation of dehydrolinalool to linalool, a perfumery chemical	MAC	Parshall and Nugent (1988)
16. Oxidation of indole to indigo, a dye, in the presence of cumyl peroxide	MAC	Yamamoto et al. (1995), Otake (1995)

[a]MAC = mechanically agitated contactor, BCR = bubble column reactor, PCR = plate column reactor. See Chapter 17 for a description of these reactors.

As already noted, these reactions are characterized by hyperbolic rate forms. Empirical power law models can also be used, but their applicability is restricted to the ranges of parameter values used in their formulation. Note in the table that mechanically agitated contactors (MACs) and bubble-column reactors (BCRs) are the most commonly used reactors. The design of such reactors is considered in Chapter 16.

Chapter 15

LIQUID-LIQUID AND SOLID-LIQUID REACTIONS

Chapter 14 was concerned with gas-liquid reactions, although it was mentioned at the beginning of the chapter that the same principles would apply equally if phase 1 were a liquid or a solid. The premise for such a statement was that phase 1 comes into the picture only to the extent that it supplies the solute A that diffuses into phase 2 and reacts with B in the film, bulk, or both, in that phase only. Because the state of phase 1 is immaterial in such a situation, the theories are equally applicable when the phase is a liquid or a solid.

However, an additional factor comes into play when phase 1 is also a liquid. Component B from phase 2 can have a finite solubility in phase 1, diffuse into that phase and react there with A. Thus reaction can occur in both phases. This is equally true when phase 1 is a solid, but the mechanism of diffusion and reaction in a solid is different. It is also possible for a gas and a solid to simultaneously dissolve and react in a liquid, but as three phases are involved here, it is considered in Chapter 17.

In this chapter we consider the following two classes of reactions: liquid-liquid and solid-liquid. One encounters reactions belonging to these classes quite frequently in organic chemical technology. We shall give instances of these while dealing with the individual systems.

LIQUID-LIQUID REACTIONS

Hydrolysis and saponification of esters and fats (Jeffreys et al., 1961; Donders et al., 1968; Sharma and Nanda, 1968), sulfonation and nitration of aromatic compounds (Albright and Hanson, 1969, 1975; Hanson and Ismail, 1976; Barona and Prengle, 1973), alkylation of isobutane, toluene, and phenols with

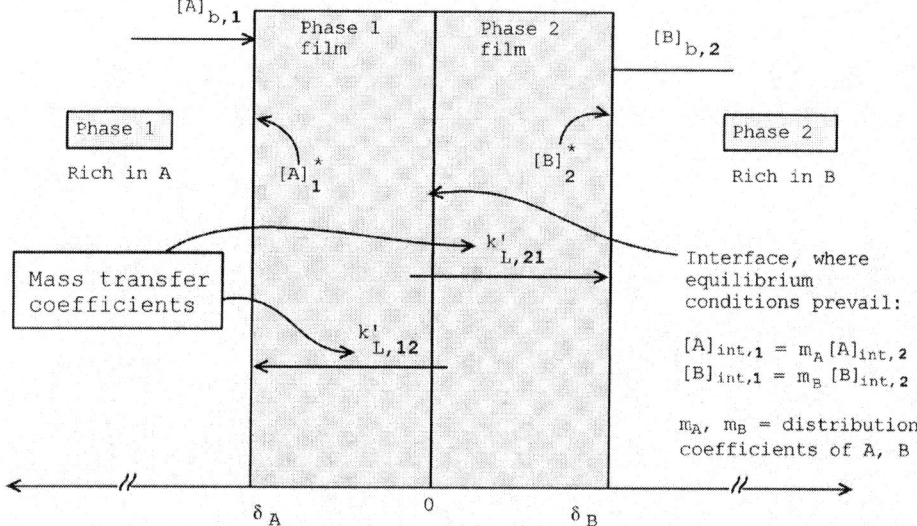

Figure 15.1 General representation of liquid-liquid reactions.

isobutylene (Jernigan et al., 1965; Mosby and Albright, 1966; Sprow, 1969; Tiwari and Sharma, 1977; Zaldivar et al., 1996), and oximation of cycloyhexanone (Rod, 1974, 1976) are some of the liquid-liquid reactions in which reaction occurs in both phases. The reaction last mentioned between cyclohexanone (which constitutes the organic phase) and aqueous hydroxylamine sulfate (containing ammonium sulfate) to give cyclohexanone oxime comprises the first step in the manufacture of caprolactam, an important reactant in the synthesis of Nylon. The reaction occurs in the aqueous phase at all levels of conversion, but at higher conversions it also occurs in the organic phase (Doraiswamy and Sharma, 1984).

As in gas-liquid reactions, here also we consider various regimes and their combinations but focus on their qualitative behavior. For the more important systems, we also summarize in tabular form the conditions to be satisfied for the different regimes and the corresponding rate equations.

Because two-liquid films are present here, two terms appear for each component, one for phase 1 and the other for phase 2.[1] The distinction between them is explained in Figure 15.1. $[A]_{b,1}$ is the concentration of A in phase 1, and $[A]_{\text{int},1}$ the concentration of A in phase 1 at the interface; $D_{A,2}$ refers to the diffusion of A in the phase containing B, phase 2, and $D_{B,1}$ to the diffusion of B in the phase containing A, phase 1; m_A and m_B are the distribution coefficients of A and B, respectively, between phases 1 and 2; and $k'_{L,12}$ and $k'_{L,21}$ are the mass transfer coefficients shown in the figure.

---

[1] Although all terms are defined under Notation, we specifically explain a few here in the interest of clarity.

470   Fluid-Fluids and Fluid-Fluid-Solids

Liquid-liquid reactions, like their gas-liquid counterparts, are affected by the physical and chemical characteristics of the system and also by the mechanical features of the equipment. Dispersion phenomena such as the coalescence and breakage of droplets, drop size distribution, and phase inversion will also affect both conversion and selectivity (see, e.g., Sprow, 1967; Laddha and Degaleesan, 1976; Delichatsios and Probstein, 1976; Tavlarides and Stamatoudis, 1981; Backes et al., 1990; Hernandez et al., 1993; Kumar et al., 1993; Brooks and Richmond, 1994). Analysis of dispersion in reactors is a vast field in itself and will not be considered in this book, except as a feature of the equipment.

Very Slow and Slow Reactions

In both cases, reaction occurs in the bulk. There are no gradients in either film for very slow reactions (regime 1), as shown in Figure 15.2a. On the other hand, for slow reactions (regime 2), diffusion of A in phase 2 and of B in phase 1 with no accompanying reaction in either case are the controlling steps. As shown in Figure 15.2b, the bulk concentrations of A in phase 2 and of B in phase 1 drop to zero. The conditions for this regime are the converse of those for regime 1; both are listed in Table 15.1 along with the corresponding rate equations. It is a relatively simple matter to extend the analysis when phase 1 is in regime 1 and phase 2 is in regime 2, or vice versa.

Fast Reactions

The presence of two liquid films leads to several combinations of regimes in the two films. In view of the importance of different regimes being operative in the two phases, we illustrate the approach by considering two combinations: (1) fast reaction in one phase (reaction-diffusion in the film of this phase) and slow reaction in the other (reaction in the bulk of this phase with mass transfer in the film), and (2) fast reaction in both phases. The treatment can be readily extended to other combinations.

Fast reaction in one phase only (say, phase 2)

This is analogous to that of a fast reaction in gas-liquid reactions discussed in Chapter 14, but is not exactly identical. Referring to Figure 15.3, there is a finite flux of B at the interface toward phase 1. No such flux exists in gas-liquid reactions except where B desorbs from phase 2 and diffuses into phase 1, a rather rare occurrence. This flux of B leads to a situation where the interfacial concentration of B is different from the bulk concentration, that is,

$$[B]_{\text{int},2} \neq [B]_{\text{b},2} \tag{15.1}$$

The conditions under which the above inequality is valid and the corresponding rate equations for no depletion of B in phase 2 are included in Table 15.1.

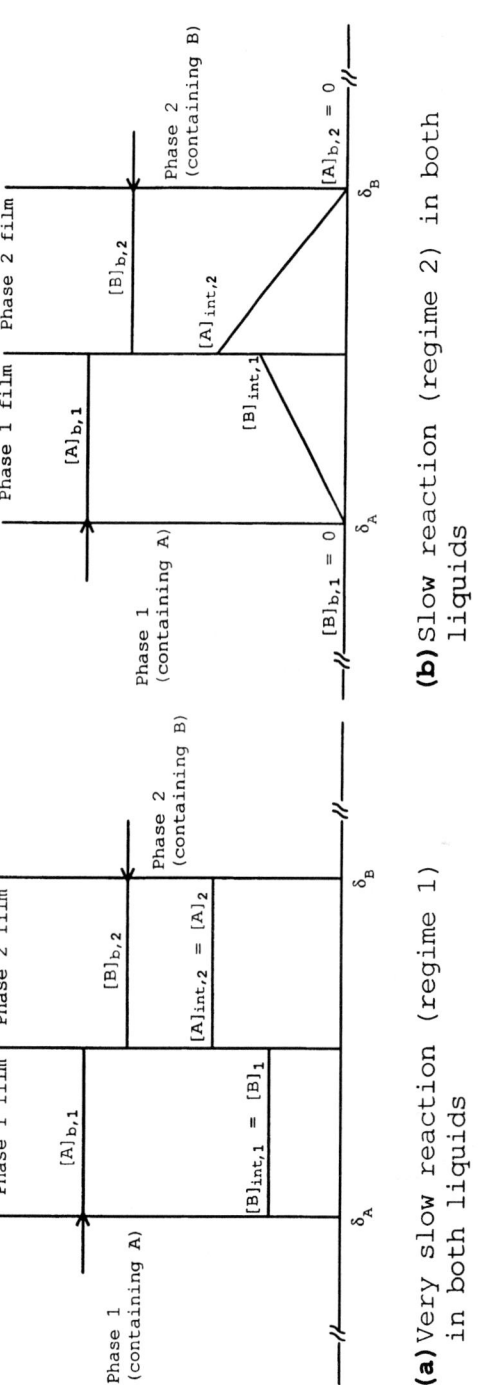

(a) Very slow reaction (regime 1) in both liquids

(b) Slow reaction (regime 2) in both liquids

Figure 15.2 Liquid-liquid reaction in bulk.

Table 15.1 Reaction in both phases of liquid-liquid reactions[a]

Regime	Conditions	Rate equations (sum of rates in the two phases)
1. Very slow (regime 1) in both phases	For reaction in phase 1: $k'_{L,12}a_L[A]_{int,2} \gg bk_{mn,2}[A]^m_{int,2}[B]^n_{b,2}$ For reaction in phase 2: $k'_{L,21}a_L[B]_{int,1} \gg (1-b)k_{mn,1}[A]^m_{b,1}[B]^n_{int,1}$	$r'_A = \left(\dfrac{b}{a_L}k_{mn,2}[A]^m_{int,2}[B]^n_{b,2}\right) + \left(\dfrac{1-b}{a_L}k_{mn,1}[A]^m_{b,1}[B]^n_{int,1}\right)$
2. Slow (regime 2) in both phases	Reverse of above for both phases plus: $\dfrac{\left(\dfrac{2}{m+1}D_{A,2}k_{mn,2}[B]^n_{b,2}[A]^{m-1}_{int,2}\right)^{1/2}}{k'_{L,12}} \ll 1$ $\dfrac{\left(\dfrac{2}{m+1}D_{B,1}k_{mn,1}[A]^m_{b,1}[B]^{n-1}_{int,1}\right)^{1/2}}{k'_{L,21}} \ll 1$ (these ensure no reaction in film)	$r'_A = k'_{L,12}[A]_{int,2} + \dfrac{k'_{L,21}[B]_{int,1}}{\nu_B}$
3. Fast (regime 3) in phase 2 only	$[B]_{int,2} \neq [B]_{b,2}$ (i.e., there is a finite flux of B toward phase 1) if $\sqrt{M} = \dfrac{\left[\left(\dfrac{2}{m+1}\right)D_{A,2}k_{mn,2}[B]^n_{b,2}[A]^{m-1}_{int,2}\right]}{k'_{L,12}}$ $\approx \dfrac{[B]_{b,2}}{\nu_B[A]_{int,2}}$	$r'_A = \eta k'_{L,12}[A]_{int,2}$ $= k'_{L,11}([A]_{b,1} - [A]_{int,1}$ Note: If LHS $\ll$ RHS in the condition, there is no flux of B toward phase 1, that is, $[B]_{int,2} \approx [B]_{b,2}$, and the system behaves as a gas-liquid reaction

4. Fast (regime 3) in both phases; no depletion (pseudo-nth-order in phase 1 and pseudo-mth-order in phase 2)

For reaction in phase 1:

$$3 < \sqrt{M_{\text{B-A}}} = \frac{\left(\dfrac{2}{n+1} D_{\text{B,1}} k_{\text{mn,1}} [A]_{\text{b,1}}^{\text{m}} [B]_{\text{int,1}}^{\text{n}-1}\right)^{1/2}}{k'_{\text{L,21}}}$$

$$\ll q_{\text{B-A}} = \frac{\nu_{\text{B}} [A]_{\text{b,1}}}{[B]_{\text{b,2}} m_{\text{B}}}$$

For reaction in phase 2:

$$3 < \sqrt{M_{\text{A-B}}} = \frac{\left(\dfrac{2}{m+1} D_{\text{A,2}} k_{\text{mn,2}} [B]_{\text{b,2}}^{\text{n}} [A]_{\text{int,2}}^{\text{m}-1}\right)^{1/2}}{k'_{\text{L,12}}}$$

$$\ll q_{\text{A-B}} = \frac{m_{\text{A}} [B]_{\text{b,2}}}{\nu_{\text{B}} [A]_{\text{b,1}}}$$

5. Fast (regime 3) in both phases; depletion

The last two terms in the two conditions given in item (4) are comparable.

6. Instantaneous (regime 4)

If $\delta_{\text{A}} = \delta_{\text{B}}$,

$$\frac{[B]_{\text{b,2}}}{\nu_{\text{B}} [A]_{\text{b,1}}} \sqrt{\frac{D_{\text{B,2}}}{D_{\text{A,1}}}} \qquad \text{(penetration theory)}$$

> 1, reaction in phase 1
< 1, reaction in phase 2
= 1, reaction at interface

$$r'_{\text{A}} = m_{\text{B}} [B]_{\text{b,2}} \left(\frac{2}{n+1} D_{\text{B,1}} k_{\text{mn,1}} [A]_{\text{b,1}}^{\text{m}} [B]_{\text{int,1}}^{\text{n}-1}\right)^{1/2}$$

$$+ \frac{[A]_{\text{b,1}}}{m_{\text{A}}} \left(\frac{2}{m+1} D_{\text{A,2}} k_{\text{mn,2}} [B]_{\text{b,2}}^{\text{n}} [A]_{\text{int,2}}^{\text{m}-1}\right)^{1/2}$$

Analytical expression not possible. See Scriven (1961), also Lee (1968), Rod (1974), and Sada et al. (1977).

When reaction occurs in phase 2,

$$r'_{\text{A}} = k'_{\text{L,12}} [A]_{\text{b,1}} \frac{1 + q_2}{(1 + q_2) m_{\text{A}} + (1 + q_2) \left(\dfrac{D_{\text{A,2}}}{D_{\text{A,1}}}\right)}$$

where $q_2 = \dfrac{D_{\text{B,2}} [B]_{\text{b,2}}}{\nu_{\text{B}} D_{\text{A,1}} [A]_{\text{b,1}}}$

When reaction occurs in phase 1,

$$r'_{\text{A}} = \frac{k'_{\text{L,21}} [B]_{\text{b,2}}}{\nu_{\text{B}}} \frac{1 + q_1}{(1 - q_1)/m_{\text{B}} + (1 + q_1) \left(\dfrac{D_{\text{B,1}}}{D_{\text{B,2}}}\right)}$$

where $q_1 = \dfrac{\nu_{\text{B}} D_{\text{A,1}} [A]_{\text{b,1}}}{D_{\text{B,2}} [B]_{\text{b,2}}}$

---

[a] $[A]_{\text{int,2}} = [A]_{\text{b,1}}/m_{\text{A}}$; $[B]_{\text{int,1}} = [B]_{\text{b,2}} m_{\text{B}}$; and $[A]_{\text{int,1}} = [A]_{\text{b,2}}/m_{\text{A}} = [A]_{\text{b,2}}/m_{\text{A}}$; m_{A} and m_{B} are distribution coefficients of A and B between phases **1** and **2**.

474 Fluid-Fluids and Fluid-Fluid-Solids

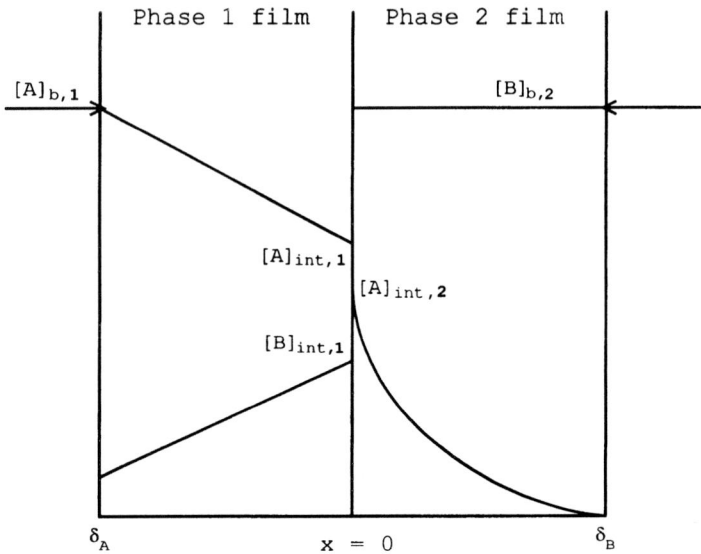

Figure 15.3 Fast reaction in phase 2 only.

Fast reaction in both phases

Here, we consider the case where diffusion and reaction occur simultaneously in both films. First, we examine the case sketched in Figure 15.4a, where the main components of the two phases are not depleted in their respective films. A is not depleted in the film of phase 1, and B is not depleted in the film of phase 2. The concentration profiles for the diffusion of A accompanied by a fast reaction in phase 2 and of B accompanied by a fast reaction in phase 1 are shown in the figure. As already mentioned, allowance should be made for a finite flux of B into

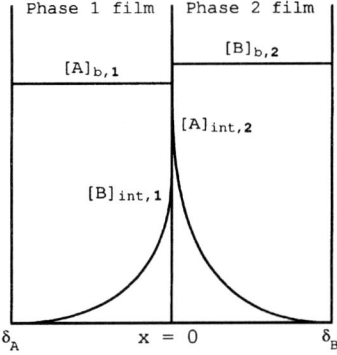 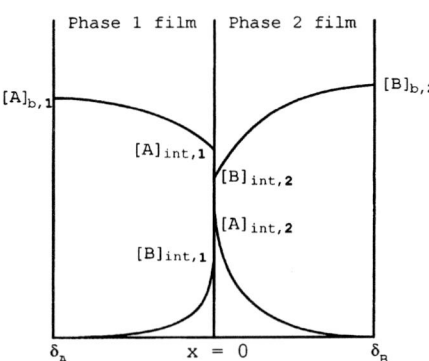

(a) No depletion in either phase   **(b)** Depletion in each phase

Figure 15.4 Fast reaction in both phases.

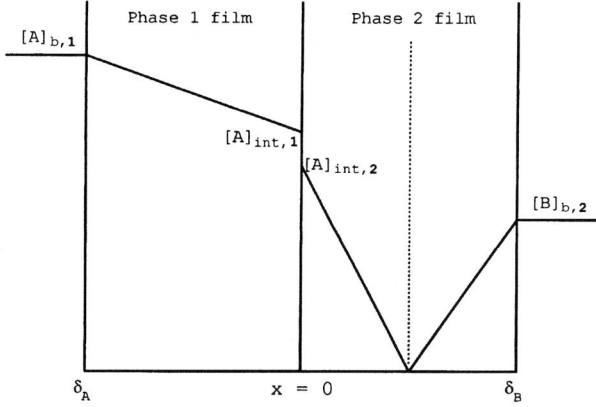

Figure 15.5 Instantaneous reaction in phase 2 film.

phase A. The conditions for this regime and the corresponding rate equations are included in Table 15.1.

Now we consider the case where there is depletion of A in the phase 1 film and of B in the phase 2 film, leading to the profiles sketched in Figure 15.4b. The conditions for this regime along with analytical solutions for the rate are included in Table 15.1. It would, however, be best to use collocation techniques to obtain the rate because the analytical expression tends to be quite approximate—although it is good enough as a first-order approximation. One can also visualize depletion in one phase and no depletion in the other and write equations for this case based on the previous two.

Instantaneous Reaction

The chief features of an instantaneous reaction have already been described in Chapter 14. However, in a liquid-liquid reaction with its two liquid films, the reaction plane can be located within either one of these films. Thus a criterion is needed for determining the phase in which it is located. Figure 15.5 illustrates a reaction in phase 2. Such a criterion has been derived by Scriven (1961) from the penetration theory and by Mhaskar and Sharma (1975) from the simpler film theory. The application of this criterion tells when the plane is located in phase 1, when in phase 2, and when at the interface of the two liquids. Table 15.1 gives this criterion along with the rate equations.

EXAMPLE 15.1

Operating regimes in the production of *m*-chloroaniline by liquid-liquid reduction of *m*-nitrochlorobenzene with disodium sulfide

A batch of *m*-chloroanaline (MCA) is to be produced by reducing *m*-nitrochlorobenzene (MNCB) with an aqueous solution of sodium disulfide (Na_2S_2) in a fully baffled mechani-

cally agitated batch contactor at 105 °C. Given the following data, find the operating regimes of this reaction:

dispersed phase: MNCB (phase 1)
continuous phase: Aqueous Na_2S_2 solution (phase 2)
reaction occurs only in phase 2
initial concentration of Na_2S_2, $([B]_{b,2})_i = 3 \times 10^{-3}$ mol/cm^3
initial concentration of MNCB in the organic phase

$$([A]_{b,1})_i = 8.454 \times 10^{-3} \text{ mol/cm}^3$$

conversion of MNCB = 99%
final concentration of MNCB in the organic phase,

$$([A]_{b,1})_f = 8.454 \times 10^{-5} \text{ mol/cm}^3$$

distribution coefficient, $m_A = 8.56 \times 10^{-4}$
diffusivity of MNCB in aqueous sodium disulfide,

$$D_{A,2} = 3.12 \times 10^{-5} \text{ cm}^2/\text{s}$$

second-order rate constant (first-order each in A and B),

$$k_{2,2} = 3.46 \times 10^3 \text{ cm}^3/\text{mol s}$$

true continuous-phase mass transfer coefficient,

$$k'_{L,12} = 5 \times 10^{-3} \text{ cm/s}$$

stoichiometric coefficient, $\nu_B = 1$

SOLUTION

The two parameters that determine the regime of a reaction are $\sqrt{M}$ and q. In this case, from Table 15.1

$$\sqrt{M_{A-B}} = \frac{\sqrt{\frac{2}{m+1} D_{A,2} k_{2,2} [B]_{b,2}^n [A]_{int}^{m-1}}}{k'_{L,12}} \tag{E15.1.1}$$

or

$$\sqrt{M_1} = \frac{\sqrt{D_{A,2} k_{1,2} [B]_{v,2}}}{k'_{L,12}} \tag{E15.1.2}$$

and

$$q_{A-B} = \frac{[B]_{b,2}}{\nu_B [A]_{b,1} m_A} \tag{E15.1.3}$$

where $m = n = 1$, $k_{1,2}$ (pseudo-first-order rate constant in phase 2) $= k_{2,2}[B]_{b,2}$, and M_1 represents M for a pseudo-first-order reaction.

Now we find $\sqrt{M_1}$ and q_{A-B} for the initial and final conditions (represented by i and f, respectively).

$$(\sqrt{M_1})_i = \frac{\sqrt{3.12 \times 10^{-5} \times 3.46 \times 10^3 \times 3 \times 10^{-3}}}{5 \times 10^{-3}} = 3.6$$

$$(q_{A-B})_i = \frac{3 \times 10^{-3}}{1 \times 8.45 \times 10^{-3} \times 8.56 \times 10^{-4}} = 414.6$$

Hence

$$3 < (\sqrt{M_1})_i \ll (q_{A-B})_i$$

clearly indicating that the reaction falls in regime 3.

Similar calculations for the final conditions give [with $([B]_{v,2})_f = 3 \times 10^{-5}$ mol/cm^3]

$$(\sqrt{M_1})_f = 0.358, \quad (q_{A-B})_f = 414.6$$

Hence

$$(\sqrt{M_1})_f < 1, \text{ but not} \ll 1$$

indicating that the reaction falls between regimes 1 and 2.

CONCLUSION

This reaction starts out in regime 3 but shifts to regime 1–2 at the end. In other words, a fast reaction slows down as it progresses to the end.

In Chapter 16 we shall see how this information can be used in designing a reactor.

Laboratory Reactors

A good reactor to use for laboratory experiments is a batch reactor with a qiescent interface. This is particularly useful in obtaining rate data under different controlled conditions of mass transfer. A simple way of varying the mass transfer. A simple way of varying the mass transfer effect is to float various sizes of plastic balls at the liquid-liquid interface, thus controlling the area of mass transfer. Reactors in which droplets of one liquid are introduced into the other can also be used.

SOLID-LIQUID REACTIONS

Solid-liquid reactions can be classified into three categories depending on the solubility of the solid in the liquid:

1. highly soluble
2. sparingly soluble
3. insoluble

When the solid is highly soluble, mass transfer considerations are unimportant, and the reaction rate can be expressed exclusively in terms of its chemical kinetics, that is, the reaction occurs in regime 1. On the other hand, when the

solid is sparingly soluble, the rate of dissolution becomes an important factor and must be considered in the formulation of rate equations. As in liquid-liquid reactions, it is also possible that part of the reaction occurs within the solid. The nature of this reaction depends essentially on the behavior of the diffusing liquid-phase component within the solid matrix and calls for a different approach to the analysis. When the solid is insoluble, the reaction occurs exclusively in the solid phase.

Mass Transfer Correlations

A number of studies have been reported on solid-liquid mass transfer in different kinds of contactors, and they have been periodically reviewed, for example Miller (1971), Nienow (1975), Wen and Fan (1975), Briens et al. (1993). Of these, only the mechanically agitated contactor is used with or without a gas phase. It is the only truly two-phase solid-liquid contactor. The other types of contactors, such as the bubble-column contactor (usually), the trickle-bed reactor, and the three-phase fluidized-bed reactor, all involve three phases and are considered in Chapter 17 on multiphase reactions.

A large number of correlations covering a variety of systems and operating conditions have been reported for MAC, and the most important of these have been tabulated by Doraiswamy and Sharma (1984b). One is due to Levins and Glastonbury (1972a,b).

$$Sh = 2 + 0.44 \left(\frac{d_p u_{es} \rho_L}{\mu_L} \right)^{0.5} Sc^{0.38} \tag{15.2}$$

where u_{es} is an effective slip velocity which is the resultant of three types of slip velocity u_1, u_2 and u_3—that is, $u_{es} = (u_1^2 + u_2^2 + u_3^2)^{0.38}$. The original reference may be consulted for the definitions and estimation of the three velocities. Other recommended correlations are:

$$Sh = \frac{k'_{SL} d_p}{DS_F} = 0.4 (ed_p^4 \rho_L^3)^{1/4} Sc^{1/3} \quad \text{(Sano et al., 1974)} \tag{15.3}$$

$$Sh = \frac{k'_{SL} d_p}{D} = 2.0 + Re^{1/2} Sc \quad \text{(Cussler, 1997)} \tag{15.4}$$

where S_F is the shape factor and e is the energy supplied per unit mass of the slurry.

Solid Sparingly Soluble

Consider the general reaction

$$A + \nu_B B(l) \rightarrow \text{products} \qquad [15.1]$$

The analysis is similar to that for gas-liquid reactions involving the same four regimes, but where A is a dissolving solid instead of a gas. One major theoretical point of difference must be noted. In liquid films in gas-liquid systems, either the film theory, the penetration theory, or the surface renewal theory is used. On the

other hand, where a solid surface is involved, the film theory and the boundary layer theory are the most frequently used (although several other hydrodynamic theories have also been proposed). However, it has been shown (Vieth et al., 1963) that almost the same values of enhancement are obtained even when the turbulent boundary layer theory is used for solid-liquid reactions. All theories give exactly the same result:

$$\text{for } \sqrt{M} > 3, \quad \eta = \sqrt{M} \tag{15.5}$$

Regimes 1, 2, and 3

The reaction of terephthalic acid (as suspended solid particles) with ethylene oxide dissolved in *n*-butanol to give bis(2-hydroxyethyl) terephthalate (which is the monomer in the manufacture of polyethylene terephthalate) occurs in regime 1 (Bhatia et al., 1976). Nitration of solid aromatic compounds such as naphthalene and acetylated *p*-anisidine proceeds in regime 3.

Regime between 2 and 3

Here, we are concerned with solid dissolution accompanied by reaction, part of which occurs in the liquid film and the rest in the bulk. It will be recalled that gas-liquid reactions were analyzed using essentially the film theory. In view of Equation 15.5, we use the film theory here also.

REACTION IN SOLIDS

One can visualize a situation where component B diffuses from the liquid phase and reacts with A on the solid surface. Example 15.2 illustrates the approach to such a situation, as a prelude to the case of an insoluble solid considered in a later section.

EXAMPLE 15.2

Discerning the controlling regime in reactive dyeing

Dyeing of textiles with reactive dyes is a very important industrial process. Unlike in usual cases, there is no dissolution of reactant A from the solid phase, the fiber, followed by reaction in solution, but the reactive dye B from the solution is taken up by the fiber where it reacts with the cellulosate ions to form a uniform chemically bound dye (Figure 15.6).

Let us consider a typical reactive dye Procion Brilliant Red 2B:

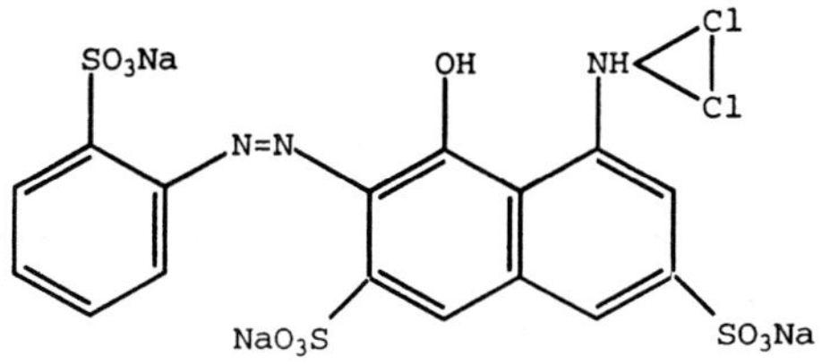

where

<Cl|Cl stands for —C(=N—C(Cl)=N—C(Cl)=N) (triazinal residue)

In the dyeing process, one —Cl or <Cl|Cl reacts with cellolose O⁻ and sometimes also with OH⁻.

To apply the diffusion-reaction theory to reactive dyeing, we use the fact that the overall behavior is generally similar to that of regime 2–3. The following equation based on the penetration theory (Danckwerts, 1970) can be derived for regime 2–3:

$$Q' = [B]^* \sqrt{D_B k_1} \left(t + \frac{1}{2k_1} \right) \qquad (E15.2.1)$$

where Q' is the amount of dye reacted per unit area and k_1 is the first-order rate constant. Doraiswamy and Sharma (1984) used reported experimental data (see Sumner and Weston, 1963; Rattee and Breuer, 1974; Peters, 1975) to prove that the equation does apply and that the process indeed involves diffusion and reaction—not mere diffusion as would be the case with the more common nonreactive dye. The reported experiments were carried out in a flow system where a freshly prepared mixture of dye and alkali was allowed to flow past a viscose film, and the data were expressed as the rate of dye uptake versus time. Equation E15.2.1 shows that a plot of Q' versus t should give a straight line, and this is indeed the case. Thus dyeing with a reactive dye occurs in the regime between 2 and 3.

The establishment of the regime of operation provides clear guidelines for efficient dyeing. Thus two factors must be considered: (1) the rate of dyeing which determines whether dyeing can be completed in an economical time period and (2) the efficiency of reaction (given by Equation E15.2.1) which is an obvious measure of the extent of dyeing. Because $k_1 = k_2[O^-]$, where k_2 is the second order-rate constant, it varies from dye to dye,

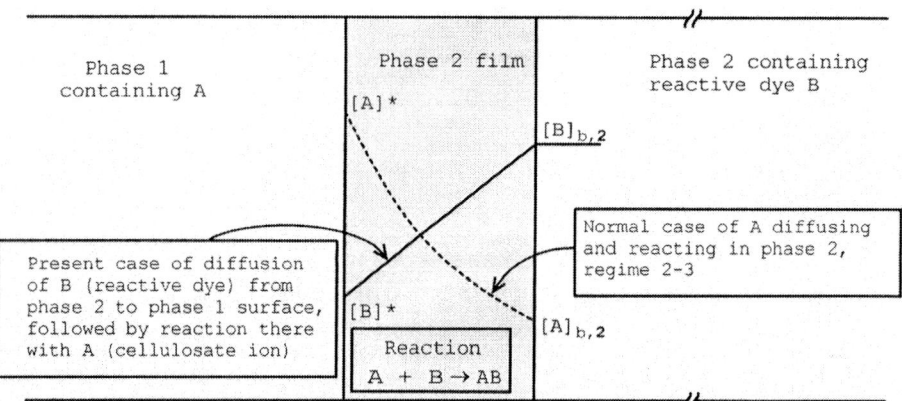

Figure 15.6 Illustration of reactive dyeing for the reaction $A + B \to AB$, where A is the cellulosic ion (O⁻) and B is the dye.

and D_B is a function of the ratio $[B]^*/[B]_{b,2}$ referred to in the dyeing trade as the *substantivity ratio*, both of which influence Q'. For a given dye, Q' is influenced largely by the substantivity ratio which in turn can be increased by using neutral salts.

This example demonstrates the power of the general methods of analysis presented here to extract meaningful conclusions from simple rate data even for a process such as dyeing practiced for long as an empirical art (and regarded as outside the reach of analysis). Thus it enables a clearer understanding of the effect of operating variables in such processes.

Regime between 3 and 4

In this regime the depletion regime, the enhancement factor for gas-liquid reactions is given by Equation 14.29. However, the following expression for η_a in that equation should be used instead of Equation 14.24 (Doraiswamy and Sharma, 1984):

$$\eta_a = 1 + \left(\frac{[B]_b}{\nu_B[A]^*}\right)\left(\frac{D_B}{D_A}\right)^{2/3} \tag{15.6}$$

Regime 4

This is a regime in which the diffusion coefficients of A and B in the liquid are the controlling parameters, and chemical reaction plays practically no part. Thus it has frequently been used to compare various theories of mass transfer to and from solid surfaces. The main conclusion is that the value of the exponent p in $(D_B/D_A)^p$ is different for different theories. The value of n for the boundary layer theory is 2/3. Recalling the values for the film and penetration theories,

$p = 1$ for the film theory,
$p = 1/2$ for the penetration theory, and
$p = 2/3$ for the boundary layer theory.

It does not seem possible at this stage to single out any particular theory as the most acceptable. In any case, because the diffusivities do not differ greatly and also considering the error in their estimation, the exact value of p is not crucial. It is often adequate to use the film theory and then apply the necessary correction by using the appropriate value of p provided that the diffusivities are accurately known. Thus the methods and equations given in Chapter 14 should be applicable.

Solid Insoluble

In systems of this kind, because the solid is insoluble, no reaction can occur between A and B in the liquid phase. Hence the component from the liquid phase must diffuse into the solid phase and react there. The most important example of such a system in organic technology is the manufacture of organometallic compounds, for example, the Grignard reagent. We shall discuss the Grignard reaction at some length later in this section. Another example is the production of

acetylene, one of the most important building blocks of organic technology, by reaction between water and insoluble calcium carbide suspended in it:

$$CaC_2 + H_2O \rightarrow C_2H_2 + CaO \qquad [15.2]$$

There are many instances where reaction between an insoluble solid and liquid occurs in the presence of a gas. The gas dissolves in the liquid and then diffuses into the solid where reaction occurs. This is a three-phase system and therefore will be deferred to Chapter 17.

The following steps are involved in the overall reaction between a solid and a fluid (the fluid can be a liquid or a gas): (1) diffusion through the fluid film adjacent to the solid and (2) diffusion and reaction in the solid according to several possible models. The analysis is very similar to that for gas-solid reactions (Levenspiel, 1972, 1993, 1998; Szekely et al., 1976; Ramachandran and Doraiswamy, 1982; Doraiswamy and Sharma, 1984 (vol. 1); Sahimi et al., 1990; Bhatia and Gupta, 1993).

Gas-solid reactions are ubiquitous in inorganic technology but rather rare in organic technology and even more rare in fine chemicals synthesis. However, the methods of analysis developed for these reactions are equally applicable to solid-liquid reactions. Hence we refer to such reactions in this chapter by the more general term *fluid-solid reactions*. The liquid-phase component can also be supplied by a dissolving gas (and not always by a solute previously dissolved in the liquid, as envisaged above). In such a case, the resistance of the liquid film surrounding the gas would also be a factor; this is considered separately in Chapter 17.

Figure 15.7 is a schematic representation of a typical fluid (f)-solid (s) reaction:

$$A(f) + B(s) \rightarrow R(g) + S(s) \qquad [15.3]$$

Figure 15.7a shows reaction at the interface between a solid reactant (B) and a product (S) after the fluid has diffused through an inwardly advancing shell of the product (*ash*). There is no reaction in either the *ash* layer or in the body of the reactant B (the core), but only at the surface of B. This is the *shrinking core* or *sharp interface* model and represents perhaps the most common mechanism of gas-solid reactions for nonporous solids. The overall reaction can be controlled

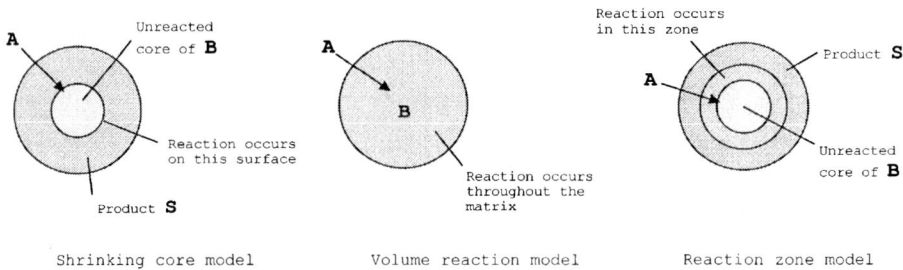

Figure 15.7 Three simple fluid-solid raction models

by diffusion through the hydrodynamic film adjacent to the solid, diffusion through the product R, reaction at the interface, or a combination of them. Different equations for conversion as a function of time result for these cases (Mazet, 1992; Levenspiel, 1999), but their use is not always warranted in the solid-liquid reactions normally encountered in organic synthesis.

An unusual variation of the shrinking core model is the *expanding core* model in which the reaction starts at the center (see Doraiswamy and Gokarn, 1988). The product—"ash core" instead of the reactant core in normal behavior—moves outward till it completely envelops the particle. The controlling mechanisms described for the shrinking core model apply equally here, but the equations are different. We shall discuss this at some length later in this section while describing a model for the disproportionation of potassium terephthalate.

Two other traditional models are the *volume reaction* and *zone models.* In the volume reaction model (Figure 15.7b), applicable largely to highly porous solids, the fluid reactant diffuses relatively quickly into the solid matrix, so that now reaction occurs at every point within this matrix. If the diffusion is very rapid, the rate of reaction is the same at every point, but if it is slow enough to establish a concentration gradient of A within the solid, the reaction rate will be a function of position. In the zone model (Figure 15.7c), developed mostly for solids of slightly lower porosity, reaction occurs in a moving reaction zone sandwiched between the advancing product and the retreating reactant. Several other models have been proposed for gas-solid reactions. Although some of them may be germane to specific solid-liquid reactions in organic technology, in general most of them are not. A detailed overview of these theories may be found in the reviews of Ramachandran and Doraiswamy (1982), Sahimi et al. (1990), and Bhatia and Gupta (1993).

EXAMPLE 15.3

Hankel reaction: disproportionation of potassium benzoate to potassium terephthalic acid

Terephthalic acid finds extensive application in the manufacture of synthetic polyester-type fibers (notably Dacron and Terylene). Commercially, it is produced by the liquid phase oxidation of *p*-xylene. Thermal disproportionation of the potassium salt of benzene carboxylic acid (obtained from benzene) in the presence of cadmium as a catalyst, followed by hydrolysis of the potassium terephthalate, is another potentially attractive route.

Several mechanisms have been suggested for the thermal decomposition reaction, such as a bimolecular reaction, carboxylation-decarboxylation, and the formation of an intermediate phenyl ion. Of these, the most acceptable is the last (Ratusky and Sorm, 1959).

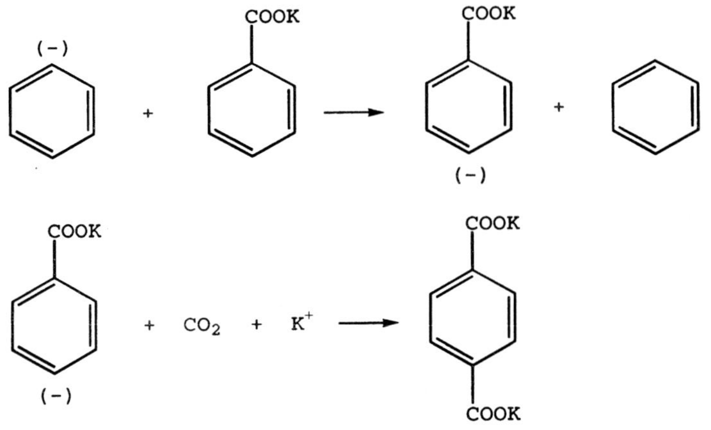

[Reaction scheme E15.3.1: benzene anion + potassium benzoate → potassium benzoate anion + benzene; and potassium benzoate anion + CO₂ + K⁺ → dipotassium terephthalate]

The first two steps are very fast. Hence, we can write

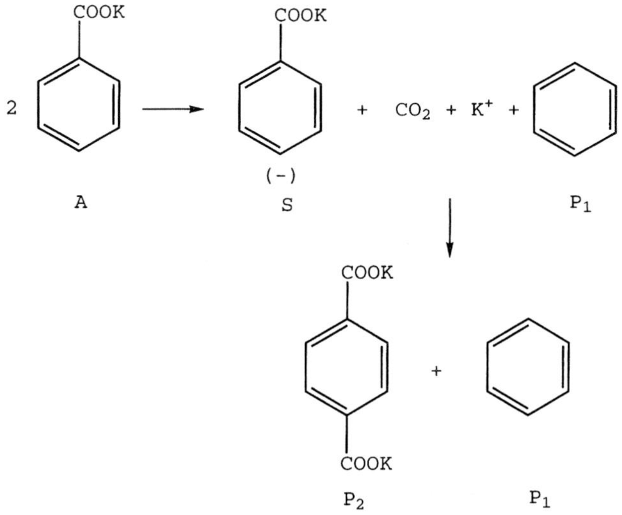

[Reaction scheme showing: 2 potassium benzoate (A) → potassium benzoate anion (S) + CO₂ + K⁺ + benzene (P₁), then → dipotassium terephthalate (P₂) + benzene (P₁)]

$$2A(s) \rightarrow S(s) + R(g) \rightarrow P_1(g) + P_2(s) \quad [E15.3.2]$$

where R is a mixture of CO_2 and benzene (P_1) from step 1; note that benzene comes out unreacted from step 2.

With the chemical steps identified, detailed experimental studies were carried out by Doraiswamy and collaborators (Gokhale et al., 1975; Kulkarni and Doraiswamy, 1980; Revankar et al., 1987, Revankar and Doraiswamy, 1992) to elucidate the mechanism by which the solid particles of potassium benzoate behave during the reaction. The results, represented in Figure 15.8, reveal rather unusual behavior. The reaction starts at the center and proceeds outward till it is complete, thus conforming to the expanding core behavior mentioned earlier.

To explain such behavior, let us first consider a simple first-order decomposition of solid A present as a pellet:

$$A(s) \rightarrow R(g) + S(s) \quad [E15.3.3]$$

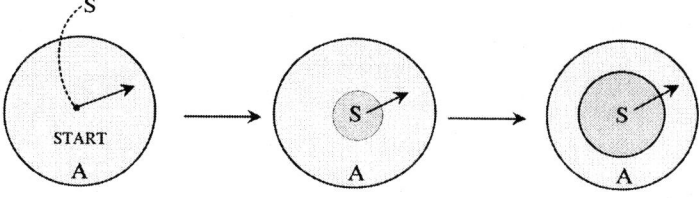

A = reactant, S = product

Figure 15.8 The expanding core or "rotting apple" (because the apple rots from the center) model for fluid-solid reactions.

The transport of gas R to the surface of the pellet is governed by an unsteady-state diffusion equation. By writing such an equation and performing certain mathematical manipulations, we get (Kulkarni and Doraiswamy, 1980)

$$[\text{Gas}] = \frac{2}{\pi^2} \phi^2 \sum_{n=1}^{\infty} -(-1)^n \frac{\exp(-\alpha t)}{n} \tag{E15.3.1}$$

where ϕ is the Thiele modulus defined as $R\sqrt{k[A]/D_{eR}}$, R is the radius of the solid, D_{eR} is the effective diffusivity of R in the solid, and α is a constant. This equation shows that the concentration is directly proportional to the square of the Thiele modulus. Thus higher values of the Thiele modulus (corresponding to lower diffusivities) lead to higher concentrations of gas within the solid. Although this is not a new observation in the context of diffusion-reaction theories, the result is particularly appropriate to the reaction system at hand, as we shall see below.

For the simple decomposition, reaction E15.3.3, the results indicate that a large portion of the gas is "trapped" inside the pellet in view of its inability to diffuse fast enough to the surface (particularly because the effective diffusivities are very low, on the order of 10^{-6}–10^{-7} cm^2/s). Thus a certain maximum pressure is built up inside the pellet, but this is seldom large enough to cause any instabilities or particle damage.

However, if the analysis is extended now to the complex disproportionation reaction, the second reaction corresponding to step 2 of the scheme would start at the center where the pressure (or concentration) of the generated gas is maximum (Figure 15.9).

The main postulate of this *delayed diffusion* model is that phenyl anion II (designated as S), formed as the solid intermediate, reacts with the generated gas R after a certain minimum pressure is reached to give the final product P_2. Based on this hypothesis, plots have been prepared for the rate of outward advance (as against inward advance in the normal shrinking core model) of the second solid phase as a function of position within

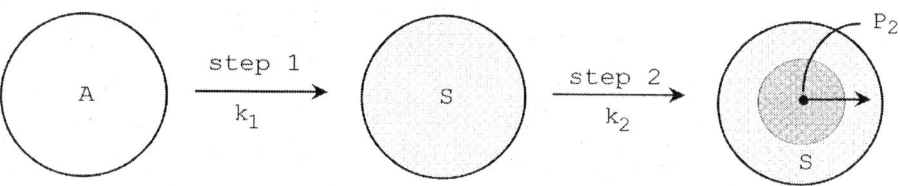

Figure 15.9 Schematic of the outwardly advancing interface in potassium benzoate disproportionation (the rotting apple model).

EXAMPLE 15.4

Grignard reaction: relative roles of reaction and mass transfer

The Grignard reagent is formed by the reaction of an alkyl halide with metallic magnesium. The reaction is carried out by suspending the metal in a solution of the halide under specified conditions (which vary from system to system). A simplified reaction pathway is shown in reaction scheme E15.4.1 for the formation of the Grignard reagent along with by-products from alkyl halides.

$$\begin{array}{c} R\text{-}X + Mg(s) \longrightarrow {}^\cdot R\text{-}X^- + {}^\cdot Mg^+(s) \\ \downarrow \\ {}^\cdot R + X\text{-}Mg^\cdot(s) \longrightarrow R\text{-}Mg\text{-}X \\ R^\cdot \diagup \quad \diagdown R^\cdot \\ R\text{-}R \qquad R(H) + R(\text{-}H) \end{array}$$ [E15.4.1]

The desired product is R-Mg-X, but note that side reactions also occur.

Reported data (see Rogers et al., 1980a,b; Sergeev et al., 1983) indicate that the rate-determining chemical step is usually the abstraction of an electron from magnesium, step 1 of the mechanism shown in scheme E15.4.1. With the controlling chemical step identified, now we can write the following mass balances for the consumption of the substrate RX represented by A:

$$\left(\frac{dN_A}{dt}\right) = -k'_{SL} a_p ([A] - [A]_s) V_r = -k'_I a_p [A]_s V_r = -k'_{ov} a_p [A]_b V_r \qquad (E15.4.1)$$

where

$$\frac{1}{k'_{ov}} = \frac{1}{k'_L} + \frac{1}{k'_I} \qquad (E15.4.2)$$

k'_{ov} is an overall rate constant and a_p is the interfacial area of magnesium per unit volume. The overall constant can be readily determined as a function of agitation (rpm). The asymptotic value of k'_{ov} gives the true rate constant in the absence of film diffusion effects, $k'_{ov} = k'_I$.

On the other hand, when film mass transfer controls, the overall coefficient would be equal to the mass transfer coefficient, $k'_{ov} = k'_{SL}$, and the value obtained should match the estimated value from established correlations. Even if the reaction is only partially influenced by mass transfer, the use of Equation E15.4.2 along with a value of k'_L should give the reaction rate constant.

True values of the mass transfer coefficient can be obtained only by using equi-accessible surfaces, those with a uniform boundary layer which therefore are equally accessible at all points. One such surface is the rotating disk. Thus the best way to determine precise values of k'_{SL} is to measure the coefficient using a rotating disk of

magnesium in a solution of the halide. Then these values can be compared with k'_{SL} estimated from available correlations for mass transfer from rotating disks.

Depending on whether the flow is laminar or turbulent, the following correlations can be used:

Laminar (Levich, 1954):

$$Sh = 0.62 Re^{0.5} Sc^{0.33} \tag{E15.4.3}$$

Turbulent (Cornet et al., 1969):

$$Sh = 0.0198 Re^{0.8} Sc^{0.33} \tag{E15.4.4}$$

where the dimensionless numbers have the usual definitions, but note that Re is defined as $Re = 2R_{disk}^2 (\text{rps})/\mu'$ where R_{disk} is the radius of the disk, μ' is the kinematic viscosity, and rps is the number of revolutions per second.

These correlations are for a smooth disk. In the Grignard reaction, because the product is formed by "removing" magnesium from the disk, the surface becomes rougher. Therefore, the exponent on Re in these equations must be modified to take this into account. Because such modified correlations are not available, the exponent should first be determined experimentally for the system at hand.

Hammerschmidt and Richarz (1991) determined the mass transfer coefficients using a rotating disk of $12.6 \, \text{cm}^2$ area and an electrochemical reaction (these reactions are usually very fast; hence mass transfer dominates). The mass transfer coefficients varied from 3.5×10^{-4} to 40×10^{-4} cm/s at 25 °C with increasing stirrer rpm. Then they studied the kinetics of the Grignard reaction of bromocyclopentane with magnesium and used the overall rate constant thus obtained to get the chemical reaction rate constant from Equation E15.4.2.

$$k_I = 55 \times 10^{-4} \, \text{s}^{-1} \text{ at } 25 \,°\text{C}$$

They also measured the true reaction rate constant directly by carrying out experiments at very high rotating speeds. The value

$$k_I = 52 \times 10^{-4} \, \text{s}^{-1} \text{ at } 25 \,°\text{C}$$

so obtained compares very well with that noted earlier (which was extracted from the overall constant at a lower speed).

The chief conclusion from this study is that except at very high rates of agitation, the reaction tends to be greatly influenced by mass transfer to the magnesium surface. Another important conclusion is that the selectivity of the reaction is also greatly influenced by mass transfer limitations. In the absence of mass transfer effects the selectivity is higher, thus making it important to eliminate these effects.

Laboratory Reactors

A small scale batch reactor, usually a 100–200 ml stirred flask, is all that is often required. Although the reactor would be thermostatted, isothermal conditions are difficult to maintain bewcause of the lower surface-to-volume ratio than, for example, for an ampule (see Figure 4.15). Hence it may often become necessary to operate at low conversions for reactions involving a reasonable heat effect.

A Rigorous Strategy for Modeling Solid-Liquid Reactions

The two examples given above clearly suggest the need for considering the solid-phase reaction in solid-liquid reacting systems and for modeling it along the lines of gas-solid reactions (which have been extensively modeled). However, a complete modeling strategy must account for reaction in both the liquid and solid phases for systems where the solid is not insoluble. Thus both of the methods outlined in the two previous sections should be considered and the total reaction obtained as the sum of the two. Such a strategy, used by Hagenson and Doraiswamy (1998) for an ultrasound-enhanced reaction (Chapter 22), is shown in Figure 15.10. As it turned out, in the reaction studied by them, the liquid phase reaction was negligible.

Use of Solvents

It is sometimes convenient to first dissolve a solid reactant in a solvent and then react it with the liquid phase reactant. Several methods have been proposed for a quick screening of solvents and the more important of these have been summarized by Frank et al. (1999). These include: a chart (Robbins, 1980; Robbins and Cusack, 1997), the UNIFAC approach (Gmehling et al., 1978, 1992; Hansen et al., 1991), Hansen's solubility parameters (Hansen, 1969, 1999; Barton, 1993), and the extended Hansen method (Ashton et al., 1983, 1991). The original articles may be referred for details of their use. Robbin's method is particularly useful as it is more general, being applicable to liquid-liquid systems as well. It is based on an evaluation of hydrogen-bonding and electron donor-acceptor interactions for 900 binary systems. It includes 12 general classes of functional groups, divided into three main categories: hydrogen bond donors, hydrogen

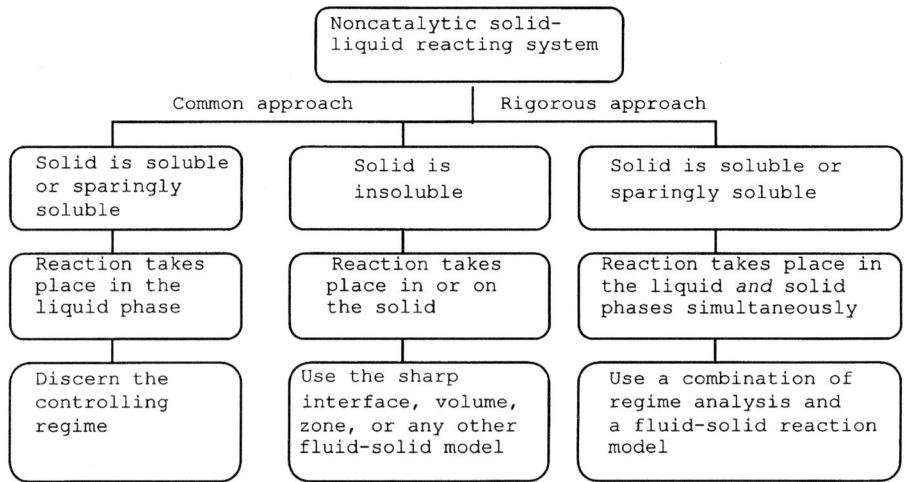

Figure 15.10 Flowchart of modeling approaches when using noncatalytic solid-liquid reaction systems (Hagenson and Doraiswamy, 1998).

Table 15.2 Examples of solid-liquid reactions

Reaction	Reference
Solid slightly soluble	
1. Reaction of cyanuric chloride with amines to give reactive dyes	Venkatraman (1972)
2. Sulfonation and nitration of naphthalene at low temperatures	Groggins (1958)
3. Reaction between terephthalic acid and ethylene oxide in a solvent in the presence of dissolved catalysts	Hydrocarbon Proc. Petrol. Refiner (1971); Bhatia et al. (1976)
4. Manufacture of diisocyanate by reaction between phosgene and hydrochlorides of amines	
5. Alkaline hydrolysis of solid esters such as di-β-chloroethyl oxalate and nitrobenzoic acid esters	Sharma and Sharma (1969, 1970a,b)
Solid insoluble	
6. Generation of acetylene by reaction of CaC_2 with water	
7. Manufacture of organometallic compounds (Grignard reagent)	Horak et al. (1975)
8. Reaction between dialkyl maleate and sodium bisulfite to give the Na salt of sulfosuccinic ester	Garrett (1972)
9. Reaction between RCl and sodium cyanate to give isocyanates	Twitchett (1974)
10. Sulfonation and chloromethylation of styrene–DVB copolymer for the manufacture of ion-exchange resins	Pepper et al. (1953)
11. Iron acid reduction of nitro compounds	Groggins (1958)

bond acceptors, and nonhydrogen bonding groups. based on whether the interactions are repulsive (position deviation from ideality), attractive (negative deviation from ideality), or zero (ideal), a quick approximate selection can be made.

Examples of Solid-Liquid Reactions

Several studies have been reported on solid-liquid reactions, some of which are listed in Table 15.2. However, there are only a few that have attempted to model these reactions along the lines just outlined. Where the authors have indicated whether the solid is slightly soluble or insoluble, this fact is mentioned in the table; but in many cases such information is also not available. It is a reasonable conclusion from these observations that solid-liquid reactions, particularly those involving reaction in the solid phase, need to be studied more rigorously.

Chapter 16

Gas-Liquid and Liquid-Liquid Reactor Design

The design equation for any reactor expresses its output as a function of the input. For a gas-liquid reactor, the output may be expressed as

[Output] = f[input]

$$= f \begin{bmatrix} \text{(flow rates of reactants), (regime of} \\ \text{operation and the corresponding rate} \\ \text{equation), (mixing characteristics of} \\ \text{the gas and liquid phases as determined} \\ \text{by the type of contactor)} \end{bmatrix} \quad (16.1)$$

In Chapter 14, we formulated a number of regimes with corresponding conditions and governing rate equations. In the present chapter, we recast the rate equations in a general form that indicates the relative roles of reaction, liquid film diffusion, and gas film diffusion. Then we briefly discuss the design principles of the more common classes of fluid-fluid reactors. Detailed treatments of design may be found in the books of Astarita (1967), Danckwerts (1970), Shah (1979), Levenspiel (1972, 1993), Doraiswamy and Sharma (1984b), Bisio et al. (1985), and Kastánek et al. (1993).

A GENERALIZED FORM OF EQUATION FOR ALL REGIMES

The most convenient way to define a rate process is by the ratio

$$\text{Rate} = \frac{\text{driving force}}{\text{resistance}} = \left[\frac{\text{driving force}}{1/\text{rate coefficient}} \right] \quad (16.2)$$

The rate coefficient can be for reaction (k), for any of the mass transfer steps (k_G or k_L), or for combinations thereof. Therefore, Equation 16.2 can be written as

$$\text{Rate} = \frac{\text{driving force}}{\text{resistance}} = \frac{\text{driving force}}{\sum \frac{1}{k_i}} \tag{16.3}$$

where k_i represents any rate coefficient. Thus overall rate equations for each of the regimes considered in the previous chapter can be recast in this form. It is often convenient to express gas-phase concentrations in terms of partial pressures. Additionally, if the gas phase is not a pure gas but a mixture (i.e., A is mixed with an inert), then gas-film resistance should also be included.

Regime 1: Very Slow Reaction

Because the reaction is kinetically controlled throughout in this regime, Equation 16.3 becomes

$$-r_A = b \frac{p_A [B]_b}{\frac{H'_A}{k_2}}, \tag{16.4}$$

where the concentrations are expressed in terms of partial pressures, k_2 is the second-order reaction rate constant with units of m^3/mol s, and H'_A is the reciprocal of Henry's law constant H_A (as defined in this work).

Regime 2 and Regime between 1 and 2: Diffusion in the Film with and without Reaction in the Bulk

The controlling resistance in regime 2 is that due to the liquid film. In addition, as already stated, there can also be a gas-film resistance, resulting in the equation

$$-r_A = \frac{p_A}{\frac{1}{k_{GP}} + \frac{H'_A}{k_L}} = \frac{p_A}{\frac{1}{k'_{GP} a_L} + \frac{H'_A}{k'_L a_L}} \tag{16.5}$$

If reaction in the bulk is also a contributing resistance, we add this resistance from Equation 16.4 to this expression, with the result

$$-r_A = \frac{p_A}{\frac{1}{k'_{GP} a_L} + \frac{H'_A}{k'_L a_L} + \frac{H'_A}{b k_2 [B]_b}} \tag{16.6}$$

This corresponds to the regime between 1 and 2 described by Equation 14.9 with an added resistance for the gas film.

Regime 3: Fast Reaction

Referring to Equation 14.18, the reaction in the film (which occurs simultaneously with diffusion in this regime) can be written in terms of pure diffusion in the liquid film multiplied by the enhancement factor:

$$-r_A = (k'_L a_L [A]^*)\eta = \left[k'_L a_L \left(\frac{p_A}{H'_A} \right) \right] \eta \tag{16.7}$$

When this rate is added to the rate of gas-film diffusion,

$$-r_A = \frac{p_A}{\dfrac{1}{k'_{GP} a_L} + \dfrac{H'_A}{k'_L a_L \eta}} \tag{16.8}$$

Regime between 2 and 3

In this regime, reaction occurs in both the film and the bulk. A simple way of accounting for the reaction in the bulk would be to add the resistance corresponding to this additional reaction to Equation 16.8, giving

$$-r_A = \frac{p_A}{\dfrac{1}{k'_{GP} a_L} + \dfrac{H'_A}{k'_L a_L \eta} + \dfrac{H'_A}{k_2 b [B]_b}} \tag{16.9}$$

A more rigorous approach that considers regimes 2 and 3 simultaneously (Table 14.2) leads to the following form:

$$-r_A = \frac{p_A}{\dfrac{1}{k'_{GP} a_L} + \dfrac{H'_A}{a_L \sqrt{D_A k_1 + k'^2_L}}} \tag{16.10}$$

where k_1 is the pseudo-first-order rate constant (1/s).

Regime between 1, 2, and 3

Simultaneous consideration of all three regimes, 1, 2, and 3, gives

$$-r_A = \frac{p_A}{\dfrac{1}{k'_{GP} a_L} + \dfrac{H'_A}{a_L \sqrt{D_1 k_1 + k'^2_L}} + \dfrac{H'_A}{b k_2 [B]_b}} \tag{16.11}$$

Regime 4: Instantaneous Reaction

In Chapter 14 we saw that the enhancement factor for this regime is an asymptotic value because it corresponds to the extreme case of enhancement due to an instantaneous reaction. The enhancement for no reaction is obviously one and that corresponding to any other regime will lie between these two asymptotes.

From Equations 14.18 and 14.24, the reaction rate within the liquid film may be written as

$$-r_A = \frac{[A]^*}{\dfrac{1}{k'_L a_L \eta_a}} \tag{16.12}$$

where

$$\eta_a = 1 + \frac{D_B[B]_b}{D_A \nu_B [A]^*} = 1 + \frac{D_B[B]_b H'_A}{D_A \nu_B p_A} \tag{16.13}$$

Adding the gas-film resistance to the rate equation gives

$$-r_A = \frac{p_A}{\dfrac{1}{k'_{GP} a_L} + \dfrac{H'_A}{k'_L a_L \eta_a}} \tag{16.14}$$

Note that because the reaction occurs instantaneously and completely in the film, there is none in the bulk. Hence Equation 16.14 does not contain a term for reaction in the bulk.

A special case of instantaneous reaction is when it occurs entirely at the gas-liquid interface. Clearly, when reaction occurs so fast as to prevent any penetration of A into the film, the controlling step is the rate at which gas A is supplied to the interface. In other words, the reaction is gas-film controlled, and the rate is given by

$$-r_A = k'_{GP} a_L p_A = \frac{p_A}{\dfrac{1}{k'_{GP} a_L}} \tag{16.15}$$

CLASSIFICATION OF GAS-LIQUID CONTACTORS

There are two broad methods of classifying gas-liquid contactors: (1) based on the manner of achieving contact between gas and liquid; and (2) based on the manner of delivering power for dispersion of one phase into the other.

Classification 1 (Based on the Manner of Phase Contact)

Using classification 1, most industrial contactors may be classified in three broad categories with a number of variations in each (Van Krevelen, 1950):

A. Contactors in which the liquid flows as a thin film
   a. packed columns
   b. thin film reactors
   c. disk reactors
B. Contactors with dispersion of gas into liquids
   a. plate columns (including controlled-cycle contactors)
   b. mechanically agitated contactors
   c. bubble columns
   d. packed bubble columns

e. sectionalized bubble columns
   f. two-phase horizontal cocurrent contactors
   g. coiled reactors
   h. vortex reactors
 C. Contactors where liquid is dispersed in the gas phase
   a. spray columns
   b. venturi scrubbers

The title of each category clearly describes the nature of the contact, and the subtitles indicate the manner of achieving such contact.

Classification 2 (Based on the Manner of Energy Delivery)

The rate and manner of delivering energy influence such important fluid-fluid parameters as the dispersed phase holdup, the interfacial area, the mass transfer coefficient (Calderbank, 1958; Nagel et al., 1972, 1973; Kastánek, 1976; Zahradnik et al., 1982; Oldshue, 1983). Thus an energy-based classification would be both useful and appropriate. In this classification, the contactors are essentially divided into three broad groups:

1. Contactors in which energy is supplied through the gas phase
2. Contactors in which energy is supplied through the liquid phase
3. Contactors in which energy is delivered through mechanically agitated parts

This classification is not always unambiguous. For example, in reactors belonging to groups 2 and 3, some energy is also delivered by the gas phase.

The principal features of the major contactors, irrespective of their classification, have been described at considerable length by many authors, for example Laddha and Degaleesan (1976), Schügerl (1977, 1980, 1983), Shah et al. (1982), Doraiswamy and Sharma (1984), Shah and Deckwer (1985), and Kastánek et al. (1993). A few common designs are sketched in Figure 16.1.

Mass Transfer Coefficients and Interfacial Areas of Some Common Contactors

Values of the mass transfer coefficients and interfacial areas for the more common contactors (packed columns, plate columns, bubble columns, mechanically agitated contactors, and static mixers) are usually known, or can be estimated from correlations published in the literature, or are supplied by the manufacturers. Typical values are given in Table 16.1.

Role of Backmixing in Different Contactors

A particularly important consideration in designing contactors for gas-liquid reactions is the validity of the ideal flow assumption (plug or mixed flow) normally used in the design. Countless studies have been reported on the role

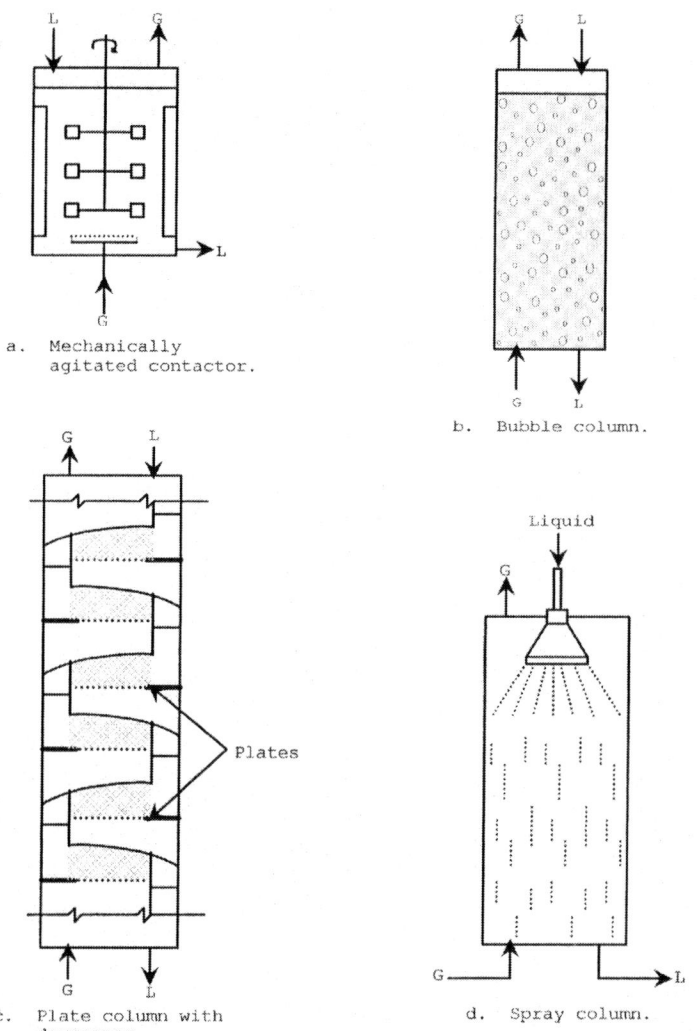

Figure 16.1 Sketches of some common gas-liquid reactor designs.

of nonideality (i.e., backmixing) in gas-liquid contactors, and based on them, some qualitative guidelines can be formulated. Such an attempt has been made in several reviews and books, for example, Doraiswamy and Sharma (1984), Shah and Deckwer (1985), and more exhaustively by Kastánek et al. (1993).

REACTOR DESIGN FOR GAS-LIQUID REACTIONS

The Overall Strategy

Knowledge of the regime of operation, the location of the reaction region within the gas-liquid system, is important in selecting the type of reactor to be used.

Table 16.1 Flow and mass transfer features of the more common gas-liquid contactors (adapted from Doraiswamy and Sharma, 1984)

Type of contactor	Typical range of superficial gas velocities (cm/s)	Mixing characteristics		k'_L (cm/s)	a_L (cm²/cm³)	Liquid holdup
		Gas phase	Liquid phase			
1. Packed column	10–100	PF	PF	0.3–2.0	0.2–3.5	0.05–0.1
2. Bubble column	1–30	PF	MF	1–4	0.25–10	0.6–0.8
3. Packed bubble column	1–20	PF	PMF[a]	1–4	1–3	0.5–0.7
4. Plate column (without downcomers)	50–300	PF	MF[b]	1–4	1–2	0.5–0.7
5. Mechanically agitated contactor	0.1–2.0	MF (or PMF)	MF	1–5	2–10	0.5–0.8
6. Spray column	5–300	PMF[a]	PF	0.5–1.5	0.2–1.5	0.05

[a] PMF = partial mixed flow.
[b] Good mixing on plates, no mixing between plates.

Depending on whether the reaction occurs in the bulk or in the film, the "effective amount of liquid" or liquid holdup will be different and forms an important parameter in reactor choice.

For purposes of design, the reaction regime can be determined from laboratory experiments carried out within the framework of the analytical methods described in Chapters 14 and 15. Table 14.2 summarizes the criteria and rate equations for a simple irreversible reaction in different regimes, and Tables 14.4–14.6 summarize similar information for more complicated systems. The basic objectives of design are then to minimize the reactor volume and the energy required for a given production rate. The overall strategy can be stated as follows:

1. Make a preliminary list of reactors and flow arrangements (plug or mixed for the two phases in countercurrent or cocurrent flow) that deserve to be considered
2. Calculate the reactor volume for each of the preselected reactors
3. Using the parameter values available from calculations in step 2, make a more rational choice of a reactor based on volume minimization
4. Assess each reactor for its energy requirement and select that with the minimum requirement. It is important to note that the minimum energy criterion serves more as a basis for reactor selection to achieve a given reaction rate than as a design procedure.

Calculation of Reactor Volume

For the calculations in step 2, it is convenient to assume that each phase is either in plug or mixed flow. One can readily visualize several combinations of these flows (Table 16.2 and Figure 16.2). Reactor types that closely approximate these combinations are also included in the table.

The first step in design (for volume calculation) is a mass balance (differential for plug flow and overall for mixed flow) written in terms of the transfer of a

Table 16.2 Reactors with ideal combinations of plug and mixed flows

Type of flow			
Liquid	Gas	Type of operation	Type(s) of reactor[a]
PF	PF	Steady-state	Bubble column ($H_r/d_r > 10$), packed column
MF	MF	Steady-state	Mechanically agitated contactor, packed column ($H_r/d_r < 3$), spray column
MF	PF	Steady-state	Plate column
PF	MF	Steady-state	Spray column
Uniform concn	MF	Unsteady-state	Batch reactor

[a] H_r = reactor height; d_r = reactor diameter.

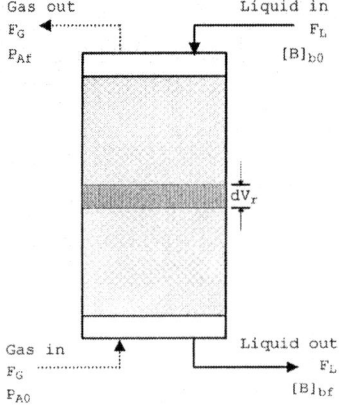
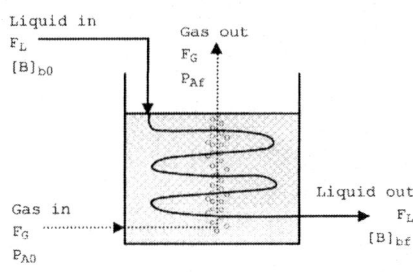

(a) Plug flow G - plug flow L

(b) Plug flow G - mixed flow L

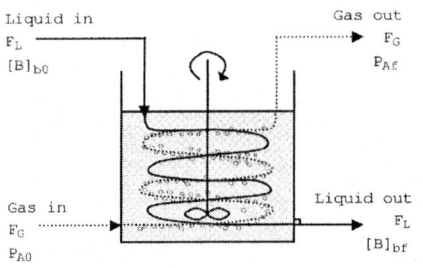
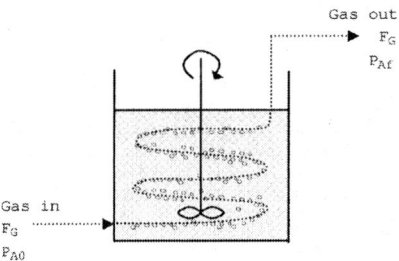

(c) Mixed flow G - mixed flow L

(d) Mixed flow G - batch L

Figure 16.2 Four ideal flow reactor configurations.

selected component (say, A) from one phase to the other, followed by reaction. Thus for the reaction

$$A(g) + \nu_B B(l) \rightarrow \text{Products} \quad [16.1]$$

$$\begin{bmatrix} \text{rate of loss} \\ \text{of } A \text{ by the} \\ \text{gas} \end{bmatrix} = \begin{bmatrix} \text{rate of loss} \\ \text{of } B \text{ by the} \\ \text{liquid} \end{bmatrix} = \begin{bmatrix} \text{rate of reaction} \\ \text{of } A \text{ in the} \\ \text{liquid} \end{bmatrix} \quad (16.16)$$

The equation can be recast to produce the now-familiar $1/-r_A$ versus X_A plots. Then the volume can be determined graphically or numerically. The limitation of Equation 16.16 should be noted, however. It assumes that all of A from phase 1 that enters phase 2 reacts in the latter phase without any accumulation. Thus it cannot account for the regime between 1 and 2 or the regime overlapping 1, 2, and 3 where A exists in a finite concentration in phase 2 or is carried away by the liquid.

Now we consider reaction 16.1 again and formulate design equations for all five cases presented in Table 16.2 (see Levenspiel, 1993, 1999, for a more detailed discussion).

Case 1: plug gas, plug liquid, countercurrent steady state

Because both gas and liquid are in plug flow, we write the following differential balance (Figure 16.2a):

$$F_{GU} dy'_A = -\frac{F_{LU} dz'_B}{\nu_B} = (-r_A) dV_r$$

$$= (-r'_A a_L) dV_r \tag{16.17}$$

where y'_A and z'_A represent the mole ratios of A to the inert in the gas and liquid phases, respectively:

$$y'_A = \frac{p_A}{p_U}, \quad z'_A = \frac{[A]}{[U]_G}, \quad z'_B = \frac{[B]}{[U]_L} \tag{16.18}$$

and F_{GU} and F_{LU} are the flow rates (mol/s) of inerts in the gas (U_G) and liquid (U_L) phases, respectively. Integration of the equation leads to

$$V_r = F_{GU} \int_{y'_{A1}}^{y'_{A2}} \frac{dy'_A}{(-r'_A) a_L} = \frac{F_{LU}}{\nu_B} \int_{z'_{B2}}^{z'_{B1}} \frac{dz'_B}{(-r'_B) a_L} \tag{16.19}$$

with

$$F_{GU}(y'_{A2} - y'_{A1}) = \frac{F_{LU}}{\nu_B} (z'_{B1} - z'_{B2}) \tag{16.20}$$

Now it only remains to introduce the appropriate equation for the rate in Equation 16.19 and solve it for the volume.

Case 2: same as case 1 but with cocurrent flow

The mass balance equations are the same as those for case 1, but the signs of the flow terms F_{GU} and F_{LU} will change. If both streams flow down the column F_{GU} becomes $-F_{GU}$, and when both streams flow up the column F_{LU} becomes $-F_{LU}$.

Case 3: plug gas, mixed liquid, steady state

In this case, sketched in Figure 16.2b, a differential balance must be written for the gas phase because it is in plug flow, but an overall balance for the liquid phase because it is in mixed flow. Thus for the gas phase, by equating the A lost by gas to the amount disappearing by reaction,

$$F_{GU} dy'_A = (-r'_A)_{\text{liq exit}} a_L dV_r \tag{16.21}$$

Then for the liquid phase, by equating the A lost by gas to the B lost by liquid,

$$F_{GU}(y'_{A0} - y'_{Af}) = \frac{F_{LU}}{\nu_B} (z'_{B0} - z'_{Bf}) \tag{16.22}$$

Integrating Equation 16.21 for the gas phase and combining with Equation 16.22 leads to

$$V_r = F_{GU} \int_{y'_{Af}}^{y'_{A0}} \frac{dy'_A}{(-r')_{\text{liq exit}} a_L} \qquad (16.23)$$

Because the liquid phase is mixed, Equation 16.23 can be used directly to calculate V_r provided that the exit conditions are known. Alternatively, if V_r is known, y'_{Af}, and therefore z'_{Bf} (from Equation 22) can be found by repeating the calculation for several assumed values of y'_{Af} or z'_{Bf} (and therefore of the final liquid phase concentration $[B]_{bf}$) till the given value of V_r is matched.

Case 4: mixed gas, mixed liquid, steady state

This case, sketched in Figure 16.2c, corresponds to mixed conditions in the entire reactor. Thus the following mass balance can be written for the vessel as a whole:

$$F_{GU}(y'_{A0} - y'_{Af}) = \frac{F_{LU}}{\nu_B}(z'_{B0} - z'_{Bf}) = (-r'_A)_{\text{at liq gas exit}} a_L V_r \qquad (16.24)$$

The solution of this equation for V_r is straightforward.

Case 5: mixed gas, batch liquid, unsteady state

In this semibatch operation a batch of liquid is taken into a reactor and a stream of gas passed through it (Figure 16.2d). The problem here is usually one of calculating the time needed for a given conversion in a reactor of known volume.

The following material balance can be written:

$$F_{GU}(y'_{A0} - y'_{Af}) = -\frac{V_L}{\nu_B}\frac{d[B]_b}{dt} = (-r'_A)a_L V_r \qquad (16.25)$$

An iterative procedure must be used for calculating the time needed for a given conversion. The following steps are involved:

1. assume a value of $[B]_b$,
2. guess a value of y'_{Af} (i.e., p_{Af}) for this $[B]_b$,
3. calculate the rate from the appropriate rate equation for the reaction.
4. compare the first and third terms of Equation 16.25, and iterate till they match,
5. repeat these above steps for another value of $[B]_b$, and prepare a plot of reciprocal rate versus $[B]_b$ and
6. integrate to get the time:

$$t = \frac{V_L}{\nu_B} \int_{[B]_{bf}}^{[B]_{b0}} \frac{d[B]_b}{(-r'_A)a_L} \qquad (16.26)$$

Comments

The following observations should serve as useful guidelines in assessing the applicability of the design equations.

1. In general, mixed flow (MF) of any phase is not desirable. However, this should be of no consequence when the reaction is controlled by diffusion or diffusion accompanied by reaction in the film and is completed within it, that is, in regimes 2 and 3.
2. In writing Equation 16.17, plug flow (PF) was assumed for both the gas and liquid phases. Thus this equation should be valid for packed columns.
3. Mechanically agitated contactors are characterized by complete backmixing of both liquid and gas phases; thus Equation 16.24 should be applicable to these contactors. Deviation from MF is possible at low stirrer speeds. If the reaction is in regime 2, accurate values of k_L are needed. Because predictive correlations for this are unreliable, the values should be obtained by conducting experiments in a geometrically similar laboratory reactor 20–30 cm diameter. To obtain such similarity, it is recommended that the ratio of tip speed to tank diameter should be kept constant (Juvekar, 1976).
4. In the widely used bubble column, PF for the gas phase and MF for the liquid phase are passable assumptions. Thus Equations 16.21–16.23 should normally be valid for these reactors. This should, however, be reviewed in the light of more detailed information on backmixing.
5. Design equations can be developed for partial mixing in either or both phases. The first step is to experimentally determine the residence time distribution for each phase. But the procedures tend to be quite involved, and are outside the scope of this treatment (see Kastánek et al., 1993, for a comprehensive exposition).

EXAMPLE 16.1

Design of a semibatch reactor for the oxidation of *n*-butyraldehyde

It is proposed to oxidize a batch of 1000 liters of *n*-butyraldehyde dissolved in *n*-butyric acid in the presence of 0.1% manganese acetate as a catalyst at essentially atmospheric pressure in a mechanically agitated contactor by passing air continuously through it. Given the following data, calculate the time required to obtain a conversion corresponding to a drop in the aldehyde concentration $[B]_i$ from 6.0 to 0.85×10^{-3} mol/cm^3:

REACTION

$$O_2 + 2CH_3CH_2CH_2CHO \rightarrow 2CH_3CH_2CH_2COOH$$
(A)   (B)   →   (R)   [E16.1.1]

m (order in O_2) = 1, n (order in aldehyde) = 2, $\nu_B = 2$

Temperature = 20 °C
Initial concentration of *n*-butyraldehyde,

$$[B]_{bi} = 6 \times 10^{-3} \, \text{mol/cm}^3$$

Final concentration of n-butyraldehyde,

$[B]_{bf} = 0.85 \times 10^{-3}$ mol/cm^3

Diameter of contactor = 100 cm
Diameter of impeller = 35 cm
Rate of inert (nitrogen) feed $F_N [= F_{GU}$ of the previous section] 0.13 mol/s
Gas-liquid interfacial area $a_L = 12.0$ cm^2/cm^3 of liquid
True liquid side mass transfer coefficient $k'_L = 2 \times 10^{-2}$ cm/s

The following average properties may be used at 20 °C: Henry's law constant of oxygen in the liquid,

$H_A = 1.043 \times 10^{-5}$ mol/cm^3 atm

Diffusivity of oxygen in the liquid, $D_A = 1.8 \times 10^{-5}$ cm^2/s
Third-order constant for the reaction

$k_3 = 4 \times 10^6$ (cm^3/mol)^2/s (Ladhabhoy and Sharma, 1969).

SOLUTION

It is assumed that the gas phase is completely backmixed and losses of n-butyraldehyde due to vaporization are negligible. (Although there will be a significant change in the viscosity of the liquid when it is oxidized, for the sake of illustration, average property values are assumed.)

The regime of the reaction is not known. Hence we shall assume a relatively common regime, 2–3, and verify its applicability as part of the calculations. The rate equation for this regime (Table 14.2) is given by

$$-r'_A a_L = a_L [A]^* \sqrt{D_A k_3 [B]_b^2 + k'^2_L}, \quad \text{(with } k = k_3 [B]_b^2\text{)} \tag{E16.1.1}$$

The conditions to be satisfied are

$$\frac{\sqrt{D_A k_3 [B]_b^2}}{k'_L} \cong 1 \tag{E16.1.2}$$

and

$$\frac{\sqrt{D_A k_3 [B]_b^2}}{k'_L} \ll \frac{[B]_b}{\nu_B [A]^*} \sqrt{\frac{D_B}{D_A}} \tag{E16.1.3}$$

Because the gas is completely mixed, the solubility of oxygen in the liquid should be based on the outlet partial pressure of oxygen in air,

$$[A]^* = H_A P \left(\frac{y'_{Af}}{1 + y'_{Af}} \right) \tag{E16.1.4}$$

where y'_{Af} is the molar ratio of oxygen to nitrogen in air. If $y'_{Af} \ll 1$, then Equation (E16.1.4) can be approximated as

$$[A]^* = H_A P y'_{Af} \tag{E16.1.5}$$

A material balance on oxygen for this semibatch reactor gives (see Equation 16.25).

$$F_N(y'_{A0} - y'_{Af}) = -\frac{V_r}{\nu_B}\frac{d[B]_b}{dt} = (-r'_A)a_L V_r \quad (E16.1.6)$$

From Equations E16.1.1, E16.1.5, and E16.1.6,

$$F_N(y'_{A0} - y'_{Af}) = H_A P y'_{Af} V_r a_L \sqrt{D_A k_3 [B]_b^2 + k'^2_L} \quad (E16.1.7)$$

The outlet mole ratio y'_{Af} can be obtained from the inlet ratio y'_{A0} by recasting Equation (E16.1.7) as

$$y'_{Af} = \frac{y'_{A0}}{1 + \alpha'\sqrt{[B]_b^2 + \beta'^2}} \quad (E16.1.8)$$

where

$$\alpha' = \frac{a_L H_A P V_r \sqrt{D_A k_3}}{F_N}, \quad \beta' = \frac{k'_L}{\sqrt{D_A k_3}} \quad (E16.1.9)$$

Then, by substituting Equation E16.1.8 for y'_{Af} in Equation E16.1.6, we get

$$-\frac{\gamma'}{y'_{A0}}\frac{d[B]_b}{dt} = \frac{\alpha'\sqrt{[B]_b^2 + \beta'^2}}{1 + \alpha'\sqrt{[B]_b^2 + \beta'^2}} \quad (E16.1.10)$$

where

$$\gamma' = \frac{V_r}{\nu_B F_N} \quad (E16.1.11)$$

Integration of Equation E16.1.10 gives the semibatch time as

$$t_{SB} = \frac{\gamma'}{y'_{A0}\alpha'}\left[\ln\left(\frac{[B]_{bi} + \sqrt{[B]_{bi}^2 + \beta'^2}}{[B]_{be} + \sqrt{[B]_{be}^2 + \beta'^2}}\right) + \alpha'([B]_{bi} - [B]_{be})\right] \quad (E16.1.12)$$

Now we check to see if conditions E16.1.2 and E16.1.3 for the validity of the assumed rate equation are satisfied at both the beginning and end of the operation.

AT THE BEGINNING

$$\frac{\sqrt{D_A k_3 [B]_{bi}^2}}{k'_L} = 2.55 \cong 1$$

To apply the second condition E16.1.3, we combine it with E16.1.5 to give

$$\frac{\sqrt{D_A k_3 [B]_{bi}^2}}{k'_L} \ll \frac{[B]_{bi}}{\nu_B H_A P(y'_{Af})_i}\sqrt{\frac{D_B}{D_A}} \quad (E16.1.13)$$

Because $(y'_{Af})_i$, the outlet mole ratio at initial condition, is unknown, we calculate it from Equation (E16.1.8). The parameter values to be used are $\alpha' = 8.169 \times 10^3$,

$\beta' = 2.35 \times 10^{-3}$, $y'_{A0} = 0.266$, and $[B]_{bi} = 6 \times 10^{-3}\,\text{mol/cm}^3$. Thus we obtain $(y'_{Af})_i = 0.00496$, giving

$$\frac{[B]_{bi}}{\nu_B H_A P (y'_{Af})_i} = 5.8 \times 10^4$$

Hence both the conditions are satisfied.

AT THE END

By repeating these calculations for the exit mole ratio at the final condition [i.e., for $y'_{Af} = (y'_{Af})_e$], it can be shown that the two conditions continue to be valid.

Therefore, we may conclude that the rate equation, with the corresponding regime 2–3, is valid for the entire duration of the run. Thus Equation E16.1.12 derived from this rate equation can be used for calculating t_{SB}.

Substituting the values of the relevant parameters in Equation E16.1.12, $F_N = 0.13\,\text{mol}$ N_2/s, $V_r = 10^6\,\text{cm}^3$, $\alpha' = 8.169 \times 10^3$, $\beta' = 2.35 \times 10^{-3}$, $\gamma' = 3.846 \times 10^6$, $y'_{A0} = 0.266$, we get

semibatch time $t_{SB} = 21.32\,\text{h}$

REACTOR CHOICE FOR GAS-LIQUID REACTIONS

The Criteria

As must be evident from a previous section on classification, gas-liquid reactions can be carried out in a large number of reactor types. This is also true of other multiphase reactions in which a liquid phase is involved. For other reactions such as gas-solid, catalytic or noncatalytic, the choice of reactor is confined to a lesser number of variations. Therefore, although reactor choice is an important consideration for all reactions, particularly heterogeneous reactions, it is more so for gas-liquid, liquid-liquid, and slurry systems, all of which are widely used in industrial organic synthesis. We discuss below the cost minimization criteria for a rational choice of reactors for gas-liquid reactions.

Minimization of reactor volume is clearly one such criterion. In addition to directly reducing the reactor cost, this also lowers the energy cost. The latter derives from the reduction in energy for agitating or pumping the fluids. A second criterion (actually a set of two criteria) can also be formulated based exclusively on minimization of the energy supply. Here, two situations can be visualized: minimization of the overall energy input to the reactor to initiate and sustain interfacial contact and minimization of energy delivered to the system for maximizing mass transfer rates. These criteria are summarized in Table 16.3.

Volume Minimization Criterion

Any criterion for volume minimization must be based on the extent to which reaction occurs in the film. The criterion for a reaction to occur entirely in the film is given by the inequality

$$\sqrt{M} \gg 1 \qquad (16.27)$$

Table 16.3 Classification of criteria for cost minimization

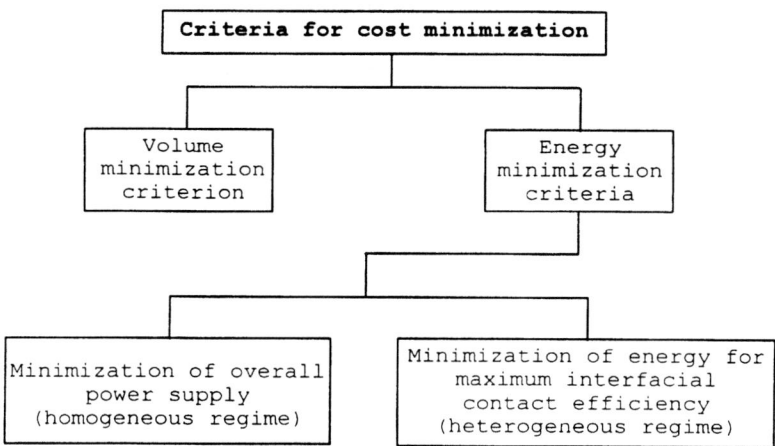

(which is the same as inequality 14.11), and for a reaction to occur entirely in the bulk, by the reverse of this criterion.

A more elaborate criterion (Kastánek et al., 1993) can also be used based on a parameter N defined as

$$N = [(M/q) + M - (M/q)(a + \eta)]^{1/2} \qquad (16.28)$$

where

$$a = \frac{[A]_0}{[A]^*}, \quad q = \frac{[B]_b D_B}{\nu_B [A]^* D_A} \text{ (for the film model)}$$

$$\text{or } \frac{[B]_b}{\nu_B [A]^*} \left(\frac{D_B}{D_A}\right)^{1/2} \text{ (for the penetration model)} \qquad (16.29)$$

The magnitude of N provides a valuable guideline for reactor selection. Table 16.4 presents a broad indication of the types of contactors recommended for

Table 16.4 Guidelines for reactor selection (adapted from Kastánek et al., 1993)

Criterion based on value of N	Regime	Type of reactor
$N \ll 0.03$	1, 2, 1–2	Bubble column
$0.03 \ll N \ll 1.3$	3, 2–3, 1–2–3	Packed column
		Plate column
		Mechanically agitated contactor
$N \gg 1.3$	3, 4	Plate column with shallow liquid layers on plates
		Spray column
		Wetted wall column

values of N ranging from $N \ll 0.03$ for reaction in the bulk to $N \gg 1.3$ for reaction in the film. These correspond to $\sqrt{M} \ll 0.03$ and $\sqrt{M} > 3$.

Limitations of volume minimization

It is useful to note that certain types of reactors are automatically eliminated for certain situations, such as the packed-bed reactor for dust-laden gas. Further, reactor selection is usually straightforward for extreme situations such as reaction entirely in the bulk or in the film. In both cases, the quantity of bulk liquid is determined by the rate of production, but the extent to which the bulk phase is utilized for the reaction is minimal when the reaction is confined to the film and maximal when it occurs fully in the bulk.

On the other hand, when the reaction occurs in both bulk and film, the reactor should combine good interfacial area generation with efficient bulk liquid utilization. A measure of the *efficiency of bulk liquid utilization* is the parameter γ_{bu} given by (Kramers and Westerterp, 1963)

$$\gamma_{bu} = \frac{1}{\sqrt{M}\psi} \left[\frac{\sqrt{M}(\psi - 1) + \tanh\sqrt{M}}{\sqrt{M}(\psi - 1)\tanh\sqrt{M} + 1} \right] \quad (16.30)$$

where

$$\psi = \frac{V_c/V}{a_L \delta}$$

The dimensionless group ψ is a direct measure of the ratio of the continuous phase fraction V_c/V to the film volume fraction $a\,\delta$. In other words, it represents the relative roles of reactor volume and interfacial area in the reaction. Equation 16.30 is graphically displayed in Figure 16.3 as plots of γ_{bu} versus ψ for different values of $\sqrt{M}$.

The unshaded regions in the figure represent clear cases of reaction in the bulk (regime 1) and the film (regimes 3 and 4). For large values of $\sqrt{M}$ corresponding to regime 3 or 4, reactors that can generate large interfacial areas are desirable. Further, liquid-phase holdup is negligibly small in such cases. On the other hand, for regime 1, reaction in the bulk, a large holdup is a primary requirement with no emphasis on area generation.

Reactor selection tends to be complicated in the range $0.03 < \sqrt{M} < 0.1$ where large values of both interfacial area and liquid holdup are required. This is also true of regime 2, despite its very low value of $\sqrt{M}$, because pure mass transfer through the film is the controlling resistance, so that interfacial area generation becomes an important consideration. In all such cases, the minimum volume criterion for reactor selection becomes inadequate, and the choice has to be based only on the minimum energy criterion to be discussed in the next section.

Steps in volume minimization

Procedures for calculating the reactor volume for various combinations of plug and mixed flow of the liquid and gas phases were described in an earlier section.

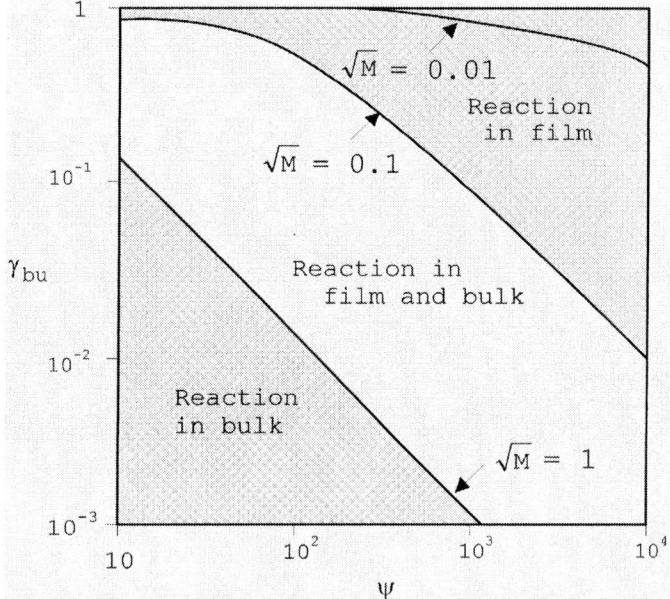

Figure 16.3 Efficiency of bulk-phase utilization γ_{bu} as a function of the composite parameter $\psi = f_c/a_L\delta$ and Hatta number $\sqrt{M}$ (adapted from Kramers and Westerterp, 1963).

However, they constitute only one step in the overall strategy for the minimization of reactor volume for a given rate of production. The full strategy involving a number of steps is outlined in Table 16.5. Note in the table that volume minimization does not always lead to a unique optimal design. A second set of criteria based on energy minimization may frequently be necessary.

Energy Minimization Criteria

The energy minimization criteria depend on the regime of operation. We broadly classify these as *homogeneous* and *heterogeneous* regimes 2(a) and 2(b).

2(a). minimization of overall power supply to the contactor
2(b). maximization of interfacial contact efficiency

Criterion 2(a): homogeneous regime (regime 1)

Regime 1 pertains to reactions that are so slow that they can be regarded as homogeneous reactions. For such reactions, the only energy consideration is the minimum energy required to establish contact between phases for reaction to occur, irrespective of the efficiency of this contact. In other words, criterion 2(a) will hold. Information on the minimum specific energy required P_a (i.e., power input per unit volume of reactor) for each type of contactor has been reported by Sittig and Heine (1977) and Viesturs et al. (1986). Table 16.6 lists the values based on the data reported in these studies. Because some of the energy in bubble

Table 16.5 Steps in the minimization of gas-liquid reactor volume

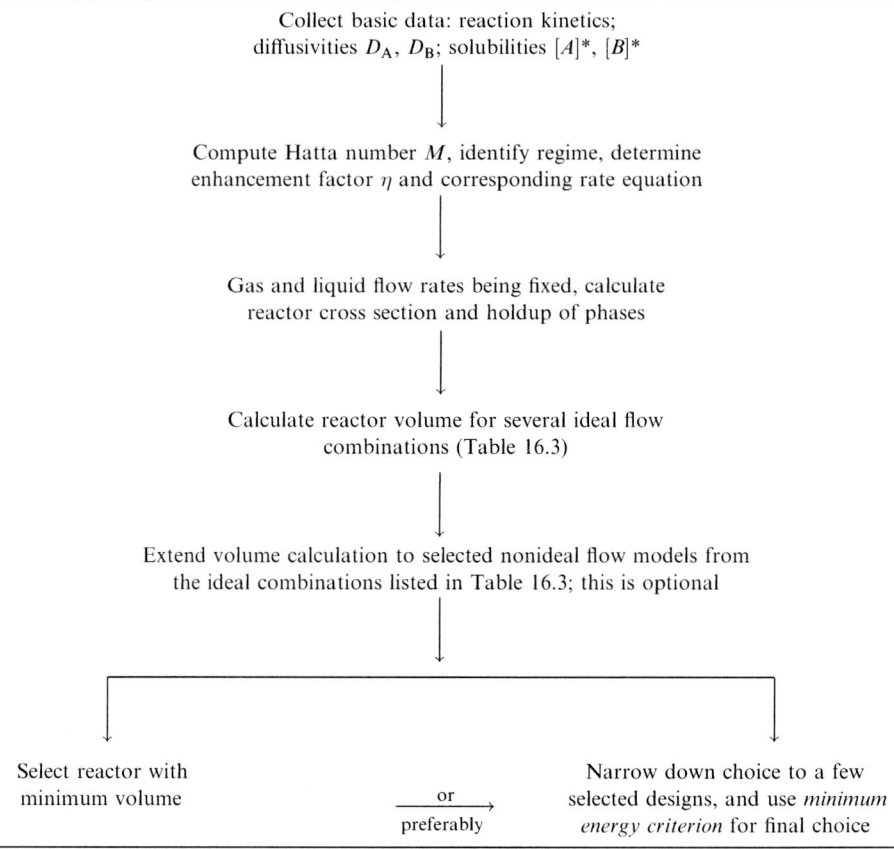

columns is supplied by the natural movement of the bubbles themselves, the specific energy requirement is minimum. Agitated contactors clearly require much more energy—which further increases as more agitators are added.

Note that to decide on the most favorable reactor, the specific energy P_a should be multiplied by the reactor volume as calculated in the previous section,

$$P = P_a V_r \tag{16.31}$$

Criterion 2(b): heterogeneous regime (regimes 2–4)

In all regimes other than regime 1, the efficiency of interfacial contact becomes an important consideration. Thus the quantity needed to be minimized is not the overall energy used for homogeneous reactions but the energy required per unit area of interfacial contact generated or material transported. This includes (1) energy losses in motors and moving devices such as pumps and agitators, (2) losses due to pressure drops in pipings and distributor elements, and (3) loss of

Table 16.6 Comparison of energy minimization criteria 2(a) and 2(b) (adapted from Kastánek et al., 1993)

Reactor type	Criterion 2(a): minimum specific energy required to achieve contact, P_a (kW/m^3)	Criterion 2(b): interfacial area generated per unit volume per unit energy input, Ω (m^3/kWh)
1. Agitated contactor with multiple stirrers	10	20
2. Agitated contactor with multiple stirrers and a central draft tube	11	20
3. Bubble column	2.5	60–80
4. Bubble column with forced liquid circulation	4	50
5. Multistage bubble column	3.5	100–300
6. Tower reactor with forced gas supply	5	140

energy at the place of dispersion formation. The last of these is obviously the most important and depends on the mechanism of dispersion. A detailed treatment of these energy losses is outside the scope of this effort. Items (1) and (2) are usually "external" to the actual energy evaluation of a specific system. However, they can often be quite important. For example, proper plate design in a plate column can minimize energy losses due to transport and pressure drop in the column, or a properly designed gas distributor device can equally enhance energy utilization efficiency (Bohmer and Blenke, 1972; Zahradnik et al., 1982; Henzler, 1982; Mann, 1983; Rylek and Zahradnik, 1984; Zahradnik et al., 1985). However, the "internal" source represented by item (3) constitutes the main element of criterion 2(b), whose final effect is reflected in the value of the interfacial area generated per unit power input per unit reactor volume. Thus a parameter

$$\Omega = \frac{k'_L a_L}{P/V_r} = \frac{k'_L a_L}{P_a} \tag{16.32}$$

can be used as a measure of the energy effectiveness of a given contactor.

Obviously, the mass transfer coefficient $k'_L a_L$ is the most important parameter in determining energy effectiveness, and several correlations have been reported for estimating it, for example, Meister et al. (1979), Botton (1980), Heijnen and Van't Riet (1984). Direct correlations have also been proposed for interfacial area a_L as a function of energy dissipation rate, for example Nagel et al. (1972), Bucholz et al. (1983), Bisio and Kabel (1985), Sisak and Hung (1986). Typical value ranges for k'_L and a_L for the more common reactors are listed in Table 16.1. An important conclusion from all of these studies is that the energy effectiveness of a contacting device decreases with increasing demand on the intensity of interfacial contact (Kastánek et al., 1993). In other words, each additional increase of mass flow rate requires a disproportionately larger increase of the specific energy dissipation rate.

Comparison of criteria

Values of Ω for different contactors, again calculated from the data of Sittig and Heine (1977), are included in Table 16.6. It is clear from the table that the two criteria 2(a) and 2(b) lead to two entirely different choices of contactors. As already noted, criterion 2(b) is applicable to a much wider range of reactions than criterion 2(a) which is essentially restricted to reactions in the kinetic regime.

LIQUID-LIQUID CONTACTORS

As already noted, the volume of a liquid-liquid reactor for a given output is calculated using the same principles as outlined for gas-liquid reactors. One major factor to be considered in liquid-liquid reactions, however, is the much smaller density difference between the two phases compared to gas-liquid reactions. Although in general, this is conducive to creating greater interfacial area, too small a difference is likely to cause problems of phase separation. This often leads to emulsion formation and the consequent attention to designs that promote emulsion breaking.

Classification of Liquid-Liquid Reactors

Liquid-liquid extractors can be classified as follows (adapted from Treybal, 1963; Warwick, 1973):

 A. Gravity operated extractors
  a. no mechanical moving parts
   spray columns
   packed columns
   sieve plate columns
  b. mechanically agitated contactors
   mixer-settlers; inert gas-agitated contactors
   columns agitated with rotating stirrers
   pulsed columns
 B. Pipeline (horizontal) contactors; static mixers
 C. Centrifugal extractors

Note that contactors in category A are common to gas-liquid and liquid-liquid reactions. The main difference stems from the need for phase separation in liquid-liquid reactions. A few common designs are shown in Figure 16.4. The design of liquid-liquid extractors in general has been well treated by Laddha and Degaleesan (1976) and the general subject of mass transfer by Taylor and Krishna (1993).

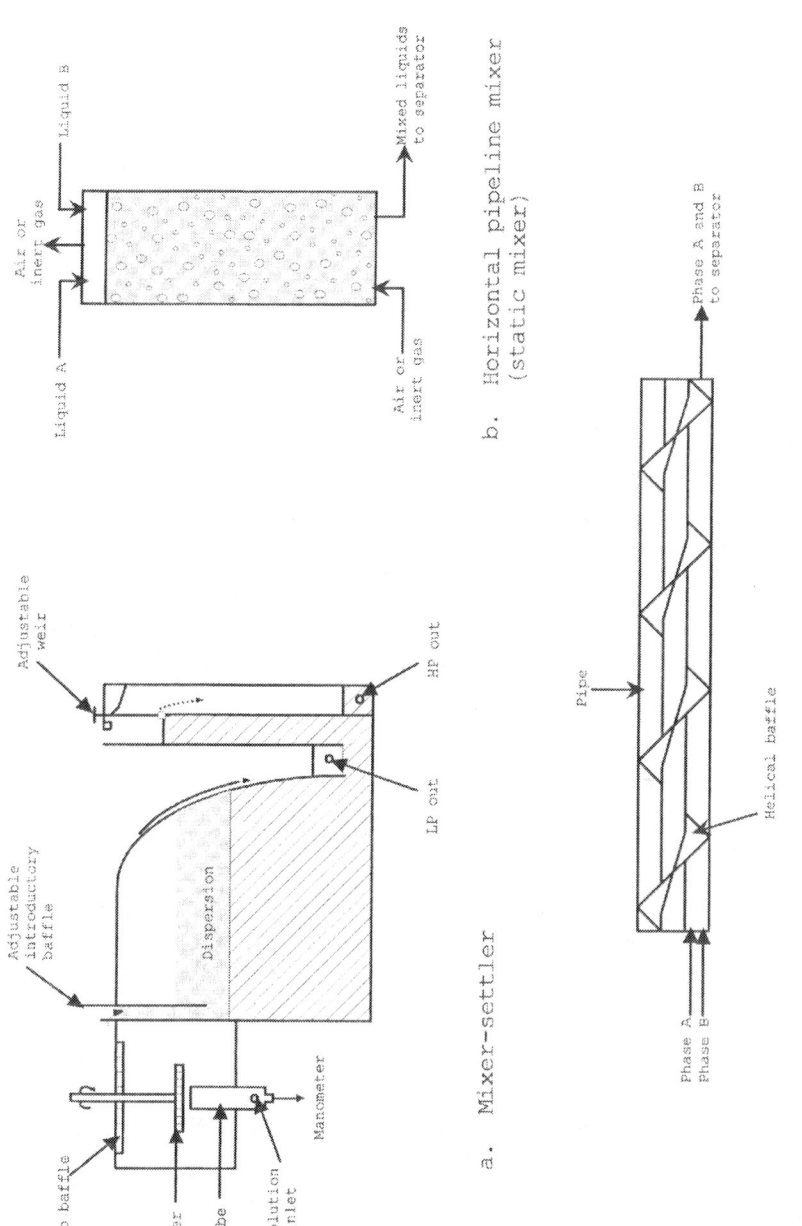

Figure 16.4 Sketches of some common liquid-liquid reactor designs. HP = heavy phase, LP = light phase.

Table 16.7 Flow and mass transfer features of the more common liquid-liquid contactors (adapted from Doraiswamy and Sharma, 1984)

Type of contactor	Mixing characteristics		Residence time of continuous phase	Fractional dispersed phase holdup	k'_L (cm/s)	a_L (cm^2/cm^3)
	Continuous phase	Dispersed phase				
1. Mechanically agitated contactor	MF	PMF	Variable over a wide range	0.05–0.40	0.3–1.0	1–800
2. Packed column	PF	PF	Limited	0.05–0.10	0.3–1.0	1–10
3. Spray column	MF	PF	Limited	0.05–0.10	0.1–1.0	1–10
4. Air (inert gas)-agitated contactor	MF	PMF	Variable over a wide range	0.05–0.30	0.1–0.3	10–100
5. Horizontal mixer	PF	PF	Limited	0.05–0.20	0.1–1.0	1–25

Values of Mass Transfer Coefficients and Interfacial Areas for Different Contactors

Because k'_L and a_L are important parameters in contactor selection, approximate values of them for the more important contactors are listed in Table 16.7, along with some other features of these contactors.

Calculation of Reactor Volume/Reaction Time

The calculation procedures are essentially similar to those for gas-liquid reactions, except that reaction in both phases must be considered. As may often happen, reaction may well be confined to a simple phase, but such an assumption can only be made after experimental verification, or if the parameter values involved distinctly rule out reaction in the other phase. We illustrate a typical procedure in Example 16.2.

EXAMPLE 16.2

Design of a batch reactor for *m*-chloroaniline by liquid-liquid reduction of *m*-nitrochlorobenzene with disodium sulfide.

A batch of *m*-chloroaniline is to be produced by reducing m-nitrochlorobenzene (MNCB) with an aqueous solution of sodium disulfide (Na_2S_2) in a fully baffled 2000-liter mechanically agitated batch contactor at 105 °C. The reaction is known to be restricted to the continuous aqueous phase, phase 2. The regime of this reaction (for 100% conversion with stoichiometric quantities of reactants), as shown in Example 1 of Chapter 15, shifts from regime 3 (fast) at the start to regime 1–2 at the end. Using this information and the following data, calculate the reaction time.

DATA

Amount of sodium disulfide, Na_2S_2 = stoichiometric requirement for 100% conversion of MNCB

Final concentration of aqueous sodium disulfide, $[B]_{f,2} = 2.96 \times 10^{-5}\,mol/cm^3$

Volume of continuous phase, V_2 or $V_c = 8.805 \times 10^5\,cm^3$

Volume of dispersed phase, V_1 or $V_d = 3.125 \times 10^5\,cm^3$

Effective interfacial area, $a_L = 58.95\,cm^2/cm^3$ of dispersion

Holdup, $b = 0.738$

SOLUTION

Because the regime shifts from 3 to 1–2 and because this shift would probably be quite gradual, it is reasonable to consider that the reaction occurs in regime 1–2–3. The reaction is also known to be restricted to phase 2, and hence the equations given in Chapter 14 can be used to calculate the reactor volume for a given conversion or the reaction time for a given volume for the same conversion.

Equation 16.11 gives the reaction rate for regime 1–2–3. This equation can be recast as follows for a liquid-liquid reaction, where the dispersed phase (liquid phase 1) replaces the gas phase:

$$-r_A = \frac{k_{2,c}[B]_{b,c} a_L [A]_d^* \sqrt{D_{A,c} k_{2,c}[B]_{b,c} + k_L'^2}}{k_{2,c}[B]_{b,c} + \left(\dfrac{a_L}{b}\right)\sqrt{D_{A,c} k_{2,c}[B]_{b,c} + k_L'^2}} \qquad (E16.2.1)$$

Note that there is no gas-phase resistance involved here. Hence the term $1/k'_{GP} a_L$ drops out. The various other terms are as defined in Notation, except that for phase 1, the symbol d is used instead of 1 and for phase 2, the symbol c is used instead of 2. For example, $k_{2,c}$ represents the second-order rate constant in phase 2 (with $k_{1,c} = [B]_{b,c} k_{2,c}$), the continuous phase.

The material balance for this case may be written as

$$-V_d \frac{d[A]_{b,d}}{dt} = -\frac{V_c}{\nu_B}\frac{d[B]_{b,c}}{dt} \qquad (E16.2.2)$$

which on integration gives

$$[B]_{b,c} = ([B]_{b,c})_i - \frac{\nu_B V_d}{V_c}[([A]_{b,d})_i - [A]_{b,d}] \qquad (E16.2.3)$$

where $([A]_{b,d})_i$ represents the initial concentration of A (MNCB) in the dispersed phase and $([B]_{b,c})_i$ that of B (Na_2S_2) in the continuous phase. Also,

$$-V_d \frac{d[A]_{b,d}}{dt} = (-r_A') a_L V_c \qquad (E16.2.4)$$

Further, the interfacial concentration of A can be written as

$$[A]_d^* = m_A [A]_{b,d} \qquad (E16.2.5)$$

Combining Equations E16.2.1, E16.2.3, E16.2.4, and E16.2.5 and rearranging results in

$$-\frac{d[A]_{b,d}}{dt} = \frac{\alpha''\rho''(\gamma'' + [A]_{b,d})\beta''[A]_{b,d}\sqrt{\gamma'' + [A]_{b,d} + \delta''}}{\alpha''(\gamma'' + [A]_{b,d}) + \eta''\sqrt{\gamma'' + [A]_{b,d} + \delta''}} \qquad (E16.2.6)$$

where

$$\alpha'' = \frac{k_{2,c}\nu_B V_d}{V_c}, \quad \beta'' = a_L m_A \sqrt{\frac{D_{A,c}k_{2,c}\nu_B V_d}{V_c}}$$

$$\gamma'' = \frac{V_c([B]_{b,c})_i}{\nu_B V_d} - ([A]_{b,d})_i$$

$$\delta'' = \frac{k_L'^2 V_c}{D_{A,c}k_{2,c}\nu_B V_d}, \quad \eta'' = \frac{a_L}{b}\sqrt{\frac{D_{A,c}k_{2,c}\nu_B V_d}{V_c}} \qquad (E16.2.7)$$

$$\rho'' = \frac{V_c}{V_d}$$

The initial and final conditions are

$$\begin{aligned} \text{at } t &= 0, \quad [A]_{b,d} = ([A]_{b,d})_i \\ \text{at } t &= t_B, \quad [A]_{b,d} = ([A]_{b,d})_f \end{aligned} \qquad (E16.2.8)$$

In addition, because the reactants are present in stoichiometric amounts, the term γ'' becomes zero, and Equation E16.2.6 reduces to

$$-\frac{d[A]_{b,d}}{dt} = \frac{\rho''\alpha''\beta''[A]_{b,d}^2\sqrt{[A]_{b,d} + \delta''}}{\alpha''[A]_{b,d} + \eta''\sqrt{[A]_{b,d} + \delta''}} \qquad (E16.2.9)$$

Integration of this equation with boundary conditions E16.2.8 gives

$$t_B = \frac{1}{\rho''\beta''\sqrt{\delta''}}\left[\ln\left(\frac{M_f + \sqrt{\delta''}}{M_f - \sqrt{\delta''}}\right) - \ln\left(\frac{M_i + \sqrt{\delta''}}{M_i - \sqrt{\delta''}}\right)\right]$$

$$+ \frac{\eta''}{\rho''\alpha''\beta''}\left(\frac{1}{([A]_{b,d})_f} - \frac{1}{([A]_{b,d})_i}\right) \qquad (E16.2.10)$$

where

$$M_f = \sqrt{([A]_{b,d})_f + \delta''}$$

$$M_i = \sqrt{([A]_{b,d})_i + \delta''}$$

The numerical values of the various parameters can be calculated using the given data. Thus

$$\alpha'' = 1.228 \times 10^3, \quad \beta'' = 9.88 \times 10^{-3}$$
$$\delta'' = 6.53 \times 10^{-4}, \quad \eta'' = 15.635$$
$$\rho'' = 880.5 \times 10^3 / 312.5 \times 10^3 = 2.818,$$
$$([A]_{b,d})_i = 8.454 \times 10^{-3}, \quad ([A]_{b,d})_f = 8.454 \times 10^{-5}.$$

Substituting these in Equation E16.2.10 gives the batch time as
$$t_B = 9485\,\text{s} = 2.63\,\text{h}$$

STIRRED TANK REACTOR: SOME PRACTICAL CONSIDERATIONS

The mechanically agitated contactor, referred to also as the stirred tank reactor (STR), is the most commonly used reactor for small and medium volume production in organic synthesis. Therefore, we shall summarize the important practical features of this reactor, which should also be applicable to CSTR.

1. Several types of impellers are used. The most common are the simple marine impeller and the highly efficient double-motion stirrer. The reactors are always suitably baffled.
2. An important consideration is mixing or dispersion (see point 5). As a rule of thumb, for homogeneous reactions or for reactions in the kinetic regime of fluid-fluid reactions, a tip speed of 2.5–3.3 m/s is required for good mixing, and for fluid-fluid reactions in the fast regime a tip speed of 5–6 m/s is necessary. The power input for homogeneous reactions is on the order of $0.1\,\text{kW/m}^3$, and for fluid-fluid systems, it is $2.0\,\text{kW/m}^3$.
3. For reactions with a sizable heat effect, heat is abstracted in several ways: reflux condenser, internal coil, external heat exchanger, cooling jacket, and half-round pipes wound on the reactor body. The overall heat transfer coefficient with water in the tank for a water-cooled or steam-heated jacketed vessel is 0.15–1.7, for a tank with water-cooled half-round pipe is 0.3–0.9, and for a tank with water-cooled internal coil is 0.5–$1.2\,\text{kW/m}^2\text{K}$ (Rose, 1981).
4. Reactors are usually available only in standard sizes. It is much cheaper to buy an oversized tank that conforms to the nearest standard than to have a reactor of actual calculated dimensions specially fabricated.
5. *Scale-up criteria*. In addition to design by modeling, as discussed in previous sections, it is also possible to use simple scale-up criteria based on laboratory results. For this purpose, all of the available criteria are consolidated in a common form in Table 16.8, with different values of the exponent c for the different criteria used. This kind of scale-up should be acceptable for the scales of production normally encountered in fine chemicals and intermediates synthesis but could be misleading for bulk chemicals production.

Table 16.8 Scale-up criteria for a stirred reactor

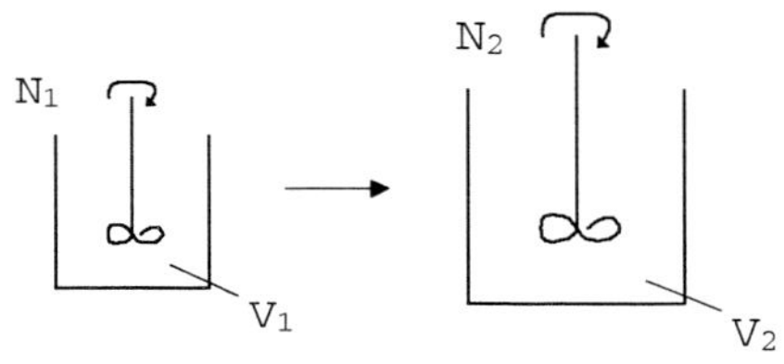

Criterion	$\dfrac{N_2}{N_1} = \left(\dfrac{V_2}{V_1}\right)^c$	Remarks
Constant stirrer power/unit volume	$c = -\dfrac{2}{9}$	To be used when the main object is mixing; criterion 1
Constant heat transfer coefficient	$c = -0.15$	To be used when heat removal is the main problem; additional heat exchange capacity must be added when needed
Constant tip speen	$c = -\dfrac{1}{3}$	Usually for a gas-liquid reaction; maintains the same gas distribution in the two reactors; satisfies criterion 2(b)
Constant pumping rate/unit volume	$c = 0$ (i.e., $N_1 = N_2$)	

Chapter 17

Multiphase Reactions and Reactors

The first three chapters of this part dealt with two-phase reactions. Although catalysts are not generally present in these systems, they can be used in dissolved form in the liquid phase. This, however, does not increase the number of phases. On the other hand, there are innumerable instances of gas-liquid reactions in which the catalyst is present in solid form. A popular example of this is the slurry reactor so extensively employed in reactions such as hydrogenation and oxidation. There are also situations where the solid is a reactant or where a phase-transfer catalyst is immobilized on a solid support that gives rise to a third phase (see Chapter 20). A broad classification of three-phase reactions and reactors is presented in Table 17.1 (not all of which are considered here). This is not a complete classification, but it includes most of the important (and potentially important) types of reactions and reactors.

The thrust of this chapter is on reactions and reactors involving a gas phase, a liquid phase, and a solid phase which can be either a catalyst (but not a phase-transfer catalyst) or a reactant, with greater emphasis on the former. The book by Ramachandran and Chaudhari (1983) on three-phase catalytic reactions is particularly valuable. Other books and reviews include those of Shah (1979), Chaudhari and Ramachandran (1980), Villermaux (1981), Shah et al. (1982), Hofmann (1983), Crine and L'Homme (1983), Doraiswamy and Sharma (1984), Tarmy et al. (1984), Shah and Deckwer (1985), Chaudhari and Shah (1986), Kohler (1986), Chaudhari et al. (1986), Hanika and Stanek (1986), Joshi et al. (1988), Concordia (1990), Mills et al. (1992), Beenackers and Van Swaaij (1993), and Mills and Chaudhari (1997). Doraiswamy and Sharma (1984) also present a discussion of gas-liquid-solid noncatalytic reactions in which the solid is a reactant.

Table 17.1 Classification of three-phase reactions and reactors. G = gas, L = liquid, S = solid, TPC = triphase catalysis

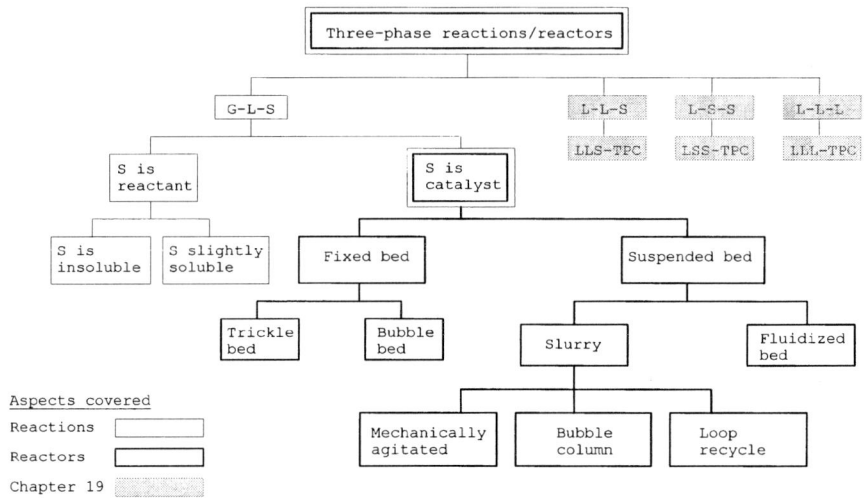

ANALYSIS OF THREE-PHASE CATALYTIC REACTIONS

The Basis

In Chapter 7 we saw how Langmuir–Hinshelwood–Hougen–Watson (LHHW) models are normally used to describe the kinetics of gas-solid (catalytic) or liquid-solid (catalytic) reactions, and in Chapters 14 to 16 we saw how mass transfer between gas and liquid phases can significantly alter the rates and regimes of these two-phase reactions. A noteworthy feature of the analyses presented in those chapters is the difference in the two approaches. In two-phase catalytic reactions within a solid, diffusion is viewed as an intruder, and its effect is expressed in terms of an effectiveness factor ε. On the other hand, in (two-phase) gas-liquid reactions, diffusion is regarded as the main phenomenon, and the role of reaction is expressed through an enhancement factor η. In reactions involving all three phases, one might choose either approach. However, because the reaction on the catalyst is the chief focus in most three-phase reactions, it has been the accepted practice to use the effectiveness factor concept.

In two-phase catalytic reactions (gas-solid or liquid-solid, where the catalyst is the solid phase), all concentrations are expressed in terms of concentrations in the liquid bulk. Because concentrations at the catalyst surface are used to formulate the rate equation and these are seldom known, they are related to the corresponding bulk concentrations (which are observable values) through appropriate mass transfer coefficients. In extending this procedure to a three-phase reaction, it must be noted that more mass transfer steps are involved here, as shown in Figure 17.1. Thus let us consider a gas-liquid reaction such as

$$A(g) + \nu_B B(l) \xrightarrow{\text{solid catalyst}} R \qquad [17.1]$$

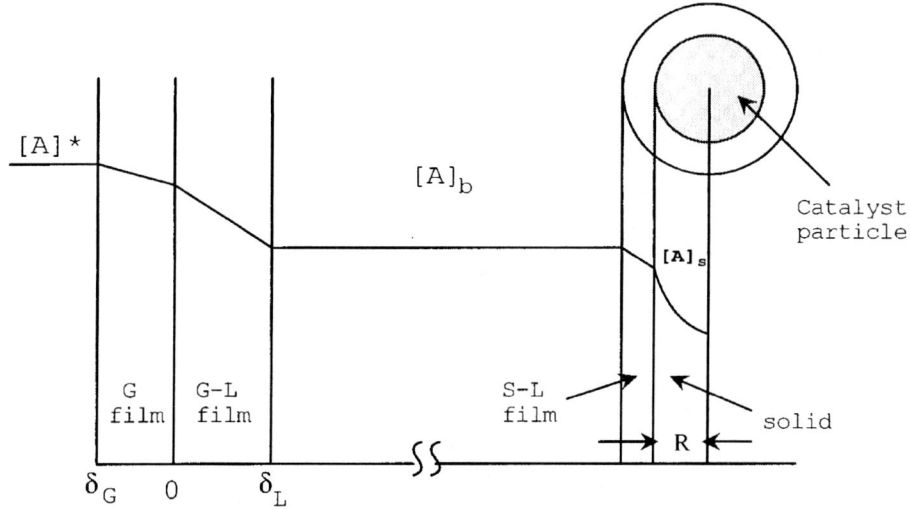

Figure 17.1 Concentration profiles in a slurry reactor. G = gas, L = liquid, R = particle radius, δ_G = gas film thickness, δ_L = liquid film thickness.

in which the reaction between A in the gas phase and B in the liquid occurs on the surface of a catalyst dispersed as a slurry in the liquid. The measurable concentration is clearly the equilibrium concentration $[A]^*$ of the gas in the liquid (which is related to the partial pressure of A by Henry's law). When the effectiveness factor is based on $[A]^*$, the resulting correction factor is referred to as an *overall effectiveness factor E* for the three-phase reaction. Thus

$$E = \frac{\text{actual rate with all the diffusional resistances}}{\text{rate based on a uniform concentration of } [A]^*} = \frac{r_{Aa}}{r_A([A]^*)} \qquad (17.1)$$

All of our analyses will be based on the use of Equation 17.1 and a *generalized Thiele modulus*. We consider both power law and LHHW kinetics.

Power Law Kinetics

Referring to Figure 17.1 and Equation 17.1, it is clear that an overall mass transfer coefficient is needed that accounts for mass transfer at both the gas-liquid interface ($k'_{G,A}$ for the gas-side and and $k'_{L,A}$ for the liquid-side films) and the solid-liquid interface ($k'_{SL,A}$). Such a coefficient is given by

$$-r_A = \bar{k}_A([A]^* - [A]_s) \qquad (17.2)$$

where

$$\bar{k}_A = \left(\frac{1}{k'_{GL,A} a_L} + \frac{1}{k'_{SL,A} a_p}\right)^{-1} \qquad (17.3)$$

The coefficient $k'_{GL,A}$ represents the overall effect of gas- and liquid-side mass transfer and is given by

$$\frac{1}{k'_{GL,A}a_L} = \frac{1}{H'_{A,C}k'_{G,A}a_L} + \frac{1}{k'_{L,A}a_L} \left(= \frac{1}{k_{GL,A}} \right) \qquad (17.4)$$

where $H'_{A,C}$ is the Henry's law constant in units of (mol/m^3 gas)/(mol/m^3 liquid). If the mass transfer coefficient $k'_{G,A}$ is expressed in units of mol/m^2 atm s (i.e., as $k'_{GP,A}$), then the Henry's law constant will be H'_A (atm m^3/mol).

The corresponding equation for component B at the solid-liquid interface is simply

$$\bar{k}_B = k'_{SL,B} a_p \qquad (17.5)$$

because there is no gas-phase diffusion here.

It is important to note that only regimes 1 and 2 of gas absorption accompanied by a simple reaction (Chapter 14) are operative here. Because the reaction occurs only on the solid surface (or within it), regimes 3 and 4 for reaction in the liquid film do not apply. An exception would be when the solid enters the gas-liquid film as a microphase. The behavior of microphases in general is analyzed in Chapter 23.

Consider reaction 17.1 again, and assume that the concentration $[B]_b$ of the liquid-phase reactant remains constant (i.e., the reaction is pseudo-order). The rate equation in the presence of diffusional resistance in the catalyst can be written as

$$-r_A = \varepsilon w k_{wm}[A]_s^m \qquad (17.6)$$

where w is the weight of catalyst per unit volume of reactor (i.e., $\rho_c f_c$), ε is the catalyst effectiveness factor, and k_{wm} is the weight-based rate constant [mol/kg cat s) (m^3/mol)^m] given by

$$-r_{wA} = k_{wm}[A]_s^m, \quad \text{mol/kg cat s} \qquad (17.7)$$

The term $1/w k_{wm}$ represents the resistance due to the reaction (see Equation 14.3). For a first-order reaction, we add this resistance to the mass transfer resistances of Equation 17.2 to give

$$-r_A = [A]^* \left(\frac{1}{\bar{k}_A} + \frac{1}{w k_{w1}} \right)^{-1} \qquad (17.8)$$

The simple relationship given by Equation 17.8 does not hold for nonlinear kinetics. For an mth-order reaction, the following implicit expression can be derived:

$$-r_A = \bar{k}_A \left[[A]^* - \left(\frac{-r_A}{w k_m} \right)^{1/m} \right] \qquad (17.9)$$

Using the rate equations for power law kinetics already given, the general expression for the effectiveness factor given by Equation 17.1, and an extension of the procedure outlined in Chapter 7, specific equations can be derived for a

general Thiele modulus $\bar{\phi}$ and an overall effectiveness factor E for a reaction of any kinetics. The equations for an mth-order reaction are listed in Table 17.2, and plots of E versus ϕ for different values of σ_A (defined in both the table and the figure) for a first-order reaction ($m = 1$) are presented in Figure 17.2. Note that the equations are implicit in the two parameters.

Expressions can also be derived (Ramachandran and Chaudhari, 1983) for the case where the liquid-phase reactant concentration $[B]_b$ does not remain constant, that is, diffusion of both A and B is limiting, and the reaction is second order. This case is also included in Table 17.2, along with the conditions for B limiting. It is a relatively straightforward matter to extend the equation to the general case of an mth order reaction (Mills and Chaudhari, 1997). The general expression for the reaction is then: $(-r_A) = E k_{w,mn}[A]^{*m}[B]_b^n$.

EXAMPLE 17.1

Hydrogenation of glucose

The hydrogenation of glucose to sorbitol comprises the first step in the commercial synthesis of vitamin C. It has been reported (Brahme and Doraiswamy, 1976; see also Levenspiel, 1993) that the kinetics of this reaction

$$H_2(g) + C_6H_{12}O_6(l) \xrightarrow[\text{nickel}]{\text{Raney}} C_6H_{15}O_6(l)$$
$$(A) \qquad (B) \qquad\qquad\qquad (R)$$

may be expressed as

$$-r_{wA} = -r_{wB} = k_{w2}[A]^{0.6}[B] \qquad (E17.1.1)$$

It is desired to carry out this reaction in an available 2-m^3 reactor at a feed rate Q_B of 0.01 m^3/s and a hydrogen feed rate Q_A of 0.2 m^3/s at 200 atm and 150 °C. Given the following data, calculate the conversion that can be achieved: catalyst: $d_p = 10\,\mu\text{m}$, $\rho_c = 8900\,\text{kg/m}^3$, $f_c = 0.056$; $[B]_{b0} = 2000\,\text{mol/m}^3$; $H'_A = 2.776 \times 10^5\,\text{Pa m}^3/\text{mol}$; $D_{eA} = 2.1 \times 10^{-9}\,\text{m}^3/\text{s}$; $k_{GL,A} = 0.05\,\text{s}^{-1}$; $k'_{SL,A} = 10^{-3}\,\text{m/s}$; $k_{w2} = 5.96 \times 10^{-6}\,(\text{mol/kg s})/(\text{m}^3/\text{mol})^{1.6}$.

SOLUTION

The reaction order with respect to the liquid-phase reactant B is 0.6 and with respect to hydrogen is 1.0. To establish whether it can be regarded as a pseudo-order reaction, we compare the values of $[A]^*$ and $[B]_{b0}$:

$$[A]^* = [A]_0 = \frac{p_A}{H'_A} = \frac{200 \times 101{,}325\,\text{Pa mol}}{277{,}600\,\text{Pa m}^3}$$

$$= 73.0\,\text{mol/m}^3$$

$$\therefore \quad [B]_{b0} = 2000\,\text{mol/m}^3 \gg [A]_0 = 73.0\,\text{mol/m}^3$$

Hence the rate equation can be written as

$$-r_{wA} = k_{w1}[A]^{0.6} \qquad (E17.1.2)$$

Table 17.2 Equations for estimating overall effectiveness factors for selected situations in the reaction $A + \nu_B B \rightarrow R$

Reaction kinetics	Generalized Thiele modulus[a]	Overall effectiveness factor[a]
1. mth order in A, $[B]$ constant (i.e., A limiting). Rate equation: $-r_{wA} = \varepsilon w k_{wm}[A]_s^m$	$\bar{\phi} = \phi \alpha_A^{(m-1)/2}$ (a) where $\phi = \dfrac{R}{3}\left[\left(\dfrac{m+1}{2}\right)\left(\dfrac{\rho_c k_{wm}([A]^*)^{m-1}}{D_{eA}}\right)\right]^{1/2}$ (b)	$E = \dfrac{\alpha_A^{(m+1)/2}}{\bar{\phi}}\left[\coth(3\bar{\phi}) - \dfrac{1}{3\bar{\phi}}\right]$ (c) See Ramachandran and Chaudhari (1983) for solution
2. Both A and B limiting, but diffusion of A or B in solid can be controlling as shown by (1) and (2) below. 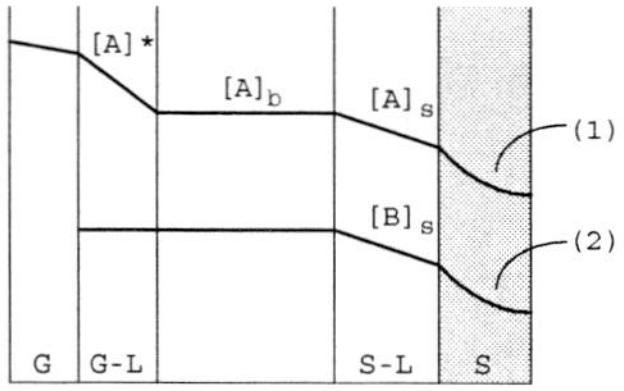	Approximate condition for nonuniform concentration gradient of liquid phase reactant B (i.e. for B also limiting): $\dfrac{D_{eB}[B]_s}{\nu_B D_{eA}[A]_s} < 10$ With this overall condition: (1) Diffusion of A in catalyst controlling: $\bar{\phi} = \dfrac{\phi \alpha_B}{\left(\alpha_B - \dfrac{\beta \alpha_A}{3}\right)^{1/2}}$ (d) Conditions: $\dfrac{D_{eB}[B]_b \alpha_B}{\nu_B D_{eA}[A]^* \alpha_A} > 1$	$E = \dfrac{1}{\bar{\phi}}\left(\coth 3\bar{\phi} - \dfrac{1}{3\bar{\phi}}\right)\alpha_A \alpha_B$ (f) *Solution.* Equation (d), or (e), and (f) are solved simultaneously for E for different assumed values of $\bar{\phi}$. *Note.* Conditions can only be tested at the end since α_A and α_B depend on E (see footnote) which is sought as the solution.

Rate equation: $-r_A = \varepsilon w k_{w11}[A]_s[B]_s$
(first order in both A and B)

(2) Diffusion of B in catalyst controlling:

$$\bar{\phi} = \frac{\phi\alpha\Delta A}{\left(\alpha_A - \dfrac{\beta\alpha_B}{3}\right)^{1/2}} \quad (e)$$

Condition:

$$\frac{D_{eB}[B]_b \alpha_B}{\nu_B D_{eA}[A]^* \alpha_A} < 1$$

where

$$\beta = \frac{\nu_B D_{eA}[A]^*}{D_{eA}[B]_b}, \quad \phi = \frac{R}{3}\left(\frac{\rho_c k_{w11}[B]_h}{D_{eA}}\right)$$

(see footnote for α_A, α_B)

---

[a] $\alpha_A = [1 - (E/\sigma_A)]$, $\alpha_B = [1 - (E/\sigma_B)]$; $\sigma_A = \bar{k}_A/(wk_{wm}[A]^{*m-1})$ in Equations (a) and (c); $\sigma_A = \bar{k}_A/(wk_{w11}[B]_b)$, $\sigma_B = \bar{k}_B/(\nu_B w k_{w11}[A]^*)$ in Equations (d), (e), and (f).

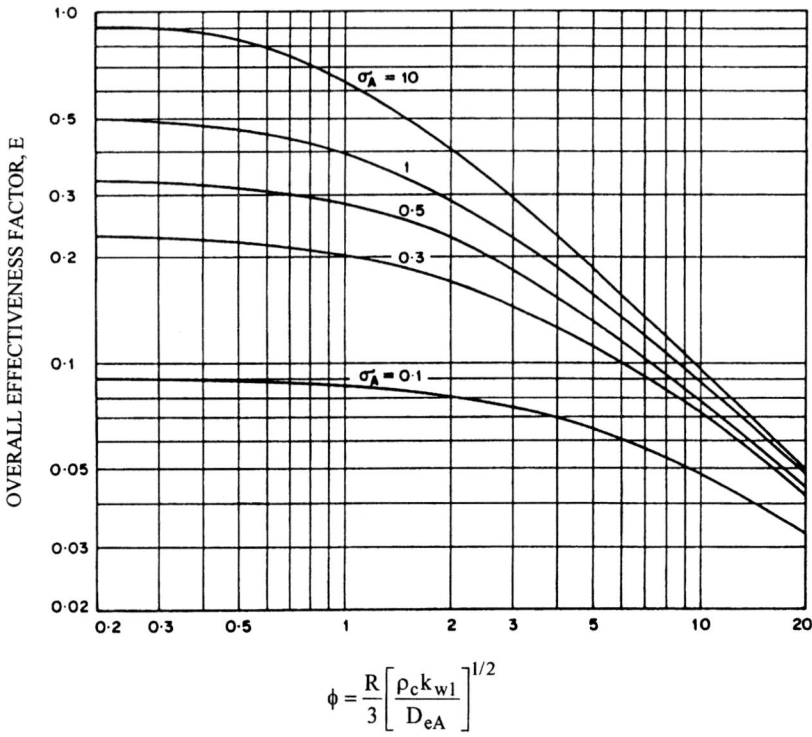

Figure 17.2 Overall effectiveness factor plots for a first-order reaction (Chaudhari and Ramachandran, 1980): $\sigma_A = \bar{k}_A/wk_{w1}$.

where

$$k_{w1} = k_{w2}[B]_{bf} \tag{E17.1.3}$$

Because the reactor is fully mixed, the concentration of B falls to $[B]_{bf}$ and remains constant thereafter. To find $[B]_{bf}$, we use the material balance

$$Q_B([B]_{b0} - [B]_{bf}) = (-r_A)V_r \tag{E17.1.4}$$

Now we need to calculate $(-r_A)$ to solve Equation (E17.1.4).

USING LINEAR EQUATION 17.8

Here, we force a first-order fit by writing the rate equation as

$$-r_{wA} = k_{w1}[A] \tag{E17.1.5}$$

where now

$$k_{w1} = k_{w2}[B]_{bf}[A]^{-0.4} \tag{E17.1.6}$$

Then we use Equation 17.8 to get

$$-r_A = \frac{[A]}{\dfrac{1}{\bar{k}_A} + \dfrac{1}{\varepsilon k_{w2} w [B]_{bf}[A]^{-0.4}}} \tag{E17.1.7}$$

where

$$\frac{1}{\bar{k}_A} = \frac{1}{k'_{GL,A} a_L} + \frac{1}{k'_{SL,A} a_p} \tag{E17.1.8}$$

$$\frac{1}{k'_{GL,A} a_L} \left(= \frac{1}{k_{GL,A}}\right) = \frac{1}{H'_A k'_{GP,A} a_L} + \frac{1}{k'_{L,A} a_L} = \frac{1}{0.05}$$

$$w = f_c \rho_c = 0.056 \times 8900 = 498.4 \, \text{kg cat/m}^3 \, \text{reactor}$$

$$a_p = \frac{6 f_c}{d_p} = \frac{6 \times 0.056}{1 \times 10^{-5}} = 3.36 \times 10^4 \, \text{m}^{-1}$$

$$k'_{SL,A} a_p = 1 \times 10^{-3} \times 33600 = 33.6 \, \text{s}^{-1}$$

and ε is the effectiveness factor which can be estimated by first calculating the Thiele modulus ϕ_1 from

$$\phi_1 = L \sqrt{\left(\frac{n+1}{2}\right) \frac{k_{w2} \rho_c [A]^{-0.4}}{D_{eA}}} [B]_{bf}^{1/2}$$

$$= 0.00326 [B]_{bf}^{1/2}$$

Even with $[B]_{bf}^{1/2} = [B]_{bo}^{1/2} = 2000^{1/2}$, ϕ_1 is very small, giving $\varepsilon = 1$. Hence there is no internal diffusional resistance in the pellet.

Substituting numerical values,

$$\bar{k}_A = 0.05$$

$$-r_A = \frac{1}{0.274 + \dfrac{25.66}{[B]_{bf}}} \tag{E17.1.9}$$

Combining Equations E17.1.4 and E17.1.9 and solving the resulting binomial,

$$[B]_{bf} = 1318.5 \, \text{mol/m}^3$$

(the negative value, -141.81, is rejected). Thus

$$1 - X_B = \frac{1318.5}{2000} = 0.659$$

giving

conversion $X_B = 34.1\%$

Note: Because $[B]_{bf} = 1318.5 \gg [A]_0 = 73.0$, the assumption of pseudo-order kinetics is correct.

USING NONLINEAR EQUATION 17.9

Substituting the values $[A]^* = 73.0$, $w = 498.4$, $m = 0.6$, and $\bar{k}_A = 0.05$ in this implicit equation and solving for the rate give

$$-r_A = 3.604 \, \text{mol/m}^3 \text{s}$$

Using this value of the rate in the material balance of Equation E17.1.4 and solving for $[B]_{bf}$,

$$[B]_{bf} = 1279.2 \, \text{mol/m}^3$$

Then the conversion is calculated as

$$1 - X_B = \frac{1279.2}{2000} = 0.640$$

giving

$$X_B = 36\%$$

We may conclude from the results of the linear and nonlinear models that the two cases lead to practically the same results. Hence the use of the true kinetic model is not always necessary.

LHHW Kinetics

One of the earliest LHHW models to be developed for a slurry reaction was for the hydrogenation of glucose to sorbitol, considered in Example 17.1. Models for several other three-phase catalytic reactions have since been developed, many of which are listed in Table 17.3.

An effectiveness factor ε can be defined and incorporated in any LHHW model as, for example, in a typical single-site surface-reaction model, giving

$$-r_A = \frac{\varepsilon w k_w [A]}{1 + K_A [A]} \tag{17.10}$$

Using Bischoff's approximation (1965), implicit expressions can then be derived for the generalized Thiele modulus $\bar{\phi}$ and overall effectiveness factor E. Based on these, plots can be prepared of E versus $\bar{\phi}$ (the familiar Thiele modulus for the catalyst) for different values of $\sigma_A = \bar{k}_A[(1 + K_A[A]^*)/wk_w]$ and $K_A[A]^*$. A few representative plots are shown in Figure 17.3.

DESIGN OF THREE-PHASE CATALYTIC REACTORS

The Approach

Our emphasis in this chapter will be on slurry reactors (various designs of which are extensively used), but other types of reactors will also be briefly considered.

Three-phase reactors are operated in either the semibatch or continuous mode, and batch operation is almost never used because the gas phase is invariably continuous. The general principles of design are the same for all types of reactors for a given mode of operation, semibatch or continuous. They differ with respect to their hydrodynamic features, particularly mass and heat transfer. Thus, for simple first-order reactions, Equation 17.8 is valid for any reactor. The rate constant k_{w1} would be the same for all of the reactors, but specific to each reactor type is the mass transfer term $\bar{k}_A$. Hence we consider first the design of

Table 17.3 Examples of LHHW models for gas-liquid-solid catalytic reactions in organic synthesis (from Mills and Chaudhary, 1997)

	Reaction System	Catalyst	Kinetic Model[a]	Reference
1.	Ethynylation of formaldehyde	Cu_2C_2-silica gel	$-r_B = \dfrac{wk_w[B]}{(1+K_B[B]+K_E[E]^*)}$	Kale et al. (1981)
2.	Oxidative carbonylation of aniline	Pd-C/NaI	$-r_B = \dfrac{wk_w[A]^*[E]^*[B]^2}{(1+K_A[A]^*+K_B[B]+K_E[E]^*)^2}$	Gupte and Chaudhari (1992)
3.	Hydrodesulfurization of o-aminobenzyl sulfides	$Co_xMo_yO_z$	$-r_B = \dfrac{wk_w(p_A/H_A')[B]}{(1+\sqrt{K_A p_A/H_A})^2(1+K_B[B])}$	Tremont et al. (1988)
4.	Hydroformylation of 1-hexene and allyl alcohol	$HRh(CO)(PPh_3)_3$	$-r_A = \dfrac{k([A]^*)^{1.52}[B][C][E]^*}{(1+K_B[E]^*)^m(1+K_B[B])^n}$	Deshpande and Chaudhari (1988; 1989a,b)
5.	Hydrogenation of glucose	Raney Ni	$-r_A = \dfrac{wk_w[A]^*[B]}{(1+K_A[A]^*)}$	Brahme and Doraiswamy (1976)
6.	Hydrogenation of o-nitroanisole	Pd/C	$-r_A = \dfrac{wk_w[A]^*[B]}{(1+K_A[A]^*)(1+K_B[B])}$	Chaudhari et al. (1984)
7.	Hydrogenation of m-nitrochlorobenzene	Pt-S/C	$-r_B = \dfrac{wk_w[A]^*[B]}{(1+K_B[B])}$	Rode et al. (1994)
8.	Hydrogenation of 2-ethylhexenal (first step: to 2-ethylhexanal)	Pd-SiO_2-monolithic	$-r_B = \dfrac{wk_w K_A K_B[A][B]}{(1+\sqrt{K_A[A]^*}+K_B[B]+K_P[P])^3}$	Irandoust and Anderson (1988)

(*continued*)

Table 17.3 (Continued)

Reaction System	Catalyst	Kinetic Model[a]	Reference
9. Hydrogenation of 2,4-dinitrotoluene	Pd/Al_2O_3	$-r_i = \dfrac{wk_{w,i}K_H K_i (p_H/R_g T)^{0.5}[i]}{[1+K_H(p_H/R_g T)^{0.5}]\left[1+K_i[i]+\sum_{j\neq i} K_j[j]\right]}$	Molga and Westerterp (1992)
10. Hydrogenation of p-isobutyl acetophenone	Ni/Y-Zeolite	$-r_i = \dfrac{wk_{w,i}[A^*][B]}{(1+K_A[A^*])}$	Rajasekharam and Chaudhari (1996)
11. Hydrogenation of 12-ethyl, 5, 6, 7, 8-tetrahydroanthraquinone (first step)	Pd/Support	$-r_i = \dfrac{wk_{w,i}[A][B]}{(1+\sqrt{K_A[A^*]})^2}$	Santacesaria et al. (1994)
12. Hydrogenation of 1, 5, 9-cyclododecatriene	Pd/Al_2O_3	$-r_i = \dfrac{wk_{w,i}K_j([A^*])^{\alpha}[j]}{\left(1+\sum_{j=1}^{3} K_j[j]\right)}$	Benaissa et al. (1996)

[a] A = gas-phase reactant other than CO (usually hydrogen); B = liquid-phase reactant (subscript b for bulk is omitted); E = CO (gas phase reactant); C = catalyst; H_A' = Henry's law constant, atm m^3/mol; p_A = partial pressure of A, atm; $k_{w,i}$ = rate constant for the ith step ($i=1$ for the first step), m^3/kg s (assuming first order dependence on surface coverage by concerned reactant); R_g = gas law constant, m^3 atm/mol K; i = any substrate (including intermediate products); j = any substrate other than i, P = product.

Notes: (1) original references must be consulted for details of equations; (2) reactions 8–12 are complex, and the rate equation in each case represents the rate of single-step conversion of species i to j.

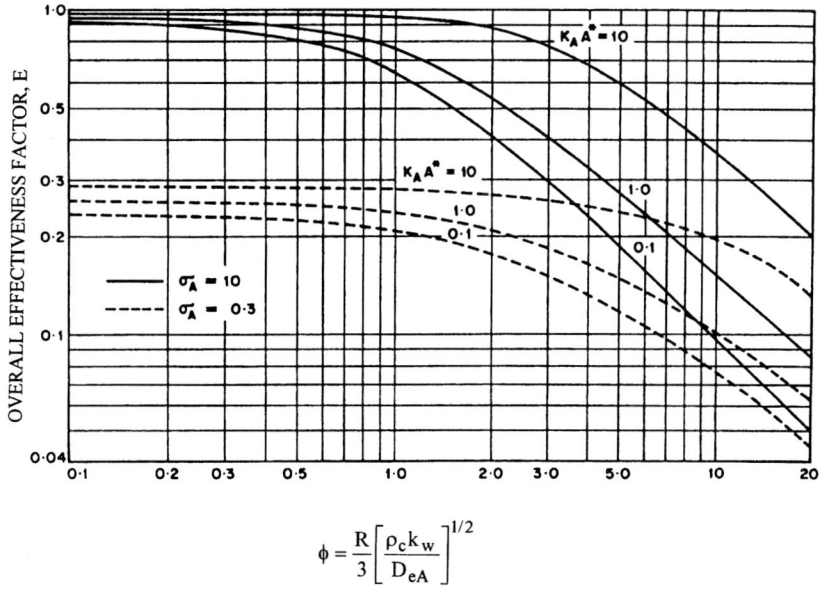

$$\phi = \frac{R}{3}\left[\frac{\rho_c k_w}{D_{eA}}\right]^{1/2}$$

Figure 17.3 Overall effectiveness factor plots for LHHW kinetics (Chaudhari and Ramachandran, 1980). $\sigma_A = (1 + k_A[A]^*)/wk_w$. A^* in figure $= [A]^*$.

semibatch and continuous reactors in general, and then briefly outline the mass transfer and other significant features of the more common types of reactors listed in Table 17.1.

Semibatch Reactors: Design Equations for (1,0)- and (1,1)-Order Reactions

Consider reaction 17.1, for which the rate equation for both cases can be written as

$$-r_A = k_1 [A]^* \tag{17.11}$$

where k_1 is the true first-order rate constant for the $(1,0)$-order reaction, but for the $(1,1)$ case is equal to $k_{11}[B]_b$ and hence is a function of $[B]_b$. The final equations for semibatch time for $(1,0)$- and $(1,1)$-order reactions are (Ramachandran and Chaudhari, 1980a)

(1,0)-ORDER REACTION

$$t_{SB} = \frac{[B]_{bi} X_B}{\nu_B [A]^*} \left(\frac{1}{\bar{k}_A} + \frac{1}{\varepsilon w k_{wl}}\right) \tag{17.12}$$

where

$$\varepsilon \text{ (for a sphere)} = \frac{1}{\phi}\left[\coth(3\phi) - \frac{1}{3\phi}\right] \quad (17.13)$$

$$\phi = \frac{R}{3}\left(\frac{\rho_c k_{w1}}{D_{eA}}\right)^{1/2} \quad (17.14)$$

(1, 1)-ORDER REACTION

$$t_{SB} = \frac{[B]_{bi} X_B}{\nu_B [A]^*}\left(\frac{1}{k_A}\right) + \frac{[B]_{bi} R^2 \rho_c I}{3w\nu_B[A]^* D_{eA}} \quad (17.15)$$

$$I = \frac{2}{\phi_i^2} \ln\left[\frac{\phi_i \cosh\phi_i - \sinh\phi_i}{\phi_i\sqrt{1-X_B}\cosh(\phi_i\sqrt{1-X_B}) - \sinh(\phi_i\sqrt{1-X_B})}\right] \quad (17.16)$$

$$\phi_i = \frac{R}{3}\left(\frac{\rho_c k_{w11,i}[B]_{bi}}{D_{eA}}\right)^{1/2} \quad (17.17)$$

Equations 17.15 and 17.16 can be simultaneously solved to predict the semibatch time as a function of conversion and the Thiele modulus ϕ_i (for initial conditions) defined by Equation 17.17. Subscript i represents the initial condition in those cases.

Continuous Reactors

Three-phase reactions can also be run continuously in several types of reactors, for example packed-bed reactors, fluidized-bed reactors, bubble-column slurry reactors, and continuous stirred tank or sparged slurry reactors. These reactors differ mainly in the flow patterns of the individual phases, and the performance equations will differ accordingly. However, a general equation applicable to all types of flow of the two phases can be developed if the gas-phase concentration is assumed to be constant (thus eliminating gas-phase hydrodynamics from consideration) and a suitable mixing model, such as the dispersion model (see Chapters 12 and 13) is used to characterize liquid-phase flow. Then the equation can be reduced to plug or mixed flow of liquid by appropriate simplification. The assumption of constant gas-phase concentration is justified when pure gas is used or when high rates of gas flow are used thus minimizing concentration changes. The first assumption is usually valid for hydrogenation, carbonylation, or hydroformylation reactions, whereas for oxidation reactions (where air is normally used), conditions must be adjusted to validate the second assumption. Equations can also be developed for varying gas-phase concentrations (Goto and Smith, 1978) but will not be attempted here.

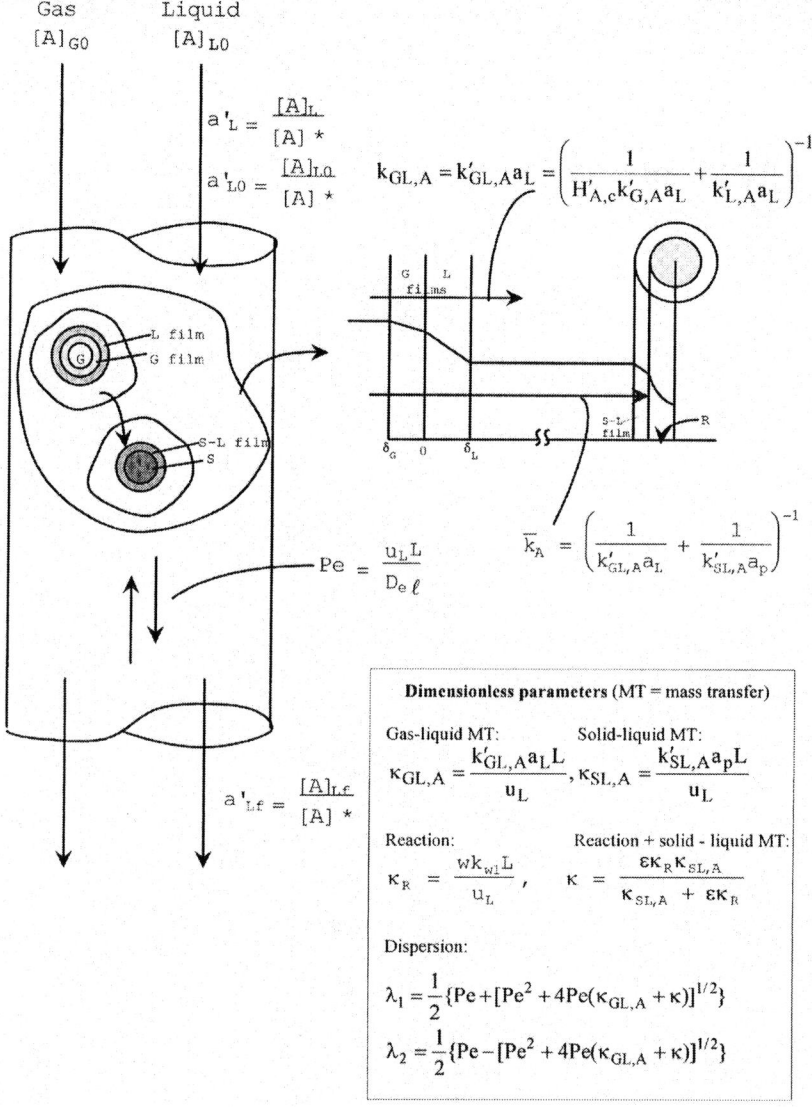

Figure 17.4 The variables and dimensionless groups used in the analysis of continuous three-phase slurry reactors.

The variables affecting continuous reactor operation and their dimensionless forms are explained in Figure 17.4. Considering an isothermal reaction with pseudo-first-order kinetics (Equation 17.6 with $m = 1$) and accounting for backmixing by using the dispersion model characterized by the Peclet number

$$Pe = \frac{u_L L}{D_{el}} \qquad (17.18)$$

the following material balance equation can be written:

$$\begin{pmatrix} \text{moles of } A \\ \text{reacted} \end{pmatrix} = \begin{pmatrix} \text{moles of } A \\ \text{in entering} \\ \text{liquid} \end{pmatrix} - \begin{pmatrix} \text{moles of } A \\ \text{in leaving} \\ \text{liquid} \end{pmatrix}$$

$$+ \begin{pmatrix} \text{moles of } A \text{ transferred} \\ \text{from the gas phase} \end{pmatrix} \qquad (17.19)$$

$$V_r(-r_A) = Q_L[A]_{L0} - Q_L[A]_{Lf} + A_c \int_0^L k'_{GL,A} a_L ([A]^* - [A]_L) dl \qquad (17.20)$$

This equation can be recast in terms of a *reactor efficiency* E_R defined as

$$E_R = \frac{\begin{pmatrix} \text{actual rate of reaction of } A \text{ over} \\ \text{the entire catalyst in the react} \end{pmatrix}}{\begin{pmatrix} \text{rate based on the assumption that the} \\ \text{entire catalyst in the reactor is} \\ \text{exposed to a uniform concentration of } [A]^* \end{pmatrix}} \qquad (17.21)$$

to give

$$E_R = \frac{a'_{L0} - a'_{Lf}}{\kappa_R} + \frac{\kappa_{GL,A}}{\kappa_R} \int_0^1 (1 - a'_L) dz \qquad (17.22)$$

where $a'_{L0} = [A]_{L0}/[A]^*$, $a'_{Lf} = [A]_{Lf}/[A]^*$, and $z = l/L$. The dimensionless quantities in this equation are defined in Figure 17.4.

The only single variable in Equation 17.22 is the reduced concentration a'_L, and equations for this can be developed for the PF, MF, and dispersion models. Then the following final expressions can be derived for E_R for the three models (Ramachandran and Chaudhari, 1983):

PLUG-FLOW MODEL

$$E_R = \frac{\kappa}{\kappa_R(\kappa_{GL,A} + \kappa)} \left[\kappa_{GL,A} + \left(a'_{L0} - \frac{\kappa_{GL,A}}{\kappa_{GL,A} + \kappa} \right) \right] [1 - e^{(\kappa_{GL,A} + \kappa)}] \qquad (17.23)$$

MIXED-FLOW MODEL

$$E_R = \frac{\kappa(a'_{L0} + \kappa_{GL,A})}{\kappa_R(1 + \kappa_{GL,A} + \kappa)} \qquad (17.24)$$

DISPERSION MODEL

$$E_R = \frac{\kappa}{\kappa_R(\kappa_{GL,A} + \kappa)} \left[\kappa_{GL,A} + \left(a'_{L0} - \frac{\kappa_{GL,A}}{\kappa_{GL,A} + \kappa} \right) \right] \left[\left(1 - \frac{Pe(\lambda_1 - \lambda_2)e^{Pe}}{\lambda_1^2 e^{\lambda_1} - \lambda_2^2 e^{\lambda_2}} \right) \right]$$

$$(17.25)$$

Definitions of the dimensionless quantities appearing in Equations 17.23–17.25 are included in Figure 17.4.

An indication of the effect of the dimensionless variables that represent mass transfer at the gas-liquid and solid-liquid interfaces ($\kappa_{GL,A}$, $\kappa_{SL,A}$), chemical reaction (κ_R), and axial diffusion (Pe) can be had from the simulation results of Goto et al. (1976) (see also Shah and Paraskos, 1975; Shah et al., 1976; Goto and Smith, 1978). These may be summarized as follows:

1. Backmixing has very little effect, and at high values of the groups that represent mass transfer at the two interfaces [$(\kappa_{GL,A} + \kappa_{SL,A} > 3)$], the predictions of the two extreme models are almost identical, and
2. Representing the group [$\kappa_{GL,A}/(\kappa_{GL,A} + \kappa)$] by ξ, the following conclusions may be drawn: for $a'_{L0} < \xi$, MFR performs better; for $a'_{L0} > \xi$, PFR performs better; and for $a'_{L0} = \xi$, the models predict identical results; this is a significant conclusion because PFR is always superior to MFR in two-phase reactors.

The fractional conversion of B can be predicted from

$$X_B = q\kappa_R E_R \tag{17.26}$$

where q is defined as $\nu_B[A]^*/[B]_{bi}$.

TYPES OF THREE-PHASE REACTORS

Of the reactors listed in Table 17.1, the four most important are the mechanically agitated slurry reactor (MASR), the bubble-column slurry reactor (BCSR), the loop-recycle slurry reactor (LRSR), and the trickle-bed reactor (TBR). The first three, sketched in Figure 17.5, are slurry reactors, whereas the fourth is a fixed-bed reactor. The features most relevant to a preliminary design of these reactors in organic synthesis and technology are briefly described here.

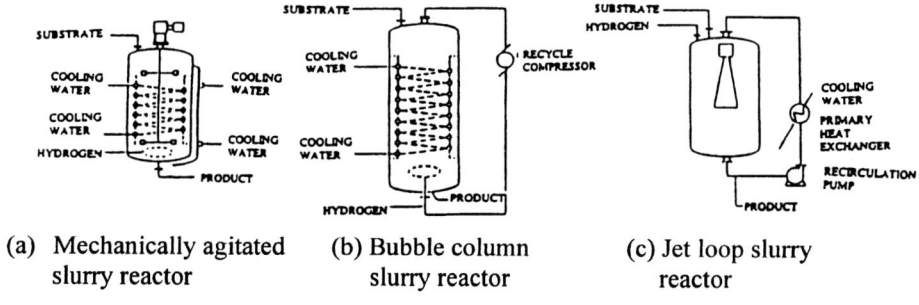

(a) Mechanically agitated slurry reactor
(b) Bubble column slurry reactor
(c) Jet loop slurry reactor

Figure 17.5 Schematic diagrams of different types of three-phase slurry reactors.

Mechanically Agitated Slurry Reactors

This class of reactors is the most commonly used in organic synthesis. A cascade of stirred reactors can also be used. For example, a cascade with a column configuration, as shown in Figure 17.6, is used in the production of the vitamin, carbol (Wiedeskehr, 1988).

Mass transfer

Extensive literature is available on the gas-liquid and solid-liquid mass transfer features of MASRs, for example, Levins and Glastonbury (1972a,b), Sano et al. (1974), Yagi and Yoshida (1975), Bern et al. (1976), Teshima and Ohashi (1977), Boon-Long et al. (1978), Patil and Sharma (1984). Mills and Chaudhari (1997) recommend the following correlations of Yagi and Yoshida (1975) (Equation 17.27) and Bern et al. (1976) (Equation 17.28) for gas-liquid mass transfer:

$$\frac{k'_L a_L d_s^2}{D} = 0.06 \left(\frac{d_s^2 N \rho_L}{\mu_L}\right)^{1.5} \left(\frac{d_s N^2}{g}\right)^{0.19} \left(\frac{\mu_L}{\rho_L D}\right)^{0.5} \left(\frac{\mu_L u_G}{\sigma_L}\right)^{0.6} \left(\frac{N d_s}{u_G}\right)^{0.32} \quad (17.27)$$

$$k'_L a_L = 1.1 \times 10^{-2} N^{1.16} d_s^{1.98} u_G^{0.32} V_L^{-0.52} \text{ (cgs)} \quad (17.28)$$

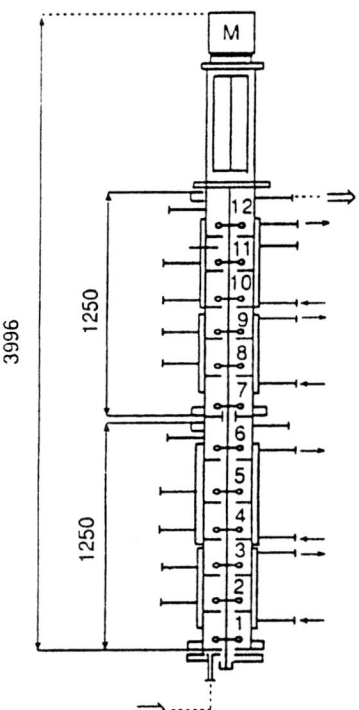

Figure 17.6 A cascade reaction column with multiple agitators for the synthesis of the vitamin "carbol" (after Wiedeskehr, 1988).

An important relatively recent achievement is the development of the so-called liquid entrained reactor (LER) for liquid phase methanol synthesis. To assess the performance of these reactors, much simultaneous work has been done on MARs, comparing the performance of the two types of reactors for this reaction. A good discussion of this can be found in a paper by Vijayaraghavan et a. (1993). An equation developed for the overall gas-liquid mass transfer coefficient (see Ko, 1987; Lee et al., 1988; Parameswaran et al., 1991) is

$$Sh = 1.054 \times 10^{-6} Re^{1.42} Sc^{0.5} G_f^{-0.38} \left(\frac{T}{T_0}\right)^{11.5} \tag{17.29}$$

where G_f is a flow group given by $\mu_L u_G / \sigma$. A rather disturbing feature of this correlation is the exponent of 11.15 for the dimensionless temperature term.

Liquid-liquid mass transfer depends on whether the transfer is from the continuous to the dispersed phase or vice versa. The liquid-liquid interfacial area a_{LL} can be estimated from $a_{LL} = 6h_L/d_0$ where h_L is the dispersed liquid phase holdup and d_0 is the average size of the dispersed droplets which can be determined from a correlation given by Okufi et al. (1990).

Solid-liquid mass transfer coefficients can be calculated from the following correlation proposed by Sano et al. (1974):

$$\frac{k'_{SL} d_p}{D S_F} = 2 + 0.4 \left(\frac{e d_p^4 \rho_L^3}{\mu_L^3}\right)^{1/4} \left(\frac{\mu_L}{\rho_L D}\right)^{1/3} \tag{17.30}$$

where $e = P/\rho_L V_L$ (P = power), energy supplied per unit mass of slurry by agitation, m²/s³.

This correlation is also recommended for liquid-liquid systems.

Minimum speed for complete suspension

It is necessary to ensure complete suspension of catalyst (see Joshi et al., 1982; Beenackers and Van Swaaij, 1993). The most important single parameter that influences complete suspension is the speed of agitation. This can be calculated from the early but still valid correlation of Zweitering (1958). In practice, the minimum agitation speed for uniform distribution of catalyst in the common loading range of $5 < w' < 30 \,\text{g}/100\,\text{g}$ liquid and particle size range of $10 < d_p < 150\,\mu\text{m}$ is 150–600 rpm.

Gas holdup

Knowledge of gas holdup is important because it is an indication of the gas residence time. A useful correlation is that of Yung et al. (1979):

$$h_G = 6.8 \times 10^{-3} \left(\frac{Q_G}{N d_s^3}\right)^{0.5} \left(\frac{\rho_L N^2 d_s^3}{\sigma_L}\right)^{0.65} \left(\frac{d_s}{d_T}\right)^{1.4} \tag{17.31}$$

Controlling regimes in an MASR

One can identify at least five major controlling regimes in an MASR: gas-liquid mass transfer, liquid-solid mass transfer of A, liquid-solid mass transfer of B, pore diffusion, and surface reaction. Table 17.4 lists the effects of different variables on the reaction for all of these regimes of control. These effects have been classified as major and minor effects.

Bubble-Column Slurry Reactors

The bubble column slurry reactor (BCSR) is used mainly for large volume productions, but in a few cases it is also used for specialty chemicals, particularly where a gaseous product is produced and corrosion problems prohibit use of agitated systems. The extent of backmixing can be controlled by using a draft tube. Such sparged reactors are particularly useful for chlorination, sulfonation, and phosgenation reactions.

Regimes of flow

The principal operating feature of a BCSR is that the catalyst is kept in suspension by the turbulence induced by the gas flow. Depending on the gas and liquid velocity ranges, BCSR can operate in different regimes of flow, as shown in Figure 17.7. In the homogeneous (bubbly) flow regime observed at very low velocities, gas bubbles of uniform size are equally distributed in both the axial and radial directions. At higher velocities, bubbles tend to coalesce, leading to nonuniform sizes. This regime of flow is termed heterogeneous churn-turbulent flow regime. Reactor scale-up is difficult in this regime, and hence should be avoided. As the velocity is further increased, the so-called heterogeneous slug flow regime is observed—especially in small diameter columns. This regime is also not of practical interest.

Mass transfer

A number of correlations have been proposed for both gas-liquid and solid-liquid mass transfer, for example, by Sano et al. (1974), Lemcoff and Jameson (1975), Hikita et al. (1981), Patil and Sharma (1983), Arters and Fan (1986, 1990), and Kikuchi et al. (1988, 1995). The recommended correlations are as follows:

GAS-LIQUID MASS TRANSFER COEFFICIENT (Akita and Yoshida 1973, 1974)

$$k'_L a_L = 0.6 D^{0.5} \left(\frac{\mu_L}{\rho_L}\right)^{-0.12} \left(\frac{\sigma_L}{\rho_L}\right)^{-0.62} d_T^{0.17} g^{0.93} h_G^{1.1} \quad \text{(cgs)} \qquad (17.32)$$

SOLID-LIQUID MASS TRANSFER COEFFICIENT

Equation 17.30 for MASR is especially valid for BCSR, but with $e = u_G g$. Kikuchi et al (1995) give a similar equation but with a slightly different definition

Table 17.4 Controlling regimes in slurry reactors: effect of pertinent variables on r_A (Doraiswamy and Sharma, 1984)

Controlling resistance	Variables whose influence is		
	Major	Minor	Negligible
Gas–liquid mass transfer, $k'_L a'_L$ (for A)	$[A]^*$, type of impeller and stirring speed (mechanically agitated contactor), gas velocity (sparged contactor)	Temperature	$[B]_b$, d_p, w
Liquid–solid mass transfer, $k'_{SL} a_p$ (for A)	d_p, w, $[A]^*$, stirring speed (mechanically agitated contactor, turbulent regime)	Temperature, gas velocity (sparged reactor, fine particles)	$[B]_b$
Liquid–solid mass transfer, $k'_{SL} a_p$ (for B)	d_p, w, $[B]_b$, stirring speed (mechanically agitated contactor, turbulent regime)	Temperature, gas velocity (sparged reactor, fine particles)	$[A]^*$
Surface reaction (pore diffusion negligible)	d_p, w, temperature, $[A]^*$, $[B]_b$; independent of $[A]$ for $m = 0$; independent of $[B]_b$ for $n = 0$		Type of impeller and stirring speed (mechanically agitated contactor), gas velocity (sparged contactor)
Surface reaction with pore diffusion	d_p, w, $[A]^*$, $[B]_b$, temperature	Pore structure	Type of impeller and stirring speed (mechanically agitated contactor), gas velocity (sparged contactor)

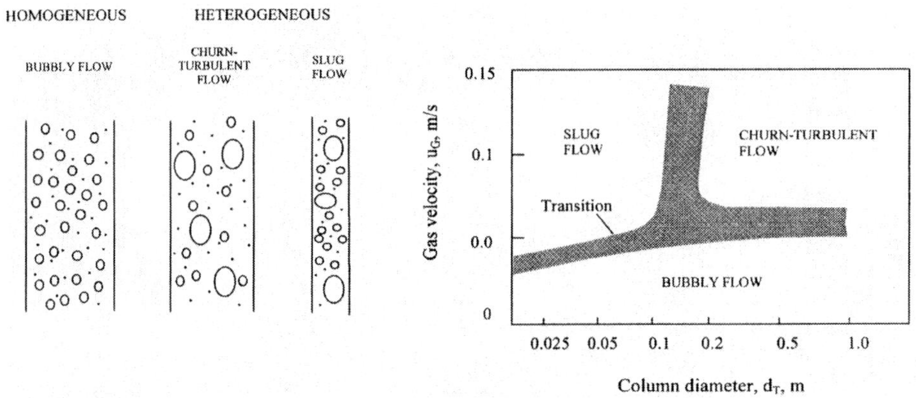

Figure 17.7 Flow regimes and transition gas velocities in a bubble column slurry reactor (adapted from Shah et al., 1982).

of the power group and with different values of the exponents of the various dimensionless groups.

Unlike in MASRs, where liquid mixing is always considered complete, in this case allowance must be made for partial mixing. Thus it may often be necessary to use the dispersion model given by Equation 17.25. The liquid-phase axial diffusion coefficient for estimating the Peclet number in this equation may be calculated from the correlations of Hikita and Kikukawa (1975) or Mangartz and Pilhofer (1981).

Minimum velocity for complete solids suspension

The following correlation proposed by Roy et al. (1964) for the maximum amount of solids that can be kept in complete suspension per unit volume of slurry (w_{max}) may be used:

$$\frac{w_{max}}{\rho_L} = 6.8 \times 10^{-4} \frac{C_\mu d_T u_G \rho_G}{\mu_G} \left(\frac{\sigma_L h_G}{u_G \mu_L}\right)^{-0.23} \left(\frac{h_G u_{TS}}{u_G}\right)^{-0.18} \omega_F^{-3} \qquad (17.33)$$

where C_μ is the viscosity correlation factor which may be estimated from $C_\mu = 0.23 - 0.179(\log \mu_L) + 0.103 \log \mu_L^2$, u_{TS} is the terminal velocity of the particles (see Roy et al., 1964, for equations for estimating u_{TS}), and ω_F is the wettability factor (which can usually be assumed to be one).

Gas holdup

Of the many correlations proposed (see Hikita et al., 1980a,b, for a review), those of Akita and Yoshida (1973) and Hikita et al. (1980b) are useful. One of

the first correlations is that of Akita and Yoshida (1973), which continues to be the one recommended:

$$\frac{h_G}{(1-h_G)^4} = 0.2 \left(\frac{gd_T^2 \rho_L}{\sigma_L}\right)^{1/8} \left(\frac{gd_T^3 \rho_L^2}{\mu_L^2}\right)^{1/12} \left(\frac{u_G}{\sqrt{gd_T}}\right) \qquad (17.34)$$

A more useful but slightly less accurate correlation is that of Hikita et al. (1980b).

$$h_G = 0.672 g^{-0.131} u_G^{0.578} \rho_G^{0.062} \rho_L^{0.069} \mu_G^{0.107} u_L^{-0.053} \sigma_T^{-0.185} \quad \text{(cgs)} \qquad (17.35)$$

EXAMPLE 17.2

Reactor choice for hydrogenation of aniline

Aniline can be hydrogenated to cyclohexylamine in the presence of Raney nickel according to the reaction

$$3H_2(g) + C_6H_5NH_2(l) \rightarrow C_6H_{11}NH_2(l)$$
$$\text{(A)} \qquad \text{(B)} \qquad \qquad \text{(R)}$$

It is desired to choose between a MASR and a BCSR for carrying out the reaction on a large scale. To enable rational choice, calculate the conversion of aniline in a 50-cm bench-scale reactor with a liquid height of 50 cm. In addition, agitation speed is also a variable for MASR. The following data of Ramachandran and Chaudhari (1983) may be used:

$k_{wl} = 51.49 \text{ cm}^3/\text{g s}$
$\rho_G = 6.05 \times 10^{-4} \text{ g/cm}^3$, $\rho_L = 1 \text{ g/cm}^3$, $\rho_C = 1.75 \text{ g/cm}^3$
$\mu_G = 1.1 \times 10^{-4} \text{ g/cm s}$, $\mu_L = 5 \times 10^{-3} \text{ g/cm s}$
$\sigma_T = 30 \text{ dyne/cm}^2$, $[A]^* = 4.46 \times 10^{-6} \text{ mol/cm}^3$
$D = 1.16 \times 10^{-4} \text{ cm}^2/\text{s}$, $[B]_{bi} = 1 \times 10^{-3} \text{ mol/cm}^3$
$D_{eA} = 1.93 \times 10^{-5} \text{ cm}^2/\text{s}$, $d_p = 0.01 \text{ cm}$, $d_s = 25 \text{ cm}$
$u_G = 0.5 \text{ cm/s}$, $w = 3.0 \times 10^{-2} \text{ g/cm}^3$, $a'_{LC} = 0$,
$u_L = 0.2 \text{ cm/s}$

SOLUTION

The various dimensionless groups to be considered are (see Figure 17.4):

$$\kappa_{GL,A} = \frac{k'_{GL,A} a_L L}{u_L}, \quad \kappa_{SL,A} = \frac{k'_{SL,A} a_p L}{u_L}$$

$$\kappa_R = \frac{w k_{wl} L}{u_L}, \quad \kappa = \frac{\varepsilon \kappa_R \kappa_{SL,A}}{\kappa_{SL,A} + \varepsilon \kappa_R} \qquad (E17.2.1)$$

The conversion can be calculated from the relationship

$$X_B = q \kappa_R E_R \qquad (E17.2.2)$$

where

$$q = \frac{\nu_B [A]^*}{[B]_{bi}} \qquad (E17.2.3)$$

Calculations for MASR

GAS-LIQUID MASS TRANSFER COEFFICIENT

Liquid volume in the reactor, $V_L = 98000 \text{ cm}^3$. From Equation 17.28

$$k'_L a_L = 1.1 \times 10^{-2} \times 10^{1.16} \times 25^{1.98} \times 0.50^{0.32} \times 98000^{-0.52}$$
$$= 0.187 \text{ s}^{-1}$$

LIQUID-SOLID MASS TRANSFER COEFFICIENT

Assume $e = 3.033 \times 10^5 \text{ erg/s}$ (calculated from the correlations given by Ramachandran and Chaudhari, 1983). Then, from Equation 17.30, $k'_{SL,A} = 0.226 \text{ cm/s}$, and because

$$a_p = \frac{6w}{\rho_c d_p} = \frac{6 \times 3 \times 10^{-2}}{1.75 \times 0.01} = 10.28 \text{ cm}^{-1}$$

$$k'_{SL,A} a_p = 2.324 \text{ s}^{-1}$$

DIMENSIONLESS MASS TRANSFER COEFFICIENT AND REACTION RATE PARAMETERS

$$\kappa_{GL,A} = 467.5, \quad \kappa_{SL,A} = 5810, \quad \kappa_R = 3862$$

The catalyst effectiveness factor ε is calculated from

$$\varepsilon = \frac{1}{\phi}\left(\coth 3\phi - \frac{1}{3\phi}\right)$$

where the Thiele modulus ϕ for a first-order reaction is obtained from

$$\phi = \frac{d_p}{6}\left(\frac{k_1}{D_{eA}}\right)^{0.5} = \frac{0.01}{6}\left(\frac{51.49 \times 1.75}{1.93 \times 10^{-5}}\right)^{0.5} = 3.6$$

giving $\varepsilon = 0.252$, and $\kappa = 833.5$.

CONVERSION

Overall reactor efficiency:

Assuming full mixing

$$E_R = \frac{833.5(0 + 467.5)}{3861.75(1 + 467.5 + 833.5)} = 0.0775$$

Assuming plug flow

$$E_R = \frac{833.5}{3861.75(467.5 + 833.5)}$$
$$\times \left[467.5 + \left(0 - \frac{467.5}{467.5 + 833.5}\right)\{1 - \exp[-(467.5 + 833.5)]\}\right]$$
$$= 0.0775$$

Both fully mixed and plug-flow calculations result in the same overall reactor efficiency.
Concentration ratio q:

$$q = \frac{1/3 \times 4.46 \times 10^{-6}}{1 \times 10^{-3}} = 1.487 \times 10^{-3}$$

Conversion of aniline (from Equation 17.26):

$$X_B = 1.487 \times 10^{-3} \times 3861.75 \times 0.0775 = 0.445, \text{ that is, } 44.5\%.$$

Calculations for BCSR

Calculate the gas-liquid mass transfer coefficient from Equation 17.32 and the solid-liquid mass transfer coefficient from Equation 17.30 with $e = u_G g$.

Knowledge of the gas holdup is needed for calculating the gas-liquid mass transfer coefficient. This can be calculated from Equation 17.34 or 17.35.

The following sample calculations are given for the conditions: gas velocity $u_G = 0.5$ cm/s, liquid velocity $u_L = 0.2$ cm/s, and catalyst loading $w = 0.05$ g/cm^3.

GAS HOLD-UP

From Equation 17.35

$$h_G = 0.672 \times 981^{-0.131} \times 0.5^{0.578} \times (6.05 \times 10^{-4})^{0.062} \times 1^{0.069}$$
$$\times (1.1 \times 10^{-4})^{0.107} \times (5 \times 10^{-3})^{-0.053} \times 30^{-0.185} = 0.0307$$

GAS-LIQUID MASS TRANSFER COEFFICIENT

From Equation 17.32

$$k'_L a_L = 0.6 \times (1.16 \times 10^{-4})^{0.5} \times \left(\frac{5 \times 10^{-3}}{1}\right)^{-0.12} \times \left(\frac{30}{1}\right)^{-0.62}$$
$$\times 50^{0.17} \times 981^{0.93} \times 0.0307^{1.1}$$
$$= 0.0378 \text{ s}^{-1}$$

LIQUID-SOLID MASS TRANSFER COEFFICIENT

$e = 0.5 \times 981 = 490.5 \text{ cm}^2/\text{s}^3$, and $a_P = 10.28 \text{ cm}^{-1}$. Then, from Equation 17.30, $k'_{SL,A} a_p = 0.657 \text{ s}^{-1}$.

DIMENSIONLESS MASS TRANSFER COEFFICIENT AND REACTION RATE PARAMETERS

These are calculated in the same manner as for MASR. The results are as follows:

$$\kappa_{GL,A} = 94.5, \quad \kappa_{SL,A} = 1642.3, \quad \kappa_R = 3861.75, \quad \kappa = 611$$

CONVERSION

Calculations are identical to those for MASR.
Overall reactor efficiency:
   Assuming fully mixed condition, $E_R = 0.0212$
   Assuming plug flow condition, $E_R = 0.0212$
Concentration ratio: $q = 1.487 \times 10^{-3}$
Aniline conversion for both fully mixed and plug flow:

$X_B = 0.122$, that is, 12.2%.

Clearly, under the conditions given, MASR performs better than BCSR.

Loop Slurry Reactors

Types of loop reactors

There are essentially three types of loop slurry reactors: jet loop slurry reactor, gas lift loop slurry reactor, and propeller loop slurry reactor. Of these, the jet loop reactor is the most commonly used in organic synthesis and technology (Figure 17.5c) for the following reasons: higher heat and mass transfer rates, rapid dissipation of heat leading to precise temperature control, controlled residence time in the mixing nozzle where most of the reaction occurs, and easier scale-up to commercial size for mass transfer controlled reactions (see Chaudhari and Shah, 1986). In actual operation, the liquid phase that contains the catalyst particles is injected at very high velocity (>20 m/s) through a nozzle from either the top or bottom of the reactor, and hence an important factor in the scale-up of this reactor is the nozzle configuration. Typically, a loop slurry reactor generates an interfacial area of 2000–3000 m^2/m^3 with bubble diameters in the range of 3–6 mm.

Many catalytic hydrogenations traditionally carried out in agitated slurry reactors have been switched to jet loop reactors (Leuteritz, 1973). Table 17.5 lists typical products made in loop slurry reactors along with remarks on the corresponding performance of mechanically agitated reactors.

Mass transfer

In a highly turbulent system like the jet loop reactor, the value of k'_L is not strongly dependent on hydrodynamics. Indeed, it has been found (Blenke and Hirner, 1974) that k'_L has an average value of 4.6×10^{-2} cm/s over a wide range of gas velocities (0.4 to 6.5 cm/s). The interfacial area can be calculated from the following correlation (Mills and Chaudhari, 1997):

$$a_L = 5.4 \times 10^3 u_G^{0.4} \left(\frac{P}{V_L}\right)^{0.66} \tag{17.36}$$

Table 17.5 Comparison of loop slurry reactor performance with mechanically agitated reactor (MASR) performance under corresponding conditions (adapted from Leuteritz, 1973)

Hydrogenation of	Catalyst concentration relative to that in MASR	Hydrogenation time and operating conditions relative to those in MASR
2,5-Dichloronitrobenzene	No basis for comparison because of different catalysts	Hydrogenation time about one-sixth as long; milder conditions
3,4-Dichloronitrobenzene	10% less	Hydrogenation time twice as long
p-Chloronitrobenzene	No basis for comparison as type of catalyst changed	Shorter time and milder conditions
o-Chloronitrobenzene	25% less	Shorter time
p-Nitroaniline	25% less	Shoter time
p-Nitroxylene	Same	Shorter time
o-Nitroethylbenzene	Same	Shorter time
o-Nitroaniline	25% less	Shorter time
Bisphenol A	25% more	Longer time
o-Nitroanisole	30% less	Shorter time

Trickle-Bed Reactors (with Cocurrent Down Flow)

As indicated in Table 17.1, there are essentially three main classes of three-phase fixed-bed catalytic reactors. The class of reactors characterized by cocurrent downflow of gas and liquid is called the trickle bed reactor (TBR). We shall be concerned here only with these reactors, for they are more commonly used in organic technology than the other two variations.

Regimes of flow

These reactors are in many ways similar to packed-bed absorption columns but operate at much lower gas and liquid velocities, usually 0.1–2.0 and 10–300 cm/s, respectively. Depending on the flow rates of the individual phases, four regimes of operation can be identified (Charpentier and Favier, 1975): *trickle (film) flow*, where the liquid flows at a low rate as a laminar film over the packing in the continuous gas phase; *pulse flow*, corresponding to higher gas and liquid rates, with alternate gas-rich and liquid-rich elements passing through the column; *bubble flow*, corresponding to even higher rates, where the liquid phase becomes the continuous phase and the gas flows through it in the form of bubbles; and *spray flow*, where the liquid is dispersed in the form of fine droplets in the continuous gas phase, but a part of it continues to flow as a film over the packing.

Several flow maps have been proposed in which the various regimes are demarcated, for example, those by Sato et al., 1973b; Charpentier and Favier,

1975; Specchia and Baldi, 1977; Chou et al., 1977; Fukushima and Kusaka, 1977; Talmor, 1977). These are useful in understanding TBR performance.

Mass transfer

The recommended correlations for gas-liquid and solid-liquid mass transfer for the different regimes are summarized in Table 17.6. The effect of liquid backmixing is usually unimportant.

Table 17.6 Correlations for trickle-bed reactors

Flow regime	Correlation[a]	Reference
Gas-liquid mass transfer		
Trickle flow	$\dfrac{k'_L a_L d_p^2}{D(1-h_L/f_B)} = 2\left(\dfrac{S_e}{d_p^2}\right)^{0.2} Re_L^{0.73} Re_G^{0.2} \left(\dfrac{\mu_L}{\rho_L D}\right)\left(\dfrac{d_p}{d_T}\right)^{0.2}$	Fukushima and Kusaka (1977)
Pulse flow	$\dfrac{k'_L a_L d_p^2}{D(1-h_L/f_B)} = 0.11 Re_L Re_G^{0.4} \left(\dfrac{\mu_L}{\rho_L D}\right)^{0.5} \left(\dfrac{d_p}{d_T}\right)^{-0.3}$	Fukushima and Kusaka (1977)
Gas-liquid interfacial area		
Trickle flow	$\dfrac{a_L d_p}{1-h_L/f_B} = 3.4 \times 10^{-1} \left(\dfrac{S_e}{d_p^2}\right)^{-1.0} Re_L^{0.4}$	Fukushima and Kusaka (1977)
Pulse flow	$\dfrac{a_L d_p}{1-h_L/f_B} = 5.9 \times 10^{-2} \left(\dfrac{S_e}{d_p^2}\right)^{-0.3} Re_L^{0.67} Re_G^{0.2}$	Fukushima and Kusaka (1977)
Dispersed bubble flow	$\dfrac{a_L d_p}{1-h_L/f_B} = 4.5 \times 10^{-4} \left(\dfrac{S_e}{d_p^2}\right)^{-0.9} Re_L^{1.8}$	Fukushima and Kusaka (1977)
Solid-liquid mass transfer		
Trickle flow	$\dfrac{k'_{SL} d_p}{D} \dfrac{a_w}{a_p} = 0.815 Re_L^{0.82} Sc_L^{0.33}$	Satterfield et al. (1978)
	$\dfrac{Sh^*}{Sc^{0.33}} = (1.19 + 0.0072 Re_G^*) Re_l^{*0.494} Ga^{-0.22}$	Burghardt et al. (1995)
Pulse flow	$\dfrac{k'_{SL} d_p}{D} \dfrac{a_w}{a_p} = 0.334 \left(\dfrac{e_L \rho_L^2 d_p^4}{h_L \mu_L^3}\right)^{0.2} Sc_L^{0.3}$	Satterfield et al. (1978)
	$\dfrac{Sh^*}{Sc^{0.33}} = 2.269 Re_L^{*0.494} Re_G^{0.178} Ga^{-0.278}$	Burghardt et al. (1995)
Liquid holdup		
Static holdup	$\dfrac{h_{Ls}}{f_B} = 3.7 \times 10^{-2} Bd^n$ where $n = -0.07$ for $Bd < 1$ $= -0.65$ for $Bd > 1$	Mersmann (1972)
Dynamic holdup	$\dfrac{h_{Ld}}{f_B} = 3.86 Re^{0.545} Ga^{-0.42} \left(\dfrac{a_p d_p}{f_B}\right)^{0.65}$, $3 > Re_L < 470$	Specchia and Baldi (1977)
	$h_{Ld} = 1.125 \left(\dfrac{a_b d_p}{f_B}\right)^{0.3} (Ga^*)^{-0.5} (Re_G + 2.28)^{-0.1}$ $\tanh[48.9(Ga^*)^{-1.16} Re_L^{0.41}]$	Burghardt et al. (1995)

[a] $Re = d_p u \rho / \mu$; $Sc = \mu/\rho D$; $Sh = k'_{SL} d_p / D$; Ga (Galileo number) $= d_p^3 \rho_L (g\rho_L + \Delta P')/\mu_L^2$; Ga^* (modified Ga) $= d_p (g\rho_L^2/\mu_L^2)^{1/3}$; Re^* (modified Re) $= G/a_B \mu$; Sh^* (modified Sh) $= (k'_{LS}/D)(\mu_L^2/g\rho_L)^{1/3}$; Bd (Bond number $= \rho_L g / \sigma_L a_p^2$; $e_L = u_L \Delta P'$ (kg/m s^3).

Controlling regimes in trickle-bed reactors

In addition to the flow regimes characteristic of trickle-bed reactors, there are also the usual controlling regimes, as described previously for MASRs. We summarize in Table 17.7 the effects of different variables on trickle-bed reactor performance in these regimes.

Other Reactors: A Brief Overview

Multiphase reaction engineering has developed into a very active field of research, with international symposia held at frequent intervals. We have only considered a few common reactor types currently in use. A few others of potential importance in relatively large volume intermediates production are the gas-liquid-solid fluidized beds, liquid entrained reactors, and rotating drum reactors.

Gas-liquid-solid fluidized bed reactors

The performance of these reactors is greatly influenced by (1) axial, radial, and global distribution of liquid and solids in the bed and (2) changes in bubble size, velocity, breakup, and coalescence. The second set of factors leads to an enhancement in the rates of heat and mass transfer. This happens because each particle (assumed to be spherical) is surrounded by a gas-liquid mixture of low pseudohomogheneous density. Consequently, the particle terminal velocity increases, which in turn has a positive effect on the mass transfer coefficient. A number of papers have been published (e.g., Arters and Fan, 1986, 1990; Fan, 1989; Nikov and Delmas, 1992; Boskovic et al., 1994; Kikuchi et al., 1995) on mass transfer in these reactors.

Liquid entrained reactors

These reactors were developed essentially to increase the productivity of methanol per unit mass of catalyst in the liquid phase process for its manufacture. Their main feature is that the catalyst particles in powder form are uniformly suspended in the liquid and the slurry is continuously circulated by a pump through a PFR. The energy cost for pumping the slurry is much less than for agitation in an MASR. The slurry concentration is about 50% by mass as against only 30% in an MASR (see Lee et al., 1988; Parameswaran *et al.*, 1991). A Sherwood number correlation for overall gas-liquid mass transfer has been developed (Vijayaraghavan *et al.*, 1993) which contains a Froude group defined as $Fr = \mu_L u_G / \sigma_L$. This reactor is unlikely to be used in industrial organic synthesis for a long time, but it has the potential for use in relatively large scale intermediates production.

Rotating drum reactor

These reactors are simple to construct, provide good control of solids retention time, and consume less power than other reactors. Mass transfer correlations of the usual form, Sh as a function of Re and Sc, are available (e.g., Machon and Linek, 1974). In a more recent study, Fr has been used as a corresponding

Table 17.7 Controlling mechanisms in trickle-bed reactors (film flow regime): effect of pertinent variables on r_A (Doraiswamy and Sharma, 1984)

Controlling resistance	Variables whose influence is		
	Major	Minor	Negligible
Gas-liquid mass transfer, $k_L' a_L$	$[A]^*$, d_p, superficial liquid velocity	Temperature	$[B]_0$, superficial gas velocity (at lower values), replacement of active by inactive catalyst particles
Liquid-solid mass transfer, $k_{SL}' a_p$ (for A)	d_p, $[A]^*$, superficial liquid velocity	Temperature	$[B]_b$, superficial gas velocity (at lower values), replacement of active by inactive catalyst particles
Liquid-solid mass transfer of $k_{SL}' a_p$ (for B)	d_p, $[B]_b$, superficial liquid velocity	Temperature	$[A]^*$, superficial gas velocity[a]
Surface reaction (pore diffusion negligible)	d_p, $[A]^*$, $[B]_b$ temperature, replacement of active by inactive catalyst particles	Superficial liquid velocity (above certain minimum)	Superficial gas velocity[a]
Surface reaction with pore diffusion	d_p, $[A]^*$, $[B]_b$, temperature,[b] replacement of active by inactive catalyst particles	Superficial liquid velocity (above certain minimum)	Superficial gas velocity[a]

[a] It is assumed that this does not cause any change in the flow regime.
[b] The effect of temperature on r_A is less pronounced in this case than for surface reaction where pore diffusion is negligible.

parameter (Gray et al., 1993). it would be useful to explore the use of these reactors in the production of organic intermediates.

COLLECTION AND INTERPRETATION OF LABORATORY DATA FOR THREE-PHASE CATALYTIC REACTIONS

Experimental Methods

Of the different types of reactors that can be used, the stirred basket reactor is the most amenable to manipulation in terms of regimes (see Kenney and Sedriks, 1972). Such a reactor for gas-solid (catalytic) reactions was considered in Chapter 7 (Figure 7.4). Typically, to operate under chemical control conditions, say in a fully baffled 15-cm diameter reactor provided with an 8-cm turbine agitator, the speed of agitation should be in the range of 1000–5000 rpm (corresponding to an impeller tip speed of 24,000–120,000 cm/min).

Effect of Temperature

The overall effect of temperature is the resultant of the increasing effect on mass transfer and the decreasing effect on solubility (see Chapter 14). However, this general conclusion should be viewed with caution because there are exceptions. An important one is hydrogen, whose solubility increases with temperature (e.g., the solubility in soybean oil increases by 60% as the temperature is raised from 20 to 100 °C (Doraiswamy and Sharma, 1984). Thus the overall effect in hydrogenation, perhaps the most frequent "user" of the slurry reactor, will be that the rate will increase significantly with an increase in temperature.

Interpretation of Data

Consider a simple first-order reaction. The total resistance to reaction can be expressed as the sum of resistances as in Equation 17.3, which then can be combined with Equation 17.6 for a first-order reaction ($m = 1$) to give

$$-r_A = [A]^* \left(\frac{1}{k'_{GL} a_L} + \frac{1}{k'_{SL} a_p} + \frac{1}{\varepsilon w k_{w1}} \right)^{-1} \quad (17.37)$$

If it is assumed that the catalyst particles and the gas bubbles are spherical, the interfacial areas a_L and a_p can be expressed as

$$a_L = \frac{6h_G}{d_b(1 - h_G)}, \quad a_p = \frac{6w}{\rho_c d_p} \quad (17.38)$$

Substituting these in Equation (17.37) gives

$$\frac{[A]^*}{-r_A} = \left(\frac{d_b(1 - h_G)}{6h_G k'_{GL}} \right) + \frac{\rho_c d_p}{6w} \left(\frac{1}{k'_{SL}} + \frac{6}{\varepsilon \rho_c d_p k_{w1}} \right) \quad (17.39)$$

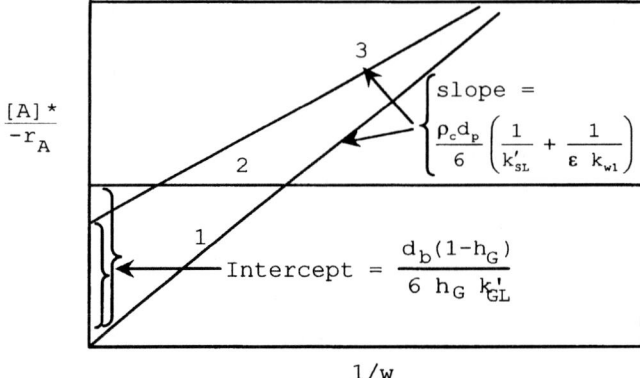

Figure 17.8 Interpretation of data for linear kinetics.

Thus a plot of $[A]^*/(-r_A)$ versus $1/w$ gives straight lines with slope and intercept as shown in lines 1, 2, and 3 of Figure 17.8. An intercept of zero, as in line 1, indicates negligible gas-liquid mass transfer resistance, a slope of zero (line 2) indicates gas-liquid mass transfer control, and finite values of slope and intercept (line 3) indicate combined control by all three resistances, gas-liquid, solid-liquid, and reaction.

By operating under conditions of high agitation where gas-liquid mass transfer effects are absent, line 1 is produced. The rate constant can be obtained from the slope and estimated value of k'_{SL}. Alternatively, if the rate constant is known, the correct value of k'_{SL} for the system at hand can be extracted from the slope. Line 2 is produced at very low agitation, and k'_{GL} may be obtained from the intercept.

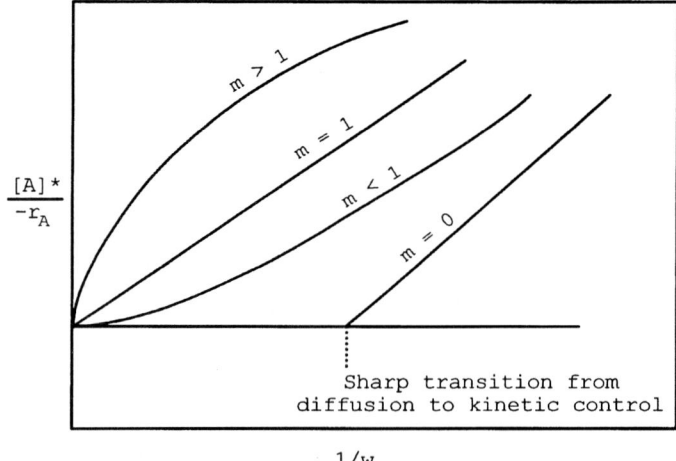

Figure 17.9 Plots of $[A]^*/(-r_A)$ for different orders (nonlinear kinetics).

If the kinetics is nonlinear, the simple additivity of the resistances principle used in this analysis will not apply. As a result, the $[A]^*/(-r_A)$ versus $1/w$ plots will no longer be straight as in Figure 17.8, but will be curves, as sketched in Figure 17.9, for different orders. Even in these cases it is possible to obtain $k'_L a_L$ by extrapolating the curves to $1/w = 0$, but a large number of experimental points very close to zero would be required. The value thus obtained, though not very accurate, should be adequate for most calculations.

EXAMPLES OF THREE-PHASE CATALYTIC REACTIONS IN ORGANIC SYNTHESIS AND TECHNOLOGY

Three-phase catalytic reactions are among the most common in organic synthesis and technology. They find application in such diverse areas as polymers, organic intermediates ranging from large to small volume chemicals, perfumery chemicals, pesticides, dyes, drugs, pharmaceuticals, vitamins, and specialty chemicals in general. Selected examples of these wide ranging products are listed in Table 17.8. Because their application is most common in the pharmaceutical industry, we list separately in Table 17.9 several examples of three-phase catalytic reactions used in the manufacture of drugs, pharmaceuticals, and vitamins, along with the type of reactor employed in each case, usually MASR, BCSR, or JLSR.

THREE-PHASE NONCATALYTIC REACTIONS

Here, we consider reaction between solid $[B]$ suspended in a liquid and gas A bubbled through it. If solid B is slightly soluble in the liquid, reaction occurs between dissolved A and B in the liquid phase. If B is insoluble, dissolved A diffuses and reacts with B within the solid. The situation represented by the first case is similar to gas-liquid reactions but with provision for solid dissolution. The second case is similar to that considered in Chapter 15 for reaction between an insoluble solid and a liquid.

Solid Slightly Soluble

This is an interesting situation in organic technology and synthesis. Examples are the dissolution of acetylene in an aqueous slurry of CuCl as a step in the manufacture of propylene oxide and the alkylation of naphthalene with ethylene in a liquid medium in the presence of BF_3-phosphoric acid as a dissolved catalyst. Note that the presence of the dissolved catalyst does not in any way alter the physical features of the system.

Table 17.8 Examples of gas-liquid-solid catalytic reactions in organic synthesis (from Mills and Chaudhari, 1997)

Reaction	Catalyst	Product (Application)	Reference
1. Isomerization of propylene oxide	Li_3PO_4 suspended in diphenyl	Allyl alcohol, an intermediate	Sittig (1967), Sukhdev (1972)
2. Hydrogenation of furfurol and its derivatives	Cu-Cr oxide	Important intermediates	Heidegger et al. (1971)
3. Amination of monoethanolamine	Undisclosed	Ethylenediamine, versatile chemical	Kohn (1978)
4. Hydrogenation of glucose	Raney Ni	Sorbitol, intermediate in pharmaceuticals	Brahme and Doraiswamy (1976)
5. Hydrogenation of o-, m-, and p-nitro chlorobenzenes	Pt/C-sulfided	Chloroanilines, intermediates in pharmaceuticals and dyes	Kosak (1980)
6. Hydrogenation of o-nitroanisole	Pd/C	o-Anisidine, intermediate in dyes and fine chemicals	Chaudhari et al. (1984)
7. Hydrogenation of butynediol	Pd-Zn/$CaCO_3$	cis-Butenediol, intermediate for vitamin A and Endosulfan, a pesticide	Chaudhari et al. (1985)
8. Hydrogenation of p-isobutyl acetophenone	Pd/C	p-Isobutyl phenyl ethanol, intermediate for ibuprofen, a pharmaceutical	Elango et al. (1990)
9. Hydrogenation of 2,4-dinitrotoluene	Pd/Alumina or Raney Ni	Toluenediamine, intermediate for TDI, used in fine chemicals	Westerterp et al. (1992)
10. Oxidation of isobutylene glycol	Pt/C	α-Hydroxybutyric acid, an intermediate	*Chemtech* (1975)

11.	Air oxidation of ethanol	Pd-based catalyst	Acetic acid, a bulk chemical	Klassen and Kirk (1955), Hsu and Ruether (1978)
12.	Hydrogenation of adiponitrile	Raney Ni	Hexamethylenediamine (HMDA), intermediate for Nylon 6,6; also for specialty chemicals	Mathieu et al. (1992)
13.	Hydrogenation of 1,5,9-cyclododecatriene	Pd/Al$_2$O$_3$	Cyclododecene, intermediate for 12-lauroactam, a pharmaceutical	Stuber et al. (1995) Benaissa et al. (1996)
14.	Hydrogenation of cinnamaldehyde	Pt-Co/C or Pt-Ru/C	Cinnamyl alcohol used in fine chemicals, perfumery	Fouilloux (1988)
15.	Hydrogenation of 3-hydroxypropanal	Ni-support	1,3 Propanediol, used in fine chemicals	Valerius et al. (1996)
16.	Hydrodesulfurization of o-aminobenzyl sulfides, 2-methyl (trimethyl-6-tri-fluoromethylaniline)	Co-Molybdate on Al$_2$O$_3$	2-Methyl-6-trifuloro-methyl aniline (MTMA), a preemergent herbicide intermediate	Tremont et al. (1988)
17.	Ethnylation of formaldehyde	Cu-acetylide-Bi-silica gel	Butynediol, intermediate in fine chemicals and pharmaceuticals	Kale et al. (1981)
18.	Oxidation of glucose	Pd-Bi/C	Gluconic acid, used in foods, detergents, and pharmaceuticals	Nikov and Paev (1995)

Table 17.9 Examples of three-phase reactions in the pharmaceutical industry

Reaction	Catalyst	Application	Reactor type
Hydrogenation of glucose to sorbitol	Raney Ni or supported Ni	Vitamin C	MASR or BCSR
Hydrogenation of butynediol to 2-butene-1,4 diol	Cu-Ni/Silica or Pd/CaCO$_3$	Vitamin B$_6$, Vitamin A	MASR or JLSR
Hydrogenation of nitrobenzene to p-aminophenol	Pt/C or PtO with aq. H$_2$SO$_4$	Paracetomol	MASR
Hydrogenation of m-nitrochlorobenzene	Sulfided Pt/C	Chloramphenicol	MASR or JLSR
Hydrogenation of p-isobutyl acetophenone to p-isobutyl phenyl ethyl alcohol	Pd/Al$_2$O$_3$ or Ni/Al$_2$O$_3$	Ibuprofen	MASR or JLSR
Hydrogenation of fluoromethyl-acetylene derivative of methyl phenyl sulfone	Pd/CaCO$_3$	Florfenicol	MASR
Hydrogenation of 2,8-dichloro-adenosine to adenosine	Pd/BaSO$_4$	Adenosine, a neuroregulator drug	MASR
Reductive amination of 4-chloroacetyl catechol to adrenaline	Pd/support	Adrenaline	MASR
Hydrogenation of 4-aminoacetylphenol to octopamine	Pd/support	Octopamine	MASR
Oxidation of phenol to catechol	Zeolite	Intermediate in drugs	MASR

Such reactions can typically be represented as

$$A(g) \rightarrow A(l)$$
$$B(s) \rightarrow B(l) \qquad [17.2]$$
$$A(l) + \nu_B B(l) \rightarrow \text{products}$$

Clearly, two liquid films are involved here, one surrounding the gas (which we designate F_1) and the other surrounding the solid (F_2). We consider two cases, one with negligible dissolution and the other with significant dissolution in F_2.

Negligible dissolution of solid in the gas-liquid film

This represents a relatively simple situation where no solid dissolves in the gas-liquid film F_1. The following steps are involved: diffusion of gas A through the film; dissolution of solid B; and diffusion and simultaneous reaction of B with dissolved A in film, F_1. The last step can occur in regime 3 (fast reaction, pseudo-order), regime 3–4 (fast reaction with depletion), or regime 4 (instantaneous

reaction). The first two cases are sketched in Figure 17.10a and the third in 17.10b. The conditions and rate equations for the three cases are summarized in Table 17.10.

Significant dissolution of solid in the gas-liquid film

If solid dissolution in the gas-liquid film is significant, the analysis becomes more involved. If the reaction occurs in the fast pseudo-first-order regime in film F_1, the bulk and film concentrations of B are the same. Thus dissolution of B in film F_1 will not make any difference to the reaction. On the other hand, an instantaneous reaction in film F_1 will be further enhanced by this dissolution. The ultimate result is that the reaction plane approaches the gas-liquid interface (i.e., $\lambda \to 0$). The condition and rate equation for this situation are included in Table 17.10.

Solid Insoluble

A typical example is the Kolbe–Schmitt carbonation of the sodium salt of β-naphthol in an inert liquid medium in the production of β-oxynaphthoic acid (commonly known as BON acid), a useful intermediate in the manufacture of dyes and other chemicals. Chlorination of wood pulp suspended in water is another example.

The only additional features in the analysis compared to that for solid-liquid reactions are the mass transfer resistances (if any) associated with the gas-liquid films. We demonstrate the procedure by considering the *Kolbe–Schmitt* reaction modeled by Phadtare and Doraiswamy (1965, 1969).

EXAMPLE 17.3

Kolbe–Schmitt carbonation of β-naphthol in an inert medium

β-Oxynaphthoic acid (BON acid), a useful organic intermediate used mainly in the manufacture of dyes, is produced by the *Kolbe–Schmitt* carbonation of sodium naphthenate prepared by reacting naphthalene with sodium hydroxide. The carbonation step is normally carried out in a dry atmosphere, but the possibility of conducting the reaction in a liquid medium such as kerosene has also been explored. The reaction scheme may be represented as

naphthol-OH + NaOH ⟶ naphthol-ONa + H₂O   [E17.3.1]

2 naphthol-ONa →(carbonation with CO₂) naphthol(ONa)(COONa) [Na salt of BON acid] + naphthol-ONa   [E17.3.2]

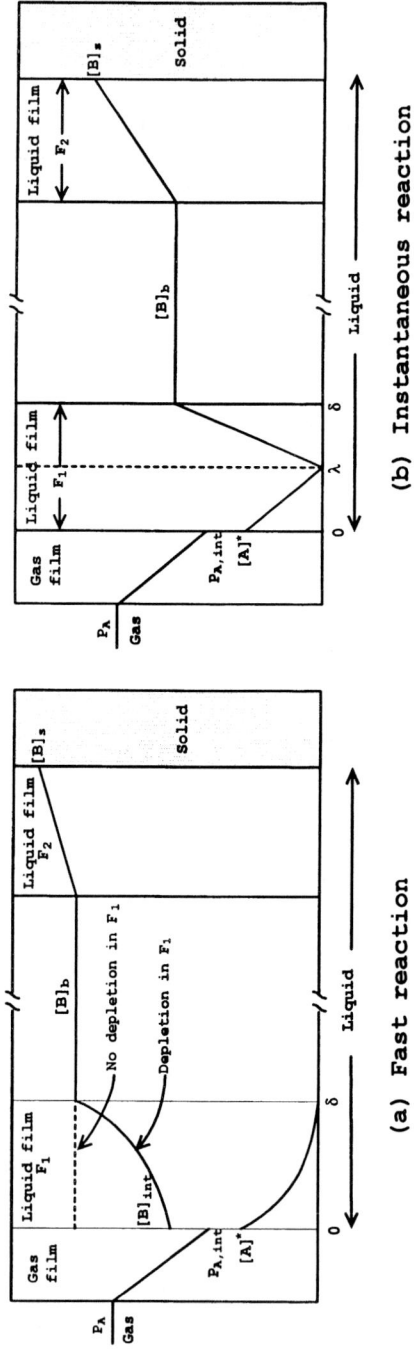

Figure 17.10 Gas-liquid-solid reaction: no solid dissolution in film (F_1) next to gas

Table 17.10 Controlling regimes in gas-liquid-solid noncatalytic reactions: $A(g) \to A(l)$, $B(s) \to B(l)$, $A(l) + \nu_B B(l) \to R$

Regime	Condition(s)	Rate equation[a]
1. Negligible solid dissolution in liquid film F_1 next to gas	$\dfrac{k'_{SL} a_p}{4 k'^2_L} \dfrac{D_A^2}{D_B} \ll 1$ $d_p \gg \delta$	
a. Fast reaction, no depletion		$-r_A = \dfrac{a''_L H_A p_A \sqrt{D_A k_2 [B]_b}}{1 + \dfrac{H_A \sqrt{D_A k_2 [B]_b}}{k'_G}}$ where $[B]_b = [B]_s - \dfrac{\nu_B(-r_A)}{k'_{SL} a_p}$
b. Fast reaction, depletion		$-r_A = \dfrac{a''_L H_A p_A \sqrt{D_A k_2 [B]_{int}}}{1 + \dfrac{H_A \sqrt{D_A k_2 [B]_{int}}}{k'_G}}$ where $[B]_{int} = [B]_s + \dfrac{\nu_B D_A}{D_B} H_A p_A$ $- \nu_B r_A \left[\left(\dfrac{H_A}{k'_{GP} a''_L} + \dfrac{1}{k'_L a''_L} \right) \dfrac{D_A}{D_B} + \dfrac{1}{k'_{SL} a_p} \right]$
c. Instantaneous reaction	$p_{A,int} \geq 0$ $[B]_b \geq 0$	$r_A = \dfrac{H_A p_A + \dfrac{D_B}{D_A} \dfrac{[B]_s}{\nu_B}}{\dfrac{H_A}{k'_{GP} a''_L} + \dfrac{1}{k'_L a''_L} + \dfrac{D_B}{D_A} \dfrac{1}{k'_{SL} a_p}}$ where $p_{A,int} = p_A - \dfrac{(-r_A)}{k'_{GP} a''_L}$ $[B]_b = [B]_s - \dfrac{\nu_B(-r_A)}{k'_{SL} a_p}$
d. Gas film control	$p_{A,int} \to 0$, $[B]_b > 0$	$-r_A = k'_{GP} a''_L p_A$
e. Solid dissolution control	$p_{A,int} > 0$, $[B]_b \to 0$	$-r_A = \dfrac{k'_{SL} a_p [B]_s}{\nu_B}$
2. Significant solid dissolution in liquid film F_1 next to gas	$\dfrac{k'_{SL} a_p}{4 k'^2_L} \dfrac{D_A^2}{D_B} \gg 1$ $d_p < \dfrac{\delta}{5}$	
a. Instantaneous reaction at gas-liquid	$\dfrac{\sqrt{D_A k_2 [B]_s}}{k'_L} \gg$ $\dfrac{[B]_s}{\nu_B [A]^*} \left(1 + \dfrac{k'_{SL} a_p}{4 k'^2_L} \dfrac{D_A^2}{D_B} \right)$	Solve following two equations for $-r'_A$ by assuming various value of λ and δ till they are satisfied: $-r'_A = \dfrac{D_A[A]^*}{\lambda} + k'_{SL} a_p [B]_s \dfrac{\lambda}{\nu_B}$ (1) $-r'_A = \dfrac{[B]_s}{\nu_B} \sqrt{D_B k'_{SL} a_p} \coth\left[(\delta - \lambda) \sqrt{\dfrac{k'_{SL} a_p}{D_B}}\right]$ $+ k'_{SL} a_p [B]_s \dfrac{\lambda}{\nu_B}$ (2)
b. Instantaneous reaction at gas-liquid interface ($\lambda \to 0$)	$\sqrt{\dfrac{k'_{SL} a_p}{D_B}} \delta > 5$ $\dfrac{[B]_s}{\nu_B} \gg [A]^*$	$-r'_A = \dfrac{[B]_s}{\nu_B} \sqrt{D_B k'_{SL} a_p}$ i.e., rate $\propto \sqrt{a_p} \propto \sqrt{(\text{particle loading}, w)}$

[a] a''_L = gas-liquid interfacial area, m^2/m^3 clear liquid; k_2 = second-order rate constant, m^3/mol s.

556   Fluid-Fluid and Fluid-Fluid-Solid

When carried out in a kerosene medium, the carbonation occurs in a three-phase system in which CO_2 first dissolves in the liquid and then diffuses to the solid naphthenate as a particulate suspension in the liquid. Because the solid is insoluble in kerosene, reaction occurs only in the solid phase. Using the data of Phadtare and Doraiswamy (1965, 1969), reproduced in Table 17.11, formulate a model to predict the conversion of sodium naphthenate to BON acid as a function of time.

SOLUTION

As briefly outlined in of Chapter 15, the shrinking core models are applicable to solids of very low porosity, and may be represented as shown in Figure 15.6.

The controlling resistance can be one of the following: (1) diffusion through the kerosene film surrounding the solid, (2) diffusion through an increasing layer of solid product (known as ash in gas-solid literature, but it is not an appropriate term for the solid product of a solid-liquid reaction); (3) chemical reaction on the surface of the receding reactant core; or (4) combinations of the three.

Table 17.11 Kolbe–Schmitt carbonation of β-naphthol: experimental time–conversion data

Reaction time t (min)	Total conversion, X_B		
	30 psig	70 psig	100 psig
	Temperature, 230 °C		
5	0.1798	0.1930	0.4733
15	0.4036	0.3610	0.3220
30	0.5544	0.5184	0.4500
40	0.6580	0.5920	0.5360
60	0.7180	0.6562	0.6220
120	0.6600	0.6890	0.6750
180	0.7360	0.7490	0.7390
	Temperature, 250 °C		
5	0.2308	0.1771	0.2500
15	0.3430	0.4200	0.4102
30	0.4998	0.5740	0.5616
40	0.5960	0.6800	0.6480
60	0.6750	0.702	0.7360
120	0.7260	0.7690	0.7530
180	0.8070	0.8044	0.8140
	Temperature, 270 °C		
5	0.3020	0.3720	0.3620
15	0.4580	0.4720	0.4580
30	0.6400	0.6572	0.6100
40	0.7332	0.7650	0.7140
60	0.7728	0.7990	0.7760
120	0.7200	0.7860	0.7840
180	0.7800	0.8260	0.8590

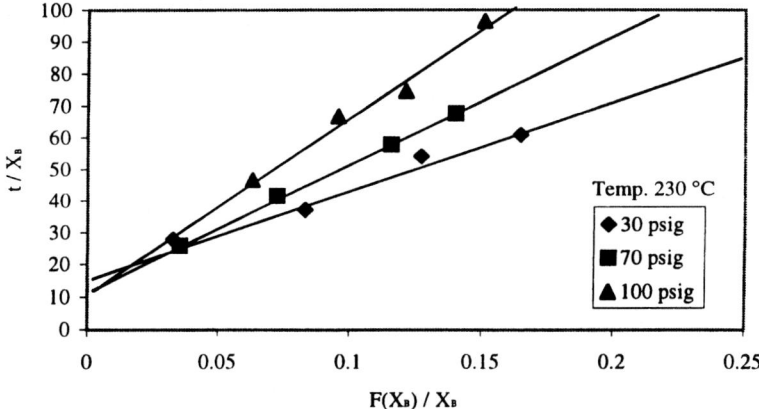

Figure 17.11 Validation of the mass transfer model (Equation E17.3.3) for the Kolbe–Schmitt carbonation of β-naphthol.

In solid-liquid reactions, the effective diffusivity of the fluid reactant through the solid product may be defined as

$$D_{eA} = \alpha D_{bA} \tag{E17.3.1}$$

where α is a factor that accounts for tortuosity and the density of particles in the product shell. Because D_{eA} cannot be estimated, unlike in gas-solid catalytic reactions (Chapter 7) or even noncatalytic reactions, we express it entirely as a function of D_b and use Equation E17.3.1 in the analysis. The constant α is an unknown quantity and can be determined only by curve fitting as part of a mass transfer group (as described here).

First, we examine the possibility of chemical control. This would require a linear dependence of the function $[1 - (1 - X_B)^{1/3}]$ on time. The actual plot (not shown) is, however, distinctly nonlinear. Hence we can conclude that the reaction is not kinetically controlled. Thus now we consider the case where the controlling resistances are diffusion through the kerosene film and the solid product. Because the particles are nearly spherical and the porosity of the solid reactant is very low, we can relate the solid conversion to the shrinking core radius by the equation

$$(1 - X_B) = \begin{pmatrix} \text{reaction of solid} \\ \text{unreacted} \end{pmatrix} = \frac{4/3(\pi r_c)^3}{4/3(\pi R^3)}$$

giving

$$X_B = 1 - \left(\frac{r_c}{R}\right)^3 \tag{E17.3.2}$$

Then, by writing rate equations for the diffusion of CO_2 through the kerosene film and the product shell, the following expression can be derived for the present reaction (Phadtare and Doraiswamy, 1965, 1969):

$$\frac{t}{X_B} = \frac{A_M}{3} + \frac{A_M k_M}{3} \left[\frac{F(X_B)}{X_B}\right] \tag{E17.3.3}$$

Table 17.12 Comparison of experimental values of model parameters with those from independent mass transfer studies

Temperature (°C)	Pressure, (psig)	A_m		k_m
		From kinetic data	From mass transfer studies	From kinetic data
230	30	54.48	55	14.48
	70	36.86	35	32.19
	100	35.97	13	45.74
250	30	17.15	—	82.76
	70	58.02	39	11.82
	100	6.38	20	183.33
270	30	34.67	—	18.99
	70	41.08	74	13.47
	100	10.71	26	84.03

where

$$A_M = \frac{2\rho_s R}{k_L'[C]_0}, \quad k_M = \frac{R}{\delta\alpha}, \quad F(X_B) = [1.5 - X_B - 1.5(1 - X_B)] \tag{E17.3.4}$$

Thus a plot of t/X_B versus $F(X_B)/X_B$ should give a straight line with

$$\text{slope} = \frac{A_M k_M}{3}$$

$$\text{intercept} = \frac{A_M}{3} \tag{E17.3.5}$$

Now Equation E17.3.3 can be tested by plotting the data of Table 17.11 in the manner mentioned. Plots at 230 °C and different pressures are shown in Figure 17.11. The values of A_M and k_M obtained from the slopes and intercepts of these plots for all of the temperatures are given in Table 17.12. The values of A_M were also determined by Phadtare and Doraiswamy (1969) from an independent series of mass transfer experiments and are included in the table. They are quite close to the kinetically determined constants at some temperatures and pressures and are of the same order of magnitude at others.

Notation to Part IV

A, A_1, A_2	Chemical species.
A_c	Cross sectional area of reactor, m^2.
A_M	Parameter defined by Equation E.17.3.4.
$[A]^*$	Equilibrium concentration of A in liquid, mol/m^3.
a_b	External area of bed per unit volume, m^2/m^3.
a_L (a_{GL} in specific cases)	Gas-liquid interfacial area, m^2/m^3.
a_{LL}	Liquid-liquid interfacial area, m^2/m^3.
a_P	External surface area of particle per unit volume, m^2/m^3.
a_w	Wetted external area of particles per unit volume of reactor, m^2/m^3.
a'	Ratio $[A]/[A]^*$.
a_L''	Gas-liquid interfacial area in gas-liquid-solid systems, m^2/m^3 of clear liquid.
B, B_1, B_2	Chemical species.
b	Liquid phase holdup.
C_μ	Viscosity correlation factor in Equation 17.33.
c	Scale-up index as defined in Table 16.8.
D	General notation for diffusivity, m^2/s.
D_b	Bulk diffusivity, m^2/s.
D_{bi}	Bulk diffusivity of i, m^2/s.
$D_{e\ell}$	Effective axial diffusivity, m^2/s.
D_{ei}	Effective diffusivity of i, m^2/s.
D_i	Diffusivity of species i, m^2/s.
$D_{i,1}, D_{i,2}$	Diffusivity of i in phases 1 and 2, m^2/s.
d	General symbol for diameter, m.
d_b	Bubble diameter, m.
d_0	Average diameter of dispersed droplets, m.

d_P	Particle diameter, m.
d_r	Reactor diameter, m.
d_s	Stirrer diameter, m.
d_T	Tank diameter, m.
E	Overall effectiveness factor defined by Equation 17.1.
E_R	Reactor efficiency given by Equation 17.21.
e	Power group defined by Equation 17.30, m^2/s^3; energy supplied per unit mass of slurry.
e_L	Energy dissipation term $(\mu_L \Delta p')$, kg/ms^3.
g	Acceleration due to gravity, m/s^2.
F_{GU}	Flow rate of inerts in the gas phase, mol/s.
F_{LU}	Flow rate of inerts in the liquid phase, mol/s.
F_N	Flow rate of inert (nitrogen), mol/s.
f_c	Fraction of reactor volume occupied by the continuous phase.
Ga, Ga^*	Galilio and modified Galilio numbers defined in Table 17.6.
G_f	Flow group defined by $\mu_L u_G / \sigma$.
g	Acceleration due to gravity, m/s^2.
H, H_r	Total reactor height, m.
H_A	Henry's law constant of A, mol/m^3 atm.
H'_A	Henry's law constant of A, atm m^3/mol.
$H'_{A,C}$	Henry's law constant of A, (mol A in gas/m^3)/(mol A in liquid/m^3).
h	Total holdup.
h_d	Height of dispersed phase, m.
h_G	Fractional gas phase holdup.
h_L	Fractional liquid phase holdup.
h_{Ld}	Dynamic liquid phase holdup.
h_{Ls}	Static liquid phase holdup.
$[i]_b$	Concentration of i in liquid bulk, mol/m^3.
$[i]_{b,c}$	Concentration of i in continuous liquid phase, mol/m^3.
$[i]_{b,d}$	Concentration of i in dispersed liquid phase, mol/m^3.
$[i]_{bf}$	Final concentration of i in liquid, mol/m^3.
$[i]_{bi}$	Initial concentration of i in liquid, mol/m^3.
$[i]_{b0}$	Inlet concentration of i, mol/m^3.
$[i]_{b,1}, [i]_{b,2}$	Concentrations of i in liquid bulk 1, 2.
$[i]_G$	Concentration of i in gas.
$[i]_{int}$	Concentration of i at interface, mol/m^3.
$[i]_{int,1}, [i]_{int,2}$	Concentrations of i at phase 1 and 2 interfaces.
$[i]_L$	Concentration of i in liquid.
$[i]_s$	Concentration of i at the catalyst surface, mol/m^3.
j	Chemical species.
$K_A, K_B, \ldots, K_i$	Adsorption equilibrium constants of $A, B, \ldots$, i, m^3/mol.
k	General notation for volume-based reaction rate constant, (m^3/mol)$^{n-1}$ 1/s (1/s for $n = 1$).
k_I, k_{II}	Rate constants for steps 1 and 2, m^3/mol s.
k_a, k_b	Parameters defined by Equation 14.36.
$\bar{k}_A, \bar{k}_B$	Overall mass transfer coefficient (gas-liquid + solid-liquid) of A, B given by Equations 17.3 and 17.5, m/s.
k_G ($k_{G,A}$ specifically for A)	Volume-based gas-side mass transfer coefficient, 1/s.

Notation to Part IV

k_{GL} ($k_{GL,A}$ specifically for A)	Total volumetric mass transfer coefficient of A at the gas-liquid interface, 1/s.
k_{GP} ($k_{GP,A}$ specifically for A)	Gas-side mass transfer coefficient in partial pressure units, mol/m^3 s atm.
k_i	General notation for rate constant for reaction of species, 1/s.
k_M	Parameter defined by Equation E17.3.4.
k_n	Rate constant for an nth order reaction, (m^3/mol)$^{n-1}$/s.
$k_{n,c}$	nth order rate constant for reaction in continuous phase, (m^3/mol)$^{n-1}$ 1/s.
k_p	Reaction rate constant in partial pressure units, mol/m^3 s atm.
k_{SL} ($k_{SL,A}$ specifically for A)	Volume-based solid-liquid mass transfer coefficient, 1/s.
k_w	General notation for weight-based rate constant, appropriate units.
k_{wm}, k_{w1}	Reaction rate constant for a slurry reaction of order m, (mol/kg cat s) (m^3/mol)^m (=m^3/kg cat s for k_{w1}).
k_{wmn}	Reaction rate constant for a slurry reaction of orders m, n in gas and liquid phase reactants, (mol/g cat s) (m^3/mol)$^{m+n}$.
$k_{1,c}$	First-order rate constant in continuous phase.
k_1, k_2, k_m, k_{mn}	Volume-based rate constants for reactions of order 1, 2, m, mn, (m^3/mol)$^{m-1}$ 1/s or (m^3/mol)$^{m+n-1}$ 1/s (=1/s for $m=1$).
$k_{2,2}$	Second-order rate constant in phase 2, m^3/mol s.
k'	General notation for surface-based reaction rate constant, (m^3/mol)^n (mol/m^2 s), (m/s for $n=1$).
k'_G	Surface-based gas-side mass transfer coefficient at the gas-liquid interface, m/s.
k'_{GL} ($k'_{GL,A}$ specifically for A)	Total surface-based mass transfer coefficient of A at the gas-liquid interface, m/s.
k'_{GP} ($k'_{GP,A}$ specifically for A)	Gas-side mass transfer coefficient in partial pressure units, mol/m^2 s atm.
k'_L ($k'_{L,A}$ specifically for A)	Surface-based liquid-side mass transfer coefficient at the gas-liquid interface, m/s.
k'_{LR}	Overall coefficient defined by Equation 14.10.
$k'_{L,12}$	Surface-based liquid mass transfer coefficient for transport from phase 1 to 2, m/s.
$k'_{L,21}$	Surface-based liquid mass transfer coefficient for transport from phase 2 to 1, m/s.
k'_{ov}	Overall surface-based mass transfer coefficient, m/s.
k'_{SL} ($k'_{SL,A}$ specifically for A)	Surface-based solid-liquid mass transfer coefficient at the solid-liquid interface, m/s.
$k'_1, k'_2, k'_m, k'_{mn}$	Surface-based rate constant for reactions of order 1, 2, m, mn, (mol/m^2 s)(m^3/mol)^m or (mol/m^2 s) (m^3/mol)$^{m+n}$ (= m/s for $m=1$).
L	Height of reactor, m.
L_{wire}	Length of wire, m.
M (or $\sqrt{M}$)	General parameter defined by Equation 14.7, which represents the extent of reaction in a film.
$\sqrt{M_{A-B}}, \sqrt{M_{B-A}}$	Groups defined in Table 15.1.
$\sqrt{M_i}$	Film-to-bulk reaction ratio for species i.
$\sqrt{M_{LH}}$	Film-to-bulk reaction ratio for Langmuir–Hinshelwood model.

$\sqrt{M'_{LH}}$	Generalized definition of $\sqrt{M_{LH}}$.
$\sqrt{M_m}$	Film-to-bulk reaction ratio for a pseudo-mth-order reaction.
$\sqrt{M_{mn}}$	Film-to-bulk reaction ratio for an m, nth-order reaction.
$\sqrt{M_1}$	Film-to-bulk reaction ratio for a pseudo-first-order reaction.
m, m'	Reaction order.
m_i	Distribution coefficient of i, $[i]_{org}/[i]_{aq}$.
m'_i	Distribution coefficient of i, $[i]_{aq}/[i]_{org}$.
N	Stirrer speed, rpm; parameter defined by Equation 16.28.
N_i	Number of moles of i.
N_m	Minimum agitation for complete suspension of solids, rpm.
n, n'	Reaction order; n is also any number.
P	Pressure, atm; power for gas-liquid agitation; total power $P_a V_r$.
P_a	Minimum specific energy required to achieve contact, kW/m^3.
Pe	Peclet number.
P_0	Power consumption for agitation of a gas-free liquid.
P_1, P_2	Chemical species.
$\Delta P'$	Pressure drop per unit bed height, dynes/cm^3 or kg/m^2 s^2.
p	Exponent of (D_A/D_B).
p_i	Partial pressure of component i, atm.
$p_{i,int}$	Partial pressure of component i at interface, atm.
Q'	Amount of dye reacting per unit area, m^3/m^2.
Q_A	Group defined as $[B]_b D_B / \nu_B [A]^* D_A$.
Q_G	Gas flow rate, m^3/s.
Q_L	Liquid flow rate, m^3/s.
$Q_{L,wier}$	Liquid flow rate, m^3/s m weir.
q	Parameter defined by Equation 16.29 and Equation 17.26.
q_{A-B}, q_{B-A}	Parameters defined in Table 15.1.
$q_{A_1}, q_{A_2}, q_{A_1 A_2}$	Groups defined by Equation 14.32.
R	Chemical species; radius, m.
R_{disk}	Radius of disk, m.
Re	Reynolds number, $2R_{disk}^2 (rps)/\mu'$.
r	Radial coordinate, m; general symbol for rate.
r_c	Shrinking core radius, m.
r_i	Volume-based reaction rate of i, mol/m^3 s.
r'_i	Surface-based reaction rate of i, mol/m^2 s.
r_{ia}	Actual rate of reaction of i, mol/m^3 s.
rps	Number of revolutions per second.
r_{wA}	Rate of reaction of A, mol/kg cat s.
S	Species S.
S_e	External surface area of particles, m^2 ($= 4\pi R^2$ for spherical particles).
S_F	Shape factor.
Sc	Schmidt number, $\mu/D\rho$.
Sh	Sherwood number, $D/d_p k'$, where k' is any mass transfer coefficient (m/s).
t	Time, s.
t_B	Batch time, s.
t_{SB}	Semibatch time, s.
u_b	Bubble velocity in free rise, m/s.

u_{es}	Effective slip velocity, m/s.
u_G	Gas velocity, m/s.
u_L	Liquid velocity, m/s.
u_{TS}	Terminal settling velocity of particles, m/s.
V	Volume, m^3.
V_c	Volume of continuous phase, m^3.
V_d	Volume of dispersed phase, m^3.
V_L	Volume of liquid, m^3.
V_r	Volume of reactor, m^3.
w	Weight of catalyst per unit volume of liquid or catalyst loading, kg/m^3.
w'	Solids loading, %.
x	Distance parameter, m.
X_B	Conversion of B.
$(y'_{Ab})_i$	Initial mole ratio of A in the liquid phase.
$(y'_{Ab})_f$	Final mole ratio of A in the liquid phase.
y'_i	Mole ratio of i to inerts in the gas phase.
z	Dimensionless distance in the reactor.
z'_i	Mole ratio of i to inerts in the liquid phase, mol/s.

Greek

α	Any constant.
α', β'	Constants defined by Equation E16.1.9.
$\alpha'', \beta'', \gamma'', \delta'', \eta'', \rho''$	Groups defined by Equation E16.2.7
δ	Film thickness, m.
ε	Catalyst effectiveness factor.
ϕ	Thiele modulus for a spherical catalyst defined as $R\sqrt{k[A]^{n-1}/D_{eA}})$.
γ'	Constant defined by Equation E16.1.11
γ_{bu}	Group representing efficiency of bulk liquid utilization.
η	Enhancement factor.
η_a	Asymptotic enhancement factor.
$\kappa, \kappa_R, \kappa_{GL,A}, \kappa_{SL,A}$	Dimensionless groups defined in Figure 17.4.
λ	Distance within the film, m.
λ_1, λ_2	Parameters defined in Figure 17.4.
ψ	Measure of the relative roles of reactor volume and interfacial area.
ℓ	Axial distance, m.
μ	Viscosity, kg/m s.
μ'	Kinematic viscosity, m^2/s.
ν_i	Stoichiometric coefficient of i.
ρ	Density, kg/m^3.
ρ_c	Catalyst density, kg/m^3.
σ	Surface tension, dynes/m.
Ω	Minimum energy, kWh/m^3.
ω_F	Wettability factor.

Subscripts/Superscripts

Subscripts

b	Liquid bulk in a fluid-fluid system.

c	Continuous phase; catalyst.
d	Dispersed phase.
e	Final condition (at $t = t_{\text{final}}$).
f	Outlet condition.
G	Gas.
i	Initial condition (at $t = 0$).
L	Liquid.
mn	Order mn (i.e., $mn = 2$ when $m = n = 1$).
mn, 1	Order mn in phase 1, i.e., 2, 1 = overall second order in phase 1.
mn, 2	Order mn in phase 2, i.e., 2, 2 = overall second order in phase 2.
0	Inlet condition.
U	Inert.
1	Phase 1.
2	Phase 2.

Superscripts

*	Interfacial or saturation value.
int	Interface

References to Part IV

Akita, K., and Yoshida, F. *Ind. Eng. Chem. Proc. Des. Dev.,* **13**, 84 (1974); **12**, 76 (1973).
Albright, L.F., and Hanson, C. *Loss Prev.,* **3**, 26 (1969); *Industrial and Laboratory Nitration* (Am. Chem. Soc. Symp. Ser. No. 22), *Am. Chem. Soc.,* Washington, D.C., 1975.
Albright, L.F., and Smith, C.S. *AIChE J.,* **14**, 325 (1968).
Arters, D.C., and Fan, L.S. *Chem. Eng. Sci.,* **41**, 107 (1986); **45**, 965 (1990).
Ashton, N.F., McDermott, C., and Brench, A., *Chemistry of Extraction of Nonreacting Soluts,* Chap. 1 in *Handbook of Extraction* (eds. Lo, T.C., Baird, M.H.I., and Hansen, C. M. Wiley, New York, 1983; reprinted by Kieger, Malabar, FL, 1991.
Astarita, G. *Mass Transfer with Chemical Reactions,* Elsevier, Amsterdam, 1967.
Backes, H.M., Ma, J.J., Bender, E., and Maurer, G. *Chem. Eng. Sci.,* **45**, 275 (1990).
Balasubramanian, S.N., Rihani, D.N., and Doraiswamy, L.K. *Ind. Eng. Chem. Fundam.,* **5**, 185 (1966).
Barona, N., and Prengle, H.W., Jr. *Hydrocarbon Process. Petrol. Refiner,* **52**(12), 73 (1973).
Barton, A.F.M., *Handbook of Solubility Parameters and Other Cohesion Parameters,* 2nd ed., CRC Press, Boca Raton, Fl., 1983.
Beenackers, A.A.C.M., and Van Swaaij, W.P.M. *Chem. Eng. Sci.,* **48**, 3109 (1993).
Benaissa, M., Carillo Le Roux, G., Joulia, X., Chaudhari, R.V., and Delmas, H. *Ind. Eng. Chem. Res.,* 1996 (in press).
Bern, L., Lidefelt, J.O., and Schoon, N.H. *J. Am. Oil Chem. Soc.,* **53**, 463 (1976).
Bhatia, S., Gopala Rao, M., and Rao, M.S. *Chem. Eng. Sci.,* **31**, 427 (1976).
Bhatia, S., and Gupta, J.S. *Rev. Chem. Eng.,* **8**, 177 (1993).
Bhattacharya, A., and Chaudhari, R.V. *Can. J. Chem. Eng.,* **65**, 1018 (1987); *Ind. Eng. Chem. Res.,* **29**, 317 (1990).
Bischoff, K.B. *AIChE J.,* **11**, 351 (1965).
Bisio, A., and Kabel, R.L. *Scale Up of Chemical Processes; Conversion from Laboratory Scale Test to Successful Commercial Scale Design,* Wiley, New York, 1985.
Blenke, H., and Hirner, W. *VDI Ber.,* **218**, 549 (1974).

Bohmer, K., and Blenke, H. *Verfahrenstechnik,* **6**, 50 (1972).
Boon-Long, S., Laguerie, C., and Couderc, P.J. *Chem. Eng. Sci.,* **33**, 813, (1978).
Botteghi, C., Ganzerla, R., Lenarda, M., and Moretti, G. *J. Mol. Cat.,* **40**, 129 (1987).
Bošković, N., Grbavčić, Z.B., Vuković, D.V., and Marković-Grabčić, M. *Powder Technol.,* **79**, 217 (1994).
Botton, R. *Chem. Eng. J.,* **20**, 87 (1980).
Brahme, P.H., and Doraiswamy, L.K. *Ind. Eng. Chem. Proc. Des. Dev.,* **15**, 130 (1976).
Briens, C.L., Del Pozo, M., and Chiu, K. *Chem. Eng. Sci.,* **48**, 973 (1993)
Brooks, B.W., and Richmond, H.N. *Chem. Eng. Sci.,* **49**, 1065 (1994).
Buchholz, R., Tsepetonides, J., Steinemannn, J., and Onken, U. *Ger. Chem. Eng.,* **6**, 105 (1983).
Burghardt, A., Bartelmus, G., Jaroszynski, M., and Kolodziej, A. *Chem. Eng. J.,* **58**, 83 (1995).
Calderbank, P.H. *Trans. Inst. Chem. Eng.,* **36**, 443 (1958).
Carlson, K.D., Sohns, V.E., Perkins, R.B., and Huffman, E. L. *Ind. Eng. Chem. Proc. Des. Dev.,* **16**, 95 (1977).
Chabala, J.C., and Fischer, M.H. U.S. Patent, 4 199,569, 1980.
Chandalia, S.B. *Indian J. Technol.,* **6**, 88 (1968a); **6**, 249 (1968b).
Charpentier, J.C., and Favier, M. *AIChE J.,* **21**, 1213 (1975).
Chaudhari, R.V. *Frontiers in Chemical Reaction Eng.,* **1**, 291 (1984).
Chaudhari, R.V., and Doraiswamy, L.K. *Chem. Eng. Sci.,* **29**, 675 (1974).
Chaudhari, R.V., Parande, M.G., Ramachandran, P.A., and Brahme, P.H. *ISCRE-8, Pergamon,* **205**, 1984.
Chaudhari, R.V., Parande, M.G., Ramachandran, P.A., Brahme, P.H., Vadgaonkar, H.G., and Jaganathan, R. *AIChE J.,* **31**, 1891 (1985).
Chaudhari, R.V., and Ramachandran, P.A., *AIChE J.,* **26**, 177 (1980).
Chaudhari, R.V., and Shah, Y.T. In *Concepts and Design of Chemical Reactors* (Eds., Whitaker, S., and Cassano, A.E.), Gordon and Breach, New York, 1986.
Chaudhari, R.V., Shah, Y.T., and Foster, N.R., *Cat. Rev. Sci. Eng.,* **28**, 431 (1986).
Chemtech, **5**, 189 (1975).
Chou, T.S., Worley, F.L., Jun., and Luss, D. *Ind. Eng. Chem. Proc. Des. Dev.,* **16**, 424 (1977).
Chua, Y.H., and Ratcliffe, J.S. *Mech. Chem. Eng. Trans.,* **7**, 6, 11, 17 (1971).
Concordia, J.J. *Chem. Eng. Prog.,* 50 (1990).
Cornet, I., Lewis, W.N., and Kappesser, R. *Trans. Inst. Chem. Eng.,* **47**, T222 (1969).
Crine, M., and L'Homme, G.A. In *Mass Transfer with Chemical Reaction in Multiphase Systems* (Ed., Alper, E.), Vol. II, Martinus Nijhoff, The Hague, 1983.
Cussler, E.L. *Diffusion: Mass Transfer in Fluid Systems,* 2nd ed., Cambridge University Press, Cambridge, UK, 1997.
Dake, S.B., Kohle, D.S., and Chaudhari, R.V. *J. Mol. Catal.,* **24**, 99 (1984).
Danckwerts, P.V. *Gas-Liquid Reactions,* McGraw-Hill, New York, 1970.
Deckwer, W.D. *Reaktionstechnik in Blasensäulen,* Otto Salle Verlag, Frankfurt am Main, 1985.
Delichatsios, M.A., and Probstein, R.F. *Ind. Eng. Chem. Fundam.,* **15**, 134 (1976).
Deshpande, R.M., and Chaudhari, R.V. *Ind. Eng. Chem. Res.,* **27**, 1996 (1988); *J. Mol. Catal.,* **57**, 177 (1989a); *J. Catal.,* **115**, 326 (1989b).
Donders, A.J.M., Wijffels, J.B., and Rietema, K. *Proc. 4th European Symp. Chem. React. Eng.* (Suppl. To *Chem. Eng. Sci.*), 1968.
Doraiswamy, L.K., and Gokarn, A.N. *Chem. Eng. Technol.,* **11**, 438 (1988).

Doraiswamy, L.K., and Sharma, M.M. *Heterogeneous Reactions—Analysis Examples and Reactor Design*, Vol. 2, Wiley, New York, 1984.
Elango, V., et al., *Eur. Patent*, **400**, 892 (C1.C07C57/30), 1990.
Fan, L.S. *Gas Liquid Fluidization Engineering*, Butterworth, MA, 1989.
Fouilloux, P. *Stud. Surf. Sci. and Catal.*, **59**, 245 (1988).
Frank, T.C., Downey, J.R., and Gupta, S.K., *Chem. Eng. Prog.*, **95**, 41 (Dec. 1999).
Friend, L., Wender, L., and Yarze, J.C. *Adv. Chem. Ser.*, **70**, 168 (1968).
Fukushima, S., and Kusaka, K. *J. Chem. Eng. Jpn*, **10**, 461 and 468 (1977).
Garrett, H.E. *Surface Active Chemicals*, Pergamon, Oxford, 1972.
Gehlawat, J.K., and Sharma, M.M. *J. Appl. Chem. Biotechnol.*, **21**, 141 (1971).
Gmehling, J.G., Anderson, T.F., and Prausnitz, J.M., *Ind. Eng. Chem. Fundam.*, **17**, 269 (1978).
Gmehling, J.G., Li, J., and Schiller, M., *Ind. Eng. Chem. Res.*, **32**, 178 (1992).
Gokhale, M.V., Naik, A.T., and Doraiswamy, L.K. *Chem. Eng. Sci.*, **10**, 1409 (1975).
Goto, S., and Smith, J.M. *AIChE J.*, **24**, 294 (1978).
Goto, S., Watabe, S., and Matsubara, M. *Can. J. Chem. Eng.*, **54**, 551 (1976).
Gray, M.R., Mehta, B., and Masliyah, J.H. *Chem. Eng. Sci.*, **48**, 3442 (1993).
Groggins, P.H. Ed. *Unit Processes in Organic Synthesis*, 5th ed., McGraw-Hill, New York, 1958.
Gupte, S.P., and Chaudhari, R.V. *Ind. Eng. Chem. Res.*, **31**, 2069 (1992).
Hagenson, L.C., and Doraiswamy, L.K. *Ind. Eng. Chem. Res.*, **53**, 131 (1998).
Haidegger, E., Hodossy, L., Kriza, D., and Peter, I. *Chem. Proc. Eng.*, **52**(8), 39 (1971).
Hammerschmidt, W.W., and Richarz, W., *Ind. Eng. Chem. Res.*, **30**, 82 (1991).
Hanika, J., and Stanek, V. In *Handbook of Heat and Mass Transfer* (Ed. Cheremisinoff, N.P.), Vol. 2, Gulf, Houston, Tex., 1986, p. 1029.
Hansen, C.M., *Ind. Eng. Chem. Prod. Res. Dev.*, **8**, 2 (1969); Hansen's Solubility Parameters: *A User's handbook*, CRC Press, Boca Raton, FL, 1999.
Hansen, H.K., Rasmussen, A., Fredenslund, M., Schiller, M., and Gmehling, J.G., *Ind. Eng. Chem. Res.*, **30**, 2352 (1991).
Hanson, C., and Ismail, H.A.M. *J. Appl. Chem. Biotechnol.*, **26**, 111 (1976).
Hatch, L.F. *Hydrocarbon Process. Petrol. Refiner.*, **49**(3), 101 (1970).
Heijnen, J.J., and Van't Riet, K. *Chem. Eng. J.*, **B21**, 42 (1984).
Henzler, H.J. *Chem. Ing. Tech.*, **54**, 8 (1982).
Hernàndez, H., Zalvidar, J.M., and Barcons, C. *Comput. Chem. Eng.*, **17S**, 45 (1993).
Hikita, H., and Asai, S. *J. Chem. Eng. Japan*, **2**, 77 (1964).
Hikita, H., Asai, S., Ishikawa, H., and Uku, J. *Chem. Eng. Commun.*, **5**, 315 (1980a).
Hikita, H., Asai, H., Tanigawa, K., Segawa, K., and Kitao, M. *Chem. Eng. J.*, **20**, 59 (1980b).
Hikita, H., Asai, S., Kikuwa, H., Zaike, T., and Masahiko, O. *Ind. Eng. Chem. Proc. Des. Dev.*, **20**, 540 (1981).
Hikita, H., and Kikukawa, H. *Chem. Eng. J.*, **8**, 412 (1975).
Hofmann, H. In *Mass Transfer with Chemical Reaction in Multiphase Systems*, Ed., Alper, E., Vol. II, Three-Phase Systems, NATO-ASI Series, Series E, Applied Sciences No. 72–73, Nijhoff, Hingam, Mass, 1983.
Horak, M., Palm, V., and Soogenbits, U. *Reaktsii Sposobu. Organ. Soedin.*, **11**(3), 709 (1975); *Chem. Abstr.*, **84**, 73330 (1976).
Hsu, S.H., and Ruether, J.A. *Ind. Eng. Chem. Proc. Des. Dev.*, **17**, 524 (1978).
Hydrocarbon Proc. Petr. Refiner, **50**(11), 226 (1971).
Irandoust, S., and Anderson, B. *Chem. Eng. Sci.*, **43**, 1983 (1988).
Jeffreys, G.V., Jenson, V.G., and Miles, F.R. *Trans. Inst. Chem. Eng.*, **39**, 389 (1961).

Jernigan, E.C., Gwyn, J.E., and Claridge, E.L. *Chem. Eng. Prog.,* **61**(11), 94 (1965).
Joshi, J.B. *Chem. Eng. J.,* **24**, 313 (1982).
Joshi, J.B., Shertukde, P.V., and Godbole, S.P. *Rev. Chem. Eng.,* **5**, 71 (1988).
Juvekar, V.A. *Studies in Mass Transfer in Gas-Liquid and Gas-Liquid-Solid Systems,* Ph.D. (Tech.) Thesis, University of Bombay, India, 1976.
Kale, S.S., Chaudhari, R.V., and Ramachandran, P.A. *Ind. Eng. Chem. Proc. Des. Dev.,* **20**, 309 (1981).
Karlsson, H.T., and Bjerle, I. *Trans. Inst. Chem. Eng.,* **58**, 138 (1980).
Kastánek, F. *Coll. Czech. Chem. Commun.,* **41**, 3709 (1976).
Kastánek, F., Zahradník, J., Kratochvíl, J., and Cermák, J., *Reactors for Gas-Liquid Systems,* Ellis Horwood, New York, 1993.
Kelkar, A.A., Kohle, K.S., and Chaudhari, R.V. *J. Organomet. Chem.,* **430**, 111 (1992).
Kenney, C.N., and Sedriks, W. *Chem. Eng. Sci.,* **27**, 2029 (1972).
Kikuchi, K., Takashi, H., and Sugarawa, T. *Can. J. Chem. Eng.,* **73**, 313 (1995).
Kikuchi, K., Tadakuma, Y., Sugawara, T., and Ohashi, H. *J. Chem. Eng. Japan,* **20**, 134 (1988).
Klassen, J., and Kirk, R.S. *AIChE J.,* **1**, 488 (1955).
Ko. M.K. *Mass Transfer Analysis of the Liquid Phase methanol Synthesis process,* Ph.D. dissertation, University of Skron, Akron, OH, 1987.
Kogami, K., and Kumanotami, J. *Bull. Chem. Soc. Jpn.,* **46**, 3562 (1973).
Kohler, M.A., *Appl. Cat.,* **22**, 21 (1986).
Kohn, P.M. *Chem. Eng.,* **85**(7), 90 (1978).
Kosak, J.R. *Ann. N.Y. Acad. Sci.,* **172**, 175 (1980).
Kramers, H., and Westerterp, K.R. *Elements of Chemical Reactor Design and Operation,* Chapman and Hall, London, 1963.
Kulkarni, B.D., and Doraiswamy, L.K. *AIChE J.,* **21**, 501 (1975); *Chem. Eng. Sci.,* **35**, 817 (1980).
Kumar, S., Kumar, R., and Gandhi, K.S. *Chem. Eng. Sci.,* **48**, 2025 (1993).
Laddha, G.S., and Degaleesan, T.E. *Transport Phenomena in Liquid Exctraction,* McGraw-Hill, New Delhi, India, 1976.
Ladhabhoy, M.E., and Sharma, M.M. *J. of Appl. Chem.,* **19**, 267 (1969).
Lee, E.S. *AIChE J.,* **14**, 490 (1968).
Lee, S., Parameswaran, E.R., and Sawant, A.V. *Mass Transfer in the Liquid Phase Methanol Synthesis (LPMeOH^{TM}) Process,* Ap-5758, Electric Power Research Institute, Palo Alto, CA, 1988.
Lemcoff, N.O., and Jameson, G.J. *Chem. Eng. Sci.,* **30**, 363 (1975); *AIChE J.,* **21**, 730 (1975).
Leuteritz, G., Process Eng., **54**, 62 (1973).
Levenspiel, O. *Chemical Reaction Engineering,* 2nd ed., Wiley, New York, 1972; 3rd ed., 1999; *Chemical Reactor Omnibook,* Oregon State University Bookstore, Corvallis, Ore., 1993.
Levenspiel, O., and Godfrey, J.H. *Chem. Eng. Sci.,* **29**, 1723 (1974).
Levich, V.G. *Solution of the Equation for Convective Diffusion to the Surface of a Rotating Disk,* In *Physicochemical Hydrodynamics,* Prentice Hall, Englewood Cliffs, NJ. 1962.
Levins, D.M., and Glastonbury, J.R. *Chem. Eng. Sci.,* **27**, 537 (1972a); *Trans. Inst. Chem. Eng.,* **50**, 132 (1972b).
Machon, V., and Linek, V. *Chem. Eng. J.,* **8**, 53 (1974).
Mangartz, K.H., and Pilhofer, TH. *Verfahrenstechnik,* **14**, 40 (1980).
Mann, R. *Gas-Liquid Contacting in Mixing Vessels,* The Instn. Chem. Engrs., Rugby, 1983.

Mathieu, C., Dietrich, E., Delmas, H., and Jenck, J. *Chem. Eng. Sci.*, **47**, 2289 (1992).
Mazet, M. *Int. Chem. Eng.*, **32**, 395 (1992).
McIver, R.G., and Ratcliffe, J.S. *Trans. Inst. Chem. Eng.*, **52**, 276 (1974).
Meister, D., Post, T., Dunn, I.J., and Bourne, J.R. *Chem. Eng. Sci.*, **34**, 1367 (1979).
Mersmann, A. *Verfahrenstechnik*, **6**, 203 (1972).
Mhaskar, R.D., and Sharma, M.M. *Chem. Eng. Sci.*, **30**, 811 (1975).
Miller, D.N. *Ind. Eng. Chem. Proc. Des. Dev.*, **10**, 365 (1971).
Miller, S.A. *Chem. Proc. Eng.*, **47**(6), 268 (1966).
Mills, P.L., and Chaudhari, R.V. *Catal. Today*, **37**, 367 (1997).
Mills, P.L., Ramachandran, P.A., and Chaudhari, R.V. *Rev. Chem. Eng.*, **8**, 1 (1992).
Molga, E.J., and Westerterp, K.R. *Chem. Eng. Sci.*, **47**, 1733 (1992).
Mosby, J.F., and Albright, L.F. *Ind. Eng. Chem. Proc. Des. Dev.*, **5**, 183 (1966).
Nagel, O., Kurten, H., and Hegner, B. *Chem. Ing. Tech.*, **45**, 913 (1973).
Nagel, O., Kurten, H., and Sinn, R. *Chem. Ing. Tech.*, **44**, 899 (1972).
Nienow, A.W. *Chem. Eng. J.*, **9**, 153 (1975).
Nikov, I., and Delmas, H. *Chem. Eng. Sci.*, **47**, 673 (1992).
Nikov, I., and Paev, K. *Catal. Today*, **24**, 41 (1995).
Okufi, S., Perez de Ortiz, E.S., and Sawistowski, H. *Can. J. Chem. Engr.*, **68**, 400 (1990).
Oldshue, J.Y. *Fluid Mixing Technology*, McGraw-Hill, New York, 1983.
Onda, K., Sada, E., Kobayashi, T., and Fujine, M. *Chem. Eng. Sci.*, **25**, 1023 (1970a); **25**, 761 (1970b).
Otake, M., *Chem. Tech.*, 36 (1995).
Parameswaran, V.R., Gogate, M.R., Lee, B.G., and Lee, S. *Fuel Sci. technol. Instl.*, **9**, 695 (1991).
Parshall, G.W., and Nugent, W.A. *Chem. Tech.*, **18**, 184 March 1988.
Patil, V.K., and Sharma, M.M. *Chem. Eng. Res. Des.*, **61**, 21 (1983); **62**, 247 (1984).
Pepper, K.W., Paisley, H.M., and Young, M.A. *J. Chem. Soc.*, 4097 (1953).
Peters, R.H. Textile Chemistry, Vol. 3, *The Physical Chemistry of Dyeing*, Elsevier, Amsterdam, 1975, p. 598.
Phadtare, P.G., and Doraiswamy, L.K. *Ind. Eng. Chem. Proc. Des. Dev.*, **4**, 274 (1965); **8**, 165 (1969).
Prengle, H.W., Mauk, C.E., Legan, R.W., and Hewes, C.G. *Hydrocarbon Proc. Petrol. Refiner*, **54**(10), 82 (1975).
Rajasekharam, M.V., and Chaudhari, R.V. *Chem. Eng. Sci.*, **51**, 1663 (1996).
Ramachandran, P.A. *Chem. Eng. Sci.*, **26**, 349 (1971).
Ramachandran, P.A., and Chaudhari, R.V. *Chem. Eng. J.*, **20**, 75 (1980a); *Three Phase Catalytic Reactors*, Gordon and Breach, New York, 1983.
Ramachandran, P.A., and Doraiswamy, L.K. *AIChE J.*, **28**, 881 (1982).
Ramachandran, P.A., and Sharma, M.M. *Trans. Inst. Chem. Eng.*, **49**, 253 (1971).
Rattee, I.D., and Breuer, M.M. *The Physical Chemistry of Dye Adsorption*, Academic Press, London, 1974, p. 244.
Ratusky, J., and Sorm, F. *Coll. Czech. Chem. Commun.*, **24**, 2553 (1959).
Revankar, V.V.S., and Doraiswamy, L.K. *Ind. Eng. Chem. Research*, **31**, 781 (1992).
Revankar, V.V.S., Kulkarni, B.D., and Doraiswamy, L.K. *Ind. Eng. Chem. Res.*, **26**, 1018 (1987).
Robbins, L.A., *Chem. Eng. Prog.*, **76**, 58 (Oct 1980).
Robbins, L.A., and Cusack, R.W., *Liquid-Liquid Extraction Operations and Equipment*, Sect. 15 in *Perry's Chemical Engineers' Handbook*, 7th ed., eds. Perry, R.H. and Green, D.W., McGraw-Hill, New York, 1997.

Rod, V. *Chem. Eng. J.*, **7**, 137 (1974); *Proc. 4th Int./6th European Symp. Chem. React. Eng.*, 1976.
Rode, C.V., Gupte, S.P., Chaudhari, R.V., Pirozhkov, C.D., and Lapidus, A.L. *J. Mol. Cat.*, **91**, 195 (1994).
Rogers, H.R., Deutsch, J., and Whitesides, G.M. *J. Am. Chem. Soc.*, **102**, 217 (1980a).
Rogers, H.R., Hill, C.L., Fujiwara, Y., Rogers, R.J., Lee Mitchell, H., and Whitesides, G.M. *J. Am. Chem. Soc.*, **102**, 217 (1980b).
Rose, L. M. *Chemical Reactor Design in Practice*, Elsevier, Amsterdam, 1981.
Roy, N.K., Guha, D.K., and Rao, M.N. *Chem. Eng. Sci.*, **19**, 215 (1964).
Rylek, M., and Zahradnik, J. *Coll. Czech. Chem. Commun.*, **49**, 1939 (1984).
Sada, E., Kumuzawa, H., and Butt, M.A. *Can. J. Chem. Eng.*, **54**, 421 (1976); **55**, 475 (1977).
Saeki, G., Weng, P.K.J., and Johnson, A.I. *Can. J. Chem. Eng.*, **50**, 730 (1972).
Sahimi, M., Gavalas, G.R., and Tsotsis, T.T. *Chem. Eng. Sci.*, **45**, 1443 (1990).
Sano, Y., Yamaguchi, N., and Adachi, T. *J. Chem. Eng. Jpn.*, **7**, 255 (1974).
Santacesaria, E., Di Serio, M., Velotti, R., and Leone, U. *Ind. Eng. Chem.*, **33**, 277 (1994).
Satterfield, C.N., Van Eck, M.W., and Bliss, G.S. *AIChE J.*, **24**, 709 (1978).
Sato, Y., Hirose, T., and Ida, T. *Kagaku Kogaku*, **38**, 543 (1973b).
Schügerl, K. *Chem. Ing. Tech.*, **49**, 605 (1977); **52**, 951 (1980).
Schügerl, K. In *Mass Transfer with Chemical Reaction in Multiphase Systems* (Ed., Alper, E.), Vol. 1, Martin Nijhoff, The Hague, 1983.
Scriven, L.E. *AIChE J.*, **7**, 524 (1961).
Sergeev, G.B., Zagorsky, V.V., and Badaev, F.Z. *J. Organomet. Chem.*, **243**, 123 (1983).
Shah, Y.T. *Gas Liquid Solid Reactor Design*, McGraw-Hill, New York, 1979.
Shah, Y.T., and Deckwer, W.D. In *Scale-up of Chemical Processes Conversion from Laboratory Scale Tests to Successful Commercial Size Design* (Eds., Bisio, A., and Kabel, R.L.), Wiley, New York, 1985.
Shah, Y.T., Kelkar, B.G., Godbole, S.P., and Deckwer, W.D. *AIChE J.*, **28**, 353 (1982).
Shah, Y.T., Mhaskar, R.D., and Paraskos, J.A. *Ind. Eng. Chem. Proc. Des. Dev.*, **15**, 400 (1976).
Shah, Y.T., and Paraskos, J.A. *Chem. Eng. Sci.*, **30**, 465 (1975).
Shah, Y.T., Stiegel, G.J., and Sharma, M.M. *AIChE J.*, **24**, 369 (1978).
Sharma, M.M., and Nanda, A.K. *Trans. Inst. Chem. Eng.*, **46**, T44 (1968).
Sharma, R.C., and Sharma, M.M. *J. Appl. Chem.*, **19**, 162 (1969); *Bull. Chem. Soc. Japan*, **43**, 642 (1970a).
Sharma, R.C., and Sharma, M.M. *Bull. Chem. Soc. Japan*, **43**, 1282 (1970b).
Sisak, C., and Hung, J. *Ind. Chem.*, **14**, 39 (1986).
Sittig, M. *Organic Chemical Process Encyclopedia*, Noyes, Park Ridge, N.J., 1969.
Sittig, W., and Heine, H. *Chem. Ing. Tech.*, **49**, 595 (1977).
Specchia, V., and Baldi, G. *Chem. Eng. Sci.*, **32**, 515 (1977).
Sprow, F.B. *Chem. Eng. Sci.*, **22**, 435, (1967); *Ind. Eng. Chem. Proc. Des. Dev.*, **13**, 433 (1969).
Stinson, S.C. *Chem. Eng. News*, June and September, 1986.
Stroh, R., Ebersberger, J., Haberland, H., and Hahn, W. *Angew. Chem.*, **69**, 124 (1957).
Stuber, F., Benaissa, M., and Delmas, H. *Catal. Today*, **24**, 95 (1995).
Sukhdev, J. *Sci. Ind. Res.* (India), **31**, 60 (1972).
Sumner, H.H., and Weston, C.D. *Am. Dyestuff Rep.*, **52**, 442 (1963).
Szekely, J., Evans, J.W., and Sohn, H.Y. *Gas-Solid Reactions*, Academic Press, New York, 1976.
Talmor, E. *AIChE J.*, **23**, 868 (1977).

Tarmy, B., Chang, M., Coulaloglou, C., and Ponzi, P. *Chem. Eng.*, **407**, 18 (1984).
Tavlarides, L.L., and Stamatoudis, M. *Advances in Chemical Engineering*, Vol. 11 (Eds., Drew, T.B., Cokelet, G.R., Hoopes, J.W., Jr. and Vermeulen, T.), Academic Press, New York, 1981.
Taylor, R., and Krishna, R. *Multicomponent Mass Transfer*, Wiley, New York, 1993.
Teshima, H., and Ohashi, Y. *J. Chem. Eng. Jpn.*, **10**, 70 (1977).
Throckmorton, P.E., and Pryde, E.H. *J. Am. Oil. Chem. Soc.*, **49**, 643 (1972).
Tiwari, R.K., and Sharma, M.M. *Chem. Eng. Sci.*, **32**, 1253 (1977).
Tremont, S.J., Mills, P.L., and Ramachandran, P.A. *Chem. Eng. Sci.*, **43**, 2221 (1988).
Treybal, R.E. *Liquid Extraction*, 2nd ed., McGraw-Hill, New York, 1963; *Mass Transfer Operations*, McGraw-Hill, New York, 1968.
Twitchett, H.J. *Chem. Soc. Rev.*, **3**(2), 209 (1974).
Valerius, G., Zhu, X., Hofmann, H., Amtz, D., and Haas, T. *Chem. Eng. & Proc.*, **35**, 11 (1996).
Van Krevelen, D.W. *Research*, **3**, 106 (1950).
Venkatraman, K. Ed., *The Chemistry of Synthetic Dyes*, Vol. 6, *Reactive Dyes*, Academic Press, New York, 1972.
Viesturs, U.E., Kuznecov, A.M., and Samenkov, V.V. *Fermentation Systems* (in Russian), Zinatne, Riga, 1986.
Vieth, W.R., Porter, J.H., and Sherwood, T.K. *Ind. Eng. Chem. Fundam.*, **2**, 1 (1963).
Vijayaraghavan, P., Kulik, C.J., and Lee, S. *Fuel Sci. Technol. Intl.*, **11**, 1577 (1993).
Villermaux, J. In *Multiphase Chemical Reactors VI. Fundamentals (Eds., Rodriguez, A.E., Calo, J.M., and Sweed, N.H., Sijthoff et Noordhoff, Alphen aan den Rijn, 1981, p. 285.*
Warwick, G.C.I. *Chem. Ind.*, 5 May, 882 (1973).
Wen, C.Y., and Fan, L.T. *Models for Flow Systems and Chemical Reactors*, Marcel Dekker, New York, 1975.
Westerterp, K.R., Janssen, H.J., and Van der Kwast, H.J. *Chem. Eng. Sci.*, **47**(15), 4179 (1992).
Wiederkehr, H. *Chem. Eng. Sci.*, **43**, 1783 (1988).
Yagi, H., and Yoshida, F. *Ind. Eng. Chem. Proc. Des. Dev.*, **14**, 488 (1975).
Yamamoto, Y., Inoue, Y., and Suzuki, H. *Shokubai (Catalyst)*, **37**, 179 (1995).
Yung, C.N., Wong, C.W., and Chang, C.L. *Can. J. Chem. Eng.*, **57**, 672 (1979).
Zahradnik, J., Kastànek, F., and Kratochvíl, J. *Coll. Czech. Chem. Commun.*, **15**, 27 (1982).
Zahradnik, J., Kratochvil, and Rylek, M. *Coll. Czech. Chem. Commun.*, **50**, 2535 (1985).
Zaldivar, J.M., Molga, E., Alós, M.A., Hernandez, H., and Westerterp, K.R. *Chem. Eng. of Proc.*, **35**, 91 (1996).
Zwietering, T.N. *Chem. Eng. Sci.*, **8**, 244 (1958).

PART V

STRATEGIES FOR ENHANCING THE RATES OF ORGANIC REACTIONS

It takes a great deal of elevation of thought to produce a very little elevation of life.

Ralph Waldo Emerson

Chapter 18

Biphasic Reaction Engineering

The chemical equilibria of many industrially important organic reactions in aqueous solutions are often displaced in the direction of the reactants, leading to very low conversions. Therefore, there is a need for an environmentally friendly strategy that will shift the equilibria toward the products, resulting in enhanced conversions. A particularly effective technique is to add a second phase, appropriately termed *biphasing*.

In general, biphasing is the intentional addition of an immiscible phase to a reaction mixture to increase the yield of the desired product or to facilitate separation of product from (say) catalyst. Much of the effort till recently has been on adding a *water-immiscible* organic solvent in enzyme-catalyzed organic reactions in the aqueous phase. Although strictly the term biphasing should apply only to soluble catalysts, thus preserving the purity of its definition, in practice it also includes insoluble catalysts such as immobilized enzymes (which would constitute a third phase).

Biphasing received an exciting stimulus around 1984 when it was used to overcome the inherent and perhaps the most telling deficiency of homogeneous catalysis (see Chapter 8). By biphasing with an aqueous phase (unlike in enzymatic catalysis where the biphasing liquid is an organic solvent), the catalyst was fully retained in that phase, whereas the product (and unused reactant) remained in the organic phase. The consequent easy separation of catalyst from product added a new dimension to homogeneous catalysis that gives it a decided edge over its heterogeneous counterpart for many reactions. Yet another dimension to biphasing was added in the last decade when it was found that both phases could be aqueous. This variant of traditional biphasing has many obvious advantages. Although still in its infancy, its enormous potential is not difficult to visualize. The chief advantages and disadvantages of biphasing are listed in Table 18.1.

Table 18.1 Advantages and disadvantages of biphasic systems

Biphastic Reactions
Advantages
Thermodynamically unfeasible reactions become possible
High reactant and product concentrations in the organic phase
Reduction of reactant and product inhibition
Enzyme remains in the aqueous phase
Ease of biocatalyst and product separation
Higher yield possible than in either aqueous or organic phase alone
Disadvantages
Likely catalyst deactivation or denaturing by organic solvent
Mass transfer limitations across interfacial boundaries

We begin our treatment of biphasing by developing the theoretical foundation for predicting an *apparent* or *effective equilibrium constant* for a biphasic reaction. This will be done specifically for enzyme-catalyzed reactions, but it can be extended to straight organic synthesis. Several important aspects of these biphasic systems, such as solvent selection and the role of mass transfer, will be discussed. This will be followed by a similar treatment of biphasing in organic synthesis. However, as no new engineering concepts are involved here (beyond that presented in Chapter 15), the treatment will focus on some general examples followed by a brief discussion of water-soluble ligands and their use in organic technology, particularly in homogeneous catalysis.

EQUILIBRIA IN BIPHASIC REACTIONS

Consider the reaction

$$A + B \xrightleftharpoons{K} R + S \qquad [18.1]$$

If R is the desired product, its concentration at equilibrium is

$$[R] = K\left(\frac{[A][B]}{[S]}\right) \qquad (18.1)$$

Clearly, an increase in $[R]$ is possible by decreasing $[S]$. This can be accomplished by combining various types of separation with reaction, including using a second liquid phase (extractive reaction) or a zeolite. The latter was briefly mentioned in Chapter 6, and the former will comprise the contents of Chapter 25.

The basis of the treatment discussed here is (1) postulation of an apparent or effective equilibrium constant for the reaction that occurs simultaneously and interactively in both phases and (2) finding a relationship between this and the true equilibrium constants in the individual phases. Such an approach (Martinek et al., 1977, 1980, 1981a,b; Martinek and Semenov, 1981a,b; Semenov et al.,

1980, 1981, 1987) should prove highly useful in preevaluating the effectiveness of biphasing with different solvents.

Ionization of reactants will obviously have an important bearing on the performance of a biphasic reaction. Thus, after first considering the effect of biphasing with nonionizing reactants, we extend the treatment to include the influence of ionization.

Equilibrium Shift with No Ionization

First we analyze a simple unimolecular reaction and then the more important bimolecular scheme.

Unimolecular reaction

Consider the simple unimolecular reaction

$$A \leftrightarrow R \qquad [18.2]$$

Now if a solvent is added so that both reactant and product partition between the two phases, then the following reaction-partition scheme results:

$$\begin{array}{c} [A \underset{}{\overset{K_w}{\rightleftarrows}} R]_W \\ m_A \updownarrow \qquad \updownarrow m_B \\ [A \underset{K_S}{\rightleftarrows} R]_S \end{array} \qquad [18.3]$$

where K_W and K_S are the equilibrium constants of the reaction in the aqueous and organic (solvent) phases, respectively, and m_A and m_R are the partition coefficients:

$$K_W = \frac{[R]_W}{[A]_W} \quad \text{and} \quad K_S = \frac{[R]_S}{[A]_S} \qquad (18.2)$$

$$m_A = \frac{[A]_S}{[A]_W} \quad \text{and} \quad m_R = \frac{[R]_S}{[R]_W} \qquad (18.3)$$

The material balances for the total reaction volume are

$$[A]_t(V_W + V_S) = [A]_W V_W + [A]_S V_S$$
$$[R]_t(V_W + V_S) = [R]_W V_W + [R]_S V_S \qquad (18.4)$$

Combining Equations 18.1–18.4 and using the definitions of the partition coefficients, the apparent equilibrium constant K_{biphasic} for the biphasic system can be expressed as

$$K_{\text{biphasic}} = K_W \frac{1 + \alpha m_R}{1 + \alpha m_A} = K_S \frac{1 + \frac{1}{\alpha} m_A}{1 + \frac{1}{\alpha} m_R} \qquad (18.5)$$

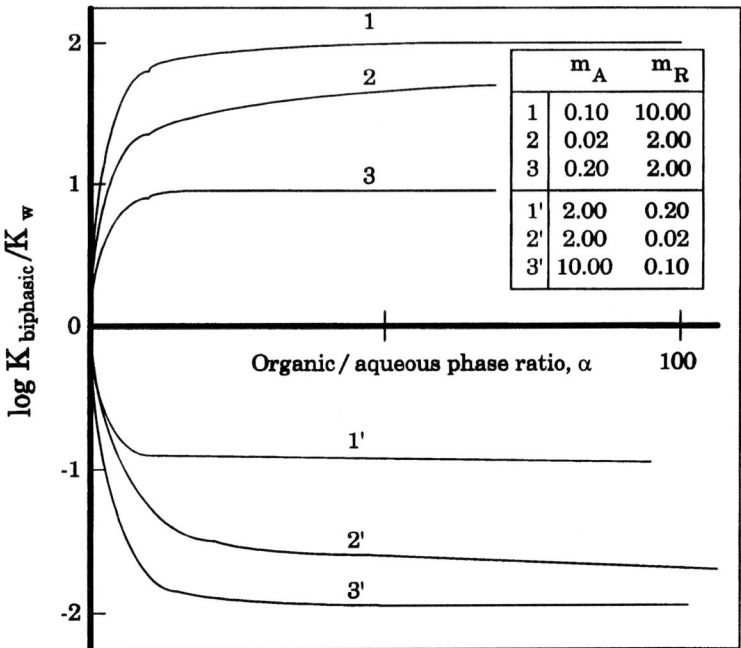

Figure 18.1 Effective equilibrium constant for the reaction $A \leftrightarrow R$ as a function of the volume ratio α (from Martinek et al., 1981a).

where

$$\alpha = \frac{V_S}{V_W} \qquad (18.6)$$

is the volume ratio of organic to aqueous phases. It is evident from Equation 18.5 that if the equilibrium constant in either phase alone is known and also the partition coefficients of the reagents and products between the two phases, then K for the biphasic system can be predicted.

An important observation is that $K_{biphasic}$ is a strong function of the phase volume ratio α (Figure 18.1). Another is that $K_{biphasic}$ increases when $m_R > m_A$, and decreases when $m_R < m_A$.

Bimolecular reaction

Consider the general aqueous phase bimolecular reaction

$$[A + B \leftrightarrow R + S]_W \qquad [18.4]$$

for which the reaction–partition scheme in a biphasic medium can be written as

$$\begin{array}{c} [A + B \xrightleftharpoons[]{K_W} R + S]_W \\ m_A \Vert\ m_B \Vert \qquad \Vert m_R \Vert m_S \\ [A + B \xrightleftharpoons[K_S]{} R + S]_S \end{array} \qquad [18.5]$$

Then the following expression can be developed for the apparent equilibrium constant for the biphasic mixture:

$$K_{\text{biphasic}} = K_W \frac{(1+\alpha m_R)(1+\alpha m_S)}{(1+\alpha m_A)(1+\alpha m_B)} \tag{18.7}$$

where

$$K_W = \frac{[R][S]}{[A][B]} \tag{18.8}$$

As the volume of the organic phase increases (i.e., as α increases), this function may have an extremum. This can be determined by examining the first derivative $(\partial K_{\text{biphasic}}/\partial \alpha)$, based on which it can be shown that it is monotonic under the following conditions:

FOR $K_{\text{biphasic}} > K_W$

$$m_R + m_S \geq m_A + m_B$$

and

$$\frac{1}{m_R} + \frac{1}{m_S} < \frac{1}{m_A} + \frac{1}{m_B} \tag{18.9}$$

FOR $K_{\text{biphasic}} < K_W$

$$m_R + m_S \leq m_A + m_B$$

and

$$\frac{1}{m_R} + \frac{1}{m_S} > \frac{1}{m_A} + \frac{1}{m_B} \tag{18.10}$$

The two cases are graphically displayed as curves 1 and 1' in Figure 18.2.
However, when

$$m_R + m_S > m_A + m_B$$

and

$$\frac{1}{m_R} + \frac{1}{m_S} > \frac{1}{m_A} + \frac{1}{m_B} \tag{18.11}$$

a maximum appears for K_{biphasic} as a function of α (Figure 18.2, curves 2, 3, and 4), and when

$$m_R + m_S < m_A + m_B$$

and

$$\frac{1}{m_R} + \frac{1}{m_S} < \frac{1}{m_A} + \frac{1}{m_B} \tag{18.12}$$

a minimum appears (Figure 18.2, curves 2', 3', and 4'). Clearly, it is possible that the value of K_{biphasic} can be higher or lower than that of either K_W or K_S,

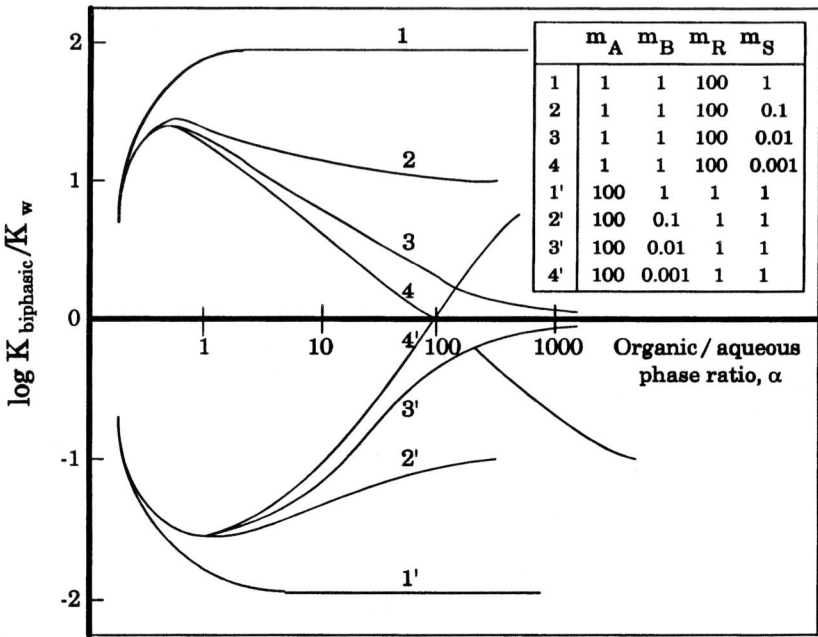

Figure 18.2 Effective equilibrium constant of the reaction $A + B \leftrightarrow R + S$ as a function of the volume ratio α (from Martinek et al., 1981a).

depending on the partition coefficients. Therefore, it is important to be able to predict the equilibrium shift of the reaction using this model before any laboratory scale testing is undertaken.

Equation 18.5 or 18.7 can be recast in terms of the ratio K_{biphasic}/K_W, which is called the *biphasic enhancement factor* (BEF). Thus for Equation 18.7, one can write

$$\text{BEF} = \frac{K_{\text{biphasic}}}{K_W}$$

$$= \frac{K \text{ in the presence of organic solvent}}{K \text{ in the absence of organic solvent}} = \frac{(1 + \alpha m_R)(1 + \alpha m_S)}{(1 + \alpha m_A)(1 + \alpha m_B)} \quad (18.13)$$

Then this can be used as a measure of the effectiveness of biphasing.

BEF AS A CRITERION FOR BIPHASING

BEF > 1.0:   Reaction is enhanced with addition of organic solvent.
BEF = 1.0:   Organic solvent has no effect.
BEF < 1.0:   Organic solvent inhibits the reaction and should not be used.

EXAMPLE 18.1

Equilibrium in the α-chymotrypsin synthesis of N-benzoyl-L-phenylalinine ester from its constituent acid and ethanol

One of the products of this reaction is water, so its partition coefficient is low (it does not exceed 0.01 in most organic solvents). The reaction can be expressed as

$$\text{N–Bz–Phe} + \text{EtOH} \xrightleftharpoons{\text{enzyme}} \text{N–Bz–Phe–OEt} + \text{H}_2\text{O} \qquad [\text{E}18.1.1]$$

First, the equilibrium constant for the reaction in the aqueous phase was determined, and it was found that the equilibrium of the reaction was shifted almost entirely toward the left ($K_W = 2 \times 10^{-3}$). The yield of the ester obtained for reaction in the aqueous phase was as low as 0.01%. Next, four different organic solvents (chloroform, benzene, carbon tetrachloride, and diethyl ether) were selected; the partition coefficient was determined for each of the four solvent–water binaries. Then the biphasic equilibrium constant was found from Equation 18.7 and plotted as a function of α (Figure 18.3).

As predicted by the theoretical model, K_{biphasic} exhibits a maximum for each of the solvents used. The physical explanation of the maximum is that the apparent equilibrium constant of the biphasic reaction can be higher than that for either of the pure phases, organic or aqueous.

This point is more effectively brought out in Figure 18.4 through a plot of K_{biphasic} as a function of α using experimentally determined values for a biphasic system that consists of chloroform and water. The lower and upper lines are the equilibrium constants for the aqueous and organic phases, respectively. It is evident that the apparent equilibrium constant for the biphasic system is higher than that for either phase alone.

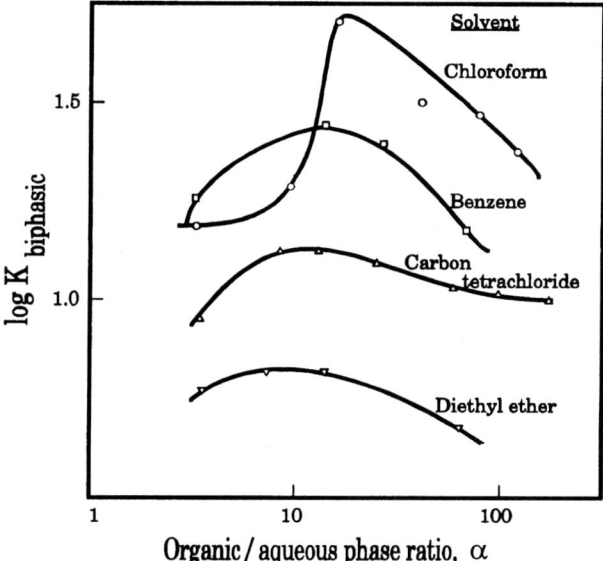

Figure 18.3 Experimental K_{biphasic} as a function of the volume ratio α for the α-chymotrypsin-catalyzed synthesis of the ethyl ester (N-benzoyl-L-phenylalanine) for four different solvents (redrawn from Martinek and Semenov, 1981b).

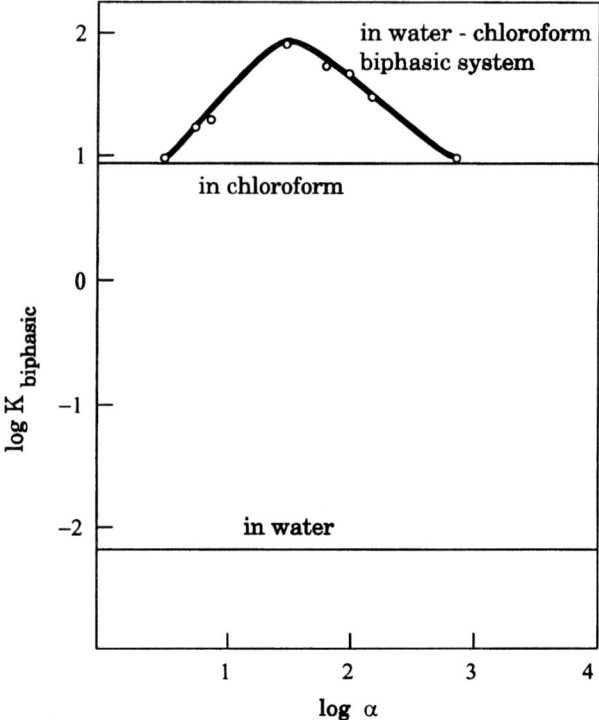

Figure 18.4 Equilibrium constant as a function of α for water, chloroform, and the water–chloroform biphasic system for the α-chymotrypsin-catalyzed synthesis of the ethyl ester of N-benzoyl-L-phenylalanine (from Semenov et al., 1987).

Equilibrium Shift with Ionization

A number of organic chemicals dissociate in water. This dissociation is often the basis for acid/base catalysis. In such a case, if one of the components of a reaction dissociates in one of the phases of a biphasic medium, further complexity is added to the analysis just presented. Then we must consider one of the following two dissociation equilibria, depending on whether the reactant is an acid or a base:

$$HA + B \xrightleftharpoons{K} R + S$$

where

$$HA \xrightleftharpoons{K_a} H^+ + A^- \qquad [18.6a]$$

$$A + B \xrightleftharpoons{K} R + S$$

where

$$B + H^+ \xrightleftharpoons{K_b} BH^+ \qquad [18.6b]$$

and K_a and K_b are the acid and base dissociation equilibrium constants, respectively. The effect of biphasing on reactions of these types is the resultant of two factors: the effect on the reaction equilibrium constant which was already considered and the effect on the degree of ionization at equilibrium.

The basic equations

If the ionization reaction 18.6a is carried out in a biphasic medium, it can be recast as

$$\left[HA \xrightleftharpoons{K_a} H^+ + A^- \right]_W$$
$$\Big\updownarrow m_{HA} \tag{18.7}$$
$$[HA \longrightarrow \text{No ionization}]_S$$

assuming that HA does not dissociate in the organic solvent.[1] Expressing K_a as pK_a, it can be shown that $pK_{a,\text{biphasic}}$ is related to the intrinsic pK_a by the following equation (Martinek and Semenov, 1981a; Semenov and Martinek, 1980):

$$pK_{a,\text{biphasic}} = pK_a + \log(1 + \alpha m_{HA}) \tag{18.14}$$

Similarly, for reaction 18.6b (i.e., protonation of B), the following expression can be derived:

$$pK_{b,\text{biphasic}} = pK_b - \log(1 + \alpha m_B) \tag{18.15}$$

It can be seen from Equations 18.14 and 18.15 that the apparent value of pK_a increases for an acid by $\log(1 + \alpha m_{HA})$ and decreases for a base by $\log(1 + \alpha m_B)$. Because the partition coefficient m_{HA} or m_B is fixed for a given system, the apparent value of pK increases with the volume fraction of solvent α for an acid and decreases with α for a base.

Reactions with one ionizing reactant

Consider a bimolecular reaction with one ionogenic group such as esterification which involves ionization of the acid:

$$\begin{bmatrix} \text{RCOOH} + \text{R}'\text{OH} \xrightleftharpoons{K_W, \text{nonionic}} \text{RCOOR}' + \text{H}_2\text{O} \\ K_a \Big\updownarrow \\ \text{RCOO}^- + \text{H}^+ \end{bmatrix}_W \tag{18.8}$$

When this reaction is allowed to take place in a biphasic medium, one has to account for the simultaneous occurrence of the main nonionic reactions in the two phases and for acid ionization. Thus the scheme represented in 18.8 is modified to scheme 18.9 in a biphasic medium.

---

[1] This assumption is by no means universal but has been made to illustrate the principle with minimum mathematical manipulation.

$$\left[\text{RCOOH} + \text{R}'\text{OH} \xrightleftharpoons{K_{S,\text{nonionic}}} \text{RCOOR}' + \text{H}_2\text{O} \right]_S$$

$$\updownarrow m_{\text{acid}} \quad \updownarrow m_{\text{alc.}} \quad \updownarrow m_{\text{ester}} \quad \updownarrow m_{\text{H}_2\text{O}}$$

$$\begin{bmatrix} \text{RCOOH} + \text{R}'\text{OH} \xrightleftharpoons{K_{W,\text{nonionic}}} \text{RCOOR}' + \text{H}_2\text{O} \\ K_a \updownarrow \\ \text{RCOO}^- + \text{H}^+ \end{bmatrix}_W \quad [18.9]$$

Based on this scheme, the following expression for K_{biphasic} can be derived:

$$K_{\text{biphasic}} = \frac{K_{W,\text{nonionic}} \left(\dfrac{1 + \alpha m_{\text{ester}}}{1 + \alpha m_{\text{acid}}} \right) \left(\dfrac{1 + \alpha m_{\text{H}_2\text{O}}}{1 + \alpha m_{\text{alc.}}} \right)}{1 + \dfrac{K_{a,\text{biphasic}}}{[\text{H}^+]}} \quad (18.16)$$

where $K_{W,\text{nonionic}}$ is the equilibrium constant for reaction 18.8 obtained from the following expression for its apparent equilibrium constant:

$$K_{W,\text{app}} = \frac{[\text{RCOOR}'][\text{H}_2\text{O}]}{[\text{RCOOH} + \text{RCOO}^-][\text{R}'\text{OH}]} = \frac{K_{W,\text{nonionic}}}{1 + \dfrac{K_a}{[\text{H}^+]}} \quad (18.17)$$

and $K_{a,\text{biphasic}}$ can be determined by using Equation 18.14.

Reactions where both reactants ionize

Here, we consider a nonionic biomolecular reaction with two ionogenic groups, exemplified by the synthesis of an amide:

$$\begin{bmatrix} \text{RCOOH} + \text{R}'\text{NH}_2 \xrightleftharpoons{K_{W,\text{nonionic}}} \text{RCONHR}' + \text{H}_2\text{O} \\ \updownarrow K_a \quad \updownarrow K_b \\ \text{RCOO}^- \quad \text{R}'\text{NH}_3^+ \end{bmatrix}_W \quad [18.10]$$

Note that no ionic reaction is considered. When conducted in a biphasic medium, the partition of all of the reactants and products in the two phases must be taken into account. Thus, now the reaction scheme becomes

$$\left[\text{RCOOH} + \text{R}'\text{NH}_2 \xrightleftharpoons{K_{S,\text{nonionic}}} \text{RCONHR}' + \text{H}_2\text{O} \right]_S$$

$$\updownarrow m_{\text{acid}} \quad \updownarrow m_{\text{amine}} \quad \updownarrow m_{\text{amine}} \quad \updownarrow m_{\text{H}_2\text{O}}$$

$$\begin{bmatrix} \text{RCOOH} + \text{R}'\text{NH}_2 \xrightleftharpoons{K_{W,\text{nonionic}}} \text{RCONHR}' + \text{H}_2\text{O} \\ \updownarrow \quad \updownarrow \\ \text{RCOO}^- \quad \text{R}'\text{NH}_3^+ \end{bmatrix}_W \quad [18.11]$$

Clearly, the partition of the unionized reagents between the two phases affects the extent of ionization of the acid and the amine and the equilibrium of amide synthesis from the nonionic forms of the reactants. Next we give the set of equations needed for calculating K_{biphasic}:

$$K_{\text{biphasic}} = \frac{[\text{amide}]_t [\text{H}_2\text{O}]_t}{[\text{RCOOH} + \text{RCOO}^-]_t [\text{R}'\text{NH}_2 + \text{R}'\text{NH}_3^+]}$$

$$= \frac{K_{\text{nonionic,biphasic}}}{1 + \dfrac{[\text{H}^+]}{K_{\text{b,biphasic}}} + \dfrac{K_{\text{a,biphasic}}}{[\text{H}^+]} + \dfrac{K_{\text{a,biphasic}}}{K_{\text{b,biphasic}}}} \quad (18.18)$$

where subscript t represents the total concentration in the two phases taken together; $K_{\text{a,biphasic}}$ and $K_{\text{b,biphasic}}$ are given by Equations 18.14 and 18.15, respectively; and

$$K_{\text{nonionic,biphasic}} = K_{\text{W,nonionic}} \frac{(1 + \alpha m_{\text{amide}})(1 + \alpha m_{\text{water}})}{(1 + \alpha m_{\text{amine}})(1 + \alpha m_{\text{acid}})} \quad (18.19)$$

where $K_{\text{W,nonionic}}$ is given by Equation 18.17.

An inspection of Equation 18.18 and of those defining the various parameters therein shows that K_{biphasic} is a function of two operating parameters, pH (i.e., pK_a or pK_b) and α. In fact, it can be shown that at realistic values of α ($10^2 - 10^3$) the equilibrium constant can increase up by two to three orders of magnitude.

Reactions with ionization accompanied by both nonionic and ionic reactions

This represents a coupled situation that defines the system completely. However, it leads to lengthy and cumbersome equations and is not attempted here. Furthermore, it is merely an algebraic extension of the analysis just presented.

SOLVENT SELECTION

The solvent should be immiscible with the aqueous phase. This is particularly important in biocatalytic systems to prevent denaturing the biocatalyst (see, e.g., Martinek and Berezin, 1977; Butler, 1979; Playne and Smith, 1983). In addition, it should have high capacity for the reactants and products. Ideally, the substrates and products should partition entirely into the organic phase, which allows easier product separation after the reaction is completed. This also prevents the products from inhibiting the reaction and from undergoing further hydrolysis themselves. Table 18.2 contains a list of solvents typically used in laboratory studies of biphasic reactions (Lilly, 1982). It is obvious that some of these solvents are unacceptable for use in industrial processes today due to increasingly stringent waste and emission regulations. It is also important to consider the flammability of the solvent. For example, diethyl ether is highly flammable and may pose a fire hazard if the reaction requires constant bubbling of oxygen through the mixture to maintain high catalytic activity of the enzyme.

Table 18.2 Commonly used organic solvents in biphasing

Benzene	Diethyl ether	Hexadecane
Butyl acetate	Di-isopropyl ether	Hexane
Carbon tetrachloride	Dipentyl ether	Methylene chloride
Chloroform	Ethyl acetate	Petroleum ether
Cyclohexane	Ethylene chloride	Toluene
Dibutyl ether	Heptane	Trichloroethane

Guidelines for Solvent Selection

From the theoretical equations presented in the foregoing section, we note that BEF is a function of the nature of the solvent used, as well as the ratio α of its volume to that of water. It is always tempting to formulate rules for selecting the "best" solvent from both of these points of view.[2] This is seldom possible, however, and resort to trial-and-error selection seems unavoidable. Even so, some useful guidelines can be suggested.

A criterion based on solvophobic interactions

An important criterion is the solvent's capacity for solvophobic interactions, of which hydrophobic interactions are a specific case (Ray, 1971; see also Cramer, 1977; Klibanov et al., 1978; Tanford, 1978; Kauzmann, 1979; Hildebrand, 1979). According to this criterion, solvents can be grouped in the following three classes in decreasing order of solvolytic interactions, that is, of biphasic effectiveness:

Class 1 Water, glycerol, ethylene glycol, amino ethanol, and formamide
Class 2 Methyl formamide and dimethyl formamide
Class 3 Methanol, ethanol, and toluene

Class 1 solvents are most probably effective because they contain at least two centers capable of forming hydrogen bonds. Extensive evaluation of this criterion was reported by Klibanov et al. (1978) (see also Martinek and Semenov, 1981b) using the catalytic activity of a number of enzymes in several water–solvent biphasic media. A typical plot from their results is produced in Figure 18.5. Class 1 solvents, glycerol and ethylene glycol, deactivate the catalyst to a lesser extent than dimethyl formamide, a class 2 solvent, or ethanol, a class 3 solvent, the last two of which fall on a single curve.

In addition to these criteria, mentioned above, certain guidelines specific to immobilized enzymes have emerged (Laane et al., 1985; Brink and Tramper, 1985). These guidelines can also be extended to free enzymes if it is remembered that they are more fragile than their immobilized counterparts and may experience denaturing at the aqueous–organic interface. The following four guidelines are noteworthy.

---

[2] This is also true for the case of reaction extraction considered in Chapter (25).

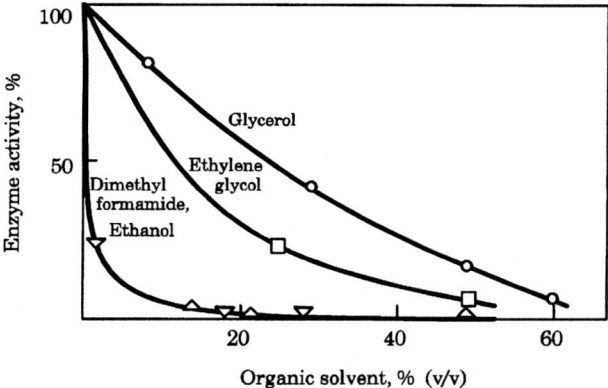

Figure 18.5 The relative catalytic activities of penicillin amidase in water–organic systems as a function of concentration (v/v) of the organic component (Martinek and Semenov, 1981b).

1. A weak, negative correlation exists between the polarity of the solvent and the retention of activity by the enzymes. This indicates that the less polar the solvent, the more suitable it would be for the biphasic system.
2. Although no specific relationship exists between the size of the solvent molecule and activity retention, there is little doubt that the size of the solvent molecule is important. The reason behind this is that the larger the solvent molecule, the more susceptible it is to steric hindrances that prevent it from diffusing into the enzyme support. This protects the enzyme from denaturing by the solvent molecule.
3. A strong correlation exists between the activity of enzymes and the logarithm of the partition coefficient of the solvent in the standard octanol–water system (Laane et al., 1985). The partition coefficient can be determined experimentally or can be calculated from hydrophobic fragmental constants (see Chapter 3). When the molecular weight of the solvent molecule is also taken into consideration, more than 70% of the enzymatic activity is retained with organic solvents that have a log m value greater than 4. Solvents whose log m is less than 2 deactivate the enzymes altogether.
4. It is advisable to use solvents with high partition coefficients when enzymes are inhibited by a high substrate or product concentration in the aqueous phase (Carrea, 1984).

In general, however, solvent selection is an exercise in compromise. For example, a solvent that partially destabilizes and inhibits an enzyme but has higher substrate solubility is actually better suited than one that does not affect the enzyme but has low substrate solubility.

ROLE OF AQUEOUS PHASE VOLUME

The volume ratio of organic to aqueous phases can greatly affect the reaction. The majority of reported studies recommend using only enough water to dissolve

or activate the catalyst. A large amount of water increases the thickness of the protective aqueous layer surrounding the catalyst, with a corresponding increase in the diffusion path of the reactants and products to and from the catalyst. Therefore, the volume of the aqueous phase typically needs to be optimized.

Once the organic solvent has been selected and the optimum volume ratio of organic to aqueous phases is determined, the water content of the solvent must be considered. There are several different reasons for this. First, using the same solvent from different suppliers leads to different results due to dissimilar water contents of the solvents (Yokozeki et al., 1982). To remedy the situation, water-saturated solvent can be used. It has also been found that when solvents are poorly hydrated, the solutes may "drag" water molecules into the organic phase (Tsai et al., 1993; Fan et al., 1994). A theoretical analysis of this "water-dragging effect" in biphasic systems has also been attempted.

When the products in the organic phase are susceptible to hydrolysis, the activity of the water content must be reduced while maintaining the activity of the catalyst. This can be accomplished either by immersing a desiccant in the organic solvent to lower the water activity of the liquid phase or by placing it in the headspace of the reactor to reduce the activity of the water vapor.

MATCHING THE pH OPTIMA FOR REACTION EQUILIBRIUM AND CATALYST ACTIVITY

We have seen how the pH of the medium can influence the activity of an enzyme, as well as the thermodynamic conversion of a reaction. Usually, the pH optima for the two are far removed from each other. Thus optimization with respect to one can be done only at the expense of the other. For instance, for β-lactam antibiotics, the optimum pH is in the range of 3–5 (Margolin et al., 1980), for p-nitroanilides the optimum pH is 2–3, and for N-substituted amino acid esters the optimum pH is 3 (Kozlov and Ginodman, 1965; D'Yachenko et al., 1971). But most of the enzymes that catalyze these reactions, for example, penicillin amydase and tripsin, are active and stable only at neutral and near neutral pH.

A practical way of matching the disparate pH values for equilibrium synthesis and maximum enzyme activity is to use a biphasic medium. Thus, consider the case of α-chymotripsin-catalyzed synthesis of N-benzoyl-L-phenylalanine ethyl ester:

$$\text{R-COOH} + \text{R'OH} \xrightleftharpoons{K} \text{R-COO-R'} + \text{H}_2\text{O}$$
$$\Updownarrow K_a \qquad\qquad\qquad\qquad\qquad\qquad\qquad [18.12]$$
$$\text{R-COO}^-$$

The pH dependence of the equilibrium constant for this reaction in water is shown in the lower curve of Figure 18.6 (Kozlov and Ginodman, 1965). Clearly, the equilibrium constant at pH $\approx$ 7, at which the activity of α-chymotripsin is maximum, is very low. Hence it is highly uneconomical to carry out the reaction in water. However, when the reaction is conducted in a biphasic medium, the

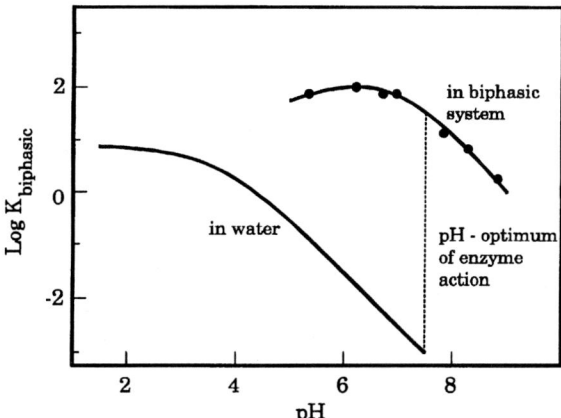

Figure 18.6 pH dependence of the apparent equilibrium constant for the synthesis of N-benzoyl-L-phenylalanine ethyl ester in water and in a biphasic system consisting of chloroform and 5% (v/v) water (redrawn from Martinek et al., 1980, 1981b)

effective pK_a of the system is raised considerably (see Equation 18.14), and the pH dependence of the equilibrium constant shifts toward neutral (and alkaline) pH. The upper curve of Figure 18.6 shows this dependence for a chloroform–water medium consisting of 5% (v/v) water (Semenov and Martinek, 1980; Martinek et al., 1981b). It can be seen that the pH thermodynamically favorable for the reaction has been made to coincide with the pH corresponding to the maximum α-chymostripsin stability and activity.

MODELING OF BIPHASIC BIOCATALYTIC SYSTEMS

The general principles of liquid-liquid reactions were described in Chapter 15. Although these principles are equally applicable to biphasic biocatalytic systems, an additional feature must be considered in these reactions, the special role of surface reaction. In regular liquid-liquid reactions there is only one set of intrinsic reactions, whether at the surface or in the bulk. In biocatalytic systems, however, the intrinsic reaction rate at the surface is elevated which complicates the analysis. First we classify the various types of biocatalytic biphasic reactions. Then, considering the simplest case, we present a brief analysis that simultaneously accounts for the elevated rate at the surface and absorption accompanied by reaction in the aqueous film and/or bulk. In regular nonenzymatic liquid-liquid systems, only the latter effect, as analyzed in Chapters 15 and 16, is operative.

Classification of Reaction Systems

Reaction in a liquid-liquid biocatalytic system is usually confined to the aqueous phase and occurs by mass transfer of the reactant from the organic phase to the

enzyme-containing aqueous phase. The actual extent of reaction and the ease of product removal, however, depend on the distribution of the reactant(s) and the product(s) of reaction between the two phases. Again this depends on the type of reaction involved. Even for the simplest case of the type $A \rightarrow R$, four distinct types of distribution are possible:

1. Both A and R are predominantly in the aqueous phase, that is, the distribution coefficients m_A and m_R are close to zero.
2. Both A and R are predominantly in the organic phase, that is, both m_A and m_R are close to infinity.
3. A is predominantly in the organic phase and R in the aqueous phase, that is, m_A is close to infinity, and m_R to zero.
4. A is predominantly in the aqueous phase and R in the organic phase, that is, m_A is close to zero and m_R to infinity.

Simple reasoning will show that nine distributions are possible for reactions of the type $A + B \rightarrow R + S$. Table 18.3 (adapted from Lilly and Woodley, 1985) lists examples of four types of reactions with a clear indication of the phase(s) in which the different components are present in each case, that is, whether the distribution coefficient m toward the solvent is infinity or zero. For instance, in reaction 1 of the table, reactant A is in the organic phase, and product R is in the aqueous phase. The methods of analysis for all these cases are similar to those described in Chapter 15, but each case will have its own set of material balance equations that must be incorporated in the overall methodology of regime analysis and reactor design described in that chapter. A fuller consideration of them is outside the scope of this treatment.

Because two liquid phases are involved, any one of them can be the dispersed phase, and the other the continuous phase. However, in biocatalytic systems, the volume of the aqueous phase is always much less than that of the organic phase. Hence the aqueous phase is normally the dispersed phase in these systems. One can visualize three situations:

1. dissolved catalyst, $V_w < V_s$
2. insoluble catalyst, $V_w < V_s$
3. insoluble catalyst, $V_w \ll V_s$.

The physical representations and concentration profiles for all of these cases are shown in Table 18.4. Case 1 involving a soluble catalyst is the simplest, and the profiles are similar to that of any liquid-liquid reaction described in Chapter 15. When the catalyst is insoluble in the aqueous phase (case 2), an increase in the concentration of A can be noticed at the aqueous–biocatalyst interface because the reactant is more soluble in the catalyst than in the aqueous phase (a situation not observed in straight liquid-liquid systems). This system is further complicated if diffusional effects within the catalyst are considered, as discussed in Chapter 7 on solid catalysts. The profile shown for case 3 is characteristic of systems in which the catalyst is insoluble in the aqueous phase and the volume of the aqueous phase is so small that there is no aqueous bulk but only an aqueous film surrounding each solid particle, which can be any supported catalyst, an

Table 18.3 Examples of biphasic reactions of different classes

Type of reaction	Distribution coefficient[a]				Reaction	Reference
	m_A	m_B	m_R	m_S		
1. $A \xrightarrow{E} R$	∞	NA	0	NA	Synthesis of benzene-cis-glycol	Ballard et al. (1983)
	∞	NA	∞	NA	Dehydrogenation of steroid Δ'	Yamane et al. (1979)
2. $A \xrightarrow{E} R + S$	0	NA	∞	∞	Hydrolysis of methyl acetate	Brookes (1984)
3. $A + B \xrightarrow{E} R$	0	∞	∞	NA	Synthesis of fat	Semenov et al. (1981)
	0	NA	0	∞	Synthesis of phenylalanine	Mitsubishi Co (1982)
4. $A + B \xrightarrow{E} R + S$	0	∞	0	∞	Reduction of cortisone	Cremonesi et al. (1975)

[a] NA = not applicable.

immobilized enzyme, or a suspended cell. The modeling of these systems becomes progressively more difficult as one moves from case 1 to 3. We examine case 1 below.

Reaction and Mass Transfer in Soluble Enzyme Systems

Consider a soluble enzyme E reacting with a substrate A from the organic phase (Figure 18.7) to give product R according to the reaction

$$A + E \rightarrow AE \rightarrow R + E \qquad [18.13]$$

First A must diffuse through the organic film where no reaction is assumed to occur. Then it dissolves in the aqueous phase at the interface and diffuses through the aqueous film into the bulk, with or without reaction in the film. Then the rest of the reaction (or the entire reaction) is completed in the aqueous bulk. The concentration profiles of A, E, and R are shown in the figure. It is assumed that there is no depletion of E in the film, even if reaction should occur there. Mathematical analyses based on this representation (and its variants) have been attempted, for example, by Pereira et al. (1987), Woodley et al. (1991), Miyake et al. (1991), Kamat et al. (1992), Chae and Yoo (1997). The following treatment is based essentially on the rigorous analysis of Miyake et al. (1991).

Total rate

The situation described previously conforms to regimes 1, 2, and 3, or 2–3 of Chapter 14. There is, however, a major difference in the formulation of the overall rate. In this case, because of the surface active nature of the reactants, the reaction rate at the surface would be higher than the intrinsic rate in the

Table 18.4 Physical representations and concentration profiles for systems involving dispersed aqueous phase with dissolved or insoluble catalyst

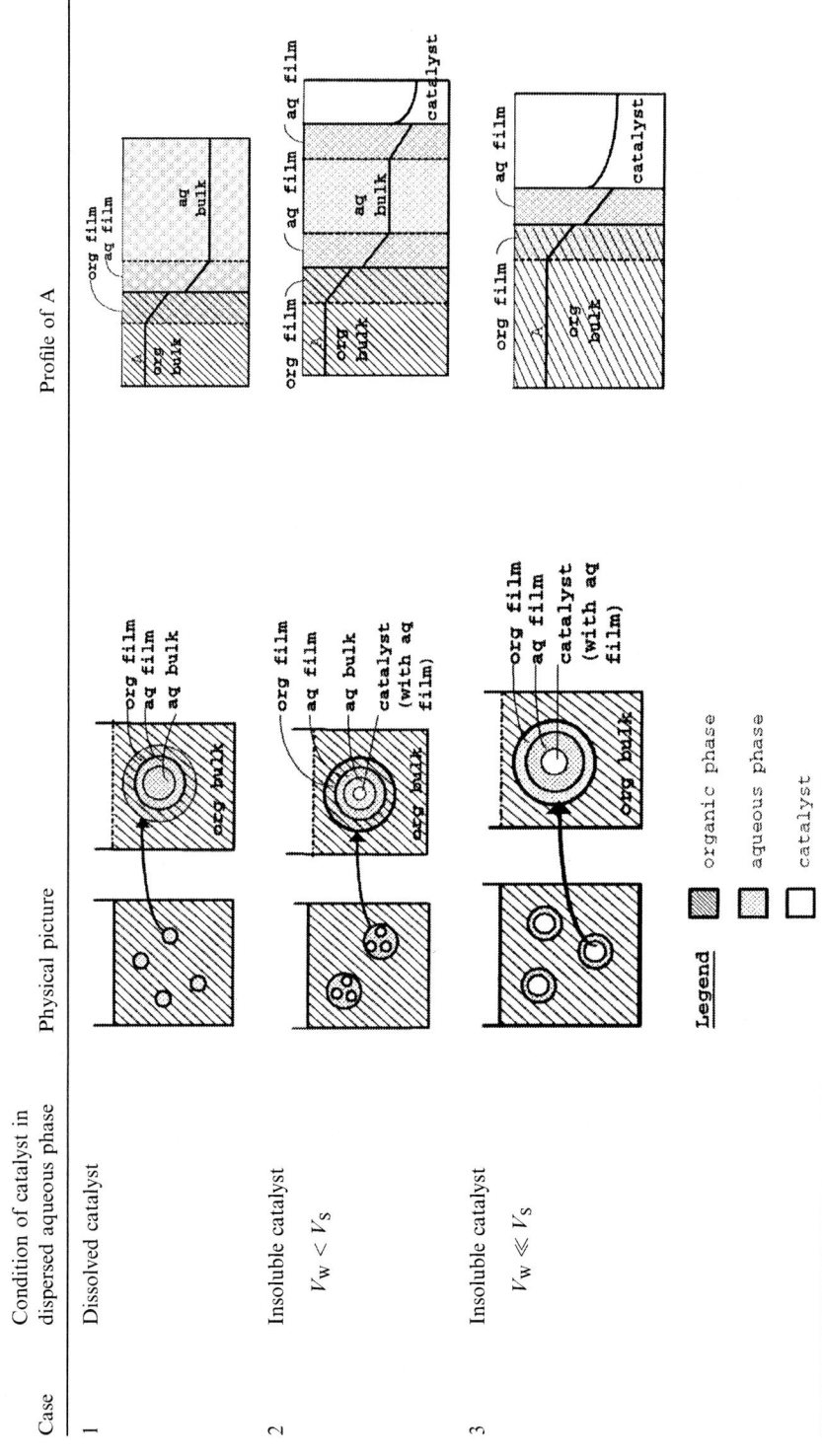

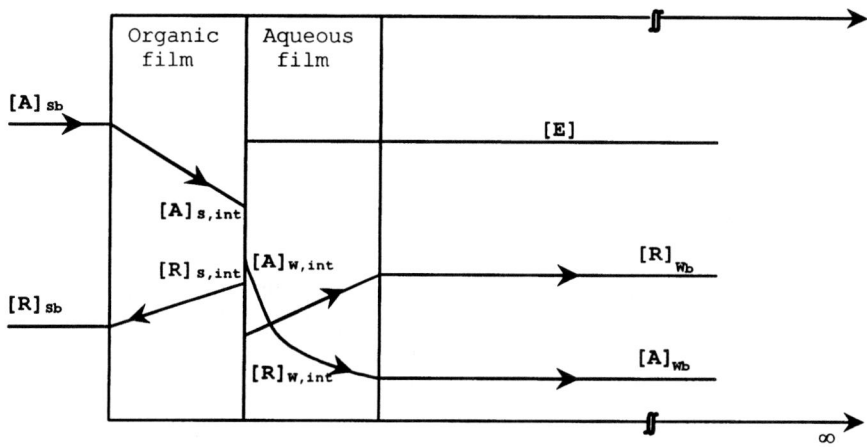

Figure 18.7 Reaction-mass transfer model for the enzymatic reaction $A \xrightarrow{E} R$.

aqueous bulk. Thus,

$$(\text{total rate}) = \begin{bmatrix} \text{liquid-phase} \\ \text{reaction in film} \\ \text{and/or bulk, i.e.,} \\ \text{regime 1, 2, 3,} \\ \text{or 2-3} \end{bmatrix} + \begin{bmatrix} \text{surface} \\ \text{reaction} \\ \text{at the} \\ \text{interface} \end{bmatrix} \quad (18.20)$$

The total reaction can be followed by monitoring the reactant or product concentration as a function of time. Then the following overall expression can be derived for the total product concentration $[R]$ as a function of time (Miyake et al., 1991):

$$[R](t) = [R]_1(t) + [R]_2(t) \quad (18.21)$$

where $[R]_1$ and $[R]_2$ represent the contributions to the total concentration of R from the reactions in the aqueous film/bulk and at the interface respectively. The equations for R_1 and R_2 are cumbersome but easy to use and are not given here (see Miyake et al., 1991).

Aqueous phase reaction (film and bulk)

The aqueous phase contribution $[R]_1$ is a function of the interfacial concentration $[A]_{\text{int}}$, mass transfer coefficient k'_L, partition coefficients m_A and m_R, and rate constant k_b in the aqueous phase. These can be experimentally determined; m_A, m_R, and k'_L can also be determined from available correlations (Chapters 3 and 17). The interface concentration $[A]_{\text{int}}$ can be taken as the solubility of A.

Once $[R]_1$ is known, $[R]_2$ can be calculated from Equation 18.21. If one of these is less than 10% of the total, it can usually be neglected and the rate based on the other. If $[R]_1$ is controlling, then the analysis becomes identical to that for liquid–liquid reactions, and the controlling regime can be determined as

described in Chapter 15. If R_2 is controlling, an appropriate Michaelis–Menton equation must be established based on actual surface values.

Interfacial reaction

In formulating an expression for the rate of this reaction, we must remember that the Michaelis–Menten equation given in Chapter 20 is applicable only to reaction in the liquid (aqueous) phase and can occur in all the regimes simultaneously with diffusion. Thus, although it can be used as the rate equation for calculating the bulk phase contribution (the second term of Equation 18.20), it has to be modified for calculating the surface rate by allowing for the enhanced surface concentrations of A and E. This is done by incorporating an excess adsorption quantity in calculating the surface concentrations of A and E. Thus the actual or effective concentrations of reactant and enzyme are used in the Michaelis–Menten equation instead of the normal liquid-phase concentrations to give (Harada and Miyake, 1986)

$$-r'_{A,e} = \frac{k_e[E]_e[A]_e}{K_{M,e} + [A]_e} \tag{18.22}$$

In this equation $K_{M,e}$ and k_e are the new effective constants of the Michaelis–Menten equation, and

$$[E]_e = [E] + \frac{\Gamma_E}{\delta'} \tag{18.23}$$

$$[A]_e = [A]_{int} + \frac{\Gamma_A}{\delta'} \tag{18.24}$$

where Γ is the excess quantity and δ' is the thickness of the actual interfacial zone where the reaction occurs (to be distinguished from the film thickness δ). The excess adsorption quantities are typically expressed by Langmuir type equations,

$$\Gamma_E = \frac{\Gamma_E^* K_E[E]_W}{1 + K_E[E]_W + K_A[A]_{int}} \tag{18.25}$$

$$\Gamma_A = \frac{\Gamma_A^* K_A[A]_{int}}{1 + K_E[E]_W + K_A[A]_{int}} \tag{18.26}$$

The adsorption constants of these equations can be obtained from interfacial tension measurements (Matijevic, 1969).

The calculation procedure consists of the following steps:

1. Estimate $[R]_1(t)$ at time t and then obtain $[R]_2(t)$ from Equation 18.21. These values give the fraction of the concentration corresponding to the surface reaction.
2. From experimental values of $[A]_s$ (assumed equal to $[A]_{int}$) and $[E]_W$, values of Γ_A and Γ_E obtained from Equations 18.25 and 18.26 using interfacial tension measurements, and estimated values of δ'

(Mallamace et al., 1981), solve Equation 18.22 to find the constants k_e and $K_{M,e}$ of that equation.
3. If one of the contributions (R_1 or R_2) is less than 10% of the total, ignore it, and base the rate exclusively on the other.

USE OF BIPHASING IN ORGANIC SYNTHESIS

The theoretical developments presented previously are quite general and are equally applicable to organic reactions without enzymes. However, in such cases, biphasing is only one of a number of strategies in which reaction is combined with a variety of separative techniques, for example, extraction, distillation, and crystallization. Therefore, it will be considered along with other strategies in Chapter 25 and will be concerned largely with methods based on the principles of acid–base reactions.

In this section we describe a strategy generally called chemical protection by biphasing and another in which biphasing is used to dramatically enhance the scope of homogeneous catalysis through development and use of novel catalysts characterized by their water solubility. The potential of the latter appears to be enormous.

Chemical Protection by Biphasing

Theory

By chemical protection we mean protecting a chemical species produced by a reaction from undergoing further conversion to an unwanted product by removing it from the scene of the reaction. For example, in the reaction

$$HO(CH_2)_6OH + HBr \rightarrow HO(CH_2)_6Br + H_2O \qquad [18.14]$$

the second OH group can be protected from undergoing further reaction to $Br(CH_2)_6Br$ by removing it from the phase in which the reaction is carried out (aqueous phase in this case) by biphasing the system, that is, by adding a hydrocarbon that preferentially extracts the monobromide. A completely general analysis of such a situation for any two liquid phases is difficult, but it is possible to draw some broad conclusions for systems where one of the phases is water. Thus, if we designate the compound to be protected as A and water as B, we can generalize this strategy by considering the following three possible ways in which A can react: (1) with water dissolved in the organic phase, (2) with water in the aqueous phase, and (3) with water at the interface.

Usually, the solubility of water in the organic solvents normally used in biphasing is very low. Thus good protection can usually be achieved in the organic phase. In the aqueous phase, however, A is totally unprotected. Because reaction is slow in the organic phase, a measure of protection can be achieved by using a solvent with a high distribution coefficient for A and by further increasing it by "salting out" A with salts such as NaCl, K_2CO_3, and NaOH (Brändström, 1983). The greatest amount of reaction occurs by the third

method, reaction at the interface. By assuming that the surface concentrations are the same as in their respective phases, we can write

$$\text{Rate} = k_a [A]_s [B]_W \tag{18.27}$$

The observed rate constant k_a can be approximated by

$$k_a = k \frac{V_{int}}{V_s} \tag{18.28}$$

where V_s and V_{int} are the volumes of the organic and interface regions, respectively. Thus an important requirement for protecting A is that the interfacial volume and/or rate constant should be low.

Examples

There are essentially two methods of protecting a desired product by biphasing: (1) removing the product itself from the reaction phase by layer separation, and (2) removing by layer separation the "harmful" reagent that has the potential to react with it further.

PREPARATION OF OPTICALLY ACTIVE EPIBROMOHYDRIN

An example of category 1 is the preparation of optically active epibromohydrin by base-catalyzed hydrolysis and dehydrogenation of the monoester of phthalic acid, as shown in reaction 18.15.

$$\text{[phthalic acid monoester with COOCH}_2\text{CHBrCH}_2\text{Br and COOH]} \xrightarrow{\text{pH 12}} \text{H}_2\text{C}\underset{O}{-}\text{CHCH}_2\text{Br} \qquad [18.15]$$

To protect the product from further reaction, it is extracted from the reaction mixture by adding a hydrocarbon or dichloromethane.

PREPARATION OF DIMETHOATE

Another example of category 1 is the manufacture of dimethoate (O,O-dimethyl-(S)-methylcarbomoylmethyl phosphorodithioate), an antiparasitic with low toxicity to warm blooded animals. It is prepared by reacting an alkali metal salt of O,O-dimethyl phosphorodithioate with N-methylchloroacetamide in a homogeneous medium like acetone. However, the yield can be considerably improved by bringing the nucleophile and the substrate into reactive contact in water in the presence of a second organic solvent immiscible with the first and capable of dissolving the reaction product (Gonzalez-Velasco et al., 1996). Thus the product is immediately removed from the scene of the reaction in the aqueous phase, preventing any impurity from being formed and dispensing with the need for purification.

SCHOTTEN–BAUMANN ACYLATION

Ester formation is a typical example of category 2. Although other methods such as the use of distillation column reactors (see Chapter 25) are preferred, layer separation can also be used by adding dichloromethane to remove the water. A well-known example of biphasing in organic synthesis belongs to this class, the Schotten–Baumann acylation reaction

$$RCOCl + R'NH_2 \rightarrow RCONHR' + HCl \qquad [18.16]$$

in which the HCl is removed by an aqueous sodium hydroxide layer.

REPLACEMENT OF HAZARDOUS TECHNOLOGY: HOFMANN REARRANGEMENT

A particularly attractive feature of biphasing is that it can replace the use of hazardous chemicals like phosgene. The Hofmann reaction for converting an amide (e.g., 2-ethylhexanoic acid amide) to the corresponding isocyanate with aqueous sodium hypochlorite is a good example because a variety of isocyanates are used in the manufacture of fine chemicals. The Hofmann reaction involves a haloamide as an intermediate and rearrangement of the amide to the isocyanate group

$$R-C(=O)-NH_2 \longrightarrow R-C(=O)-NBr^{(-)} \longrightarrow \boxed{R-N=C=O \xrightarrow{H_2O} R-NH_2} \qquad [18.17]$$

The isocyanate group adds on water (boxed step) to give carbamic acid (not shown) which is decarboxylated to RNH_2 under alkaline conditions. However, by biphasing with a selected second liquid phase that has a high distribution coefficient for the isocyanate, the reaction can be practically arrested at the isocyanate stage. Thus the usual method of making isocyanates from phosgene can be circumvented.

USE OF WATER AS EXTRACTANT

In the previous examples, an organic solvent was used to extract and thus protect the desired compound. It is also possible to use water to extract the desired product from an organic phase. An interesting example (Brändström, 1983) is the epoxidation of olefinic compounds with metachloroperoxybenzoic acid, which is converted to metachloroxybenzoic acid (MCBA) in the process. Because MCBA undergoes further reaction with the epoxy compound in most organic solvents, it can be protected by biphasing with a buffered aqueous medium at pH 7, when more than 99% of MCBA is extracted into the aqueous phase.

Biphasing can also be advantageously used to improve hydrolytic efficiencies. This is particularly relevant to the manufacture of drug intermediates and relies heavily on the principle of reactive extraction. The biphasic efficiencies of these reactions are largely due to differences in the pK_a values. This aspect of biphasic reactions is considered at some length in Chapter 25.

TRANSFER OF IONS: A PRELUDE TO PHASE-TRANSFER CATALYSIS

A particularly attractive way of biphasing is to force the reactant from one phase into another where it can react with the reactant present in that phase. This is not exactly a method of protection but of forcing a reaction between components in two immiscible phases. It is always preferable to force ions into the organic phase than neutral molecules into the aqueous phase. This is best achieved by a technique known as phase-transfer catalysis. In view of the rapidly increasing importance of phase-transfer catalysis in organic technology/synthesis, it is considered separately in the next chapter.

Biphasing in Homogeneous Catalysis: "Mobile Immobilization"

Biphasing emerged as an attractive strategy for enhancing the industrial acceptability of homogeneous catalysis with the successful implementation in 1984 of the aqueous biphasic oxo technology for the hydroformylation of propylene. In this process (Kuntz, 1987), the reactants and products remain in the same phase, and the highly water-soluble catalyst (solubilized by using a water-soluble ligand) remains in the aqueous phase. As a result, the expensive processing steps of earlier technologies to separate product from catalyst are eliminated. In effect, the catalyst may be considered "heterogenized by a mobile support (water)" (see Kalck and Monteil, 1992; Herrmann and Kohlpainter, 1993; Haggin, 1994; NATO Advanced Research Workshop, 1994; Lubineau et al., 1994; Wiebus and Cornils, 1994, 1995; Cornils, 1995; Cornils and Wiebus, 1996).

Although water-soluble homogeneous catalysts have great potential in organic synthesis in general, developmental work so far has largely been restricted to hydroformylation of olefins (see, e.g., Morel, 1980; French Pats. 1981, 1983; Buhling et al., 1995,; Herrmann et al., 1995; Monflier et al., 1995, 1997; Ritter et al., 1996; Papadogianakis et al., 1997; Chen and Alper, 1997). Shell's Higher Olefins Process (SHOP) also uses a biphasic homogeneous catalyst but is unique in that the second liquid is also an organic solvent (Moulijn et al., 1993). A recent study also reports a novel class of versatile (nonaqueous) solvents for the hydrogenation, isomerization, and hydroformylation of alkenes (Chauvin et al., 1995a,b). These involve using room-temperature ionic liquids based on 1-n-butyl-3-methylimidazolium (BMI^+) salts, for example,

$$\left[Me\text{-}N\overset{+}{\frown}N\text{-}nBu \right] A^-, \quad A^- = BF_4^-, \ PF_6^-, \ SbF_6^-$$

Figure 18.8 lists several examples of the more successful water-soluble ligands used in forming biphasic homogeneous catalysts.

Because of its uniquely benign environmental quality, the use of water per se as the biphasing liquid has clearly been an important factor in the success of water-soluble homogeneous catalysts. It is equally important to note, however,

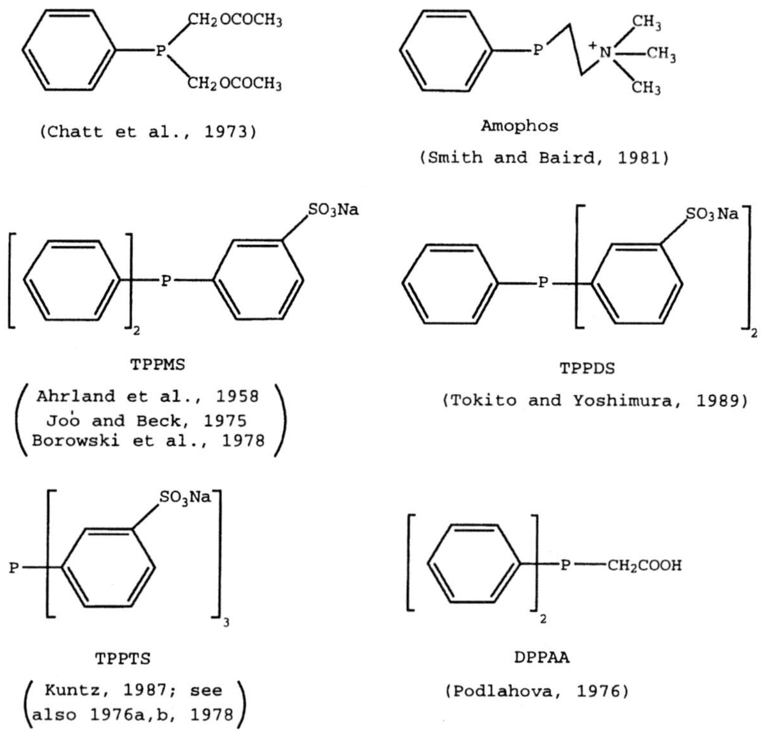

Figure 18.8 Examples of some successful water-soluble ligands. Amphos = (2-diphenylphosphinoethyl)-trimethylammonium cation; TPP = triphenylphosphine; MS, DS, TS = mono-, di-, and trisubstituted; DPPAA = diphenylphosphinoacetic acid (adapted from Cornils and Wiebus, 1995).

that the solubility of the starting organic reactants is usually very low in water, leading to correspondingly low reaction rates. This problem can be overcome by adding a surfactant, a solvating agent, or a cosolvent; but any such "foreign" additive leads to purification problems of its own, which may often be as severe as those associated with the very processes it seeks to replace. A potential solution is to use a "promoter ligand" which is soluble exclusively in the organic phase. Chaudhari et al. (1995) showed that the hydroformylation (with [HRh(CO)(TPPTS)$_3$] as catalyst) of a highly water-insoluble substrate such as 1-octene can be enhanced 10–50 times simply by adding triphenylphosphine to the octene. This strategy also suffers from the fact that again a foreign substance is introduced. Nevertheless, it opens up the possibility of developing ligands with carefully balanced hydrophilic and hydrophobic features.

Aqueous-Aqueous Biphasing

In the previous section we considered the case of biphasing an organic phase with an aqueous phase. An even more interesting technique that is steadily

gaining currency is where both the liquid phases are aqueous (see Walter and Johansson, 1994). This has unlimited potential, particularly in view of the increasing demand for environmentally benign technologies. The major techniques under investigation are aqueous biphasic systems (ABS), cloud point extraction (CPE), extractions using thermosetting polymer systems (TPS), and micellar extraction (ME) or, more appropriately, micellar enhanced ultrafiltration (MEUF). A recent review by Huddleston et al. (1999) attempts to put all of them on a common footing. They all involve several common features: they all solubilize otherwise insoluble hydrophobic species; no organic volatile compounds (VOCs) are involved; all are formed by addition of polymers to water; and all rely on the structuring properties of liquid water for separating it into two (or more) phases. Although it is doubtful if these types of extraction can be called reactive extractions as they are being used today, they can be (initially) used for systems of the type covered earlier in this chapter.

Aqueous biphasic systems (ABS)

ABS is generally used for extracting polymeric macromolecules from biological systems. Here, two or more water-soluble polymers, or a polymer and a salt, are added to an aqueous solution above a critical concentration. The single aqueous phase then separates into two distinct aqueous phases. Several theories have been advanced for this type of biphasing (see, e.g., Van Oss et al., 1989; Abbott et al., 1990; Huddleston et al., 1990). A particularly attractive theory is that the hydrated polymers cause differently structured water molecules, leading to their separation into two (or more) phases, all aqueous. Phase separation is enhanced by an increase in temperature (de Belval et al., 1998) and depends on the molecular weight of the polymer. A potential application is in the extraction of organic hydrocarbon species (Rogers et al., 1998; Willauer et al., 1999). It is interesting that the partitioning of organic species in the two aqueous phases follows the partitioning scheme for octanol-water systems outlined in Chapter 3 (see also Valsaraj et al., 1989; Zaslavsky et al., 1990; Willauer et al., 1999]. Typical examples of polymer-polymer and polymer-salt systems used in ABS are (Johansson et al., 1998; Huddleston et al., 1999): polymer-polymer systems: dextran with poly(ethylene glycol) or poly(propylene glycol), poly(ethylene glycol)-poly(propylene glycol), and poly(vinyl alcohol)-dextran); polymer-salt systems: Poly(ethylene glycol) with NaOH, KOH, Na_2CO_3, K_2CO_3, Na (formate) (univalent ions), $MgSO_4, Na_2WO_4, (NH_4)_2SO_4, Na_3S$ (divalent ions), $Na_3PO_4, (NH_4)_3$(citrate) (trivalent ions), Na_4SiO_4, Na_4(HEDPA) (tetravalent ions).

Cloud point extraction (CPE)

CPE is a technique in which an aqueous solution, upon addition of a surfactant and raising the temperature to its cloud point, becomes turbid and amenable to splitting into two phases: a surfactant phase and a dilute aqueous phase (Hinze and Pramauro, 1993). A *consolute curve* (i.e., cloud point versus concentration curve) may be determined, separating the one-phase and two-phase regions of

the system. This technique is particularly useful for the extraction of metals, preconcentration, or purification, but no great prospects are foreseeable in organic synthesis. Among the more common surfactants used for cloud point extraction are poly(oxyethylene glycol) monoethers (trade name: Brij), poly(oxyethylene methyl-*n*-alkyl ethers (trade name: Emulgen), and poly(oxyethylene) nonylphenyl ethers (trade name: Triton).

Thermosetting polymer systems (TPS)

TPS may be regarded as an extension of ABS in that it substitutes thermosetting polymers for the polymer-polymer or polymer-salt systems of the latter (see, e.g., Johansson et al., 1998; Persson et al., 1998). Examples of the polymers used are poly(ethylene glycol) (PEG), poly(vinyl alcohol) (PVA), PEO-PPO block polymers (Ploronic) and PEO-PPO random copolymers (UCON) (Hurter et al., 1995). This technique is particularly relevant in organic synthesis since it can be used in the production of pharmaceuticals (Alexandridis et al., 1994; Lopes and Loh, 1998).

Micellar enhanced ultrafiltration (MEUF)

In this type of extraction, micellar structures are retained by correctly selecting the ultrafiltration (UF) membrane (Scamehorn et al., 1988). Hydrophobic species are solubilized within the micelles, but surfactant monomers in equilibrium with the micelles can penetrate the membrane along with the free solutes in equilibrium with those solubilized in the micelles. Whereas several uses for this technique have been suggested, such as the collection of radioactive uranium and plutonium present in acid wastes during nuclear plant decommissioning, from our point of view its principal use is in enantiomeric separation (Overdevest et al., 1998).

EXAMPLES OF BIPHASING

From the discussion presented in the previous sections it would be clear that biochemical processes have been the more extensive beneficiaries of biphasing than has straight organic synthesis. Although the recent trend has been to use this highly effective tool in organic reactions in increasing measure, including aqueous-aqueous biphasing, the largest number of available examples is still in the area of bichemical biphasing. Many examples of biphasic reactions in organic synthesis/technology were presented in sections devoted to this area of biphasing. In Table 18.5 we give a number of examples of the use of biphasing in its conventional environment, that is, in biochemical processes.

Table 18.5 Examples of biphasic biochemical reactions

Reaction	Catalyst	Organic solvent	Enhancement	Reference
Enzymatic oxidation of ethanol and butanol	Alcohol oxidase	Toluene	Conversion increases with increasing volume fraction of organic phase	Hidaka et al. (1995)
Synthesis of poly(2,6-dimethyl-1,4-phenylene oxide)	Cuprous chloride, also used a surface active ligand to promote the reaction	Toluene and chloroform	Chloroform slightly more effective than toluene: polymer molecular weight on the order of 50,000 and yield in excess of 80%	Dautenhahn and Lim (1992)
		n-Heptane and diethyl ether	Not effective	
Hydrolysis of 2-naphthyl acetate	Lipase from *Rhizopus delemar*	n-Heptane	Assuming film thickness $\delta = 10^{-7}$ cm, enhancement is 10^6 times that in aqueous phase alone	Miyake et al. (1991)
Hydrolysis of olive oil	*Candida rugosa* lipase	Isooctane	No enhancement information available Equilibrium conversions higher than 98% obtained	Tsai et al. (1991)
Hydrolysis of tributyrin	Lipase of *Candida cylindracea*	n-Heptane	No enhancement information available	Ucar et al. (1989)
Converting olive oil to an interesterified fat (1- and 3-positional specific interesterification)	Immobilized lipase from *Rhizopus delemar*	Water saturated n-hexane	No comparison of yields in aqueous and biphasic system available	Yokozeki et al. (1982)
Synthesis of N-benzoyl-L-phenylalanine ethyl ester	α-Chymotrypsin	None	Yield as low as 0.01%	Martinek et al. (1981b)
		Diethyl ether	% yield: 26	
		Carbon tetrachloride	62	
		Benzene	64	
		Chloroform	80	

Process	Biocatalyst	Solvent	Notes	Reference
Synthesis of N-acetyl-1-tryptophanyl-L-leucine amide from N-acet-trp and L-Leu-NH$_2$	α-Chymotrypsin	Ethyl acetate	Yield is almost 100% compared with about 0.1% in water alone	Semenov et al. (1981)
Stereoselective hydrolysis of dl-menthyl succinate	*Rhodotorula minuta* var. *texensis* cells	Two-phase system of 20 mM potassium phosphate buffer and solvent (1 : 5)	The following are the % yields of L-menthol: t-Butyl acetate 0 Benzene 22.4 Benzene-n-heptane (1 : 1) 24.9 n-Heptane 72.6	Omata et al. (1981)
Synthesis of a dipeptide from an N-substituted aspartic acid and a lower alkyl ester of phenylalanine	Immobilized metalloproteinase	Recommend several different immiscible organic solvents	No enhancement information given Give 17 different reactions as examples (patent)	Oyama et al. (1980)
Transformation of steroids	Hydroxysteroid dehydrogenases β-HSDH, 20β-HSDH and 3α-HSDH free and immobilized on CNBr-activated Sepharose (comparative study)	Ethyl acetate	Behaves as a weak but competitive inhibitor for the substrate	Carrea et al. (1979)
Stoichiometric oxidation of 4-androstene-3,17-dione (4-AD) to androst-1,4-diene-3,17-dione (ADD) in the presence of phenazine methosulfate (PMS)	*Nocardia rhodocrous*, either free or immobilized by entrapment in a hydrophilic gel (H-gel) or lipophilic gel (L-gel)	Benzene–n-heptane (1 : 1 by volume)	The disappearance rate of 4-AD is increased by about 180 times when the solvent is added to the system	Yamane et al. (1979)

(*Continued*)

Table 18.5 (Continued)

Reaction	Catalyst	Organic solvent	Enhancement	Reference
Conversion of a hydroxyl group of a steroid to an oxo group	Enzymatic oxidation catalyst which is a microorganism or an extracted enzyme from the *Mycobacterium*, *Nocardia*, *Brevibacterium*, or *Arthrobacter*	CCl_4 is preferred 50% by volume	Enhancement information not available. Several examples of steroid conversion are given	Dunnill and Lilly (1976)
Synthesis of N-acetyl-L-tryptophan ethyl ester	α-Chymotrypsin	None Chloroform	Yield as low as 0.01% Yield almost 100% Equilibrium constant in biphasic system exceeds that in the aqueous system by more than three orders of magnitude	Martinek and Klibanov (1977)
Reduction of ketosteroids coupled with the catalyzed dehydrogenation of ethanol	20β-Hydroxysteroid 20β-(HSDH) and alcohol dehydrogenase (ADH)	None *n*-Hexane Isooctane Carbon tetrachloride Chlorobenzene Diethyl ether Ethyl acetate Butyl acetate	% cortisone conversion: 0 < 5 < 5 < 5 15 10 90 100	Cremonesi et al. (1975)

Conversion of cholesterol to cholest-4-ene-3-one	*Nocardia* sp.	Toluene Hexadecane Carbon tetrachloride Ether Benzene Petroleum ether	% conversion: 24 22 21 18 15 10	Buckland et al. (1975)
Dehydrogenation of testosterone coupled to pyruvate reduction	Enzyme 1: β-hydroxysteroid dehydrogenase Enzyme 2: lactate dehydrogenase	3 ml aqueous buffer plus 2.5 ml organic solvent: Acetone Ethanol Carbon tetrachloride Ethyl acetate Butyl acetate	% conversion of testosterone into androstenedione: 2 2 5 90 100	Cremonesi et al. (1974) Cremonesi et al (1973)
Oxidation of steroid hormones	Fungal laccase A from *Polyporus versicolor*, a copper-containing oxidase	Discusses several different solvents	No enhancement information available	Lugaro et al. (1973)

Chapter 19

Phase-Transfer Reaction Engineering

There are many situations in organic synthesis where it is desirable to bring about reaction between reactants present in two (or more) immiscible phases. Agents known as phase-transfer catalysts are used for this purpose. Their role in initiating or accelerating such reactions has been proven extensively since the early seventies, and the principles of their operation are being increasingly understood [see Weber and Gokel, 1977; Reuben and Sjoberg, 1981; Frechet, 1984; Freedman, 1986; Goldberg, 1992 (English translation); Dehmlow and Dehmlow, 1993; Starks et al., 1994; Yufit, 1995; Sasson and Neumann, 1997; Naik and Doraiswamy, 1998]. To date, an estimated 500 different commercial chemical processes (mostly for small volume chemicals) using about 5–25 million pounds per annum of phase-transfer catalysts have been reported (Starks et al., 1994), and well over 6,500 compounds have been synthesized in the laboratory using PTC (Keller, 1979, 1986). A large number of industrial applications of phase-transfer catalysis are found in the pharmaceutical, agrochemical, and fine chemicals industries. Additionally, it is now being increasingly used in processes related to the environment, in process modifications for eliminating the use of solvents, and in reactions related to the treatment of poisonous effluents. Not surprisingly, then, there has been a constant stream of publications and patents every year.

Phase-transfer catalysis (PTC) is an area that has largely been the province of the preparatory organic chemist (defined broadly to include organometallic and polymer chemists)[1]. It is only since the early eighties that the engineering aspects

---

[1] According to Tundo et al (1989), PTC was unconsciously used by organic chemists long before it was recognized as a distinct technique. They often catalyzed a nucleophilic reaction with a tertiary amine, which in the presence of the alkyl halide produced a quaternary ammonium salt, a typical PTC.

of phase-transfer catalysis are being explored, including such traditional features as mass and heat transfer and reactor design. Our main objective is to present a brief but coherent engineering analysis of PTC, following an introduction to its basic principles.

GENERAL FEATURES

Basic Principles

When two reactants are present in two different, immiscible liquid phases (usually one aqueous and the other organic), they can often be brought together by addition of a solvent that is both water-like and organic-like (e.g., ethanol, which derives its hydrophilic nature from its hydroxyl group and its lipophilicity from the ethyl group). However, the rate enhancement tends to be limited due to excessive solvation of the nucleophile. On the other hand, dipolar aprotic solvents such as dimethylformamide or acetonitrile that have lipophilic methyl groups and polar functional groups can be more effective because, in the absence of hydroxyl groups, they do not solvate anions. However, these solvents suffer from disadvantages such as a high boiling point, high cost, and difficulty in recovery.

An elegant and effective method developed during the last twenty five years is the use of phase-transfer agents in catalytic amounts (the origin of this term is attributed to Starks 1971; see also Makosza, 1975; Brändström, 1977). Any substance that can ion-pair with the anion of a nucleophile in the aqueous or solid phase (e.g., quaternary ammonium salts) or complex with its cationic half (e.g., crown ethers), thus extracting the nucleophile into the immiscible organic phase and then activating it for a reaction to occur there, can function as a phase-transfer (PT) catalyst.

Consider the reaction between an organic substrate RX and a nucleophilic reagent MY present in an immiscible aqueous solution or an insoluble solid phase. The addition of a quaternary ammonium salt, traditionally represented as QX, leads to the following reactions:

$$QX_{\text{org/aq}} + MY_{\text{S/L}} \rightarrow QY_{\text{org}} + MX_{\text{S/L}} \text{ (ion exchange)} \qquad [19.1a]$$

$$QY_{\text{org}} + RX_{\text{org}} \rightarrow RY_{\text{org}} + QX_{\text{org/aq}} \text{ (organic reaction)} \qquad [19.1b]$$

where subscript org represents the organic phase and S/L indicates that the nucleophile can be a solid or a liquid. In the ion-exchange reaction, the catalyst (QX) reacts with the nucleophilic reagent (MY) to give the active form of the PT catalyst (QY), and the second reaction in which the PT catalyst (QX) is regenerated occurs entirely in the organic phase. The cycle repeats itself.

Another important factor is the likelihood that the catalyst (QX, QY) will partition between the aqueous and organic phases in liquid-liquid systems. Thus electrostatic interactions and mass transport tend to govern much of the thermodynamics and kinetics of the PTC cycle.

PT Catalysts Commonly Used

Quaternary *onium* (ammonium, phosphonium) salts, often referred to as quats, were the first PT catalysts to be used; these ion-paired compounds continue to be the dominant class. Traditionally, quaternary onium salts, crown ethers, cryptands, and open chain polyethers such as polyethylene glycols (PEGs) have been used as phase-transfer catalysts. Various other specialized PT catalysts have also been reported in the literature, such as polymeric analogs of the dipolar aprotic solvents dimethyl sulfoxide, N,N-dimethylformamide, N-methyl-2-pyrrolidone, and tetramethylurea, in both soluble and immobilized forms (Kondo et al., 1994 and references therein); chiral PT catalysts based on the optically active amines ephedrine, chinine, or other cinchona alkaloids (Bhattacharya et al., 1986); TDA-1 (tris(3,6-dioxaheptyl) amine) synthesized by Rhone-Poulenc; a novel high-temperature PT catalyst (EtHexDMAP, N-alkyl salt of 4-dialkylaminopyridine) (Brunelle, 1987); and some special multisite PT catalysts (Idoux and Gupton, 1987; Balakrishnan and Jayachandran, 1995). The main features of the more important classes of catalysts are summarized in Table 19.1, and a few selected structures are given in Figure 19.1.

Classification of PTC Reactions

PTC reactions can be broadly (but not rigidly) classified into two main classes: *soluble PTC* and *insoluble PTC* (Figure 19.2). Further classification is possible

Table 19.1 Main features of the more important classes of PT catalysts

Catalyst	Cost	Stability and activity	Use and recovery of catalyst
Ammonium salts	Cheap	Moderately stable under basic conditions and up to 100 °C, decomposition by Hofmann elimination under basic conditions, moderately active	Widely used; recovery is relatively difficult
Phosphonium salts	Costlier than ammonium salts	More stable thermally than ammonium salts, although less stable under basic conditions	Widely used; recovery is relatively difficult
Crown ethers	Expensive	Stable and highly active catalysts both under basic conditions and at higher temperatures even up to 150–200 °C	Often used; recovery is difficult and poses environmental problems due to their toxicity
Cryptands	Expensive	Stable and highly reactive except in the presence of strong acids	Used sometimes despite high costs and toxicity, due to higher reactivity
PEG	Very cheap	More stable than quaternary ammonium salts, but lower activity	Often used; can be used when larger quantities of catalyst cause no problems; relatively easy to recover

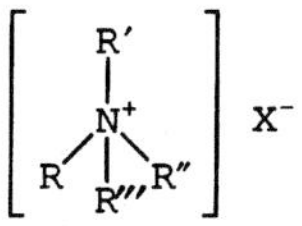

onium salt
N = N, P or Ar
R = alkyl

e.g. 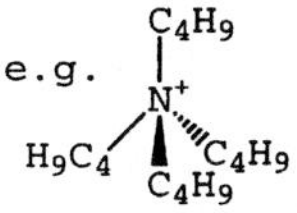

Tetrabutylammonium bromide

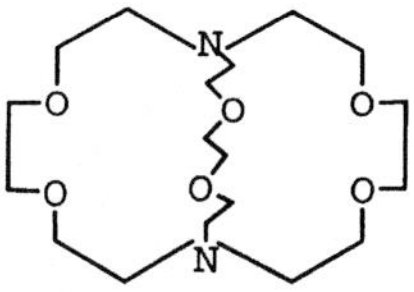

Cryptate
(Cryptofix 222)

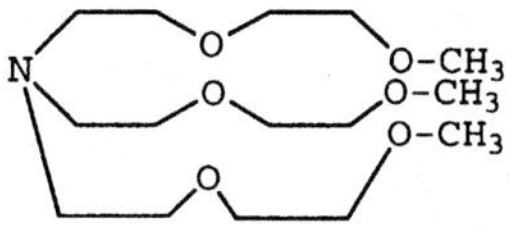

Tris(3,6-dioxaheptyl)amine
(TDA-1, open chain cryptate)

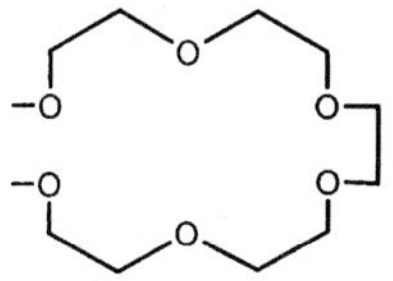

Crown ether
(18-crown-6)

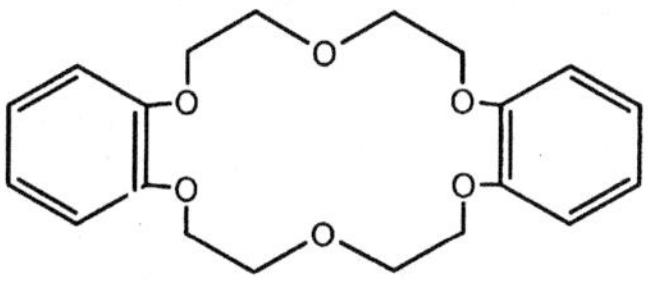

Podand
(Pentaethylene glycol
dimethyl ether)

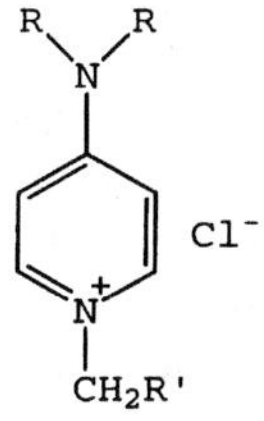

Thermally stable
trisdialkylamino pyridinium quat salt

(R, R' = alkyl)

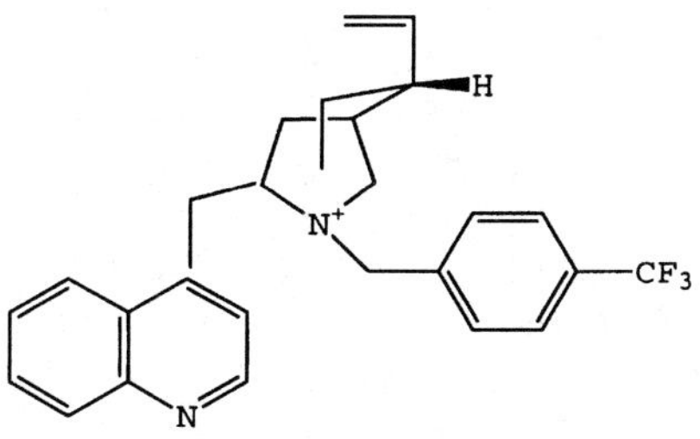

Br⁻ (8*R*),(9*S*)-*N*-(*p*-Trifluoromethyl benzyl)
cinchonium bromide

Figure 19.1 Structure of a few selected PT catalysts

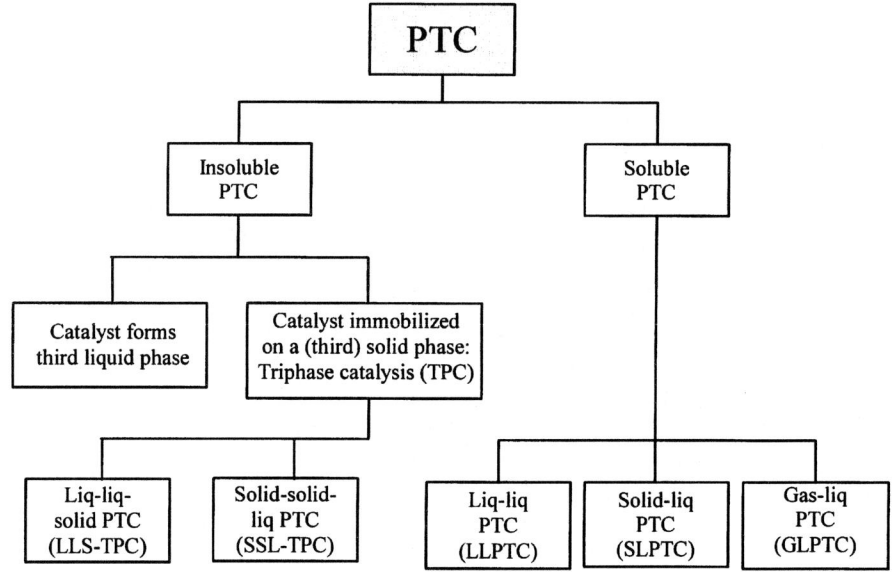

Figure 19.2 Classification of PTC reactions

within each class, depending on the actual phases involved. Thus in soluble PTC, we have liquid-liquid PTC (LLPTC), solid-liquid PTC (SLPTC), and gas-liquid PTC (GLPTC). Similarly, in insoluble PTC also, we have two subcategories: immobilized PTC where the catalyst is rendered insoluble by immobilizing on a solid support, and catalysis by a third liquid phase formed by the catalyst. Again, in immobilized PTC, often referred to as triphase catalysis (TPC) because of the presence of three phases, the two reactant phases can be liquids, giving rise to liquid-liquid-solid triphase catalysis (LLS-TPC), or one of them can be a solid, giving rise to solid-solid-liquid triphase catalysis (SSL-TPC). This classification, which is further discussed here, is not mutually exclusive because the catalyst in LLPTC can also be insoluble.

Soluble PTC

LIQUID-LIQUID AND SOLID-LIQUID SYSTEMS

In LLPTC, reaction occurs between a reagent in an aqueous phase and another in the organic phase. On the other hand, in SLPTC, reaction occurs between a solid nucleophile (MY) or a solid base (such as K_2CO_3) and an organic substrate usually in an organic solvent in which the solid nucleophile is insoluble or sparingly soluble. There is sometimes a distinct advantage in operating in the solid-liquid mode because the elimination of the aqueous phase lowers the degree of hydration of the ion pair and leads to an increase in its reactivity. On the other hand, traces of water are invariably needed in SLPTC to enhance the rate of the PTC cycle by allowing solid dissolution in a

thin aqueous film called the *omega phase* formed around the solid surface (Liotta et al., 1987).

GAS-LIQUID SYSTEMS

Gas-liquid PTC (GLPTC) is much less studied than SLPTC or LLPTC (Tundo et al., 1989; Tundo and Selva, 1995, and references therein). In this case, the organic substrate is present in gaseous form and is passed over a solid bed of the inorganic reagent (Tundo and Venturello, 1979) or an inert inorganic support (Angeletti et al., 1984), both of which are coated with a PT catalyst in its molten state. This last feature of these reactions is perhaps responsible for its somewhat misleading name "gas-liquid PTC."

Insoluble PTC

TRIPHASE CATALYSIS

A vexing problem with PTC is the difficulty and cost of separating the catalyst from the product. This can be overcome by immobilizing it on a suitable solid support, making separation very easy and economical—often requiring no more than a simple filtration (see Ford, 1984; Ford and Tomoi, 1984). However, the supported forms tend to be much more expensive and less active (due to diffusional limitations) than their soluble analogs and hence have not found much industrial acceptance despite their operational ease. In general, a common basic mechanism is applicable to both soluble PTC and immobilized PTC [or triphase catalysis (TPC)[2]], although the latter includes some additional steps to account for diffusion and mass transfer within the support phase.

CATALYSIS BY A THIRD (LIQUID) PHASE

In some liquid-liquid systems, the PT catalyst forms a separate (third) liquid phase, and this variant of PTC is generally referred to simply as insoluble PTC. The sudden increase in reaction rate that is sometimes observed when the quat concentration is increased beyond a critical value is attributed to the formation of a third phase. This phase is rich in the PT catalyst and contains little organic solvent or the aqueous nucleophile and only small quantities of water. Thus, ideally, it is possible to have a rapid rate of reaction, followed by separation of the organic and aqueous phases, with the third (catalyst-rich) phase amenable to easy recovery for further use (Wang and Weng, 1988; Ido et al., 1995a, 1995b).

Inverse PTC

A nontypical variant of PTC is inverse PTC (IPTC). This refers to a class of heterogeneous reactions similar to traditional PTC systems but based on the use of a phase-transfer agent to transfer species from the organic to the aqueous phase. Thus in this case, the main reaction occurs in the aqueous phase. Commonly used inverse PT catalysts include 4-diaminopyridine based compounds, pyridine, pyridine-N-oxides, and different cyclodextrin derivatives.

---

[2] We shall use immobilized PTC and triphase catalysis interchangeably in this chapter.

Noncatalytic Phase-Transfer Reactions

Another technique similar to phase-transfer catalysis is ion-pair extraction (Brändström, 1977) wherein lipophilic quaternary onium salts are used in stoichiometric amounts rather than catalytic amounts. Ion-pair extraction is particularly useful for leaving groups with high lipophilicity such as I^- and $ArSO_3^-$. Quaternary salts of highly hydrophilic anions such as HSO_4^- are used to transfer the anionic reactant from the aqueous phase into the organic phase. Although interesting, this technique has found limited application.

KINETICS AND MODELING OF SOLUBLE PTC

The most general model of PTC needs simultaneous consideration of all steps operative in the PTC cycle and their relative speeds and contributions to the overall rate. The actual kinetics of the reactions (with which the chemist is concerned) is part of this larger picture. Because this has been dealt with in detail by many authors (principally Starks et al., 1994), we restrict the treatment here largely to modeling with short detours into kinetics where needed.

Modeling of LLPTC

A common mechanism of LLPTC action for a nucleophilic substitution reaction is sketched in Figure 19.3a. According to this mechanism, the quat must necessarily dissolve in the aqueous phase, that is, it must partition between the phases to pick up the nucleophile. On the other hand, if the catalyst is insoluble in either one of the phases, the ion-exchange reaction is confined to the interfacial area between the two liquids. The resulting surface-oriented PTC cycle is illustrated in Figure 19.3b.

Factors that affect PTC reaction

The PTC cycle consisting of reactions 19.1a and 19.1b is influenced by several factors. These relate to both diffusion and reaction and are briefly discussed here.

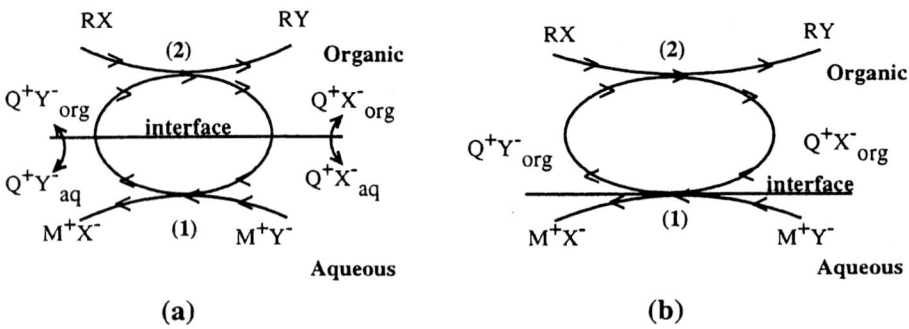

Figure 19.3 Mechanisms of LLPTC action: (a) quat is soluble in the aqueous phase; (b) quat is insoluble in the aqueous phase

DISTRIBUTION COEFFICIENT

The coefficients for the distribution of the quat species between the organic and aqueous phases are defined as

$$m_{QX} = \frac{[QX]_{org}}{[QX]_{aq}}, \quad m_{QY} = \frac{[QY]_{org}}{[QY]_{aq}} \tag{19.1}$$

These are often assumed to be constant during reaction, although they sometimes vary with aqueous phase ionic strength (Asai et al., 1991, 1992) and temperature (Wang and Yang, 1991a).

DISSOCIATION CONSTANT IN AQUEOUS PHASE

Although it is usually assumed that the quat is completely ionized in the aqueous phase to form free ions and is present as an ion pair in the organic phase, in general in any solvent there can be free ions (thermodynamically distinct entities) that coexist in equilibrium with the ion pairs. Thus it is important, especially in the aqueous phase, to define a dissociation constant as

$$K_{QX} = \frac{[Q^+]_{aq}[X^-]_{aq}}{[QX]_{aq}} \tag{19.2}$$

The value of K_{QX} can be derived from Bjerrum's theory for ion pairs (Brändström, 1977). In organic solvents with high dielectric constants such as acetone, the dissociation can be high enough for the anion formed to play an active role in the reaction.

MASS TRANSFER COEFFICIENTS FOR QY and QX

The transfer of QX from the aqueous phase to the organic phase (or of QY from the organic phase to the aqueous phase) can be quantified by a differential equation (Wang and Yang, 1991a), which when solved gives a simple correlation for calculating the overall mass transfer coefficient. For example, for QY we can derive

$$\ln\left[\frac{[QY]_{org}}{[QY]_{org,b}} + \frac{m_{QY}V_{org}}{V_{aq}}\left(\frac{[QY]_{org}}{[QY]_{org,c}} - 1\right)\right]\left(\frac{1}{m_{QY}} + \frac{V_{org}}{V_{aq}}\right)^{-1} = -(k_{QY}a)t \tag{19.3}$$

Thus, using the concentration profile with time and distribution coefficient values, a simple linear regression gives us the values of the overall mass transfer coefficient for QY.

INTRINSIC KINETICS OF THE ION-EXCHANGE REACTION

The intrinsic kinetics of the ion-exchange reaction can be found by carrying out the PTC reaction in the absence of the organic substrate (but with the organic phase included). Having independently calculated the mass transfer coefficients, we can calculate the intrinsic kinetics of QY generation via ion exchange by tracking the concentration of QY in the organic phase. Wang and Yang (1991a) report differential equations for the dynamics of QY in

both the aqueous and organic phases in two-phase reactions with no organic substrate added. Then the ion-exchange reaction rate constants can be calculated by solving the equations using the experimental data.

INTRINSIC KINETICS OF THE ORGANIC REACTION

By carrying out a homogeneous reaction in the organic phase with the organic substrate and a known quantity of QY (instead of using MY), the intrinsic rate constant of the organic reaction can be found.

Model development

Using the information just given, the overall kinetics of the PTC cycle in a two-phase system can be determined. Considering the complexity of the systems, several approaches to LLPTC modeling have been taken (Evans and Palmer, 1981; Lele et al., 1983; Chen et al., 1991; Wu, 1993, 1996; Bhattacharya, 1996). All these are based essentially on the classification of fluid-fluid reactions into four regimes, as described in Chapters 14 and 15.

If the reaction occurs in regime 1, the organic phase reaction controls the overall rate, and the only information needed is the kinetics of this reaction. On the other hand, if we assume that there is film transfer resistance but the reaction still occurs in the liquid bulk, regime 2 is operative. Much of the reported modeling of LLPTC is for regime 2 (although no specific statements to that effect are made). We use an approach for this regime that provides a self-consistent framework for analyzing PTC reactions in general, including SLPTC reactions considered in the next section (see Wu, 1993). Then we present a preliminary analysis of PTC in regime 3 where the reaction occurs in the film.

REGIME 1

The role of mass transfer of Q^+X^- from the organic to the aqueous phase and of Q^+Y^- from the aqueous to the organic phase can be evaluated by defining the following dimensionless groups (Wu, 1993):

$$\psi_{QY} = \frac{[QY]_{aq} V_{aq} m_{QY}}{[QY]_{org} V_{org}}, \qquad \psi_{QX} = \frac{[QX]_{org} V_{org}}{[QY]_{aq} V_{aq} m_{QX}} \qquad (19.4)$$

If the PT catalyst in both the phases is in extractive equilibrium (i.e., there is equilibrium distribution of catalyst) and mass transfer resistances are absent or neglected completely, then ψ_{QY} and ψ_{QX} are each equal to 1. In other words, the reaction is in regime 1, and no further modeling besides determining the intrinsic kinetics is required.

REGIME 2

Assuming extractive equilibrium and complete dissociation of MY and MX in the aqueous phase, the following sets of rate equations and mass balances can be written:

RATE EQUATIONS

$$-\frac{[RX]_{\text{org}}}{dt} = k_2[RX]_{\text{org}}[QY]_{\text{org}}$$

$$-\frac{d[QX]_{\text{org}}}{dt} = k_2[RX]_{\text{org}}[QY]_{\text{org}} + k'_{QX}a([QX]_{\text{org}} - m_{QX}[QX]_{\text{aq}})$$

$$-\frac{d[QY]_{\text{org}}}{dt} = k_2[RX]_{\text{org}}[QY]_{\text{org}} - k'_{QY}a(m_{QY}[QY]_{\text{aq}} - [QY]_{\text{org}}) \quad (19.5)$$

$$-\frac{d[QX]_{\text{aq}}}{dt} = k_1[MY]_{\text{aq}}[QX]_{\text{aq}} - k_{-1}[MX]_{\text{aq}}[QY]_{\text{aq}} - k'_{QX}a([QX]_{\text{org}} - m_{OX}[QX]_{\text{aq}})$$

$$-\frac{d[QY]_{\text{aq}}}{dt} = k'_{QY}a(m_{QY}[QY]_{\text{aq}} - [QY]_{\text{org}}) - k_1[MY]_{\text{aq}}[QX]_{\text{aq}} + k_{-1}[MX]_{\text{aq}}[QY]_{\text{aq}}$$

MASS BALANCES

$$q_0 = V_{\text{org}}([QY]_{\text{org}} + [QX]_{\text{org}}) + V_{\text{aq}}([QX]_{\text{aq}} + [QY]_{\text{aq}})$$

$$N_{MY,0} = V_{\text{org}}([QY]_{\text{org}} + [RX]_{\text{org},0} - [RX]_{\text{org}}) + V_{\text{aq}}([Y^-]_{\text{aq}} + [QY]_{\text{aq}}) \quad (19.6)$$

$$N_{MX,0} + V_{\text{org}}[RX]_{\text{org}} + q_0 = V_{\text{org}}([QX]_{\text{org}} + [RX]_{\text{org}}) + V_{\text{aq}}([X^-]_{\text{aq}} + [QX]_{\text{aq}})$$

Using these equations, an effectiveness factor defined as

$$\eta_{\text{org}} = \frac{\text{actual rate of reaction}}{\text{rate of reaction with all of the catalyst present as } QY} \quad (19.7)$$

can be calculated from the equation (Wu, 1996)

$$\eta_{\text{org}} = \frac{(V_{\text{org}}/V_{\text{aq}})m_{QY} + 1}{(V_{\text{org}}/V_{\text{aq}})m_{QY}} + \left[\frac{Da_{QY}}{(V_{\text{org}}/V_{\text{aq}})m_{QY}} + Da_{QX}\right]$$

$$+ [(V_{\text{org}}/V_{\text{aq}})m_{QY} + 1]\left[\frac{Da_{QY} + 1}{(V_{\text{org}}/V_{\text{aq}})m_{QY}} \frac{1}{K} + \frac{k_2}{k_1}\frac{V_{\text{org}}}{V_{\text{aq}}}\right] \quad (19.8)$$

where Da_{QY} and Da_{QX} are the Damköhler numbers for the catalyst species QY and QX, respectively.

Recently, Bhattacharya (1996) developed a simple, general framework for modeling PTC reactions in liquid–liquid systems. The main feature of this analysis is that it can model complex multistep reactions in both aqueous and organic phases and is thus applicable to both normal and inverse PTC reactions. It does not resort to the commonly made pseudo-steady-state assumption nor does it assume extractive equilibrium.

Models for LLPTC become even more complicated for special cases such as PTC systems that involve reactions in both aqueous and organic phases, systems with a base reaction even in the absence of PT catalyst, or other complex series-parallel multiple reaction schemes. For example, Wang and Wu (1991) studied the kinetics and mass transfer implications of a sequential reaction using PTC

that involved a complex reaction scheme with six sequential S_N2 reactions in the organic phase, along with interphase mass transfer and ion exchange in the aqueous phase. Similarly, another complicated PTC system was reported by Wang and Chang (1991a, 1991b) involving the allylation of phenoxide with allyl chloride in the presence of PEG as a PT catalyst.

Modeling of LLPTC reactions in the presence of bases is also complicated due to the complex nature of the mechanism. Although a few studies of individual systems have been reported (Wang and Wu, 1990; Asai et al., 1992; Wang and Chang, 1994, 1995), there is yet no general model for such reactions.

REGIME 3

Almost all of the reported studies of PTC have been concerned with slow reactions that occur in the bulk with or without mass transfer resistance in the liquid film. The general reasoning for this has probably been that PTC helps only slow reactions and that fast reactions would be relatively unaffected. Detailed experimental studies have shown, however, that PTC can also enhance the rates of fast reactions, those in regime 3 (Lele et al., 1983). Using the alkaline hydrolysis of different formate and acetate esters with Aliquot 336 and cetyltrimethylammonium bromide as PT catalysts and after confirming that the reactions were in regime 3 (by using the appropriate equations from Chapter 15), it was found that the reactions were considerably speeded up, with enhancement factors varying from 20 to more than 200. No modeling studies have been reported, however. The possibility of reaction in regime 3 has been considered by Naik and Doraiswamy (1997) for SLPTC, and a similar approach should hold for LLPTC.

THE RIGID SPHERE MODEL

Another approach to regime 3 is to regard LLPTC as a dispersion of droplets of one liquid (usually the organic phase) into a continuous phase of the other (usually the aqueous phase). If now it is assumed that the droplets behave as rigid spheres, then the diffusion-reaction model developed in Chapter 7 can be used (see Yang, 1998). Thus it should be possible to account for all of the three major influences involved: reaction, diffusion within the sphere, and diffusion to the surface of the sphere. In a sense, the whole sphere becomes a "film," with reaction occurring either in the entire film or in a part of it (corresponding to the outer regions of the sphere). The only difference from the approach outlined in Chapter 7 for a catalyst would be with respect to the nature of the reaction itself. Scheme a of Figure 19.3 will now be applicable, and separate equations must be written for the organic and aqueous phases along with appropriate initial and boundary conditions. Since the catalyst QX is transformed in the aqueous phase to the active QY form, and is therefore involved in a cyclic process of use and regeneration, an equation for its conservation must also be written. All these equations (see Table 19.2) can then be nondimensionalized and solved in terms of a dimensionless time (since this is a dynamic process), an appropriately defined Thiele modulus, and a Biot number. Note, however, that the equations listed in Table 19.2 are in their native dimensional form.

Table 19.2 Equations for the rigid sphere model of LLPTC (after Yang, 1998)

Organic phase:

$$\frac{\partial [RX]_{org}}{\partial t} = \frac{D_{RX}}{r^2} \frac{\partial}{\partial r}\left(r^2 \frac{\partial [RX]_{org}}{\partial r}\right) - k_{org}[RX]_{org}[QY]_{org}$$

$$\frac{\partial [QX]_{org}}{\partial t} = \frac{D_{QX}}{r^2} \frac{\partial}{\partial r}\left(r^2 \frac{\partial [QX]_{org}}{\partial r}\right) + k_{org}[RX]_{org}[QY]_{org}$$

$$\frac{\partial [QY]_{org}}{\partial t} = \frac{D_{QY}}{r^2} \frac{\partial}{\partial r}\left(r^2 \frac{\partial [QY]_{org}}{\partial r}\right) - k_{org}[RX]_{org}[QY]_{org}$$

Initial and boundary conditions:

$t = 0$: $[RX]_{org} = [RX]_{org,0}$, $[QY]_{org} = [QX]_{org} = 0$

$r = 0$: $\dfrac{\partial [RX]_{org}}{\partial r} = \dfrac{\partial [QX]_{org}}{\partial t} = \dfrac{\partial [QY]_{org}}{\partial t} = 0$

$r = R$: $\dfrac{3 D_{RX}}{R} \dfrac{\partial [RX]_{org}}{\partial R} = -k_{org}[RX]_{org}[QY]_{org}$

$\dfrac{3 D_{QX}}{R} \dfrac{\partial [QX]_{org}}{\partial r} = -\dfrac{3 k'_{QX}}{R}\left([QX]_{org} - m_{QX}[QX]_{aq}\right) + k_{org}[RX]_{org}[QY]_{org}$

$\dfrac{3 D_{QY}}{R} \dfrac{\partial (QY)_{org}}{\partial r} = \dfrac{3 k'_{QY}}{R}\left([QY]_{org} - \dfrac{1}{m_{QY}}[QY]_{org}\right) - k_{org}[RX]_{org}[QY]_{org}$

Aqueous phase:

$$\frac{d[QY]_{aq}}{dt} = k_{aq}[MY]_{aq}[QX]_{aq} - k'_{QY}\alpha\left(\frac{V_{org}}{V_{aq}}\right)\left([QY]_{aq} - \frac{1}{m_{QY}}[QY]_{org,s}\right)$$

$$\frac{d[QX]_{aq}}{dt} = k'_{QX}\alpha\left(\frac{V_{org}}{V_{aq}}\right)[QX]_{org,s} - m_{QX}[QX]_{aq} - k_{aq}[MY]_{aq}[QX]_{aq}$$

$$\frac{d[MY]_{aq}}{dt} = -k_{aq}[MY]_{aq}[QX]_{aq}$$

Initial conditions:

$t = 0$: $[QY]_{aq} = 0$, $[MY]_{aq} = [MY]_0$, $[QX]_{aq} = [QX]_0$

Catalyst conservation:[a]

$$V_{aq}[QX]_0 = V_{aq}([QX]_{aq} + [QY]_{aq}) + V_{org}(\overline{[QY]}_{org} + \overline{[QX]}_{org})$$

where

$$\overline{[i]}_{org} = \frac{3}{R^3}\int_0^R r^2 [i]_{org}\, dr$$

Conversion of RX in organic phase:

$$X = 1 - \frac{\overline{[RX]}_{org}}{[RX]_{org,0}}$$

---

[a] $\overline{[i]}_{org}$ represents the average concentration of i within an organic droplet.

The results of simulation of these equations show that the effects of Thiele modulus and Biot number are similar to those for a solid catalyst. Thus for low values of $\phi (< 1)$, the reaction is chemically controlled, while for higher values $(\phi > 2)$ it is diffusion controlled within the sphere. External mass transfer is not important at high Biot numbers. The concentration of the active form of catalyst within the sphere is an imporant determinant of the rate. High values of ϕ lead to lower reaction rates, with the active catalyst concentrated close to the surface. More important, for a given set of parameter values, there appears to be (in most cases) an optimum time at which the active catalyst concentration within the sphere reaches a maximum.

Modeling of SLPTC

In dealing with SLPTC, two types of mechanisms must be recognized, homogeneous and heterogeneous solubilization. We present the outlines of a reasonably rigorous model for each.

Homogeneous and heterogeneous solubilization

In SLPTC, the quat (Q^+X^-) approaches the solid surface and undergoes ion exchange with the solid nucleophilic salt at or near the solid interface (or in some cases within the solid) to form Q^+Y^-, followed by reaction of Q^+Y^- with the organic substrate RX. The organic reaction takes place only in the liquid (organic) phase. Depending on the location and mechanism of the ion-exchange reaction and on the solubility of the solid in the organic phase, two general mechanisms for SLPTC have been proposed (Figure 19.4): homogeneous and heterogeneous solubilization. These may be regarded as SLPTC counterparts of the LLPTC mechanisms shown in Figure 19.3.

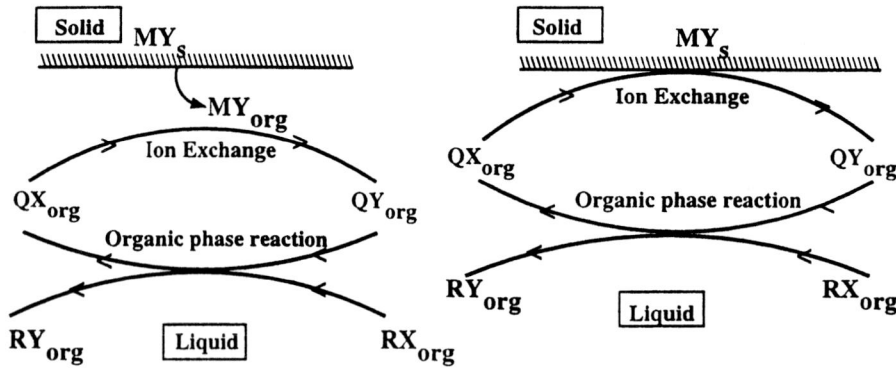

Figure 19.4 Mechanisms of SLPTC action: homogeneous and heterogeneous solubilization

Homogeneous solubilization requires that the nucleophilic solid have some finite solubility (though low) in the organic phase and involves the dissolution of the inorganic salt in the organic phase, followed by ion exchange of the quat in the liquid phase with dissolved MY. Heterogeneous solubilization occurs by transfer of the nucleophilic anion by a PT catalyst from the surface of the solid crystalline lattice to the organic phase. In this case, the ion-exchange reaction takes place at the solid surface or within the solid phase. A new kinetic model, the ternary adsorption complex model (Esikova and Yufit, 1991a,b), has recently been proposed, in which a series of adsorption steps at the surface replaces the ion-exchange reaction. There is no substantive evidence so far, however, to favor this theory over the common ion-exchange reaction. Therefore, the following development is based entirely on the ion-exchange mechanism.

Reaction models based on homogeneous solubilization

It would be useful to develop a stepwise approach, with incremental additions of complexity, to modeling SLPTC for both homogeneous and heterogeneous solubilizations, and to obtain expressions for an effectiveness factor limited to the organic phase reaction and another for an overall effectiveness factor. For homogeneous solubilization, a comprehensive model accounting for the different possible controlling steps, solid dissolution, ion exchange, mass transfer of quat, and the organic reaction, can be developed on the basis of the film theory of mass transfer (Naik and Doraiswamy, 1997). Figure 19.5 shows a schematic of models that progressively account for these different steps, with one or more controlling at a time. It should be noted that the ion-exchange reaction is permitted to be in

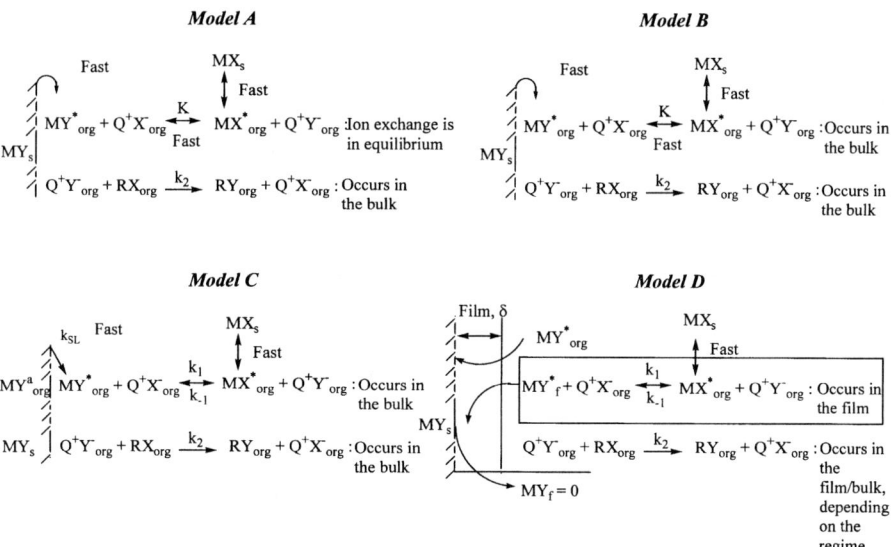

Figure 19.5 Models for homogeneous solubilization (Naik and Doraiswamy, 1997)

Table 19.3 Operating regimes of the four homogeneous solubilization SLPTC models shown in Figure 19.5

Model	Step(s) contributing to overall rate	Characteristic features of the model	Regime of ion exchange	Regime organic reaction
Model A	Organic phase reaction	Constant Q^+Y^- concentration; pseudo-first-order rate	(fast equilibrium)	1
Model B	Organic phase and ion-exchange reactions	Changing Q^+Y^- concentration	1	1
Model C	Organic phase reaction, ion exchange, and solid dissolution	Rate of solid dissolution important at low values of $k_{SL}a$	Between 1 and 2	1
Model D	Organic phase reaction, ion exchange, solid dissolution, and mass transfer steps	Mass transfer limitations slow down reaction	3	1, 2, or 3

different regimes, unlike in all previous studies (LLPTC or SLPTC) in which it was assumed to be in equilibrium. Also, the possibility that the main (organic) reaction occurs in regime 1, 2, or 3 has been considered. The main features of the models are summarized in Table 19.3, and the corresponding equations are given in Table 19.4.

The equations for models A and B are quite straightforward. For model C, because solid dissolution is also a contributing resistance, the interfacial area of mass transfer a_p changes with time. Hence an equation for $\delta a_p / \delta t$ is included in the model equations. Model D incorporates the effects of transport of QY between the film and the organic bulk. Because the ion-exchange step is usually faster than the organic reaction, we assume that it occurs completely in the film (regime 3). On the other hand, the organic phase reaction is slower and can occur in regime 1, 2, or 3. The equations given in the table are for regime 3. Because film concentrations are involved in these equations, they are represented as $[i]_f$.

Figure 19.6 presents a comparison of the results for all of these models. Clearly, assuming a simple model that accounts for only the organic phase reaction can lead to a gross overestimate of the reaction rate.

Reaction model based on heterogeneous solubilization (accounting for reaction within the solid)

Ion exchange within the solid phase can play an important role in determining the kinetics of the PTC cycle. In such a situation, ion exchange in the solid phase can be one of the rate-controlling steps because access to the anions in the solid can be restricted by mechanical hindrances due to the lattice structure and deposition of the product MX. For a reactive solid, transient conditions prevail within the solid, thus complicating the analysis. Structural changes in the solid can also, occur with reaction progress, which can lead to a continual shift in the controlling regime with time. In the general case, the controlling steps can be either the liquid-phase transfer steps (external mass transfer), diffusion steps within the reactive solid, adsorption-desorption steps (if any), surface (ion-exchange) reaction, liquid-phase organic reaction, or combinations of them. Thus, as brought out in Figure 15.7, in any reaction involving a solid, a complete analysis must allow for reaction within the solid in addition to that in the liquid.

Depending on the porosity of the solid and other factors, different models can be operative for the solid phase ion-exchange reaction. The more common models are sharp interface model, volume reaction model, and reaction zone model (see Chapter 15). Naik and Doraiswamy (1997) recently developed a general model incorporating many of these considerations (Figure 19.7). Using the volume reaction model for the ion-exchange reaction in the solid, the following equations were developed:

Table 19.4 Equations for the four homogeneous solubilization models shown in Table 19.3 (adapted from Naik and Doraiswamy, 1997)

Model A

$$\frac{-d[RX]_{\text{org}}}{dt} = k_2[QY]_{\text{org}}[RX]_{\text{org}} = k_2 \frac{Kq_0}{K+a}[RX]_{\text{org}}$$

where $k_2 \dfrac{Kq_0}{K+a} = $ constant (say k_1), gives pseudo first-order kinetics $\dfrac{-d[RX]_{\text{org}}}{dt} = k_1[RX]_{\text{org}}$, with I.C.: $t = 0$; $[RX]_{\text{org}} = [RX]_{\text{org},0}$ giving $[RX] = [RX]_0 \exp(-k_1 t)$

Model B

$$\frac{-d[RX]_{\text{org}}}{dt} = k_2[RX]_{\text{org}}[QY]_{\text{org}}$$

$$\frac{-d[QY]_{\text{org}}}{dt} = k_2[RX]_{\text{org}}[QY]_{\text{org}} + k_{-1}[QY]_{\text{org}}[MX]^*_{\text{org}} - k_1[QX]_{\text{org}}[MY]^*_{\text{org}}$$

I.C. $t = 0$; $[RX]_{\text{org}} = [RX]_{\text{org},0}$, $\quad [QY]_{\text{org}} = 0$

Model C

$$\frac{d[MY]_{\text{org}}}{dt} = k'_{\text{SL}} a_p[[MY]^*_{\text{org}} - [MY]_{\text{org}}] + k_{-1}[QY]_{\text{org}}[MX]_{\text{org}} - k_1[QX]_{\text{org}}[MY]_{\text{org}}$$

$$\frac{-d[RX]_{\text{org}}}{dt} = k_2[RX]_{\text{org}}[QY]_{\text{org}}$$

$$\frac{-d[QY]_{\text{org}}}{dt} = k_2[RX]_{\text{org}}[QY]_{\text{org}} + k_{-1}[QY]_{\text{org}}[MX]^*_{\text{org}} - k_1[QX]_{\text{org}}[MY]_{\text{org}}$$

$$\frac{da_p}{dt} = -\frac{R^3 \rho_s^3 a_p^4 C_{RX,0}}{81\theta w_s^3} \frac{d[RX]_{\text{org}}}{dt}$$

I.C. $t = 0$, $[RX]_{\text{org}} = [RX]_{\text{org},0}$, $[MY]_{\text{org}} = 0$, $[QY]_{\text{org}} = 0$, $a_p = a_{p,0} = \dfrac{3w_s}{\rho_s R}$

where

$$a_p = \frac{3w_s}{R\rho_s} \qquad \theta = \frac{\text{moles of } MY \text{ taken}}{\text{moles of } RX \text{ taken}}$$

Model D

$$\frac{\partial [QY]_{\text{org}}}{\partial t} = D_{QY}\frac{\partial^2[QY]_{\text{org}}}{\partial x^2} - k_2[RX]_{\text{org}}[QY]_{\text{org}} + k_1[MY]_{\text{f}}[QX]_{\text{org}} - k_{-1}[MX]_{\text{f}}[QY]_{\text{org}}$$

$$\frac{d[RX]_{\text{org}}}{dt} = -k_2[RX]_{\text{org}}[QY]_{\text{org}}$$

$$\frac{\partial C_{MY_{\text{f}}}}{\partial t} = D_{MY}\frac{\partial^2[MY]_{\text{f}}}{\partial x^2} - k_1[MY]_{\text{f}}[QX]_{\text{org}} + k_{-1}[MX]_{\text{f}}[QY]_{\text{org}}$$

I.C. $t = 0$, $[RX]_{\text{org}} = [RX]_{\text{org},0}$; $[QY]_{\text{org}} = [MY]_{\text{f}} = 0$

B.C. $x = 0$, $D_{MY}\dfrac{d[MY]_{\text{f}}}{dx} = -k'_{\text{SL}}[MY]^*_{\text{org}}$

$x = \delta$, $[MY]_{\text{f}} = 0$; $\quad x = 0$, $\dfrac{d[QY]_{\text{org}}}{dx} = 0$

Dimensionless groups

$$\tau = \frac{D_{RX} t}{R^2}, \quad \phi = R\sqrt{\frac{k_{\text{org}}[QX]_0}{D_{RX}}}$$

$$Bi_{QY} = \frac{k'_{QY} R}{D_{QY}}, \quad Bi_{QX} = \frac{k'_{QX} R}{D_{QX}}$$

Phase-Transfer Reaction Engineering   623

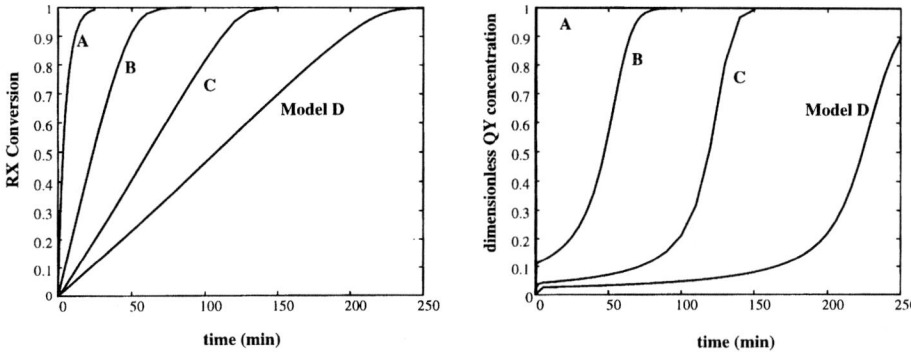

Figure 19.6 Plots of conversion and catalyst concentration as functions of time for homogeneous solubilization models A–D (Naik and Doraiswamy, 1997)

$$\frac{\partial [QX]_{sol}}{\partial t} = \frac{D_{e,Q}}{r^2}\frac{\partial}{\partial r}\left(r^2 \frac{\partial [QX]_{sol}}{\partial r}\right) - k_S\left([QX]_{sol} - \frac{[QY]_{sol}}{K}\right)$$

$$\frac{\partial [QY]_{sol}}{\partial t} = \frac{D_{e,Q}}{r^2}\frac{\partial}{\partial r}\left(r^2 \frac{\partial [QY]_{sol}}{\partial r}\right) - k_S\left([QX]_{sol} - \frac{[QY]_{sol}}{K}\right) \quad (19.9)$$

$$\frac{d[QX]_b}{dt} = k_2[RX]_b[QY]_b - k_Q a([QX]_b - [QX]_S)$$

$$\frac{d[RX]_b}{dt} = -k_2[RX]_b[QY]_b$$

where $D_{e,Q}$ is the effective diffusivity of the quat (assumed to be the same for QX and QY) in the solid, and subscripts sol, s, and b refer to the solid, surface, and bulk, respectively. $[QY]_b$ may be calculated from a quat species balance as

$$[QY]_b = q_0 - [QX]_b - [QX]_{sol,ave} - [QY]_{sol,ave} \quad (19.10)$$

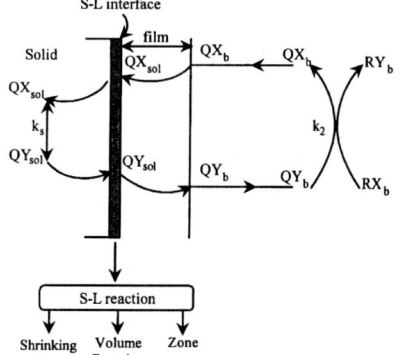

Steps involved in the PTC cycle are:

1. Liquid-phase mass transfer: diffusion of Q^+X^- to the solid surface
2. Diffusion of Q^+X^- within the solid
3. Possible adsorption of Q^+X^- on the solid surface
4. Reaction (ion exchange) at the solid surface, forming Q^+Y^-
5. Desorption of Q^+Y^-
6. Diffusion of Q^+Y^- out of the solid
7. Liquid-phase mass transfer: diffusion of Q^+Y^- to the liquid bulk
8. Organic reaction of Q^+Y^- with RX, regenerating Q^+X^-

Figure 19.7 A comprehensive model for SLPTC with provision for including any existing model for fluid-solid reactions (such as the sharp interface, volume reaction, or zone model) (Naik and Doraiswamy, 1997)

where $[QX]_{\text{sol,ave}}$ and $[QY]_{\text{sol,ave}}$ are, respectively, the volume averaged concentrations of QX and QY in the solid.

INITIAL CONDITIONS

$$t = 0, [RX]_b = [RX]_{\text{sol}}, [QX]_{\text{sol}} = [QY]_{\text{sol}} = [QY]_b = 0, [QX]_b = q_0 \quad (19.11)$$

BOUNDARY CONDITIONS

$$r = 0, \frac{d[QX]_{\text{sol}}}{dr} = \frac{d[QY]_{\text{sol}}}{dr} = 0$$

$$r = R, D_{e,Q} \frac{d[QX]_{\text{sol}}}{dr} = k'_Q([QX]_b - [QX]_{\text{sol,S}}) \quad (19.12)$$

$$r = R, D_{e,Q} \frac{d[QY]_{\text{sol}}}{dr} = k'_Q([QY]_b - [QY]_{\text{sol,S}})$$

As in effective diffusivity, here also we assumed that the two forms of the catalyst QX and QY have the same solid-liquid mass transfer coefficient k'_Q.

AN IMPORTANT CONCLUSION

It is possible to nondimensionalize these equations in terms of the Thiele modulus ϕ, the Biot number for mass transfer Bi_m, and nondimensional time and distance. An important conclusion emerges from the solution of these nondimensional equations (Figure 19.8): the solid phase ion-exchange reaction has a significant effect on the organic phase reaction. Although a typical fluid-solid reaction that occurs within a solid (in this case the ion-exchange reaction) is greatly influenced by reactant diffusion into the solid, heterogeneous solubiliza-

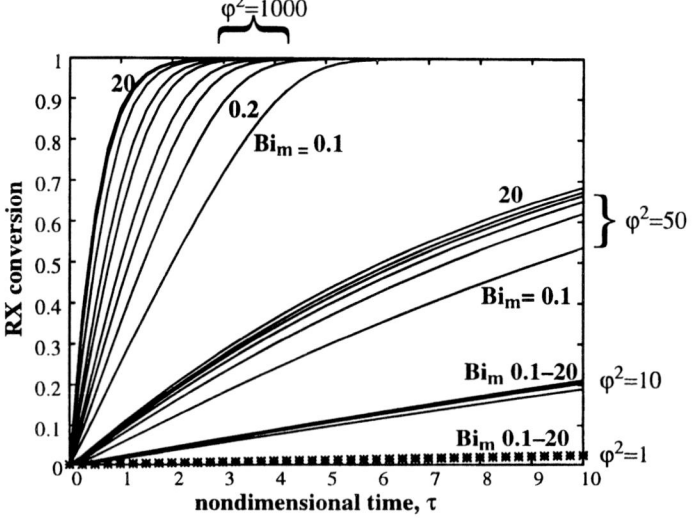

Figure 19.8 Effectiveness factor for an SLPTC reaction as a function of the solid-phase and organic liquid-phase reaction parameters (Naik and Doraiswamy, 1997)

tion additionally influences the main organic phase reaction (which is different from the solid phase reaction). Indeed, we are primarily interested in this effect of diffusion, which is not normally considered in the analysis of traditional catalytic reactions (Chapter 7). It can be noticed in Figure 19.8 that at low ϕ, the overall organic reagent conversion is low because the ion-exchange reaction within the solid is too slow to deliver sufficient quantities of the quat species QY into the liquid phase. Film transfer can become important at higher values of ϕ, where the rate of the ion-exchange reaction is fast compared to the diffusional effect within the solid and its neighborhood (i.e., the liquid film at the solid-liquid interface). It is significant that higher values of ϕ, indicative of more severe diffusional limitation, give higher overall RX conversions. This happens because ion exchange is much faster than the diffusional effect and significant amounts of QY are generated for consumption in the organic liquid-phase reaction.

IMMOBILIZED PTC

Support Materials

Both inorganic and organic (mostly polymeric) supports have been used.

Inorganic supports

It has been reported that adsorption of an organic substrate on an inorganic support tends to increase its local concentration on the support surface, thereby leading to greater reactivity, for example, in the reduction of ketones with sodium borohydride and in the synthesis of primary alkyl chlorides from primary alcohols (Tundo and Venturello, 1979). Thus inorganic solids such as alumina (Tundo et al., 1982), silica gel (Tundo and Venturello, 1981), and silica (Arrad and Sasson, 1990) have been used to support quaternary ammonium salts, with the PT catalyst either physically adsorbed (Tundo et al., 1982) or chemically bonded via long spacer groups (Tundo and Venturello, 1979) to the surface of the solid support. Because alumina by itself can also catalyze various heterogeneous reactions (Quici and Regen, 1979; Pradhan and Sharma, 1992) and is also more stable in alkali than silica gel, it is often used as an effective support for PTC reactions (Tundo and Badiali, 1995). Clays (Lin and Pinnavaia, 1991), and more so zeolites, that have well-defined chemical and morphological characteristics, also have great potential in the synthesis of highly stereoselective and chiral products. Their negatively charged aluminum silicate structure and strong electrostatic interactions governing adsorption can be used to physically or chemically adsorb quaternary onium salts, but some special techniques must be used to enhance catalyst loading (see Akelah et al., 1994).

Organic (polymeric) supports

The most commonly used organic support is polystyrene (crosslinked with DVB) in its microporous (1–2% crosslinking) form, although it has also been used in its macroporous and popcorn forms (Ford et al., 1982; Shan et al., 1989). Various other polymeric catalysts have also been used, such as polyvinylpyridine resins,

commercial ion-exchange resins (Arrad and Sasson, 1989), modified dextran anion exchangers (Kise et al., 1981), and macroporous glycidyl methacrylate–ethylene dimethacrylate resins (Hradil and Svec, 1984).

Kinetics and Mechanism of Immobilized PTC

Liquid-liquid-solid triphase catalysis (LLS-TPC)

Let us first consider LLS-TPC. What is involved here is not just a planar phase boundary as in classical two-phase systems. Instead, one visualizes a volume element that incorporates the catalytically active sites, as well as the two phases with the reagents. Diffusion of the aqueous phase (for ion exchange to take place) and the organic phase (for the main reaction to take place) into the solid support is important. It is a reasonable assumption that only one of the phases (depending on the support microenvironment) fills the pores of the solid catalyst, whereas the second phase is dispersed as globules within the support matrix and must diffuse to the active sites through the first phase. The fact that the aqueous phase and the organic phase interact differently with different solid supports further complicates the issue. A realistic mechanism (Tomoi and Ford, 1981; Hradil et al., 1988) involves the collision of droplets of the organic (or aqueous) phase with solid catalyst particles dispersed in a continuous aqueous (or organic) phase. The nature of the solid support (hydrophilic or hydrophobic) determines which phase fills the pores of the catalyst and acts as the continuous phase. A similar model for transport of the organic reagent from the bulk phase through water to the catalyst particle has been developed for emulsion polymerization (Svec, 1988). Free migration of the ion pairs between phases is not possible because the cation is part of the solid support. Therefore, it is necessary that the immobilized PT catalyst be just at the boundary of the two phases or oscillate between the two. The presence of spacer chains is believed to help such an oscillation of the catalyst between the aqueous and organic phases, assuming that both these phases are present within the pores of the solid support. The novelty of the concept of a spacer chain with a head group of the catalyst oscillating between the two liquid phases has attracted many studies (e.g., Hradil et al. 1987; Ruckenstein and Park, 1988; Ruckenstein and Hong, 1992).

A LHHW TYPE MODEL FOR LLS-TPC REACTIONS

It is commonly assumed that the reaction rate of a TPC system follows pseudo-first-order kinetics—that is, $[QY]_s$ is constant and the rate is simply a function of $[RX]$. This behavior is usually observed in reaction systems in which the inorganic nucleophile is in excess and, more important, where the leaving anion from the organic phase does not interfere with the overall reaction. Such reactions are limited only by the organic-phase reaction step. However, the pseudo-first-order model might not be valid for a TPC system whose overall reaction rate is limited by the ion-exchange reaction or by both the organic phase and ion-exchange reactions. In such a reaction system, the leaving anion is strongly coordinated with the catalyst cation compared to the inorganic

nucleophile. Therefore, the leaving ion tends to interfere with the ion-exchange step by creating a competition with the inorganic ion in coordinating with the catalyst. This leads to a decrease in the concentration of the inorganic ion transferred to the organic phase as the reaction progresses.

A general mechanistic model for substitution nucleophilic reactions has been developed by Satrio et al. (2000) by incorporating the effect of the leaving anion on the ion-exchange step. This model is based on the Langmuir-Hinshelwood and Eley-Rideal mechanisms, which were considered in Chapter 7 under the general category of LHHW models. Its derivation is outlined in Table 19.5. The model can be used to determine whether a triphase catalytic reaction is limited by the ion-exchange step, organic-reaction step, or by a combination of the two steps.

Solid-solid-liquid triphase catalysis (SSL-TPC)

In SSL-TPC, Kondo et al. (1994) suggest direct solid-solid interaction that leading to the formation of a complex between the solid nucleophilic reagent and the solid polymeric support by the cooperative coordination of active sites in the polymer and the alkali metal ion of the reagent. On the other hand, a simple extraction mechanism similar to classical PTC has been suggested by Arrad and Sasson (1991), in which reaction rates are controlled either by mass transfer of the inorganic reagent by small amounts of water or by the organic phase chemical reaction. Ion exchange is believed to proceed by transportation of the nucleophile into the solid catalyst pores via dissolution in the traces of water which are always present and form a fourth saturated phase in the system. In the absence of water, direct interaction of the nucleophile with the catalyst surface can lead to small amounts of reaction. More importantly, however, traces of water increase the rate of reaction, possibly by increasing the mass transfer of the nucleophile (MacKenzie and Sherrington, 1981).

Modeling of Immobilized PTC (LLS-TPC)

SSL-TPC systems involve both liquid-liquid and solid-solid interfaces, and it is not possible at this stage to model such systems with any degree of certainly. Therefore, we restrict the treatment to LLS-TPC.

To model triphase catalysis, one has to develop equations that consider diffusion and reaction in two immiscible liquid phases within the solid phase (the equations of Chapter 7 for a single fluid phase would not be adequate). Thus a realistic model for triphase catalysis has to account for mass transfer of both organic and aqueous phase species from their respective bulk phases to the surface of the catalyst, diffusion of the two phases through the pores of the solid catalyst, the intrinsic kinetics of reactions at the immobilized catalyst sites, and the diffusion of products back to the catalyst surface and into the bulk solutions. This overall scheme is shown in Figure 19.9. Because the catalyst support is usually lipophilic in a typical triphase system, the organic phase can be taken as the continuous phase, with the dispersed aqueous phase droplets diffusing through the continuous organic phase to come in contact with the quat species

Table 19.5 Derivation of a mechanistic model for nucleophilic substitution reactions in the presence of a triphase catalyst (Satrio et al., 2000)[a]

TPC reaction system:

$$(RX)_{org} + (M^+Y^-)_{aq} \leftrightarrow (RY)_{org} + (M^+X^-)_{aq} \qquad [1]$$

Overall reaction rate:

$$-r_{org} = k_{org}[RX]_{org}[Q^+Y^-]_{org} \qquad (1)$$

Mechanistic steps
  Ion-exchange step:

$$(Q^+X^-)_s + (Y^-)_{aq} \leftrightarrow (Q^+Y^-)_s + (X^-)_{aq} \qquad [2]$$

  Organic phase reaction step:

$$(Q^+Y^-)_s + (RX)_{org} \rightarrow (Q^+X^-)_s + (RY)_{org} \qquad [3]$$

The reversible ion-exchange reaction is expressed by using the traditional notation of heterogeneous catalysis and assuming the formation of a transitional site $X^- \cdot S^+ \cdot Y^-$ between the forward and reverse reaction steps, where S^+ is the "site" equivalent of the triphase catalyst's cation Q^+.

$$Y^- + S^+ \cdot X^- \leftrightarrow X^- \cdot S^+ \cdot Y^- \leftrightarrow X^- + S^+ \cdot Y^- \qquad [4]$$

Note that the steps are treated as two separate equilibrium attachment/detachment steps. The forward reaction step is seen as the "attachment/detachment" of Y^- anion to the inactive site $S^+ \cdot X^-$,

$$Y^- + S^+ \cdot X^- \leftrightarrow X^- \cdot S^+ \cdot Y^- \qquad [5]$$

and the reverse reaction step as the "attachment/detachment" of X^- anion to the active site $S^+ \cdot Y^-$,

$$X^- + S^+ \cdot Y^- \leftrightarrow X^- \cdot S^+ \cdot Y^- \qquad [6]$$

By assuming that the rates of attachment and detachment of each step are in equilibrium and the transition sites are transformed instantaneously to either active sites or inactive sites, the following equations are obtained:

$$\theta_{Y^-} = K_{Y^-}[Y^-]_{aq}(1 - \theta_{Y^-} - \theta_{X^-}) \qquad (2)$$

$$\theta_{X^-} = K_{X^-}[X^-]_{aq}(1 - \theta_{Y^-} - \theta_{X^-}) \qquad (3)$$

By combining the expressions for θ_{Y^-} and θ_{X^-}, a hyperbolic equation for the fraction of active TPC sites is obtained. Expressed in terms of catalyst concentration, the equation is

$$[S^+ \cdot Y^-] = [S^+]_t \frac{K_{Y^-}[Y^-]_{aq}}{1 + K_{Y^-}[Y^-]_{aq} + K_{X^-}[X^-]_{aq}} \qquad (4)$$

Finally, by combining Equations 1 and 4, the following expression for the overall rate of reaction is obtained:

$$-r_{org} = k_{org}[RX]_{org}[S^+]_t \frac{K_{Y^-}[Y^-]_{aq}}{1 + K_{Y^-}[Y^-]_{aq} + K_{X^-}[X^-]_{aq}} \qquad (5)$$

[a] Notation: $S^+ = Q^+$ regarded as a site in the language of heterogeneous catalysis; $S^+ \cdot Y^-$ = active form of the site; $S^+ \cdot X^-$ = inactive form of the site; $X^- \cdot S^+ \cdot Y^-$ = transitional phase catalytic site; $[S^+]_t$ = concentration of active sites (Q) expressed as total Q per total liquid volume; θ_{Y^-} = fraction of total number of TPC cations attached to Y^- anions; θ = fraction of total number of TPC cations attached to X^- anions. See Notation for the rest.

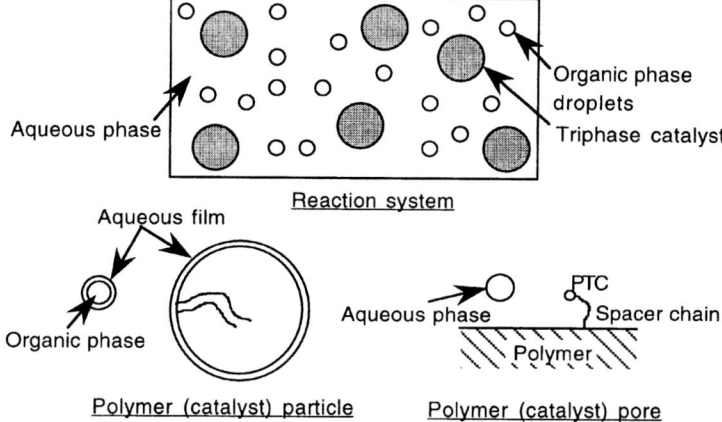

Figure 19.9 A schematic representation of triphase catalysis

immobilized at the solid surface. Mass transfer resistances for the aqueous phase reactant include the aqueous film resistance at the aqueous-organic interface and the organic film resistance at the organic liquid-solid particle interface. For the organic phase reactant, film diffusion is restricted to that at the organic liquid-solid particle interface. In addition to the film resistances, internal diffusional limitations within the pores of the solid catalyst are also to be considered for both phases. Such a model has been developed by Wang and Yang (1991b, 1992) for an irreversible organic phase reaction and generalized by Desikan and Doraiswamy (1995) to account for the reversibility of the ion-exchange reaction. The relevant equations are as follows:

MASS BALANCE OF QUAT WITHIN CATALYST

$$\frac{d[QX]_c}{dt} = -k_1[Y]_{aq,c}[QX]_c + k_{-1}[X]_{aq,c}[QY]_c + k_2[RX]_{org,c}[QY]_c \qquad (19.13)$$

MASS BALANCE OF ORGANIC SUBSTRATE WITHIN CATALYST

$$\varepsilon_{org}\frac{\partial[QY]_{org,c}}{\partial t} = \frac{D_{e,RX}}{r^2}\frac{\partial}{\partial r}\left[r^2\frac{\partial[RX]_{org,c}}{\partial r}\right] - \rho_c k_2[RX]_{org,c}[QY]_c \qquad (19.14)$$

MASS BALANCE OF INORGANIC SPECIES WITHIN CATALYST

$$\varepsilon_{aq}\frac{\partial[Y]_{aq,c}}{\partial t} = \frac{D_{e,Y}}{r^2}\frac{\partial}{\partial r}\left[r^2\frac{\partial[Y]_{aq,c}}{\partial r}\right] - \rho_c k_1\left([Y]_{aq,c}[QX]_c - \frac{1}{K}[QY]_c[X]_{aq,c}\right) \qquad (19.15)$$

IONIC SPECIES BALANCE IN THE AQUEOUS PHASE

$$[X]_{aq,c} = [Y]_{aq,0} - [Y]_{aq,c} \qquad (19.16)$$

QUAT BALANCE IN THE PARTICLE

$$[QY]_c = [QX]_{c,0} - [QX]_c \tag{19.17}$$

with the initial and boundary conditions

$$t = 0, \quad [RX]_{\text{org,c}} = [Y]_{\text{aq,c}} = 0, \quad [QX]_c = [QX]_{c,0}$$

$$r = 0, \quad \frac{\partial [RX]_{\text{org,c}}}{\partial r} = \frac{\partial [Y]_{\text{aq,c}}}{\partial r} = 0$$

$$r = R, \quad D_{e,RX}\frac{\partial [RX]_{\text{org,c}}}{\partial r} = k'_Y([RX]_{\text{org,c}} - [RX]_{\text{org,s}}) \tag{19.18}$$

$$D_{e,Y}\frac{\partial [Y]_{\text{aq,c}}}{\partial r} = k'_Y([Y]_{\text{aq,c}} - [Y]_{\text{aq,s}})$$

where k'_{RX} and k'_Y represent the mass transfer coefficients of RX and Y in the bulk phase, respectively.

In addition to these equations, the following bulk phase mass balances are also needed to keep track of changes in the organic and aqueous reagent concentrations:

$$-V_{\text{org}}\frac{d[RX]_{\text{org}}}{dt} = k'_{RX}aV_{\text{cat}}([RX]_{\text{org,b}} - [RX]_{\text{org,s}}) \tag{19.19}$$

$$-V_{\text{aq}}\frac{d[Y]_{\text{aq}}}{dt} = k_YaV_{\text{cat}}[Y]_{\text{aq,b}} - [Y]_{\text{aq,s}}) \tag{19.20}$$

with the initial conditions

$$t = 0, \quad [RX]_{\text{org,b}} = [RX]_0, \quad [Y]_{\text{org,b}} = [Y]_0 \tag{19.21}$$

This set of equations can be nondimensionalized in terms of reduced concentrations (with respect to initial values) and other physically relevant parameters like the Thiele modulus (ϕ), Biot number for mass transfer (Bi_m), and the dimensionless time and radial position,

$$\phi = R\left(\frac{\rho_c k_2 [QX]_{c,0}}{D_{e,RX}}\right)^{1/2}, \quad [Bi]_m = \frac{k_{RX}R}{D_{e,RX}}, \quad \tau = \frac{D_{e,RX}t}{(\varepsilon_{\text{org}} - \varepsilon_{\text{aq}})R^2}, \quad \omega = \frac{r}{R} \tag{19.22}$$

A few other (more specific) parameters also appear in the equations (see Desikan and Doraiswamy, 1995, for details; also Example 1).

An expression for the catalyst effectiveness factor based on the organic reaction can be obtained as

$$\eta = \frac{\int_V k_2 [RX]_{\text{org,c}}([QX]_{c,0} - [QX]_c)dV}{k_2 [RX]_{\text{org,s}}([QX]_{c,0} - [QX]_s)V} \tag{19.23}$$

which in dimensionless form becomes

$$\eta = \frac{3 \int_0^1 [R\hat{X}]_{\text{org,c}}(1 - [\hat{Q}X]_c)\omega^2 d\omega}{[R\hat{X}]_{\text{org,s}}(1 - [Q\hat{X}]_s)} \tag{19.24}$$

where $[R\hat{X}]$ represents the concentration of RX normalized with respect to its initial concentration. The overall effectiveness factor for the entire cycle is given by

$$\eta_{\text{overall}} = \frac{\eta}{1 + \eta(\phi^2/Bi_m)} \tag{19.25}$$

Figure 19.10 shows a typical plot of the intraparticle effectiveness factor versus dimensionless time for different values of the Thiele modulus. It should be noted that the reversibility of the aqueous phase ion-exchange reaction leads to lower effectiveness factors than those for irreversible reactions. However, the figure also shows that as the reaction becomes increasingly diffusion controlled, reversibility can actually produce a favorable effect. Note that, in addition to the two effects mentioned, simulation can also give the concentrations of the various components as functions of the radial position, time, and the Thiele modulus.

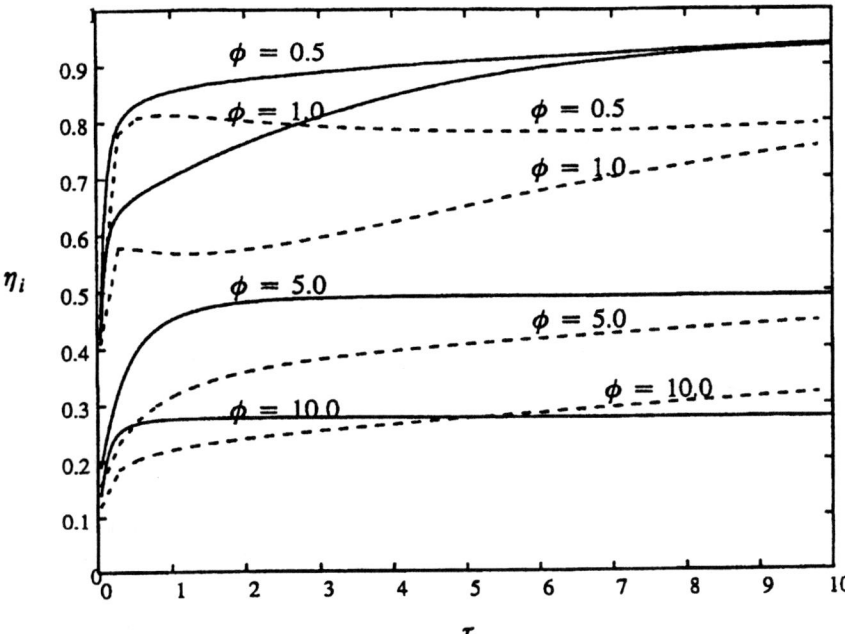

Figure 19.10 Triphase catalysis: effectiveness factor as a function of dimensionless time for various values of the Thiele modulus (Desikan and Doraiswamy, 1995). Solid lines represent irreversible reactions ($K \to \infty$) and broken lines reversible reactions ($K = 0.1$). Note: $\eta_i = \eta$ of Equation 19.24

They can provide useful guidelines for conducting a triphase reaction (see Example 19.1).

Comparison of Soluble and Immobilized Forms of PT Catalysts

It will be recalled that immobilized PT catalysts have not found favor with the industry because of the diffusional resistance of the solid that leads to reduced reaction rates. However, if an immobilized version of a soluble catalyst can be prepared that at least matches the performance of the soluble catalyst, it should be favored because of its other advantages. Quaternary ammonium salts are the most commonly used catalysts along with copolymers of polystyrene cross-linked to divinyl benzene as the solid supports. Use of PEG as a support has also been reported (MacKenzie and Sherrington, 1980; Kimura and Regen, 1983; Hradil and Svec, 1984).

Note that when a PT catalyst is immobilized on a polymer support, it reacts with a functional group in the polymer resulting in a structure that can differ from that of the soluble form used in immobilization. To make a proper comparison, therefore, the soluble catalyst used in the comparative studies must have the same structure as that in the bound form. Desikan and Doraiswamy (2000) made such a comparison for the model reaction between benzyl chloride in the organic phase and sodium acetate in the aqueous phase to form benzyl acetate. If methyltributylammonium chloride (MTBAC) is used as the soluble catalyst, its immobilized form would be benzyltributylammonium chloride (BTBAC), as shown by the structures below.

$CH_3(C_4H_9)_3N^+Cl^-$ (MTBAC)

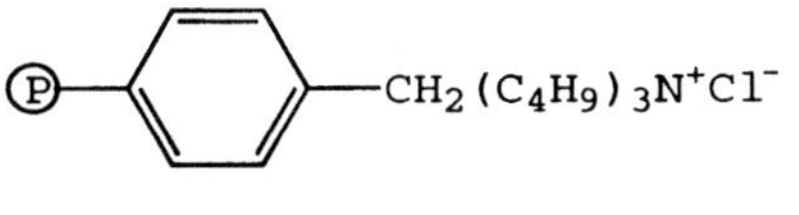

(P)—⟨benzene ring⟩—$CH_2(C_4H_9)_3N^+Cl^-$

(Polymer bound MTBAC)

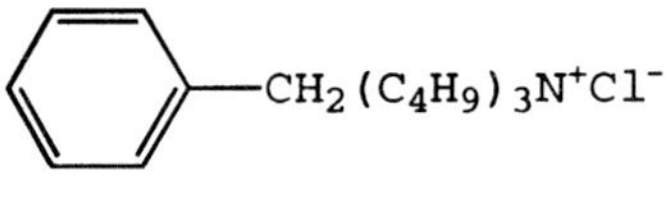

⟨benzene ring⟩—$CH_2(C_4H_9)_3N^+Cl^-$

(BTBAC)

Thus comparisons were made between soluble and immobilized MTBAC and also the "theoretical" native form of the immobilized version BTBAC.

The results, shown in Figure 19.11, are quite revealing. The immobilized catalyst gave the highest conversion. More detailed studies at different temperatures gave activation energies of 28.0 and 21.5 kcal/mol, respectively, for the soluble and immobilized versions of MTBAC, providing additional evidence of the superiority of the immobilized catalyst. The main reason for this is the increased lipophilicity of the catalyst due to its polymer backbone. If this were true, use of a nonpolymer solid such as silica gel must give a lower conversion. Experimental data confirmed this behavior. Indeed, the conversion was even

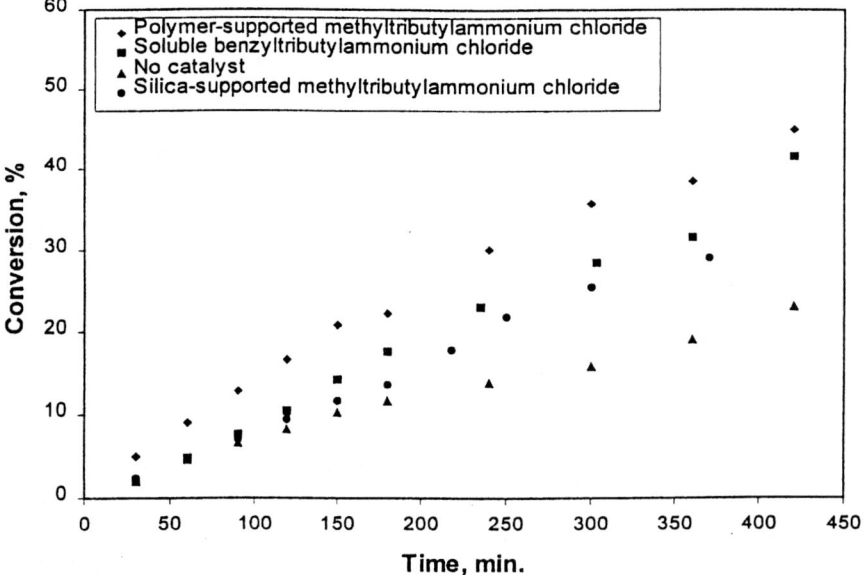

Figure 19.11 Comparison of soluble and immobilized forms of a PT catalyst: benzyltributylammonium chloride in the esterification of benzyl chloride with sodium acetate (redrawn from Desikan and Doraiswamy, 2000)

lower than that with the soluble catalyst—because of the diffusional effect. Clearly, therefore, use of a polymeric support leads to two opposing effects, an enhancing polymer effect and a retarding diffusional effect. Based on this observation, it should be possible to prepare an immobilized catalyst in which the enhancing effect more than offsets the retarding effect.

A second reason could be the microphase action of the solid particles (see Chapter 25). However, studies with the support material without the catalyst showed no enhancement in conversion. Therefore, we may conclude that increased lipophilicity is the only reason for enhanced activity in this case.

Effect of Nonisothermicity

To evaluate the role of nonisothermicity in TPC reactions, it is necessary to add energy balances to the material balances represented by Equations 19.13 to 19.21. The equations, in both their dimensional and dimensionless forms, have been developed by Desikan and Doraiswamy (1999) (see also Desikan, 1997), and their use is illustrated in Example 19.1.

EXAMPLE 19.1

Application of the model equations for TPC to a specific reaction, both for isothermal and nonisothermal situations

It is desired to conduct the transesterification of benzyl chloride with sodium acetate using polymer-supported tributylmethylammonium chloride as a triphase catalyst. The reactor is maintained at isothermal conditions, but temperature gradients can exist within the pellets. Using the following data, generate plots of reactant concentration as a function of time for both isothermal and nonisothermal pellets.

DATA

Volume of organic phase = 500 mL; volume of aqueous phase = 500 mL; initial concentration of organic reactant = 3 mol/ℓ; initial concentration of inorganic reactant = 5 mol/ℓ; catalyst concentration = 5 g immobilized catalyst/ℓ; reactor temperature = 60 °C; catalyst radius = 200 microns, porosity = 0.3, density = 0.9 g/mL, thermal conductivity = 0.8 W/m K; rate constant for the forward exchange reaction $k_1 = 5.2 \times 10^{-5}$/s (Tomoi et al., 1983); the rate constant of the organic reaction is given by

$$k_{app} = k_2 \rho_c [QX]_{c,0} = 2.77 \times 10^{10} \exp(-11136/T) - 1.305 \times 10^{15} \exp(-15200/T), 1/s$$

where $[QX]_{c,0}$ is the initial concentration of active catalyst in the solid in mmol/g and the equilibrium constant of the exchange reaction $K = 20.5$.

SOLUTION

The mass balance equations to be used are the same as Equations 19.13–19.21. The basis of the heat balance equation is the Prater formulation

$$\lambda_e (T - T_s) = D_{e,RX}([RX]_{org,s} - [RX]_{org,c}) \tag{E19.1.1a}$$

where λ_e is the effective thermal conductivity of the pellet. This equation describes the relation between T and $[RX]$ at any point in the catalyst. We use it to define a Thiele modulus based on the (measurable) surface temperature. The result is the following final expression for the nonisothermal Thiele modulus (Desikan and Doraiswamy, 1999):

$$\phi^2 = \phi_s^2 \exp\left[\alpha_s \beta_m \frac{([R\hat{X}]_{org,s} - [R\hat{X}]_{org,c})}{1 + \beta_m([R\hat{X}]_{org,s} - [R\hat{X}]_{org,c})}\right] \tag{E19.1.1b}$$

The expression for effectiveness factor is then

$$\eta_{noniso} = \frac{3 \int_0^1 \exp\left[\alpha_s \beta_m \frac{[R\hat{X}]_{org,s} - [R\hat{X}]_{org,c}}{1 + \beta_m([R\hat{X}]_s - [R\hat{X}]_c)}\right] [R\hat{X}]_{org,c} (1 - [Q\hat{X}]_{org,c}) \omega d\omega}{[R\hat{X}]_{org,s}(1 - [Q\hat{X}]_s)}$$

(E19.1.2)

In these equations, ϕ is the Thiele modulus at any point within the catalyst, ϕ_s is the modulus at the surface, and the parameters α_s and β_m are, as in any fluid–solid (catalytic) reaction (Chapter 7), the additional parameters necessary to characterize nonisothermal operation.

Table 19.6 Dimensionless groups and their values for Example 1.1

$$\alpha_1 = \frac{k_1}{k_2} = 10, \quad \alpha_2 = \frac{k_1}{k_2 K} = 0.01, \quad \gamma_Y = \frac{[Y]_0}{\rho_c [QX]_{c,0}} = 0.5$$

$$\gamma_{RX} = \frac{[RX]_0}{\rho_c [QX]_{c,0}} = 0.25 \quad \phi_s^2 = \frac{R^2 \rho_c k_2 [QX]_0}{D_{e,RX}} = 0.1,$$

$$\chi = \frac{k'_y}{k'_{RX}} = 1.0, \quad \psi = \frac{D_{e,Y}}{D_{e,RX}} = 1.0$$

$$\theta_{QX} = \frac{1}{\varepsilon_{org} + \varepsilon_{aq}} = 1.0, \quad \theta_{org} = \frac{\varepsilon_{org}}{\varepsilon_{org} + \varepsilon_{aq}} = 0.5,$$

$$\theta_{aq} = \frac{V_{aq}}{V_{org}} \frac{1}{\varepsilon_{org} + \varepsilon_{aq}} = 0.5, \quad \theta'_{org} = \frac{V_{org}}{V_{cat}} \left(\frac{1}{\varepsilon_{org} + \varepsilon_{aq}} \right) = 500,$$

$$\alpha_s = \frac{E}{R_g T_s} = 10, 20 \quad \beta_m = \frac{D_{e,RX}(-\Delta H_r) \rho_c [QX]_{c,0} \gamma_{RX}}{\lambda_e T_s} = 0.2, 0.05, 0.0, -0.05, -0.2$$

To solve these equations, values of the various diffusion coefficients and the heat of reaction of the organic phase reaction are necessary. These were estimated by the methods outlined in Chapters 3 and 7, and the values are (see Desikan and Doraiswamy, 1999):

D_{RX} (benzyl chloride in toluene) = 2.49×10^{-5} cm²/s,
D_Y (sodium acetate in water) = 1.925×10^{-5} cm²/s,
$D_{e,RX}$ (benzyl chloride in catalyst) = D_{RX}, $f_c/\tau = 2.49 \times 10^{-6}$ cm²/s,
$D_{e,Y}$ (sodium acetate in catalyst) = D_{RX}, $f_c/\tau = 1.925 \times 10^{-6}$ cm²/s,
$(-\Delta H_{org}) = 334.5$ kJ/mol.

It will also be noticed from the equations that a number of dimensionless groups are involved. These were calculated using the parameter values listed above and are summarized in Table 19.6. Mass transfer coefficients for three-phase systems cannot easily be estimated (although an approximate equation is given later on). In the absence of this information, simulation studies were confined to a very high Biot number (100), so that film transfer resistances could be ignored.

Simulation results are plotted as RX concentration versus time in Figure 19.12 for both isothermal and nonisothermal situations (it is only necessary to ignore the heat transfer groups to simulate the isothermal condition). Clearly, there is some effect at higher reaction times, but it is not significant. From the values of the Thiele modulus ($\phi_S = 0.33$), we can assume that the reaction is kinetically controlled. The low value of β_m also justifies the small effect of nonisothermicity.

Another interesting feature is shown in Figure 19.13, which is a plot of the effectiveness factor versus time for different values of the Thiele modulus at fixed values of α_s and β_m. Clearly, there is an optimum time at which the effectiveness factor is maximum. These are physically realizable values for which a catalyst can be tailored. Recall that an optimum time (for maximum active catalyst concentration, and therefore for maximum rate) was also observed for LL-PTC.

Note: although this example has been designed primarily to illustrate the effect of nonisothermicity by using Equations 19.13–19.21 alone important parameters (such as

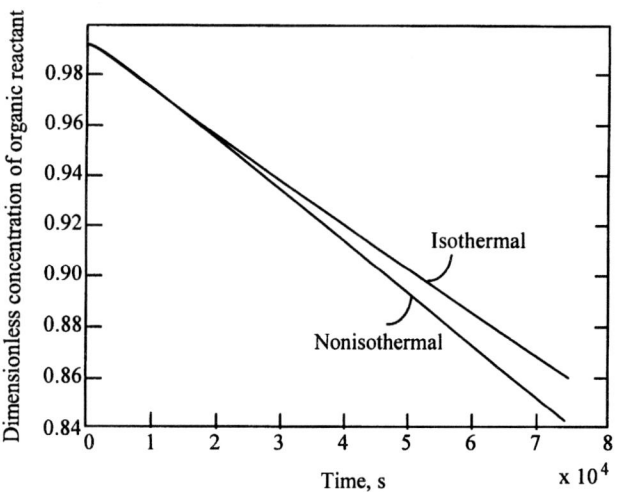

Figure 19.12 Concentration vs. time plots for isothermal and nonisothermal operations in Example 19.1 (Desikan and Doraiswamy, 1999)

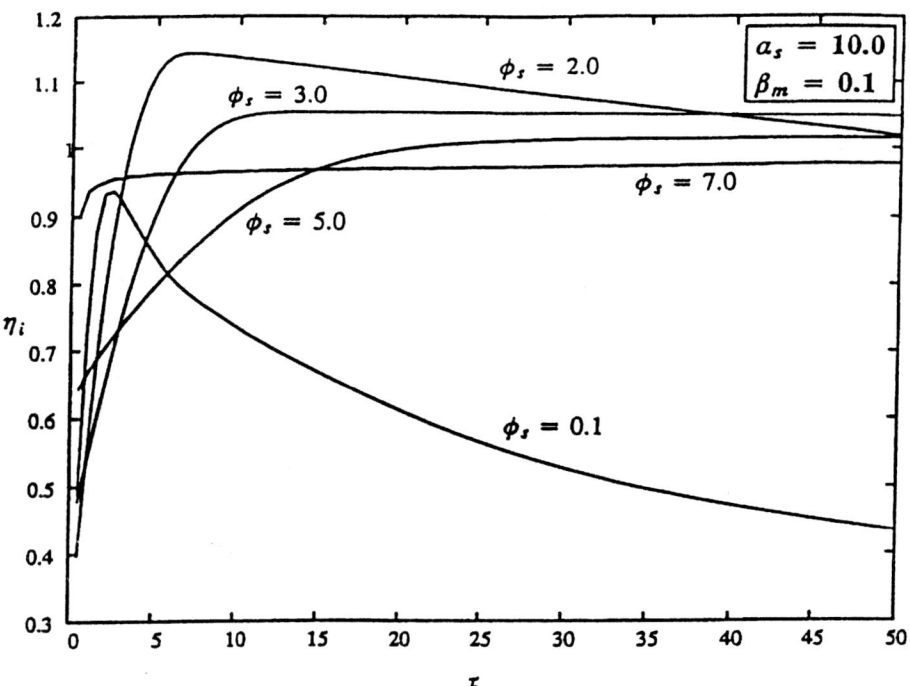

Figure 19.13 Effectiveness factor vs dimensionless time plots for different values of ϕ_s at fixed values of α_s and β_m for Example 19.1 (Desikan and Doraiswamy, 1999)

Autocatalysis in Triphase *PT* Reactions

In experiments carried out on the hydrolysis of octyl acetate with sodium hydroxide as a base and benzyltributylammonium chloride as a PT catalyst, the conversion–time curves were *S*-shaped, as shown in Figure 19.14 (Glatzer and Doraiswamy, 2000a). Such curves are typical of autocatalysis and have important implications in PTC.

When the reaction starts, there is no product alcohol, and because the organic phase has very low polarity, the catalyst is prevented from crossing the interface. Thus the reaction is restricted to the organic–aqueous interfacial boundary layer. The concentration of the catalyst in this layer is very high; so is the concentration of activated hydroxide ions. This explains the observed enhancement due to the homogeneous catalyst even at the commencement of the run (when there is no product). As the reaction progresses, increasing amounts of the product (alcohol with an ability to solubilize the catalyst) are formed. Thus the PT catalyst pairs with the hydroxide ions and crosses the interface into the organic phase, enhancing the rate of reaction in that phase. As more alcohol is formed, the rate increases further (instead of falling as in any normal reaction) and then levels off.

The same reaction was performed with a much more lipophilic catalyst, Aliquat 336. The result was a fast pseudo-first-order reaction with no auto-

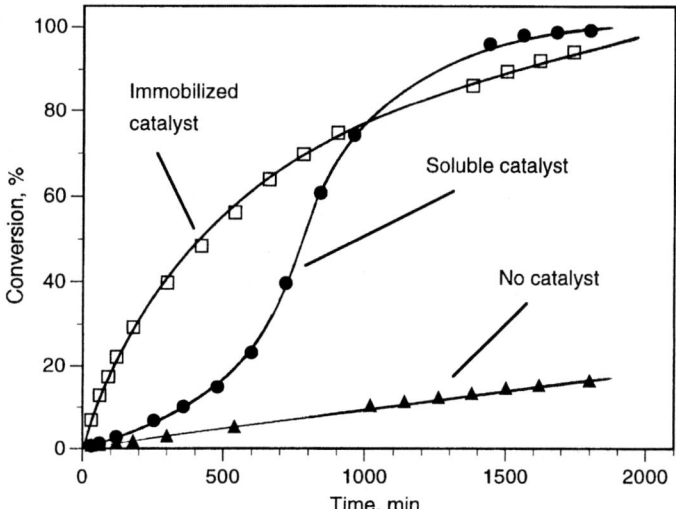

Figure 19.14 Autocatalysis in PTC: conversion vs. time curves in the hydrolysis of octyl acetate under basic conditions (using sodium hydroxide) with benzyltributylammonium chloride as a PT catalyst (Glatzer, 1999; also Glatzer and Doraiswamy, 2000a)

catalytic enhancement. Similar behavior was reported for the same catalyst in another system (Asai et al., 1992). Thus it appears that autocatalysis occurs only when a PT catalyst that is sparingly soluble in nonpolar organic media is used and changes in polarity allow the catalyst to dissolve in increasing measure as the reaction progresses.

A rate expression was derived based on the earlier studies of Tamhankar et al. (1981) (see Example 5.5), that accounts for both the noncatalytic and catalytic components of the overall reaction which represents the data with less than 2% error (Glatzer and Doraiswamy, 2000a).

Mass Transfer in Triphase Catalysis

In the SLPTC models previously developed, film mass transfer coefficients for the organic and aqueous phases are important parameters. In Chapter 14 we saw how the coefficient for gas-liquid systems can be determined. The same method can be used for liquid-liquid systems. In all of these cases, mass transfer rates are calculated using contactors with known interfacial areas. Where a solid phase is involved, the constant area criterion can be met by using a rotating disk of solid reactant or catalyst, as the case may be (Melville and Goddard, 1985, 1988; Hammerschmidt and Richarz, 1991).

In liquid-liquid-solid (three phase) systems such as triphase catalysis, mass transfer in the aqueous and organic phases occurs serially in a single PTC cycle, that is, ion exchange in the aqueous phase followed by the main reaction in the organic phase. The design of a contactor for getting these values simultaneously has to be quite different from that for two-phase systems.

Rotating disk contactor

In a rotating disk contactor (RDC) developed recently (Glatzer et al., 1998), provision was made for such a determination. As shown in Figure 19.15, a disk carrying the supported PT catalyst is rotated about its horizontal axis located exactly at the level of the liquid-liquid interface. Perpendicular orientation of the disk is necessary to ensure equal exposure of the solid-bound catalyst to both bulk phases during one revolution. By taking samples from the two phases during a run, the mass transfer coefficient in each phase can be determined. The construction features of this RDC have been fully described by Glatzer et al. (1998).

A correlation for Sherwood number (based on RDC data)

Many correlations have been suggested to estimate mass transfer coefficients in two-phase systems for various types of physical scenarios. Oldshue (1983) gives a summary of available correlations for liquid-solid, liquid-liquid, and gas-liquid mass transfer, as well as an estimate of the mass transfer rate of a G-L-S system—namely, the oxidation of sodium sulfite in water. Cussler (1997) gives

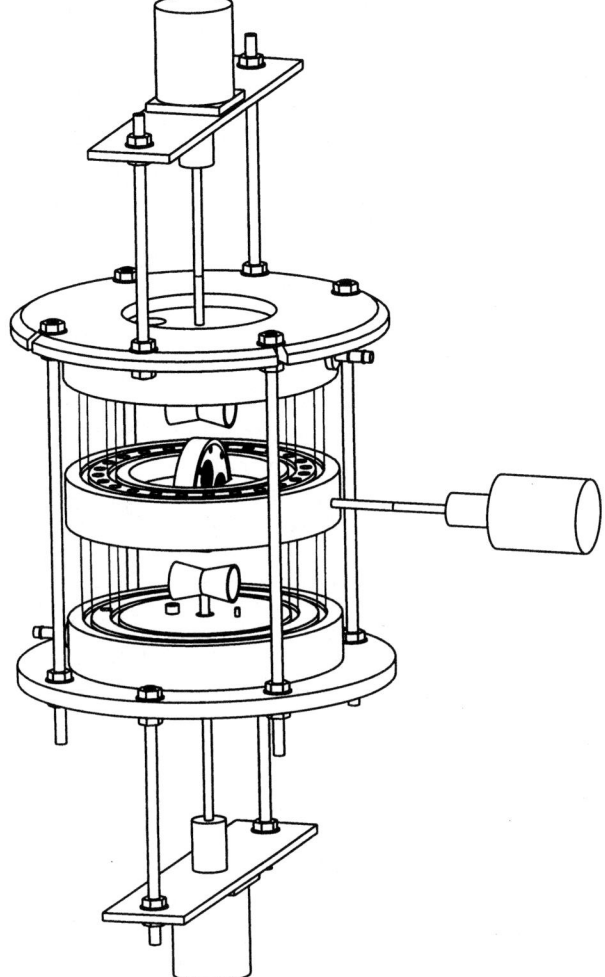

Figure 19.15 Rotating disk contactor for measuring mass transfer in liquid-liquid-solid systems (Glatzer et al., 1998)

selected mass transfer correlations for different physical situations for fluid-fluid and fluid-solid interfaces. Of these, the following is particularly noteworthy:

$$k'_L \left(\frac{\rho}{\mu g}\right)^{1/3} = 0.0051 \left(\frac{u\rho}{a\mu}\right)^{0.67} \left(\frac{D\rho}{\mu}\right)^{0.50} (ad_P)^{0.4} \tag{19.26}$$

This equation is used to estimate the mass transfer coefficient at liquid-solid interfaces in packed towers, and it might be the best available correlation for liquids. Note that in this equation the reciprocal Schmidt number is raised to a positive fractional power.

Despite the limitations of available data with RDC (three different TPC systems were investigated), it was possible to suggest a first generalization (Glatzer and Doraiswamy, 2000b). The key idea for a general correlation for TPC systems is to break up the three-phase system (L-L-S) into two L-S systems and correlate the data for each "side" of the three-phase system separately.

The following equation correlating the Sherwood number with the Reynolds and Schmidt numbers was found to predict the organic phase mass transfer coefficient well:

$$\frac{k'_L d_S}{D} = 0.005 \left(\frac{d_S^2 \omega' \rho}{2\mu}\right)^{3/4} \left(\frac{D\rho}{\mu}\right)^{1/3} \tag{19.27}$$

where d_S is the stirrer diameter and ω' the disk rotation in radians/time. Equation 19.27 has some similarities with Equation 19.26: the factors are almost the same (0.0051 and 0.005), and both equations raise the reciprocal Schmidt number to a positive fractional power. Note that in the RDC used, the liquid circulates around the particles held in a single layer on the disk. Hence the system approaches slurry-bed behavior and the equation can approximately predict mass transfer in slurry reactors. It is capable of predicting the Sherwood number to within 20% error. However, in most cases the prediction is significantly better. This is commonly an accepted level of uncertainty in the prediction of Sherwood number in these types of correlations.

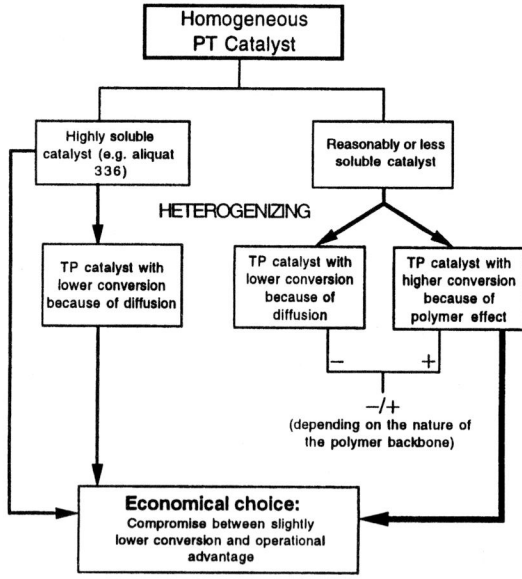

Figure 19.16 A methodology for choice between soluble and immobilized forms of a catalyst for a given reaction (Glatzer, 1999; also Glatzer and Doraiswamy, 2000a)

Choice between Soluble and Immobilized Forms of PT catalysts

Triphase catalysts are difficult and more expensive to prepare than their soluble counterparts. Doubts also persist about their ability ever to compete with catalysts such as Aliquat 336 that are highly soluble in an organic phase. Therefore, it is necessary to construct a viable methodology involving comparisons at various stages. We propose such a methodology in Figure 19.16 which is self-explanatory.

INDUSTRIAL APPLICATIONS OF PTC

The most extensive beneficiaries of PTC are the pharmaceutical, pesticide, and other small and medium volume chemical industries. Its use in the pharmaceutical industry with particular emphasis on synthetic penicillins and drugs derived by *O*- and *N*-alkylation of various heterocycles has been reviewed by Lindblom and Elander (1980). Some examples are morphine, azapine, erythromycin A, pyrimidine diones, and vincamines. Examples of its applications in the pesticide industry have also been given, among others, by Slaoui (1982), Neumann and Sasson (1984), Reinick and Sheldon (1983), and Yoshikowa et al. (1984). Typical pesticides prepared using PTC are fenvalerate and cypermethrin (both belong to the general class of synthetic pyrethroids). A recent review by Sharma (1997) and another by Naik and Doraiswamy (1998) list several other examples of the use of PTC. The former is particularly useful for current and potential applications. A few selected ones, drawn from different sectors of the chemical industry, are listed in Table 19.7. A representative list of the more common pesticides made by PTC-assisted reactions appears in Table 19.8.

The extensive use of PTC in the pharmaceutical and pesticide industries notwithstanding, chiral synthesis has superseded them as its most successful application. It is always the industry's hope, the chemist's challenge, and the engineer's lot to make chiral synthesis of a desired "handed" compound (usually a pharmaceutical) technically feasible. Economically, it seldom fails to be superior to the commonly used method of separation from a racemic mixture. In this connection it is important to note that an optically pure PTC (such as a crown ether), if present in a chirally discriminating tight ion pair in the transition state, can yield the desired pure product in more than 90% enantiomeric excess. This can be accomplished either by using a prochiral substrate or by kinetic resolution of a racemic mixture. The PTCs normally used in drug synthesis are derived from natural alkaloids (Dolling, 1986; Dolling et al., 1987). A PTC belonging to this category, $(8R),(9S)$-N-(p-trifluoromethyl benzyl) cinchonium bromide, is included in Figure 19.1. Asymmetric induction has also been accomplished for a number of other reactions by using suitably handed PTCs, for example, alkylation, oxidation, epoxidation, Michael addition, and Darzens reaction (see Freedman, 1986; Dehmlow and Dehmlow, 1993; Starks et al., 1994).

Table 19.7 Selected examples of industrially important reactions

Application	Reaction	Reference
Chiral synthesis using cinchinidium derived optically active PT catalyst		
Synthesis of indacrinone, a diuretic drug candidate	C-alkylation of indanone derivatives and oxindoles using cinchona alkaloids	Dolling et al. (1987) Bhattacharya et al. (1986)
Synthesis of chiral α-amino acids	Alkylation of imines, glycine derivatives, and Schiff base derivatives	O'Donnel (1993) O'Donnel et al. (1989)
Agrochemicals		
Synthesis of an antidote for herbicides	N-alkylation of hexamethylenetetramine with chloromethyl ketones	Smith (1990)
Synthesis of herbicides and insecticides	Selective O-alkylation and O-phosphorylation of ambient pyridimates	Cutie and Halpern (1992)
Synthesis of insecticideal pyrethroid and insect pheremones	Wittig reaction of aliphatic aldehydes and alkenyl alcohols with 50% NaOH or solid K_2CO_3	Deng et al. (1989)
Synthesis of naturally occurring pellitorine, possessing insecticidal activity	PTC vinylation of (E)-1-iodo-1-heptene with vinyl acetate	Jeffery (1988)
Synthesis of a herbicide	N-alkylation of substituted phenyl N-hydroxyurea with dimethylsulfate	Fujita et al. (1982)
Perfumery and fragrance chemicals		
Enhancement and augmentation of aroma of perfumes	Alkylation of acetophenone moiety with allyl chloride	Sprecker and Hanna (1982)
Intermediate step in the synthesis of a fragrance from furfural	C-alkylation of propanal and butanal by 2-chloromethylfuran	Norwicki and Gora (1991)
Synthesis of phenylacetic acid, an intermediate in the perfumery industry	Carbonylation of benzyl chloride in the presence of a palladium-based catalyst	Cassar et al. (1976)
Compounds with biological activity		
One-pot synthesis of carboxamides and peptides	Reaction of a free acid or a carboxylic ester with an amine with KOH/K_2CO_3 and a phenylphosphonate coupling agent	Watanabe and Mukiyama (1981a)

One-pot synthesis of benzofuran derivatives with wide ranging biological activities	Reaction of o-chloronitrobenzenes with sodium azide	Ayyangar et al. (1987)
Synthesis of aminopyrroles, intermediates in synthesis of biologically active compounds like pyrrolytriazenes	N-alkylation of N-unsubstituted 3-aminopyrrole with TDA-1 as PT catalyst	Almerico et al. (1989)
Pharmaceuticals		
Synthesis of various drugs such as dicyclonine, phenoperidine, oxaladine, ritaline, etc.	Alkylation of phenylacetonitrile using NaOH, instead of expensive sodium ethoxide	Lindblom and Elander (1980)
Synthesis of (R)-fluorenyloxyacetic acid, useful in the treatment of brain odema	Use of nonionic surfactant, Triton X, with a cinchonidinium based PT catalyst to accelerate the alkylation step	Dolling (1986)
Synthesis of the commercial antibiotic chloramphenicol	Aldol condensation in the presence of NaOH and a PT catalyst	Koch and Magni (1985)
Synthesis of chlorpromine and imipramine antidepressants	N-alkylation of carbazones, phenothiazines, acridanone, and indoles using alkyl ahalides and aqueous NaOH/solid K_2CO_3	Schmolka and Zimmer
Synthesis of lysergic acid based pharmaceuticals and other molecules with the indole skeleton	Facile and selective monoalkylation of the indole nitrogen using PTC, instead of using K azide in liquid ammonia at $-40\,°C$	Lindblom and Elander (1980)
Synthesis of calcitriol derivatives	O-alkylation using t-butylbromoacetate	Neef and Steinmeyer (1991)
Other specialty chemicals		
Synthesis of allyltribromophenol, a flame retardant polymer	Etherification of hindered tribromophenol with allyl bromide	Wang and Yang (1990)
Synthesis of dialkyl sulfides (additives for lubricants, stabilizers for photographic emulsions)	Reaction of sodium sulfide with benzyl chloride	Pradhan and Sharma (1992) Hagenson et al. (1994)
Synthesis of β-lactams	Reaction of amino acids and methanesulfonyl chloride	Watanabe and Mukiyama (1981b)
Synthesis of dichlorovinyl carbazole, used in preparation of photoconductive polymers	Dichlorovinylation of carbazole in solid-liquid system	Pielichowski and Czub (1995)

Table 19.8 Some industrial examples of PTC assisted pesticides synthesis

Type of pest control chemical	Brief description	Reactions
Ethion (insecticide)	Reaction of dialkyl phosphorodithioic acids with a halo compound in the presence of PTC and alkali	$2(CH_3O)_2PS\text{-}S^-Na^+ + Cl\text{-}CH_2\text{-}CH_2\text{-}Cl \xrightarrow{PTC}$ $(CH_3O)_2P\text{-}S\text{-}S\text{-}CH_2\text{-}CH_2\text{-}S\text{-}S\text{-}P(OCH_3)_2$ Ethion
Cypermethrin (insecticide)	PTC condensation of 3-(2,2-dichloroethenyl)-2,2-dimethylcyclopropanecarboxylic acid chloride with m-phenoxybenzaldehyde cyanohydrin PTC: Benzyltriethylammonium chloride (TEBAC)	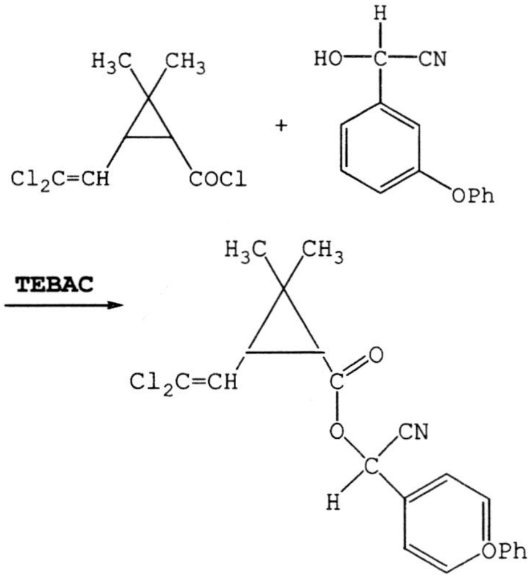 (structures shown) Cypermethrin

Fenvalerate (insecticide)	PTC condensation of 2-(4-chlorophenyl)-3-methyl-butyryl chloride with *m*-phenoxybenzaldehyde cyanohydrin. Note: 2-(4-chlorophenyl)-3-methyl-butyryl chloride also synthesized using PTC.	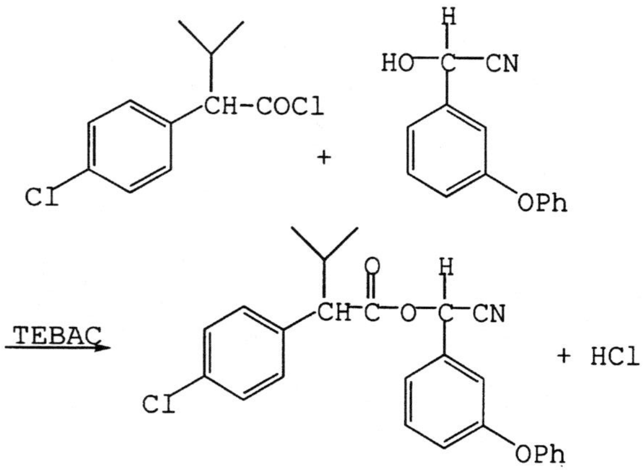 Fenvalerate
Metamitron (herbicide)	Reaction of benzoyl chloride with NaCN to yield corresponding acyl cyanide (intermediate). PTC: Tetrabutylammonium bromide.	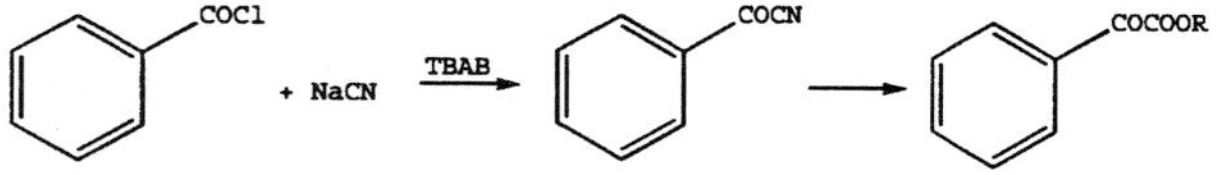 Metamitron
	Carbene reaction under PTC conditions. Reaction of dichlorocarbene generated from chloroform and alkali with benzaldehyde to yield mandelic acid. Subsequent hypochlorite oxidation of methyl mandelate and esterification to form metamitron.	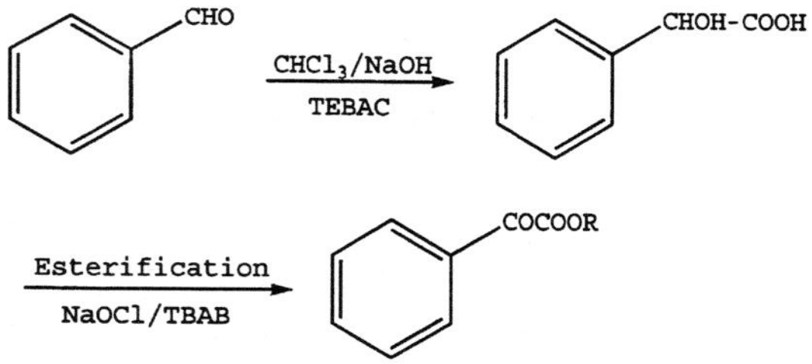

(*Continued*)

Table 19.8 (Continued)

Type of pest control chemical	Brief description	Reactions
Triforine (fungicide)	PTC condensation of piperazine with $CCl_3CHClNHCHO$ in presence of PTC and potassium carbonate. triforine = N,N'-bis-(1-formamido-2,2,2-trichloroethyl) piperazine.	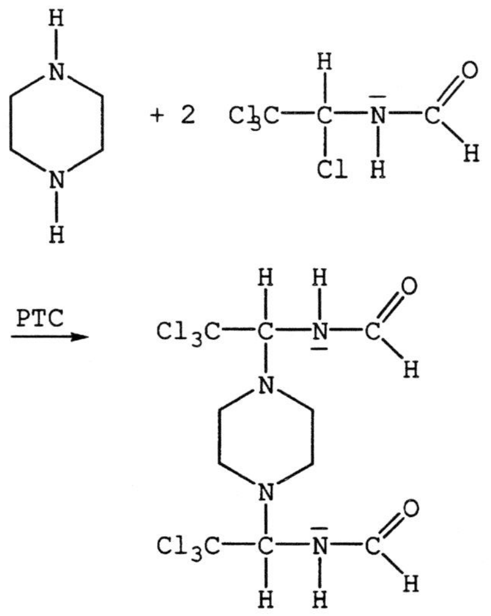 piperazine + 2 Cl_3C-CHCl-NH-CHO $\xrightarrow{PTC}$ N,N'-bis(1-formamido-2,2,2-trichloroethyl)piperazine (Triforine)
Thiols (intermediates for agrochemicals)	Reaction of alkyl halides with ammonium or alkali metal hydrogen sulfide in presence of PT catalyst.	$RX + NaSH_{(aq)} \xrightarrow[TBAB]{PTC} RSH + NaX$ Thiol

Chapter 20

Bioorganic Synthesis Engineering

Biological processes, from the simplest to the most complex, can broadly be classified as those caused by the catalytic action of living entities known as *microorganisms* or *microbes*, and those promoted and catalyzed by "lifeless substances" produced by microorganisms, known as *enzymes*. The two together are often referred to as *biocatalysts*. *The microbial kingdom* of living entities consists of all living things with a very simple biological organization. Both microbes and enzymes can be used to promote or selectively achieve a wide range of chemical transformations.

Indeed, biocatalysts occupy a unique position in the wide spectrum of catalysts used in organic technology and synthesis. One of the chief beneficiaries of the rising emphasis on environmentally friendly processes is the enzyme, for it is being increasingly pressed into service to generate technologies that are both highly selective and pollution free. As catalysts, enzymes accelerate the rates of reactions at milder conditions, are highly selective, are biodegradable, and can be used in "free" solution form or as immobilized heterogeneous catalysts. The last feature, their use in immobilized form, has been a major factor in the movement of the enzyme from laboratory to industry.

Two main shortcomings of the conventional enzyme that have limited its application in organic synthesis are its restriction to reactions in the aqueous phase and to very mild temperatures and pressures. Research in the last few years has "released" the enzyme from these restrictions (see Govardhan and Margolin, 1995; Adams et al., 1995). Thus now it is possible to use enzymes in aqueous solutions containing water-miscible organic cosolvents, aqueous organic biphasic mixtures, and anhydrous organic solvents. Research has also uncovered microorganisms from a variety of unconventional *habitats* such as the biosphere and the depths of the oceans that have the unique ability to

accomplish chemical transformations at extreme conditions covering a wide range of temperatures, pressures, and salt concentrations. Hence it seems almost certain that enzymes will play an increasingly important role in industrial organic synthesis.

Several books and reviews have appeared on the industrial applications of biocatalysts, for example, Trevan (1980), Chibata et al. (1982), Dixon and Webb (1984), Tramper (1985), Wiseman (1985), Chibata (1978, 1987), Scott (1987), Peppler and Reed (1987), Cheetham (1987), Nakamura et al. (1988), Dordick (1991, 1992), Tanaka et al. (1993), Govardhan and Margolin (1995), Adams et al. (1995), and also on the design and analysis of biocatalytic reactors, for example, Bailey and Ollis (1986), Van't Riet and Tramper (1991), Shuler and Kargi (1992), Peter (1992), Dunn et al. (1992),[1] Tanaka et al. (1983), Vieth and Wolf (1994), Nielsen and Villadsen (1994).

The objectives of this chapter are to outline the basic aspects of cell and enzyme action, review the methods of immobilization, outline the principles of bioreaction analysis, briefly describe the methods of bioreactor design, and conclude by giving several examples of the use of cells and enzymes in organic technology and synthesis, including enantioselective synthesis.

MICROBES AND ENZYMES

All living matter is composed of cells and their products. Thus all classes of microbes are described in terms of cells or their specific functional areas known as *organelles* that have discrete but intricately coordinated roles. Despite their great diversity, there is one underlying entity that controls their action, DNA. This feature enables the use of the continually evolving methods of *genetic engineering* to manipulate cell behavior. Our treatment will be based on the optimal use of available species and will not consider the genetic engineering to generate new ones or their discovery from harsher *habitats* (which can often lead to a variety of new drugs and other useful chemicals).

Microbes

There are two classes of microbes, *procaryotes* and *eucaryotes*. Procaryotic cells are usually characterized by simple unicellular structures. Exceptions are also known such as the cyanobacteria, which are often multicellular. The only procaryotes are bacteria which are also the most abundant form of life on earth. Eucaryotes encompass all other forms of life, including animal and plant.

Eucaryotic organisms can be either unicellular (e.g. *protozoa*) or more complex structures. They all contain organelles such as *mitochondria*, *ribosomes* (also found in procaryotes), *lysosomes*, and *plastids*. Yeasts are unicellular fungi

---

[1] This book is particularly useful to those without a strong background in chemical engineering and gives a very clear exposition (with easily understandable mathematics) of biokinetics, dynamic differential balances, and modeling of a number of reactions and reactor types normally encountered in bioorganic synthesis.

known largely for their use in alcoholic beverage production. *Aspergillus* and *Penicillium*, which are used the world over for the production of antibiotics and biological catalysts, are filamentous fungi.

Enzymes

General features of enzymes

All enzymes are proteins. They are isolated from microbial fermentations and animal and plant by-products. The one feature of enzymes that places them above all other forms of catalysts is their remarkable specificity to a desired product. Enzymes are essentially large molecules whose molecular weights range from 10,000 to several million and whose cross-sectional dimensions extend up to about 10 nm. To develop catalytic action, however, many enzymes require the cooperation of low molecular weight nonprotein substances known as *coenzymes* or *cofactors*. The protein moiety in such cases is specifically referred to as *apoenzyme*. There are essentially six classes of common enzymes, as listed in the first part of Table 20.1.

Two distinguishing features of enzyme catalysis are noteworthy. There is usually an optimum pH value at which the enzyme activity is maximal. This is substantiated by many systems, for example, the enzyme-catalyzed oxidation of glucose (Figure 20.1). The second is the existence of an optimum temperature for maximum activity (see Johnson et al., 1954).

Table 20.1 Classes of enzymes and extremozymes

Enzyme(s)	Function
Common Enzymes	
Oxidoreductases, e.g., oxidases, dehydrogenases	Oxidation of hydroxyl, ketone, amino groups; and reduction of $-CH_2CH_3-$
Transferases, e.g., glucotranferases, transaminases	Transfer of methyl, formyl, alkyl, sulfate groups between molecules
Hydrolases, e.g., esterases, lipases, proteases, glucosidases, sulfatases	Hydrolysis of esters, phosphates, glycosides
Lyases, e.g., ketolases, decarboxylases, hydratases	Cleave C–C, C–N, C–O bonds to produce double bonds, or add groups to double bonds
Isomerases, e.g., racemases, epimerases	Intramolecular isomerization
Ligases, e.g., carboxylases, synthetases	Linking of molecules to form C–O, C–N, C–C, C–S bonds
Extremozymes	
From *Pyrococcus furiosus*	High temperatures, up to 100 °C
From *Bacillus TA41*	Low temperatures, down to 4 °C
From *Methanococcus janaschii*	High pressures, up to 250 atm; and high temperatures up to 85 °C
From *Clostridium paradoxam*	High pH, 10.1; and temperatures up to 56 °C
From *Metallasphaera sedula*	Low pH, 2.0; and temperatures up to 75 °C
From *Halobacterium halobium*	High salt, 4–5 M NaCl

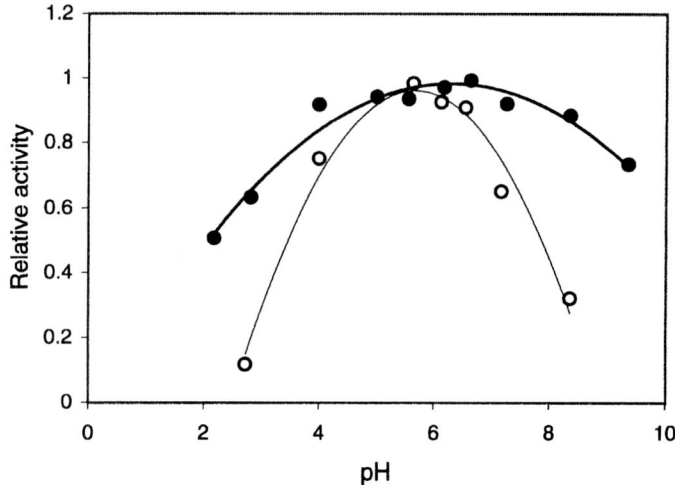

Figure 20.1 The pH dependence of glucose oxidase activity for soluble (○) and immobilized (●) forms (Cho and Bailey, 1978)

Enzymes that function under extreme conditions (extremozymes) and in organic media

Microorganisms have been discovered that grow under extreme conditions such as at 0–140 °C, pressures up to 250 atm, and salt contents as high as 4–5 M NaCl. These are referred to generally as *extremophiles* and specifically according to the *habitats* where they were discovered. For example, those found in hot springs are called *thermophiles*, in Antarctic sea waters *psychrophiles*, in deep sea hydrothermal vents *arophiles*, in alkaline lakes *alkalophiles*, in hot sulfurous springs *acidophiles*, and in salt lakes *halophiles*.

The enzymes produced by these extremophiles, known as *extremozymes*, can function under extreme conditions. An illustrative list along with an indication of the extreme environments in which they can function is included in Table 20.1 (sources: Kushner, 1978; Jones et al., 1983; Huber et al., 1989; Li et al., 1993; Davail et al., 1994; Adams et al., 1995). Enzymes extracted from these microorganisms have been tested for a variety of reactions and optimum temperatures have been found. Examples are enzymes from *Pyrococcus furiosus* for α- and β-glucosidase, α-amylase, protease, and hydrogenase activities (Bryant and Adams, 1989; Costantino et al., 1990; Blumentals et al., 1990; Kegen et al., 1993; Laderman et al., 1993).

Almost all of the enzymes discovered till recently function only in aqueous media. The inherent disadvantage of this feature is that the solubility of many organic compounds in water is low, which leads to low reaction rates. This has led to research that has resulted in enzymes that can function efficiently in organic solvents (Klibanov, 1986; Margolin, 1991), water–solvent mixtures (Dordick, 1989; Margolin, 1991), and in supercritical solvents (Randolph et al., 1988a,b) (see Chapter 26). A significant finding is that the activity of enzymes

suspended in organic solvents can be enhanced by adding carbohydrates (Yamane et al., 1990; Dabulis and Klibanov, 1993), polymers (Ottolina et al., 1992), organic buffers (Strika-Alexopoulos and Freedman, 1993), and salts such as KCl (Khmelnitsky et al., 1994). The enhancement due to salt addition is particularly dramatic.

Nature is not the only source of extremozymes. They are increasingly being designed in the laboratory by protein engineering methods (Govardhan and Margolin, 1995). A new class of enzymes known as CLECs (cross-linked enzyme crystals) has also been produced by growing them as crystals out of a crude mixture of commercially available enzymes and then chemically cross-linking the molecules within the crystals (St. Clair and Navia, 1992; Walters, 1997). These microcrystalline enzymes are much more stable than traditional enzymes and outperform them in many ways.

Immobilization of Enzymes

The binding forces in an enzyme are relatively weak and lead to its facile deactivation. However, if the enzyme were attached to a firm solid surface, it would inherit the inflexibility of its support and usually (but not always) show greater stability. This general technique known as immobilization was considered in Chapter 7, and is particularly relevant to enzymes because of their structural flexibility. These *immobilized enzymes* can withstand harsher conditions than "free" or soluble enzymes. The resulting catalytic action of such enzymes is truly heterogeneous—with all its advantages and disadvantages.

Methods of immobilization

Methods of immobilizing biocatalysts can be broadly classified as shown in Table 20.2. This classification covers most of the prevalent methods. Recent developments in immobilization are based as much on a judicious combination of the methods listed in the table as on developing new strategies. For example, adsorption and subsequent cross-linking of enzymes, which avoids the disadvantages of the individual methods and exploits their advantages, is a new strategy that is being increasingly used. Detailed descriptions of the various techniques can be found in many of the books and reviews listed earlier.

Effects of immobilization

Immobilization of an enzyme within a solid matrix causes a significant change in the environment in which the enzyme functions. This change is attributable to several factors, the most significant being the diffusional resistance offered by the solid. This is considered in a later section.

The Intact Cell versus the Enzyme in Organic Synthesis

It is the enzyme that acts as the catalyst, whether as an intact microorganism or as an isolated enzyme. Although structurally complex (polyamides constructed

Table 20.2 Methods of immobilization

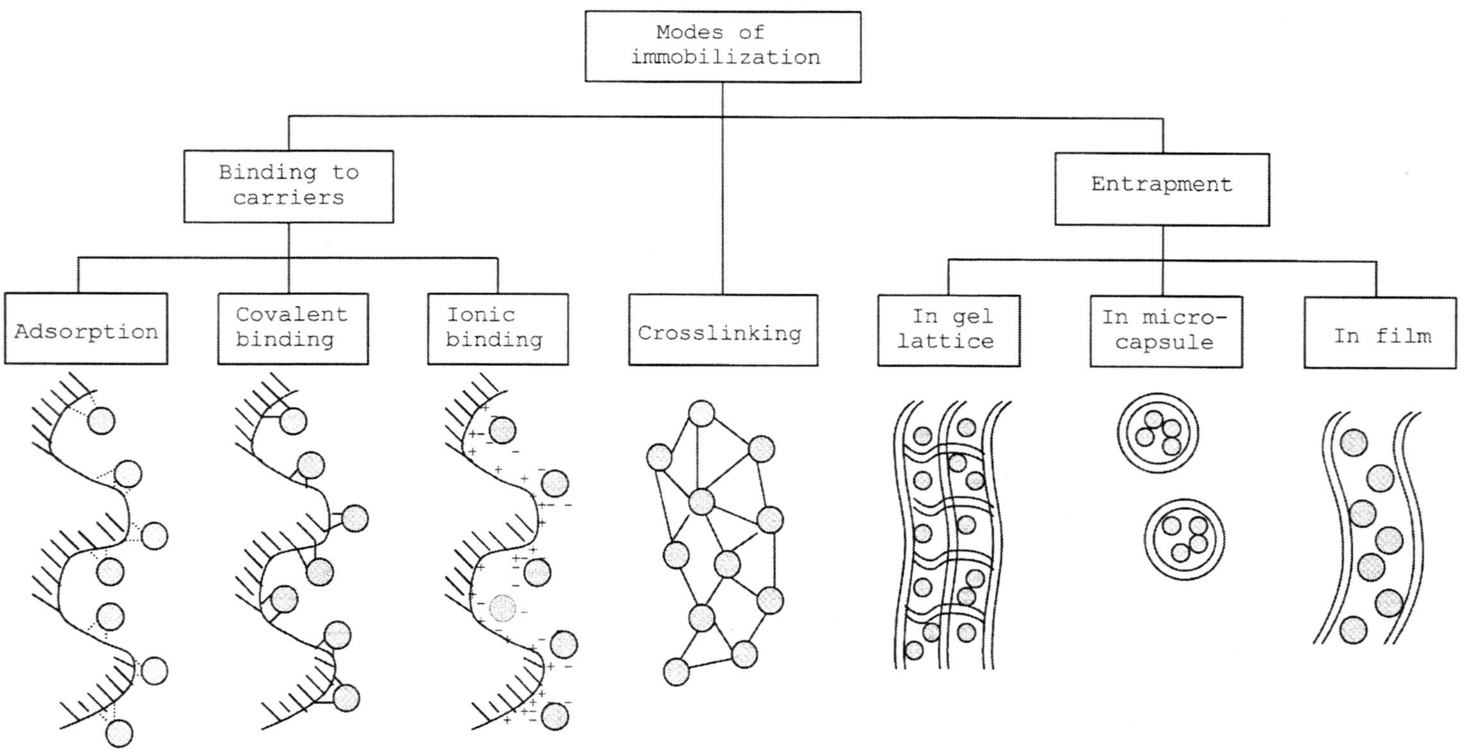

```
                        Modes of
                     immobilization
                            |
        ┌───────────────────┼───────────────────┐
   Binding to           Crosslinking         Entrapment
   carriers                                      |
        |                                        |
   ┌────┼────┐                         ┌─────────┼─────────┐
Adsorption Covalent Ionic           In gel    In micro-   In film
          binding  binding          lattice   capsule
```

from amino acids), the enzyme is a well-defined chemical entity. On the other hand, the cell is a complex assembly of several thousand enzymes and other constituents such as cofactors. Its catalytic role is not attributable to a single chemically defined enzyme but is due to the cooperative action of various cell constituents (not necessarily all of them). The cells are preferable from a practical point of view because they can be used directly. On the other hand, if the constituent enzyme is to be used, first it must be isolated by disintegrating the cells (often by passing them from a region of very high pressure to one of atmospheric pressure and sometimes by using ultrasound), thus adding to the total cost.

KINETICS AND MODELING OF BIOREACTIONS

The reactions considered here are those in which an organic feed is converted into a product by the catalytic action of an enzyme or a microbe. These reactions can be represented as follows:

ENZYMATIC REACTION

$$\begin{pmatrix} A \\ \text{organic} \\ \text{reactant} \end{pmatrix} \xrightarrow{E \text{ (enzyme)}} \begin{matrix} R \\ \text{(product)} \end{matrix} \qquad [20.1]$$

MICROBIAL REACTION

$$\begin{pmatrix} A \\ \text{organic} \\ \text{reactant} \end{pmatrix} \xrightarrow{C \text{ (cells)}} \begin{matrix} C \\ \text{(more cells)} \end{matrix} + \begin{matrix} R \\ \text{(product)} \end{matrix} \qquad [20.2]$$

The main distinction between them is that in an enzymatic reaction, the enzyme, a lifeless chemical entity, does not reproduce. On the other hand, in microbial transformation, the reaction occurs within a living cell where it is catalyzed by the enzyme produced by it (just as in an enzymatic reaction). However, the cell also reproduces itself, generating more enzyme in the process. Thus the modeling of microbial reactions must take this growth process into account. Further, these reactions can produce three possible results, as shown in Table 20.3. The reaction can be tailored specifically to produce any one of these results. Our concern in this chapter is with result 1, the production of a specific product R.

Enzymatic Reactions

Michaelis–Menten and Briggs–Haldane models

Enzymatic reactions can typically be represented as

$$A + E \underset{k_{-1}}{\overset{k_1}{\rightleftarrows}} AE \xrightarrow{k_2} R + E \qquad [20.3]$$

Table 20.3 Products of microbial fermentation

Reaction:	C +	A	⟶	C +	R
	(cell)	(reactant, which is food for cell growth)		(more cells)	(product)

Desired result		
1. Product R	2. Cells C	3. Breakdown of A
("Waste" left after cell growth)	Used in the production of *single cell protein* for animal feed	For use in waste water treatment
Used in the production of antibiotics such as *penicillin*		

where AE is the substrate–enzyme complex. An enzyme balance gives

$$[E]_0 = [E] + [AE] \tag{20.1}$$

To develop a model for such a scheme, the first assumption we make is that a fraction of the enzyme remains attached to the substrate, that is, $[AE] \neq 0$. Following this, we can make either of two further assumptions: a rate-determining step is present or the steady-state approximation holds for the intermediate. Depending on which set of assumptions is made, we have either of two celebrated models, the Michaelis–Menten (MM) model or the Briggs–Haldane (BH) model, although the latter appears to be favored (Chance, 1943).

The following single form of equation can be derived for the two models:

$$-r_A = r_R = \frac{k_2[E]_0[A]}{K_M + [A]} = \frac{V_m[A]}{K_M + [A]} \tag{20.2}$$

where

FOR THE MICHAELIS–MENTEN MODEL

$$K_M = \frac{k_{-1}}{k_1} \tag{20.3}$$

FOR THE BRIGGS–HALDANE MODEL

$$K_M = \frac{k_{-1} + k_2}{k_1} \tag{20.4}$$

In Equation 20.2, the term $k_2[E]_0$ represents the maximum possible reaction rate and is usually denoted by V_m.

The parameters of the MM or BH model can be estimated by the statistical methods briefly referred to in Chapter 7. However, preliminary values can be obtained by using different linearized forms of the model equation known by different names. The most important are summarized in Table 20.4 along with the corresponding plots.

Table 20.4 Linearized forms of the Michaelis-Menten equation

Name of plot	Linearized form	Plot
1. Lineweaver–Burke	$\dfrac{1}{-r_A} = \left(\dfrac{K_M}{V_m}\right)\dfrac{1}{[A]} + \dfrac{1}{V_m}$	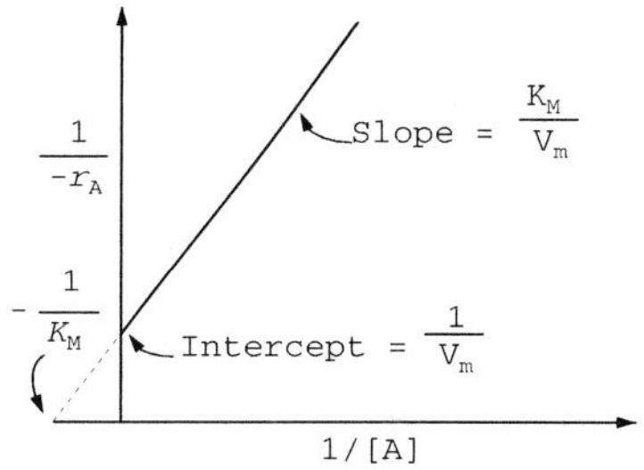 Plot of $\dfrac{1}{-r_A}$ vs $1/[A]$; Slope $= \dfrac{K_M}{V_m}$, Intercept $= \dfrac{1}{V_m}$, x-intercept $= -\dfrac{1}{K_M}$
2. Eadie–Hofstee	$-r_A = V_m - \dfrac{K_M(-r_A)}{[A]}$	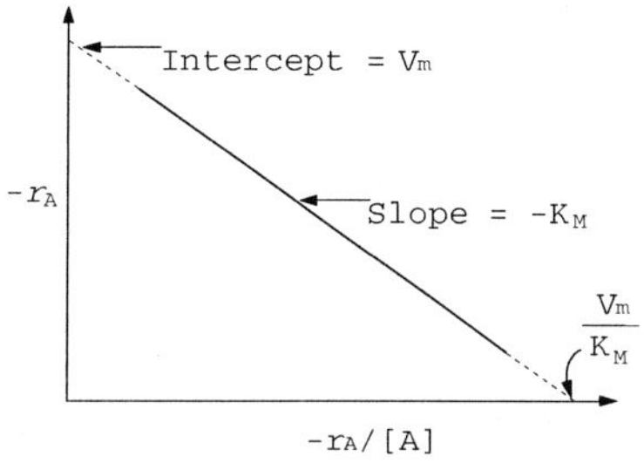 Plot of $-r_A$ vs $-r_A/[A]$; Intercept $= V_m$, Slope $= -K_M$, x-intercept $= \dfrac{V_m}{K_M}$
3. Hanes — multiply (1) by $[A]$	$\dfrac{[A]}{-r_A} = \dfrac{K_M}{V_m} + \dfrac{[A]}{V_m}$	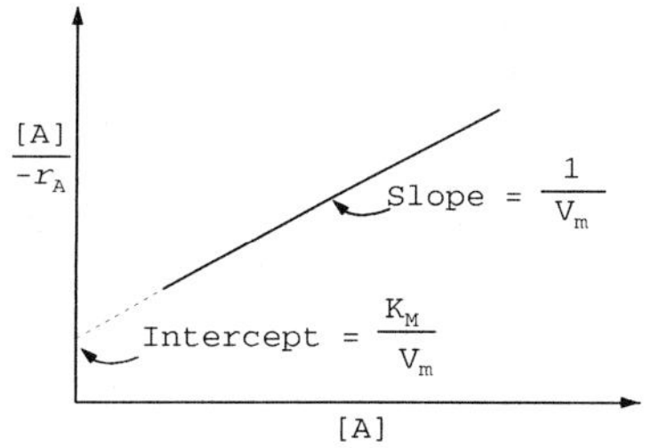 Plot of $[A]/-r_A$ vs $[A]$; Slope $= \dfrac{1}{V_m}$, Intercept $= \dfrac{K_M}{V_m}$

Inhibition

Inhibitors are substances that tend to decrease the rates of enzyme-catalyzed reactions. A study of these is obviously important in mitigating their effect or in accounting for it in reactor design. Another reason is that certain enzymes are harmful to the human system, and their action can be blocked by using a suitable inhibitor, which is usually an organic chemical that may have to be specially synthesized.

Inhibitors can be grouped broadly in two categories, reversible and irreversible. Reversible inhibitors are weakly bonded to the surface and therefore can be removed relatively easily (by dialysis or simple dilution). Irreversible inhibitors are essentially those that cannot be easily removed. Several categories of reversible enzymes are possible; the most important are competitive, noncompetitive, and substrate. Given in Table 20.5 are brief descriptions of these types of inhibition along with their effects on the constants of the MM model, as reflected in the original MM equation (see Levenspiel, 1993, for derivations).

Table 20.5 Main features of competitive, noncompetitive, and substrate inhibition[a]

Type	Competitive	Noncompetitive	Substrate inhibition
Description	B adsorbs competitively at the binding site of A 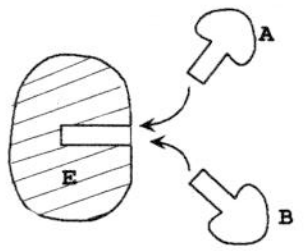	A and B adsorb on different sites, but adsorption of B stops action of A 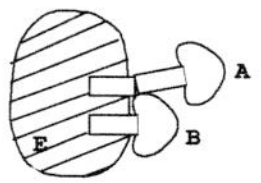	Reactant itself can sometimes inhibit if used in excess or too strongly adsorbed
Scheme	$A + E \underset{k_{-1}}{\overset{k_1}{\rightleftarrows}} AE \xrightarrow{k_2} R + E$ $B + E \underset{k_{-3}}{\overset{k_3}{\rightleftarrows}} BE$	$A + E \underset{k_{-1}}{\overset{k_1}{\rightleftarrows}} AE \xrightarrow{k_2} R + E$ $B + E \underset{k_{-3}}{\overset{k_3}{\rightleftarrows}} BE$ $B + AE \underset{k_{-4}}{\overset{k_4}{\rightleftarrows}} ABE$	$A + E \underset{k_{-1}}{\overset{k_1}{\rightleftarrows}} AE \xrightarrow{k_2} R + E$ $A + AE \underset{k_{-3}}{\overset{k_3}{\rightleftarrows}} AAE$ (complex AAE does not contribute to reaction)
Model equation	$r_R = \dfrac{V_m'[A]}{K_M' + [A]}$ where $V_m' = V_m = k_2[E]_0$ $K_M' = K_M[1 + [B]_0 K_3]$	$r_R = \dfrac{V_m''[A]}{K_M'' + [A]}$ where $V_m'' = \dfrac{V_m}{[1 + [B]_0 K_4]}$ $K_M'' = K_M \left[\dfrac{1 + [B]_0 K_3}{1 + [B]_0 K_4} \right]$	$r_R = \dfrac{V_m'''[A]}{K_M''' + [A] + [A]^2 K_3}$ where $V_m''' = V_m$ $K_M''' = K_M$
Graphical representation of effect on r_R vs. [A] plot	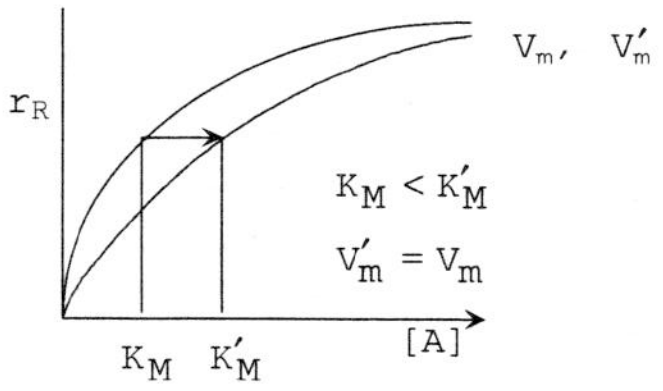 $K_M < K_M'$ $V_m' = V_m$	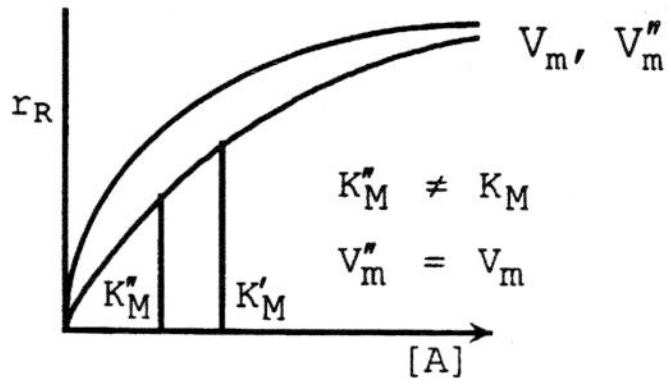 $K_M'' \ne K_M$ $V_m'' = V_m$	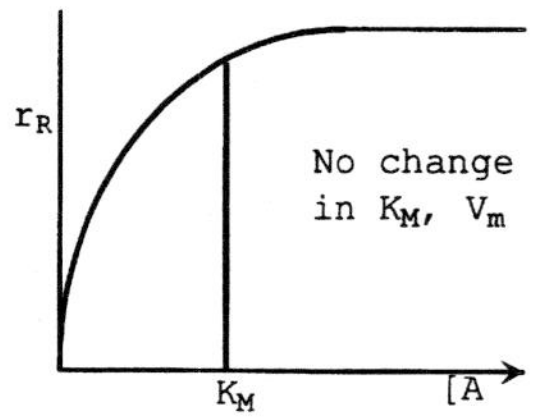 No change in K_M, V_m

[a] $K_1 = \dfrac{k_1}{k_{-1}}$, $K_3 = \dfrac{k_3}{k_{-3}}$, $K_4 = \dfrac{k_4}{k_{-4}}$. In mixed behavior, both K_M and V_m are affected

Immobilized enzymes (IME)

We consider here the two features normally associated with all solid catalysts (Chapter 7), internal diffusion and external film transport. An additional feature peculiar to enzymes is the relatively large effect of pH.

EFFECT OF INTERNAL DIFFUSION

Although the internal diffusion effect is common to all solid-catalyzed reactions, it is far more significant in IME reactions than in conventional solid-catalyzed reactions because the substrate in IME reactions is a liquid and molecular diffusivities in liquids (particularly aqueous solutions) are lower than in gas-solid systems. Two other contributing factors are: (1) the catalytic activity of enzymes is usually higher than that of solid catalysts and (2) the size of the species being transported is usually larger in enzyme-catalyzed reactions than in conventional catalytic reactions, leading to greater diffusional resistance.

In analyzing the role of diffusion in IME reactions, we assume that the reaction follows single-substrate MM kinetics and that there is no appreciable change in temperature or pH. Then the following nondimensional mass balance can be written for a single spherical pellet:

$$\frac{d^2[\hat{A}]_p}{d\hat{R}^2} + \frac{2}{\hat{R}} \frac{d[\hat{A}]_p}{d\hat{R}} = \phi_S^2 \left(\frac{[\hat{A}]_p}{1 + \beta[\hat{A}]_p} \right) \quad (20.5)$$

with the boundary conditions

$$\hat{R} = 0, \quad \frac{d[\hat{A}]_p}{d\hat{R}} = 0; \quad \hat{R} = 1, \quad [\hat{A}]_p = 1 \quad (20.6)$$

where (assuming $[A]_s = [A]_b$)

$$[\hat{A}]_p = \frac{[A]_p}{[A]_b}, \quad \hat{R} = \frac{r}{R} = \frac{2r}{d_p}, \quad \beta = \frac{[A]_b}{K_M}$$

$$\phi_s = \begin{pmatrix} \text{Thiele modulus} \\ \text{for a sphere} \end{pmatrix} = R\sqrt{\frac{V_m}{K_M D_{eA}}} \quad (20.7)$$

As in conventional solid catalysts (Chapter 7), the Thiele modulus can be generalized for any shape by using the ratio of volume to area (Λ_0) of the shape in place of R (e.g., R becomes $R/3$ for a sphere), thus replacing ϕ_s by ϕ (see Marsh et al., 1973; Engasser and Horvath, 1973; Blanch and Dunn, 1974).

Again, as in gas-solid catalytic reactions, we define an effectiveness factor which, here, may be expressed in dimensionless form as

$$\varepsilon = \frac{\left(\dfrac{d[\hat{A}]_p}{d\hat{R}} \right)_{\hat{R}=1}}{3\phi^2 \left(\dfrac{1}{1+\beta} \right)} \quad (20.8)$$

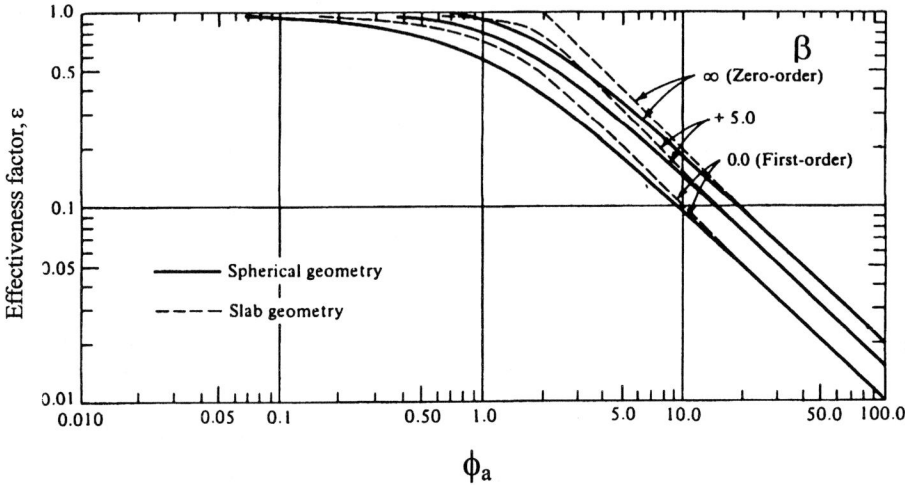

Figure 20.2 Effectiveness factor vs. the observable modulus ϕ_a for immobilized enzyme catalysts with Michaelis–Menten intrinsic kinetics ($\beta = [A]_b/K_M$) (Bailey and Ollis, 1986)

Note that the Thiele modulus used is the generalized Thiele modulus ϕ. A shortcoming of the modulus defined by Equation 20.8 is that values of V_m and K_M may not always be available. To overcome this practical deficiency, we turn to the Weisz modulus ϕ_a of Chapter 7 based on the *observed reaction rate* r_{Aa} which has been widely used in the analysis of conventional catalytic reactions. Using this modulus, plots of ε versus ϕ_a for various values of β can be obtained, as shown in Figure 20.2 (Bailey and Ollis, 1986). Clearly, $\beta = 0$ conforms to first-order and $\beta = 1$ to zero-order kinetics. Intermediate values conform to MM kinetics.

Lee and Reilly (1981) defined a more rigorous form of the Thiele modulus based on the generalized modulus of Bischoff and Aris (see Chapter 7) which is particularly useful in analyzing the role of diffusion in deactivation. Their analysis shows that, in reactant-independent deactivation, the presence of a strong diffusional limitation lowers the rate of deactivation to half the diffusion-free value. Thus, surprisingly, diffusion seems to have a favorable effect on the performance of a deactivating immobilized enzyme catalyst.

EFFECT OF EXTERNAL DIFFUSION

The theoretical treatment of the effect of external diffusion follows the same lines as for conventional catalytic reactions except again that the rate term is different. Thus

$$-r_A = k_L([A]_b - [A]_s) = \frac{V_m[A]_s}{K_M + [A]_s} \tag{20.9}$$

which can be expressed in nondimensional form as

$$\frac{1 - [\hat{A}]_{sb}}{Da} = \frac{[\hat{A}]_{sb}}{1 + \beta[\hat{A}]_{sb}} \qquad (20.10)$$

where

$$[\hat{A}]_{sb} = \frac{[A]_s}{[A]_b} \qquad (20.11)$$

and Da is a Damköhler number given by

$$Da = \frac{V_m}{k_L[A]_b} = \frac{\text{maximum reaction rate}}{\text{maximum mass transfer rate}} \qquad (20.12)$$

The value of k_L in Da can be found from any of the many Sherwood number or j_d correlations (for k'_L) reported (see Chapter 7). A few common correlations specific to immobilized biocatalysts are listed in Table 20.6.

From the value of $[\hat{A}]_{sb}$ thus obtained from Equation 20.10 and that of $[A]_b$, which is the experimentally observed concentration of A, $[A]_s$ can be calculated from Equation 20.11. Hence either the observed rate (based on the bulk value) or the true rate (based on the surface value) can be found from the rate equation. Several other significant studies (experimental and theoretical) have been reported in this area, for example, Rovito and Kittrell, 1973; Brams and McLaven, 1974; Horvath and Engasser, 1974; Toda, 1975; Greenfield et al., 1975; Lee et al., 1979; and Patwardhan and Karanth, 1982).

COMBINED EFFECTS OF INTERNAL AND EXTERNAL DIFFUSION

The analysis is similar to that described in Chapter 7, except that MM kinetics replaces the first-order equation. However, the general trend of the effect is similar to that for a first-order reaction in a flat plate and is given by

$$\frac{1}{\varepsilon_a} = \frac{1}{\varepsilon} + \frac{\phi^2}{Bi_m} \qquad (20.13)$$

Table 20.6 Selected correlations for liquid-side mass transfer coefficient in IME systems

Correlation[a]	Reference
1. Porous glass beads: $j_d = 1.625(Re')^{-0.507}$	Rovito and Kittrell (1973)
2. Collagen-enzyme chips: $j_d = 5.7(Re')^{-0.78}$	Davidson et al. (1974)
3. Spherical microencapsulated enzymes: $k'_L = \dfrac{0.000464 G^{1/3}}{f_B d_p^{2/3}}$	Mogensen and Vieth (1973)

[a] $Re' = d_p u \rho / \mu$; $G = g/cm^2 \, s$; $d_p = cm$, $j_d = (Sh)(Sc)/Re'$.

where ε_a is the actual effectiveness factor in the combined presence of internal and external effects, ϕ is given by Equation 20.7 with Λ_0 replacing R, and Bi_m is a Biot number for mass transfer defined by

$$Bi_m = \frac{k'_L(V_p/A_p)}{D_{eA}} \tag{20.14}$$

APPARENT MICHAELIS–MENTEN CONSTANTS

In the developments just presented, the constants of the MM equation are the intrinsic constants of the IME catalyst which are independent of particle size and liquid flow rate. Another approach is to extract apparent values of the constants for any given flow rate and particle size and correlate them as functions of these parameters. Several methods of extracting these apparent constants have been proposed (e.g., Lilly et al., 1968; Shiraishi, 1993, 1995; Miyakawa and Shiraishi, 1997). It is important to note, however, that even the intrinsic constants have different values for the free and immobilized forms of an enzyme, but this can often be ignored.

EFFECT OF pH AND TEMPERATURE

It has been shown (Ollis, 1972; Karanth and Bailey, 1978; Reilly and Lee, 1981) that the effect of slow diffusion in general is to increase the width of the pH versus apparent stability curve. This is illustrated in Figure 20.1 for glucose oxidase activity (Cho and Bailey, 1978).

COMPLEX SYSTEMS

Many enzymatic reactions involve two substrates or one substrate and a cofactor. Atkinson and Lester (1974) presented a general theoretical treatment of two-substrate kinetics along with an asymptotic approximation. Immobilized systems of two enzymes that catalyze a reaction sequence can have enhanced efficiency (Goldman and Katchalski, 1971; Lin, 1977). Nonuniform distribution of the biocatalysts can develop during immobilization or can be deliberately forced to enhance the effectiveness (Buchholz and Gödelmann, 1978; Carleysmith et al., 1980).

Microbial Reactions

The growth of microbial cells is affected by intracellular and environmental factors. The most important intracellular factor is the RNA content. Among the environmental factors are pH, temperature, reactor geometry, and rheological properties.

Clearly, a complete description of the growth kinetics of a culture must recognize two facts: each cell of the culture is *structured*, and the cells are *segregated* (i.e., they differ from one another). The *structured segregated* model is obviously the most realistic description of cell growth. By the same token, models in which structure and segregation are both absent are the simplest. These are the *unstructured nonsegregated* models where the culture is treated as an unstruc-

tured continuum. In any unstructured model, the cell composition is fixed, that is, it does not change in response to new growth conditions. This is referred to as *balanced growth*. Between the two extremes, one can have *structured nonsegregated* and *unstructured segregated* models.

Unstructured nonsegregated models

The basic unstructured nonsegregated model is the Monod model. Several modifications of this model have also been proposed.

THE MONOD MODEL

The overall microbial growth rate r_C can be quantified in terms of the *specific growth* rate μ (1/s) and the cell concentration [C]. The specific growth rate itself can be predicted by the same form of kinetic equation as the Michaelis–Menten equation for enzyme-catalyzed reactions,

$$\mu = \frac{\mu_m[A]}{K_m + [A]} \tag{20.15}$$

where K_m is the substrate concentration at which the cells multiply at half the maximum rate and is called the *saturation constant* or the *half-velocity constant*, but more commonly the Monod constant, and μ_m is the maximum rate (1/s) when $[A] \gg K_m$. Thus r_C may be expressed as the product

$$r_C = \frac{d[C]}{dt} = \mu[C] = \frac{\mu_m[C][A]}{K_m + [A]} \tag{20.16}$$

or

$$\mu = \frac{1}{[C]}\frac{d[C]}{dt} \tag{20.17}$$

Known after Monod (1950), this expression is best understood in terms of the representation shown in Table 20.7 along with the Michaelis–Menten model for comparison (see Levenspiel, 1993). It can be readily seen that

$$r_C = Y_{C/A}(-r_A) = Y_{C/R}(r_R) \tag{20.18}$$

where $Y_{C/A}$ and $Y_{C/R}$ are the yield coefficients defined by

$$Y_{C/A} = \frac{\text{moles C formed}}{\text{moles A converted}} \tag{20.19}$$

$$Y_{C/R} = \frac{\text{moles C formed}}{\text{moles R formed}} \tag{20.20}$$

MODIFICATIONS OF THE MONOD MODEL

The simple Monod model does not account for several factors that tend to complicate the system, for example, maintenance, which refers to the consumption of energy for all processes other than growth, inhibition by substrate or product, the presence of two substrates, the growth of microorganisms at the

Table 20.7 Comparison of models for enzyme (Michaelis–Menten) and microbial (Monod) reactions

Michaelis–Menten model	Monod model[a]
Mechanism: $$A + E \underset{k_{-1}}{\overset{k_1}{\rightleftarrows}} AE$$ $$AE \xrightarrow{k_2} R + E$$	Mechanism: $$A + C_{\text{resting}} \underset{k_{-1}}{\overset{k_1}{\rightleftarrows}} C_{\text{pregnant}}$$ $$C_{\text{pregnant}} \xrightarrow{k_2} 2C_{\text{resting}} + R$$
Enzyme balance: $[E]_0 = [E] + [AE]$	Cell balance: $[C]_{\text{total}} = [C]_{\text{resting}} + [C]_{\text{pregnant}}$
Equation: $$r_R = \frac{k_2[E]_0[A]}{K_M + [A]}$$	Equation: $$r_R = \frac{1}{Y_{C/R}} \frac{k_2[C][A]}{K_m + [A]}$$

[a] The notation used to represent the Monod equations is that of Levenspiel (1993).

solid-liquid interface, and the product formation kinetics. A discussion of these is outside the scope of this treatment (see Shuler and Kargi, 1992).

Structured models

The basic feature of a structured model is that the cell is permitted to have a number of mutually interacting subcomponents. The simplest structured model would obviously have two subcomponents, but the more sophisticated models can have as many as 40. The kind of model to be developed will depend on the mechanism of cell growth and the cell model assumed. The degrees of structure and segregation will depend on the number of subpopulations into which the total cell population is divided, each population modeled as a single cell. Hence, no generalized structured model is possible. However, certain common guidelines for deriving such models have been formulated (Bailey and Ollis, 1986; Shuler and Kargi, 1992).

BIOREACTORS

In view of the essential differences between enzyme and microbial actions as explained previously, the analysis and design of the two classes of bioreactors are quite different.

Enzyme Reactors

We consider the usual three types of reactors, batch, plug-flow, and mixed-flow.

Batch reactors

Assuming a simple one-substrate enzyme-catalyzed reaction, the following familiar mass balance equation results:

$$-\frac{d[A]}{dt} = -r_A \tag{20.21}$$

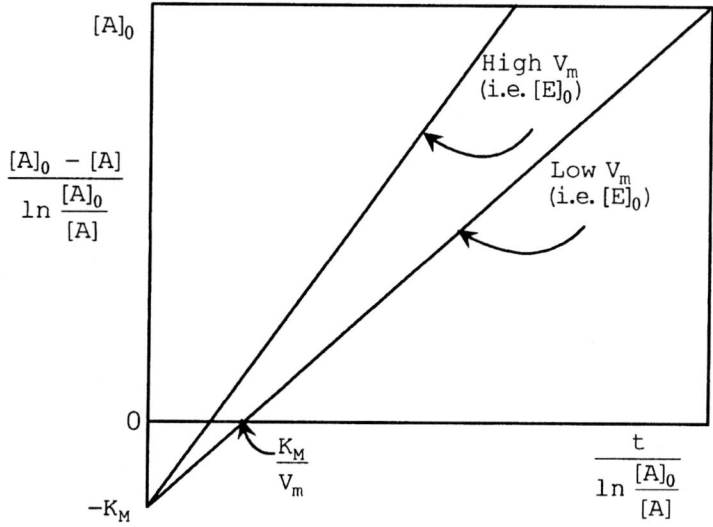

Figure 20.3 Plot of Equation 20.22 for a PFR (or BR)

Upon substituting Equation 20.2 for the rate and integrating, we get

$$\frac{[A]_0 - [A]}{\ln([A]_0/[A])} = -K_m + V_m \left[\frac{t}{\ln([A]_0/[A])}\right] \quad (20.22)$$

which then can be plotted as shown in Figure 20.3 to obtain K_M and V_m directly.

Expressions similar to Equation 20.22, and written in terms of conversion, can also be derived for other simple kinetic equations such as substrate inhibition. These are identical with the corresponding expressions for PFR considered later and listed in Table 20.8, except that $\bar{t}$ is replaced by t and $(1 - f_B)/f_B$ is unity. For more complicated rate expressions, the governing balance equation has to be integrated numerically.

Plug-flow tubular reactor

Restricting our attention to a one-dimensional plug-flow reactor, the variation of substrate concentration (as against time in a batch reactor) is given by

$$-u\frac{d[A]}{d\ell} = -r_A \quad (20.23)$$

Because it is customary to employ this type of reactor for immobilized enzymes, it is useful to generalize Equation 20.23 to include both soluble enzyme and packed immobilized enzyme versions of a PFR. This can be done by allowing for bed porosity f_B as follows:

$$-u\frac{d[A]}{d\ell} = \frac{(1-f_B)}{f_B}(-r_A) \quad (20.24)$$

For soluble enzymes, the term $(1 - f_B)/f_B$ becomes unity.

Table 20.8 Design equations for enzyme-catalysed reactions in a PFR (or a BR with $\bar{t}$ replaced by t) and a CSTR (from Messing, 1975)[a]

Kinetic expression for $(-r_A)$	PFR design equation for $V_m \bar{t} \left(\dfrac{1-f_B}{f_B} \right)$ (with $t = \bar{t}$ and $(1-f_B)/f_B = 1$ for BR)	CSTR design equation for $V_m \bar{t}$

Michaelis–Menten:

$$\dfrac{V_m [A]}{K_M + [A]}$$

PFR: $[A]_0 X_A - K_M \ln(1 - X_A)$

CSTR: $X_A \left(\dfrac{K_M}{1 - X_A} + [A]_0 \right)$

Reversible Michaelis–Menten:

$$\dfrac{V_m \left([A] - \dfrac{[R]}{K} \right)}{K_M + [A] + \dfrac{K_M [R]}{K_R}}$$

PFR: $[A]_0 \left(1 - \dfrac{K_M}{K_R} \right) \left[\dfrac{X_A}{a} + \dfrac{1}{a^2} \ln(1 - aX_A) \right] - \left(K_M + [A]_0 + \dfrac{K_M [R]_0}{K_R} \right) \dfrac{1}{a} \ln(1 - aX_0)$

where $a = \dfrac{K+1}{K}$

CSTR: $\dfrac{X_A \left[K_M + [A]_0 - X_A [A]_0 + \dfrac{K_M [R]_0 + [A]_0 X_A}{K_R} \right]}{1 - X_A \left(1 + \dfrac{1}{K} \right)}$

Competitive product inhibition:

$$\dfrac{V_m^2 [A]}{[A] + K_M \left(1 + \dfrac{[R]}{K_i} \right)}$$

PFR: $[A]_0 \left(1 - \dfrac{K_M}{K_i} \right) [X_A + \ln(1 - X_A)] - \left(K_M + [A]_0 + \dfrac{K_M [R]_0}{K_i} \right) \ln(1 - X_A)$

CSTR: $\dfrac{X_A \left[K_M + [A]_0 - X_A [A]_0 + \dfrac{K_M ([R]_0 + [A]_0 X_A)}{K_i} \right]}{1 - X_A}$

Substrate inhibition:

$$\dfrac{V_m}{1 + \dfrac{K_M}{[A]} + \dfrac{[A]}{K_i}}$$

PFR: $[A]_0 X_A - K_M \ln(1 - X_A) + \dfrac{[A]_0^2}{K_i} \left(X_A - \dfrac{X_A^2}{2} \right)$

CSTR: $X_A [A]_0 \left[1 + \dfrac{K_M}{[A]_0 (1 - X_A)} + \dfrac{(1 - X_A)[A]_0}{K_i} \right]$

[a] K_R, K_i, are constants.

Defining the mean residence time as

$$\bar{t} = \frac{\ell(1 - f_B)}{f_B u} \tag{20.25}$$

we can obtain integrated expressions relating substrate conversion X_A to system parameters for different rate forms. Some of these are presented in Table 20.8.

The development can be readily extended to more complicated models, as described in Chapter 12.

Continuous Stirred Tank Reactor

The mass balance for an ideal CSTR can be written as

$$Q([A]_0 - [A]) = (-r_A)V_r \tag{20.26}$$

Defining the mean residence time as $\bar{t} = V_r/Q$, we can write the following expression for the usual MM kinetics:

$$[A] = -K_M + V_m \left(\frac{[A]\bar{t}}{[A]_0 - [A]} \right) \tag{20.27}$$

which then can be plotted as shown in Figure 20.4 to give K_M and V_m.

Expressions can also be derived for other forms of the kinetic equation, including some types of inhibition, and these are included in Table 20.8.

Fluidized-bed reactors

The essential features of fluidized bed reactors were discussed in Chapter 12. When using these reactors for immobilized enzymes, we must remember that we are normally concerned here with liquid fluidized beds. Thus in general, fixed-bed reactors are preferred to fluid-bed reactors for immobilized enzyme systems

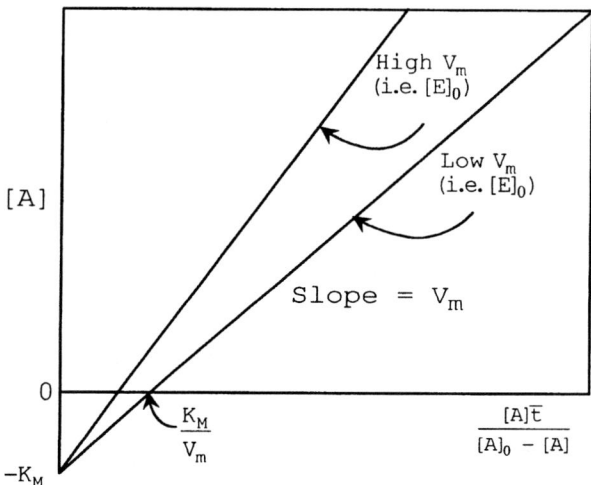

Figure 20.4 Plot of Equation 20.27 for a CSTR

involving anaerobic fermentation. However, in aerobic fermentation, gas-liquid-solid contacting becomes important, and this is better achieved in a fluidized-bed reactor than in a fixed-bed reactor. Where a deactivating catalyst is involved, the fluidized-bed may also often be a better alternative. In the following example we highlight the difference between fixed- and fluidized-bed reactor performances for a deactivating IME catalyst.

EXAMPLE 20.1

Comparison of fixed- and fluidized-bed reactor performances for a deactivating IME catalyst: reaction of NADH and sodium pyruvate to give NAD and sodium lactate on lactate dehydrogenase immobilized on DE-81 cellulose membrane.

First we develop the performance equations based on the equations for conventional fixed- and fluidized-bed reactors with a deactivating catalyst listed in Table 12.7 (see also Sadana and Doraiswamy, 1971) and then use these equations to compare the two reactor types for the present IME reaction [Doraiswamy and Sharma, 1984; Sadana, 1980, 1991 (see acknowledgments); also see Chapter 12].

DEVELOPMENT OF EQUATIONS

In the reactor equations developed in Chapter 12, the reaction rate was represented by a simple power law model of order m, and the deactivation was expressed in terms of the decrease in the rate constant with time according to

$$k = k_0 e^{-k_d t_p} \tag{E20.1.1}$$

The situation here is slightly different in that the catalyst is an enzyme for which the rate equation is given by

$$-r_A = \frac{k[E]_0[A]}{K_M + [A]} \tag{E20.1.2}$$

Assuming exponential deactivation of the enzyme, the following equation can be written:

$$[E] = [E]_0 e^{-k_d t_p} \tag{E20.1.3}$$

where k_d is the deactivation constant.

The next step is to combine these equations with the reactor equations for fixed- and fluidized-bed reactors developed in Chapter 12.

FIXED-BED REACTOR

Let t_{pl} be the total time for a predetermined ultimate level of decay, say, 10% of the initial activity. If the liquid transit time within the reactor is much less than the decay time t_{pl}, Equation 12.49 can be recast as

$$\frac{dy_A}{dz} = -B'' e^{\lambda \hat{t}} \left(\frac{y_A}{\hat{K}_M + y_A} \right) \tag{E20.1.4}$$

where

$$\hat{K}_M = \frac{K_M}{[A]_0}, \quad B'' = \frac{k[E]_0 V_r}{[A]_0 Q} = \frac{k[E]_0}{[A]_0 S_v}, \quad S_v = \frac{Q}{V_r}$$

$$\lambda = k_d t_{pl}, \quad \hat{t} = \frac{t_p}{t_{pl}}$$

(E20.1.5)

The solution to Equation E20.1.4 with boundary condition $y_A(0) = 1$ is

$$\hat{K}_M \ln y_A + (y_A - 1) + B'' e^{-\lambda \hat{t}} = 0 \tag{E20.1.6}$$

and the time-averaged conversion at the reactor exit $(z = 1)$ is

$$\bar{X}_A = 1 - \int_0^1 y_A \, d\hat{t} \tag{E20.1.7}$$

Plots of the time-averaged conversion for different representative values of $\hat{K}_M$ and λ are given in Figure 20.5a.

FLUIDIZED-BED REACTOR

We assume here that the reactor is fully mixed. Thus Equation E20.1.3 for enzyme decay should be replaced by the following relationship for the average concentration (Weekman, 1968):

$$[\bar{E}] = \frac{E_0}{1 + \lambda} \tag{E20.1.8}$$

Now Equation 12.55 for a conventional fluidized-bed reactor with a deactivating catalyst becomes

$$\frac{dy_A}{dz} = -\frac{B''}{1 + \lambda}\left(\frac{y_A}{\hat{K}_M + y_A}\right) \tag{E20.1.9}$$

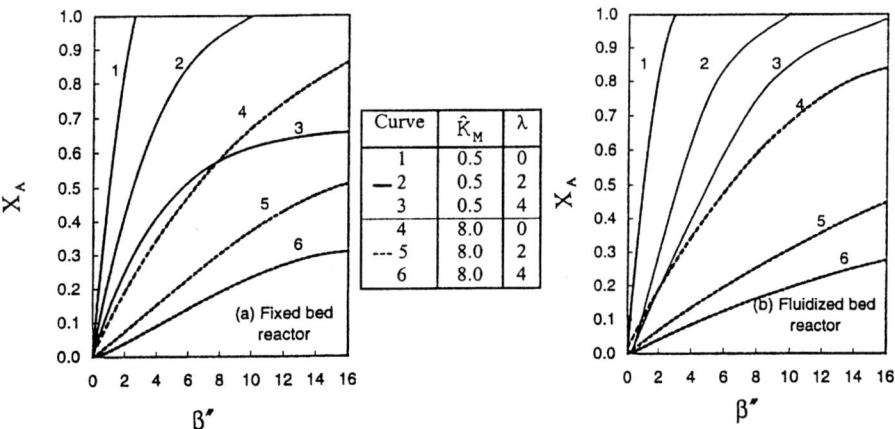

Figure 20.5 Conversion vs. B'' plots for different values of λ and K_M for an immobilized-enzyme reaction in a (a) fixed-bed reactor and (b) fluidized-bed reactor (adapted from Sadana, 1991).

The solution to this equation with boundary condition $y_A(0) = 1$ is

$$\hat{K}_M \ln y_A + (y_A - 1) + \left(\frac{B''}{1+\lambda}\right)z = 0 \qquad (E20.1.10)$$

and the conversion at the reactor exit may be calculated from

$$X_A = (1 - y_A) \qquad (E20.1.11)$$

Representative plots of X_A versus B'' are included in Figure 20.5b.

COMPARISON OF REACTOR PERFORMANCES FOR THIS REACTION

The reaction may be represented by

$$\text{NADH} + \text{Na pyruvate} \leftrightarrow \text{NaD} + \text{Na lactate} \qquad [E20.1.1]$$

We use the following data given by Wilson et al. (1968):

$k_d = 0.345 \text{ min}^{-1}$

$t_{pl} = 20 \text{ min}$

$K_M = 1.2 \times 10^{-4} \text{ mol}/\ell$

$Q = 2.82 \text{ cm}^3/\text{min}$

$[A]_0 = 8.3 \times 10^{-5} \text{ mol}/\ell$

A parameter C known as *reactor capacity* has been defined and a value of 0.49 μmol/min given for the flow rate used.

The parameters $\hat{K}_M$, B'', and λ may be calculated as

$$\hat{K}_M = \frac{1.2 \times 10^{-4}}{8.3 \times 10^{-5}} = 1.445$$

$$B'' = \frac{C}{Q[A]_0} \quad \begin{pmatrix} \text{a form equivalent to Equation E20.1.5} \\ \text{because } C = k[E]_0 V_r \text{ moles/time} \end{pmatrix}$$

$$= \frac{4.9 \times 10^{-7}}{2.82 \times 8.3 \times 10^{-8}} = 2.08$$

$$\lambda = k_d t_{pl} = 0.345 \times 20 = 6.9$$

Substituting these values in Equations E20.1.6 and E20.1.7 for the fixed-bed and in E20.1.10 and E20.1.11 for the fluidized-bed reactors, we get

    conversion in the fixed-bed reactor: 49.3%

    conversion in the fluidized-bed reactor: 42.7%

For the low $\hat{K}_M$ and B'' values characteristic of this system, the conversion obtained in the fixed-bed reactor is higher than that in the fluidized-bed reactor.

Role of internal diffusion in IME reactors

Reactor design in the presence of internal diffusion involves simultaneous solution of the reactor and pellet equations (see Chapter 12). These are given in Table 20.9 along with the boundary conditions in both dimensional and nondimensional forms.

Table 20.9 Continuity equations for a tubular reactor packed with pellets of IME. (Normalization has been done with $[A]_0$ both for the reactor and the pellet, and not with $[A]_s$ for the pellet)

Equation	Reactor equation	Pellet equation for estimating ε in the reactor equations (spherical pellet)
Dimensional form	$-u \dfrac{d[A]}{d\ell} = \varepsilon(1-f_B) \dfrac{V_m[A]}{K_M + [A]}$	$D_{eA}\left(\dfrac{d^2[A]_p}{dr^2} + \dfrac{2}{r}\dfrac{d[A]_p}{dr}\right) = \dfrac{V_m[A]_p}{K_M + [A]_p}$
Boundary conditions	$[A] = [A]_0$ at $\ell = 0$	$\dfrac{d[A]_p}{dr} = 0$ at $r=0$, $[A] = [A]_s$ at $r=R$
Dimensionless form	$\dfrac{da}{dz} = \bar{t}\varepsilon(1-f_B)\dfrac{V_m a}{K_M + a}$	$\dfrac{d^2 a_p}{d\hat{R}^2} + \dfrac{2}{\hat{R}}\dfrac{da_p}{d\hat{R}} = \phi_s^2 \dfrac{a_p}{\hat{K}_M + a_p}$ where $\phi_s^2 = \dfrac{V_m R^2}{D_{eA}}$
Boundary conditions	$a = a_0 = 1$ at $z = 0$	$\dfrac{da_p}{d\hat{R}} = 0$ at $\hat{R} = 0$, $a_p = a$ at $\hat{R} = 1$
Notation	$a = \dfrac{[A]}{[A]_0}$, $z = \dfrac{\ell}{L}$, $\hat{K}_M = \dfrac{K_M}{[A]_0}$	$a_p = \dfrac{[A]_p}{[A]_0}$, $\hat{R} = \dfrac{r}{R}$, $a_{ps} = \dfrac{[A]_{ps}}{[A]_0}$
		($= a$ at z in reactor equation)

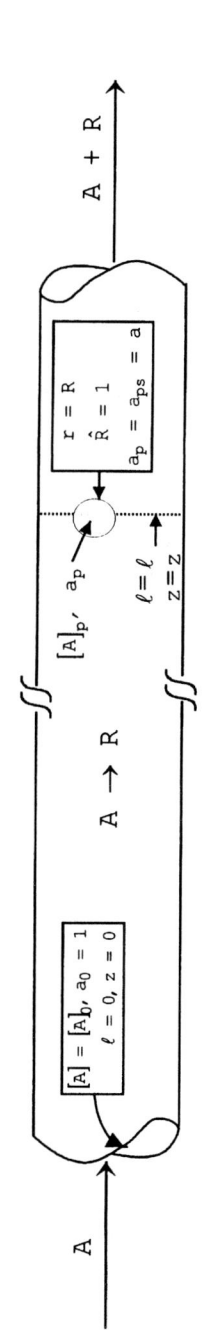

Role of external mass transfer

The effect of external mass transfer is incorporated exactly as described in Chapter 12. The boundary condition at the pellet surface (at $\hat{R} = 1$) is changed to account for the fact that $[A]_{ps} \neq [A]_b$ at any axial position z in the reactor. This is done through a Biot number with k'_L estimated from any of the equations listed in Table 20.6.

Liquid-liquid-solid three-phase fixed-bed reactors

Liquid-liquid-solid reactors are commonly used for biphasic reactions catalyzed by immobilized phase-transfer catalysts (which form the third, solid phase). Certain basic aspects of such reactors were considered in Chapter 19. Three-phase reactions of this type are also encountered in biological reactions, for example, the enzymatic synthesis of amino acid esters in polyphasic media (Vidaluc et al., 1983), and the production of L-phenylalanine utilizing enzymatic resolution in the presence of an organic solvent (Dahod and Empie, 1986). Predictably, the performance of these reactors is influenced by the usual kinetic and mass transfer aspects of heterogeneous systems (see Lilly, 1982; Chen et al., 1982; Woodley et al., 1991a,b). Additionally, performance is also influenced by the complex hydrodynamics associated with the flow of two liquids past a bed of solids (Mitarai and Kawakami, 1994; Huneke and Flaschel, 1998). It is noteworthy, for instance, that phase distribution within the reactor is different from that in the feed and is also a function of position within the reactor and within the voids of each pellet in the bed. More intensive research is needed before these reactors can be rationally designed.

Microbial Reactors

Biological reactors, like their enzymatic counterparts, can also be classified into batch, plug-flow and mixed-flow reactors. However, it is the semibatch reactor that is perhaps the most useful for microbial systems.

Batch reactors

An important consideration in microbial reactions is cell growth. Hence we first describe the main features of cell growth in a batch reactor before taking up the modeling of this reactor.

GROWTH CYCLES IN BATCH SEPARATION

The batch cultivation process is characterized by four phases. The first is the *initial lag phase* during which there is no significant increase in the number of cells. This is followed by a rapid growth stage normally known as the *exponential growth phase*. Growth ceases when the nutrient is exhausted or, less often, when toxic products increase to an intolerable level. Eventually, cells begin to die due to the accumulation of toxin and/or depletion of food, and this phase is termed the *death phase*. Figure 20.6 represents the different phases just described. It is difficult to develop a generalized kinetic model for the lag and death phases, but

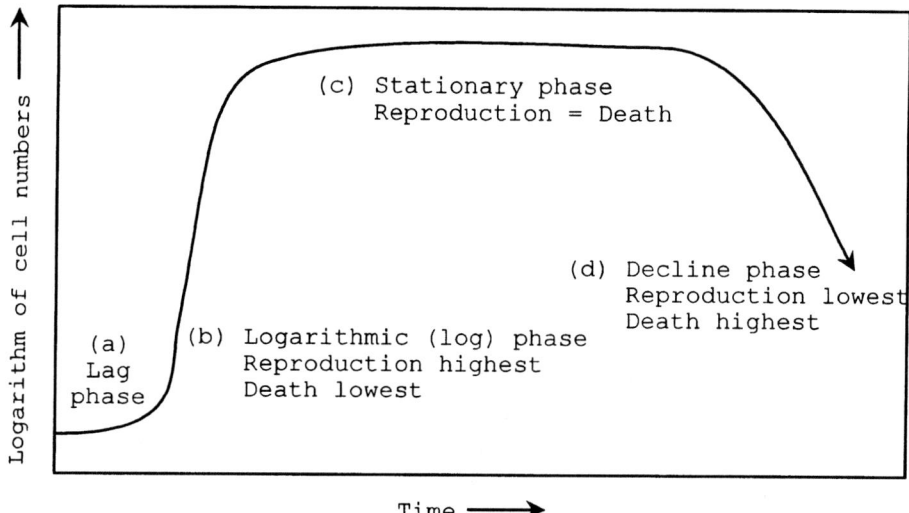

Figure 20.6 Different phases of a cell's growth cycle in batch culture

governing equations for the exponential growth phase have been developed using the Monod model or its extensions (see Bailey and Ollis, 1986).

MATHEMATICAL MODEL BASED ON MONOD KINETICS

Batch growth of the cell population is one of the most extensively used operations of the pharmaceutical industry. Once the culture is introduced, the concentrations of nutrients, biomass, and product vary with time as growth proceeds. The material balance for the growth of cells can be written as follows:[2]

$$r_C = \frac{d[C]}{dt} = \frac{k[C][A]}{K_m + [A]} \qquad (20.28)$$

in which μ_m of Equation 20.16 is replaced by the more familiar k. This can be recast as

$$kt = \int_{[C]_0}^{[C]} \frac{K_m + [A]}{[C][A]} d[C] \qquad (20.29)$$

Integration of this equation gives the cell concentration as a function of time expressed more conveniently as $kt = f([C])$.

$$kt = \left[\frac{K_m}{[A]_0 + ([C]_0/Y_{C/A})} + 1\right] \ln \frac{[C]}{[C]_0} - \left[\frac{K_m}{[A]_0 + ([C]_0/Y_{C/A})}\right] \ln \frac{[A]}{[A]_0} \qquad (20.30)$$

---

[2] Strictly, this is not valid for batch growth because r_C lags behind any change in [A]. The validity increases at higher values of [A], that is, as the reaction becomes increasingly zero order in A.

The substrate balance is given by

$$-\frac{d(V[A])}{dt} = \frac{1}{y_{C/A}} r_C V \tag{20.31}$$

For a constant culture volume, this becomes

$$-\frac{d[A]}{dt} = \frac{r_C}{y_{C/A}} = \frac{1}{y_{C/A}} \frac{d[C]}{dt} \tag{20.32}$$

giving

$$[A]_0 - [A] = \frac{1}{y_{C/A}}([C] - [C]_0) \tag{20.33}$$

Now this equation can be combined with Equation 20.30 to eliminate the cell concentration $[C]$ and give the substrate concentration $[A]$ as a function of time.

Tubular reactor

A cell balance over a differential element of a PFR gives

$$r_C = u\frac{d[C]}{d\ell} = \frac{k[C][A]}{K_m + [A]} \tag{20.34}$$

which can be recast as

$$\frac{d[C]}{d\bar{t}} = \frac{k[C][A]}{K_m + [A]} \tag{20.35}$$

Similarly, by analogy with Equation 20.32 the equation for the substrate can be written as

$$-\frac{d[A]}{d\bar{t}} = \frac{1}{Y_{C/A}}\frac{d[C]}{d\bar{t}} \tag{20.36}$$

The relationship between substrate consumption and cell formation is given by Equation 20.33. Then Equation 20.36 can be integrated and combined with 20.33 to give $[A]$ as a function of $\bar{t}$.

Note that we cannot run the PFR with a sterile feed (i.e., with $[C]_0 = 0$) because plug flow would prevent the fluid from being inoculated by cells. Whatever cells were initially introduced into the reactor get washed out without recycle. This phenomenon of cell *washout* is considered next. It can be circumvented by introducing a recycle so that the input stream is inoculated before entering the reactor.

Continuous stirred tank reactor (chemostat)

The CSTR is known as a *chemostat* in biochemical language, but we shall continue with the term CSTR or MFR. Because this is a continuous mixed reactor, the question arises: Should there be a continuous introduction of cells along with the reactant, that is, should $[C]_0$ be zero or a finite value. We address this question briefly without going into mathematical details.

CASE 1: $[C]_0 = 0$

The chief conclusion is that cell production and therefore the microbial reaction can be sustained (i.e., washout can be avoided) provided that the dilution rate $\mathcal{D} = Q/V$ (s^{-1}) is less than the maximum value given by

$$\mathcal{D}_{\text{for max cell output}} = k\left(1 - \sqrt{\frac{K_m}{K_m + [A]_0}}\right) \tag{20.37}$$

In other words, the space time $\bar{t} = 1/\mathcal{D}$ (s) must be maintained above a certain minimum value.

Using the basic CSTR equation

$$\bar{t} = \frac{\Delta[i]}{r_i} \quad i = A, C, \text{ or } R \tag{20.38}$$

the performance equations in terms of [C] and [A] can be written as (Novick and Szillard, 1950)

$$[A] = \frac{K_m}{k\bar{t} - 1}, \quad \text{for } k\bar{t} > 1 \tag{20.39}$$

$$[C] = Y_{C/A}\left([A]_0 - \frac{K_m}{k\bar{t} - 1}\right), \quad \text{for } k\bar{t} > 1 \tag{20.40}$$

CASE 2: $[C]_0 \neq 0$

To avoid the possibility of washout altogether and place no restrictions on space time, it is best to operate at $[C]_0 \neq 0$. Thus we operate the reactor as a conventional CSTR with a steady input of reactant and cells. Again, we can use Equation 20.38, substitute the Monod equation for the rate term, and rearrange the resulting expression to give, finally (Levenspiel, 1993)

$$k\bar{t} = \frac{([A]_0 - [A])([A] + K_m)}{(1/Y_{C/A})[C]_0[A] + [A]([A]_0 - [A])} \tag{20.41}$$

To obtain an expression for the highest production rate, we write Equation 20.38 as

$$r_C = \frac{[C] - [C]_0}{\bar{t}} = \text{maximum} \tag{20.42}$$

leading to

$$k\bar{t}_{\text{opt}} = \frac{P}{P-1}\left[\frac{[A]_0 - (P-1)K_m}{[A]_0 + [C]_0/Y_{C/A} - (P-1)K_m}\right] \tag{20.43}$$

where

$$P = \left(1 + \frac{[A]_0 + ([C]_0/Y_{C/A})}{K_m}\right)^{1/2} \tag{20.44}$$

A chemostat fitted with some arrangement for feedback can increase the biomass concentration in the reactor much above the theoretical limit of the simple chemostat (Herbert, 1961; Pirt and Kurowski, 1970).

Fed-batch reactor

It often becomes necessary in biochemical reactions to continuously add one (or more) substrate(s), a nutrient, or any regulating compound to a batch reactor, from which there is no continuous removal of product. A reactor in which this is accomplished is conventionally termed the semibatch reactor (Chapter 4) but is referred to as a *fed-batch reactor* in biochemical language. The fed-batch mode of operation is very useful when an optimum concentration of the substrate (or one of the substrates in a multisubstrate system) or of a particular nutrient is desirable. This can be achieved by imposing an optimal feed policy.

From the design point of view, the procedure outlined in Chapter 10 is applicable. From Table 10.2 (but using the Monod model for the rate), the following equation can be constructed for reactant A (which is continuously fed to an initial batch):

$$-\frac{d[A]}{dt} = \frac{([A]_0 - [A])}{\bar{t}} - \frac{1}{Y_{C/A}}\left(\frac{k[C][A]}{K_m + [A]}\right) \tag{20.45}$$

Clearly, the feed rate $Q(t)$ can be manipulated to give an optimal feed policy (with respect to time) to maximize performance. Usually, a constant or exponential addition policy is the most practical and the simplest. Yet, in realistic situations, the feed rate must be judiciously varied so that a robust *optimization criterion* can be met. Thus first a suitable objective function must be evolved. Many such objective functions are possible, for example, maximization of yield and minimization of production cost.

EXAMPLES OF THE USE OF BIOCATALYSTS IN ORGANIC SYNTHESIS

It is estimated that until the advent of the 1990s more than 3,000 enzymes had been characterized, of which over 300 were available commercially for various applications (Sigma Chemical Catalog, 1989). The number used in organic synthesis is clearly considerably lower. Still, these figures do provide a measure of perspective to biocatalysis in organic synthesis, for the number of regular solid catalysts in use (as can be gleaned from Chapter 6) is considerably higher. Several perceived and real difficulties, Table 20.10 (see Dordick, 1991), in the use of enzymes are rapidly disappearing, so that the stage appears to be set for a rapid expansion of the biocatalytic base in organic synthesis.

Industrial Production of Organic Chemicals

Several industrial applications of biocatalysts (enzymes and cells) are listed in Table 20.11, many of which are used in immobilized form. Although in some cases the enzymatic process involved is a single-step process that gives the final product directly from the basic reactant, in most industrial processes this constitutes one step (sometimes more) in a multistep chemical synthesis, but cuts down the total number of steps and thus the cost of production. It also imparts greater selectivity and purity to the product.

Table 20.10 Real and perceived difficulties in exploiting enzymes as commercial catalysts

Robustness	Enzymes are perceived as too fragile for commercial use. They normally denature above 60 °C, and the rate of deactivation is considerably higher than the rate of activation. However, enzymes have been isolated from thermophilic organisms that are very stable (Kelly and Deming, 1988). Further, in many instances, the reaction conditions are mild enough that the fragility is not a major problem.
Restriction of action to aqueous solutions	This might have been a major restriction some years ago, but it is much less so now (Butler, 1979; Lilly, 1982; Klibanov, 1986; Halling, 1987; Dordick, 1989; Adams et al., 1995).
Reaction speed vs. selectivity	There is frequently much confusion between rate and selectivity. Although enzymes are highly selective, because of their high molecular weight (50,000 on an average), they occupy much more space, thus reducing the density of active sites. Hence the productivity is lower than for a chemical catalyst, although the selectivity is remarkably high. The choice between chemical and enzyme catalysts is therefore determined by the importance of speed relative to selectivity.
Specificity vs. general applicability	An enzyme is generally perceived to be too specific to be of use for any other reaction. This is not always true, for many enzymes are known to catalyze a wide range of reactions, for example, lipases and proteases catalyze the hydrolysis of a number of esters and proteins (Inouye et al., 1979; Posorske, 1984; Neuberger and Brocklehurst, 1987), and peroxidases catalyze a wide variety of substrates (Saunders et al., 1964; Musso, 1967; Dordick et al., 1987).
Susceptibility to product inhibition	This is a serious shortcoming but can be overcome by the use of novel reactors such as membrane reactors (Matson and Quinn, 1986; see also Chapter 24), nonaqueous media (Dordick, 1989; Rethwisch et al., 1990), and recombinant DNA technology (Arbige and Pitcher, 1989).
Difficulty of isolation	This is not a problem for immobilized enzymes but is perceived to be one for soluble enzymes. However, this is not always true because in most cases separation of the enzyme into crude mixtures is adequate (Goldstein, 1987).

A particularly significant example is acrylamide, one of the most important commodity chemicals used in the production of a variety of polymers. In the conventional chemical synthesis of acrylamide, acrylonitrile is hydrated in the presence of a copper salt as the catalyst. This process, however, is beset with difficulties arising from the complex nature of catalyst preparation and regeneration and product separation and purification. Nagasawa and Yamada (1989) describe a process developed by a combined group of Kyoto University and Nitto Chemical Industry (Watanabe et al., 1979; Asano et al., 1982; Watanabe and Okumura, 1986) which is based on the use of a nitrile hydratase of *Corynebacterium* N-774 and *Pseudomonas chlororaphis* B23. By adding the nitrile gradually in a semicontinuous feed, it was possible to overcome inhibition of the hydratase activity and obtain over 99% conversion to acrylamide (hence avoiding completely the formation of acrylic acid as a side product) and thus doing away with the need for separation and recycling. In actual industrial production (Nakai et al., 1988), the cells are entrapped in a cationic acrylamide-based poly-

Table 20.11 Selected applications of immobilized enzymes or cells in organic synthesis[a]

Chemical	Biocatalyst	Source of biocatalyst
Glucose (from starch)	Glucoamylase (Amyloglucosidase)	*A. niger, Rhizopus niveus*
Aspartame (sweetener) (from phenylalanine)	Fructosyl transferase	*B. subtilis, Candida spp.*
	Thermolysin	*B. thermoproteolyticus*
High fructose corn syrup (from corn)	Glucose isomerase	*B. coagulans, Acetomyces, Lactobacillus brevis*
L-dopa (from L-tyrosine)	β-Tyrosinase	Mushroom
L-Aspartate (from fumarate)	Aspartase	*E. coli*
L-Alanine (from L-aspartate)	L-Aspartate-4-decarboxylase	*P. decumhae*
Acrylamide (from acrylonitrile)	Nitrilase	*Carynebacterium sp.*
Propylene glycol (from propylene)	Chloroperoxidase	
L-Phenylalanine (from *trans*-cinnamic acid) (from acetamidocinnamic acid)	Phenylalanine ammonia lyase	*Rhodotorula rubra*
		B. sphaericus
p-Hydroxylphenyl glycine [from DL-5-(*p*-hydroxyphenyl)-hydantoin]	DL-Hydantoinase	*Bacillus brevis* cells
6-Aminopenicillanic acid (6-APA) (precursor to semisynthetic antibiotics—from penicillin G or V)	Penicillin amidase	*Basodiomycete spp.*
L-Lysine (from DL-aminocaprolactam)	L-Aminocaprolactam hydrolase	*Candida humicola*
L-Malate (from fumarate)	Fumarase	*Brevibacterium ammonia* genes
Gibberellic acid (from cheese whey)		*Fusarium moniliform* cells
2,3-Butanediol (from whey permeate)		*Klebsiella pneumoniae* cells
L-Glutamic acid		*Brevibacterium flavum* (multienzyme)
α-Aminocaproic acid	Cyclic dimer hydrolase	*Achromobacter gluttatus*
Natural flavors (derived from natural substrates, e.g., cheese: upgrading of the sesquiterpene, valencene, to a higher sesquiterpene; of ethanol to acetaldehyde, etc.)		Fungi (e.g., *Aspergillus niger*), bacteria (e.g., *Bacillus subtilis*) and yeasts (e.g., *Candida lipolytica*) (not immobilized)

[a] Sources: Arima (1965), Fukumara (1974), Kinoshita et al. (1975), Yamamoto et al. (1980), Constantinides et al. (1981), Jandel et al. (1982), Godfrey and Reichelt (1983), Lee and Maddox (1986), Kahlon and Malhotra (1986), Armstrong and Yamazaki (1986), Dordick (1991).

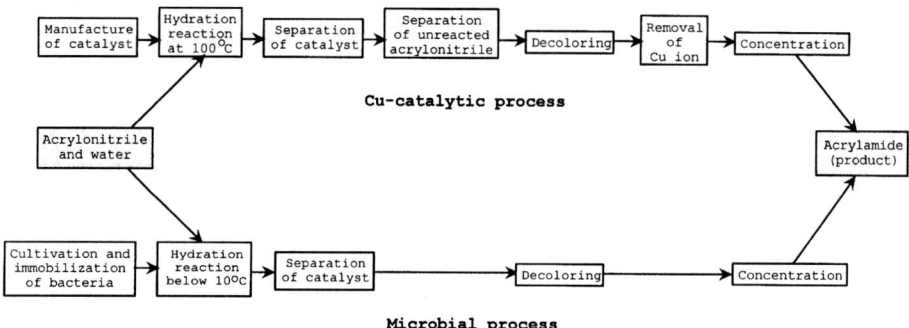

Figure 20.7 Comparison of microbial and conventional processes for acrylamide

mer gel (Watanabe, 1987), and because the conversion is at 100% no separation is involved. Also, the vexing problem of freeing the product from copper ions no longer exists. A comparison of the chemical and biocatalytic routes appears in Figure 20.7.

Biocatalysis in Asymmetric Synthesis (Bioasymmetric Synthesis)

While discussing asymmetric synthesis in Chapter 9 we noted that enzymes are among the best candidates for producing a desired enantiomer with remarkable selectivity. A few specific examples of microbiologically mediated asymmetric synthesis were given to emphasize the attractiveness of this route.

Examples of common classes of bioasymmetric reactions

Examples involving four common synthetic steps, oxidation, reduction, addition, and epoxidation, are given in Figure 20.8 to illustrate the more general role of biocatalysts (microbial and enzymatic) in asymmetric synthesis. Table 20.11 contains several other examples of bioasymmetric synthesis. The classical application of this route is the well-known total synthesis of penicillins starting from various building blocks using a variety of *Penicillium* strains.

EXAMPLE 20.2

Development of a process for the bioasymmetric synthesis of L-aspartic acid via immobilized aspartase

The use of immobilized aspartase or whole cells for L-aspartic acid production has been described by many investigators (Chibata, 1978; Fusee et al., 1981; Smith et al., 1982; Wood and Calton, 1984). The complete development of such a process consists of two major steps, reportedly carried out as follows (Hamilton et al., 1985).

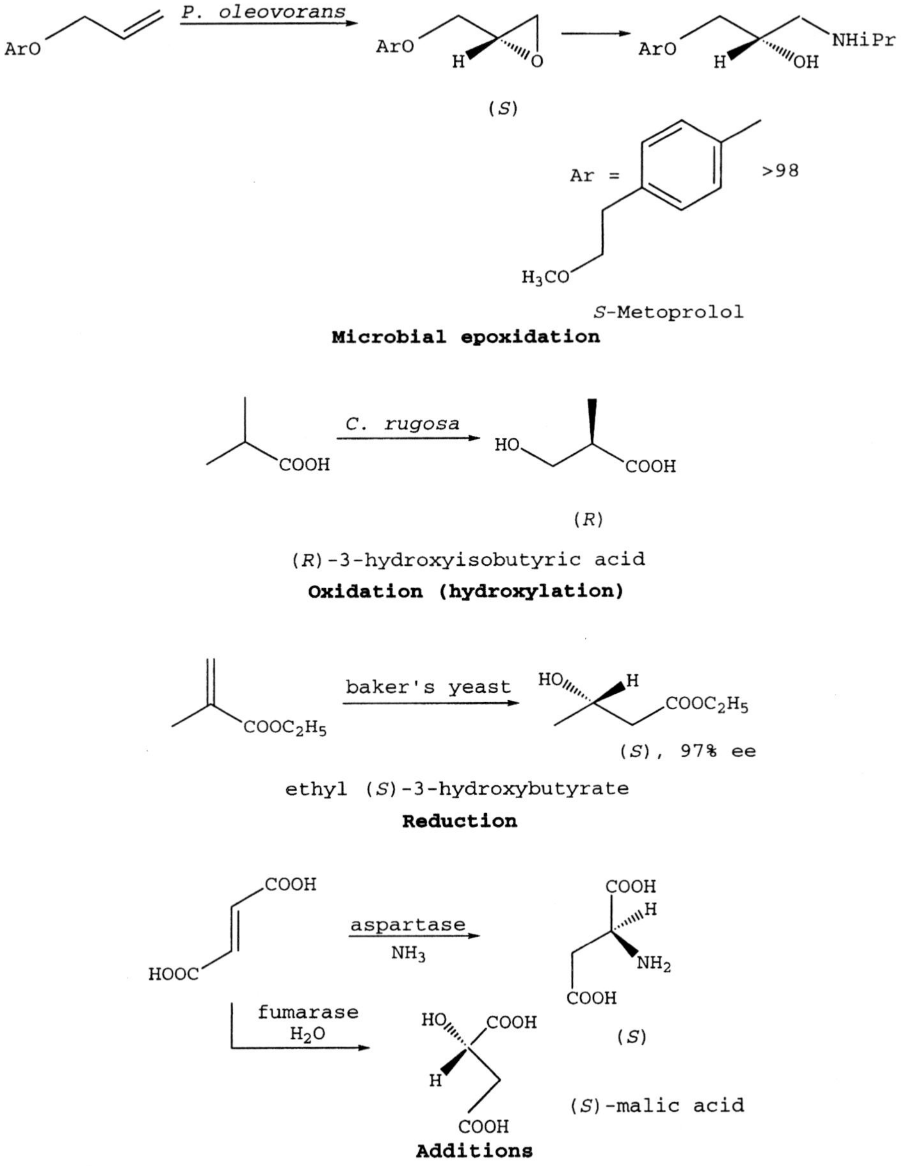

Figure 20.8 Examples of common classes of reactions in enzyme- and microbe-mediated stereochemical synthesis

IMMOBILIZATION OF WHOLE CELLS CONTAINING ASPARTASE

A strain of *E. coli* expressing high aspartase activity was grown in an aerated fermentor at 37 °C for 12 h, and the cells were recovered by centrifugation. Then they were made into a paste and immobilized on a vermiculite carrier at a very high density: 500 g wet cell paste/ℓ

of bioreactor volume. Of the several possible techniques of immobilization described in Table 20.2, entrapment was chosen as the most appropriate, and a special entrapment technology was developed and patented (Swann, 1983).

BIOREACTOR CHOICE AND OPERATION

A fixed-bed reactor was selected in view of the many uncertain features of the fluidized-bed reactor (see Chapter 12). The feed used was a 1.8 M aqueous solution of ammonium fumarate containing 1 mM magnesium sulfate. This solution (pH 8.5) was passed through the bioreactor at a rate of 2.1 $\ell/\text{hr}/\ell$ reactor volume, giving,

$$\frac{V_r}{Q} = \frac{1}{2.1} = 0.476 \text{ h}$$

L-Aspartic acid was formed at the rate of 237 g/ℓ with almost 100% conversion of the fumarate. This corresponded to a productivity of 0.5 kg/h/ℓ reactor volume (3.7 mol/h/ℓ). The L-aspartic acid was recovered from the mixture issuing from the reactor by precipitation with acid followed by filtration.

Enantioselectivity by enzymatic resolution

The use of enzymes in the resolution of racemates was briefly discussed in Chapter 9. In this section, we develop equations for the kinetic analysis of resolution. Lipases have been the most commonly used enzymes for this purpose, such as in the resolution of racemic ibuprofen (Mustranta, 1992), racemic naproxen (Tsai et al., 1996), and 3-hydroxy esters (Capewell et al., 1996).

The basic principle of this method is to react one of the enantiomers preferentially with an enzyme (free or immobilized, largely the latter). Expressions can be developed for the enantiomeric excess of the desired enantiomer (the one left unreacted) as a function of the extent of conversion of the racemate as a whole and the parameters of the Michaelis–Menten models for the two enantiomers (see Chen et al., 1982; Rakels et al. 1994; Wu, 1997).

Considering an irreversible reaction with no product inhibition, the following kinetic mechanism would usually be valid for a mixture of enantiomers R and S:

$$\begin{array}{c} R \xrightleftharpoons[k_{-1R}]{k_{+1R}} ER \xrightarrow{k_{2R}} EP_1 \xrightarrow{k_{3R}} E + P_1 \quad \boxed{\text{FAST}} \\ E \\ S \xrightleftharpoons[k_{-1S}]{k_{+1S}} ES \xrightarrow{k_{2S}} EP_2 \xrightarrow{k_{3S}} E + P_2 \quad \boxed{\text{SLOW}} \end{array} \quad [20.4]$$

If we define the *specificity constant* α of an enzymatic reaction as the ratio V_m/K_M for each enantiomer, then the ratio of the specificity constants for the two reactions can be written as

$$E_c = \frac{\alpha_R}{\alpha_S}, \quad \alpha_R = \frac{V_{m,R}}{K_{M,R}}, \quad \alpha_S = \frac{V_{m,S}}{K_{M,S}} \tag{20.46}$$

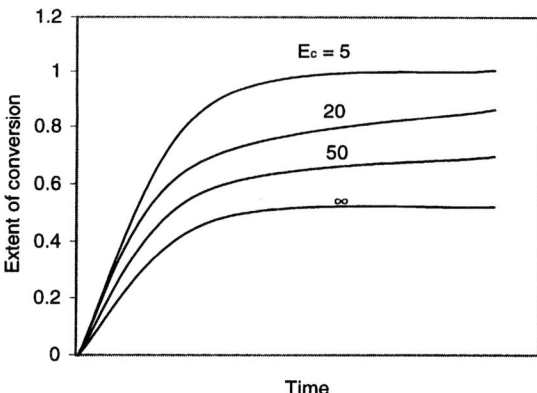

Figure 20.9 Effect of enantiomeric ratio on progress curve

If S is the slow reacting enantiomer left largely unreacted after treatment with the enzyme, then E_c can also be written in terms of the overall conversion X of $R + S$ and the enantiomeric excess of enantiomer S, ee (S). The final expression obtained is

$$E_c = \frac{\ln[(1-X)[1-ee(S)]]}{\ln[(1-X)[1+ee(S)]]} \tag{20.47}$$

This equation is graphically displayed in Figure 20.9. The *progress curves* given therein are very useful because they give the enantiomeric excess directly as a function of conversion for fixed Michaelis–Menten parameters of the two reactions (as determined by E_c).

Estimation of E_c for use in Equation 20.47 requires knowledge of the parameters of the MM models α_R and α_S for the two enantiomers. These parameters can be determined by experimentally measuring the rate of disappearance of the racemate (i.e., the overall conversion) and preparing a progress curve as shown in Figure 20.9. Although the general shape of the curve will be as shown in the figure, the exact curve for a given system will depend on the kinetic parameters V_m and K_M for the two reactions. The total rate of reaction is given by (Rakels et al., 1994)

$$-\frac{d([R]+[S])}{dt} = \frac{\alpha_R[E][R] + \alpha_S[E][S]}{1 + \frac{[R]}{K_{M,R}} + \frac{[S]}{K_{M,S}}} + k_h([R]+[S]) \tag{20.48}$$

and the rate of enzyme deactivation by

$$-\frac{d[E]}{dt} = k_d[E] \tag{20.49}$$

where k_h and k_d are the first-order autohydrolysis and deactivation constants for the system. The constants of the equations can be estimated by standard statistical methods.

The chief drawback of this method is the uncertainty associated with parameter estimation. To overcome this problem, the reaction can be carried out in a chromatographic reactor packed with beads of the immobilized enzyme and the amounts of the enantiomers eluted directly measured. An unsteady-state analysis of such a reaction leads to the following simple expression (Wu, 1998):

$$E_c = \frac{\ln(1 - X_R)}{\ln(1 - X_S)} \tag{20.50}$$

The only data required are the conversions of the two enantiomers. Because a transient method is used, complications due to enzyme deactivation do not arise. Thus the method does away with the need for allowing for it, as in Equation 20.49. Further, it enables calculation of E_c from just one experimental measurement of the enantiomer eluted.

Chapter 21

Electroorganic Synthesis Engineering

Historically, electrochemical processes have been limited to the production of inorganic compounds, and commercial processes based on electroorganic synthesis have found only limited application. It appeared to be an "odious truth" (Fry, 1972) that electrochemical techniques were ignored in organic synthesis. But the past 25 years have witnessed the introduction of a fairly large number of new electroorganic processes with attendant advances in electrochemical process analysis. The most remarkable has been Monsanto's highly successful electrochemical route for the production of adiponitrile. A particularly notable advance is the electrosynthesis of fine chemicals and natural products. Combinations of electrosynthesis with other strategies of rate or selectivity enhancement such as catalysis by PTC and by enzymes (Chapters 19 and 20) are also adding exciting possibilities to organic synthesis. Simultaneously, fundamental understanding of the principles of organic electrochemistry, electrode kinetics, and transport processes in electrochemical systems has grown rapidly in the last decade.

A number of books and reviews have appeared on electroorganic chemistry during this period, for example, Eberson and Schafer (1971), Fry (1972), Beck (1974), Perry and Chilton (1976), Rifi and Covitz (1975, 1980), Weinberg (1974, 1990), Swann and Alkire (1980), Kyriacou (1981), Pletcher (1982), Baizer and Lund (1983), Baizer (1973, 1984), Shono (1984), Pletcher and Walsh (1990), Little and Weinberg (1991), Bowden (1997), Bockris (1998), Hamann (1998). This period also saw the emergence of *electrochemical reaction engineering* as a distinct discipline of chemical reaction engineering, as evidenced by a number of books and reviews on the subject, for example, Picket (1979), Udupa (1979), Danly (1980, 1984), Alkire and Beck (1981), Weinberg et al. (1982), Alkire and Chin (1983), Fahidy (1985), Hine (1985), Goodridge et al. (1986), Rousar et al.

(1986), Heitz and Krysa (1986), Ismail (1989), Scott (1991), Prentice (1991), Goodridge and Scott (1995).

Electroorganic synthesis offers opportunities for performing many of the conventional organic reactions at controlled rates and greater product selectivities without the addition of any catalyst. The processes almost always employ milder conditions and are characterized by greatly reduced air and water pollution. Further, there are a number of syntheses that can only be carried out electrochemically, such as the Kolbe synthesis and electrochemical perfluorination.

As in all other chapters connected with different methods of organic synthesis, we briefly outline in this chapter the basic principles of electroorganic chemistry, discuss the modeling of electrochemical reactions and reactors, and give several examples of electroorganic synthesis.

BASIC CONCEPTS AND DEFINITIONS

The Electrochemical Reactor (or Cell)

In electrochemical terminology, the vessel in which electrochemical transformation occurs is called the cell. To bring the terminology in line with that of chemical reaction engineering, we refer to it as the *electrochemical reactor* (ECR). An ECR is one in which electrical energy is converted into chemical energy leading to chemical change.

Consider the electrochemical reactor shown in Figure 21.1. The reactor consists of two electrodes immersed in a conducting solution known as an *electrolyte*. The terminals of the electrodes are connected to an external dc power supply. The positive electrode, anode, and the negative electrode, cathode, are separated from each other by a cell separator thereby dividing the cell into two compartments. The electrolytes in the anode and cathode compartments are known as the *anolyte* and *catholyte*, respectively. Two identical reference electrodes are used to measure the individual electrode potentials. When a voltage of sufficient magnitude is applied across the two electrodes, electron transfer occurs between each electrode and its adjacent electrolyte resulting in current flow in the external loop and hence to *electron transfer reactions* at the electrodes. This phenomenon is known as *electrolysis*.

The electrode at which the desired reaction occurs is called the *working electrode*, and the other electrode is termed the *counterelectrode*. The overall cell voltage E for the reactor in Figure 21.1 and the anode and cathode potentials E_a (or E^+) and E_c (or E^-) are very important parameters in electroorganic synthesis. The second electrode is necessary to complete the cycle. On the other hand, it can also be the site for unwanted side reactions. The presence of these side reactions makes any analysis quite complicated. Hence this treatment will largely be restricted to single electrode reactions on the working electrode.

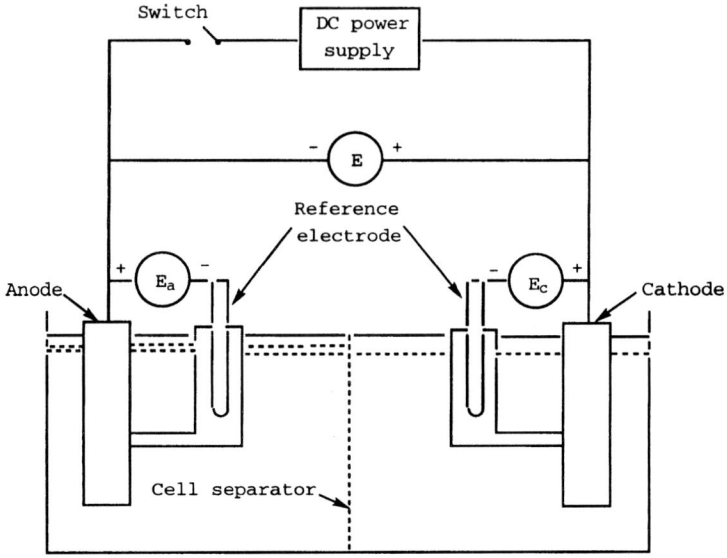

Figure 21.1 A typical setup for conducting an electrochemical reaction

Examples of Difficult/Selective Organic Reactions Carried Out Electrochemically

Since the success of the Monsanto process for adiponitrile, several challenging syntheses have been accomplished electrochemically. An important feature of many of these syntheses is that they cover particularly difficult or selective organic reactions. These include such classes of reactions as oxidation, reduction, carboxylation, dehalogenation, cyanation, etc., several specific examples of which are listed in Table 21.1) (see Swann and Alkire, 1980, for more examples).

Faraday's Law

The basic relationship applicable to all electrochemical reactions is Faraday's law that relates the amount of a substance reacted (kg) at the electrode surface to the charge passed:

$$W = \frac{ItM}{nF} \tag{21.1}$$

where I is the current passed (amperes, A) in t seconds, M is the molecular weight, n is the number of electrons transferred per mole of reaction, and F is the Faraday (96,500 coulombs or A s/mol). Because all of the current passed does not lead to the desired product, we define *current efficiency* (C.E) as

$$C.E = \frac{W_a}{W} \times 100 \tag{21.2}$$

where W_a is the actual weight obtained of a particular product and W is the quantity theoretically expected by Faraday's law.

Table 21.1 Examples of difficult/selective electroorganic syntheses (from Jansson, 1984)

Type of reaction	Example
1. Difficult oxidation	$HOCH_2C \equiv CCH_2OH \xrightarrow{-8e^-} HOOCC \equiv CCOOH$
2. Difficult reduction	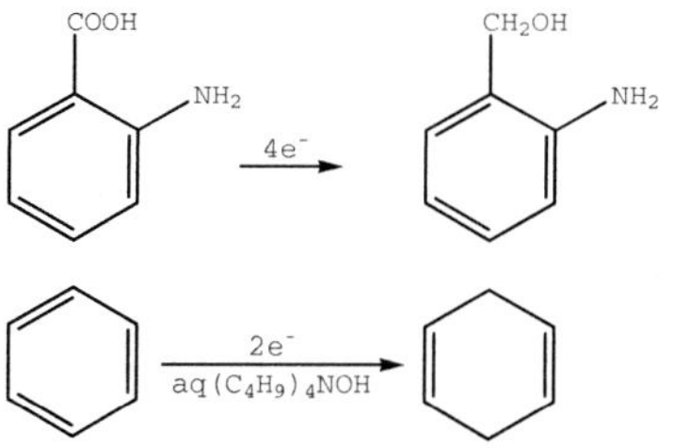 o-(COOH)(NH$_2$)C$_6$H$_4$ $\xrightarrow{4e^-}$ o-(CH$_2$OH)(NH$_2$)C$_6$H$_4$ benzene $\xrightarrow[aq(C_4H_9)_4NOH]{2e^-}$ cyclohexadiene
3. Selective reduction	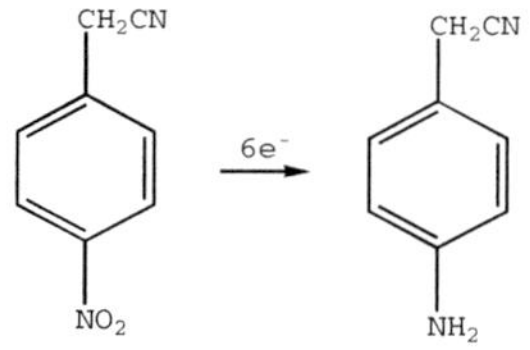 p-(CH$_2$CN)(NO$_2$)C$_6$H$_4$ $\xrightarrow{6e^-}$ p-(CH$_2$CN)(NH$_2$)C$_6$H$_4$
4. Oxidative coupling	$O_2S(NR^-)_2 \xrightarrow{-2e^-} RN=NR + SO_2$
5. Carboxylation	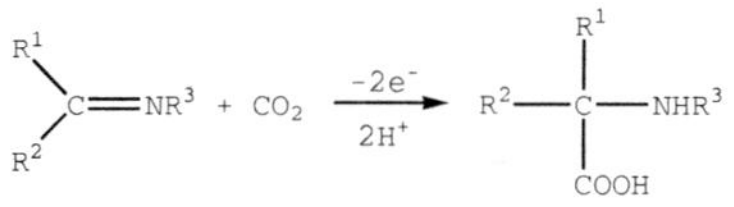 $R^1R^2C=NR^3 + CO_2 \xrightarrow[2H^+]{-2e^-} R^2\text{--}C(R^1)(NHR^3)\text{--}COOH$
6. Anodic decarboxylation	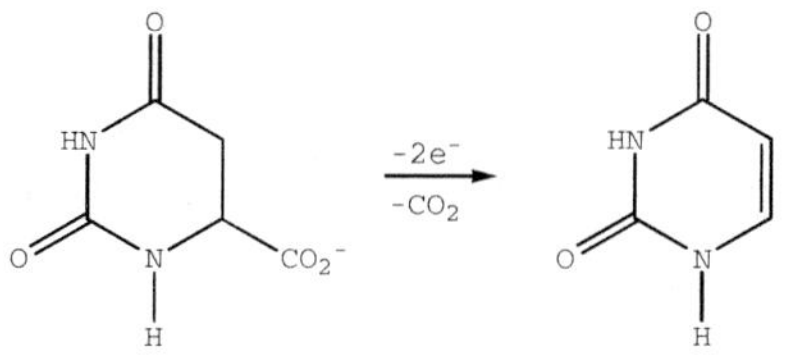 dihydroorotate $\xrightarrow[-CO_2]{-2e^-}$ orotate (uracil)
7. Selective dehalogenation	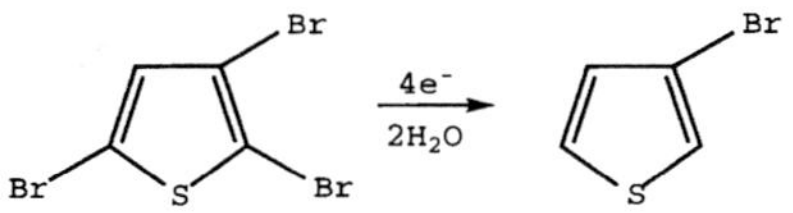 2,3,5-tribromothiophene $\xrightarrow[2H_2O]{4e^-}$ 3-bromothiophene
8. Desulfurization	$ArCH_2S^+(CH_3)_2 \xrightarrow[H^+]{2e^-} ArCH_3 + (CH_3)_2S$
9. Dehydrodimerization	$2CH_2(COOC_2H_5)_2 \xrightarrow[-2H^+]{-2e^-} (C_2H_5OCO)_2CHCH(COOC_2H_5)_2$

(*Continued*)

686 Strategies for Enhancing Rate Reactions

Table 21.1 (Continued)

Type of reaction	Example
10. Direct cyanation	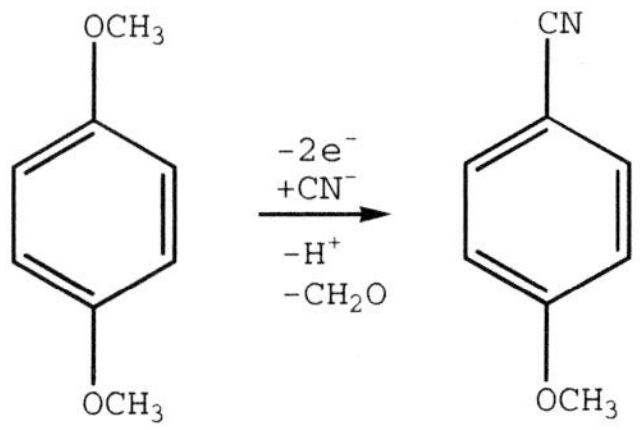 p-dimethoxybenzene $\xrightarrow[-H^+, -CH_2O]{-2e^-, +CN^-}$ 4-cyanoanisole
11. Fluorination	$C_nH_{2n+1}COF \xrightarrow[\text{oxidation}]{HF} C_nF_{2n+1}COF$
12. Anodic substitution	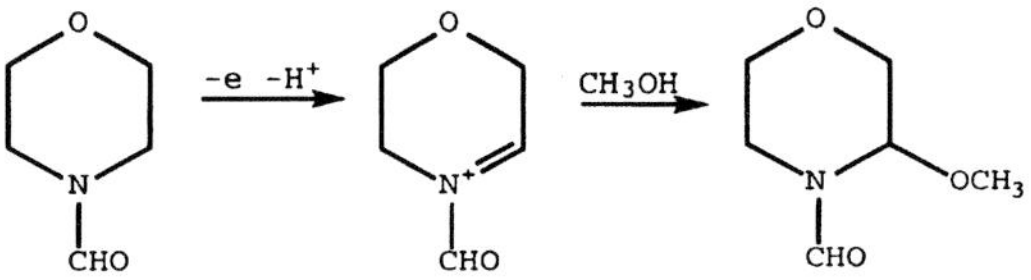 N-formylmorpholine $\xrightarrow{-e, -H^+}$ iminium $\xrightarrow{CH_3OH}$ α-methoxy-N-formylmorpholine
	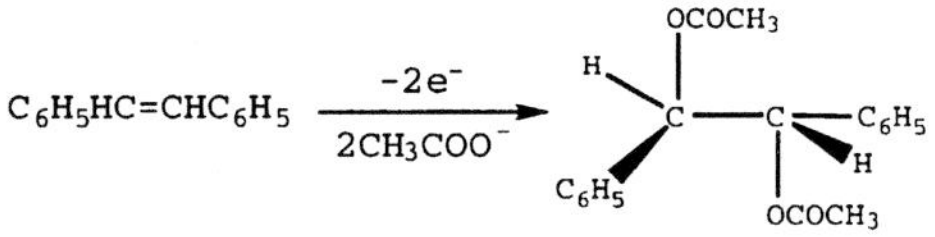 $C_6H_5HC=CHC_6H_5 \xrightarrow[2CH_3COO^-]{-2e^-}$ 1,2-diacetoxy-1,2-diphenylethane
13. Pinacolization	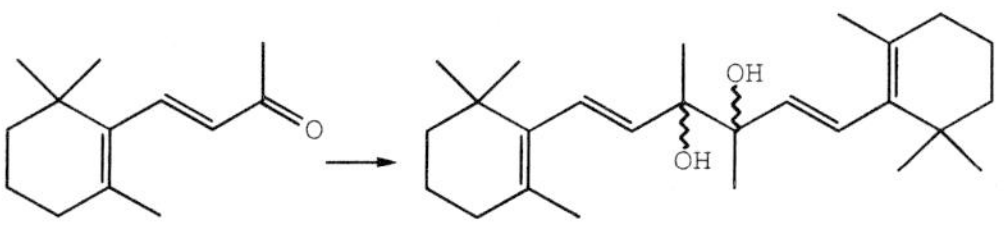 β-ionone derivative → pinacol dimer

Another important parameter associated with electrosynthesis is the amount of energy consumed. This refers to the electrical power in kWH required to produce a unit weight (kg) of the desired product and is given by

$$\text{Energy consumption} = \frac{nFE}{36M(C.E)}, \quad \text{kWh/kg} \tag{21.3}$$

where E is the cell voltage in volts.

Electrochemical Measure of Reaction Rate

The parameters that define the rate of a chemical reaction are determined generally by the broad area of the reaction, that is, chemical, photochemical, electrochemical, etc. In electrochemical reactions, the characteristic measurable parameter is the current density i which is the current in A/m² of the electrode, and the parameter that converts it to molar rate is the Faraday. Thus the reaction rate (for the disappearance of reactant A) can be expressed as

$$-r_A = \frac{V_r}{A_c}\frac{d[A]}{dt} = \frac{i_A}{nF}, \quad \text{mol/m}^2\,\text{s} \tag{21.4}$$

Because n (for a given system) and F are constants, the current density i is often used as a direct measure of the rate.

Open Circuit and Standard Open Circuit Voltages

A typical electrode reaction can be represented as follows:

CATHODIC

$$\nu_A A + \nu_B B + \cdots + ne \rightarrow \nu_R R + \nu_S S + \ldots \qquad [21.1]$$

ANODIC

$$\nu_A A + \nu_B B + \cdots - ne \rightarrow \nu_R R + \nu_S S + \cdots \qquad [21.2]$$

A term commonly used in electrochemistry for characterizing any such reaction is *open circuit voltage* E_0 ($= E^- - E^+$) with zero current flowing between the electrodes through the electrolyte. In theory, E_0 should obviously be zero. In practice, however, it has a finite value which, for a reversible reaction, is related to the Gibbs free energy change by

$$(\Delta G)_{T,P} = -nFE_0 \qquad (21.5)$$

Another commonly used term is *standard open circuit voltage* E^0 which is related to the standard free energy change by

$$\Delta G^0 = -nFE^0 \qquad (21.6)$$

Then, using the well-known relationship between free energy change and the activities (a_j) of the reaction components, the following relationship between E_0 and E^0 can be derived:

$$E_0 = E^0 - \frac{R_g T}{nF} \sum \nu_j \ln a_j \qquad (21.7)$$

Thus E_0, the cell voltage at zero current, can be estimated from the standard free energy change or from the standard electrode potentials, values of which are available in the literature (Perry and Chilton, 1976).

Polarization and Overpotentials

The relationship just presented, which provides a means of computing the theoretical minimum cell voltage, has been derived by assuming that the system is reversible and in equilibrium. However, industrial electroorganic reactions are mostly irreversible, and hence any such reaction requires an applied emf or cell voltage $E > E_0$ to proceed in one direction. A net current passes in the system, and the electrodes are said to be *polarized*. This irreversible phenomenon of *polarization* is the basis of electrode kinetics.

The overall picture

Because all electrochemical reactions involve anodic and cathodic reactions, polarization will have components for both reactions. As will be explained later, the electrode potentials have two terms for each electrode: *surface overpotential* η_a or η_c and *concentration overpotential* η_{con}. Apart from these overpotentials, electrical energy will also be expended due to the electrical resistance of the cell components such as electrolyte, diaphragm, busbar, etc. Thus the practical cell voltage E_t, when a net current is flowing through the cell, is the sum

$$E_t = E_0 + |\eta_c + \eta_{con}|_c + |\eta_a + \eta_{con}|_a + [IR_t] \tag{21.8}$$

where R_t is the sum of all internal resistances in the cell system and subscripts c and a refer to the cathode and anode, respectively.

The sum of the surface and concentration overpotentials is called the total *overpotential* or *overvoltage* η of the electrode which can be measured by standard reference electrodes, as already shown in Figure 21.1. All electrode potentials are related to the standard hydrogen electrode (SHE), the potential of which is set by definition at zero at all temperatures. The saturated calomel electrode (SCE), $Hg/Hg_2Cl_2/Cl^-$, is the most widely used reference electrode for potential measurement. Other electrode systems used are $Hg/Hg_2SO_4/SO_4^{2-}$, $Hg/HgO/OH^-$, and $Ag/AgCl/Cl^-$.

Regimes of control

As in any heterogeneous reaction, two major controlling regimes are also possible in electrochemical reactions, surface reaction and external mass transfer, referred to specifically as surface overpotential (or *charge transfer*) and concentration overpotential, respectively, in electrochemical terminology.

Charge transfer control (surface overpotential)

Consider the electrode reaction

$$A + ne \leftrightarrow R \tag{21.3}$$

At or near equilibrium conditions, the reduction of A to R and the oxidation of R to A will take place at equal rates resulting in no net chemical change. Because current density represents the rate of reaction in an electrochemical system, $i_- = i_+$, where i_- (or i_c) and i_+ (or i_a) are, respectively, the current densities (A/m²) associated with the forward (cathodic) and reverse (anodic) reactions. The value of i_- or i_+ at equilibrium is called the *exchange current density* i_0. Thus, at equilibrium,

$$i_-(\text{or } i_c) = i_+(\text{or } i_a) = i_0 \tag{21.9}$$

Now if an external potential E (also called electrode potential) is applied to the system, the rates of the forward and reverse reactions as well as the concentrations of A and R will shift from their equilibrium values. The extent of this shift, uninfluenced by physical factors such as external mass transfer or any of

the other factors contained in Equation 21.8, gives the surface overpotential η directly:

$$\eta = E - E_0 \tag{21.10}$$

It can be shown that

$$i_- = i_0 \exp\left(-\alpha\eta_c \frac{nF}{R_g T}\right) \tag{21.11}$$

and

$$i_+ = i_0 \exp\left((1-\alpha)\eta_a \frac{nF}{R_g T}\right) \tag{21.12}$$

where

$$i_0 = k'_0 nF[A]_b \exp\left(-\frac{\alpha E nF}{R_g T}\right) = k'_a nF[R]_b \exp\left(\frac{(1-\alpha)E nF}{R_g T}\right) \tag{21.13}$$

k'_c and k'_a are, respectively, the rate constants for the forward (cathodic) and reverse (anodic) reactions, and α is known as the *transfer coefficient*. Note that Equation 21.13 is similar (but not equivalent) in form to a first-order chemical reaction with the usual Arrhenius type dependence of the rate constant on temperature. The net current i is given by

$$i = i_- - i_+$$

$$= i_0 \left[\exp\left(-\alpha\eta_c \frac{nF}{R_g T}\right) - \exp\left((1-\alpha)\eta_a \frac{nF}{R_g T}\right)\right] \tag{21.14}$$

This is the well-known Butler–Volmer equation for electrode kinetics in the absence of mass transfer effects between electrode and electrolyte.

If the reverse step is negligible, that is, $i_- \gg i_+$, then the Butler–Volmer equation for a cathodic process reduces to the more familiar form

$$i = i_c = i_0 \exp\left(-\alpha\eta_c \frac{nF}{R_g T}\right) \tag{21.15a}$$

or

$$\ln(i) = \ln(i_c) = \ln(i_0) - \frac{\alpha\eta_c nF}{R_g T} \tag{21.15b}$$

commonly known as the Tafel equation. These equations are frequently expressed in the form

$$\eta_c = \left(\frac{R_g T}{\alpha nF}\right)\ln(i_0) - \left(\frac{R_g T}{\alpha nF}\right)\ln(i_c) \tag{21.16a}$$

or in the empirical form

$$\eta_c = a + b\ln(i_c) \tag{21.16b}$$

where

$$a = \frac{R_g T}{\alpha nF}\ln(i_0), \quad b = -\frac{R_g T}{\alpha nF} \quad \text{(Tafel slope)} \tag{21.17}$$

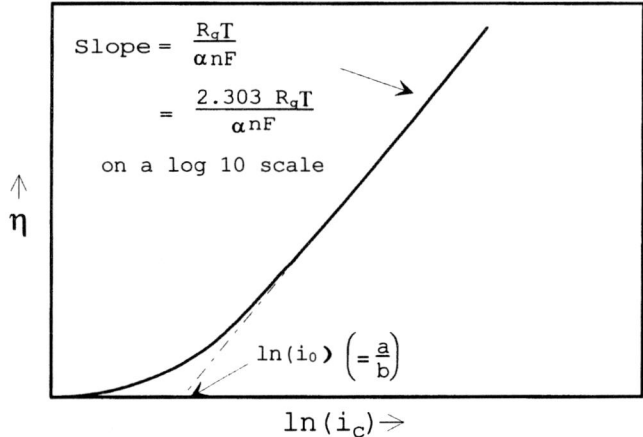

Figure 21.2 A qualitative illustration of the Tafel plot

Equation 21.16b is equivalent to 21.16a when $b = -(R_g T/\alpha nF)$. A plot of the cathodic overpotential η_c against $\ln(i_c)$ will give a straight line of slope $(-R_g T/\alpha nF)$ and intercept $\ln(i_0)$ on the $\ln(i_c)$ axis. This kind of plot is referred to as a *Tafel plot*, and the slope, usually denoted by the letter b as in Equation 21.17, is a *Tafel slope*. For anodic reactions, $i_+ \gg i_-$ and the slope is $R_g T/(1-\alpha)nF$. A typical Tafel plot is shown in Figure 21.2.

Mass transfer control (concentration overpotential)

Mass transport in an electrochemical reactor occurs by three mechanisms: migration in the electrical field, film diffusion, and convection. The first of these is a special feature of electrochemical reactions, whereas the other two are common to all reactions that have a solid phase. However, where an inert-supporting electrolyte is used, the effect of migration can be neglected. With this assumption, let us consider a single electrode reaction given by reaction 21.3. When a finite current is passed through the cell and conditions are perfectly reversible, the concentration overpotential can be expressed as (Pickett, 1979)

$$\eta_{con} = \frac{R_g T}{nF} \ln\left(\frac{[A]_s[R]_b}{[R]_s[A]_b}\right) \qquad (21.18)$$

where (see Figure 21.3) subscript s refers to the electrode surface. Subscript b denotes the liquid bulk (and is assumed when not specifically used).[1] Now we apply this equation to mass transfer control. For this we invoke the basic equation for mass transfer appropriately recast for the electrochemical reaction at the electrode as

$$i = nFk'_L([A]_b - [A]_s) \qquad (21.19)$$

---

[1] For example, concentrations such as $[A]$, $[A]_0$, $[A]_f$, $[A]_1$, etc. that appear throughout this chapter all refer to values at bulk conditions.

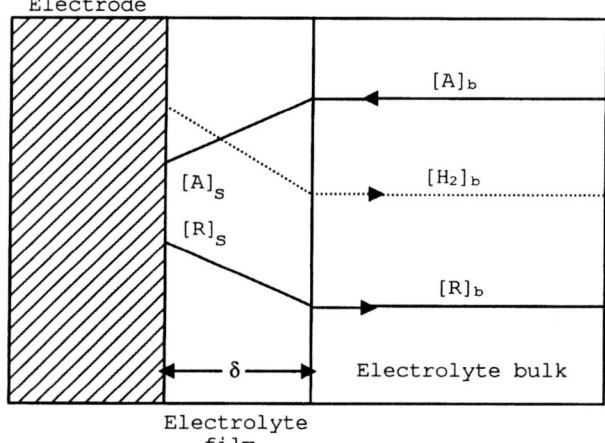

Figure 21.3 Concentration profiles in the primary (———) and loss (· · · · ·) reactions under mass transfer control

When applied to the limiting condition of mass transfer control ($[A]_s = 0$), this becomes

$$i_{\lim} = nFk'_L[A]_b \qquad (21.20)$$

The limiting current density $i_{\lim}$ represents the maximum rate, at 100% current efficiency, at which a given electrode reaction can occur (Selman and Tobias, 1978). The values in many practical situations are usually lower than this limiting value.

Substituting Equation 21.19 for $[A]_s$ in Equation 21.18,

$$\eta_{\text{con}} = \frac{R_g T}{nF} \ln\left[\frac{(i_{\lim} - i)}{i_{\lim}} \frac{[R]_b}{[R]_s}\right] \qquad (21.21)$$

If diffusion of the product plays no part in the controlling step, $[R]_b = [R]_s$ and Equation 21.21 reduces to

$$\eta_{\text{ccn}} = \frac{R_g T}{nF} \ln\left(\frac{[A]_s}{[A]_b}\right) = \frac{R_g T}{nF} \ln\left[\frac{(i_{\lim} - i)}{i_{\lim}}\right] \qquad (21.22)$$

Other equivalent forms are

$$i = i_{\lim}\left[1 - \exp\left(\eta_{\text{con}} \frac{nF}{R_g T}\right)\right] \qquad (21.23a)$$

$$\frac{[A]_b}{[A]_s} = \frac{i_{\lim}}{i_{\lim} - i} \qquad (21.23b)$$

A plot of this equation appears in Figure 21.4.

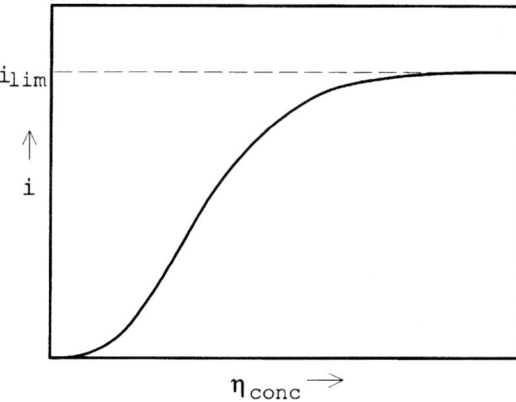

Figure 21.4 Plot of Equation 21.23a

Combined charge transfer and mass transfer control

When charge transfer and mass transfer proceed at comparable rates, the net cathodic current i_c is given by

$$i_c = i_0 \left(\frac{[A]_s}{[A]_b}\right) \exp\left[-\alpha\eta \frac{nF}{R_g T}\right] \qquad (21.24)$$

Note the use of η, the total overpotential, in this equation. Also, the term $[A]_s/[A]_b$ is introduced to replace $[A]_b$ in the definition of i_0 (Equation 21.13) by $[A]_s$. Combining this equation with Equation 21.19 gives

$$i_c = \frac{1}{\dfrac{1}{nFk'_L[A]_b} + \dfrac{1}{i_0 \exp\left(-\alpha\eta \dfrac{nF}{R_g T}\right)}} \qquad (21.25)$$

Equation 21.25 is an important general equation for current density that accounts for both mass and charge transfer resistances. In the absence of mass transfer resistance, it reduces to the Tafel equation (Equation 21.15a).

Correlations for Mass Transfer at Electrodes

The usual Sherwood number or j_d correlations have been used to correlate mass transfer coefficients with system properties and flow characteristics. Scores of such correlations have been reported covering a wide range of electrode-electrolyte configurations. A few of these that represent mass transfer in the more common electrochemical reactors are listed in Prentice (1991).

Current Distribution

In the design of ideal batch and continuous stirred tank reactors, we assume uniform current distribution at the electrodes to simplify the mathematical

formulations. Lack of uniform current distribution can lead to loss of current efficiency and product selectivity (see Newman, 1973, and the text by Scott, 1991, for details). Depending on the degree of simplification, current distribution can be classified into three categories: (1) *primary*, when it is entirely determined by electrode configuration; (2) *secondary*, when it takes into account the surface overpotential, that is, electrode kinetics; and (3) *tertiary*, when it takes into account all the overpotentials, ohmic, concentration, and surface. A discussion of these will not be attempted here.

MODELING OF ELECTROORGANIC REACTIONS AND REACTORS

The modeling of ECR systems involves, as that of any other reaction system, the development of a suitable reaction model for subsequent use in reactor modeling. The reaction model for an ECR is an expression for the dependence of current density on reaction parameters such as reactant concentration, electrode potential, rate constants, pH, temperature, etc. The reactor model relates the reactor parameters to performance criteria. The objective is to evolve suitable expressions for the computation of the electrode area required for a desired conversion, batch time, etc. We devote the next section to developing reaction models for simple electrochemical systems and proceed to reactor modeling in the following section.

REACTION MODELING

Consider the simple electrode reaction 21.3. Usually, any such primary reaction is accompanied by a side reaction that leads to a reduction of current efficiency. Such a reaction is referred to as a *loss reaction* in electrochemical terminology. The major loss reaction is the hydrogen evolution reaction

$$2H^+ \xrightarrow{2e} H_2 \qquad [21.4]$$

Reactions 21.3 and 21.4 consist of the following steps:

1. Transport of A from bulk of electrolyte to electrode surface
2. Conversion of reactant A to product R
3. Transport of R from electrode surface to bulk.

In the loss reaction, hydrogen must be transported from the electrode surface (site of evolution) to the bulk. Now we develop a typical model for these reactions.

Assuming that the main reaction, (reaction 21.3), is governed by a simple Tafel type expression, we combine Equations 21.10, 21.11 and 21.13 to get

$$\frac{i_A}{nF} = k_A''[A]_b = k_A' e^{-b_A E}[A]_b \qquad (21.26)$$

where

$$k_A'' = k_A' \exp(-b_A E) \tag{21.27}$$

$$b_A = \frac{\alpha n F}{R_g T} \tag{21.28}$$

In these equations, the terms k_A' and $\exp(-b_A E)$ have been combined to give a composite rate constant k_A''. This has been done merely to emphasize the formal similarity between the conventional and electrochemical rate equations.

A material balance on A gives

$$\begin{bmatrix} \text{rate of} \\ \text{change in} \\ \text{concentration} \\ \text{at electrode} \\ \text{surface} \end{bmatrix} = \begin{bmatrix} \text{rate of} \\ \text{transport to} \\ \text{electrode} \end{bmatrix} - \begin{bmatrix} \text{rate of} \\ \text{consumption} \\ \text{due to} \\ \text{reaction} \end{bmatrix} \tag{21.29}$$

$$\frac{d[A]_s}{dt} = k_L' a([A]_b - [A]_s) - k_A' a[A]_s e^{-b_A E} \tag{21.30}$$

At steady state, $d[A]_s/dt = 0$, and hence

$$\frac{i_A}{nF} = k_L'([A]_b - [A]_s) = k_A'[A]_s e^{-b_A E} \tag{21.31}$$

Eliminating $[A]_s$,

$$\frac{i_A}{nF} = \frac{[A]_b}{\dfrac{1}{k_L'} + \dfrac{1}{k_A' \exp(-b_A E)}} = k'[A]_b \tag{21.32}$$

where

$$\frac{1}{k'} = \frac{1}{k_L'} + \frac{1}{k_A' \exp(-b_A E)} \tag{21.33}$$

In addition to the primary reaction, the secondary reaction should also be modeled. Assuming that the hydrogen evolution reaction on the electrode under consideration is also of the Tafel type but zero order in hydrogen (because the solvent composition is practically constant during electrolysis), we can write

$$\frac{i_H}{2F} = k_H' e^{-b_H E} \tag{21.34}$$

where

$$b_H = \frac{2\alpha' F}{R_g T} \tag{21.35}$$

and i_H and α' are the current density and transfer coefficient associated with hydrogen evolution, respectively, and k_H' and b_H are the electrochemical parameters of the hydrogen evolution reaction.

In view of the assumption that the reaction is zero order in hydrogen, any mass transfer effect with respect to hydrogen transport would be of no conse-

quence. Hence an overall rate constant corresponding to k' of Equation 21.32 is not involved, and the rate constant k'_H can be used as such to compute i_H.

The total current i_t assuming no other side reaction is given by

$$i_t = i_A + i_H \tag{21.36}$$

Thus the current efficiency for the primary reaction is

$$C.E. = \frac{i_A}{i_t} \tag{21.37}$$

These equations can be solved if the kinetic parameters b_A, b_H, and k'_A, k'_H, k'_L are known. Thus the reaction model enables the theoretical computation of current efficiency for the applied potential if the kinetic parameters are known [see Prentice, 1991; Goodridge and Scott, 1995, for details].

REACTOR MODELING

Electrochemical reactors are heterogeneous by their very nature. They always involve a solid electrode, a liquid electrolyte, and an evolving gas at an electrode. Electrodes come in many forms, from large-sized plates fixed in the cell to fluidizable shapes and sizes. Further, the total reaction system consists of a reaction (or a set of reactions) at one electrode and another reaction (or set of reactions) at the other electrode. The two reactions (or sets of reactions) are necessary to complete the electrical circuit. Thus, although these reactors can, in principle, be treated in the same manner as conventional catalytic reactors, detailed analysis of their behavior is considerably more complex. We adopt the same classification for these reactors as for conventional reactors, batch, plug-flow, mixed-flow (continuous stirred tank), and their extensions.

Approaches to Design

The electrochemical reactor usually has a reservoir for the electrolyte along with the reactor. A comparison of the more important conventional chemical reactors with their corresponding electrochemical counterparts is shown in Table 21.2.

To simplify the mathematical analysis, we assume first-order kinetics. This is justified because most electrochemical reactions are first order, and those that are not can be assumed to be so without serious loss of accuracy.

Potentiostatic and Galvanostatic Operations

Electrochemical reactors can be operated under conditions of constant electrode potential, constant current, or constant cell voltage. The first two are referred to as *potentiostatic* and *galvanostatic* modes of operation, respectively. The potentiostatic method is characterized by constant values of the kinetic parameters and hence enables integration of the differential equations describing the different reactors. On the other hand, galvanostatic operation is characterized by an inevitable change of electrode potential with time, leading to variations in the kinetic parameters. Hence we restrict our treatment to potentiostatic operation.

Table 21.2 Analogy between electrochemical and conventional chemical reactors

Mode of operation	Conventional chemical reactor[a]	Electrochemical reactor
Batch	Reactor: $|A| + |B| \rightarrow R$	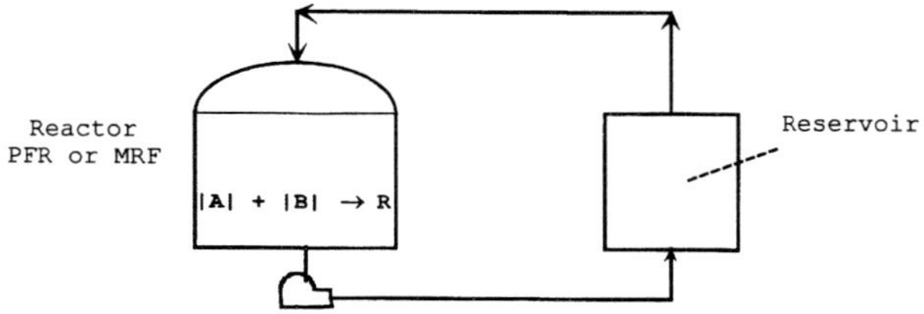 Reactor PFR or MRF: $|A| + |B| \rightarrow R$; Reservoir
Semibatch	Reactor with B added: $|A| + B \rightarrow R$	Reactor with B added: $|A| + B \rightarrow R$; Reservoir
Continuous	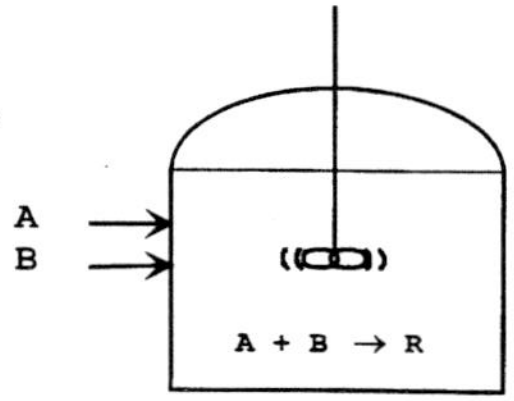 Reactor with A and B added: $A + B \rightarrow R$	Reactor with A and B added: $A + B \rightarrow R$; Reservoir

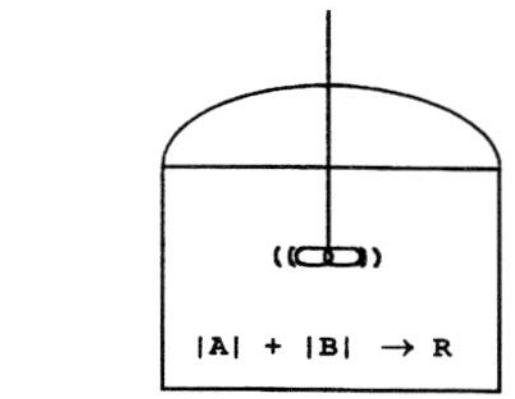

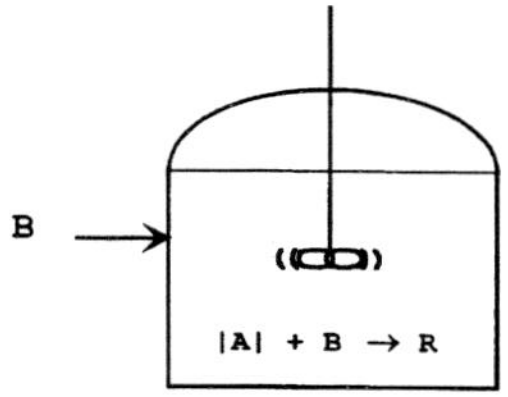

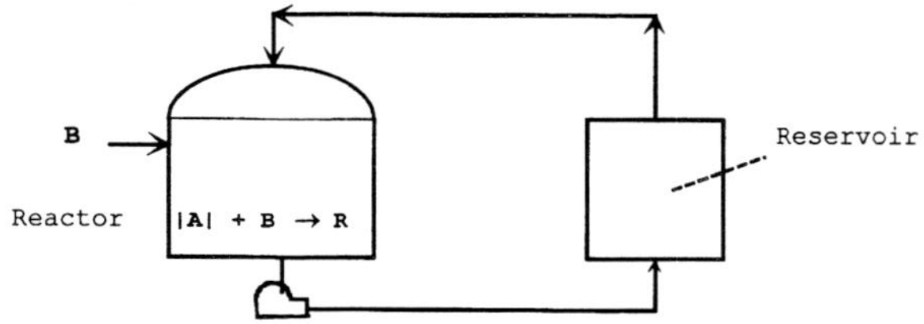

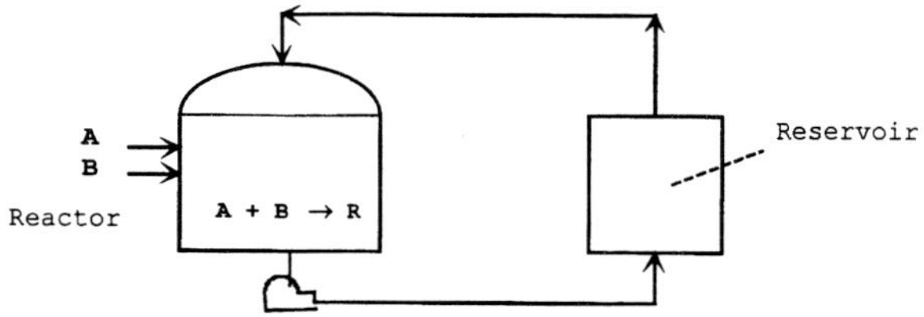

[a] $|A|$ or $|B|$ indicates that the reactant is initially present in the reactor and not added during reaction.

Galvanostatic operation is commonly adopted, however, because potentiostats are not available for large-scale operations and are also much costlier.

Batch Reactor

The simple batch reactor was considered in Chapter 4. A modified version in which the electrolyte is recirculated, however, is the preferred mode of operation in the electrochemical industry because it provides flexible batch volume and also enhances the mass transfer characteristics of the cell due to circulation. Further, with recirculation, the reactor can be operated either in the plug-flow or mixed-flow mode. We consider all three cases here along with a few other common modes of operation.

The simple batch reactor (without recirculation)

This reactor (Figure 21.5) provides a logical link with the conventional stirred batch reactor and a starting point for establishing a framework for more useful cell designs. For a single cathodic reaction that occurs in the reactor with 100% current efficiency, the following mass balance can be written:

$$\frac{i_A}{nF} = -\left(\frac{V_r}{A_c}\right)\left(\frac{d[A]_b}{dt}\right) \qquad (21.38)$$

where V_r is the reactor volume. Assuming Tafel-like behavior for the reaction, from Equations 21.32 and 21.38

$$\frac{i_A}{nF} = k'[A]_b = -\frac{V_r}{A_c}\left(\frac{d[A]_b}{dt}\right) \qquad (21.39)$$

Integrating Equation 21.39 between the limits of initial ($[A] = [A]_0, t = 0$) and final ($[A] = [A]_f, t = t_B$) conditions leads to

$$A_c = \frac{V_r}{t_B}\int_{[A]_f}^{[A]_0} \frac{1}{k'}\frac{d[A]_b}{[A]_b} \qquad (21.40)$$

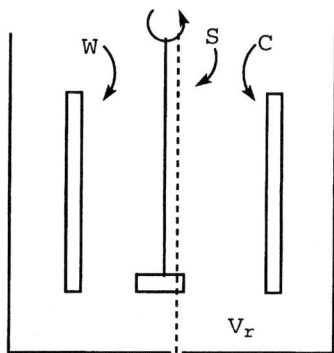

Figure 21.5 Conventional type batch electrochemical reactor (with no recirculation). W = working electrode, C = counterelectrode, S = separator

Equation 21.32 is the reaction model that must be solved simultaneously with the reactor Equation 21.40 to obtain the required electrode area as a function of process parameters.

If the current efficiency is less than 100% due to the loss reaction (reaction 21.4), this reaction must also be considered. Its current density is given by Equation 21.34. Thus from a current balance, $i_t = i_A + i_H$, giving

$$i_t = nFk'[A]_b + 2Fk'_H e^{-b_H E} \tag{21.41}$$

where k' and k'_H are given by Equations 21.33 and 21.34, respectively.

Equations 21.40 and 21.41 can be combined and solved to obtain A_c as a function of conversion, batch time, total current, and other system parameters such as $[A]_0, k'_L, k'_A, b_A,$ and t. Remembering that $X_A = ([A]_0 - [A])/[A]_0$, the final equations obtained in terms of conversion are

$$i_t = nFk'[A]_0(1 - X_A) \tag{21.42}$$

$$A_c = \frac{V_r}{t_B} \int_0^{X_{Af}} \frac{1}{k'} \frac{dX_A}{(1 - X_A)} \tag{21.43}$$

Equations 21.42 and 21.43 along with Equation 21.33 can be solved to provide an expression for A_c as a function of the process parameters.

It is clear from the foregoing derivations that the current efficiency is a practically important parameter. This can be obtained from the ratio (i_A/i_t), that is, Equation 21.37, where i_A is the current density for the main reaction and i_t is the total current density given by Equation 21.41.

Batch operation with recirculation through a plug-flow reactor

Now we consider a reactor that is operated with total recirculation of electrolyte, but plug flow prevails within the reactor itself (Figure 21.6). A material balance on A gives

$$-Qd[A] = \frac{i_{A\ell} A_c}{nF} \frac{d\ell}{L} \tag{21.44}$$

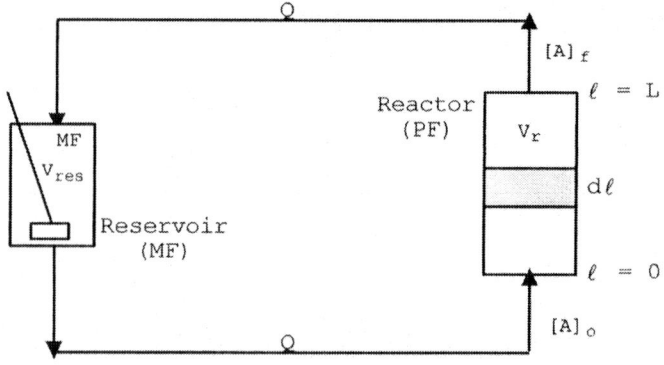

Figure 21.6 Batch plug-flow reactor

which on combining with the reaction model, Equation 21.32, and integrating gives

$$[A]_f = [A]_0 e^{-k'A_c/Q} \tag{21.45}$$

This is the design equation for a PFR with total reflux and, not surprisingly, is reminiscent of the conventional PFR equation where k' is defined as a combined rate constant (Equation 21.33) and the electrode area A_c appears as a parameter.

Similarly, by drawing a material balance over the reservoir, the following equation for reactant concentration (or conversion) as a function of batch time can be derived:

$$[A]_f = [A]_0 \exp\left\{ -\frac{t_B}{\bar{t}_{res}} \left[1 - \exp\left(-\frac{k'A_c}{Q} \right) \right] \right\} \tag{21.46a}$$

where $\bar{t}_{res} = V_{res}/Q$. This can also be recast as an expression for A_c as

$$A_c = \frac{Q}{k'} \ln\left[\frac{1}{1 - \frac{\bar{t}_{res}}{t_B} \ln\left(\frac{[A_0]}{[A]_f}\right)} \right] = \frac{Q}{k'} \ln\left[\frac{1}{1 - \frac{\bar{t}_{res}}{t_B} \ln\left(\frac{1}{1-X_A}\right)} \right] \tag{21.46b}$$

Equations 21.46a,b constitute the reactor model for the present configuration. Either of these can be combined with the reaction model, Equation 21.42, and with Equation 21.33 (the defining equation for k') to give the electrode area as a function of conversion, current density, and other system parameters.

Batch operation with circulation through a stirred tank reactor

This reactor is sketched in Figure 21.7. A material balance over the reactor alone gives

$$Q([A]_1 - [A]_2) = (-r'_A)A_c \tag{21.47}$$

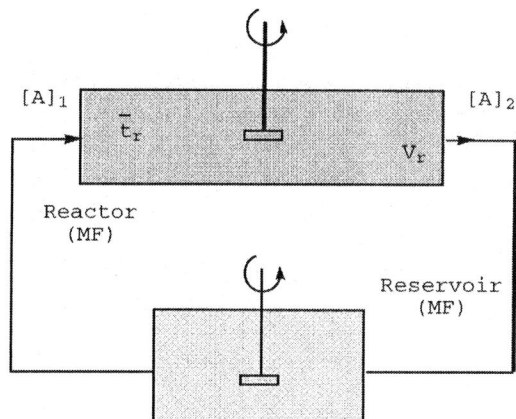

Figure 21.7 Batch mixed-flow reactor

Assuming Tafel-like behavior via Equation 21.32, that is, $-r'_A = k'[A]_2$,

$$[A]_2 = \frac{[A]_1}{\left(1 + \dfrac{k'A_c}{Q}\right)} \qquad (21.48)$$

Similarly, an instantaneous material balance on the reservoir combined with Equation 21.48 gives

$$V_{res}\frac{d[A]_1}{dt} + Q[A]_1 = Q[A]_2 = \frac{Q[A]_1}{\left(1 + \dfrac{k'A_c}{Q}\right)} \qquad (21.49)$$

Solution of this equation with boundary conditions

$$[A]_1 = [A]_0 \quad \text{at } t = 0, \qquad [A]_1 = [A]_f \quad \text{at } t = t_B \qquad (21.50)$$

results in

$$[A]_f = [A]_0 \exp\left\{-\frac{t_B}{\bar{t}_{res}}\left[1 - \frac{1}{\left(1 + \dfrac{k'A_c}{Q}\right)}\right]\right\} \qquad (21.51a)$$

or

$$A_c = \frac{Q}{k'}\left\{\frac{1}{\left[1 - \dfrac{\bar{t}_{res}}{t_B}\ln\left(\dfrac{1}{1 - X_{Af}}\right)\right]} - 1\right\} \qquad (21.51b)$$

Equation 21.51a or b is the design equation for a stirred tank batch reactor with recirculation of electrolyte. It can be solved by combining it with the reaction model given by Equation 21.41 or 21.42 to obtain A_c as a function of conversion, current density, and other system parameters. The current efficiency can be calculated from i_A and i_t (as in Example 21.1).

Cascade Reactor

When a number of cells are connected in series as a cascade, reaction occurs stepwise in each of the reactors resulting in high overall conversion. Both CSTR and PFR models can be analyzed as cascades. This reactor system is identical to that developed in Chapter 10 but with appropriate electrochemical parameters. Thus for a system of N CSTRs in a cascade, the design equation is

$$\frac{[A]_N}{[A]_0} = \frac{1}{\left(1 + \dfrac{k'A_c}{Q}\right)^N} \qquad (21.52a)$$

or

$$A_c = \frac{Q}{k'}\left[\left(\frac{[A]_0}{[A]_N}\right)^{1/N} - 1\right] \qquad (21.52b)$$

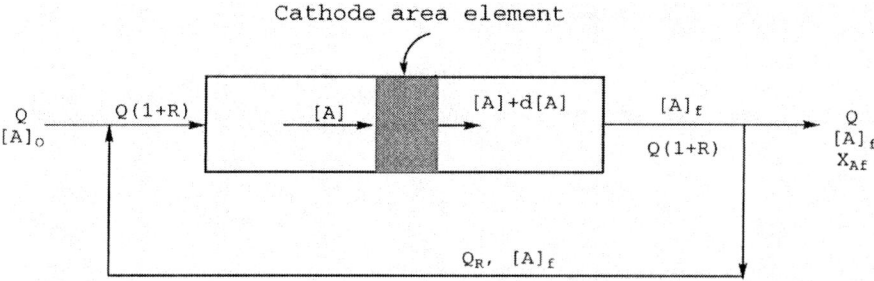

Figure 21.8 Recycle-flow reactor

Reactor with Recycle

Here again the development of the design equation proceeds along the same lines as for the recycle reactor described in Chapter 10. Thus, considering Figure 21.8, the following mass balance can be written over a differential cathode area element dA_c:

$$i_c dA_c = -nFQ(1+R)d[A] \tag{21.53}$$

Assumption of Tafel-like behavior permits combining this equation with Equation 21.32 to give

$$A_c = Q(1+R) \int_{[A]_0'}^{[A]_f} -\frac{1}{k'} \frac{d[A]}{[A]} \tag{21.54}$$

where

$$[A]_0' = \frac{[A]_0 + R[A]_f}{1+R} \tag{21.55}$$

Equation 21.54 can be solved simultaneously with the reaction model, Equation 21.42, to obtain the electrode area A_c in terms of conversion and other parameters of the system.

Now we illustrate the application of some of the reactor modeling methods outlined to a typical electroorganic process, the electrochemical production of p-anisidine, using the parameter values given by Goodridge and Scott (1995) (see also Clark et al., 1988).

EXAMPLE 21.1

Electrolytic reduction of nitrobenzene to p-anisidine in a methanol medium

p-Anisidine can be prepared by the electrolytic reduction of nitrobenzene in an acid medium. The reaction is carried out in a methanol medium and proceeds via phenyl-hydroxylamine as an intermediate. p-Aminophenol is formed as a significant by-product and the other by-products are aniline (which is formed by the reduction of phenylhydroxylamine) and o-anisidine (reaction E21.1.1). However, to simplify the mathematical development of the model, we shall refer to p-aminophenol, o-anisidine, and p-anisidine collectively as S.

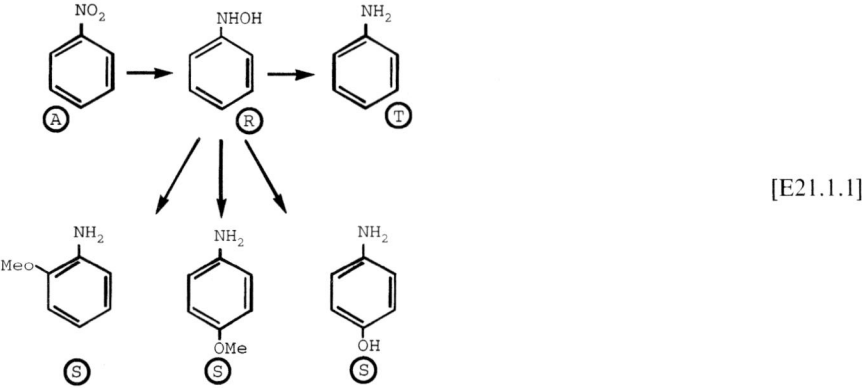

[E21.1.1]

DEVELOPMENT OF MODEL EQUATIONS

Let us construct a reactor model that will predict the effects of conversion, current density, and electrolyte circulation rate on the chemical yield and run time of a cell for the production of *p*-anisidine (see Clark et al., 1988 for details). The electrolytic cell used will be assumed to be of the narrow gap filter press type with total recirculation of electrolyte.

The reaction scheme can be represented as follows:

MAIN REACTION

$$A_s \xrightarrow[-r'_A]{2e} R_s \xrightarrow[r'_R]{2e} T_s$$

$$N_A \uparrow \quad \downarrow N_R \quad \downarrow$$

$$A_b \qquad R_b \qquad T_b$$

$$\downarrow$$

$$S$$

[E21.1.2]

SIDE REACTION

$$2H^+ \xrightarrow[r_H]{2e} H_2 \qquad [E21.1.3]$$

Assuming that there is no accumulation of A and R on the electrode surface and that Tafel behavior is applicable, we can obtain the following relationship for the current density of A:

$$\frac{i_A}{4F} = \frac{[A]_b}{\dfrac{1}{k'_L} + \dfrac{1}{k'_A \exp(-b_A E)}} \qquad (E21.1.1)$$

Assuming further that the rate of diffusion of R away from the surface is fast compared to the rate of the chemical step $R \to S$, as a first approximation we can neglect the buildup of R in the bulk phase. With Tafel behavior again, the current density of R can be written as

$$\frac{i_R}{2F} = k'_R [R]_s \exp(-b_R E) \qquad (E21.1.2)$$

From a coulombic balance,

$$i_A = 4Fr_A \quad \text{and} \quad i_R = 2Fr_R \tag{E21.1.3}$$

and from a mole balance of species R

$$r'_A = k'_L[R]_s + r_R \tag{E21.1.4}$$

The first term on the RHS of this equation represents the number of moles of R diffusing away from the surface of the electrode assuming that accumulation of R in the bulk is negligible.

Equations E21.1.2–E21.1.4 can be combined to give

$$\frac{i_A}{4F} = k'_L[R]_s + k'_R[R]_s \exp(-b_R E) \tag{E21.1.5}$$

from which

$$[R]_s = \frac{i_A}{4F[k_L + k'_R \exp(-b_R E)]} \tag{E21.1.6}$$

Substituting this expression for $[R]_s$ in Equation E21.1.2, the current density for R becomes

$$i_R = \frac{i_A k'_R \exp(-b_R E)}{2[k'_L + k'_R \exp(-b_R E)]} \tag{E21.1.7}$$

Recall that for the hydrogen reaction on the surface,

$$\frac{i_H}{2F} = k'_H \exp(-b_H E) \tag{E21.1.8}$$

Thus the total current density can be expressed as

$$i_t = i_A + i_R + i_H \tag{E21.1.9}$$

or

$$i_t = \frac{4F[A]_b}{\dfrac{1}{k'_L} + \dfrac{1}{k'_A \exp(-b_A E)}} \left\{ 1 + \frac{k'_R \exp(-b_R E)}{2[k'_L + k'_R \exp(-b_R E)]} \right\} + 2Fk'_H \exp(-b_H E) \tag{E21.1.10}$$

Equation E21.1.5 for the current density for the reaction is the electrochemical equivalent of the rate equation for a conventional reaction. We now solve this equation simultaneously with the reactor equation for the selected reactor, for example, Equation 21.46 for a conventional BR with circulation operated as a PFR, or Equation 21.51 for a BR with circulation operated as a CSTR. Thus the electrode area A_c can be estimated as a function of conversion $X_{Af}, [A]_f$, current density i_A, and reaction time $t_B, \bar{t}$. All electrochemical parameters of the equations can be experimentally determined from polarization studies. The reactor efficiency can then be obtained from i_A and i_t.

PARAMETER VALUES

Experimental methods of determining the various parameters are described in standard electrochemical engineering texts. We use the values of Goodridge and Scott (1995):

$b_A = 0.0339 \text{ (mV}^{-1}\text{)}, \quad b_R = 0.019 \text{ (mV}^{-1}\text{)},$

$k'_A = 5.622 \times 10^{-10} \text{ A/m}^2 \text{ per kmol/m}^3,$

$k'_R = 5.337 \times 10^{-9} \text{ A/m}^2 \text{ per kmol/m}^3,$

$k'_L = 1.0 \times 10^{-5} \text{ m/s}, \quad b_H = 0.037 \text{ mV}^{-1},$

$k'_H = 4.9 \times 10^{-10} \text{ A/m}^2 \text{ per kmol/m}^3$

SOLUTION OF EQUATIONS

We shall assume that a conventional batch reactor with circulation operated as a PFR is used. The corresponding equation is 21.46b. Then, this equation can be solved simultaneously with Equation 21.1.5 for an assumed Q to give A_c as a function of voltage E for fixed values of X_A or as a function of voltage E for fixed values of i_A (and hence i_t).

The results are plotted in Figures 21.9a and 21.9b. Note the large effect on electrode area of the total current density and voltage at the lower end of their scales and at higher conversions. At a fixed low value of voltage < 175 mV or current density < 3000 A/m^2, the electrode area increases exponentially for higher conversions.

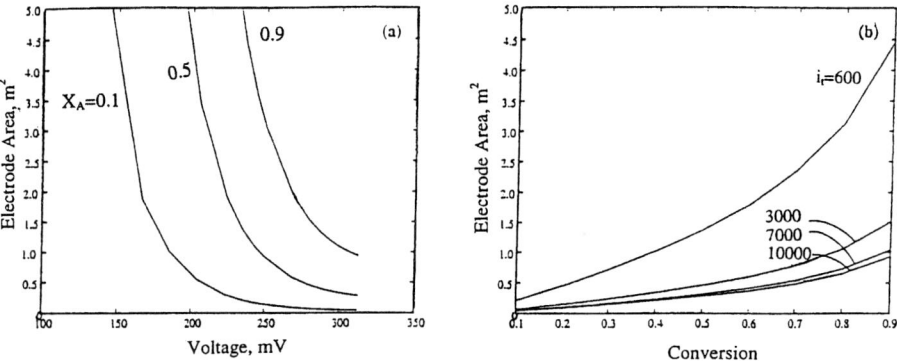

Figure 21.9 Solutions to equations for the *p*-anisidine problem: (a) Electrode area vs. voltage for given conversions. (b) Electrode area vs. conversion for given current densities. Assumption: $t_B/\bar{t}_{res} = 1$

SCALE-UP OF ELECTROCHEMICAL REACTORS

As in conventional chemical reactors, similarity criteria are also employed in the scale-up of electrochemical reactors. Apart from similarities such as geometric similarity, kinematic similarity, and chemical similarity, electrical similarity is a unique criterion in the scale-up of electrochemical cells. It is defined as the condition where geometrically, kinematically, and chemically similar cells have identical cell voltages and current distributions inside the cells. It is also important to note that, although the principles of similarity criteria are the same for chemical and electrochemical reactors, suitable modifications have to be made to obtain electrochemical similarity.

Criteria for Chemical Similarity

The following criteria hold for batch cells operated at constant current conditions:

BATCH

$$\left[\frac{(C.E)It}{nF[A]_0 V_r}\right]_{\text{cell I}} = \left[\frac{(C.E)It}{nF[A]_0 V_r}\right]_{\text{cell II}} \tag{21.56}$$

CSTR

$$\left[\frac{(C.E)I}{nF[A]_0 Q}\right]_{\text{cell I}} = \left[\frac{(C.E)I}{nF[A]_0 Q}\right]_{\text{cell II}} \tag{21.57}$$

PFR

$$\left[\frac{(C.E)i_c a' L}{nF[A]_0 Q}\right]_{\text{cell I}} = \left[\frac{(C.E)i_c a' L}{nF[A]_0 Q}\right]_{\text{cell II}} \tag{21.58}$$

where a' = electrode area per unit length, L = electrode length, and C.E = current efficiency.

Criteria for Electrical Similarity

The criterion for electrical similarity can be stated as follows: two similar cells will have identical current distributions if one of the following three parameters is held constant.

1. Dimensionless limiting current density N_{LCD} (for mass transfer controlled reactions)

$$N_{\text{LCD}} = \frac{i_{\text{lim}} FL}{\kappa R_g T} \tag{21.59}$$

2. Dimensionless exchange current density N_{ECD} (for reactions that are activation controlled in the linear region)

$$N_{\text{ECD}} = \frac{i_0 FL}{\kappa R_g T} \tag{21.60}$$

3. Dimensionless average current density N_{ACD} (for the Tafel region)

$$N_{\text{ACD}} = \frac{i_{\text{av}} FL}{\kappa R_g T} \tag{21.61}$$

Equations 21.59–21.61 can be used, depending on the controlling regime. To achieve the same cell voltage for two different size cells operated at the same current density, the anode–cathode spacing, thickness, and conductivity of the separator should be identical.

Scale-up based on these similarity criteria is widely practiced in the development of industrial cells. An example of their use in the scale-up of a frequently used cell is given below.

EXAMPLE 21.2

Scale-up of a rotating electrode cell

A laboratory cell of 25-mm-diameter rotating electrode is operated at 1500 rpm. It is required to scale up the cell to commercial size with a 200-mm-diameter rotating cylindrical electrode using the same electrolyte and operating conditions such as current density, temperature, reactant concentration, pH, etc.

The rotating electrode cell employs a working electrode which is rotated about its axis. Because the electrolyte is stirred in the vicinity of the electrode–electrolyte interface, convective mass transfer occurs, for which the following correlation can be employed:

$$Sh = 0.079 Re^{0.7} Sc^{0.356} \tag{E21.2.1}$$

The influence of the change in diameter on peripheral velocity or rpm of the rotating electrode can be computed directly by mathematical simplification of Equation E21.2.1.

FOR CELL 1

$$\frac{k'_L d_1}{D} = 0.079 \left(\frac{d_1 u_1 \rho}{\mu}\right)^{0.7} (Sc)^{0.356} \tag{E21.2.2}$$

FOR CELL 2

$$\frac{k'_L d_2}{D} = 0.079 \left(\frac{d_2 u_2 \rho}{\mu}\right)^{0.7} (Sc)^{0.356} \tag{E21.2.3}$$

Combining Equations E21.2.2 and E21.2.3,

$$\frac{u_2}{u_1} = \left[\frac{d_2}{d_1}\right]^{0.429} \tag{E21.2.4}$$

where 1 and 2 denote values for the laboratory and scaled-up cells, respectively, and u is the linear velocity of the rotating electrode ($\pi d_{rot.elect.} N$ where N is the rotation rate per second).

By substituting the relevant values, we get $u_2 = 4.78$ m/s, which corresponds to a rotation rate of 457 rpm for the larger cell. Hence the correlation given by Equation (E21.2.4) is useful for computing design data for the larger cell. However, in employing these correlations, suitable modifications should be made if there is a variation in hydrodynamics and the geometric factors of the cells.

INDUSTRIAL ELECTROCHEMICAL CELL COMPONENTS
AND CONFIGURATIONS

Cell Components

In choosing the reactor configuration for an electrosynthetic process, decisions will have to be made regarding the selection of a suitable cell system and its

principal components, electrolyte, electrodes, and cell separators. These have been extensively described and will not be discussed here, except for two major developments, one with respect to electrodes and the other cell separators.

1. Most metals are stable in operation as cathodes but unstable as anodes. Satisfactory stability is at least 1,000 hours of continuous operation and preferably 10,000 hours for commercial operations. Of the many categories of electrodes developed, four merit special attention for use in electrosynthesis in different media: *chemically modified electrodes* (CMEs), *dimensionally stable anodes* (DSAs), *conducting polymers*, and catalytic cathodes (especially of Raney nickel).
2. Common diaphragm materials used in commercial electrolytic cells are porous ceramics, asbestos, and microporous plastics, but they all suffer from poor mechanical strength and availability in limited sizes and shapes. It was only after the availability of the highly stable *perfluorinated ion-exchange membranes* that the problems of cell separators for industrial production were solved. These ion-exchange membranes allow only cation or anion (with water) transport between electrode chambers.

Types of Electrochemical Reactors

The choice of an optimum cell design is critical to the commercial success of an electrochemical process. Several review articles have been published on industrial cell designs that have been developed and are in use (Tomilov and Fioshin, 1971; Goodridge and King, 1982; Danly, 1984; Goodridge et al., 1986; Scott, 1991; Goodridge and Scott, 1995). Factors such as the desired productivity, adequate mass transfer rates, and reaction environment will have to be considered in the choice of a suitable cell design. A few typical reactor types are shown in Figure 21.10 (see Goodridge and Scott, 1995, for concise descriptions).

SELECTED EXAMPLES OF INDUSTRIAL ELECTROORGANIC SYNTHESIS

Clearly, electroorganic synthesis is a methodology for carrying out a wide range of organic reactions. It encompasses convential reactions such as oxidation or reduction of suitable functional groups and also a wide range of reactions such as anodic or cathodic substitution, addition, coupling, bond cleavage, etc. It is also an excellent method for generating intermediates such as anion radicals, dianions, cation radicals, and dications which subsequently engage in radical, electron transfer, nucleophilic, or acid-base reactions. Some of the difficult/selective organic reactions that have been carried out electrochemically have already been listed in Table 21.1.

Though hundreds of syntheses have been reported on a laboratory scale (see Swann and Alkire, 1980; Little and Weinberg, 1991), only a small fraction of them have been commercialized, and some are still in the development stage. Several R & D organizations in both the private and public sectors in the USA,

(a). A general purpose cell (ICI)-section.

(b). Capillary gap cell (BASF).

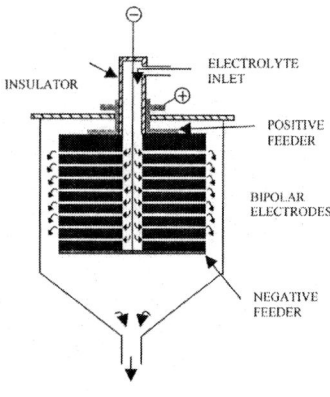

(c). Rotating cylindrical cell.

(d). A typical electrochemical reactor for gas-liquid reactions.

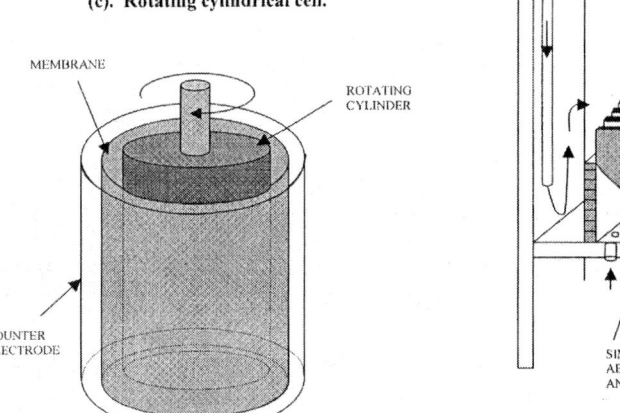

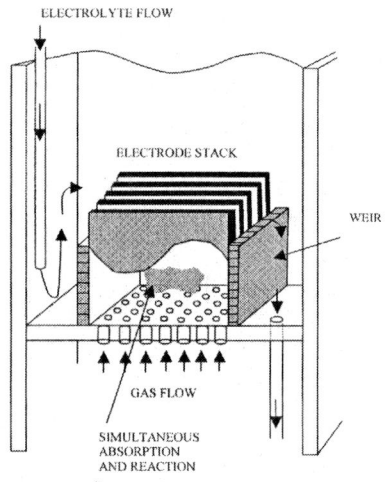

Figure 21.10 Some representative electrochemical reactors (redrawn from Goodridge and Scott, 1995)

Germany, UK, India, Japan, and Russia have been working on the development of electroorganic processes and have also successfully commercialized a few of them. Table 21.3 provides a partial list of reactions that have been developed on bench, pilot, and commercial scales in different countries. More recent laboratory-scale studies on electrochemical synthesis of certain fine chemicals and natural products are also included in the table.

Table 21.3 A representative list of electroorganic syntheses that have been commercialized or are under development[a]

No.	Reaction	Status	Developer	Reference
1	Nitrobenzene to p-aminophenol	P/C	Bayer, Mills Lab. CECRI	Jayaraman et al. (1977)
2	Furan to 2,5-dimethoxydihydrofuran	C	BASF	Krishnan et al. (1979); Degner (1982)
3	Acrylonitrile to adiponitrile	C	Monsanto, BASF, Asahi, UCB	Danly (1981); Danly and Campbell (1982)
4	Formaldehyde to ethylene glycol	B/P	Electrosynthesis Co.	Weinberg et al. (1982)
5	Acetone to pinacol	P	BASF, Bayer	
5	Benzene to hydroquinone	P	Tennessee Eastman	Feess and Wendt (1982)
6	Diacetone sorbose to diketogluconic acid	C	Hoffmann-LaRoche	Seiler and Robertson (1982)
7	Reduction of phthalic acid to dihydrophthalic acid at cathode and oxidation of acetylenic alcohol to the acid at anode	L	Germany	Degner (1982)
8	Oxidation of organic compounds such as alcohols to aldehydes and acids at nickel hydroxide electrode	L	Germany	Kaulen and Schafer (1982)
9	2-Methylindole to 2-methylindoline	C	BASF	Beck et al. (1985)
10	p-Methoxy benzaldehyde to p-methoxy benzyl alcohol	C	BASF	Beck et al. (1985)
11	Anthranilic acid to o-aminobenzyl alcohol	P/C	BASF	Beck et al. (1985)
12	Nitrobenzene to p-anisidene	P/C	BASF	Beck et al. (1985)
13	Furfuryl alcohol to maltol	C	Otsuka	Beck et al. (1985)
14	Butene to methyl ethyl ketone	P	Exxon	Beck et al. (1985)
15	Octanoyl chloride to perfluoro-octanoic acid	C	3M, Dai Nippon, Asahi Glass, Bayer	Drakesmith and Hughes (1986) (see also Simons, 1950)
16	Glucose to gluconic acid/calcium gluconate	C	CECRI, Sandoz	Subbiah et al. (1988)
17	Electrosynthesis of trifluoromethylbenzaldehyde (a pesticide intermediate) from p-chlorotrifluoromethyl benzene	L	—	Saboureau (1989)
18	Synthesis of disilyl compound, an intermediate in the synthesis of Profenid, an anti-inflammatory drug	L	—	Biran and Bourdeau (1993)

(Continued)

Table 21.3 (Continued)

No.	Reaction	Status	Developer	Reference
19	Electrochemical addition of carboanions of nitrocompounds to levoglucosenone, a compound with a chiral carbohydrate skeleton	L	—	Laikhter (1993)
20	Thiocyanation of aminobenzene compounds	L	Switzerland	Sandoz (1996)
21	Reduction of 1,3-disubstituted 4,6-dinitrobenzene 4,6-diaminoresorcinol and its salts	L	Japan	Nissan Chemicals (1996)
22	Synthesis of thioethers for pharmaceutical use	L	Spain	Spanish Pat. (1996)
23	Preparation of tetraalkyl 1,2,3,4-butane tetracarboxylates	L	USA	Bagley et al. (1997)
24	Derivatives of 2,7-diaminosuberic acid and 2,4-diaminoadipic acid by Kolbe reaction	L	—	Biebl et al. (1998)
25	Indirect methoxylation with a chlorine mediator for the preparation of chiral cationic glycine equivalents	L	—	Kardassis et al. (1998)
26	Electrocarboxylation reactions with sacrificial Al or Mg anodes (a step in the production of naproxen)	B/P	SNPE France	—

[a] B = bench scale; C = commercial; L = laboratory scale; P = pilot plant scale; CECRI = Central Electrochemical Research Institute, India. It is not clear whether the processes developed on a pilot plant scale were subsequently commercialized or dropped.

Chapter 22

Sonoorganic Synthesis Engineering

Ultrasonics or *ultrasound* refers to sound waves beyond the audible range of the human ear. The normal human hearing range is 16–16,000 cycles per second. The accepted terminology for one cycle per second is the Hertz (or Hz), and hence the hearing range is expressed as 16 Hz to 16 kHz. Ultrasound is normally considered to lie approximately in the range of 15 kHz to 10 MHz, that is, 15×10^3 to 10000×10^3 cycles per second, with acoustic wavelengths of 10 to 0.01 cm. Like any sound wave, ultrasound is propagated through a medium in alternating cycles of *compression* and stretching or *rarefaction*. These produce certain effects in the medium that can be usefully exploited. One such application is in the field of synthetic organic chemistry, first reported by Richards and Loomis (1927) and designated *sonochemistry*. The most appealing feature of sonochemistry is its ability to enhance reaction rates, often to remarkably high levels under environmentally benign conditions. Despite this potential, economic considerations have precluded the use of sonochemical processes. It is noteworthy, however, that a change in perspective appears to be emerging, as evidenced by the fact that a pilot plant is currently being funded by a French company to sonochemically oxidize cyclohexanol to cyclohexanone, and developmental work is underway in Germany to produce 4 tons of Grignard reagent per year (Ondrey et al., 1996).

A number of books and reviews covering mostly the chemical aspects of sonochemistry have appeared over the years, for example, Suslick, 1988, 1989, 1990a,b; Ley and Low, 1989; Mason, 1986, 1990a,b, 1991; Mason and Lorimer, 1989; Price, 1992; Bremner, 1994; Low, 1995; Luche, 1998. A recent review by Thompson and Doraiswamy (1999) covers both the chemical and engineering aspects of sonochemistry and another by Keil and Swamy (1999) examines the present state of our understanding of sonoreactor design.

Sonochemical enhancement of reaction rates is caused by a phenomenon called *cavitation*. Therefore, we largely confine the treatment in this chapter to the chemical and reaction engineering (scale-up) aspects of cavitation and its associated effects (see Shah et al., 1999, for a detailed treatment). An alternative means of achieving the same result is by mimicking the ultrasonic effect by inducing "hydrodynamic cavitation." Because of the practical importance of this technique, we conclude the chapter by outlining its main features.

DEVICES FOR PRODUCING ULTRASOUND

Ultrasonic energy is produced in a medium by using transducers which are devices for converting one form of energy into another, in this case electrical or mechanical energy into sound energy. Transducers can be (1) fluid-driven or (2) electromechanical. In the first category, high-velocity gas or liquid is forced across a vibrating plate, thus sonicating the fluid. Liquid-driven transducers are often referred to as whistles. The second category, however, is the more commonly used. Electromechanical (or *electrosonic*) devices convert electrical to sound energy by utilizing the inverse of the piezoelectric effect observed in certain crystals. The piezoelectric effect is the potential difference produced across the opposite faces of a crystal when it is subjected to sudden compression. The inverse phenomenon is the effect produced when a rapidly alternating potential is applied across the faces of a piezoelectric crystal. This effect induces dimensional changes in the crystal resulting in the production of vibrational (sound) energy. The chemical effects of ultrasound are usually realized in the frequency range of 20–100 kHz and can be produced in three ways:

1. By directly immersing a vibrating metal plate or a horn into the reaction medium (direct sonication)
2. By placing the reaction vessel in a tank containing a coupling fluid (usually water) which is sonicated either by a metal plate or a horn (indirect sonication); ultrasonic waves travel through this fluid before contacting the reaction vessel
3. By in situ generation of ultrasonic power in the reacting liquid by forcing it across a vibrating blade (the whistle referred to earlier).

Several variations of these are possible, particularly for large-scale operation. Figure 22.1 shows schematic sketches of different types of devices used to generate ultrasound (Berlan and Mason, 1992). By using any of these transducer designs, it is possible to introduce *power ultrasound* (as it is normally called) into the reacting system. A reaction conducted under the influence of ultrasound is usually represented as

$$A + B \xrightarrow{)))} R + S \qquad [22.1]$$

A number of sonicators are commercially available, for example, the Suslick sonicator and a design recently marketed by the Belgian company Undatim. Figure 22.2 is a sketch of a typical setup for conducting a sonochemical reaction.

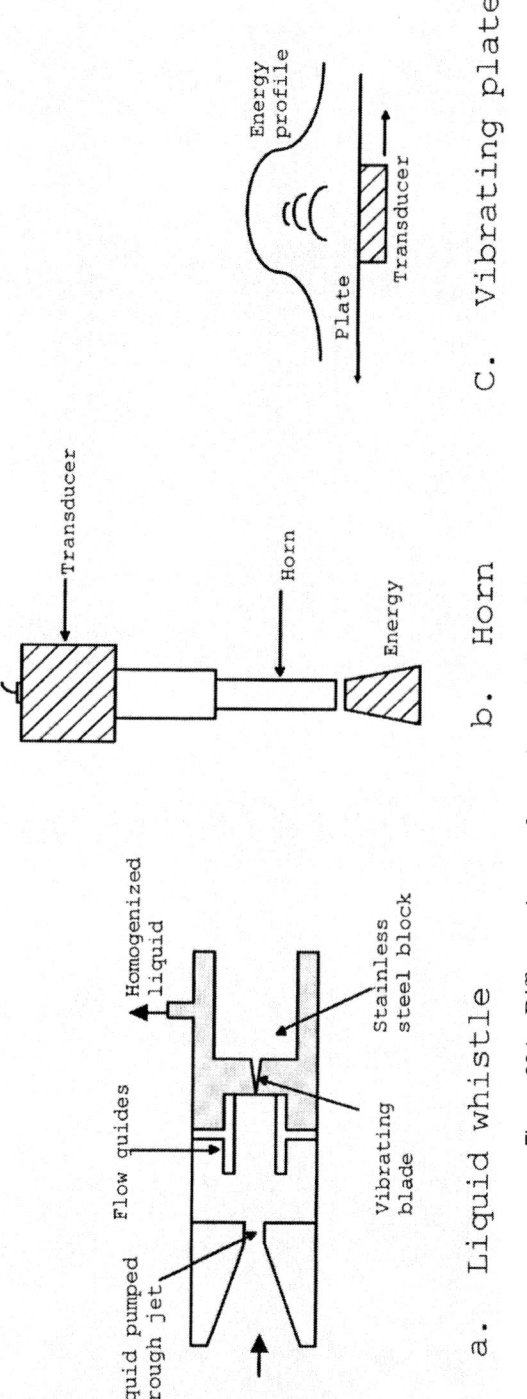

Figure 22.1 Different classes of transducers (adapted from Berlan and Mason, 1992)

714 Strategies for Enhancing Rate Reactions

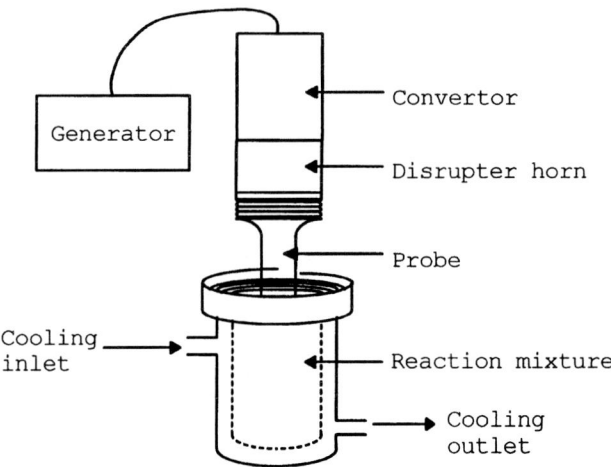

Figure 22.2 A typical setup for conducting a sonochemical reaction

Sonochemistry has developed essentially around reactions that can be carried out in a liquid medium. This medium can be a homogeneous liquid phase or a heterogeneous medium in which at least one phase is liquid—to serve as the vehicle for transmitting ultrasonic power. We shall consider the effects of ultrasound in both systems.

HOMOGENEOUS REACTIONS

In this section, we cover the important basic aspects of sonochemistry, including such features as cavitation and the roles of frequency, pressure, solvent, and dissolved gases.

Ultrasound: Some Basic Considerations

Sound is transmitted through a medium in the form of a sine wave characterized by an acoustic pressure P_a (also called the applied pressure)

$$P_a = P_A \sin \omega t \tag{22.1}$$

where

$$\omega = 2\pi f \tag{22.2}$$

P_A is the amplitude of the ultrasound wave, f is the frequency of the wave, and ω is the angular frequency.

An important parameter in sonochemistry is the pressure P_∞ away from the sonication zone on the downstream side, which may be expressed as the difference between the hydrostatic (or barometric) pressure P_h that already exists in the system and the acoustic pressure given by Equation 22.1,

$$P_\infty = P_h - P_A \sin \omega t = P_h - P_a \tag{22.3}$$

Another characteristic property of the wave is the intensity I, the energy transmitted per unit time per unit area of the fluid, defined as

$$I = \frac{P_A^2}{2\rho_L u_s} \tag{22.4}$$

where ρ_L is the density of the medium (liquid) and u_s is the velocity of sound in that medium. This energy, however, does not remain constant as the wave travels through the medium but is attenuated due to frictional and thermal losses. The intensity at a distance ℓ may be expressed in terms of the initial (undamped) intensity I_0 as

$$I = I_0 \exp(-2\alpha\ell) \tag{22.5}$$

where α is the *absorption* or *attenuation constant*, whose magnitude is determined by the nature of the medium and the frequency of the wave.

Cavitation: Its Physical Basis and Characterization

The dimensions associated with ultrasound are not on the molecular scale. Thus the chemical effects of ultrasound cannot be attributed to any direct interaction of the acoustic wave with chemical species on a molecular level. Indeed, it is well known now that these effects are the result of the physical processes associated with ultrasonic waves. The most important of these is cavitation.

Sound waves passing through a medium in accordance with Equation 22.1 alternately compress and stretch the molecular structure of the medium. In fact, during the rarefaction cycle, the (negative) stretching force is so strong that the molecules are pulled apart beyond the critical distance separating them. This leads to the creation of voids or microbubbles between molecules, referred to as *cavities*. In the absence of any gas or vapor in the medium, these cavities would enclose a vacuum. In reality, however, there is considerable vaporization of the liquid, and in addition there is often a localized inclusion of gas (dissolved in the liquid). Thus the microbubbles (or cavitation bubbles) usually enclose gas or a mixture of gas and vapor. The behavior of these bubbles depends on whether they are *transient* or *stationary* bubbles:

1. Transient bubbles are small bubbles whose lifetimes are 10^{-8}–10^{-9} s (Vogel et al., 1989), are produced at ultrasonic intensities higher than 10 W/cm^2, and collapse within one acoustic cycle. Very high temperatures (up to 5000 K) and pressures (in excess of 500 atmospheres) are generated during such collapse, resulting in the fracture of the chemical species inside and the formation of free radicals. The nature of a typical radius–time profile of a transient bubble is shown in Figure 22.3.
2. Stable bubbles are relatively large, are produced at intensities less than 3 W/cm^2, and oscillate for several acoustic cycles, almost in unison with the acoustic wave, before collapsing, or never collapse at all. Such cavitation can also lead to chemical reaction via free radical formation.

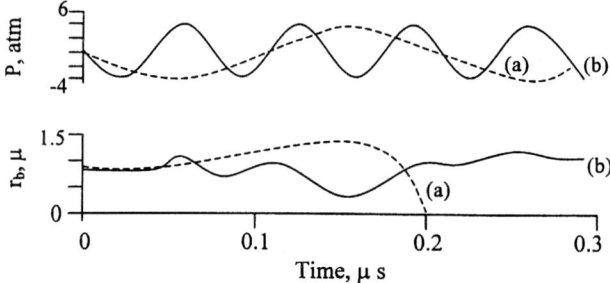

Figure 22.3 Radius–time and pressure–time profiles of a transient cavity (or microbubble) (adapted from Lorimer and Mason, 1991)

Therefore, sonochemistry can be described as the result of acoustic cavitation consisting of three events: creation, expansion, and implosive collapse of bubbles in ultrasonically irradiated liquids (Apfel, 1981; also see Suslick, 1986). The microbubble, or cavitational, field is characterized by spherical bubbles partially filled with noncondensable gases. The radius of the bubble r_b at any time deviates from its initial value r_{b0} in a periodic fashion. This dynamic behavior of the bubble is well described by the Raleigh–Plesset equation given by

$$r_b \frac{d^2 r_b}{dt^2} + \frac{3}{2}\left(\frac{dr_b}{dt}\right)^2 = \frac{1}{\rho}\left[\left(P_0 + \frac{2\sigma}{r_{b0}} - P_v\right)\left(\frac{r_{b0}}{r_b}\right)^{3\gamma} - \frac{2\sigma}{r_b} - \frac{4\mu}{r_b}\left(\frac{dr_b}{dt}\right)\right.$$
$$\left. - (P_0 - P_\infty)\right] \qquad (22.6)$$

where dr_b/dt and d^2r_b/dt^2 are the velocity and acceleration of bubble wall motion, respectively; and P_0 is the pressure in the bulk liquid in the absence of ultrasound (often equal to the hydrostatic pressure P_h). The use of this equation is illustrated in Example 22.2.

Effect of Frequency

The cavitational effect tends to be lower at higher frequencies (Mason, 1991) because increased frequencies result in shortened rarefaction half-periods, leading to a reduction in the extent of cavitation. Then the power or amplitude of irradiation will have to be increased considerably to maintain cavitation. This is illustrated in Figure 22.4. Thus lower frequencies in the range of 20–30 kHz are normally selected for sonochemical applications.

Some contradictory results have also been reported, however. For instance, in the decomposition of carbon disulfide (Entezari et al., 1997), frequency has no apparent effect, whereas in reactions such as oxidation higher frequencies may lead to enhanced reaction rates (Entezari and Kruus, 1994). The latter is probably due to the high rate of hydroxyl radical formation at higher frequencies (Petrier et al., 1992; Entezari and Kruus, 1996; Mason et al., 1994).

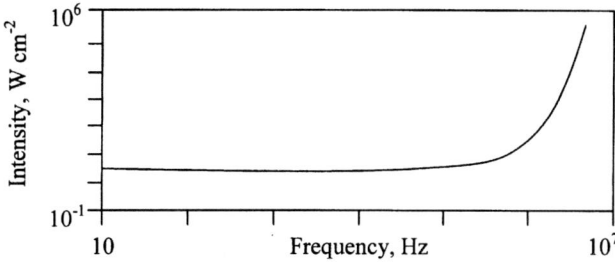

Figure 22.4 Relationship between ultrasonic frequency and power input (adapted from Mason, 1990b)

Effect of Ambient Pressure

It can be shown that cavitation probably does not occur at the equilibrium bubble size at atmospheric pressure. However, the bubble can be driven to cavitation by reducing the pressure. Conversely, higher pressures require greater ultrasonic energy to induce cavitation. This leads to a larger intensity of cavitational collapse and therefore to a considerably enhanced ultrasonic effect, as demonstrated in the oxidation of indane to indane-1-one using potassium permanganate (Cum et al., 1988).

However, if the pressure is too high, cavitation may not occur at all, as observed in the hydrogenation of soybean oil (Moulton et al., 1983; see also Moulton et al., 1987). Thus, for any given system, an optimum operating pressure most likely exists (Berlan, 1994).

Effect of Ambient Temperature

When a reacting medium is irradiated, part of the sonochemical energy is dissipated as heat which raises the temperature of the medium as a whole. Although this rise in temperature is usually not more than 10–15 °C, in some cases it can be considerably higher (40–50 °C), particularly when sonicating small volumes. Therefore, it is necessary to thermostat the sonicator by removing the heat, preferably by circulating a liquid or using an ice bath. This rather mild temperature rise should be distinguished from the dramatic rise at cavitation centers due to the implosive collapse of bubbles.

Increasing the ambient reaction temperature invariably increases the rates of conventional reactions. However, when using ultrasound, this is often not the case. Consider a liquid medium in which a sonochemical reaction is occurring at a temperature not far from its boiling point. In the presence of an ultrasonic wave during its rarefaction cycle, there will be a considerable reduction in pressure. This will tend to bring the liquid to its boiling point at that pressure. As a result, any cavitational bubbles formed during the rarefaction cycle will immediately suck in the solvent vapor. When these bubbles collapse during the subsequent compression cycle, they will tend to implode less violently because the vapor is more dense and effectively absorbs some of the impact. This results

in much lower temperatures and pressures than associated with implosions when the solvent is not at its boiling point. It may also happen (Mason, 1990a, 1991) that the bubbles will not collapse at all due to excessive cushioning by the enclosed vapor.

Our concern is primarily with the transient bubble because it is such a bubble that produces the highest temperature. The maximum temperature T_{max} and pressure P_{max} of a transient bubble at its moment of collapse, assuming that it contains an ideal gas and no vapor and neglecting the effect of surface tension and fluid viscosity, are given by

$$T_{max} = \frac{T_0 P_m (\gamma - 1)}{P_n} \tag{22.7}$$

$$P_{max} = P_n \left[\frac{P_m (\gamma - 1)}{P_n} \right]^{\gamma/\gamma - 1} \tag{22.8}$$

where T_0 is the bulk temperature of the system, P_m is the pressure in the liquid at the moment of collapse

$$P_m = P_h + P_a \approx P_h + P_A \tag{22.9}$$

and P_n is the pressure in the bubble at the moment of collapse, that is, $P_n = P_v + P_G \approx P_v$. Note that the pressure P_G within the gas can be neglected as a result of the assumption that the transient bubble has grown so fast that there is practically no influx of gas during the short period of its existence; hence P is approximated as the vapor pressure of the liquid P_v.

The process of cavitation has since been the subject of several intensive studies, for example, Cum et al., 1990, 1992; Atchley et al., 1988; Alippi et al., 1992; Leighton, 1995. One of the theories is that cavitation is the result of events that occur within a cloud of bubbles and not just within a single bubble (as assumed in the above equations), but reaction engineering approaches taken so far have been based on the classical single-bubble theories.

Now to determine the effect of solvent vapor pressure (or, equivalently, that of temperature), we write the Arrhenius equation for the two widely different temperatures T_0 and T_{max}:

$$\ln k = \ln k^0 - \frac{E}{R_g T_0} \tag{22.10}$$

$$\ln k_{et} = \ln k_{et}^0 - \frac{E}{R_g T_{max}} \tag{22.11}$$

On substituting Equation 22.7 for T_{max}, the elevated constant becomes

$$\ln k_{et} = \ln k_{et}^0 - \frac{E}{R_g (\gamma - 1) P_m} \left(\frac{P_v}{T_0} \right) \tag{22.12}$$

A plot of Equation 22.12 as rate versus P_v for the bleaching of 2,2-diphenyl-1-picrylhydrazyl (DPPH) in a number of solvents appears in Figure 22.5 (Suslick et al., 1983a). The linearity of the plot validates Equation 22.12 and also brings

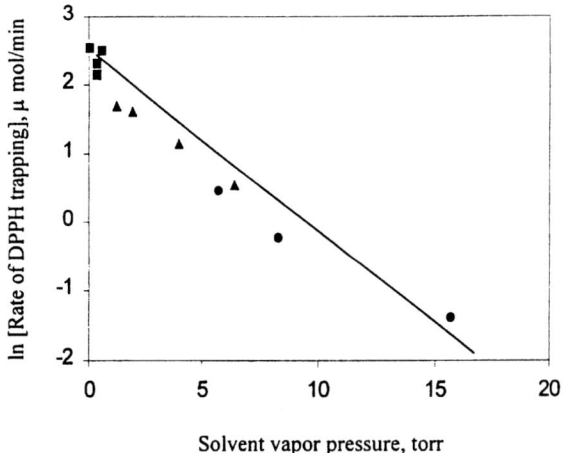

Figure 22.5 Plot of ln (rate) vs. P_v for 2,2-diphenyl-1-picrylhydrazyl (DPPH) (redrawn from Suslick et al., 1983a)

out a more fundamental point. *The reaction rate increases with a decrease in solvent partial pressure*, that is, with a decrease in temperature. This is contrary to normal behavior where the effect of increased temperature is to increase the reaction rate (Figure 22.6). Lorimer et al. (1991) modified Equation 22.12 to give the following expression for the ratio of the sonicated and "silent" reaction rates:

$$\ln\left(\frac{k_{us}}{k} - 1\right) = \ln F + \frac{E}{R_g T_0} - \frac{E}{R_g P_m (\gamma - 1)} \frac{P_v}{T} \qquad (22.13)$$

where the terms k_{us} and k are the rate constants for reactions with and without ultrasound, respectively; and F accounts for the volume of the reaction mixture and the time of its exposure to the extreme temperature of the cavitational event. Equation 22.13 has been successfully used, in a rearranged form, to describe the

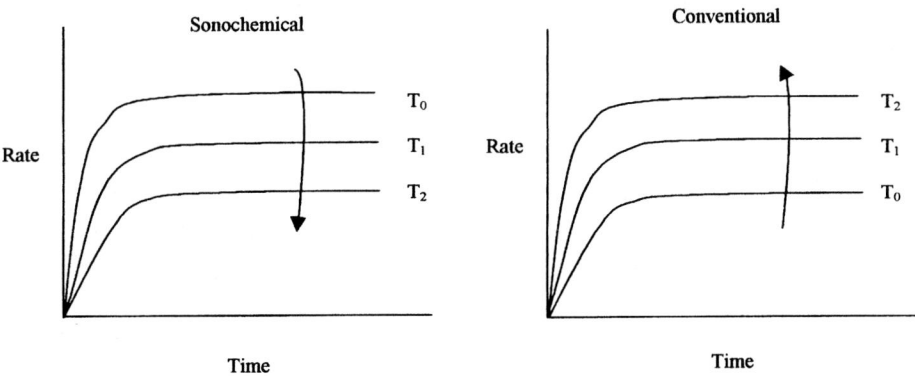

Figure 22.6 Plots to illustrate the unusual effect of temperature on a sonochemical reaction

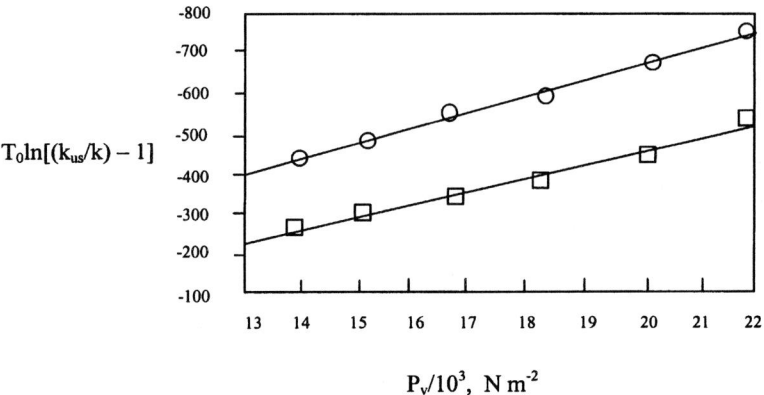

Figure 22.7 Plot of $T_0 \ln\left(\frac{k_{us}}{k} - 1\right)$ vs. P_v for $K_2S_2O_8$ (redrawn from Lorimer et al., 1991)

behavior of the sonochemically enhanced decomposition of aqueous $K_2S_2O_8$ as shown in Figure 22.7 (see Lorimer et al., 1991, for further details).

Therefore, it is desirable to conduct a sonochemical reaction at as low a temperature as practically feasible when the solvent vapor pressure is dominant, consistent with a reasonable overall reaction time. A simple means of achieving this is to use solvents that have low vapor pressure.

Effect of Acoustic Intensity

Acoustic intensity has a profound effect on the rates of sonochemical reactions. Below a certain value, the amplitude of the wave would be too small to cause nucleation or bubble growth. As the intensity is increased beyond that required for incipient cavitation, the effective size of the liquid zone undergoing cavitation increases and with it the sonochemical rate (Mason, 1990b). This increase is not sustained, however, because amplitudes beyond a certain point lead to excessive cavitation. As a result, there is a reduction in the penetration of the wave into the liquid and therefore of any sonochemical enhancement.

Effect of Solvent

We have already seen how an increase in solvent vapor pressure can lead to a seemingly paradoxical reduction of the sonochemical rate (Figure 22.5). Thus it is necessary to choose a solvent whose vapor pressure is neither too low nor too high.

If cavitation occurs at a low intensity, the effectiveness of cavitation will be much reduced. The intramolecular forces within a solvent determine the level of sonication that will drive a bubble to cavitation. Thus it is necessary to choose a solvent that has strong structural cohesion, such as through hydrogen bonding, and high surface tension. Water is clearly the best candidate from this point of

view. Inertness is another major consideration. In general, a compromise in properties has to be accepted. For example, one might have to settle for a hydrocarbon such as xylene instead of water, even though it cavitates at a much lower intensity of sonication.

Effect of Dissolved Gases

We have already stated that usually gases other than the solvent vapor are also present. When a cavitational bubble is initiated during the rarefaction cycle because of the sudden release of pressure, any dissolved gas in the liquid will be "sucked" out. Then it forms the nucleus of a cavitational bubble. Even a gas that is highly soluble is subject to this sucking action because of the very high negative pressures generated during rarefaction. Equations 22.7 and 22.8 show that the maximum temperature and pressure generated during cavitational collapse are strong functions of the polytropic ratio $\gamma = C_p/C_v$ in the ambient gas. This is to be expected because collapse occurs in the short time of approximately 3.5 μs (Prasad Naidu et al., 1994) and can therefore be assumed to be adiabatic. The polytropic ratio is a measure of the heat released during the adiabatic compression of the gas and is responsible for raising the temperature. The dramatic nature of the polytropic effect may be illustrated by considering cavitation in the presence of a monotonic gas and a typical Freon. The polytropic ratios of the two are 1.66 and 1.1, respectively, and from Equation 22.7 the maximum temperature for the monotonic gas is about seven times that for Freon. Sometimes, however, the effect of solubility outweighs that of the polytropic ratio as in carbon disulfide decomposition. The reaction rate in the presence of He is higher than in the presence of Ar, even though $\gamma_{Ar} > \gamma_{He}$ because of the greater solubility of helium than that of argon (Entezari et al., 1997). When hydroxyl radicals are generated in the reaction, oxygen accelerates the reaction more than a polytropic gas such as argon (see Hart and Henglein, 1985; Berlan, 1994).

In general, however, best results are obtained when a monotonic gas, usually argon, is continuously bubbled through the irradiated medium. This helps to counteract the degassing of the system by the passage of ultrasound.

Miscellaneous Effects (Mainly of Viscosity)

In addition to the important factors previously considered which influence sonochemical reactions, a few others need to be considered as well. Among these are surface tension, viscosity, and solubility. The viscosity of a liquid increases as the pressure is increased or the temperature is decreased. Solvents with higher viscosity require higher amplitudes (or power) for cavitation to occur. In other words, cavitation becomes difficult to induce in high viscosity liquids. This is a situation that enhances the ultrasonic effect, and hence higher viscosities should normally lead to greater rate enhancements.

Both vapor and noncondensable gases can cushion an implosion. Gases that are extremely soluble may redissolve before cavitation is initiated or the bubble size may become so large (because of easy penetration of the gas into it) that the

bubbles may float to the surface. Hence there is considerable uncertainty about the cushioning effect of highly soluble gases.

HETEROGENEOUS REACTIONS

Rates of heterogeneous reactions, such as those between two immiscible liquids, between a gas and a liquid, or between a solid and a liquid, are also enhanced by using ultrasound. Slurry reactions in which gas, liquid, and solid are all present are also known to benefit remarkably by sonication. Indeed, considering the involvement of at least one mass transfer resistance in all these reactions, the ultrasonic effect is even more dramatic than in homogeneous reactions because of its enhancing effect on mass transfer rates. Once again, cavitation is the reason for intrinsic rate enhancements. However, implosive collapse can also occur by asymmetrical cavitation (in addition to the usual symmetrical cavitation) in the presence of a solid-liquid surface. In the field of organometallic chemistry, in particular, where a solid phase is present, the use of sonication has been most extensive (Luche, 1987; Suslick, 1988; Ley and Low, 1989; Mason et al., 1990a; Price, 1992).

Solid-Liquid Reactions

Where solid particles are used, they can either be consumed in the reaction or function as a catalyst. Enhancement by ultrasound can be due to several reasons.

1. *The cleaning action of ultrasound*: A common example is the use of magnesium in the preparation of a Grignard reagent (Tuulmets et al., 1995). Sonication plays an important role in that it continually exposes fresh surface to the reaction. In addition, it causes pitting of the surface, thus exposing new surface to the reaction.
2. *The sweeping action of ultrasound*: In many cases, ultrasound also sweeps away from the vicinity of the surface both the product and any intermediates that may be formed.
3. *The size reduction action of ultrasound*: In reactions involving dispersion of a solid in a liquid, the use of ultrasound leads to fragmentation of the solid with consequent increase in surface area.

The most pertinent effects of ultrasound in solid-liquid reactions are mechanical, which are attributed to *symmetrical* and/or *asymmetrical cavitation*. Symmetrical cavitation (the type encountered in homogeneous systems) leads to localized areas of high temperatures and pressures and also to shock waves that can create microscopic turbulence (Elder, 1959). As a result, mass transfer rates are considerably enhanced. For example, Hagenson and Doraiswamy (1998) observed a twofold increase in the intrinsic mass transfer coefficient in the reaction between benzyl chloride (liquid) and sodium sulfide (solid). In addition, a decrease in particle size and therefore an increase in the interfacial surface area appears to be a common feature of ultrasound-assisted solid-liquid reactions (Suslick et al., 1987; Ratoarinoro et al., 1992, 1995; Hagenson and Doraiswamy, 1998).

In any reaction involving a solid reactant, the solid surface may be several orders of magnitude larger than that of the cavitating bubble (see Suslick, 1990b). When this occurs, cavitation is induced, but now it occurs asymmetrically on the surface (Neppiras, 1980; Lauterborn and Vogel, 1984; Crum, 1995;). In such a situation, microjets of solvent perpendicular to the solid surface are formed as the bubble collapses and impinge on the surface. This leads to the well-known cleaning affects of ultrasound which can enhance the rates of certain reactions, for example, the synthesis of organosilicon hydrides using magnesium hydride particles in a liquid medium (Klein et al., 1995), in which surface inhibiting hydrides are removed from the particles, thus increasing their reactivity. When the solid is a metallic powder, the situation gets complicated due to the tendency of the degraded metal particles to agglomerate (and melt) with prolonged sonication, leading to localized melting (Doktycz and Suslick, 1990) (a conclusion that has been contested by Margulis, 1992). On the other hand, the catalytic activity of powders such as those of Ni (so commonly used in the hydrogenation of organic compounds) is considerably enhanced by sonication as a result of the removal of inhibitory coatings (Suslick and Casadonte, 1987).

Liquid-Liquid Reactions

In liquid-liquid reactions, one liquid is dispersed in the other. Under the influence of ultrasound, mixing becomes so severe that the dispersion soon turns into an emulsion. This leads to an enormous rise in the interfacial area of contact and therefore in the reaction rate constant. It is not clear whether it also enhances the specific rate coefficient in liquid-liquid reactions.

Modeling of Heterogeneous Reactions

Reaction zones

Spin trapping and other studies of volatile and nonvolatile solutes have established the existence of three zones of sonochemical activity, as shown in Figure 22.8. Estimates of the relative sizes of the three zones have been made by Suslick

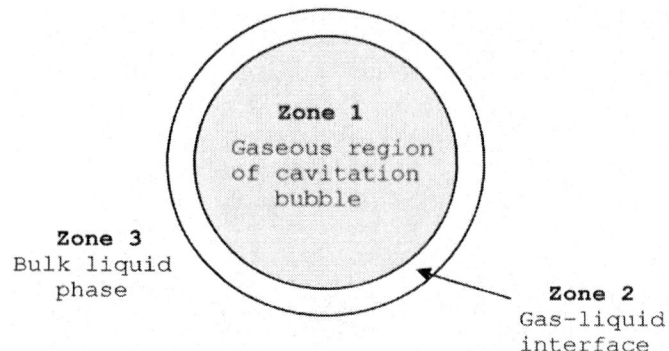

Figure 22.8 Three zones of sonochemical activity

and Hammerton (1986). Zone 1 corresponds to the inside of the cavitating bubble characterized by extremely high temperatures and pressures, leading to breakage of molecules (Suslick, 1990a; Misik and Riesz, 1996a,b). Reactions of nonvolatile solutes or those with low volatility occur in zone 2 (interfacial region), for example, degradation of thymine (Sehgal and Wang, 1981; Kondo et al., 1988). Investigation of the sonolysis of several organic compounds of low volatility has shown that, if the reactant is not volatile enough to enter zone 2, sonochemical enhancement does not occur (Griffing, 1952; Misik and Riesz, 1996b; Ando et al., 1996). Reaction can occur in zone 3 only if the radicals involved escape the implosion. When that happens, the kinetics is similar to that of radiation chemistry (Kondo et al., 1988; Fuchs and Heusinger, 1995).

Kinetics

Studies on ligand substitution of metal carbonyls (useful in homogeneous catalysis) (see Suslick et al., 1983b; Suslick and Hammerton, 1986) and primary radical formation from *n*-alcohols (Misik and Riesz, 1996a) suggest that the rate constant decreases linearly with reactant vapor pressure according to Equation 22.12.

However, in these studies, the total pressure was allowed to increase with an increase in vapor pressure. When the experiments were modified to keep the total pressure constant (Suslick and Hammerton, 1986), it was found that the observed rate constant increased linearly with reactant vapor pressure—and additionally produced a nonzero intercept. From this it was concluded that the linear dependence indicated reaction in zone 1 and the intercept suggested reaction in zone 2 (see Suslick, 1990b). Based on these studies, as well as those of Misik et al. (1995), it seems likely that zones 1 and 2 become indistinguishable from one another during and immediately following bubble collapse. This would result in a system that is comparable to a point source at a very high temperature and pressure, with steep temperature gradients to the bulk liquid at ambient temperature and pressure, as shown in Figure 22.9 (Thompson and Doraiswamy, 2000).

A practically useful approach is to compare the rate constants of a reaction with and without ultrasound. Such studies (Lorimer et al., 1991) show that the rate constant with ultrasound k_{us} is not equal to the sum of the rate constants without ultrasound k and the rate constant associated with cavitational collapse k_{bub}, that is,

$$k_{us} \neq k + k_{bub} \tag{22.14}$$

This absence of additivity (at a given temperature) results from the much higher temperature associated with k_{bub} for bubble collapse. Allowance can be made for this by introducing two more parameters, F_V for the fraction of reaction mixture volume occupied by bubbles and F_t for the cycle time (compression phase) during which reaction occurs at the elevated temperature of the cavitational zone. Using these parameters, the following expression can be developed:

$$\left(\frac{k_{us}}{k} - 1\right) = F_V F_t \left(\frac{k_{et}}{k} - 1\right) \tag{22.15}$$

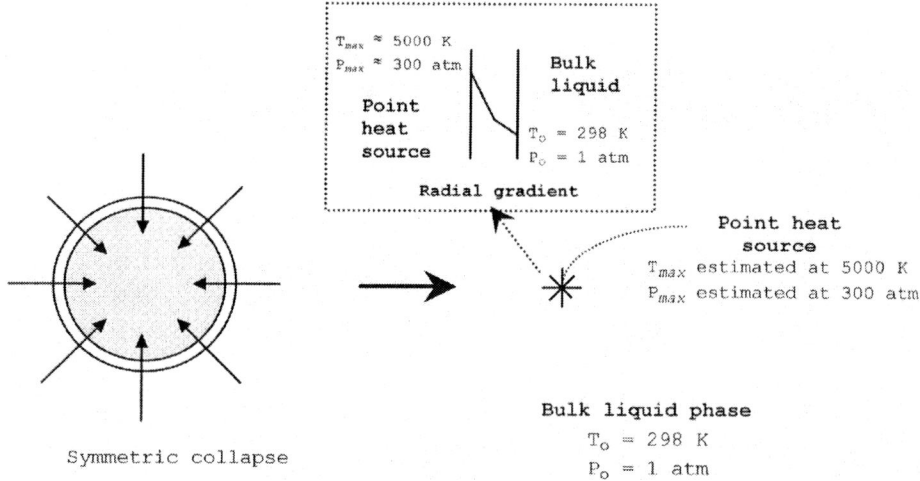

Figure 22.9 Symmetric collapse resulting in localized hot spot (Thompson and Doraiswamy, 2000)

where k_{et} is the rate constant of the reaction at the elevated temperature within the cavitational bubble and is given by Equation 22.11. Now if we express T_{max} by Equation 22.7 and substitute it in Equation 22.15, we recover Equation 22.13 but with $F = F_v F_t$. This expression explains the solvolysis of 2-chloro-2-methylpropane in ethanol–water solution. Further, it can be extended to nonaqueous systems by assuming that $k_{us} \gg k$. A shortcoming of this expression is the assumption that $E_{et} = E$, which may not be true (Tatsumoto and Fuji, 1987; Mills et al., 1995; Whillock and Harvey, 1997).

Mass Transfer

Film transfer and effective diffusivity

As in any solid-liquid reaction, when the solid is sparingly soluble, reaction occurs within the solid by diffusion of the liquid-phase reactant into it across the liquid film surrounding the solid. Thus two diffusion parameters are operative, the solid-liquid mass transfer coefficient k_{SL} and the effective diffusivity D_e of the reactant in the solid. A reaction in the solid can occur by any of several mechanisms. The simpler and more common of these were briefly explained in Chapter 15. For reactions following the sharp interface model, ultrasound can enhance either or both these constants. Indeed, in a typical solid-liquid reaction such as the synthesis of dibenzyl sulfide from benzyl chloride and sodium sulfide ultrasound enhances k_{SL} by a factor of 2 and D_e by a factor of 3.3 (Hagenson and Doraiswamy, 1998). Similar enhancement in k_{SL} was found for a Michael addition reaction (Ratoarinoro et al., 1995) and for another mass transfer-limited reaction (Worsley and Mills, 1996).

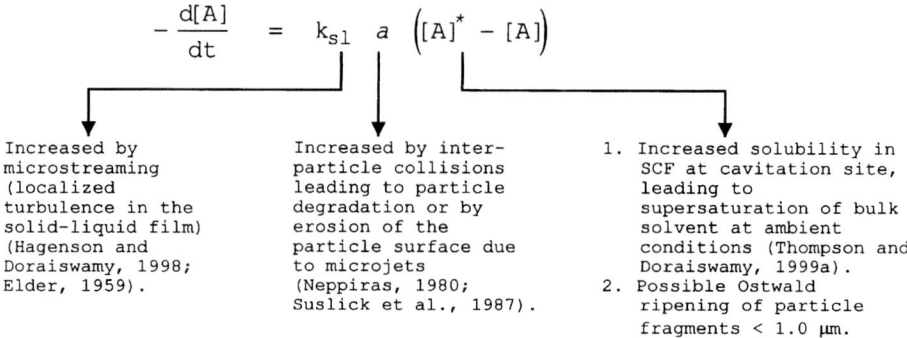

Figure 22.10 Sketch illustrating the three ways in which mass transfer can be enhanced (Thompson and Doraiswamy, 1999)

Dissolution

Ultrasound increases the rates of dissolution of sparingly soluble solids in a variety of solid-liquid systems (see Geier, 1989; Booth et al., 1997). In all of these cases, it is the surface area or the mass transfer coefficient that is enhanced. A significant observation is that ultrasound can also induce supersaturation of the solid, thus adding another dimension to rate enhancement by increasing the driving force for mass transfer (Thompson and Doraiswamy, 2000). Figure 22.10 illustrates the three ways in which the rate of mass transfer can be enhanced. Ultrasound can enhance any or all of these simultaneously, that is the mass transfer coefficient, the interfacial area, and the driving force. Although the first two are, in a sense, expected, the third places ultrasound on a distinctive footing. Thompson and Doraiswamy's results on the concentration of Na^+ as a function of time for both the silent and sonicated cases are shown in Figure 22.11. The supersaturation level appears to cycle and reaches a minimum every 20 minutes. This cycling may be due to random fluctuations in the average or to competing dissolution and precipitation processes. Similar supersaturation was observed in the dissolution of calcium citrate in water. This phenomenon has great potential in processes where the reaction occurs in the bulk liquid phase.

ULTRASOUND IN ORGANIC SYNTHESIS

Since the advent of sonochemistry as a distinct branch of ultrasonics, it is being used increasingly in organic synthesis. We treat the chemistry part in this section. Like most reactions involving a liquid phase (see Chapters 14–17), heterogeneous sonochemical reactions can also be divided into three broad categories. Including homogeneous reactions, the categories are as follows:

1. homogeneous reactions
2. liquid-liquid reactions
3. liquid-solid reactions
4. gas-liquid-solid reactions.

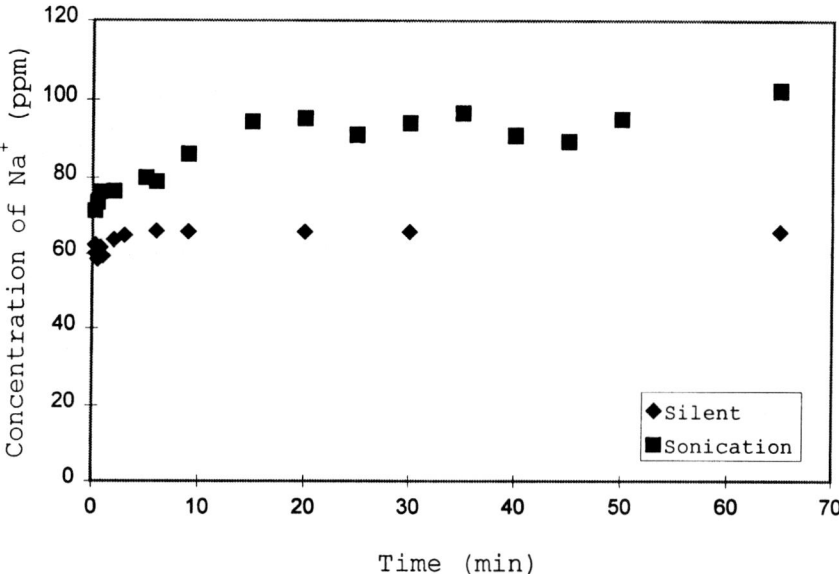

Figure 22.11 Concentration of Na^+ as a function of time in the reaction for both the silent and sonicated cases (Thompson and Doraiswamy, 2000)

In each of these categories, four broad types of the ultrasonic effect on organic reactions can be identified:

A. reactions that are initiated by ultrasound
B. reactions that are enhanced by ultrasound
C. reactions whose pathways are changed by ultrasound.
D. reactions that are unaffected by ultrasound.

Table 22.1 shows the effect of ultrasound on selected classical organic reactions such as the Grignard reaction, the Barbier reaction, the Michael addition, the Baltz-Schieman reaction, the Strecker synthesis, etc. (see Thompson and Doraiswamy, 1999, for a more extensive list). Other reactions of interest are those in which forcing conditions are replaced by much milder conditions, those in which the reactions are rendered user-friendly, and those in which the use of a phase-transfer catalyst can be avoided, although it will be shown later (Chapter 26) that the combined use of PTC and ultrasound can produce more striking enhancements. Many reactions are carried out under conditions where safety becomes a major problem and important precautions become necessary. The use of ultrasound in such cases can significantly reduce hazard in addition to enhancing the reaction. Two examples of this are (1) the preparation of the highly reactive "Rieke" metal powders and (2) the Barbier reaction.

Rieke powders are normally prepared by reducing the corresponding metal halides with potassium metal in refluxing tetrahydrofuran, which is clearly a hazardous procedure. Much of the hazard can be avoided if lithium metal is used in place of potassium and the reaction is carried out at room temperature

Table 22.1 Examples of reactions enhanced, modified, or unaffected by ultrasound

Homogeneous (H)/Heterogeneous Reactions

Type	Reaction	Traditional conditions/results	Ultrasonic conditions/results	Reference
B	Diels–Alder cyclization (H) 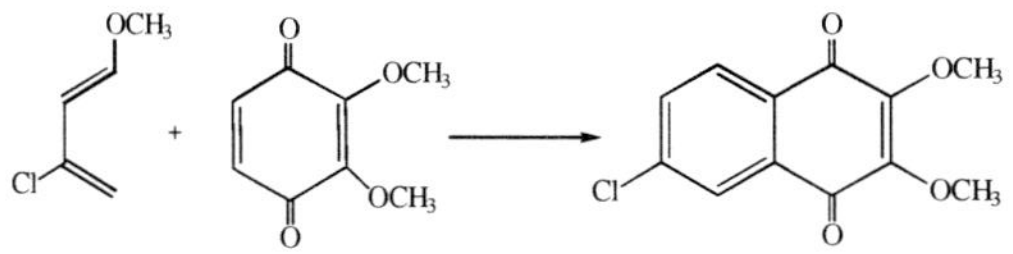 Solvent: toluene	Stirring conditions not specified Rxn time: 35.0 h Yield: 77.9%	Probe system Freq: 20 kHz Temp: 25 °C Rxn time: 3.5 h Yield: 97.3%	Javed et al. (1995)
B	Oxidation of indane to indan-1-one (H) 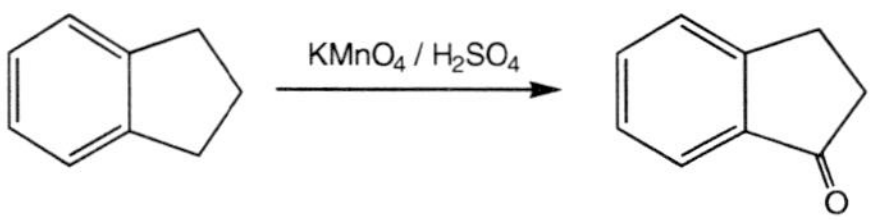	Stirring: 500 rpm Temp: 25 °C Press: 760 torr Rxn time: 3 h Yield: $\leq 27\%$ $k_{obs} = 5.12 \times 10^{-5}\,\text{s}^{-1}$	Probe system Freq: 21.5 kHz Power: 90 W Temp: 25 °C Press: 760 torr Rxn time: 3 h Yield = 73% $k_{obs} = 2.96 \times 10^{-4}\,\text{s}^{-1}$	Cum et al. (1988)
D	Acid catalyzed ketalization of acetophenone (H) 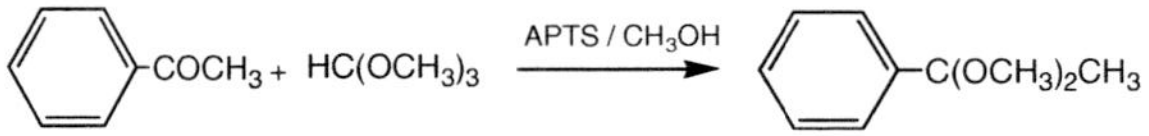 Follows an ionic mechanism	Stirring Temp: 12 °C Rxn time: 30 min Yield: 44%	Bath system Temp: 12 °C Rxn time: 30 min Yield: 48%	Einhorn et al. (1990)
		Rxn time: 3.5 h Yield: 94%	Rxn time: 3.5 h Yield: 94%	

A Reduction of methoxysilane

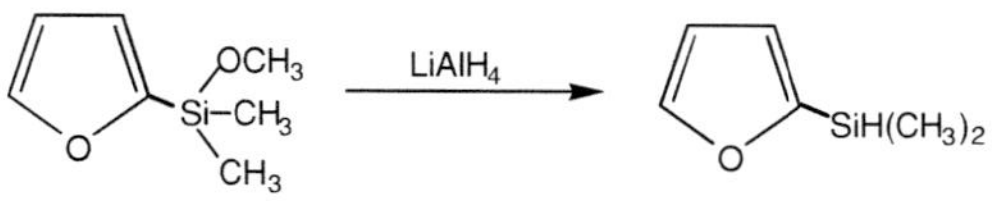

[structure: 2-furyl-Si(OCH$_3$)(CH$_3$)$_2$] $\xrightarrow{\text{LiAlH}_4}$ [structure: 2-furyl-SiH(CH$_3$)$_2$]

Solvent: pentane

Stirring
No reaction

Bath system
Temp: 35 °C
Rxn time: 3 h
Yield: 100%

Bremner (1986)

B Epoxidation of long chain unsaturated fatty esters

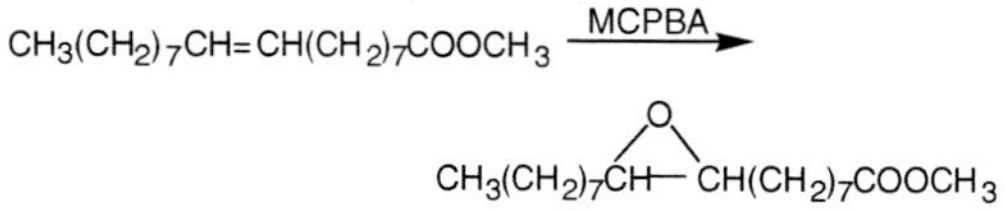

CH$_3$(CH$_2$)$_7$CH=CH(CH$_2$)$_7$COOCH$_3$ $\xrightarrow{\text{MCPBA}}$ CH$_3$(CH$_2$)$_7$CH—CH(CH$_2$)$_7$COOCH$_3$ (epoxide)

MCPBA = m-chloroperoxybenzoic acid

Stirring
Conditions not specified
Rxn time: 2 h
Yield: 48%

Probe system
Freq: 20 kHz
Temp: 20 °C
Rxn time: 15 min
Yield: 92%

Lie Ken Jie (1995)

B Oxidation of arylalkanes

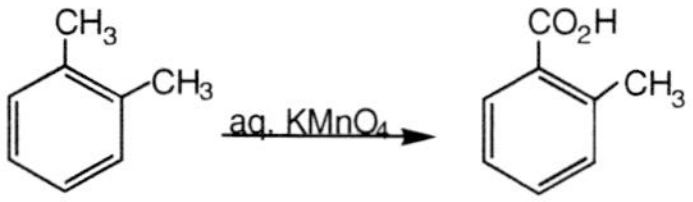

[structure: o-xylene] $\xrightarrow{\text{aq. KMnO}_4}$ [structure: o-methylbenzoic acid]

Solvent: water

Stirring: 150 rpm
Temp: 30–35 °C
Rxn time: 4 h
Yield: 12%

Bath system
Freq: 23 kHz
Power: 120 W
Temp: 30–35 °C
Rxn time: 4 h
Yield: 80%

Soudagar and Samant (1995)

B Michael addition of nitroalkanes to monosubstituted α,β-unsaturated esters

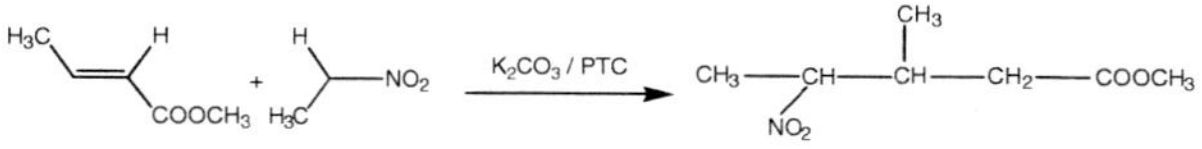

H$_3$C–CH=CH–COOCH$_3$ + H$_3$C–CH$_2$–NO$_2$ $\xrightarrow[\text{PTC}]{\text{K}_2\text{CO}_3}$ CH$_3$–CH(NO$_2$)–CH(CH$_3$)–CH$_2$–COOCH$_3$

PTC: Aliquat® 336

Temp: 40 °C
Rxn time: 2 days
Yield: 85%

Bath system
Freq: 60 kHz
Power: 80–160 W
Temp: 25 °C
Rxn time: 2 h
Yield: 90%

Jouglet et al. (1991)

(continued)

Table 22.1 (Continued)

Homogeneous (H)/Heterogeneous Reactions

Type	Reaction	Traditional conditions/results	Ultrasonic conditions/results	Reference
B	Permanganate oxidation of 2-octanol 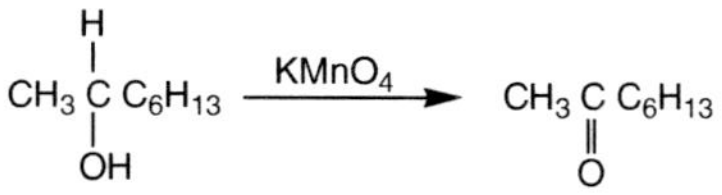 $CH_3\text{-}\overset{H}{\underset{OH}{C}}\text{-}C_6H_{13} \xrightarrow{KMnO_4} CH_3\text{-}\overset{\parallel}{\underset{O}{C}}\text{-}C_6H_{13}$ Solvent: hexane	Stirring Temp: 50 °C Rxn time: 5 h Yield: 3%	Bath system Temp: 50 °C Rxn time: 5 h Yield: 93%	Ando and Kimura (1990)
B	Synthesis of chalcones by Claisen–Schmidt codensation 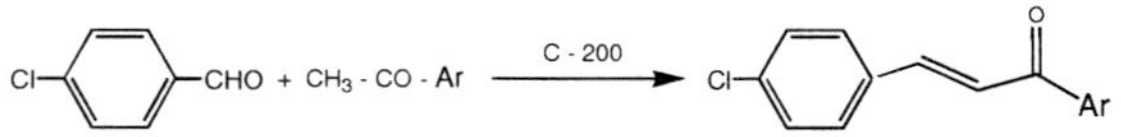 $Cl\text{-}C_6H_4\text{-}CHO + CH_3\text{-}CO\text{-}Ar \xrightarrow{C\text{-}200} Cl\text{-}C_6H_4\text{-}CH=CH\text{-}CO\text{-}Ar$ Solvent: ethanol (86%) Catalyst: activated barium hydroxide C-200	Stirring Temp: 25 °C Rxn time: 60 min Yield: 5% Catalyst wt: 1.0 g	Bath system Temp: 25 °C Rxn time: 10 min Yield: 76% Catalyst wt: 0.1 g	Fuentes et al. (1987)
B	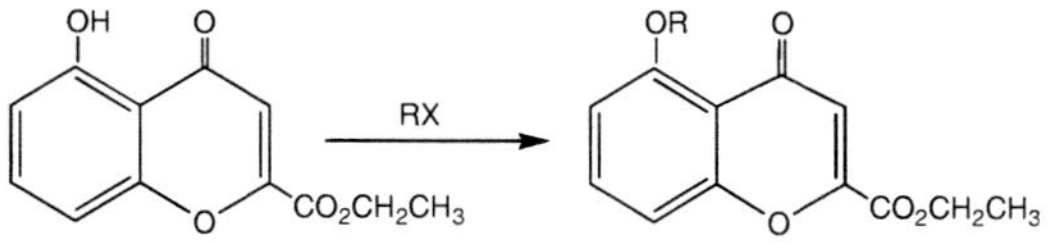 The o-alkylation of 5-hydroxy chromones 5-hydroxychromone-2-CO$_2$CH$_2$CH$_3$ + RX → 5-OR-chromone-2-CO$_2$CH$_2$CH$_3$	Stirring Temp: 65 °C Rxn time: 105 min Yield: 48%	Probe system Temp: 65 °C Rxn time: 60 min Yield: 79%	Mason et al. (1990b)

RX = benzyl bromide
Solvent: N-methylpyrrolidinone (NMP)
Other examples are given in article

B	Strecker synthesis of α-aminonitriles using an alumina support PhCHO + KCN + NH$_4$Cl ⟶ PhCH(CN)(NH$_2$) + H$_2$O + KCl Support: Al$_2$O$_3$ (improves the selectivity of PhCH(CN)NH$_2$) Solvent: acetonitrile	Stirring Temp: 50°C Rxn time: 24 h Yield: 64%	Bath system Freq: 45 kHz Power: 100 W Temp: 50°C Rxn time: 24 h Yield: 90%	Hanafusa et al. (1987); see also Menendez et al. (1986)
B	Ullmann coupling of 2-iodonitrobenzene 2-O$_2$N-C$_6$H$_4$-I ⟶(Cu)⟶ 2,2'-dinitrobiphenyl Mole ratio of Cu/substrate = 4 Copper powder presonicated for 15 minutes	Stirring Temp: 63 ± 1°C Rxn time: 2 h Yield < 1.5%	Microtip probe system Freq: 20 kHz Power: 135 W Temp: 63 ± 1°C Rxn time: 2 h Yield: 70.4%	Lindley et al. (1987)
B	Barbier reaction: retention of optical activity from S(+) 2-octyl halides 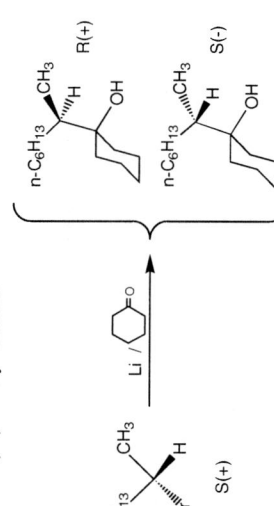 Solvent: tetrahydrofuran (THF)	Stirring Temp: 0°C Rxn time: 7 h Yield: 50% Configuration: R(+) % e.e: 6	Probe system Temp: 0°C Rxn time: 30 min Yield: 59% Configuration: R(+) % e.e: 10	de Souza-Barboza et al. (1987); see also de Souza-Barboza et al. (1988)

(continued)

Table 22.1 (Continued)

Homogeneous (H)/Heterogeneous Reactions

Type	Reaction	Traditional conditions/results	Ultrasonic conditions/results	Reference
B	Reformatsky reaction 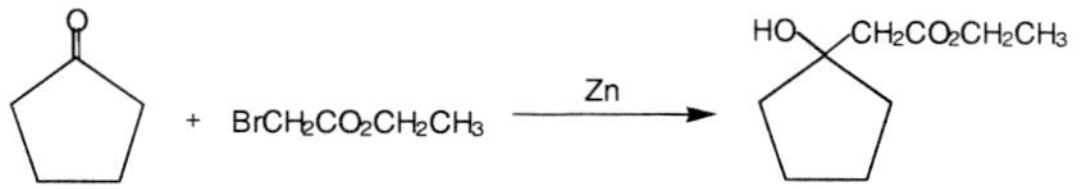 ⬡=O + BrCH₂CO₂CH₂CH₃ —Zn→ (cyclopentane)-OH, CH₂CO₂CH₂CH₃ Various solvents and types of zinc powders were tested	Stirring Temp: 80 °C Rxn time: 12 h Yield: 50%	Bath system Freq: 50/60 Hz Power: 150 W Temp: 25–30 °C Rxn time: 30 min Yield: 98%	Han and Boudjouk (1982), see also Boudjouk and Han (1984)
C	Change in pathway from Friedel–Crafts reaction to nucleophilic substitution 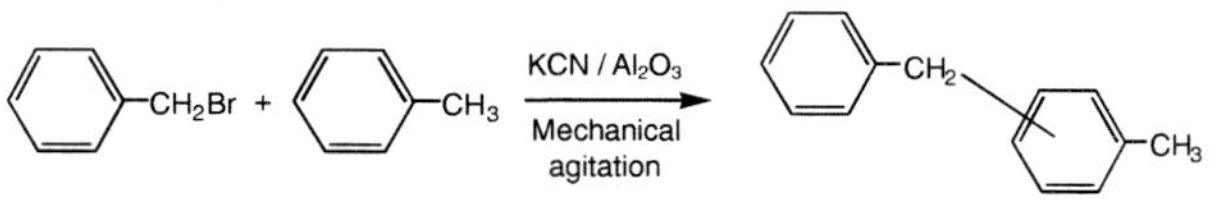 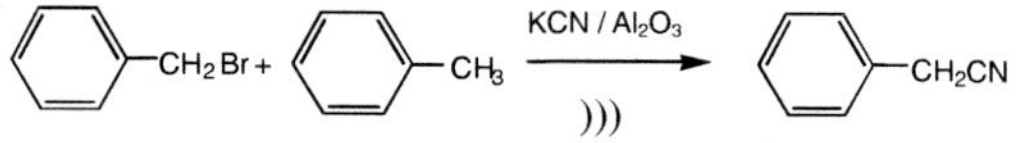 Ph-CH₃ + Ph-CH₂Br —KCN/Al₂O₃, Mechanical agitation→ Ph-CH₂-C₆H₄-CH₃ Ph-CH₃ + Ph-CH₂Br —KCN/Al₂O₃,)))→ Ph-CH₂CN	Stirring Temp: 50 °C Product: o- and p-benzyltoluenes Yield: 75%	Bath system Freq: 45 kHz Power: 200 W Product: benzyl cyanide Yield: 71%	Ando et al. (1984a)

[a] A = initiated; B = enhanced; C = changed the pathway; D = no effect.

with ultrasound (Mason, 1990a). Typically, in the preparation of Rieke copper, the reflux time is reduced from 8 hours to 40 minutes. The reactions involved may be represented as

$$MX_n + nA \rightarrow M^* + nAX \qquad [22.2]$$

(MX = metal halide, A = Li, Na, K)

The Barbier reaction, extensively studied by Luche and co-workers (see Luche et al., 1990), is similar to the Grignard synthesis except that it proceeds in one step. In a recent variation that has now become common practice, lithium is used instead of magnesium. The scope of the reaction, rather restricted so far, can be widened by conducting it in an ultrasonic bath. No special precautions or equipment are needed, and even impure solvents can be used.

SCALE-UP OF SONOCHEMICAL REACTORS

The various advantages of using ultrasound (as can be inferred from Table 22.1) notwithstanding, sonochemistry continues to be a laboratory curiosity with few industrial applications. The chief reasons are the lack of a scale-up rationale and economics.

Large-Scale Reactors

Large-scale batch sonochemical reactors can be designed on the basis of the performance of conventional laboratory sonicators if it is assumed that there are no serious scale-up factors. These are the "cleaning bath" reactors (indirect sonication) and reactors with immersible transducers or sonic probes (direct sonication). Continuous reactors use either wall-mounted transducers (indirect sonication) or sonic probes (direct sonication).

Several designs are shown in Figure 22.12: (a) common batch "cleaning bath" reactor with wall mounted transducers; (b) batch reactor with immersible transducers; (c) batch reactor with sonic probe; (d) continuous flow tubular reactor with wall-mounted transducers; (e) the Harwell sonochemical reactor; and (f) a shell-and-tube reactor. A number of other designs are discussed by Thompson and Doraiswamy (1999b).

The cleaning bath and immersible transducer reactors

The usual cleaning bath reactor (Figure 22.12a) suffers from the fact that there are air gaps between the transducer (usually flat) and the reactor wall (usually circular) leading to inefficient energy transmission to the liquid inside the reactor. This problem can often be overcome by using a hexagonal cross-sectional reactor, in which contact occurs between flat faces.

Reactors with immersible transducers (Figure 22.12b) give higher intensities and can be retrofitted easily. On the other hand, a mechanical stirrer cannot easily be used with certain types of transducers. As a result, the acoustic (and

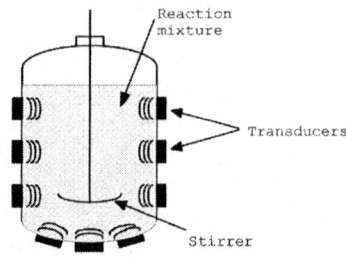
(a) 'Cleaning bath' sonoreactor with wall mounted transducers

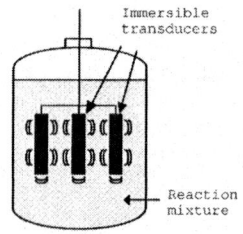
(b) Sonoreactor with immersible transducers

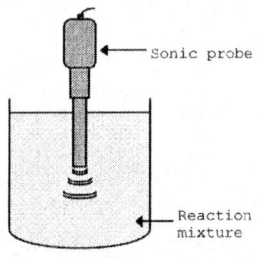
(c) Sonoreactor with sonic probe

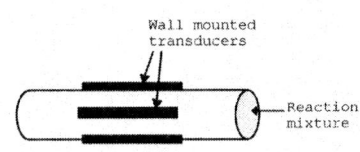
(d) Tubular flow sonoreactor with wall mounted transducers

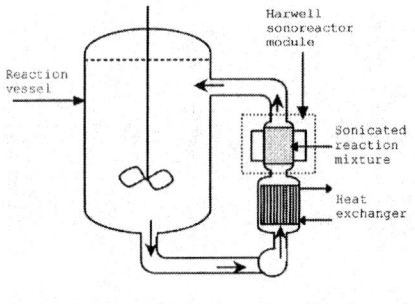
(e) Harwell sonochemical reactor

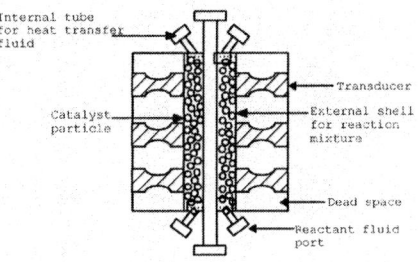
(f) Shell-and-tube sonoreactor (Ragaini, 1992)

Figure 21.12 Sketches of a few common designs of sonochemical reactors

therefore the temperature) field inside the reactor tends to be nonuniform, thus detracting from reproducible operation.

Reactor with sonic probe

Now let us consider the sonic probe reactor (Figure 22.12c). Here, only the probe is inserted, not the entire assembly with the transducer. The probe generates a high intensity field around its tip and can be inserted easily into the liquid

through a suitable port at the top of the vessel. These two features make the sonic probe reactor much more attractive than the other reactors. The major disadvantage is that the intensity tends to be unevenly distributed, especially when the vessel is large. Thus any industrial-scale reactor must use a large number of suitably placed probes to overcome this problem.

EXAMPLE 22.1

Approximate design of a sonic probe reactor

Based on what is known so far, the volume of any sonochemical reactor can only be fixed "arithmetically" to meet a given production capacity. It is not possible to compute the reaction time for a given volume, or vice versa, even assuming that the Arrhenius parameters are known, because the actual temperature and the cavitational field within the reactor would be unknown. An approximate procedure is suggested here. Attempts to map the cavitational and energy fields are briefly discussed in a later section.

As a first step, based on laboratory-scale data at different times, a rate equation can be developed and then the batch reactor performance equation

$$t = [A]_0 \int_0^{X_A(t)} \frac{dX_A}{-r_A} \tag{E22.1.1}$$

used to predict the reaction time. The dubious assumption here is that the average temperature and the sonication field in the scaled-up reactor are identical to those in the laboratory reactor.

The next step is the design of the ultrasound part of the reactor system. For this, we consider a 10-m^3 vessel of 3 m in diameter and 1.5 m high. The calculation proceeds as follows:

1. The largest available sonic probe has a rating of 2.5 kW from a 5-cm diameter probe tip. Thus the ultrasonic intensity is

   $$I = 2500/[\pi(5/2)^2] \approx 120 \text{ W/cm}^2$$

2. Assuming that the probe will irradiate 1 liter or 0.001 m^3 of liquid (based on available information on a number of systems), this will account for 1 in 10,000 parts of the reactor liquid. Thus roughly 10,000 probes will be needed. Assuming that the probes are available in arrays of 10, about 1000 arrays are needed.
3. The approximate cost of these arrays is $600,000.

Computations show (Goodwin, 1990) that this cost is much less than that for the cleaning bath reactor or for the reactor with immersible transducers. A practical problem with this reactor is that the tips are exposed to a very small fraction of the liquid, hence a very efficient method of circulation is necessary.

Tubular flow reactor

This reactor (Figure 22.12d) is identical to the tubular (plug-flow) reactor described in Chapter 4, except that now suitable transducers are attached to the outer wall of the reactor. The size of the irradiated zone can be controlled by the number of transducers used. Like all reactors with externally mounted

Strategies for Enhancing Rate Reactions

transducers, this reactor is also noninvasive. Further, transducers can be fixed to any existing tubular reactor but preferably to one with flat edges. Most importantly, because the exposed volume at any time is small, a relatively small number of transducers is required.

The Harwell and shell-and-tube sonoreactors

The chief merit of the Harwell reactor, shown in Figure 22.12e, is that the ultrasound intensity is focused at the core and therefore the walls of the larger of the reactors are saved from damage due to cavitational collapse in their vicinity. The efficiency of this reactor increases with increasing throughput, and the modularity of the active region is a distinct advantage. One major disadvantage is that the active species (activated by sonication) may be very short-lived. Another useful design is the shell-and-tube sonoreactor shown in Figure 22.12f, which is very similar to the conventional shell-and-tube reactor.

Reactions with Special Features

It is often necessary to consider the special features of a reaction before undertaking the design of a reactor for commercial operation. These can be ascertained only by experiments carried out on a relatively large scale. Unfortunately, there are too few reported results to enable any generalizations.

One reaction that has been studied in some detail is the Simmons–Smith reaction (Repic and Vogt, 1982):

$$CH_2I_2 + Zn \rightarrow ICH_2ZnI$$

$$\underset{CH_3(CH_2)_7}{\overset{H}{\diagdown}}C=C\underset{H}{\overset{H}{\diagup}} + ICH_2ZnI \rightarrow CH_3(CH_2)_7-\underset{\underset{CH_2}{\diagdown \diagup}}{C-C} \qquad [22.3]$$

which is carried out in the presence of zinc. By using sonochemically activated zinc (i.e., presonicated zinc) the yield could be raised to 91% (by continuing sonication during reaction) from the 51% obtained by the traditional method which uses iodine or lithium for activation. Zinc must be in powder form in this activation.

In a scaled-up version, a 22-liter vessel was placed in a 50-gallon ultrasonic bath (Repic et al., 1984). The zinc was cast in two 800-gram lumps using conical flasks as molds. About 593 grams of methyl oleate was used along with 1.3 liters of diiodomethane and 2.7 liters of dimethoxyethane to give the product in 82% yield. This is about 9% lower than the laboratory-scale yield of 91% with presonicated zinc. Therefore, a scale-up problem exists which has yet to be addressed.

The main advantages of this method are that the zinc can be in any form (usually rod or foil) and the reaction can be controlled by removing the metal from the scene of the reaction.

DESIGN OF CAVITATION REACTORS

As noted earlier, the modeling of any reactor with a separate field of its own involves coupling the equations describing this field with the other reactor equations. In a sonochemical reactor, the concerned field develops due to cavitation. It is important to realize that cavitation bubbles begin to emerge at a threshold pressure of 1 Mpa (Yount et al., 1984; Prosperetti and Commander, 1989; Ye and Ding, 1995). Depending on the amplitude of the pressure, the volume fraction of these bubbles can vary from 10^{-4} to 10^{-1} (Dahnke et al., 1999a,b). Additionally, there is a density distribution function associated with the bubbles, as well as a bubble radius distribution function. So, what we essentially have is a sound wave propagating in a bubbly liquid consisting of a mixture of liquid and gas bubbles of varying densities and radii. This can have a profound influence on the behavior of the sound wave and its effect on the reacting system. For example, the velocity of sound, which is about 340 m/s in water, can be annihilated down to about 20 m/s in a bubbly liquid (Dähanke et al., 1999b).

Complications in modeling such a situation arise due to the following reasons: (1) the relevant wave parameters like velocity of sound in the medium and the attenuation constants are functions of the bubble density distribution and (2) the distribution itself is a function of the sound amplitude and frequency. This complex interdependence necessitates a special algorithm to calculate the pressure field in a sonochemical reactor. A number of models of varying degrees of complexity have been developed by Keil and coworkers (e.g., Keil and Dähnke, 1997a,b; Dähnke and Keil, 1998a,b; Dähnke et al., 1999a,b). In view of the highly mathematical nature of these models, we limit the treatment to a brief semiqualitative description.

The starting point in all these models is the calculation of the time-dependent pressure field using the following three-dimensional homogeneous wave equation (Junger and Feit, 1986):

$$\Delta P(\mathbf{r}, t) = \frac{1}{u_s} \frac{\partial^2 P(\mathbf{r}, t)}{\partial t^2} \tag{22.16}$$

where $P(\mathbf{r}, t)$ is the spatial and time-dependent pressure, $\mathbf{r}$ is the coordinate vector, and Δ is the Laplace operator. This equation has to be modified as follows to allow for the effect of pressure damping:

$$\frac{\partial^2 P(\mathbf{r}, t)}{\partial t^2} + 2\alpha' \frac{\partial P(\mathbf{r}, t)}{\partial t} = \left(\frac{\alpha'^2 + \omega^2}{\mathbf{k}^2}\right) \Delta P(\mathbf{r}, t) \tag{22.17}$$

where α' represents the time-dependent damping coefficient, and $\mathbf{k}$ is the wave vector in an undamped fluid. Note that the wave vector for a mixture of liquid and gas ($\mathbf{k}_{mix}$) actually consists of a real part and an imaginary part. Equation 22.17 considers only the real part $\mathbf{k}$. The second term on the left-hand side represents the damping term, while the principal factor on the right-hand side accounts for the variation of velocity (Dähnke et al., 1999). The damping coefficient α must be distinguished from the space-dependent damping coefficient which in one-dimensional form is given by Equation 22.5.

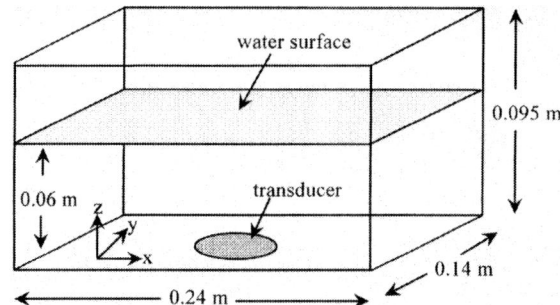

Figure 21.13 The simple reactor considered for modeling (adapted from Dähnke et al., 1999b)

Equation 22.17 can now be solved to give the pressure as a function of time for the entire three-dimensional interior of the reactor, provided that pressure values at two points in time are known. For a reactor sketched in Figure 22.13, the calculated pressure distribution in a plane at a height of 0.02 m above the transducer is shown in Figure 22.14 (Dähnke et al., 199b). The complex nature of the pressure distribution is easy to see. The conversion in any reactor will also be a function of space and time, and there is no method known yet for calculating it. A reasonable method would be to calculate the volume fraction of the bubbles and relate it to conversion empirically. Even this method is beset with uncertainties since only a part of the cavitational bubbles leads to sonochemical events. Much research needs to be done before a rational method of correlating conversion with cavitation can be established.

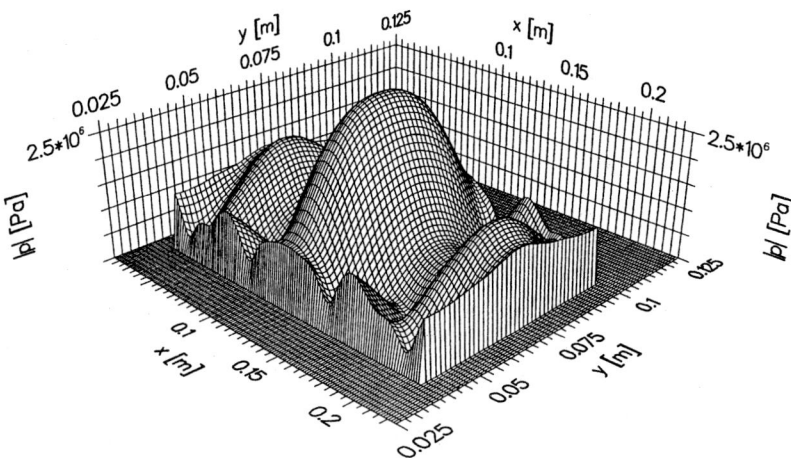

Figure 22.14 Calculated pressure field distribution in the plane at a height of 0.02 m above the transducer shown in Figure 22.13 (Dähnke et al., 1999b). The figure in original was kindly supplied for reproduction by Professor Keil

In another (somewhat different) approach, a probability density function (PDF) has been proposed (Moholkar and Pandit, 1997). This is used to map the cavity dynamics in the reaction medium covering all three phases of a cavity's lifetime: growth, oscillation, and collapse. An ultrasonic reactor is considered highly efficient if the PDF shows peaks in the collapse regime at all of the locations in the cavitation field. This is an indication that pressure pulses exist throughout the medium and are not restricted to just a few locations. In other words, the cavitational intensity is uniformly distributed. If peaks occur in the growth and collapse regimes, it is desirable to place the reactor inside the sonicated medium at a location where the maximum probability of collapse is indicated.

SONOCHEMICAL EFFECT WITHOUT ULTRASOUND: HYDRODYNAMIC CAVITATION

Rate enhancements obtained by sonication, as described previously, are all based on cavitation. If cavitation can be induced by an alternative method without the complexities of sonication, reactor scale-up can become more facile. A method that is becoming increasingly attractive is *hydrodynamic cavitation*.

Hydrodynamic cavitation is achieved by throttling a valve downstream from a pump (see Figure 22.15). When the pressure downstream from the orifice (used to throttle the flow) falls below the vapor pressure of the medium, partial vaporization of the medium occurs. This results in the generation of vapor, gas, or vapor/gas bubbles that oscillate and collapse some distance downstream as the pressure is recovered. Such a phenomenon is referred to as hydrodynamic cavitation.

A striking example is the hydrolysis of fatty oils (Pandit and Joshi, 1993) in an ultrasonic generator or in a flow loop. Cavitation in a flow loop was created by throttling a valve downstream from the pump, leading to local temperatures of the order of 300 °C and pressures of the order of 10 atm (normally used in such hydrolyses). This means that reactions normally carried out at relatively high temperatures and pressures can now be carried out not only by using reactors

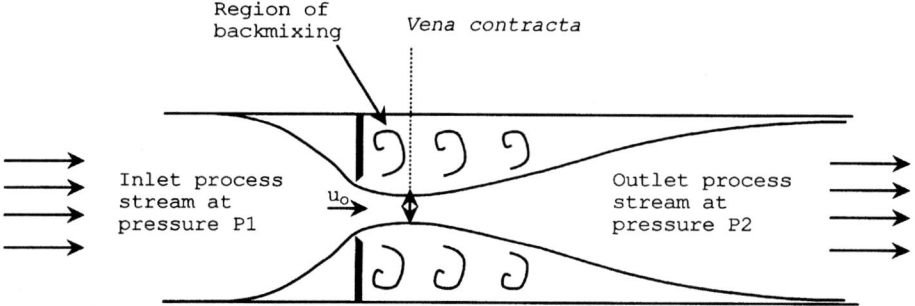

Figure 22.15 Hydrodynamic cavitation reactor

insonated by a regular sonic horn, as described in the earlier sections, but even by such a simple technique as downstream throttling to generate cavitation.

Inception of Hydrodynamic Cavitation

The parameters that are most important in hydrodynamic cavitation are the downstream pressure P_2, the vapor pressure P_v and density ρ of the medium, and the velocity of the fluid through the orifice. These are grouped together into a *cavitation number* defined as

$$C_N = \frac{(P_2 - P_v)}{(1/2)\rho_L u_{or}^2} \tag{22.18}$$

It has been shown (Yan et al., 1988) that C_N is independent of velocity but depends on orifice size, and that it increases linearly with an increase in the ratio of orifice-to-pipe diameters. Cavitation is induced at values of C_N in the range of 1.0–2.0 (usually closer to 1), depending on the size and shape of the constriction and the change in external pressure from the vena contracta to the point where it is fully recovered.

Reactor Design and Scale-Up

An important consideration is the level of turbulence downstream of throttling. Including this effect, the parameters that can be externally controlled to influence reactor performance are the discharge pressure of the pump (usually referred to as the *cavitating pump*), the ratio of orifice-to-pipe diameters, and the downstream pipe size. In hydrodynamic cavitation, the bubbles flow along with the main stream. Thus the longer the bubble life, the more the effect of cavitation on the liquid downstream from the orifice. Remembering this fact, the effect of changes in these variables on maximum bubble size, cavitational volume, and bubble life (the most important aspects of cavitating bubbles) can be summarized as follows:

1. A rise in the discharge pressure and hence in the final recovered pressure results in an increase in the active cavitational volume downstream from the orifice, with increased cavitational intensity.
2. An increase in the orifice-to-pipe ratio [without altering the maximum bubble (i.e., cavity) size] leads to a considerable increase in the life of the cavity and hence in the cavitational effect.
3. An increase in pipe size downstream from the orifice does not significantly alter the life of the cavity, but it does enhance the maximum size reached before an implosive collapse.

Perhaps the most significant feature of hydrodynamic cavitation is that as the flow changes from nonturbulent to turbulent, bubble behavior changes from oscillatory to transient, and the latter resembles acoustic cavitation. Thus features similar to those of acoustic cavitation can be replicated in hydrodynamic flow by manipulating the level of turbulence.

The following example illustrates how the three parameters just identified can be used to produce an effective design.

EXAMPLE 22.2

Design of a hydrodynamic cavitation reactor

It is proposed to design a hydrodynamic cavitation reactor using the following major design parameters: recovered discharge pressure $P_2 = 3$ atm, pipe size downstream from the orifice $d_P = 2$ in, and orifice-to-pipe diameter ratio of 0.5. Find the maximum volume that can be treated per unit time and also the volume of the insonated zone (i.e., the active volume) downstream from the orifice.

SOLUTION

The solution consists of four major steps: (1) calculation of orifice and pipe velocities and pressure recovery time, (2) modeling of the turbulence downstream of the orifice, (3) solution of the bubble dynamic equation to obtain the important parameters of the reactor as functions of time, and (4) use of the results obtained in step 3 to design the reactor.

Hydrodynamic parameters

The first two steps involve purely hydrodynamic procedures and hence will only be very briefly illustrated here. But the parameter values obtained therefrom will be used in the third step. They are as follows:

1. cavitation number $C_N = 1$ (assumed)
2. orifice velocity $u_{or} = 24.08$ m/s, obtained from Equation 22.18 with $C_N = 1$ and known values of P_2 (assumed as 3 atm) and solvent vapor pressure P_v. The equation has to be modified for a vaporous cavity, or a cavity with zero noncondensible content (Moholkar and Pandit, 1997).
3. pipe velocity $u_P = 6.12$ m/s, obtained from the orifice-to-pipe diameter ratio
4. pressure recovery time $\tau = 0.0257 s$, obtained from the typical pressure recovery profile downstream from the orifice.
5. length scale of the eddy $\ell_e = 3.048 \times 10^{-3}$ m, obtained from $\ell_e = 0.08 (d_{or} + d_p)/2$
6. power input for unit mass of the cavitating liquid $P_M = 15855$ W/kg, obtained from

$$P_M = \frac{(\Delta P)_{\text{permanent}}(\text{volumetric liquid flow rate}), \text{m}^3/\text{s}}{\rho \,(\text{volume of pipe over which pressure recovery occurs})}$$

7. turbulent fluctuating velocity $u' = 3.39$ m/s, obtained from $u' = (P_M \times \ell_e)^{1/3}$
8. frequency of turbulence pressure variations $f_t = 1.12$ kHz, obtained from (u'/ℓ_e).

The instantaneous local velocity u_i of the fluid in the cavitating zone is an important hydrodynamic parameter. This is obtained from the equation

$$u_i = u_t + u' \sin(2\pi f_t t) \tag{E22.2.1}$$

where u_t is the local mean velocity and u' is the fluctuating component superimposed on it. Using this instantaneous velocity, the instantaneous static pressure can be calculated from the following form of Bernoulli's equation:

$$P_i = P_v + (1/2)\rho u_{or}^2 - (1/2)\rho u_i^2 \qquad (E22.2.2)$$

Then this instantaneous pressure, which incorporates the effect of fluctuations due to turbulence, is used in place of P_∞ in the bubble dynamics equation given by Equation 22.6, assuming that linear pressure recovery does not collapse the cavity (Moholkar and Pandit, 1999).

Solution of the bubble dynamics equation

This is a nonlinear, second-order, second-degree differential equation, which can be readily solved by the substitution:

$$\frac{dr_b}{dt} = \beta \qquad (E22.2.3)$$

giving

$$\frac{d\beta}{dt} = \frac{1}{\rho r_b}\left[\left(P_0 - P_v + \frac{2\sigma}{r_{b0}}\right)\left(\frac{r_{b0}}{r_b}\right)^{3\gamma} - \frac{2\sigma}{r_b} - (P_0 - P_i) - \frac{4\mu\beta}{r_b}\right] - \frac{3\beta^2}{2r_b} \qquad (E22.2.4)$$

and with the initial conditions

$$t = 0, \quad r_b = r_{b0}, \quad \beta = \frac{dr_b}{dt} = 0 \qquad (E22.2.5)$$

Clearly, an important parameter in the solution is the initial bubble radius r_{b0}. Yan et al. (1988) measured the distribution of bubble sizes and found that it is in the range of 1–100 μm. We select sizes 2, 5, and 20 μm.

The solution gives us three parameters as functions of time: bubble radius, bubble wall velocity, and the pressure pulse P_L, which represents the pressure in the vicinity of the bubble (felt in the liquid just outside the bubble surface in zone 2 of the reaction) and is given by

$$P_L = P_b - \frac{2\sigma}{r_b} - \frac{4\mu}{r_b}\frac{dr_b}{dt} \qquad (E22.2.6)$$

where P_b is the pressure inside the bubble given by

$$P_b = P_{b0}\left(\frac{r_{b0}}{r_b}\right)^{3\gamma} \qquad (E22.2.7)$$

The computer program can be terminated when the wall velocity reaches the velocity of sound in water, 1500 m/s. Beyond this value, the liquid is no longer incompressible and therefore Equation 22.6 is no longer valid.

Implications of the solution for reactor design

The results obtained from the solution of the Raleigh–Plesset equation can be used to estimate the volume and intensity of the "active zone" downstream from the orifice. The volume of the zone can be calculated by multiplying the bubble lifetime (i.e., the total time used in solving the equation or the time of bubble collapse where the bubble wall velocity

Table 22.2 Results of simulations for three initial bubble sizes

Initial bubble size r_{bo} (μ)	Bubble life time t_L (ms)	Distance the bubble travels downstream of orifice (mm)	Volume of isonated zone downstream of orifice (cm^3)	Pressure pulse obtained after complete collapse of bubble (atm)
2	0.982	24.83	50.3	69.21
5	1.026	25.90	52.0	40.67
10	1.038	26.25	53.0	12.41

equals the velocity of sound) by the orifice velocity. Thus the intensity (or strength) of the cavitational zone is defined as the average pressure pulse P_L experienced by the reactant as a result of cavity collapse.

The results of the computations are summarized in Table 22.2 for the three bubble sizes selected. From them, one can determine whether the design is capable of producing the desired intensity, that is, of generating a pressure pulse of the magnitude required to bring about the necessary physical or chemical transformation, and one can also estimate the residence time of the reactants in the active zone based on the cavity life time t_L and the average fluid bulk velocity in the pipe downstream from the orifice. If the cavitational intensity and/or residence time are/is not sufficient, the following modifications are suggested:

1. To extend the active zone, increase the orifice-to-pipe diameter ratio.
2. To increase the magnitude of the cavitation intensity, that is, to increase the magnitude of the pressure pulse obtained from cavitational collapse, increase the pipe size downstream from the orifice; this results in faster pressure recovery and correspondingly more violent cavity collapse.
3. Increase the recovered discharge pressure P_2 by providing a secondary orifice of larger diameter.

Reference was made previously to three zones of reaction. The pressure P_b inside the bubble determines the zone in which the reaction occurs. High P_b values correspond to reaction in zone 1, and high P_L values to reaction in zone 2. For reaction to occur in zone 3 (the Weissler reaction, for instance), P_b should be high, and also the residence time of the radicals generated should not be so high in zones 1 and 2 that they react among themselves and neutralize any enhancing effect.

Chapter 23

Microphase-Assisted Reaction Engineering

A relatively recent concept in organic reaction engineering is the use of submicron particles to enhance the rate of a reaction. These are usually microparticles of solids, but can also be microdroplets of liquids, or even microbubbles of gases. They can be external agents, participating reactants, or precipitating solids. In this chapter, we cover the role of small particles as a whole, which may be regarded as constituting an additional colloid-like phase normally referred to as the *microphase*. We begin by classifying the microphase in terms of its mode of action and then proceed to an analysis of the following categories of importance in organic technology: microslurry of (1) catalyst or adsorbing particles in a reactive mixture; (2) solid reactant particles in a continuous phase of the second reactant; and (3) solid particles precipitating from reaction between two dissolved reactants, one of which can be a solid dissolving and reacting simultaneously with the other reactant. The microphase in the first case is externally added, whereas that in the last two cases is a reactant or a product. The field is still developing (with many unproven theories), and hence we restrict the treatment to a simple analysis of selected situations based on reasonable assumptions (thus avoiding often unjustified complexity).

CLASSIFICATION OF MICROPHASES

A microphase can be described as an assemblage of very small dispersed phase particles with average size (d_p) much less than the diffusional length scale of the solute. Usually $d_p < 10\,\mu m$, compared to the diffusional length scale which is of the order of 50–60 μm. Although the microphase is a distinct phase, the phase in which it is present is commonly regarded as pseudohomogeneous. In a stricter sense, however, it should be regarded as a *microheterogeneous* phase. Indeed,

several studies have been reported on modeling heterogeneous microphase systems (Holstvoogd et al., 1986, 1988; Yagi and Hikita, 1987). In view of the ability of the particles of such a system, pseudohomogeneous or pseudoheterogeneous, to get inside the fluid film, they can enhance the transport rate of the solute through the film. Experimental observations in typical gas-liquid and slurry systems have clearly demonstrated (see Ramachandran and Sharma, 1969; Uchida et al., 1975; Sada et al., 1977a,b, 1980; Alper et al., 1980; Pal et al., 1982; Bruining et al., 1986; Bhaskarwar et al., 1986; Bhagwat et al., 1987; Mehra et al., 1988; Mehra and Sharma, 1988a; Hagenson et al., 1994) the enhancing role of a microphase made up of fine particles. The case of a second liquid phase acting as a microphase or of a solid product performing a similar function has also been studied and found to enhance the reaction rate (Janakiraman and Sharma, 1985; Mehra and Sharma, 1985, 1988b; Anderson et al., 1998). Mehra et al. (1988) and Mehra (1990a,b, 1996) presented a detailed account of the role of different types of microphases in rate enhancement. In all these cases, either a microphase is separately introduced or one of the reactants or products acts as a microphase.

Based on these studies, the following classes of microphase action can be identified:

1. Here, as shown in Figure 23.1a, the microphase particles (or droplets in the case of a liquid microphase) exhibit a kind of affinity for solute A and transport it from the interface to the bulk of phase 2 where reaction occurs. Reaction occurs on the microphase, and A is not discharged into the bulk of phase 2 for reaction to occur there. Because the reactant "disappears" into the microphase, the latter is referred to in this case as a "sink."
2. The microphase is a sink as in case (1) but is catalytically active (Pal et al., 1982; Wimmers et al., 1984; Karve and Juvekar, 1990; Nagy and Moser, 1995). Examples are micelles, carbon particles, and catalyst particles in a slurry reactor. Here, too, reaction occurs within or on the particles of the microphase, and Figure 23.1a applies equally. The efficiency of a slurry reactor, so commonly used in organic synthesis, can be greatly enhanced if the catalyst is present as a microslurry.
3. The microphase is present as a microslurry of a sparingly soluble salt reacting with a diffusing gas (Ramachandran and Sharma, 1969; Uchida et al., 1975; Sada et al., 1980; Alper et al., 1980; Alper and Deckwer, 1981; Sada and Kumazawa, 1982; Yagi and Hikita, 1987). The reactant microparticles enter the liquid film around the gas bubbles and enhance mass transfer (Figure 23.1b). Because in this case the microphase provides the reactant to the bulk, it is referred to as a "source."
4. One of the products acts as the microphase that transports A from the interface to the bulk (Mehra and Sharma, 1988b; Pradhan et al., 1992; Saraph and Mehra, 1994; Anderson et al., 1998a,b). Then reaction occurs between A and B after B has diffused on to the microphase. Alternatively, the microphase discharges its load into the bulk, where further reaction occurs. Figure 23.1c depicts this situation. From the

746   Strategies for Enhancing Rate Reactions

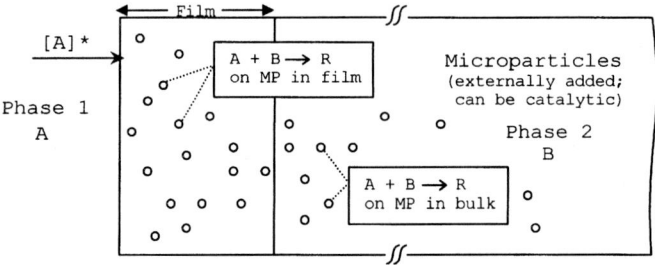

a. Microphase as catalytic or physical "sink."

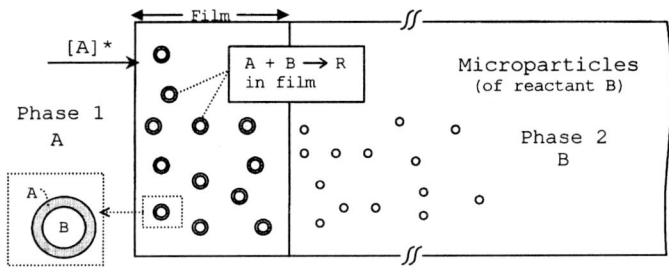

b. Microphase as "source," i.e. a reactant acts as microphase.

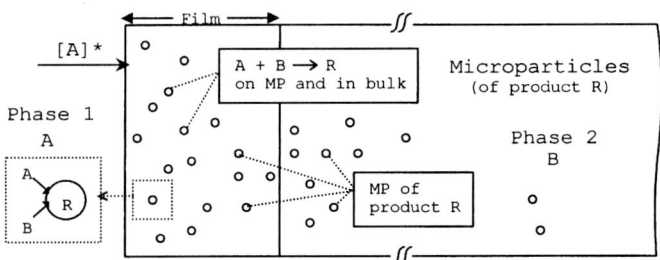

c. Microphase as autocatalyst, i.e. a product acts as microphase

Figure 23.1   Categories of microphase action

practical point of view, this constitutes a very important class of microphase-assisted reactions in organic technology.

ANALYSIS OF MICROPHASE ACTION

Using the pseudohomogeneous model (surprisingly upheld by a number of experiments) as the fundamental basis, we list here some commonly used

terms as well as other concepts and assumptions pertinent to the analysis to follow.

Basic Concepts and Assumptions

1. Figure 23.2 is a general representation of microphase action. Reactant A from phase 1 diffuses into phase 2 where it reacts with B to form products. The microphase enhances the reaction by picking up more of A from the vicinity of the macrointerface and transporting it to the bulk of phase 2. The concentration gradients of A for pure mass transfer, mass transfer with reaction, and mass transfer with reaction accompanied by microphase action are shown in the figure. Notice the progressive steepening of the gradient corresponding to increasing reaction rate.
2. Because the holdup of the microphase in the continuous phase is very low, a requirement for the pseudohomogeneous model, the macrointerface—or the interface between the two major phases—is assumed to be physically undisturbed by the presence of the microphase.
3. Although the microparticles circulate in the fluid bulk as a result of external agitation, it is assumed that the particles contained in the fluid elements in transient contact with the interface remain stationary. Thus it is possible to assume a Sherwood number of 2, a generally accepted value for mass transfer to and from a spherical particle in an infinite medium.
4. *Macro- and Microregimes.* Now we modify the theory of mass transfer accompanied by chemical reaction presented in Chapter 14 to include the effect of a microphase. This effect can manifest itself as a shift in a controlling regime. Thus a reaction that is, say, in regime 2 in the absence of microphase (curve I of Figure 23.3) can very well shift to regime 3 in the presence of one (curve II). Although there is still no

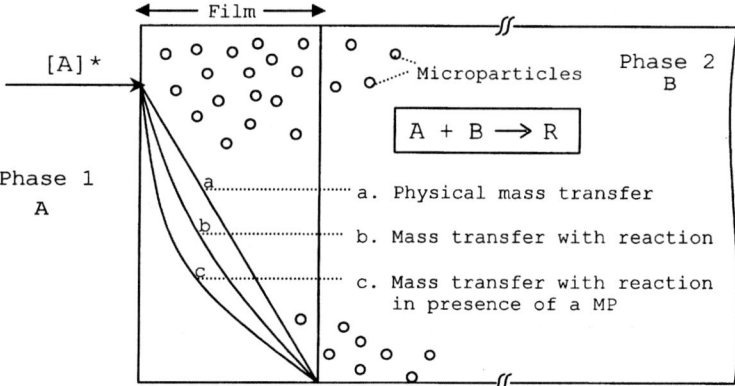

Figure 23.2 Sketch illustrating rate enhancement due to microphase action (based on film theory).

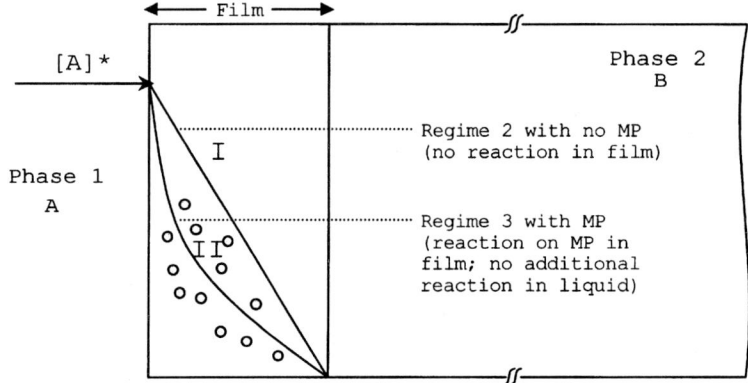

Figure 23.3 Film theory representation of microphase action

reaction in the liquid within the film and in a strict sense therefore the reaction continues in regime 2, the fact is that reaction does occur on the microphase in the film. Hence it makes practical sense to identify the macroregimes in the presence of a microphase in terms of the overall behavior of the liquid–microphase system and not just the liquid phase.

Now let us look at the regime within the microenvironment corresponding to the microphase. Different controlling regimes for reaction in (or on) the microphase are possible within each of the macroregimes, depending on the mechanism of solute uptake by the microphase. Thus we can have kinetic, mass transfer, or mixed control, and the regimes corresponding to them are referred to as *microregimes*. A classification of these regimes appears in Table 23.1, and Figure 23.4 illustrates the situation for macroregime 2.

Table 23.1 Classification of regimes in the presence of a microphase

Macroregime	Microregime		
	Kinetic	Mass transfer	Mixed
Regime 2 (slow reaction)	2 (a)	2 (b)	2 (c)
Regime between 2 and 3	2–3 (a)	2–3 (b)	2–3 (c)
Regime 3 (fast reaction)	3 (a)	3 (b)	3 (c)
Regime 4 (instantaneous reaction)	4 (a)	4 (b)	4 (c)

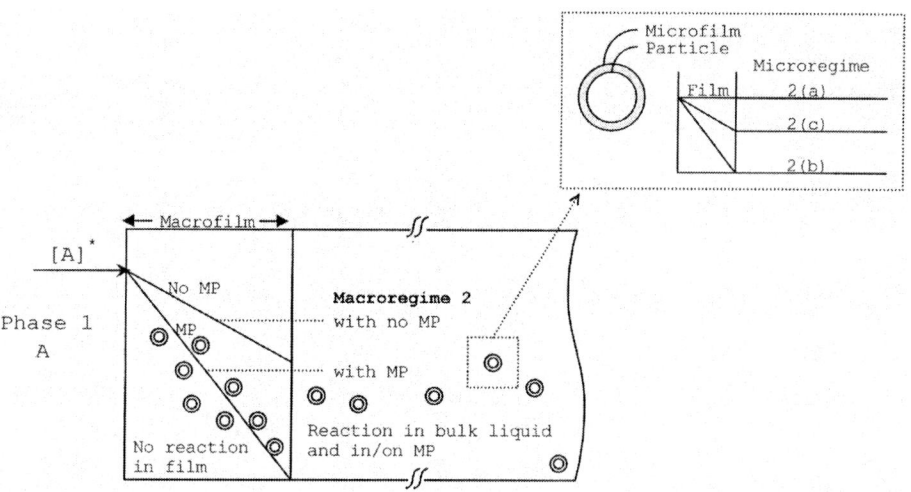

Figure 23.4   Classification of microregimes for macroregime 2

MICROPHASE AS A CHEMICAL (CATALYTIC) OR PHYSICAL SINK

When the microphase is an externally introduced catalyst, as in a slurry reactor or a physically adsorbing solid such as microfine carbon particles, the solute is consumed within (or on) this phase. Thus there is no accumulation of A in the microphase, resulting in steady-state operation. This enables the use of the simple film theory that is restricted to steady-state transport and is inherently inapplicable to time-dependent situations (see chapter 4).

When a microphase is added to the system, the effect is to increase the specific reaction rate by increasing the overall mass transfer rate. This leads to different effects in different regimes. Regime 1 being kinetically controlled is little influenced by fluctuations in the transport rate, particularly when this rate increases, as in the present case. Regime 4 is rather distinctive and involves complete depletion of B in the film (and is considered separately). Thus we are largely concerned with slow and fast reaction regimes (2 and 3).

Where liquid-liquid systems are concerned, it is conceivable that the microphase is present in both liquid phases. The analysis of such systems is more complicated and will not be attempted here (see Janakiraman and Sharma, 1985). The developments that follow are restricted to systems where the microphase is present only in the continuous phase.

Slow Reaction Regime (No Reaction in Film Liquid)

The condition for the slow reaction regime where no reaction occurs in the film and the rate equation for this case are given in Table 14.2. This rate increases in the presence of a microphase as a result of the uptake of A from phase 1.

The uptake coefficient

To formally extend the analysis to include this enhancing effect of the microphase, we write the transport equation for this case as

$$D_A \frac{d^2[A]}{dx^2} = \text{reaction in microphase in the film} \tag{23.1}$$
$$= h_0 k_0 [A] = h_0 k_0' a_{MP} [A]$$

where h_0 is the microphase holdup (volume of microphase/total volume of liquid + microphase), k_0 (with units of 1/s) is the *uptake coefficient* based on the transport rate expressed as moles per unit time per unit volume $(-r_A)$:

$$-r_A = k_0 [A] \tag{23.2}$$

k_0' (with units of m/s) is the coefficient based on the rate r_A' expressed as moles per unit time per unit interfacial area between the microphase and the bulk liquid:

$$-r_A' = k_0' [A] \tag{23.3}$$

and a_{MP} is the interfacial area per unit volume of the microphase.

We have expressed the interfacial area as area per unit volume of microphase, although the reaction is expressed in terms of the volume of the liquid phase. An alternative, and often more useful, way of expressing the microphase interfacial area is per unit volume of the liquid $a_{L,MP}$. Thus $a_{L,MP} = h_0 a_{MP}$, and

$$h_0 k_0 = h_0 a_{MP} k_0' = a_{L,MP} k_0' \tag{23.4}$$

Therefore, Equation 23.1 can also be written as

$$D_A \frac{d^2[A]}{dx^2} = a_{L,MP} k_0' [A] = h_0 a_{MP} k_0' [A] = h_0 k_0 [A] \tag{23.5}$$

Notice that a new constant has been introduced in these equations, the uptake coefficient expressed as k_0 or k_0'. Now we shall examine the physical implications of this coefficient before proceeding with the solution of Equation 23.5 and its interpretation. As classified in Table 23.1, three microphase regimes are possible within each main macroregime, slow, fast, and mixed.

Kinetic control: regime 2(a)

Here, the overall rate must be equal to the reaction rate within the microphase, with $k_0 = k_{MP}$ (to be distinguished from k for reaction in the bulk). Note that $k_{MP,1}$ (or simply k_{MP}) is a pseudo-first-order rate constant given by $k_{MP,2}[B]_0$ where $k_{MP,2}$ is the second-order rate constant. Because only a fraction of A is within (or on) the microphase, the extent of pickup of A by the microphase should be known. This is given by the distribution coefficient m_A of A between the microphase and phase 2:

$$m_A = \frac{\text{concentration of } A \text{ in microphase}}{\text{concentration of } A \text{ in phase 2}} \tag{23.6}$$

Now the equation for this regime can be written by assuming that the mass transfer rate to the microphase is so high as to ensure that it is locally saturated with A:

$$k_0 = m_A k_{MP}, \quad k_0' = m_A k_{MP}' \tag{23.7}$$

Film diffusion control: regime 2(b)

Here, the coefficient k_0 must be equal to the coefficient for mass transfer across the liquid film surrounding the microphase $k_{L,MP}$:

$$k_0 = k_{L,MP}' a_{MP}, \quad \text{or} \quad k_0' = k_{L,MP}' \tag{23.8}$$

where $k_{L,MP}'$ is the area-based mass transfer coefficient for the film surrounding the microphase (m/s).

To use Equation 23.8, $k_{L,MP}'$ should be known. This can be obtained from knowledge of the Sherwood number $(k_{L,MP}' d_{p,MP}/D_A)$. It is reasonable to assume the well-known limit of $Sh = 2$ for stagnant flow, so that

$$\text{Sherwood No. } (Sh) = \frac{k_{L,MP}' d_{p,MP}}{D_A} = 2; \quad k_{L,MP} = \frac{2 D_A}{d_{p,MP}} \tag{23.9}$$

Combining this with Equation 23.8 and noting that $a_{MP} = 6/d_{P,MP}$, the following final expression for k_0 results:

$$k_0 = \frac{12 D_A}{d_{p,MP}^2} \tag{23.10}$$

Mixed control: regime 2(c)

Because both mass transfer and reaction are simultaneously operative in this regime, k_0 can be obtained by summing the resistances of the two:

$$k_0 = \frac{1}{1/k_{L,MP}' a_{MP} + 1/m_A k_{MP}} \tag{23.11}$$

In this case the concentration of A in the microphase is between zero and saturation.

Expressions for the reaction rate

Having formulated the equations for the uptake coefficient for different regimes, now we write the following boundary conditions for solving Equation 23.5:

$$\begin{aligned} x &= 0, \quad [A] = [A]^* \\ x &= \delta, \quad [A] = [A]_b \end{aligned} \tag{23.12}$$

We also write a mass balance at the end of the film $x = \delta$. This can be easily done because at steady state the entire amount of A passing the film at $x = \delta$ has to be consumed by reaction both in the bulk and in the microphase in the bulk.

$$\begin{pmatrix} \text{amount} \\ \text{passing film} \\ \text{at } x = \delta \end{pmatrix} = \begin{pmatrix} \text{amount} \\ \text{reacting} \\ \text{in bulk} \end{pmatrix} + \begin{pmatrix} \text{amount} \\ \text{reacting on} \\ \text{microphase} \end{pmatrix} \quad (23.13)$$

$$-D_A a_L \left[\frac{d[A]}{dx}\right]_{x=\delta} = k[A]_b + k_0 h_0 [A]_b$$

$$= k[A]_b + k_0' a_{L,MP}[A]_b$$

The solution of Equation 23.5 along with Equations 23.12 and 23.13 leads to a rather lengthy general expression for the specific rate (not given here). Simpler, and often more useful, forms of this expression can be obtained by reducing it to specific situations. For this purpose, we use the magnitude of the parameter

$$\sqrt{M_{MP}} = \frac{\sqrt{D_A k_0 h_0}}{k_L'} = \frac{\sqrt{D_A k_0' a_{L,MP}}}{k_L'} \quad (23.14)$$

as a measure of the extent of reaction in the microparticles in the film (recall from Chapter 14 that $\sqrt{M}$ is the extent of reaction in the liquid film). Note that k_L' in the above definition is the usual mass transfer coefficient for the liquid macrofilm. The difference between the two mass transfer coefficients k_L' and $k_{L,MP}'$ is brought out in Figure 23.5. Now we consider three cases, depending on the contribution of the microphase to reaction in the film. Regime 2 in which no reaction occurs in the liquid phase within the film is involved here.

CASE 1

$$\sqrt{M_{MP}} < 0.3, \qquad \frac{r_A'}{k_L [A]^*} \leq 1.0 \quad (23.15)$$

Here, the contribution of the microphase to reaction in the film is negligible. The reaction, which occurs exclusively in the bulk, is given by

$$(-r_A') a_L = [A]^* \left[\frac{1}{k_L' a_L} + \frac{1}{(k + k_0 h_0)}\right]^{-1} \quad (23.16)$$

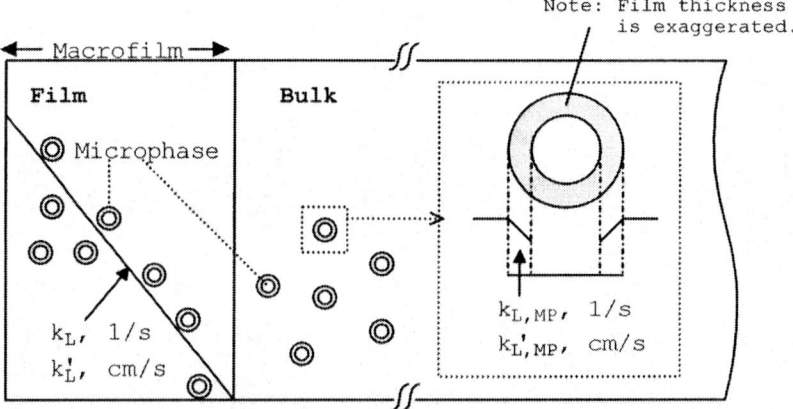

Figure 23.5 The mass transfer coefficients involved in the analysis of microphase action

Equation 23.16 makes no assumptions as to the relative rates of reaction and mass transfer with respect to reaction in the microphase, and therefore represents the general case of mixed control regime 2(c). If $k'_L a_L \gg (k + k_0 h_0)$, the reaction is in the kinetic regime, regime 2(a), and for the reverse situation in the mass transfer regime, regime (2b).

CASE 2

$$0.3 < \sqrt{M_{MP}} < 3.0, \quad \frac{r'_A}{k'_L[A]^*} > 1.0 \tag{23.17}$$

The situation here is one where part of the reaction occurs in/on the microphase in the film and the rest in the bulk, that is, both terms on the right-hand side of Equation 23.13 are operative. Therefore the reaction is in regime 2–3, and depending on the relative magnitudes of $k'_{L,MP} a_{MP}$ and $m_A k_{MP}$, it can be in microregime 2–3(a), 2–3(b), or 2–3(c). The specific rate is given by

$$-r'_A = k'_L[A]^* \frac{\sqrt{M_{MP}}}{\tanh\sqrt{M_{MP}}} \tag{23.18}$$

CASE 3

$$\sqrt{M_{MP}} > 3.0 \tag{23.19}$$

Here, the reaction occurs entirely in the film, regime 3. In other words, the effect of the microphase shifts the reaction from the slow to the fast macroregime, as shown by the concentration profiles of A in Figure 23.3. Depending on the relative magnitudes of $k'_{L,MP} a_{MP}$ and $m_A k_{MP}$, the reaction can be in microregime 3(a), 3(b), or 3(c). Then the specific rate of reaction is

$$-r'_A = k'_L[A]^* \sqrt{M_{MP}} \tag{23.20}$$

This equation holds only if there is no depletion of B, for which the condition is

$$\sqrt{M_{MP}} < \frac{[B]_b D_B}{\nu_B [A]^* D_A} \tag{23.21}$$

Fast Reaction Regime (Reaction in Film Liquid)

In writing the mass balance for the fast reaction regime, we consider two reaction pathways: (1) reaction of A, diffusing into the liquid film, with B; and (2) reaction of A, picked up by the microphase, with B that diffuses onto this phase. The second constitutes the additional term introduced due to the presence of the microphase. The concentration profiles of A without the microphase and with the microphase would be similar to those shown in Figure 23.2. The profile for pure mass transfer is also included as curve A for comparison. The specific rate is obviously highest in the presence of a microphase, for which the following mass balance can be written

$$D_A \frac{d^2[A]}{dx^2} = k[A] + h_0 k_0 [A] \tag{23.22}$$

The boundary conditions at the phase 1–film and phase 2 (bulk)–film interfaces are

$$x = 0, \quad [A] = [A]^*$$
$$x = \delta, \quad [A] = [A]_b = \text{(usually)} \tag{23.23}$$

These equations can be solved to give the specific rate r'_A:

$$-r_A = -D_A \left[\frac{d[A]}{dx}\right]_{x=0} = \frac{k_L [A]^* (M')^{1/2}}{\tanh(M')^{1.2}} \tag{23.24}$$

where $\sqrt{M'}$ is a dimensionless group defined as

$$\sqrt{M'} = \frac{\sqrt{(D_A(k_1 + h_0 k_0))}}{k'_L} \tag{23.25}$$

This group is usually designated as a modified Hatta number Ha' to distinguish it from the original Hatta number Ha (see Chapter 14) which does not include the microphase term $h_0 k_0$.

Now we can define a microphase-assisted enhancement factor much in the manner of the factor defined in Chapter 14, but modified specifically to reflect microphase action:

$$\eta' = \frac{\text{rate with a microphase}}{\text{rate without a microphase}} \tag{23.26}$$

The rate without a microphase can be found by solving Equation 23.22 with the microphase term $h_0 k_0 [A] = 0$. Then, from Equation 23.26,

$$\eta' = \left(\frac{M'}{M}\right)^{1/2} \frac{\tanh(M)^{1/2}}{\tanh(M')^{1/2}} \tag{23.27}$$

As in a slow reaction, k_0 is given by Equation 23.7, 23.10, or 23.11. The appropriate expression may be introduced in Equation 23.25, which then may be solved to obtain an expression for the specific rate.

Instantaneous Reaction Regime

Here, reaction occurs within the film (or the fluid element) on a plane where the concentrations of both A and B approach zero (see Figure 14.4). The penetration theory accounts for reaction plane movement toward the bulk without a microphase, but the presence of one reverses the direction. We shall not pursue this problem further.

EXAMPLE 23.1

Oximation of cyclododecanone: A problem to illustrate the analysis of regimes for a reaction with an external microphase.

Oximation of cyclododecane, a solid-liquid reaction, was studied by Janakiraman and Sharma (1985) by reacting finely ground particles of cyclododecanone (CD) with aqueous hydroxylamine sulfate (HAS) in a mechanically agitated contactor, with and without microparticles of carbon. The carbon particles are not catalysts but strongly adsorb the solid reactant (which is otherwise only sparingly soluble in the aqueous phase) and transport it to the aqueous bulk, where reaction occurs at an enhanced rate. The reaction was carried out at 28 °C, at an agitator speed of 1200 rpm, and a ketone holdup of 0.04 g/cm^3. Carbon microparticles of two sizes, 1.7 µ and 4.33 µ, were used, and volumetric rates (mol/s cm^3 aqueous phase) were determined at different microphase loadings for each size. These are presented in Table 23.2.

The values of the system parameters used in the computations were either measured experimentally or estimated from available correlations and are as follows: holdup of ketone = 0.4 g/cm^3, [HAS] = 0.4 M, average initial particle size = 157 µ, a_L = 15.44 cm^2/cm^3, $k'_L = 1.2 \times 10^{-3}$ cm/s, $m_A = 0.022$, $a_{L,ave} = 15.44$ cm^2/cm^3 aqueous phase, $D_A = 5.8\ 10^{-6}$ cm^2/s, solubility of cyclododecanone in 0.4 M HAS solution $[A]^* = 1.44 \times 10^{-7}$ mol/cm^3, carbon density $\rho_{MP} = 0.5$, and carbon particle size $d_{MP} = 1.7 \times 10^{-4}$ cm.

Using the data and results in the table, it is desired to identify the operating macro- and microregimes of this system and also to determine if the microphase merely hastens the reaction in a given regime or whether it causes a shift in regime.

SOLUTION

Macroregime without a microphase

Because most equations use the surface-based rate $-r'_A$, we first calculate it from $-r'_A = -r_A/a_{L,ave}$ using a value of $a_{L,ave} = 15.44$ cm^2/cm^3 (see original article). Then we determine the macroregime by calculating the ratio of the rates of reaction and mass transfer.

$$\frac{(-r'_A)}{k'_L[A]^*} = \frac{1.1 \times 10^{-10}}{1.2 \times 10^{-3} \times 1.44 \times 10^{-7}} = 0.64$$

Because this ratio is less than 1, the reaction is slow and occurs exclusively in the bulk. The rate is given by Equation 23.16 with $k_0 h_0 = 0$. Thus we obtain

$$k = 0.033\ \text{s}^{-1}$$

Because k and $k_L a_{L,ave}\ (= 0.0185\ \text{s}^{-1})$ are of comparable magnitudes, the reaction falls between regimes 1 and 2.

Macroregime with a microphase

Let us again calculate the ratio $-r'_A/k'_L[A]^*$, this time for run 2 with a carbon loading of 0.25. The value obtained is 1.1. Because this value is greater than 1, the reaction is in macroregime 3. Thus it has shifted from macroregime 1–2 without carbon to macroregime 3 (fast reaction) with it. This is true for all loadings (see table).

Table 23.2 Rates of oximation of cyclododecanone in the presence of microparticles of carbon (from Janakiraman and Sharma, 1985)

w (g/cm^3 × 100)	Carbon particle size (μ)	$-r'_A$ (mol/cm^2 s × 10^{10})	ϕ	$-r'_A/[A]*k_L$	$\sqrt{M_{MP}}$	$k'_0 a_{MP}$ (1/s)
0	—	1.10	1.0	0.64	0	0
0.25	1.7	1.90	1.73	1.10	0.58	0.08
0.50	1.7	2.58	2.34	1.49	1.30	0.42
0.75	1.7	3.17	2.88	1.83	1.73	0.74
1.00	1.7	3.65	3.32	2.11	2.05	1.04
1.50	1.7	4.37	3.97	2.53	2.53	1.59
0.50	4.33	1.83	1.66	1.06	0.43	0.05
0.75	4.33	2.18	1.98	1.26	0.91	0.20
1.00	4.33	2.59	2.35	1.50	1.30	0.42
1.50	4.33	3.19	2.91	1.85	1.73	0.74
2.00	4.33	3.80	3.46	2.20	2.15	1.15

Microregime

The specific rate in macroregime 2–3 is given by Equation 23.18. This equation is also valid for regime 3, but the denominator vanishes (when the value of $\sqrt{M_{MP}}$ exceeds 3.0). Knowing the specific rate, we calculate $\sqrt{M_{MP}}$ from this equation. Then, from the definition of $\sqrt{M_{MP}}$ given by Equation 23.14, we calculate the microphase contribution $k_0 h_0$, or $k'_0 a_{L,MP}$. Thus for run 2, $k'_0 a_{L,MP} = 0.08 \, \text{s}^{-1}$.

By a similar procedure, we calculate $k'_0 a_{L,MP}$ for all of the other runs with microphase, and the results are included in Table 23.2. Then we get k'_0 from (taking run 2)

$$k'_0 = \frac{0.08}{a_{L,MP}} = \frac{0.08}{(6w/\rho_{MP} d_{p,MP})} = \frac{0.08 \times 0.5 \times 1.7 \times 10^{-4}}{6 \times 0.25}$$

$$= 4.5 \times 10^{-6} \, \text{cm/s}$$

To determine the regime of the microphase reaction (i.e., the microregime), we compare k'_0 with $k'_{L,MP}$ for the mass transfer limit and with $k'_{MP} m_A$ for the kinetic limit. To obtain the mass transfer coefficient $k'_{L,MP}$, we use the relationship

$$Sh = \frac{k'_{L,MP} d_{P,MP}}{D_A} = 2$$

giving $k'_{L,MP} = 2 \times 5.8 \times 10^{-6} / 1.7 \times 10^{-4} = 0.068 \, \text{cm/s}$. To get the kinetic limit $k'_{MP} m_A$, we assume that the rate constant for reaction in the microparticle is approximately equal to that in the bulk, $k \cong k_{MP}$. This gives $k'_{MP} m_A = m_A k_{MP} / a_L \cong 0.022 \times 0.033 \times (1/15.44) = 4.7 \times 10^{-5} \, \text{cm/s}$. Thus

$$k'_0 = 4.5 \times 10^{-6} \, \text{cm/s}$$

$$k'_{L,MP} = 6.8 \times 10^{-2} \, \text{cm/s}$$

$$k'_{MP} m_A = 4.7 \times 10^{-5} \, \text{cm/s}$$

From the magnitudes of the various rate constants, we see that $k'_0 \ll k'_{L,MP}, k'_0 \ll k'_{MP} m_A$. We may conclude that the reaction is in the kinetically controlled microregime 3(a).

Similar calculations for 4.33 μ particles show that the reaction is also in regime 3(a) for this case. Hence we conclude that

regime without microphase:	regime 1–2
regime with microphase:	regime 3
microregime:	regime 3(a)
overall regime classification:	regime 3(a)

MICROPHASE AS A PHYSICAL CARRIER

Now if we extend the analysis to a microphase that acts merely as a physical carrier of the solute (Figure 23.1), there would be an accumulation of A in the microphase by solubilization or adsorption within (or on) this phase. As a result, the steady-state equations would no longer be applicable. The extent of accumulation is determined by the distribution coefficient of A between the microphase and the continuous phase. In such a situation, an unsteady-state model must be used. Intentional addition of a noncatalytic microphase as

MICROPHASE AS A SOURCE—REACTANT AS MICROPHASE

Two important situations can be identified in this category: (1) reactant B is a sparingly soluble solid and (2) reactant A is present both as large and fine (micro) bubbles of gas. The latter involves dissolution of gaseous A in phase 2 in the form of macrobubbles accompanied by enhancement of the rate by microbubbles of A and is not considered here. Case 1 is developed below.

Consider Figure 23.6 in which phase 2 contains a sparingly soluble reactant B, which provides B in the bulk for reaction with A diffusing from phase 1. Because the solubility of B is inherently low in this system, the reaction rate is also correspondingly low. Therefore, to achieve a nontrivial degree of enhancement in the specific rate, B should be consumed as fast as possible near the macrointerface, thus leading to its depletion in that region and creation of a driving force for the dissolution of more B. Rapid dissolution of solid B can be accomplished by using microparticles of the solid which can enter the film. Because the depletion of B would be maximum in the instantaneous regime, one would primarily be concerned with this regime when the microphase acts as a source. Therefore, our treatment of this situation will, be restricted to the instantaneous regime (Ramachandran and Sharma, 1969; see Uchida et al., 1975, for a more elaborate treatment, and also Sada et al., 1980).

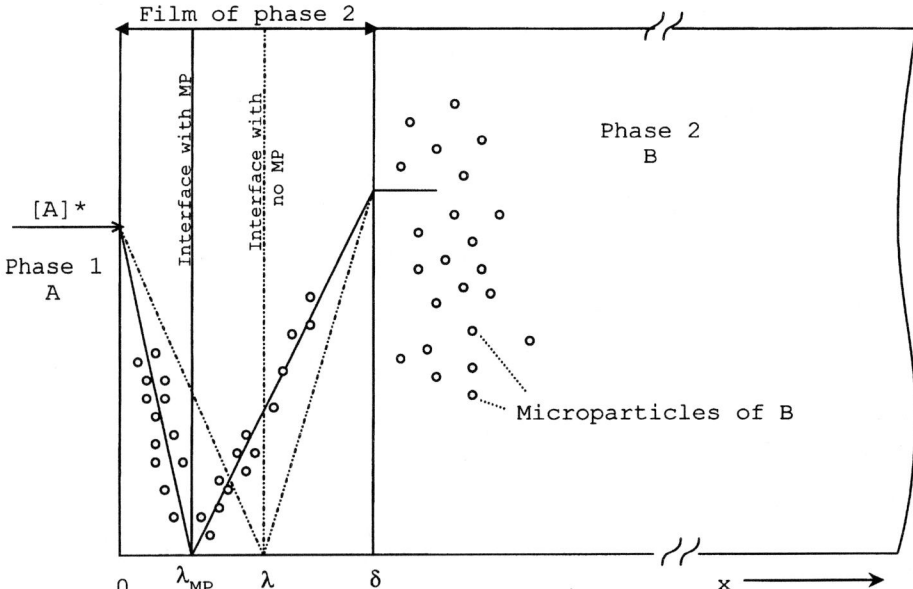

Figure 23.6 Representation of reaction between solid B dissolving in phase 2 film and diffusing gas A

In the instantaneous regime, A and B diffuse from opposite directions to a plane in the film where they are consumed fully by reaction. Thus in this case, balance equations for species A and B must be written, remembering that A disappears progressively in the region $0 < x < \lambda$ and B in the region $\lambda < x < \delta$ (Figure 23.6), and must be solved under appropriate boundary conditions. The final equation obtained is

$$(-r'_A) = \frac{D_A[A]^*}{\lambda} + \frac{h_0 k_0 \lambda [B]_b^*}{2\nu_B}$$

$$= \left(\frac{[B]_b^*}{\nu_B}\right)(D_B h_0 k_0)^{1/2} \coth\left(\frac{h_0 k_0}{D_B}\right)^{1/2}(\delta - \lambda) + \left(\frac{h_0 k_0 [B]_b^*}{\nu_B}\right) \quad (23.28)$$

where $[B]_b^*$ is the saturation concentration (solubility) of B in the liquid bulk, and k_0 for this case is $k'_{L,MP} a'_{MP}$.

Because the solubility of B is very low in the situation considered, the reaction plane tends to coincide with the macrointerface ($x = 0$). This implies that now the rate is controlled only by the diffusion and dissolution of B. For this to happen, the condition

$$\left(\frac{h_0 k_0}{D_B}\right)^{1/2} \delta > 5.0 \quad (23.29)$$

should be satisfied, and Equation 23.28 simplifies to

$$(-r'_A) = \frac{[B]_b^*}{\nu_B}(D_B h_0 k_0)^{1/2} \quad (23.30)$$

When solid dissolution is unimportant, we have a situation where absorption of A may be deemed to occur in a solution saturated with B. This is identical to regime 14.4 described in Chapter 14, and Equation 14.24 is recovered.

SOLID PRODUCT AS A MICROPHASE—MICROPHASE-ASSISTED AUTOCATALYSIS

In this section we shall apply the basic principles of microphase action developed earlier to a product acting as a microphase. Because a product of reaction provides the microphase for rate enhancement, we refer to this as *microphase-assisted autocatalysis*. Besides the conceptual distinctiveness of this system, its major additional features are the crystallization and growth kinetics of the product. Clearly, this would call for a detailed analysis of the entire process of crystallization, but we shall not attempt it here. Our approach is to let the basic features of the theory unfold as we apply it to a specific system of industrial importance, production of citric acid.

EXAMPLE 23.2

Microphase-assisted autocatalysis in the production of citric acid: Analysis of rate enhancement through a study of controlling regimes and crystallization kinetics

The prevalent manufacturing process for citric acid is fermentation followed by downstream purification. A common method of purification is to treat the products of fermentation with lime slurry followed by reaction of the solid calcium citrate formed with aqueous sulfuric acid to give an aqueous solution of pure citric acid and a precipitate of calcium sulfate. Our interest is in the final purification step which may be written as

$$Ca_3(C_6H_5O_7)_2(s) + 3H_2SO_4(l) \rightarrow 2C_6H_8O_7(l) + 3CaSO_4(s) \quad \text{[E23.2.1]}$$
$$(A) (B)$$

Note that this represents a system in which a sparingly soluble solid (calcium citrate) reacts with a liquid to give a liquid product (citric acid) and a solid precipitate (calcium sulfate).

METHODOLOGY OF MODEL DEVELOPMENT AND VALIDATION

The film theory representation of the reaction is similar to that shown in Figure 23.1c. If the R formed acts as a microphase, the rate of mass transfer of A from the solid to the liquid phase is enhanced. Then the "additional" A picked up by R is discharged into the bulk where it reacts with B. Alternatively, B is transported onto the microphase where it reacts with A. Thus a set of events occurs in the macrofilm associated with A, and another with the microfilm surrounding a microparticle of product R. The construction of a model for this situation will require the precise identification of these macro- and microregimes (as in Example 23.1), formulation of appropriate equations, and coupling of these equations with those for crystallization. This coupling, absent in the cases considered so far, is crucial when a product acts as a microphase because it accounts for the changing condition of the microphase through factors such as rates of crystal nucleation and growth.

Experimental verification of autocatalytic enhancement

Experimental data, plotted in Figure 23.7 as conversion versus time, show a sudden rise in conversion at 85% when precipitation starts. This clearly suggests "autocatalytic" action by the precipitating solid.

Mass transfer and intrinsic kinetics

The mass transfer coefficient across the macrointerface, determined by the usual concentration versus time data, is plotted in Figure 23.8 as a function of microphase loading. The reaction rate parameters in the absence of microphase were also determined by the usual method from the rate equation

$$(-r_A)_{aq} = k_{mn}[A]^m[B]^n \quad (E23.2.1)$$

Other parameters such as the interfacial area, diffusion coefficient, etc. were estimated using well-known methods. The values at 15 °C are $m = n = 0.5$, $-r_A = 0.025$ mol/ℓs, $D_A = 1.65 \times 10^{-5}$ cm^2/s, mass transfer rate $= 8.3 \times 10^{-4}$ mol/ℓs, $a_L = 50$ cm^2/cm^3, $[A]^* = 0.014$ mol/ℓ, $[B]_b = 0.01$ mol/ℓ, $m_A = 5000$, $m_B = 5000$, $k'_L = 1.93 \times 10^{-2}$ cm/s, and $k_{mn} = 3.9$ s^{-1} (see also Anderson, 1996).

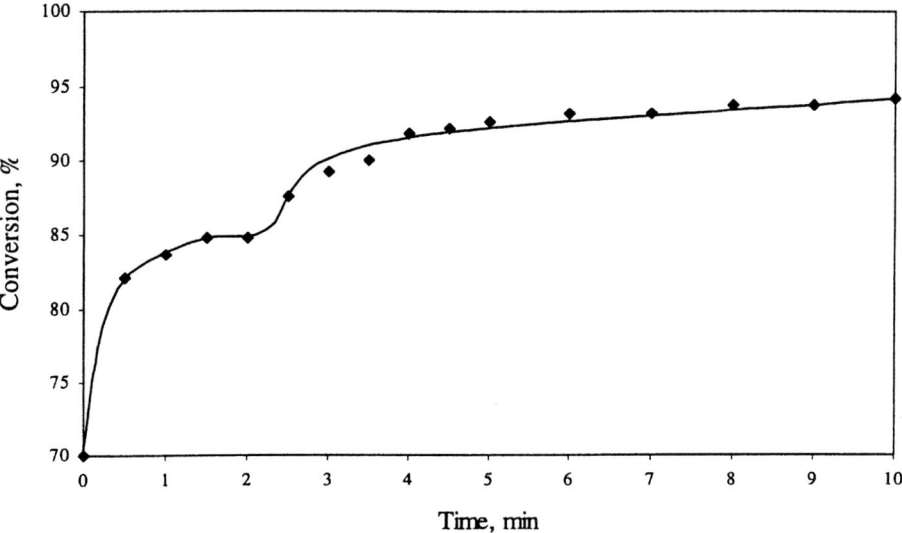

Figure 23.7 Autocatalytic behavior due to the crystallized product (calcium sulfate) acting as a microphase in the citric acid reaction (Anderson et al., 1998b)

Macroregime without a microphase

A comparison of the reaction and mass transfer rates enables identification of the controlling macroregime. Based on the parameter values given, comparison of the two rates for this reaction shows that the mass transfer rate is much lower than the reaction rate. Thus the reaction is probably in regime 2 (3 and 4 cannot also be ruled out). A more selective test (see Chapter 14) is the magnitude of the Hatta number: it should be less than one if there is no reaction in the bulk, that is,

$$Ha = \sqrt{M} = \frac{\sqrt{\frac{2}{m+1} D_A k_{mn} [A]^{*m-1} [B]_b^n}}{k'_L} < 1 \qquad (\text{E}23.2.2)$$

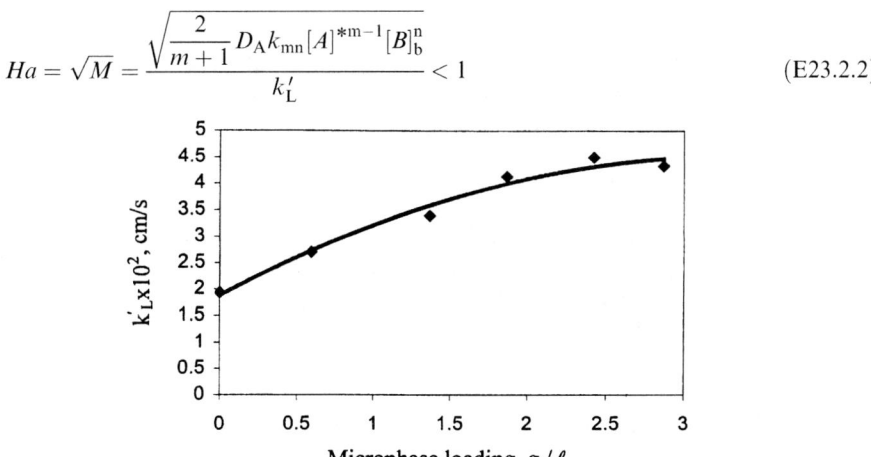

Figure 23.8 Effect of microphase (calcium sulfate) loading in the citric acid reaction (Anderson et al., 1998b)

Using the values listed before,
$$Ha = 0.44 < 1$$
Thus there is negligible reaction in the film. In other words, regimes 3 and 4 may be ruled out. Hence we may conclude that the reaction falls in regime 2, slow reaction.

Macroregime with a microphase

The total reaction rate with a microphase is equal to the homogeneous reaction rate in the aqueous phase plus the reaction rate on/in the microphase, and is given by
$$(-r_A)_t = k_{mn}[A]^m[B]^n + k_m([A]m_A[B]m_B) \tag{E23.2.3}$$
The first term on the right represents the homogeneous reaction rate in the aqueous phase and is not affected by the microphase. Clearly, the reaction rate on the microphase (or any additional reaction in the aqueous phase due to the microphase) can be obtained by subtracting the observed aqueous phase rate without a microphase from the observed (total) rate with a microphase.

Experimental data show that for an initial microphase loading of, say, $0.57 \text{ g}/\ell$ (based only on particles less than 10 μm from the full amount of R added; 10 μm is the assumed cutoff point for microphase action), the experimental conversion rate was 1.76 times without a microphase. Now let us assume that the observed increase in the mass transfer coefficient due to the presence of microphase, as shown in Figure 23.8 (2.61×10^{-2} at a loading of $0.57 \text{ g}/\ell$ as against 1.82×10^{-2} cm/s without a microphase) is reflected entirely in an increase in the aqueous phase concentration of the citrate. When this increased concentration ($0.0045 \text{ mol}/\ell$) is substituted in Equation E23.2.1, the rate is
$$-r_A = 3.9(0.0045)^{0.5}(0.01)^{0.5} = 0.026 \text{ mol}/\ell s$$
which closely matches the experimentally determined value with the microphase ($0.025 \text{ mol}/\ell s$). Therefore, the assumption made above is correct, and the increased reaction rate is due entirely to an increase in the aqueous phase concentration.

The Hatta number with microphase can be calculated by using the increased value of $k'_L(2.61 \times 10^{-2}$ cm/s) for the mass transfer coefficient and of $[A]^*$ equal to $0.014 \text{ mol}/\ell$. Thus
$$Ha = 0.36 < 1$$
The enhanced mass transfer rate is calculated as $2.61 \times 10^{-2} \times 50 \times (0.014 - 0.008) = 6.8 \times 10^{-3} \text{ mol}/\ell s$. Because Ha continues to be less than one and the enhanced mass transfer rate is still less than the reaction rate, we conclude that the reaction continues to be in regime 2 despite the increase in rate due to microphase action.

Microregime

Because there is no reaction on the microphase, mass transfer of calcium citrate between the aqueous phase and the microphase during its pickup in the neighborhood of the solid (calcium citrate) or during its subsequent discharge in the aqueous bulk is the only possible rate-limiting step to be considered.

Reaction equations

WITHOUT A MICROPHASE

Because the reaction occurs in regime 2, the unsteady-state mass balance equation for A is given by
$$D_{bA} \frac{\partial^2[A]}{\partial x^2} = \frac{\partial[A]}{\partial t} \tag{E23.2.4}$$

with initial and boundary conditions

I.C. $[A]$ = finite value

B.C. $x = 0, \quad [A] = [A]^*$ (E23.2.5)

$x = \delta, \quad [A] = 0$

These equations can be solved to obtain the concentration in the film as a function of time (see Astarita, 1967; Doraiswamy and Sharma, 1984).

The rate of consumption of sulfuric acid is given by the mass transfer rate of calcium citrate,

$$-\frac{d[B]}{dt} = k'_L a_L [A]^*$$ (E23.2.6)

It is important to be able to calculate this rate because experimental results are often more easily obtained in terms of the rate of consumption of the liquid-phase reactant (sulfuric acid in this case).

WITH A MICROPHASE

The reaction with a microphase is far more difficult to model. Equations for both the phases and crystallization must be developed and solved simultaneously. Because all of these processes are time-dependent in a batch reaction such as this, the penetration model (see Chapter 4) is more appropriate for describing mass transfer.

The continuous phase unsteady-state mass balance equation is similar to that for the base reaction given by Equation E23.2.4, except that an additional term is needed to account for the uptake of the microphase:

$$D_A \frac{\partial^2 [A]}{\partial x^2} = \frac{\partial [A]}{\partial t} + h_0 k_0 \left([A] - \frac{[A]_{MP}}{m_A} \right)$$ (E23.2.7)

where the uptake coefficient k_0 is expressed by Equation 23.11. The conditions given by Equation E23.2.5 apply equally to this case.

The rate of consumption of sulfuric acid given by Equation E23.2.6 also has an additional term now to account for the uptake to (or discharge from) the microphase. Thus

$$-\frac{d[B]}{dt} = k'_L a_L [A]^* + h_0 k_0 \left([A] - \frac{[A]_{MP}}{m_A} \right)$$ (E23.2.8)

It is apparent from Equations E23.2.7 and E23.2.8 that the concentration on the microphase as a function of time is an important requirement of the model. We address this problem next.

The general mass balance equation for A on the microphase can be written as

$$\frac{d[A]_{MP}}{dt} = h_0 k_0 \left([A] - \frac{[A]_{MP}}{m_A} \right) - k_{MP} [A]_{MP}^m [B]_{MP}^n$$ (E23.2.9)

by assuming that the diffusion of microparticles themselves is negligible. The second term, denoting reaction on the microphase, is zero in this case.

Equations E23.2.8 and E23.2.9 can be solved using the appropriate initial and boundary conditions to compute the concentrations of A in the continuous and microphases as functions of time. However, these model equations depend on the average particle diameter, surface area, and volume of the microphase. Because they are constantly changing as more crystals nucleate and grow, complete knowledge of the crystallization kinetics of calcium sulfate is necessary to solve the equations.

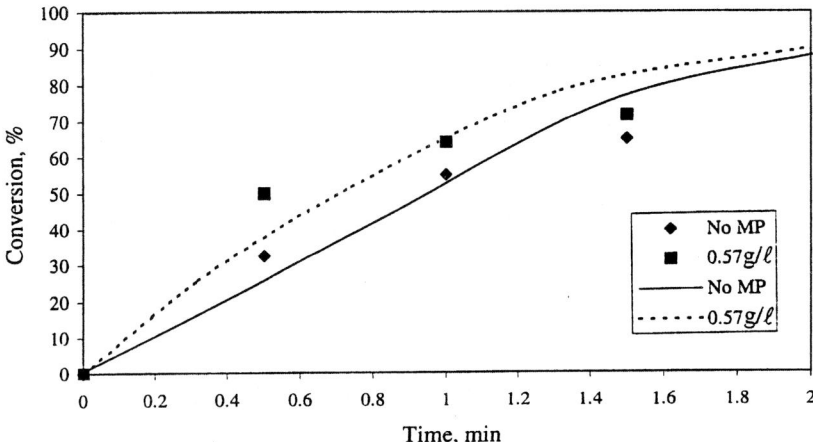

Figure 23.9 Experimental test of the proposed autocatalytic model (Anderson et al., 1998b)

CRYSTALLIZATION KINETICS

In modeling normal crystallization, the entire range of crystal sizes comprising the precipitate must be considered (Randolph and Larson, 1988). On the other hand, when modeling the role of a microphase, only the size range corresponding to the microphase must be considered, for the rest does not influence the reaction. It is here that the greatest uncertainty lies, for specific numbers should be assigned to the upper and lower limits of the microphase that are active in enhancing the rate. The best approach would be to make several assumptions, solve the equations for each of them, and compare the values with the experimental results. This would mean a very large number of trials. Therefore, we present only the final results obtained on the assumption that the upper and lower limits of the microphase are 10 μm and 1 μm, respectively (see Anderson et al., 1998a,b, for details of crystallization kinetics).

MODEL VERIFICATION

The model results without a microphase and with 0.57 g of microphase are shown in Figure 23.9 with the corresponding experimental data. The average deviation is approximately 12%. This is caused by several factors, some of which probably are that there is (as already noted) no accurate way of determining the particle size cutoff points for microphase action and the values used are arbitrary; the value of the diffusion coefficient used is no more than a good estimate; and the distribution coefficient may not be independent of concentration, as assumed in the model.

This example (supplemented by the additional results of Anderson et al., 1998a,b) shows that a microphase can greatly influence the nucleation and growth rates of crystals, which in turn can drastically alter the crystal size distribution of the precipitate. In other words, the average precipitate size can be controlled to enhance the precipitation and therefore the reaction rate. A parallel example is the use of microemulsion media to obtain monodispersed calcium carbonate particles via carbonation of calcium phenate (Marsh, 1987).

Chapter 24

Membrane-Assisted Reactor Engineering

Like zeolites that combine shape selectivity with catalysis (Chapter 6), membranes combine separation with catalysis to enhance reaction rates. The dual functionality of zeolites derives from the nature of the *catalytic material*, whereas that of membranes derives from the nature of *the reactor material*. The catalyst in the membrane reactor can be a part of the membrane itself or be external to it (i.e., placed inside the membrane tube). The chief property of a membrane is its ability for selective *permeation* or *permselectivity* with respect to certain compounds.

Organic membrane reactions are best carried out in reactors made of *inorganic membranes*, such as from palladium, alumina, or ceramics. Good descriptions of these reactions and the membranes used are available in many reviews, for example, Gryaznov (1986, 1992), Stoukides (1988), Armor (1989), Govind and Ilias (1989), Bhave (1991), Zaspalis and Burggraaf (1991), Hsieh (1989, 1991), Shu et al. (1991), Shieh (1991), Gellings and Bouwmeister (1992), Tsotsis et al. (1993b), Harold et al. (1994), Saracco and Specchia (1994), Sanchez and Tsotsis (1996). A recent trend has been to develop polymeric-inorganic composite type membranes formed by the deposition of a thin dense polymeric film on an inorganic support (Kita et al., 1987; Rezac and Koros, 1994, 1995; Zhu et al., 1996). Another class of membranes under development for organic synthesis is the *liquid membrane* (Marr and Kopp, 1982; Eyal and Bressler, 1993). The permselective barrier in this type of membrane is a liquid phase, often containing a dissolved "carrier" or "transporter" that selectively reacts with a specific permeate to enhance its transport rate through the membrane. Our main concern in this chapter will be with inorganic membrane reactors.

We commence our treatment with an introduction to the exploitable features of membrane reactors (with no attempt to describe membrane synthesis). Then

we describe the main variations in design and operating mode of these reactors, develop performance equations for the more important designs, and compare the performances of some important designs with those of the traditional mixed- and plug-flow reactors. Finally, we present a summary of the applications of membrane reactors in enhancing the rates of organic reactions.

GENERAL CONSIDERATIONS

Inorganic Membranes for Organic Reactions/Synthesis

Inorganic membranes can be grouped under two broad classes, dense and porous. Membranes made of dense palladium or its alloys and of porous glass are the most commonly used membranes for organic reactions. For reactions involving hydrogen, the dense membranes are typically metallic and are made of Pt or Pd (Gryaznov et al., 1986). These can be in the form of hollow tubes or foils or thin films deposited on porous supports prepared by various methods (Gryaznov et al., 1993; Shu et al., 1993; Li et al., 1993). These membranes have low permeability but high permselectivity.

Porous membranes are usually characterized by mesopores, with transport in the Knudsen regime. Typical examples are Vycor glass and γ-Al_2O_3. Thin film ceramic membranes also belong to this category and are prepared by sol-gel coating techniques (Leenaars et al., 1984, 1985; Uhlhorn et al., 1987, 1992a,b) or vapor deposition (CVD) (Gavalas et al., 1989; Lin and Burggraaf, 1992). Porous membranes have high permeability but relatively low permselectivity.

Polymeric–inorganic composites have been developed mainly for the separation of organic compounds through pervaporation (PV) and vapor permeation (VP). However, they can also be advantageously used for such reactions as esterification and condensation.

Potentially Exploitable Features of Membranes

Several attractive features of membranes derive from their unique characteristics and amenability to novel modes of operation. These are sketched in Figure 24.1 and briefly described here.

EQUILIBRIUM SHIFT

The most important feature of a membrane reactor that gives it a pronounced advantage over other reactors is its ability to remove a product selectively by letting it permeate out of the reactor through its inert (Figure 24.1a) or catalyst-containing (Figure 24.1b) membrane wall. Thus one can "beat the equilibrium" and achieve conversions beyond the limits of equilibrium.

CONTROLLED ADDITION OF REACTANTS

Controlled dosing of one of the reactants (usually hydrogen or oxygen) is often an important consideration in partial hydrogenation and oxidation reactions. The reduction of selectivity in the inlet region of a fixed-bed reactor can be

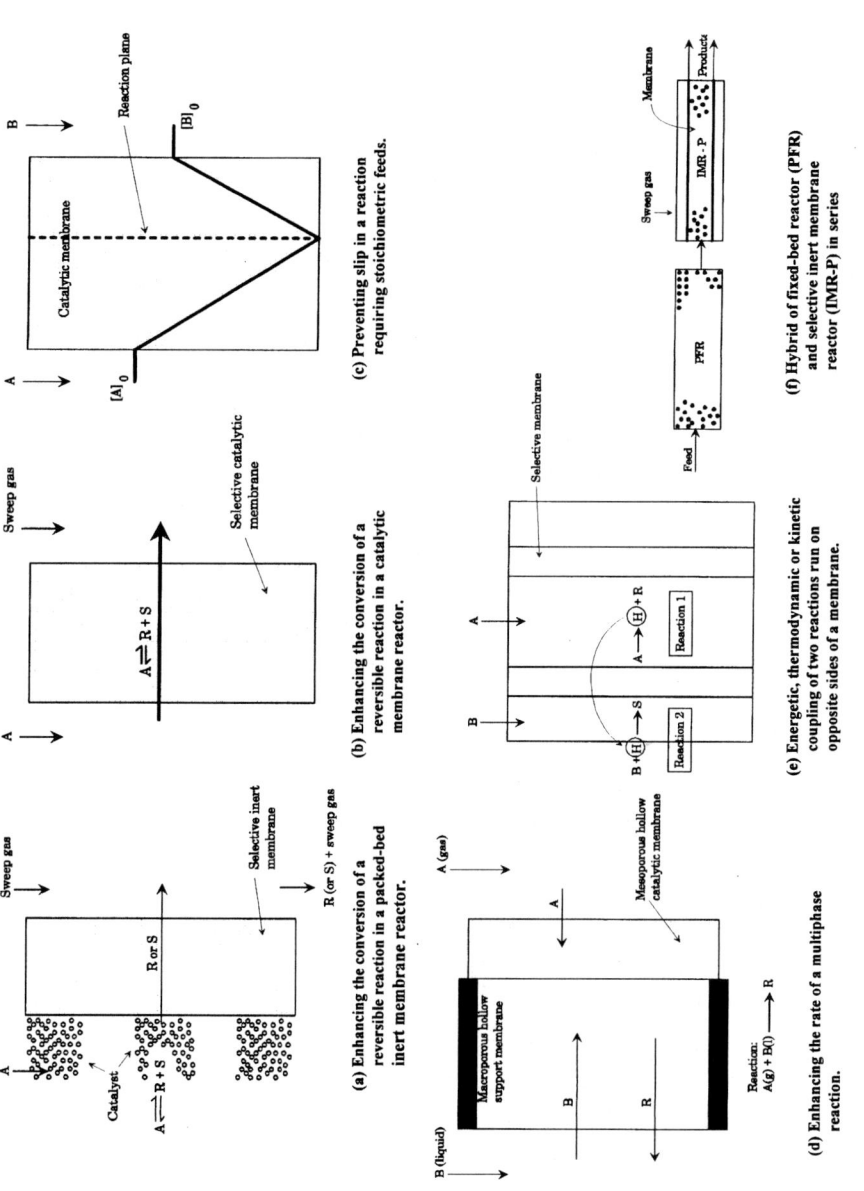

Figure 24.1 Exploitable features of membrane reactors

avoided by maintaining a uniformly low reactant concentration by controlled supply of the second reactant across the membrane wall of the entire reactor length.

PREVENTING EXCESS REACTANT "SLIP" IN REACTIONS REQUIRING STRICT STOICHIOMETRIC FEEDS

Certain reactions that involve more than one reactant require introduction of the reactants in precisely stoichiometric proportions. Thus in a reaction such as

$$\nu_A A + \nu_B B \rightarrow \nu_R R + \nu_S S \qquad [24.1]$$

any feed of A (or B) in excess of the ratio ν_A/ν_B (or ν_B/ν_A) "slips" into the product stream. Such a slip, generally undesirable, would be particularly unacceptable if the reactant concerned happened to be a toxic pollutant.

If A and B are allowed to diffuse into the membrane from opposite sides and react instantaneously and completely at a plane whose location is determined by stoichiometry (Figure 24.1c), then there would be no slip on either side (Zaspalis et al., 1991; Sloot et al., 1992). Two other noteworthy features of this concept are that permselectivity of the membrane is not essential and permeability is usually low.

MIMICKING TRICKLE-BED OPERATION WITH IMPROVED PERFORMANCE

Another use of nonpermselective membranes is in multiphase organic reactions involving trickle-bed type reactors (Harold et al., 1989, 1994; Cini and Harold, 1991). The reactor consists essentially of a hollow macroporous membrane tube coated on the inside with a hollow mesoporous catalyst layer, as shown in Figure 24.1d. Liquid and gas are allowed to flow on opposite sides of the membrane. Because the gas comes into *direct contact* with the liquid-filled catalyst, it resembles a trickle-bed reactor. However, because there is no separate liquid film to hamper the supply of gas to the catalyst sites, it performs better than the traditional trickle-bed reactor.

COUPLING OF REACTIONS

A particularly attractive application of membrane reactors is in coupling two reactions carried out on the opposite sides of a membrane. A product from reaction 1 on one side can serve as a reactant for reaction 2 on the other side, while the exothermic heat from reaction 2 supplies the endothermic heat for reaction 1 (Gryaznov et al., 1986). This is illustrated in Figure 24.1e.

HYBRIDIZATION

An operational variation reminiscent of combined MT and TM reactors is the use of a plug-flow, fixed-bed reactor followed in series by a packed inert membrane reactor, as shown in Figure 24.1f (Wu and Liu, 1992).

Major Types of Membrane Reactors

The many attractive features of membrane reactors described in the previous section underscore the potential of these reactors in organic chemical technology and synthesis. A broad classification of these reactors is given in Table 24.1 and sketches of a few specific ones are given in Figure 24.2.

Figure 24.2a shows a design consisting of a permselective membrane tube placed coaxially inside an outer shell. Reaction occurs in the inner tube which is filled with catalyst. One (or more) of the products from the inner tube permeates through the catalytically inert membrane wall into the outer shell where it is swept away by an inert gas (usually argon). When the catalyst is a packed bed as shown in Figure 24.2a, the reactor is designated as a *packed-bed inert selective*[1] *membrane reactor* (IMR-P).

A useful variation of this design, shown in Figure 24.2b, consists of three concentric tubes (Oertel et al., 1987). The inner of the two annular spaces formed is filled with the catalyst, and selective permeation of products to the central (product) tube is achieved by placing a number of tubular membranes inside this packed volume. Therefore, it is called the *packed-bed inert selective multimembrane reactor* (IMMR-P). In yet another version of an IMR-P, the membrane is supported on the inner surface of a hollow fiber membrane tube and the catalyst is loaded around the hollow fiber (Figure 24.2c).

When the catalyst in design (a) is fluidized, it is designated as a *fluidized-bed inert selective*[1] *membrane reactor* (IMR-F). It is, however, more common to have the fluidized bed in the shell and sweep the products out through the membrane tubes, usually placed horizontally, as shown in Figure 24.2d.

Figure 24.2e shows a design that is similar to Figure 24.2a, but the catalyst is supported on the membrane wall (and not placed inside the membrane tube). In some cases, the membrane itself is the catalyst. In this kind of reactor, as the reactant passes through the inner tube, it permeates across the catalytic membrane, and diffusion and reaction occur simultaneously inside the membrane wall. We refer to this reactor as the *catalytic selective*[1] *membrane reactor* (CMR-E). When it is packed with catalyst, the result is a *packed-bed catalytic selective membrane* reactor (CMR-P) (Figure 24.2f). Modified designs have been used by Gryaznov and coworkers (Gryaznov, 1986; Gryaznov et al., 1974a,b) in which the reactor geometry provides simultaneously for a high degree of utilization of the vessel space and extended catalyst surface.

A schematic of a reactor made from a nonselective membrane for preventing the slip of an excess reactant is shown in Figure 24.2g. The principle of this reactor was outlined before. In the particular design shown, one of the reactants (B) is continuously recirculated on one side of the membrane so that complete conversion of A can be achieved on the opposite side without any slip. We refer to such a *catalytic nonselective*[2] *membrane reactor* without packing as a CNMR-E. When packed, it is referred to as a CNMR-P. Another nonselective

---

[1] Use of the word "selective" denoting the permselective nature of the membrane is optional.

[2] The word "nonselective" (abbreviated to N) should be retained to emphasize its difference from the permselective membrane.

Table 24.1 Broad classification of membrane reactors[a]

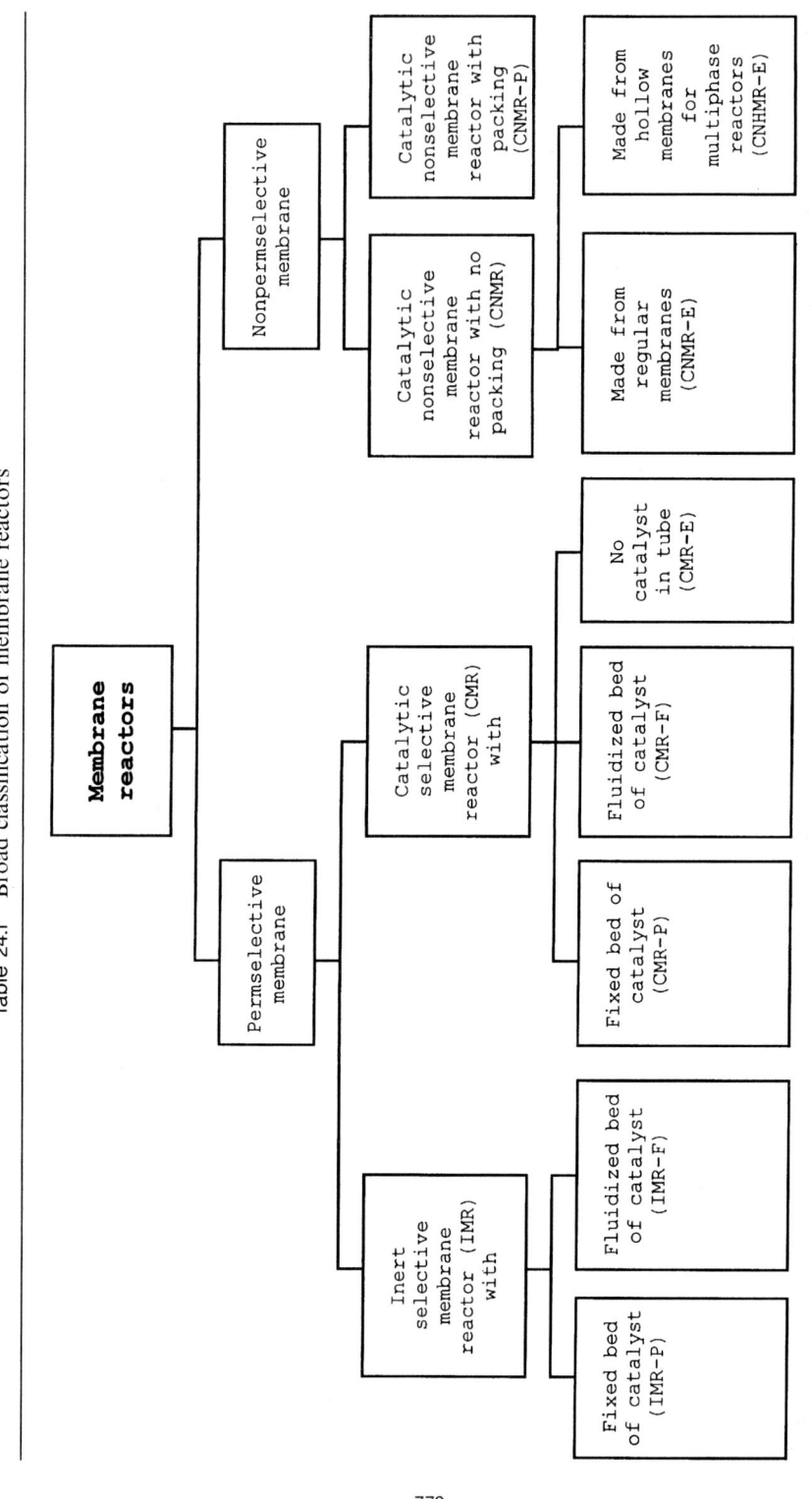

[a] E = empty, F = fluidized bed, P = packed bed.

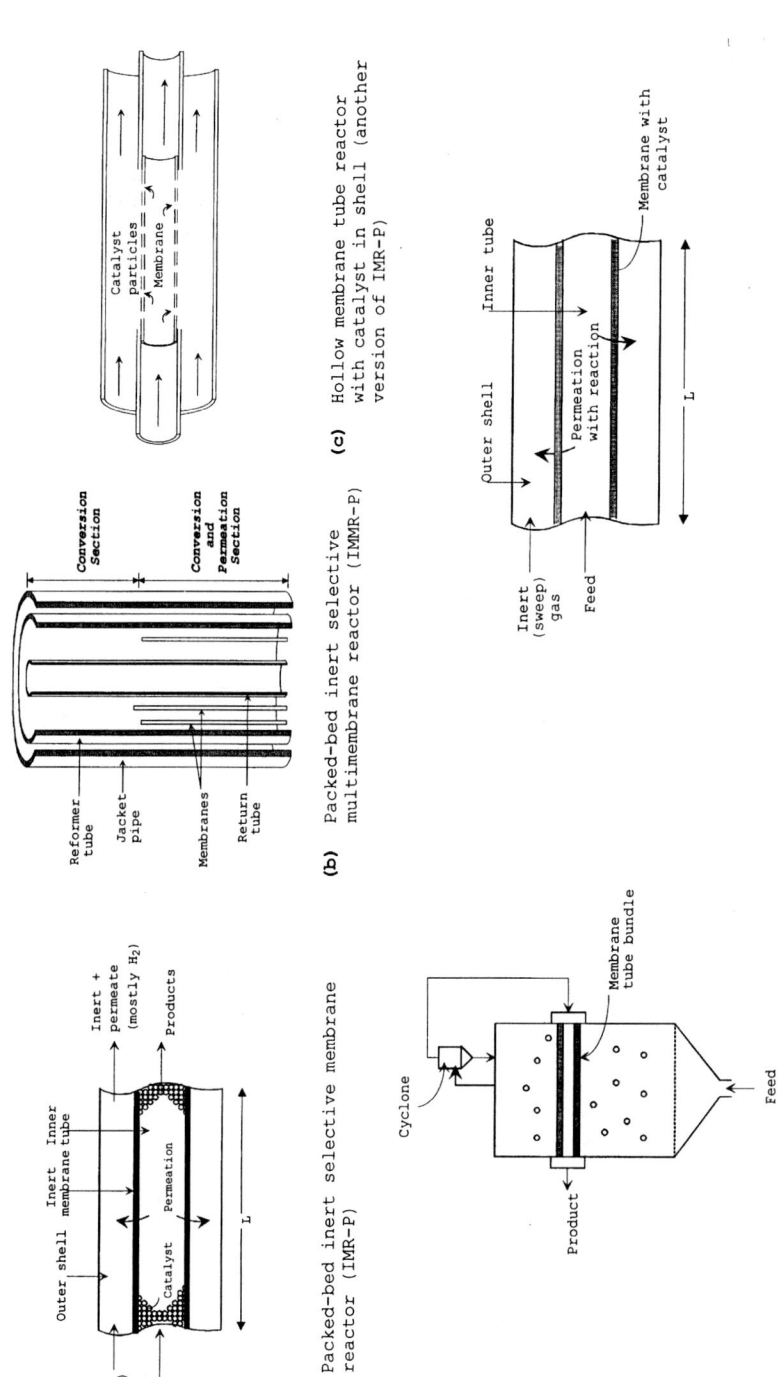

Figure 24.2 Types of reactors

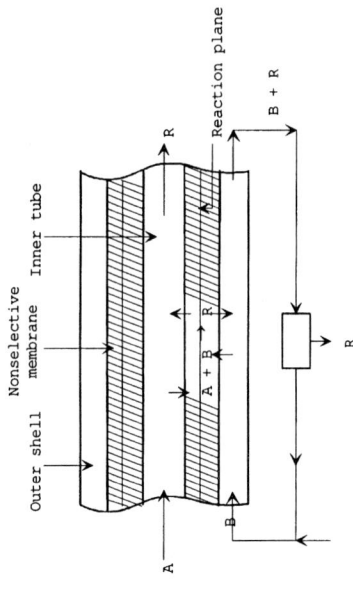

(f) Packed-bed catalytic selective membrane reactor (CMR-P)

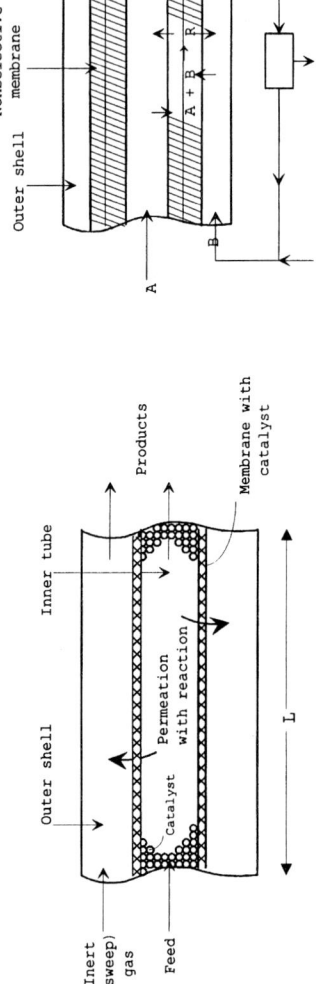

(h) Catalytic nonselective hollow membrane reactor (CNHMR-E) for multiphase reactions: G = gas, L = liquid

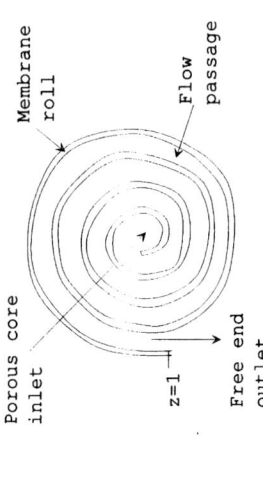

(g) Catalytic nonselective membrane reactor (CNMR-E)

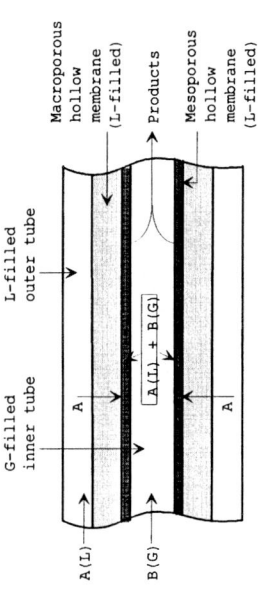

(i) Immobilized-enzyme membrane reactor (IEMR) (Shao et al., 1989)

Figure 24.2 (Continued)

membrane reactor with considerable potential is the *catalytic nonselective[2] hollow membrane reactor* (CNHMR-E) for multiphase reactions. A useful design based on the principle explained in Figure 24.1d is sketched in Figure 24.2h.

A design used in enzyme-catalyzed synthesis is shown in Figure 24.2i (Shao et al., 1989). This consists essentially of an enzyme immobilized in a sheet of microporous plastic (such as PVC) which is then rolled into a spiral and placed in a reactor vessel. Like the Gryaznov reactor referred to earlier, this *immobilized-enzyme membrane reactor* (IEMR) provides for maximum utilization of reactor space and extension of catalyst surface.

MODELING OF MEMBRANE REACTORS

We outline here the important features of the two most basic types of membrane reactors, IMR-P and CMR-E, supplemented by a brief discussion of some of their variations.

Packed-Bed Inert Selective Membrane Reactor with Packed Catalyst (IMR-P)

Hydrogenation and dehydrogenation reactions have benefited most from the use of membrane reactors, and a number of studies have been reported on the modeling of these systems, for example, Itoh et al. (1985, 1990), Mohan and Govind (1986), Itoh and Govind (1989a,b), Govind and Mohan (1988), Tsotsis et al. (1992, 1993). Thus consider the following typical dehydrogenation reaction:

$$A \leftrightarrow R + \nu_H H \qquad [24.2]$$

where H is hydrogen. Figure 24.3 is a sketch of this reaction conducted in an isothermal IMR-P. It will be assumed that the flow in both the inside and outside tubes is tubular.

Model equations

The general mass balance can be written as

$$\begin{pmatrix} \text{rate of change of} \\ \text{component } i, \text{ mol/m}^3\text{ s} \end{pmatrix} = \begin{pmatrix} \text{rate of formation} \\ \text{or disappearance, mol/m}^3\text{ s} \end{pmatrix}$$
$$- \begin{pmatrix} \text{rate of permeation,} \\ \text{mol/m}^3\text{ s} \end{pmatrix} \qquad (24.1)$$

The last term is merely the specific rate of permeation Π_i (mol/m^2 s) multiplied by the area a per unit volume, which for a tubular reactor is $2/R_1$ (1/m). The product $\Pi_i a$ gives the rate of permeation in units of mol/m^3 s.

INNER (FEED) TUBE

The mass balance for this case becomes

$$\frac{dG_i^T}{d\ell} = r_i - \frac{2\Pi_i}{R_1} \qquad (24.2)$$

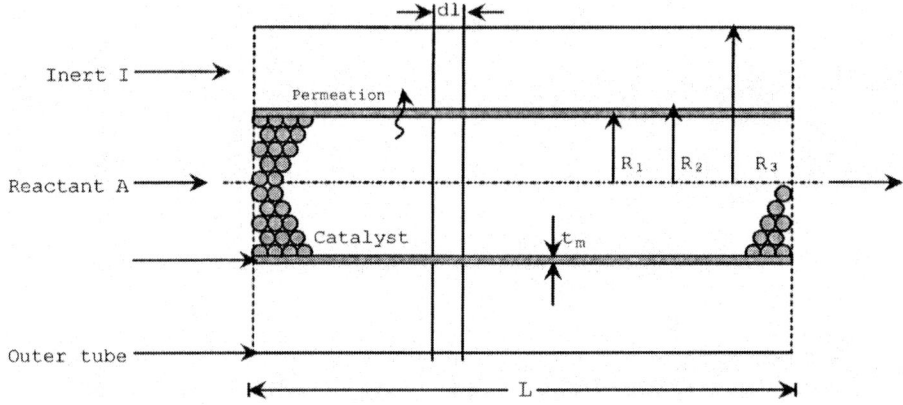

Figure 24.3 Packed-bed inert selective membrane reactor (IMR-P)

where G_i^T is the specific molar flow rate of species i on the tube (feed) side in mol/m² s; R_1 is the inside radius of the tube; and r_i is the rate in mol/m³ s, with $r_i' < 0$ for a reactant, > 0 for a product, and $= 0$ for an inert. The initial conditions are

$$\ell = 0, \quad G_{i=R,H}^T = 0, \quad G_A^T = G_{A0}^T, \quad G_I^T = \alpha G_{A0}^T \tag{24.3}$$

where α is the ratio of sweep gas to reactant gas rates:

$$\alpha \text{ (sweep ratio)} = \frac{G_{I0}^S}{G_{A0}^T} = \frac{Q_{I0}^S}{Q_{A0}^T} \tag{24.4}$$

OUTER (SHELL) SIDE

In writing the equations for the shell side, note that the reaction rate term in Equation 24.1 is zero. Thus the only term on the right is the rate at which each component permeates into the annulus and adds on to the flow from the upstream side. In formulating the permeation term, we must also remember that the permeation area per unit volume of the outer tube is given by

$$a = \frac{\text{surface area of the inner tube}}{\text{volume of the annular space}}$$

$$= \frac{2R_1}{(R_3^2 - R_2^2)} \tag{24.5}$$

Thus the equations can be written in compact form as

$$\frac{dG_i^S}{d\ell} = \left(\frac{2R_1}{R_3^2 - R_2^2}\right) \Pi_i \tag{24.6}$$

where G_i^S is the specific molar flow rate in the outer shell (mol/m² s) and R_2 and R_3 are the outside radius of the inner tube and the inside radius of the outer tube, respectively (Figure 24.3).

The initial conditions are

$$\ell = 0, \quad G^S_{i \neq I} = 0, \quad G^S_I = G^S_{I0} = \alpha G^T_{A0} \tag{24.7}$$

PERMEATION

Because the membrane is assumed to be inert, the following simple mass balance can be written for the rate of permeation:

$$\Pi_i = \frac{D_i(p_i^T - p_i^S)}{t_m} \tag{24.8}$$

where D_i is the diffusivity, referred to as permeability in the membrane literature, of component i in units of mol/m atm s, p_i^T and p_i^S are the partial pressures of i in the inner tube and outer shell, respectively, and t_m is the membrane thickness.

THE OVERALL CONVERSION EQUATION

The overall balance for reactant A in the inner and outer tubes leads to the following expression for its conversion:

$$\begin{aligned} X_A &= \frac{\pi R_1^2 (G^T_{A0} - G^T_A) - \pi(R_3^2 - R_2^2) G^S_A}{\pi R_1^2 G^T_{A0}} \\ &= 1 - \frac{G^T_A}{G^T_{A0}} - \frac{G^S_A}{G^T_{A0}} \frac{(R_3^2 - R_2^2)}{R_1^2} \end{aligned} \tag{24.9}$$

For no permeation (i.e., $G^S_A = 0$), this reduces to

$$G^S_A = 0, \quad X_A = 1 - (G^T_A / G^T_{A0}) \tag{24.10}$$

We now demonstrate the use of these equations in simulating an IMR-P for a specific reaction.

EXAMPLE 24.1

Simulation of an IMR-P for the dehydrogenation of cyclohexane to benzene

The reaction

$$\text{(A)} \rightleftharpoons \text{(R)} + 3H_2 \tag{E24.1.1}$$

is carried out in a porous Vycor membrane reactor with A = cyclohexane, R = benzene, and H = hydrogen (Figure 24.3). Cyclohexane is passed through the inner tube and sweep gas argon (I) through the outer tube.

It is desired to determine if conversions beyond the equilibrium limit corresponding to the temperature and pressure of the reaction can be achieved. Because membrane thickness controls the permeation rate, it is also desired to optimize the thickness for maximum performance. The input data for the simulation are given in Table 24.2.

Table 24.2 Input parameters for simulation (adapted from Itoh et al., 1985)

Reactor dimensions (cm)	$R_1 = 0.7,\ R_2 = 0.85,\ R_3 = 2.0,\ L = 20$
Gas flow rate at inlet, mol/m²/s	$G_{IO}^T = 0.0797,\ G_{AO}^T = 0.0152,\ G_{IO}^S = 0.111$
Reaction temperature, K	$T = 483$
Reaction pressure, Pa	$P_t = 1.013 \times 10^5$
Kinetic parameters (from some previous studies)	$k^0 = 1.42 \times 10^{-5}\ \text{mol m}^3\ \text{s/Pa},\ E = 67.7\ \text{kJ/mol}$

RIGOROUS SOLUTION (ISOTHERMAL OPERATION)

Governing Equations

Assuming plug flow, the material balance for the inner tube is given by Equation 24.2 which can be expanded into the following set:

$$\frac{dG_A^T}{d\ell} = -(-r_A) - \frac{2\Pi_A}{R_1} \tag{E24.1.1}$$

$$\frac{dG_R^T}{d\ell} = (-r_A) - \frac{2\Pi_R}{R_1}, \quad [r_R = (-r_A)] \tag{E24.1.2}$$

$$\frac{dG_H^T}{d\ell} = 3(-r_A) - \frac{2\Pi_H}{R_1}, \quad [r_H = 3(-r_A)] \tag{E24.1.3}$$

$$\frac{dG_I^T}{d\ell} = -\frac{2\Pi_I}{R_1} \tag{E24.1.4}$$

In these equations the rate refers to the net rate given by Equation 24.1.10, which includes the reverse rate. Similarly, the material balance for the outer tube, Equation 24.6, can be expanded to give

$$\frac{dG_A^S}{d\ell} = \frac{2R_2\Pi_A}{(R_3^2 - R_2^2)} \tag{E24.1.5}$$

$$\frac{dG_R^S}{d\ell} = \frac{2R_2\Pi_R}{(R_3^2 - R_2^2)} \tag{E24.1.6}$$

$$\frac{dG_H^S}{d\ell} = \frac{2R_2\Pi_H}{(R_3^2 - R_2^2)} \tag{E24.1.7}$$

$$\frac{dG_I^S}{d\ell} = \frac{2R_2\Pi_I}{(R_3^2 - R_2^2)} \tag{E24.1.8}$$

The initial conditions are

$$G_A^T = G_{A0}^T$$
$$G_R^T = G_R^S = 0$$
$$G_I^T = \alpha G_{A0}^T \tag{E24.1.9}$$
$$G_A^S = G_R^S = G_H^S = 0$$
$$G_I^S = G_{I0}^S$$

PARAMETER VALUES

The rate equation is given by

$$(-r_A) = k^0 \exp\left(\frac{E}{R_g T}\right)\left(p_A - \frac{p_R p_H^3}{K}\right) \tag{E24.1.10}$$

The values of k^0, E, K, and of the other parameters of the system are listed in Table 24.2.

Now we have all of the data needed to solve Equations E24.1.1–E24.1.9 along with E24.1.10. This can be easily done by any of the well-known numerical methods.

Discussion of Results

Figure 24.4a (recalculated results of Itoh et al., 1985) shows the mole fraction profiles of cyclohexane, benzene, and hydrogen in the inner (reaction) tube. As anticipated earlier, the reactant (cyclohexane) profile shows a continuously falling trend, whereas the profiles of the products benzene and hydrogen show maxima. The latter occur because of competition between the rate of reaction and that of the permeation of hydrogen. One dominates the entrance region and the other the exit region. As for benzene, because the membrane is more selective to hydrogen, the permeation rate of benzene is slower than that of hydrogen, resulting in a slower rate of decrease of its concentration in the reaction tube. On the other hand, as expected, the mole fractions of all three components rise steadily in the outer separation tube (Figure 24.4b).

The effect of membrane thickness is shown in Figure 24.5. Opposing influences are exerted by reaction and permeation, resulting in an optimum thickness of 0.1 cm.

A SIMPLIFIED APPROACH

A simplified approach would be to straightaway postulate that only hydrogen gas permeates through the membrane and that the rate of permeation is given by the simple relationship

$$\Pi_H a = k_{\text{per}}[H], \quad \text{mol/m}^3 \text{ s} \tag{E24.1.11}$$

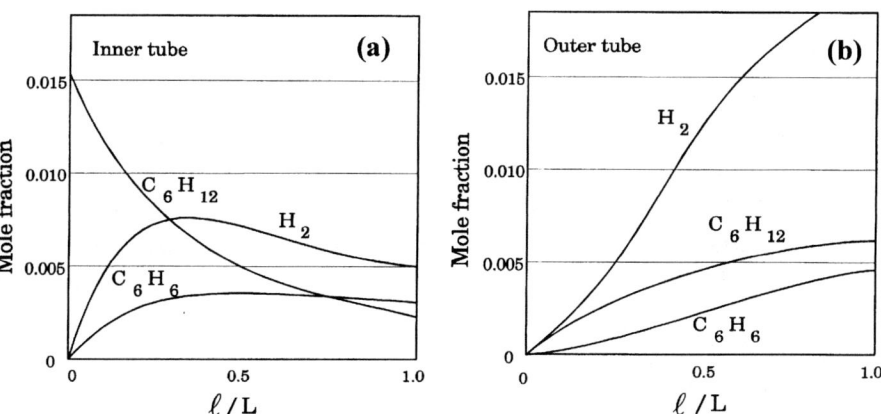

Figure 24.4 Dehydrogenation of cyclohexane: computed profiles in the (a) tube and (b) shell sides (from the results of Itoh et al., 1985)

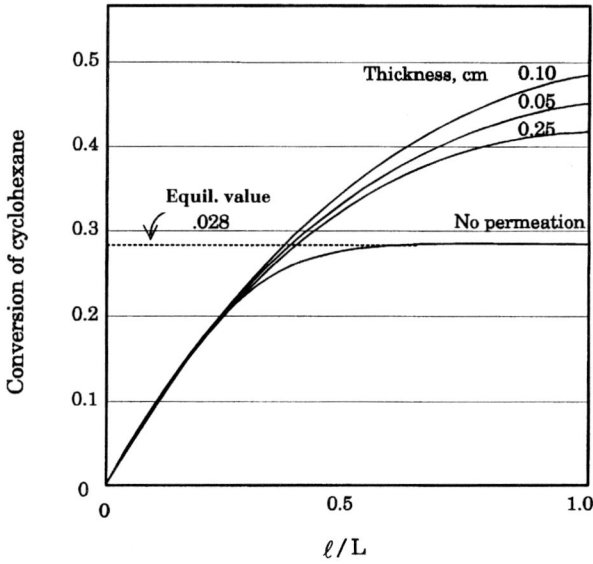

Figure 24.5 Dehydrogenation of cyclohexane: effect of membrane thickness on conversion (from the results of Itoh et al., 1985)

Then, using the material balance of Equation 24.1 and restricting the analysis to the inner tube (thus dropping the superscripts T and S), the following equations can be readily written:

$$\frac{dF_A}{dV} = -(-r_A), \quad \frac{dF_H}{dV} = 3(-r_A) - \Pi_H a$$

or

$$\frac{dF_A}{d\bar{t}} = -(-r_A)Q_0$$

$$\frac{dF_R}{d\bar{t}} = (-r_A)Q_0 \tag{E24.1.12}$$

$$\frac{dF_H}{d\bar{t}} = 3(-r_A)Q_0 - Q_0 k_{\text{per}}[H]$$

where $\bar{t} = V/Q_0 = L/u_0$. The initial conditions are

$$\bar{t} = 0, \quad F_A = F_{A0}, \quad F_R = F_H = 0, \quad Q = Q_0 \tag{E24.1.13}$$

Noting that concentration $= F/Q$, these equations can be solved to obtain the exit flow rates F_A, F_R, F_H as functions of the residence time $\bar{t}$.

Extension to consecutive reactions

It seems likely that the use of IMR-P for a consecutive reaction can have beneficial effects on selectivity. Indeed, using the partial oxidation of methane to formaldehyde as the test reaction, the selectivity was found to be higher than

in a PFR when the IMR-P was operated in the conventional way with the usual assumptions: sweep gas in the shell side under isothermal plug-flow conditions (Lund, 1992; Agarwalla and Lund, 1992).

Fluidized-Bed Inert Selective Membrane Reactor (IMR-F)

The *fluidized-bed inert selective membrane reactor* (IMR-F) sketched in Figure 24.2d is a useful configuration. Work reported in the last few years (Elnashaie and Adris, 1989; Wagialla and Elnashaie, 1991; Elnashaie et al., 1991, 1992, 1993; Adris et al., 1991; Abdalla and Elnashaie, 1995) has thrown some interesting light on the performance of the IMR-F. An advantage of the fluidized-bed reactor, inadequately emphasized in the literature, is the fact that in reactions such as the dehydrogenation of ethylbenzene to styrene the bubbles of hydrogen act as "natural membranes" which remove the hydrogen formed in the dense phase.

The use of an IMR-F improves the selectivity for the desired product in a complex reaction and also can even increase it beyond that of the fixed bed. For instance, in the dehydrogenation of ethylbenzene to styrene given by the abbreviated scheme (Sheel and Crowe, 1969, for full scheme)

$$C_6H_5CH_2CH_3 \leftrightarrow C_5H_5CH = CH_2 + H_2$$
$$C_6H_5CH_2CH_3 \rightarrow C_6H_6 + C_2H_4$$
$$C_6H_5CH_2CH_3 + H_2 \rightarrow C_6H_5CH_3 + CH_4$$
$$2H_2O + C_2H_4 \rightarrow 2CO + 4H_2$$
$$H_2O + CH_4 \rightarrow CO + 3H_2$$
$$H_2O + CO \rightarrow CO_2 + H_2$$

[24.3]

the removal of hydrogen from the scene of the reaction in the dense phase by both transport of hydrogen bubbles into the dilute phase (where little reaction occurs) and by the permselective action of the membrane tube wall (Figure 24.2d) prevents the conversion of styrene to toluene by reaction 24.3, thus improving styrene selectivity.

Catalytic Selective Membrane Reactor (CMR-E)

In analyzing the performance of a CMR-E (Figure 24.6), we make the same assumptions as we did for an IMR-P. There is a major difference in the equations, however. Because the reaction occurs entirely within the membrane, it is bounded by the internal surface of the membrane wall, corresponding to $r = R_1$ for flow in the inner tube, and by the outer surface of the same wall, corresponding to $r = R_2$, for flow in the shell side. The resulting material balance equations for reaction 24.2 are formulated here for flow in the two tubes and for reaction in the membrane.

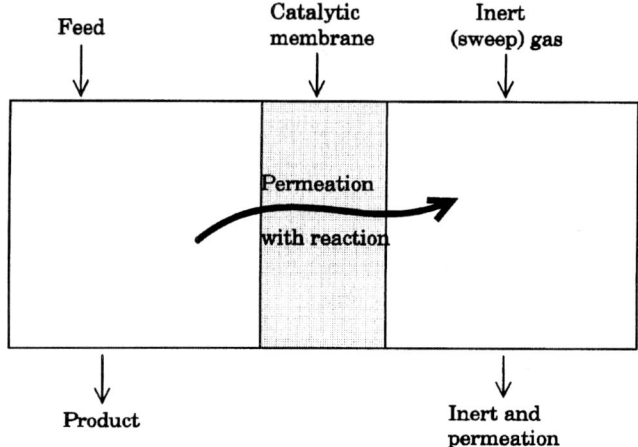

Figure 24.6 Catalytic selective membrane reactor

Model equations

INNER (FEED) TUBE

$$\frac{dF_i^T}{d\ell} = \frac{2\pi D_i}{R_g T} \left[r \left(\frac{dp_i}{dr} \right) \right]_{r=R_1} \tag{24.11}$$

with

$$\ell = 0, \quad F_i^T = F_{i0}^T = Q_0^T \left(\frac{y_{i0}^T P_t^T}{R_g T} \right) \tag{24.12}$$

OUTER (SHELL SIDE) TUBE

$$\frac{dF_i^S}{d\ell} = -\frac{2\pi D_i}{R_g T} \left[r \left(\frac{dp_i}{dr} \right) \right]_{r=R_1} \tag{24.13}$$

with

$$\ell = 0, \quad F_i^S = F_{i0}^S = Q_0^S \left(\frac{y_{i0}^S P_t^S}{R_g T} \right), \quad Q_I^S = \alpha Q_{A0}^T \tag{24.14}$$

where α is the sweep ratio given by Equation 24.4.

INSIDE THE MEMBRANE (WITH REACTION)

$$\frac{D_i}{R_g T} \frac{1}{r} \frac{d}{dr} \left[r \left(\frac{dp_i}{dr} \right) \right] = -\nu_i r_i$$
$$= -\nu_i k \left(p_A - \frac{p_R p_S}{K} \right) \tag{24.15}$$

The boundary conditions are

$$r = R_1, \quad p_i = y_i^T P_t^T = y_i^T P_t$$
$$r = R_2, \quad p_i = y_i^S P_t^S = y_i^S P_t \quad (24.16)$$

The last terms in these conditions are the result of the assumption $P_t^T = P_t^S =$ constant(P_t) along the length of the reactor.

Main features of CMR-E

Equations 24.11–24.16 can either be solved as such or can be nondimensionalized and then solved. Solutions can be obtained by the IMSL subroutine DBVFD along with a third-order Runge–Kutta technique. Experimental data on the dehydrogenation of ethane (Champagnie et al., 1992) reasonably uphold the predicted ethane profiles in both the tube and shell sides.

Further experimental data on the same reaction confirm trends typical of these reactions. For example, Figure 24.7 (Tsotsis et al., 1993) shows the variation of conversion with the sweep ratio α for an inner tube pressure of 2 atm and outer shell pressure of 1 atm. It can be seen that the actual conversion obtained is significantly higher than the equilibrium conversion. In conformity with the nature of the reaction, that is, an increase in volume, the equilibrium conversion on the shell side (which is at a lower pressure) is higher than in the inner tube. It increases with increasing α because the concentration of R on the shell side is reduced with increasing α, resulting in an increase in the rate of depletion of R in the reaction tube. This leads in turn to a further increase in the rate of reaction on the catalyst in the inner tube.

A vexing problem with the inert sweep gas is the need to separate it from product. In reactions such as dehydrogenation of butane (Gobina and Hughes, 1996) and of cyclohexane (Wang et al., 1992), for example, this has been overcome by using oxygen in admixture with an inert such as CO after eliminating

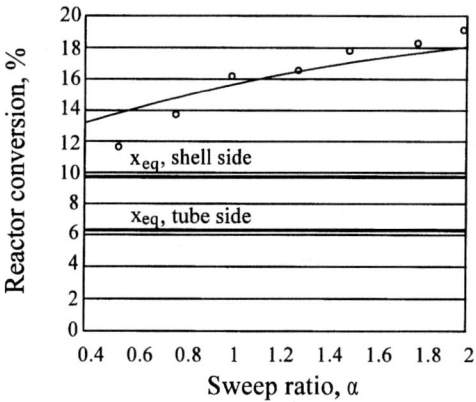

Figure 24.7 Conversion vs. sweep ratio in the dehydrogenation of ethane at fixed values of other parameters (redrawn from Tsotsis et al., 1993)

the potential formation of an oxide layer by prior high-temperature reduction with hydrogen. Such reaction-assisted transport can substantially increase conversions.

As in an IMR, here, too, the thickness of the membrane is an important adjustable parameter that must be optimized. It appears as the boundary values R_1 and R_2, where the difference $(R_2 - R_1)$ is the membrane thickness. Experimental results on ethane dehydrogenation (Champagnie, et al., 1992) clearly indicate the existence of an optimum thickness for best performance.

Packed-Bed Catalytic Selective Membrane Reactor (CMR-P)

This reactor obviously has two catalystic zones, the packed bed and the membrane itself. As already stated, although theoretically daunting, it offers the best configuration for a complete analysis of membrane reactors. Experimental studies on the dehydrogenation of ethane showed considerable enhancement over both tube and shell side equilibrium conversions (Tsotsis et al., 1992). Further improvement was possible with an increase in the sweep ratio.

Catalytic Nonselective Membrane Reactor (CNMR-E)

This reactor, sketched in Figure 24.2g, was developed especially to prevent slip in reactions required to be strictly stoichiometric. Modeling of a CNMR-E has been attempted for both fast irreversible and reversible reactions (Sloot et al., 1990, 1992; Zaspalis et al., 1991; Veldsink et al., 1992), notably the Claus reaction $2H_2S + SO_2 \rightarrow (3/8)S_8 + 2H_2O$. This concept can also be used in partial oxidations in organic technology, for example, partial oxidation of ethylene to acetaldehyde (Harold et al., 1992). The stoichiometry of this reaction can be represented as

$$A + \nu_B B \rightarrow \nu_{R1} R$$
$$A + \nu_{R2} B \rightarrow \nu_S S$$
[24.4]

where R is the desired product. Experiments in which the active side of the membrane was exposed to a $(C_2H_4 + He)$ mixture and the support side to a $(O_2 + He)$ mixture gave significantly higher selectivities than in experiments in which the active layer was exposed to a $(C_2H_4 + O_2 + He)$ mixture and the support side to pure He.

Catalytic Nonselective Hollow Membrane Reactor for Multiphase Reactions (CNHMR-MR)

The tubular multiphase hollow membrane wall reactor briefly described before and sketched in Figure 24.1h is a multiphase reactor design very similar to the trickle-bed reactor. In a regular trickle-bed reactor, the liquid flows over a partially wetted pellet as a thin film and supplies the liquid-phase reactant to the catalyst pores. This action, however, has the effect of hindering pore access to

the gas, thus reducing the reaction rate. On the other hand, in the multiphase membrane reactor, the liquid-filled membrane is directly accessible to the gas flowing in the inside tube. Thus mass transfer in this reactor is considerably more efficient than in the conventional trickle-bed reactor.

It is also possible to have the liquid side fully mixed. This would mean that the external surface of the hollow tube would be exposed to a liquid of uniform composition.

One can develop governing equations for the two cases mentioned and compare their performances with the conventional trickle-bed reactor modeled as a string of suspended spherical pellets contacted by cocurrent flow of gas and liquid. Based on such a comparison, the following observations can be made (Harold and Cini, 1989; Harold et al., 1989; Cini et al., 1991):

1. Direct supply of gas to the catalyst pores without an intervening liquid film greatly improves the reactor performance for very active catalysts.
2. For catalysts of low or moderate activity, the tube walls should (and can) be made as thin as possible to improve internal transport. Such flexibility does not exist in the conventional trickle-bed reactor because the particle size can be reduced only at the cost of increased pressure drop.

Immobilized-Enzyme Membrane Reactor

Immobilized-enzyme reactors were considered in Chapter 20. It is easy to extend the concept to immobilization on the walls of a membrane tube. What is even more practical is to immobilize the enzyme in the usual manner on solid particles such as silica and encapsulate the particles in a ribbed sheet of a microporous plastic such as PVC. Then this sheet can be rolled in a jelly-roll configuration inside a spiral reactor (Figure 24.2i). The consequent large surface area of immobilized enzyme available per unit volume of reactor space makes such a spiral reactor an attractive choice.

A reactor with such a configuration was used in the clarification of fruit juice by elimination of pectin by the enzyme pectinase. The pectin, which is present in colloidal form, aggregates in the presence of the enzyme and settles, leading to easy physical clarification (Shao et al., 1989).

OPERATIONAL FEATURES

Several operational features of membrane reactors were listed in an earlier section. We consider a few important ones here.

Combining Exothermic and Endothermic Reactions

A particularly useful feature of a membrane is that it integrates reaction and separation into a single process, thereby increasing the conversion beyond equilibrium. If this concept can be extended to an integration of two reactions,

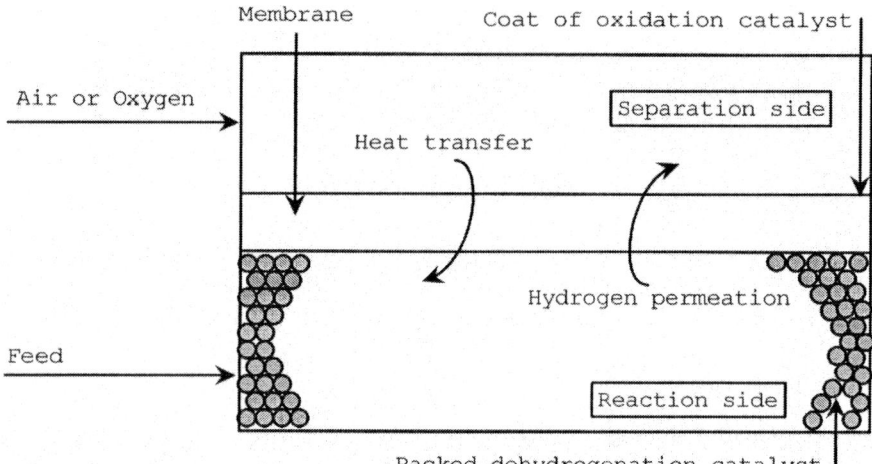

Figure 24.8 Coupling of endothermic (dehydrogenation) and exothermic (oxidation) reactions

one exothermic and the other endothermic, carried out on the opposite sides of the membrane, then we would have a thermally self-sustaining reaction. To expand this concept further (Figure 24.8), if hydrogen from an endothermic dehydrogenation reaction in the inner tube is permselectively transported to the shell side where, instead of being physically swept from the system, it is oxidized (exothermally) by oxygen over a catalyst there, we would have an interactive thermal effect superimposed on the membrane's predisposition to separation. In other words, the heat liberated during oxidation in this "separation side" flows through the membrane into the "reaction side," thus providing the heat required for this endothermic reaction.[3]

Dehydrogenation and hydrogenation are obvious choices for such coupling. The main disadvantage is that there is seldom a complete energetic, kinetic, or thermodynamic matching of the two reactions (Basov and Gryaznov, 1985). Thus it may often be necessary to supplement the "reactant" generated on one side or heat on the other by direct additional supply of the deficient quantity to the opposite side.

One can set up equations based on this model for transport of mass, as well as energy, axially through the reactor which also includes transport across the membrane. But we restrict our treatment here to a few important qualitative observations (see Itoh and Govind, 1989a, for a quantitative model) when the reactor is operated adiabatically.

The reaction side temperature (solid lines of Figure 24.9) registers a fall in the vicinity of the entrance at low values of the heat transfer rate ($\Gamma = UA/C_{pA}F_{A0}$), as indicated by curve A_1 for $\Gamma = 1$. This fall disappears as the heat transfer rate

---

[3] It seems more appropriate to call these "reaction 1 side" and "reaction 2 side" instead of reaction and separation sides, but to avoid conflict with the existing literature we continue with the original nomenclature.

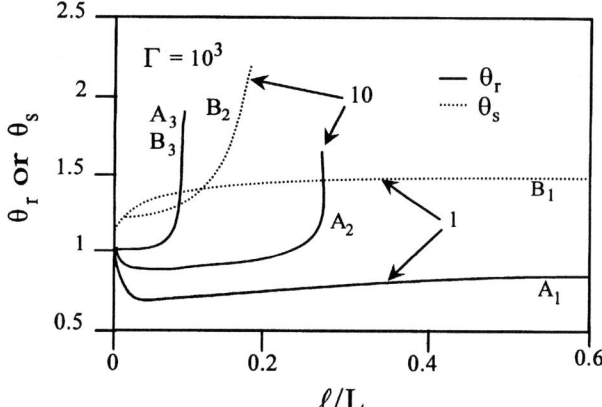

Figure 24.9 Temperature profiles in adiabatic operation of coupled reactions (redrawn approximately from Itoh and Govind, 1989a); $\theta_s = T^s/T_0$, $\theta_r = T^T/T_0$, $\Gamma = UA_m/C_{PA/F_A}$.

rises, and reaches a very high value within the first small fraction of the reactor length (curves A_2 and A_3). At the separation side (broken lines), because of the exothermicity of the reaction, there can be no minimum in temperature, but the trends beyond the initial region are similar to those for the reaction side (the two curves A_3 and B_3, corresponding to a very high tranfer rate, almost completely coinciding with each other). Similar trends are observed for the conversion (with no dip at any point in its value).

Controlled Addition of One of the Reactants in a Bimolecular Reaction Using an IMR-P

A reference to this method of operation was made earlier. There are several instances of industrial organic reactions that are bimolecular and exothermic. An important example is the production of chloromethanes. The temperature rise can be controlled by axially distributed addition of chlorine at several discrete points into a packed-bed, fluidized-bed, or empty tube reactor through which methane is passed (Doraiswamy et al., 1975). The membrane is an ideal choice for such reactions because now it can be allowed to permeate over the entire length of the membrane from the shell side into the inner tube or vice versa.

Experimental results on the oxidative dehydrogenation of ethane to ethylene (Tonkovich et al., 1995) clearly demonstrate the superior performance of an IMR-P in relation to that of a PFR. With ethane in the tube and air in the outer shell, the ratio of ethane to air (β) is an important parameter in determining the performance of IMR-P. At high values of β, the amount of air introduced into the shell is relatively small, and hence its permeation into the inner tube is also correspondingly small. Thus the contact time of ethane with the catalyst does not change significantly through the reactor. Because plug-flow conditions can be assumed within the reactor and permeation does little to alter the

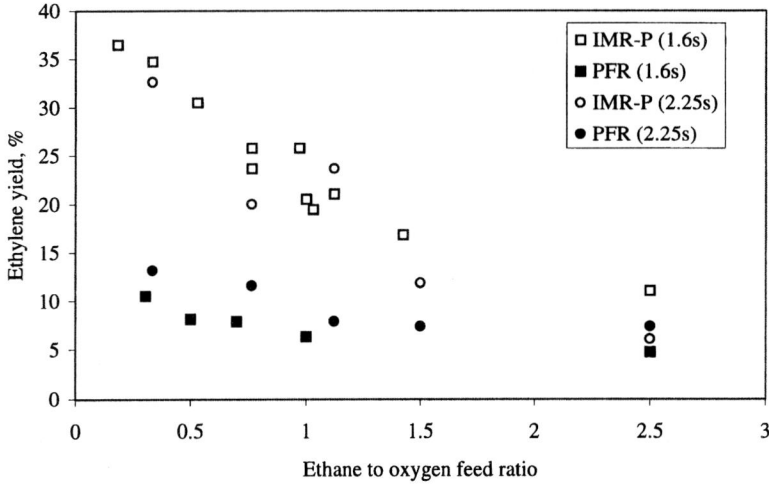

Figure 24.10 Ethylene yields in IMR-P and PFR runs (redrawn from Tonkovich et al., 1995)

situation, the performance of an IMR-P will be similar to that of a PFR. This is brought out in the high β runs plotted in Figure 24.10.

For low β, on the other hand, the conditions at the entrance correspond to long contact times, which are not favorable for high selectivity of the intermediate (ethylene), but the low value of β favors high selectivity. As the ethane moves downstream, the situation is reversed. The rate of permeation decreases (and hence β increases), and at the same time the contact time decreases. It will, therefore, be seen that the residence time changes favorably down the reactor, whereas the ethane–air ratio changes unfavorably. The overall effect appears to be a very large enhancement in selectivity over the PFR values at low ratios. In fact, there is a threefold enhancement over a PFR at $\beta \approx 0.3$, as shown in the figure.

The plug-flow reactor is generally accepted as the most favorable for intermediate selectivity in series-parallel reactions. The results of this example clearly show that the membrane reactor can significantly outperform the PFR.

Effect of Tube and Shell Side Flow Conditions

In the developments just presented, cocurrent plug-flow was assumed in both the tube and shell sides of the reactor. It would be instructive to analyze the effect of countercurrent flow, as well as different combinations of plug and mixed flow on the two sides of the membrane. Countercurrent flow can be achieved merely by changing the direction of sweep gas flow. However, this results in a split boundary value problem because the conditions on the shell side, unlike those on the tube side, are specified at the outlet instead of at the inlet. Substitution of mixed flow for plug flow is straightforward because one has only to use uniform concentrations everywhere in the region.

Five models are possible, as listed in Figure 24.11. The results of simulation using these models have shown that (Itoh et al., 1990) model (a), the counter-

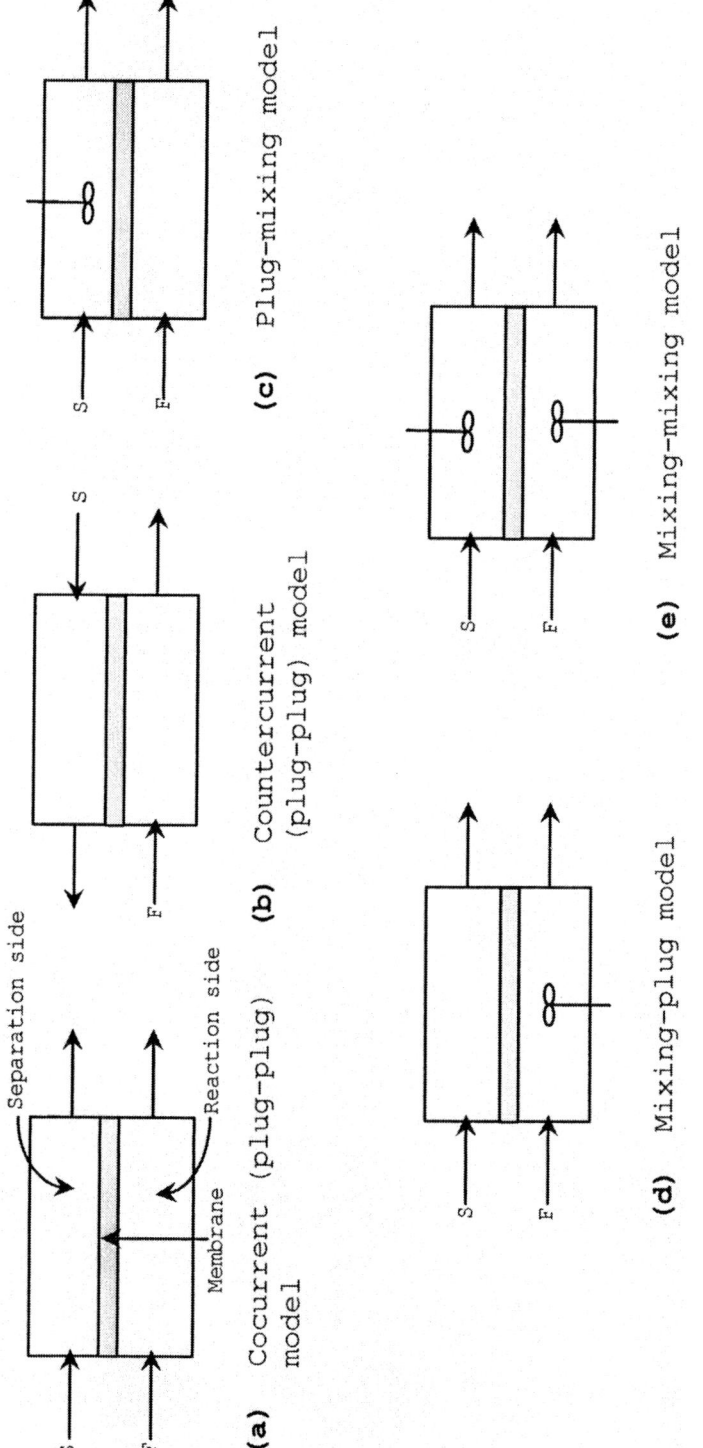

Figure 24.11 Schematic representation of ideal flow configurations. F = feed, S = sweep gas

current plug flow model, is clearly the best, whereas model (e), the mixing-mixing model, gives the poorest performance. Changes in parameter values do not change the order of these extreme models but do alter the sequence of the intermediate models.

COMPARISON OF REACTORS

It would be instructive to compare the performances of an IMR-P and a CMR-E themselves and also with those of a PFR and an MFR. We shall do so by making the following assumptions: the reaction is isothermal, and fluids on each side of the membrane are fully mixed. The latter is a simplifying assumption that permits one to assign a specific pressure to each side. In comparing the performances of these reactors, we must consider the effects due to the difference in pressure between the tube and shell sides [effect (1)] and of selective separation [effect (2)] (Sun and Khang, 1990).

EFFECT 1

Because reaction occurs simultaneously with permeation, the equilibrium conversion on the separation side will depend on whether P^S is lower than, equal to, or higher than P^T. This in turn will depend on whether there is a decrease, no change, or an increase in the number of moles. Thus we have Case 1: $\Delta\nu > 0$, Case 2: $\Delta\nu = 0$, Case 3: $\Delta\nu < 0$.

EFFECT 2

This is the characteristic separation effect of the membrane.

COMBINED EFFECT

A qualitative grading of the performance of an IMR-P and a CMR-E, along with those of a PFR and an MFR both operated at the tube and shell side pressures, is presented in Table 24.3 [adapted from Sun and Khang (1990)]. This table may be used as a preliminary guide to reactor selection.

Table 24.3 Order of performance of different reactors on a scale of 1 to 6

Reaction: Any reaction such as $\nu_A A + \nu_B B \rightarrow \nu_R R + \nu_S S$, $\Delta\nu = (\nu_R + \nu_S) - (\nu_A + \nu_B)$

Reactor type	Low space time			High space time		
	$\Delta\nu > 0$	$\Delta\nu = 0$	$\Delta\nu < 0$	$\Delta\nu > 0$	$\Delta\nu = 0$	$\Delta\nu < 0$
IMR-P	4	2	2	4	1	1
CMR-E	6	4	6	3	2	4
PFR (P^T)	3	1	1	5	3	2
PFR (p^S)	1	—	4	1	—	5
MFR (P^T)	5	3	3	6	4	3
MFR (P^S)	2	—	5	2	1	6

EXAMPLES OF THE USE OF MEMBRANE REACTORS IN ORGANIC TECHNOLOGY AND SYNTHESIS

So far, membrane reactors in organic technology/synthesis have not been industrially exploited. However, they hold promise for commercial application in the near future. Many of the laboratory-scale studies reported are on hydrogenation, dehydrogenation, and hydrogenolysis reactions involving medium to large volume chemicals. Most of the modeling studies considered in the earlier sections were also based on these reactions. The majority of these investigations have used dense platinum-based membranes. Extensive lists of these studies (many as patents with inadequate information) are presented by Shu et al. (1991), Hsieh (1991), and Saracco and Specchia (1994). Studies of small- and medium-volume chemicals, more germane to our purpose, are relatively few, and are listed in Table 24.4. The more recent attempts to produce esters in membrane reactors are also noteworthy and are included in the table. Details of two processes, for vitamin K and linalool, are outlined here.

Table 24.4 Examples of organic reactions in membrane reactors

Reaction	Membrane material	Reference
Dehydrogenation of cyclohexanediol to pyrocatechol	Pd/Ni	Sarylova et al. (1970); Mishchenko et al. (1977)
Hydrodemethylation of methyl- and dimethylnaphthalenes	Pd/Mo	Lebedeva (1981)
Hydrogenation of nitrobenzene to aniline	Pd/Ru	Mishchenko et al. (1979)
Hydrogenation of quinone and acetic anhydride to 2-methyl-1,4-diacetoxynaphthalene	Pd/Ni	Gryaznov and Karvanov (1979)
Synthesis of vitamin K_4	Pd alloys	Gryaznov (1986); Gryaznov et al. (1986)
Hydrogenation of an acetylenic to an ethylenic alcohol in the synthesis of the fragrance linalool	Pd/Ru	Gryaznov et al. (1982); Karavanov and Gryaznov (1984)
Hydrogenation of p-carboxylbenzaldehyde to toluic acid		Gryaznov et al. (1981)
Hydrogenation of 2,4-dinitrophenol to 2,4-diaminophenol	Pd/Ru	Mishchenko and Gryaznov (1981)
Dehydrogenation of borneol to camphor simultaneously with hydrogenation of cyclopentadiene to cyclopentene	Pd/Ni	Smirnov et al. (1981a,b)
β-Ionol to dehydro-β-ionone	Pd/Ni	Smirnov et al. (1978)
Esterification of ethanol by acetic acid	Polymer (poly-etherimide)-coated ceramic	Zhu et al. (1996)
Esterification of oleic acid	Polymer-coated ceramic	Okamoto et al. (1993)
Esterification of carboxylic acid with ethanol	Polymer-coated ceramic	Kita et al. (1988)
Phenol-acetone condensation	Polymer-coated ceramic	Okamoto et al. (1991)

Small and Medium Volume Chemicals

Vitamin K

Conventional production of vitamin K consists of four steps: hydrogenation of 2-methylnaphthoquinone-1,4 to 2-methylnaphthohydroquinone-1,4 in a solvent in the presence of Raney nickel; separation of the product from the catalyst by filtration; evaporation of the solvent; and boiling with acetic anhydride. Because the anhydride is highly corrosive, it tends to attack the nickel, and hence complete separation of the catalyst is necessary. On the other hand, use of a palladium alloy membrane reactor eliminates corrosion and makes it possible to complete the whole process in a single step (Gryaznov et al., 1986). The overall reaction is

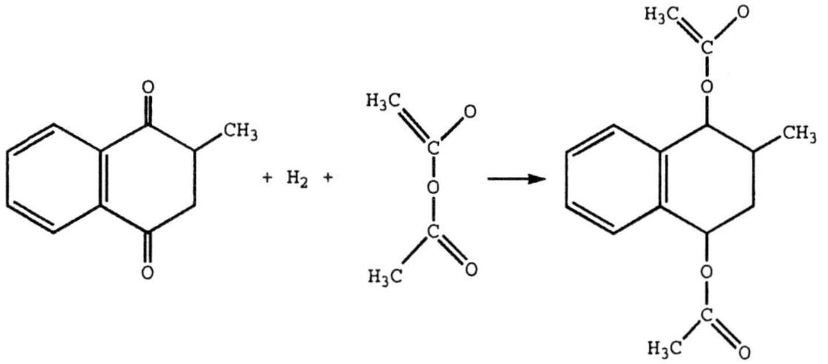

[24.5]

A 95% conversion was obtained on a pilot plant scale compared to 80% by the conventional process.

Linalool (a fragrance)

Another successful application of palladium-based membranes is the hydrogenation of an acetylenic alcohol to an ethylenic alcohol in the synthesis of the perfume linalool with a 98% yield. A palladium-ruthenium catalytic membrane was used with high selectivity toward triple-bond to double-bond hydrogenation. We have already noted the importance of this kind of selectivity in organic synthesis (Chapter 6). Membranes can give higher selectivities than even the best hydrogenation catalysts in straight catalysis. The production of linanol shown here (Gryaznov et al., 1982, 1983) is a pointer to further industrial applications of catalytic membranes.

$$CH_3-\underset{\underset{CH_3}{|}}{C}=CH-CH_2-CH_2-\underset{\underset{CH_3}{|}}{\overset{\overset{OH}{|}}{C}}-C\equiv CH + H_2 \longrightarrow$$

$$CH_3-\underset{\underset{CH_3}{|}}{C}=CH-CH_2-CH_2-\underset{\underset{CH_3}{|}}{\overset{\overset{OH}{|}}{C}}-CH=CH_2 \qquad [24.6]$$

Membrane Reactors for Economic Processes (Including Energy Integration)

One way of using a membrane reactor is to selectively feed hydrogen from a hydrogen-rich stream into the reactor. Hydrogen-rich gases are available from refineries, ammonia plants, and many large-scale dehydrogenation plants. They cannot be used directly to produce high purity chemicals such as pharmaceuticals and fragrances because of the often unwanted impurities present. Hence expensive pure hydrogen is almost always used. Membrane reactors make it possible to use these hydrogen-rich gases directly.

Reactions used to convert triple C≡C bonds to double C=C bonds, nitro groups to amino, quinones to hydroquinones, and aldehydes to acids, as well as those used to replace multistep syntheses, have been carried out with better results than without membranes (Gryaznov, 1986). An attractive feature of many of these reactions, for example, the production of vitamin K, is that corrosive conditions are avoided. Often in single-step reactions, the same catalysts have been used as in the corresponding conventional processes. A reaction such as selective hydrogenation of butynediol to butenediol (Chapter 6), for which new catalysts are under constant development, can benefit greatly by using membrane reactors. In fact, most reactions involving selective saturation of a triple bond to a double bond and partial oxidation are promising candidates.

Another major advantage of membrane processes is the energy integration that can be introduced by combining an exothermic reaction on one side of the membrane with an endothermic process on the other. Shu et al. (1991) and Saracco and Specchia (1994) give a number of examples of such combined reactions, for example, the dehydrogenation of borneol to camphor combined with the hydrogenation of cyclopentadiene to cyclopentene (Smirnov et al., 1981a,b); the dehydrogenation of isopropanol to acetone combined again with the hydrogenation of cyclopentadiene to cyclopentene (Mikhalenko and coworkers, 1977, 1978; Gryaznov et al., 1981); and the dehydrogenation of cyclohexanol to cyclohexanone combined with the hydrogenation of phenol to cyclohexanone (Basov and Gryaznov, 1985). The last example is particularly attractive as it produces the same product, cyclohexanone, in both reactions (Figure 24.12).

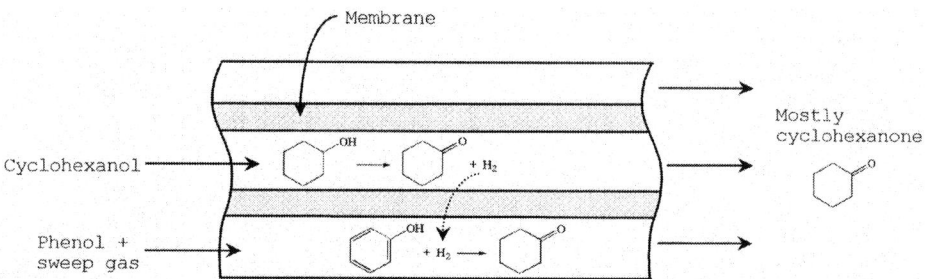

Figure 24.12 Coupling of reactions to produce the same product (cyclohexanone)

Chapter 25

Multifunctional Reactor Engineering

Looking back to the progress achieved in the areas of conventional separation and reactor design, it seems that major advances have now been made. This has led to increasing research into methods in which reaction and separation are combined in a single unit. The equipment in which this dual function is carried out is sometimes referred to as the *combo reactor*.

Combo reactors can be of two types, separation oriented and reaction oriented. In the first, reaction is used to achieve efficient separation, such as in the separation of *p*-cresol from its mixture with *m*-cresol. This method need not necessarily be restricted to separation, for it can also be attractive from the reaction point of view in the following scenario: the required product from a process comes out with a by-product of low value in a very difficultly separable mixture of the two. If the by-product can be converted to a useful coproduct in an easily separable mixture with the primary product, then we would have an attractive process. Alternatively, there can be two unimportant by-products in a difficultly separable mixture which can be converted to an easily separable mixture of useful products. An interesting example of this is illustrated later in this chapter.

In the second, separation is used to enhance the reaction. An example of this is the intentional biphasing of a homogeneous liquid-phase reaction by addition of a second liquid phase. The second phase can act in several ways to enhance the productivity of the system. Another example is the large-scale version of the chemist's apparatus in which a reflux column condenser is connected to a batch reactor (usually a round-bottomed flask) as shown in Figure 25.1. The product and the heat of reaction are continuously removed, and the reactant is returned to the reactor.

Modeling of type 1 systems leads to equations for the *separation factor* of a mixture enhanced by reaction. For type 2 systems, on the other hand, equations

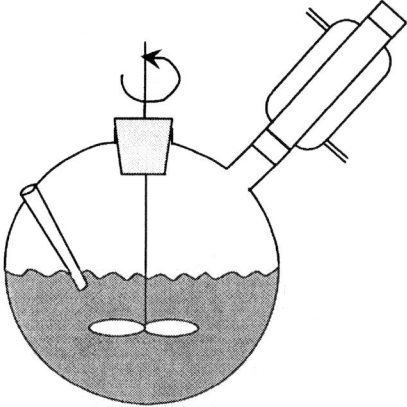

Figure 25.1 A simple laboratory setup for reaction with distillation

are obtained for the conversion or yield enhanced by separation. If the component separated is the desired product, it is really immaterial which definition is used. The reaction itself is unimportant in type 1 systems, however, except as a vehicle for enhancing the separation.

Irrespective of whether reaction or separation is of primary concern, three types of reaction-separation techniques are commonly used: *reaction-extraction*, *reaction-distillation*, and *reaction-crystallization*. Each of these can in theory be either reaction or separation oriented. Among other, less conventional, methods of combining reaction with separation are biphasing and the use of membrane reactors. These were considered in earlier chapters.

Table 25.1 lists several combinations of reaction and separation. The sequencing of the two in the nomenclature of the different combinations clearly reveals their orientations. This chapter is primarily concerned with reactive extraction (also termed dissociation-extraction), extractive reaction, reactive distillation (or dissociation-extractive-distillation), and distillative reaction (or distillation column reactors). Crystallization is almost always used for separation and seldom for enhancing a reaction. A notable exception is when one of the reactants is a sparingly dissolving solid and the size of the crystallizing solid is less than the thickness of the film surrounding the reactant. Then the crystallizing microphase enhances the rate of dissolution and hence the rate of reaction, a situation that was considered in Chapter 23.

REACTION-EXTRACTION

First we explain the basic principles of reactive (i.e., dissociation) extraction and then outline the more economical regenerative dissociation-extraction technique. We follow this up with a brief description of extractive reaction as an extension of what was described in Chapter 18 on biphasing.

Table 25.1 A broad classification of reaction-separation techniques and strategies

```
                          REACTION-SEPARATION
    ┌──────────┬──────────────┬──────────────┬──────────────┐
Use of      Reaction -    Reaction -    Reaction -     Reaction -
Zeolites    Permeation    Extraction    Distillation   Crystallization
            (Membrane
            Reactors)
  │              │         ┌────┴────┐   ┌─────┴─────┐   ┌──────┴──────┐
Separation                Reactive  Extractive  Reactive          Reactive
  │                       Extraction Reaction   Distillation      Crystallization/
Reaction                                                          Precipitation
  │                                                                   │
Separative       Separation                    Distillative       Precipitative
Reaction             │                         Reaction           Reaction
                 Separative                    (Distillation
                 Reaction                      Column Reactor)
```

Dissociation-Extraction (Separation Mode)

The basic principle

Dissociation-extraction involves equilibrating a mixture of acids (or bases) dissolved in an organic solvent with an aqueous phase containing a neutralizing agent in a stoichiometrically deficient amount just sufficient to react with the stronger component of the mixture. This *stoichiometric deficiency* of the neutralizing agent with respect to the total acids (or bases) leads to a competition between the components to react with the neutralizing agent. Thus the aqueous phase becomes enriched in the stronger component as its salt, and the organic phase is enriched in the weaker component.

As an illustration of this principle, consider a mixture of the weak organic acids, *p*-cresol and *m*-cresol. These have normal boiling points differing only by a fraction of a degree, 201.8 and 202 °C, and thus cannot be separated by distillation. The difference in their molecular structures, however, leads to different acid strengths. The acid dissociation constants are 7.9×10^{-11} and 5.24×10^{-11} for *m*-cresol and *p*-cresol, respectively. This difference can be exploited to separate a mixture by dissociation-extraction. If the mixture is partially neutralized by a strong base such as aqueous sodium hydroxide, there will be a competition between these two organic acids to react with the alkali. The stronger acid, *m*-cresol, having the higher dissociation constant, will react preferentially to form an ionized salt in the aqueous phase. This cresylate salt is insoluble in the organic phase. The weaker acid, *p*-cresol, will be retained in the organic phase.

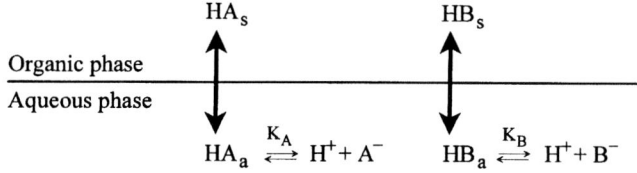

Figure 25.2 Principle of dissociation-extraction

Dissociation-extraction has been used to separate various isomeric and non-isomeric organic acids and bases, particularly those occurring in coal tar, such as m-cresol/p-cresol (Walker, 1950; Karr and Scheibel, 1954; Yamamoto, 1957; Ellis and Gibbon, 1963; Kostyuk et al., 1969; Anwar et al., 1971a; Kafarov et al., 1976), chlorophenols (Milnes, 1974; Laddha and Sharma, 1978; Anwar et al., 1979; Wadekar and Sharma, 1981a; Gaikar and Sharma, 1984), N-substituted anilines (Jagirdar and Sharma, 1981), C-substituted anilines (Wadekar and Sharma, 1981b), nitrocresols (Sasaki et al., 1977), o-/p-nitroanilines (Kemula et al., 1961), quinoline and isoquinoline (Yamamoto et al., 1958), monohydric alkyl phenols, 2,4-/2,5-xylenols (Coleby, 1971, Kiseleva et al., 1971). It has been commercially exploited in the separation of dichlorophenols, penicillin acids, and benzoic acid derivatives. For some of these systems, dissociation-extraction is the only practical method of separation.

Theory of dissociation-extraction

When a mixture of two acids HA and HB (HA is the weaker of the two) dissolved in a water-immiscible organic solvent is contacted with an aqueous phase containing the neutralizing agent in a stoichiometric deficiency, the following equilibria are established, as shown in Figure 25.2 (see Anwar et al., 1971b).

1. Distribution of the undissociated organic acids between the phases.

$$[HA]_s \xrightarrow{m_A} [HA]_a \quad \text{and} \quad [HB]_s \xrightarrow{m_B} [HB]_a \qquad [25.1]$$

where

$$m_A = \frac{[HA]_s}{[HA]_a} \qquad (25.1)$$

$$m_B = \frac{[HB]_s}{[HB]_a} \qquad (25.2)$$

and subscripts a and s represent the aqueous and organic phases, respectively.

2. Dissociation of the organic acids in the aqueous phase.

$$HA \xrightleftharpoons{K_A} H^+ + A^- \qquad [25.2]$$

$$HB \xrightleftharpoons{K_B} H^+ + B^- \qquad [25.3]$$

where

$$K_A = \frac{[H^+]_a [A^-]_a}{[HA]_a} \frac{\gamma_{H^+} \gamma_{A^-}}{\gamma_{HA}} \qquad (25.3)$$

$$K_B = \frac{[H^+]_a [B^-]_a}{[HB]_a} \frac{\gamma_{H^+} \gamma_{B^-}}{\gamma_{HB}} \qquad (25.4)$$

3. Competition between HA and HB leading to the following key equilibrium reaction in the aqueous phase (see Figure 25.2):

$$HB + A^- \xrightleftharpoons{K} HA + B^- \qquad [25.4]$$

$$K = \frac{[HA]_a [B^-]_a}{[HB]_a [A^-]_a} \cdot \frac{\gamma_{HA} \gamma_{B^-}}{\gamma_{HB} \gamma_{A^-}} \qquad (25.5)$$

From Equations 25.3–25.5

$$K = \frac{K_B}{K_A} \qquad (25.6)$$

and from Equations 25.1, 25.2, 25.5 and 25.6

$$\frac{[B^-]_a}{[A^-]_a} = \frac{K_B m_A [HB]_s}{K_A m_B [HA]_s} \cdot \frac{\gamma_{HB} \gamma_{A^-}}{\gamma_{HA} \gamma_{B^-}} \qquad (25.7)$$

Assuming the following equalities for activity coefficients:

$$\gamma_{BH} = \gamma_{AH}$$

and

$$\gamma_{A^-} = \gamma_{B^-} \qquad (25.8)$$

Equation 25.7 simplifies to

$$\frac{[B^-]_a}{[A^-]_a} = \frac{K_B m_A [HB]_s}{K_A m_B [HA]_s} \qquad (25.9)$$

Because the total organic acids are in a stoichiometric excess over the extracting agent of concentration N, this concentration can be expressed as

$$N = [A^-]_a + [B^-]_a \qquad (25.10)$$

Now we define a separation factor α to quantify the extent of separation:

$$\alpha = \frac{[HA]_s / ([HA]_a + [A^-]_a)}{[HB]_s / ([HB]_a + [B^-]_a)} \qquad (25.11)$$

Substituting for $[HA]_s, [HB]_s, [HA]_a, [HB]_a, [A^-]_a,$ and $[B^-]_a$, the expression for the separation factor becomes

$$\alpha = \frac{m_A K_B}{m_B K_A} \left\{ \frac{N(\delta+1) + [AB]_s[1/m_B + (K_A/K_B)(\delta/m_A)]}{N(\delta+1) + [AB]_s[(K_B/K_A)(1/m_B) + \delta/m_A]} \right\} \quad (25.12)$$

where $[AB]_s$ and δ are the total concentration and molar ratio of acids in the organic phase, respectively; and $[AB]_a$ is the total concentration of organic acids in the aqueous phase:

$$[AB]_s = [HA]_s + [HB]_s$$
$$[AB]_a = [HA]_a + [HB]_a + [A^-]_a + [B^-]_a \quad (25.13)$$
$$\delta = \frac{[HA]_s}{[HB]_s}$$

An approximate estimate of the separation factor can be obtained from the ratio of distribution coefficients and of the dissociation constants:

$$\alpha = \frac{m_A K_B}{m_B K_A} \quad (25.14)$$

Although the analysis presented is for a mixture of acids, it is equally applicable to bases.

Prediction of separation factors

In using Equation 25.12 to predict the separation factor for given acidic/basic mixtures, values of m_A, m_B, K_A and K_B are necessary and are easily obtained experimentally. However, the dissociation constants can also be predicted from available correlations (Chapter 3).

Regenerative dissociation-extraction: general considerations

Although dissociation-extraction processes have been employed commercially, their wider application is hampered by high operational costs. The major factor is the cost of the extracting and neutralizing agents required to recover the bases (or acids) from the extract phase. Another factor is the strong interaction between a compound and the reagent. In an ideal separation process, this interaction must be strong enough to give the desired separation but weak enough to be broken down without the need for excessive chemical or thermal energy.

Anwar et al. (1974, 1979) used a weak extracting agent followed by secondary extraction of the aqueous extract with a sufficiently polar solvent, thus setting up a regenerative process (Figure 25.3). Consider again the separation of *m*- and *p*-cresols. If the cresols dissolved in an organic solvent of low to moderate polarity, say, a paraffinic or aromatic solvent, are contacted with an aqueous solution of a weak base such as trisodium phosphate, the competition between the cresols for reaction with the phosphate leads to separation. The more acidic *m*-cresol reacts preferentially forming a dissociated salt in the aqueous phase, and the less acidic *p*-cresol remains predominantly in the organic phase. The

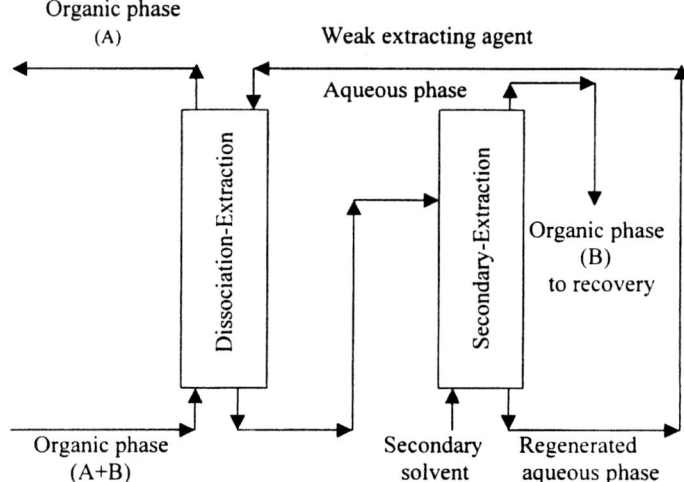

Figure 25.3   Regenerative dissociation-extraction using a weak extracting agent

aqueous phase containing sodium *m*-cresylate is contacted with an organic solvent with high affinity for cresols, say chloroform, whereupon the reaction is reversed and trisodium phosphate is regenerated. Then the cresol is recovered from the organic solvent by distillation. However, a limitation of regenerative processes is that the amount of solute transferred to the aqueous phase may be low.

Selection of solvent in dissociation-extraction

Dissociation-extraction exploits the difference in the dissociation constants and also the difference in the distribution coefficients of the components of a mixture. By changing the solvent, it is often possible to change the difference in distribution coefficients. Thus a judicious selection of solvent is important in realizing efficient separation. The relative interactions of different components with a solvent can determine the difference in distribution coefficients.

Organic solvents can be classified according to their ability to interact with phenolic and basic substances. Aliphatic hydrocarbons such as *n*-heptane and cyclohexane constitute a class of inert solvents with negligible tendency to interact with –OH or –NH_2 groups. These are followed by aromatic hydrocarbons, halogenated hydrocarbons, ethers and ketones, esters, and finally alcohols, in order of increasing tendency to interact with solutes. Alcohols can form strong hydrogen bonds with various hydrogen bond donors and acceptors. Knowledge of these interactions coupled with structural information about the molecules can help in selecting a proper solvent.

To have better selectivity in dissociation-extraction, it is essential that the weaker component is retained in the organic phase and the stronger component does not specifically interact with the solvent.

On the basis of solvent–solute interactions and steric hindrance to these interactions, it seems reasonable to use

1. an inert solvent if the stronger component has a relatively free solvophilic group; and
2. a polar solvent if the weaker component has a relatively free solvophilic group.

EXAMPLE 25.1

Choice of solvent in reactive extraction

6-Chloro-o-cresol (6COC) is an intermediate in the production of herbicides and pharmaceuticals. Its manufacture yields a mixture of 6COC and unreacted o-cresol (OC). Because both OC and 6COC boil in the temperature range of 191–192°C, distillation is not a practical option for separating this mixture.

Dissociation-extraction may be considered as a feasible alternative. If benzene is chosen as the solvent for this technique, practically no separation can be achieved despite a large difference in the pK_a values. This results from the very high distribution coefficient difference acting in the opposite direction. 6COC has a much higher distribution coefficient than o-cresol because of the presence of the bulky, hydrophobic –Cl group. However, with a polar solvent such as dibutyl ether and n-octanol, the separation factors are 4 and 9, respectively, even though the distribution coefficient difference (278 for COC and 28.2 for OC) still acts in the opposite direction. This is because OC has a comparatively free hydroxyl group that can interact with these polar solvents and thus is retained in the organic phase. There is no equivalent change in the distribution coefficient of 6COC because of the steric hindrance provided to the –OH group by the –Cl group.

Now let us consider the separation of 2,6-xylenol/p-cresol mixtures. Here, the situation is reversed. The distribution coefficient of 2,6-xylenol ($pK_a = 10.62$) is higher than that of p-cresol ($pK_a = 10.28$) because of the close proximity of the two methyl groups to the hydroxyl group. For such a mixture, the use of a polar solvent will be detrimental because a free –OH group on p-cresol can interact with polar solvents. Thus an inert solvent is a better choice. Indeed, separation factors are about 6 in dibutyl ether, 13 in benzene, and 35 in n-heptane (Gaikar and Sharma, 1985).

EXAMPLE 25.2

Use of dissociation-extraction in the manufacture of a useful product from a difficultly separable by-product mixture

In the manufacture of m-nitrochlorobenzene, a mixture of o-nitrochlorobenzene and m-nitrochlorobenzene is obtained as a by-product. One possible way of using this mixture is by subjecting it to methanolysis followed by reduction:

By this process, the *o*-isomer is converted to *o*-ansidine (ANS), and *m*-nitrochlorobenzene gives *m*-chloroaniline (MCA). The mixture of ANS and MCA thus obtained can be separated by dissociation-extraction.

The pK_as of ANS and MCA are 4.52 and 3.52, respectively, and their distribution coefficients are 41.5 and 62.4 between chlorobenzene and water. The protonation constants of ANS and MCA are 3.31×10^{-10} and 3.31×10^{-11}, respectively. Thus ANS can be extracted into the aqueous phase containing a mineral acid as a neutralizing agent in a stoichiometrically deficient amount.

The equations derived for phenolic mixtures can be suitably modified to treat basic mixtures with an acid neutralizing agent. The equilibration of a 1.0 kmol/m³ solution of a mixture of MCA/ANS in the ratio of 0.25:1 with an aqueous 1.0 kmol/m³ solution of hydrochloric acid gives the following compositions in the organic and aqueous phases:

$[MCA]_s = 0.121$ $\quad [ANS]_s = 0.076$

$[MCA]_a = 0.002$ $\quad [ANS]_a = 0.002$

$[MCA - H^+]_a = 0.096$ $\quad [ANS - H^+]_a = 0.904$

$\alpha(\text{calc.}) = 14.77$

$\alpha(\text{exptl.}) = 14.56$

Differences in both the basic strength and the distribution coefficient act in the same direction. Thus they supplement each other to give a high separation factor.

This example demonstrates a particularly useful feature of organic synthesis engineering. Any by-product mixture from a process that cannot be easily separated can be converted by reaction to a more useful mixture, whose components can then be separated relatively easily by dissociation-extraction.

Extractive Reaction (Reaction Mode)

Biphasing by deliberate addition of a second liquid phase was considered in Chapter 18. We extend the treatment in this section to aspects of biphasing that involve extractive reaction. Thus we focus the discussion on extractive reactions that depend on the dissociation equilibria of the components involved, and we restrict the treatment to a few instructive examples.

1. *Epoxidation of olefinic compounds with metachloroperbenzoic acid.* An interesting example (Brändström, 1983) is the epoxidation of olefinic

compounds with metachloroperbenzoic acid (MCPBA) to give metachlorobenzoic acid (MCBA). Because MCBA undergoes further reaction with MCPBA in most organic solvents, its yield can be increased if it can be appropriately protected. This can be done by exploiting the large difference in the pK_a values of the two compounds and a reasonable difference in their distribution coefficients between dichloromethane and water: pK_a = 3.82 and m = 1.2 for MCBA, and pK_a = 7.3 and m = 1.8 for MCPBA. Clearly, MCPBA is a much weaker acid and is more lipophilic than MCBA. Thus if a solution of the two acids in dichloromethane is stirred in a buffered aqueous medium at pH 7, more than 99% of MCBA can be extracted into the aqueous phase and only 1% of MCPBA.

A practical method of carrying out the reaction is to dissolve the reactant in dichloromethane along with the epoxidizing agent MCPBA, add a layer of aqueous sodium hydrogen carbonate, and carry out the extractive oxidation in this two-phase system. Almost the entire product MCBA is found in the aqueous phase. An alternative method is to use any aqueous solution (in place of sodium hydrogen carbonate) whose pH is electronically controlled at 7–8.

2. *Extractive hydrolysis of formate drug intermediates to the corresponding alcohols.* Extractive reaction can be advantageously used to improve hydrolytic efficiencies and is particularly relevant to the manufacture of drug intermediates. A good example of this is the reaction

[25.5]

which is an intermediate step in the synthesis of the broad spectrum antibiotic primaxin (King et al., 1985). This step, in which the formate portion of (1) is hydrolyzed to the alcohol portion of (2), suffers from the disadvantage that impurities are formed to the extent of 75% of the solid mass. A strategy to overcome this deficiency is to separate (1) from the impurities by selectively manipulating the "other portion" of (1). This can be done by titration with an aqueous base when the β-ketoester portion is converted to the enolate salt which is soluble in water (the impurities are not). Because the main reaction leading to (2) is very fast, it is clear that (1) can be efficiently converted to (2) by a two-phase extractive hydrolysis.

3. *Hydroformylation reactions.* Propylene can be hydroformylated by using a homogeneous Rh catalyst complexed to tris(m-sulfophenyl)phosphine.

The catalyst is soluble in water and can be extracted, but its sodium salt cannot be. A situation of this kind is clearly an attractive candidate for extractive reaction. Because a variety of pharmaceuticals and their intermediates are made via hydroformylation (see Chapter 8), this strategy should prove useful in the pharmaceutical industry.

DISTILLATION-REACTION

Distillation combined with reaction has been successfully used for separating close boiling mixtures. When used in this *separation mode*, the technique is frequently referred to *as dissociation-extractive distillation*. It can also be used in the *reaction mode* by continuous separation of the reaction products from the reactants. The equipment used in the latter case is often referred to as a distillation column reactor (DCR). The chief advantage of this method is that the reactants can be used in stoichiometric quantities, with attendant elimination of recycling costs.

Dissociation-Extractive Distillation (Separation Mode)

The basic principle

The mole fraction of the ith component (y_{Gi}) in the vapor phase in equilibrium with a liquid mixture with a mole fraction y_{Li} at a total pressure P (not much higher than atmospheric) is given by

$$y_{Gi} = \frac{\gamma_i y_{Li} P_i}{P_T} \qquad (25.15)$$

where γ_i is the activity coefficient of i. By definition, the relative volatility of a mixture of components i and j is given by

$$\beta_{ij} = \frac{y_{Gi}/y_{Li}}{y_{Gj}/y_{Lj}} \qquad (25.16)$$

or

$$\frac{y_{Gi}}{y_{Gj}} = \frac{y_{Li}\gamma_i P_i}{y_{Lj}\gamma_j P_j} \qquad (25.17)$$

It is evident that the relative volatility can be manipulated through the γs by adding suitable reacting/complexing agents.

Theory

Acid-base reactions, which are typically very fast and reversible, can take advantage of steric and/or acidity differences of the components of a mixture, usually isomers. Consider a mixture of two bases B_1 and B_2 that are to be separated by imposing a chemical reaction with an acid introduced in a stoichiometric

deficiency. The reactions involved are

$$B_1 + HA \xrightleftharpoons{K_{B_1}} B_1H^+A^- \qquad [25.6]$$

$$B_2 + HA \xrightleftharpoons{K_{B_2}} B_2H^+A^- \qquad [25.7]$$

The complexes formed are assumed to be nonvolatile. In other words, the chemical reactions are assumed to be confined to the liquid phase. Now the following competitive reaction occurs between the bases and the complexes:

$$B_1H^+A^- + B_2 \xrightleftharpoons{K_{12}} B_2H^+A^- + B_1 \qquad [25.8]$$

where

$$K_{12} = \frac{[B_2H^+A^-]_L[B_1]_L}{[B_2]_L[B_1H^+A^-]_L} = \frac{K_{B_2}}{K_{B_1}} \qquad (25.18)$$

Liquid-vapor equilibrium is also established between the free bases, giving

$$\beta_{12} = \frac{[B_1]_G[B_2]_L}{[B_2]_G[B_1]_L} \qquad (25.19)$$

Two steps are involved in the protonation of the bases, ion-pair formation and dissociation of the ion pair into charged species as shown here.

$$B_i + HA \xrightleftharpoons{K_{PT}} B_iH^+A^- \xrightleftharpoons{K_D} B_iH^+ + A^- \qquad [25.9]$$

where K_{PT} is the protonation constant and K_D the ion-pair dissociation constant. The solvent dielectric constant ε plays a role in the ion-pair dissociation step (the attractive force between ions is inversely proportional to ε). Thus with weak acids in nonaqueous media, the formation of free ions should be very low. On the other hand, the higher stability of the anion A^- of a strong acid allows ion dissociation, in which case the free ions should be formed in greater amount if the dielectric constant of the solvent is high. Moreover, the K_{12} value should be higher with free ions than with ion pairs. Normally, K_{12} depends on the difference between the pK_as of the acid and bases involved in this acid-base reaction except for *ortho*-substituted compounds because of the steric effects. The value of K_{12} must be known from experiments with and without the reacting components.

If the entire amount of acid is consumed in the complexation because of stoichiometric deficiency, then one can assume that

$$N = [B_1H^+A^-]_L + [B_2H^+A^-]_L \qquad (25.20)$$

Now we define an apparent relative volatility as

$$\beta_a = \frac{[[B_1]_G/([B_1]_L + [B_1H^+A^-]_L)]}{[[B_2]_G/([B_2]_L + [B_2H^+A^-]_L)]}$$

$$= \left(\frac{[B_1]_G [B_2]_L}{[B_2]_G [B_1]_L}\right)\left(\frac{1 + [B_2H^+A^-]_L/[B_2]_L}{1 + [B_1H^+A^-]_L/[B_1]_L}\right)$$

$$= \beta_{12}\left(\frac{1 + [B_2H^+A^-]_L/[B_2]_L}{1 + [B_2H^+A^-]_L/[B_1]_L}\right) \quad (25.21)$$

Elimination of $[B_1H^+A^-]$ and $[B_2H^+A^-]$ through Equations 25.18 and 25.20 leads to the following expression for relative volatility:

$$\beta_a = \beta_{12}\left[1 + \frac{N(\delta' + 1)(K_{12} - 1)}{N(\delta' + 1) + K_{12}[B_{12}]_L + \delta'[B_{12}]_L}\right] \quad (25.22)$$

where δ' and $[B_{12}]_L$ are two new parameters defined as

$$\delta' = \frac{[B_1]_L}{[B_2]_L} \quad (25.23)$$

$$[B_{12}]_L = [B_1]_L + [B_2]_L$$

The value of K_{12} depends on the relative strengths of the two bases. For $K_{12} > 1$, β_a will always be greater than β_{12}.

Clearly, competitive reactions can increase the relative volatility of the mixture. For most close boiling substances, especially isomers, β_{12} is close to unity. Thus the increase in relative volatility results from the bracketed quantity which can be manipulated by varying the reacting component and thereby changing K_{12}. Figure 25.4 shows the variation in β_a/β_{12} with K_{12} for a set of selected parameter values.

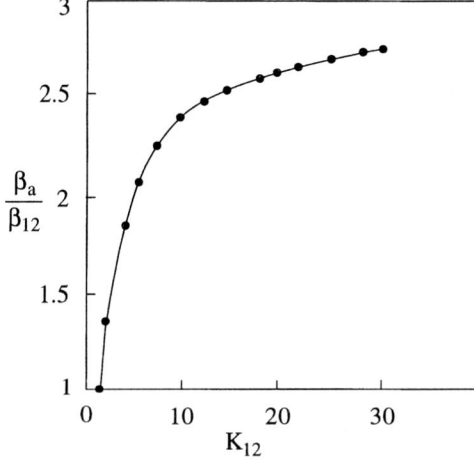

Figure 25.4 Enhancement in relative volatility due to competitive reaction 25.8

Examples

An acid–base reaction can be used to enhance separation by taking advantage of the difference in the pK_a values of the components to be separated (Duprat and Gau, 1991). For example, in the close boiling 3-/4-picolines mixture, addition of trifluoroacetic acid in stoichiometric deficiency results in preferential complexation of 4-picoline with a selectivity of about 2 in formamide as the solvent. Because the pyridinium salts are nonvolatile, the vapor phase is enriched with respect to 3-picoline. A near complete separation can be expected by repetition of this enrichment by countercurrent staging. 3-Picoline will leave the column as the distillate at the top, and 4-picoline will leave as a liquid phase complex with the acid as the bottoms product. 4-Picoline can be regenerated from the complex by adding a stronger base.

Acid–base reactions can also be exploited by taking advantage of the steric and/or acidity differences of alcohol isomers by conducting the distillation of these mixtures in the presence of amines (Gassend et al., 1985). More recently, organic bases have been used in the reactive distillation of close boiling phenolic substances (Mahapatra et al., 1988).

Distillative Reaction (Reaction Mode)

The basic principle

Consider a reversible reaction of the type

$$A + B \leftrightarrow R + S \qquad [25.10]$$

If one of the products can be continuously removed by carrying out the reaction simultaneously with distillation, then the reaction will be driven further forward thus increasing the conversion. The reaction can be carried out either in a simple batch reactor or in a continuous distillation column. The continuous column can be either a plate column (with variations in design) or a packed column. Several studies have been reported on modeling such units, for example, Marek (1954), Nelson (1971), Suzuki et al. (1971), Komatsu and Holland (1977), Holve (1977), Komatsu (1977), Sawistowski and Pilavakis (1979, 1988), Grosser et al. (1987), Steppan et al. (1987), Barbosa and Doherty (1987a,b; 1988a,b,c), DeGarmo et al. (1992), Doherty and Buzad (1992), Masamoto and Matsuzaki (1994), Kolah et al. (1996), and Chopade and Sharma (1997).

Although plate columns are more common for large volume products, packed columns are preferred for the relatively small volume products involved in intermediates and fine chemicals manufacture. However, the latter can also be used for relatively large volume productions, as in the manufacture of methyl tertiary butyl ether (MTBE), an important antiknock agent that replaces tetraethyl lead in gasoline. Hence we consider the following two cases: (1) a batch reactor with continuous removal of one of the products and (2) a packed-column reactor where the packing is also the catalyst. A brief qualitative discussion of DCRs in general along with industrial examples is also presented.

Batch reactor with continuous removal of product

Let reaction 10 be carried out in a batch reactor as shown in Figure 25.1, with an attached column for separating R. We assume that stoichiometric quantities of A and B are present initially and consider two cases: (1) there is no accumulation of S because it is vaporized as soon as it is formed, and (2) there is an accumulation of S because only a fraction of it is vaporized.

CASE 1: NO ACCUMULATION OF S

$$N_A = N_{A0}(1 - X_A), \qquad N_S = N_{A0}X_A - \int_0^t F_s \, dt \tag{25.24}$$

The rate equation may be written as

$$-r_A = k_2[A][B] - k_{-2}[R][S] \tag{25.25}$$

where

$$[A] = [B] = \frac{N_{A0}(1 - X_A)}{V}, \qquad [R] = \frac{N_{A0}X_A}{V} \tag{25.26}$$

Thus

$$-r_A V = N_{A0}\frac{dX_A}{dt} = \frac{k_2 N_{A0}^2 (1 - X_A)^2}{V} - \frac{k_{-2} N_{A0} X_A}{V}\left(N_{A0}X_A - \int_0^t F_s \, dt\right) \tag{25.27}$$

An overall mass balance gives

$$\frac{d(\rho V)}{dt} = -F_s M_s \tag{25.28}$$

which, for constant density, becomes

$$\frac{dV}{dt} = -\left(\frac{M_s}{\rho}\right) F_s = -\alpha' F_s \tag{25.29}$$

where M_s is the molecular weight of S and $\alpha' = M_s/\rho$.

Because it is assumed that the product S evaporates as quickly as it is formed,

$$N_{A0}X_A = \int_0^t F_s \, dt \tag{25.30}$$

giving

$$F_s = N_{A0}\frac{dX_A}{dt} = \frac{k_2 N_{A0}^2 (1 - X_A)^2}{V} \tag{25.31}$$

Substituting in Equation 25.29,

$$\frac{dV}{dt} = -\alpha' N_{A0}\frac{dX_A}{dt}$$

or

$$\frac{dV}{dX_A} = -\alpha' N_{A0} \tag{25.32}$$

giving

$$V = V_0(1 + \varepsilon_{L1} X_A) \tag{25.33}$$

where

$$\varepsilon_{L1} = -\alpha'[A]_0 \tag{25.34}$$

Combining Equations 25.31 and 25.33,

$$\frac{dX_A}{dt} = \frac{k_2 N_{A0}(1-X_A)^2}{V} = k_2[A]_0 \frac{(1-X_A)^2}{(1+\varepsilon_{L1}X_A)} \tag{25.35}$$

This equation can be solved analytically to give the time required for a specific conversion X_A:

$$t = \frac{1}{k_2[A]_0}\left[\frac{(1+\varepsilon_{L1})X_A}{1-X_A} - \varepsilon_{L1}\ln\frac{1}{1-X_A}\right] \tag{25.36}$$

CASE 2: S IS NOT COMPLETELY VAPORIZED

Let us assume that a fraction of product S formed is lost by vaporization. This fraction β' depends on the vapor–liquid equilibrium and the heating policy used.

$$N_s = N_{A0} X_A (1 - \beta') \tag{25.37}$$

where

$$\beta' = \text{reaction-separation parameter} = \frac{\int_0^t F_s dt}{N_{A0} X_A} \tag{25.38}$$

The parameter β' can be physically interpreted as the ratio of the moles of S removed by vaporization to the moles formed by reaction and may therefore be regarded as a reaction-separation parameter. Equation 25.33 for this case will be modified as follows:

$$V = V_0 - \alpha' \int_0^t F_s dt = V_0 - \alpha' \beta' N_{A0} X_A \tag{25.39}$$

$$= V_0(1 + \varepsilon_{L2} X_A)$$

where

$$\varepsilon_{L2} = \alpha'\beta'[A]_0 = \frac{M_s \int_0^t F_s dt}{\rho N_{A0} X_A}[A]_0 \tag{25.40}$$

Then, by combining Equations 25.25, 25.26, 25.37, and 25.39, we finally get

$$\frac{dX_A}{dt} = \frac{k_2[A]_0(1-X_A)^2}{(1+\varepsilon_{L2}X_A)} - \frac{k_{-2}[A]_0 X_A^2}{(1+\varepsilon_{L2}X_A)}(1-\beta') \tag{25.41}$$

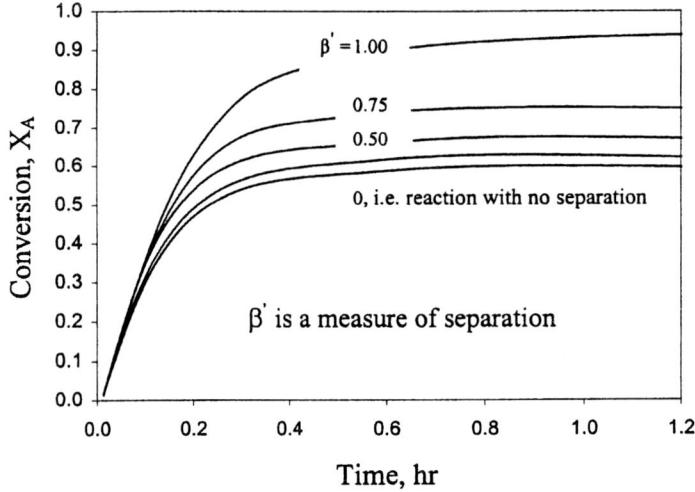

Figure 25.5 Performance of a distillation column reactor for the reaction $A + B \leftrightarrow R + S$. Conversion–time profiles for different values of the extent of product (S) removal β'

Equation 25.41 can be solved numerically for different values of β' corresponding to the extent to which the product is removed. Some results of numerical integration are shown in Figure 25.5. Clearly, the conversion increases with an increase in β'. The maximum conversion is obtained when the product is instantaneously removed from the reaction mixture. At $\beta' = 0$, the conversion approaches the value corresponding to the limiting condition of reaction with no separation.

Packed distillation column reactor

A packed DCR has the advantage that it speeds up the reaction in the column and also supplies a packing surface for mass transfer at the vapor–liquid interface. As in any packed-bed reactor, the principal aim of modeling a packed DCR is to obtain the concentration (or mole fraction) profiles of the different components along the reactor. A basic requirement for doing this is an equation for mass flux at the surface that incorporates the effect of chemical reaction. This is not needed in modeling plate columns. In addition, the packing usually acts also as a catalyst. Hence the large number of models developed for the plate column are not applicable to the packed column.

Consider Figure 25.6 which represents a differential element of a packed column reactor. The following component mass balance can be readily written:

$$d(F_G y_{Gj}) = -N_j' a A_c d\ell = d(F_L y_{Lj}) - r_j h_0 A_c d\ell \tag{25.42}$$

where F_G and F_L are the gas and liquid molal flow rates, respectively, A_c is the cross-sectional area of the column, h_0 is the liquid holdup in the packing per unit volume of the packed bed, N_j' is an overall mass flux that includes the effect of

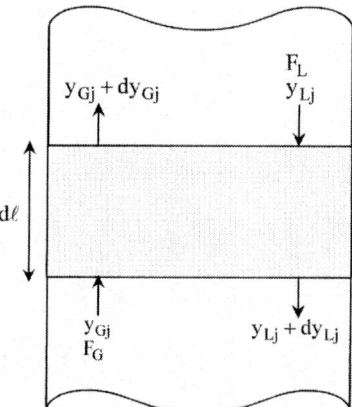

Figure 25.6 Flows in a differential element of a packed distillation column reactor

reaction, and a is the area per unit volume. The following expression for N_j' can be used (Sawistowski and Pilavakis (1979, 1988)):

$$N_j' = \left[\frac{1}{k_{Gj}^0} + \left(\frac{K_j}{k_{Lj}^0}\right)\left(\frac{1+\varepsilon_G}{1+\varepsilon_L/2}\right)\right]^{-1}\left[\left(\frac{1+\varepsilon_G+\varepsilon_L/2}{1+\varepsilon_L/2}\right)y_{Gj}^* - y_{Gj}\right] \quad (25.43)$$

where the ks represent the mass transfer coefficients of species j in the gas and liquid phases (mol/m^2 s), respectively, for pure distillation, K_j is the equilibrium constant for gas–liquid equilibrium at the surface (i.e., $y_{Gj,s} = K_j y_{Lj,s}$), and ε_G and ε_L are constants defined by

$$\varepsilon_G = \frac{(q_s/\Delta H_v)}{k_{Gj}^0}, \quad \varepsilon_L = \frac{(q_s/\Delta H_v)}{k_{Lj}^0} \quad (25.44)$$

Based on Equation 25.42, the following overall and component derivatives with respect to column height can be written:

$$\frac{dF_G}{d\ell} = -\left(\frac{q_s}{\Delta H_v}\right)aA_c \quad (25.45)$$

$$\frac{dF_L}{d\ell} = -\left(\frac{q_s}{\Delta H_v}\right)aA_c \quad (25.46)$$

$$\frac{dy_{Gj}}{d\ell} = \frac{\left[-y_{Gj}\left(\dfrac{dF_G}{d\ell}\right) - N_j a A_c\right]}{F_G} \quad (25.47)$$

$$\frac{dy_{Lj}}{d\ell} = \frac{\left[r_j h_0 A_c - y_{Lj}\left(\dfrac{dF_L}{d\ell}\right) - N_j a A_c\right]}{F_L} \quad (25.48)$$

810   Strategies for Enhancing Rate Reactions

The term q_s corresponds essentially to the heat of reaction (heat of dilution and heat loss are negligibly small in comparison) which can be calculated from

$$q_s = \frac{r_j h_0 \Delta H_v}{a} \qquad (25.49)$$

The set of Equations 25.42–25.48 can be solved provided the following information is available: vapor–liquid equilibrium data, for example, the ternary equilibrium data for a typical esterification reaction; mass and enthalpy balances around the feed point, reflux inlet, and reboiler to account for the flow rates, compositions, and thermal conditions of the external streams; mass transfer coefficients in the absence of reaction (either by experimental determination or estimation from available correlations); liquid holdup (usually from available correlations); and an expression for the reaction rate. Then the equations can be solved by any convenient method, preferably the Runge–Kutta routine, to get the mole fraction of each component as a function of height.

OVERALL EFFECTIVENESS FACTOR IN A PACKED DCR

An important feature of packed DCRs is the need to pack the catalyst in a special way to ensure good flow, mass transfer, and contact characteristics. An example of this is the use of an ion-exchange resin catalyst (Amberlyst 15) in methyl tertiary butyl ether (MTBE) manufacture. The bed consists of bags made in the form of a cloth belt with narrow pockets sewn across it (Figure 25.7). The pockets are filled with catalyst granules, and the belt is twisted into a helical form, referred to as a bale (see Smith, 1980, for details). Clearly, each pocket represents a closely packed bed of unconsolidated particles, and the pocket and the individual particles exhibit, respectively, their own distinctive macro- and microdiffusional features. This is broadly similar to the particle–pellet model of a catalyst pellet (see Chapter 7) but with distinctly different "pellet" behavior. Therefore, it is necessary to define an overall effectiveness factor which takes this unique feature into account. An attempt to do this was recently reported (Xu

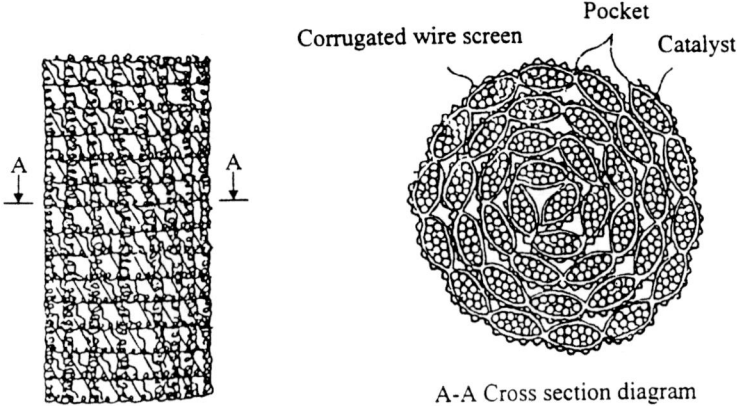

Figure 25.7   Structure of the catalyst bale used in MTBE production (Xu et al., 1995)

et al., 1995) for the MTBE reaction, for which the overall effectiveness factor was found to vary from 0.2 to 0.9 depending on the conditions of operation.

Discussion with examples

A common example of distillative reaction is esterification such as Eastman Kodak's process for methyl acetate:

$$\text{Methanol} + \text{acetic acid} \leftrightarrow \text{methyl acetate} + \text{water} \qquad [25.11]$$

in which the conversion is limited by thermodynamics. By continuously removing water or methyl acetate from the reaction mixture, the reaction equilibrium can be forced completely to the product side without using an excess of any reactant.

It is also possible to suppress undesirable chemical reactions, such as in the alkylation of isobutane by butene in the manufacture of isooctane. In the presence of butene, isooctane can undergo further alkylation, thus reducing the selectivity. Use of distillative reaction removes isooctane continuously from the column, thus enhancing the selectivity.

Even for highly exothermic reactions where the heat release can significantly affect the conversion rate, by conducting the reaction in a reactive distillation column, the heat of reaction can be used to remove product continuously from the reaction mixture. Thus chemical reactions that exhibit either an unfavorable reaction equilibrium or significant heat of reaction can benefit from reactive distillation column technology.

A few other industrial applications of DCRs are noteworthy. MTBE, TAME (tertiary amyl methyl ether), and ETBE (ethyl tertiary butyl ether) are clean burning octane enhancers that are replacing lead in gasoline as antiknock agents. MTBE is made from methanol and isobutene on a strongly acidic ion-exchange resin with excess methanol to prevent dimerization of isobutene. In the new process using a DCR (referred to earlier), most of the reaction is completed in a prereactor, and the near-equilibrium reaction mixture is fed to the DCR in the catalytic zone in the middle of the reactor (see Smith and Huddleston, 1982). Isobutene is completely converted in this zone of the reactor. The stripping section separates MTBE from the reactants which are continuously forced up into the reaction zone. Almost all new MTBE plants use this technology now.

Methylal synthesis is another outstanding example of distillative reaction (see Masamoto and Matsuzaki, 1994). Formaldehyde is conventionally produced by methanol oxidation as a 55% aqueous solution which is the maximum achievable concentration. Any other use of the aldehyde in higher concentrations is rendered impractical due to the high cost of concentration. In a process commercialized by Asahi Chemical Co., a much more concentrated solution is obtained by oxidizing methylal instead of methanol ($CH_3OCH_2OCH_3 + O_2 \rightarrow 3CH_2O + H_2O$), since only one mole of water is formed for every three moles of formaldehyde. Methylal itself is produced by acid-catalyzed reaction between methanol and aqueous formaldehyde solution, a by-product in the purification of formaldehyde. The reaction is carried out in a DCR to produce 70%

formaldehyde, which is then directly used as a reactant in acetal homopolymer and copolymer units.

Some important features of DCRs

The design of DCRs has developed into a highly specialized area, and no attempt will be made to cover this in any detail here. However, some of the theoretical foundations are qualitatively discussed, based essentially on the studies of Barbosa and Doherty, 1987a,b; 1988a,b,c; Buzad and Doherty, 1994; and Venimadhavan et al., 1994.

The design equations would include, in addition to the usual heat and mass balances and vapor-liquid equilibria, equations for chemical equilibria and/or reaction kinetics. The occurrence of a chemical reaction can severely restrict the allowable ranges of temperatures and phase compositions by virtue of the additional equations for chemical equilibrium/kinetics. This effect can be quantitatively analyzed by constructing a *residue curve map* (RCM). It explicitly shows the shifting of distillation boundaries in the presence of reaction and defines the limits of feasible distillation column operation. We illustrate this (Venimadhavan et al., 1994) by considering the reaction

$$A + B \leftrightarrow 2R \qquad [25.12]$$

with a temperature-independent equilibrium constant of 2. We assume ideal vapor-liquid behavior of the system, and also that A and B have constant volatilities of 5 and 3 relative to R. The RCM for a nonreactive mixture of this three-component system is shown in Figure 25.8a. If now reaction occurs and the system reaches equilibrium instantaneously, a maximum boiling azeotrope can occur, as it does in this case (Figure 25.8b). Phase compositions corresponding to this point are called *reactive azeotropes*. At the reactive azeotropy point, a temperature is reached where the rate of vaporization (or condensation) and the rate of reaction of each species are such that vaporization occurs without change of composition in either phase. If the reaction proceeds at a finite rate (with a constant vapor flow rate), the behavior is as depicted in Figure 25.8c. Note that the chemical equilibrium limit is recovered at long times. The behavior becomes more complicated as the time is further increased.

It is clear from Figure 25.8 that the condition for azeotropy is not equality of compositions in the liquid and vapor phases, as in straight distillation. Instead, the following equality has been proposed (Barbosa and Doherty (1988a,b,c):

$$\frac{y_{L1} - y_{G1}}{\nu_1 - \nu_t y_{L1}} = \frac{y_{Li} - y_{Gi}}{\nu_i - \nu_t y_{Li}}, \quad i = 2, \ldots, c$$

$$\nu_t = \sum_0^\infty \nu_i \qquad (25.50)$$

where c is the number of components and ν_i is the stoichiometric coefficient of component i.

Equation 25.50 shows that composition is not a convenient measure of azeotropy in reactive distillation. For equilibrium reactive mixtures, it is more

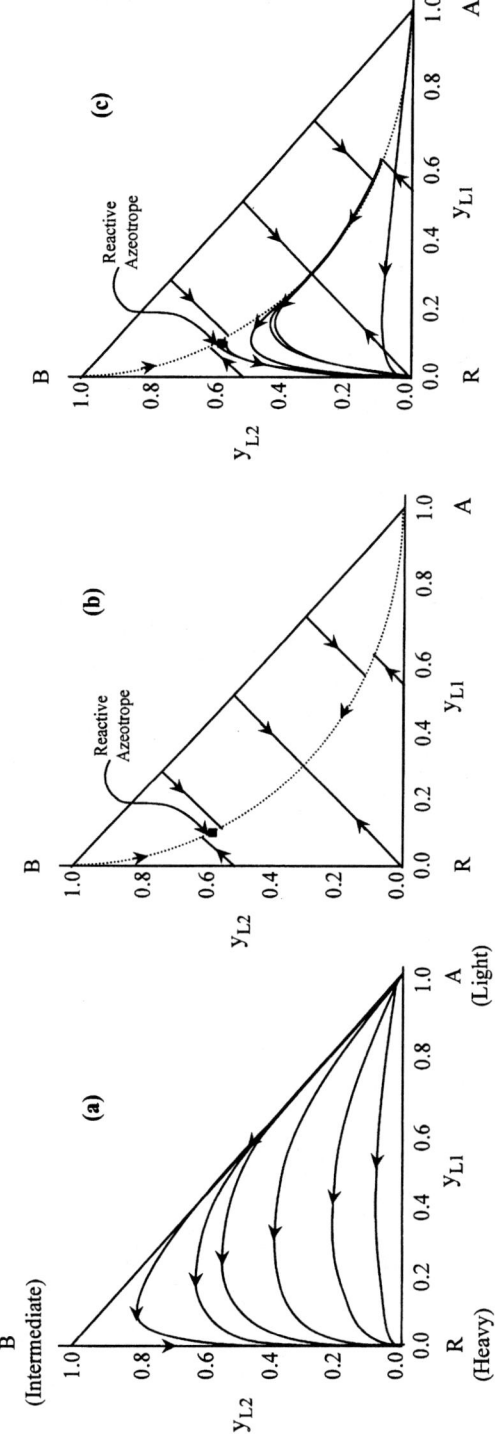

Figure 25.8 Residue curve maps for different situations in a distillation column reactor (from Venimadhavan et al., 1994)

convenient to use transformed variables which represent the equivalent amounts of reactants present in the equilibrium mixture. Thus for the reaction

$$\nu_A A + \nu_B B \leftrightarrow \nu_R R \qquad [25.13]$$

these variables are defined as

$$Y_{LA} = \frac{y_{LA} - (\nu_A/\nu_R)y_{LR}}{1 - (\nu_t/\nu_R)y_{LR}}$$

$$Y_{LB} = \frac{y_{LB} - (\nu_A/\nu_R)y_{LR}}{1 - (\nu_t/\nu_R)y_{LR}} \qquad (25.51)$$

The transformed composition variables have two convenient properties: they have the same numerical value before and after reaction, and their sum is unity:

$$Y_{LA} + Y_{LB} = 1 \qquad (25.52)$$

The condition for reactive azeotropy when expressed in terms of these transformed variables takes the familiar form of compositional equality, and may be written as

$$Y_{Li} = Y_{Gi} \qquad (25.53)$$

The design equations for a single-feed distillation column using these variables have been formulated by Barbosa and Doherty (1988a,b,c) (see also Rev, 1994). The procedure has been extended to multiple reactions by several authors (see, e.g., Ung and Doherty, 1994a,b,c, and Kolah et al., 1996). The appearance of multiple steady states in the solution of the DCR equations has also been considered (Huan et al., 1995) in methyl-*tert*-butyl ether production.

Chapter 26

Other Important (and Some Lesser Known) Strategies of Rate Enhancement

The literature contains examples of several strategies of rate enhancement not covered in the previous chapters. Many of these are essentially strategies for individual reactions with little general appeal. On the other hand, a few are very important, and several others combine two or more strategies. Of these, photochemical and micellar enhancements are as important as the strategies considered earlier in this part. However, in photochemical enhancement, recent studies have shown that the basis of scale-up used so far is questionable (Cassano et al., 1995), and designs based on newer concepts are still in their infancy. In micellar catalysis, despite the advances made, there are few industrial applications. As a result, these are included in this chapter on other strategies. Hydrotropes and supercritical fluids, although "old" with respect to other uses, are emerging as strong contenders for rate enhancement and ease of processing. Hence these two strategies are considered at some length in this chapter. Also included are the use of microwaves and several combinatorial strategies such as PTC with electrochemical, enzymatic, or sonochemical techniques; the use of supercritical fluids in similar combinations; enzymatic reactions in micelles; and PTC reactions in supercritical fluids or membrane reactors.

PHOTOCHEMICAL ENHANCEMENT

Interaction of light with a chemical species can initiate or enhance a chemical reaction. Reactions of this type are known as *photochemical reactions*. Of the many distinctive features of photochemistry, the following is particularly noteworthy: in thermal excitation processes, all three forms of energy, electronic, transational, and rotational, are raised to higher levels. In contrast, photoexcitation raises only the electronic energy level which leads to higher selectivity, as

exemplified by the photochlorination of the methyl group of toluene without any ring chlorination. Further, photochemical reactions are ecologically clean and require much less aggressive methods than conventional syntheses. Examples of reactions initiated or enhanced by light are many, and a small number are in industrial use, particularly in the production of halogenated hydrocarbons, alkane sulfates, and fine organic chemicals, including vitamins and fragrances. But the potential is enormous.

Of the many books, reviews, and leading articles available on the subject, the following are specially pertinent to our purpose: Cassano et al. (1967), Suppan (1972), Borrell (1973), Fischer (1978), Roger and Villermaux (1979, 1983), Bloomsfield and Owsley (1982), De Bernardez and Cassano (1982, 1985, 1986), Pfoertner (1984, 1986, 1990, 1991), Yue (1985a,b,c), André et al. (1986, 1988), Alfano et al. (1986a,b,c), Cassano et al. (1986, 1994, 1995), De Bernardez et al. (1987), Braun (1988), Alfano and Cassano (1988a,b), Cassano and Alfano (1991), Cabrera et al. (1991a,b,c, 1994), Bajic et al. (1995)[1].

Basic Energy Considerations

Visible light constitutes a small fraction of the total radiation and is often referred to as white light. If u_ℓ is the velocity of light ($= 2.998 \times 10^8 \, \text{m s}^{-1}$),

$$\lambda \nu = u_\ell \tag{26.1}$$

The frequency ν is usually expressed as hertz (Hz) in cycles per second (s^{-1}). A frequently used parameter is the *wave number* (m^{-1}) which is the reciprocal of the *wavelength* λ, the distance between the crests of a wave.

Light may also be regarded as a collection of photons or "packets of energy." A photon has no mass but has a specific energy which is proportional to the frequency of the radiation:

$$q(\text{energy of a quantum}) = h\nu = \frac{hu_\ell}{\lambda}, \quad e = N_0 q \tag{26.2}$$

where q is in joules or ergs per molecule, e is the energy in einsteins (joules or ergs per mole), and h is Planck's constant ($= 6.6265 \times 10^{-34}$ joule-seconds).

The basic feature of a photochemical reaction is interaction between *one* molecule or atom of a chemical species A and *one* photon (q ergs) to produce a molecule A^* in an *excited state*. The overall reaction in which n photons of light are absorbed may be written as

$$A + nq \rightarrow R \tag{26.1}$$

Photochemical reactions also include those in which light is not absorbed by the reactant directly, but by an added sensitizer. These are referred to as *photosensitized reactions*. Ketones such as acetophenone, α-substituted aceto-

---

[1] This is a useful review that covers photochemical reactors and auxiliaries such as light sources, filters, solvents, etc., particularly relevant to preparative photochemistry.

phenones, and benzophenone are among the more commonly used photosensitizers.

The wavelength of light absorbed from an incident beam by an atom or a molecule essentially influences the primary step of its photochemical transformation. The secondary reactions are usually thermal, but they also tend to be influenced by the wavelength of the absorbed light.

Quantum and Synthetic Quantum Yields

In photochemistry, the most important reaction parameter is *quantum yield* defined as

$$\phi_\lambda = \left(\frac{\text{moles of product formed}}{\text{number of photons absorbed by reactant}}\right)_\lambda \quad (26.3)$$

which represents the efficiency of utilization of light of a given wavelength λ. Note that this is not the yield with respect to the material reactant A as in any normal reaction but with respect to the "photoreactant"—the photon. The yield that is practically more important is the *synthetic quantum yield* ϕ' defined as

$$\phi' = \frac{\text{moles of product formed/kW h}}{\text{einsteins/kW h}} \quad (26.4)$$

which represents the quantum yield over a range of wavelengths rather than at a specific wavelength and can be quite easily estimated by measuring the moles formed per kilowatt-hour of lamp input.

Simple Thermodynamic and Kinetic Considerations

There are two basic laws of photochemistry. The first law leads to the well-known *Beer–Lambert law*, which gives the following expression for the attenuation of the intensity of light with distance:

$$I = I_0 \exp(\alpha_a [A] s') = I_0 \exp(\mu s') \quad (26.5)$$

where I_0 and I are the incident and transmitted light intensities (e/m² s), respectively; s' is the distance (m) travelled in the medium; α_a is the *absorptivity* or the *attenuation constant* (m²/mol); and μ is the *linear absorption coefficient* given by the product $\alpha_a [A]$, (m^{-1}). The light absorbed is the difference $(I_0 - I)$.

A number of deactivating events (e.g., collision of an activated molecule with an inactive one) occur when a photon is absorbed in an organic molecule, and the second law provides an energetic description of these deactivation processes. Thus, if the volumetric rate of light absorption is given by

$$(\text{Rate of absorption}) = k_i I_v \quad (26.6)$$

where I_v is the amount absorbed per unit time per unit reactor volume (e/m³ s), k_i will be independent of λ (i.e., $k_i = 1$) only if I_v is exclusively for the primary process that causes reaction. The quantity I_v is a very important parameter in

photoreactor design. Knowledge of the radiation field is obviously necessary to obtain an average value of this quantity.

Equation 26.6 can be extended to explicitly account for the photochemical nature of a reaction by expressing the rate of formation of product R as a function of both reactant concentration and the volumetric rate of absorption of light:

$$\frac{d[R]}{dt} = kI_v^m[A]^n, \qquad m = 1 \text{(usually)} \tag{26.7}$$

where k is the rate constant with units of $(m^3)^{m+n-1} s^{m-1}/(mol^{n-1} e^m$ $[= (m^3)^n/e$ $(mol)^{n-1}$ for $m = 1$, or m^3/e for $m = n = 1]$.

It is clear from these equations that a standard method of measuring light is needed for any quantitative assessment of a photochemical process. For this purpose, the intensity of light falling on a sample must be precisely defined in photons per second, so that it can be related to the quantum yield. This is done by using a reference sample, commonly termed the *actinometer*, of known quantum yield. The method itself is known as *chemical actinometry* (see Borrell, 1973, for details).

Photochemical Reactors

Approaches to design

Photochemical reactor design involves simultaneous solution of the mass, energy, and momentum balance equations (as in normal reactors) along with equations for the radiation field and energy source (which are specific to photochemical reactors). Two approaches are possible: (1) the intensity of the incident light, irrespective of the source, is used as the inlet boundary condition (*incidence models*); (2) the emission from the source itself is part of the mathematical description (*emission models*). The first approach has been extensively used but suffers from the weakness that the incident light is a function of scale, and hence a priori design from laboratory scale data tends to be uncertain. The second approach is formally correct, and involves no such uncertainty.

It is important to note that the near-classical first approach is increasingly being discarded by the theoretician, whereas the second (with a growing literature) is highly mathematical and largely untested. Further, the design of heterogeneous systems is yet to be effectively addressed. This is particularly true of reactors with a particulate phase (such as slurry and fluidized-bed reactors) which involve scattering of light from solid particles, a phenomenon that further adds to the complexity of the emission model (see, e.g., Maruyama and Nishimoto, 1992; Alfano et al., 1994). Thus we do not address the question of rigorous photochemical reactor design in this book. On the other hand, we outline certain basic aspects of the first approach as an order-of-magnitude guide to reactor development and at the end of the section give an example of photochemical reactor development based on certain logical practical considerations that will continue to be valid in industrial practice.

Reactor configurations

One of the most important features of photocemical reactors is the reactor configuration, the lamp–reactor geometry. Of the many configurations possible, a few typical ones are sketched in Figure 26.1 (see Doede and Walker, 1955; Yue, 1985a; Braun, 1988, for several more designs): a cylindrical (tubular) reactor with an elliptical reflector, an annular reactor, a film-type reactor with a specially designed reflector, a single-lamp multitubular continuous flow reactor, and a cylindrical reactor irradiated from the bottom.

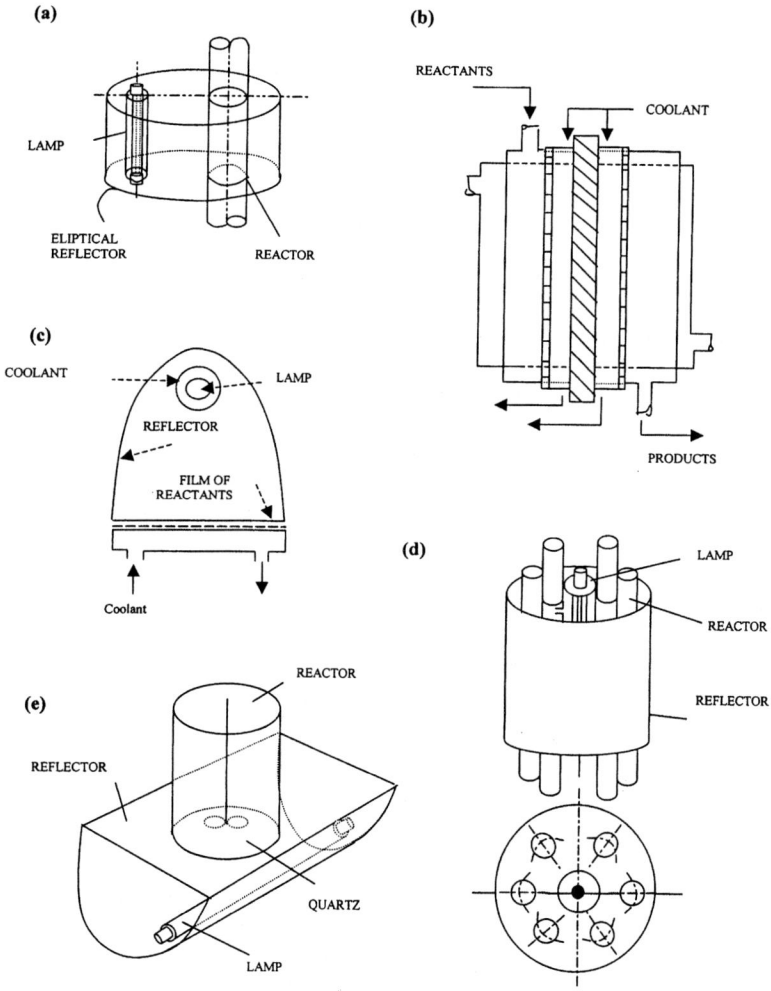

Figure 26.1 Examples of basic photochemical reactors (some adapted from Cassano et al., 1995): (a) tubular photoreactor inside a cylindrical reflector of elliptical cross section; (b) annular photoreactor; (c) film-type photoreactor; (d) single-lamp multitube continuous photoreactor; (e) perfectly-mixed semibatch cylindrical photoreactor irradiated from the bottom by a tubular source and a parabolic reflector

The principal features of the cylindrical, annular, and parallel plate configurations (using the incidence model) are summarized in Table 26.1. They are based on the further simplification that the radiation field can be completely described by the radial transmission of light from the wall to the axis of the reactor (thus ignoring its diffusion from all directions within the reactor).

Optical thickness

An important consideration in photochemical reactor design is the rapid attenuation of light intensity with the distance from the wall. It defines the zone in which much of the reaction occurs and is called the *optical thickness*. A useful expression was developed (Fischer, 1978) for light intensity as a function of thickness d for 99% absorption from a lamp of radius r_{l0}.

$$\log \frac{I_0}{I} = \log \frac{r_1}{r_{10}} + 2.3\mu(r_1 - r_{10}), \quad (r_1 - r_{10}) = d \tag{26.8}$$

where r_1, is a specific radial position less than r_{10}. From Equation 26.5 we see that high values of α_a and/or $[A]_0$ can reduce the thickness considerably usually to micrometer levels. Despite this deficiency, high values of $[A]_0$ are invariably preferred in industry to avoid excessive costs associated with solvent losses and recovery. Thus it seems that low optical thicknesses are almost unavoidable. To overcome this problem, the incident radiation is widened to include less absorbing parts of the chosen absorption spectrum. Then the reaction zone (i.e., the optical thickness) is likely to be extended significantly (Coyle, 1986). Designs to accomplish this have been suggested by Tiedt (1968) and Braun et al. (1986).

Reactor equations

Consider the simple reaction

$$A \xrightarrow{h_\nu} R \tag{26.2}$$

The steady-state continuity equation for A is given by (see Chapter 12).

$$u\frac{\partial [A]}{\partial \ell} + D_r \left[\frac{1}{r}\frac{\partial}{\partial r}\left(r\frac{\partial [A]}{\partial r}\right) \right] - (-r_A) = 0 \tag{26.9}$$

Because we are adopting the first approach, we assume knowledge of the incident light and use the following equation for light intensity as a function of the radial and axial positions of monochromatic light (Harano and Smith, 1968):

$$I_{r,\ell} = \frac{2I_0 R}{r}\left\{\exp\left[\alpha_a \int_0^R [A](r',\ell)dr'\right]\right\} \cosh\left[\alpha_a \int_0^r [A](r',\ell)dr'\right] \tag{26.10}$$

where $I_{r,\ell}$ is the absorbed light intensity (e/m^2 s) at position (r, ℓ) in the reactor and r' is a dummy radial variable of integration. This equation can be simultaneously solved with the continuity Equation 26.9.

Table 26.1 Radial intensity equations for parallel plate, cylindrical, and annular configurations (based on the approximate incidence model) (adapted from Roger and Villermaux, 1986)

Configuration	Parallel plate	Cylindrical	Annular
Sketch	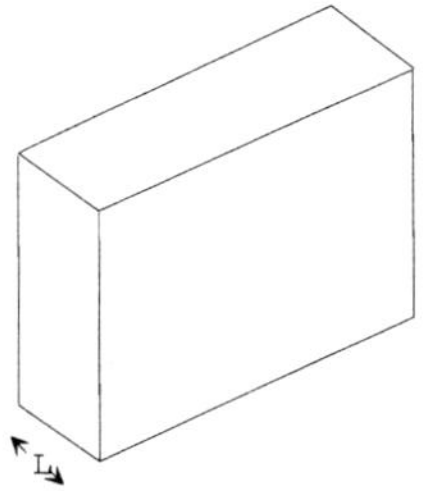	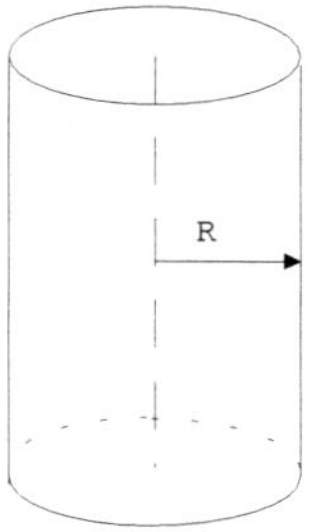	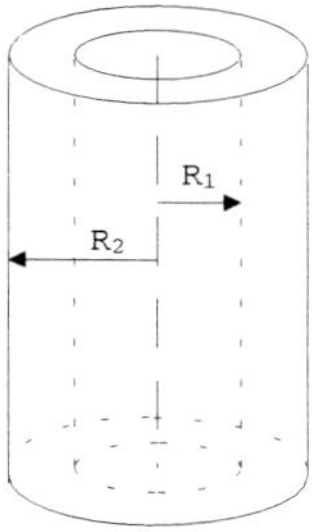
Characteristic dimension, d	L	$2R$	$R_2 - R_1$
Photochemical power, P	$\dfrac{I_0}{L}$	$\dfrac{2I_0}{R}$	$\dfrac{2I_0 R_1}{R_2^2 - R_1^2}$
Intensity, I	$I_0 \exp(-\mu \ell)$	$\dfrac{I_0 R}{r}\{\exp[-\mu(R-r)] + \exp[-\mu(R+r)]\}$	$\dfrac{I_0 R_1}{r}\exp[-\mu(r - R_1)]$
Mean intensity, $\bar{I}$	$\dfrac{P}{\mu}[1 - \exp(-\mu L)]$	$\dfrac{2I_0}{\mu R}[1 - \exp(-2\mu R)] = \dfrac{P}{\mu}[1 - \exp(-2\mu R)]$	$\dfrac{2I_0 R_1}{\mu(R_2^2 - R_1^2)}\{1 - \exp[-\mu(R_2 - R_1)]\}$ $= \dfrac{P}{\mu}\{1 - \exp[-\mu(R_2 - R_1)]\}$
Mean volumetric intensity, $\bar{I}_v$	$P[1 - \exp(-\mu L)]$	$P[1 - \exp(-2\mu R)]$	$P\{1 - \exp[-\mu(R_2 - R_1)]\}$
Absorption efficiency, η	$1 - \exp(-\mu L)$	$1 - \exp(-2\mu R)$	$1 - \exp[-\mu(R_2 - R_1)]$

If we assume plug flow, $D_r = $ infinity, and the solution yields (Ragonese and Williams, 1971)

$$\check{R} = \int_0^{X_A} \frac{dX_A}{(1 - X_A')^n \{1 - \exp[-\sigma(1 - X_A')]\}} \tag{26.11}$$

where X_A' is a dummy conversion variable; σ and $\check{R}$ are the optical density and a reaction group, respectively, both dimensionless:

$$\sigma = \alpha_a [A]_0 d_T$$

$$\check{R} = \frac{kE_t}{Q[A]^{1-n}} \tag{26.12}$$

in which α_a is the absorptivity, m^2/mol; E_t is the total incident energy, e/s; Q is the volumetric flow rate, m^3/s; d_T is the tube (or reactor) diameter; and k is the rate constant $(m^3)^n/e \, (mol)^{n-1}$. A comparison of this equation with experiment is shown in Figure 26.2.

If the reactor considered is operated at a large recycle, it will approach CSTR performance, and the expression for $\check{R}$ becomes

$$\check{R} = \frac{X_A}{(1 - X_A)^n \{1 - \exp[-\sigma(1 - X_A)]\}} \tag{26.13}$$

This is valid only if the wavelength range used is very narrow. Given this restriction, Equation 26.13 can be used for scale-up of a fully mixed reactor (a CSTR or a PFR with a large recycle) for a single-step photochemical reaction. A comparison of this equation with experimental data is included in Figure 26.2.

Note: In addition to the limitations of the above equations already noted, it is equally important to note that the reaction itself is usually a chain reaction involving initiation, propagation, and termination steps.

Examples of Photochemical Synthesis

The most important applications of photochemical reactions in organic technology/synthesis are in the field of free radical reactions such as halogenation (mainly chlorination), sulfochlorination, sulfoxidation, and nitrosation. There is an increasing trend now toward development of nonradical photoreactions, particularly in the synthesis of vitamins, drugs, and fragrances. Table 26.2 lists representative syntheses of commercial value that have been carried out with light (see, among others, Coyle, 1986; Pfoertner, 1991, for more examples).

The importance of asymmetric synthesis in organic technology was emphasized in Chapter 9. It is also possible to introduce chirality through a photochemical reaction by transferring it from a chiral auxiliary attached to the reacting molecule. If a bimolecular photoreaction is involved, the chiral auxiliary may be attached to any one of the reactants. Examples of such asymmetric induction are cyclobutanes, oxetanes, and cyclohexenes by [4 + 2] photocycloaddition (Pfoertner, 1990).

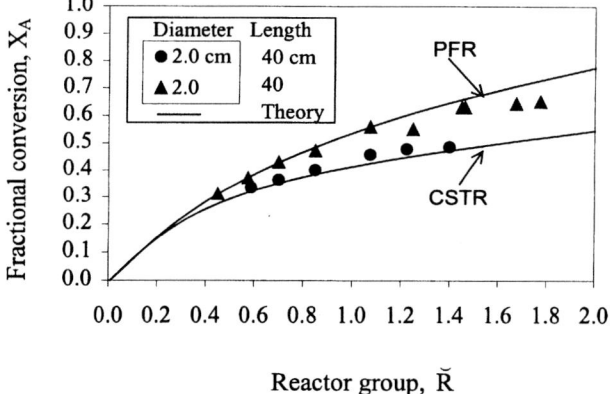

Figure 26.2 Comparison of the theoretical predictions of Equation 26.11 for a PFR and Equation 26.13 for a CSTR with experimental results (modified from Ragonese and Williams, 1971). Reaction: photolysis of tetrachloroplatinic acid with parameter values giving an optical density of 2.9

EXAMPLE 26.1

Photochemical synthesis of previtamin D: A simple design procedure

Vitamin D3 is used mainly as an additive in livestock feeds, and to a small extent in the prevention and cure of rickets in children. Industrially, it is produced from cholesterol extracted from wool grease. The cholesterol is converted to 7-dehydrocholesterol (1) in four steps. Then photochemical ring opening of (1) gives previtamin D3 (2) which on heating undergoes a hydrogen shift to give the final vitamin (3) (see Fragner, 1964; Sanders et al., 1969; Greenbaum, 1970).

[E26.1.1]

This reaction is particularly valuable in illustrating the roles of several aspects of photochemical synthesis covered in the earlier sections, which often dictate the course of photochemical process development.

Table 26.2 Examples of photoorganic syntheses

Reaction	Comments	Reference(s)
I. Free radical reactions (e.g., chlorination, sulfoxidation, nitrosation; mostly for large volume chemicals		
1. Benzyl chloride (1), benzal chloride (2) and benzotrichloride (3) from toluene 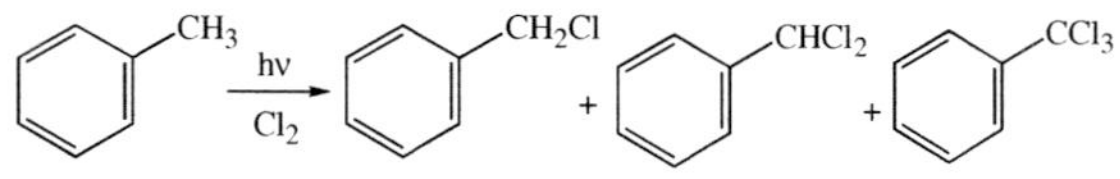	(1) gives benzyl alcohol, a well-known fragrance; (2) gives benzaldehyde; and (3) benzoic acid derivatives	Haring and Knoll (1964, 1965),
2. Sulfoxidation of *n*-alkanes to alkane sulfonates (4) $$RH + SO_2 + 1/2 O_2 \xrightarrow{h\nu} RSO_2OH$$ $$RSO_2OH - H_2O + Na \xrightarrow[\text{decomp.}]{\text{radical}} RSO_3Na + H_2SO_4$$	Immersed lamp reactor	Hartig (1975)
3. Photonitrosation of cyclohexane to the oxime 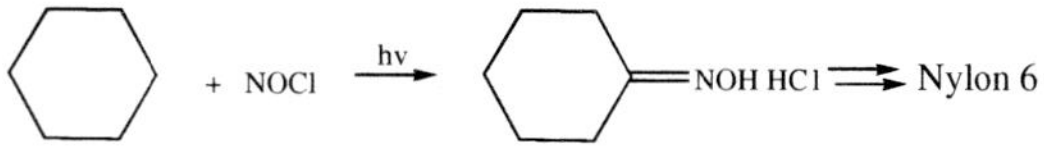		Hutson and Logan (1972)
II. Drugs, vitamins, fragrances; mostly small volume chemicals from non-free radical reactions		
4. Synthesis of rose oxide (5) by photooxygenation of citronellal as the first step 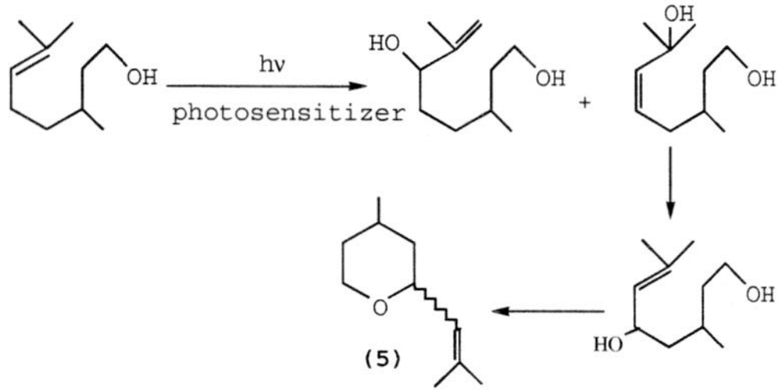	Rose Bengal is used as a photosensitizer	Ohloff et al. (1961)

5. Synthesis of a mixture of isomers (6), (7) (a fragrance) from isophorone and cyclohexane

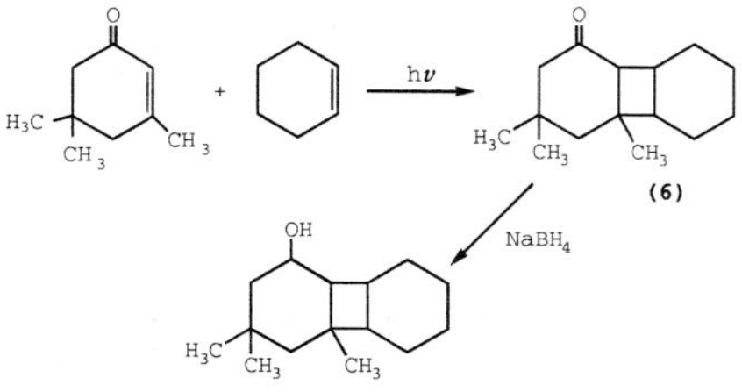

An olefin is used as solvent with a 7.5-kW lamp; $\phi' = 0.25$ with Corning glass filtered light

Bloomsfield and Owsley (1977)

6. Photoisomerization of 11-*cis*-vitamin A acetate (8) to all-*trans* form (9)

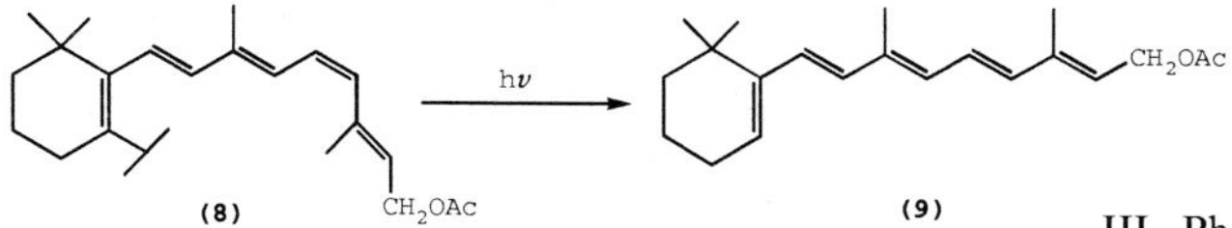

Wittig synthesis produces (8) and (9)

Fischer (1978)

III. Photosensitized reactions

7. (E/Z) Isomerization of tachysterol
   See Example E26.1, Reaction E26.1.2.

Improves the selectivity of vitamin D synthesis

Hoffman-La Roche (1985).
Stevens (1985)

8. Synthesis of a plant growth regulator (10) from 3-sulfolene and maleic anhydride (by cycloaddition)

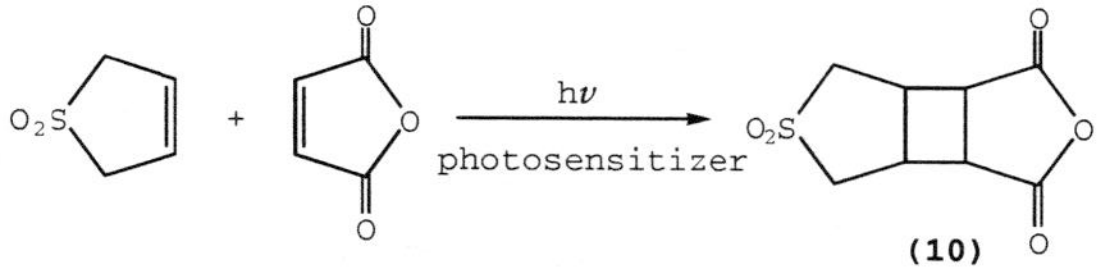

A high value chemical from cheap raw materials; acetophenone used as a sensitizer

Bloomsfield (1975)

Note: Several combinations with other strategies have been used. Another area of great importance is photocatalysis by solids.

REACTANT CONCENTRATION AND OPTICAL THICKNESS

7-Dehydrocholesterol has a very high absorbance for light. It has been used in concentrations varying from as low as 1 wt% in ethanol or ether to as high as 5% in *t*-butyl methyl ether (Pfoertner, 1991). Although a high concentration is obviously desirable for a high reaction rate, it results (as already noted) in a much reduced output. The solution to this problem lies in selecting a proper reactor.

REACTOR CHOICE AND SCALE-UP

The falling film reactor would be ideally suited because it has the merit of being able to handle the entire reactant solution in film form. The best design would be one where the reaction is allowed to proceed to the desired conversion at the end of the film by adjusting the flow rate.

Scaling up of production by using a higher power lamp is not recommended because the length of the light-emitting part of a lamp is almost independent of power, and any increased power would lead to overirradiation and corresponding loss of selectivity. Thus scale-up can be accomplished only by using several lamps of 20, 40, or 60 kW capacity. The question now is whether the conversion from each lamp and therefore its capacity can be increased by any means.

SELECTIVITY AND CONVERSION CONSIDERATIONS

It is known that the photochemical reaction E26.1.1 is not a straightforward sequential reaction leading to the final product (3). As the photoreaction of (1) progresses, photons are also absorbed by the previtamin (2) giving the undesirable trans isomeric E-triene, tachysterol (4).

(E-form)

(4)

As a first step in minimizing this unwanted reaction, we notice the crucial fact that the spectra of (1) and (2) (see Fischer, 1978) overlap almost completely, thus precluding the use of specific wavelengths to promote one reaction over the other. The only way to avoid excessive loss of 7-dehydrocholesterol by the wasteful reaction is to arrest the main reaction at a certain conversion, but this reduces the reactor throughput. Thus a second photochemical reaction was developed (Hoffmann-La Roche, 1985) in which the E-form was converted back to the desired Z-form. This so-called E/Z isomerization is carried out with a photosensitizer, as shown in reaction E26.1.2.

$$(4) \text{ (E-form)} \xrightarrow[\text{visible light}]{\text{Sensitizer}} (3) \text{ (Z-form)} \qquad [\text{E26.1.2}]$$

LAMP EFFICIENCY AND COST CONSIDERATIONS

Referring again to the UV spectra of (1) and (2) (Fischer, 1978), it is clear that the emission from a mercury lamp is not particularly good for irradiating this reaction. The relatively weak emission lines at 28, 289, and 297 nm constitute only a small part of the output of the lamp, amounting to only 8.5 einsteins/hour for a 40-kW lamp, as against a total emission of 174 einsteins/hour (Fischer, 1978). The quantum yield without the corrective reaction E26.1.2 is reported to be 0.31. Thus a single 40-kW bulb gives $8.5 \times 0.31 = 2.64$ mol/hr of the previtamin. Based on this value, the number of 40 kW lamps can be calculated for a given rate of production.

ENHANCEMENT BY MICELLES

The Nature of Micelles

Micelles are aggregates of surfactants (also called detergents) that contain hydrophobic (or apolar) "heads" and hydrophilic (or nonpolar) "tails" in aqueous solution. The tail which usually consists of a long hydrocarbon chain extends into the interior of the aggregates, while the polar head groups collect at the interface with water. Because of their hydrocarbon-like interior, micelles can solubilize large quantities of hydrophobic substances that are otherwise only sparingly soluble in water. Micelles can also incorporate ionic reagents due to the presence of water around them. Because of the ability of micelles to take up both aqueous and nonaqueous phase reactants, the rates of reactions in micellar solutions can be considerably higher than in the absence of micelles when they are limited due to the low solubility of hydrophobic reagents in aqueous media.

Several books and reviews are available, of which the following are particularly relevant to our presentation: Mukherjee and Mysels (1970), Cordes (1973), Fendler and Fendler (1975), Mittal (1977), Fendler (1982), Bunton (1979, 1991), Moroi (1992).

The fundamental property of a surface active agent, as mentioned before, is that it contains both polar and nonpolar moieties in its structure. This property is termed *amphiphilicity* or *amphipathicity*, and the substances that possess it are called *amphiphiles*. An amphiphile can be anionic or cationic, depending on whether its hydrophobic moiety is an anion or a cation. A zwitterion is an ion that possesses both anionic and cationic groups on the hydrophobic moiety and can behave either as an anionic, cationic or neutral species. An example of an ionic type is sodium dodecyl sulfate; a cationic: dodecyltrimethylammonium bromide; a zwitterionic: *N*-dodecyl-3-aminopropionic acid; and a nonionic: *N*, *N*-dimethyldodecylamine oxide.

Unlike the aqueous micells just described (often referred to simply as micelles), there can also be micelles in which the apolar groups are in contact with the solvent at the exterior and the polar or ionic groups are associated with water in the interior. These are referred to as *reverse* or *inverse micelles*. Reverse micelles are particularly noted for their accelerating or retarding effect on biochemical reactions.

Aqueous Micelles

Expanding on the definition of micelles, they are formed when a sufficiently large number (50 to several thousands) of amphiphilic molecules or ions aggregate in aqueous solution to a concentration (usually a narrow range of concentrations) above a certain level called the *critical micellization concentration* (CMS) (see Elworthy et al., 1968; Shinoda et al., 1963; Mittal, 1977). Any quantity beyond CMC only increases the concentration of the micellized surfactant. Many properties of aqueous surfactant solutions, most importantly, their ability to solubilize a reactant, suddenly change at the CMC due to micelle formation. Several observations concerning CMC have been documented (see, for example, Mukherjee and Mysels, 1970; Cordes and Gitler, 1973; Fendler and Fendler, 1975; Fisher and Oakenfull, 1977). In general, a positively charged substrate is solubilized only by an anionic micelle, and a negatively charged substrate only by a cationic micelle. Thus electrostatic interactions play a crucial role in solubilization.

When micelles are formed from surfactants with functional head groups that contain an anionic nucleophilic moiety such as an amino, hydroxyl, thionate, iodosobenzoate, imidazole, or gem-diolate group, they are often referred to as reactive micelles. Such surfactants can either be micellized by themselves or comicellized with common inert surfactants (Fendler and Fendler, 1975; Sudholter et al., 1979; Fendler, 1982; Romsted, 1984). It can generally be assumed that aqueous-phase reaction is negligible (i.e., all of the substrate is micellar-bound), and the micellar head groups are completely neutralized by counterions (Bunton, 1991).

Nonaqueous (Reverse) Micelles

Reverse micelles are normally used in enzyme-catalyzed reactions. The water in the core of the micelle is called the *water pool*. At a constant surfactant concentration, the amount of water introduced determines the micellar size. The nature of the entrapped water in reverse micelles has been a subject of considerable debate. At low amounts of water, it is thought that most of it is bound, leading to low enzyme activity. At higher amounts, the water becomes more "free" with a resultant increase in enzymatic activity.

As with aqueous micelles, catalytic activity in reversed micelles is due to substrate solubilization. The concentrations in a reverse micellar system can be based on either the total reaction volume or the micellar aqueous pool. The former is obviously simpler though somewhat less realistic.

Reaction acceleration in reverse micelles may reach factors of 10^6 (or more), as against about 10^2 in normal micelles, and is caused, as in aqueous micellar catalysis, by local concentration effects and orientations and interactions in the microenvironment of the micellar cavity. Because reverse micelles are normally used for enzymatic reactions, we defer a treatment of this combinatorial strategy until later.

Kinetic Modeling of Micellar Reactions (Without Transport Limitations)

A number of models have been proposed to explain micellar reactions. Of these, the *pseudophase models* are the most relevant to the organic reactions normally encountered. They are based on the assumption that the aqueous and micellar phases, though present as two distinct phases, can be regarded as two pseudophases acting as a single pseudohomogeneous phase with no mass tranfer resistances between them. The chief parameters of these models are molecularity and order of reaction (as in any chemical reaction), reactant distribution among micelles (uniform or nonuniform), and size distribution (or dispersity) of the micelles (mono- or polydispersed). It is assumed that the system is isothermal. What is expected of each of these models is an expression for the overall *observed rate constant* k_{an} (with no transport intrusions) in terms of the fundamental parameters of the system (mainly the individual rate constants in the aqueous and micellar phases, the distribution coefficient of reactant between micelle and water, and the micellar concentration). Table 26.3 lists a few such models along with the corresponding reaction schemes and final expressions.

In addition to the relatively simple pseudophase models just described, several other models have also been proposed (e.g. Almgren and Rydholm, 1979; Quina and Chaimovich, 1979; Quina et al., 1980; Gunnarsson et al., 1980; Romsted, 1984, 1985; Rodenas and Vera, 1985), but they will not be considered here.

Mass Transfer Effects

There are two broad categories of mass transfer effects involved in micellar reactions, mass transfer in pseudophase models and mass transfer in heterogeneous models.

Mass transfer in pseudophase models

Assume that reactant A is dissolved in an aqueous micellar solution. One part of it reacts with B in the aqueous bulk, and the other is solubilized in the micelles and reacts with B concentrated near the polar groups in the micelles. For these reactions to occur, mass transfer of A and B must occur between the phases. Taking all of these processes into account, a model for the overall reaction rate was formulated by Asai et al. (1996a), which was later specifically applied to the micellar oximation of cyclohexanone by sodium dodecyl sulfate (Asai et al., 1996b).

Mass transfer in heterogeneous models

The fact that micelles are molecular aggregates of sizes typically in the radius range of 10–30 Å suggests that they may be treated as microparticles that would enhance mass transfer rates in liquid-liquid or solid-liquid systems. Rate enhancements by microparticles in general has been treated in some detail in Chapter 23. Similar situations can arise in micellar reactions, where the micelles themselves act as microparticles (see Janakiraman and Sharma, 1985).

Table 26.3 Some models of micellar reactions[a]

Model	Scheme	Molecularity, order	Reactant dist. among phases	Dispersity of micelles	Equation
1	$M + A \underset{k_W}{\overset{K_1}{\rightleftharpoons}} MA \xrightarrow[k_{M1}]{+B} P$	Unimolecular, first order	Uniform	Mono-dispersed	$k_{a1} = \dfrac{k_W + k_{M1} K_1 [M]}{1 + K_1 [M]}$
2	Same as 1	Unimolecular, nth order	Uniform	Mono-dispersed	$k_{an} = \dfrac{k_W + k_{Mn} m_A^n V_D [M]}{1 + m_A [M] V_D}$
3	A exists as cation AH_2^+, neutral compound AH, or anion A^-. $AH_2^+ \underset{+B}{\overset{K_1}{\rightleftharpoons}} AH \underset{+B}{\overset{K_2}{\rightleftharpoons}} A^-$ → Products	Bimolecular, first order in A and B	Uniform	Mono-dispersed	Equation for k_{a2} involves several other equations, and is therefore not given here (see Martinek et al., 1975)
4	Reactions of these occur both in water and micellar phases. $M + A \xrightarrow{K_1} AH \xrightarrow{k_{M1}} M + P$; $k_W \to P$; $M_1 + A \xrightarrow{K_2} MA_2 \xrightarrow{k_{M2}} MA_1 + P$; $k_W \to P$; $\vdots$ $MA_{n-1} + A \xrightarrow{K_n} MA_n \xrightarrow{k_{Mn}} MA_{n-1} + P$; $k_W \to P$	Unimolecular	Nonuniform (Poisson dist.)	Mono-dispersed	$k_a = \dfrac{k_W + k_{Mi} K_1 [M] \exp \bar{N}_A}{1 + K_1 [M] \exp \bar{N}_A}$ or $\dfrac{(k_a - k_W)}{(k_{M1} - k_a)} = K_1 [M_t]$

[a] M = Micelle; $\bar{N}_A$ = average number of reactant molecules per micelle = $([A_t] - [A])/M_t$; M_t = total number of micelles; k_{an} = observed nth order rate constant; $[M]$ = concentration of micelles = $[D] - CMC$; D = detergent; V_D = molar volume of D; $m_A = [A]_M/[A]_W$; $k_{Mi} = k_M$ for ith step. See Notation for the rest.

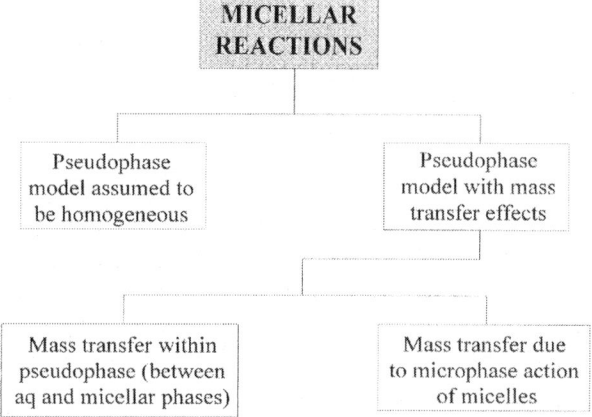

Figure 26.3 Mass transfer in micellar reactions: the total picture

Consider the reaction between a hydrophobic reactant A in phase 1, the organic phase, and a hydrophilic reactant B in phase 2, the aqueous phase:

$$A + \nu_B B \rightarrow \text{Products} \qquad [26.3]$$

If this reaction were to occur only in the aqueous bulk, the film has no role to play, the entire system is pseudohomogeneous, and a true kinetic analysis as outlined earlier is possible. If mass transfer effects are present between the pseudophases, the analysis outlined before for such systems would apply. However, if reaction occurs in the film, two situations can arise: (1) reaction occurs only in the micelles present in the film and not in the rest of the film, and (2) reaction occurs in both the micellar and aqueous phases in the film. The analysis of both the situations is very similar to that for microphase action described in Chapter 23.

The Total Picture

Based on the analyses presented in the foregoing section, a total picture of micellar action emerges covering both chemical and engineering considerations. This is shown in Figure 26.3.

MICROWAVE-ASSISTED ORGANIC SYNTHESIS

Microwaves and Their Heating Effect

During the past decade, there has been an increasing interest in the potential of microwave energy as a source of rate enhancement in organic synthesis. Since the initial reports on the use of microwaves (Gedye et al., 1986; Giguere et al., 1986), different classes of reactions have been carried out using microwave heating.

Many reviews (e.g., Gedye et al., 1991; Mingos and Baghurst, 1991; Abramovich, 1991; Whitaker and Mingos, 1994; Caddick, 1995) listing the chemical reactions studied and discussing the rate enhancements achieved have been written. We present here some of the applications of microwaves in organic and organometallic synthesis and the engineering challenges associated with the scale-up of microwave reactors.

The microwave region of the electromagnetic spectrum corresponds to wavelengths of 1 cm to 1 m (frequencies of 30 GHz to 300 MHz, respectively). Of these, the wavelengths in the range of 1–25 cm are extensively used for radar transmissions and the rest for telecommunications. In order not to interfere with these operations, domestic and commercial microwave heaters are assigned the following two wavelengths: 12.2 cm (2.45 GHz) and 33.3 cm (900 MHz). Most domestic microwave ovens operate at 2.45 GHz.

The heating effect produced by microwaves is greatly influenced by the dielectic constant ε' of the medium, which is a measure of the ability of the molecules to be polarized by an external electrical field. An additional important factor is *dielectric loss* ε'', which is a measure of the efficiency with which electromagnetic energy can be converted into heat. The dielectric loss for a molecule goes through a maximum as the dielectric constant is decreased. The ratio of dielectric loss to dielectric constant is defined as the *dielectric loss tangent*, tan δ, which is an important measure of characterizing microwave heating.

Microwave Effect in Organic Synthesis

Microwave irradiation has been widely used in organic synthesis on a laboratory scale during the last few years. Several categories of reactions, for example, Diels–Alder reactions, *ortho*-Claisen condensations, ene-reactions, oxidations, esterification of carboxylic acids with alcohols, and hydrolysis of esters and amides to carboxylic acids, have been successfully carried out in conventional microwave ovens. Table 26.4, adapted from Majetich and Hicks (1994), provides representative examples of some of these reactions. It can be seen that the conditions under which the conventional and microwave reactions are carried out differ significantly for most reactions. Hence, in comparing reactions with microwaves with conventional syntheses with reflux, it is necassary to make the comparisons based on the total amount of energy input to the reaction volume.

Some of the early reports claiming specific microwave effects have been discounted in recent years as no more than thermal effects (e.g., Westaway and Gedye, 1995; Stuerga and Gaillard, 1996a,b). It is likely that many of the reported rate enhancements with microwaves can be attributed to the heating of solvents above their boiling points due to an increase in the operating pressure. A wide range of polar organic molecules interact with microwaves at 2–45 GHz and undergo rapid (1–2 °C/s) heating. This high energy input to the reaction mixture leads to superheating, which in the absence of rapid local mixing leads to regions of superheated liquids. Temperatures 20–50 °C higher than the conventional boiling points have been observed. These effects are magnified in high pressure vessels, and solutions have been known to be maintained

Table 26.4 Examples of microwave-assisted organic syntheses (adapted from Majetich and Hicks, 1994)

Synthesis reaction		Conventional synthesis	Microwave synthesis
Diels–Alder reaction		Solvent: DMF Reaction temp: reflux (153 °C) Yield: 82% Time: 120 min	Solvent: DMF Reaction temp: 192–197 °C (20 psi) Yield: 75% Time: 12 min
ortho-Claisen rearrangement		Solvent: DMF Reaction temp: reflux (153 °C) Yield: 23% Time: 80 h	Solvent: DMF Reaction temp: 193–198 °C (30 psi) Yield: 80% Time: 5 h
		Solvent: DMF Reaction temp: reflux (153 °C) Yield: 83% Time: 4 days	Solvent: DMF Reaction temp: 192–197 °C (17 psi) Yield: 91% Time: 20 min
Ene reaction		Solvent: DMF Reaction temp: reflux (153 °C) Yield: 14% Time: 40 h	Solvent: DMF Reaction temp: 179–184 °C (90 psi) Yield: 49% Time: 20 min

(Continued)

Table 26.4 (Continued)

Synthesis reaction		Conventional synthesis	Microwave synthesis
Esterification 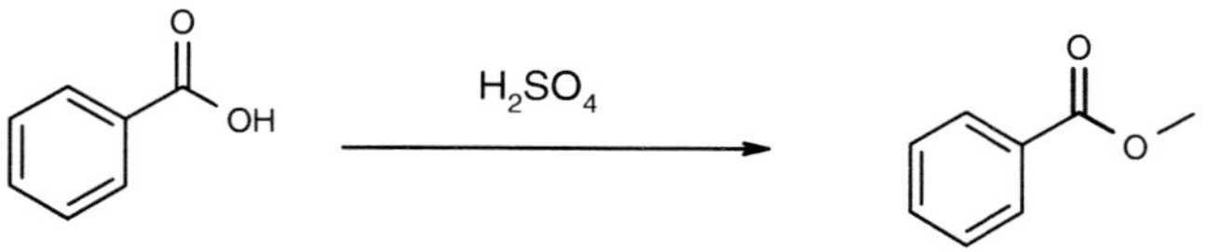	PhCOOH →(H₂SO₄) PhCOOMe	Solvent: Methanol Reaction temp: reflux (65 °C) Yield: 92% Time: 80 min	Solvent: Methanol Reaction temp: 120 °C (50 psi) Yield: 92% Time: 1 min
Hydrolysis 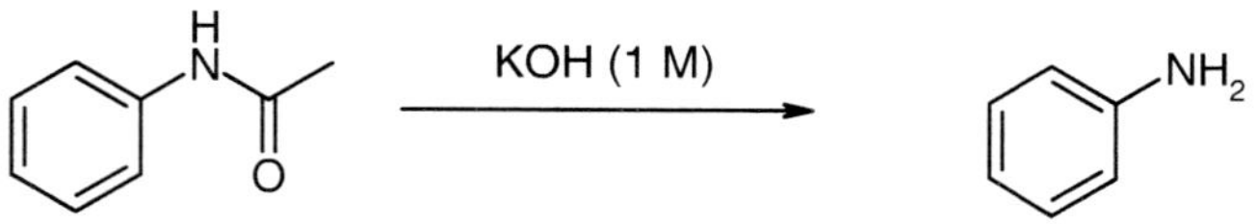	PhNHCOCH₃ →(KOH (1 M)) PhNH₂	Solvent: Methanol–water Reaction temp: reflux (79 °C) Yield: 60% Time: 36 h	Solvent: Methanol–water Reaction temp: 130–136 °C (90 psi) Yield: 83% Time: 45 min
Williamson ether synthesis 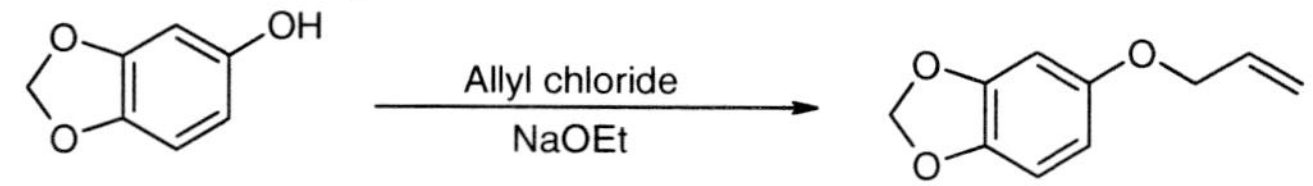	ArOH →(Allyl chloride / NaOEt) ArOCH₂CH=CH₂	Solvent: Ethanol Reaction temp: reflux (78 °C) Yield: 89% Time: 60 min	Solvent: Ethanol Reaction temp: 130 °C (85 psi) Yield: 96% Time: 1 min
Beckmann rearrangement 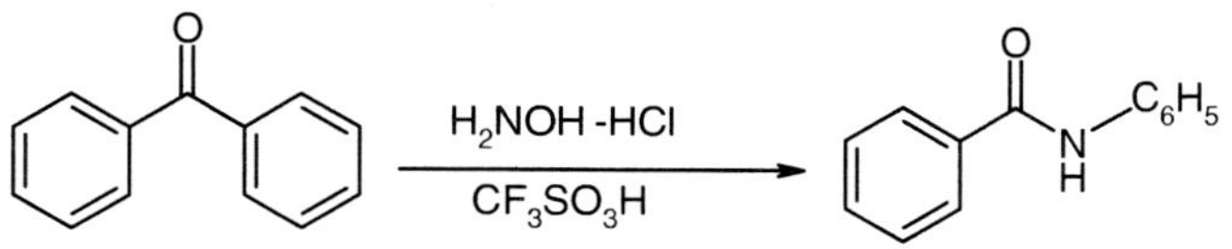	Ph₂C=O →(H₂NOH·HCl / CF₃SO₃H) PhCONHC₆H₅	Solvent: Formic acid Reaction temp: reflux (101 °C) Yield: 96% Time: 80 min	Solvent: Formic acid Reaction temp: 171 °C (90 psi) Yield: 99% Time: 3 min

at 50–150 °C above their conventional reflux temperatures. The increase in pressure (and hence temperature) in the reaction vessel depends on the amount of microwave energy absorbed, which in turn depends on the power level for a fixed amount of reaction mixture. The volume of solvent used also affects the pressure and temperature rise at a fixed power level. When the volume of the solvent is small, the pressure increases as the volume of the solvent is increased. However, as the volume is further increased, a point is reached when the increased volume can no longer be heated more rapidly. The volume for a maximum heating rate depends on the molecular weight and heat capacity of the solvent used.

One of the main problems in comparing microwave-assisted reactions with conventionally heated reactions is that it is difficult to accurately measure the internal temperature of microwave-assisted reactions. For most liquid-phase reactions, xylene thermometers are used instead of mercury or alcohol because the latter are microwave-active. Xylene, on the other hand, is inert, nonconducting, and has a low dielectric loss tangent. Other complicated but currently expensive methods such as gas pressure thermometry, infrared pyrometry, and fluoroptic thermometry using fiber optic cables are also available. In heterogeneous or solid media, analysis of the microwave effect is even further complicated because it is almost impossible to measure the actual temperature at the surface of the solids with current technology. However, it is expected that any enhancing effect by microwaves is probably caused in this case too by high thermal effects at the solid surface.

There are a few instances where beneficial nonthermal effects of microwave irradiation have been reported, for example, mutarotation of α-D-glucose (Pagnotta et al., 1993). Equilibrium was reached much more rapidly using microwave irradiation than conventional heating. Interestingly, organic reactions carried out under "dry" (solvent-free) conditions have also been widely reported to be enhanced by microwaves, with or without synergy with phase-transfer catalysis (see, Loupy et al., 1993, and references therein). Microwaves also enhance reactions using solid reagents supported on inorganic mineral supports such as alumina, silica gel, or clay, under solvent-free conditions (see, Villemin and Labiad, 1990). Another area where microwaves are extensively used is in the synthesis of semiconductors and other solid-state reactions (Baghurst and Mingos, 1992a).

Thus, although it is still not clear whether specific microwave effects exist, the indisputable observation that a variety of reactions can be carried out under microwave conditions is of significant environmental and commercial value. Reaction times are drastically reduced, resulting in significant economies in electricity and heating costs. Solvent handling problems are also eliminated in microwave-assisted solid-state reactions and organic reactions carried out under dry conditions.

Microwave Reactors

Radiation in a microwave oven is generated by a magnetron—a microwave tube which acts by the interaction of electrons with a magnetic field. The resulting microwaves are guided into a microwave cavity by a waveguide, reflected by the

walls of the cavity, and then absorbed by the material in the cavity. There are a variety of methods for carrying out microwave-assisted organic reactions using domestic and commercial microwave ovens.

Domestic (laboratory-scale) microwave ovens

The relatively low cost of domestic microwave ovens makes them an attractive option for carrying out reactions on a laboratory scale. A common modification is the addition of a port to allow the insertion of reaction tubes or flasks into the cavity. The dimensions should be such as to prevent radiation leakage by attenuation of the microwaves. Details of port construction are documented in several reviews (e.g., Mingos and Whittaker, 1997). The experimental procedure can be extremely simple, and reactions are typically carried out in open Erlenmeyer flasks subjected to short periods of irradiation (or in some cases in sealed Teflon vessels). Adequate safety precautions are always necassary while carrying out microwave-assisted reactions to avoid the risk of explosion due to the rapid increase in pressure (in sealed vessels) and temperature in the vessel. In most microwave-assisted reactions, one of the components, usually the solvent, is microwave-active. Polar solvents such as water, methanol, dimethyl formamide (DMF), ethyl acetate, acetone, chloroform, acetic acid, and dichloromethane absorb microwaves, leading to heating when irradiated with microwaves. Nonpolar solvents such as hexane, toluene, carbon tetrachloride, diethyl ether do not couple with microwaves and hence are not heated by irradiation. DMF is often the preferred solvent for carrying out microwave-assisted reactions. Mixtures of solvents consisting of microwave active/inactive components are also sometimes used.

There are a number of published methods for modifying domestic microwave ovens for safer operation (e.g. Giguere et al., 1986, Baghurst and Mingos, 1992b). One of them is a simple technique commonly used in conventional reactions. A reflux condenser is attached to the reactor vessel, and the vapors from the reaction are condensed back into the vessel. If reactions have to be carried out at elevated pressures, the Parr autoclave (Parr Instrument Company) is a good choice because it can withstand pressures up to 80 atm and temperatures up to 250 °C. The reactor is provided with a rotating carousel, so that a number of reaction vessels can be irradiated and agitated at the same time. Recently, microwave reactors suitable for organic reactions and kinetic studies in the laboratory have been developed (Constable et al., 1992; Raner et al., 1995). A combination reactor for carrying out both ultrasound and microwave reactions in the same reactor has also been developed (Chemat et al., 1996).

Scale-up of microwave reactors

As in most other cases of reactor design with distributed properties in the reaction field (e.g., distributed sonic intensity, distributed photo energy), the reaction field in a microwave reactor is also a distributed system (with respect to temperature, through which alone the microwave effect seems to be manifested in most cases). The results from a domestic microwave reactor can

give only single point data in this distributed field. Including this, the most important problems in microwave reactor scale-up are (1) energy efficiency, (2) uniformity of heating in the irradiated zone, and (3) safe monitoring and control of the reaction.

ENERGY AND POWER EFFICIENCY

Power transferred to a medium by microwaves is given as

$$P_d = (\lambda + 2\pi\nu\varepsilon')|E|^2 \tag{26.14}$$

where P_d is the power transferred in a unit volume of material with conductivity λ and dielectric constant ε' and E is the local electric field intensity of the microwave field of frequency ν. Typical applications for which microwaves have been used so far are dielectric materials for which conductive heat transfer is negligible. In reactions involving catalysts, however, the conductive heat transfer may be significant. Because, in general, the power transferred in a microwave reactor is utilized to heat many materials besides providing the heat of reaction, the total power P_t supplied can be divided into two components:

$$P_t = P_d + \sum \frac{\Delta T}{\Delta t} C_p \rho \tag{26.15}$$

The summation in the second term indicates that there are many materials that are heated by the microwave. As we can see, the goal of the microwave engineer is to minimize the second term. It can be done, for instance, by appropriate choice of insulation and other materials. Materials with lower heat capacity and density exhibit lower absorption of microwave energy.

FIELD UNIFORMITY AND PENETRATION OF MICROWAVES

As already noted, spatial nonuniformities are a major concern in the scale-up of microwave reactions. Traditionally, turntables and mode mixers in domestic microwave ovens have been used to reduce spatial nonuniformity. There are two types of nonuniformities in the microwave heating process. The first can be termed the *standing wave effect*, which is the repeated field intensity variation within the microwave applicator. This intensity typically follows a half sine wave pattern. Due to this standing wave effect, the field intensity can change from the maximum to zero in a distance of one-quarter of the operating wavelength. The second type of nonuniformity is related to the penetration depth of microwaves. An approximate relationship for the penetration depth ℓ_{pd} for small dielectric heat loss is given by

$$\ell_{pd} \propto \lambda_{mw} \sqrt{\left(\frac{\varepsilon'}{\varepsilon''}\right)} \tag{26.16}$$

where λ_{mw} is the wavelength of the microwave radiation. From this expression, we see that the depth to which the microwave can penetrate depends largely on the material properties and the wavelength used. Hence a reaction carried out in an Erlenmeyer flask in a domestic microwave oven, as is often done in studying

microwave chemistry, may not proceed to any appreciable extent when carried out in a larger microwave reactor. This classical problem in reactor scale-up has yet to be seriously addressed for microwave-assisted reactions.

The problem of spatial nonuniformity with microwave heating can largely be overcome by using mechanical stirrers to mix its contents—as in any conventional reactor. Another way to overcome this limitation is to use microwaves of different wavelengths. However, this is not possible because regulations require that only microwaves of 2.45 GHz frequency be used. Perhaps the best way to overcome this problem is to use continuous reactors.

USE OF CONTINUOUS REACTORS

A novel continuous microwave reactor system (Figure 26.4) was recently described (Strauss and Trainor, 1995). Because only a small volume of the reaction solution is exposed to microwave irradiation, the power needed for heating is low. The design also overcomes the penetration depth problem discussed earlier. Liquids flowing through a small tube are likely to be heated uniformly in the radial direction while passing through the microwave field. The main factor in the design of such reactors is the length of the reaction section, which depends on the residence time of the reaction mixture. The residence time, in turn, depends on the desired reactor capacity and conversion.

A variety of other continuous flow systems has also been proposed, for example, a microwave-transparent PFA (polyfluoroalkyl Teflon) coil in a microwave oven (Cablewski et al., 1994), and a more recent one suitable for homogeneous reactions and also for heterogeneous catalytic reactions (Chemat et al., 1995, 1998).

Choice of Frequencies Other Than Microwaves

The use of microwaves is essentially a method of using electromagnetic radiation to provide the energy needed for a chemical reaction. Radio-frequency waves, which are electromagnetic radiation at lower frequencies, have been used for industrial heating for a few decades. The most common commercial radio fre-

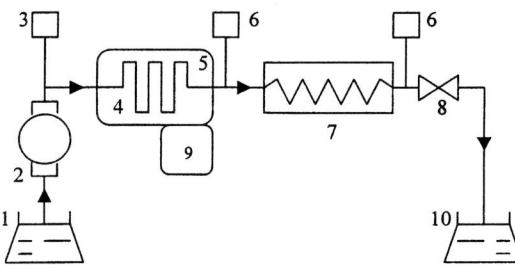

Figure 26.4 A novel continuous microwave reactor system (redrawn from Strauss and Trainor, 1995): 1: reactants for processing; 2: metering pump; 3: pressure transducer; 4: microwave cavity; 5: reaction coil; 6: temperature sensor; 7: heat exchanger; 8: pressure regulator; 9: microprocessor controller; 10: product vessel

quencies are 13, 27, and 40 MHz. For various reasons, however, their use in chemical reactions has not found favor with the chemical industry.

ORGANIC SYNTHESIS IN SUPERCRITICAL FLUIDS

Supercritical Fluids

Supercritical fluids (SCFs) were first used in the organic chemical industry to extract ingredients in the production of pharmaceuticals and food chemicals. The scope of their usage has since expanded enormously and today includes an array of homogeneous and heterogeneous chemical reactions. Quite clearly, the chief reasons are the "cleanness" of the fluids, mainly supercritical (sc) CO_2, and the complete absence of any environmental fallout. Apart from CO_2, a large number of other gaseous compounds such as ethylene and perchlorinated and fluorinated hydrocarbons, and liquid compounds such as acetone, hexane, and pentane have been introduced as SCFs. Several reviews on the subject are available, for example, Subramaniam and McHugh (1986), Savage et al. (1995), Dinjus et al. (1997).

Properties of SCFs

The phase diagram of the most commonly used sc fluid, CO_2, is shown in Figure 26.5. It can be divided into three regions, gas, liquid, and solid. Every binary transition is represented by a line, whereas all three phases exist in equilibrium at only one point, the triple point (TP). The critical point, that point at which gas and liquid become indistinguishable from each other, is reached as the fluid is heated along the gas-liquid boundary (the boiling point line). When the fluid is compressed and heated beyond the critical point, it is referred to as a supercritical fluid. Therefore, the critical point is an important property of an SFC. The value can be as low as 31 °C (for CO_2) and as high as 374 °C (for water). At temperatures above T_c, the transition to the liquid state cannot occur regardless of the pressure applied.

From an operational point of view, the greatest advantage of an SCF is that the density at any given temperature can be varied, without any discontinuity, from the density of a vapor to that of a liquid, merely by varying the pressure (Figure 26.6). This opens up the possibility of *engineering a solvent* for a given requirement. In particular, when both the reduced temperature and pressure are in the vicinity of unity, the density is very sensitive to pressure. This enables conducting the reactions over a wide range of property values within a narrow pressure range and constitutes a major advantage in parameter optimization. In effect, this *tunable density* of an SCF provides an additional parameter with which to optimize or "tune" chemical reactions.

The use of SCFs in conducting organic (and related) reactions has undergone explosive growth in recent years in such diverse areas as homogeneous and heterogeneous catalysis, electrochemistry, conversion of biomass to fuels and chemicals, and enzyme-catalyzed organic reactions. Indeed, *solvent engineering*

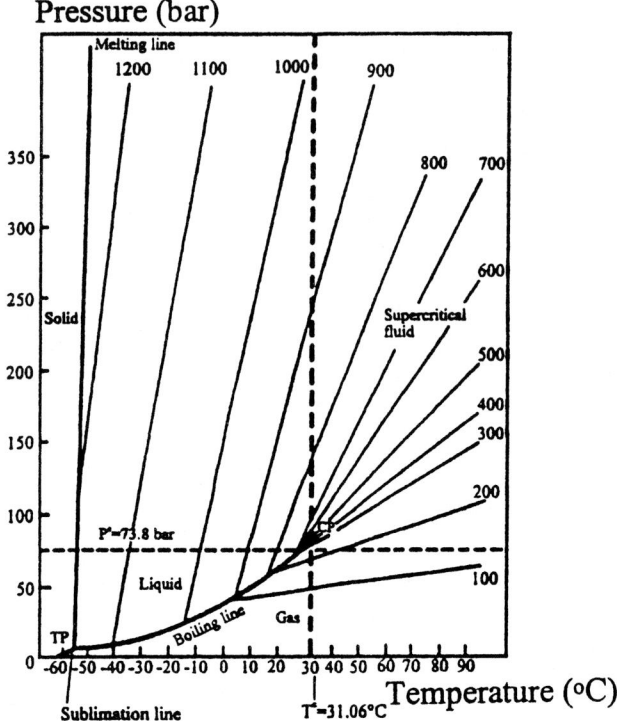

Figure 26.5 Phase diagram of CO_2, a typical supercritical fluid (Dinjus et al., 1997)

by using SCFs to tune enzymatic reactions may be regarded as the "next frontier" in this important area (Hammond et al., 1985; Kamat et al., 1995), and will be discussed later.

Use of SCF in Organic Synthesis

Diels–Alder reactions

Considering the importance of the Diels–Alder reaction in organic synthesis and its proven sensitivity to solvent, a number of studies have been carried out using

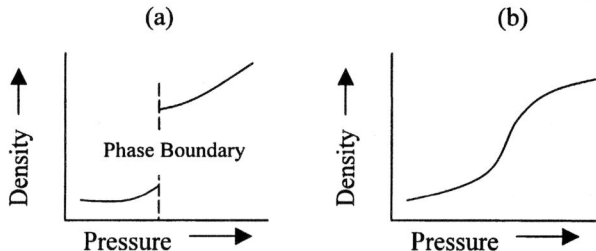

Figure 26.6 Variation of density with pressure for (a) normal and (b) supercritical fluids

SCFs as solvents for this reaction, for example, Paulaitis and Alexander (1987), Kim and Johnston (1987), Ikushima et al. (1990, 1991, 1992), Isaacs and Keating (1992), Kaupp (1994). The results of these and other studies confirm the expectation that there is no discontinuity in the reaction rate in the transition from fluid to sc CO_2 at constant density, but at the same time there is no significant rise in the reaction rate either. An example of a reaction that shows a reasonable increase in the rate constant is the cycloaddition of *p*-benzoquinone and cyclopentadiene in sc CO_2 (Isaacs and Keating, 1992). The rate constant increases with pressure and is about 20% higher than in diethyl ether.

An interesting result is that the selectivity in a parallel Diels-Alder reaction can be controlled by pressure, as demonstrated in the cycloaddition of methyl acrylate and cyclopentadiene in sc CO_2 (Kim and Johnston, 1987).

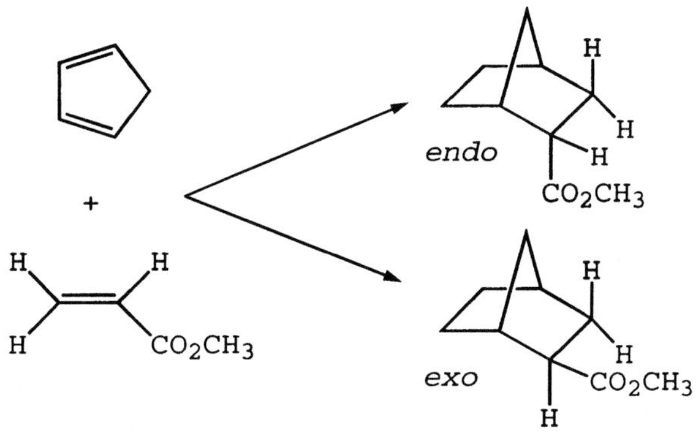

[26.4]

The *endo/exo* selectivity increased with pressure, although the maximum increase was only about 2.5%.

Heterogeneous (catalytic) reactions

The usefulness of an SCF in reactions catalyzed by solids stems from the fact that the carbon deposited on the catalyst in most organic reactions can be dissolved in the SCF, thus preventing catalyst deactivation (Tiltscher and co-workers, 1981, 1984, 1987). However, experimental data on the isomerization of 1-hexene on Al_2O_3 and Pt/Al_2O_3 catalysts show some intriguing results (Saim and Subramaniam, 1988, 1990, 1991; Saim et al., 1989; Manos and Hofmann, 1991; Baptist-Nguyen and Subramaniam, 1992; Ginosar and Subramaniam, 1994). Coke deposits are indeed dissolved by the fluid under supercritical conditions, but this does not seem to arrest catalyst deactivation. In fact, the rate of deactivation increases due to the liquid-like nature of the supercritical fluid, which results in a higher diffusional resistance in the pores than in the case of subcritical gaseous fluids. This is clearly brought out in Figure 26.7 which shows a distinct reduction in catalyst activity beyond an optimal fluid density. Thus

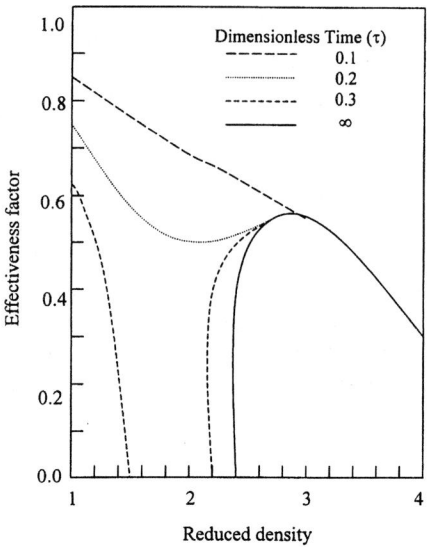

Figure 26.7 Plot showing a distinct reduction in catalytic activity beyond an optimal density of a supercritical fluid (Baptist-Nguyen and Subramaniam, 1992)

an optimal combination of solvent and diffusional properties of the system is necessary to obtain maximum rates.

As in any heterogeneous catalytic reaction (Chapter 7), it is necessary to use a gradientless reactor to obtain precise kinetic data. An internal recycle reactor was recently proposed by Bertucco et al. (1997) for this purpose. The data should be obtained at times shorter than needed for a perceptible onset of deactivation.

The usefulness of SCFs in organic synthesis as already outlined notwithstanding, their rate-enhancing ability is most evident in combination with some of the other methods of synthesis discussed in earlier chapters—for example, bio-organic, PTC, photochemical, and electroorganic syntheses. We consider these in subsequent reactions along with other combinatorial strategies.

Reactions where a reactant itself is a supercritical fluid

The use of SCFs as solvents gives rise to the logical question: how would the reaction behave if, instead of the solvent being an SCF, a reactant itself were one? An important reaction of this type is the hydrogenation of sc CO_2 to formic acid in the presence of homogeneous catalysts such as Wilkinson's rhodium-based catalyst (Inoue et al., 1975) and the more recent ruthenium-based catalyst of Noyori and collaborators (Jessop et al., 1994a,b, 1995, 1996). Because CO_2 is both a solvent and a reactant, the situation gets quite complicated, and a much deeper study than reported so far is needed to understand such reactions. Noyori's catalyst has been found very active largely because of the presence of CO_2 in supercritical form with its extremely high miscibility with hydrogen (Tsang and Streett, 1981; Howdle and Poliakoff, 1989) and excellent mass transfer characteristics.

A remarkable application of reactions of this type would be if carbonylation with CO is replaced by reaction with CO_2. An example of this is the use of sc CO_2 in the catalytic production of dimethylformamide (DMF), a very important organic solvent. The present industrial production of DMF is based on the carbonylation of dimethylamine in the presence of methanol. Recent studies have shown that DMF can also be produced by reacting dimethylamine, sc CO_2, and hydrogen in the presence of a homogeneous ruthenium catalyst (Jessop et al., 1994a). Reactions of this type can add a new dimension to the more conventional carbonylation by CO in the presence of a homogeneous catalyst (see Chapter 8).

USE OF HYDROTROPES (HYDROTROPIC SOLUBILIZATION)

General Features of Hydrotropes

Aqueous solutions of certain salts, such as those of benzoic, salicylic, benzenesulfonic, naphthoic, and various hydroaromatic acids, possess the power of dissolving certain substances otherwise not soluble in water. These substances, known as *hydrotropes*, are structurally similar to surfactants in that they have hydrophilic and hydrophobic moieties in the same molecule, but the alkyl chain is shorter. A few typical hydrotropes are listed in Figure 26.8. When solubilization

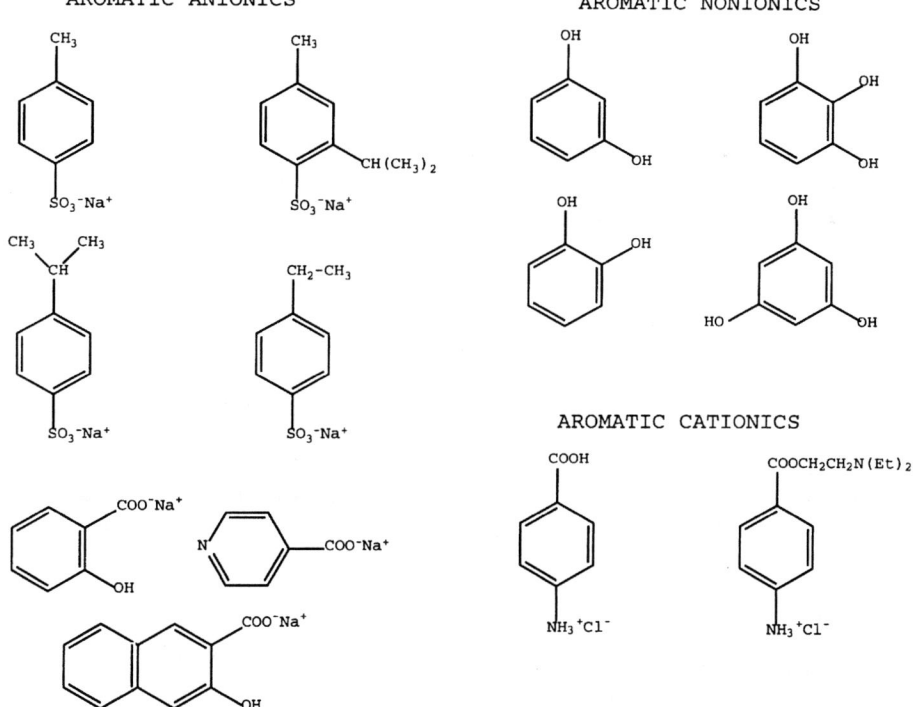

Figure 26.8 Examples of some common hydrotropes

is caused or increased by a hydrotrope as the additive, it is specifically referred to as *hydrotropic solubilization*. Clearly, such an increase in solubilization can greatly enhance the rates of reaction in aqueous solutions.

Hydrotropes have many features in common with micelles. The most important is the presence of a *minimum hydrotrope concentration* (CHC) analogous to the minimum micellar concentration (CMC) described earlier (Balasubramanian et al., 1989). The most important difference is that in hydrotropes, the dissolved solute is precipitated on dilution, whereas with surfactants dilution leads to emulsification with consequent problems of separation. Another difference is that surfactants show solubility enhancements at low concentrations, usually in the millimolar range, whereas hydrotropic solubilization occurs in the molar concentration range. Yet another difference is that, unlike micellar solubilization which is general and nonselective, hydrotropes do not solubilize all hydrotropic substances and are hence selective. This is obviously an advantage where reactant selectivity is important.

Possible Mechanisms of Hydrotropic Action: Similarity to Salting-In

A great amount of theoretical work has been reported on the behavior of nonelectrolytes in aqueous electrolyte solutions, but very little information is available on hydrotropy, probably because of the lack of a clear distinction between hydrotropes and salts, and the resulting tendency to group hydrotropes with *salting-in* substances. A salting-in substance is one that enhances the solubility of a nonelectrolyte in a solvent, that is, the solubility is greater in the salt solution than in the corresponding pure solvent. A more common occurrence, however, is one where the converse is true: the solubility of a nonelectrolyte in a salt solution is lower than in the corresponding solvent, a situation referred to generally as the *salting-out* effect. Assuming that hydrotropes behave in a similar fashion, the following equation can be used to predict the solubility of an organic compound in a hydrotropic solution:

$$\log \frac{S}{S_H} = K_S[H] \tag{26.17}$$

where K_S is the Setschenow constant, H is the hydrotrope, and S_H and S are the solubilities of an organic solute in water with and without the hydrotrope, respectively. For a positive hydrotropic effect corresponding to a salting-in process, K_S is positive, and for a negative hydrotropic effect corresponding to a salting-out process, K_S is negative. This empirical equation, a good predictive correlation for the salt effects, is applicable to hydrotropes only in a limited concentration range. A more accurate correlation is that of Schneider (1976) (see also Pandit and Sharma, 1989; Agarwal, 1993).

Applications of Hydrotropy in Organic Synthesis

Any technique that enhances the solubility of a sparingly soluble solute in a liquid finds immediate application in organic synthesis. Thus research over the

last several years has shown that hydrotropes can indeed enhance the rates of a large number of organic reactions. An important factor in accelerating these reactions in water is the hydrophobic effect, which is the tendency of nonpolar species to aggregate in aqueous solution thus decreasing the hydrocarbon–water interfacial area (Rideout and Breslow, 1980). Such hydrophobic packing of reactant molecules is perhaps the main reason for the observed rate acceleration in water. The most significant industrial advantages of using water are that the conditions are usually mild and the use of hazardous solvents is avoided. Now we cite several instances of hydrotrope-assisted organic syntheses in water.

In the production of certain common dyes, for example, Metanil Yellow or Orange 1V, the reactant (diphenylamine) has to be at least slightly soluble. Ordinarily, it is dissolved in strong sulfuric acid, and this objectionable acid is later neutralized with alkali. When diphenylamine is dissolved in a hydrotropic solution and diazotized metanilic acid and NaOH are added, brilliant yellow crystals of Metanil Yellow separate in excellent yield.

[26.5]

When sulfanilic acid is used instead of metanilic acid, Orange 1V is likewise obtained in equally good yield.

Aromatic alcohols are useful intermediates in a variety of chemical reactions. The reduction of aldehydes without α-hydrogen can be easily accomplished through the Cannizarro reaction. Typically, the use of PT catalysts does not seem to have any significant effect on the Cannizarro conversion of benzaldehyde to the acid (Gokel et al., 1976). On the other hand, with cymenesulfonate as a hydrotrope, the conversion was 72.5% as against 4.9% with no hydrotrope. The main factor against the industrial application of the Cannizarro reaction is that it involves simultaneous oxidation and reduction, leading to a mixture of the corresponding acid and alcohol. This can be overcome by adding formaldehyde to initiate a cross-Cannizarro reaction, thus leading to a preponderance of the alcohol in the final product. Consequently, in the reaction

[26.6]

the yield of alcohol was 75% in the presence of PEG 200 as a hydrotrope (Sane and Sharma, 1987), and the overall rate went up by two orders of magnitude.

The increase in rate, though very large, is to be expected, but that in the yield is surprising.

The Claisen–Schmidt condensation of p-substituted benzaldehydes with p-substituted acetophenones gives the corresponding α, β-ketones (chalcones) in hydrotropic media in the presence of NaOH as catalyst (Sadvilkar, 1995).

[26.7]

It was found that the rate increased with the hydrophobicity of the alkyl group of the hydrotrope.

Dihydropyridines are a group of drugs used as calcium channel blockers in cardiovascular disorders (Gilman et al., 1985). Nitrendipine is an important member of this class. This drug is synthesized by the reaction of m-nitrobenzaldehyde with methyl acetoacetate and ethyl acetoacetate in the presence of aqueous ammonia. Excellent results have been reported with more than 90% selectivity for the asymmetrical dihydropyridines in hydrotropic solutions (Latha, 1998; see also Sadvilkar et al., 1995).

A particularly attractive application of hydrotropy in organic synthesis arises when the product is bulkier than the reactant, with the result that it has lower solubility than the reactant in the hydrotrope solution. Consequently, it selectively precipitates out of the reaction mixture and can be easily filtered out. Then the hydrotropic solution can be recycled, thus minimizing the environmental hazards associated with waste disposal. An important example is the synthesis of Diels–Alder adducts that act as flame retardants for polymer blends and formulations. One of these is also used in the manufacture of the pesticide Endosulfan. The reaction involves a diene such as hexachloropentadiene or anthracene and a dienophile such as p-benzoquinone or maleic anhydride. The following typical reaction carried out by Sadvilkar (1995) gave excellent results:

[26.8]

Hydrotropes can also be used to specifically enhance the selectivity for a desired reaction in a parallel scheme. An example of this is the reduction of α, β-unsaturated aldehydes and ketones by agents such as sodium borohydride and sodium dithionite, leading to various products via parallel 1,2 and 1,4-attacks.

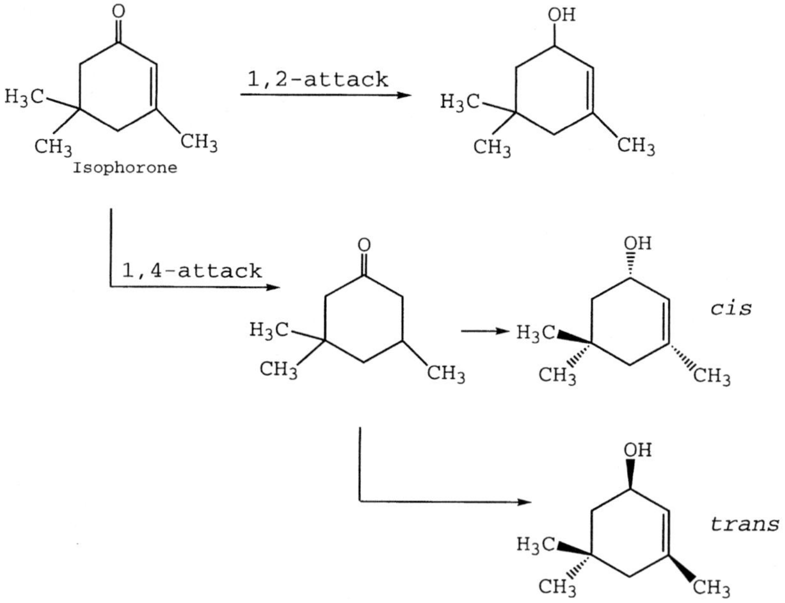

[26.9]

Experiments showed (Laxman and Sharma, 1990) that the 1,4-reduction is much more favored in the presence of a hydrotrope such as sodium butyl monoglycol sulfate than in its absence. This is because the likelihood of an H ion attacking the 4-position decreases when the electron-withdrawing tendency of the carbonyl group is weakened and the addition of hydrotrope increases the hydrophobic microdomains in the aqueous phase thus favoring 1,4-reduction.

COMBINATORIAL STRATEGIES

The various strategies of rate enhancement mentioned previously are usually employed one at a time. However, it has been found advantageous to use two (or in rare cases, three) of these strategies simultaneously for a given reaction. Quite clearly, a number of such binary combinations are possible—although not every combination may be feasible. Based on available studies, it seems reasonable to divide them into three main categories: combinations of PTC with other strategies; enzymatic, photochemical, and electrochemical syntheses in sc fluids; and a combination of ultrasound with electrochemistry (sonoelectrochemistry). Combinations not covered in these groups will be discussed under miscellaneous combinations.

Combinations of PTC with Other Strategies

PTC has been used along with a number of other rate-enhancing techniques such as sonochemistry, microwaves, electrochemistry, microphases, photochemistry, and supercritical fluids. Other (less known and only marginally combinatorial) strategies include the use of cocatalysts (Dolling, 1986; Dehmlow et al., 1985, 1988) and dual PT catalysts (Szabo et al., 1987; Tsanov et al., 1995; Savelova and

Table 26.5 Examples of binary combinations of PTC with other rate enhancement strategies (adapted from Naik and Doraiswamy, 1998)

Combination of PTC with	Illustrative example	Reference
Photochemistry	Photocyanation of aromatic hydrocarbons	Beugelmans et al. (1978)
	PTC carbonylation of aryl and vinyl halides under UV irradiation	Brunet et al. (1983)
	Reduction of nitrobenzenes to the corresponding oximes or quinones using viologens	Tomioka et al. (1986)
	Photohydrogenation of acetylenic groups with viologen, Pt or Pd, and a photosensitizer	Maidan and Willner (1986)
	Photochemically induced polymerization of methyl methacrylate	Shimada et al. (1989, 1990)
Microwaves	Alkylation of ethylphenyl sulfonyl acetate, diethyl malonate, anions derived from active methylene	Wang and Jiang (1992), Wang et al. (1995)
	Dihalocyclopropanation of substituted olefins under LLPTC and SLPTC conditions	Villemin and Labiad (1992)
	Ethoxylation of o,p-nitrochlorobenzene	Yuan et al. (1992a)
	Reactions of carboxylic acids with halides	Yuan et al. (1992b)
Ultrasound	Esterifications, etherification and hydrolysis of esters	Davidson et al. (1987)
	Michael addition involving the addition of chalcone to diethyl malonate in SLPTC mode	Ratoarinoro et al. (1992)
		Contamine et al. (1994)
	Synthesis of benzyl sulfide by reaction of solid sodium sulfide with benzyl chloride	Hagenson et al. (1994)
	Synthesis of fluvenes from phenylacetylene	Wang and Zhao (1996)
Electroorganic synthesis	Electroinitiated copolymerization of α-methyl styrene and isoprene	Akbulut et al. (1986)
	Chlorination of substituted naphthalenes	Forsyth et al. (1987)
	Oxidation of toluene and aromatic hydrocarbon using Ce^{4+}/Ce^{3+} as the redox mediator	Pletcher and Valdes (1988a,b)
	Electrodecarboxylation of carboxylate salts (Kolbe reaction)	Walton et al. (1990)
	Anthracene oxidation to anthraquinone using Mn^{3+}/Mn^{2+} as the redox mediator	Chou et al. (1992)
	Electrooxidation of pyrrole	Benahcene et al. (1995)
	Reduction of methylviologen and ferricyanide	Benahcene et al. (1995)
	Electroreduction of benzaldehyde and benzoquinone	Durant et al. (1996)
	Electrodehalogenation of 3- and 4-bromobenzophenone and o-nitrobenzene	Compton et al. (1996b)
Microphases	Synthesis of carboxylic acid with halides	Hagenson et al. (1994)

Vakhiova, 1995; Jagdale et al., 1996). Table 26.5 lists selected examples of the use of many of these combinations (see Naik and Doraiswamy, 1998).

PTC in ultrasound-assisted reactions

The chemical effects of ultrasound are generally attributed to intense local conditions involving the formation of free radicals. However, in PTC reactions following the ionic mechanism, rate enhancements are typically due to mechanical effects, mainly through an enhancement of mass transfer. The importance of water in SLPTC systems has already been pointed out. Ultrasound can be expected to perform a function similar to that of traces of water by weakening the crystal lattice structure of the solid reagent, thereby enabling the PT catalyst to easily ion exchange at the solid surface. It also appears that the PTC-ultrasound combination is superior to either technique alone (Davidson et al., 1987; Jouglet et al., 1991). In such cases, the PT catalyst initiates the reaction by transfer of species across the interface, and ultrasound merely facilitates the transfer, probably by increasing the interfacial transfer area.

PTC in electroorganic syntheses

The use of PTC in electroorganic oxidation and reduction reactions is widespread because it involves in situ generation and regeneration of oxidizing and reducing agents. Most applications are in liquid-liquid systems, but it can also be used in solid-liquid PTC systems (Chou et al., 1992). Recently, Do and collaborators extensively studied the electrochemical oxidation of benzyl chloride in the presence of a PT catalyst (soluble and immobilized) in both batch (Do and Chou, 1989, 1990, 1992) and in continuous (Do and Do, 1994a,b,c) electrochemical reactors. Mathematical models for both types of reactors were also proposed.

PTC in supercritical fluids

Reactions using a PT catalyst can also be carried out in supercritical fluids (usually, sc CO_2). In one of the few studies of this kind reported so far (Dillow et al., 1996), the two phases involved in the reaction were sc CO_2 and a solid phase. The mechanism of PTC remains the same as in conventional PTC involving transfer of the reactant ion from the solid phase to the supercritical fluid by a quaternary ammonium salt or a crown ether. However, the choice of catalyst is restricted in this case by its solubility in the SCF phase. A polar or protic cosolvent is often necessary, and even small amounts of a solvent such as acetone can greatly increase the solubility. Thus, compared to traditional PTC, SCF-PTC requires much lower amounts of solvents. This combinatorial strategy has great potential for industrial application.

PTC with ultrasound and microphase

The use of ultrasound or microphase in combination with PTC enhances the reaction rate (Hagenson et al., 1994). What is more significant is that when all three strategies are used simultaneously, the enhancement is higher than with

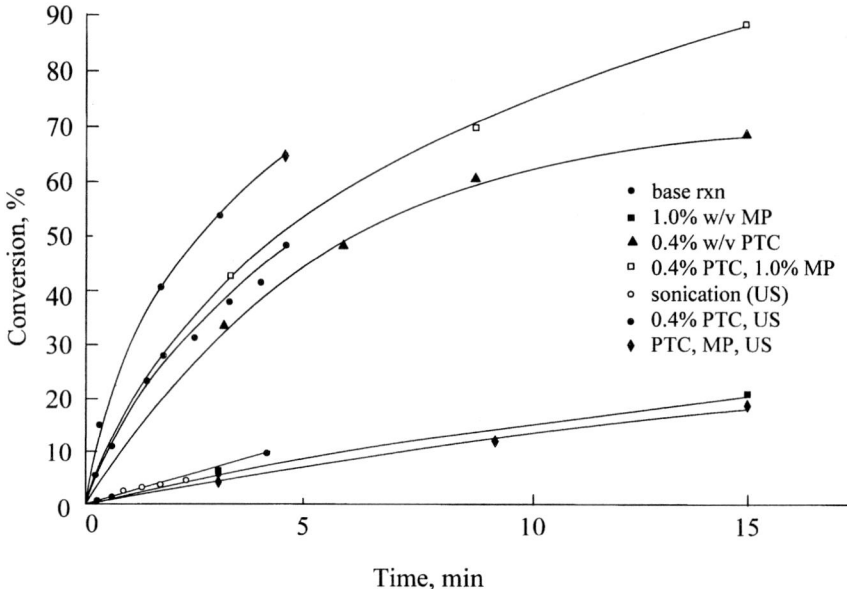

Figure 26.9 Enhancements by PTC, ultrasound, microphase, and combinations thereof (redrawn from Hagenson et al., 1994)

any one or any binary combination of the three strategies. The enhancement, however, is not additive, as shown in Figure 26.9.

Enzymatic, Photochemical, and Electrochemical Syntheses in Supercritical Fluids

The use of SCFs has been extended to include enzymatic, photochemical, and electrochemical reactions. We give a few selected examples to illustrate the growing importance of combining these less conventional methods of organic synthesis with an unconventional one (use of SCFs).

Enzymatic reactions in supercritical fluids

Conducting an enzymatic reaction in a supercritical fluid has the distinct advantage that enzymes are usually insoluble in SCFs. Thus the recovery and reuse of enzymes become easier and much more economical, and there are similar attendant advantages in processing the reaction products. Other advantages are the high solubility of oxygen and hydrogen in SCFs and the higher diffusivities of substrates due to the lower viscosities of SCFs (Russell and Beckman, 1991; Kamat et al., 1995). When these advantages are combined with the normal characteristic of SCFs that they can be readily removed without residue after the reaction by simple decompression, the overall attractiveness of conducting enzymatic reactions in SCFs becomes obvious. Supercritical CO_2 is the most extensively used SCF (Aaltonen and Rantakylä, 1991). Several other fluids have also been used as SCFs (Hammond et al. 1985; Kamat et al. 1992,

1993; Chaudhary et al. 1995; Russell et al., 1994). Important reviews covering the various enzymatic reactions carried out in supercritical fluids are those of Noritomi et al. (1995), Gunnlaughsdottir and Sivik (1995), Knez et al. (1995), and Kamat et al. (1995).

It is usually desirable that the substrate be soluble in the supercritical medium, but this is not a requirement. For example, in the hydrolysis of *p*-nitrophenylphosphate in sc CO_2, the substrate is not soluble in the SCF, and it is postulated that the reaction occurs in the aqueous film between sc CO_2 and the enzyme (Randolph et al., 1985). Apparently, the film is formed by the adsorption of water dissolved in sc CO_2 onto the enzyme.

The literature contains the results of a number of reactions carried out in supercritical fluids. A good example is the equilibrium shift between the *trans* and *gauche* rotamers of $(CF_3)_2CHOH$ in cs CHF_3. Increasing density favors the gauche form (Kazarian and Poliakoff, 1995) because of the increase in dielectric constant that accompanies the increase in density. When the dielectric constant remains unchanged with an increase in density, as with sc SF_6, there is no equilibrium shift. In some cases, such as the enantioselective esterification of ibuprofen using immobilized *Mucor michei*, the rate can be slightly increased by varying the pressure, but the enantioselectivity remains unchanged (Rantakylä and Aaltonen, 1994). It is also important to note that in some cases there is no enhancement of the rate and the only advantage of an SCF over a conventional solvent is the much easier separation afforded by it. In general, a minimum amount of water must be present for the enzyme to have any activity at all (Dordick, 1989). However, there is an optimum water content beyond which a hydration sphere is formed that hinders the diffusion of the hydrophobic substrate onto the enzyme. Further, an excess of water can denature the enzyme (Dumont et al., 1993).

On the other hand, immobilized enzymes seem to perform much better in SCFs. Thus in the esterification of glycidol with butyric acid on porcine pancreatic lipase, the immobilized form of the enzyme (with 20–25% hydration) gives three times higher rates than the optimized free enzyme containing 10% water (Martins et al., 1994).

[26.10]

The yield and selectivity in sc CO_2 are comparable to those with the most favorable organic solvents.

Photochemical reactions in supercritical fluids

A particularly striking example of the use of supercritical fluids is the photochemical dimerization of isophorone in sc CO_2 and in sc CHF_3

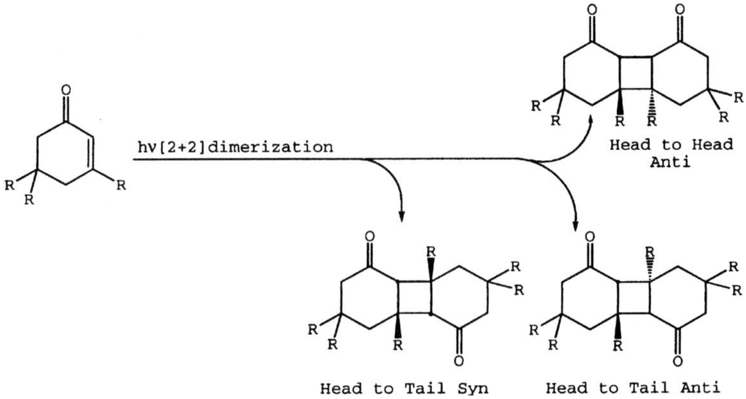

[Scheme 26.11: hv [2+2] dimerization of a cyclohexenone giving Head to Head Anti, Head to Tail Syn, and Head to Tail Anti dimers.]

The tunable density feature of an SCF has been used to best advantage in this reaction to change the relative yields of the three stereochemical dimers (Hrnjez et al., 1989). This change has been attributed to the fact that the three transition states involved in the reaction have significantly different dipole moments and hence interact with the solvent to different extents. Any such interaction depends on the solvent density and hence can be optimized by tuning the density.

The example cited is significant, for it illustrates the role of an SCF in influencing product distribution. Most other studies illustrate its role in changing the rate constant of the reaction. A good example of the latter is the conversion of hermalones (α-acids) in a mixture with lupidones (β-acids) to the *trans-iso* form via an oxidative π-methane rearrangement (André et al., 1985). Both acids are extracted from hop and contribute to the bitterness of beer, but the α-acids must be converted to the *iso-α* form before addition to the beer.

Electrochemical reactions in supercritical fluids

In almost all of the work reported so far on electrochemical reactions in supercritical fluids, the emphasis has not been on preparative organic synthesis. One of the principal questions tackled has been whether the Stokes–Einstein relationship between diffusion and viscosity continues to be valid for an SCF, and the conclusion is that it probably is. In synthesis, the only significant study is the production of dimethyl carbonate from CO and methanol using sc CO_2 as a cosolvent with methanol (Dombro et al., 1988). The use of sc CO_2 in place of water enabled the reaction to be carried out near the critical point, that is, at a lower temperature. Tetrabutylammonium bromide (TBAB) in a concentration of 1–5% was used as an electrolyte and bromide source. The reactions occurring at the two electrodes are

Anode: $2(n\text{-Bu})_4\text{N}^+ + 2\text{Br}^- \longrightarrow 2(n\text{-Bu})_4\text{N}^+ + \text{Br}_2 + e^-$

$$\text{Br}_2 + \text{CO} \longrightarrow \text{COBr}_2$$

$$\text{COBr}_2 + 2\text{CH}_3\text{OH} \longrightarrow \text{CH}_3-\text{O}-\underset{\underset{\text{O}}{\|}}{\text{C}}-\text{O}-\text{CH}_3 \qquad [26.12]$$

Cathode: $2\text{H}^+ + 2e^- \longrightarrow \text{H}_2$

$$2\text{CH}_3\text{OH} + \text{CO} \longrightarrow \text{CH}_3-\text{O}-\underset{\underset{\text{O}}{\|}}{\text{C}}-\text{O}-\text{CH}_3$$

Sonoelectrochemistry

The combination of ultrasound with electrochemistry is a rapidly emerging field called *sonoelectrochemistry* (Compton et al., 1997). Table 26.5 includes illustrative examples of electroorganic synthesis with ultrasound.

Scale-up of electrochemical processes for industrial exploitation often suffers due to the limitation of mass transfer to the electrode surface and fouling of the electrode surface. Application of ultrasound in electrochemical systems overcomes these limitations by increasing the rate of mass transfer to the electrode, raising the limiting current density, and reducing the diffusion layer thickness. In addition to enhancing mass transfer, ultrasound can have a chemical effect in some homogeneous systems that involve highly reactive radical species. Thus dual activation by simultaneous passing of current and ultrasound through a reaction mixture is possible.

Important factors in sonoelectrochemistry

The asymmetric collapse of imploding cavitational bubbles near the electrode–solution interface promotes turbulence and leads to microstreaming at the electrode surface and thereby to enhanced limiting currents. Thus ultrasonic irradiation tremendously decreases the depletion of electroactive species near the electrode. Typical values of the diffusion layer thickness under sonication are around 5 μm, compared to normal values of approximately 500 μm. Because the limiting current is given by

$$i_{\lim} = \frac{nFDa[A]}{\delta} \qquad (26.18)$$

this leads to a substantial increase in the limiting current in most cases (see Chapter 23 for notation).

The distance that separates the electrode disk and the ultrasound horn tip is an important variable because the limiting current enhancement decreases with increasing distance of separation. The frequency of ultrasound at constant intensity has no effect on the limiting current (Lorimer et al., 1996). However, a change in intensity at constant frequency does affect the limiting current. Using a rotating disk electrode with well-defined hydrodynamics, it is possible

to compare silent and sonicated reactions by calculating the theoretical speed at which the disk would have to rotate under silent conditions to achieve the same transport limit as found under sonication. Such a comparison shows that ultrasound leads to limiting current conditions well beyond those attainable in practice by mere rotation.

Electrochemical reactor with ultrasound: the sonotrode

To combine the use of ultrasound with electrochemical reactions, usually a powerful immersion ultrasound horn is introduced into a thermostatted conventional three-electrode cell. In a recent study, the tip of the sonic horn itself was electrically insulated and used as the electrode—called the "sonotrode" (Reisse et al., 1994; Durant et al.,1996; Compton et al., 1996a). Combining the working electrode with the ultrasound tip leads to a rather different perturbation at the electrode surface. Using the titanium horn itself as the electrode leads only to moderate currents and no anodic currents due to the presence of a semiconducting TiO_2 layer. However, very high limiting currents (corresponding to diffusion layer thicknesses of less than 1 µm) are obtained at relatively low ultrasound intensities by embedding a platinum electrode directly into the horn tip.

Miscellaneous Combinations

In addition to the combinations described, examples of less formal combinations clearly provide additional evidence of the growing importance of combinatorial strategies in general. We outline some of them here.

Enzymatic reactions in reverse micelles

The principle of reverse micelles was outlined earlier. The advantages of low-water environment organic media include the following:

1. Greater solubility (specific solvation effects) and greater concentrations of certain reagents in organic media. Many reactions are thermodynamically possible only in specific organic solvents.
2. For reactions involving water as a by-product, the main product synthesis is favored in the low water environment of an organic solvent. Hydrolytic side reactions are also minimized.
3. Often, one of the products can be partitioned from the enzyme microenvironment, leading to a favorable shift in equilibrium and completion of the reaction (see Chapter 18).
4. Substrate specificity of the enzyme may be potentially controlled by the choice of solvents.
5. Greater thermostability is achieved.

The classical Michaelis-Menten equation is also obeyed in reverse micelles for the simple case of unimolecular reactions. Because the enzyme concentration is low compared to reverse micelle concentration, model 1 of Table 26.3 should be

applicable. The corresponding classical form of the enzymatic reaction is

$$A + E \leftrightarrow EA \rightarrow E + \text{Products} \tag{26.19}$$

based on which the final equation developed is (Khmelnitsky et al., 1990)

$$-r_A = \frac{k_{aM}[E]_{M0}[A]_{M0} V_M}{K_{aMM} + [A]_{M0}} \tag{26.20}$$

where k_{aM} and K_{aMM} are the apparent micellar phase and Michaelis–Menten constants[2] and V_M is the micellar volume fraction. The apparent Michaelis–Menten constant is related to the true constant by

$$K_{aMM} = K_{MM} \frac{1 + V_M(m_{MA} - 1)}{m_{MA}} \tag{26.21}$$

Clearly, if the substrate is solubilized exclusively in the micellar phase, $K_{aMM} \rightarrow K_{MM} V_M$, and if solubilized completely in the organic bulk, $K_{aMM} \rightarrow K_{MM}(1 - V_M)/m_{MA}$.

Use of hydrotropes in electroorganic synthesis

If the solubility of poorly soluble substances in electrochemical reactions can be increased by using hydrotropes, remarkable enhancements of the rate can be achieved. Most of the hydrotropes withstand cathodic reduction much better than anodic attack, and permanent loss of hydrotrope may occur due to anodic attack. Two early examples are the reduction of nitro compounds such as nitrophenol to aminophenol in quantitative yield and the oxidation of benzyl alcohol or benzaldehyde to benzoic acid.

PTC-micellar catalysis and ultrasound-microwave combinations

A few other specific combinations besides the more researched ones cited above have also been reported. Among them, the two potentially important combinations are PTC-micellar catalysis, and ultrasound-microwave.

USE OF MICROMIXING AND FORCED UNSTEADY-STATE OPERATION

The strategies described so far are essentially chemistry based in the sense that different environments are created for a reaction to be initiated or enhanced. It is also possible to achieve enhancement by physically manipulating the same environment. One way of accomplishing this is by introducing controlled micromixing or optimizing the sequence of reactant additions. Another way is to operate under forced unsteady-state conditions by periodic reversals of the direction of flow. The main aspects of these strategies were considered in Chapter 13.

---

[2] Note that in this chapter we use the subscript MM instead of M, as in Chapter 20, to denote Michaelis–Menten kinetics to avoid confusion with the notation for micellar reactions.

Notation to Part V

Unlike in other parts, the notation is given chapterwise for part V. This is because of the distinctive nature of each of the chapters comprising this part.

CHAPTER 18

A	Reactant.
$[A]_{int}$	Effective interfacial concentration of A, mol/m^3.
$[A]_e$	Effective concentration of A, mol/m^3.
B	Reactant.
BEF	Biphasic enhancement factor defined by Equation 18.13.
E	Enzyme.
$[E]_b$	Bulk concentration of enzyme, mol/m^3.
$[E]_e$	Effective concentration of enzyme, mol/m^3.
$[i]$	Concentration of species 1, mol/m^3.
$[i]_S$, $[i]_W$	Concentration of i in solvent and water phases, mol/m^3.
K	Equilibrium constant.
K_a, K_b	Acid and base dissociation constants.
K_A, K_E	Adsorption constants of A and E, m^3/mol.
$K_{M,e}$	Effective Michaelis–Menton constant, mol/m^3.
K_S, K_W	Equilibrium constants in the organic and aqueous phases.
$K_{W,app}$	Apparent equilibrium constant in water.
$K_{a,biphasic}$, $K_{b,biphasic}$	Acid and base dissociation constants in a biphasic system.
$K_{biphasic}$	Equilibrium constant in a biphasic system.
$K_{nonionic,biphasic}$	Equilibrium constant for nonionic reaction in a biphasic system.

$K_{S,\text{nonionic}}, K_{W,\text{nonionic}}$	Equilibrium constants for nonionic reactions in the organic and aqueous phases.
k_a	Observed reaction rate constant, 1/s.
k_b	Rate constant in the aqueous phase, 1/s.
k_e	Effective constant of the Michaelis–Menten equation, 1/s.
k'_L	Liquid-side mass transfer coefficient, m/s.
m_i	Distribution coefficient of i between solvent (organic phase) and water.
R	Product.
$[R]_1, [R]_2$	Contributions to the total concentration of R from reactions in the liquid film/bulk and at the interface.
$-r_{A,e}$	Effective rate of disappearance of A, mol/m^3 s.
S	Product.
V_{int}	Volume of interface region, m^3.
V_S, V_W	Volumes of organic and water phases, m^3.

Greek

α	Volume ratio of phases V_s/V_W.
δ	Thickness of film, m.
δ'	Thickness of the interface zone, m.
Γ_i	Excess adsorption quantity of species i, mol/m^2.
$\Gamma*_i$	Excess adsorption quantity of species i at saturation, mol/m^2.

Subscripts/Superscripts

Subscripts

b	Bulk.
e	Effective
int	interface.
s	Solvent (organic) phase.
t	Total.
w	Water.

Superscripts

*	Saturation.

CHAPTER 19

a	Interfacial area, m^2/m^3.
a_p	Interfacial area of particle, m^2/kg.
D	Liquid-phase or bulk diffusivity, m^2/s.
$D_{e,Q}$	Effective diffusion coefficient of catalyst species within the solid or catalyst phase, m^2/s.

$D_{e,i}$	Effective diffusion coefficient of species i, m²/s.
d_p	Particle diameter, m.
d_s	Stirrers diameter, m.
g	Acceleration due to gravity, m/s²
i	Any species i.
$[i]$	Concentration of species i.
$[\hat{i}]$	Dimensiolless concentration $[i]/[i]_0$.
$[\hat{i}]_s$	Dimensioless surface concentration, $[i]_s/[i]_0$.
K	Equilibrium constant for ion exchange.
K_i	Dissociation constant for species i.
k_a	Observed pseudo-first-order rate constant, 1/s.
k_{mn}	Organic reaction mnth-order rate constant.
k_{SL}	Solid dissolution mass transfer coefficient, m/s.
k_1	Forward reaction rate constant for ion exchange, m³/mol s.
k_{-1}	Reverse reaction rate constant for ion exchange, m³/mol s.
k_2	Organic reaction second-order constant, m³/mol s.
k_i'	Overall mass transfer coefficient for species i, m/s.
k_L'	Mass transfer coefficient, m/s
$k_{L,i}'$	Mass transfer coefficient for species i, m/s.
MX	Product of ion-exchange reaction.
MY	Aqueous phase reactant (solid or liquid).
m_i	Distribution coefficient for quat between organic and aqueous phases.
N_i	Number of moles of species i, kmol.
Q	Any quaternary salt (or quat)
QX	PT catalyst.
$[QX]_c$	Catalyst concentration in the solid support, mol/kg.
$[QX]_{sol}$	Concentration of QX in the solid reactant, mol/m³.
QY	Activated form of PT catalyst.
Q^+	Cation of catalyst.
q_0	Initial concentration of added quat, mol/m³.
R	Radius of catalyst (solid) particle, m.
RX	Organic-phase reactant.
$[R\hat{X}]$	Dimensionless concentration, $[RX]/[RX]_0$.
$[R\hat{X}]_s$	Dimensionless surface concentration, $[RX]_s/[RX]_0$.
RY	Product of organic-phase reaction.
r	Rate of volume reaction, mol/m³ s; radial coordinate, m.
r_i	Rate of reaction of species i, mol/m³ s.
r	Rate of surface reaction, mol/m² s.
t	Time, s.
V_{org}	Volume of organic phase, m³.
V_{aq}	Volume of aqueous phase, m³.
V_{cat}	Volume of catalyst phase, m³.
X^-	Anion.
Y^-	Anion.

Greek

δ	Film thickness, m.
ε_i	Volume fraction of i phase in support pores.
η	Effectiveness factor.
ρ_c	Density of solid support, kg/m^3.
ω'	Disk rotation in radians.

Dimensionless Groups

Da, Da_i	Damköhler number; Da for species i.
$\alpha_1, \alpha_2, \alpha_s, \beta_m, \gamma_Y,$ $\gamma_{RX}, \psi, \phi_s, \theta_{QX}, \theta_{org},$ $\theta_{aq}, \theta'_{org}, \chi$	See Table 19.6.
$(Bi)_m, \phi, \tau, \omega$	See Equation 19.22.
ψ_1	See Equation 19.4.

Subscripts/Superscripts

Subscripts

a	Actual (observed).
aq	Aqueous phase.
b	Bulk.
c	Catalyst.
f	Film.
i	Species i.
int	Interface.
L	Liquid.
MX	Product of ion-exchange reaction.
MY	Aqueous-phase reactant.
noniso	Nonisothermal
0	Initial conditions.
org	Organic phase.
Q	Catalyst.
QX	PT catalyst.
QY	Activated form of PT catalyst.
RX	Organic-phase reactant.
RY	Product of organic-phase reaction.
SL	Solid-liquid.
Sol	Solid.
s	Surface.

Superscripts

*	Saturation.
overbar	Average value within a droplet.

CHAPTER 20

A	Rectant (substrate).
A_p	Area of pellet, m^2/kg.
$[A]_v$	Concentration of A in the bulk, mol/m^3.
$[A]_p$	Concentration of A within an immobilized enzyme pellet, mol/m^3.
$[A]_s$	Concentration of A at the pellet surface, mol/m^3.
$[\hat{A}]_p$	Normalized concentration of A within the pellet, $[A]_p/[A]_s$.
$[\hat{A}]_{sb}$	Normalized surface concentration, $[A]_s/[A]_b$.
AE	Substrate-enzyme complex.
a	Normalized concentration in of A in the reactor, $([A]/[A]_0)$.
a_p	Normalized concentration of A in pellet, $([A]_p/[A]_o)$.
a_{ps}	Normalized concentration of A at the pellet surface, $([A]_s/[A]_o)(= a$ in the absence of external diffusion).
B''	Parameter defined by Equation E20.1.5.
Bi_m	Biot number.
C	Cells; reactor capacity, $\mu mol/min$ (Example 20.1).
D	Diffusion coefficient, m^2/s.
Da	Damköhler number.
D_{eA}	Effective diffusion coefficient of A, m^2/s.
d_p	Diameter of pellet, m.
E	Enzyme.
$[\bar{E}]$	Average concentration of enzyme, mol/m^3.
E_c	Ratio of specificity constants of the enantiomers, α_R/α_S.
ee	Enantiomeric excess.
f_B	Bed voidage.
G	Mass flow rate, $kg/m^2 s$.
$[i]$	Concentration of species i, mol/m^3.
j_d	Mass transfer group defined in Table 20.6.
K_M	Michaelis–Menten constant, mol/m^3.
$\hat{K}_M$	Dimensionless Michaelis–Menten constant, $K_M/[A]_0$.
$K_{M,R}, K_{M,S}$	Michaelis–Menten constants for R and S.
K_m	Monod constant, mol/m^3.
k	General notation for rate constant, s^{-1} (for a first-order reaction).
k_0	Rate constant defined specifically to indicate absence of deactivation, 1/s.
k_1, k_2	Forward reaction rate constants, 1/s for first order, $m^3/mol\ s$ for second order.
k_{-1}, k_{-2}	Reverse reaction rate constants, 1/s for first order, $m^3/mol\ s$ for second order.
k_d	Deactivation constant, 1/s for first order.
k_h	Autohydrolysis constant, 1/s for first order.

Notation for Part V

k'_L	Liquid-side mass transfer coefficients, m/s.
k_L	Liquid-side mass transfer coefficient, 1/s.
L	Length, m.
Q	Volumetric flow rate, m³/s.
R	Product; radius of pellet, m.
R	Enantiomer R.
$\hat{R}$	Normalized radial coordinate, r/R.
r_C	overall rate of cell growth, kg/m³s.
$r_i(-r_A)$	Rate of reaction (disappearance) of species i, mol/m³s.
S	Enantiomer S.
S_V	Space velocity (Q/V_r), s⁻¹.
t	Time, s.
t_p	Reaction or decay time, s.
t_{P1}	Total time for a predetermined ultimate level of decay, s.
$\bar{t}$	Mean residence time (V_r/Q), s.
$\hat{t}$	Dimensionless time defined by Equation E20.1.5.
u	Velocity, m/s.
V	Volume (also reactor volume), m³.
V_m	Maximum possible rate of an enzymatic reaction $(E_0 k)$, mol/m³ s.
$V_{m,R}, V_{m,S}$	V_m written specifically for R and S, mol/m³ s.
V_p	Volume of pellet, m³.
V_r	Reactor volume, m³.
X_A	Conversion of A.
$\bar{X}_A$	Time-averaged conversion of A.
y_A	Mole fraction of A.
$y_{C/A}, y_{C/R}$	Yield coefficients defined by Equations 20.19 and 20.20.
z	Dimensionless axial distance ℓ/L.

Greek

α	Specificity constant V_m/K_M.
α_R, α_S	Defined as $V_{m,R}/K_{M,R}$ and $V_{m,S}/K_{M,S}$.
β	Defined as $[A]_s/K_M$.
ε	Catalyst effectiveness factor.
ε_a	Effectiveness factor in the combined presence of internal and external diffusion.
ϕ	Generalized Thiele modulus; general notation for Thiele modulus.
ϕ_a	Observable Thiele modulus based on the rate.
ϕ_s	Thiele modulus for a sphere defined by Equation 20.7.
λ	Dimensionless deactivation constant defined as $k_d t_{p1}$.
μ	Specific growth rate of cells, s⁻¹.
μ_m	Maximum growth rate of cells, s⁻¹.
μ_L	Viscosity of liquid, kg/ms.
ℓ	Length coordinate.

862   Strategies for Enhancing Rate Reactions

ρ_L	Density of liquid, kg/m^3.
$\mathcal{D}$	Dilution rate, s^{-1}.
Λ	Length coordinate, m.
Λ_0	Characteristic length of a shape, m.
$\bar{\Lambda}$	Normalized general shape coordinate, Λ/Λ_0.

Subscripts/Superscripts

Subscripts

0	Entrance or initial conditions.
R	R enantiomer.
S	S enantiomer.
s	Pellet surface.
b	Bulk.
p	Inside the pellet.

CHAPTER 21

A	Chemical species.
A_c	Cathodic area, m^2.
a	Parameter defined by Equation 21.17; area per unit volume, 1/m.
a'	Area per unit length of electrode, m.
a_j	Activity of species j.
B	Chemical species.
b	Parameter defined by Equation 21.17.
b_A	Parameter defined by Equation 21.28, V^{-1}.
b_H	Parameter defined by Equation 21.35, V^{-1}.
C.E.	Current efficiency.
D	Diffusivity, m^2/s.
d	Diameter, m.
E	Cell voltage, V.
E^+, E_A	Anodic voltage, V.
E^-, E_c	Cathodic voltage, V.
E_0	Open circuit voltage, V.
E^0	Standard open circuit voltage, V.
e	Electron.
F	Faraday, coloumbs, A s/mol.
ΔG	Gibbs' free energy change, kJ/mol.
ΔG^0	Standard free energy change, kJ/mol.
I	Current, A.
i	Current density, A/m^2; species i.
i_0	Exchange current density, A/m^2.
i_-, i_c	Cathodic current density, A/m^2.
i_c, i_a	Anodic current density, A/m^2.
i_j	Current density associated with species j, A/m^2.

Notation for Part V

i_{lim}	Limiting current density, A/m^2.
i_t	Total current density, A/m^2.
k'	Overall rate constant defined by Equation 21.33, m/s.
k_A	nFk'_A for nitrobenzene reduction.
k'_A	Parameter defined by Equation 21.26, representing the rate constant for the reaction of A, m/s.
k''_A	Parameter defined by Equation 21.27, m/s.
k'_a	Rate constant for the anodic reaction m/s.
k'_c	Rate constant for cathodic reaction, m/s.
k'_H	Rate constant for the loss reaction, m/s.
k'_L	Mass transfer coefficient, m/s.
k_R	nFk'_R for the reaction of R.
k'_R	Rate constant for the reaction of R, m/s.
L	Length, m.
M	Molecular weight, kg/mol.
N	Total number of CSTRs in a cascade.
$N_{LCD}, N_{ECD}, N_{ACD}$,	Criteria parameters defined by Equations 21.59–21.61.
n	Number of electrons.
Q	Volumetric flow rate, m^3/s.
R	Chemical species; recycle ratio.
r_A	Volumetric rate of reaction, mol/m^{3}s.
r'_A	Rate of surface reaction, mol/m^{2}s.
Re	Reynolds number, $du\rho/\mu$.
R_g	Gas constant, m^3 atm/mol K.
S	Chemical species.
Sc	Schmidt number, $\mu/\ell D$.
Sh	Sherwood number, $k'_L d/D$.
T	Temperature, K.
t	Time, s.
t_B	Batch time, s.
$\bar{t}_{res}$	Residence time in reservoir, V_{res}/Q, s.
u	Linear velocity of rotating electrode, m/s.
V, V_r	Reactor volume, m^3.
V_{res}	Volume of reservoir, m^3.
W	Theoretical weight, kg.
W_a	Actual weight, kg.
X_A	Conversion of A.

Greek

α	Transport coefficient for the main reaction.
α'	Transport coefficient for the loss reaction.
η	Overvoltage (overpotential), V.
κ	Specific electrical conductivity, Siemens (S).
ℓ	Length coordinate, m.
$\left.\begin{array}{l}\mu_{con}, \mu_a \\ \mu_c\end{array}\right\}$	Concentration, anodic, and cathodic overpotentials, V.

864   Strategies for Enhancing Rate Reactions

μ	Viscosity, kg/m s.
ν_i	Stoichiometric coefficient of i.
ρ	Density, kg/m^3.

Subscripts/Superscripts

Subscripts

0	Initial.
a	Anode.
b	Bulk.
c	Cathode.
f	Final.
lim	Limiting.
s	Surface.
t	Total.

CHAPTER 22

A	Reactant.
$[A]_0$	Initial concentration of A, mol/m^3.
B	Reactant.
C_N	Dimensionless cavitation number.
C_p, C_v	Heat capacities at constant pressure and constant volume, kcal/kg K.
D_e	Effective diffusivity, m^2/s.
d_{or}	Diameter of orifice, m.
d_p	Diameter of pipe, m.
E	Activation energy, kcal/mol.
E_{et}	Activation energy at elevated temperature, kcal/mol.
F	$F_v F_t$.
F_t	Cycle time at which bubble phase reaction occurs, s.
F_v	Volume of reaction mixture occupied by bubbles, m^3.
f	Frequency ($= \omega/2\pi$), 1/s.
f_T	Frequency of turbulent pressure variations, 1/s.
I	Acoustic intensity, kW/m^2.
I_0	Initial (undamped) acoustic intensity, kW/m^2.
k	Rate constant, (m^3/mol)$^{n-1}$ 1/s.
k^0	Arrhenius parameter, (m^3/mol)$^{n-1}$ 1/s.
k_{bub}	Rate constant associated with bubble collapse, (m^3/mol)$^{n-1}$ 1/s.
k_{et}	Rate constant of reaction occurring within the bubble, (m^3/mol)$^{n-1}$ 1/s.
k_{et}^0	Arrhenius parameter of reaction occurring within the bubble, units of k_{et}.
k_{SL}	Solid-liquid mass transfer coefficient, 1/s.

Notation for Part V

k_{us}	Rate constant in the presence of ultrasound, $(m^3/mol)^{n-1}$ 1/s.
$\mathbf{k}$	Real part of $\mathbf{k}_{mix}$ (which has also an imaginary component).
$\mathbf{k}_{mix}$	Wave vector in a mixture of gas and liquid.
$P(\mathbf{r}, t)$	Spatial and time-dependent pressure, atm.
P_0	Pressure in the bulk liquid in the absence of ultrasound, atm.
P_2	Pressure in the fluid after it passes through the orifice, atm.
P_∞	Pressure far from the bubble in an infinite fluid, atm.
P_A	Pressure corresponding to the peak amplitude of the wave, atm.
P_G	Pressure in the gas portion of the sonicated system, atm.
P_L	Pressure pulse, atm.
P_M	Power input per unit mass of the cavitating liquid, kW/kg.
P_a	Acoustic pressure, atm.
P_b	Pressure inside the bubble, atm.
P_{b0}	Initial gas pressure inside the bubble, atm.
P_h	Hydrostatic pressure, atm.
P_i	Instantaneous pressure, atm.
P_m	Liquid pressure at transient collapse, atm.
P_{max}	Maximum pressure developed at moment of bubble collapse, atm.
P_{mean}	Mean static pressure at any point downstream of the orifice, atm.
P_n	Pressure in the bubble at the moment of collapse, atm.
P_v	Vapor pressure of liquid, atm.
R	Product.
R_g	Universal gas constant, m^3 atm/mol K.
$-r_A$	Rate of disappearance of A, mol/m^3 s.
r_b	Bubble radius, m.
r_{b0}	Initial bubble radius, m.
$\mathbf{r}$	Coordinate vector
T	Temperature, K.
T_0	Ambient temperature, K.
T_{max}	Maximum temperature developed at moment of bubble collapse, K.
t	Time, s.
t_L	Bubble lifetime, s.
u'	Turbulent fluctuating velocity, m/s.
u_i	Instantaneous local velocity, m/s.
u_{or}	Velocity of the fluid at the orifice, m/s.
u_p	Velocity of the fluid in the pipe, m/s.
u_s	Velocity of sound in the liquid medium, m/s.
u_t	Mean local velocity, m/s.
X_A	Conversion of species A.

Greek

α	Attenuation coefficient of the medium.
α'	Time-dependent attenuation coefficient of the medium.
β	Velocity at the wall, $dr_b/d\ell$, m/s.
γ	Polytropic ratio of specific heats of bubble mixture.
λ	Wave number.
ℓ	Distance from the source of ultrasound source, m.
ℓ_e	Length scale of eddy, m.
ω	Angular frequency, s^{-1}.
ρ_L	Density of the bulk liquid medium, kg/m^3.
σ	Surface tension of the bulk liquid medium, dyne/m.
τ	Pressure recovery time, s.
μ	Viscosity of the cavitating liquid, kg m/s

CHAPTER 23

A	Reactant.
a_L	Interfacial area between macrophases, m^2/m^3 aqueous phase.
$a_{L,MP}$	Microphase interfacial area per unit volume of liquid, m^2/m^3.
a_{MP}	Microphase interfacial area per unit volume of microphase, m^2/m^3.
D_i	Diffusivity of species i, m^2/s.
$d_{P,MP}$	Particle diameter of microphase, m.
h_0	Microphase holdup, m^3 microphase/m^3 liquid plus microphase.
$[i]$	Concentration of i, mol/m^3.
$[i]_b$	Concentration of i in the liquid bulk, mol/m^3.
$[i]_b^*$	Saturation concentration (solubility) of i in the liquid bulk, mol/m^3.
k	General symbol for rate constant for an nth-order reaction, $(m^3/mol)^{n-1}(1/s)$.
k_0	Uptake coefficient, 1/s.
$k_{L,MP}$	Mass transfer coefficient for film around microparticles, 1/s.
k_{mn}	Rate constant for a reaction mth-order in A and nth-order in B, $(m^3/mol)^{m+n-1}(1/s)$.
k_{MP}	Reaction rate constant for an nth-order reaction on microphase, $(m^3/mol)^{n-1}(1/s)$.
k_0'	Uptake coefficient, m/s.
$k_{L,MP}'$	Mass transfer coefficient for film around microparticles, m/s.
$\sqrt{M}$ or Ha	Hatta number (measure of the extent of reaction in the film).
$\sqrt{M'}$ or Ha'	Modified Hatta number (with a microphase) defined by Equation 23.25.

$\sqrt{M_{MP}}$	Measure of the extent of reaction in the microparticles in the film.
m	Reaction order.
m_A, m_B	Distribution coefficients of A, B, defined by Equation 23.6.
R	Solid product.
$-r_A$	Rate of disappearance of A, mol/m^{3}s.
$-r'_A$	Rate of disappearance of A, mol/m^{2}s.
Sh	Sherwood number, $k'_{L,MP} d_{P,MP}/D_A$.
t	Time, s.
w	Microphase holdup, kg/m^3.
x	Distance parameter, m.

Greek

δ	Film thickness, m.
η'	Ratio of the rates with and without a microphase.
ν_i	Stoichiometric coefficient of species i.
ρ	Density, kg/m^3.
λ	Distance within the film, m.

Subscripts/Subscripts

Subscripts

b	Liquid bulk.
L	Macrointerface with liquid.
L,MP	Interface between liquid and microphase.
MP	Microphase.
n	Reaction order.
0	Initial value.
p	Particle.
t	Total.

Superscripts

*	Saturation.

CHAPTER 24

A	Reactant.
A_m	Area of membrane, m^2.
a	Area per unit volume, 1/m.
C_{pA}	Heat capacity of reactant, kcal/mol K.
D_j	Diffusivity of species j, m^2/s or mol/m atm s.
d_M	Molecular diameter, m.
F_j	Flow rate of species j, mol/s.
F_j^T, F_j^S	Flow rates of species j in inner tube and outer shell, respectively, mol/s.

G_j^T, G_j^S	Flow rates per unit area of species j in inner tube and outer shell, respectively, mol/m²s.
H	Product (hydrogen).
I	Inert (sweep) gas, usually argon.
K	Reaction equilibrium constant.
k	Rate constant, $(m^3/mol)^{n-1}(1/s)$.
k^0	Arrhenius frequency factor, same units as the rate constant.
k_{per}	Percolation rate constant, appropriate units.
L	Reactor length, m.
M_j	Molecular weight of species j.
N_0	Avogadro number.
P^S, P^T	Pressures on the shell and tube sides.
p_j^S, p_j^T	Shell and tube side partial pressures of species j.
Q_j	Volumetric flow rate of species j, m³/s.
R	Reactant; radius, m.
R_1	Inner radius of membrane tube, m.
R_2	Outer radius of membrane tube, m.
R_3	Inner radius of shell, m.
R_g	Gas constant, kcal/mol K.
r_j	Rate of reaction of species j, mol/m³s.
S	Product.
$\bar{t}$	Residence time (V/Q_o), s.
t_m	Thickness of membrane wall, m.
T	Temperature, K.
U	Heat transfer coefficient, kcal/m² K s.
u	Velocity, m/s.
V	Reactor volume, m³
X_A	Conversion of A.
X_e^S, X_e^T	Equilibrium conversion on the shell and tube sides.

Greek

α	Ratio of sweep gas to feed gas flow rates at inlet, Equation 24.4.
β	Ratio of ethane to oxygen.
λ	Mean free path, m.
ℓ	Length parameter, m.
$\Delta \nu$	Volumetric difference between products and reactants.
θ_r	Dimensionless temperature T^T/T_0.
θ_s	Dimensionless temperature T^S/T_0.
ν_j	Stoichiometric coefficient of species j.
Π_j	Specific rate of permeation of species j, mol/m²s.
Γ	Heat transfer rate, $UA_m/C_{pA}F_{A0}$.

Subscripts/Superscripts

Subscripts

0	Initial/entrance conditions.
e	Equilibrium.

Superscripts

T	Tube side.
S	Shell side.

CHAPTER 25

A, B	Acid, reactant.
a	Interfacial area, 1/m.
$[AB]_a, [AB]_s$	Total concentration of acids (HA + HB) in the aqueous and organic phases, mol/m³.
A_c	Cross sectional area, m².
$[A^-]_a, [B^-]_a$	Concentrations of anions A^-, B^- in the aqueous phase, mol/m³.
B, B₁, B₂	Bases, reactants.
$[B_1]_s, [B_2]_s$	Concentrations of bases B_1 and B_2 in the organic phase, mol/m³.
F_G	Molal flow rate of gas, mol/s.
F_L	Molal flow rate of liquid, mol/s.
F_j	Molal flow rate of species j, mol/s.
$[HA]_a, [HB]_a$	Concentrations of acids HA, HB, HX in undissociated form in the aqueous phase, mol/m³.
$[HA]_s, [HB]_s$	Concentrations of acids HA and HB in the organic phase, mol/m³.
ΔH_r	Heat of reaction, kcal/mol.
ΔH_v	Heat of vaporization, kcal/mol.
h_0	Liquid holdup in the packing per unit volume of the packed bed.
K_A, K_B	Dissociation constants of acids HA and HB (Equations 25.3–25.4).
K_{B_1}, K_{B_2}	Equilibrium constants for complexation of bases B_1 and B_2 with an acid (reactions 25.6–25.7).
K	Equilibrium constant for reaction 25.4 (Equations 25.5–25.6).
K_{12}	Equilibrium constant for reaction 25.8.
K_D	Equilibrium constant of ion-pair dissociation.
K_{PT}	Equilibrium constant for protonation of a base in the organic phase.
k	General notation for rate constant (with any units).
k_2, k_{-2}	Second-order rate constants for forward and reverse reactions.

k_{Gj}^0	Gas-phase mass transfer coefficient of j for pure distillation, mol/m²s.
k_{Lj}^0	Liquid-phase mass transfer coefficient of j for pure distillation, mol/m²s.
M_j	Molecular weight of species j.
m_A, m_B	Distribution coefficients of free acids HA and HB.
N	Concentration of neutralizing agent, mol/m³.
N_j	Number of moles of j.
N_j'	Overall mass flux of species j, including the effect of reaction, mol/m² s.
pK_a	-Log (acid dissociation constant).
P_T	Total pressure of the system, bar or atm.
p_i, p_j	Vapor pressure of components i and j at the system temperature, bar or atm.
q_s	Heat flux, kcal/m² mol.
r_j	Rate of reaction of j, mol/m³ s.
R	Product.
S	Product.
t	Time, s.
V	Volume, m³.
X_A	Conversion of A.
$[X^-]_a$	Concentration of anion X⁻ in the aqueous phase, mol/m³.
Y_{Li}, Y_{Gi}	Transformed variables defined by Equation 25.51 for Y_{Li}
y	Mole fraction.
y_{Li}, y_{Gi}	Mole fractions of i in the liquid and vapor phases.

Greek

α	Separation factor.
α'	Defined by $(-M_s/\rho)$ (Equation 25.29).
β_{ij}	Relative volatility of components i and j in a mixture of the two.
β_a	Apparent relative volatility of the components of a mixture.
β'	Reaction-separation parameter defined by Equation 25.38.
δ	Molar ratio of free acids $[HA]_s/[HB]_s$ in the organic phase at equilibrium in dissociation-extraction.
δ'	Molar ratio of free bases $[B_1]/[B_2]$ in the liquid phase in dissociation-extractive distillation.
$\varepsilon_G, \varepsilon_L$	Constants defined by Equation 25.44.
$\varepsilon_{L1}, \varepsilon_{L2}$	Defined by Equations 25.34 and 25.40.
ℓ	Length (height) coordinate, m.
ν_i	Stoichiometric coefficient of i.
γ	Activity coefficient.
ρ	Density, mol (or kg)/m³.

Notation for Part V   871

Subscripts/Superscripts

Subscripts

a	Aqueous phase.
G	Gas (vapor) phase.
i	Ion-pair
L	Liquid phase.
0	Initial condition.
s	Solvent (organic) phase.

CHAPTER 26

A	Reactant.
$[A]$	Concentration of A, mol/m^3
a_c	Area per unit volume, 1/m.
B	Reactant.
C_p	Heat capacity, kcal/kg °C.
CMC	Critical micellar concentration, mol/m^3.
D	Detergent; diffusivity, m^2/s.
D_r	Radial diffusivity, m^2/s.
d	Thickness of layer surrounding lamp, m.
d_T	Diameter of tube, m.
E	Local electrical field intensity of microwave field of frequency ν; energy; enzyme.
E_t	Total incident energy, einstein/s.
e	Einstein, energy of a molecule, kW (erg).
F	Faraday, coloumbs (amp s/mol).
h	Planck's constant, 6.6265×10^{-34} Joule s.
I	Intensity of transmitted light, einstein/m^2 s.
I_0	Intensity of incident light, einstein/m^2 s.
$I_{r,\ell}$	Intensity of absorbed light at (r,ℓ), einstein/m^{2}s.
I_v	Light absorbed per unit volume, einstein/m^{3}s.
$K_{a,MM}$	Apparent Michaelis–Menten constant.
K_{MM}	Michaelis–Menten constant.
K_n	Equilibrium constants for the steps shown in Table 26.3.
K_s	Setschenow constant.
k	Rate constant of a photochemical reaction with orders m and n, (m^3)$^{m+n-1}$ s^{m-1}/(mol)$^{n-1}$ e^m.
k_a	Observed (apparent) rate constant, appropriate units.
k_{an}	Observed (apparent) pseudo-nth-order rate constant, (m^3/mol)$^{n-1}$(1/s)
k_{Mn}	Pseudo-nth-order rate constant for reactions 1, 2, ..., n (of Table 26.3) in the micellar phase.
k_W	Rate constant for reaction in the aqueous phase, appropriate units.
L	Length, m.

M	Micellar.
m	Reaction order for I_v.
m_{MA}	Distribution coefficient of A between micellar and water phases.
N_0	Avogadro's number.
n	Reaction order for the organic reactant; any number.
P	Radiant power of lamp, W; product.
P_d	Power transferred in a unit volume of material with conductivity λ and dielectric constant ε'.
P_t	Total power supplied per unit volume, W/m^3 (or cal/m^3 s).
Q	Feed rate, m^3/s.
q	Energy of a quantum, erg/molecule.
R	Product; radius.
$\check{R}$	Reaction group defined by Equation 26.12.
r	Radial coordinate, m.
$-r_A$	Rate of disappearance of A, mol/m^3 s.
r'	Dummy variable.
r_{10}	Radius of lamp, m.
S	Solubility of an organic solute in water.
S_H	Solubility of an organic solute in water in the presence of a hydrotrope.
SCF	Supercritical fluid.
sc	Supercritical.
s'	Distance, m.
T_c	Critical temperature, K.
t	Time, s.
u	Velocity, m/s.
u_ℓ	Velocity of light, m/s.
V_D	Volume of detergent, m^3.
V_M	Volume of micellar phase, m^3.
V_r	Reactor volume, m^3.
X_A	Conversion.
X_A'	Dummy variable.
z	Normalized length, ℓ/L.

Greek

α_a	Absorptivity, m^2/mol.
δ	Diffusion layer thickness, m.
ε'	Dielectric constant.
ε''	Dielectric loss.
H	Hydrotrope.
ϕ_λ	Quantum yield defined by Equation 26.3.
ϕ'	Synthetic quantum yield defined by Equation 26.4.
λ	Wavelength of light, m.
λ_{mw}	Wavelength of microwave radiation, m.

ℓ	Length parameter, m.
ℓ_s	Length of the sample, m.
ℓ_{pd}	Penetration depth, m.
μ	Linear absorption coefficient, $\alpha_a[A]$, m^{-1}.
ν	Frequency of light, Hz.
ρ	Density, kg/m^3.
σ	Optical density defined by Equation 26.12.

Subscripts/Superscripts

Subscripts

A	Reactant.
a	Actual (apparent).
E	Enzyme.
M	Micellar phase.
MM	Michaelis–Menten.
T	Tube.
W	Water phase.

References to Part V

Unlike in other parts, references are given chapterwise for part V. This is because of the distinctive nature of each of the chapters comprising this part.

CHAPTER 18

Abbott, N.L., Blankschtein, D., and Hatton, T.A. *Bioseparation*, **1**, 191 (1990).
Ahrland, S., Chatt, J., Davies, N.R., and Williams, A.A.J. *J. Chem. Soc.*, 276 (1958).
Alexandridis, P., Holzwarth, J.F., and Hatton, T.A. *Macromolecules*, **27**, 2414 (1994).
Ballard, D.H.G., Courtis, A., Shirley, I.M., and Taylor, S.C. *J. Chem. Soc. Chem. Commun.*, 954 (1983).
Borowski, A.F., Cole-Hamilton, D.J., and Wilkinson, G. *Nouv. J. Chim.*, **2**, 137 (1978).
Brändström, A. *J. Mol. Catal.*, **20**, 93 (1983).
Brink, L.E.S., and Tramper, J. *Biotechnol. Bioeng.*, **27**, 1258 (1985).
Brookes, I.K., Ph.D. Thesis, University of London 1984.
Buckland, B.C., Dunnill, P., and Lilly, M.D. *Biotechnol. Bioeng.*, **17**, 815 (1975).
Buhling, A., Kamer, P.C.J., and Van Leeuwen, P.W.N.M. *J. Mol. Catal.*, **98**, 69 (1995).
Butler, L.G. *Enzyme Microb. Tech.*, **1**, 253 (1979).
Carrea, G., Colombi, F., Mazzola, G., Cremonesi, P., and Antonini, E. *Biotechnol. Bioeng.*, **21**, 39 (1979).
Carrea, G. *Trends Biotechnol.*, **2**, 102 (1984).
Chae, H-J-, and Yoo, Y-J- *J. Technol. Biotechnol.*, **70**, 163 (1997).
Chatt, J., Leigh, G.J., and Leigh, R.M. *J. Chem. Soc., Dalton Trans.*, **19**, 2021 (1973).
Chaudhari, R.V., Bhanage, B.M., Deshpande, R.M., and Delmas, H. *Nature*, **373**, 501 (1995).
Chauvin, Y., Einloft, S., and Olivier, H. *Ind. Eng. Chem. Res.*, **34**, 1149 (1995a).
Chauvin, Y., Mussmann, L., and Olivier, H. *Angew. Chem. Int. Ed.*, **34**, 2698 (1995b).
Chen, J., and Alper, H. *J. Am. Chem. Soc.*, **119**, 893 (1997).

Cornils, B., and Wiebus, E. *Chemtech*, 33 (1995); *Recl. Trav. Chim. Pays-Bas*, **115**, 211 (1996).
Cornils, B. *Angew. Chem., Int. Ed. Eng.*, **34**, 1575 (1995).
Cramer, R.D. *J. Amer. Chem. Soc.*, **99**, 5408 (1977).
Cremonesi, P., Carrea, G., Ferrara, L., and Antonini, E. *Biotechnol. Bioeng.*, 17, 1101 (1975).
Cremonesi, P., Carrea, G., Ferrara, L., and Antonini, E. *Eur. J. Biochem.*, **44**, 401 (1974).
Cremonesi, P., Carrea, G., Sportoletti, G., and Antonini, E. *Arch. Biochem. Biophys.*, **159**, 7 (1973).
D'Yachenko, E.D., Kozlov, L.V., and Antonov, V.K. *Biokhimiya*, 36, 981 (1971).
Dautenhahn, P.C., and Lim, P.K. *Ind. Eng. Chem. Res.*, 31, 463–469 (1992).
de Belval, S., le Breton, B., Huddleston, J.G., and Lyddiatt, A. *J. Chromatogr. B*, **711**, 19 (1998).
Dunnill, P., and Lilly, M.D. In *Methods in Enzymology*, Vol. XLIV, (Ed., Mosbach, K.), Academic Press, New York, 1976.
Fan, W., Tsai, R.S., Tayar, E.N., Carrupt, P.A., and Testa, B. *J. Phys. Chem.*, **98**, 329 (1994).
French Patent. 2,505,322, 1981; Rhône Poulenc Ind; 2,541,675, 1983; Rhône Poulenc Ind.
Gonzalez-Velasco, J.R., Gonzalez-Marcos, J.A., Celaya, J., and Gutierrez-Ortiz, M.A. *Ind. Eng. Chem. Res.*, 35, 4389 (1996).
Haggin, J. *Chem. Eng. News*, Oct. 10, 1994.
Harada, M., and Miyake, Y. *J. Chem. Eng. Jpn.*, **19**, 196 (1986).
Herrmann, W.A., and Kohlpaintner, C.W. *Angew. Chem., Int. Ed. Engl.*, **32**, 1524 (1993).
Herrmann, W.A., Albanere, G.P., Manetsberger, R.B., Lapp, P., and Bahrmann, H. *Angew. Chem., Int. Ed. Engl.*, **34**, 811 (1995).
Hidaka, N., Matsumoto, T., and Morroka, S. *Ind. Eng. Chem. Res.*, **34**, 2272 (1995).
Hildebrand, J.H. *Proc. Nat. Acad. Sci. USA*, **76**, 194 (1979).
Hinze, W.L., and Pramauro, E. *CRC Crit. Rev. Anal. Chem.*, **24**, 133 (1993).
Huddleston, J.G., and Lyddiatt, A. *Appl. Biochem. Biotechnol.*, **26**, 249 (1990).
Huddleston, J.G., Willauer, H.D., Griffin, S.T., and Rogers, R.D. *Ind. Eng. Chem. Res.*, **38**, 2523 (1999).
Hurter, P.N., Alexandridis, P., and Hatton, T.A. In *Solubilization in Surfactant Aggregates of Surfactant Science series* (Eds. Christian, S.D., and Scamehorn, J.F.), Vol 55, Marcel Dekker, New York, 1995.
Johansson, H.-O., Karlstrom, G., Tjerneld, F., and Hayes, C. *J. Chromatogr. B.*, **711**, 3 (1998).
Joó, F., and Beck, M.T. *React. Kinet. Catal. Lett.*, **2**(3), 257 (1975).
Kalck, P., and Monteil, F. *Adv. Organomet. Chem.*, **34**, (1992).
Kamat, S., Beckman, E.J., and Russell, A.J. *Enzyme Microb. Technol.*, **14**, 265 (1992).
Kauzmann, W. In *Advances in Protein Chemistry* (Eds., Anfinsen, C.B., Jr., Anson, M.L., Baily, K., and Edsall, J.T.), Vol. 14, Academic Press, New York, 1979.
Klibanov, A.M., Semenov, A.N., Samokhin, G.P., and Martinek, K. *Bioorgan. Khim.*, **4**, 82 (1978).
Kozlov, L.V., and Ginodman, L.M. *Biokhimiya*, **30**, 1051 (1965).
Kuntz, E. French Patent 2,349,562, 1976a, Rhône Poulenc Ind.; French Patent 2,366,237, 1976b, Rhône Poulenc Ind.; Ger. Patent 2,733,516, 1978, Rhône Poulenc Ind.; *Chemtech*, **17**, 570 (1987).
Laane, C., Boeren, S., and Vos, K. *Trends Biotech.*, **3**, 251 (1985).
Lilly, M.D. *J. Chem. Techn. Biotechnol.*, **32**, 162 (1982).
Lilly, M.D., and Woodley, J.M. *Stud. Org. Chem.*, **22**, 179 (1985).

Lopes, J.R., and Loh, W. *Langmuir*, **14**, 750 (1998).
Lubineau, A., Augé, J., and Queneau, Y. *Synthesis*, **8**, 741 (1994).
Lugaro, G., Carrea, G., Cremonesi, P., Casellato, M.M., and Antonini, E. *Arch. Biochem. Biophys.*, **159**, 1 (1973).
Mallamace, F., Migliardo, P., Vasi, C., and Wanderlingh, F. *Phys. Chem. Liquids*, **11**, 47 (1981).
Margolin, A.L., Svedas, V.K., and Berezin, I.V. In *Future Directions for Enzyme Engineering* (Eds., Wingard, L.B., Jr., Berezin, I.V., and Klyosov, A.A.), Plenum, New York, 1980.
Martinek, K., and Berezin, I.V. *J. Solid-Phase Biochem.*, **2**, 343 (1977).
Martinek, K., Klibanov, A.M., Samokhin, G.P., Semanov, A.N., and Berezin, I.V. *Bioorgan. Khim.*, **3**, 696 (1977).
Martinek, K., and Semenov, A.N. *Biochim. Biophys. Acta*, **658**, 90 (1981a); *J. Appl. Biochem.*, **3**, 93 (1981b).
Martinek, K., Semenov, A.N., and Berezin, I.V. *Dokl. Akad. Nauk SSSR*, **253**, 358 (1980); *Biochim. Biophys. Acta*, **658**, 76 (1981a); *Biotechnol. Bioeng.*, **23**(4), 1115 (1981b).
Matijevic, E. *Surface and Colloid Science*, Vol. 1, Wiley, New York, 1969.
Mitsubishi Oils and Fats Co. Ltd., Japan Patent 59-114796, 1982.
Miyake, Y., Ohkubo, M., and Teramoto, M. *Biotechnol. Bioeng.*, **38**, 30 (1991).
Monflier, E., Fremy, G., Castanet, Y., and Mortreux, A. *Angew. Chem., Int. Ed.*, **34**, 2269 (1995).
Monflier, E., Tilloy, S., Bertoux, F., Castanet, Y., and Mortreux, A. *New J. Chem.*, **21**, 857 (1997).
Morel, D. French Patent 2,486,525, 1980; Rhône Poulenc Ind.
Moulijn, J.A., Van Leeuwen, P.W.N.M., and van Santen, R.A., Eds. *Catalysis, Elsevier*, Amsterdam, 1993.
NATO Advanced Research Workshop *Aqueous Organometallic Chemistry and Catalysis*, Debrecen, Hungary, Aug. 29–Sept. 1, 1994; Preprint.
Omata, T., Iwamoto, N., Kimura, T., Tanaka, A., and Fukui, S. Eruo. *J. Appl. Microb. Biotech.*, **11**, 199 (1981).
Overdevest, P.E., Keurentjes, J.T.F., Van der Padt, A., and Van't Riet, K. *Abstracts of Papers for the 215th ACS National Meeting*, Dallas, 1998 (Am. Chem. Soc., Washington, D.C., 1998, I&EC 001).
Oyama, K., Nishimura, S., Nonaka, Y., Hasimoto, T., and Kihara, K. U.K. Patent Appl. No. 8011159, April 2, 1980.
Papadogianakis, G., Verspui, G., Maat, L., and Sheldon, R.A. *Catal. Lett.*, **47**, 43 (1997).
Pereira, V., Tigli, H., and Gryte, C.C. *Biotechnol. Bioeng.*, **30**, 505 (1987).
Persson, J., Nystrom, L., Ageland, H., and Tjerneld, F. *J. Chromatogr. B*, **711**, 97 (1998).
Playne, M.J., and Smith, B.R. *Biotechnol. Bioeng.*, **25**, 1251 (1983).
Podlahova, J. *J. Inorg. Nucl. Chem.*, **38**(1), 125 (1976).
Ray, A. *Nature*, **231**, 313 (1971).
Ritter, U., Winkhofer, N., Schmidt, H.G., and Roesky, H.W. *Angew. Chem., Int. Ed.*, **35**, 524 (1996).
Rogers, R.D., Willauer, H.D., Griffin, S.T., and Huddleston, J.G. *J. Chromatogr. B.*, **711**, 255 (1998).
Scamehorn, J.F., and Harwell, J.H. In *Surfactants in Chemical/Process Engineering* (Eds. Wasan, D.T., Ginn, M.E., and Shah, D.O., *Surfactants Science Series*) Marcel Dekker, New York, vol. 28, 1988.
Semenov, A.N., and Martinek, K. *Bioorgan. Khim.*, **6**, 1559 (1980).
Semenov, A.N., Berezin, I.V., and Martinek, K. *Biotechnol. Bioeng.*, **23**, 355 (1981).

Semenov, A.N., Khmelnitski, Y.L., Berezin, I.V., and Martinek, K. *Biocatal.*, **1**, 3 (1987).
Semenov, A.N., Martinek, K., and Berezin, I.V. *Bioorgan. Khim.*, **6**, 600 (1980).
Smith, R.T., and Baird, M.C. *Transit. Metal. Chem.*, **6**, 187 (1981).
Tanford, C. *Science*, 200, 1012 (1978).
Tokitoh, Y., and Yoshimura, N. U.S. Patent 4,808,756 (to Kuraray Corp.), 1989.
Tsai, R.S., Fan, W., Tayer, N.E., Carrupt, P.A., Testa, B., and Kier, L.B. *J. Am. Chem. Soc.*, **115**, 9632 (1993).
Tsai, S.W., Wu, G.H., and Chiang, C.L. *Biotechnol. Bioeng.*, **38**, 761 (1991).
Ucar, T., Ekiz, H.I., and Caglar, M.A. *Biotechnol. Bioeng.*, **33**, 1213 (1989).
Valsaraj, K.T., and Thibodeaux, L. *Water Res.*, **23**, 183 (1989).
Van Oss, C.J., Chaudhury, M.K., and Good, R.J. *Sep. Sci. Technol.*, **24**, 15 (1989).
Walter, H., and Johansson, G., (eds.), *Aqueous Two-Phase Systems*, in *Methods in Enzymology*, vol. 228, Academic Press, San Diego, 1994.
Wiebus, E., and Cornils, B. *Chem. Ing. Tech.*, **66**, 916 (1994); *Chemtech.*, **25**, 33 (1995).
Willauer, H.D., Huddleston, J.G., Griffin, s.T., and Rogers, R.D. *Sep. Sci. Technol.*, 1999 (in press).
Woodley, J.M., Brazier, A.J., and Lilly, M.D. *Biotechnol. Bioeng.*, **37**, 133 (1991).
Yamane, T., Nakatani, H., Sada, E., Omata, T., Tanaka, A., and Fukui, S. *Biotechnol. Bioeng.*, **21**, 2133 (1979).
Yokozeki, K., Yamanaka, S., Takinami, K., Hirose, Y., Tanaka, A., Sonomoto, K., and Fukui, S. *Eur. J. Appl. Microb. Biotech.*, **14**, 1 (1982).
Zaslavsky, B. Yu., Gulaeva, N.D., Djafarov, S., Masimov, E.A., and Mihheva, L.M. *J. Colloid Interface Sci.*, **137**, 147 (1990).

CHAPTER 19

Akelah, A., Rehab, A., Selim, A., and Agag, T. *J. Mol. Catal.*, **94**, 311 (1994).
Almerico, A.M., Cirrincione, G., Aiellio, E., and Daltolo, G. *J. Heter. Chem.*, **26**, 1631 (1989).
Angeletti, E., Tundo, P., Venturello, P., and Trotta, F. *British Polym. J.*, **16**, 219 (1984).
Arrad, O., and Sasson, Y. *J. Org. Chem.*, **54**, 4493 (1989); **55**, 2952 (1990); *J. Chem. Soc., Perkins Trans.*, **2**, 457 (1991).
Asai, S., Nakamura, H., and Furuichi, Y. *J. Chem. Eng. Jpn.*, **24**, 653 (1991); *AIChE J.*, **38**, 397 (1992).
Ayyangar, N.R., Madan Kumar, S., and Srinivasan, K.V. *Synthesis*, 616 (1987).
Balakrishnan, T., and Jayachandran, J.P. *J. Chem. Soc., Perkins Trans.*, **2**, 2081 (1995).
Bhattacharya, A. *Ind. Eng. Chem. Res.*, **35**, 645 (1996).
Bhattacharya, A., Dolling, U., Grabowski, E.J., Krady, S., Ryan, K.M., and Weinstock, L.M. *Angew. Chem., Int. Ed. Engl.*, **25**, 476 (1986).
Brändström, A. *Adv. Phys. Org. Chem.*, **15**, 267 (1977).
Brunelle, D.J. In *Phase-Transfer Catalysis: New Chemistry, Catalysts, and Applications* (ed., Starks, C.M.), American Chemical Society Symposium Series No. 326, American Chemical Society, Washington, D.C., 1987.
Cassar, L., Foa, M., and Gardano, A. *J. Organomet. Chem.*, **121**, C55 (1976).
Chen, C.T., Hwang, C., and Yeh, M.Y. *J. Chem. Eng. Jpn.*, **24**, 284 (1991).
Cussler, E.L. *Diffusion: Mass Transfer in Fluid Systems*, Cambridge, New York, 1997.
Cutie, Z.G., and Halpern, M. U.S. Patent 5,120,846 (Chem. Abstr. 117:90510c), 1992.
Dehmlow, E.V., and Dehmlow, S.S. *Phase Transfer Catalysis*, 3rd ed., Verlag Chemie, Weinheim, 1993.
Deng, Y.N., Li, D., and Xu, H.S. *Chin. Sci. Bull.* (CA 114:23571y), **34**, 203 (1989).

Desikan, S., Ph.D. thesis, *Studies in Immobilized Phase-Transfer Catalysis*, Iowa State University, Ames. Iowa, 1997.
Desikan, S., and Doraiswamy, L.K. *Ind. Eng. Chem. Res.*, **34**, 3524 (1995); **38**, 2634 (1999); *Chem. Eng. Sci.* (2000), in press.
Dolling, U.H. U.S. Patent 4,605,761 (Chem. Abstr. 106:4697n) (1986).
Dolling, U.H., Hughes, D.L., Bhattacharya, A., Ryan, K.M., Karady, S., Weinstock, L.M., Grenda, V.J., and Grabowski, E.J.J. In *Phase-Transfer Catalysis: New Chemistry, Catalysts, and Applications* (Ed., Starks, C.M.), ACS Symposium Series (326), American Chemical Society, Washington D.C., 1987.
Esikova, I.A., and Yufit, S.S. *J. Phys. Org. Chem.*, **4**, 149 (1991a); **4**, 336 (1991b).
Evans, K.J., and Palmer, H.J. *AIChE Symp. Ser.*, **77**(202), 104 (1981).
Ford, W.T. *Chemtech*, **436** (1984).
Ford, W.T., Lee, J., and Tomoi, M. *Macromolecules*, **15**, 1246 (1982).
Ford, W.T., and Tomoi, M. *Adv. Polym. Sci.*, **55**, 49 (1984).
Frechet, J.M.J. In *Crown Ethers and Phase Transfer Catalysis in Polymer Science* (Eds., Mathias, L.J., and Carraher, C.E., Jr.), Plenum, New York, 1984.
Freedman, H.H. *Pure Appl. Chem.*, **58**, 857 (1986).
Fujita, F., Itaya, N., Kishida, H., and Takemoto, I. (Sumitomo Chemical Co. Ltd.), Eur. Pat. Appl. 3835 (Chem. Abstr. 92:110707t), U.S. Patent 4,328,166 (1982).
Glatzer, H.J. Triphase Catalysis: Mass Transfer and Kinetic Studies Ph.D. Thesis, Iowa State University, Ames, Iowa, 1999.
Glatzer, H.J., and Doraiswamy, L.K. *Chem. Eng. Sci.* (2000a,b), in press.
Glatzer, H.J., Desikan, S., and Doraiswamy, L.K. *Chem. Eng. Sci.*, **53**(13), 2431 (1998).
Goldberg, Y. *Phase Transfer Catalysis, Selected Problems and Applications*, Gordon and Breach, Philadelphia, 1992.
Hagenson, L.C., Naik, S.D., and Doraiswamy, L.K. *Chem. Eng. Sci.*, **49**, 4787, (1994).
Hammerschmidt, W.W., and Richarz, W. *Ind. Eng. Chem. Res.*, **30**, 82 (1991).
Hradil, J., and Svec, F. *Poly. Bull.*, **11**, 159 (1984).
Hradil, J., Svec, F., and Frechet, J.M.J. *Polymer*, **28**, 1593 (1987).
Hradil, J., Svec, F., Konak, C., and Jurek, K. *Reactive Polym.*, **9**, 81 (1988).
Ido, T., Yamaguchi, K., Itoh, H., and Goto, S. *Kagaku Kogaku Ronbushu*, **21**, 715 (1995a).
Ido, T., Yamaguchi, K., Yamamoto, T., and Goto, S. *Kagaku Kogaku Ronbushu*, **21**, 804 (1995b).
Idoux, J.P., and Gupton, J.T. In *Phase-Transfer Catalysis: New Chemistry, Catalysts, and Applications* (Ed., Starks, C.M.), *Am. Chem. Soc. Symp. Ser.* No. 326, Washington, D.C., 1987.
Jeffery, T. *Synth. Commun.*, **18**, 77 (1988).
Keller, W.E. *Compendium of Phase Transfer Reactions and Related Synthetic Methods*, Fluka, Switzerland, 1979; *Phase Transfer Reactions*, Fluka Compendium, Vol. 1, Georg Thieme Verlag, 1986.
Kimura, Y., and Regen, S.L. *J. Org. Chem.*, **48**(3), 385 (1983).
Kise, H., Araki, K., and Seno, M. *Tetrahedron Lett.*, **22**, 1017 (1981).
Koch, M., and Magni, A. (Gruppo Lepetit), U.S. Patent 4,501,919 (Chem. Abstr. 102:204296k) (1985).
Kondo, S., Takesue, M., Suzuki, M., Kunisada, H., and Yuki, Y. *J. M. S. - Pure Appl. Chem.*, **A31**, 2033 (1994).
Lele, S.S., Bhave, R.R., and Sharma, M.M. *Chem. Eng. Sci.*, **38**, 765 (1983).
Lin, C.L., and Pinnavaia, T.J. *Chem. Mater.*, **3**, 213 (1991).
Lindblom, L., and Elander, M. *Pharmaceutical Tech.*, **4**, 59 (1980).

Liotta, C.L., Burgess, E.M., Ray, C.C., Black, E.D., and Fair, B.E., *Am. Chem. Soc.* Ser. No. 326, Washington, D.C., 15, (1987).
Makosza, M. *Pure Appl. Chem.*, **43**, 439 (1975).
McKenzie, W.M., and Sherrington, D.C. *Polymer*, **22**, 431 (1981).
Melville, J.B., and Goddard, J.D. *Chem. Eng. Sci.*, **40**, 2207 (1985); *Ind. Eng. Chem. Res.*, **27**, 551 (1988).
Naik, S.D., and Doraiswamy, L.K. *Chem. Eng. Sci.*, **52**, 4533 (1997); *AIChE J.*, **44**, 612 (1998).
Neef, G., and Steinmeyer, A. *Tetrahedron Lett.*, 5073 (1991).
Neumann, R., and Sasson, Y. *J. Org. Chem.*, **49**, 3448, (1984).
Norwicki, J., and Gora, J. *Pol. J. Chem.*, **65**, 2267 (1991).
O'Donnell, M.J. In *Catalytic Asymmetric Synthesis* (Ed., Ojima, I.), VCH, New York, 1993.
O'Donnell, M.J., Bennet, W.D., and Wu, S. *J. Am. Chem. Soc.*, **111**, 2353 (1989).
Oldshue, J.Y. *Fluid Mixing Technology*, McGraw-Hill, New York, 1983.
Pielichowski, J., and Czub, P. *Synth. Commun.*, **25**, 3647 (1995).
Pradhan, N.C., and Sharma, M.M. *Ind. Eng. Chem. Res.*, **31**, 1610 (1992).
Quici, S., and Regen, S.L. *J. Org. Chem.*, **44**, 3436 (1979).
Reinick, A., and Sheldon, R.A. U.S. Patent 4,175,094 (1983).
Reuben, B., and Sjoberg, K. *Chemtech*, **315** (1981).
Ruckenstein, E., and Hong, L. *J. Catal.*, **136**, 378 (1992).
Ruckenstein, E., and Park, J.S. *J. Poly. Sci., Part C: Polym. Lett.*, **26**, 529 (1988).
Sasson, Y., and Neumann, R. Eds. *Handbook of Phase Transfer Catalysis*, Blackie Academic of Professional, London, 1997.
Satrio, J.A.B., Glatzer, H.J., and Doraiswamy, L.K. *Chem. Eng. Sci.* (2000), in press.
Schmolka, S.J., and Zimmer, H. *Synthesis*, **29** (1984).
Shan, Y., Kang, R.H., and Li, W. *Ind. Eng. Chem. Res.*, **28**, 1289 (1989).
Sharma, M.M. In *Handbook of Phase Transfer Catalysis* (Eds., Sasson, Y., and Neumann, R.), Blackie Academic of Professional, London, 1997.
Slaoui, S. *Tetrahedron Lett.*, **23**, 1681 (1982).
Smith, L.R. U.S. Patent 4,962,212 (Chem. Abstr. 114:143116a) (1990).
Sprecker, M.A., and Hanna, M.R. Br. U.K. Patent Appl. GBB2139222, U.S. Appl. 487045 (Chem. Abstr. 102:P220567e) (1982).
Starks, C.M. *J. Amer. Chem. Soc.*, **93**, 195 (1971).
Starks, C.M., Liotta, C.L., and Halpern, M. *Phase-Transfer Catalysis: Fundamentals, Applications, and Industrial Perspectives*, Chapman and Hall, New York, 1994.
Svec, F. *Pure Appl. Chem.*, **60**, 377 (1988).
Tamhankar, S.S., Gokarn, A.N., and Doraiswamy, L.K. *Chem. Eng. Sci.*, **36**, 1365 (1981).
Tomoi, M., and Ford, W.T. *J. Am. Chem. Soc.*, **103**, 3821 (1981).
Tomoi, M., Hosokawa, Y., and Kakiuchi, H. *Macromol. Chem. Rapid Commun.*, **4**, 227 (1983).
Tundo, P., and Badiali, M. *Reactive Poly.*, **10**, 55 (1995).
Tundo, P., and Selva, M. *Chemtech*, **25**(5), 31 (1995).
Tundo, P., and Venturello, P. *J. Am. Chem. Soc.*, **101**, 6606, (1979); **103**, 856 (1981).
Tundo, P., Moraglio, G., and Trotta, F. *Ind. Eng. Chem. Res.*, **28**, 881 (1989).
Tundo, P., Venturello, P., and Angeletti, E. *J. Am. Chem. Soc.*, **104**, 6551 (1982).
Wang, D.H., and Weng, H.S. *Chem. Eng. Sci.*, **43**, 209 (1988).
Wang, M.L., and Chang, S.W. *Can. J. Chem. Eng.*, **69**, 340 (1991a); *Ind. Eng. Chem. Res.*, **30**, 2378 (1991b); *Ind. Eng. Chem. Res.*, **33**, 1606 (1994); **34**, 3696 (1995).

Wang, M.L., and Wu, H.S. *J. Org. Chem.*, **55**, 2344 (1990); *Chem. Eng. Sci.*, **46**, 509 (1991).
Wang, M.L., and Yang, H.M. *Ind. Eng. Chem. Res.*, **29**, 522 (1990); *Chem. Eng. Sci.*, **46**, 619 (1991a); *Ind. Eng. Chem. Res.*, **30**, 2384 (1991b); *Ind. Eng. Chem. Res.*, **31**, 1868 (1992).
Watanabe, Y., and Mukiyama, T. *Chem. Lett.*, 285 (1981a); 443 (1981b).
Weber, W.P., and Gokel, G.W. In *Reactivity and Structure Concepts in Organic Chemistry*, 4, Springer Verlag, Berlin, 1977.
Wu, H.S. *Ind. Eng. Chem. Res.*, **32**, 1323 (1993); *Chem. Eng. Sci.*, **51**, 827 (1996).
Yang, H.-M. *Ind. Eng. Chem. Res.*, **37**, 398 (1998).
Yoshikawa, H., Shono, T., and Eto, M. *Chem. Abstr.*, **103**, 87984 R (1984).
Yufit, S.S. *Russ. Chem. Bull.*, **44**, 1989 (1995).

CHAPTER 20

Adams, M.W.W., Perler, F.B., and Kelly, R.M. *Biotechnology*, **13**, 662 (1995).
Arbige, M.V., and Pitcher, W.H. *Trends Biotechnol.*, **7**, 330 (1989).
Arima, K. In *Global Impacts of Applied Microbiology* (Ed., Starr, M.P.), Wiley, 1965.
Armstrong, D.W., and Yamazaki, H. *Trends Biotechnol.*, **4**, 264 (1986).
Asano, Y., Yasuda, T., Tani, Y., and Yamada, H. *Agric. Biol. Chem.*, **46**, 1183 (1982).
Atkinson, B., and Lester, D.E. *Biotech. Bioeng.*, **16**, 1321 (1974).
Bailey, J.E., and Ollis, D.F. *Biochemical Engineering Fundamentals*, McGraw-Hill, New York, 1986.
Blanch, H.W., and Dunn, I.J. *Adv. Biochem. Eng.*, **3**, 127 (1974).
Blumentals, I.I., Robinson, A.S., and Kelly, R.M. *Appl. Environ. Microbiol.*, **56**, 1992 (1990).
Brams, W.H., and McLaven, A.D. *Soil Biol. Biochem.*, **6**, 183 (1974).
Bryant, F.O., and Adams, M.W.W. *J. Bio. Chem.*, **264**, 5070 (1989).
Buchholz, K., and Gödelmann, B. *Biotechnol. Bioeng.*, **20**, 1201 (1978).
Butler, L.G. *Enzyme Microb. Technol.*, **1**, 253 (1979).
Capewell, A., Wendel, V., Bornscheuer, U., Meyer, H.H., and Scheper, T. *Enzyme Microb. Technol.*, **19**, 181 (1996).
Carleysmith, S.W., Eanes, M.B.L., and Lilly, M.D. *Biotech. Bioeng.* **22**, 957 (1980).
Chance, B. *J. Biol. Chem.*, **151**, 553 (1943).
Cheetham, P.S.J. *Enzyme Microb. Technol.*, **9**, 194 (1987).
Chen, C.-S., Fujimoto, Y., Girdaukas, G., and Sih, C.J. *J. Am. Chem. Soc.*, **104**, 7294 (1982).
Chibata, I. *Immobilized Enzymes*, Halstead Press, New York, 1978; In *Biotechnology*, Vol. 7a (Ed. Kennedy, J.F.), VCH, New York, 1987.
Chibata, I., Fukui, S., and Wingard, L.B. Eds. *Enzyme Engineering*, Plenum, New York, 1982.
Cho, Y.K., and Bailey, J.E. *Biotechnol. Bioeng.*, **20**, 1651 (1978).
Constantinides, A., Bhatia, D., and Vieth, W.R. *Biotechnol. Bioeng.*, **23**, 899 (1981).
Constantino, H.R., Brown, S.H., and Kelly, R.M. *J. Bacteriol.*, **172**, 3654 (1990).
Dabulis, K., and Klibanov, A.M. *Biotechnol. Bioeng.*, **41**, 566 (1993).
Dahod, S.K., and Empie, M.W. In *Biocatalysis in Organic Media* (Eds., Laane, C., Tramper, J., and Lilly, M.D.), Elsevier, Amsterdam, 1986.
Davail, S., Feller, G., Narinx, E., and Gerday, C. *J. Bio. Chem.*, **269**, 7448 (1994).
Davidson, B., Veith, W.R., Wang, S.S., Zwiebel, S., and Gilmore, R. *AIChE Symp. Ser.*, **144**, 182 (1974).

Dixon, M., and Webb, E.D. *Enzymes*, 3rd ed., Academic Press, New York, 1984.
Dordick, J.S. *Enz. Microb. Technol.*, **11**, 194 (1989); In *Biocatalysts for Industry* (Ed., Dordick, J.S.), Plenum, New York, 1991; *Trends in Biotechnol.*, **10**, 287 (1992).
Dordick, J.S., Marletta, M.A., and Klibanov, A.M. *Biotechnol. Bioeng.*, **30**, 31 (1987).
Doraiswamy, L.K., and Sharma, M.M. *Heterogeneous Reactions: Analysis, Examples and Reactor Design*, Vol. 1: *Gas-Solid and Solid-Solid Reactions*, Wiley, New York, 1984.
Dunn, I.J., Heinzle, E., Ingham, J., and Prenosil, J.E. *Biological Reaction Engineering: Principles, Applications, and Modelling with PC Simulation*, VCH, New York, 1992.
Engasser, J.M., and Horvath, C. *Biochemistry*, **13**, 3845 (1974).
Fukumura, T. Japan Patent 74-15795 (1974).
Fusee, M.C., Swann, W.E., and Calton, G.J. *Appl. Environ. Microbiol.*, **42**, 672 (1981).
Godfrey, T., and Reichelt, J. *Ind. Enzymol.*, The Nature Press, New York, 1983.
Goldman, R., and Katchalski, E. *J. Theor. Biol.*, **32**, 243 (1971).
Goldstein, W.E. In *Basic Biotechnology No. 9* (Eds., Bu'Lock, J., and Kristiansen, B.), Academic Press, London, 1987.
Govardhan, C.P., and Margolin, A.L. *Chem. Ind.*, **17**, 689 (1995).
Greenfield, P.F., Kittrell, J.R., and Laurence, R.L. *Anal. Biochem.*, **65**, 109 (1975).
Halling, P.J. *Biotechnol. Adv.*, **5**, 47 (1987).
Hamilton, B.K., Hsiao, H.-Y., Swann, W.E., Anderson, D.M., and Delente, J.J. *Trends Biotechnol.*, **3**, 64 (1985).
Herbert, D. *Symp. Soc. Gen. Microb.*, **11**, 391 (1961).
Horvath C., and Engasser, J.M. *Biotechnol. Bioeng.*, **16**, 909 (1974).
Huber, R., Spinler, C., Gambarcorta, A., and Stetter, K.O. *Syst. Appl. Microbiol.*, **12**, 32 (1989).
Huneke, F.U., and Flaschel, E. *J. Chem. Technol. Biotechnol.*, **71**(3), 213 (1998).
Inouye, K., Watanabe, K., Morihara, K., Tochino, Y., Kanaya, T., Emura, J., and Sakakibara, S. *J. Am. Chem. Soc.*, **101**, 751 (1979).
Jandel, A.-S., Husted, H., and Wandrey, C. *Eur. J. Appl. Microbiol. Biotechnol.*, **15**, 59 (1982).
Johnson, F.H., Eyring, H., and Plissar, M.J. *The Kinetic Basis of Molecular Biology*, Wiley, New York, 1954.
Jones, W.J., Leigh, J.A., Mayer, F., Woese, C.R., and Wolfe, R.S. *Arch. Microbiol.*, **132**, 254 (1983).
Kahlon, S.S., and Malhotra, S. *Enzyme Microb. Technol.*, **8**, 613 (1986).
Karanth, N.G., and Bailey, J.E. *Biotechnol. Bioeng.*, **20**, 1817 (1978).
Kegen, S.W.M., Luesink, E.J., and Stams, A.J.B. *Eur. J. Biochem.*, **213**, 305 (1993).
Kelly, R.M., and Deming, J.W. *Biotechnol. Prog.*, **4**, 47 (1988).
Khmelnitsky, Y.L., Welsh, S.H., Clark, D.S., and Dordick, J.S. *J. Am. Chem. Soc.*, **116**, 2647 (1994).
Kinoshita, S., Muranaka, M., and Okada, H. *J. Ferment. Technol.*, **53**, 223 (1975).
Klibanov, A.M. *Chemtech*, **16**, 354 (1986).
Kushner, D.J. In *Microbial Life in Extreme Environments* (Ed., Kushner, D.J.), Academic Press, London, 1978.
Laderman, K.A., Davis, B.R., Krutzch, H.C., Lewis, M.S., Griko, Y.V., Privalov, P.L., and Anfinsen, C.B. *J. Biol. Chem.*, **268**, 4394 (1993).
Lee, G.K., and Reilly, P.J. *Chem. Eng. Sci.*, **36**, 1967 (1981).
Lee, H.K., and Maddox, I.S. *Enzyme Microb. Technol.*, **8**, 409 (1986).
Lee, S.B., Kim, S.M., and Ryu, D.D.Y. *Biotechnol. Bioeng.*, **21**, 2023 (1979).
Levenspiel, O. *The Chemical Reactor Omnibook*, Oregon State University Bookstore, Corvallis, Oe., 1993.

Li, Y., Mandelco, L., and Wiegel, J. *Int. J. Sysm. Bacteriol.*, **43**, 450 (1993).
Lilly, M.D. *J. Chem. Tech. Biotechnol.*, **32**, 162 (1982).
Lilly, M.D., Hornby, W.E., and Crook, E.M. *Biochem. J.*, **100**, 718 (1968).
Lin, S.H. *Biophys. Chem.*, **7**, 229 (1977).
Margolin, A.L. *Chemtech*, **21**, 160 (1991).
Marsh, D.R., Lee, Y.Y., and Tsao, G.T. *Biotechnol. Bioeng.*, **15**, 483 (1973).
Matson, S.L., and Quinn, J.A. *Ann. N.Y. Acad. Sci.*, **469**, 152 (1986).
Messing, R.A. *Immobilized Enzymes for Industrial Reactors*, Academic Press, New York, 1975.
Mitarai, H., and Kawakami, K. *J. Chem. Technol. Biotechnol.*, **60**, 375 (1994).
Miyakawa, H., and Shiraishi, F. *J. Chem. Technol. Biotechnol.*, **69**, 456 (1997).
Mogensen, A.O., and Vieth, W.R. *Biotechnol. Bioeng.*, **15**, 467 (1973).
Monod, J. *Annales de l'Institut Pasteur*, **79**, 390 (1950).
Musso, H. In *Oxidative Coupling of Phenols* (Eds., Taylor, W.I., and Battersby, A.R.), Dekker, New York, 1967.
Mustranta, A. *Appl. Microbiol. Biotechnol.*, **38**, 61 (1992).
Nagasawa, T., and Yamada, H. *Trends Biotechnol.*, **7**, 153 (1989).
Nakai, K., Watanabe, I., Sato, Y., and Enomoto, K. *Nippon Nogeikagaku Kaishi*, **62**, 1443 (1988).
Nakumara, K., Aizawa, M., and Miyawaki, O. *Electroenzymology, Coenzyme Regeneration*, Springer Verlag, New York, 1988.
Neuberger, A., and Brocklehurst, K., Eds. *Hydrolytic Enzymes*, Elsevier, Amsterdam, 1987.
Nielsen, J.H., and Villadsen, J. *Bioreaction Engineering Principles*, Plenum, New York, 1994.
Novick, A., and Szillard, L. *Science*, **112**, 715 (1950).
Ollis, D.F. *Biotechnol. Bioeng.*, **14**, 871 (1972).
Ottolina, G., Carrea, G., Riva, S., Sartore, L., and Veronese, F.M. *Biotechnol. Lett.*, **14**, 947 (1992).
Patwardhan, V.S., and Karanth, N.G. *Biotechnol. Bioeng.*, **24**, 763 (1982).
Peppler, H.J., and Reed, G. In *Biotechnology*, Vol. 7a (Ed. Kennedy, J.F.), VCH, New York, 1987.
Peter, G. Ed. *Enzyme Engineering: Immobilized Biosystems*, Ellis Harwood, New York, 1992.
Pirt, S.J., and Kurowski, W.M. *J. Gen. Micro.*, **63**, 357 (1970).
Posorske, L.H. *J. Am. Oil Chem. Soc.*, **61**, 1758 (1984).
Rakels, J.L.L., Borein, B., Straathof, A.J.J., and Heijnen, J.J. *Biotechnol. Bioeng.*, **43**, 411 (1994).
Randolph, T.W., Blanch, H.W., and Prausnitz, J.M. *AIChE J.*, **34**, 1354 (1988a).
Randolph, T.W., Clark, D.S., Blanch, H.W., and Prausnitz, J.M. *Science*, **238**, 387 (1988b).
Reilly, P.J., and Lee, G.K. *Chem. Eng. Commun.*, **12**, 195 (1981).
Rethwisch, D.G., Subramanian, A., Yi, G., and Dordick, J.S. *J. Am. Chem. Soc.*, **112**, 1649 (1990).
Rovito, B.J., and Kittrell, J.R. *Biotechnol. Bioeng.*, **15**, 143 (1973).
Sadana, A. Personal communication of solution (worked out with Doraiswamy, L.K.), National Chemical Laboratory, Pune (c1980); *Biocatalysis: Fundamentals of Enzyme Deactivation Kinetics*, Prentice-Hall, Englewood Cliffs, N.J., 1991.
Sadana, A., and Doraiswamy, L.K. *J. Catal.*, **32**, 384 (1971).

Saunders, B.C., Holmes-Siedle, A.G., and Stark, B.P. *Peroxidase*, Butterworth, Washington, D.C., 1964.
Scott, C.D. *Enzyme Microb. Technol.*, **9**, 66 (1987).
Shiraishi, F. *Enzyme Microb. Technol.*, **15**, 150 (1993); *J. Ferment. Bioeng.*, **79**, 373 (1995).
Shuler, M.L., and Kargi, F. *Bioprocess Engineering: Basic Concepts*, Prentice-Hall, Englewood Cliffs, N.J., 1992.
Sigma Chemical Catalog, 1989.
Smith, W.J., Inloes, D.S., Martin, A., Robertson, C.R., and Cohen, S.N. *SIM News 32*, Abst. P30 (1982).
St. Clair, N. L., and Navia, N. M. *J. Am. Chem. Soc.*, **114**, 7314 (1992).
Strika-Alexopoulos, E., and Freedman, R.B. *Biotechnol. Bioeng.*, **41**, 887 (1993).
Swann, W.E. Luxembourg Patent No. 84,920 (1983).
Tanaka, A., Tosa, T., and Kobayashi, T. Eds. *Industrial Application of Immobilized Biocatalysts*, Dekker, New York, 1993.
Toda, K. *Biotechnol. Bioeng.*, **17**, 1729 (1975).
Tramper, J. *Trends Biotechnol.*, **3**, 45 (1985).
Trevan, M.D. *Immobilized Enzymes: An Introduction and Applications in Biotechnology*, Wiley, New York, 1980.
Tsai, S.-W., Liu, B.-Y., and Chang, C.-H. *J. Chem. Tech. Biotechnol.*, **65**, 156 (1996).
Van't Riet, K., and Tramper, J. *Basic Bioreactor Design*, Dekker, New York, 1991.
Vidaluc, J.L., Baboulene, M., Speziale, V., Lattes, A., and Monsan, P. *Tetrahedron*, **39**, 269 (1983).
Vieth, W.R., and Wolf, R. *Bioprocess Engineering: Kinetics, Mass Transport, Reactors, and Gene Expression*, Wiley, New York, 1994.
Walters, -L. *Chem. Ind.*, 412, June 2 (1997).
Watanabe, I. *Methods Enzymol.*, **136**, 523 (1987).
Watanabe, I., and Okumura, M. Japan Patent No. 162,193, 1986.
Watanabe, I., Sato, T., and Kuono, T. Japan Patent No. 129,190, 1979.
Weekman, V.W., Jr. *Ind. Eng. Chem. Proc. Des. Dev.*, **7**, 90 (1968).
Wilson, R.J.H., Kay, G., and Lilly, M.D. *Biochem. J.*, **108**, 845 (1968).
Wiseman, A. Ed. *Handbook of Enzyme Biotechnology*, 2nd ed., Ellis Horwood, New York, 1985.
Wood, L.L., and Calton, G.J., U.S. Patent 4,436,813, 1984.
Woodley, J.M., Cunnah, P.J., and Lilly, M.D. *Biocatalysis*, **5**, 1 (1991a).
Woodley, J.M., Bracier, A.J., and Lilly, M.D. *Biotechnol. Bioeng.*, **37**, 133 (1991b).
Wu, J.-Y. *AIChE J.*, **44**, 474 (1998).
Yamamoto, K., Tosa, T., and Chibata, I. *Biotechnol. Bioeng.*, **22**, 2045 (1980).
Yamane, T., Ichurya, T., Nagata, M., Ueno, A., Shimizu, S. *Biotechnol. Bioeng.*, **36**, 1063 (1990).

CHAPTER 21

Alkire, R.C., and Beck, T.R. Eds. Tutorial Lectures in Electrochemical Engineering and Technology, *AIChE*, **77**, No. 204 (1981).
Alkire, R.C., and Chin, D.-T. Eds. Tutorial Lectures in Electrochemical Engineering and Technology, *AIChE*, **9**, No. 229 (1983).
Baizer, M.M. Ed. *Organic Electrochemistry*, Dekker, New York, 1973.
Baizer, M.M., and Lund, H. Eds. *Organic Electrochemistry: An Introduction and a Guide*, 2nd ed. Dekker, New York, 1983.
Baizer, M.M. *Tetrahedron*, **40**, 935 (1984).

Bagley, M.R., Dutton, M.C., and Dennis, J. U.S.Patent WO 26,389 [Cl.C.25B3/10], to Monsanto, U.S.A., 1997.
Beck, F. *Electroorgansche Chemie: Grundlagen and Anwendungen*, Verlag Chemie, Weinbein, 1974.
Beck, F., Goldacker, H., Kreysa, G., Vogt, H., and Wendt, H. Eds. *Ullmann's Encyclopedia of Industrial Chemistry*, Weinheim, Germany, 1985, p. 240.
Biran, C., and Bourdeau, N. *Synth. Commun.*, **23**, 1727 (1993).
Bockris, J.O.M. *Modern Electrochemistry*, Plenum, New York, 1998.
Bowden, M.E. *Chemistry is Electric!* Chemical Heritage Foundation, Philadelphia, 1997.
Clark, J.M.T., Goodridge, F., and Plimley, R.E. *J. Appl. Electrochem.*, **18**, 899 (1988).
Danly, D.E., and Campbell, C.R. In *Techniques of Electroorganic Synthesis* (Eds., Weinberg, N.L., and Tilak, B.V.), Part III, Wiley, New York, 1982, p. 283.
Danly, D.E. *Chem. Tech.*, **10**, 302 (1980).
Danly, D.E. *Emerging Opportunities for Electroorganic Processes*, Dekker, New York, 1984.
Danly, D.E. *Hydrocarbon Process.*, **60**, No. 4, 141 (1981).
Degner, D. In *Electrosynthesis, From Laboratory, To Pilot, To Production* (Eds., Genders, J.D., and Pletcher, D.), Electrosynthesis Co., New York, 1982.
Drakesmith, F.G., and Hughes, D.A. *J. Fluorine Chem.*, **32**, 103 (1986).
Eberson, L., and Schafer, H. *Organic Electrochemistry*, Springer Verlag, Berlin, 1971.
Fahidy, T.Z. *Principles of Electrochemical Reactor Analysis*, Elsevier, New York, 1985.
Feess, H., and Wendt, H. In *Techniques of Electroorganic Synthesis* (Eds., Weinberg, N.L., and Tilak, B.V.), Part III, Wiley, New York, 1982.
Fry, A.J. *Synthetic Organic Chemistry*, Harper & Row, New York, 1972.
Goodridge, F., and King, C.J.H. *Techniques of Electroorganic Synthesis*, Part III, Wiley, New York, 1982, p. 58.
Goodridge, F., and Scott, K. *Electrochemical Process Engineering*, Plenum, New York, 1995.
Goodridge, F., Mamoor, G.M., and Plimley, R.E. *Inst. Chem. Eng. Symp. Ser.*, No. 98, Electrochemical Engineering, 1986, p. 61.
Hamann, C.H. *Electrochemistry*, Wiley-VCH, New York, 1998.
Heitz, E., and Kreysa, G. *Principles of Electrochemical Engineering*, VCH, New York, 1986.
Hiebl, J., Blanka, M., Guttman, A., Kollmann, H., Leitner, K., Mayrhofer, G., Rovenszky, F., and Winkler, K. *Tetrahedron*, **54**, 2059 (1998).
Hine, F. *Electrode Processes and Electrochemical Engineering*, Plenum, New York, 1985.
Ismail, M.I. Ed., *Electrochemical Reactors: Their Science and Technology*, Elsevier, New York, 1989.
Jansson, R. *Chem. Eng. News*, **62** (Nov. 19), 44 (1984).
Jayaraman, K., Udupa, K.S., and Udupa, H.V.K. *Trans. SAEST*, **12**, 143 (1977).
Kardassis, G., Brunge, P., and Steckhan, E. *Tetrahedron*, **54**, 3471 (1998).
Kaulen, J., and Schafer, H.J. *Tetrahedron*, **38**, 3299 (1982).
Krishnan, V., Muthukumaran, A., Mahalingam, N., and Udupa, H.V.K. *Trans. SAEST*, **14**, 1 (1979).
Kyriacou, D.K. *Basics of Electroorganic Synthesis*, Wiley Interscience, New York, 1981.
Laikhter, A.L. *Tetrahedron Lett.*, **34**, 4465 (1993).
Little, R.D., and Weinberg, N.L. Eds. *Electroorganic Synthesis*, Dekker, New York, 1991.
Newman, J.S. *Electrochemical Systems*, Prentice-Hall, Englewood Cliffs, N.J., 1973.
Nissan Chemicals Ltd., Japan Patent 08,283,975, 1996.

Perry, R.H., and Chilton, C.H. *Chemical Engineers Handbook*, McGraw-Hill, New York, 1976.
Pickett, D.J. *Electrochemical Reactor Design*, Elsevier, New York, 1979.
Pletcher, D. *Industrial Electrochemistry*, Chapman and Hall, New York, 1982.
Pletcher, D., and Walsh, F.C. *Industrial Electrochemistry*, Chapman and Hall, London, 1990.
Prentice, G. *Electrochemical Engineering Principles*, Prentice-Hall, Englewood Cliffs, N.J., 1991.
Rifi, M.R., and Covitz, F.H. *Introduction to Organic Electrochemistry*, Dekker, New York, 1975, 1980.
Rousar, I., Micka, K., and Kimla, A. *Electrochemical Engineering*, Elsevier, New York, 1986.
Saboureau, C. *J. Chem. Soc., Chem. Commun.*, 885 (1989).
Sandoz Ltd. Switzerland, U.S. Patent 5,573,653 [Cl.205], 1996.
Scott, K. *Electrochemical Reaction Engineering*, Academic Press, London, 1991.
Seiler, P., and Robertson, P.M. *Chimia*, **36**, 305 (1982).
Selman, J.R., and Tobias, C.W. *Adv. Chem. Eng.*, **10**, 212 (1978).
Shono, T. *Tetrahedron*, **40**, 811 (1984).
Simons, J.H. U.S. Patent 2,519,983 (1950).
Spanish Patent 9,638,601, 1996.
Subbiah, P., Jayaraman, K., Seshadri, C., Thirunavukkarasu, P., and Udupa, K.S. *Electrochem.*, **4**(2), 149 (1988).
Swann, S., Jr., and Alkire, R. *Bibliography of Electroorganic Syntheses*, Port City Press, Baltimore, MD., 1980.
Tomilov, A.P., and Fioshin, M.Ya. *Brit. Ch. Eng.*, **16**, 154 (1971).
Udupa, H.V.K. *AIChE Symp. Ser.* 185, **75**, 26 (1979).
Weinberg, N.L. Ed. *Technique of Electroorganic Synthesis, in Techniques in Chemistry Series*, Vol. 5 (Ed., Weissberger, A.), Part I, 1974; Part III, Wiley, New York, 1975.
Weinberg, N.L. Tilak, B.V., and Sarangapani, S. Eds. *Techniques of Electroorganic Synthesis, Scale-Up and Engineering Aspects, in Techniques in Chemistry Series* (Ed., Weissberger, A.), Vol. 5, Part III, Wiley, New York, 1982.
Weinberg, N.L. In *Electrosynthesis, From Laboratory, To Pilot, To Production* (Eds., Genders, J.D. and Pletcher, D.), Electrosynthesis Co., New York, 1990, Chapter 5.

CHAPTER 22

Alippi, A., Cataldo, F., and Galbato, A. *Ultrasonics*, **30**(3), 148 (1992).
Ando, T., Fujita, M., Kimura, T., Leveque, J.M., Luche, J.L., and Sohmiya, H. *Ultrasonics Sonochemistry*, **3**, S223 (1996).
Ando, T., and Kimura, T. *Ultrasonics*, **28**, 326 (1990).
Ando, T., Sumi, D., Kawate, T., Ichihara, J., and Terukiyp, H. *J. Chem. Soc., Chem. Commun.*, 439 (1984a).
Apfel, R.E. *J. Acoust. Soc. Am.*, **69**(6), 1624 (1981).
Atchley, A.A., Frizzell, L.A., Apfel, R.E., Holland, C.K., Madanshetty, S., and Roy, R.A. *Ultrasonics*, **26**, 280 (1988).
Berlan, J. *Ultrasonics Sonochemistry*, **1**(2), S97 (1994).
Berlan, J., and Mason, T.J. *Ultrasonics*, **30**(4), 203 (1992).
Booth, J., Compton, R.G., Hill, E., Marken, F., and Rebbitt, T.O. *Ultrasonics Sonochemistry*, **4**, 1 (1997).
Boudjouk, P.R., and Han, B.H. U.S. Patent 4,466,879, 1984.

Bremner, D.H. *Ultrasonics Sonochemistry*, **1**(2), S119 (1994); *Chem. Britain*, **22**, 633 (1986).
Crum, L.A. *Ultrasonics Sonochemistry*, **2**(2), S147 (1995).
Cum, G., Galli, G., Gallo, R., and Spadaro, A. *Il Nuovo Cimento*, **12**(10), 1423 (1990); *Ultrasonics*, **30**(4), 267 (1992).
Cum, G., Gallo, R., and Spadaro, A. *J. Chem. Soc., Perkins Trans II*, 375 (1988).
Dähnke, S., and Keil, F.J. *Chem. Eng. Technol.*, **21**, 873 (1998a); *Ind. Eng. Chem. Res.*, **37**, 848 (1998b); *Chem. Eng. Sci.*, **54**, 2865 (1998).
Dähkne, S., Swamy, K.M., and Keil, F.J. *Ultrasonics Sonochemistry*, **6**, 31 (1999a); 221 (1999b).
de Souza-Barboza, J.C., Petrier, C., and Luche, J.L. *J. Org. Chem.*, **53**, 1212 (1988).
de Souza-Barboza, J.C., Luche, J.L., and Pertrier, C. *Tetrahedron Lett.*, **28**(18), 2013 (1987).
Doktyz, S.J., and Suslick, K.S. *Science*, **247**, 1067 (1990).
Einhorn, C., Einhorn, J., Dickens, M.J., and Luche, J.L. *Tetrahedron Lett.*, **31**(29), 4129, 1990.
Elder, S.A. *J. Acoust. Soc. Am.*, **31**, 54 (1959).
Entezari, M.H., Kruus, P., and Otson, R. *Ultrasonics Sonochemistry*, **4**, 49 (1997).
Entezari, M.H., and Kruus, P. *Ultrasonics Sonochemistry*, **1**(2), S75 (1994); **3**, 19 (1996).
Fuchs, R., and Heusinger, H. *Ultrasonics Sonochemistry*, **2**(2), S105 (1995).
Fuentes, A., Marinas, J.M., and Sinisterra, J.V. *Tetrahedron Lett.*, **28**(39), 4541 (1987).
Geier, G.E. U.S. Patent 4,845,125 1989.
Goodwin, T.J. In *Sonochemistry: The Uses of Ultrasound in Chemistry* (Ed., Mason, T.J.), Royal Society of Chemistry, Cambridge, 1990.
Griffing, V. *J. Chem. Phys.*, **20**(6), 939 (1952).
Hagenson, L.C., and Doraiswamy, L.K. *Chem. Eng. Sci.*, **53**(1), 131 (1998).
Han, B.H., and Boudjouk, P. *J. Org. Chem.*, **47**, 5030 (1982).
Hanafusa, T., Ichihara, J., and Ashida, T. *Chem. Lett.*, 687 1987).
Hart, E.J., and Henglein, A. *J. Phys. Chem.*, **89**, 4342 (1985).
Javed, T., Mason, T.J., Phull, D.D., Baker, N.R., and Robertson, A. *Ultrasonics Sonochemistry*, **2**(1), S3 (1995).
Jouglet, B., Blanco, L., and Rousseau, G. *Synlett*, 907 (1991).
Junger, M.C., and Feit, D. *Sound, Structures and Their Interactions*, MIT Press, London, 1986.
Keil, F.J., and Dähnke, S. *Hungarian J. Ind. Chem.*, **25**, 71 (1997a); *Periodica Polytechnica Ser. Chem. Eng.*, **41**, 41 (1997b).
Keil, F.J., and Swamy, K.M. *Rev. Chem. Eng.*, **15**, 85 (1999).
Klein, K.D., Knott, W., and Koerner, G. U.S. Patent 5,455,367, 1995.
Kondo, T., Krishna, C.M., and Riesz, P. *Int. J. Radiat. Biol.*, **53**(6), 891 (1988).
Lauterborn, W., and Vogel, A. *Annu. Rev. Fluid Mech.*, **16**, 223 (1984).
Leighton, T.G. *Ultrasonics Sonochemistry*, **2**(2), S123 (1995).
Ley, S.V., and Low, C.M.R. *Ultrasound in Synthesis*, Springer Verlag, Berlin, 1989.
Lie Ken Jie, M.S.F. and Lam, C.K. *Ultrasonics Sonochemistry*, **2**(1), S11 (1995).
Lindley, J., Lorimer, J.P., and Mason, T.J. *Ultrasonics*, **25**, 45 (1987).
Lorimer, J.P., Mason, T.J., and Fiddy, K. *Ultrasonics*, **29**(4), 338, 1991.
Low, C.M.R. *Ultrasonics Sonochemistry*, **2**(2), S153 (1995).
Luche, J.L. *Ultrasonics*, **25**, 40 (1987); *Synthetic Organic Sonochemistry*, Plenum, New York, 1998.
Luche, J.L., Einhorn, C., Einhorn, J., de Souza-Barboza, J.C., Petrier, C., Dupuy, C., Delair, P., Allavena, C., and Tuschl, T. *Ultrasonics*, **28**(5), 316 (1990).

Margulis, M.A. *Ultrasonics*, **30**(3), 152 (1992).
Mason, T.J. *Ultrasonics*, **24**, 245 (1986); *Chemistry with Ultrasound*, Critical Reports on Applied Chemistry, Vol. 28, Elsevier Applied Science, 1990a; *Sonochemistry: The Uses of Ultrasound in Chemistry*, Royal Society of Chemistry, Cambridge, 1990b; *Practical Sonochemistry: User's Guide to Applications in Chemistry and Chemical Engineering*, Ellis Horwood, Chichester, England, 1991.
Mason, T.J., and Lorimer, J.P. *Endeavour, New Series*, **13**(3), 123 (1989).
Mason, T.J., Lorimer, J.P., Bates, D.M., and Zhao, Y. *Ultrasonics Sonochemistry*, **1**(2), S91 (1994).
Mason, T.J., Lorimer, J.P., and Walton, D.J. *Ultrasonics*, **28**(5), 333 (1990a).
Mason, T.J., Lorimer, J.P., Paniwnyk, L., Harris, A.R., Wright, P.W., Bram, G., Loupy, A., Ferradou, G., and Sansoulet, J. *Synth. Commun.*, **20**, 3411 (1990b).
Menendez, J.C., Trigo, G.G., and Sollhuber, M.M. *Tetrahedron Lett.*, **27**(28), 3285 (1986).
Mills, A., Li, X., and Meadows, G. *Ultrasonics Sonochemistry*, **2**(1), S39 (1995).
Misik, V., Miyoshi, N., and Riesz, P. *J. Phys. Chem.*, **99**, 3605 (1995).
Misik, V., and Riesz, P. *Ultrasonics Sonochemistry*, **3**, 25 (1996a); S173(1996b).
Moholkar, V.S., and Pandit, A.B. *AIChE J.*, **43**, 1641 (1997); *Ultrasonics Sonochemistry*, **6**, 53 (1999).
Moulton, K.J., Koritala, S., and Frankel, E.N. *J. Am. Oil Chem. Soc.*, **60**(7), 1257 (1983).
Moulton, K.J., Koritala, S., Warner, K., and Frankel, E.N. JAOCS, **64**(4), 542 (1987).
Neppiras, E.A. *Phys. Rep.*, **61**(3), 159 (1980).
Ondrey, G., Kim, I., and Parkinson, G. *Chem. Eng.*, 39 (1996).
Pandit, A.B., and Joshi, J.B. *Chem. Eng. Sci.*, **48**(19), 3440 (1993).
Petrier, C., Jeunet, A., Luche, J.L., and Reverdy, G. *J. Am. Chem. Soc.*, **114**(8), 3148 (1992).
Prasad Naidu, D.V., Rajan, R., Kumar, R., Gadhi, K.S., Arakeri, V.H., and Chandrasekaran, S. *Chem. Eng. Sci.*, **49**(6), 877 (1994).
Price, G.J. *Current Trends in Sonochemistry*, Royal Society of Chemistry, Cambridge, 1992.
Prosperetti, A., and Commander, K.W. *J. Acoust. Soc. Am.*, **85**, 732 (1989).
Ragaini, V. U.S. Patent 5,108,654, 1992.
Ratoarinoro, N., Contamine, F., Wilhelm, A.M., Berlan, J., and Delmas, H. *Chem. Eng. Sci.*, **50**(3), 554 (1995).
Ratoarinoro, N., Wilhelm, A.M., Berlan, J., and Delmas, H. *Chem. Eng. J.*, **50**, 27 (1992).
Repic, O., Lee, P.G., and Giger, U. *Org. Preps. and Proc. Int.*, **16**(1), 25 (1984).
Repic, O., and Vogt, S. *Tetrahedron Lett.*, **23**(27), 2729 (1982).
Richards, W.T., and Loomis, A.L. *J. Am. Chem. Soc.*, **49**, 3086 (1927).
Sehgal, C.M., and Wang, S.Y. *J. Am. Chem. Soc.*, **103**, 6606 (1981).
Shah, Y.T., Moholkar, V.S., and Pandit, A.B. *Cavitation Reaction Engineering*, Plenum, New York, 1999.
Soudagar, S.R., and Samant, S.D. *Ultrasonics Sonochemistry*, **2**(1), S15 (1995).
Suslick, K.S. *Adv. Organometallic Chem.*, **25**, 73 (1986); *Ultrasound: Its Chemical, Physical, and Biological Effects*, VCH, New York, 1988; *Scientific American*, **260**, 80 (1989); *Science*, **247**, 1439 (1990a); *New Scientist*, **125**, 50 (1990b).
Suslick, K.S., and Casadonte, D.J. *J. Am. Chem. Soc.*, **109**, 3459 (1987).
Suslick, K.S., Casadonte, D.J., Green, M.L.H., and Thompson, M.E. *Ultrasonics*, **25**, 56 (1987).
Suslick, K.S., Gawlenowski, J.J., Schubert P.F., and Wang, H.H. *J. Phys. Chem.*, **87**, 2299 (1983a).

Suslick, K.S., Goodale, J.W., Schubert, P.F., and Wang, H.H. *J. Am. Chem. Soc.*, **105**, 5781 (1983b).
Suslick, K.S., and Hammerton, D.A. *IEEE Trans. Ultrasonics, Ferroelectrics and Frequency Control*, UFFC-33(2), 143 (1986).
Tatsumoto, N., and Fujii, S. *J. Acoust. Soc. Jpn.*, **E8**(5), 191 (1987).
Thompson, L.H., and Doraiswamy, L.K. *Ind. Eng. Chem. Res.*, **38**, 1215 (1999); *Chem. Eng. Sci.*, **55**, 3085 (2000).
Tuulmets, A., Kaubi, K., and Heinoja, K. *Ultrasonics Sonochemistry*, **2**(2), S75 (1995).
Vogel, A., Lauterborn, W., and Timm, R. *J. Fluid Mech.*, **206**, 299 (1989).
Whillock, G.O.H., and Harvey, B.F. *Ultrasonics Sonochemistry*, **4**, 23 (1997).
Worsley, D., and Mills, A. *Ultrasonics Sonochemistry*, **3**, S119 (1996).
Yan, Y., Thorpe, R.B., and Pandit, A.B. In *Proceedings of the International Symposium on Flow Induced Vibration and Noise*, Chicago, ASME, New York, 1988.
Ye, Z., and Ding, L. *J. Acoust. Soc. Am.*, **98**, 1629 (1995).
Yount, D.E., Gillary, E.W., and Hoffman, D.C. *J. Acoust. Soc. Am.*, **76**, 1511 (1984).

CHAPTER 23

Alper, E., and Deckwer, W.D. *Chem. Eng. Sci.*, **36**, 1097 (1981).
Alper, E., Wichtendahl, B., and Deckwer, W.D. *Chem. Eng. Sci.*, **35**, 217 (1980).
Anderson, J.G. Microphase-Assisted Autocatalysis, Ph.D. thesis, Iowa State University, Ames, Iowa, 1996.
Anderson, J.G., Doraiswamy, L.K., and Larson, M.A. *Chem. Eng. Sci.*, **53**, 2451 (1998a).
Anderson, J.G., Larson, M.A., and Doraiswamy, L.K. *Chem. Eng. Sci.*, **53**, 2459 (1998b).
Astarita, G. *Mass Transfer with Chemical Reaction*, Elsevier, Amsterdam, 1967.
Bhagwat, S.S., Joshi, J.B., and Sharma, M.M. *Chem. Eng. Commun.*, **58**, 311 (1987).
Bhaskarwar, A.N., and Kumar, R. *Chem. Eng. Sci.*, **41**, 399 (1986).
Bruining, W.J., Joosten, G.E.H., Beenackers, A.A.C.M., and Hofman, H. *Chem. Eng. Sci.*, **41**, 1873 (1986).
Doraiswamy, L.K., and Sharma, M.M. *Heterogeneous Reactions: Analysis, Examples and Reactor Design*, Vol. 2: *Fluid-Fluid and Fluid-Fluid-Solid Reactions*, Wiley, New York, 1984.
Hagenson, L.C., Naik, S.D., and Doraiswamy, L.K. *Chem. Eng. Sci.*, **49**, 4787 (1994).
Holstvoogd, R.D., Ptasinski, K.J., and van Swaaij, W.P.M. *Chem. Eng. Sci.*, **41**, 1943 (1986).
Holstvoogd, R.D., van Swaaij, W.P.M., and Dierendonck, L.L., *Chem. Eng. Sci.*, **43**, 2181 (1988).
Janakiraman, B., and Sharma, M.M. *Chem. Eng. Sci.*, **40**, 235 (1985).
Karve, S., and Juvekar, V.A. *Chem. Eng. Sci.*, **45**, 587 (1990).
Marsh, J.F. *Chem. Ind.*, **14**, 470 (1987).
Mehra, A. In *Handbook of Heat and Mass Transfer* (Ed., Cheremissingoff, N.P.), Vol. 4, Chap. 16, p. 677, Gulf, Houston, Tex, 1990a; *Chem. Eng. Sci.*, **45**, 1525 (1990b); *Chem. Eng. Sci.*, **51**, 461 (1996).
Mehra, A., Pandit, A.B., and Sharma, M.M. *Chem. Eng. Sci.*, **43**, 913 (1988).
Mehra, A., and Sharma, M.M. *Chem. Eng. Sci.*, **40**, 2382 (1985); *Chem. Eng. Sci.*, **43**, 2541 (1988a); *Chem. Eng. Sci.*, **43**, 1071 (1988b).
Nagy, E., and Moser, A. *AIChE J.*, **41**, 23 (1995).
Pal, S.K., Juvekar, V.A., and Sharma, M.M. *Chem. Eng. Sci.*, **37**, 327 (1982).
Pradhan, N.C., Mehra, A., and Sharma, M.M. *Chem. Eng. Sci.*, **47**, 493 (1992).
Ramachandran, P.A., and Sharma, M.M. *Chem. Eng. Sci.*, **24**, 1681 (1969).

Randolph, A.D., and Larson, M.A. *Theory of Particulate Processes: Analysis and Techniques of Continuous Crystallization*, 2nd ed., Academic Press, New York, 1988.
Sada, E., and Kumazawa, H. *Chem. Eng. Sci.*, **37**, 945 (1982).
Sada, E., Kumazawa, H., and Butt, M.A. *Chem. Eng. Sci.*, **32**, 1165 (1977a); 1499 (1977b); **35**, 771 (1980).
Saraph, V.S., and Mehra, A. *Chem. Eng. Sci.*, **49**, 949 (1994).
Uchida, S., Koide, K., and Shindo, M. *Chem. Eng. Sci.*, **30**, 644 (1975).
Wimmers, O.J., Paulussen, R., Kermeulen, D.P., and Fortuin, J.M.H. *Chem. Eng. Sci.*, **39**, 1415 (1984).
Yagi, H., and Hikita, H. *Chem. Eng. J. Biochem. Eng.* **36**, 169 (1987).

CHAPTER 24

Abdalla, B.K., and Elnashaie, S.S.E.H. *Membr. Sci.*, **101**, 31 (1995).
Adris, A.M., Elnashaie, S.S.E.H., and Hughes, R. *Can. J. Chem. Eng.*, **69**, 1061 (1991).
Agarwalla, S., and Lund, C.R.F. *J. Membr. Sci.*, **70**, 129 (1992).
Armor, J.N. *Appl. Catal.*, **49**, 1 (1989).
Basov, N.L., and Gryaznov, V.M. *Membr. Katal.*, 117 (1985).
Bhave, R.R. *Inorganic Membranes Synthesis, Characteristics and Applications*, Van Nostrand Reinhold, New York, 1991.
Champagnie, A.M., Tsotsis, T.T., Minet, R.C., and Wagner, E. *J. Catal.*, **134**, 713 (1992).
Cini, P., and Harold, M.P. *AIChE J.*, **37**, 997 (1991).
Cini, P., Blaha, S.R., Harold, M.P., and Venketeraman, K. *J. Membr. Sci.*, **55**, 199 (1991).
Doraiswamy, L.K., Krishnan, G.R.V., and Sadasivan, N. National Chemical Laboratory (India) Report, 1975.
Elnashaie, S.S.E.H., Abdalla, B.K. and Hughes, R. *Ind. Eng. Chem. Res.*, **32**, 2537 (1993).
Elnashaie, S.S.E.H., and Adris, A.M. In *Fluidization VI* (Grace, G.R., Shemilt, L.W., and Bergougnou, M.A., Eds.), Engineering Foundation, New York, 1989.
Elnashaie, S.S.E.H., Wagialla, K.M., and Helal, A.M. *Math. Comput. Modelling*, **15**, 43 (1991).
Elnashaie, S.S.E.H., Wagialla, K.M., and Abashar, M.E.E. *Math. Comput. Modelling*, **16**, 35 (1992).
Eyal, A.M., and Bressler, E. *Biotech. Bioeng.*, **41**(3), 287 (1993).
Gavalas, G.R., Megiris, C.E., and Nam, S.W. *Chem. Eng. Sci.*, **44**, 1829 (1989).
Gellings, P.J., and Bouwmeister, H.J.M. *Catal. Today*, **12**, 1 (1992).
Gobina, E., and Hughes, R. *Appl. Catal. A*, **137**, 119 (1996).
Govind, R., and Ilias, S. *AIChE Symp. Ser.*, **85**(268), 18 (1989).
Govind, R., and Mohan, K. *AIChE J.*, **34**(9), 1493 (1988).
Gryaznov, V.M. *Plat. Met. Rev.*, **30**(2), 68 (1986); *Plat. Met. Rev.*, **36**(2), 70 (1992).
Gryaznov, V.M., and Karvanov, A.N. *Khim.-Farm. Zh.*, **13**, 74 (1979).
Gryaznov, V.M., Karavanov, A.N., Belosljudova, T.M., Ermolaev, A.M., Maganjuk, A.P., and Sarycheva, I.K. Br. Patent 2,096,595 1982; U.S. Patent 4,388,479, 1983.
Gryaznov, V.M., Mishchenko, A.P., Smirnov, V.A., Kashdan, M.V., Sarylova, M.E., and Fasman, A.B., French Patent 2,595,092, 1986.
Gryaznov, V.M., Serebryannikova, O.S., Serov, Y.M., Ermilova, M.M., Karavanov, A.N., Mischenko, A.P., and Orekhova, N.V. *Appl. Catal.*, **96**, 15 (1993).
Gryaznov, V.M., Smirnov, V.S., Mischenko, A.P., and Aladyshev, S.I. British Patent 1,342,869, 1974a; U.S. Patent 3,849,076, 1974b.

Gryaznov, V.M., Smirnov, V.S., and Slin'ko, M.G. In *Proc. 7th Int. Cong. Catal.*, Part A, (Eds., Seiyama, T., and Tanabe, K.), 1981.
Harold, M.P., Lee, C., Burggraaf, A.J., Keizer, K., Zaspalis, V.T., and de Lange, R.S.A. *Mat. Res. Soc. MRS Bulletin*, **29**, 34 (1994).
Harold, M.P., Cini, P., Patanaude, B., and Venkatraman, K. *AIChE Symp. Ser.*, **85**, 26 (1989).
Harold, M.P., and Cini, P. *AIChE Symp. Ser.*, **85**, 26 (1989).
Harold, M.P., Zaspalis, V.T., Keizer, K., and Burggraaf, A.J. 5th NAMS Meeting, Lexington, K., May 1992.
Hsieh, H.P. *AIChE Symp. Ser.*, **85**(268), 53 (1989); *Catal. Rev. – Sci. Eng.*, **33**, 1 (1991).
Itoh, N., and Govind, R. *AIChE Symp. Ser.*, **85**, 10 (1989a); *Ind. Eng. Chem. Res.*, **28**, 1554 (1989b).
Itoh, N., Shindo, Y., and Haraya, K. *J. Chem. Eng. Jpn.*, **23**, 420 (1990).
Itoh, N., Shindo, Y., Haraya, K., Obata, K., Hakuta, T., and Yoshitome, H. *Int. Chem. Eng.*, **25**(1), 139 (1985).
Karavanov, A.N., and Gryaznov, V.M. *Kinet. Catal.*, **25**, 56 (1984).
Kita, H., Sasaki, S., Tanaka, K., Okamoto, K., and Yamamoto, M. *Chem. Lett.*, 2025 (1988).
Lebedeva, V.I. *Met. I Splavy Kak Membran. Katalizatory M.*, **112** (1981); *Chem. Abstract*, **98**(9), 71615P (1983).
Leenaars, A.F.M., Keizer, K., and Burggraaf, A.J. *J. Mater. Sci.*, **19**, 1077 (1984); *J. Coll. Inter. Sci.*, **105**, 27 (1985).
Li, Z.Y., Maeda, H., Kusakabe, K., Morooka, S., Anzai, H., and Akiyama, S. *J. Membr. Sci.*, **78**, 247 (1993).
Lin, Y.S., and Burggraaf, A.J. *AIChE J.*, **38**, 444 (1992).
Lund, C.R.F. *Catal. Lett.*, **12**, 395 (1992).
Marr, R., and Kopp, A. *Int. Chem. Eng.*, **22**(1), 44 (1982).
Mikhalenko, N.N., and Tabares, C. *Sovrem. Zadachi V Tochn. Naukakh*, 126 (1977).
Mikhalenko, N.N., Khrapova, N.E.V., and Gryaznov, V.M. *Neftekhimiya*, **18**, 354 (1978).
Mishchenko, A.P., and Gryaznov, V.M. German Patent 3,013,799, 1981.
Mishchenko, A.P., Gryaznov, V.M., Smirnov, V.S., Senina, E.D., Parbuzina, I.L., Roshan, N.R., Polyakova, V.P., and Savitsky, V.M. U.S. Patent 4,179,470, 1989; German Patent 2,816,279, 1979.
Mishchenko, A.P., Saryalova, M.E., Gryaznov, V.M., Smirnov, V.S., Roshan, N.R., and Polyakova, V.P. *Izv. Acad. Nauk. SSSR. Ser. Khim.*, **7**, 1620 (1977).
Mohan, K., and Govind, R. *AIChE J.*, **32**(12) 2083 (1986).
Oertel, M., Schmitz, J., Welrich, W., Jendryssek-Neumann, D., and Schulten, R. *Chem. Eng. Technol.*, **10**, 248 (1987).
Okamoto, K., Semoto, R., Tanaka, K., and Kita, H. *Chem. Lett.*, 167–170 (1991).
Okamoto, K., Yamamoto, M., Otoshi, Y., Semoto, T., Yano, M., Tanaka, K., and Kita, H. *J. Chem. Eng. Jpn*, **26**, 475–481 (1993).
Rezac, M.E., and Koros, W.J. *J. Membr. Sci.*, **93**, 193–201 (1994); *Ind. Eng. Chem. Res.*, **34**, 862–868 (1995).
Sanchez, J., and Tsotsis, T. In *Fundamentals of Inorganic Membrane Science and Technology* (Eds., Burggraaf, A.J., and Cot, L.), Elsevier, Amsterdam, 1996.
Saracco, G., and Specchia, V. *Catal. Rev. – Sci. Eng.*, **36**, 302 (1994).
Sarylova, M.E., Mishchenko, A.P., Gryaznov, V.M., and Smirnov, V.S. *Izv. Akad. Nauk SSSR. Ser. Khim.*, **190**, 144 (1970).
Shao, X., Xu, S., and Govind, R. *AIChE Symp. Ser.*, **268**, 1 (1989).

Sheel, H.P. and Crowe, C.M. *Can. J. Chem. Eng.*, **47**, 183 (1969).
Shu, J., Grandjean, B.P.A., Ghali, E., and Kaliaguine, S. *J. Membr. Sci.*, **77**, 181 (1993).
Shu, J., Grandjean, B.P.A., Van Neste, A., and Kaliaguine, S. *Can. J. Chem. Eng.*, **69**(10), 1036 (1991).
Sloot, H.J., Smolders, C.A., Van Swaaij, W.P.M., and Versteeg, G.F. *AIChE J.*, **38**, 887 (1992).
Sloot, H.J., Versteeg, G.F., and Van Swaaij, W.P.M. *Chem. Eng. Sci.*, **45**(8), 2415 (1990).
Smirnov, V.S., Gryaznov, V.M., Ermilova, M.M., Orekhova, N.V., Roshan, N.R., Polyakova, V.P., and Savitskii, E.M. German Patent 3,003,993, 1981a; Soviet Patent 870,393, 1981b.
Smirnov, V.S., Gryaznov, V.M., Niropol'skaya, M.A., Samokhvalov, G.I., Kozlovskaya, N.K., Ermilova, M.M., and Mishchenko, A.P. Soviet Patent 437,743, 1978.
Stoukides, M. *Ind. Eng. Chem. Res.*, **27**, 1745 (1988).
Sun, Y.M., and Khang, S.J. *Ind. Eng. Chem. Res.*, **29**(2), 232 (1990).
Tonkovich, A.L.Y., Secker, R.B., Reed, E.L., Roberts, G.L., and Cox, J.L. *Sep. Sci. Technol.*, **30**, 1609 (1995).
Tsotsis, T.T., Champagnie, A.M., Vasileiadis, S.P., Ziaka, Z.D., and Minet, R.C. *Chem. Eng. Sci.*, **47**(9/11), 2903 (1992); *Sep. Sci. Technol.*, **28**, 397 (1993a).
Tsotsis, T.T., Minet, R.G., Champagnie, A.M., and Liu, P.K.T. In *Computer-Aided Design of Catalysts* (Eds., Becker, E.R., and Pereira, C.J.), Dekker, New York, 1993b.
Uhlhorn, R.J.R., Huis in't Veld, M.B.H.J., Keizer, K., and Burggraaf, A.J. *Sci. Ceram.*, **14**, 55 (1987).
Uhlhorn, R.J.R., Keizer, K., and Burggraaf, A.J. *J. Membr. Sci.*, **66**, 259 (1992a); *J. Membr. Sci.*, **66**, 271 (1992b).
Veldsink, J.W., Damme, R.M.J., Versteeg, G.F., and Van Swaaij, W.P.M. *Chem. Eng. Sci.*, **47**, 2939 (1992).
Wagialla, K.M., and Elnashaie, S.S.E.H. *Ind. Eng. Chem. Res.*, **30**, 2298 (1991).
Wang, A.W., Reich, B.A., Johnson, B.K., and Foley, H.C. *Symp. Octane and Cetain Enhanc. Proc. Red. Emissions Motor Fuels*, *Am. Chem. Soc.*, San Francisco Meeting, 5–10 April, 1992.
Wu, J.C.S., and Liu, P.K.T. *Ind. Eng. Chem. Res.*, **31**(1), 322 (1992).
Zaspalis, V.T., and Burggraaf, A.F. In *Inorganic Membranes Synthesis, Characteristics and Applications* (Ed., Bhave, R.R.), Van Nostrand Reinhold, New York, 1991.
Zaspalis, V.T., Van Praag, W., Keizer, K., Van Ommen, J.G., Ross, J.R.H., and Burggraaf, A.J. *Appl. Catal.*, **74**, 249 (1991).
Zhu, Y., Minet, R.G., and Tsotsis, T.T. *Chem. Eng. Sci.*, **51**, 4103 (1996).

CHAPTER 25

Anwar, M.M., Cook, S.T.M., Hanson, C., and Pratt, M.W.T. *Proc. Int. Solvent Extraction Conf.*, ISEC-1974, Soc. Chem. Ind., London, Vol. 1, 1974, p. 895; *Proc. Int. Solvent Extraction Conf.*, ISEC1977, Can. Inst. Mining and Metal., Vol. 2, 1979, p. 679.
Anwar, M.M., Hanson, C., and Pratt, M.W.T. *Proc. Int. Solvent Extraction Conf.* ISEC1971, Soc. Chem. Ind., London, Vol. 1, 1971a, p. 119; *Trans. Inst. Chem. Eng.*, **49**, 95 (1971b).
Barbosa, D., and Doherty, M.F. *Chem. Eng. Sci.*, **43**, 529 (1988a); **43**, 541 (1988b); **43**, 1523 (1988c); *Proc. R. Soc.*, **A413**, 443 (1987a); **A413**, 459 (1987b).
Brändström, A. *J. Mol. Catal.*, **20**, 93 (1983).

Buzad, G., and Doherty, M.F. *Chem. Eng. Sci.*, **49**, 1994.
Chopade, S.P., and Sharma, M.M. *React. Funct. Poly.*, **32**, 53 (1997).
Coleby, J. *Recent Advances in Liquid-Liquid Extraction* (Ed., Hanson, C.) Pergamon, Oxford, 1971, p. 124.
DeGarmo, J.L., Parulekar, V.N., and Pinjala, V. *Chem. Eng. Prog.*, 43, March (1992).
Doherty, M.F., and Buzad, G.B. *The Chemical Engineer*, Supplement, 17, August (1992).
Duprat, F., and Gau, G. *Can. J. Chem. Eng.*, **69**, 1320 (1991).
Ellis, S.R.M., and Gibbon, J.D. In *The Less Common Means of Separation* (Ed., Pirie, J.M.), The Institution of Chemical Engineers, London, 1963.
Gaikar, V.G., and Sharma, M.M. *J. Sep. Proc. Technol.*, **5**, 53 (1984); *Solv. Extn. Ion Exchange*, **3**, 679 (1985).
Gassend, R.G., Duprat, F., and Gau, G. *Nouv J. Chim.*, **19**, 703 (1985).
Grosser, J.H., Doherty, M.F., and Malone, M.F. *Ind. Eng. Chem. Res.*, **26**, 983 (1987).
Holve, W.A. *Ind. Eng. Chem. Fundam.*, **16**, 56 (1977).
Huan, S., Hertzberg, T., and Lein, K.M. *Comput. Chem. Eng.*, **19**, Suppl., S327 (1995).
Jagirdar, G.C., and Sharma, M.M. *J. Sep. Proc. Technol.*, **2**, 37 (1981).
Kafarov, V.V., Vygon, V.G., Chulok, A.I., and Kostyuk, V.A. *Russ. J. Phys. Chem.*, **50**, 1618 (1976).
Karr, A.E., and Scheibel, E.G. *Ind. Eng. Chem.*, **46**, 1583 (1954).
Kemula, W., Brzozowski, S.H., and Pawlowski, W. *Roczniki. Chem.*, **35**, 711 (1961); *Chem. Abstr.*, **55**, 23391 (1961).
King, M.L., Forman, A.L., Orella, C., and Pines, S.H. *Chem. Eng. Prog.*, 36, Ma7 (1985).
Kiseleva, E.N., Belyaeva, V.A., and Gel'prin, N.I. *Khim. Prom.*, **47**, 178 (1971).
Kolah, A.K., Mahajani, S.M., and Sharma, M.M. *Ind. Eng. Chem. Res.*, **35**, 3707 (1996).
Komatsu, H., and Holland, C.D. *J. Chem. Eng. Jpn.*, **10**, 292 (1977).
Komatsu, H. *J. Chem. Eng. Jpn.*, **10**, 200 (1977).
Kostyuk, V.A., Mikhailova, G.S., Grigor'ev, S.M., Chernomordik, E.Ya., and Markus, G.A. *Koks. Khim.*, **12**, 48 (1968); *Chem. Abstr.*, **70**, 70035 (1969).
Laddha, S.S., and Sharma, M.M. *J. Chem. Tech. Biotechnol.*, **28**, 69 (1978).
Mahapatra, A., Gaikar, V.G., and Sharma, M.M. *Sep. Sci. Technol.*, **23**, 429 (1988).
Marek, J. *Coll. Czech. Chem. Commun.*, **19**, 1055 (1954).
Masamoto, J., and Matsuzaki, K. *J. Chem. Eng. Jpn*, **27**, 1 (1994).
Milnes, M.M. *Proc. Int. Solvent Extraction Conf.*, ISEC1974, Soc. Chem. Ind., London, Vol. 1, 1974, p. 983.
Nelson, P.A. *AIChE J.*, **17**, 1043 (1971).
Rev, E. *Ind. Eng. Chem. Res.*, **33**, 2174 (1994).
Sasaki, M., Nodera, K., Mukai, K., and Yoshioka, H. *Bull. Chem. Soc. Jpn.*, **50**, 276 (1977).
Sawistowski, H., and Pilavakis, P. A. *Int. Symp.* Distillation, London, *Inst. Chem. Eng. Symp. Series*, **56**, 42 (1979); *Chem. Eng. Sci.*, **43**, 355 (1988).
Smith, J.F. *Fluid Phase Equilibria*, **78**, 219 (1992).
Smith, L.A., Jr. U.S. Patent 4,242,530 1980.
Smith, L.A., Jr., and Huddleston, M.N. *Hydrocarbon Process.*, **61**, 121 (1982).
Steppan, D.D., Doherty, M.F., and Malone, M.F. *J. Appl. Polym. Sci.*, **33**, 2333 (1987).
Suzuki, I., Yagi, H., Komatsu, H., and Hirata, M. *J. Chem. Eng. Jpn.*, **4**, 26 (1971).
Ung, S., and Doherty, M.F. *Chem. Eng. sci.*, **50**, 3201. (1994a); *Ind. Eng. Chem. Res.*, **34**, 2555 (1994b); **34**, 3195 (1994c).
Venimadhavan, G., Buzad, G., Doherty, M.F., and Malone, M.F. *AIChE J.*, **40**, 1814 (1994).

Wadekar, V.V., and Sharma, M.M. *J. Chem. Tech. Biotechnol.*, **31**, 279 (1981a); *J. Sep. Proc. Technol.*, **2**, 1 (1981b).
Walker, A.C. *Ind. Eng. Chem.*, **42**, 1226 (1950).
Xu, X., Zheng, Y., and Zheng, G. *Ind. Eng. Chem. Res.*, **34**, 2232 (1995).
Yamamoto, A. Japan Patent 1517, 1956; *Chem. Abstr.*, **51**, 8809, 1957.
Yamamoto, A., Arakawa, H., Higuchi, H., and Koji, Y. Japan Patent 5220 (1957); *Chem. Abstr.*, **52**, 9572 (1958).

CHAPTER 26

Aaltonen, O., and Rantakylä, M. *Chemtech*, 240 (1991).
Abramovitch, R.A. *Org. Prep. Proced. Int.*, **23**(6), 685 (1991).
Agarwal, M. Separation Process, Ph.D. (Tech) Thesis, University of Bombay, 1993.
Akbulut, U., Toppare, L., and Yurttas, B. *Polymer*, 803 (1986).
Alfano, O.M. *Ind. Eng. Chem. Res.*, **34**, 488 (1995); *Chem. Eng. Sci.*, **49**, 5327 (1994).
Alfano, O.M., and Cassano, A.E. *Ind. Eng. Chem. Res.*, **27**, 1087 (1988a); **27**, 1095 (1988b).
Alfano, O.M., Romero, R.L., and Cassano, A.E. *Chem. Eng. Sci.*, **41**, 421 (1986a); **41**, 1137 (1986b); **41**, 1155 (1986c).
Alfano, O.M., Cabrera, M.I., and Cassano, A.E. *Chem. Eng. Sci. (Proc. 13th Intl. Symp. Chem. Reaction Eng.)*, **49**, (1994).
Almgren, M., and Rydholm, R. *J. Phys. Chem.*, **83**, 360 (1979).
André, J.C., Said, A., and Viriot, M.L. Fr. Patent 851,7436, 1985.
André, J.C., Viriot, M.L., and Said, A. *J. Photochem. Photobiol. A: Chem.*, **42**, 383 (1988).
André, J.C., Viriot, M.L., and Villermaux, J. *Pure Appl. Chem.*, **58**, 907 (1986).
Asai, S., Nakamura, H., Kimura, Y., and Sakamoto, M. *Trans. IchemE*, **74**(A3), 240 (1996a).
Asai, S., Nakamura, H., Yamaguchi, H., and Shimizu, H. *Trans. IChemE*, **74**(A), 349 (1996b).
Baghurst, D.R., and Mingos, D.M.P. *J. Organomet. Chem.*, 384 C57 (1990); *J. Chem. Soc., Dalton Trans.*, **115**, 1 (1992); *British Ceramic Trans.*, **91**, 124 (1992a); *J. Chem. Soc., Dalton Trans.*, 1151 (1992b).
Bajic, M., Bajic, Ma., and Karminski-Zamola, G. *Preparative Organic Photochemistry: Experimental Techniques*, Allured, Carol Stream, Il., 1995.
Balasubramanian, D., Srinivas, V., Gaikar, V.G., and Sharma, M.M. *J. Phys. Chem.*, **93**, 3865 (1989).
Baptist-Nguyen, S., and Subramaniam, B. *AIChE J.*, **38**, 1027 (1992).
Benahcene, A., Petrier, C., Reverdy, G., and Labbe, P. *New J. Chem.*, **19**, 989 (1995).
Bertucco, A., Canu, P., Devetta, L., and Zwahlen, A.G. *Ind. Eng. Chem. Res.*, **36**, 2626 (1997).
Beugelmans, R., Ginsburg, A., Lecas, A., LeGoff, M.T., and Roussi, G. *Tetrahedron Lett.*, **35**, 3271 (1978).
Bloomsfield, J.J. (to Monsanto). U.S. Patent 3, 873, 568 1975, 1975.
Bloomsfield, J.J., and Owsley, D.C. U.S. Patent 4,004,057 1977a; 4,051,076 1977b; in *Kirk and Othmer's Encyclopedia of Chemical Technology*, 3rd ed., Vol. 6, Wiley, New York, 1982.
Borrell, P. *Photochemistry: A Primer* (Eds., Stokes, B.J., and Malpas, A.J.), Edward Arnold, London, 1973.
Braun, A.M., *Photocatalysis and Environment* (Ed., Schiavello, M.), Kluwer Academic, New York, 1988.

Braun, A., Maurette, M.T., and Oliveros, E. *Technologie Photochemique*, 1st ed., Presses Polytechniques Romandes, Lausanne, 1986.
Brunet, J.J., Sidot, C., and Caubere, P. *J. Org. Chem.*, **48**, 1166 (1983).
Bunton, C.A. *Catal.Rev.-Sci. Eng.*, **20**(1), 1 (1979); in *Micellar Rate Effects upon Organic Reactions in Kinetics and Catalysis in Microheterogeneous Systems* (Eds., Graetzel, M., and Kalyanasundaram, K.), Dekker, New York, 1991.
Cablewski, T., Foux, A.F., and Strauss, C.R. *J. Org. Chem.*, **59**, 3408 (1994).
Cabrera, M.I., Alfano, O.M., and Cassano, A.E. *Chem. Eng. Commun.*, **107**, 95 (1991a); *Chem. Eng. Commun.*, **107**, 123 (1991b); *AIChE J.*, **37**, 1471 (1991c); *Ind. Eng. Chem. Res.*, **33**, 3031 (1994).
Caddick, S. *Tetrahedron*, **51**(38), 1403 (1995).
Cassano, A.E. *Ind. Eng. Chem. Res.*, **34**, 2155 (1995).
Cassano, A.E, Alfano, O.M., and Romero, R.L. In *Concepts and Design of Chemical Reactors* (Eds., Whitaker, S., and Cassano, A.E.), Gordon and Breach, New York, 1986.
Cassano, A.E., and Alfano, O.M. In *Photochemical Conversion and Storage of Solar Energy* (Eds., Pellizzetti, E., and Schiavello, M.), Kluwer, Dordrecht, 1991.
Cassano, A.E., Martin, C.A., and Alfano, O.M. *Trends Photochem. Photobiol.*, **3**, 55–79 (1994).
Cassano, A.E., Martin, C.A., Brandi, R.J., and Alfano, O.M. *Ind. Eng. Chem. Res.*, **32**, 2155 (1995).
Cassano, A.E., Silveston, P.L., and Smith, J.M. *Ind. Eng. Chem.*, **59**, 18 (1967).
Chaudhary, A.K., Beckman, E.J., and Russell, A.J. *J. Am. Chem. Soc.*, **117**, 3728 (1995).
Chemat, F., Esveld, D.C., Poux, M., and DiMartino, J.L. *J. Microwave Power and Electromagnetic Energy*, **33**, 88 (1998).
Chemat, F., Poux, M., DiMartino, J.L., and Berlan, J. *Chem. Eng. Technol.*, **19**, 420 (1995).
Chemat, F., Poux, M., DiMartino, J.L., and Berlan, J. *J. Microwave Power and Electromagnetic Energy*, **31**, 19 (1996).
Chou, T.C., Do, J.S., and Cheng, C.H. In *Modern Methodology in Organic Synthesis* (Ed., Shono, T.), VCH, New York, 1992.
Compton, R.G., Eklund, J.C., Marken, F., Rebbitt, T.O., Akkermans, R.P., and Waller, D.N. *Electrochimica Acta*, **42**, 2919 (1997).
Compton, R.G., Eklund, J.C., Marken, F., and Waller, D.N. *Electrochimica Acta*, **41**, 315 (1996a).
Compton, R.G., Marken, F., and Rebbitt, T.O. *J. Chem. Soc., Chem. Commun.*, 1017 (1996b).
Constable, D., Raner, K., Somlo, P., and Strauss, C. *J. Microwave Power and Electromagnetic Energy*, **27**, 195 (1992).
Contamine, F., Faid, F., Wilhelm, A.M., Berlan, J., and Delmas, H. *Chem. Eng. Sci.*, **49**, 5865 (1994).
Cordes, E.S. Ed., *Reaction Kinetic in Micelles*, Plenum, New York, 1973.
Cordes, E.H., and Gitler, C. *Prog. Bioorg. Chem.*, **2**, 1 (1973).
Coyle, J.D., Ed., *Photochemistry in Organic Synthesis*, special publication No. 57, The Royal Society of Chemistry, London, 1986.
Davidson, R.S., Safdar, A., Spencer, J.D., and Robinson, B. *Ultrasonics*, **25**, 35 (1987).
De Bernardez, E.R., and Cassano, A.E. *Lat. Am. J. Heat Mass Transfer*, **6**, 333 (1982); *J. Photochem.*, **30**, 285 (1985); *Ind. Eng. Chem. Proc. Des. Dev.*, **25**, 601 (1986).
De Bernardez, E.R., Claria, M.A., and Cassano, A.E. In *Chemical Reaction and Reactor Engineering* (Eds., Carberry, J.J., and Varma, A.), Dekker, New York, 1987.

Degrand, C., Prest, R., and Compagnon, P.L. *J. Org. Chem.*, **52**, 5229 (1987).
Dehmlow, E.V., Raths, H.C., and Soufi, H. *J. Chem. Res., Synop.*, 334 (1988).
Dehmlow, E.V., Thieser, R., Sasson, Y., and Pross, Z. *Tetrahedron*, **41**, 2927 (1985).
Dillow, A.K., Yun, S.L., Suleiman, D., Boatright, D.L., Liotta, C.L., and Eckert, C.A. *Ind. Eng. Chem. Res.*, **35**, 1801 (1996).
Dinjus, E., Fornika, R., and Scholz, M. In *Chemistry Under Extreme or Non-Classical Conditions* (Eds., Van Eldik, R., and Hubbard, C.D.), Wiley, New York, 1997.
Do, J.S., and Chou, T.C. *J. Appl. Electrochem.*, **19**, 922 (1989); **20**, 978 (1990); **22**, 966 (1992).
Do, J.S., and Do, Y.L. *Electrochimica Acta*, **39**, 2037 (1994a); 2299 (1994b); 2311 (1994c).
Doede, C.M., and Walker, C.A., *Chem. Eng.*, **62**, 159 (1955).
Dolling, U.H. *Chem. Abstr.*, 106:4697n (1986).
Dombro, R.A., Prentice, G.A., and McHugh, M.A. *J. Electrochem. Soc.*, **135**, 2219 (1988).
Dordick, J.S. *Enzyme Microb. Technol.*, **11**, 194 (1989).
Dumont, T., Barth, D., and Perrut, M. *J. Supercrit. Fluids*, **6**, 85 (1993).
Durant, A., Francois, H., Reisse, J., and DeMesmaker, A.K. *Electrochimica Acta*, **41**, 277 (1996).
Elworthy, P.H., Florence, A.T., and MacFarlane, C.B. *Solubilization by Surface Active Agents*, Chapman and Hall, London, 1968.
Fendler, J.H., *Membrane Mimetic Chemistry*, Wiley Interscience, New York, 1982.
Fendler, J.H., and Fendler, E.J. *Catalysis in Micellar and Macro Molecular Systems*, Academic Press, New York, 1975.
Fischer, M. *Angew. Chem., Int. Ed. Engl.*, **17**, 16 (1978).
Fisher, L.F., and Oakenfull, D. *Q. Rev. Chem.*, **2**, 1 (1977).
Forsyth, S.R., Pletcher, D., and Healy, K.P. *J. Appl. Electrochem.*, **17**, 905 (1987).
Fragner, J. *Vitamine. Gustav Fischer Verlag, Jena*, **1**, 613, (1964).
Gedye, R., Smith, F., and Westaway, K.C. *J. Microwave Power and Electromagnetic Energy*, **26**, 3 (1991).
Gedye, R., Smith, F., Westaway, K.C., Ali, H., Baldisera, L., Laberge, L., and Rousell, J. *Tetrahedron Lett.*, **27**(3), 279 (1986).
Giguere, R.J., Bray, T.L., Duncan, S.M., and Majetich, G. *Tetrahedron Lett.*, **27**(41), 4945 (1986).
Gilman, A.G, Goodman, L.S., Rall, T.W., and Murad, F. *The Pharmacological Basis of Therapeutics*, 7th ed., Macmillan Publishing, New York, 1985.
Ginosar, D.M., and Subramaniam, S. *Proc. Int. Symp. on Catalyst Deactivation* (Eds., Delmon, B., and Froment, G.F.), Elsevier, Amsterdam, 1994.
Gokel, G.W., Gerdes, H.M., and Robert, N.W. *Tetrahedron Lett.*, **653** (1976).
Greenbaum, S.B. In *Kirk-Othmer: Encyclopedia of Chemical Technology*, 2nd ed., Interscience, Vol. 21, New York, 1970.
Gunnarsson, G., Jonsson, B., and Wennerstom, H. *J. Phys. Chem.*, **84**, 3114 (1980).
Gunnlaughsdottir, H., and Sivik, B. *J. Am. Oil Chem. Soc.*, **72**, 399 (1995).
Hagenson, L.C., Naik, S.D., and Doraiswamy, L.K. *Chem. Eng. Sci.*, **49**, 4787 (1994).
Hammond, D.A., Karel, M., Klibanov, A.M., and Krukonis, V. *J. Appl. Biochem. Biotechnol.*, **11**, 393 (1985).
Harano, Y., and Smith, J.M. *AIChE J.*, **14**, 584 (1968).
Haring, H.G., and Knoll, H.W. *Chem. Process Eng.*, **45**, 560, 619, 690 (1964); **46**, 38 (1965).
Hartig, H. *Chem.-Ztg.*, **99**, 179 (1975).

Hofmann-La Roche (to Pfoertner, K.H., and Hanson, H.J.), U.S. Patent 4,551,214 (1985).
Howdle, S.M., and Poliakoff, M. *J. Chem. Soc., Chem. Commun.*, 1099 (1989).
Hrnjez, B.J., Metha, A.J., Fox, M.A., and Johnston, P.K. *J. Am. Chem. Soc.*, **111**, 2662 (1989).
Hutson, T., Jr., and Logan, R.S. *Chem. Eng. Prog.*, **68**, 76 (1972).
Ikushima, Y., Ito, S., Asano, T., Yokoyama, T., Saito, N., Hatakeda, K., and Goto, T. *J. Chem. Eng. Jpn*, **23**, 96 (1990).
Ikushima, Y., Saito, N., and Arai, M. *J. Phys. Chem.*, **96** 2293 (1992); *Bull. Chem. Soc. Jpn*, **64**, 282 (1991).
Inoue, Y., Sasaki, Y., and Hashimoto, H. *J. Chem. Soc., Chem. Commun.*, 718 (1975).
Issacs, N., and Keating, N. *J. Chem. Soc., Chem. Commun.*, 876 (1992).
Jagdale, S.J., Patil, S.V., and Salunke, M.M. *Synth. Commun.*, **26**, 1747, (1996).
Janakiraman, B., and Sharma, M.M. *Chem. Eng. Sci.*, **40**, 235 (1985).
Jessop, P.G., Hsiao, Y., Ikariya, T., and Noyori, R. *J. Am. Chem. Soc.*, **116**, 8851 (1994a); **118**, 344 (1996).
Jessop, P.G., Ikariya, T., and Noyori, R. *Nature*, **368**, 231 (1994b); *Science*, **269**, 1065 (1995).
Jouglet, B., Blanco, L., and Rousseau, G. *Synlett*, **907** (1991).
Kamat, S.V., Barrera, J., Beckman, E.J., and Russell, A.J. *Biotechnol. Bioeng.*, **40**, 158 (1992).
Kamat, S.V., Beckman, E.J., and Russell, A.J. *Crit. Rev. Biotechnol.*, **15**, 41 (1995).
Kamat, S.V., Iwaskewycz, B., Beckman, E.J., and Russell, A.J. *Proc. Natl. Acad. Sci. U.S.A.*, **90**, 2940 (1993).
Kaupp, G. *Angew. Chem., Int. Ed. Engl.*, **33**, 1452 (1994).
Kazarian, S.G., and Poliakoff, M. *J. Phys. Chem.*, **99**, 8624 (1995).
Khmelnitsky, Yu.L., Neverova, I.N., Poyakov, V.I., Grinberg, V.Ya., Levashov, A.V., and Marinek, K. *Eur. J. Biochem.*, **190**, 155 (1990).
Kim, S., and Johnston, K.P. (Eds. Squires, T.G., and Paulaitis, M.E.), *ACS Symposium Series*, **329**, 42 (1987).
Knez, Z., Rizner, V., Habulin, M., and Bauman, D. *J. Am. Oil Chem. Soc.*, **72** (1995).
Latha, V. *Studies in Surfactant Systems and Hydrotropy*, Ph.D. Thesis, University of Mumbai, 1998.
Laxman, M., and Sharma, M.M. *Synth. Commun.*, **20**, 111, 1990.
Lorimer, J.P., Pollet, B., Phull, S.S., Mason, T.J., Walton, D.J., and Geissler, U. *Electrochimica Acta*, **41**, 2737 (1996).
Loupy, A., Petit, A., Ramdani, M., Yvanaeff, C., Majdoub, M., Labaid, B., and Villemin, D. *Can. J. Chem.*, **71**, 90 (1993).
Maidan, R., and Willner, I. *J. Am. Chem. Soc.*, **108**, 1080 (1986).
Majetich, G., and Hicks, R. *Res. Chem. Intermed.*, **20**(1), 61 (1994).
Manos, G., and Hofmann, H. *Chem. Eng. Technol.*, **14**, 73 (1991).
Martinek, K., Osipov, A.P., Yatsimirski, A.K., and Berezin, I.V. *Tetrahedron*, **31**, 709 (1975).
Martins, J.F., Borges de Carvalho, I., Correa de Sampaio, T., and Barreiros, S. *Enzyme Microb. Technol.*, **16**, 785 (1994).
Maruyama, T., and Nishimoto, T. *Chem. Eng. Commun.*, **117**, 111 (1992).
Mingos, D.M.P., and Baghurst, D.R. *Chem. Soc. Rev.*, **20**, 1 (1991).
Mingos, D.M.P., and Whittaker, A.G. In *Chemistry Under Extreme or Non-Classical Conditions* (Eds., Van Eldik, R., and Hubbard, C.D.), Wiley, New York, 1997.

Mittal, K.L. Ed., *Micellization, Solubilization and Microemulsion*, Vols. 1 and 2, Plenum, New York, 1977.
Moroi, Y. *Micelles: Theoreotical and Applied Aspects*, Plenum, New York, 1992.
Mukherjee, P. and Mysels, K.J. *Critical Micelle Concentrations of Aqueous Surfactant Systems*, National Bureau of Standards, U.S. Government Printing Office, Washington D.C., 1970.
Naik, S.D., and Doraiswamy, L.K. *AIChE J.*, **44**, 612 (1998).
Noritomi, H., Miyata, M., Kato, S., and Nagahama, K. *Biotechnol. Lett.*, 17 (1995).
Ohloff, G., Klein, E., and Schenck, G.O. *Angew. Chem.*, **73**, 578, (1961).
Pagnotta, M., Pooley, C.L.F., Gurland, B., and Choi, M. *J. Phys. Org. Chem.*, **6**, 407 (1993).
Pandit, A.B., and Sharma, M.M. *Chem. Eng. Sci.*, **44**, 218 (1989).
Paulaitis, M.E., and Alexander, G.C. *Pure. Appl. Chem.*, **59**, 61 (1987).
Pfoertner, K.H. *J. Photochem.*, **25**, 91 (1984); in *Photochemistry in Organic Synthesis* (Ed., Coyle, J.D.), Royal Society of Chemistry, London, 1986; *J. Photochem. Photobiol. A: Chemistry*, **51**, 81 (1990); in *Ullmann's Encyclopedia of Industrial Chemistry* (Ed., Elvers, B.), VCH, Weiheim, Germany, 1991.
Pletcher, D., and Valdes, E.M. *Electrochimica Acta*, **33**, 499 (1988a); **33**, 509 (1988b).
Quina, F.H. and Chaimovich, H. *J. Phys. Chem.*, **83**, 1844 (1979).
Quina, F.H., Politi, M.J., Cuccovia, I.M., Baumgarten, E., Martins-Franchetti, S.M., and Chainovich, H. *J. Phys. Chem.*, **84**, 361 (1980).
Ragonese, F.P., and Williams, J.A. *AIChE J.*, **17**, 1352 (1971).
Randolph, T.W., Blanch, H.W., Prausnitz, J.M., and Wilke, C.R. *Biotechnol. Lett.*, **7**, 325 (1985).
Raner, K.D., Strauss, C.R., Trainor, R.W., and Thorn, J.S. *J. Org. Chem.*, **60**, 2456 (1995).
Rantakylä, M., and Aaltonen, O. *Biotechnol. Lett.*, **16**, 824, (1994).
Ratoarinoro, N., Wilhelm, A.M., Berlan, J., and Delmas, H. *Chem. Eng. J.*, **50**, 27 (1992).
Reisse, J., Francois, H., Vandercammen, J., Fabre, O., Mesmaeker, A.K., Maerschalt, C., and Delplancke, J.L. *Electrochimica Acta*, **39**, 37 (1994).
Rideout, D.C., and Breslow, R. *J. Am. Chem. Soc.*, **102**, 7816 (1980).
Rodenas, E., and Vera, S. *J. Phys. Chem.*, **89**, 513 (1985).
Roger, M., and Villermaux, J. *AIChE J.*, **21**, 1207 (1975); *Chem. Eng. J.*, **17**, 219 (1979); **26**, 85, (1983).
Romsted, L.S. In *Micellization, Solubilization and Microemulsion*, Vol. 2 (Ed., Mittal, K.L.), Plenum, New York, 1977; in *Surfactants in Solution*, Vol. 2 (Eds., Mittal, K.L., and Lindman, B.), Plenum, New York, 1984; *J. Phys. Chem.*, **89**, 5107, 5113 (1985).
Russell, A.J., and Beckman, E.J. *Enzyme Microb. Technol.*, **13**, 1007 (1991).
Russell, A.J., Beckman, E.J, and Chaudhary, A.K. *Chemtech*, 33 (1994).
Sadvilkar, V.G. *Aqueous Hydrotrope Solutions as Novel Reaction Media*, Ph.D. (Science) Thesis, University of Bombay, 1995.
Sadvilkar, V.G., Khadilkar, B.M and Gaikar, V.G. *J. Chem. Tech. Biotech.*, **63**, 33, 1995.
Saim, S., Ginosar, D.M., and Subramaniam, B. (Ed., Johnston, K.P.), *ACS Symposium Series*, **406**, 301 (1989).
Saim, S., and Subramaniam, B. *Chem. Eng. Sci.*, **43**, 1837 (1988); *J. Supercrit. Fluids*, **3**, 214 (1990); *J. Catal.*, **131**, 445 (1991).
Sanders, G.M., Pot, J., and Havinga, E. *Fortschr. Chem. Org. Naturst.*, **27**, 131 (1969).
Sane, P.V., and Sharma, M.M. *Synth. Commun.*, **17**, 1331 (1987).

Savage, P.E., Gopalan, S., Mizam, T.I., Martino, C.J., and Brock, E.E. *AIChE J.*, **41**, 1723 (1995).

Savelova, V.A., and Vakhiova, L.N. *Russ. Chem. Bull.*, **44**, 2012 (1995).

Schneider, H. In *The Selective Solvation of Ions in Mixed Solvents, in Solute-Solvent Interactions*, Vol. 2 (Eds., Coetzee, J.F., and Ritchie, C.D.), Dekker, New York, 1976.

Shimada, S., Nakagawa, K., and Tabuchi, K. *Polym. J.*, **21**, 275 (1989).

Shimada, S., Obata, Y., Nakagawa, K., and Tabuchi, K. *Polym. J.*, **22**, 777 (1990).

Shinoda, K., Nakagawa, T., Tamamushi, B.T., and Isemura, T. *Colloidal Surfactants*, Academic Press, New York, 1963.

Stevens, R.D.S. (to Solarchem, Brolor Investments, Canada). U.S. Patent 4,686,023, 1985.

Strauss, C.R., and Trainor, R.W. *Aust. J. Chem.*, **48**(10), 1665 (1995).

Stuerga, D.A.C., and Gaillard, P. *J. Microwave Power and Electromagnetic Energy*, **31**, 87, (1996a); **31**, 101, (1996b).

Subramaniam, B., and McHugh, M.A. *Ind. Eng. Chem. Proc. Des. Dev.*, **25**, 1, (1986).

Sudholter, E.J.R., Van der Langkruis, and Engberts, J.B.F.N. *Recl. Trav. Chim. Pas.-Bas Beig*, **99**, 73 (1979).

Suppan, P. *Principles of Photo-Chemistry, Monographs for Teachers*, No. 22, The Chemical Society, Burlington House, London, 1972.

Szabo, T., Aranyosi, K., Csiba, M., and Toke, L. *Synthesis*, 565 (1987).

Tiedt, W. *Chem. Ztg. Chem. Appar.*, **90**, 813 (1966); **91**, 17, 299, 569, 968 (1967); **92**, 76 (1968).

Tiltscher, H., and Hofmann, H. *Chem. Eng. Sci.*, **42**, 959 (1987).

Tiltscher, H., Wolf, H., and Schelchshorn, J. *Angew. Chem., Int. Ed. Engl.*, **20**, 892 (1981); *Ber. Bunsenges. Phys. Chem.*, **88**, 897 (1984).

Tomioka, H., Ueda, K., Ohi, H., and Izawa, Y. *Chem. Lett.*, 1359 (1986).

Tsang, C.Y., and Streett, N.B. *Chem. Eng. Sci.*, **36**, 993 (1981).

Tsanov, T., Vassiler, K., Stamenova, R., and Tsvetonov, C. *J. Polym. Sci., Part A: Polym. Chem.*, **33**, 2623 (1995).

Villemin, D., and Labiad, B. *Synth. Commun.*, **20**, 3213, 3333 (1990); **22**, 2043 (1992).

Walton, D.J., Chyla, A., Lorimer, J.P., and Mason, T.J. *Synth. Commun.*, **20**, 1843 (1990).

Wang, J.X., and Zhao, K. *Synth. Commun.*, **26**, 1617 (1996).

Wang, Y., Deng, R., Aiqiao, M., and Jiang, Y. *Synth. Commun.*, **25**, 1761 (1995).

Wang, Y., and Jiang, Y. *Synth. Commun.*, **22**, 2287 (1992).

Westaway, K.C., and Gedye, R.N. *J. Microwave Power and Electromagnetic Energy*, **30**, 219 (1995).

Whitaker, A.G., and Mingos, D.M.P. *J. Microwave Power and Electromagnetic Energy*, **29**, 195 (1994).

Yuan, Y., Gao, D., and Jiang, Y. *Synth. Commun.*, **22**, 2117 (1992a).

Yuan, Y., Jiang, Y., and Gao, D. *Synth. Commun.*, **22**, 3109 (1992b).

Yue, P.L. In *Photoelectrochemistry, Photocatalysis, and Photoreactors* (Ed., Schiavello, M.), Reidel, Dordrecht, Holland, 1985a; 1985b; 1985c.

A PROCESS OVERVIEW

We saw in the previous chapters how chemistry and chemical engineering should be combined to generate a technology which, given all operating constraints, may be regarded as the best at the time of the assessment. What we did not consider was the roles of the two in devising the final scheme for a new product. These roles, distinct and exploitable as they are for any chemical, become particularly important for intermediate and small volume chemicals where chemistry and engineering are more closely linked. The old concept that "the smaller the volume of production, the smaller the role of the engineer" is rapidly giving place to a more balanced interaction between the two. Thus, in concluding this book, devoted largely to an exposition of these roles, we formulate a methodology that can be of practical use in any organic process development activity. For this, we start with the "process vision" concept proposed by Basu et al. (1999)[1] for the pharmaceutical industry and modify it for industrial organic synthesis in general.

Like a corporate vision statement or a university vision statement, a process must also be regarded in its entirety and should be entitled to a vision statement. It is not restricted to common requirements such as maximizing yield and purity, but also includes such factors as safety, waste minimization, environmental requirements, cost, operability, and time. Therefore, it is neither a chemist's vision nor an engineer's vision. It even transcends the combined vision of the two, and as Basu et al. (1999)[1] point out, "it is a vision of the process that all stakeholders in development, manufacturing, and marketing can share and own."

[1] Basu, P. K., Mack, R. A., and Vinson, J. M. *Chem. Eng. Prog.*, **95**, 82, August (1999).

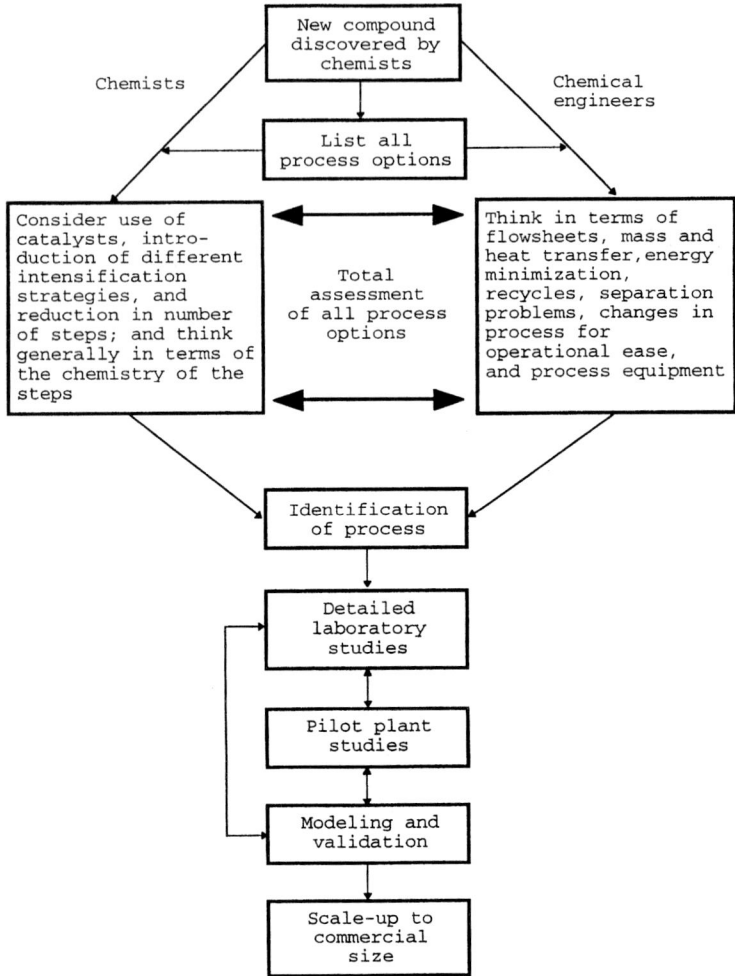

The chemist-chemical engineer interactive network: a feature of the "process vision."

Thus, starting with the discovery of a new product (attributed essentially to the chemist), the subsequent development of the final process depends on constant interaction between chemists, engineers (of all denominations), and marketing people. A scheme for such an interactive approach (involving mainly chemists and chemical engineers) is sketched in the above figure. The basic feature of this scheme is that the chemical engineer is involved from a very early stage in the development process. He (or she) is usually not a member of the discovery team but is not necessarily excluded from it. The roles of the chemists and the engineers are clearly indicated in the methodology. It will be noticed that there is scope here for including catalysis in its various forms in one

or more steps of the process and for considering the possibility of using one of the many strategies for rate enhancement described in Part V.

In closing, I would like to mention that in recent years a number of reactor configurations have been developed for intermediate or small volume chemicals, such as (1) use of a single reactor for many operations, (2) combining two or more chemical steps in a single operation, (3) use of novel designs such as the rotating disk reactor,[2] and (4) using biphasing or reaction-separation combinations to achieve optimal results. What the book gives is a treatment of the use of catalysts in a variety of forms and situations and of the various strategies individually. What it does not attempt is an integration of these various aspects of a process. This can be best done by people devoted to such an overall interactive enterprise which requires a group of diverse talents (including engineers from a number of other disciplines) working in unison. Further, the book is restricted to the reactor part of a process, or to the reaction-separation part if one is involved.

---

[2] See, for example, Oxley, P., Brechtelsbauer, C., Ricard, F., Lewis, N., and Ramshaw, C. *Ind. Eng. Chem. Res.*, **39**, 2175 (2000).

EPILOGUE

As will be evident to the reader, the book is formally addressed to chemical engineers, but chemistry is so inextricably woven into it that it should also be accessible, albeit to a lesser degree, to chemists and practicing technologists. Indeed, as already stressed in the preface, the chief motivation of the book has been to bring chemist and chemical engineer together by providing a theme, a list of topics, and a format of treatment that, one hopes, will benefit both in some measure.

The fact that so many strategies, preceded by basic general information on the required tools of analysis and design, have been included might provoke a reminder of the common but rather unsubtle adage: "A little knowledge is dangerous."

I hope even the semblance of such an interpretation can be effectively quelled by the more telling—and somewhat euphemistic—rejoinder of T.H. Huxley: "If a little knowledge is dangerous, where is the man who has so much as to be out of danger?"

In the final analysis, what has emerged is what was intended: a broad picture of the various general strategies of rate enhancement, leading to a reasoned choice between competing alternatives. This has been aided by an adequate explanation of the principles involved, followed by a large number of examples of their use. Extensive references have been provided to supplement the presentation—by enabling the more probing reader to acquire a vaster knowledge of a chosen set of promising possibilities.

Acknowledgments for Figures and Tables

CHAPTER 2

Figure 2.3.   With permission from Dover Publishers.
Figure 2.4.   With permission from Dover Publishers.

CHAPTER 4

Table 4.3.   With permission from Prentice-Hall.

CHAPTER 5

Figure 5.3.   With permission from American Chemical Society.
Figure 5.5a,b. With permission from Prentice-Hall.
Table 5.1.   With permission from Elsevier Science.
Table 5.3.   With permission from American Chemical Society.

CHAPTER 6

Table 6.4.   With permission from American Chemical Society.
Figure 6.4.   With permission from Academic Press.

CHAPTER 7

Figure 7.3.   Dover Publishers; with permission from the author, R. Aris.
Figure 7.5.   With permission from Elsevier Science.
Figure 7.6.   With permission from Elsevier Science.
Figure 7.8.   With permission from Academic Press.
Figure 7.9.   With permission from McGraw-Hill.
Figure 7.10.   With permission from American Chemical Society.
Table 7.6.   With permission from McGraw-Hill.

CHAPTER 8

Figure 8.1.	With permission from American Chemical Society.
Figure 8.5.	With permission from University Science Books.
Figure 8.6.	With permission from Elsevier Science.
Figure 8.7.	With permission from American Chemical Society.
Figure 8.8.	With permission from American Chemical Society.
Figure 8.9.	With permission from American Chemical Society.
Table 8.1.	With permission from University Science Books.
Table 8.2.	With permission from University Science Books.

CHAPTER 9

Table 9.8.	With permission from Elsevier Science.

CHAPTER 10

Figure 10.13.	With permission from American Chemical Society.
Table 10.1.	With permission from Prentice-Hall.

CHAPTER 11

Table 11.2.	Book out of print. Rights reverted to author (L.M. Rose), but author could not be contacted.

CHAPTER 12

Table 12.4.	With permission from Elsevier Science.

CHAPTER 14

Figure 14.13.	With permission from Elsevier Science.

CHAPTER 15

Figure 15.10.	With permission from Elsevier Science.

CHAPTER 17

Figure 17.2.	With permission from Gordon and Breach.
Figure 17.3.	With permission from Gordon and Breach.
Figure 17.6.	With permission from Elsevier Science.
Table 17.3.	With permission from Elsevier Science.
Table 17.4.	With permission from John Wiley.
Table 17.7.	With permission from John Wiley.
Table 17.8.	With permission from Elsevier Science.

CHAPTER 18

Figure 18.1.	With permission from Elsevier Science.
Figure 18.2.	With permission from Elsevier Science.
Figure 18.3.	With permission from Academic Press.
Figure 18.4.	With permission from Gordon and Breach.
Figure 18.5.	With permission from Academic Press.
Figure 18.6.	With permission from Academic Press.

Acknowledgments for Figures and Tables 907

CHAPTER 19

Figure 19.5.	With permission from Elsevier Science.
Figure 19.6.	With permission from Elsevier Science.
Figure 19.7.	With permission from Elsevier Science.
Figure 19.8.	With permission from Elsevier Science.
Figure 19.10.	With permission from American Chemical Society.
Figure 19.11.	With permission from Elsevier Science.
Figure 19.12.	With permission from American Chemical Society.
Figure 19.13.	With permission from American Chemical Society.
Figure 19.14.	With permission from Elsevier Science.
Figure 19.15.	With permission from Elsevier Science.
Figure 19.16.	With permission from Elsevier Science.

CHAPTER 20

Figure 20.1.	With permission from John Wiley and sons.
Figure 20.2.	With permission from McGraw-Hill.
Example 20.1.	With permission from Prentice-Hall Pearson Education.

CHAPTER 21

Figure 21.10.	Redrawn with permission from Plenum.

CHAPTER 22

Figure 22.5.	With permission from American Chemical Society.
Figure 22.7.	with permission from Elsevier Science.
Figure 22.9.	With permission from American Chemical Society.
Figure 22.10.	With permission from American Chemical Society.
Figure 22.11.	With permission from Elsevier Science.
Figure 22.14.	With permission from Elsevier Science.

CHAPTER 23

Figure 23.7.	With permission from Elsevier Science.
Figure 23.8.	With permission from Elsevier Science.
Figure 23.9.	With permission from Elsevier Science.
Table 23.2.	With permission from Elsevier Science.

CHAPTER 24

Figure 24.7.	With permission from Marcel Dekker.
Figure 24.10.	Redrawn with permission from Marcel Dekker.

CHAPTER 25

Figure 25.7.	With permission from American Chemical Society.
Figure 25.8.	With permission from American Institute of Chemical Engineers.

CHAPTER 26

Figure 26.5.	With permission from John Wiley.
Figure 26.7.	With permission from American Institute of Chemical Engineers.
Figure 26.9.	With permission from Elsevier Science.

INDEX

absolute configuration, 246
absorption of gas in liquid accompanied by reaction. See diffusion of gas in liquid accompanied by
absorptivity, 817
adiabatic reactor, 368–376
   design by dynamic programming, 374
   strategies for heat exchange in, 374
   trajectories in, 370–371
additivity principle, 29–30. See also group contributions
adsorption, 173
anolyte, 683
apoenzyme, 649
aprotic solvents, dipolar, 607
aqueous-aqueous biphasing. See biphasing
asymmetric catalysis, 260–279
   heterogeneous, 276–279
   homogeneous, 260–276
asymmetric synthesis, 243–279
   by biocatalysis (see biocatalysis)
   by chromatographic resolution, 255–256
   classification of, 259–260
   methods of generating chirality in, 258–259
   traditional synthesis followed by resolution, 248–258
asymptotic enhancement factor, 440
attenuation constant. See absorptivity
autocatalytic enhancement, 760

batch reactor, 59–69

   in electroorganic synthesis (see electrochemical reactors)
   nonisothermal operation of, 66–69
   optimum operating policies for, 69
   rate constants from, 61–66
   for reactions with no volume change, 59–60
   for reactions with volume change, 61
biocatalysis
   in asymmetric synthesis, 677–681
   in organic synthesis, 674–681
biorganic synthesis engineering, 647–681
bioreactors, 662–674
biphasic biocatalytic systems, 589–595
   classification of, 589–591
   mass transfer in (see mass transfer)
biphasic reactions, 578–591
   equilibria (equilibrium shift) in, 576–585
   matching pH optima in, 588–589
   role of aqueous phase volume, 587–588
   solvent selection in, 585–587
biphasic reaction engineering, 575–605
biphasing
   aqueous-aqueous, 599–601
   chemical protection by, 595–598
   in dimethoate preparation, 596
   examples of, 601–605
   in Hofmann rearrangement, 597
   in homogeneous catalysis (see mobile immobilization)
   in Schotten-Baumann acylation, 597
   use of water as extractant in, 597

Bjerrum's theory of ion pairs. *See* ion pairs
Briggs–Haldane model, 653–656
Brönsted constant/relationship, 30–31
bubble column slurry reactors
  gas holdup in, 538–539
  mass transfer in (*see* mass transfer)
  minimum velocity in, 538
Butler-Volmer equation, 689

carbonylation, 232–234
  for ibuprofen, 242
  of methanol, 232–234, 240–241
cascade reactor. *See* electrochemical reactors
catalysis
  asymmetric (*see* asymmetric catalysis)
  clays in, 145–148
  by cyclodextrins, 154–155
  footprint, 279
  homogeneous (*see* homogeneous catalysis)
  by ion-exchange resins, 152–154
  layered vs. porous catalysts, 146
  role of solvents in, 168–170
  by solids, 125–170
  by titanates, 155–156
  by zeolites (*see* zeolites)
catalysts
  bimetallic, 142–143
  dilution of, 211–212
  heteropolyacid, 143–145
  metallic glass, 150–152
  modified forms of, 127–129
  solid base, 149–150
  solid superacid, 149
  water-soluble, 275–276
catholyte, 685
cavitation, 717–718. *See also* hydrodynamic cavitation
cavitational reactors, design of. *See* sonochemical reactors
cell voltage, overall, 683
charge transfer control, 688–690
chemical protection. *See* biphasing
chemyzymes, 268
chiral catalysis/catalysts
  in PTC, 608
  solid-supported, 276–278
  supported aqueous-aqueous (SAP), 278–279
  zeolite-supported, 278
chiral pools, 258–259
chirality, importance of, 247–248
cloud point extraction, 600–601
coenzymes, 651
cofactors. *See* coenzymes
combinatorial strategies, 847–855

electroorganic synthesis in hydrotropes, 855
enzymatic reactions in reverse micelles, 854–855
PTC with other strategies, 847–850: in electroorganic synthesis, 849; in scf, 849; in ultrasound-assisted reactions, 849; in ultrasound and microphase-assisted reactions, 849–850
sonoelectrochemistry, 853–854
use of scf in: electrochemical reactions, 852–853; enzymatic reactions, 850–851; photochemical reactions, 851–852
combined charge and mass transfer control, 692
combo reactors, 792
complex reactions, 85–112
  analytical solutions for, 97–102
  determination of rates in, 94–95
  extent of reaction in, 92–94
  independent reactions in (*see* independent reactions)
  mathematical representation of, 88–89
  multiple and multistep reactions in, 85
  multistep reactions in, 105–110
  reactor design for (*see* reactor design for complex reactions)
  selectivity and yield in, 95–97, 106
  simultaneous homogeneous and catalytic reactions in, 110–112
  stoichiometry of, 86–88
concentration overpotential. *See* mass transfer
continuous stirred tank reactor (CSTR). *See* mixed-flow reactor (MFR)
coordination number, geometry. *See* transition-metal catalysis
coordinative unsaturation. *See* transition-metal catalysis
crystallization kinetics, 766
criteria
  for chemical similarity in electrochemical reactors, 705
  for electrical similarity, 705
  for eliminating transport disguises, 209
  for energy minimization in gas-liquid reactions, 507
  for volume minimization in gas-liquid reactors, 505–507
CSTR. *See* mixed-flow reactor
current
  distribution, 692–693
  density, 686–687
  efficiency, 684

deactivating catalyst, reactor design for, 389–395
  comparison of reactor performances, 393–394
  in fixed-bed reactor, 390–391
  in fluidized-bed reactor, 391–393
dielectric loss, 832
diffusion
  effect of, in solid catalyzed reactions: external transport, 201–203; internal and external transport, 204; internal diffusion, 183–200 (see also effectiveness factor)
  multicomponent, 194–195
  in pellets, modes of, 183–186
  relative roles of internal and external transport, 205–206
diffusion of gas(es) in liquid accompanied by
  a complex reaction, 443–448
  a simple irreversible reaction, 433–443: discerning the controlling regimes in, 443; effect of temperature in, 441–443; in regimes 1, 2, 3, 4, 433–441
  reaction of two absorbing gases, 448–456: with liquid phase reactant, 452–455; with each other, 448–452; accompanied by a complex reaction, 455–456; a reverse reaction, 455–456
  reaction with complex rate models (LHHW type) involving: one gas and one liquid, 462–463; two gases and a liquid, 463–464
  reaction with two liquid phase reactants, 457–458
dissociation constant. See properties
dissociation-extraction, 794–802
  extractive reaction (reaction mode), 800–802
  regenerative, 797–798
  selection of solvent in, 798–799
  separation mode, 794–800
  theory of, 795–797
distillation column reactor. See distillation-reaction
distillation-reaction, 802–814
  dissociation-extractive distillation (separation mode), 802–805
  distillative reaction (reaction mode), 805–814
  methyl acetate synthesis, 811
  methylal synthesis, 811
  multiple steady states in, 814
  packed column reactor, 808–811
  overall effectiveness factor in, 810–811
  reaction-separation parameter, 807–808

residue curve map (RCM), 812
transformed composition variables, 814
distribution coefficient
  estimation of (see properties)
  of quat species, 615
dynamic programming. See adiabatic reactor

effectiveness factor, 186–194
  in bimolecular reactions, 191–193
  in complex reactions, 196–197
  generalized, 188–190
  in immobilized PT catalysts (see immobilized PTC)
  isothermal, 186–188
  in LHHW kinetics, 196
  for liquid phase reactions, 197–200
  nonisothermal, 193–194
  in supported/immobilized catalysts, 200–201
  in triphase catalysis (see immobilized PTC)
electrical similarity. See critera
electrochemical (electroorganic) cell (reactor), 683
  industrial cell components of, 706–707
  measure of reaction rate in, 686–687
electrochemical reactions. See combined charge and mass transfer control; charge transfer control; mass transfer, concentration overpotential, overpotential
electrochemical reactors, 695–707
  batch without recirculation, 697–698
  batch with recirculation through a PFR, 698–699
  batch with recirculation through an STR, 699–700
  cascade, 700
  with recycle, modeling of, 701
  reduction of nitrobenzene, 701–704
  scale-up of, 704–706
  types of, 707
electrode, counter- and working, 683
electrolyte, 683
electron rules. See transition metal catalysis
electroorganic chemistry, 682
eletroorganic reactions, modeling of, 693–695
electroorganic reactors. See electrochemical reactors
electroorganic synthesis engineering, 682–709
enantiomeric excess, 247–248
enantioselectivity, by enzymatic resolution, 679–681
enhancement factor, 439

enzymes, 649-651
  comparison of reactors for, 666-668
  reactors for, 662-666
enantiomers, pure, methods of preparing by, 248-258
  asymmetric catalysis, 258
  biocatalysis, 256-258
  chromatographic resolution, 255-256
  diastereomer crystallization, 251-253
  directed crystallization, 255
  kinetic resolution, 253-255
epoxidation
  by heteropolyacids, 145
  by homogeneous catalysis, 236
  by TS-1 catalyst, 141
equilibria, extrathermodynamic approach to, 17, 28-30
equilibrium composition
  in complex reactions, 24-26
  in simple reactions, 21-23
equilibrium constant, 20
  temperature dependence of, 21
equilibrium shift. See membrane reactors
ethylation of aniline
  batch reactor equations for, 334-335, 338-339
  independent reactions in (see independent reactions)
  MFR (CSTR) equations for, 340-343
expanding core model, 483
extractive reaction. See dissociation-extraction
extremophiles, 650
extremozymes, 649-651

Faraday's law, 684
fast reactions. See mixing
fed-batch reactor, 674. See also semibatch reactor
fixed-bed reactor, 357-368
  cell model of, 358
  for a deactivating catalyst, 390-392
  heterogeneous models of, 360, 364-365
  parametric sensitivity in, 365-368
  pseudohomogeneous models of, 358
  quasicontinuous models of, 358
  reactor choice, 375
  runaway in (see fixed-bed reactors, parametric sensitivity)
  variations in design of, 377
fluidized-bed reactor, 377-389
  bed properties of, 384
  bubble rise velocity in, 380
  bubbling bed model of, 379-385
  bubbling velocity, 378
  classification of (see Geldart's classification)
  conversion in, 383-385
  end region models of, 385-386
  equivalent diameter of, 387
  gas-liquid-solid, 545
  grid (or jet) region in, 385-386
  heat transfer in, 383
  liquid-liquid-solid, 670
  mass transfer in (see mass transfer)
  minimum fluidization velocity in, 377
  optimization of, 389
  slugging velocity in, 378
  solids distribution in, 382
forced unsteady-state operation (FUSO), 414-416
free energy change, 18
  medium and substituent effects on, 27-28
  standard, 20, 45

gas-liquid contactors (reactors), 493-510. See also criteria; mass transfer
  backmixing in, 494-495
  calculation of reactor volume, 497-501
  classification-1 and -2, 493-494
  generalized form of equation for, 491-493
  interfacial areas in, 494
  reduction of general equation to regimes 1-4, 491-493
gas-liquid reactions. See diffusion of gas in liquid accompanied by
Geldart's classification of fluidizer beds, 378
Gibb's free energy. See free energy change
Grignard reaction. See solid-liquid reactions
group contributions, 29-30, 37

haloform reaction
  analysis and optimization of, 102-105
  reactor design for, 335-337
Hammett relationship, 31-32
Hankel reaction. See solid-liquid reactions
haptophilicity, 168
Hatta number, 436
  modified (with microphase), 756, 764
heat, of combustion, formation, and reaction, 18-19
heat transfer, 80-81
heterogeneous solubilization. See SLPTC
heterogenized "homogeneous" catalysts, 163-167
homogeneous catalysis, 213-242
  addition reactions in, 224-225
  basic reactions of, 223-226
  coordination reactions in, 224

916   Index

heat capacity of liquid, 40–41
heat of combustion, 43
heat of formation, standard, 43–44
heat of vaporization, 39–40
heats of reaction, formation, and
    combustion, 18–19
  by law of corresponding states, 37
  reaction, 43–46
  surface tension, 51
  thermal conductivity of gas, 50–51
  thermal conductivity of liquid, 51
  thermodynamic and equilibrium, 38–43
  transport, 46–51
  ultrasonic velocity in liquids, 51
  vapor pressure, 38–39
  viscosity of gas, 46–47
  viscosity of liquid, 47–48
pseudohomogeneous models for
  fixed-bed reactors, 358–360
  microphase action, 746–748
PTC reactions
  cycle of, 607, 612
  insoluble, 611
  inverse, 611
  liquid-liquid (LLPTC), 612–618
  liquid-liquid-solid (LLS–TPC)
    (see immobilized PTC)
  solid-liquid (see SLPTC)
  solid-solid-liquid (SSL–TPC) (see
    immobilized PTC)
  soluble, 610–611

quantum yield, 817
quaternaty ammonium salts, 608

racemic mixtures, resolution of, 246–256
rarefaction, 711
rate constants
  from batch reactor (BR) data, 61–66
  from mixed-flow reactor (MFR) data, 76
  from plug-flow reactor (PFR) data, 72
  in reactions with an autocatalytic step,
    110–112
rate-determining step, 173–174
reaction rates, 52–58
  basic rate equation, 54–56
  different definitions, 53–54
  stoichiometry of, 56–58
reaction-separation parameter, 807–808
reaction zone model, 484–485
reactive azeotrope, 812
reactive azeotropy point, 812
reactive dying, 479–481
reaction-extraction.
    See dissociation-extraction

reactive extraction.
    See dissociation-extraction
reactors. See batch reactor; mixed-flow
    reactor; plug-flow reactor; semibatch
    reactor
  continuous, in three-phase catalytic
    reactions, 530–533
  integral, differential, mixed, 210–212
  laboratory, 81–84
reactor design for catalytic reactions. See
    fixed-bed reactors, fluidized-bed
    reactors, gas-liquid contactors, three-
    phase reactors
reactor design for complex reactions, 333–356
  batch reactor, 334–339
  mixed-flow reactor (MFR or CSTR),
    340–343
  optimum temperatures in a CSTR, 348–349
  optimum temperature profiles in a PFR
    (and BR), 349–355
  reactor choice for: parallel-consecutive
    reactions, 334–339; parallel reactions,
    343–345
reactor efficiency. See three-phase catalytic
    reactors
recycle-flow reactor. See PFR with recycle
regimes of control, 190
residue curve map (RCM), 812–813
Reynolds number, 202
Rieke metal powder, 727
rotating drum reactor, 545–547

salting-in, salting-out, 844
Schmidt number, 202
segregated flow. See mixing
selectivity
  extrathermodynamic approach to, 32–33
  maximization of, 102–105
  in multiple reactions, 96–105
  in multistep reactions, 105–109
semibatch reactor, 315–326
  constant volume, 316–317
  general expression for, 320–321
  for menadione, 321–325
  in three-phase catalytic reactions, design of,
    531–532
  variable volume, 317–321
Setschenow constant, 844
sharp interface model. See shrinking core
    model
Sharpless synthesis, 266–268
Sherwood number, 202
shrinking core model, 482–483
sink, microphase
  as chemical (catalytic), 749–751

domestic ovens, 836
scale-up of, 836–838
mixed-flow reactor (MFR or CSTR), 73–77
   basic (and design) equations for, 74–76
   in electrochemical synthesis (see
      electrochemical reactors, batch with
      recirculation through an STR)
   in series: design equations, 309–310;
      minimization of volume in, 310–311;
      nonisothermal operation of, 76–77
   rate parameters from, 76
mixing
   addition sequence in, 407–409
   in fast reactions, 405–406
   partial, 402–405
   practical implications of, 406–409
   role of, 396–409
   segregated flow, 401
mobile immobilization, 598–599
multifunctional reactor engineering, 792–814
   reaction-distillation, 802–814
   reaction-extraction, 793–802
multiphase reactions and reactors. See three-phase catalytic reactors; three-phase noncatalytic reactions
multiple solutions, 409–414
multiple steady states
   in an adiabatic CSTR, 410–411
   stability of the steady state, 411–414

nucleophile (main places), 607, 610
nucleophilic reagent (main place), 607
Nusselt number, 202

omega phase, 611
onium salts, 608
optimization of HMB production, 102–105
optimum temperature profiles in a PFR
   consecutive reactions, 351
   extension to batch reactor, 355–356
   parallel-consecutive reactions, 352–353
   parallel reactions, 349–351
   reversible reactions, 351–352
   summary of, 355
optimum temperatures in a CSTR (or MFR), 348–349
overall effectiveness factor. See three-phase catalytic reactions
overpotential, 687–692. See also charge transfer control; mass transfer, at electrodes
oxidation
   comparison of catalysts for menadione, 158–159
   in homogeneous catalysis, 236
   partial oxidation catalysts, 156–158
   state, 215–217
oxidation state. See oxidation, state
oximation of cyclododecanone, 755–757

partial molal properties. See thermodynamics
partial oxidation. See oxidation
PFR. See plug-flow reactor
phase-transfer
   catalysis (see phase transfer, reaction engineering)
   catalysts, 608–609
   reaction engineering, 606–646
   reactions, industrial applications of, 641–646
piezoelectric effect, 712
photochemical enhancement, 815–827
   kinetic considerations in, 817–818
   thermodynamic considerations in, 817–818
photochemical reactors, 818–822
   approaches to design of, 818
   configurations of, 819–820
   equations for, 820–822
   optical thickness in, 820
photochemical synthesis
   examples of, 822, 824–825
   of previtamin D, 825–827
plug-flow reactor (PFR), 69–73
   basic (and design) equations for, 70–72
   in electrochemical synthesis (see
      electrochemical reactors, batch reactor
      with recirculation through a PFR)
   nonisothermal operation of, 72–73
   rate parameters from, 72
   with recycle, 307–310
Polanyi relation, 30–31
polarization, 687–692
power ultrasound, 712
Prandtl number, 202
properties, estimation
   acid and base dissociation constants, 45–46
      (see also Hammett relationship)
   basic, 37–38
   boiling point, 38
   critical, 37–38
   density of gas (vapor), 41
   density of liquid, 41–42
   diffusivity in gas, 48
   diffusivity in liquid, 48–50
   dissociation constant, acid and base, 45–46
   distribution coefficient, 42–43
   entropy of ideal gas, 44–45
   free energy of formation, standard, 45
   by group contributions, 37
   heat capacity of ideal gas, 40

914　Index

macrofilm mass transfer coefficient, 752
macromixing, 399–402
　partial, 402–404
macroregime, 747, 755, 761–762
mass transfer/mass transfer coefficients (in)
　biphasic biocatalytic systems, 591–595
　bubble colum slurry reactors, 536–538
　electrochemical reactions, 690–691
　(at) electrodes (concentration overpotential), 690–692
　film theory of, 79
　fluid-bed reactors, 381–382
　gas-liquid contactors, 494
　immobilized PTC (triphase catalysis), 638–640
　liquid-liquid contactors, 512
　loop slurry reactors, 542
　measurement of, in gas-liquid systems, 464–465
　mechanically-agitated slurry reactors, 534–535
　micellar reactions, 829–831
　penetration theories of, 79–80
　solid-liquid reactions, 478
　sonochemical reactions, 725–726
　surface renewal theory of (see penetration theories of)
　trickle-bed reactors, 544
　two-film theory of, 80–81
mechanically agitated slurry reactor, 534–536
　controlling regimes in, 536
　gas holdup in, 535
　mass transfer in (see mass transfer)
　minimum speed (velocity) in, 535
membranes
　inorganic, 766
　potentially exploitable features of, 766–768
membrane reactors
　catalytic nonselective and nonselective hollow, 782
　catalytic selective, 779–782
　combining/coupling of reactions in, 768, 783–785
　comparison of, 788
　controlled addition of reactants in, 766–768, 785–786
　for economic processes, 791
　effect of tube and shell side flows in, 786–788
　equilibrium shift in, 766
　fluidized–bed inert selective, 769, 779
　immobilized enzyme, 773, 783
　major types of, 769–773
　modeling of, 773–783

operational features of , 783–788
in organic synthesis, use of, 789–790
packed-bed catalytic selective, 769, 773–779, 786–792
preventing slip in, 768
trickle-bed operation, mimicking of , 768
membrane-assisted reactor engineering, 765–791
micellar enhanced ultrafiltration, 601
micellar reactions, 829–831
　kinetic modeling of, 829
　mass transfer in (see mass transfer)
MFR. See mixed-flow reactor
micelles
　aqueous, 828
　enhancement by, 827–831
　nonaqueous, 828
　reverse/inverse (see nonaqueous)
Michaelis-Menten (MM) constant (K_{MM}), 654
　effect of inhibition on K_{MM}, 655–656
Michaelis-Menten (MM) model, 653–656
microbes, 648–649
microbial reactions, 660–662
microbial reactors, 670–674
microbubbles, 744, 758
microdroplets, 744
microemulsion media, 764
microheterogeneous phase, 744
micromixing, 399–402
　partial, 404
　policy, 401–402
microparticles, 744, 758
microphase
　action, analysis of, 746–748
　as a chemical or physical sink, 749–757
　as a physical carrier, 757–758
　as a source, 758–759
　classification of, 744–746
　solid product as, 759–764
microphase-assisted autocatalysis. See microphase, solid product as
microphase-assisted reaction engineering, 744–764. See microphase
microregime, 747–748, 755–757, 762
microslurry, 744
microwave
　choice of frequencies other than, 838–839
　effect in organic synthesis, 832–835
　heating effect of, 831–832
　penetration depth of, 837
microwave-assisted organic synthesis, 831–839
microwave reactors, 835–838
　continuous reactors, 838

elimination in, 226
heterolytic addition in, 224
homolytic addition in, 224–225
in gas-liquid reactions (extension to complex rate models), 458–464
insertion in, 225
operational scheme of, 222
oxidation catalysts in (*see* oxidation)
oxidative addition in, 225
role of diffusion in, 240
Wilkinson catalysts in (*see* hydrogenation)
homogeneous solubilization. *See* SLPTC
hydrodynamic cavitation, 739–743
  inception of, 740
  reactor design for, 740–743
hydroformylation, 234–236
hydrogenation
  of aniline, 539–542
  by asymmetric catalysis, 260–265
  enantioselective, 269
  general model for (in homogeneous catalysis), 231
  of glucose, 521–526
  by homogeneous catalysis, 229–231
  of nitrobenzene in: adiabatic reactor, 371–373; fixed–bed reactor, 363–364; fluidized-bed reactor, 387–389
  solid catalysts for, 159–163
  by Wilkinson's catalyst, 229–231: catalyst cycle in, 229; kinetics and modeling of, 230–231
hydrotropes, 843–847
  applications in organic synthesis, 844–847
  general features of, 843–844
  possible mechanisms of action of, 844

immobilization of enzymes, 651–653
  effect of external diffusion, 658–659
  effect of internal diffusion, 657–658
  effect of internal and external diffusion, 659–660
  methods of, 651
immobilized PTC, 625–637
  autocatalysis in, 637–640
  comparison with soluble PTC, 632–633
  effectiveness factor for, 630–632
  effect of nonisothermicity in, 633–637
  inorganic supports for, 625
  kinetics and mechanism of LLS-TPC and SSL-TPC, 626–627
  mass transfer in (*see* mass transfer)
  modeling of LLS–TPC, 627–632
  organic supports for, 625–626
  Sherwood number correlation for, 638–640
independent reactions, 89–92

in cyclization of ethylenediamine, 89–90
in ethylation of aniline, 91–92
internal and external transport (in catalyst pellets)
  combined role of heat and mass, 204
  relative roles of heat and mass, 205–206
inverse PTC, 611
ion pairs, Bjerrum's theory of, 613
isomerization
  of aromatics, 133–135
  by homogeneous catalysis, 228

j factors (j_d, j_h), 202

Kolbe-Schmitt carbonation. *See* three-phase noncatalytic reactions

laminar flow reactors, 404–405
Langmuir isotherm, 173
Lewis acid catalysts, zeolites as, 130–135
LHHW models, 172–183
  experimental pitfalls in, 180–181
  a generalized approach to, 181–182
  influence of surface nonideality in, 182
  model selection, 178–180
  paradox of, heterogeneous kinetics, 182–183
  placebo effect, 182–183
  theoretical pitfalls in, 179–180
ligands
  charges and coordination numbers of, 217
  role in transition metal catalysis, 218–220
linear absorption coefficient, 817
lipophilicity, 607
liquid-entrained reactor, 545
liquid-liquid contactors, 510–515
  calculation of reactor volume, 512–515
  classification of, 510–511
  interfacial areas in, 512
  mass transfer coefficients in (*see* mass transfer)
liquid-liquid PTC (LLPTC)
  modeling of, 612–618
  rigid sphere model of, 616–618
liquid-liquid reactions, 468–477
  fast, 470–475
  instantaneous, 475
  reduction of m-nitrochlorobenzene, 475–477
  very slow and slow, 470
liquid phase reactions, thermodynamic implications of, 19
loop slurry reactor, 542–543
  mass transfer in (*see* mass transfer)
  types of, 542

Index   917

as physical, 749–751
SLPTC
  heterogeneous solubilization models of, 621–625
  homogeneous solubilization models of, 619–621
  modeling of, 618–625
solid-liquid reactions, 477–489
  Grignard reaction, 486–487
  Hankel reaction, 483–486
  laboratory reactors for, 487
  mass transfer in (see mass transfer)
  solid insoluble, 481–487
  solid sparingly soluble, 478–481
  strategy for modeling of, 488
sonochemical reactions
  cavitation, devices for producing (see ultrasound)
  effect of: acoustic intensity, 720; ambient pressure, 717; ambient temperature, 717–720; dissolved gases, 721; frequency, 717–718; solvent, 720–721; viscosity, 721–722
  heterogeneous, 722–726
  homogeneous, 714–722
  mass transfer in (see mass transfer)
sonochemical reactors
  approximate design of, 735
  design of cavitation reactors, 737–739
  for reactions with special features, 736
  scale-up of, 733–736
sonoelectrochemistry. See combinatorial strategies
sonoorganic synthesis engineering, 711–743
stirred tank reactor (STR), practical considerations, 515–516
supercritical fluid(s) (scf), 839–843
  catalytic reactions in, 841–842
  Diels-Alder reactions in, 840–841
  electrochemical, enzymatic, photochemical syntheses in (see combinatorial strategies)
  properties of, 839–840
  reactant itself as scf, 842–843
  use in organic synthesis, 840–843
synthetic quantum yield, 817

Tafel equation, 689
thermodynamics
  basic relationships in, 17–28
  of Gattermann-Koch reaction, 23–24
  medium and substituent effects, 27–28
  partial molal properties, 26
  of reactions in solution, 26–28
Thiele modulus, 187

three-phase catalytic reactions, 517–526
  analysis of, 518–526
  basis, 518–521
  examples in organic synthesis, 549–551
  LHHW kinetics in, 526
  overall effectiveness factor in, 519
  power law kinetics in, 519–521
three-phase catalytic reactors, 526–547
  continuous, 530–533
  dispersion model of, 532
  laboratory data: interpretation of, 547–549; effect of temperature on, 547
  mixed-flow model of, 532
  plug-flow model of, 532
  reactor efficiency, 532
  semibatch, 529–530
three-phase noncatalytic reactions
  Kolbe-Schmitt carbonation of β-naphthol, 553–558
  solid insoluble, 553
  solid slightly soluble, 549–553
transition-metal catalysis
  16-18-electron rule, 221–222
  18-electron rule, 220–221
  coordination number, geometry, 217–218
  coordinative unsaturation, 217–218
  formalisms in, 214–222
  hydride complex $ReH_7(PMe_2Ph)_2$, 214–215
  oxidation state in (see oxidation)
  role of ligands in, 218–220
trickle-bed reactors, 543–545
  controlling regimes in, 545
  mass transfer in (see mass transfer)
  regimes of flow in, 543–544
triphase catalysis. See immobilized PTC

ultrasonics, 711
ultrasonic velocity. See properties
ultrasound, 711
  devices for producing, 712–714
uptake coefficient, 750

variable volume reactor, 326–330
  measures of mixing in, 330–331
  semibatch operation of, 327–330
viscosity. See properties
voltage
  cell, 683
  open circuit, 687
  standard open circuit, 687
volume reaction model, 483

Weisz modulus, 190
whistles, 714–715

Wilkinson catalyst, 229–231
  catalytic cycle in, 229–230
  kinetics and modeling of reactions using, 230–231

yield
  in multiple reactions, 96–105
  in multistep reactions, 105–109
  vs. number of steps in a multistep scheme, 109

zeolites, 129–142
  applications in organic synthesis, 131
  as Brönsted and Lewis acid catalysts, 132–137
  as basic catalysts, 139
  in hybrid catalysis, 141–142
  in oxidation reactions, 139–141
  noncatalytic uses of, 131
  structure and shape selectivity of, 129–130
  supported, in chiral catalysts (*see* chiral catalysis)